PRAISE FROM ACROSS THE NATION
FOR THE JOBBANK SERIES...

"A superior series of job-hunt directories."
-**Cornell University Career Center's** *Where to Start*

"Help on the job hunt ... Anyone who is job-hunting in the New York area can find a lot of useful ideas in a new paperback called *The Metropolitan New York JobBank* ..."
-**Angela Taylor,** *New York Times*

"One of the better publishers of employment almanacs is Adams Media Corporation ... publisher of *The Metropolitan New York JobBank* and similarly named directories of employers in Texas, Boston, Chicago, Northern and Southern California, and Washington DC. A good buy ..."
-*Wall Street Journal's*
National Business Employment Weekly

"For those graduates whose parents are pacing the floor, conspicuously placing circled want ads around the house and typing up resumes, [*The Carolina JobBank*] answers job-search questions."
-*Greensboro News and Record*

"A timely book for Chicago job hunters follows books from the same publisher that were well received in New York and Boston ... [*The Chicago JobBank* is] a fine tool for job hunters ..."
-**Clarence Peterson,** *Chicago Tribune*

"Job-hunting is never fun, but this book can ease the ordeal ... [*The Los Angeles JobBank*] will help allay fears, build confidence, and avoid wheel-spinning."
-**Robert W. Ross,** *Los Angeles Times*

"Job hunters can't afford to waste time. *The Minneapolis-St. Paul JobBank* contains information that used to require hours of research in the library."
-**Carmella Zagone**
Minneapolis-based Human Resources Administrator

"*The Florida JobBank* is an invaluable job-search reference tool. It provides the most up-to-date information and contact names available for companies in Florida. I should know -- it worked for me!"
-**Rhonda Cody, Human Resources Consultant**
Aetna Life and Casualty

"*The Boston JobBank* provides a handy map of employment possibilities in greater Boston. This book can help in the initial steps of a job search by locating major employers, describing their business activities, and for most firms, by naming the contact person and listing typical professional positions. For recent college graduates, as well as experienced professionals, *The Boston JobBank* is an excellent place to begin a job search."

-Juliet F. Brudney, Career Columnist
Boston Globe

"*The Phoenix JobBank* is a first-class publication. The information provided is useful and current."

-Lyndon Denton
Director of Human Resources and Materials Management
Apache Nitrogen Products, Inc.

"*The Seattle JobBank* is an essential resource for job hunters."

-Gil Lopez, Staffing Team Manager
Battelle Pacific Northwest Laboratories

"Through *The Dallas-Ft. Worth JobBank,* we've been able to attract high-quality candidates for several positions."

-Rob Bertino, Southern States Sales Manager
CompuServe

"*The San Francisco Bay Area JobBank* ... is a highly useful guide, with plenty of how-to's ranging from resume tips to interview dress codes and research shortcuts."

-A.S. Ross, *San Francisco Examiner*

"[*The Atlanta JobBank* is] one of the best sources for finding a job in Atlanta!"

-Luann Miller, Human Resources Manager
Prudential Preferred Financial Services

"Thanks to our listing in *The Detroit JobBank*, we have received a large number of professional referrals."

-Michael A. Solonika, Human Resources Manager
Prestige Stamping, Inc.

"*The Metropolitan Washington DC JobBank* is a valuable resource for anyone who likes to thoroughly research a company before making a commitment to the interview process. This book provides us with an informed candidate."

-Mel Rappleyea, Human Resource Director
Hit or Miss, Inc.

What makes
The JobBank Guide to Computer & High-Tech Companies
the nation's best resource for getting a job
in the computer industry?

With vital employment information on thousands of employers across the nation, *The JobBank Guide to Computer & High-Tech Companies* is the most comprehensive directory of its kind available today.

The book provides information on **one of the hottest industries in the country**, with the primary employer listings providing contact information, telephone numbers, addresses, a summary of the firm's business, and in many cases Websites and e-mail addresses, fax numbers, descriptions of the firm's typical professional job categories, the principal educational backgrounds sought, internships, and the fringe benefits offered.

In addition to the **detailed primary employer listings**, *The JobBank Guide to Computer & High-Tech Companies* gives telephone numbers and addresses for **thousands of additional employers**, as well as information about executive search firms and permanent placement agencies.

All of the reference information contained in this book is as up-to-date and accurate as possible. The company profiles have been thoroughly researched and verified by mail and by telephone. Adams Media Corporation publishes **more employment guides more often** than any other publisher of career directories.

In addition, *The JobBank Guide to Computer & High-Tech Companies* features current information about the job scene -- **forecasts on which industry segments are hiring, overviews of economic trends, online resources for jobseekers**, and **lists of professional associations** so you can get your job hunt started off right.

A condensed, but thorough, review of the entire job search process is presented in the chapter **The Basics of Job Winning**, a feature which has received many compliments from career counselors. In addition, the book includes a section on **resumes and cover letters** the *New York Times* has acclaimed as "excellent."

The JobBank Guide to Computer & High-Tech Companies gives job hunters the most comprehensive, timely, and accurate career information, organized and indexed to facilitate the job search. It's a must-have for anyone trying to find work in the fast-growing computer industry today.

Published by Adams Media Corporation
260 Center Street, Holbrook, MA 02343

Manufactured in the United States of America.

Because addresses and telephone numbers of smaller companies change rapidly, we recommend you call each company and verify the information before mailing to the employers listed in this book. Mass mailings are not recommended.

While the publisher has made every reasonable effort to obtain and verify accurate information, occasional errors are inevitable due to the magnitude of the database. Should you discover an error, or if a company is missing, please write the editors at the above address so that we may update future editions.

"This publication is designed to provide accurate and authoritative information with regard to the subject matter covered. It is sold with the understanding that the publisher is not engaged in rendering legal, accounting, or other professional advice. If legal advice or other expert assistance is required, the services of a competent professional person should be sought."
      --From a *Declaration of Principles* jointly adopted by a Committee of the American Bar Association and a Committee of Publishers and Associations

The appearance of a listing in the book does not constitute an endorsement from the publisher.

ISBN: 1-55850-740-X

*This book is available at quantity discounts for bulk purchases.
For information, call 800/872-5627.*

Visit our exciting job and career site at http://www.careercity.com

The JobBank Guide to

Computer

& High-Tech Companies

Managing Editor:	Steven Graber
Assistant Managing Editor:	Jennifer J. Pfalzgraf
Editor:	Marcie DiPietro
Associate Editors:	William P. McNeill Matthew P. Moran Heidi E. Sampson
Researchers:	Andy Richardson Michelle Roy Ashlee K. Stokes

Adams Media Corporation
HOLBROOK, MASSACHUSETTS

Top career publications from Adams Media Corporation

The JobBank Series:
each JobBank book is $16.95

The Atlanta JobBank, 1997
The Austin/San Antonio JobBank, 1st Ed.
The Boston JobBank, 1997
The Carolina JobBank, 4th Ed.
The Chicago JobBank, 1997
The Cincinnati JobBank, 1st Ed.
The Cleveland JobBank, 1st Ed.
The Dallas-Fort Worth JobBank, 1997
The Denver JobBank, 9th Ed.
The Detroit JobBank, 7th Ed.
The Florida JobBank, 1997
The Houston JobBank, 1997
The Indianapolis JobBank, 1st Ed.
The Las Vegas JobBank, 1st Ed.
The Los Angeles JobBank, 1997
The Minneapolis-St. Paul JobBank, 1997
The Missouri JobBank, 1st Ed.
The Northern New England JobBank, 1st Ed.
The New Mexico JobBank, 1st Ed.
The Metropolitan New York JobBank, 1997
The Upstate New York JobBank, 1st Ed.
The Greater Philadelphia JobBank, 1997
The Phoenix JobBank, 6th Ed.
The Pittsburgh JobBank, 1st Ed.
The Portland JobBank, 1st Ed.
The Salt Lake City JobBank, 1st Ed.
The San Francisco Bay Area JobBank, 1997
The Seattle JobBank, 1997
The Tennessee JobBank, 3rd Ed.
The Virginia JobBank, 1st Ed.
The Metropolitan Washington DC JobBank, 1997
The Wisconsin JobBank, 1st Ed.

The JobBank Guide to Computer & High-Tech
    Companies, $16.95

The National JobBank, 1997
    (Covers the entire U.S.: $295.00 hc)

The JobBank Guide to Employment Services,
    1996-1997
    (Covers the entire U.S.: $160.00 hc)

Other Career Titles:

The Adams Cover Letter Almanac ($10.95)
The Adams Electronic Job Search Almanac, 1997
    ($9.95)
The Adams Job Interview Almanac ($12.95)
The Adams Jobs Almanac, 1997 ($15.95)
The Adams Resume Almanac ($10.95)
America's Fastest Growing Employers, 2nd Ed.
    ($16.00 pb, $30.00 hc)
Career Shifting ($9.95)
Cold Calling Techniques, 3rd Ed. ($7.95)
The Complete Resume & Job Search Book for
    College Students ($9.95)
Cover Letters That Knock 'em Dead, 2nd Ed.
    ($10.95)
Every Woman's Essential Job Hunting & Resume
    Book ($10.95)
The Harvard Guide to Careers in the Mass Media
    ($7.95)
High Impact Telephone Networking for Job
    Hunters ($6.95)
How to Become Successfully Self-Employed, 2nd
    Ed. ($9.95)
How to Get a Job in Education, 2nd Ed. ($15.95)
The Job Hunter's Checklist ($5.95)
The Job Search Handbook ($6.95)
Knock 'em Dead, The Ultimate Jobseeker's
    Handbook, 1997 ($12.95)
The Lifetime Career Manager ($20.00 hc)
The MBA Advantage ($12.95)
The Minority Career Book ($9.95)
The National Jobline Directory ($7.95)
The New Rules of the Job Search Game ($10.95)
Outplace Yourself ($15.95 pb)
Over 40 and Looking for Work? ($7.95)
Reengineering Yourself ($12.95)
The Resume Handbook, 3rd Ed. ($7.95)
Resumes That Knock 'em Dead, 2nd Ed. ($10.95)
Richard Beatty's Job Search Networking ($9.95)
300 New Ways to Get a Better Job ($7.95)

Almost 20,000 potential job opportunities at the click of a button...

Adams JobBank FastResume Suite CD-ROM contains almost 20,000 employer listings, 1,800 executive search firms, 1,100 employment agencies, 600 professionally written sample resumes, 600 sample cover letters, and 100 important interview questions, with powerful answers and expert advice. Pinpoint potential employers by state, industry, or positions commonly offered. Then review employer profiles that typically include contact names for professional hiring, address, phone number, company description, common professional positions, educational backgrounds sought, and even fringe benefits offered. Also included are profiles of executive search firms and employment agencies. You can produce multiple letters for mass mailings. After writing a cover letter, select contact names and just click on "print letter." Keep track of it all with the built-in contact log. Windows CD-ROM.

To order books or software, please send check or money order (including $4.50 for postage) to:
Adams Media Corporation, 260 Center Street, Holbrook MA 02343
(Foreign orders please call for shipping rates.)

Ordering by credit card? Just call 800/USA-JOBS (In Massachusetts, call 617/767-8100).
Please check your favorite retail outlet first

TABLE OF CONTENTS

INTRODUCTION

HOW TO USE THIS BOOK

Right now, you hold in your hands one of the most effective job-hunting tools available anywhere. In *The JobBank Guide to Computer & High-Tech Companies*, you will find a wide array of valuable information to help you launch or continue a rewarding career. But before you open to the book's employer listings and start calling about current job openings, take a few minutes to learn how best to use the resources presented in *The JobBank Guide to Computer & High-Tech Companies*.

The JobBank Guide to Computer & High-Tech Companies will help you to stand out from other jobseekers. While many people looking for a new job rely solely on newspaper help-wanted ads, this book offers you a much more effective job-search method -- direct contact. The direct contact method has been proven to be twice as effective as scanning the help-wanted ads. Instead of waiting for employers to come looking for you, you'll be far more effective going to them. While many of your competitors will use trial and error methods in trying to set up interviews, you'll learn not only how to get interviews, but what to expect once you've got them.

In the next few pages, we'll take you through each section of the book so you'll be prepared to get a jump-start on your competition.

The Computer Industry: An Overview

To get a feel for the state of the computer industry, read the introductory section called *The Computer Industry*. In it, we'll recap the industry's recent performance. Even more importantly, you'll learn where the industry is headed. What are the prospects for the industry that forms one of the fastest-growing sectors of the nation's economy? Which occupations have the most potential growth and which ones are laying off? Are there any companies or geographic areas that are especially hot?

To answer these questions for you, we've pored over business journals and newspapers and interviewed business leaders and labor analysts. Whether you are new to the job market and need a source of industry information, or are a long-time member of the computer world looking for a new job, you'll find this section to be a concise thumbnail sketch of the industry as a whole.

This type of information is potent ammunition to bring into an interview. Showing that you're well versed in current industry trends helps give you an edge over job applicants who haven't done their homework.

The Basics of Job Winning

Preparation. Strategy. Time-management. These are three of the most important elements of a successful job search. *The Basics of Job Winning* helps you address these and all the other elements needed to find the right job.

One of your first priorities should be to define your personal career objectives. What qualities make a job desirable to you? Creativity? High pay? Prestige? Use *Basics of Job Winning* to weigh these questions. Then use the rest of the chapter to design a strategy to find a job that matches your criteria.

In *The Basics of Job Winning,* you'll learn which job-hunting techniques work, and which don't. We've reviewed the pros and cons of mass mailings, help-wanted ads, and direct contact. We'll show you how to develop and approach contacts in your field; how to research a prospective employer; and how to use that information to get an interview and the job.

Also included in *The Basics of Job Winning*: interview dress code and etiquette, the "do's and don'ts" of interviewing, sample interview questions, and more. We also deal with some of the unique problems faced by those jobseekers who are currently employed, those who have lost a job, and college students conducting their first job search.

Resumes and Cover Letters

The approach you take to writing your resume and cover letter can often mean the difference between getting an interview and never being noticed. In this section, we discuss different formats, as well as what to put on (and what to leave off) your resume. We review the benefits and drawbacks of professional resume writers, and the importance of a follow-up letter. Also included in this section are sample resumes and cover letters which you can use as models.

The Best and Worst Ways to Find a Job

In this section we take a quick look at the different methods of conducting a job search. We help define what works best for you, and what doesn't.

CD-ROM Job Search

Jobseekers who are looking for an edge against the competition may want to check out these CD-ROM products.

Online Job Search

The Information Superhighway grows larger every day and offers a multitude of services for the job hunter. This section offers a list of jobseeking sites with employer profiles, company biographies, help wanted advertisements, and more. You can also post your resume on certain sites making them available to nationwide, and in some cases worldwide, employers.

The Employer Listings

Employers are listed by the field they specialize in, and within each field, alphabetically by company names. When a company does business under a person's name, like "John Smith & Co.," the company is usually listed by the surname's spelling (in this case "S").

The JobBank Guide to Computer & High-Tech Companies is broken down into different segments of the computer industry. Each company profile is assigned to one of the chapters listed below.

Software Developers/Programming Services
Hardware Manufacturers
Computer Components Manufacturers
Semiconductor Manufacturers
Networking and Systems Services
Computer Consultants
Internet Providers
Online Services
Miscellaneous Computer Products and Services
Computer Repair/Rental and Leasing
Computer Distributors, Resellers, and Wholesalers

Many of the company listings offer detailed company profiles. In addition to company names, addresses, and phone numbers, these listings also include contact names or hiring departments, and descriptions of each company's products and/or services. Many of these listings also feature a variety of additional information including:

E-mail and World Wide Web addresses - The addresses of how to reach companies electronically is essential, especially for those highly-competitive positions where getting your resume to a company a few days before the U.S. postal service can make all the difference. Also make sure to check out a company's home page to get background notes and other useful information to really show off your knowledge about the company in your cover letter or during an interview.

Common positions - A list of job titles that the company commonly fills when it is hiring, organized in alphabetical order. Note: Keep in mind that *The JobBank Guide to Computer & High-Tech Companies* is a directory of major employers in the industry, not a directory of openings currently available. Many of the companies listed will be hiring, others will not. However, since most professional job openings are filled without the placement of help-wanted ads, contacting the employers in this book directly is still a more effective method than browsing the Sunday papers.

Educational backgrounds sought - A list of educational backgrounds that companies seek when hiring.

Benefits - What kind of benefits packages are available from these employers? Here you'll find a broad range of benefits, from the relatively common (medical insurance) to those that are much more rare (health club membership; child daycare assistance).

Special programs - Does the company offer training programs, internships, or apprenticeships? These programs can be important to first-time jobseekers and college students looking for practical work experience. Many employer profiles will include information on these programs.

Parent company - If an employer is a subsidiary of a larger company, the name of that parent company will often be listed here. Use this information to supplement your company research before contacting the employer.

Number of employees - The number of workers a company employs.

Companies may also include information on other U.S. locations and any stock exchange the firm may be listed on.

Because so many job openings are with small and mid-sized employers, or at a small office of a big company, we've also included the addresses and phone numbers of such employers. While none of these listings include any additional hiring information, many of them do offer rewarding career opportunities. These companies are found under each industry segment heading. Within each industry segment, they are organized by the type of product or service offered.

A note on all employer listings that appear in *The JobBank Guide to Computer & High-Tech Companies*: This book is intended as a starting point. It is not intended to replace any effort that you, the jobseeker, should devote to your job hunt. Keep in mind that while a great deal of effort has been put into collecting and verifying the company profiles provided in this book, addresses and contact names change regularly. Inevitably, some contact names listed herein have changed even before you read this. We recommend you contact a company before mailing your resume to ensure nothing has changed.

Employment Services

Immediately following the employer listings section of this book are listings of employment services firms. Many jobseekers supplement their own efforts by contracting headhunters and other employment search firms to generate potential job opportunities.

This section is a comprehensive listing of such firms, arranged alphabetically by state. Within each state, Executive Search Firms are listed first alphabetically, followed by Permanent Employment Agencies. Each listing includes the firm's name, address, telephone number, and contact person. Most listings also include the industries the firm specializes in, the type of positions commonly filled, and the number of jobs filled annually.

Associations

We have included a directory of other industry-specific resources to help you in your job search. These include professional and industrial associations, many of which can provide employment advice and job search help.

Indexes

The JobBank Guide to Computer & High-Tech Companies has two indexes: one organized geographically by state, and the other a straight alphabetical listing.

THE COMPUTER INDUSTRY: AN OVERVIEW

These days, it seems you can do almost anything with a computer, without ever having to leave the comfort of your own home: researching any subject known to humankind, checking the balance of your savings account, shopping for everything from groceries to automobiles -- most of which can be delivered right to your door. And that's just on the consumer side. Companies by the thousands are creating intranets and extranets to quickly and efficiently share information with their employees, customers, and vendors. In the highly competitive game of computers, companies are doing all they can to make their products better, faster, and more user-friendly. And with technology improving at an astronomical rate, jobseekers who are computer-savvy are in high demand across the country.

Silicon Valley is one of the best places to be if you're looking for work in the computer industry. *Time* magazine states that 50,000 jobs are being created there each year by software companies alone! Programmers that can speak Java and other emerging languages arc increasingly sought after. The area also boasts some of the highest salaries in the nation. Other good bets in California are computer artists and animators for the film industry.

Boston is another hot spot for computer professionals, thanks in part to the area along Route 128 known as America's Technology Highway. Software employment there grew from 95,000 in 1994 to 130,000 in 1996 -- an increase of almost 37 percent. More double-digit growth is expected through the end of the century, according to the Massachusetts Software Council.

Time published a list of the 15 hottest places for jobs in 1997, among which Albuquerque, Austin, Boise, Phoenix, Raleigh, and Seattle all boasted strong high-tech employment. Also making an appearance was Portland for semiconductor manufacturing, which is predicted to have a growth rate of 4.5 percent. Orlando is a great locale for software designers, as well as for film animators working in the city's expanding entertainment industry.

Software

Formerly a segment of the computer industry controlled by the relatively few (**Microsoft**, in particular), the software world is seeing some changes in the late '90s. Credit goes to the Internet. The January 1996 release of **Sun Microsystems**'s Java began this shift, as the software allows programmers to write code that can be transported via the Web and run on any type of operating system.

This change of focus to cross-platform development has been swept up by many big Internet players. **Netscape**'s last two versions of Navigator are Java-enabled browsers. So is Microsoft's Internet Explorer 3.0. And that's just the tip of the iceberg. Microsoft's Internet software division alone now totals 2,500 people. The company bought out 20 startups in 1996, and most of those purchases were designed to increase its presence in the Internet software arena.

Between mid-1995 and the end of 1996, the company hired 250 people, many as a direct result of acquisitions. Most of these new hires were software developers.

According to The Yankee Group Inc., a market researcher, companies bought $500 million worth of software tools in 1996 just for developing Internet applications. An additional $400 million was spent on software for intranets. In 1997, Internet software sales were expected to reach $300 billion. Sales of software to run intranet servers will total over $4 billion in 1997, and increase further to $8 billion by 1998, according to Zona Research Inc. Jobseekers who can create, use, or sell Internet and intranet software will be in big demand.

Systems analysts and computer engineers are good professions to be in as well. Both occupations made *Time* magazine's list of the "15 Hottest Fields" in 1997. Jobs for programmers who only use C++ are on the decline, as companies move toward using newer development tools and are hiring applicants who are multilingual. Those with knowledge of the latest languages will have an easier time finding work, as Java has sparked the demand for new applications.

Jobseekers experienced with object-oriented software tools will be in demand through the end of the century. The development of software for network computers will also provide many opportunities.

Hardware

The excellent sales growth that the personal computer industry experienced in the early and mid '90s has begun to taper off. PC sales had shot up 22 percent in 1995, but in 1996 International Data Corporation calculated the sales increase at only 16 percent. Market researcher Dataquest expected that increase to hit 18 percent by the end of 1997. *Business Week* still predicts annual sales of PCs will reach 132 million by the end of the century.

By the year 2000, 35 percent of computers sold annually will be laptops, according to Giga Information Group. Another hot growth area is servers, whose sales could climb over 30 percent in 1997, believes *Business Week*. Mainframes are still on the rise, with Unix-based machines doing particularly well. Sales of minicomputers and workstations are likely to increase only slightly in the coming years.

To keep sales figures up, companies are starting to explore international markets, primarily Japan, other parts of Asia, and Latin America. However, Sony, already a force to be reckoned with in the consumer electronics marketplace, could become a worthy foreign competitor.

Despite the slowing growth in personal computer sales, there are still plenty of job opportunities in hardware manufacturing. Jobseekers with a background in working with multimedia-based machines or with the latest hardware technology will have some of the best prospects. Network computers will also be big players in the computer game through the end of the decade, as more and more companies are establishing LANs and intranets. As a result, network administrators and MIS managers should have even more opportunities.

The Internet and the World Wide Web

The original Internet, Arpanet, was created by the Defense Department in 1969 as a basic system of e-mail to share military research being done on incompatible computers at universities across the country. When Arpanet connected to other networks in the 1970s, the Internet was born.

Since that time, technology on the Internet has increased exponentially -- and so has the number of users. According to market researcher Jupiter Communications, over 15 million households in North America alone had some form of access to the Internet at the end of 1996. By the end of the century, that number is expected to hit 38 million.

But all this growth doesn't come without its share of problems. In August 1996, **America Online** experienced technical difficulties and its 6 million customers were locked out of the Internet for 19 hours. This rapid expansion was more than even the nation's largest online service could handle. By the end of the year, many customers were frustrated and angry. In January 1997, AOL offered these customers one month's free airtime, or a two month refund.

Companies that provide direct access to the Internet are sprouting up all over the country, and giving online service giants AOL and **CompuServe** some very tough competition. Forrester Research Inc. predicts that 32 million people will be customers of Internet service providers by the year 2000, versus only 12.7 million for online service providers. Even telecommunications leader AT&T has joined in. February 1996 saw the birth of WorldNet, AT&T's Internet access service.

Of course, there wouldn't even be such a thing as "online" if the World Wide Web hadn't come along. The Web was created as a way to combine graphics, text, audio, and video and send them over telephone lines in the form of a page, using Internet technology. Since its inception in 1990, the Web has experienced tremendous growth. As of January 1, 1996, the World Wide Web consisted of 20,000 sites containing close to 20 million pages of information.

The birth of the Web has literally changed the way Corporate America does business. It has brought about the Electronic Publishing Age, and become the foundation for an emerging consumer marketplace. Forrester Research projects that $6.6 billion worth of goods will be purchased online by the end of the century! Revenues from advertising alone on the Web are expected to climb as high as $700 million in 1998, and $5 billion by the year 2000, according to Jupiter Communications. The biggest adsites: Netscape, **InfoSeek**, and **Yahoo!**.

Market researcher Hambrecht & Quist predicts sales of Web-authoring tools to be $300 million by the end of the decade, up from only $2 million in 1995. And this all adds up to good news for the Web-savvy. Corporations nationwide are scrambling to get connected to the Internet, and the more you know about getting online and staying competitive there, the better. If you can speak HTML, you're in luck. The creation and maintenance of company home pages and Websites, both for the Internet and intranets, is an excellent area for jobseekers.

Semiconductors

From the invention of its first microprocessor in 1971 to its latest MMX technology, which will execute multimedia functions faster than ever before, **Intel** has been at the heart of the semiconductor game. *Business Week* estimates Intel to be worth almost $100 billion, as much as the Big Three U.S. automakers combined. Much of the company's recent success can be attributed to the Pentium Pro, a speedy little chip comprised of 5.5 million transistors that can process 400 million instructions per second.

Market researcher Dataquest, Inc. estimated that 60 million of the approximately 95 million microprocessors expected to be sold to the computer world in 1997 will be Pentium Pros and compatibles. However, competition could be stiff from the likes of **Motorola**, whose PowerPC chips originally designed for use in Apple machines are now being used in other applications, such as network and NT servers. **Advanced Micro Devices**, whose chips are less expensive than Intel's and are also IBM-compatible, is another company to watch. Makers of less costly PCs could likely choose lower-priced components, such as those made by AMD, potentially giving Intel a run for its money.

According to *Business Week*, 40 percent of all semiconductors and two-thirds of all memory chips manufactured are for use in personal computers. This direct link between the prosperity of chipmakers and personal computer sales has caused the industry to seek other markets for their chips, such as the dozens of upcoming consumer electronics goods, many of which are expected to hit stores in 1998.

What does this all mean for the semiconductor industry? The sector as a whole experienced a 10.5 percent decrease in revenues in 1996, the steepest drop since 1985. This was largely a result of overestimated demand, which led to a lot of costly expansion. DRAM (dynamic random-access memory) chips saw a 75 percent drop in price. The Semiconductor Industry Association expected revenues from DRAMs to decline by 13.9 percent, from $24.6 billion in 1996 to $21.1 billion in 1997. However, the industry as a whole was expected to improve by 7.4 percent in 1997 with sales of $139 billion.

Long-term prospects for jobseekers looking to get into this highly cyclical industry are favorable. Growth rates are expected to be anywhere from 15 to 25 percent in 1998 through the turn of the century.

THE JOB SEARCH

THE BASICS OF JOB WINNING:
A CONDENSED REVIEW

This chapter is divided into four sections. The first section explains the fundamentals that every jobseeker should know, especially first-time jobseekers. The following three sections deal with special situations faced by specific types of jobseekers: those who are currently employed, those who have lost a job, and college students.

THE BASICS:
Things Everyone Needs to Know

Career Planning   The first step to finding your ideal job is to clearly define your objectives. This is better known as career planning (or life planning if you wish to emphasize the importance of combining the two). Career planning has become a field of study in and of itself.

If you are thinking of choosing or switching careers, we particularly emphasize two things. First, choose a career where you will enjoy most of the day-to-day tasks. This sounds obvious, but most of us have at one point or another been attracted by a glamour industry or a prestigious job title without thinking of the most important consideration: Would we enjoy performing the everyday tasks the position entails?

The second key consideration is that you are not merely choosing a career, but also a lifestyle. Career counselors indicate that one of the most common problems people encounter in job-seeking is that they fail to consider how well-suited they are for a particular position or career. For example, some people, attracted to management consulting by good salaries, early responsibility, and high-level corporate exposure, do not adapt well to the long hours, heavy travel demands, and constant pressure to produce. Be sure to ask yourself how you might adapt to not only the day-to-day duties and working environment that a specific position entails, but also how you might adapt to the demands of that career or industry choice as a whole.

Choosing Your Strategy   Assuming that you've established your career objectives, the next step of the job search is to develop a strategy. If you don't take the time to develop a strategy and lay out a plan, you may find yourself going in circles after several weeks of randomly searching for opportunities that always seem just beyond your reach.

The most common job-seeking techniques are:

- following up on help-wanted advertisements
- using employment services
- relying on personal contacts
- contacting employers directly (the Direct Contact method)

Many professionals have been successful in finding better jobs using each one of these approaches. However, the Direct Contact method boasts twice the success rate of the others. So unless you have specific reasons to believe that other strategies would work best for you, Direct Contact should form the foundation of your job search.

If you prefer to use other methods as well, try to expend at least half your effort on Direct Contact, spending the rest on all of the other methods combined. Millions of other jobseekers have already proven that Direct Contact has been twice as effective in obtaining employment, so why not benefit from their experience?

With your strategy in mind, the next step is to work out the details of **Setting** your search. The most important detail is setting up a schedule. Of course, **Your** since job searches aren't something most people do regularly, it may be **Schedule** hard to estimate how long each step will take. Nonetheless, it is important to have a plan so that you can monitor your progress.

When outlining your job search schedule, have a realistic time frame in mind. If you will be job-searching full-time, your search could take at least two months or more. If you can only devote part-time effort, it will probably take at least four months.

You probably know a few currently employed people who seem to spend their whole lives searching for a better job in their spare time. Don't be one of them. If you are presently working and don't feel like devoting a lot of energy to job-seeking right now, then wait. Focus on enjoying your present position, performing your best on the job, and storing up energy for when you are really ready to begin your job search.

> **The first step in beginning your job search is to clearly define your objectives.**

Those of you who are currently unemployed should remember that job-hunting is tough work physically and emotionally. It is also intellectually demanding work that requires you to be at your best. So don't tire yourself out by working on your job campaign around the clock. At the same time, be sure to discipline yourself. The most logical way to manage your time while looking for a job is to keep your regular working hours.

If you are searching full-time and have decided to choose several different contact methods, we recommend that you divide up each week, designating some time for each method. By trying several approaches at once, you can evaluate how promising each seems and alter your schedule accordingly. But be careful -- don't judge the success of a particular technique just by the sheer number of interviews you obtain. Positions advertised in the newspaper, for instance, are likely to generate many more interviews per opening than positions that are filled without being advertised.

If you are searching part-time and decide to try several different contact methods, we recommend that you try them sequentially. You simply won't have enough time to put a meaningful amount of effort into more than one method at once. Estimate the length of your job search, and then allocate so many weeks or months for each contact method, beginning with Direct Contact.

And remember that all schedules are meant to be broken. The purpose of setting a schedule is not to rush you to your goal but to help you periodically evaluate how you're progressing.

The Direct Contact Method Once you have scheduled your time, you are ready to begin your search in earnest. If you decide to begin with the Direct Contact method, the first step is to develop a checklist for categorizing the types of firms for which you'd like to work. You might categorize firms by product line, size, customer type (such as industrial or consumer), growth prospects, or geographical location. Your list of important criteria might be very short. If it is, good! The shorter it is, the easier it will be to locate a company that is right for you.

Now you will want to use this *JobBank* book to assemble your list of potential employers. Choose firms where *you* are most likely to be able to find a job. Try matching your skills with those that a specific job demands. Consider where your skills might be in demand, the degree of competition for employment, and the employment outlook at each company.

Separate your prospect list into three groups. The first 25 percent will be your primary target group, the next 25 percent will be your secondary group, and the remaining names you can keep in reserve.

After you form your prospect list, begin work on your resume. Refer to the Resumes and Cover Letters section following this chapter to get ideas.

Once your resume is complete, begin researching your first batch of prospective employers. You will want to determine whether you would be happy working at the firms you are researching and to get a better idea of what their employment needs might be. You also need to obtain enough information to sound highly informed about the company during phone conversations and in mail correspondence. But don't go all out on your research yet! You probably won't be able to arrange interviews with some of these firms, so save your big research effort until you start to arrange interviews. Nevertheless, you should plan to spend several hours researching each firm. Do your research in batches to save time and energy. Start with this book, and find out what you can about each of the firms in your primary target group. Contact any pertinent professional associations that may be able to help you learn more about an employer. Read industry

> **The more you know about a company, the more likely you are to catch an interviewer's eye. (You'll also face fewer surprises once you get the job!)**

publications looking for articles on the firm. (Addresses of associations and names of important publications are listed after each industrial section of employer listings in this book.) Then try additional resources at your local library. Keep organized, and maintain a folder on each firm.

If you discover something that really disturbs you about the firm (they are about to close their only local office), or if you discover that your chances of getting a job there are practically nil (they have just instituted a hiring freeze), then cross them off your prospect list. If possible,

---

DEVELOPING YOUR CONTACTS: NETWORKING

Some career counselors feel that the best route to a better job is through somebody you already know or through somebody to whom you can be introduced. These counselors recommend that you build your contact base beyond your current acquaintances by asking each one to introduce you, or refer you, to additional people in your field of interest.

The theory goes like this: You might start with 15 personal contacts, each of whom introduces you to three additional people, for a total of 45 additional contacts. Then each of these people introduces you to three additional people, which adds 135 additional contacts. Theoretically, you will soon know every person in the industry.

Of course, developing your personal contacts does not work quite as smoothly as the theory suggests because some people will not be able to introduce you to anyone. The further you stray from your initial contact base, the weaker your references may be. So, if you do try developing your own contacts, try to begin with as many people that you know personally as you can. Dig into your personal phone book and your holiday greeting card list and locate old classmates from school. Be particularly sure to approach people who perform your personal business such as your lawyer, accountant, banker, doctor, stockbroker, and insurance agent. These people develop a very broad contact base due to the nature of their professions.

---

supplement your research efforts by contacting individuals who know the firm well. Ideally you should make an informal contact with someone at that particular firm, but often a direct competitor, or a major supplier or customer, will be able to supply you with just as much information. At the very least, try to obtain whatever printed information the company has available -- not just annual reports, but product brochures and any other printed materials that the firm may have to offer, either about its operations or about career opportunities.

Getting the Interview Now it is time to arrange an interview, time to make the Direct Contact. If you have read many books on job-searching, you may have noticed that most of these books tell you to avoid the personnel office like the plague. It is said that the personnel office never hires people; they screen candidates. Unfortunately, this is often the case. If you can identify the appropriate manager with the authority to hire you, you should try to contact that person directly. However, this will take a lot of time in each case, and often you'll be bounced back to personnel despite your efforts. So we suggest that initially you begin your Direct Contact campaign through personnel offices. If it seems that the firms on your prospect list do little hiring through personnel, you might consider some alternative courses of action.

The three obvious means of initiating Direct Contact are:

- Showing up unannounced
- Mail (postal or electronic)
- Phone calls

Cross out the first one right away. You should never show up to seek a professional position without an appointment. Even if you are somehow lucky enough to obtain an interview, you will appear so unprofessional that you will not be seriously considered.

Mail contact seems to be a good choice if you have not been in the job market for a while. You can take your time to prepare a letter, say exactly what you want, and of course include your resume. Remember that employers receive many resumes every day. Don't be surprised if you do not get a response to your inquiry, and don't spend weeks waiting for responses that may never come. If you do send a letter, follow it up (or precede it) with a phone call. This will increase your impact, and because of the initial research you did, will underscore both your familiarity with and your interest in the firm.

Another alternative is to make a "cover call." Your cover call should be just like your cover letter: concise. Your first statement should interest the employer in you. Then try to subtly mention your familiarity with the firm. Don't be overbearing; keep your introduction to three sentences or less. Be pleasant, self-confident, and relaxed. This will greatly increase the chances of the person at the other end of the line developing the conversation. But don't press. If you are asked to follow up with "something in the mail," this signals the conversation's natural end. Don't try to prolong the conversation once it has ended, and don't ask what they want to receive in the mail. Always send your resume and a highly personalized follow-up letter, reminding the addressee of the phone conversation. *Always* include a cover letter if you are asked to send a resume.

> **Always include a cover letter if you are asked to send a resume.**

Unless you are in telephone sales, making smooth and relaxed cover calls will probably not come easily. Practice them on your own, and then with your friends or relatives.

If you obtain an interview as a result of a telephone conversation, be sure to send a thank-you note reiterating the points you made during the

---

DON'T BOTHER WITH MASS MAILINGS OR BARRAGES OF PHONE CALLS

Direct Contact does not mean burying every firm within a hundred miles with mail and phone calls. Mass mailings rarely work in the job hunt. This also applies to those letters that are personalized -- but dehumanized -- on an automatic typewriter or computer. Don't waste your time or money on such a project; you will fool no one but yourself.

The worst part of sending out mass mailings, or making unplanned phone calls to companies you have not researched, is that you are likely to be remembered as someone with little genuine interest in the firm, who lacks sincerity -- somebody that nobody wants to hire.

HELP WANTED ADVERTISEMENTS

Only a small fraction of professional job openings are advertised. Yet the majority of jobseekers -- and quite a few people not in the job market -- spend a lot of time studying the help wanted ads. As a result, the competition for advertised openings is often very severe.

A moderate-sized employer told us about their experience advertising in the help wanted section of a major Sunday newspaper:

It was a disaster. We had over 500 responses from this relatively small ad in just one week. We have only two phone lines in this office and one was totally knocked out. We'll never advertise for professional help again.

If you insist on following up on help wanted ads, then research a firm before you reply to an ad. Preliminary research might help to separate you from all of the other professionals responding to that ad, many of whom will have only a passing interest in the opportunity. It will also give you insight about a particular firm, to help you determine if it is potentially a good match. That said, your chances of obtaining a job through the want ads are still much smaller than they are with the Direct Contact method.

conversation. You will appear more professional and increase your impact. However, unless specifically requested, don't mail your resume once an interview has been arranged. Take it with you to the interview instead.

Preparing for the Interview

Once the interview has been arranged, begin your in-depth research. You should arrive at an interview knowing the company upside-down and inside-out. You need to know the company's products, types of customers, subsidiaries, parent company, principal locations, rank in the industry, sales and profit trends, type of ownership, size, current plans, and much more. By this time you have probably narrowed your job search to one industry. Even if you haven't, you should still be familiar with the trends in the firm's industry, the firm's principal competitors and their relative performance, and the direction in which the industry leaders are headed.

Dig into every resource you can! Read the company literature, the trade press, the business press, and if the company is public, call your stockbroker (if you have one) and ask for additional information. If possible, speak to someone at the firm before the interview, or if not, speak to someone at a competing firm. The more time you spend, the better. Even if you feel extremely pressed for time, you should set aside several hours for pre-interview research.

> You should arrive at an interview knowing the company upside-down and inside-out.

If you have been out of the job market for some time, don't be surprised if you find yourself tense during your first few interviews. It will probably happen every time you re-enter the market, not just when you seek your first job after getting out of school.

Tension is natural during an interview, but knowing you have done a thorough research job should put you more at ease. Make a list of questions that you think might be asked in each interview. Think out your answers carefully and practice them with a friend. Tape record your responses to the problem questions. If you feel particularly unsure of your interviewing skills, arrange your first interviews at firms you are not as interested in. (But remember it is common courtesy to seem enthusiastic about the possibility of working for any firm at which you interview.) Practice again on your own after these first few interviews. Go over the difficult questions that you were asked.

Interview Attire

How important is the proper dress for a job interview? Buying a complete wardrobe of Brooks Brothers pinstripes or Donna Karan suits, donning new wing tips or pumps, and having your hair styled every morning are not enough to guarantee you a career position as an investment banker. But on the other hand, if you can't find a clean, conservative suit or won't take the time to wash your hair, then you are just wasting your time by interviewing at all.

Top personal grooming is as important as finding appropriate clothes for a job interview. Careful grooming indicates both a sense of thoroughness and self-confidence. This is not the time to make a statement -- take out the extra earrings and avoid any garish hair colors not found in nature. Women should not wear excessive makeup, and both men and women should refrain from wearing any perfume or cologne (it only takes a small spritz to leave an allergic interviewer with a fit of sneezing and a bad impression of your meeting). Men should be freshly shaven, even if the interview is late in the day, and men with long hair should have it pulled back and neat.

Men applying for any professional position should wear a suit, preferably in a conservative color such as navy or charcoal gray. It is easy to get away with wearing the same dark suit to consecutive interviews at the same company; just be sure to wear a different shirt and tie for each interview.

Women should also wear a businesslike suit. Professionalism still dictates a suit with a skirt, rather than slacks, as proper interview garb for women. This is usually true even at companies where pants are acceptable attire for female employees. As much as you may disagree with this guideline, the more prudent time to fight this standard is after you land the job.

SKIRT VS. PANTS:
An Interview Dilemma

For those women who are still convinced that pants are acceptable interview attire, listen to the words of one career counselor from a prestigious New England college:

I had a student who told me that since she knew women in her industry often wore pants to work, she was going to wear pants to her interviews. Almost every recruiter commented that her pants were "too casual," and even referred to her as "the one with the pants." The funny thing was that one of the recruiters who commented on her pants had been wearing jeans!

The final selection of candidates for a job opening won't be determined by dress, of course. However, inappropriate dress can quickly eliminate a first-round candidate. So while you shouldn't spend a fortune on a new wardrobe, you should be sure that your clothes are adequate. The key is to dress at least as formally or slightly more formally and more conservatively than the position would suggest.

What to Bring   Be complete. Everyone needs a watch, a pen, and a notepad. Finally, a briefcase or a leather-bound folder (containing extra, *unfolded*, copies of your resume) will help complete the look of professionalism.

Sometimes the interviewer will be running behind schedule. Don't be upset, be sympathetic. There is often pressure to interview a lot of candidates and to quickly fill a demanding position. So be sure to come to your interview with good reading material to keep yourself occupied and relaxed.

The Interview   The very beginning of the interview is the most important part because it determines the tone for the rest of it. Those first few moments are especially crucial. Do you smile when you meet? Do you establish enough eye contact, but not too much? Do you walk into the office with a self-assured and confident stride? Do you shake hands firmly? Do you

---

BE PREPARED:
Some Common Interview Questions

Tell me about yourself...

Why did you leave your last job?

What excites you in your current job?

Where would you like to be in five years?

How much overtime are you willing to work?

What would your previous/present employer tell me about you?

Tell me about a difficult situation that you
faced at your previous/present job.

What are your greatest strengths?

What are your greatest weaknesses?

Describe a work situation where you took initiative
and went beyond your normal responsibilities.

Why do you wish to work for this firm?

Why should we hire you?

make small talk easily without being garrulous? It is human nature to judge people by that first impression, so make sure it is a good one. But most of all, try to be yourself.

Often the interviewer will begin, after the small talk, by telling you about the company, the division, the department, or perhaps, the position. Because of your detailed research, the information about the company should be repetitive for you, and the interviewer would probably like nothing better than to avoid this regurgitation of the company biography. So if you can do so tactfully, indicate to the interviewer that you are very familiar with the firm. If he or she seems intent on providing you with background information, despite your hints, then acquiesce.

But be sure to remain attentive. If you can manage to generate a brief discussion of the company or the industry at this point, without being forceful, great. It will help to further build rapport, underscore your interest, and increase your impact.

Soon (if it didn't begin that way) the interviewer will begin the questions, many of which you will have already practiced. This period of the interview usually falls into one of two categories (or somewhere in between): either a structured interview, where the interviewer has a prescribed set of questions to ask; or an unstructured interview, where the interviewer will ask only leading questions to get you to talk about yourself, your experiences, and your goals. Try to sense as quickly as possible in which direction the interviewer wishes to proceed. This will make the interviewer feel more relaxed and in control of the situation.

> **The interviewer's job is to find a reason to turn you down; your job is to not provide that reason.**
>
> -John L. LaFevre, author, *How You Really Get Hired*
>
> Reprinted from the 1989/90 *CPC Annual,* with permission of the National Association of Colleges and Employers (formerly College Placement Council, Inc.), copyright holder.

Remember to keep attuned to the interviewer and make the length of your answers appropriate to the situation. If you are really unsure as to how detailed a response the interviewer is seeking, then ask.

As the interview progresses, the interviewer will probably mention some of the most important responsibilities of the position. If applicable, draw parallels between your experience and the demands of the position as detailed by the interviewer. Describe your past experience in the same manner that you do on your resume: emphasizing results and achievements and not merely describing activities. But don't exaggerate. Be on the level about your abilities.

The first interview is often the toughest, where many candidates are screened out. If you are interviewing for a very competitive position, you will have to make an impression that will last. Focus on a few of your greatest strengths that are relevant to the position. Develop these points carefully, state them again in different words, and then try to summarize them briefly at the end of the interview.

Often the interviewer will pause toward the end and ask if you have any questions. Particularly in a structured interview, this might be the one chance to really show your knowledge of and interest in the firm. Have a list prepared of specific questions that are of real interest to you. Let your questions subtly show your research and your knowledge of the firm's activities. It is wise to have an extensive list of questions, as several of them may be answered during the interview.

> **Getting a job offer is a lot like getting a marriage proposal. Someone is not going to offer it unless they're pretty sure you're going to accept it.**
>
> -Marilyn Hill,
> Associate Director,
> Career Center,
> Carleton College

Do not turn your opportunity to ask questions into an interrogation. Avoid reading directly from your list of questions, and ask questions that you are fairly certain the interviewer can answer (remember how you feel when you cannot answer a question during an interview).

Even if you are unable to determine the salary range beforehand, do not ask about it during the first interview. You can always ask about it later. Above all, don't ask about fringe benefits until you have been offered a position. (Then be sure to get all the details.)

Try not to be negative about anything during the interview (particularly any past employer or any previous job). Be cheerful. Everyone likes to work with someone who seems to be happy.

Don't let a tough question throw you off base. If you don't know the answer to a question, simply say so -- do not apologize. Just smile. Nobody can answer every question -- particularly some of the questions that are asked in job interviews.

Before your first interview, you may be able to determine how many rounds of interviews there usually are for positions at your level. (Of course it may differ quite a bit even within the different levels of one firm.) Usually you can count on attending at least two or three interviews, although some firms are known to give a minimum of six interviews for all professional positions. While you should be more relaxed as you return for subsequent interviews, the pressure will be on. The more prepared you are, the better.

Depending on what information you are able to obtain, you might want to vary your strategy quite a bit from interview to interview. For instance, if the first interview is a screening interview, then be sure a few of your strengths really stand out. On the other hand, if later interviews are primarily with people who are in a position to veto your hiring, but not to push it forward, then you should primarily focus on building rapport as opposed to reiterating and developing your key strengths.

If it looks as though your skills and background do not match the position the interviewer was hoping to fill, ask him or her if there is another division or subsidiary that perhaps could profit from your talents.

Write a follow-up letter immediately after the interview, while it is still fresh in the interviewer's mind (see the sample follow-up letter format found in the Resumes and Cover Letters chapter). Then, if you haven't heard from the interviewer within a week, call to stress your continued interest in the firm, and the position, and request a second interview.

After the Interview

THE BALANCING ACT:
Looking for a New Job While Currently Employed

For those of you who are still employed, job-searching will be particularly tiring because it must be done in addition to your normal work responsibilities. So don't overwork yourself to the point where you show up to interviews looking exhausted and start to slip behind at your current job. On the other hand, don't be tempted to quit your present job! The long hours are worth it. Searching for a job while you have one puts you in a position of strength.

If you're expected to be in your office during the business day, then you have additional problems to deal with. How can you work interviews into the business day? And if you work in an open office, how can you even call to set up interviews? As much as possible you should keep up the effort and the appearances on your present job. So maximize your use of the lunch hour, early mornings, and late afternoons for calling. If you keep trying, you'll be surprised how often you will be able to reach the executive you are trying to contact during your out-of-office hours. You can catch people as early as 8 a.m. and as late as 6 p.m. on frequent occasions.

Making Contact

Your inability to interview at any time other than lunch just might work to your advantage. If you can, try to set up as many interviews as possible for your lunch hour. This will go a long way to creating a relaxed atmosphere. But be sure the interviews don't stray too far from the agenda on hand.

Scheduling Interviews

Lunchtime interviews are much easier to obtain if you have substantial career experience. People with less experience will often find no alternative to taking time off for interviews. If you have to take time off, you have to take time off. But try to do this as little as possible. Try to take the whole day off in order to avoid being blatantly obvious about your job search, and try to schedule two to three interviews for the same day. (It is very difficult to maintain an optimum level of energy at more than three interviews in one day.) Explain to the interviewer why you might have to juggle your interview schedule -- he/she should honor the respect you're

> **Try calling as early as 8 a.m. and as late as 6 p.m. You'll be surprised how often you will be able to reach the executive you want during these times of the day.**

showing your current employer by minimizing your days off and will probably appreciate the fact that another prospective employer is interested in you.

References   What do you tell an interviewer who asks for references? Just say that while you are happy to have your former employers contacted, you are trying to keep your job search confidential and would rather that your current employer not be contacted until you have been given a firm offer.

IF YOU'RE FIRED OR LAID OFF:
Picking Yourself Up and Dusting Yourself Off

If you've been fired or laid off, you are not the first and will not be the last to go through this traumatic experience. In today's changing economy, thousands of professionals lose their jobs every year. Even if you were terminated with just cause, do not lose heart. Remember, being fired is not a reflection on you as a person. It is usually a reflection of your company's staffing needs and its perception of your recent job performance and attitude. And if you were not performing up to par or enjoying your work, then you will probably be better off at another company anyway.

> **Be prepared for the question "Why were you fired?" during job interviews.**

A thorough job search could take months, so be sure to negotiate a reasonable severance package, if possible, and determine what benefits, such as health insurance, you are still legally entitled to. Also, register for unemployment compensation immediately. Don't be surprised to find other professionals collecting unemployment compensation -- it is for everyone who has lost their job.

Don't start your job search with a flurry of unplanned activity. Start by choosing a strategy and working out a plan. Now is not the time for major changes in your life. If possible, remain in the same career and in the same geographical location, at least until you have been working again for a while. On the other hand, if the only industry for which you are trained is leaving, or is severely depressed in your area, then you should give prompt consideration to moving or switching careers.

Avoid mentioning you were fired when arranging interviews, but be prepared for the question "Why were you fired?" during an interview. If you were laid off as a result of downsizing, briefly explain, being sure to reinforce that your job loss was not due to performance. If you were in fact fired, be honest, but try to detail the reason as favorably as possible and portray what you have learned from your mistakes. If you are confident one of your past managers will give you a good reference, tell the interviewer to contact that person. Do not to speak negatively of your past employer and try not to sound particularly worried about your status of being temporarily unemployed.

Finally, don't spend too much time reflecting on why you were let go or how you might have avoided it. Think positively, look to the future, and be sure to follow a careful plan during your job search.

THE COLLEGE STUDENT:
How to Conduct Your First Job Search

While you will be able to apply many of the basics covered earlier in this chapter to your job search, there are some situations unique to the college student's job search.

Perhaps the biggest problem college students face is lack of **Gaining** experience. Many schools have internship programs designed to give **Experience** students exposure to the field of their choice, as well as the opportunity to make valuable contacts. Check out your school's career services department to see what internships are available. If your school does not have a formal internship program, or if there are no available internships that appeal to you, try contacting local businesses and offering your services -- often, businesses will be more than willing to have any extra pair of hands (especially if those hands are unpaid!) for a day or two each week. Or try contacting school alumni to see if you can "shadow" them for a few days, and see what their day-to-day duties are like. Either way, try to begin building experience as early as possible in your college career.

THE GPA QUESTION

You are interviewing for the job of your dreams. Everything is going well: You've established a good rapport, the interviewer seems impressed with your qualifications, and you're almost positive the job is yours. Then you're asked about your GPA, which is pitifully low. Do you tell the truth and watch your dream job fly out the window?

Never lie about your GPA (they may request your transcript, and no company will hire a liar). You can, however, explain if there is a reason you don't feel your grades reflect your abilities, and mention any other impressive statistics. For example, if you have a high GPA in your major, or in the last few semesters (as opposed to your cumulative college career), you can use that fact to your advantage.

What do you do if, for whatever reason, you weren't able to get experience directly related to your desired career? First, look at your previous jobs and see if there's anything you can highlight. Did you supervise or train other employees? Did you reorganize the accounting system, or boost productivity in some way? Accomplishments like these demonstrate leadership, responsibility, and innovation -- qualities that most companies look for in employees. And don't forget volunteer activities and school clubs, which can also showcase these traits.

On-Campus Recruiting

Companies will often send recruiters to interview on-site at various colleges. This gives students a chance to get interviews at companies that may not have interviewed them otherwise, particularly if the company schedules "open" interviews, in which the only screening process is who is first in line at the sign-ups. Of course, since many more applicants gain interviews in this format, this also means that many more people are rejected. The on-campus interview is generally a screening interview, to see if it is worth the company's time to invite you in for a second interview. So do everything possible to make yourself stand out from the crowd.

The first step, of course, is to check out any and all information your school's career center has on the company. If the information seems out of date, call the company's headquarters and ask to be sent the latest annual report, or any other printed information.

Many companies will host an informational meeting for interviewees, often the evening before interviews are scheduled to take place. DO NOT MISS THIS MEETING. The recruiter will almost certainly ask if you attended. Make an effort to stay after the meeting and talk with the company's representatives. Not only does this give you an opportunity to find out more information about both the company and the position, it also makes you stand out in the recruiter's mind. If there's a particular company that you had your heart set on, but you weren't able to get an interview with them, attend the information session anyway. You may be able to convince the recruiter to squeeze you into the schedule. (Or you may discover that the company really isn't suited for you after all.)

Try to check out the interview site beforehand. Some colleges may conduct "mock" interviews that take place in one of the standard interview rooms. Or you may be able to convince a career counselor (or even a custodian) to let you sneak a peek during off-hours. Either way, having an idea of the room's setup will help you to mentally prepare.

Be sure to be at least 15 minutes early to the interview. The recruiter may be running ahead of schedule, and might like to take you early. But don't be surprised if previous interviews have run over, resulting in your 30-minute slot being reduced to 20 minutes (or less). Don't complain; just use whatever time you do have as efficiently as possible to showcase the reasons *you* are the ideal candidate.

LAST WORDS

A parting word of advice. Again and again during your job search you will be rejected. You will be rejected when you apply for interviews. You will be rejected after interviews. For every job offer you finally receive, you probably will have been rejected a multitude of times. Don't let rejections slow you down. Keep reminding yourself that the sooner you go out and get started on your job search, and get those rejections flowing in, the closer you will be to obtaining the job you want.

RESUMES AND COVER LETTERS

When filling a position, a recruiter will often have 100-plus applicants, but time to interview only a handful of the most promising ones. As a result, he or she will reject most applicants after only briefly skimming their resumes.

Unless you have phoned and talked to the recruiter -- which you should do whenever you can -- you will be chosen or rejected for an interview entirely on the basis of your resume and cover letter. Your cover letter must catch the recruiter's attention, and your resume must hold it. (But remember -- a resume is no substitute for a job search campaign. *You* must seek a job. Your resume is only one tool.)

RESUME FORMAT:
Mechanics of a First Impression

The Basics Recruiters dislike long resumes, so unless you have an unusually strong background with many years of experience and a diversity of outstanding achievements, keep your resume length to one page. If you must squeeze in more information than would otherwise fit, try using a smaller typeface or changing the margins.

Keep your resume on standard 8-1/2" x 11" paper. Since recruiters often get resumes in batches of hundreds, a smaller-sized resume may get lost in the pile. Oversized resumes are likely to get crumpled at the edges, and won't fit easily in their files.

First impressions matter, so make sure the recruiter's first impression of your resume is a good one. Print your resume on quality paper that has weight and texture, in a conservative color such as white, ivory, or pale gray. Use matching paper and envelopes for both your resume and cover letter.

Getting it on Paper Modern photocomposition typesetting gives you the clearest, sharpest image, a wide variety of type styles, and effects such as italics, bold-facing, and book-like justified margins. It is also much too expensive for many jobseekers. And improvements in laser printers mean that a computer-generated resume can look just as impressive as one that has been professionally typeset.

A computer or word processor is the most flexible way to type your resume. This will allow you to make changes almost instantly and to store different drafts on disk. Word processing and desktop publishing systems also offer many different fonts to choose from, each taking up different amounts of space. (It is generally best to stay between 9-point and 12-point font size.) Many other options are also available, such as bold-facing for emphasis, justified margins, and the ability to change and manipulate spacing.

The end result, however, will be largely determined by the quality of the printer you use. You need at least "letter quality" type for your resume. Do not use a "near letter quality" or dot matrix printer. Laser printers will generally provide the best quality.

Household typewriters and office typewriters with nylon or other cloth ribbons are *not* good enough for typing your resume. If you don't have access to a quality word processor, hire a professional who can prepare your resume with a word processor or typesetting machine.

Don't make your copies on an office photocopier. Only the personnel office may see the resume you mail. Everyone else may see only a copy of it, and copies of copies quickly become unreadable. Either print out each copy individually, or take your resume to a professional copy shop, which will generally offer professionally-maintained, extra-high-quality photocopiers and charge fairly reasonable prices.

Proof with Care
Whether you typed it yourself or paid to have it produced professionally, mistakes on resumes are not only embarrassing, but will usually remove you from further consideration (particularly if something obvious such as your name is misspelled). No matter how much you paid someone else to type, write, or typeset your resume, *you* lose if there is a mistake. So proofread it as carefully as possible. Get a friend to help you. Read your draft aloud as your friend checks the proof copy. Then have your friend read aloud while you check. Next, read it letter by letter to check spelling and punctuation.

If you are having it typed or typeset by a resume service or a printer, and you can't bring a friend or take the time during the day to proof it, pay for it and take it home. Proof it there and bring it back later to get it corrected and printed.

> The one piece of advice I give to everyone about their resume is: Show it to people, show it to people, show it to people. Before you ever send out a resume, show it to at least a dozen people.
>
> -Cate Talbot Ashton,
> Associate Director,
> Career Services,
> Colby College

If you wrote your resume on a word processing program, also use that program's built-in spell checker to double-check for spelling errors. But keep in mind that a spell checker will not find errors such as "to" for "two" or "wok" for "work." It's important that you still proofread your resume, even after it has been spell-checked.

Types of Resumes
The two most common resume formats are the functional resume and the chronological resume (examples of both types can be found at the end of this chapter). A functional resume focuses on skills and de-emphasizes job titles, employers, etc. A functional resume is best if you have been out

of the work force for a long time and/or if you want to highlight specific skills and strengths that your most recent jobs don't necessarily reflect.

Choose a chronological format if you are currently working or were working recently, and if your most recent experiences relate to your desired field. Use reverse chronological order. To a recruiter your last job and your latest schooling are the most important, so put the last first and list the rest going back in time.

Your name, phone number, and a complete address should be at the **Organization** top of your resume. Try to make your name stand out by using a slightly larger font size or all capital letters. Be sure to spell out everything -- never abbreviate St. for Street or Rd. for Road. If you are a college student, you should also put your home address and phone number at the top.

Next, list your experience, then your education. If you are a recent graduate, list your education first, unless your experience is more important than your education. (For example, if you have just graduated from a business school, have some computer experience, and are applying for a job in the computer industry, you would list your computer experience first.)

Keep everything easy to find. Put the dates of your employment and education on the left of the page. Put the names of the companies you worked for and the schools you attended a few spaces to the right of the dates. Put the city and state, or the city and country, where you studied or worked to the right of the page.

This is just one suggestion that may work for you. The important thing is simply to break up the text in some way that makes your resume visually attractive and easy to scan, so experiment to see which layout works best for your resume. However you set it up, stay consistent. Inconsistencies in fonts, spacing, or tenses will make your resume look sloppy. Also, be sure to use tabs to keep your information vertically lined up, rather than the less precise space bar.

RESUME CONTENT:
Say it with Style

You are selling your skills and accomplishments in your resume, so it **Sell Yourself** is important to inventory yourself and know yourself. If you have achieved something, say so. Put it in the best possible light. But avoid subjective statements, such as "I am a hard worker" or "I get along well with my coworkers." Just stick to the facts.

While you shouldn't hold back or be modest, don't exaggerate your achievements to the point of misrepresentation. Be honest. Many companies will immediately drop an applicant from consideration (or fire a current employee) if inaccurate information is discovered on a resume or other application material.

Keep it Brief   Write down the important (and pertinent) things you have done, but do it in as few words as possible. Your resume will be scanned, not read, and short, concise phrases are much more effective than long-winded sentences. Avoid the use of "I" when emphasizing your accomplishments. Instead, use brief phrases beginning with action verbs.

While some technical terms will be unavoidable, you should try to avoid excessive "technicalese." Keep in mind that the first person to see your resume may be a human resources person who won't necessarily know all the jargon -- and how can they be impressed by something they don't understand?

Also, try to keep your paragraphs at six lines or shorter. If you have more than six lines of information about one job or school, put it in two or more paragraphs. The shorter your resume is, the more carefully it will be examined. Remember: Your resume usually has between eight and 45 seconds to catch an employer's eye. So make every second count.

Job Objective   A functional resume may require a job objective to give it focus. One or two sentences describing the job you are seeking can clarify in what capacity your skills will be best put to use.

Examples: An entry-level position in the computer industry.
A challenging position in software development requiring analytical thought and excellent programming skills.

Don't include a job objective in a chronological resume. Even if you are certain of exactly what type of job you desire, the presence of a job objective might eliminate you from consideration for other positions that a recruiter feels are a better match for your qualifications. But even though you may not put an objective on paper, having a career goal in mind as you write can help give your resume a sense of direction.

Work Experience   Some jobseekers may choose to include both "Relevant Experience" and "Additional Experience" sections. This can be useful, as it allows the jobseeker to place more emphasis on certain experiences and to de-emphasize others.

Emphasize continued experience in a particular job area or continued interest in a particular industry. De-emphasize irrelevant positions. Delete positions that you held for less than four months (unless you are a very recent college grad or still in school). Stress your results, elaborating on how you contributed in your previous jobs. Did you increase sales, reduce costs, improve a product, implement a new program? Were you promoted? Use specific numbers (i.e., quantities, percentages, dollar amounts) whenever possible.

Mention all relevant responsibilities. Be specific, and slant your past accomplishments toward the position that you hope to obtain. For example, do you hope to supervise people? If so, then state how many people, performing what function, you have supervised.

Keep it brief if you have more than two years of career experience. **Education** Elaborate more if you have less experience. If you are a recent grad with two or more years of college, you may choose to include any high school activities that are directly relevant to your career. If you've been out of school for awhile, list post-secondary education only.

Mention degrees received and any honors or special awards. Note individual courses or research projects you participated in that might be relevant for employers. For example, if you are a recent college graduate applying for a job that doesn't directly translate to your major, be sure to mention any courses you took that are relevant to the position sought.

USE ACTION VERBS

How you write your resume is just as important as *what* you write. The strongest resumes use short phrases beginning with action verbs. Below are a few action verbs you may want to use. (This list is not all-inclusive.)

achieved	developed	integrated	purchased
administered	devised	interpreted	reduced
advised	directed	interviewed	regulated
analyzed	discovered	invented	reorganized
arranged	distributed	launched	represented
assembled	eliminated	maintained	researched
assisted	established	managed	resolved
attained	evaluated	marketed	restored
budgeted	examined	mediated	restructured
built	executed	monitored	revised
calculated	expanded	negotiated	scheduled
collaborated	expedited	obtained	selected
collected	facilitated	operated	served
compiled	formulated	ordered	sold
completed	founded	organized	solved
computed	generated	participated	streamlined
conducted	headed	performed	studied
consolidated	identified	planned	supervised
constructed	implemented	prepared	supplied
consulted	improved	presented	supported
controlled	increased	processed	tested
coordinated	initiated	produced	trained
created	installed	proposed	updated
designed	instituted	provided	upgraded
determined	instructed	published	wrote

Highlight Impressive Skills   Be sure to mention any computer skills you may have. You may wish to include a section entitled "Additional Skills" or "Computer Skills," in which you list any software programs you know. An additional skills section is also an ideal place to mention fluency in a foreign language.

Personal Data   This section is optional, but if you choose to include it, keep it very brief (two lines maximum). A one-word mention of hobbies such as fishing, chess, baseball, cooking, etc., can give the person who will interview you a good way to open up the conversation. It doesn't hurt to include activities that are unusual (fencing, bungee jumping, snake-charming) or that somehow relate to the position or the company you're applying to (for instance, if you are a member of a professional organization in your industry). Never include information about your age, health, physical characteristics, marital status, or religious affiliation.

References   The most that is needed is the sentence, "References available upon request," at the bottom of your resume. If you choose to leave it out, that's fine.

HIRING A RESUME WRITER:
Is it the Right Choice for You?

If you write reasonably well, it is to your advantage to write your own resume. Writing your resume forces you to review your experience and figure out how to explain your accomplishments in clear, brief phrases. This will help you when you explain your work to interviewers.

If you write your resume, everything will be in your own words -- it will sound like you. It will say what you want it to say. If you are a good writer, know yourself well, and have a good idea of which parts of your background employers are looking for, you should be able to write your own resume better than anyone else can. If you decide to write your resume yourself, have as many people review and proofread it as possible. Welcome objective opinions and other perspectives.

> **Those things [marital status, church affiliations, etc.] have no place on a resume. Those are illegal questions, so why even put that information on your resume?**
>
> -Becky Hayes, Career Counselor
> Career Services, Rice University

When to Get Help   If you have difficulty writing in "resume style" (which is quite unlike normal written language), if you are unsure of which parts of your background you should emphasize, or if you think your resume would make your case better if it did not follow one of the standard forms outlined either here or in a book on resumes, then you should consider having it professionally written.

There are two reasons even some professional resume writers we know have had their resumes written with the help of fellow professionals. First, they may need the help of someone who can be objective about their background, and second, they may want an experienced sounding board to help focus their thoughts.

The best way to choose a writer is by reputation -- the **If You Hire** recommendation of a friend, a personnel director, your school placement **a Pro** officer, or someone else knowledgeable in the field.

Important questions:
- "How long have you been writing resumes?"
- "If I'm not satisfied with what you write, will you go over it with me and change it?"
- "Do you charge by the hour or a flat rate?"

There is no sure relation between price and quality, except that you are unlikely to get a good writer for less than $50 for an uncomplicated resume and you shouldn't have to pay more than $300 unless your experience is very extensive or complicated. There will be additional charges for printing.

Few resume services will give you a firm price over the phone, simply because some resumes are too complicated and take too long to do for a predetermined price. Some services will quote you a price that applies to almost all of their customers. Once you decide to use a specific writer, you should insist on a firm price quote before engaging their services. Also, find out how expensive minor changes will be.

COVER LETTERS:
Quick, Clear, and Concise

Always mail a cover letter with your resume. In a cover letter you can show an interest in the company that you can't show in a resume. You can also point out one or two skills or accomplishments the company can put to good use.

The more personal you can get, the better. If someone known to the **Make it** person you are writing has recommended that you contact the company, **Personal** get permission to include his/her name in the letter. If you have the name of a person to send the letter to, address it directly to that person (after first calling the company to verify the spelling of the person's name, correct title, and mailing address). Be sure to put the person's name and title on both the letter and the envelope. This will ensure that your letter will get through to the proper person, even if a new person now occupies this position. But even if you don't have a contact name and are simply addressing it to the "Personnel Director" or the "Hiring Partner," definitely send a letter.

Type cover letters in full. Don't try the cheap and easy ways, like using a computer mail merge program, or photocopying the body of your letter and typing in the inside address and salutation. You will give the impression that you are mailing to a host of companies and have no particular interest in any one.

Cover letter do's and don'ts

- *Do* keep your cover letter brief and to the point.
- *Do* be sure it is error-free.
- *Don't* just repeat information verbatim from your resume.
- *Don't* overuse the personal pronoun "I."
- *Don't* send a generic cover letter -- show your personal knowledge of and interest in that particular company.
- *Do* accentuate what you can offer the company, not what you hope to gain from them.

FUNCTIONAL RESUME
(Prepared on a word processor
and laser printed.)

BRIAN RAYMOND BETHEL
5 Scarlet Avenue
Charlotte, NC 28214
(704) 555-0605

BACKGROUND
Extensive and diversified computer hardware and software knowledge in personal computers. Expertise in prototype computer testing. Excellent investigative and research skills. Self-taught in many programming languages, word processors, operating systems, database and spreadsheet software applications.

TECHNICAL QUALIFICATIONS
Programming Languages - Borland C, QBasic, FORTRAN, WordBasic
Operating Systems - MS-DOS, Windows, Novell Netware
Database - FoxPro, askSam
Spreadsheets - Lotus, Excel
Word Processors - WordStar, WordPerfect, Word for Windows, Word for DOS

SUMMARY OF ACCOMPLISHMENTS
- Organized preproduction testing of computer prototype
- Researched, wrote, and edited test procedures that were used for testing of computer prototypes
- Correlated and verified technical information and communicative style in test procedures written by engineers/technical writers
- Co-authored Maxmillian (MDS) Microsoft DOS 5.0 software test plan and was presented with plaque for acknowledging outstanding efforts on development of DOS 5.0
- Wrote database application to track problems found during developmental stages of product, and to generate reports on outstanding issues found during product development
- Designed storage/retrieval and tracking system and cost analysis for manufactured products
- Developed computer engineering test tools
- Analyzed, developed, and implemented new work-related quality processes to improve product testing
- Created software applications that automated work-related processes such as generating status and engineering change request reports

CAREER HISTORY
Maxmillian Data Systems, **Systems Engineer,** Charlotte, NC
Gilford College, **Instructor,** Greensboro, NC

EDUCATION
BGS, Duke University, Durham, NC
Majors: Industrial and Systems Engineering, Computer Computational Mathematics, Mathematics

CHRONOLOGICAL RESUME

(Prepared on a word processor
and laser printed.)

WENDY M. KNOX
319 Governor Street
Albuquerque, NM 87104
(505) 555-1472

CAREER SUMMARY

An experienced professional with expertise in the design and development of multi-user database management systems running on a Local Area Network. Skilled in LAN management and USER training.

BUSINESS EXPERIENCE

JEFFERSON MANUFACTURING CORP., Albuquerque, NM   1990 - Present
Documentation Development Coordinator

Analyze, develop, and maintain application software for engineering LAN. Provide training and user support for all applications to LAN users. Maintain departmental PC workstations including software installation and upgrades. Reduced data entry errors and process time by developing an online program which allowed program managers to submit model number information. Replaced time-consuming daily review board meetings by developing a program which allowed engineers to review and approve model and component changes online. Developed an online program which reduced process time, standardized part usage, and allowed engineers to build part lists for new products and components.

Computer Systems Analyst   1986-1990

Analyzed and designed database management systems; maintained and repaired workstations, and managed LAN. Reduced process time and purchasing errors by developing an online program which allowed the purchasing department to track the status of all purchasing invoices. Developed a purchase order entry program for the purchasing department which improved data entry speed and reduced the number of data entry errors.

LAFAYETTE GROUP, INC., Albuquerque, NM   1981-1986
Engineering Technician III

Prototyped and tested new PC products, drawing schematics and expediting parts. Designed and coded multi-user database management software for engineering use. Expedited parts for 25 or more telecommunications terminal prototypes. Built, troubleshot, and transferred those prototypes to various departments for testing.

EDUCATION

Associates in Electronics Engineering Technology, University of Notre Dame 1981
Continuing education training courses: Advanced Digital Electronics, C Language Hands-On
    Workshop, Visual BASIC Programming, and Structured Analysis and Design Methods.

COMPUTER EXPERIENCE

IBM PCs, Tape Backup Systems, LANs, MS-DOS, Lotus 1-2-3, dBase III, DBXL\ Quicksilver, Clipper, C Language, Netware 3.11, MS Windows, Visual BASIC, and SQL Language.

FUNCTIONAL RESUME
(Prepared on an office
quality typewriter)

DANTE JOSEPH CADELLA
32 Snake Lane
Olympia, WA 98505
(206) 555-9107

Employment Experience:
Eden Data Systems, Inc. - July 1987 to Present.
Olympia, WA

Pre-Sales Technical Support:
Support the Eastern Area sales staff, which averages about
five reps. Formulate and implement sales strategies with
sales reps. Work with prospects to understand business
problems and propose appropriate technical solutions. Write
and deliver technical product presentations. Develop and
deliver customer product demos. Design proposed system
configurations and write proposals.

Key Achievements:
Managed large data migration effort. Designed and
implemented an open-systems migration plan that replaced a
large Eden Data mainframe with a Silicon Graphics compute
server and a network of thirteen SGI workstations. This
effort included: system installation and configuration,
development of file migration utilities, conversion of in-
house FORTRAN code, conversion of over 50,000 files, and
developing and delivering system-administrator and end-user
training.

Education:
Washington State University, Pullman, WA
B.S. in Graphic Communications, 1987

Programming Languages:
FORTRAN, C, VAX Assembler

Operating Systems:
UNIX, MS-DOS, NOS, NOS/VE, AOS/VS

Hardware Experience:
Eden Data mainframes, Silicon Graphics servers and
workstations, Sun workstations, DEC workstations, Hewlett-
Packard workstations, PCs, Macintosh

Professional Awards:
Eden Data Professional Excellence Award, 1989
Eden Data Sales Analyst of the Month, 1992

CHRONOLOGICAL RESUME
(Prepared on a word processor
and laser printed)

MADELINE T. GREENE
912 Edwards Court
Cambridge, MA 02139
(617) 555-5555

TECHNICAL SUMMARY
Hardware Platforms: PC's, Workstations, Mainframes, Servers
Operating Systems: VMS, MVS, DOS, ULTRIX/UNIX, CP/M
Networks: DECNET, TCP/IP
Approaches: OODB, Client Servers, Remote Data Access, RDB,   DBMS, CAD/CAM,
         EDI, CASE, 4GL's

EMPLOYMENT HISTORY
NILES CORPORATION (1984 to Present)
Ultrix Systems and Software, MIS Manager (1993 to Present)
Developed and implemented an overall information architecture for the manufacturing and engineering operational units of the low-end business. Directed the efforts focused on utilizing client/server technology for manufacturing plants in the United States and Canada. Managed 24 professionals and an annual budget of $2.7 million, as well as a computer resource group that serviced 1,200 customers. Managed downsizing effort which resulted in budget reductions.

Consultant & Information Systems Marketing (1990 to 1993)
Major responsibility was to be the MIS subject matter expert for the development of courses designed to educate senior sales and software service people on MIS functions, its problems, and Digital's solutions. Designed and implemented various seminars and symposiums for the purpose of educating customers. All these activities resulted in increased sales and the establishment of new accounts.

Field Service Logistics (FLS) IS Manager (1988 to 1990)
Directed the design, development and implementation of the New Business Process. Directed and managed a staff of senior MIS professionals which implemented the MRPII application in multiple locations. The result was a 25% savings in inventory and expense reductions of $10 million.

MIS Manager, Low-End Business Center (1987 to 1988)
Directed MIS activities in support of order processing business unit with high-volume activity. In order to stabilize the operations, analyzed the operation, hired staff, proposed investments, and drove solutions. The result was smoother operations, quicker response to customers' orders, and reduced indirect labor costs.

Manager, Personal Data Systems (1984 to 1987)
Managed and directed the PDS Department in the pursuit of its long-range plan, the main focus of which was to deliver a distributed data base system. Resulted in reduced labor costs at headquarters and quicker response to employees.

EDUCATION
Massachusetts Institute of Technology (1988). Master of Science in Computer Science
Boston University (1974). Bachelor of Business Administration (Accounting)
Currently enrolled in the Certified Financial Planning program

GENERAL MODEL
FOR A COVER LETTER

Your mailing address
Date

Contact's name
Contact's title
Company
Company's mailing address

Dear Mr./Ms. _____:

Immediately explain why your background makes you the best candidate for the position that you are applying for. Describe what prompted you to write (want ad, article you read about the company, networking contact, etc.). Keep the first paragraph short and hard-hitting.

Detail what you could contribute to this company. Show how your qualifications will benefit this firm. Describe your interest in the corporation. Subtly emphasizing your knowledge about this firm and your familiarity with the industry will set you apart from other candidates. Remember to keep this letter short; few recruiters will read a cover letter longer than half a page.

If possible, your closing paragraph should request specific action on the part of the reader. Include your phone number and the hours when you can be reached. Mention that if you do not hear from the reader by a specific date, you will follow up with a phone call. Lastly, thank the reader for their time, consideration, etc.

Sincerely,

(signature)

Your full name (typed)

Enclosure (use this if there are other materials, such as your resume, that are included in the same envelope)

SAMPLE COVER LETTER

568 North Main Street
Decorah, IA 52101
September 12, 1997

Ms. Leah Conroy
Personnel Director
BRG Corporation
1140 Main Street
Cedar Rapids, IA 52402

Dear Ms. Conroy:

As a graduate student completing a Master of Science degree in Computer Science in December of this year, a challenging application software or system-oriented position interests me very much.

I believe that my two years' experience as a Data Base Designer, as well as my six months' experience as an intern at Salinger Corporation, gives me the necessary capabilities to successfully handle the responsibilities of a Computer Software Designer. I have considerable experience with DBMS packages such as Oracle, Ingres, DB2, Foxpro, and OS/2 Data Manager.

My academic training has provided valuable experience with C and many languages in the Unix/Windows environment. My education has exposed me to extensive programming assignments, wherein I could work individually or as part of a team. With competency in Unix, C, SAS, Pascal, and a variety of other programming languages, I am comfortable in SUNOS, DOS, and VAX operating systems. I have effectively used a variety of graphics, spreadsheet, database, desktop publishing, word processing, and telecommunication applications.

I have a high degree of initiative, have the ability to communicate effectively, and grasp new concepts quickly. Further, I believe that my analytical skills and experience would positively contribute to the design department of your organization.
I would appreciate the opportunity to discuss my qualifications in a personal interview. I will call your office next Monday to follow up on my inquiry. I look forward to learning more about the software design projects at BRG Corporation.

Sincerely,

Andrew Markson

Andrew Markson

Enclosure

GENERAL MODEL FOR A
FOLLOW-UP LETTER

Your mailing address
Date

Contact's name
Contact's title
Company
Company's mailing address

Dear Mr./Ms._____:

Remind the interviewer of the reason (i.e., a specific opening, an informational interview, etc.) you were interviewed, as well as the date. Thank him/her for the interview, and try to personalize your thanks by mentioning some specific aspect of the interview.

Confirm your interest in the organization (and in the opening, if you were interviewing for a particular position). Use specifics to re-emphasize that you have researched the firm in detail and have considered how you would fit into the company and the position. This is a good time to say anything you wish you had said in the initial meeting. Be sure to keep this letter brief; a half-page is plenty.

If appropriate, close with a suggestion for further action, such as a desire to have an additional interview, if possible. Mention your phone number and the hours that you can be reached. Alternatively, you may prefer to mention that you will follow up with a phone call in several days. Once again, thank the person for meeting with you, and state that you would be happy to provide any additional information about your qualifications.

Sincerely,

(signature)

Your full name (typed)

THE BEST AND WORST WAYS TO FIND A JOB

You may be surprised to learn that some of the most popular job search methods are not successful for the majority of the people who use them. This chapter offers a chance to take a look at the real value of the services and techniques at your disposal.

EMPLOYMENT AGENCIES

Employment agencies are commissioned by employers to find qualified candidates for job openings. However, the agency's main responsibility is to meet the needs of the employer -- not necessarily to find a suitable job for the candidate.

This is not to say that employment agencies should be ruled out altogether. There are employment agencies specializing in specific industries that can be useful for experienced professionals. However, employment agencies are not always a good choice for entry-level jobseekers. They often try to steer inexperienced candidates in an unwanted direction and frequently place overqualified candidates in clerical positions.

If you decide to register with an agency, your best bet is to find one that is recommended by a friend or associate. Barring that, names of agencies can be found in *The JobBank Guide to Employment Services* (Adams Media Corporation), found in your local public library. Or you may contact:

National Association of Personnel Services (NAPS)
3133 Mount Vernon Avenue
Alexandria VA 22305
703/684-0180

Be aware that there are an increasing number of bogus employment service firms, usually advertising in newspapers and magazines. These companies promise even inexperienced jobseekers top salaries in exciting careers -- all for a sizable fee. Others use expensive 900 numbers that jobseekers are encouraged to call. Unfortunately, most people find out too late that the jobs they have been promised do not exist.

As a general rule, most legitimate employment agencies will never guarantee a job and will not seek payment until after the candidate has been placed. Even so, every agency you are interested in should be checked out with the local chapter of the Better Business Bureau (BBB). Find out if the agency is licensed and has been in business for a reasonable amount of time. See if any valid complaints have been registered against the agency recently and if the firm has been responsive to these complaints.

If everything checks out with the BBB, call the firm to find out if it specializes in your area of expertise and how it will go about marketing your qualifications. After you have selected a few agencies (three to five is best), send each one a resume with a cover letter. Make a follow-up phone call a week or two later, and try to schedule an interview. Be prepared to take a battery of tests at the interview.

Above all, do not expect too much. Only a small percentage of all professional, managerial, and executive jobs are listed with these agencies, so they are not a terrific source of opportunities. Use them as an addition to your job search campaign, but focus your efforts on other, more promising methods.

EXECUTIVE SEARCH FIRMS

Also known as headhunters, these firms are somewhat similar to professional employment agencies. They seek out and screen candidates for high-paying executive and managerial positions and are paid by the employer. Unlike employment agencies, they typically approach viable candidates directly, rather than waiting for candidates to approach them. Many prefer to deal with already employed candidates and will not accept "blind" inquires from job hunters.

These organizations are not licensed, so if you decide to go with an executive search firm, make sure it has a solid reputation. Names of search firms can be found in *The JobBank Guide to Employment Services* or by contacting the following:

Association of Executive Search Consultants (AESC)
500 5th Avenue
Suite 930
New York NY 10110
212/398-9556

American Management Association (AMA)
Management Services Department
135 West 50th Street
New York NY 10020
212/586-8100

As with employment agencies, do not let an executive search firm become a critical element of your job search campaign -- no matter how encouraging it may sound.

NEWSPAPER CLASSIFIED ADS

Contrary to popular belief, newspaper classified ads are *not* a good source of opportunities for job hunters. Few people find jobs, although many spend a tremendous amount of time and effort poring through newspaper after newspaper.

Only a small percentage of all job openings are advertised in classified ads. Unfortunately, so many applicants respond to these ads that the competition is extremely fierce. Even if your qualifications are good, your chances of getting an interview are not. If you use classified ads as a source for seeking employment, be sure to focus only a small portion of your job search efforts in this direction.

"BLIND" ADS

"Blind" ads are newspaper advertisements that do not identify the employer. Jobseekers are usually instructed to send their resumes to a post office box number. Although they may seem suspicious, "blind" ads can be a source of legitimate job opportunities. A firm may choose to run "blind" advertisements because it does not wish to be deluged with phone calls, or it may be trying to replace someone who has not been terminated yet.

JOB COUNSELING SERVICES

Most states have an employment service (sometimes called the Job Service), which provides career counseling and job placement assistance. Your local city or town may also offer job counseling services. Check your local phone book for the nearest office. Or try writing:

National Board for Certified Counselors
3D Terrace Way
Greensboro NC 27403
910/547-0607

You may also wish to consult *The JobBank Guide to Employment Services*, which lists career counseling services, as well as executive search firms, employment agencies, and temporary help services located throughout the U.S.

NONPROFIT AGENCIES

It may be worthwhile to find out what services are provided by nonprofit agencies in the local community. Many of these organizations offer counseling, career development, and job placement services to specific groups such as women and minorities. Also, try contacting the national organizations below for information on career planning, job training, and other employment-related topics.

For Women:

U.S. Department of Labor
Women's Bureau
200 Constitution Avenue NW
Washington DC 20210
202/219-6606

Wider Opportunities for Women
815 15th Street NW
Suite 916
Washington DC 20005
202/638-3143

For Minorities:

NAACP
Attn: Job Services
4805 Mount Hope Drive
Baltimore MD 21215-3197
410/358-8900

National Urban League
Employment Department
500 East 62nd Street
New York NY 10021
212/310-9000

National Urban League
Washington Operations
1111 14th Street NW
6th Floor
Washington DC 20005
202/898-1604

For the Blind:

Job Opportunities for the Blind Program
National Federation for the Blind
1800 Johnson Street
Baltimore MD 21230
800/638-7518

For the Physically Challenged:

President's Committee on Employment of People with Disabilities
1331 F Street NW
Third Floor
Washington DC 20004-1107
202/376-6200

COLLEGE CAREER CENTERS

If you're a recent college graduate, your school's career center is a great place for you to start your job search. Your college career center can help you identify and evaluate your interests, work values, and skills. Most offer workshops on topics like job search strategy, resume writing, letter writing, and effective interviewing, or job fairs for meeting with prospective employers. Yours should also have a career resource library where you can find job leads and an alumni network to help you get started on networking.

Even if you've long since graduated, your college's career center may be able to provide you with valuable networking contacts. You should call to find out if your alma mater offers this or other services to alumni.

ELECTRONIC DATABASES

Electronic databases compile information on jobseekers to be scanned and searched by hiring companies. Databases may be the new wave for jobseekers of the future, but right now they are primarily used by college students and recent graduates, often with limited results.

These services usually charge between $30 and $50 to list a resume for one year. If you decide to go this route, be sure to make it only one small part of your job search plan.

DIRECT CONTACT

One of the best ways to find a job is by direct contact. Direct contact means introducing yourself to potential employers, usually by way of a resume and cover letter, without a prior referral. This type of "cold contact" can be very effective if done the right way. For more information, turn to the chapter entitled "The Basics of Job Winning."

NETWORKING

Another excellent method of finding work is through networking, a strategy that focuses primarily on developing a network of insider contacts. This is a great approach to use even if you do not have any professional contacts. Networking takes many forms; with a little skill and a lot of effort, it can be a very productive tool.

CD-ROM JOB SEARCH

Jobseekers who are looking for any edge they can find may want to check out the following selected CD-ROM products:

ADAMS JOBBANK FASTRESUME SUITE
260 Center Street
Holbrook MA 02343
800/872-5627
The CD-ROM version of the best-selling *JobBank* series contains almost 20,000 detailed profiles of companies in all industries, 1,800 executive search firms, and 1,100 employment agencies. For most companies, you will find the name, address, company description, and key contact name. The database also lists commonly held positions and information on benefits for most companies. You can search the database by company name, state, industry, and job title. Calling itself a "total job search package," the CD-ROM also creates personalized resumes and cover letters and offers advice on job interviews, including over 100 sample interview questions and answers. *Adams JobBank FastResume Suite* CD-ROM is for Windows® and Windows®95.

AMERICAN BIG BUSINESS DIRECTORY
5711 South 86th Circle
P.O. Box 27347
Omaha NE 68127
800/555-5211
American Big Business Directory has profiles of 160,000 privately and publicly held companies employing more than 100 people. The CD-ROM contains company descriptions which include company type, industry, products, and sales information. Also included are multiple contact names for each company, more than 340,000 in all. You can search the database by industry, SIC code, sales volume, employee size, or zip code.

AMERICAN MANUFACTURER'S DIRECTORY
5711 South 86th Circle
P.O. Box 27347
Omaha NE 68127
800/555-5211
Made by the same company that created *American Big Business Directory*, *American Manufacturer's Directory* lists over 531,000 manufacturing companies of all sizes and industries. The directory contains product and sales information, company size, and a key contact name for each company. The user can search by region, SIC code, sales volume, employee size, or zip code.

BUSINESS U.S.A.
5711 South 86th Circle
P.O. Box 27347

Omaha NE 68127
800/555-5211
Also from the makers of *American Big Business Directory* and *American Manufacturer's Directory*, this CD-ROM contains information on 10 million U.S. companies. The profiles provide contact information, industry type, number of employees, and sales volume. The profiles also indicate whether a company is public or private, as well as providing information about the company's products. There are a number of different search methods available, including key words, SIC code, geographic location, and number of employees.

CAREER SEARCH - INTEGRATED RESOURCE SYSTEM
21 Highland Circle
Needham MA 02194-3075
617/449-0312
Career Search is a database which contains listings for over 490,000 privately and publicly held companies. It has contact information, including names of human resources professionals or other executives, for companies of virtually all sizes, types, and industries. The extensive database can be searched by industry, company size, or region. This product is updated monthly.

COMPANIES INTERNATIONAL
835 Penobscot Building
645 Griswald Street
Detroit MI 48226
800/877-GALE
Produced by Gale Research Inc., this database is geared toward businesses and job hunters. The CD-ROM is compiled from *Ward's Business Directory* and the *World's Business Directory*, and contains information on more than 300,000 companies worldwide. You can find industry information, contact names, and number of employees. Also included is information on the company's products and revenues. The database can be searched by industry, company products, or geographic location.

CORPTECH DIRECTORY
12 Alfred Street, Suite 200
Woburn MA 01801-1915
800/333-8036
The *CorpTech Directory* on CD-ROM contains detailed descriptions of over 45,000 technology companies. It also lists the names and titles of nearly 155,000 executives -- CEOs, sales managers, R&D managers, and human resource professionals. Home page and e-mail addresses are also available. In addition to contact information, you can find detailed information about each company's products or services and annual sales revenues. The *CorpTech Directory* also lists both the number of current employees, and the number of employees twelve months previously. Some companies also list the number of employees they anticipate to have in one year, so you can get an idea of how much a

company expects to grow. You can search the database by type of company, geographic location, or sales revenue. Updated quarterly.

DISCOVERING CAREERS & JOBS
835 Penobscot Building
645 Griswald Street
Detroit MI 48226
800/877-GALE
Provides overviews on 1,200 career choices, 1,000 articles from trade publications, and contact information for professional associations. This CD-ROM also contains self-assessment tests, college profiles and financial aid data.

DISCOVERING CAREERS & JOBS PLUS
835 Penobscot Building
645 Griswald Street
Detroit MI 48226
800/877-GALE
This Gale Research Inc. CD-ROM gives users contact information on more than 45,000 companies, with 15,000 in-depth profiles and 1,000 company-history essays. In addition, the product also provides profiles and application procedures for all major two- and four-year U.S. colleges and universities.

DUN & BRADSTREET MILLION DOLLAR DISC PLUS
3 Sylvan Way
Parsippany NJ 07054
800/526-0651
This CD-ROM lists information on over 400,000 companies. Of these, about 90 percent are privately held, but all have at least $3 million in annual sales or at least 50 employees. The *Million Dollar Disc Plus* is a broad directory covering virtually every industry. The directory provides the number of employees, sales volume, name of the parent company, and corporate headquarters or branch locations. This product includes the names and titles of top executives, as well as biographical information, including education and career background. Searches can be done by location, industry, SIC code, executive names, or key words in the executive biographies. This directory is updated quarterly.

ENCYCLOPEDIA OF ASSOCIATIONS:
NATIONAL ORGANIZATIONS OF THE U.S.
835 Penobscot Building
645 Griswald Street
Detroit MI 48226
800/877-GALE
Contains nearly 23,000 national organizations' descriptions and contact information. You can search for data by association name, geographic location, and key words. This CD-ROM is available in both single- and multi-user formats. Gale offers four other modules in this series covering international, local, and nonprofit organizations, as well as association materials information.

GALE BUSINESS RESOURCES CD-ROM
835 Penobscot Building
645 Griswald Street
Detroit MI 48226
800/877-GALE
This contains statistics on over 200,000 businesses nationwide. You can search by company name, products, market share, and more. This two CD-ROM set contains detailed profiles on certain industries and covers the major companies in each industry. This product is available in both single- and multi-user formats. Users can search by company name, industry type, products, or by combining any of the preceding criteria.

GALE'S CAREER GUIDANCE SYSTEM
835 Penobscot Building
645 Griswald Street
Detroit MI 48226
800/877-GALE
This CD-ROM provides contact information, as well as the names of key personnel, for more than 210,000 potential employers. You can choose from among 250 job categories and 1,200 job titles for a more personalized search. The database can also be searched by SIC code. This annually updated directory also includes the full text of two federal government publications, the *Dictionary of Occupational Titles* and also the *Occupational Outlook Handbook*.

HARRIS INFOSOURCE INTERNATIONAL
2057 East Aurora Road
Twinsburg OH 44087
800/888-5900
This directory of manufacturers profiles thousands of companies. Although the majority of the companies listed are located in the United States, the directory also provides listings for some businesses overseas. Besides contact information, you can find out the number of employees, plant size, and sales revenue. The database also contains the names and titles of top executives. This CD-ROM is updated annually and can be purchased in smaller regional or state editions.

HOOVER'S COMPANY AND INDUSTRY DATABASE ON CD-ROM
1033 La Posada Drive
Suite 250
Austin TX 78752
512/454-7778
This CD-ROM contains *Hoover's Handbook of American Businesses*, *Hoover's Handbook of World Businesses*, and *Hoover's Handbook of Emerging Businesses*, as well as listings of various computer companies, media companies, and more. *Hoover's Company and Industry Database* contains 2,500 company profiles in 450 industries. This CD-ROM has three quarterly updates and is available in both single- and multi-user formats.

MOODY'S COMPANY DATA
99 Church Street
New York NY 10007
800/342-5647
Moody's Company Data is a CD-ROM which has detailed listings for over 10,000 publicly traded companies. In addition to information such as industry, company address, and phone and fax numbers, each listing includes the names and titles of its top officers, including the CEO, president, and vice president; company size; number of shareholders; corporate history; subsidiaries; and financial statements. Users can conduct searches by region, SIC codes, industry, or earnings. This CD-ROM is updated monthly.

NATIONAL FAX DIRECTORY
835 Penobscot Building
645 Griswald Street
Detroit MI 48226
800/877-GALE
A directory of fax numbers for 250,000 businesses, the *National Fax Directory* gives the user access to 170,000 fax numbers for companies, associations, and government agencies. Information can be obtained alphabetically or by subject. This CD-ROM is updated annually.

STANDARD & POOR'S REGISTER
65 Broadway
8th Floor
New York NY 10004
800/221-5277
The CD-ROM version of this three-volume desk reference provides the same information as its printed companion. The database lists over 55,000 companies, including more than 12,000 public companies. In addition to contact information, which includes the names and titles of over 500,000 executives, you can find out about each company's primary and secondary sources of business, annual revenues, number of employees, parent company, and subsidiaries. When available, the *Standard & Poor's Register* also lists the names of banks, accounting firms, and law firms used by each company. Also, the directory provides biographies of more than 70,000 top executives, which include information such as directorships held and schools attended. There are 55 different search modes available on the database. You can search geographically, by zip code, industry, SIC code, or stock symbol. You can also limit your search to only private or only public companies. This directory is updated quarterly.

WALKER'S CORPORATE DIRECTORY OF U.S. PUBLIC COMPANIES
835 Penobscot Building
645 Griswald Street
Detroit MI 48226
800/877-GALE

A directory of 9,500 publicly held U.S. corporations. The user can access each company's income, earnings per share, assets, stock exchange symbol, and more. Searches on business information are classified by company, SIC code, stock exchange, and more. This CD-ROM also allows the user to print company reports and labels.

ONLINE JOB SEARCH

Finding a job can be easier than ever thanks to the vast Information Superhighway. The Internet and the World Wide Web have put an end to combing through a maddening mass of Sunday classifieds looking for the perfect job. Newspapers, employer profiles, industry information, and more are at your fingertips with the aid of a modem and an Internet service provider. You can post your resume and do a nationwide job search in one sitting.

CAREERCITY
http://www.careercity.com
Industries: All
Search Location: Worldwide
One of the most comprehensive and helpful sites on the World Wide Web, *CareerCity* contains worldwide searches by keyword, location, and job category for thousands of job openings. Users can also search through 100 job newsgroups by keyword. Job hunters can submit a resume to *CareerCity's* resume database. *CareerCity* also includes a wealth of other beneficial job search information, including resume, cover letter, and interview tips from Adams career publications and career experts.

THE AT&T COLLEGE NETWORK
http://www.att.com/college
Industries: All
One of the most graphically pleasing and easy-to-use sites, *The AT&T College Network* offers a great deal of jobseeking data. Job listings, research source links, student services, job links, expert advice, and more make this a valuable site for the college student or job hunting professional.

AMERICA'S JOB BANK
http://www.ajb.dni.us/index.html
Industries: All
Search Location: Nationwide
Offers a vast database with over 250,000 employment opportunities.

ATLANTA COMPUTERJOBS STORE
http://www.computerjobs.com
Industries: High-Tech
Search Location: Southeast

THE BEST JOBS IN THE USA TODAY
http://www.bestjobsusa.com
Industries: All
Search Location: Nationwide

A location full of helpful information and suggestions for today's job hunter. Post your resume in the "Resume Pavilion" or shop at the "Career Store" for books, audio tapes, and videos to aid you in your job search. *The Best Jobs in The USA Today* also has *Employment Review* magazine online, which offers new employment opportunities and employment trends. The database is updated every 48 hours.

CAREER CONNECTION
http://www.connectme.com
Industries: All
Search Location: Nationwide
A career site searchable by job description.

CAREER MAGAZINE
http://www.careermag.com/careermag
Industries: All
Career Magazine offers employer profiles, a directory of career consultants, a directory of executive recruiters, a resume bank, job openings, a career forum, and news and articles, including information from *The Wall Street Journal*'s *National Business Employment Weekly*.

CAREERMART
http://www.careermart.com/main.html
Industries: All
An impressive site with an abundance of job-related information. You can click on *CareerMart*'s entrance to obtain job listings, employer profiles, company home pages, and conduct a job search by company, location, and job listing. You can browse in the "Newsstand" and access various publications such as *USA Today*, *CNN Interactive*, *Asian Week*, *Fortune*, *The Economist*, and more. You can enter the "ADvise Tent" for a "CareerChat," employment outlooks, career fair information, and more.

CAREER MOSAIC
http://www.careermosaic.com
Industries: All
Produced by Bernard Hodes Advertising, this is currently one of the hottest job search sites. *Career Mosaic* contains an online job fair, a career resource center, resume posting, and a forum for college jobseekers. Surfing this site can prove fruitful, as it is one of the choice spots for businesses to post job openings.

CAREER PARADISE
http://www.emory.edu/career
Industries: All
A career Website from Emory University offering job links and employment-related information.

CAREERPATH
http://www.careerpath.com
Industries: All
Search Location: Nationwide
An enormous site of employment advertisements from seven of the nation's top newspapers. *CareerPath* consists of the complete Sunday want ad sections for the *Boston Globe, Chicago Tribune, Los Angeles Times, Philadelphia Inquirer, San Jose Mercury News, South Florida Sun-Sentinel*, and *Washington Post.* You can search these advertisements by newspaper, job category, and other keywords.

CAREER RESOURCE CENTER
http://www.careers.org
Industries: All
Search Location: Nationwide
A career directory offering links listed by region, state, industry, and company.

CAREER SHOP
http://www.careershop.com
Industries: All
An online database of employment opportunities. Posting a resume on this site is free of charge and allows you to find out which employers have downloaded your credentials. *Career Shop* offers tips on writing your electronic resume, interview tips, and employment searches by region, company name, and industry.

CAREERWEB
http://www.cweb.com
Industries: All
Search Location: Worldwide
This site offers a database containing worldwide employers and franchisers. You can search by job, state, country, company, or keyword. *CareerWEB* allows the user to search its library of employment publications and posts listings from the *Wall Street Journal's National Business Employment Weekly.* "Jobmatch" allows you to fill out a form with your credentials and *CareerWEB* will match them with an employer seeking your qualifications.

THE CATAPULT
http://www.jobweb.org/catapult
Industries: All
Search Location: Worldwide
An employment Website operated by the National Association of Colleges and Employers with job site, regional, and international links.

COLLEGE GRAD JOB HUNTER
http://www.collegegrad.com
Industries: All

A good site for recent college graduates offering interviewing tips, resume development and posting, job search preparation, and entry-level job opportunities. You can search for entry-level positions by keywords.

COMPUTERWORLD
http://www.computerworld.com
Industries: Computers
Created by OnlineFocus and International Data Group's Interactive Services Group, this location is the computer-related newspaper's World Wide Web career site. *Computerworld* conducts a search by skills, job level (entry-level or experienced), job title, company, and city and state. One feature of this site is "CareerMail," which matches your search criteria against *Computerworld's* database once a day, with e-mail sent to you when your criteria have been met. This site also has corporate profiles, IDG's publications, an index of graduate schools, and other informative and educational resources.

DALLAS COMPUTERJOBS STORE
http://www.computerjobs.com/dallas
Industries: High-Tech
Search Location: Southwest

EMPLOYMENT EDGE
http://sensemedia.net/employment.edge
Industries: Accounting, auditing, engineering, legal, management, and computer programming
This site offers career fair calendars, salary guides, job search tips, resume posting, and more. You can browse through the employment database using keywords.

ENTRY LEVEL JOB SEEKER ASSISTANT
http://www.utsi.edu:70/students
Industries: All
A directory for jobseekers who have never had a full-time, permanent position in their field or have not yet been in the professional workforce for a year.

FLEET HOUSE CAREER MATCH
http://www.fleethouse.com/career
Industries: All
A joint effort from Fleet House Information Management Solutions and Wood and Associates, this site offers a low-cost service that matches employees with employers. Subscribers can post resumes and inquire about firms.

4WORK
http://www.4work.com
Industries: All
Search Location: Worldwide

This site lists worldwide *Fortune* 500 corporations. You can search through businesses by location, company, position, or keyword. You can also locate internship and volunteer opportunities.

HELP WANTED-USA
http://www.webcom.com/-career/hwusa.html
Industries: All
Search Location: Nationwide
A division of Gonyea & Associates, Inc., this nationwide network of employer listings allows you to search by state, industry, or keyword. *Help Wanted-USA* also offers a resume bank.

HOT JOBS
http://www.hotjobs.com
Industries: High-Tech
Search Location: Nationwide

HYPER-MEDIA RESUMES
http://www.webcom.com
An online resume management site for professionals. This site also has huge lists of career-related sites for the job hunter.

INTELLIMATCH
http://www.intellimatch.com
Industries: All
Search Location: Worldwide.
A service which matches jobseekers and employers. You can use *IntelliMatch* to create your resume, which you can then either print out or post on a worldwide database. This site also allows you to post your resume confidentially. The listings of hot jobs and the online career center include companies such as Hewlett-Packard, Mitsubishi, Oracle, Pitney Bowes, Silicon Graphics, and more.

JOB SEARCH AND EMPLOYMENT OPPORTUNITIES: BEST BETS FOR THE NET
http://www.asa.ugl.lib.umich.edu/chdocs/employment/job-guide.toc.html
A collection of Websites searchable by industry category.

JOB SOURCE
http://jobsource.com
Industries: All
Search location: Nationwide
Job Source offers job searches using keywords, industry, or location. It also has employer profiles, a resume generator, employer of the week, and a career center offering articles, links, and various job-hunting resources.

JOB TRAK
http://www.jobtrak.com
Industries: All
College students and recent graduates will enjoy this site designed just for them. Used by over 200,000 employers, this site has over 2,100 new job openings posted each day. There are career fairs, job search suggestions, resumes online, job listings, top recruiters, and a guide to graduate schools. This site works with hundreds of college and university career centers and provides a search for both students and alumni to search their alma mater for job opportunities.

JOB WEB
http://www.jobweb.com
Industries: All
The *Job Web* offers a surplus of jobseeking options, such as career planning resources, job searches, industry information, and the opportunity to have your job search questions answered by human resource and career service professionals. You can peruse the company directory, job postings, internships, federal government positions, and more. The *Job Web* search engine allows Web surfers to comb through listings by keyword, state, and location.

JOBANK
http://www.jobank.com
Industries: All
Search Location: Nationwide

JOBS FOR PROGRAMMERS ACROSS AMERICA
http://www.jfpresources.com
Industries: Computer programming
Search Location: Nationwide

KAPLAN'S CAREER CENTER
http://www.kaplan.com/career
Industries: All
Kaplan's Career Center is a great place to start or enhance your job search. This site offers trade advice with fellow job hunters, career links, career counselors, resume and interview tips, and an amusing interview game online.

THE MONSTER BOARD
http://www.monster.com
Industries: All
Search Location: Worldwide
ADION's dazzling graphics and thousands of job listings from 1,500 hot companies are found on *The Monster Board*. Whether you are going on a "Career Safari" or admiring the "Corporate Sphere," this is likely to be an enjoyable job search. The "Career Safari" offers an exhaustive career search, online resume posting, and employer profiles. The "Corporate Sphere" contains human resource information and employment success stories. Jobseekers

specify keyword searches by skill, job title, or criteria. Over 300 employer profiles are available, as well as information on job fairs and helpful job-hunting tips.

THE NATIONAL MULTIMEDIA ASSOCIATION OF AMERICA'S (NMAA) JOB BANKER

http://www.nmaa.org/jobbank.html

Industries: All

At this site you can search by keyword, location, duration, commitment, and industry. NMAA's search engine also accommodates requests for short-term or part-time placement.

ONLINE CAREER CENTER

http://www.occ.com

Industries: All

Search Location: Nationwide

Internet surfers can post a resume, and find job fair listings, electronic books and directories, and employment opportunities from over 300 United States companies.

POST A JOB

http://www.postajob.com

Industries: Information technology

Search Location: Nationwide

Post A Job utilizes technology that e-mails you when new opportunities are posted. This location also allows you to post your resume and ensure proper privacy.

THE RILEY GUIDE

http://www.jobtrak.com/jobguide

Industries: All

Search Location: Nationwide

An employment guide of free listings assembled by Margaret Riley of Worcester Polytechnic Institute.

SELECTJOBS

http://www.selectjobs.com

Industries: All

Post a resume and search the job database by region, discipline, and skills on *SelectJOBS*. Once your search criteria has been entered, this site will automatically e-mail you when a job opportunity matches your requests.

STUDENTCENTER

http://www.studentcenter.com

Industries: All

A great resource for graduates or current college students, *StudentCenter* allows the surfer to browse a database of over 35,000 companies and 1,000 industries.

This highly informative site offers internship opportunities, career advice, tips on resume and cover letter building, resume construction, a student forum, and a fun virtual interview.

TECHNOLOGY REGISTRY
http://www.techreg.com
Industries: Manufacturing, service, and technology
One of the nation's largest online employment databases serving the technology, manufacturing, and service industries. The database contains employment profiles on more than 1 million individuals and is used nationally by retained, senior-level executive search firms, venture capital funds, and corporate recruiters.

3DSITE
http://3dsite.com/3dsite
Industries: Computer graphics
Search Location: Nationwide

VIRTUAL JOB FAIR
http://www.vjf.com
Industries: All
Search Location: Nationwide
This site offers job searches, a resume center, high-technology careers, career resources, and more. *Virtual Job Fair* allows you to search by job title, industry, location, or company.

WORK ONLINE NETWORK SYSTEMS
http://www.wons.com
An employment database searchable by state, industry, or job.

ZDNET JOB DATABASE
http://www.zdnet.com/~zdi/jobs/jobs.html
Industries: High-Tech
Search Location: Nationwide
Ziff-Davis Interactive and ADION, in partnership with The Monster Board, offer this nationwide job search directory which allows you to search by state, industry, and position.

SOFTWARE DEVELOPERS/PROGRAMMING SERVICES

*O*ne thing is clear: To be a software developer, it is essential to stay current with the latest technology. One Applications Specialist recommends taking courses at a local university, especially if the company you work for will pay for them. The more programming languages a person knows, the greater the range of opportunities that will present themselves.

Demand for experienced developers remains high, and opportunities for broader responsibility certainly exist. For Rich Perkett and Tom Nickel, both former Systems Analysts, programming duties are also expanding into managerial responsibilities. A typical day can include writing and debugging code, testing a product, and coordinating the efforts of other team members.

Competition among software development companies will remain high and the battle for shelf space in retail stores continues. Some developers may be asked to find new sales channels to break through the barrier holding back their products. Programmers will have to continue to add new areas of knowledge to their skill set if they want to succeed.

ACS SOFTWARE
1155 16th NW, Washington DC 20036. 202/872-4363. **Contact:** Human Resources. **Description:** A manufacturer of phone directory disks.

A4 HEALTH SYSTEMS
5501 Dillard Drive, Cary NC 27511. 919/851-6177. **Contact:** Carla Nolan, Human Resources. **Description:** Formerly Management Systems Associates, a4 Health Systems develops software and related hardware for hospitals.

AMS
20 Waterside Drive, Farmington CT 06032. 860/677-5224. **Contact:** Jennifer Warner, Recruiting. **Description:** Develops rating software for insurance agencies.

ASA INTERNATIONAL LTD.
10 Speen Street, Framingham MA 01701. 508/626-2727. **Fax:** 508/626-0645. **Contact:** Human Resources Department. **Description:** ASA designs and develops proprietary vertical market software and installs software on a variety of computers and networks. **Common positions include:** Computer Programmer; Sales Representative; Software Engineer; Technical Support Representative. **Educational backgrounds include:** Business Administration; Computer Science; Engineering. **Benefits:** 401(k); Credit Union; Dental Insurance; Disability Coverage; Life Insurance; Medical Insurance. **Special Programs:** Internships. **Corporate headquarters location:** This Location. **Other U.S. locations:** Nashua NH; Blue Bell PA. **Subsidiaries include:** ASA Legal Systems Company, Inc.; ASA, Inc. **Operations at this facility include:** Administration; Research and Development; Sales; Service. **Listed on:** NASDAQ. **Annual sales/revenues:** $21 - $50 million. **Number of employees at this location:** 84. **Number of employees nationwide:** 187.
Other U.S. locations:
• 615 Amherst Street, Nashua NH 03063-1017. 603/889-8700.

ASA LEGAL SYSTEMS COMPANY, INC.
960-C Harvest Drive, Blue Bell PA 19422. 215/542-2100. **Contact:** Human Resources. **Description:** Develops computer applications catering to the legal industry.

ASAP
1328 East Empire Street, Bloomington IL 61701. 309/662-6100. **Contact:** Human Resources. **Description:** Develops computer software for the insurance industry.

ASD SOFTWARE
4650 Arrow Highway, Montclair CA 91763. 909/624-2594. **Contact:** Human Resources. **Description:** Markets software. ASD Software markets utility programs for Macintosh that are designed by various developers. The company also offers technical support for its products.

AW COMPUTER SYSTEMS, INC.
9000A Commerce Parkway, Mount Laurel NJ 08054. 609/234-3939. **Contact:** Human Resources. **Description:** Develops point-of-sale software.

ABRAXAS SOFTWARE
5530 SW Kelly Avenue, Portland OR 97201. 503/244-5253. **Fax:** 503/244-8375. **Contact:** Human Resources. **E-mail address:** info@abxsoft.com. **World Wide Web address:** http://www.abxsoft.com. **Description:** A developer of software tools including PCYACC, an embedded language development tool; and CodeCheck, an expert system for analyzing programming languages.

ACADIANA COMPUTER SYSTEMS, INC.
P.O. Box 90238, Lafayette LA 70509. 318/981-2494. **Contact:** Human Resources. **Description:** Designs billing software for the health care industry.

ACCENT WORLDWIDE, INC.
1401 Dove Street, Suite 470, Newport Beach CA 92660. **Toll-free phone:** 800/535-5256. **Contact:** Human Resources. **World Wide Web address:** http://www.accentsoft.com. **Description:** A developer of multilingual word processing software, offered in over 30 typing languages. **Corporate headquarters location:** This Location. **Parent company:** Accent Software International, Ltd. (Jerusalem, Israel).

ACCESS SOFTWARE
4750 Wiley Post Way, Building 1, Suite 100, Salt Lake City UT 84116. 801/359-2900. **Contact:** Human Resources. **Description:** A manufacturer of video games and interactive movies.

ACCESS SOLUTIONS INTERNATIONAL INC.
650 Ten Rod Road, North Kingston RI 02852. 401/295-2691. **Contact:** Human Resources. **Description:** Develops optical archiving computer programs.

ACCLAIM ENTERTAINMENT, INC.
One Acclaim Plaza, Glen Cove NY 11542. 516/656-5000. **Contact:** Human Resources Coordinator. **Description:** Acclaim Entertainment, Inc. publishes entertainment software for use with interactive entertainment hardware platforms, develops and publishes comic books, markets motion capture technology and studio services, distributes coin-operated arcade games, distributes of software for affiliated labels, and distributes its interactive entertainment through partnership with Tele-Communications. Acclaim publishes interactive software mainly for the Sega and Nintendo platforms. **Common positions include:** Accountant/Auditor; Advertising Clerk; Economist; Public Relations Specialist; Software Engineer; Systems Analyst. **Educational backgrounds include:** Communications; Computer Science; Engineering; Finance; Liberal Arts; Marketing. **Benefits:** 401(k); Dental Insurance; Disability Coverage; Employee Discounts; Life Insurance; Medical Insurance; Savings Plan. **Corporate headquarters location:** This Location. **Subsidiaries include:** Acclaim Comics of New York City; Iguana Entertainment in Texas; Acclaim Coin-Op in California. **Operations at this facility include:** Administration; Research and Development; Sales. **Listed on:** NASDAQ. **Number of employees at this location:** 300.

ACCUGRAPH CORPORATION
5822 Cromo Drive, El Paso TX 79912. 915/581-1171. **Contact:** Marisela Skorup, Human Resources. **Description:** Develops and supports computer-based software products and manages graphic information, focusing primarily on applications in the areas of telecommunications,

facilities management, and computer network management. Accugraph's products originally served the needs of engineers, architects, planners, and manufacturers for computer-aided design and computer-aided manufacturing software applications. The company produces and markets its proprietary software products for operation on third-party platforms. Accugraph's integrated systems are used by its customers to automate telecommunications applications and provide related services to manage computer networks and for facilities management. The company's largest customers, segmented by application, are Northern Telecom (telecommunications), BellSouth (computer network management) and Hewlett-Packard (worldwide facilities management). **Corporate headquarters location:** This Location.

ACROSCIENCE CORPORATION

1966 13th Street, Suite 250, Boulder CO 80302. 303/541-0089. **Fax:** 303/541-9056. **Contact:** Human Resources. **E-mail address:** info@acroscience.com. **World Wide Web address:** http://www.acroscience.com. **Description:** A developer of software tools for the sciences, acroScience markets Visual Science, a visual programming environment for mathematical system design and analysis.

ADAM SOFTWARE, INC.

1600 River Edge Parkway, Suite 800, Atlanta GA 30328. 770/980-0888. **Fax:** 770/955-3088. **Contact:** Human Resources. **Description:** A developer of anatomical software for medical schools, other educational institutions, and consumers.

ADOBE SYSTEMS, INC.

345 Park Avenue, San Jose CA 95110. 415/961-4400. **Contact:** Human Resources Representative. **World Wide Web address:** http://www.adobe.com. **Description:** Adobe develops, markets, and supports computer software products and technologies for Apple Macintosh, Windows, and OS/2 compatibles that enable users to create, display, print, and communicate electronic documents. The company licenses its technology to major computer and publishing suppliers, and markets software and typeface products for authoring and editing visually rich documents, including digital and video output. The company distributes its products through a network of original equipment manufacturer customers, distributors and dealers, value-added resellers, and systems integrators. The company has operations in the Americas, Europe, and the Pacific Rim. In 1994, the company acquired the Aldus Corporation. Aldus focused on three lines of business: applications for the professional print publishing, graphics, and prepress markets; applications for the general consumer market; and applications for the interactive publishing market. **Common positions include:** Electrical/Electronics Engineer; Graphic Artist; Software Engineer. **Corporate headquarters location:** This Location.

ADOBE SYSTEMS, INC.

411 1st Avenue South, Seattle WA 98104. 206/470-7000. **Contact:** Personnel Department. **World Wide Web address:** http://www.adobe.com. **Description:** This location develops several of the company's software products and provides sales and support services. Overall, Adobe develops, markets, and supports computer software products and technologies for Apple Macintosh, Windows, and OS/2 compatibles that enable users to create, display, print, and communicate electronic documents. The company licenses its technology to major computer and publishing suppliers, and markets software and typeface products for authoring and editing visually rich documents, including digital and video output. The company distributes its products through a network of original equipment manufacturer customers, distributors and dealers, and value-added resellers and systems integrators. The company has operations in the Americas, Europe, and the Pacific Rim. In 1994, the company acquired the Aldus Corporation. Aldus focused on three lines of business: applications for the professional print publishing, graphics, and prepress markets; applications for the general consumer market; and applications for the interactive publishing market. **Common positions include:** Accountant/Auditor; Computer Programmer; Financial Analyst; Software Engineer.

ADVANCED BUSINESSLINK CORPORATION

155 108th Avenue NE, Suite 210, Bellevue WA 98004. 206/455-9804. **Contact:** Human Resources. **Description:** Advanced BusinessLink Corporation develops remote access software.

ADVANCED COMPUTER SOLUTIONS
12675 Danielson Court, Suite 407, Poway CA 92064. 619/748-6800. **Contact:** Human Resources. **Description:** Designs customized computer applications. Advanced Computer Solutions creates specific motherboards and applications for their client base including businesses in various medical and industrial fields.

ADVANCED CONTROL SYSTEMS, INC.
P.O. Box 922548, Norcross GA 30092. 770/446-8854. **Contact:** Human Resources Department. **Description:** Advanced Control Systems, Inc. manufactures software for power companies.

ADVANCED DATA CONCEPTS INC.
1500 Northeast Irving, Suite 310, Portland OR 97232-4207. 503/233-1220. **Contact:** Human Resources. **Description:** Outsources to government contractors. Advanced Data Concepts Inc. develops and provides customized applications including security programs to government agencies and contractors.

ADVANCED MEDIA INC.
80 Orville Drive, Bohemia NY 11716. 516/244-1616. **Contact:** Office Manager. **Description:** Advanced Media Inc. was founded in 1993 to provide professional multimedia development products, services, and proprietary technologies to corporate accounts. Advanced Media publishes and markets a complete line of multimedia presentation, authoring, production, and database tools. The company has developed and markets two software products, Multimedia Studio and Media Master Professional. These products contain video and audio editing programs, morphing systems, and 500 MB of clip art. Advanced Media is one of the leading suppliers of active applications and services. The company has developed applications for Disney, 20th Century Fox, Turner, Columbia/Sony, Genesis, and Hallmark Entertainment.

ADVANCED PRODUCTIONS, INC.
1230 Hempstead Turnpike, Franklin Square NY 11010. 516/328-7000. **Fax:** 516/354-4015. **Contact:** Thomas Mulvey, Product Manager. **E-mail address:** tmulvey@seasoft.com. **World Wide Web address:** http://www.apigames.com. **Description:** Advanced Productions, Inc. is a publisher of video games. Founded in 1982. **NOTE:** Also at this location is Advanced Productions's sister company, Software Engineering of America. **Common positions include:** Multimedia Designer. **Educational backgrounds include:** Art/Design. **Corporate headquarters location:** This Location. **International locations:** Germany. **Listed on:** Privately held. **Annual sales/revenues:** $51 - $100 million. **Number of employees at this location:** 150. **Number of employees worldwide:** 250.

ADVANCED VETERINARY SYSTEMS
3410 Sky Park Boulevard, Eau Claire WI 54701. 715/834-0355. **Contact:** Human Resources. **Description:** Develops and sells software for veterinarians.

ADVANCED VOICE TECHNOLOGIES INC.
HOMEWORK HOTLINE
369 Lexington Avenue, 14th Floor, New York NY 10017. 212/599-2062. **Fax:** 212/697-9236. **Contact:** Jenifer Gould, Office Manager. **Description:** Advanced Voice Technologies Inc. was founded to develop custom hardware and software applications to support the voice processing industry. HOMEWORK HOTLINE was designed to meet the needs of the educational community. Over 400 primary and secondary schools have installed the HOMEWORK HOTLINE, a specialty tailored voice processing system which allows teachers and school administrators to communicate with parents and students on a daily basis.

ADVENT SOFTWARE
301 Brannan Street, Sixth Floor, San Francisco CA 94107. 415/543-7696. **Contact:** Human Resources. **Description:** Develops investment software. Advent Software supplies investment managers, financial planners, and brokerage houses with financial planning and investment applications that run on IBMs or compatibles with Windows.

AGENCY MANAGEMENT SERVICES
700 Longwater Drive, Norwell MA 02061. 617/982-9400. **Contact:** Kathe Donlan, Director of Human Resources. **Description:** Develops and markets software for insurance agents. **Common positions include:** Computer Programmer; Hardware Engineer; Insurance Agent/Broker; Manufacturing Engineer; Software Engineer. **Educational backgrounds include:** Manufacturing Management; Nursing.

AIMTECH
20 Trafalgar Square, Suite 300, Nashua NH 03063. 603/883-0220. **Contact:** Human Resources. **World Wide Web address:** http://www.aimtech.com. **Description:** A developer of cross-platform multimedia authoring systems for corporations, government, and educational institutions.

AIR GAGE CORPORATION
12170 Globe Road, Livonia MI 48150. 313/591-9220. **Contact:** Human Resources. **Description:** Develops and manufactures air gages and statistical process control software (SPCS). Distributes SPCS to manufacturers wishing to monitor, regulate, or collect data on their manufacturing processes. **Corporate headquarters location:** This Location.

ALEXANDER LAN, INC.
100 Perimeter Road, Nashua NH 03063. 603/880-8800. **Fax:** 603/880-8881. **Contact:** Human Resources. **World Wide Web address:** http://www.alexander.com. **Description:** A manufacturer of crash prevention and protection software products.

ALIAS WAVEFRONT
365 Northridge Road, Suite 250, Atlanta GA 30350. 770/393-1881. **Contact:** Human Resources. **Description:** A graphic software development company. **Parent company:** Silicon Graphics makes a family of workstation and server systems that are used by engineers, scientists, and other creative professionals to develop, analyze, and simulate complex, three-dimensional objects. Systems include Onyx, IRIS, Indigo, and Challenge, and all use RISC processors in their circuitry.

ALLEN COMMUNICATIONS
Five Triad Center, 5th Floor, Salt Lake City UT 84180. 801/537-7800. **Contact:** Human Resources. **Description:** Develops software for the multimedia industry.

ALLEN SYSTEMS GROUP INC.
750 11th Street South, Naples FL 34102. 941/263-6700. **Contact:** Human Resources. **Description:** Develops software. Allen Systems Group supplies *Fortune* 1000 companies with system management, file transfer, and help desk software.

ALLTEL HEALTHCARE INFORMATION SERVICES
200 Ashford Center North, Atlanta GA 30338. 404/847-5000. **Contact:** Human Resources. **Description:** Develops software information systems.

ALPHA SOFTWARE CORPORATION
168 Middlesex Turnpike, Burlington MA 01803. 617/229-2924. **Fax:** 617/272-4876. **Contact:** Human Resources. **World Wide Web address:** http://www.alphasoftware.com. **Description:** Develops and markets business software for IBM personal computers and compatibles. The company distributes internationally to Canada and Europe.

ALPHATRONIX, INC.
P.O. Box 13978, Research Triangle Park NC 27709. 919/544-0001. **Fax:** 919/544-4079. **Contact:** Suzanne Jones, Human Resources Administrator. **World Wide Web address:** http://www.alphatronix.com. **Description:** A provider of client/server data storage technology and an Internet quality control software developer offering comprehensive analysis of Websites. Founded in 1988. **Common positions include:** Sales Engineer; Sales Executive; Sales Manager; Sales Representative; Software Engineer. **Educational backgrounds include:** Business Administration; Computer Science; Engineering. **Benefits:** 401(k); Dental Insurance; Disability Coverage; Medical Insurance. **Corporate headquarters location:** This Location. **Number of employees at this location:** 60.

ALTA GROUP
1310 Wakarusa Drive, Lawrence KS 66049. 913/841-1283. **Contact:** Human Resources. **Description:** A computer software developer and supplier of software application tools. **Parent company:** Cadence Design Systems.
Other U.S. locations:
* 555 North Matilda Avenue, Sunnyvale CA 94086. 415/574-5800. Contact: Susan Barrett, Staffing Manager.

ALTIA INC.
5030 Corporate Plaza Drive, Suite 200, Colorado Springs CO 80919. 719/598-4299. **Fax:** 719/598-4392. **Contact:** Human Resources. **E-mail address:** info@altia.com. **World Wide Web address:** http://www.altia.com. **Description:** A provider of feature software which aids engineers and marketing professionals in developing the appearance and behavior of new products. Products include Graphics Editor, Animation Editor, Stimulus Editor, and Control Editor. **Corporate headquarters location:** This Location.

AMERICAN CYBERNETICS
1830 West University Drive, Suite 112, Tempe AZ 85281. 602/968-1945. **Fax:** 602/966-1654. **Contact:** Human Resources. **World Wide Web address:** http://www.amcyber.com. **Description:** A publisher of programmer editing software including Multi-Edit for Windows and DOS; and Evolve, an xBase programming add-on.

AMERICAN HEALTH CARE SOFTWARE
137 Iroquois Avenue, Essex Junction VT 05452. 802/872-3484. **Contact:** Marcia DeRosia, President. **Description:** Develops software for the nursing home industry and home health care agencies.

AMERICAN LASER GAMES
303 San Mateo NE, Suite 204, Albuquerque NM 87108. 505/260-9915. **Contact:** Human Resources Department. **Description:** A developer, manufacturer, and publisher of home entertainment software and video arcade games.

AMERICAN MANAGEMENT SYSTEMS INC.
14033 Denver West Parkway, Golden CO 80401. 303/215-3500. **Contact:** Director of Human Resources. **Description:** This location designs and develops software for the telecommunications industry. American Management Systems helps large organizations solve complex management problems by applying information technology and systems engineering. Industries and markets served include financial service institutions; insurance companies; federal agencies; state and local governments; colleges and universities; telecommunications firms; health care providers; and energy companies.

AMERICAN PHYSICIANS SERVICE GROUP INC. (APS)
1301 Capital of Texas Highway, Suite C300, Austin TX 78746. 512/328-0888. **Contact:** Bill Hayes, Chief Financial Officer. **Description:** American Physicians Service Group provides, through its subsidiaries, computer systems and software packages to medical clinics and medical schools; financial services that include management of malpractice insurance companies and investment services to individuals and institutions; and Spanish-language Yellow Page directories which advertise U.S. businesses and are distributed in major markets in Mexico. **Subsidiaries include:** APS Communications Corporation, 4414 Centerview, Suite 230, San Antonio TX 78228, 210/736-1200; APS Facilities Management, Inc., (located at this address); APS Financial Corporation (located at this address); APS Insurance Services, Inc. (located at this address); APS Systems, Inc. (located at this address); APS Systems, Inc./Regional Office, 10801 Executive Center Drive, Suite 400, Little Rock AR 72211, 501/225-9550; PLE Management, Inc. (located at this address); Prime Medical Services, Inc. (located at this address), 512/328-2892. **Number of employees nationwide:** 130.

AMERICAN SOFTWARE, INC.
470 East Paces Ferry Road, Atlanta GA 30305. 404/264-5599. **Fax:** 404/264-5232. **Contact:** Ryan L. Lenox, Director of Human Resources. **Description:** American Software develops, markets, and

supports integrated supply chain management and financial control systems. The company's multiplatform enterprise software applications are primarily for forecasting and inventory management, purchasing and materials control, and order processing and receivables control. The company also offers consulting and outsourcing services, providing companies with business solutions across a wide range of industries. Supply Chain Management, the primary product of American Software, allows customer orders, inventory levels, purchase orders, and other key information to flow automatically from one business to another, thus shortening cycle times, reducing inventory, enhancing customer service and loyalty, and increasing quality, productivity, profitability, and return on investment. **Common positions include:** Accountant/Auditor; Computer Programmer; Customer Service Representative; Software Engineer; Systems Analyst; Technical Writer/Editor. **Educational backgrounds include:** Accounting; Business Administration; Computer Science; Finance; Liberal Arts; Marketing; Mathematics. **Benefits:** 401(k); Dental Insurance; Disability Coverage; Life Insurance; Medical Insurance; Profit Sharing. **Corporate headquarters location:** This Location. **Operations at this facility include:** Administration; Research and Development; Sales; Service. **Listed on:** NASDAQ. **Annual sales/revenues:** $51 - $100 million. **Number of employees at this location:** 500. **Number of employees nationwide:** 600. **Number of employees worldwide:** 660.
Other U.S. locations:
- 5605 North MacArthur Boulevard, Suite 850, Irving TX 75038. 972/580-8350. This location is a regional sales office.

ANALYSIS & TECHNOLOGY, INC.
P.O. Box 220, North Stonington CT 06359-9801. 860/599-3910. **Contact:** Human Resources Department. **Description:** An engineering firm that develops software and computer systems. **Educational backgrounds include:** Biology; Chemistry; Computer Science; Data Processing; Physics.
Other U.S. locations:
- One Corporate Place, Middletown RI 02842. 401/849-5952. Contact: Sara Smith, Human Resources Representative.

ANALYSIS AND COMPUTER SYSTEMS INC.
One VandeGraaff Drive, Burlington MA 01803. 617/272-8841. **Contact:** Human Resources. **Description:** Offers software development services. Analysis and Computer Systems provides software manufacturers with support services in the Q&A and testing portions of development.

ANASAZI INC.
7500 North Dreamy Draw Drive, Suite 120, Phoenix AZ 85020. 602/870-3330. **Contact:** Human Resources. **Description:** Develops software that runs hotel reservation systems. Anasazi Inc. does business worldwide.

ANN ARBOR COMPUTER
1201 East Ellsworch Road, Ann Arbor MI 48108. 313/973-7875. **Contact:** Human Resources. **Description:** Designs software for automated inventory systems. Ann Arbor Computer produces software packages for industries or businesses utilizing warehouses with large quantities of stocked items. The company's products aid in inventory control.

ANSYS
201 Johnson Road, Houston PA 15342-1300. 412/746-3304. **Contact:** Dana DiTillo, Human Resources. **Description:** Develops finite element analysis software for engineering firms.

AONIX
595 Market Street, 10th Floor, San Francisco CA 94105. 415/543-0900. **Contact:** Human Resources. **Description:** Develops and markets computer-aided software engineering (CASE) products which allow a network of minicomputers to interact, enabling fully-integrated software environments to exist.

APERTUS TECHNOLOGIES INC.
7275 Flying Cloud Drive, Eden Prairie MN 55344. 612/828-0300. **Toll-free phone:** 800/328-3998. **Contact:** Lori Cocking, Director Of Human Resources. **World Wide Web address:**

http://www.apertus.com. **Description:** A multi-product software development company that provides organizations with cost-effective network integration solutions. The company's products include network gateways, communications software, data integration tools, and system management products. Founded in 1979. **Common positions include:** Computer Programmer; Software Engineer; Technical Writer/Editor. **Educational backgrounds include:** Computer Science. **Benefits:** 401(k); Dental Insurance; Disability Coverage; Life Insurance; Medical Insurance; Tuition Assistance. **Other U.S. locations:** El Toro CA; Atlanta GA; New York NY; Dallas TX. **International locations:** Germany; U.K. **Listed on:** NASDAQ. **Annual sales/revenues:** $21 - $50 million. **Number of employees at this location:** 165.

APIARY, INC.
500 East Markham, Suite 310, Little Rock AR 72201. 501/376-3600. **Fax:** 501/376-3666. **Contact:** Human Resources. **E-mail address:** gdowdy@apiary.com. **World Wide Web address:** http://www.apiary.com. **Description:** Develops component software and development tools. **Corporate headquarters location:** This Location.

APPLICON, INC.
4251 Plymouth Road, Suite 3200, Ann Arbor MI 48105. 313/995-6000. **Contact:** Technical Recruiter. **Description:** A software manufacturer. **NOTE:** Jobseekers should address inquiries to 6260 Lookout Road, Boulder CO 80301.

APPLIED IMAGING
2380 Walsh Avenue, Santa Clara CA 95051. 408/562-0250. **Contact:** Human Resources. **Description:** Applied Imaging develops genetic software used by hospitals and universities in detecting genetic birth defects and then provides it in bundled computer packages.

APPLIED MICROSYSTEMS CORPORATION
P.O. Box 97002, Redmond WA 98073-9702. 206/882-2000. **Contact:** Human Resources. **Description:** A manufacturer and supplier of high-quality software development tools and in-circuit emulators for embedded system development.

APPLIED QUOTING SYSTEMS INC.
625 Walnut Ridge Drive, Hartland WI 53029. 414/367-5495. **Contact:** Human Resources. **Description:** Develops software for the insurance industry.

APPLIED STATISTICS, INC.
2055 White Bear Avenue, St. Paul MN 55109. 612/481-0202. **Fax:** 612/481-0410. **Contact:** Personnel Manager. **Description:** Develops software for the Windows operating system utilizing Microsoft Visual C++. **Common positions include:** Administrative Assistant; Administrative Services Manager; Computer Programmer; Sales Engineer; Sales Manager; Software Engineer. **Educational backgrounds include:** Computer Science; Engineering. **Benefits:** 401(k); Dental Insurance; Medical Insurance; Tuition Assistance. **Corporate headquarters location:** This Location. **Listed on:** Privately held. **Annual sales/revenues:** $5 - $10 million. **Number of employees at this location:** 48.

APPLIED TECHNOLOGY ASSOCIATES
1320 Villa Street, Mountain View CA 94041. 415/965-7774. **Contact:** Human Resources. **Description:** A software developer, primarily under contract to the government. **Corporate headquarters location:** This Location.

APPLIX, INC.
112 Turnpike Road, Westborough MA 01581. 508/870-0300. **Contact:** Human Resources Department. **Description:** The company develops and markets software applications for the UNIX market. **Corporate headquarters location:** This Location.

AQUIDNECK MANAGEMENT ASSOCIATES LTD.
28 Jacome Way, Middletown RI 02842. 401/849-8900. **Contact:** Human Resources. **Description:** Develops software for government. Aquidneck Management Associates supplies local and state governments with computer applications designed to automate the court systems.

ARGENT SOFTWARE
49 Main Street, Torrington CT 06790. 860/489-5553. **Contact:** Human Resources Department. **Description:** Develops and markets systems software products.

ARKANSAS SYSTEMS INC.
17500 Chenal Parkway, Little Rock AR 72211. 501/227-8471. **Contact:** Human Resources. **Description:** Develops software for the banking industry.

ASPEN TECHNOLOGY, INC.
Ten Canal Park, Cambridge MA 02141. 617/577-0100. **Fax:** 617/577-0303. **Contact:** Human Resources. **Description:** Aspen Technology, Inc. supplies computer-aided chemical engineering software products to the process manufacturing industries, which include the chemicals, petroleum, pharmaceuticals, metals and minerals, food and consumer products, and electric power industries. This software enables chemical engineers to create mathematical models of manufacturing processes and to predict the performance of these processes under varying equipment configurations and operating conditions. Aspen Technology's modeling, simulation, and optimization software assists chemical engineers in designing more efficient and cost-effective production processes. This software also enables chemical engineers to adjust the design of production processes to comply with environmental and safety requirements. **Benefits:** 401(k). **Corporate headquarters location:** This Location. **International locations:** Belgium; England; Hong Kong; Japan. **Listed on:** NASDAQ. **Number of employees nationwide:** 300. **Other U.S. locations:**
- 14701 Saint Mary's Lane, Houston TX 77079. 281/584-1000. This location develops software.
- 2000 Technology Drive, Fairmont WV 26554. 304/368-1111. This location offers sales and testing services.

ASTROBYTE
625 East 16th Avenue, Suite 200, Denver CO 80203. 303/861-4861. **Fax:** 303/861-4876. **Contact:** Personnel Department. **E-mail address:** personnel@astrobyte.com. **World Wide Web address:** http://www.astrobyte.com. **Description:** A developer of Internet publishing tools and software. **Common positions include:** Software Engineer. **Corporate headquarters location:** This Location.

ATARI CORPORATION
455 South Matilda Avenue, Sunnyvale CA 94086. 408/328-0900. **Contact:** Personnel Director. **Description:** This location is an administrative and sales office for the video systems company. Atari manufactures and distributes video systems and video games.

ATARI GAMES CORPORATION
675 Sycamore Drive, Milpitas CA 95035. 408/434-3700. **Contact:** Human Resources. **Description:** A developer of video game software. This location is a research and development facility. **Corporate headquarters location:** This Location.

ATEX PUBLISHING SYSTEMS CORPORATION
15 Crosby Drive, Bedford MA 01730. 617/275-2323. **Fax:** 617/276-1254. **Contact:** Human Resources. **Description:** Designs, develops, and sells computer software products for the newspaper, magazine, and prepress publishing markets worldwide. Approximately one-third of the company's revenues are derived from the U.S. market, and the balance is from international sales. **Common positions include:** Accountant/Auditor; Administrative Services Manager; Blue-Collar Worker Supervisor; Budget Analyst; Buyer; Computer Programmer; Credit Manager; Customer Service Representative; Editor; Electrical/Electronics Engineer; Financial Analyst; General Manager; Human Resources Manager; Public Relations Specialist; Purchasing Agent/Manager; Quality Control Supervisor; Services Sales Representative; Software Engineer; Systems Analyst; Technical Writer/Editor. **Educational backgrounds include:** Accounting; Art/Design; Business Administration; Communications; Computer Science; Engineering; Finance; Liberal Arts; Marketing. **Benefits:** 401(k); Daycare Assistance; Dental Insurance; Disability Coverage; Employee Discounts; Life Insurance; Medical Insurance; Pension Plan; Profit Sharing; Savings Plan; Tuition Assistance. **Corporate headquarters location:** This Location. **Other U.S. locations:** Nationwide. **Parent company:** ATEX Holding International, B.V. **Operations at this facility**

include: Administration; Research and Development; Sales; Service. **Listed on:** Privately held. **Number of employees at this location:** 240. **Number of employees nationwide:** 400.

ATLAS BUSINESS SOLUTIONS, INC.

3330 Fiechtner Drive SW, Fargo ND 58103. 701/235-5226. **Fax:** 701/280-0842. **Contact:** Human Resources. **World Wide Web address:** http://www.atlasbsi.com. **Description:** A software developer and computer services firm. Product line includes Visual Staff Scheduler software, an employee scheduling system. **Corporate headquarters location:** This Location.

ATTACHMATE CORPORATION

8230 Montgomery Road, Cincinnati OH 45236. 513/745-0500. **Fax:** 513/794-8108. **Recorded Jobline:** 513/794-8555. **Contact:** Human Resources. **Description:** Develops and supports computer software.

AUDRE RECOGNITION SYSTEMS INC.

11021 Via Frontera, San Diego CA 92127-1789. 619/451-2260. **Fax:** 619/451-0267. **Contact:** Human Resources. **Description:** AUDRE develops and markets software and related systems that allow the conversion of hard copy documents into computer-accessed information. The AUDRE system transforms information resident on paper, microfilm, or as existing electronic images into smart data usable by such application systems as Computer-Aided Design, Computer-Aided Manufacturing, Database Retrieval Systems, Technical Publication Systems, and Geographic Information Systems, among others. AUDRE software enables a computer to recognize information from raster images of documents and to deliver that information to the end-user in vector form. Through the use of a document scanner, a raster image is created as a computer file. Once the image is in this format, AUDRE offers three packages of technologies and tools to address the customer's needs: the Raster Master, the Visual Vectorizer, and the Automatic Choice. **Corporate headquarters location:** This Location. **Number of employees at this location:** 60.

AURUM SOFTWARE, INC.

3385 Scott Boulevard, Santa Clara CA 95148. 408/986-8100. **Fax:** 408/654-3515. **Contact:** George Greeley, Recruiter. **Description:** A developer of software for sales force automation. **NOTE:** The company offers entry-level positions. **Common positions include:** Administrative Assistant; Applications Engineer; Attorney; Chief Financial Officer; Database Manager; Education Administrator; Finance Director; Financial Analyst; Human Resources Manager; Marketing Manager; Marketing Specialist; MIS Specialist; Project Manager; Sales Engineer; Sales Representative; Secretary; Software Engineer; Systems Manager; Teacher/Professor; Technical Writer/Editor; Webmaster. **Educational backgrounds include:** Accounting; Computer Science; Engineering; Finance; Marketing. **Benefits:** 401(k); Dental Insurance; Disability Coverage; Employee Discounts; Medical Insurance. **Listed on:** NASDAQ. **Annual sales/revenues:** $21 - $50 million. **Number of employees at this location:** 90. **Number of projected hires for 1997 - 1998 at this location:** 70.

AUTHORIZATION SYSTEMS INC.

1801 Rockville Pike, Suite 216, Rockville MD 20852. 301/770-1625. **Contact:** Human Resources. **Description:** Develops software and provides systems analysis for the government.

AUTO F/X CORPORATION

Black Point HCR 73, Box 689, Alton Bay NH 03810. 603/875-4400. **Fax:** 603/875-4404. **Contact:** Human Resources. **World Wide Web address:** http://www.autofx.com. **Description:** A developer of automated and pre-created effects and design tools for the graphic arts field.

AUTO-GRAPHICS, INC.

3201 Temple Avenue, Pomona CA 91768-3200. 909/595-7204. **Fax:** 909/595-3506. **Contact:** Human Resources. **Description:** Auto-Graphics, Inc. provides software and processing services to database and information publishers. The company's services include the computerized preparation and processing of customer-supplied information to be published in various formats including print, microform, CD-ROM, and/or online computer access. The company's customers include book, encyclopedia, dictionary, and price catalog publishing companies, and libraries across the country which publish their bibliographic information for staff and patrons in computer-generated catalogs, microform, and through CD-ROM. In addition to providing such computerized data management and typesetting services, the company markets a CD-ROM hardware and software package for access to computer generated information. **Corporate headquarters location:** This Location. **Number of employees at this location:** 115.

AUTO-TROL TECHNOLOGY CORPORATION

12500 North Washington Street, Denver CO 80241-2400. **Toll-free phone:** 800/233-2882. **Fax:** 303/252-2249. **Recorded Jobline:** 303/252-2007. **Contact:** Lisa Jayne, Manager of Human Resources. **Description:** Auto-Trol Technology is a software design and engineering firm. The company develops and markets leading-edge software in the CAD/CAM/CAE, technical illustration, network configuration, and technical information management industries. Auto-Trol Technology integrates computer hardware, operating systems, proprietary graphics software, and applications software into systems for process plant design, civil engineering, discrete manufacturing, facilities layout and design, mechanical design, technical publishing, and network configuration management. The company also offers support services. **NOTE:** Entry-level positions are offered. **Common positions include:** Computer Programmer; Mathematician; Mechanical Engineer; Software Engineer; Systems Analyst; Technical Writer/Editor. **Educational backgrounds include:** Computer Science; Engineering; Mathematics. **Benefits:** 401(k); Credit Union; Dental Insurance; Disability Coverage; Life Insurance; Medical Insurance; Paid Vacation; Tuition Assistance. **Special Programs:** Internships. **Corporate headquarters location:** This Location. **Other U.S. locations:** Nationwide. **Operations at this facility include:** Administration; Marketing; Research and Development; Sales. **Listed on:** NASDAQ. **Annual sales/revenues:** $5 - $10 million. **Number of employees at this location:** 200.

AUTODESK, INC.

111 McInnis Parkway, San Rafael CA 94903. 415/507-5000. **Contact:** Steve McMahon, Vice President/Human Resources. **Description:** Designs, develops, markets, and supports a line of computer-aided design (CAD), engineering, and animation software products for desktop computers and workstations. **Common positions include:** Computer Programmer; Quality Control Supervisor; Systems Analyst. **Educational backgrounds include:** Computer Science. **Benefits:** Dental Insurance; Disability Coverage; Life Insurance; Medical Insurance; Pension Plan; Tuition Assistance. **Corporate headquarters location:** This Location. **Operations at this facility include:** Administration; Divisional Headquarters; Manufacturing; Regional Headquarters; Research and Development; Sales; Service. **Number of employees nationwide:** 1,800.

AUTOMATION SERVICES INC.

2549 Richmond Road, Suite 400, Lexington KY 40509. 606/266-3000. **Contact:** Human Resources. **Description:** Develops software. Automation Services supplies rock quarries and cement companies with software for plant automation.

THE AVALON HILL GAME COMPANY

4517 Harford Road, Baltimore MD 21214. 410/254-9200. **Contact:** Eric Dott, President. **Description:** A developer of war games on CD-ROM for both Macintosh and IBM platforms. The Avalon Hill Game Company also manufactures board versions of its computer games.

AVANT!

1101 Slater Road, Durham NC 27705. 919/941-6600. **Contact:** Director of Human Resources. **Description:** Avant! develops, markets, and supports software products that assist IC designing engineers in performing automated design, layout, physical verification, and analysis of advanced integrated circuits. The company's proprietary hierarchical physical verification software enables design engineers to improve time to market, reduce design risk, and obtain complete and fast verification of integrated circuits. The company's VeriCheck family of design verification software was introduced in 1992 and includes hierarchical products addressing geometric and electrical verification and parasitic extraction for resistance and capacitance analysis. The company also produces a family layout products compatible with UNIX or MS-DOS operating systems and run on platforms including DEC, HP, IBM, and Sun workstations and on 386/486-based PCs. **Corporate headquarters location:** This Location. **Listed on:** NASDAQ. **Number of employees at this location:** 100.

AVANTOS PERFORMANCE SYSTEMS, INC.
5900 Hollis Street, Suite A, Emeryville CA 94608. 510/654-4600. **Fax:** 510/654-5199. **Contact:** Human Resources. **World Wide Web address:** http://www.avantos.com. **Description:** A leading developer of business management software designed to increase productivity. Products include ManagePro, an award-winning software tool; and Review Writer, used for the writing of performance appraisals.

AVID TECHNOLOGY, INC.
Metropolitan Technology Park, One Park West, Tewksbury MA 01876. 508/640-6789. **Contact:** Director of Employment. **Description:** Founded in 1987, Avid develops and markets Apple Macintosh software, specializing in programs for video and audio editing. The software works in conjunction with a hardware digital editing system. Avid markets to the TV and film industries, and general business markets. Distribution includes Canada, Europe, South America, Australia, and the Pacific Rim. **Common positions include:** Computer Programmer; Computer Scientist; Hardware Engineer; Software Engineer; Systems Analyst.

AXENT TECHNOLOGIES
2155 North Freedom Boulevard, Suite 200, Provo UT 84604. 801/224-5306. **Fax:** 801/227-3787. **Contact:** Human Resources. **Description:** Axent Technologies develops software that secures computer passwords and personal computer files.

AXENT TECHNOLOGIES
2400 Research Boulevard, Suite 200, Rockville MD 20850. 301/258-5043. **Contact:** Human Resources. **Description:** Axent Technologies develops software that secures computer passwords and personal computer files.

AXIS COMPUTER SYSTEMS
RK Executive Center, 201 Boston Post Road West, Marlborough MA 01752. 508/481-9600. **Contact:** Human Resources. **Description:** Develops software for manufacturing facilities in the metal industry.

BBN CORPORATION
150 Cambridge Park Drive, Cambridge MA 02140. 617/873-2000. **Contact:** Manager of Human Resources. **Description:** A diversified high-technology company that develops and markets statistical data analysis software used in a variety of fields, including research and development, quality control and manufacturing, and clinical data management. Products include statistical advisor systems for the design of experiments, time series, data analysis and graphics, clinical trials management, and quality control via statistical process control. **Common positions include:** Accountant/Auditor; Financial Analyst; Hardware Engineer; Software Engineer. **Corporate headquarters location:** This Location. **Subsidiaries include:** BBN Communications Corporation, a leading supplier of private, wide area network systems; BBN Software Products Corporation, a developer, marketer, and distributor of integrated statistical data analysis software products sold primarily for applications in manufacturing, engineering, and research and development; BBN Advanced Computers, Inc., a developer and marketer of high-performance applications in industry and research; and BBN Systems and Technology Corporation, engaged in acoustics and computer information, specializing in architectural acoustics and environmental technologies, physical sciences, and graphic technology.

BGS SYSTEMS INC.
One First Avenue, Waltham MA 02254. 617/891-0000. **Contact:** Recruiting Department. **Description:** BGS Systems Inc. manufactures, sells, and supports software used for MIS and business productivity. The integrated software is written for IBM, IBM compatible VAX/VMS mainframes and RS/6000 workstations, new applications, and network performance management. BGS also consults for capacity planning and related issues. The company distributes its products across North America, South America, Europe, the Middle East, and the Pacific Rim. **Common positions include:** Computer Programmer; Systems Analyst. **Number of employees at this location:** 180.

BMC SOFTWARE, INC.

2101 Citywest Boulevard, Houston TX 77042-2827. 713/918-8800. **Contact:** Mrs. Johnny Horn, Director of Human Resources. **Description:** BMC Software, Inc. develops, markets, and supports standard systems software products to enhance and increase the performance of large-scale (mainframe) computer database management systems and data communications software systems. **Common positions include:** Computer Programmer; Systems Analyst. **Corporate headquarters location:** This Location. **Number of employees nationwide:** 900.

BMC SOFTWARE, INC.

1190 Saratoga Avenue, Suite 130, San Jose CA 95129. 408/556-0720. **Contact:** Human Resources. **Description:** BMC Software, Inc. develops, markets, and supports standard systems software products to enhance and increase the performance of large-scale (mainframe) computer database management systems and data communications software systems. This location formerly operated under the name Peer Networks Inc. prior to being purchased by BMC Software. **NOTE:** Interested jobseekers should address inquiries to Mrs. Johnny Horn, Director of Human Resources, BMC Software, Inc., 2101 Citywest Boulevard, Houston TX 77042-2827. **Corporate headquarters location:** Houston TX. **Number of employees nationwide:** 900.

BANC SOFTWARE SERVICES

981 East North Tent, Abilene TX 79601. 915/672-1363. **Contact:** Human Resources. **Description:** Designs software. BANC Software Services designs, programs, and manages software packages specifically adapted for banks.

BANKERS SYSTEMS INC.

6815 Sauckview Drive, St. Cloud MN 56303. 320/251-3060. **Contact:** Human Resources Department. **Description:** Bankers Systems Inc. develops software that is used by the banking industry.

BARRA, INC.

1995 University Avenue, Suite 400, Berkeley CA 94704. 510/548-5442. **Contact:** Human Resources Department. **Description:** Barra, Inc. develops, markets, and supports application software and information services used to analyze and manage portfolios of equity, fixed income, and other financial instruments. The company serves more than 750 clients in 30 countries, including many of the world's largest portfolio managers, fund sponsors, pension and investment consultants, brokers/dealers, and master trustees. **Common positions include:** Customer Service Representative; Economist; Electrical/Electronics Engineer; Human Resources Manager; Mathematician; Payroll Clerk; Quality Control Supervisor; Receptionist; Services Sales Representative; Software Engineer; Statistician; Systems Analyst. **Educational backgrounds include:** Computer Science; Economics; Engineering; Finance; Mathematics; Physics. **Benefits:** Daycare Assistance; Dental Insurance; Disability Coverage; Employee Discounts; Home Loan Assistance; Life Insurance; Medical Insurance; Pension Plan; Profit Sharing; Savings Plan; Tuition Assistance. **Corporate headquarters location:** This Location. **Other U.S. locations:** NY. **Operations at this facility include:** Administration; Regional Headquarters; Research and Development; Sales; Service. **Number of employees at this location:** 260. **Number of employees nationwide:** 300.

BAYSTATE TECHNOLOGIES

33 Boston Post Road West, Marlborough MA 01752. 508/229-2020. **Contact:** Human Resources. **E-mail address:** webmaster@baystate.com. **World Wide Web address:** http://www.cadkey.com. **Description:** A developer of mechanical software for CAD systems. **Corporate headquarters location:** This Location.

BENDATA

1125 Kelly Johnson Boulevard, Suite 100, Colorado Springs CO 80920. 719/531-5007. **Fax:** 719/536-0620. **Contact:** Human Resources Department. **E-mail address:** postmaster@bendata.com. **World Wide Web address:** http://www.bendata.com. **Description:** Bendata is a provider of software and services for support center markets such as help desks, customer service, and MIS/IS departments.

BENEFIT TECHNOLOGY INC.
2701 South Bayshore Drive, Miami FL 33133. 305/285-6900. **Contact:** Human Resources. **Description:** Develops software for business.

BENTLEY SYSTEMS INC.
690 Pennsylvania Drive, Exton PA 19341. 610/458-5000. **Contact:** Human Resources. **Description:** Develops software. Bentley Systems supplies universities, students, and engineers with CAD software.

AM BEST COMPANY
AM Best Road, Oldwick NJ 08858. 908/439-2200. **Fax:** 908/439-3296. **Contact:** Human Resources. **World Wide Web address:** http://www.ambest.com. **Description:** A database products firm, offering software, CD-ROM, and diskette support products to the insurance industry. **Corporate headquarters location:** This Location.

BIRMY GRAPHICS CORPORATION
255 East Drive, Suite H, Melbourne FL 32904. 407/768-6766. **Contact:** Human Resources. **Description:** Develops software for printers.

BITSTREAM, INC.
215 First Street, Cambridge MA 02142. 617/497-6222. **Contact:** Human Resources. **Description:** Develops and markets digital fonts for original equipment manufacturers and end-users. Products include fonts used in graphic arts image setters, printers, and screen displays. Retail products are font packages for word processing and desktop publishing software. Products are used on both personal computer and Macintosh platforms. Founded in 1981, the company distributes its products throughout North America, Europe, and the Pacific Rim.

BLACK BELT SYSTEMS
398 Johnson Road, Building 2, Glasgow MT 59230. 406/367-5509. **Fax:** 406/367-2329. **Contact:** Human Resources. **E-mail address:** info@blackbelt.com. **World Wide Web address:** http://www.blackbelt.com. **Description:** A developer of graphics software. **Corporate headquarters location:** This Location.

BLACKBAUD INC.
4401 Belle Oaks Drive, Charleston SC 29405-8530. 803/740-5400. **Contact:** Personnel Coordinator. **Description:** Blackbaud Inc. offers computer programming services for all types of nonprofit companies. Blackbaud Inc.'s software is designed to help companies with a wide variety of activities including fundraising, administration, and organization.

BLAST SOFTWARE
P.O. Box 808, Pittsboro NC 27312. 919/542-3007. **Toll-free phone:** 800/24-BLAST. **Fax:** 919/542-0161. **Contact:** Human Resources. **E-mail address:** melinda@blast.com. **World Wide Web address:** http://www.blast.com. **Description:** BLAST (BLocked ASynchronous Transmission) develops communications software which offers cross-platform file transfer, reliable terminal emulation, and error correction. The software is designed for rugged use and features a sliding window design which, if interrupted, will continue transmission from the point of interruption rather than beginning again, ensuring a constant stream of data resulting in significant speed improvements. Founded in 1979. **NOTE:** The company's physical address is 49 Salisbury Street West, Pittsboro NC 27312.

BLUE MARBLE GEOGRAPHICS
261 Water Street, Gardiner ME 04345. 207/582-6747. **Fax:** 207/582-7001. **Contact:** Human Resources. **World Wide Web address:** http://www.bluemarblegeo.com. **Description:** A developer of mapping software. **Common positions include:** Technical Support Representative. **Corporate headquarters location:** This Location.

BLUEBIRD AUTO RENTAL SYSTEMS INC.
9 Sylvan Way, Parsippany NJ 07054. 201/984-1014. **Contact:** Human Resources. **Description:** Designs computer applications for auto rental agencies.

BLUESTONE CONSULTING INC.

1000 Briggs Road, Mount Laurel NJ 08054. 609/727-4600. **Contact:** Human Resources Department. **Description:** Develops computer technology programs and provides user training. Bluestone Consulting offers Web-based software including dynamic Web applications and GUIs. This company also offers training programs in various states around the nation.

BLYTH SOFTWARE

989 East Hillsdale Boulevard, Suite 400, Foster City CA 94404. 415/571-0222. **Contact:** Human Resources Department. **Description:** Blyth Software develops, markets, and supports software products for the development and deployment of applications for accessing multi-user databases in workgroup and enterprisewide client/server computing environments. The company's OMNIS family of products is used by corporations, system integrators, small businesses, and independent consultants to deliver custom information management applications for a wide range of uses including financial management, decision support, executive information, sales and marketing, and multimedia authoring systems. In addition to these products, Blyth provides consulting, technical support, and training to help plan, analyze, implement, and maintain applications software based on the company's technology. **Corporate headquarters location:** This Location.

BOKLER SOFTWARE CORPORATION

P.O. Box 261, Huntsville AL 35804. 205/539-9901. **Fax:** 205/882-7401. **Contact:** Human Resources Department. **E-mail address:** info@bokler.com. **World Wide Web address:** http://www.bokler.com. **Description:** Bokler Software Corporation is a developer of cryptographic software components for developers. **Corporate headquarters location:** This Location.

BOOLE & BABBAGE, INC.

3131 Zanker Road, San Jose CA 95134. 408/526-3000. **Contact:** Human Resources. **Description:** Develops automation software. **Common positions include:** Computer Programmer; Software Engineer.

BORLAND INTERNATIONAL

100 Borland Way, Scotts Valley CA 95066. 408/431-1000. **Contact:** Human Resources. **Description:** Borland International designs and markets computer software products for both businesses and software developers. **Common positions include:** Customer Service Representative; Manager of Information Systems; Public Relations Specialist; Quality Control Supervisor; Software Engineer; Technical Writer/Editor; Technician. **Number of employees nationwide:** 1,900.

BOTTLER SYSTEM INC.

8710 University Drive, Suite 310, Durham NC 27707. 919/490-8440. **Contact:** Human Resources Department. **Description:** Bottler System Inc. develops software for Coca-Cola bottling companies.

BOTTOMLINE TECHNOLOGIES

155 Fleet Street, Portsmouth NH 03801. 603/436-0700. **Contact:** Lisa Kolosey, Human Resources Manager. **Description:** Designs and manufactures software that allows users to print checks from a laser printer.

BROADWAY & SEYMOUR, INC.

128 South Tryon Street, Charlotte NC 28202-5050. 704/372-4281. **Contact:** Mark Hazzard, Recruiting Manager. **Description:** Develops software for banks and other financial institutions and offers consulting services. **Common positions include:** Computer Engineer; Computer Programmer; Computer Scientist; Customer Service Representative; Software Engineer; Systems Analyst; Technical Writer/Editor. **Educational backgrounds include:** Computer Science. **Benefits:** 401(k); Dental Insurance; Disability Coverage; Employee Discounts; Life Insurance; Medical Insurance; Stock Option; Tuition Assistance. **Corporate headquarters location:** This Location. **Other U.S. locations:** Boston MA; St. Paul MN; Raleigh NC; Worthington OH; Houston TX. **Operations at this facility include:** Administration; Research and Development; Sales; Service.

BRODERBUND SOFTWARE, INC.
500 Redwood Boulevard, Novato CA 94948. 415/382-4400. **Contact:** Human Resources. **Description:** Broderbund develops, publishes, and markets personal computer software for the home, school, and small business markets. Products are offered primarily in two consumer software categories: personal productivity and education. Since its founding in 1980, the company has sold more than 15 million units of consumer software. **Common positions include:** Computer Engineer; Computer Programmer; Design Engineer; Software Engineer. **Number of employees at this location:** 340.

BROWNS INTERACTIVE TECHNOLOGY
P.O. Box 9111, Cedar Rapids IA 52409. 319/396-3130. **Contact:** Human Resources. **Description:** Develops software. Browns Interactive Technology primarily creates interactive software for government agencies.

BUSINESS MANAGEMENT DATA, INC. (BMD)
125 Pacifica, Suite 220, Irvine CA 92618. 714/789-9100. **Contact:** Human Resources Department. **E-mail address:** resume@bmdinc.com. **Description:** Business Management Data is a software development and consulting firm providing system and software applications to the transportation, finance, health care, and insurance industries. BMD has client locations nationwide. **Common positions include:** Accountant/Auditor; Administrative Assistant; Computer Programmer; Human Resources Manager; Software Engineer; Systems Analyst. **Educational backgrounds include:** Business Administration; Computer Science; Engineering; Marketing; Mathematics. **Benefits:** Employee Discounts; Life Insurance; Medical Insurance; Profit Sharing. **Corporate headquarters location:** This Location. **Subsidiaries include:** Sriven Computer; Solutions, Ltd. (India). **Operations at this facility include:** Administration; Research and Development; Sales; Service. **Listed on:** Privately held. **Number of employees at this location:** 45. **Number of employees nationwide:** 70.

BUSINESS RECORDS CORPORATION
2901 3rd Street South, Waite Park MN 56387. 612/221-0239. **Contact:** Human Resources. **Description:** Business Records Corporation develops payroll and assessments software for governments.

BUTLER & CURLESS ASSOCIATES
7610 Falls of the News, Raleigh NC 27615. 919/847-1811. **Contact:** Human Resources Department. **Description:** Butler & Curless Associates develops and sells distribution software.

CBIS
285 International Parkway, Lake Mary FL 32746. 407/661-8000. **Contact:** Staffing Manager. **Description:** CBIS is a computer software company specializing in the telecommunications industry. **Common positions include:** Computer Programmer; Systems Analyst. **Educational backgrounds include:** Computer Science. **Benefits:** Dental Insurance; Disability Coverage; Employee Discounts; Life Insurance; Medical Insurance; Pension Plan; Tuition Assistance. **Special Programs:** Internships. **Corporate headquarters location:** Cincinnati OH. **Parent company:** Cincinnati Bell, Inc. **Operations at this facility include:** Regional Headquarters; Research and Development; Sales; Service. **Listed on:** New York Stock Exchange.

C.C.C. (COMPREHENSIVE COMPUTER CONSULTING)
7000 Central Parkway, Suite 1000, Atlanta GA 30328. 770/512-0100. **Contact:** Human Resources Department. **Description:** A technical recruiting company which provides businesses with computer programmers for network analysis and support. **Corporate headquarters location:** This Location.

CCH INC.
21250 Hawthorne Boulevard, Torrance CA 90503. 310/543-6200. **Fax:** 310/543-6544. **Contact:** Cyndi Andreu, Recruiter. **World Wide Web address:** http://www.cch.com. **Description:** One of the nation's largest developers of income tax processing software for the tax profession. The company markets its software products to tax attorneys, tax accountants, and CPAs. **Common**

positions include: Computer Programmer; Customer Service Representative; Systems Analyst; Tax Specialist; Technical Support Representative; Technical Writer/Editor. **Educational backgrounds include:** Accounting; Computer Science. **Benefits:** 401(k); Dental Insurance; Disability Coverage; Life Insurance; Medical Insurance; Pension Plan; Profit Sharing; Tuition Assistance. **Corporate headquarters location:** Riverwoods IL. **Parent company:** Wolters Kluwer. **Number of employees at this location:** 400.

CE SOFTWARE INC.
1801 Industrial Circle, West Des Moines IA 50265. 515/221-1801. **Contact:** Human Resources. **Description:** Develops and sells a wide variety of software products.

CFI PROSERVICES, INC.
400 SW Sixth Avenue, Suite 200, Portland OR 97204. 503/274-7280. **Fax:** 503/790-9293. **Contact:** Human Resources. **Description:** CFI ProServices, Inc. is a leading provider of software-based compliance products to the financial services industry. The company combines its software, banking, and legal expertise to deliver knowledge-based software that documents loan and deposit transactions in compliance with federal and state laws and customers' internal policies. CFI relies upon a nationwide network of independent legal counsels to ensure the accuracy of the compliance information built into its software solutions. The company's products are used by more than 2,500 customers in all 50 states, the District of Columbia, and Guam. The changing regulatory environment affords the company the opportunity to develop new products or acquire complementary products that respond to market needs. Due to the complexity of the banking and regulatory issues addressed by the products, CFI sells them through an experienced direct sales force located throughout the country. **Corporate headquarters location:** This Location. **Number of employees at this location:** 200.

CIS CORPORATION
Penn Center West, Building 2, Pittsburgh PA 15276. 412/787-9600. **Contact:** Human Resources. **Description:** Develops systems integration software for a variety of businesses.

CIS TECHNOLOGIES, INC.
One Warren Place, 6100 South Yale Avenue, Suite 1900, Tulsa OK 74136. 918/496-2451. **Contact:** Human Resources. **Description:** Develops software used by hospitals and other health care facilities, mainly for electronic billing.

CLR PROFESSIONAL SOFTWARE
2400 Lake Park Drive, Suite 180, Smyrna GA 30080-7608. 770/432-1996. **Contact:** Human Resources. **Description:** CLR Professional Software is a designer of tax preparation software. **Corporate headquarters location:** Dallas TX.

CMI-COMPETITIVE SOLUTIONS INC.
3940 Peninsular Drive SE, Suite 100, Grand Rapids MI 49546. 616/957-4444. **Contact:** Human Resources. **Description:** Develops software for the automotive industry.

CMS/DATA CORPORATION
101 North Monroe Street, Suite 800, Tallahassee FL 32301. 904/878-5155. **Contact:** Human Resources. **Description:** Develops financial software.

CPA SOFTWARE
One Pensacola Plaza, Suite 500, Pensacola FL 32501. 904/434-2685. **Contact:** Administrative Manager. **Description:** CPA Software sells professional software to certified public accountants. **Common positions include:** Computer Programmer; Customer Service Representative; Marketing Specialist; Public Relations Specialist; Purchasing Agent/Manager; Services Sales Representative; Systems Analyst; Technical Writer/Editor. **Educational backgrounds include:** Accounting; Business Administration; Communications; Computer Science; Marketing. **Benefits:** Dental Insurance; Disability Coverage; Life Insurance; Medical Insurance; Profit Sharing; Tuition Assistance. **Corporate headquarters location:** This Location. **Parent company:** Fenimore Software Group, Inc. **Number of employees at this location:** 80.

CPSI (COMPUTER PROGRAMMING SERVICES INC.)
6600 Wall Street, Mobile AL 36695. 334/639-8100. **Contact:** Human Resources. **Description:** Develops and installs business software for hospitals and other health care providers. CPSI's products assist in office management, electronic billing, and a wide range of other areas.

CRC INFORMATION SYSTEMS, INC.
9700 North 91st Street, Scottsdale AZ 85258. 602/451-7474. **Contact:** Human Resources. **Description:** A designer and manufacturer of computer software for printing and labeling companies.

CSK SOFTWARE
100 Saw Mill Road, Danbury CT 06810. 203/730-5300. **Contact:** Human Resources. **Description:** A developer of financial services software.

CTG (COMPUTER TASK GROUP, INC.)
1335 Gateway Drive, Suite 2013, Melbourne FL 32901. 407/725-1300. **Contact:** Human Resources. **Description:** Develops inventory control software for warehouses.

CTI INC.
1330 Braddock Place, Suite 300, Alexandria VA 22314. 703/683-1101. **Contact:** Human Resources. **Description:** Develops computer software and offers technical support.

CABLE & COMPUTER TECHNOLOGY (CCT)
1555 South Sinclair Street, Anaheim CA 92806. 714/937-1341. **Contact:** Human Resources. **Description:** A manufacturer of computer emulation products for the U.S. government.

CADENCE DESIGN SYSTEMS, INC.
555 River Oaks Parkway, San Jose CA 95134. 408/943-1234. **Fax:** 408/955-0820. **Contact:** Human Resources Department. **Description:** A producer of electronic design automation software. **Common positions include:** Design Engineer; Electrical/Electronics Engineer. **Benefits:** 401(k); Dental Insurance; Disability Coverage; Life Insurance; Medical Insurance; Tuition Assistance. **Special Programs:** Internships. **Corporate headquarters location:** This Location. **Operations at this facility include:** Research and Development; Sales. **Listed on:** New York Stock Exchange.

CADIS, INC.
1909 26th Street, Boulder CO 80302. 303/440-4363. **Fax:** 303/440-5309. **Contact:** Human Resources. **E-mail address:** hr@cadis.com. **World Wide Web address:** http://www.cadis.com. **Description:** A developer of a line of integrated software products and services for Internet and intranet information classification, publishing, storage, and retrieval. **Corporate headquarters location:** This Location.

CAERE CORPORATION
100 Cooper Court, Los Gatos CA 95030-3321. 408/395-7000. **Contact:** Human Resources. **Description:** Develops and markets information recognition software. **Common positions include:** Computer Programmer; Software Engineer.

CALYX SOFTWARE
8610 Southwestern Boulevard, Suite 616, Dallas TX 75206. 214/739-8035. **Contact:** Human Resources. **Description:** Develops software for the mortgage industry. Calyx Software designs and markets Point for Windows and Point for DOS, both of which are programs designed to help process the paperwork created by the home buying procedure.

CAMBIA NETWORKS
1851 East 1st Street, Suite 860, Santa Ana CA 92705. 714/245-1400. **Contact:** Human Resources Department. **Description:** Manufactures and develops information management software.

CAMBRIDGE TECHNOLOGY PARTNERS
304 Vassar Street, Cambridge MA 02139. 617/374-9800. **Contact:** Human Resources. **Description:** Cambridge Technology Partners provides information technology consulting and

software development services to organizations with large-scale information processing and distribution needs that are utilizing or migrating to open systems computing environments. Services include Rapid Process Implementation, Strategic Application Development, the Open Enterprise Plan, and Empowerment. Rapid Process uses technology to improve existing business processes or develop new ones. Strategic Application creates business-driven IT applications to help clients integrate information; link offices and people; manage information, time, money, and orders more efficiently; and reduce customer service cycles. The Open Enterprise Plan develops a blueprint to help clients use powerful new technologies across their entire organizations, instead of just in individual departments or functions. Empowerment provides classroom training and practical, hands-on experience in the technologies the company uses to develop strategic applications. **Corporate headquarters location:** This Location. **Other U.S. locations:** Los Angeles CA; Atlanta GA; Chicago IL; Detroit MI; Lansing MI; New York NY; Dallas TX.

CANDLE CORPORATION
2425 Olympic Boulevard, Santa Monica CA 90404. 310/829-5800. **Contact:** Human Resources. **Description:** Develops and markets systems software. **Common positions include:** Computer Programmer; Computer Scientist; Design Engineer; Electrical/Electronics Engineer; Manufacturing Engineer; Software Engineer.

CANMAX RETAIL SYSTEMS, INC.
150 West Carpenter Freeway, Irving TX 75039. 972/541-1600. **Fax:** 972/541-1155. **Contact:** Paul Dao, Technical Recruiter. **E-mail address:** postmaster@canmax.com. **World Wide Web address:** http://www.canmax.com. **Description:** Programs software for convenience stores and gas stations. Founded in 1979. **NOTE:** The company offers entry-level positions. **Common positions include:** Applications Engineer; Computer Programmer; Software Engineer; Systems Analyst; Systems Manager; Technical Writer/Editor. **Educational backgrounds include:** Computer Science; Engineering. **Benefits:** 401(k); Dental Insurance; Disability Coverage; Employee Discounts; Life Insurance; Mass Transit Available; Medical Insurance; Profit Sharing; Tuition Assistance. **Corporate headquarters location:** This Location. **International locations:** Canada. **Parent company:** International Retail Systems. **Listed on:** NASDAQ. **Stock exchange symbol:** CNMX. **Annual sales/revenues:** $21 - $50 million. **Number of employees at this location:** 150. **Number of projected hires for 1997 - 1998 at this location:** 35.

CAPSOFT DEVELOPMENT
732 East Utah Valley Drive, Suite 400, American Fork UT 84003. 801/763-3900. **Contact:** Human Resources. **Description:** Develops document assembly software.

CARLETON INC.
325 South Lafayette Boulevard, South Bend IN 46601. 219/236-4600. **Fax:** 219/236-4640. **Contact:** Human Resources. **Description:** Develops software for the consumer credit and credit insurance industries. Carleton also sells pre-programmed hand-held computers.

CARNEGIE GROUP INC.
5 PPG Place, Pittsburgh PA 15222. 412/642-6900. **Fax:** 412/642-6906. **Contact:** Human Resources. **E-mail address:** info@cgi.com. **World Wide Web address:** http://www.cgi.com. **Description:** Carnegie Group focuses on software development, systems integration, and technical consulting. This company offers knowledge based and project based solutions for a business's particular needs. Carnegie Group has strengths in databases, C++, advanced GUI, and UNIX.

CASELLE, INC.
P.O. Box 31, Spanish Fork UT 84660. 801/798-9851. **Fax:** 801/798-2097. **Contact:** Human Resources. **Description:** Caselle creates specialized accounting software for use by local government offices.

CATALYST INTERNATIONAL
8989 North Deerwood Drive, Milwaukee WI 53223. 414/377-9400. **Contact:** Human Resources. **Description:** A software manufacturer specializing in the development of warehousing systems.

CATENARY SYSTEMS, INC.
470 Belleview, St. Louis MO 63119. 314/962-7833. **Fax:** 314/962-8037. **Contact:** Human Resources. **World Wide Web address:** http://www.catenary.com. **Description:** A developer of image application software tools.

CAYENNE SOFTWARE, INC.
Eight New England Executive Park, Burlington MA 01803. 617/273-9003. **Toll-free phone:** 800/528-2388. **Fax:** 617/228-9904. **Contact:** Human Resources Manager. **E-mail address:** info@cayennesoft.com. **World Wide Web address:** http://www.cayennesoft.com. **Description:** Developers of computer-aided software engineering (CASE) tools and providers of computer design services. The company's products and services enable the rapid development and maintenance of business information systems. **Common positions include:** Applications Engineer; Marketing Specialist; Sales Executive; Sales Representative; Software Engineer; Technical Writer/Editor. **Educational backgrounds include:** Business Administration; Computer Science. **Benefits:** 401(k). **Corporate headquarters location:** This Location. **Listed on:** NASDAQ. **Annual sales/revenues:** $51 - $100 million. **Number of employees at this location:** 200. **Number of employees worldwide:** 400.

CELLULAR TECHNICAL SERVICES COMPANY (CTS)
2401 Fourth Avenue, Seattle WA 98121. 206/443-6400. **Contact:** Human Resources. **Description:** A provider, developer, and marketer of real-time information management software systems. Cellular Technical Services Company serves clients in the wireless communications industry. The company's products provide real-time data for such purposes as fraud detection, billing, customer service, and others. **Corporate headquarters location:** This Location.

CENTURA SOFTWARE CORPORATION
1060 Marsh Road, Menlo Park CA 94025. 415/321-9500. **Fax:** 415/321-5471. **Contact:** Marjorie Erwin, Staffing Specialist. **Description:** Develops software for local area networks and database management systems. Services include support, consulting, and training. **Common positions include:** Software Engineer; Technical Support Representative. **Educational backgrounds include:** Computer Science; Engineering. **Benefits:** 401(k); Dental Insurance; Disability Coverage; Employee Discounts; Flexible Benefits; Life Insurance; Medical Insurance. **Corporate headquarters location:** This Location. **Other U.S. locations:** Los Angeles CA; Washington DC; Atlanta GA; Chicago IL; New York NY; Dallas TX. **Operations at this facility include:** Administration; Research and Development; Sales; Service. **Listed on:** NASDAQ. **Number of employees at this location:** 270. **Number of employees nationwide:** 300.

CERIDIAN
300 Embassy Row, Atlanta GA 30328. 770/399-2170. **Contact:** Personnel Department. **Description:** Develops human resources/payroll software and offers payroll services. **NOTE:** Please send resumes to Personnel, 8100 34th Avenue, Bloomington MN 55440. **Corporate headquarters location:** Minneapolis MN. **Listed on:** New York Stock Exchange.

CERNER CORPORATION
2800 Rockcreek Parkway, Kansas City MO 64108. 816/221-1024. **Contact:** Human Resources. **Description:** Cerner designs, installs, and supports software systems for the health care industry worldwide, including hospitals, HMOs, clinics, physicians' offices, and integrated health organizations. All Cerner applications are structured around a single architectural design, called Health Network Architecture (HNA), which allows information to be shared among clinical disciplines and across multiple facilities. Cerner's information systems are focused in four areas: Clinical Management, Care Management, Repositories, and Knowledge. Clinical management systems include PathNet, which automates the processes of the clinical laboratory; MedNet, which supports pulmonary medicine, respiratory care, and other internal medicine departments; RadNet, which focuses on automating radiology department operations; PharmNet, which automates the processes of the pharmacy; SurgiNet, which addresses the information management needs of the operating room team; and MSMEDS, which provides information management for the pharmacy. Care management systems include ProNet, which automates the processes of patient management and registration, order communication, scheduling, and tracking. Repositories include the Open Clinical Foundation, an enterprisewide, relational database that contains information captured by

various clinical systems to form the computer-based patient record; Open Management Foundation, a repository of process-related information to support management analysis and decision making; MRNet, which automates the chart management process for the medical records department; and Open Engine, an interface engine that collates interfaces, linking systems at a single point. Cerner's knowledge system is Discern, a family of applications that provides support for improving the quality and effectiveness of care. This location develops the software. **Corporate headquarters location:** This Location.
Other U.S. locations:
- 2603 Main Street, Suite 700, Irvine CA 92614. 714/250-1024.
- Five Concourse Parkway, Suite 410, Atlanta GA 30328. 770/804-1024.
- One New England Executive Park, Burlington MA 01803. 617/270-0340.
- 28333 Telegraph Road, Suite 500, Southfield MI 48034-1903. 810/357-1818.
- 8235 Douglas, Suite 1000, Dallas TX 75225. 214/369-4210.
- 2201 Cooperative Way, Suite 301, Herndon VA 20171. 703/904-1871.
- 3245 146th Place SE, Suite 100, Bellevue WA 98007. 206/643-4433.

CHADWYCK-HEALEY INC.
1101 King Street, Suite 380, Alexandria VA 22314. 703/683-4890. **Contact:** Human Resources Department. **Description:** An electronic publisher that develops poetry CD-ROMs and Internet software.

CHAMPLAIN SOFTWARE
P.O. Box 4067, Burlington VT 05406. 802/862-1402. **Contact:** Gary Taylor, Recruiting. **Description:** Writes computer programs for higher education administration.

CHARTERHOUSE SOFTWARE CORPORATION
4195 Thousand Oaks Boulevard, Westlake Village CA 91362. 805/494-5191. **Contact:** Human Resources. **Description:** Charterhouse Software manufactures and markets their own brand of accounting software to both businesses and individuals.

CHECKFREE CORPORATION
4411 East Jones Bridge Road, Norcross GA 30092-1615. 770/441-3387. **Contact:** Mary Gentile, Human Resources Generalist. **Description:** Produces financial and related software. **Corporate headquarters location:** This Location.

CHEYENNE SOFTWARE, INC.
Three Expressway Plaza, Roslin Heights NY 11577. 516/465-4000. **Toll-free phone:** 800/243-9462. **Fax:** 516/465-5499. **Contact:** Human Resources. **E-mail address:** hr@cheyenne.com. **World Wide Web address:** http://www.cheyenne.com. **Description:** An international developer of local area network (LAN) management software. Cheyenne Software products are compatible with NetWare, Windows NT, UNIX, Macintosh, OS/2, Windows 3.1, and Windows 95 operating systems. Its line includes storage management, security, and communications products. Founded in 1983. **NOTE:** The company offers a comprehensive compensation and benefits package. Send resumes to Staffing Manager, 2000 Marcus Avenue, Lake Success NY 11042. **Common positions include:** Account Manager; Art Director; Computer Programmer; Development Manager; Internet Specialist; Marketing Specialist; Product Manager; Sales Engineer; Software Engineer; Technical Support Representative. **Benefits:** Relocation Assistance. **Special Programs:** Internships. **Corporate headquarters location:** This Location. **Listed on:** American Stock Exchange. **Stock exchange symbol:** CYE. **Annual sales/revenues:** $174 million. **Number of employees worldwide:** 1,000.

CHROMACOPY
45 Manning Road, Billerica MA 01821. 508/663-2766. **Contact:** Personnel. **Description:** Produces presentation graphics software and workstation products. Chromacopy also creates imaging systems for personal computers and Macintosh desktop packages, color electronic prepress systems, and overnight slide services.

CIM SOFTWARE CORPORATION
5735 Lindsey Street, Minneapolis MN 55422. 612/544-1752. **Contact:** Human Resources. **Description:** Develops networking software. Cim Software Corporation provides companies with system integrators to facilitate communication between computers.

CIMLINC INC.
1699 Stutz Drive, Troy MI 48084. 810/649-0240. **Contact:** Human Resources. **Description:** Cimlinc Inc. designs specialized software for industrial companies that assists in a variety of industrial processes. Clients include tool, die, and molding companies.

CINCOM SYSTEMS, INC.
2300 Montana Avenue, Cincinnati OH 45211. 513/662-2300. **Contact:** Patricia A. Sledge, Senior Staffing and Placement. **Description:** A developer and marketer of software technologies in database management systems, application development, and manufacturing and financial applications. Cincom also provides text management and project management of IBM, Digital, Hewlett-Packard, UNIX, and other platforms. **Common positions include:** Computer Programmer; Software Engineer; Systems Analyst. **Educational backgrounds include:** Computer Science; M.I.S. **Benefits:** Dental Insurance; Disability Coverage; Employee Discounts; Life Insurance; Medical Insurance; Pension Plan; Tuition Assistance. **Corporate headquarters location:** This Location. **Listed on:** Privately held.

CLARIFY INC.
2702 Orchard Parkway, San Jose CA 95134. 408/428-2000. **Fax:** 408/428-0633. **Contact:** Human Resources. **E-mail address:** hr@clarify.com. **World Wide Web address:** http://www.clarify.com. **Description:** A leader in the development of customer support management software. **Corporate headquarters location:** This Location. **Number of employees nationwide:** 150.

CLARIS CORPORATION
5201 Patrick Henry Drive, Santa Clara CA 95052. 408/987-7000. **Contact:** Human Resources. **Description:** Claris is engaged in software development and sales. Claris is the developer of ClarisWorks 3.0, a software product that combines word processing, spreadsheet, charting, graphics, database, drawing, and painting functions. **Parent company:** Apple Computer.

CLINARIUM INC.
1255 Drummers Lane, Suite 306, Wayne PA 19087. 610/995-9333. **Contact:** Cathy Becker, Recruiting. **Description:** Develops pharmaceutical software.

CODA INC.
1155 Elm Street, Manchester NH 03101. 603/647-9600. **Fax:** 603/628-7382. **Contact:** Mary Coles, Human Resources Director. **Description:** Designs, manufactures, and sells financial accounting software. **Other U.S. locations:** Los Angeles CA; San Francisco CA; Atlanta GA; Chicago IL.

COGGINS SYSTEMS, INC.
5384 Peachtree Corners East, Norcross GA 30092. 770/447-9202. **Contact:** Human Resources Department. **Description:** Designs management software for such clients as the United States government and Shell Oil Company.

COGNISOFT CORPORATION
8324 165th Street NE, Redmond WA 98052. 206/702-0800. **Contact:** Human Resources. **Description:** Cognisoft develops software for corporate databases and the Internet.

COGNOS CORPORATION
400 Interstate North Parkway, Suite 1060, Atlanta GA 30339. 770/951-0294. **Contact:** Human Resources. **Description:** Develops a variety of business software including PowerPlay.

COLUMBIA ULTIMATE, INC.
11800 Sunrise Valley Drive, Suite 1135, Reston VA 20191. 703/264-8877. **Fax:** 703/264-1987. **Contact:** Rich Irvine, General Manager. **Description:** Develops database and other software programs for various collection agencies.

COLUMBINE JDS SYSTEMS INC.
1707 Cole Boulevard, Golden CO 80401. 303/237-4000. **Contact:** Jennifer Bennett, Recruiter. **Description:** Develops software for television and radio stations.

COMMAND DATA INC.
2204 Lakeshore Drive, Suite 206, Birmingham AL 35209. 205/879-3282. **Contact:** Human Resources. **Description:** Designs business software for the construction materials industry.

COMMVAULT SYSTEMS
One Industrial Way West, Eatontown NJ 07724. 908/935-8000. **Contact:** Human Resources. **Description:** Manufactures software for computer backup systems. Commvault then sells the entire system, including hardware.

COMPACT SOFTWARE INC.
201 McLean Boulevard, Paterson NJ 07504. 201/881-1200. **Contact:** Human Resources. **Description:** Develops and distributes software. Compact Software produces their own applications focused on circuit design. This software is of a technical nature and caters to such professions as engineering.

COMPTEK FEDERAL SYSTEMS, INC.
2732 Transit Road, Buffalo NY 14224. 716/677-4070. **Contact:** Human Resources. **Description:** A developer of defense-related software.

COMPUSPEAK
15065 West 116th Street, Olathe KS 66062. 913/491-3444. **Contact:** Human Resources. **Description:** A developer of voice application software for computers.

COMPUTE RX
15 Locust Avenue, Fairmont WV 26554. 304/367-1477. **Contact:** Human Resources. **Description:** A developer of medical software which the company sells to hospitals and pharmacies.

COMPUTER ASSOCIATES INTERNATIONAL, INC.
One Computer Associates Plaza, Islandia NY 11788. 516/342-6000. **Contact:** Personnel Department. **World Wide Web address:** http://www.cai.com. **Description:** Founded in 1976, Computer Associates International is one of the world's leading developers of client/server and distributed computing software. The company develops, markets, and supports enterprise management, database and applications development, business applications, and consumer software products for a broad range of mainframe, midrange, and desktop computers. Computer Associates serves most of the world's major business, government, research, and educational organizations. **Common positions include:** Accountant/Auditor; Administrator; Commercial Artist; Computer Programmer; Customer Service Representative; Human Resources Manager; Manufacturer's/Wholesaler's Sales Rep.; Systems Analyst; Technical Writer/Editor. **Educational backgrounds include:** Accounting; Art/Design; Business Administration; Computer Science; Finance. **Benefits:** Dental Insurance; Disability Coverage; Life Insurance; Medical Insurance; Profit Sharing; Tuition Assistance. **Special Programs:** Internships. **Corporate headquarters location:** This Location. **Operations at this facility include:** Administration; Research and Development; Sales. **Listed on:** New York Stock Exchange. **Annual sales/revenues:** $3 billion. **Number of employees nationwide:** 4,000. **Number of employees worldwide:** 9,000.
Other U.S. locations:
- 8222 South 48th Street, Suite 100, Phoenix AZ 85044. 602/438-8505. Contact: Debora Farmis, Administrative Assistant. This location is a sales and support office.
- 1201 Marina Village Parkway, Alameda CA 94501. 510/769-1400. This location is a sales office.
- 9740 Scranton Road, Suite 200, San Diego CA 92121. 619/452-0170. This location develops software.
- 2525 Augustine Drive, Santa Clara CA 95054. 408/562-8800. Contact: Ms. Chris King, Manager of Human Resources. This location is engaged in the sale, research, and development of software.

- 2300 Windy Ridge Parkway, Suite 1000, Atlanta GA 30339. 770/953-9276. This location sells software and offers technical support.
- 2400 Cabot Drive, Lisle IL 60532. 630/505-6000. This location develops and sells software and offers customer service support.
- 201 University Avenue, Westwood MA 02090. 617/329-7700. This location develops and sells software.
- 2 Executive Drive, Fort Lee NJ 07024. 201/592-0009. This location sells software, offers technical support, and is home to the marketing department.
- 2000 Midlantic Drive, Mount Laurel NJ 08054. 609/273-9100. This location sells software and offers technical support.
- 7965 North High Street, Columbus OH 43235. 614/888-1775. Contact: Shelley Eyers, Human Resources. This location develops software.
- 575 Herndon Parkway, Herndon VA 20170. 703/708-3000. This location develops and supports software.
- 411 108th Avenue, Suite 600, Bellevue WA 98004. 206/688-2000. This location develops software.

COMPUTER ASSOCIATES INTERNATIONAL, INC.

909 Las Colinas Boulevard East, Irving TX 75039. 972/556-7100. **Contact:** Lavena Sipes, Personnel Manager. **World Wide Web address:** http://www.cai.com. **Description:** This location develops and sells software and offers support services. Overall, Computer Associates International develops, markets, and supports more than 300 integrated products including systems and database management, business applications, application development, and consumer solutions. Software products include systems management software, information management software, business application software, and desktop software. Computer Associates International, Inc. serves many of the world's businesses, governments, research organizations, and educational organizations. Founded in 1976, Computer Associates has since become a *Fortune* 500 company. **NOTE:** To address the challenge of finding quality new employees, Computer Associates International, Inc. looks for entry-level applicants or co-op students for the "farm system" and grooms them internally to become productive members of the Computer Associates team. The majority of promotions are given internally, and the farm system approach has provided a career path for many employees. Employees are trained in all areas of the organization to make them part of the companywide team. The company recruits at universities nationwide. **Common positions include:** Administrative Services Manager; Clerical Supervisor; Computer Programmer; Customer Service Representative; Data Processor; Financial Services Sales Rep.; Human Resources Manager; Manufacturer's/Wholesaler's Sales Rep.; Marketing Specialist; Quality Control Supervisor; Sales Representative; Services Sales Representative; Software Engineer; Support Personnel; Systems Analyst; Technical Writer/Editor. **Educational backgrounds include:** Accounting; Business Administration; Computer Science; Finance; Human Resources; Marketing. **Benefits:** 401(k); Dental Insurance; Disability Coverage; Employee Discounts; Life Insurance; Medical Insurance; Savings Plan; Smoke-free; Tuition Assistance. **Special Programs:** Internships. **Corporate headquarters location:** Islandia NY. **Other U.S. locations:** Nationwide. **Operations at this facility include:** Administration; Divisional Headquarters; Research and Development; Sales; Service. **Listed on:** New York Stock Exchange. **Annual sales/revenues:** $3 billion. **Number of employees at this location:** 350. **Number of employees nationwide:** 4,000. **Number of employees worldwide:** 9,000.

COMPUTER ASSOCIATES INTERNATIONAL, INC.

12120 Sunset Hills Road, Reston VA 20190. 703/709-4500. **Fax:** 703/709-4580. **World Wide Web address:** http://www.cai.com. **Contact:** Sean T. Friedman, Manager/Human Resources. **Description:** This location is a sales office. Founded in 1976, Computer Associates International is one of the world's leading developers of client/server and distributed computing software. The company develops, markets, and supports enterprise management, database and applications development, business applications, and consumer software products for a broad range of mainframe, midrange, and desktop computers. Computer Associates serves most of the world's major business, government, research, and educational organizations. **NOTE:** The company offers entry-level positions. **Common positions include:** Account Manager; Account Representative; Administrative Assistant; Computer Programmer; Customer Service Representative; Database Manager; Human Resources Manager; Marketing Specialist; Sales Engineer; Sales Executive;

Sales Manager; Sales Representative; Secretary; Software Engineer; Systems Analyst; Systems Manager; Technical Writer/Editor; Telecommunications Manager; Typist/Word Processor. **Benefits:** 401(k); Daycare Assistance; Dental Insurance; Disability Coverage; Employee Discounts; Financial Planning; Fitness Program; Life Insurance; Mass Transit Available; Medical Insurance; Tuition Assistance. **Corporate headquarters location:** Islandia NY. **Other U.S. locations:** Nationwide. **Operations at this facility include:** Divisional Headquarters. **Listed on:** New York Stock Exchange. **Stock exchange symbol:** CmpAsc. **Annual sales/revenues:** $3 billion. **Number of employees at this location:** 500. **Number of employees nationwide:** 4,000. **Number of employees worldwide:** 9,000.

COMPUTER CHROME
803 Transfer Road, St. Paul MN 55114. 612/646-2442. **Contact:** Human Resources. **Description:** Develops presentation graphics software.

COMPUTER CONCEPTS CORPORATION
80 Orville Drive, Bohemia NY 11716. 516/244-1500. **Contact:** Cathy Athans, Human Resources Manager. **Description:** Computer Concepts designs, markets, and supports information delivery software products, including end user data access tools for personal computers and client/server environments, and systems management software products for corporate mainframe data centers. Products include d.b.Express, which offers methods of finding, organizing, analyzing, and using information contained in databases; systems management software products, which improve mainframe system performance, reduce hardware expenditures, and enhance the reliability and availability of the data processing environment; client/server products, which develops client/server relational database administration and programmer productivity tools; Superbase, a Windows relational database technology with links to back-end server databases; MapLinx, a desktop database mapping utility for personal computers; and Perspective for Windows, a three-dimensional graphics presentation product.

COMPUTER CORPORATION OF AMERICA, INC. (CCA)
500 Old Connecticut Path, P.O. Box 9378, Framingham MA 01701-9378. 508/270-6666. **Contact:** Human Resources. **Description:** Develops mainframe database products that run on VAX systems.

COMPUTER CURRICULUM CORPORATION
1287 Lawrence Station Road, Sunnyvale CA 94089. 408/541-3308. **Contact:** Human Resources. **Description:** Programs educational software. **Common positions include:** Accountant/Auditor; Administrative Services Manager; Attorney; Budget Analyst; Buyer; Claim Representative; Clerical Supervisor; Computer Programmer; Credit Manager; Electrician; Financial Analyst; Human Resources Manager; Human Service Worker; Librarian; Mathematician; Services Sales Representative; Software Engineer; Statistician; Systems Analyst; Technical Writer/Editor. **Educational backgrounds include:** Accounting; Art/Design; Business Administration; Communications; Computer Science; Engineering; Finance; Liberal Arts; Marketing; Mathematics. **Benefits:** 401(k); Dental Insurance; Disability Coverage; Employee Discounts; Life Insurance; Medical Insurance; Savings Plan; Tuition Assistance. **Special Programs:** Internships. **Corporate headquarters location:** This Location. **Operations at this facility include:** Administration; Divisional Headquarters; Regional Headquarters; Research and Development; Sales; Service. **Listed on:** NASDAQ. **Number of employees at this location:** 400. **Number of employees nationwide:** 600.

COMPUTER DESIGN INC.
2880 East Beltline NE, Grand Rapids MI 49505. 616/361-1139. **Contact:** Human Resources. **Description:** Develops designing software for the apparel industry.

COMPUTER INFORMATION ANALYSTS INC.
3900 East Monument Street, Baltimore MD 21205. 410/522-2234. **Contact:** Human Resources. **Description:** Develops year 2000 solutions software and 4th generation technologies.

COMPUTER SCIENCES CORPORATION
4815 Bradford Drive, Huntsville AL 35805. 205/837-7200. **Contact:** Human Resources. **Description:** This location designs a wide variety of computer software. Overall, Computer

Sciences Corporation, founded in 1959, is one of the largest independent providers of information technology consulting, systems integration, and outsourcing to industry and government. One of the first providers of computer software and related services, the company has expanded its scope of services to include management consulting, education and research programs in the strategic use of information resources and the design, engineering, development, integration, installation, and operation of computer-based systems and communications systems. Computer Sciences Corporation also provides consumer credit-related services, automated systems for health care organizations, financial insurance services, and data processing services. The company's principal markets are the U.S. federal government, U.S. commercial markets, and various international markets.

COMPUTER SCIENCES CORPORATION
P.O. Box 1038, Moorestown NJ 08057-0902. 609/234-1166. **Contact:** Human Resources Department. **Description:** This location configures software. Computer Sciences Corporation primarily services the U.S. government. The four sectors of the company include the Systems Group division, the Consulting division, the Industry Services Group, and the CSC divisions. The Systems Group division designs, engineers, and integrates computer-based systems and communications systems, providing all the hardware, software, training, and related elements necessary to operate a system. The Consulting division includes consulting and technical services in the development of computer and communication systems to non-federal organizations. The Industry Services Group provides service to health care, insurance, and financial services, as well as providing large-scale claim processing and other insurance-related services. CSC Health Care markets business systems and services to the managed health care industry, clinics, and physicians. CSC Enterprises provides consumer credit reports and account management services to credit grantors. **Other U.S. locations:** Nationwide.

COMPUTER SOLUTIONS INTERNATIONAL, INC.
215 Park Avenue South, Suite 1913, New York NY 10003. 212/777-0966. **Contact:** Daniel Hubert, President. **Description:** Develops accounting software and offers support services.

COMPUTERVISION CORPORATION
100 Crosby Drive, Bedford MA 01730. 617/275-1800. **Contact:** Human Resources. **Description:** Computervision designs, produces, and markets design automation, product data management, product visualization, and other engineering workgroup software packages and related services worldwide. The design automation software includes functions that allow a customer to create and store a single solid geometric model of a particular product that can then be made available in electronic form simultaneously for review and analysis by design teams. The engineering workshop packages provide wide information access, and include software for product data management and product visualization. The company also provides a variety of value added services and hardware and network maintenance services. **Corporate headquarters location:** This Location. **Annual sales/revenues:** More than $100 million.
Other U.S. locations:
- 50 West Big Beaver Road, Troy MI 48084. 810/689-0100. This location is a sales office.

COMPUTERVISION CORPORATION
9805 Scranton Road, San Diego CA 92126. 619/587-3000. **Fax:** 619/587-3047. **Contact:** Peggy Randalls, Human Resources. **World Wide Web address:** http://www.cv.com. **Description:** This location designs CAD software. Overall, Computervision designs, produces, and markets design automation, product data management, product visualization, and other engineering workgroup software packages and related services worldwide. The design automation software includes functions that allow a customer to create and store a single solid geometric model of its product that can then be made available in electronic form simultaneously for review and analysis by design teams. The engineering workshop solutions provide wide information access, and include software for product data management and product visualization. The company also provides a variety of value-added services and hardware as well as network maintenance services. **Common positions include:** Design Engineer; Technical Writer/Editor. **Educational backgrounds include:** Computer Science; Engineering; Mathematics. **Benefits:** 401(k); Dental Insurance; Disability Coverage; Employee Discounts; Financial Planning; Life Insurance; Medical Insurance; Pension Plan; Profit Sharing; Savings Plan; Tuition Assistance. **Corporate headquarters location:**

Bedford MA. **Other U.S. locations:** Nationwide. **Listed on:** New York Stock Exchange. **Number of employees at this location:** 100. **Annual sales/revenues:** More than $100 million.

COMPUTERVISION SERVICES

2401 West Behrend Drive, Suite 17, Phoenix AZ 85027. 602/780-9411. **Contact:** Human Resources. **Description:** Computervision designs, produces, and markets design automation, product data management, product visualization, and other engineering workgroup software packages and related services worldwide. The design automation software includes functions that allow a customer to create and store a single solid geometric model of its product that can then be made available in electronic form simultaneously for review and analysis by design teams. The engineering workshop solutions provide wide information access, and include software for product data management and product visualization. The company also provides a variety of value-added services and hardware as well as network maintenance services.

COMPUTRON SOFTWARE INC.

301 Route 17 North, 12th Floor, Rutherford NJ 07070. 201/935-3400. **Contact:** Human Resources. **Description:** Develops and markets various financial software products.

COMPUWARE CORPORATION

31440 Northwestern Highway, Farmington Hills MI 48334-2564. 810/737-7300. **Contact:** Recruiting. **Description:** Develops, markets, and supports an integrated line of systems software products which improve the productivity of programmers and analysts in application program testing, test data preparation, error analysis, and maintenance. Compuware also provides a broad range of data processing professional services, including business systems analysis, design, and programming, as well as systems planning and consulting.

COMSIS CORPORATION

8737 Colesville Road, Suite 1100, Silver Spring MD 20910. 301/588-0800. **Fax:** 301/588-5922. **Contact:** Human Resources. **Description:** Develops and supports software for the transportation industry.

CONCEPTUAL REALITIES

4002 West Burbank Boulevard, Burbank CA 91505. 818/557-2911. **Contact:** Human Resources. **Description:** Conceptual Realities develops video game software.

CONCORD MANAGEMENT SYSTEMS INC.

5301 West Cypress, Suite 200, Tampa FL 33607. 813/281-2200. **Contact:** Human Resources. **Description:** Develops software for accounts receivable and accounts payable departments.

CONNECTSOFT

11130 North East 33rd Place, Suite 250, Bellevue WA 98004. 206/827-6467. **Contact:** Human Resources. **Description:** ConnectSoft develops business application software.

CONSILIUM, INC.

485 Clyde Avenue, Mountain View CA 94043. 415/691-6100. **Contact:** Human Resources. **Description:** Consilium, Inc. is the world's leading supplier of integrated manufacturing execution systems software and services. For more than 15 years, the company has been helping world class manufacturers achieve best practices in their manufacturing operations through better visibility and control of the plant floor. Consilium has helped 50,000 users in 20 countries solve critical business problems by providing solutions that optimize manufacturing performance. The company's WorkStream and FlowStream product lines provide real-time access to critical information about the production process, enabling manufacturers to identify and implement best practices in cost, quality, service, and speed. **Corporate headquarters location:** This Location.

CONSTRUCTION SYSTEMS ASSOCIATES INC.

1090 Northchase Parkway, Marietta GA 30067. 770/955-3518. **Contact:** Human Resources. **Description:** Develops software for large construction projects.

THE CONTINUUM COMPANY, INC.
9500 Arboretum Boulevard, Austin TX 78759-6399. 512/338-7591. **Fax:** 512/338-6216. **Recorded Jobline:** 512/338-7300. **Contact:** Donna Monroe, Senior Human Resources Specialist. **Description:** A leading supplier of software solutions and related services to the insurance industry worldwide. **Common positions include:** Accountant/Auditor; Buyer; Computer Programmer; Financial Analyst; Human Resources Manager; Management Analyst/Consultant; Software Engineer; Systems Analyst; Technical Writer/Editor. **Educational backgrounds include:** Accounting; Business Administration; Communications; Computer Science; Finance. **Benefits:** 401(k); Dental Insurance; Disability Coverage; Employee Discounts; Life Insurance; Medical Insurance; Retirement Plan; Tuition Assistance. **Corporate headquarters location:** This Location. **Other U.S. locations:** Nationwide. **Operations at this facility include:** Administration; Divisional Headquarters; Regional Headquarters; Research and Development; Sales; Service. **Number of employees at this location:** 650. **Number of employees nationwide:** 2,400.

CONTROL SOFTWARE GROUP INC.
3021 East Dublin Grandville Road, Columbus OH 43231. 614/882-7422. **Contact:** Human Resources. **Description:** Develops software. Control Software Group designs applications for a business's particular needs. This company specializes in creating programs for any industry that needs to follow safety compliance laws.

CORBIS CORPORATION
15395 SE 30th Place, Suite 300, Bellevue WA 98007. 206/641-4505. **Recorded Jobline:** 206/649-3431. **Contact:** Human Resources. **Description:** Develops software. Corbis Corporation produces creative CD-ROMs and maintains an archive of visual images.

COREL, INC.
567 East Timpanogos Parkway, Mailstop G24, Orem UT 84057. **Toll-free phone:** 800/772-6735. **Fax:** 801/222-4374. **Contact:** Human Resources. **E-mail address:** hr@corel.ca. **World Wide Web address:** http://www.corel.com. **Description:** A software developer. In March of 1996, Corel purchased the word-processing program WordPerfect from Novell Inc. **NOTE:** Human Resources at the Canadian headquarters address is 1600 Carling Avenue, Ottawa, Ontario, Canada, K1Z 8R7. Fax: 613/761-1146. In order to be eligible for a position at the Ottawa location, Canadian citizenship or eligibility under the North American Free Trade Agreement (NAFTA) is required. **Common positions include:** Marketing Manager; Quality Assurance Engineer; Software Engineer; Support Personnel. **Parent company:** Corel Corporation (Ottawa, Ontario, Canada). **Listed on:** NASDAQ; Toronto Stock Exchange. **Stock exchange symbol:** COSFF; COS, respectively. **Annual sales/revenues:** $205 million.

CORNET INC.
701 Main Street, Stroudsburg PA 18360. 717/420-0800. **Contact:** Human Resources. **Description:** Develops sales force automation software.

CORVALLIS MICROTECHNOLOGY INC. (CMI)
413 SW Jefferson Avenue, Corvallis OR 97333. 541/752-5456. **Contact:** Human Resources. **Description:** A developer of surveying software.

CREATIVE LABS, INC.
1901 McCarthy Boulevard, Milpitas CA 95035. 408/428-6600. **Fax:** 408/232-1284. **Contact:** Human Resources. **E-mail address:** solivas@creaf.com. **World Wide Web address:** http://www.creaf.com. **Description:** A manufacturer of computer game software. **Corporate headquarters location:** This Location.

CREATIVE MULTIMEDIA
225 SW Broadway, Suite 600, Portland OR 97205. **Fax:** 503/241-4370. **Contact:** Recruiting. **E-mail address:** recruiting@creativemm.com. **World Wide Web address:** http://www.creativemm.com. **Description:** A developer of interactive consumer multimedia information. The company has a line of over 30 CD-ROM titles. **Number of employees at this location:** 100.

CREATIVE SOLUTIONS
7322 Newman Boulevard, Dexter MI 48130. 313/426-5860. **Contact:** Human Resources Department. **Description:** Creative Solutions develops, markets, and supports accounting and tax software.

CUSA TECHNOLOGIES, INC.
986 West Atherton Drive, Salt Lake City UT 84123. 801/263-1840. **Contact:** Human Resources. **Description:** Cusa Technologies provides software support for credit unions.

CYBERMEDIA INC.
3000 Ocean Park Boulevard, Suite 2001, Santa Monica CA 90405. 310/581-4700. **Contact:** Human Resources. **Description:** A developer of software known as Oil Change, which takes inventory of all of the programs on a user's PC, and automatically downloads version upgrades from the World Wide Web as they become available.

CYBERNET SYSTEMS CORPORATION
727 Airport Boulevard, Ann Arbor MI 48108. 313/668-2567. **Fax:** 313/668-8780. **Contact:** Human Resources. **World Wide Web address:** http://www.cybernet.com. **Description:** A developer of virtual reality, force feedback, robotics, human/computer interactions, electromechanical systems, computer vision, and real-time systems. **Corporate headquarters location:** This Location.

CYBERTEK CORPORATION
244 Knollwood Drive, Bloomingdale IL 60108. 630/924-3583. **Fax:** 630/924-3581. **Contact:** Senior Recruiter. **Description:** Provides software and services to the life insurance industry. **Common positions include:** Actuary; Computer Programmer; Consultant; Insurance Agent/Broker; Systems Analyst. **Educational backgrounds include:** Computer Science; Mathematics. **Benefits:** 401(k); Dental Insurance; Disability Coverage; Employee Discounts; Life Insurance; Medical Insurance; Profit Sharing; Spending Account; Tuition Assistance. **Corporate headquarters location:** Columbia SC. **Parent company:** Policy Management Systems Corporation. **Operations at this facility include:** Regional Headquarters. **Listed on:** New York Stock Exchange. **Number of employees at this location:** 250. **Number of employees nationwide:** 4,000.
Other U.S. locations:
- 7800 North Simmons Freeway, Suite 600, Dallas TX 75247. 214/637-1540.

CYBORG SYSTEMS INC.
2 North Riverside Plaza, Suite 1200, Chicago IL 60606. 312/454-1865. **Contact:** Human Resources. **Description:** Designs software for human resource departments.

CYCARE SYSTEMS INC.
7001 North Scottsdale Road, Suite 1000, Scottsdale AZ 85253-3644. 602/596-4300. **Contact:** Human Resources. **Description:** Designs medical software. Cycare Systems supplies hospitals, universities, and doctors' offices with medical and related software.

CYLINK CORPORATION
320 Decker Drive, Irving TX 75062. 972/719-2624. **Contact:** Human Resources. **Description:** Develops software.

CYMA SYSTEMS, INC.
2330 West University, Suite 7, Tempe AZ 85281. 602/303-2962. **Contact:** Joan Nickel, Human Resources. **Description:** Cyma Systems, Inc. is engaged in the development and distribution of microcomputer-based software, specializing in accounting, medical practice management, point-of-sale, and related vertical applications. **Common positions include:** Accountant/Auditor; Administrator; Computer Programmer; Customer Service Representative; Manufacturer's/Wholesaler's Sales Rep.; Marketing Specialist; Operations/Production Manager; Public Relations Specialist; Quality Control Supervisor; Systems Analyst; Technical Writer/Editor. **Educational backgrounds include:** Accounting; Computer Science; Marketing. **Benefits:** Dental

Insurance; Disability Coverage; Life Insurance; Medical Insurance; Savings Plan; Tuition Assistance. **Corporate headquarters location:** This Location. **Operations at this facility include:** Administration; Manufacturing; Research and Development; Sales; Service. **Number of employees at this location:** 60.

DBBASICS, INC.
P.O. Box 30754, Raleigh NC 27622-0754. 919/380-7252. **Toll-free phone:** 800/646-6490. **Fax:** 919/319-9492. **Contact:** Human Resources. **World Wide Web address:** http://www.dbbasics.com. **Description:** A software development company with a focus on client/server database applications and training. The company is divided into four divisions. The Project Solutions Division, the largest division with over 40 client/server developers, develops custom databases for clients. This division has completed over 600 projects since 1988. The Contract Placement Division recruits programmers, technical writers, and database administrators for client contract assignments. The Training Division offers training courses for professional programmers. The Commercial Products Division develops products which are sold commercially including IRIS, an Internet recruiting information system. DBBasics has working relationships with a number of vendors including Microsoft, Oracle, Sybase, Novell, IBM, and Sun Microsystems among others. Founded in 1988. **NOTE:** Applications may be submitted via the DBBasics Web page under the Employment menu selection. **Corporate headquarters location:** This Location. **Listed on:** Privately held.

DCR TECHNOLOGIES INC.
8009 34th Avenue South, Bloomington MN 55425. 612/854-6109. **Contact:** Human Resources. **Description:** This location is a sales office. The company manufactures networking software, and sells the applicable hardware. DCR Technologies also offers technical assistance.

DGM&S INC.
1025 Briggs Road, Suite 100, Mount Laurel NJ 08054. 609/866-1212. **Contact:** Human Resources. **Description:** Develops middleware applications.

DHI COMPUTING SERVICE, INC.
1525 West 820 North, Provo UT 84601. 801/373-8518. **Fax:** 801/374-5316. **Contact:** Human Resources. **Description:** Develops business software and installs hardware used mainly by dairies.

DP ASSOCIATES
4900 Century, Huntsville AL 35816. 205/837-8300. **Contact:** Mr. Tommie Batts, Executive Vice President. **Description:** Performs contract computer work for the U.S. government, providing everything from programming services to systems operations.

DP SOLUTIONS
2000 East Edgewood Drive, Lakeland FL 33801. 941/666-2330. **Contact:** Human Resources. **Description:** Develops preventative maintenance software. The objective of this location is shifting towards sales operations only.

DP SOLUTIONS
4249 Piedmont Parkway, Suite 105, Greensboro NC 27410. 910/854-7700. **Contact:** Human Resources. **Description:** Develops preventative maintenance software.

DST SYSTEMS, INC.
333 West 11th Street, 3rd Floor, Kansas City MO 64105. 816/435-8614. **Toll-free phone:** 800/874-0174. **Fax:** 816/435-8618. **Contact:** Human Resources. **Description:** A software developer and transfer agent for the financial industry. **NOTE:** The company offers entry-level positions. **Common positions include:** Administrative Assistant; Computer Programmer; Customer Service Representative; Systems Analyst. **Educational backgrounds include:** Accounting; Business Administration; Computer Science; Finance; Liberal Arts; Mathematics. **Benefits:** 401(k); Dental Insurance; Employee Discounts; Life Insurance; Medical Insurance; Profit Sharing; Tuition Assistance. **Special Programs:** Internships. **Corporate headquarters location:** This Location. **Subsidiaries include:** Argus Health Systems, Inc.; NFDS. **Number of employees at this location:** 6,000.

DATA RESEARCH AND APPLICATIONS INC.
9041 Executive Park Drive, Suite 200, Knoxville TN 37923. 423/690-1345. **Contact:** Human Resources. **Description:** Develops software for data recovery and provides computer hardware. Data Research and Applications caters to businesses with Windows NT.

DATA WORKS
5910 Pacific Center Boulevard, San Diego CA 92121. 619/546-9600. **Fax:** 619/546-9777. **Contact:** Melinda DiPasquale, Recruiting Coordinator. **E-mail address:** jobs@dataworks.com. **World Wide Web address:** http://www.dataworks.com. **Description:** A developer of integrated program software for mid/size manufacturers utilizing Windows NT, Visual Basic, Powerbuilder, C++, and Systembuilder. **Common positions include:** Administrative Assistant; Computer Programmer; Software Engineer; Systems Analyst; Technical Writer/Editor. **Educational backgrounds include:** Computer Science. **Benefits:** 401(k); Dental Insurance; Disability Coverage; Employee Discounts; Life Insurance; Medical Insurance; Tuition Assistance. **Corporate headquarters location:** This Location. **Listed on:** NASDAQ. **Annual sales/revenues:** $51 - $100 million. **Number of employees at this location:** 530. **Number of projected hires for 1997- 1998 at this location:** 50.

DATA-BASICS INC.
9450 Midwest Avenue, Cleveland OH 44125. 216/663-5600. **Contact:** Human Resources. **Description:** Develops job cost accounting software for the construction and architecture industries, as well as service account management software.

DATA-CORE SYSTEMS INC.
3700 Science Center, Philadelphia PA 19104. 215/243-1990. **Fax:** 215/243-1909. **Contact:** Human Resources. **Description:** Data-Core Systems develops database applications.

DATA/WARE DEVELOPMENT INC.
9449 Carol Park Drive, San Diego CA 92121. 619/453-7660. **Contact:** Human Resources. **Description:** Designs and sells connectivity and channel emulation software.

DATABASE DESIGN INC.
716 Main Street, Hyannis MA 02601. 413/548-8896. **Contact:** Human Resources. **Description:** Develops database solutions software. Database Design also offers consulting services.

DATABEAM
3191 Nicholasville Road, Lexington KY 40503. 606/245-3500. **Fax:** 606/245-3528. **Contact:** Human Resources. **E-mail address:** resume@databeam.com. **World Wide Web address:** http://www.databeam.com. **Description:** A developer of document conferencing applications for end users. Databeam also licenses computer toolkits to third parties. Product line includes FarSite document conferencing software. **Listed on:** Privately held.

DATACAP SYSTEMS
100 New Britain Road, Chalfont PA 18914. 215/997-8989. **Contact:** Human Resources. **Description:** Develops software and point-of-sale systems.

DATACRON INC.
977 Oaklawn Avenue, Elmhurst IL 60126. 630/832-1011. **Contact:** Human Resources. **Description:** Develops and supplies software for home health agencies.

DATAFOCUS, INC.
12450 Fair Lakes Circle, Suite 400, Fairfax VA 22033. 703/803-3343. **Contact:** Human Resources. **Description:** A software manufacturer. Products include Nutcracker software.

DATAGATE INC.
6490 South McCarran Boulevard, Suite B14, Reno NV 89509-6120. 702/828-6111. **Contact:** Human Resources. **Description:** Datagate offers software support services.

DATAIR EMPLOYEE BENEFITS SYSTEMS
735 North Cass Avenue, Westmont IL 60559. 630/325-2600. **Contact:** Human Resources. **Description:** Datair develops employee benefits software for companies.

DATALOGICS INC.
101 North Wacker Drive, Chicago IL 60606. 312/266-4444. **Contact:** Human Resources. **Description:** Develops and markets software for publishing companies.

DATAMATIC INC.
2121 North Glenville Drive, Richardson TX 75082. 972/234-5000. **Contact:** Human Resources. **Description:** Offers software integration solutions. Datamatic has an engineering team and an active sales force.

DATAMEDIC CORPORATION
20 Oser Avenue, Hauppauge NY 11788. 516/435-8880. **Contact:** Human Resources. **Description:** Develops software. Datamedic specializes in developing medical applications for computers. This company offers its products to hospitals, universities, and doctors' offices.

DATASTREAM SYSTEMS, INC.
50 Datastream Plaza, Greenville SC 29605. 864/422-5001. **Toll-free phone:** 800/955-6775. **Fax:** 864/422-5000. **Contact:** Human Resources. **World Wide Web address:** http://www.dstm.com. **Description:** Develops maintenance management software incorporating implementation, training, and consulting services for businesses to reduce downtime and increase productivity. Datastream serves many major industries including government, health care, hospitality, manufacturing, and transportation. Product line includes copyrighted products such as MaintainIt and MP2 Client/Server. Datastream was listed fifth in *Business Week*'s 1996 "100 Hot Growth Companies," and 34th in *Forbes*'s 1996 list of the 200 Best Small Companies. **Common positions include:** Client/Server Specialist; Consultant; Software Developer; Technical Writer/Editor. **Corporate headquarters location:** This Location. **Listed on:** NASDAQ. **Stock exchange symbol:** DSTM. **Annual sales/revenues:** $27 million.

DATATEL INC.
4375 Fair Lakes Court, Fairfax VA 22033. 703/968-9000. **Contact:** Human Resources. **Description:** Develops and markets software. Datatel specializes in creating applications for higher education and fundraising organizations. Datatel's clientele includes universities, hospitals, and other large groups.

DATAVIZ, INC.
55 Corporate Drive, Trumbull CT 06611. 203/268-0030. **Contact:** Kathy Fetchick, Human Resources. **E-mail address:** kfetchick@dataviz.com. **World Wide Web address:** http://www.dataviz.com. **Description:** Provides software utilities for transferring files from Macintosh to PC format and vice versa. **Common positions include:** Software Engineer. **Educational backgrounds include:** Computer Science; Engineering; Mathematics; Physical Science. **Corporate headquarters location:** This Location.

DATAWARE TECHNOLOGIES INC.
222 Third Street, Suite 3300, Cambridge MA 02142. 617/621-0820. **Contact:** Human Resources. **Description:** Dataware Technologies, one of *Inc.* magazine's "100 Fastest-Growing Small Public Companies," is a leading international developer and marketer of high-performance, multiplatform, and multilingual products. Dataware provides database software products, including CD-author, CD-answer, and recorded reference books to the CD-ROM market; and provides an extensive range of services to enable CD-ROM publishers to create, update, distribute, and support their electronic information products. Dataware's software is used by 1,700 organizations worldwide. **Corporate headquarters location:** This Location.

DATAWATCH CORPORATION
234 Ballardvale Street, Wilmington MA 01887-1032. 508/988-9700. **Contact:** Human Resources. **Description:** Datawatch Corporation designs, manufactures, markets, and supports personal computer software. Datawatch's principal products are Monarch, which provides data access,

translation, and reporting capability to users of network PCs; and VIREX, which detects, repairs, and monitors virus infections for Apple Macintosh PCs. The company recently added its third principal product, marketed under the name Q-Support. This product results from a 1994 agreement with WorkGroup Systems Ltd., a software company based in the United Kingdom, for exclusive North American publishing and marketing rights to WorkGroup's Quetzal product line, which is a range of Information Center support management tools.

DAVIDSON & ASSOCIATES, INC.

19840 Pioneer Avenue, Torrance CA 90503. 310/793-0600. **Fax:** 310/793-0601. **Recorded Jobline:** 310/793-0599. **Contact:** Corporate Recruiter. **E-mail address:** jobs@davd.com. **Description:** Davidson & Associates, Inc. is a leading publisher and distributor of multimedia educational and entertainment software for both the home and school markets. The company's award-winning, best-selling consumer titles include the Blaster series, as well as products in the Kid Tools, Cool Tools, FUNdamentals, and adult self-improvement lines. Davidson also has partnerships with Fisher-Price and Simon & Schuster for the development of multimedia software based on Fisher-Price toys and Simon & Schuster books. Divisions of Davidson include Educational Resources, a leading reseller of educational software to schools; Blizzard Entertainment, a leading game developer; First Byte, a leader in text-to-speech technologies; Learningways, an educational software developer; and The Cute Company, an innovator in educational and entertainment software for children. **Common positions include:** Animator; Computer Programmer; Designer; Internet Services Manager; Systems Analyst; Video Production Coordinator. **Educational backgrounds include:** Accounting; Art/Design; Computer Science; Marketing. **Benefits:** 401(k); Dental Insurance; Disability Coverage; Employee Discounts; Life Insurance; Medical Insurance; Pension Plan; Tuition Assistance. **Special Programs:** Internships. **Corporate headquarters location:** This Location. **Other U.S. locations:** Nationwide. **Operations at this facility include:** Administration; Research and Development. **Listed on:** NASDAQ. **Stock exchange symbol:** DAVD. **Annual sales/revenues:** More than $100 million. **Number of employees at this location:** 300. **Number of employees nationwide:** 1,000.

DAY & ZIMMERMAN INFORMATION SOLUTIONS

1541 South Warner Road, Suite 322, Wayne PA 02842. 610/975-6883. **Contact:** Human Resources. **Description:** A systems engineering and software development firm engaged in systems development, large-scale systems analysis, information engineering, logistics, interactive computer-aided training, and in-service engineering of complex computer and weapon systems. **Corporate headquarters location:** Philadelphia PA. **Parent company:** Day & Zimmerman, Inc. provides a wide range of professional services including engineering design, construction, and procurement; clean room design and validation; construction management; technical services; automation and data processing consulting; mass real estate appraisal; security guard services; munitions manufacturing services; naval ship alterations; and logistical support. Customers encompass a variety of industries and government agencies.

DEALER INFORMATION SYSTEMS CORPORATION

114 West Magnolia Street, Suite 500, Bellingham WA 98225. 360/733-7610. **Contact:** Human Resources. **Description:** Develops management software for tractor and automobile dealerships.

DELORME MAPPING

P.O. Box 298, South Main Street, Freeport ME 04032. 207/865-4171. **Fax:** 207/865-7080. **Contact:** Human Resources. **E-mail address:** jobs@delorme.com. **World Wide Web address:** http://www.delorme.com. **Description:** A publisher of atlases and gazetteers, CD-ROM phone directories, and mapping software and related computer hardware. **Common positions include:** Computer Programmer; Graphic Designer; Project Manager; Sales Manager. **Corporate headquarters location:** This Location.

DELPHI INFORMATION SYSTEMS

1333 North California Boulevard, Walnut Creek CA 94596. 510/946-2850. **Contact:** Human Resources. **Description:** Manufactures agency management software for insurance agencies.

DELPHI INFORMATION SYSTEMS
3501 West Algonquin Road, Suite 500, Rolling Meadows IL 60008. 847/506-3100. **Contact:** Human Resources. **Description:** Develops and markets software. Delphi Information Systems supplies insurance companies with agency management applications.

DELPHI REDSHAW INC.
680 Andersen Drive, Foster Plaza Building #10, Pittsburgh PA 15220. 412/937-3661. **Contact:** Human Resources. **Description:** Develops software and offers technical support. Delphi Redshaw supplies insurance companies with agency management software. This location also offers technical support services to their customers.

DELTEK SYSTEMS INC.
8280 Greensboro Drive, 3rd Floor, McLean VA 22102. 703/734-8606. **Contact:** Human Resources. **Description:** Develops and markets software. Deltek Systems supplies businesses with financial accounting applications and other related packages.

DENTRIX DENTAL SYSTEMS
732 East Utah Valley Drive, Suite 500, American Fork UT 84003. 801/763-9300. **Contact:** Human Resources. **Description:** Develops clinical and practice management software for dentists.

DESIGN DATA SYSTEMS
11701 South Belcher Road, Suite 105, Largo FL 33773. 813/539-1077. **Contact:** Human Resources. **World Wide Web address:** http://www.designdatasys.com. **Description:** A computer software, consulting, and services firm. Design Data offers client/server applications for all major platforms including Windows, UNIX, Novell, and OS2, and operates in many areas including accounting, distribution, and project management. **Other U.S. locations:** Los Angeles CA; San Francisco CA; Chicago IL; New York NY.

DESIGN STRATEGY CORPORATION
600 3rd Avenue, Suite 100, New York NY 10016. 212/370-0392. **Contact:** Human Resources. **Description:** Develops and markets inventory control software.

DESKTOP ENGINEERING INTERNATIONAL, INC.
1200 MacArthur Boulevard, Mahwah NJ 07430. 201/818-9700. **Contact:** Human Resources. **Description:** A designer and producer of software for use in mechanical and structural engineering industries.

DIGITAL INSTRUMENTATION TECHNOLOGY, INC.
127 Eastgate Drive, Suite 20500, Los Alamos NM 87544. 505/662-1459. **Fax:** 505/662-0897. **Contact:** Human Resources. **World Wide Web address:** http://www.dit.com. **Description:** A developer of state-of-the-art data transfer programs that provide fast and efficient movement of data among a range of computer platforms. **Corporate headquarters location:** This Location.

DIGITAL INTERFACE SYSTEMS INC.
P.O. Box 1408, Benton Harbor MI 49023. 616/926-2148. **Contact:** Human Resources. **Description:** Designs software. Digital Interface Systems specializes in helping businesses integrate their systems including developing the applicable software.

DIGITAL TECHNOLOGY INTERNATIONAL
500 West 1200 South, Orem UT 84058. 801/226-2984. **Contact:** Human Resources. **Description:** Develops software for newspaper publishers.

DOCUMENTUM INC.
5671 Gibraltar Drive, Pleasanton CA 94588. 510/463-6800. **Fax:** 510/463-6850. **Contact:** Human Resources. **E-mail address:** job request@documentum.com. **World Wide Web address:** http://www.documentum.com. **Description:** Develops software. Documentum creates client/server software for document management.

DRAGON SYSTEMS, INC.
320 Nevada Street, Newton MA 02160. **Fax:** 617/965-8348. **Contact:** T. Rosker, Director of Human Resources. **E-mail address:** tamahr@dragonsys.com. **World Wide Web address:** http://www.dragonsys.com. **Description:** A leader in speech recognition technology for both the PC and Macintosh. Dragon Systems's voice activated products are offered in American English, British English, French, German, Italian, Spanish, and Swedish. Platforms include MS-DOS, Windows, Mac OS, UNIX, telephony, and portable systems. **Corporate headquarters location:** This Location. **Listed on:** Privately held. **Number of employees worldwide:** 200.

DYNAMAC CORPORATION
2275 Research Boulevard, Rockville MD 20850. 301/417-9800. **Contact:** Human Resources. **Description:** Dynamac develops environmental software concentrating on database applications.

DYNASTRESS INC.
55 East Mill Street, Akron OH 44308. 330/535-0102. **Contact:** Human Resources. **Description:** Dynastress creates databases for industrial use.

E.A.S. TECHNOLOGIES
4345 Security Parkway, New Albany IN 47150. 812/941-1535. **Contact:** Human Resources. **Description:** Develops time and attendence management software.

EMS (EFFECTIVE MANAGEMENT SYSTEMS, INC.)
12000 West Park Place, Milwaukee WI 53224. 414/359-9800. **Contact:** Jane Ruman, Human Resources. **Description:** EMS makes software that supports the management of manufacturing operations and has installed software modules for 700 manufacturing clients around the world. The company's software is focused around the concept that compression of time is the key to increased customer satisfaction, lower costs, and higher quality and profits for manufacturers. The software is intended to eliminate wasted time and to recommend and facilitate effective time management practices. EMS provides tactical tools to help customers achieve their strategic goals. The company's two related software systems are marketed as Time Critical Manufacturing-Shop Floor Information System (TCM-SFIS) and Time Critical Manufacturing-Enterprise Management System (TCM-EMS). TCM-SFIS provides production management, shop-floor scheduling, and operations support that may be used on a departmental or plantwide basis. TCM-EMS serves diverse applications that span customer service, production, procurement, inventory management, and finance and accounting. EMS personnel work with each new client on a step-by-step training and implementation program to assure timely and effective integration of Time Critical Manufacturing with the manufacturer's resources. The company also provides telephone, modem-based, and on-site support services as needed. **Listed on:** NASDAQ.

E3 ASSOCIATES LTD.
1800 Parkway Place, Suite 600, Marietta GA 30067. 770/424-0100. **Contact:** Human Resources. **Description:** E3 Associates develops inventory control software for businesses.

EAGLE POINT SOFTWARE
4131 Westmark Drive, Dubuque IA 52002-2627. 319/556-8392. **Contact:** Human Resources. **Description:** Eagle Point Software offers AEC and GIS software solutions.

EARLY CLOUD & COMPANY
127 Johnny Cake Hill Road, Middletown RI 02842. 401/849-0500. **Contact:** Human Resources. **Description:** Early Cloud & Company develops teleservicing software. **Parent company:** IBM.

EASTERN SOFTWARE CORPORATION
P.O. Box 856, Sharon PA 16146. 412/981-5087. **Contact:** Human Resources. **Description:** Develops and manufactures software for managing consumer lending and commercial loans.

EASTMAN KODAK COMPANY
900 Chelmsford Street, Lowell MA 01851. 508/323-7600. **Contact:** Human Resources. **Description:** Develops color management software.

EDIFY CORPORATION

2840 San Tomas Expressway, Santa Clara CA 95051. 408/982-2000. **Toll-free phone:** 800/944-0056. **Fax:** 408/982-0777. **Contact:** Human Resources. **E-mail address:** hr@edify.com. **World Wide Web address:** http://www.edify.com. **Description:** Develops software known as Electronic Workforce, which provides a company's customers with corporate information, through connections using the World Wide Web, telephone, and other interactive media. Applications include bank account inquiries, employee benefit enrollments, inventory status, and automatic shipment notifications. In May 1996, the company released Edify Electronic Banking System, which enables financial institutions to provide automated banking services over the World Wide Web. **Corporate headquarters location:** This Location. **Listed on:** NASDAQ. **Stock exchange symbol:** EDFY. **Number of employees nationwide:** 200.

EDMARK CORPORATION

P.O. Box 97021, Redmond WA 98073. 206/556-8400. **Fax:** 206/556-8998. **Contact:** Human Resources. **Description:** Edmark Corporation develops, publishes, and markets educational software and other products for the early childhood and special education markets. The company's customer base is geographically diverse and consists primarily of educational institutions, software distributors, and retailers. The company's management and development teams are led by educators who view the computer as an enabling technology that encourages children to learn in new ways. Edmark creates products that combine powerful educational principles with exciting characters, colorful graphics, and interactivity. The products employ time-honored, subtle techniques to invite exploration, teach learning strategies, encourage thoughtful risk-taking, and build self-confidence and self-esteem. **NOTE:** The company's physical address is 6727 185th Avenue NE, Redmond WA 98052. **Corporate headquarters location:** This Location. **Number of employees at this location:** 120.

J.D. EDWARDS & COMPANY

8055 East Tufts Avenue, Suite 1331, Denver CO 80237. 303/488-4000. **Contact:** Human Resources. **Description:** Develops software. J.D. Edwards & Company designs, markets, and supports a wide variety of software applications for businesses. This location is the company headquarters and home of the development processes. **Corporate headquarters location:** This Location.
Other U.S. locations:
- 2907 Butterfield Road, Oakbrook IL 60521. 630/575-8600. This location serves as a regional sales and marketing office.

ELECTRONIC ARTS, INC.

1450 Fashion Island Boulevard, San Mateo CA 94404. 415/571-7171. **Contact:** Human Resources. **Description:** Creates, markets, and distributes interactive entertainment software for use primarily on 16-bit computer platforms, including the Sega Genesis and Super Nintendo entertainment systems and floppy disk-based PCs (primarily IBM PCs and compatibles). **Number of employees nationwide:** 900.

ELECTRONIC KITS INTERNATIONAL (EKI)

178 South State Street, Orem UT 84058. 801/224-3618. **Contact:** Human Resources. **Description:** EKI develops software and hands-on electronic kits for schools.

ENCORE SYSTEMS INC.

900 Circle 75 Parkway, Suite 1700, Atlanta GA 30339. 770/612-3500. **Contact:** Human Resources. **Description:** Develops software for the hospitality industry.

ENSCO INC.

5400 Port Royal Road, Springfield VA 22151. 703/321-9000. **Contact:** Human Resources. **Description:** Develops information technologies software.

ENSCO INC.

1277 Taylor Road, Owego NY 13827. 607/687-4110. **Contact:** Human Resources. **Description:** Develops information technologies software.

ENTERPRISE SYSTEMS GROUP LTD.

427A Hayden Station Road, Windsor CT 06095. 860/688-3745. **Contact:** Human Resources. **Description:** Develops software. Enterprise Systems Group creates personal software and resells other manufacturers' titles.

ENTERPRISE SYSTEMS INC.

1400 South Wolf Road, Building 500, Wheeling IL 60090. 847/537-4800. **Contact:** Human Resources. **Description:** Develops software. Enterprise Systems produces and sells applications catering to the needs of specific medical facilities. Some of the programs include materials management, financial accounting, patient scheduling for operating rooms, and inventory control for health clinics.

ENTROPIC RESEARCH LABORATORY

600 Pennsylvania Avenue SE, Suite 202, Washington DC 20003. **Fax:** 202/546-6648. **Contact:** Personnel. **E-mail address:** personnel@entropic.com. **Description:** A developer of UNIX software tools for speech research and development. Tools are used in the areas of speech synthesis, analysis, recognition, text-to-speech synthesis, and coding.

EPIC DESIGN TECHNOLOGY, INC.

310 North Mary Avenue, Sunnyvale CA 94068-4111. 408/733-8080. **Contact:** Human Resources. **Description:** EPIC develops, markets, and supports a family of simulation and analysis software tools that helps integrated circuit (IC) designers better manage the timing and power characteristics of IC designs. The company's tools can be used by designers at different stages of IC development to help identify flaws, enhance speed, and reduce power consumption. EPIC's tools are focused on meeting the simulation and static and dynamic analysis requirements of IC designers and reducing the power consumption of complex ICs. The company's three principal products -- PathMill, PowerMill, and TimeMill -- utilize the company's core technologies to enable a designer to address timing and power requirements of an IC. **Corporate headquarters location:** This Location.

ERDAS

2801 Buford Highway NE, Suite 300, Atlanta GA 30329-2137. 404/248-9000. **Contact:** Human Resources. **Description:** Erdas is a software design company.

ERISCO

1700 Broadway, New York NY 10019. 212/765-8500. **Contact:** Human Resources. **Description:** Develops health management software. Erisco supplies its products to insurance agencies and health care providers.

ERUDITE SOFTWARE AND CONSULTING, INC.

2474 North University Avenue, Suite 100, Provo UT 84604. 801/373-6100. **Fax:** 801/373-9922. **Contact:** Human Resources. **Description:** Develops custom applications around a client's needs. Erudite Software and Consulting also offers training programs for certain Microsoft and Novell software applications.

ETAK INC.

1430 O'Brien Drive, Menlo Park CA 94025. 415/328-3825. **Contact:** Human Resources. **Description:** Develops digital mapping software. Etak supplies the automotive and transportation industry with software for digital mapping devices. These units are being developed to show drivers a computerized map of the area they are traveling through.

EVANS & SUTHERLAND COMPUTER CORPORATION

600 Komas Drive, Salt Lake City UT 84108. 801/588-1000. **Fax:** 801/588-4511. **Contact:** Mr. Pat Galleos, Human Resources Department. **World Wide Web address:** http://www.es.com. **Description:** Designs, manufactures, sells, and services special purpose 3-D computer graphics systems developed to help people solve complicated problems and help train people to perform complex tasks. Uses of this equipment include visual simulation systems for pilot training and engineering workstations for molecular modeling or engineering design. The majority of Evans & Sutherland's customers are in the aerospace and defense-related markets. **Common positions include:** Database Manager; Hardware Engineer; Software Engineer; Systems Analyst.

Educational backgrounds include: Computer Science; Engineering; Mathematics. Benefits: Dental Insurance; Disability Coverage; Life Insurance; Medical Insurance; Pension Plan; Savings Plan; Stock Option; Tuition Assistance. Corporate headquarters location: This Location. Operations at this facility include: Administration; Manufacturing; Research and Development; Sales; Service. Listed on: NASDAQ. Number of employees at this location: 900.

EX-CEL SOLUTIONS, INC.

14614 Grover Street, Suite 200, Omaha NE 68144. 402/333-6541. Fax: 402/330-6349. Contact: Nancy M. Gaeta, Human Resource Generalist. World Wide Web address: http://www.excels.com. Description: Offers programming and systems integration services. Ex-Cel also sells and services computer hardware. NOTE: The company offers entry-level positions. Common positions include: Account Manager; Account Representative; Accountant/Auditor; Administrative Assistant; Administrative Services Manager; Applications Engineer; Computer Programmer; Customer Service Representative; Electrical/Electronics Engineer; Operations/Production Manager; Production Manager; Purchasing Agent/Manager; Sales Manager; Sales Representative; Secretary; Software Engineer; Systems Analyst; Typist/Word Processor. Educational backgrounds include: Accounting; Business Administration; Communications; Computer Science; Marketing. Benefits: 401(k); Dental Insurance; Disability Coverage; Employee Discounts; Life Insurance; Medical Insurance; Pension Plan; Profit Sharing; Savings Plan; Tuition Assistance. Corporate headquarters location: This Location. Other U.S. locations: Santa Fe Springs CA; Livermore CA; Dubuque IA; Chicago IL; Blaine MN; Grandview MO; Park Ridge NJ. Listed on: Privately held. Number of employees at this location: 40.

EXCALIBUR TECHNOLOGIES CORPORATION

1959 Palomar Oaks Way, Carlsbad CA 92009. 760/438-7900. Fax: 760/438-7979. Contact: Human Resources. Description: Excalibur Technologies Corporation designs, develops, markets, and supports computer software products used for the document imaging and multimedia information retrieval marketplaces. The company also offers consulting, training, maintenance, and systems integration services in support of its customers' use of its software products. In addition, the company performs research and development under contract and licenses proprietary software products for use in office, identification, and multimedia information retrieval systems. The company distributes its products through direct sales, distributors, select resellers, and vertical market suppliers. Corporate headquarters location: This Location.

EXECUTIVE SOFTWARE INTERNATIONAL

701 North Brand Boulevard, Sixth Floor, Glendale CA 91203. 818/547-2050. Fax: 818/545-9241. E-mail address: dirpers@executive.com. World Wide Web address: http://www.execsoft.com. Contact: Director of Personnel. Description: Develops and markets systems software for the VAX computer manufactured by Digital Equipment Corporation. Executive Software International also markets, sells, and supports products for the Windows NT operating system. Common positions include: Account Representative; Computer Programmer; Software Engineer; Systems Manager; Technical Writer/Editor. Benefits: Dental Insurance; Medical Insurance. Special Programs: Training programs available. Corporate headquarters location: This Location. International locations: United Kingdom. Listed on: Privately held. Annual sales/revenues: $21 - $50 million. Number of employees at this location: 102. Number of employees worldwide: 108. Number of projected hires for 1997 - 1998 at this location: 10.

EXEMPLAR LOGIC, INC.

815 Atlantic Avenue, Suite 105, Alameda CA 94501. Contact: Human Resources. E-mail address: jobs@exemplar.com. World Wide Web address: http://www.exemplar.com. Description: A developer and marketer of EDA software tools.

EXPRESS TECHNOLOGIES CORPORATION

3753 Howard Hughes Parkway, Suite 200, Las Vegas NV 89109. 702/733-6230. Toll-free phone: 800/654-9548. Toll-free fax: 800/654-9533. Contact: Human Resources. World Wide Web address: http://www.exptech.com. Description: A developer of software called World Watch, displaying real time geographical locations throughout the world, as well as illuminated patterns which delineate regions currently experiencing daylight. Express products are also available as World Watch screen savers.

EXPRESS TECHNOLOGIES CORPORATION

107 North Broadway, De Pere WI 54115. 414/337-1640. **Toll-free phone:** 800/654-9548. **Toll-free fax:** 800/654-9533. **Contact:** Human Resources. **World Wide Web address:** http://www.exptech.com. **Description:** A developer of software called World Watch, displaying real time geographical locations throughout the world, as well as illuminated patterns which delineate regions currently experiencing daylight. Express products are also available as World Watch screen savers.

EXTENSIS CORPORATION

55 SW Yamhill, 4th Floor, Portland OR 97204. **Fax:** 503/274-0530. **Contact:** Human Resources. **E-mail address:** info@extensis.com. **World Wide Web address:** http://www.extensis.com. **Description:** A developer of graphics and desktop publishing tools.

EYRING CORPORATION

6912 South 185 West, Midvale UT 84047. 801/561-1111. **Fax:** 801/565-4697. **Contact:** Supervisor of Human Resources. **Description:** Develops computer software including Assembly Management Systems (AMS) paperless workstations. **NOTE:** The company offers entry-level positions. **Common positions include:** Accountant/Auditor; Computer Programmer; Department Manager; Electrical/Electronics Engineer; Human Resources Manager; Operations/Production Manager; Purchasing Agent/Manager; Systems Analyst; Technical Writer/Editor. **Educational backgrounds include:** Accounting; Computer Science; Engineering; Finance; Marketing. **Benefits:** Dental Insurance; Disability Coverage; Life Insurance; Medical Insurance; Profit Sharing; Savings Plan; Tuition Assistance; Vision Insurance. **Corporate headquarters location:** This Location. **Other U.S. locations:** IL; MO. **Operations at this facility include:** Administration; Manufacturing; Sales; Service. **Annual sales/revenues:** $11 - $20 million. **Number of employees at this location:** 40.

FDP CORPORATION

2140 South Dixie Highway, Miami FL 33133. 305/858-8200. **Contact:** Human Resources. **Description:** Develops financial software for insurance agencies.

FDS INTERNATIONAL

18 West Ridgewood Avenue, Paramus NJ 07652. 201/843-0800. **Contact:** Human Resources Department. **Description:** Develops freight forwarding software.

FAIRCOM

4006 West Broadway, Columbia MO 65203-0100. 573/445-6833. **Fax:** 573/445-9698. **BBS:** 573/445-6318. **Contact:** Human Resources. **E-mail address:** faircom@faircom.com. **World Wide Web address:** http://www.faircom.com. **Description:** A provider of cross-platform C development tools including client/server technology.

FARPOINT TECHNOLOGIES, INC.

133 Southcenter Court, Suite 1000, Morrisville NC 27560. 919/460-4551. **Toll-free phone:** 800/645-5913. **Fax:** 919/460-7606. **BBS:** 919/460-5771. **Contact:** Human Resources. **E-mail address:** farpoint@fpoint.com. CompuServe: 74431,1406 and GO FARPOINT. **World Wide Web address:** http://www.fpoint.com. **Description:** A developer and manufacturer of custom controls for Windows programming. Product line includes AWARE/VBX, BUTTONMAKER, TAB PRO, and TAB/VBX.

FEDERAL APD, INC.

24700 Crestview Court, Farmington Hills MI 48335. 810/477-2700. **Toll-free phone:** 800/521-9330. **Fax:** 810/473-1259. **Contact:** Julie Rockett, Human Resource Administrator. **Description:** A manufacturer of facility management system software used for parking, access, and revenue control. Products include the self-park, microprocessor-based barrier gate. **NOTE:** The company offers entry-level positions. **Common positions include:** Computer Operator; Computer Programmer; Electrical/Electronics Engineer; Project Engineer; Software Engineer; Systems Engineer. **Educational backgrounds include:** Computer Science; Engineering. **Benefits:** 401(k); Daycare Assistance; Dental Insurance; Disability Coverage; Life Insurance; Medical Insurance; Pension Plan; Profit Sharing; Savings Plan; Tuition Assistance. **Corporate headquarters location:**

Oak Brook IL. **Parent company:** Federal Signal Corporation. **Listed on:** New York Stock Exchange. **Annual sales/revenues:** $11 - $20 million.

FIDELIO TECHNOLOGIES
2640 Golden Gate Parkway, Suite 211, Naples FL 34105. 941/643-7999. **Contact:** Human Resources. **Description:** Fidelio Technologies develops software for large hotel chains.

FINANCIAL INFORMATION TECHNOLOGY INC.
900 North Woods Chapel Road, Blue Springs MO 64015. 816/229-8225. **Contact:** Al Opsal, Recruiting. **Description:** Financial Information Technology develops policy administration software packages for insurance agencies.

FINANCIAL PUBLISHING COMPANY
82 Brookline Avenue, Boston MA 02215. 617/262-4040. **Contact:** Linda Turco, Personnel Coordinator. **Description:** A developer of mortgage-backed securities software and information for the banking and financial markets. **Common positions include:** Computer Programmer; Financial Analyst; Software Engineer. **Corporate headquarters location:** This Location. **Listed on:** Privately held. **Number of employees at this location:** 85.

FINANCIAL SOFTWARE INC.
680 Engineering Drive, Suite 120, Norcross GA 30092. 770/662-0404. **Contact:** Human Resources. **Description:** Develops and supports legal and accounting software.

FIRST DATA INVESTOR SERVICES GROUP
7557 Rambler Road, Suite 1050, Dallas TX 75231. 214/891-0900. **Contact:** Sales Department. **Description:** Develops pension benefits and general retirement software. Customers include investment houses.

FISCHER INTERNATIONAL SYSTEMS CORPORATION
3506 Mercantile Avenue, Naples FL 34104. 941/643-1500. **Contact:** Human Resources. **Description:** Develops and sells software for electronic mailings and PC security.

FISERV
1818 Commerce Drive, South Bend IN 46628. 219/282-3300. **Contact:** Tracy Lee, Human Resources Manager. **Description:** Develops software for the mortgage industry.

FITECH SYSTEMS
3098 Piedmont Road NE, Suite 200, Atlanta GA 30305-2600. 404/262-2298. **Contact:** Human Resources. **Description:** Fitech Systems is a designer of software for use by credit unions.

FLAVORS TECHNOLOGY, INC.
670 North Commercial Street, Manchester NH 03101. 603/647-1270. **Contact:** Human Resources. **Description:** A software developer.

FOLIO CORPORATION
5072 North 300 West, Provo UT 84604. 801/229-6700. **Contact:** Human Resources. **Description:** Produces electronic publishing software.

FOREFRONT, INC.
4710 Table Mesa Drive, Suite B, Boulder CO 80303. 303/499-9181. **Contact:** Human Resources. **E-mail address:** info@ff.com. **World Wide Web address:** http://www.ff.com. **Description:** A software developer. Products include ForHelp, an authoring tool; ForeHTML, a Website creation tool; ForeVB, a visual basic applications process automator; OLE for Help, a help files tool; and Help Buttons, a custom button tool.

FOREST COMPUTER INC.
1749 Hamilton Road, Okemos MI 48864. 517/349-4700. **Contact:** Human Resources Department. **Description:** A software development house specializing in network software.

FOUR SEASONS SOFTWARE
2025 Lincoln Highway, Edison NJ 08817. 908/248-6667. **Contact:** Human Resources Department. **Description:** Develops and sells enterprise-level applications.

FOURGEN SOFTWARE TECHNOLOGY INC.
115 NE 100th Street, Seattle WA 98125. 206/522-0055. **Contact:** Human Resources. **Description:** Develops warehouse distribution and financial software.

FOURTH SHIFT CORPORATION
3500 Parkway Lane, Suite 200, Norcross GA 30092. 770/446-9719. **Contact:** Human Resources. **Description:** Fourth Shift Corporation is a manufacturer and retailer of manufacturing software.

FOURTH SHIFT CORPORATION
7900 International Drive, Minneapolis MN 55425. 612/851-1500. **Contact:** Human Resources. **Description:** Develops and markets manufacturing software.

FREEDOM SOFTWARE
9F Oliver Court, Pittsburgh PA 15239. 412/327-4940. **Fax:** 412/327-6518. **Contact:** Human Resources. **E-mail address:** support@freedom.lm.com. **World Wide Web address:** http://www.freedom.lm.com. **Description:** A developer of open systems technologies providing software for the open systems market.

FRONTIER TECHNOLOGIES
10201 North Port Washington Road, Mequon WI 53092. 414/241-4555. **Contact:** Human Resources. **World Wide Web address:** http://www.frontiertech.com. **Description:** A developer of Internet and intranet software.

FUNK SOFTWARE, INC.
222 Third Street, Cambridge MA 02142. 617/497-6339. **Fax:** 617/547-1031. **Contact:** Human Resources. **E-mail address:** sales@funk.com. **World Wide Web address:** http://www.funk.com. **Description:** A developer of remote access and LAN-based communications software. Product line includes Steel-Belted Radius for NetWare, a centralized authentication for dial-in users; Wanderlink, a remote access software for NetWare; Proxy, a remote control software which allows remote operation of PCs; and AppMeter, a software usage metering product. Founded in 1982. **Corporate headquarters location:** This Location. **Listed on:** Privately held.

FUTURE THREE SOFTWARE INC.
33031 Schoolcraft Road, Livonia MI 48150. 313/261-5609. **Contact:** Human Resources. **Description:** Develops software packages for large corporations.

GCC TECHNOLOGIES
209 Burlington Road, Bedford MA 01730. 617/275-5800. **Contact:** Human Resources. **Description:** Develops software for personal computers. GCC also conducts research in the areas of computer graphics, VLSI design, consumer robotics, and digital sound generation. **NOTE:** Applicants must show proven ability in one or more of the following areas: real-time programming for microprocessors; applications software for personal computers; or computer graphics. **Common positions include:** Accountant/Auditor; Administrator; Blue-Collar Worker Supervisor; Computer Programmer; Credit Manager; Customer Service Representative; Electrical/Electronics Engineer; Manufacturer's/Wholesaler's Sales Rep.; Mechanical Engineer; Technical Writer/Editor. **Educational backgrounds include:** Computer Science; Economics; Engineering; Finance; Liberal Arts; Marketing. **Benefits:** Dental Insurance; Disability Coverage; Life Insurance; Medical Insurance; Savings Plan. **Corporate headquarters location:** This Location. **Operations at this facility include:** Administration; Manufacturing; Research and Development; Sales; Service. **Number of employees at this location:** 150.

GRC INTERNATIONAL, INC.
2940 Presidential Drive, Suite 390, Fairborn OH 45324. 937/429-7773. **Fax:** 937/429-7769. **Contact:** Human Resources Department. **E-mail address:** cjobs@grci.com. **World Wide Web**

address: http://www.grci.com. **Description:** This location assists in military-based design projects and also develops some software products. GRC International is an international provider of knowledge-based professional services and technology-based product solutions to government and commercial customers. GRC International's activities encompass sophisticated telecommunications products, network systems analysis, and network software development operations for the commercial market. The company creates large-scale decision-support systems and software engineering environments; applies operations research and mathematical modeling to business and management systems; and implements advanced database technology. GRC International also provides studies and analysis capabilities for policy development and planning; modeling and simulation of hardware and software used in real-time testing of sensor, weapon, and battlefield management command, control, and communication systems; and testing and evaluation. **NOTE:** Entry-level positions are offered. **Common positions include:** Aerospace Engineer; Computer Programmer; Design Engineer; Electrical/Electronics Engineer; Industrial Engineer; Mechanical Engineer; MIS Specialist; Multimedia Designer; Operations/Production Manager; Software Engineer; Systems Analyst. **Educational backgrounds include:** Business Administration; Computer Science; Engineering. **Benefits:** 401(k); Dental Insurance; Disability Coverage; Life Insurance; Medical Insurance; Profit Sharing; Savings Plan; Tuition Assistance. **Corporate headquarters location:** Vienna VA. **Operations at this facility include:** Regional Headquarters. **Listed on:** New York Stock Exchange. **Number of employees at this location:** 50. **Number of employees worldwide:** 1, 300.
Other U.S. locations:
- 635 Discovery Drive, Huntsville AL 35806. 205/922-1941. This location provides technical support for software development and engineering.
- 985 Space Center Drive, Suite 310, Colorado Springs CO 80915. 719/596-5395. This location develops software for the U.S. government.

GSE SYSTEMS
8930 Stanford Boulevard, Columbia MD 21045. 410/312-3500. **Contact:** Human Resources. **Description:** Develops software. GSE Systems creates power simulation, nuclear simulation, and process controls programs.

GSI TRANSCOMM
1380 Old Freeport Road, Pittsburgh PA 15238. 412/963-6770. **Fax:** 412/963-6779. **Contact:** Personnel Director. **Description:** GSI Transcomm is the North American headquarters of GSI, one of the world's leading providers of business software solutions and consulting services. The company's software product, TOLAS, is for sales and customer service, product distribution, logistics and warehouse management, as well as inventory and financial control. **Common positions include:** Computer Programmer; Software Engineer; Systems Analyst. **Educational backgrounds include:** Economics. **Benefits:** 401(k); Dental Insurance; Disability Coverage; Life Insurance; Medical Insurance; Pension Plan; Tuition Assistance. **Corporate headquarters location:** This Location. **Other U.S. locations:** Huntington Beach CA; Santa Clara CA; Elmsford NY. **Parent company:** GSI (Paris, France). **Operations at this facility include:** Administration; Regional Headquarters; Research and Development; Sales. **Listed on:** Privately held. **Number of employees nationwide:** 200.

GT INTERACTIVE SOFTWARE
16 East 40th Street, New York NY 10016. 212/726-6500. **Fax:** 212/726-6590. **Contact:** Human Resources Department. **Description:** GT Interactive Software creates, wholesales, and markets a wide variety of software products. The company specializes in interactive games.

GTCO CORPORATION
7125 Riverwood Drive, Columbia MD 21044. 410/381-6688. **Fax:** 410/290-9065. **Contact:** Human Resources. **E-mail address:** webmaster@gtco.com. **World Wide Web address:** http://www.gtco.com. **Description:** GTCO (Graphics Technology Company) and its subsidiary Science Accessories are suppliers of precision 2D and 3D sonic digitizer systems for CAD, graphic arts, desktop publishing, mapping, and GIS applications. Product titles include Ultima II, Roll-Up II, QuikRuler, AccuTab, and Freepoint.

GTE DATA SERVICES INC.

P.O. Box 290152, Temple Terrace FL 33687. 813/978-4000. **Contact:** Human Resources. **Description:** Develops telecommunications software. This company is associated with GTE Corporation, one of the world's largest publicly held telecommunications companies.

GALILEO INTERNATIONAL

9700 West Higgins Road, Suite 400, Rosemont IL 60018. 847/518-4000. **Contact:** Human Resources Department. **Description:** Galileo International is a global company specializing in travel technology.

GEAC COMPUTER

61 South Paramus Road, Paramus NJ 07652. 201/843-8700. **Contact:** Human Resources. **Description:** Develops and markets business applications software in the areas of finance, human resources, inventory, materials, management, manufacturing, health care, and higher education. Among the company's products is the SmartStream series of financial software.
Other U.S. locations:
- 15621 Red Hill Avenue, Tustin CA 92780. 714/258-5800. This is a customer support location.
- 66 Perimeter Center East, Atlanta GA 30346. 404/239-2000.
- 3150 Premier Drive, Irving TX 75063. 214/490-3482. This location is a sales office.

GENELCO INC.

1600 South Brentwood Boulevard, Suite 500, St. Louis MO 63144. 314/962-2040. **Contact:** Human Resources. **Description:** Develops claim processing software for insurance agencies.

GENERAL MAGIC, INC.

420 North Mary Avenue, Sunnyvale CA 94086. 408/774-4000. **Fax:** 408/774-4024. **Contact:** Human Resources. **E-mail address:** staffing@genmagic.com. **Description:** General Magic, Inc. was founded to develop object-oriented software platform technologies for a new generation of electronic devices, networks, and services. The company has developed two platform technologies called Magic Cap and Telescript. Magic Cap software integrates electronic mail, fax, telephone, and paging capabilities, enables wireless and wireline communications, and provides an intuitive user interface for personal intelligent communicators and other communications devices. Telescript software allows the creation of network platforms by providing a new communications architecture using intelligent mobile software programs that route themselves through networks and perform various tasks on behalf of a user, from carrying messages to executing transactions. The company's technologies are open platforms intended to enable the development of communicating applications and services by independent software vendors, electronic merchants and publishers, and other third parties. **NOTE:** Indicate what position you are most interested in applying for when sending a resume.

GENERAL SCIENCES CORPORATION

6100 Chevy Chase Drive, Suite 200, Laurel MD 20707. 301/953-2700. **Fax:** 301/953-1213. **Contact:** Mrs. Pat Robinson, Personnel Manager. **Description:** Develops computer software for a wide variety of scientific studies.

GENESYS CORPORATION

6797 Dorsey Road, Suite 200, Dorsey MD 21227-6231. 410/379-5400. **Contact:** Office Manager. **Description:** Develops and distributes imaging software.

GEOWORKS

960 Atlantic Avenue, Alameda CA 94501. 510/814-1660. **Contact:** Human Resources. **Description:** Geoworks develops and markets the GEOS operating system, a key enabling technology for the consumer computing device (CCD) market. Working with leading hardware manufacturers, communications providers, and independent software vendors, Geoworks focuses on software packages under $500.

GLACI, INC.
P.O. Box 26354, Wauwatosa WI 53226. 414/475-6388. **Fax:** 414/475-7388. **Contact:** Human Resources. **E-mail address:** info@glaci.com. **World Wide Web address:** http://www.glaci.com. **Description:** A Web server software developer. Products are compatible with Novell NetWare.

GLASPAC TOTAL SOLUTIONS (GTS)
117 South 700 East, American Fork UT 84003. 801/756-3091. **Fax:** 801/756-3099. **Contact:** Human Resources. **Description:** GTS develops business and management software for glass companies of all types.

GLENAYRE
4800 River Green Parkway, Duluth GA 30136. **Fax:** 770/813-4780. **Contact:** Human Resources. **E-mail address:** imoore@atlanta.glenayre.com. **World Wide Web address:** http://www. glenayre.com. **Description:** A developer of communications software and related products.

GLOBAL VILLAGE COMMUNICATION, INC.
1144 East Arques Avenue, Sunnyvale CA 94086. 408/523-1000. **Fax:** 408/523-2407. **Contact:** Human Resources. **E-mail address:** careers@globalvillage.com. **World Wide Web address:** http://www.globalvillage.com. **Description:** A leading supplier of integrated communication solutions for personal computer uses. Global Village Communication acquired SofNet, Inc. in 1994, a leading PC software manufacturer. The company introduced a cross-platform plug-and-play Internet service for small businesses in 1995. Global Village Communication, U.K. Ltd. was the result of a 1996 acquisition and offers complete Windows-based ISDN solutions. **Benefits:** 401(k); Dental Insurance; Disability Coverage; Employee Discounts; Life Insurance; Medical Insurance; Tuition Assistance; Vision Insurance. **Corporate headquarters location:** This Location. **Other U.S. locations:** Marietta GA; Richardson TX. **Operations at this facility include:** Administration; Research and Development; Sales; Service. **Listed on:** NASDAQ. **Annual sales/revenues:** More than $100 million. **Number of employees at this location:** 180. **Number of employees nationwide:** 200. **Number of employees worldwide:** 250.

GLOBALINK, INC.
9302 Lee Highway, Suite 1200, Fairfax VA 22031. 703/273-5600. **Fax:** 703/273-3866. **Contact:** Jean Ellen Wood, Director of Human Resources. **World Wide Web address:** http://www.globalink.com. **Description:** Designs, develops, markets, and supports foreign language translation software for business and professional use, including language and applications, books, and other products for the microcomputer marketplace. In addition, the company provides professional language services through its multilingual staff and through contract arrangements with independent linguists and translators. Globalink offers a wide array of bilingual translation tools for French, German, Italian, and Spanish. The software enables users to automatically translate text at their computer. **Corporate headquarters location:** This Location. **Other U.S. locations:** San Diego CA. **Listed on:** American Stock Exchange. **Number of employees nationwide:** 90.

GNOSSOS SOFTWARE INC.
1625 K Street NW, Suite 1250, Washington DC 20006-1604. 202/463-1200. **Fax:** 202/785-9562. **Contact:** President. **E-mail address:** info@gnossos.com. **World Wide Web address:** http://www.gnossos.com. **Description:** Develops software and offers consulting services. Gnossos Software focuses on database applications for client businesses using Paradox for Windows and Delphi. The company's products include Keep in Touch: Government Relations for managing corporate affairs departments, and Keep in Touch: Contact Marketing System for sales and marketing departments. **Common positions include:** Computer Programmer; Marketing Specialist; Systems Analyst; Technical Support Representative. **Educational backgrounds include:** Art/Design; Biology; Business Administration; Computer Science; Liberal Arts; Marketing; Mathematics. **Benefits:** Life Insurance; Medical Insurance; Profit Sharing; Savings Plan; Tuition Assistance. **Corporate headquarters location:** This Location. **Listed on:** Privately held. **Number of employees at this location:** 15.

GOLDEN SOFTWARE, INC.
809 14th Street, Golden CO 80401-1866. **Toll-free phone:** 800/972-1021. **Fax:** 303/279-0909. **Contact:** Patrick Madison, Human Resources. **E-mail address:** pat@golden.com. **World Wide Web address:** http://www.golden.com/golden/. **Description:** A developer of contouring, mapping, and graphing software for Windows and DOS. **Corporate headquarters location:** This Location.

BILL GOOD MARKETING
406 West South Jordan Parkway, Suite 600, South Jordan UT 84095. 801/572-1480. **Fax:** 801/572-1496. **Contact:** Jackie Rasband, Human Resources Manager. **Description:** Develops and markets software used by stockbrokers. **Common positions include:** Computer Operator; Customer Service Representative; Database Manager; Editor; Finance Director; Graphic Artist; Human Resources Manager; Managing Editor; Marketing Manager; Marketing Specialist; Project Manager; Quality Control Supervisor; Sales and Marketing Representative; Sales Manager; Statistician; Technical Writer/Editor; Typist/Word Processor. **Educational backgrounds include:** Computer Science; Finance; Marketing. **Benefits:** 401(k); Dental Insurance; Life Insurance; Medical Insurance. **Corporate headquarters location:** This Location. **Listed on:** Privately held. **Number of employees at this location:** 67.

GRADIENT TECHNOLOGIES INC.
Two Mount Royal Avenue, Marlborough MA 01752. 508/624-9600. **Contact:** John Marino, Corporate Recruiter. **Description:** Develops distributed computing software.

GRAFTEK, INC.
6260 Lookout Road, Boulder CO 80301. 303/581-0588. **Fax:** 303/581-0566. **Contact:** Human Resources. **E-mail address:** support@graftek.com. **World Wide Web address:** http://www.graftek.com. **Description:** A developer of a line of design, drafting, analysis, and N/C programs integrated into workstations. Products include GMS, a 3D geometric modeling software; AGILE, an Algorithmic Interface Graphics LanguagE; DRAFTING, computer-aided drafting; and GMSMOTIF, a graphical user interface.

GRANITE SYSTEMS RESEARCH
296 Route 101, Suite 7, Amherst NH 03031. 603/672-3500. **Fax:** 603/672-1748. **Contact:** Human Resources. **E-mail address:** granite@granite.com. **World Wide Web address:** http://www.granite.com. **Description:** A developer of network operations software. The product line features Xper, an administration automation software package.

GRASP INFORMATION CORPORATION
725 Concord Avenue, Cambridge MA 02138. 617/499-1499. **Fax:** 617/499-1498. **Contact:** Human Resources. **E-mail address:** success@grasp.com. **World Wide Web address:** http://www.grasp.com. **Description:** A developer of software designed to access and organize information focusing on Websites, Usenet groups, CD-ROMs, and online information. **Common positions include:** Software Engineer. **Corporate headquarters location:** This Location.

GREAT PLAINS SOFTWARE, INC.
1701 38th Street SW, Fargo ND 58103. 701/281-3149. **Fax:** 701/281-6600. **Contact:** Human Resources Department. **E-mail address:** apply@gps.com. **World Wide Web address:** http://www.gps.com. **Description:** A leading developer of integrated and modular accounting and financial management software. The company offers solutions for customers ranging from small businesses to midrange corporations. Great Plains provides solutions for companies that want to take advantage of such technologies as graphical user interfaces, cross-platform connectivity, SQL, and client/server computing. More than 3,000 VARs, consultants, and independent software developers sell, install, and support Great Plains's products worldwide. Great Plains also customizes computer systems to fit individual business needs and develop vertical market software that integrates with Great Plains applications. **NOTE:** Entry-level positions and training programs are offered. **Common positions include:** Account Manager; Accountant/Auditor; Administrative Assistant; Consultant; Controller; Customer Service Representative; Database Manager; Design Engineer; Editor; Education Administrator; Electrical/Electronics Engineer; Financial Analyst; Graphic Artist; Human Resources Manager; Internet Services Manager; Librarian; Market Research Analyst; Marketing Manager; MIS Specialist; Multimedia Designer; Public Relations

Specialist; Purchasing Agent/Manager; Quality Control Supervisor; Sales Executive; Sales Manager; Sales Representative; Systems Analyst; Systems Manager; Technical Writer/Editor; Telecommunications Manager; Vice President of Marketing. **Educational backgrounds include:** Accounting; Art/Design; Business Administration; Communications; Computer Science; Economics; Engineering; Finance; Liberal Arts; Marketing; Mathematics; Public Relations. **Benefits:** 401(k); Dental Insurance; Disability Coverage; Employee Discounts; Financial Planning; Life Insurance; Medical Insurance; Pension Plan; Profit Sharing; Savings Plan; Telecommuting; Tuition Assistance. **Corporate headquarters location:** This Location. **Listed on:** Privately held. **Number of employees at this location:** 550. **Number of employees nationwide:** 600.

GRECO SYSTEMS
372 Coogan Way, El Cajon CA 92020. 619/442-0205. **Toll-free phone:** 800/234-7326. **Fax:** 619/447-8982. **Contact:** Suzanne Pierce, Director of Human Resources. **E-mail address:** grecosys@adnc.com. **Description:** A manufacturer of industrial computer software and hardware systems for communication and storage in factory automation environments. **Common positions include:** Sales Representative; Software Engineer. **Benefits:** 401(k); Dental Insurance; Disability Coverage; Life Insurance; Medical Insurance; Profit Sharing; Tuition Assistance. **Corporate headquarters location:** This Location. **Listed on:** Privately held. **Annual sales/revenues:** $5 - $10 million. **Number of employees at this location:** 60.

GREEN MOUNTAIN SOFTWARE CORPORATION
P.O. Box 700, Colchester VT 05446. 802/865-2728. **Contact:** Lou Krieg, President. **Description:** Develops software. Green Mountain Software custom designs programs to a client's particular needs.

GROUP 1 SOFTWARE, INC.
4200 Parliament Place, Suite 600, Lanham MD 20706-1844. 301/731-2300. **Contact:** Trent Lutz, Human Resources Manager. **Description:** Group 1 Software, Inc. develops, acquires, markets, and supports specialized, integrated list management, mail management, and marketing support software systems. Group 1 Software also publishes list and mail management software products. **Common positions include:** Accountant/Auditor; Administrator; Computer Programmer; Customer Service Representative; Financial Analyst; Instructor/Trainer; Marketing Specialist; Operations/Production Manager; Public Relations Specialist; Sales Executive; Software Developer; Software Engineer; Systems Analyst; Technical Writer/Editor. **Educational backgrounds include:** Accounting; Business Administration; Communications; Computer Science; Engineering; Finance; Marketing. **Benefits:** 401(k); Dental Insurance; Disability Coverage; Life Insurance; Medical Insurance; Tuition Assistance. **Corporate headquarters location:** This Location. **Other U.S. locations:** CA; GA; IL; MN; NJ; NV; TX; VA. **Parent company:** Comnet Corporation. **Operations at this facility include:** Administration; Research and Development; Sales. **Listed on:** NASDAQ. **Annual sales/revenues:** $21 - $50 million. **Number of employees at this location:** 250. **Number of employees nationwide:** 325. **Number of employees worldwide:** 365.

GROUP 1 SOFTWARE, INC.
300 Galleria Parkway, Suite 720, Atlanta GA 30339. 770/951-8581. **Contact:** Human Resources. **Description:** This is a sales office for the company. Overall, Group 1 Software, Inc. develops, acquires, markets, and supports specialized, integrated list management, mail management, and marketing support software systems. Group 1 Software also publishes list and mail management software products. **Common positions include:** Accountant/Auditor; Administrator; Computer Programmer; Customer Service Representative; Financial Analyst; Instructor/Trainer; Marketing Specialist; Operations/Production Manager; Public Relations Specialist; Sales Executive; Software Developer; Software Engineer; Systems Analyst; Technical Writer/Editor. **Educational backgrounds include:** Accounting; Business Administration; Communications; Computer Science; Engineering; Finance; Marketing. **Benefits:** 401(k); Dental Insurance; Disability Coverage; Life Insurance; Medical Insurance; Tuition Assistance. **Corporate headquarters location:** Lanham MD. **Other U.S. locations:** CA; IL; MN; NJ; NV; TX; VA. **Parent company:** Comnet Corporation. **Operations at this facility include:** Sales. **Listed on:** NASDAQ. **Annual sales/revenues:** $21 - $50 million. **Number of employees nationwide:** 325. **Number of employees worldwide:** 365.

HBO & COMPANY

700 East Gate Drive, Suite 500, Mount Laurel NJ 08054. 609/234-4041. **Contact:** Human Resources. **Description:** This location offers sales and technical support. Overall, HBO & Company provides networking solutions and software. HBO & Company supplies physicians, hospitals, and other health care facilities with network service and support. The company's products are designed to enhance quality health care by improving the information systems providers use. **NOTE:** To apply for a job at any regional HBO & Company office, it is recommended to first send resumes to the corporate headquarters. The address is: HBO & Company, Attn: HR, 301 Perimeter Center North, Atlanta GA 30346.
Other U.S. locations:
* 4720 Walnut Street, Boulder CO 80304. 303/443-9660. This location designs and installs software for the medical industry.
* 65 Main Street SE, Suite 215, Minneapolis MN 55414. 612/623-4038.
* 5 Country View Road, Malvern PA 19355. 610/296-3838.

HBS INTERNATIONAL

411 108th Avenue NE, Suite 800, Bellevue WA 98004. 206/455-2652. **Contact:** Human Resources. **Description:** HBS International provides data processing services and develops software for the health care industry.

HK SYSTEMS

P.O. Box 1512, Milwaukee WI 53201-1512. 414/860-7000. **Contact:** Human Resources. **Description:** Designs and manufactures computer-controlled machinery for manufacturing and for various warehousing processes. This location of HK Systems writes software and designs machinery. **Corporate headquarters location:** This Location.
Other U.S. locations:
* 515 East 100 South, Salt Lake City UT 84102. 801/530-4000.

HPI (HELPFUL PROGRAMS, INC.)

600 Boulevard South, Suite 106, Huntsville AL 35802. 205/880-8782. **Fax:** 205/880-8705. **Contact:** Human Resources. **E-mail address:** sales@instalit.com. **World Wide Web address:** http://www.instalit.com. **Description:** Develops software that creates installation programs for use with a variety of operating systems. **Corporate headquarters location:** This Location.

HPR, INC.

245 First Street, Cambridge MA 02142. 617/679-8000. **Contact:** Human Resources. **Description:** A developer of software that helps health care companies contain costs. **Corporate headquarters location:** This Location.

HRB SYSTEMS INC.

P.O. Box 60, State College PA 16804. 814/238-4311. **Contact:** Pam McBride, Human Resources Manager. **Description:** A computer software and electronics company. **Common positions include:** Engineer. **Parent company:** Raytheon E-Systems.

HRSOFT, INC.

16 Revere Road, Morristown NJ 07960. 201/984-6334. **Fax:** 201/984-5427. **Contact:** Human Resources. **World Wide Web address:** http://www.hrsoft.com. **Description:** A developer and provider of human resources-related business software and services.
Other U.S. locations:
* 1719 Old Farm Road, Fairfield IA 52556. 515/472-4300.

HTE INC.

390 North Orange Avenue, Suite 2000, Orlando FL 32801. 407/841-3235. **Contact:** Roxanne Walton, Human Resources. **Description:** Develops computer software for city governments.

HAGEN SYSTEMS INC.

6438 City West Parkway, Eden Prairie MN 55344. 612/944-6865. **Contact:** Bob Hanson, Human Resources. **Description:** Develops management software for the graphic arts industry.

HARBINGER EDI SERVICES
1055 Lenox Park Boulevard, Atlanta GA 30319. 404/841-4334. **Contact:** Human Resources. **Description:** Develops EDI software.

HARLAND
4700 South Syracuse Street, Suite 900, Denver CO 80237. 303/770-5190. **Toll-free phone:** 800/937-3799. **Fax:** 303/770-5155. **Contact:** Human Resources. **Description:** Develops, markets, and supports PC-based lending and platform automation software for the financial services industry. The company's software is designed to increase efficiency in financial institutions. **Common positions include:** Account Manager; Accountant/Auditor; Administrative Assistant; Administrative Services Manager; Attorney; Customer Service Representative; Database Manager; Design Engineer; Human Resources Manager; Marketing/Public Relations Manager; Project Manager; Quality Control Supervisor; Sales Executive; Sales Manager; Software Engineer; Technical Writer/Editor. **Educational backgrounds include:** Computer Science. **Benefits:** 401(k); Dental Insurance; Disability Coverage; Employee Discounts; Life Insurance; Medical Insurance; Tuition Assistance. **Corporate headquarters location:** Atlanta GA. **Parent company:** John G. Harland Company. **Listed on:** New York Stock Exchange. **Stock exchange symbol:** JH. **Annual sales/revenues:** More than $100 million. **Number of employees at this location:** 100. **Number of employees nationwide:** 6,200.

HARRIS ELECTRONIC DESIGN AUTOMATION, INC.
7796 Victor-Mendon Road, Fishers NY 14463. 716/924-9303. **Fax:** 716/924-4729. **Contact:** Cynthia Andrews, Human Resources Administrator. **E-mail address:** cynthia@harriseda.com. **World Wide Web address:** http://www.harriseda.com. **Description:** Develops, markets, and supports system-level design automation (SLDA) software for the design through manufacture of printed circuit boards, multichip modules, and ball grid arrays. The company's products complement commonly used electronic design automation tools to optimize designs, reduce costs, enhance performance, and shorten product development cycles. **Common positions include:** Accountant/Auditor; Applications Engineer; Controller; General Manager; Human Resources Manager; Marketing Manager; Sales Representative; Software Engineer; Systems Manager; Technical Writer/Editor. **Educational backgrounds include:** Computer Science. **Benefits:** 401(k); Dental Insurance; Disability Coverage; Employee Discounts; Life Insurance; Medical Insurance; Tuition Assistance. **Corporate headquarters location:** This Location. **Other U.S. locations:** Los Angeles CA; Boston MA. **International locations:** Germany; London; Paris. **Listed on:** Privately held. **Annual sales/revenues:** $11 - $20 million. **Number of employees at this location:** 60. **Number of employees worldwide:** 80. **Number of projected hires for 1997 - 1998 at this location:** 15.

HARRISDATA
611 North Barker Road, Brookfield WI 53045. 414/784-9099. **Contact:** Human Resources. **Description:** Harrisdata manufactures software that facilitates a variety of operations, including human resources, financial services, and distribution.

HARTE-HANKS DATA TECHNOLOGIES
25 Linnell Circle, Billerica MA 01821. 508/436-8981. **Fax:** 508/663-3576. **Contact:** David Lobley, Human Resources Representative. **World Wide Web address:** http://www. hartehanks.com. **Description:** Harte-Hanks Data Technologies develops proprietary software to help clients solve database problems. Founded in 1968. **NOTE:** This company offers entry-level positions, training programs, and second and third shifts. **Common positions include:** Account Manager; Account Representative; Administrative Assistant; Applications Engineer; Computer Operator; Computer Programmer; Customer Service Representative; Database Manager; Market Research Analyst; Marketing Manager; Marketing Specialist; MIS Specialist; Project Manager; Sales Engineer; Sales Executive; Sales Manager; Sales Representative; Secretary; Software Developer; Software Engineer; Systems Analyst; Systems Manager; Technical Writer/Editor. **Educational backgrounds include:** Computer Science; Economics; Engineering; Marketing; Mathematics. **Benefits:** 401(k); Daycare Assistance; Dental Insurance; Disability Coverage; Employee Discounts; Financial Planning; Job Sharing; Life Insurance; Medical Insurance; Paid Vacation; Pension Plan; Profit Sharing; Savings Plan; Telecommuting; Tuition Assistance. **Special Programs:** Internships. **Corporate headquarters location:** San Antonio TX. **Other U.S.**

locations: Nationwide. **International locations:** Asia; Australia; Canada; South America; United Kingdom. **Parent company:** Harte-Hanks Communications. **Listed on:** New York Stock Exchange. **Annual sales/revenues:** More than $100 million. **Number of employees at this location:** 425. **Number of employees nationwide:** 6,000. **Number of employees worldwide:** 7,000. **Number of projected hires for 1997 - 1998 at this location:** 250.

HEALTH CARE DATA SYSTEMS INC.
5703 Enterprise Parkway, P.O. Box 608, De Witt NY 13214-0608. 315/446-7111. **Contact:** Human Resources. **Description:** Develops a variety of software products for the health care industry.

HEALTH SYSTEMS ARCHITECTS, INC.
244 Knollwood Drive, Bloomingdale IL 60108. 630/924-3100. **Fax:** 630/924-3581. **Contact:** Al Calalang, Senior Recruiter. **Description:** A provider of computer software for the health insurance and health care industries. **NOTE:** This company offers entry-level positions. **Common positions include:** Computer Programmer; Systems Analyst. **Educational backgrounds include:** Computer Science; Health Care. **Benefits:** 401(k); Cafeteria; Dental Insurance; Disability Coverage; Employee Discounts; Life Insurance; Medical Insurance; On-Site Gym; Profit Sharing; Spending Account; Tuition Assistance. **Special Programs:** Internships. **Corporate headquarters location:** This Location. **Other U.S. locations:** Utica NY; Dallas TX. **Operations at this facility include:** Administration; Research and Development; Sales; Service. **Listed on:** Privately held. **Annual sales/revenues:** $21 - $51 million. **Number of employees at this location:** 155. **Number of employees nationwide:** 170.

HEALTH SYSTEMS INTEGRATION INC.
8009 34th Avenue South, Bloomington MN 55425. 612/851-9696. **Fax:** 612/858-7905. **Contact:** Human Resources. **E-mail address:** jgess@hsii.ccare.com. **World Wide Web address:** http://www.health-systems.com. **Description:** A designer and developer of software for the health care industry. Founded in 1977. **Common positions include:** Account Manager; Accountant/Auditor; Administrative Assistant; Chief Financial Officer; Claim Representative; Clerical Supervisor; Computer Programmer; Customer Service Representative; Database Manager; Financial Analyst; Human Resources Manager; Marketing Specialist; MIS Specialist; Operations/Production Manager; Project Manager; Sales Executive; Sales Representative; Secretary; Software Engineer; Systems Analyst; Technical Writer/Editor. **Educational backgrounds include:** Business Administration; Computer Science; Finance; Health Care; Marketing. **Benefits:** 401(k); Dental Insurance; Disability Coverage; Life Insurance; Medical Insurance. **Corporate headquarters location:** This Location. **Other U.S. locations:** Reston VA. **Parent company:** CompuCare. **Listed on:** Privately held. **Number of employees at this location:** 350.

HEWLETT-PACKARD COMPANY
SALT LAKE CITY OPERATION
348 East Winchester, Salt Lake City UT 84107. 801/265-6200. **Contact:** Bruce Betts, Office Manager. **Description:** This location produces advanced computer-aided engineering (CAE) tools. Overall, Hewlett-Packard Company designs and manufactures measurement and computation products and systems used in business, industry, engineering, science, health care, and education. Principal products include integrated instrument and computer systems (hardware and software); peripherals; and medical electronic equipment and systems. **NOTE:** Jobseekers should send resumes to: Employment Response Center, Event #2498, Hewlett-Packard Company, Mail Stop 20-APP, 3000 Hanover Street, Palo Alto CA 94304-1181. **Corporate headquarters location:** Palo Alto CA.

HILGRAEVE INC.
Genesis Center, 111 Conant Avenue, Suite A, Monroe MI 48161. 313/243-0576. **Toll-free phone:** 800/826-2760. **Fax:** 313/243-0645. **BBS:** 313/243-5915. **Contact:** Human Resources. **E-mail address:** resume@hilgraeve.com. CompuServe: GO HILGRAEVE. **World Wide Web address:** http://www.hilgraeve.com. **Description:** A developer and manufacturer of asynchronous communications software for IBM-compatible desktop and portable PCs.

HITACHI AMERICA
1601 Trapello Road, 3rd Floor, Waltham MA 02154. 617/890-0444. **Contact:** Human Resources. **Description:** A software development division of the electronics giant.

HOGAN SYSTEMS, INC.
7525 LBJ Freeway, Dallas TX 75240. 972/386-0020. **Contact:** Human Resources. **Description:** Hogan Systems, Inc. supplies integrated software products and support services to financial institutions worldwide.

HUGHES TRAINING INC.
LINK DIVISION
P.O. Box 1237, Binghamton NY 13902-1237. 607/721-5465. **Contact:** Dave Maroney, Manager of Employment. **Description:** A developer of software-based training programs. **Common positions include:** Aerospace Engineer; Computer Scientist; Electrical/Electronics Engineer.

HUNTSVILLE MICROSYSTEMS
P.O. Box 12415, Huntsville AL 35815. 205/881-6005. **Contact:** Human Resources. **Description:** Manufactures emulators for computer systems.

HYPERION SOFTWARE CORPORATION
900 Long Ridge Road, Stamford CT 06902. 203/703-3000. **Fax:** 203/322-3904. **Contact:** Cindy Crowell, Human Resources. **Description:** Hyperion Software develops, markets, and supports a family of network-based business information software products for companies with many divisions or locations. The product line provides executives, managers, and analysts with the capability to collect, process, access, and analyze critical business information in a timely manner, using networked personal computers. **Common positions include:** Accountant/Auditor; Computer Programmer; Software Engineer. **Educational backgrounds include:** Accounting; Business Administration; Computer Science; Engineering; Marketing. **Benefits:** 401(k); Dental Insurance; Disability Coverage; Life Insurance; Medical Insurance; Stock Option; Tuition Assistance. **Special Programs:** Internships. **Corporate headquarters location:** This Location. **Operations at this facility include:** Administration; Divisional Headquarters; Regional Headquarters; Research and Development; Sales; Service. **Number of employees at this location:** 500. **Number of employees nationwide:** 600.
Other U.S. locations:
- 104 Springfield Center Drive, Woodstock GA 30188. 770/591-2044. This location serves as a regional sales office.

I-CONCEPTS
2607 Walnut Hill Lane, Suite 200, Dallas TX 75229. 214/956-7770. **Contact:** Human Resources. **Description:** A developer of software for the insurance industry.

I-LOGIX, INC.
Three Riverside Drive, Andover MA 01810. 508/682-2100. **Contact:** Human Resources. **Description:** A manufacturer of software for high-technology applications.

IBM CORPORATION
3600 Carillon Point, Kirkland WA 98033. 206/803-0600. **Toll-free phone:** 800/796-9876. **Contact:** IBM Staffing Services. **Description:** This facility is a programming center. Overall, International Business Machines (IBM) is a developer, manufacturer, and marketer of advanced information processing products, including computers and microelectronic technology, software, networking systems, and information technology-related services. The company strives to offer value worldwide through its United States, Canada, Europe/Middle East/Africa, Latin America, and Asia Pacific business units by providing comprehensive and complete product choices. **NOTE:** Jobseekers should send a resume to IBM Staffing Services, Department 1DP, Building 051, P.O. Box 12195, Research Triangle Park NC 27709-2195.

ICL RETAIL
P.O. Box 631, Wake Forest NC 27588. 919/556-6721. **Contact:** Human Resources. **Description:** Designs software for a variety of businesses.

IDX SYSTEMS CORPORATION

P.O. Box 1070, Burlington VT 05402-1070. 802/862-1022. **Contact:** Administrative Manager. **Description:** Creates and develops medical software for hospitals. IDX Systems Corporation is a provider of health care information to physician groups and academic medical centers across the country. **NOTE:** The company is located at 1400 Shelburne Road, Burlington VT. **Common positions include:** Computer Programmer; Customer Service Representative; Manufacturer's/Wholesaler's Sales Rep.; Systems Analyst; Technical Writer/Editor. **Educational backgrounds include:** Computer Science; Medical Technology. **Benefits:** Dental Insurance; Disability Coverage; Life Insurance; Medical Insurance; Profit Sharing; Savings Plan. **Corporate headquarters location:** This Location. **Number of employees nationwide:** 1,200.

IDX SYSTEMS CORPORATION

116 Huntington Avenue, Boston MA 02116. 617/424-6800. **Fax:** 617/266-5419. **Contact:** Autumn Baskin, Human Resources Assistant. **E-mail address:** baskin@idx.com. **World Wide Web address:** http://www.idx.com. **Description:** Develops medical software for hospitals. IDX Systems Corporation is a provider of health care information to physician groups and academic medical centers across the country. **NOTE:** The company offers entry-level positions. **Common positions include:** Account Manager; Administrative Assistant; Applications Engineer; Computer Operator; Computer Programmer; Market Research Analyst; Systems Analyst; Technical Writer/Editor. **Educational backgrounds include:** Computer Science; Engineering; Health Care. **Benefits:** 401(k); Dental Insurance; Disability Coverage; Employee Discounts; Financial Planning; Job Sharing; Life Insurance; Medical Insurance; Pension Plan; Profit Sharing; Savings Plan; Tuition Assistance. **Corporate headquarters location:** Burlington VT. **Other U.S. locations:** CA; GA; IL; MA. **Listed on:** NASDAQ. **Annual sales/revenues:** $5 - $10 million. **Number of employees at this location:** 450. **Number of employees nationwide:** 1,200.

IFS INTERNATIONAL

Rensselear Technology Park, 185 Jordan Road, Troy NY 12180. 518/283-7900. **Fax:** 518/283-7336. **Contact:** Human Resources. **Description:** IFS is an international software integrator providing financial and retail industries with cost-effective electronic funds transfer solutions for automated teller machines and point-of-sale terminals. IFS develops solutions that can range from large international clearinghouse banks to national and regional networks to medium-sized financial or retail institutions.

IQ SOFTWARE

3295 River Exchange Drive, Suite 550, Norcross GA 30092. 770/446-8880. **Fax:** 770/446-7261. **Contact:** Human Resources Manager. **Description:** IQ Software designs, manufactures, and markets report writing software. Products include IQ (Intelligent Query) and IQ Access which are offered in Windows, Motif, and character versions and used in a wide range of industries. **Common positions include:** Accountant/Auditor; Administrative Services Manager; Clerical Supervisor; Computer Programmer; Credit Manager; Customer Service Representative; Financial Analyst; Human Resources Manager; Systems Analyst; Technical Writer/Editor. **Educational backgrounds include:** Computer Science. **Benefits:** 401(k); Dental Insurance; Disability Coverage; Employee Discounts; Life Insurance; Medical Insurance. **Corporate headquarters location:** This Location. **Other U.S. locations:** California; Michigan; New Jersey; New York; Texas. **Operations at this facility include:** Administration; Manufacturing; Research and Development; Sales; Service. **Listed on:** NASDAQ. **Number of employees at this location:** 100. **Number of employees nationwide:** 170.

ITC (INDUSTRIAL TRAINING CORPORATION)

13515 Dulles Technology Drive, Herndon VA 20171. 703/713-3335. **Toll-free phone:** 800/638-3757. **Fax:** 703/713-0065. **Contact:** Leslye Schrank, Human Resources Director. **Description:** A full-service training company specializing in the development, production, marketing, and sale of both off-the-shelf and custom multimedia training courseware for corporate, educational, and governmental organizations. The majority of ITC's products and services are multimedia programs, but the company also markets, sells, and distributes certain linear training products (primarily videotape and text-based) through its USA Training division. Standard multimedia platforms for ITC products include both laserdisc and CD-ROM. **Corporate headquarters location:** This

Location. **Subsidiaries include:** CI Acquisition Corporation. **Listed on:** NASDAQ. **Number of employees nationwide:** 75.

IAMBIC SOFTWARE
2 North First Street, Suite 212, San Jose CA 95113. 408/882-0390. **Fax:** 408/882-0399. **Contact:** Human Resources. **E-mail address:** helpwanted@iambic.com. **World Wide Web address:** http://www.iambic.com. **Description:** A developer and manufacturer of software applications for the handheld computing market. **Common positions include:** Software Engineer. **Corporate headquarters location:** This Location.

ICONTROL
P.O. Box 1503, Watertown SD 57201-1503. 605/886-3848. **Fax:** 605/886-4180. **Contact:** Human Resources. **Description:** A software development company. Products include ICONtrol Human Resources, a fully-integrated, human resources management system; ICONtrol Direct Deposit, which allows businesses using Great Plains Dynamics Payroll to electronically deposit funds directly into employees' bank accounts; and ICONtrol Manufacturing application, a program that can instantly brief the user on the status of every order, the cost of labor and raw materials, estimated time of delivery, required changes in work assignments, profitability of the order, and other useful information. **Parent company:** ANZA, Inc. is also the parent company of three other growing companies: PERSONA, Inc., a sign manufacturer; Midcom, Inc., a transformer and modem manufacturer; and Midcom Technology, Inc., an electronic assembly manufacturer.

IKOS SYSTEMS, INC.
19050 Pruneridge Avenue, Cupertino CA 95014. 408/255-4567. **Fax:** 408/366-8699. **Contact:** Human Resources. **Description:** IKOS designs, manufactures, markets, and supports logic simulation software and compatible hardware accelerators. IKOS's simulators are used in the design of Application Specific Integrated Circuits (ASIC) and ASIC-based systems. IKOS supports a direct sales operation in North America, France, and Germany, and a distributor network covering the remainder of Europe and Asia. The company's products are used by designers of electronic systems to determine whether a system design works before incurring the cost and time to build the actual system. **Corporate headquarters location:** This Location.

IMPACT TECHNOLOGIES INC.
1807 Park 270, Suite 120, St. Louis MO 63146. 314/576-2661. **Contact:** Ray Saturnino, Recruiter. **Description:** Develops software, specializing in telecommunications applications.

IMPULSE INC.
8416 Xerxes Avenue North, Minneapolis MN 55444. 612/425-0557. **Contact:** Recruiting. **Description:** Develops three-dimensional software.

IMSPACE SYSTEMS CORPORATION
2665 Ariane Drive, Suite 207, San Diego CA 92117. 619/272-2600. **Fax:** 619/272-4292. **Contact:** Human Resources. **World Wide Web address:** http://www.imspace.com. **Description:** A publisher and developer of image and multimedia cataloging software. Product line includes Kudo software for managing and publishing digital images. Imspace's products and services enhance the ability of digital photographers, desktop publishers, and multimedia producers to work with collections of digital images, video, sound, and documents. Software is used on Macintosh and Windows computers. Founded in 1991. **Corporate headquarters location:** This Location. **Listed on:** Privately held.

INFISY SYSTEMS INC.
616 FM 1960 West, Suite 550, Houston TX 77090. 281/583-1600. **Contact:** Dick Alexander, President. **Description:** A computer programming company.

INFO POINT CORPORATION
14 Lochview Drive, Windsor CT 06095. 860/285-8919. **Contact:** Neil Schulz, President. **Description:** A developer of software for general business uses and financial software for municipalities. Info Point Corporation also resells value-added products.

INFODATA SYSTEMS INC.

12150 Monument Drive, Suite 400, Fairfax VA 22033. 703/934-5205. **Fax:** 703/934-7154. **Contact:** Eva Franklin, Human Resources Generalist. **Description:** Provides mainframe-based software products and services which help business, government, and educational institutions to automate collections of documents consisting of large amounts of text. Infodata Systems also provides complete Electronic Document Management System solutions to address developing client/server technologies, through the sale of products and software integration services. In support of its proprietary software products, the company offers a range of ongoing product support services, including training, a telephone hotline service, product enhancements, and maintenance releases. **Corporate headquarters location:** This Location. **Subsidiaries include:** Infodata Systems International Inc.; Infodata Research and Development Corporation. **Listed on:** NASDAQ. **Number of employees at this location:** 60.

INFOMED

1180 SW 36th Avenue, Pompano Beach FL 33069. 954/974-0707. **Toll-free phone:** 800/463-6633. **Fax:** 954/975-3906. **Contact:** Margie L. Rosenthal, Human Resources Manager. **Description:** A developer and publisher of software for the home health care industry. Founded in 1970. **Common positions include:** Computer Programmer; Controller; Customer Service Representative; Marketing/Public Relations Manager; MIS Specialist; Quality Control Supervisor; Sales Representative; Systems Analyst; Technical Writer/Editor. **Educational backgrounds include:** Computer Science. **Benefits:** 401(k); Dental Insurance; Disability Coverage; Employee Discounts; Life Insurance; Medical Insurance; Tuition Assistance. **Corporate headquarters location:** Atlanta GA. **Other U.S. locations:** Pompano Beach FL. **Annual sales/revenues:** $11 - $20 million. **Number of employees at this location:** 100. **Number of employees nationwide:** 300.

INFORMATION ANALYSIS INC.

2222 Gallows Road, Suite 300, Dunn Loring VA 22027. 703/641-0955. **Contact:** Leslie Reed, Human Resources Director. **Description:** Information Analysis Inc. performs activities primarily related to software applications development, hardware and software consulting services, and software sales and support services. These services include systems re-engineering, feasibility and requirements analysis, systems planning analysis and design, database design and management, software development, and project management. **Corporate headquarters location:** This Location. **Subsidiaries include:** AlliedHealth; DHD Systems; Information Systems, Inc. **Number of employees nationwide:** 160.

INFORMATION BUILDERS INC.

1250 Broadway, New York NY 10001. 212/736-4433. **Fax:** 212/564-1726. **Contact:** Lila Goldberg, Personnel Manager. **Description:** A software development firm. Products include FOCUS, EDA, and LEVEL5 software for various platforms. Other services include support and sales. **Common positions include:** Accountant/Auditor; Computer Programmer; Customer Service Representative; Operations/Production Manager; Systems Analyst; Technical Writer/Editor. **Educational backgrounds include:** Accounting; Business Administration; Communications; Computer Science; Marketing. **Benefits:** 401(k); Dental Insurance; Disability Coverage; Medical Insurance; Pension Plan; Tuition Assistance. **Corporate headquarters location:** This Location. **Operations at this facility include:** Administration; Divisional Headquarters; Research and Development. **Listed on:** Privately held. **Number of employees at this location:** 800. **Number of employees nationwide:** 1,500.
Other U.S. locations:
- 224 Airport Parkway, Suite 310, San Jose CA 95110-1022. 408/453-7600.
- 1001 Fourth Avenue, Suite 810, Seattle WA 98154. 206/628-6494.

INFORMATION ENABLING TECHNOLOGIES

4001 South 700 East, #301, Salt Lake City UT 84107. 801/284-9002. **Toll-free phone:** 800/438-7933. **Fax:** 801/284-9160. **Contact:** David Warnock, Human Resources Manager. **E-mail address:** David@IFT-SCC.com. **World Wide Web address:** http://www.IFT-SCC.com. **Description:** A software development company that creates high-end, Windows-based, critical business solutions for the accounting and financial fields. Information Enabling Technologies is its own direct sales channel for its products. **Common positions include:** Account Manager; Computer Programmer;

Production Manager; Project Manager; Sales Executive; Sales Manager; Sales Representative; Technical Writer/Editor. **Educational backgrounds include:** Business Administration; Communications; Computer Science; Finance. **Benefits:** 401(k); Cafeteria Plan; Dental Insurance; Disability Coverage; Life Insurance; Mass Transit Available; Medical Insurance; Pension Plan; Profit Sharing; Savings Plan; Telecommuting; Tuition Assistance. **Corporate headquarters location:** This Location. **Listed on:** Privately held. **Annual sales/revenues:** $11 - $20 million. **Number of employees at this location:** 30. **Number of projected hires for 1997 - 1998 at this location:** 30+.

INFORMATION MANAGEMENT ASSOCIATES, INC.
One Corporate Drive, Suite 414, Shelton CT 06484. 203/925-6800. **Contact:** Patti Jarvis, Human Resources Manager. **Description:** A software development and applications company that specializes in telemarketing, marketing, sales, and customer service application software for the IBM midrange, UNIX client/server market. **Common positions include:** Account Representative; Quality Assurance Engineer. **Educational backgrounds include:** Bachelor of Arts; MBA.

INFORMATION RESOURCE ENGINEERING INC.
8029 Corporate Drive, Baltimore MD 21236. 410/931-7500. **Contact:** Human Resources. **Description:** Develops security encryption software.

INFORMATION TECHNOLOGY, INC.
1345 Old Cheney Road, Lincoln NE 68912. 402/423-2682. **Contact:** Human Resources. **Description:** Develops financial software.

INFORMEDICS, INC.
4000 Kruse Way Place, Building 3, Suite 300, Lake Oswego OR 97035. 503/697-3000. **Fax:** 503/697-7671. **Contact:** Mr. Dale Conner, Chief Financial Officer. **Description:** Develops software for the health care industry. Informedics specializes in producing blood bank management systems. The company offers its customers six product lines, including LifeLine and ClinicManager as well as ongoing support, maintenance, and programming services. The company installs software, provides training, and furnishes software and hardware upgrades. **Corporate headquarters location:** This Location. **Number of employees at this location:** 60.

INFORMIX SOFTWARE, INC.
4100 Bohannon Drive, Menlo Park CA 94025. 415/926-6300. **Fax:** 415/926-6161. **Contact:** Human Resources. **Description:** Informix Software, Inc. provides database technology to build, deploy, run, and evolve applications. Informix products include powerful distributed database management systems, application development tools, and graphical- and character-based productivity software for delivering information to every significant desktop platform. **Common positions include:** Product Engineer; Software Engineer; Technical Support Representative. **Educational backgrounds include:** Computer Science; Engineering. **Benefits:** 401(k); Dental Insurance; Disability Coverage; Life Insurance; Medical Insurance; Profit Sharing; Tuition Assistance. **Corporate headquarters location:** This Location. **International locations:** England; Singapore. **Operations at this facility include:** Administration; Research and Development; Sales. **Listed on:** NASDAQ. **Number of employees at this location:** 800. **Number of employees nationwide:** 1,300.
Other U.S. locations:
- 5299 TC Boulevard, Suite 740, Englewood CO 80111. 303/850-0210. This location serves as a regional sales office.
- 2001 Butterfield Road, Suite 300, Downers Grove IL 60515. 630/769-1600. This location serves as a regional sales office.

INFORMIX SOFTWARE, INC.
16011 College Boulevard, Lenexa KS 66219. 913/599-7100. **Contact:** Manager of Human Resources. **Description:** Informix Software, Inc. provides database technology to build, deploy, run, and evolve applications. Informix products include powerful distributed database management systems, application development tools, and graphical- and character-based productivity software for delivering information to every significant desktop platform. **Common positions include:** Computer Programmer; Customer Service Representative; Manufacturer's/Wholesaler's Sales Rep.;

Marketing Specialist; Software Engineer; Technical Writer/Editor. **Educational backgrounds include:** Computer Science; Marketing. **Benefits:** Dental Insurance; Disability Coverage; Life Insurance; Medical Insurance; Profit Sharing; Sabbatical Leave; Savings Plan; Tuition Assistance. **Special Programs:** Internships. **Corporate headquarters location:** Menlo Park CA. **International locations:** England; Singapore. **Operations at this facility include:** Divisional Headquarters; Manufacturing; Research and Development. **Listed on:** New York Stock Exchange.

INNOVATIVE TECH SYSTEMS
444 Jacksonville Road, Warminster PA 18974. 215/441-5600. **Contact:** Human Resources. **Description:** A manufacturer of business software.

INSO CORPORATION
31 St. James Avenue, 11th Floor, Boston MA 02116. 617/753-6500. **Contact:** Human Resources. **Description:** Inso Corporation develops and markets software tools for proofing, electronic reference, and information management that improve the productive use of textual information in an electronic environment. The company markets its products worldwide primarily to original equipment manufacturers of computer hardware, software, and consumer electronics products. The company's products are divided into three product lines: Proofing Tools, which are software programs and related databases for correcting errors in spelling, grammar, punctuation, capitalization, spacing, and other mistakes in documents created by users of consumer applications; Information Products, which are software search and retrieval programs for use with structured electronic reference works based on authoritative content acquired from print publishers; and Information Management Tools, which are software programs and related databases to enhance the indexing and retrieval of information stored in electronic form. **Other U.S. locations:**
* 330 North Wabash, 15th Floor, Chicago IL 60611. 312/329-0700.

INSPIRATION SOFTWARE, INC.
7412 SW Beaverton Hillside Highway, Suite 102, Portland OR 97225-2167. 503/297-3004. **Fax:** 503/297-4676. **Contact:** Human Resources. **World Wide Web address:** http://www. inspiration.com. **Description:** A developer of visual creative planning and diagramming tool software which integrates diagramming and outlining for clear communication of ideas and processes. **Corporate headquarters location:** This Location.

INSTRUCTIVISION, INC.
Three Regent Street, Suite 306, Livingston NJ 07039-1617. 201/992-9081. **Contact:** Human Resources. **Description:** A multimedia publishing company, Instructivision develops educational video and computer programs and workbooks. The company operates a full-service video production facility encompassing a production stage, an interformat digital editing suite, offline editing, 3-D animation, and audio recording equipment.

INSURANCE AUTOMATION SYSTEMS INC.
3737 Park East Drive, Suite 202, Beachwood OH 44122. 216/464-2180. **Contact:** Human Resources. **Description:** Designs software for insurance agencies.

INSURANCE SOLUTIONS
11101 Airport Drive, Hayden Lake ID 83845. 208/762-3006. **Contact:** Human Resources. **Description:** Develops software for the insurance industry.

INSURQUOTE SYSTEMS, INC.
517 East 1860 South, Provo UT 84606. 801/373-7345. **Fax:** 801/373-4017. **Contact:** Patrick Monney, Administrative Assistant. **Description:** Insurquote Systems prepares custom software for the insurance industry.

INTEGRATED DATA SYSTEMS, INC. (IDS)
6001 Chatham Center Drive, Suite 300, Savannah GA 31405. 912/236-4374. **Fax:** 912/236-6792. **Contact:** Human Resources. **World Wide Web address:** http://www.ids-net.com. **Description:** A developer of software focusing on presentation and access products for the Internet, including three-dimensional and virtual reality interfaces for the World Wide Web. The company also

develops real-time video playback interfaces and multimedia authoring packages. Founded in 1991.

INTEGRATED MEDICAL SYSTEMS, INC.
15000 West 6th Avenue, Suite 400, Golden CO 80401. 303/279-6116. **Contact:** Human Resources. **Description:** A developer of PC software for the company's nationwide Medicom networks.

INTEGRATED SERVICES
1Q242 SW Garden Place, Portland OR 97035. 503/968-8100. **Contact:** Human Resources. **Description:** Develops management software for automobile lube shops. **Number of employees at this location:** 60.

INTEGRATED SYSTEMS, INC.
201 Moffett Park Drive, Sunnyvale CA 94089. 408/542-1500. **Fax:** 408/980-0400. **Contact:** Human Resources. **Description:** Integrated Systems, Inc. is a world leader in developing design tool and operating system software for computers embedded in products ranging from everyday consumer goods to industrial and transportation sector capital equipment. Integrated Systems' products offer a complete graphical solution for rapidly designing, simulating, and testing microprocessor-based control systems. The company also provides embedded and control system application development engineering services. The company focuses on six worldwide embedded systems markets: aerospace, automotive, communications, multimedia/entertainment, office/retail automation, and industrial process control. **Corporate headquarters location:** This Location.

INTEGRATRAK
12600 SE 38th Street, Suite 250, Bellevue WA 98006. 206/401-1000. **Contact:** Human Resources. **Description:** IntegraTrak develops software for the telecommunications industry.

INTELEGRAPHICS
741 North Grand Avenue, Waukesha WI 53186. 414/574-9000. **Contact:** Human Resources. **Description:** A firm specializing in computerized utility mapping and software. **Common positions include:** Computer Programmer; Draftsperson. **Benefits:** 401(k); Dental Insurance; Disability Coverage; Life Insurance; Medical Insurance; Pension Plan; Tuition Assistance. **Corporate headquarters location:** This Location. **Operations at this facility include:** Service. **Number of employees nationwide:** 160.

INTELLESOL INC.
825 25th Street SW, Fargo ND 58103. 701/235-3390. **Contact:** Human Resources. **Description:** This location is a sales office for the accounting software developer.

INTELLICORP
1975 El Camino Real West, Suite 101, Mountain View CA 94040-2216. 415/965-5500. **Fax:** 415/965-5647. **Contact:** Human Resources. **Description:** IntelliCorp designs, develops, and markets software development tools and provides related training, customer support, and consulting services. IntelliCorp's mission is to provide its customers with object-oriented software tools for the design, development, and delivery of scalable client/server applications. The company's products include Kappa, Kappa-PC, the Object Management Workbench (OMW), and KEE. IntelliCorp utilizes its own direct sales force and authorized resellers. **Corporate headquarters location:** This Location. **Other U.S. locations:** Nationwide. **International locations:** Europe.

INTELLUTION, INC.
One Edgewater Drive, Norwood MA 02062. 617/769-8878. **Fax:** 617/769-1990. **Contact:** Human Resources. **E-mail address:** hr@intellution.com. **World Wide Web address:** http://www.intellution.com. **Description:** A developer of industrial automation solutions software. **Common positions include:** Applications Engineer; Credit Manager; Human Resources Manager; Quality Assurance Engineer; Sales Engineer; Sales Manager; Software Engineer; Technical Writer/Editor; Tool Engineer. **Parent company:** Emerson Electric (St. Louis MO).

INTERACTIVE MAGIC

P.O. Box 13491, Research Triangle Park NC 27709. 919/461-0722. **Fax:** 919/461-0723. **Contact:** Human Resources. **E-mail address:** hr@imagicgames.com. **World Wide Web address:** http://www.imagicgames.com. **Description:** A developer and publisher of strategy and simulation computer games.

INTERACTIVE SOLUTIONS INC.

377 Route 17 South, Hasbrouck Heights NJ 07604. 201/288-6699. **Contact:** Human Resources. **Description:** Develops custom applications around a client's needs. **Common positions include:** Account Representative; Applications Engineer; Computer Programmer; Database Manager; MIS Specialist; Multimedia Designer; Sales Representative; Software Engineer; Systems Analyst; Telecommunications Manager; Typist/Word Processor. **Educational backgrounds include:** Accounting; Art/Design; Computer Science; Mathematics; Physics. **Benefits:** Disability Coverage; Financial Planning; Life Insurance; Medical Insurance; Tuition Assistance. **Corporate headquarters location:** This Location. **Listed on:** Privately held. **Annual sales/revenues:** $11 - $20 million. **Number of employees at this location:** 120.

INTERCON ASSOCIATES INC.

95 Allens Creek Road, #2, Rochester NY 14618. 716/244-1250. **Contact:** Human Resources. **Description:** Develops document assembly software and produces font cartridges.

INTERFACE SYSTEMS, INC.

5855 Interface Drive, Ann Arbor MI 48103. 313/769-5900. **Fax:** 313/769-1047. **Contact:** Jane Gensheimer, Human Resources Manager. **E-mail address:** jgen@intface.com. **Description:** A computer software and hardware developer of products for communications and document management. The company's products operate on MS Windows NT, IBM's SNA, OS/2, and UNIX systems. **Common positions include:** Computer Programmer; Customer Service Representative; Sales Representative; Software Engineer; Systems Analyst; Technical Writer/Editor. **Educational backgrounds include:** Computer Science. **Benefits:** 401(k); Dental Insurance; Life Insurance; Medical Insurance. **Corporate headquarters location:** This Location. **Other U.S. locations:** Loves Park IL. **International locations:** U.K. **Operations at this facility include:** Administration; Divisional Headquarters; Manufacturing; Research and Development; Sales; Service. **Listed on:** NASDAQ. **Annual sales/revenues:** $51 - $100 million. **Number of employees at this location:** 100. **Number of employees nationwide:** 125. **Number of employees worldwide:** 175.

INTERGO COMMUNICATIONS

903 East 18th Street, Suite 230, Plano TX 75074. 972/424-7882. **Contact:** Human Resources. **Description:** InterGo Communications develops software to monitor pornographic material seen by minors on the Internet.

INTERLEAF INC.

62 Fourth Avenue, Waltham MA 02154. 617/290-0710. **Contact:** Personnel Department. **Description:** Interleaf's software and services provide organizations all over the world with innovative document management systems and solutions. **Common positions include:** Computer Programmer; Consultant; Software Engineer; Technical Writer/Editor. **Educational backgrounds include:** Computer Science. **Special Programs:** Internships. **Corporate headquarters location:** This Location.

INTERLINQ SOFTWARE CORPORATION

11255 Kirkland Way, Kirkland WA 98033. 206/827-1112. **Fax:** 206/827-0927. **Contact:** Human Resources. **Description:** INTERLINQ Software Corporation is a leading provider of PC-based software solutions for the residential mortgage lending industry. The company's MortgageWare product line is sold to banks, savings institutions, mortgage banks, mortgage brokers, and credit unions. The MortgageWare product line is a complete PC-based software system that automates all aspects of the loan origination and secondary marketing processes, from qualifying a borrower to processing, settling, closing, and selling loans. MortgageWare also includes tools to help lenders track and manage loans in their system as well as a proprietary electronic communications system that enables data to be transferred via modem between headquarters, branch offices, and laptop origination systems. **Corporate headquarters location:** This Location. **Number of employees at this location:** 180.

INTERMETRICS INC.
23 4th Avenue, Northwest Park, Burlington MA 01803. 617/221-6990. **Contact:** Human Resources. **Description:** This location develops software. Overall, Intermetrics provides software systems, services, and products to a broad base of customers around the world. Intermetrics's expertise includes language design and programmer productivity tools; digital signal processing tools and application; hardware and system simulation; computer and network security; guidance, navigation, and control; and information systems integration. The products and services are used by major corporations and government agencies for applications in such fields as transportation, financial management and decision support, automotive, communications, manned space flight, air traffic control, anti-submarine warfare, and command and control. Along with building programming tools and environments, the company responds directly to customer needs by developing software system solutions; analyzing system requirements; performing rapid prototyping; performing integration, verification, and validation testing; re-engineering systems; maintaining operational software; and providing training and support. **Corporate headquarters location:** This Location.
Other U.S. locations:
- 607 Louis Drive, Warminster PA 18974. 215/674-2913. This location engineers software.
- 2025 112th Avenue, Suite 301, Bellevue WA 98004. 206/451-1120.

INTERNATIONAL MICROCOMPUTER SOFTWARE, INC. (IMSI)
1895 Francisco Boulevard East, San Rafael CA 94901-5506. 415/257-3000. **Fax:** 415/257-3567. **Contact:** Human Resources. **World Wide Web address:** http://www.imsisoft.com. **Description:** IMSI is a leading developer of productivity software for business and home use. The company has recently established the Living Media division which focuses on CD-ROM multimedia software for learning and education. IMSI's three primary product lines are business, consumer productivity, and multimedia learning software. The majority of IMSI's revenues come from its business software that work with, and extend the functionality of Microsoft Office, as well as operate as stand alone products. IMSI's products educate and entertain users through interactive multimedia presentations, with animation, CD-quality audio, video, graphics, and text.

INTERNATIONAL RESEARCH INSTITUTE
12200 Sunrise Valley Drive, Suite 300, Reston VA 20191. 703/715-9605. **Contact:** Don Zerbe, Human Resources Director. **Description:** A software development company.

INTERNATIONAL SOFTWARE SOLUTIONS
935 North Industrial Park Drive, Orem UT 84057. 801/224-6700. **Fax:** 801/226-8990. **Contact:** Human Resources. **Description:** A software integration company that develops custom programming for its clients. The company specializes in networking solutions, e-mail systems, and the Internet.

INTERNET SYSTEMS CORPORATION
180 North Stetson, 42nd Floor, Chicago IL 60601. 312/540-0100. **Fax:** 312/540-0118. **Contact:** Gerri A. Floyd, Employee Services Analyst. **E-mail address:** gerri.floyd@isl.com. **Description:** Designs and develops systems solutions for financial institutions, including the ATLAS software product line, which is a series of financial transaction processing systems that allow financial institutions to increase productivity and reduce operating costs. **Common positions include:** Computer Programmer; Software Engineer; Speech-Language Pathologist; Webmaster. **Educational backgrounds include:** Computer Science. **Benefits:** 401(k); Dental Insurance; Disability Coverage; Life Insurance; Medical Insurance; Tuition Assistance. **Corporate headquarters location:** This Location. **International locations:** U.K. **Listed on:** Privately held. **Annual sales/revenues:** $21 - $50 million. **Number of employees at this location:** 130.

INTERSE CORPORATION
111 West Evelyn Avenue, Suite 213, Sunnyvale CA 94086. 408/732-0932. **Fax:** 408/732-7038. **Contact:** Human Resources. **E-mail address:** pstrisower@interse.com. **World Wide Web address:** http://www.interse.com. **Description:** Develops and markets Web analysis software

products that employ inference-based algorithms to reconstruct visits, users, and organizations that interact with a Website. Founded in 1994. **NOTE:** Interse maintains an employment page at its Website. **Common positions include:** Software Engineer. **Corporate headquarters location:** This Location. **Listed on:** Privately held.

INTERSOLV INC.
9420 Key West Avenue, Rockville MD 20850. 301/838-5000. **Contact:** Sue Ratcliff, Director of Human Resources. **Description:** Provides computer programming and software services.
Other U.S. locations:
* 735 SW 158th Avenue, Beaverton OR 97006. 503/645-1150.

INTRAFED
1151-A Seven Locks Road, Potomac MD 20854. 301/315-0240. **Contact:** Human Resources. **Description:** Develops software and provides systems integration services.

INTRANET, INC.
One Gateway Center, Newton MA 02158. 617/527-7020. **Contact:** Human Resources. **Description:** A manufacturer of networking software.

INTUIT, INC.
P.O. Box 7850, Mountain View CA 94039. 415/944-6000. **Contact:** Human Resources. **Description:** Intuit, Inc. develops and markets personal finance and small business accounting software and complementary supplies and services for personal computer users. Its principal product is Quicken, which allows users to organize, understand, and manage their personal finances. **Number of employees nationwide:** 500.

INTUIT, INC.
6220 Greenwich Drive, San Diego CA 92122. 619/784-4000. **Contact:** Human Resources. **Description:** Intuit, Inc. develops and markets personal finance software. At this location, Intuit engineers their software packages for small businesses and personal finance including the popular Quicken.

ISYS/BIOVATION
40 Vreeland Avenue, Totowa NJ 07512. 201/256-5858. **Contact:** Human Resources. **Description:** Develops software. ISYS/Biovation supplies hospitals, universities, and doctors' offices with computer hardware and software designed to manage laboratory information and management systems.

ISYS/ODYSSEY DEVELOPMENT, INC.
The Denver Technological Center, 8775 East Orchard Road, Suite 811, Englewood CO 80111. 303/689-9998. **Fax:** 303/689-9997. **Contact:** Human Resources. **World Wide Web address:** http://www.isysdev.com. **Description:** A manufacturer of electronic publishing and imaging products including ISYS HindSite Internet utilities; ISYS Web, for online publishing; ISYS Image, for data capture and full-text search; ISYS Electronic Publisher, a retrieval and authoring tool; and ISYS for Adobe Acrobat, a search engine with PDF files.

ITERATED SYSTEMS INC.
3525 Piedmont Road, 7 Piedmont Center, Suite 600, Atlanta GA 30305. 770/840-0310. **Contact:** Mary Reagle, Director of Human Resources. **Description:** Develops image and video compression software.

JASC INC.
P.O. Box 44997, Eden Prairie MN 55344. 612/930-9171. **Contact:** Human Resources. **E-mail address:** jobs@jasc.com. **Description:** A developer of Windows-based graphics and multimedia software. Product line includes Paint Shop Pro, JASC Media Center, and Professional Capture Systems. **Common positions include:** Computer Programmer. **Benefits:** 401(k); Dental Insurance; Life Insurance; Medical Insurance; Profit Sharing; Vision Insurance.

JBA INTERNATIONAL INC.
3701 Algonquin Road, Suite 100, Rolling Meadows IL 60008. 847/590-0299. **Contact:** Human Resources. **Description:** Develops software. JBA International works with businesses in developing inventory control, distribution, and manufacturing computer programs.

JDA SOFTWARE SERVICES INC.
11811 North Tatum Boulevard, Suite 2000, Phoenix AZ 85028. 602/404-5500. **Contact:** Human Resources. **Description:** Develops software for the retail industry.

JANIFF SOFTWARE
615 West Johnson Avenue, Cheshire CT 06410. 203/250-8210. **Fax:** 203/250-8215. **Contact:** Human Resources. **World Wide Web address:** http://www.janiff.com. **Description:** A developer of business software solutions and services and provider of computer consulting and training.

JYACC
116 John Street, 20th Floor, New York NY 10038. 212/267-7722. **Contact:** Human Resources. **Description:** Develops a variety of software products and also provides software consulting services. **Number of employees at this location:** 200.

KEANE, INC.
10 City Square, Boston MA 02129. 617/241-9200. **Contact:** Human Resources. **Description:** Keane is a software services company that designs, develops, and manages software for corporations and health care facilities. Keane's services enable clients to leverage existing information systems and develop and manage new software applications faster. The company serves clients through two operating divisions: the Information Services Division (ISD) and the Healthcare Services Division (HSD). ISD provides software design, development, and management services to corporations and government agencies with large and recurring software development needs. HSD develops, markets, and supports financial, patient care, and clinical application software for hospitals and long-term care facilities. **Common positions include:** Accountant/Auditor; Computer Programmer; Systems Analyst. **Educational backgrounds include:** Computer Science. **Benefits:** 401(k); Daycare Assistance; Dental Insurance; Disability Coverage; Life Insurance; Medical Insurance; Profit Sharing; Savings Plan; Tuition Assistance. **Corporate headquarters location:** This Location. **Operations at this facility include:** Administration. **Listed on:** American Stock Exchange. **Number of employees at this location:** 100. **Number of employees nationwide:** 4,500.
Other U.S. locations:
- 950 Ridgemont Drive NE, Cedar Rapids IA 52402. 319/393-3343. Contact: Beverly Taylor, Human Resources.
- 1000 2nd Avenue, Suite 1100, Seattle WA 98104. 206/464-1426. This location provides technical support to consumers for software.

KEANE, INC.
HEALTHCARE SERVICES DIVISION
290 Broadhollow Road, Melville NY 11747. 516/351-7000. **Toll-free phone:** 800/699-6773. **Fax:** 516/351-7115. **Contact:** Michael Labrise, Human Resources Representative. **World Wide Web address:** http://www.keane.com. **Description:** Designs, develops, and manages software for corporations and health care facilities. Through its North American network of more than 40 branch offices, Keane provides its clients with a broad range of software services, including project management and application development. Founded in 1965. **Common positions include:** Accountant/Auditor; Computer Operator; Computer Programmer; Database Manager; MIS Specialist; Operations/Production Manager; Registered Nurse; Sales Representative; Systems Analyst. **Educational backgrounds include:** Accounting; Computer Science; Health Care. **Benefits:** 401(k); Dental Insurance; Disability Coverage; Life Insurance; Medical Insurance; Pension Plan; Tuition Assistance. **Corporate headquarters location:** Boston MA. **Operations at this facility include:** Divisional Headquarters. **Listed on:** American Stock Exchange. **Stock exchange symbol:** KEA. **Annual sales/revenues:** More than $100 million. **Number of employees at this location:** 75. **Number of projected hires for 1997 - 1998 at this location:** 30.

KNOWLEDGE SYSTEMS CORPORATION

4001 Weston Parkway, Cary NC 27513. 919/481-4000. **Contact:** Human Resources. **Description:** A manufacturer of educational software and provider of consulting services.

KNOZALL SYSTEMS

375 East Elliot, Suite 10, Chandler AZ 85225. 602/545-0006. **Fax:** 602/545-0008. **Contact:** Human Resources. **World Wide Web address:** http://www.knozall.com. **Description:** A manufacturer of networking utilities for local and wide area networks. **Corporate headquarters location:** This Location.

L-SOFT INTERNATIONAL, INC.

8401 Corporate Drive, Suite 510, Landover MD 20785. 301/731-0440. **Toll-free phone:** 800/399-5449. **Fax:** 301/731-6302. **Contact:** Human Resources. **World Wide Web address:** http://www.lsoft.com. **Description:** A developer of mailing list management software. Product line includes LISTSERV, an electronic mailing list management product; LSMTP, a program allowing large quantity delivery of Internet mail; and EASE, which allows users to create mail lists on L-Soft's centrally maintained servers. Founded in 1994. **Common positions include:** Computer Programmer; Engineer; Sales and Marketing Representative. **Corporate headquarters location:** This Location.

LPA SOFTWARE INC.

290 Woodcliff Drive, Fairport NY 14450. 716/248-9600. **Contact:** Human Resources. **Description:** Develops custom software for large corporations.

LAHEY COMPUTER SYSTEMS, INC.

P.O. Box 6091, Incline Village NV 89450. 702/831-2500. **Fax:** 702/831-8123. **Contact:** Mark Reynaud, Human Resources. **E-mail address:** mreyn@lahey.com. **World Wide Web address:** http://www.lahey.com. **Description:** A software developer specializing in scientific and engineering solutions. **Common positions include:** Computer Programmer. **Corporate headquarters location:** This Location.

LANDMARK SYSTEMS CORPORATION

8000 Towers Crescent Drive, Vienna VA 22182. 703/902-8000. **Contact:** Human Resources. **Description:** Develops performance maintenance software for businesses.

LANGUAGE ENGINEERING CORPORATION

385 Concord Avenue, Belmont MA 02178. 617/489-4000. **Fax:** 617/489-3850. **Contact:** Human Resources. **E-mail address:** info@hq.lec.com. **World Wide Web address:** http://www.lec.com. **Description:** Develops software for the World Wide Web that translates English Web pages into Japanese. Language Engineering Corporation's product is called Logovista Internet.

LARI SOFTWARE INC.

207 South Elliot Road, Suite 203, Chapel Hill NC 27514. **Fax:** 919/968-0801. **Contact:** Human Resources. **E-mail address:** lari@larisoftware.com. **World Wide Web address:** http://www.larisoftware.com. **Description:** A developer of graphics software. Product line includes LightningDraw GX.

LASER DESIGN INC.

9401 James Avenue South, Suite 162, Minneapolis MN 55431. 612/884-9648. **Contact:** Human Resources. **Description:** Designs software for three-dimensional laser digitizers.

LAWSON SOFTWARE

1300 Godward Street, Minneapolis MN 55413. 612/379-2633. **Contact:** Human Resources. **Description:** Develops business applications. Lawson Software specializes in creating financial software especially for procurement departments.

LEAD TECHNOLOGIES, INC.

900 Baxter Street, Charlotte NC 28204. **Toll-free phone:** 800/637-4699. **Contact:** Human Resources. **E-mail address:** support@leadtools.com. **World Wide Web address:**

http://www.leadtools.com. **Description:** A developer of image file format, processing, and compression toolkits.

THE LEARNING COMPANY
One Athenaeum Street, Cambridge MA 02142. 617/494-1200. **Contact:** Employment Manager. **Description:** The Learning Company uses emerging technologies to create a system of highly engaging, easy-to-use software products that help build important lifelong learning and communications skills. The subject areas and design objectives for the company's products are developed with input from education and language professionals, purchasers, and end users to ensure that each new product is of the highest quality and matches an identified market need. The company has developed families of educational software products designed to help build important life-long learning and communication skills in key subject areas that are appropriate for different age groups. Most of the company's products are available in home editions, and certain products are also available in school editions, which include network, site license, and stand alone configurations. Product families include the Rabbit Family (Ages 3 to 7), the Treasure Family (Ages 6 to 8), the Writing Tools Family (Ages 8 and Up) and The Children's Writing & Publishing Center, and The Foreign Languages Family (Ages 15 and Up). **Common positions include:** Accountant/Auditor; Computer Programmer; Customer Service Representative; Quality Control Supervisor; Software Engineer; Systems Analyst; Technical Writer/Editor. **Benefits:** 401(k); Dental Insurance; Disability Coverage; Employee Discounts; Life Insurance; Medical Insurance; Tuition Assistance. **Corporate headquarters location:** This Location. **Parent company:** SoftKey International (Cambridge MA). **Listed on:** NASDAQ. **Number of employees at this location:** 250. **Number of employees nationwide:** 480.
Other U.S. locations:
* 6493 Kaiser Drive, Fremont CA 94555. 510/792-2101. Activities at this location include artwork and software development, and marketing.
* 6160 Summit Drive North, Brooklyn Center MN 55430. 612/569-1500. At this location, the company develops software and maintains a sales and marketing team.

LEXIS-NEXIS INFORMATION TECHNOLOGY
1563 Massachusetts Avenue, Cambridge MA 02139. 617/864-6151. **Contact:** Human Resources. **Description:** Designs and produces software products for law firms, legal departments, and other legal organizations.

LIANT SOFTWARE CORPORATION
5230 Pacific Concourse Drive, Suite 200, Los Angeles CA 90045. 310/643-4402. **Fax:** 310/643-4452. **Contact:** Human Resources. **World Wide Web address:** http://www.liant.com. **Description:** Liant Software Corporation's products include Relativity, an SQL relational access through ODBC to COBOL managed data for client/server Windows applications; and Open PL/I, which offers transition of PL/I mainframe and minicomputer applications from legacy systems to open, client/server environments.

LIANT SOFTWARE CORPORATION
RELATIVITY DIVISION
3006 Longhorn Boulevard, Suite 107, Austin TX 78758. 512/834-0299. **Toll-free phone:** 800/349-9222. **Fax:** 512/719-7070. **Contact:** Human Resources. **World Wide Web address:** http://www.liant.com. **Description:** Liant Software Corporation's products include Relativity, an SQL relational access through ODBC to COBOL managed data for client/server Windows applications; and Open PL/I, which offers transition of PL/I mainframe and minicomputer applications from legacy systems to open, client/server environments. **Corporate headquarters location:** Framingham MA.

LILLY SOFTWARE ASSOCIATES, INC.
239 Drakeside Road, Hampton NH 03842. 603/926-9696. **Fax:** 603/926-9698. **Contact:** Human Resources Department. **World Wide Web address:** http://mfginfo.com. **Description:** A developer of materials requirement planning (MRP) client/server software, called Visual Manufacturing 2.0, which allows managers to track all phases of production. **Corporate headquarters location:** This Location.

LOCKHEED MARTIN
P.O. Box 1570, Dahlgren VA 22448. 540/371-8175. **Contact:** Lynda Thomson, Human Resources. **Description:** This location develops defense-oriented software for the government.

LODGISTIX, INC.
1938 North Woodlawn Street, Suite 1, Wichita KS 67208. 316/685-2216. **Contact:** Human Resources. **Description:** A developer of software for the hospitality industry.

LOGICA NORTH AMERICA, INC.
32 Hartwell Avenue, Lexington MA 02173. 617/476-8000. **Contact:** Robert Eubank, Chief Administrative Officer. **Description:** A computer firm whose business activities include computer programming systems and design, and consulting services primarily in the banking, insurance, and telecommunications industries.

LOGICA, INC.
425 California Street, Suite 500, San Francisco CA 94104. 415/288-5200. **Contact:** Human Resources. **Description:** A computer firm whose business activities include computer programming systems and design, and consulting services primarily in the banking, insurance, and telecommunications industries.

LOGICAL OPERATIONS
500 Canal View Boulevard, Rochester NY 14624. 716/240-7353. **Fax:** 716/240-7465. **Contact:** Human Resources. **World Wide Web address:** http://www.logicalops.com. **Description:** Logical Operations is a leading publisher of computer coursework, produces a wide range of software products, and provides computer training. **NOTE:** The company offers entry-level positions. **Common positions include:** Budget Analyst; Buyer; Computer Programmer; Controller; Credit Manager; Customer Service Representative; Database Manager; Editor; Editorial Assistant; Electrical/Electronics Engineer; Finance Director; Financial Analyst; Human Resources Manager; Market Research Analyst; Marketing Manager; Marketing Specialist; Multimedia Designer; Online Content Specialist; Operations/Production Manager; Project Manager; Purchasing Agent/Manager; Sales Manager; Sales Representative; Systems Analyst; Systems Manager; Technical Writer/Editor. **Educational backgrounds include:** Accounting; Business Administration; Communications; Computer Science; Engineering; Finance; Liberal Arts; Marketing; Mathematics. **Benefits:** 401(k); Dental Insurance; Disability Coverage; Employee Discounts; Financial Planning; Health Club Discount; Job Sharing; Life Insurance; Medical Insurance; Profit Sharing; Sabbatical Leave; Telecommuting; Tuition Assistance. **Other U.S. locations:** Boston MA; Seattle WA. **International locations:** Montreal. **Parent company:** Ziff-Davis Publishing Company (New York NY). **Operations at this facility include:** Divisional Headquarters. **Listed on:** Privately held. **Annual sales/revenues:** $21 - $50 million. **Number of employees at this location:** 250. **Number of projected hires for 1997-1998 at this location:** 30.

LOGICON RDA
317 South Second Street, Leavenworth KS 66048. 913/651-8347. **Contact:** Human Resources. **Description:** Logicon develops state-of-the-art software technology for the United States military and government. **Parent company:** Logicon, Inc.

LOGICON STRATEGIC & INFORMATION SYSTEMS
222 West Sixth Street, San Pedro CA 90731. 310/831-0611. **Fax:** 310/521-2660. **Contact:** Human Resources Department. **Description:** Logicon Strategic & Information Systems develops state-of-the-art software technology for the United States military and government. **Common positions include:** Aerospace Engineer; Computer Programmer; Electrical/Electronics Engineer; Software Engineer; Systems Analyst. **Educational backgrounds include:** Computer Science; Engineering; Mathematics; Physics. **Benefits:** 401(k); Dental Insurance; Disability Coverage; Life Insurance; Medical Insurance; Profit Sharing; Tuition Assistance. **Corporate headquarters location:** Torrance CA. **Other U.S. locations:** Washington DC; Cocoa Beach FL; Indianapolis IN; Omaha NE; Clearfield UT. **Parent company:** Logicon, Inc. **Operations at this facility include:** Research and Development. **Listed on:** New York Stock Exchange. **Number of employees at this location:** 450.

LOGICON, INC.
3701 Skypark Drive, Suite 200, Torrance CA 90505. 310/373-0220. **Contact:** Human Resources. **Description:** Logicon is a leading contributor to state-of-the-art software technology in the United States military and government. **Corporate headquarters location:** This Location.
Other U.S. locations:
• 950 North Orlando Avenue, Winter Park FL 32789. 407/629-6010.

LOGMATICS
2140 New Market Parkway, Suite 112, Marietta GA 30067. 770/955-3446. **Contact:** Phil Angevine, President. **Description:** Manufactures software, printers, scanners, and related equipment.

LOOKING GLASS TECHNOLOGIES, INC.
100 Cambridge Park Drive, Suite 300, Cambridge MA 02140. 617/441-6333. **Toll-free phone:** 800/360-7455. **Fax:** 617/441-3946. **Contact:** Human Resources. **E-mail address:** cs@lglass.com. **World Wide Web address:** http://www.lglass.com. **Description:** A designer of 3D interactive/virtual reality computer games. **Corporate headquarters location:** This Location.

LOTUS DEVELOPMENT CORPORATION
55 Cambridge Parkway, Cambridge MA 02142. 617/577-8500. **Contact:** Human Resources. **Description:** Lotus develops, manufactures, and markets applications software and services that meet the evolving technology and business applications requirements for individuals, work groups, and entire organizations. **Number of employees nationwide:** 4,400.
Other U.S. locations:
• Onc Bush Street, Suite 600, San Francisco CA 94104-4412. 415/421-7013.

LUCAS ARTS ENTERTAINMENT COMPANY
P.O. Box 10307, San Rafael CA 94912. **Fax:** 415/444-8438. **Recorded Jobline:** 415/444-8495. **Contact:** Human Resources. **World Wide Web address:** http://www.lucasarts.com. **Description:** An international developer and publisher of entertainment software, some of which incorporate a Star Wars theme. **Corporate headquarters location:** This Location.

LUCAS BEAR
7111 Valley Green Road, Suite 230, Fort Washington PA 19034. 215/836-5161. **Fax:** 215/836-2305. **Contact:** Barbara Rufe, Director of Human Resources. **Description:** A developer and supplier of forecasting and logistics software solutions. The company's supply chain management software provides solutions for demand forecasting, inventory planning, and continuous replenishment. This software is used by manufacturers, distributors, and retailers worldwide, including *Fortune* 1000 companies. **Common positions include:** Accountant/Auditor; Administrative Services Manager; Computer Programmer; Human Resources Manager; Operations/Production Manager; Purchasing Agent/Manager; Quality Control Supervisor; Software Engineer; Systems Analyst; Technical Writer/Editor. **Educational backgrounds include:** Computer Science; Marketing; Mathematics. **Benefits:** 401(k); Dental Insurance; Disability Coverage; Life Insurance; Medical Insurance. **Corporate headquarters location:** This Location. **Operations at this facility include:** Administration; Research and Development; Sales; Service. **Listed on:** Privately held. **Annual sales/revenues:** Less than $5 million. **Number of employees at this location:** 30. **Number of employees nationwide:** 45.

LUCKMAN INTERACTIVE, INC.
1055 West Seventh Street, Suite 2580, Los Angeles CA 90017. 213/614-0966. **Fax:** 213/614-1929. **Contact:** Dena Speak, Human Resources Manager. **E-mail address:** jobs@luckman.com. **World Wide Web address:** http://www.luckman.com. **Description:** A developer of Internet client/server software including Web directories and Web editing tools.

LUGARU SOFTWARE LTD.
5824 Forbes Avenue, Pittsburgh PA 15217. 412/421-5911. **Fax:** 412/421-6371. **Contact:** Human Resources. **World Wide Web address:** http://www.lugaru.com. CompuServe: GO LUGARU. **Description:** A manufacturer of an EMACS-style programmer's text editor for Windows, DOS, and OS/2 called Epsilon Programmer's Editor. **Corporate headquarters location:** This Location.

M&I DATA SERVICES, INC.

P.O. Box 23528, Brown Deer WI 53223-0528. 414/357-2290. **Contact:** Human Resources. **Description:** Provides data processing and software services for financial institutions throughout the United States and in foreign countries. **Common positions include:** Computer Programmer; Systems Analyst; Technical Writer/Editor. **Educational backgrounds include:** Business Administration; Computer Science; Information Systems; M.I.S. **Benefits:** 401(k); Dental Insurance; Disability Coverage; Life Insurance; Medical Insurance; Pension Plan; Savings Plan; Tuition Assistance. **Special Programs:** Internships. **Corporate headquarters location:** This Location. **Parent company:** Marshall & Ilsley Corporation (Milwaukee WI). Subsidiaries of Marshall Ilsley Corporation include M&I Investment Management Corp.; Marshall & Ilsley Trust Company; M&I Marshall & Ilsley Trust Company of Arizona; Marshall & Ilsley Trust Company of Florida; M&I First National Leasing Corp.; M&I Capital Markets Group, Inc.; M&I Brokerage Services, Inc.; M&I Mortgage Corp.; Richter-Schroeder Company, Inc.; and M&I Insurance Services, Inc. **Operations at this facility include:** Administration; Divisional Headquarters; Research and Development; Sales; Service. **Listed on:** NASDAQ. **Number of employees at this location:** 1,800. **Number of employees nationwide:** 2,200.

MAI SYSTEMS CORPORATION

9501 Jeronimo Road, Irvine CA 92718. 714/580-0700. **Fax:** 714/580-2391. **Contact:** Juanita Potts, Human Resources Manager. **Description:** Develops gaming software and programs for the hotel industry. **Common positions include:** Accountant/Auditor; Buyer; Computer Programmer; Credit Manager; Customer Service Representative; Electrical/Electronics Engineer; Financial Analyst; Human Resources Manager; Purchasing Agent/Manager; Software Engineer; Systems Analyst. **Educational backgrounds include:** Accounting; Business Administration; Computer Science; Engineering; Finance; Marketing. **Benefits:** Dental Insurance; Disability Coverage; Employee Discounts; Life Insurance; Medical Insurance; Tuition Assistance. **Corporate headquarters location:** This Location. **Operations at this facility include:** Administration; Manufacturing; Research and Development; Sales; Service. **Number of employees at this location:** 150. **Number of employees nationwide:** 500.

MCS INC.

400 Penn Center Boulevard, Pittsburgh PA 15235. 412/823-7440. **Contact:** Human Resources. **Description:** Develops software for home medical equipment.

MPACT

17197 North Laurel Park Drive, Suite 201, Livonia MI 48152. 313/462-2244. **Contact:** Human Resources. **Description:** Develops computer applications, specializing in electronic data interchange software.

PAUL MACE SOFTWARE CORPORATION

400 Williamson Way, Ashland OR 97520. 541/488-2322. **Fax:** 541/488-1549. **Contact:** Human Resources Department. **World Wide Web address:** http://www.pmace.com. **Description:** Paul Mace Software Corporation is a developer, publisher, and marketer of graphic animation systems and software utilities.

MACMILLAN PUBLISHING

201 West 103rd Street, Indianapolis IN 46290. 317/581-3500. **Contact:** Human Resources. **Description:** A publisher of computer software and manuals.

MACNEAL-SCHWENDLER CORPORATION

815 Colorado Boulevard, Los Angeles CA 90041. 213/258-9111. **Contact:** Human Resources. **Description:** A leader in computer aided analysis technology, the company develops, markets, and supports software for computer aided engineering.

MACOLA SOFTWARE

333 East Center Street, Marion OH 43302. 614/382-5999. **Contact:** Human Resources. **Description:** Develops accounting, distribution, and manufacturing software.

MACROMEDIA, INC.
600 Townsend Street, Suite 310 West, San Francisco CA 94103. 415/252-2000. **Contact:** Human Resources. **Description:** Develops multimedia software for the Web.

MAGNA SOFTWARE CORPORATION
275 7th Avenue, 20th Floor, New York NY 10001. 212/691-0300. **Toll-free phone:** 800/431-9006. **Fax:** 212/691-1968. **Contact:** Sandra Booth, Manager of Recruiting. **Description:** Magna Software Corporation develops leading-edge, 3-tier, enterprise client/server open systems. The company generates applications that are fully compiled for a wide range of environments including DCE/Encina, Tuxedo, and CICS/6000. Magna Software Corporation also integrates customers' mainframes into these environments. **NOTE:** Ideal job candidates will have a strong client/server background and an entrepreneurial mindset. **Common positions include:** Account Manager; Computer Programmer; Customer Service Representative; Marketing Specialist; Software Engineer; Technical Writer/Editor; Telemarketer. **Educational backgrounds include:** Business Administration; Computer Science; Engineering; Marketing; Mathematics. **Benefits:** 401(k); Dental Insurance; Disability Coverage; Life Insurance; Medical Insurance; Profit Sharing; Stock Option. **Corporate headquarters location:** This Location. **Other U.S. locations:** Fairfax VA. **Operations at this facility include:** Administration; Sales; Service. **Listed on:** Privately held. **Annual sales/revenues:** $5 - $10 million. **Number of employees at this location:** 40.

MAINSOFT CORPORATION
1270 Oakmead Parkway, Suite 310, Sunnyvale CA 94086. 408/774-3400. **Fax:** 408/774-3404. **Contact:** Human Resources. **E-mail address:** hr@mainsoft.com. **World Wide Web address:** http://www.mainsoft.com. **Description:** Mainsoft provides Windows-to-UNIX cross-platform software development environments. **Common positions include:** Applications Engineer; Product Engineer; Software Engineer; Test Engineer. **Corporate headquarters location:** This Location. **Listed on:** Privately held. **Number of employees at this location:** 60.

MANAGEMENT CONTROL SYSTEMS
502 8th Street South, Fargo ND 58103. 701/293-8922. **Contact:** Human Resources. **Description:** Develops accounting software programs for use by businesses.

MANAGEMENT GRAPHICS INC.
1401 East 79th Street, Minneapolis MN 55425. 612/854-1220. **Contact:** Human Resources. **Description:** Develops graphic systems for computers including applicable software.

MANAGEMENT TECHNOLOGY INC. (MTI)
630 Third Avenue, 15th Floor, New York NY 10017. 212/983-5620. **Contact:** Human Resources. **Description:** Develops and markets a variety of banking and other financial software.

MANATRON INC.
2970 South 9th Street, Kalamazoo MI 49009. 616/375-5300. **Contact:** Mary Gephardt, Human Resources. **Description:** Develops software for government agencies.

MANUGISTICS, INC.
2115 East Jefferson Street, Rockville MD 20852. 301/984-5000. **Contact:** Carl DiPietro, Human Resources Manager. **Description:** Provides decision support software and services for *Fortune* 500 manufacturing, transportation, and distribution companies. **Common positions include:** Computer Programmer; Industrial Engineer; Instructor/Trainer; Mechanical Engineer; Operations/Production Manager; Statistician; Systems Analyst; Technical Writer/Editor; Transportation/Traffic Specialist. **Educational backgrounds include:** Business Administration; Computer Science; Engineering; Marketing. **Benefits:** Dental Insurance; Disability Coverage; Life Insurance; Medical Insurance; Savings Plan; Stock Option; Tuition Assistance. **Special Programs:** Internships. **Corporate headquarters location:** This Location. **Operations at this facility include:** Administration; Research and Development; Sales; Service. **Number of employees at this location:** 300.

MAPINFO CORPORATION
One Global View, Troy NY 12180. 518/285-6000. **Contact:** Human Resources. **Description:** A developer of geography software.

MARCAM CORPORATION
95 Wells Avenue, Newton MA 02159. 617/965-0220. **Contact:** Human Resources. **Description:** Marcam is a supplier of enterprise application software and services for industrial and distribution companies. The company also provides product support, implementation consulting, education, and programming services. **Corporate headquarters location:** This Location. **Number of employees nationwide:** 965.

MARCAM CORPORATION
Park 80 West, Plaza Two, Saddle Brook NJ 07663. 201/291-8400. **Contact:** Human Resources. **Description:** Develops and sells manufacturing and other business-related software. **Corporate headquarters location:** Newton MA.

MATHSOFT, INC.
101 Main Street, Cambridge MA 02142. 617/577-1017. **Fax:** 617/577-8829. **Contact:** Melisa Kahler, Human Resources Representative. **E-mail address:** hrjobs@mathsoft.com. **Description:** MathSoft, Inc. is a leading developer of mathematical software and Electronic Books for desktop computers. Founded in 1984, the company developed Mathcad, a live, interactive environment for mathematics work in a wide variety of fields, including engineering, science, and education. Mathcad is currently available for PC, Macintosh, and UNIX workstations; and in German, French, Spanish, Japanese, and Italian. MathSoft is the publisher of the Mathcad Library of Electronic Books, which includes many fully interactive titles. MathSoft also markets Maple V symbolic computation software and other third-party mathematical software. **Common positions include:** Software Engineer. **Educational backgrounds include:** Engineering; Mathematics. **Benefits:** 401(k); Dental Insurance; Disability Coverage; Life Insurance; Medical Insurance; Savings Plan; Tuition Assistance. **Corporate headquarters location:** This Location. **Other U.S. locations:** Seattle WA. **Listed on:** NASDAQ. **Stock exchange symbol:** MATH. **Annual sales/revenues:** $11 - $20 million. **Number of employees at this location:** 60. **Number of employees nationwide:** 150.

MAXTECH
1000 Cobb Place Boulevard, Building 200, Kennesaw GA 30144-3684. 770/428-5000. **Toll-free phone:** 800/582-9337. **Fax:** 770/428-5009. **Contact:** Human Resources. **E-mail address:** info@maxtech.com. **World Wide Web address:** http://www.maxtech.com. **Description:** A provider of SCO UNIX-based solutions and a software developer and reseller. Product line includes DOUBLEVISION, remote control software which allows a user to connect to another user's screen. **Corporate headquarters location:** This Location.

McAFEE ASSOCIATES, INC.
5944 Luther Lane, Suite 1007, Dallas TX 75225. 214/361-8086. **Toll-free phone:** 800/338-8754. **Fax:** 214/361-2696. **Contact:** Human Resources. **World Wide Web address:** http://www.mcafee.com. **Description:** An international supplier of enterprisewide network security and management software. McAfee products include VirusScan, an anti-virus software. The company has strategic relationships with numerous computer firms including Compaq, Hewlett-Packard, Novell, Microsoft, and AST. **NOTE:** McAfee maintains an employment opportunities page at its Website. **Common positions include:** Software Engineer. **Corporate headquarters location:** Santa Clara CA. **Other U.S. locations:** Tinton Falls NJ. **Listed on:** NASDAQ. **Stock exchange symbol:** MCAF. **Annual sales/revenues:** $90 million. **Number of employees nationwide:** 350.

McAFEE ASSOCIATES, INC.
2710 Walsh Avenue, Santa Clara CA 95051. 408/988-3832. **Toll-free phone:** 800/338-8754. **Fax:** 408/970-9727. **BBS:** 408/988-4004. **Contact:** Human Resources. **E-mail address:** andrea_nation@cc.mcafee.com. Internet FTP: ftp.mcafee.com. America Online: MCAFEE. CompuServe: GO MCAFEE. Microsoft Network: GO MCAFEE. **World Wide Web address:** http://www.mcafee.com. **Description:** An international supplier of enterprisewide network security and management software. McAfee products include VirusScan, an anti-virus software. The company has strategic relationships with numerous computer firms including Compaq, Hewlett Packard, Novell, Microsoft, and AST. **Common positions include:** Applications Engineer; Quality Assurance Engineer; Sales Representative; Software Engineer; Systems Engineer; Technician.

Corporate headquarters location: This Location. Other U.S. locations: Tinton Falls NJ; Dallas TX. Listed on: NASDAQ. Stock exchange symbol: MCAF. Annual sales/revenues: $90 million. Number of employees nationwide: 350.

MEASUREMENT TECHNIQUES
160 Old Maple Street, Stoughton MA 02072. 617/344-6230. Contact: Human Resources. Description: A developer of software that improves the performance of computer networks.

MECA SOFTWARE, LLC
115 Corporate Drive, Trumbull CT 06611. 203/452-2600. Contact: Human Resources. Description: Develops and markets financial services software including Managing Your Money, a software product that enables users to perform online banking.

MEDE AMERICA
2045 Midway Drive, Twinsburg OH 44087. 216/425-3241. Contact: Human Resources. Description: Develops computer software for the medical industry.

MEDICAL INFORMATION TECHNOLOGY, INC. (MEDITECH)
MEDITECH Circle, Westwood MA 02090. 617/821-3000. Contact: Human Resources. Description: Develops, markets, installs, and supports computer software for hospitals. Common positions include: Accountant/Auditor; Biological Scientist; Chemist; Computer Programmer; Dietician/Nutritionist; Services Sales Representative; Technical Writer/Editor. Educational backgrounds include: Accounting; Biology; Chemistry; Computer Science; Economics; Finance; Mathematics; Medical Technology; Nursing; Pharmacology. Benefits: Disability Coverage; Life Insurance; Medical Insurance; Profit Sharing; Stock Option; Tuition Assistance. Corporate headquarters location: This Location. Operations at this facility include: Administration; Research and Development; Sales; Service. Number of employees nationwide: 1,000.

MEDICODE INC.
5225 Wiley Post Way, Suite 500, Salt Lake City UT 84116. 801/536-1000. Contact: Human Resources. Description: Develops medical software including databases.

MEDICUS SYSTEMS CORPORATION
One Rotary Center, Suite 1111, Evanston IL 60201. 847/570-7500. Contact: Susan Doctors, Vice President, Human Resources. Description: Medicus Systems Corporation develops and markets specialized decision support software designed to improve the organizational and clinical effectiveness of hospitals, multi-hospital systems, academic medical centers, managed care providers, large physician groups, and other health care providers.

MEDIWARE INFORMATION SYSTEMS
7300 West 110th Street, Suite 700, Overland Park KS 66210. 913/451-6161. Contact: Human Resources. Description: Develops software for hospital pharmacies.

THE MEDSTAT GROUP, INC.
777 East Eisenhower Parkway, Suite 500, Ann Arbor MI 48108. 313/996-1180. Contact: Human Resources. Description: A health care information, software, and consulting firm which designs and builds database systems for use in analyzing health care claims and benefits for large employers, insurance companies, and the research industry. Common positions include: Computer Programmer; Consultant; Data Analyst; Database Manager; Services Sales Representative. Educational backgrounds include: Business Administration; Communications; Computer Science; Finance; Health Care; Marketing. Benefits: Dental Insurance; Disability Coverage; Flexible Benefits; Life Insurance; Medical Insurance; Profit Sharing. Corporate headquarters location: This Location. Other U.S. locations: San Francisco CA; Atlanta GA; Boston MA; Nashville TN. Operations at this facility include: Administration; Research and Development; Sales; Service. Number of employees nationwide: 450.

MEGASOFT, L.L.C.
819 Highway 33 East, Freehold NJ 07728. 908/431-5300. Contact: Christine Engstrom, Human Resources. Description: MEGASOFT is a provider of effective solutions for software

manufacturing and information distribution. The company offers a complete menu of services to address all of its customers' information delivery needs using the latest innovations in technology. Established in 1983, MEGASOFT provides a broad range of integrated software and information distribution options in multiple formats on disk, in print, and online. MEGASOFT has delivered solutions to over 14,000 customers in many industries -- including the technology, insurance, financial services, pharmaceutical, publishing, government, and transportation communities.

MENTOR GRAPHICS CORPORATION
8005 SW Boeckman Road, Wilsonville OR 97070-7777. 503/685-7000. **Recorded Jobline:** 800/554-5259. **Contact:** Staffing. **Description:** Mentor Graphics Corporation, founded in 1981, has pioneered advances in electronic design automation (EDA), the use of technically advanced computer software to automate the design, analysis, and documentation of electronic components and systems. Today, the company is one of the world's leading suppliers of EDA software and professional services, and markets its products primarily to companies in the aerospace, computer, consumer electronics, semiconductor, and telecommunications industries. Mentor Graphics products increase productivity and product quality, and reduce development costs and time to market. **Common positions include:** Software Engineer; Technical Writer/Editor. **Educational backgrounds include:** Computer Science. **Benefits:** 401(k); Daycare Assistance; Dental Insurance; Life Insurance; Medical Insurance; Tuition Assistance. **Special Programs:** Internships. **Corporate headquarters location:** This Location. **Operations at this facility include:** Administration; Research and Development. **Listed on:** NASDAQ. **Number of employees at this location:** 1,000. **Number of employees nationwide:** 1,900.
Other U.S. locations:
- 5430 LBJ Freeway, Suite 1010, Dallas TX 75240. 972/450-2300. This is a sales office.
- 155 108th Street, Suite 450, Bellevuc WA 98006. 206/451-2362. This is a regional sales office.

MERCURY INTERACTIVE CORPORATION
470 Portrero Avenue, Sunnyvale CA 94086. 408/523-9900. **Fax:** 408/523-9911. **Contact:** Human Resources. **World Wide Web address:** http://www.merc-int.com. **Description:** A provider of automated software quality (ASQ) tools for enterprise applications testing. The company's products are used to isolate software and system errors prior to application deployment. **Corporate headquarters location:** This Location.
Other U.S. locations:
- 777 South Flagler Drive, 8th Floor, West Palm Beach FL 33401. 561/820-9444. This location is a sales office.
- Five Concourse Parkway, Suite 3100, Atlanta GA 30328. 770/804-5895. This location is a sales office.
- 10255 West Higgins Road, Suite 760, Rosemont IL 60018. 847/803-3176. This location is a sales office.
- One Cranberry Hill, Lexington MA 02173. 617/860-4400. This location is a regional sales office.
- 8201 Corporate Drive, Suite 620, Landover MD 20785. 301/918-2510. This location is a sales office.
- 99 Park Avenue, 18th Floor, New York NY 10016. 212/687-4646. This location is a sales and technical support office.

MERIDIAN TECHNOLOGY
P.O. Box 4404, Chesterfield MO 63005. 314/532-7708. **Fax:** 314/532-3242. **Contact:** Human Resources. **E-mail address:** employment@meridiantc.com. **World Wide Web address:** http://www.meridiantc.com. **Description:** A developer of software focusing on networking and Internet access products. **NOTE:** The company's physical address is 11 McBride Corporate Center Drive, Suite 250, Chesterfield MO.

MESQUITE SOFTWARE, INC.
3925 West Braker Lane, Austin TX 78759. 512/305-0080. **Fax:** 512/305-0009. **Contact:** Human Resources. **World Wide Web address:** http://www.mesquite.com. **Description:** A provider of products and services to the system simulation market. Product line includes CSIM18-The

Simulation Engine, a process-oriented, general purpose simulation toolkit that allows for discrete-event simulation models. **Corporate headquarters location:** This Location.

META SOFTWARE CORPORATION
125 Cambridge Park Drive, Cambridge MA 02140. 617/576-6920. **Fax:** 617/661-2008. **Contact:** Human Resources. **E-mail address:** resumes@metasoft.com. **World Wide Web address:** http://www.metasoftware.com. **Description:** A provider of BPR tools, with a wide range of products and services aimed at improving critical business processes. Aspects of the company's services include process modeling and data capture, simulation and animation, workflow automation, and implementation. Founded in 1985. **Benefits:** 401(k); Dental Insurance; Disability Coverage; Education Assistance; Life Insurance; Medical Insurance. **Corporate headquarters location:** This Location.

METAFILE INFORMATION SYSTEMS INC.
4131 Highway 52 North, Suite B221, Rochester MN 55901. 507/286-9232. **Contact:** Human Resources. **Description:** Develops image and COLD applications.

METATOOLS INC.
6303 Carpinteria Avenue, Carpinteria CA 93013. 805/566-6200. **Contact:** Human Resources. **Description:** A developer of graphic design software known as Kai's Power GOO that allows users to manipulate and distort digital images.

METROWERKS INC.
2201 Donley Drive, Suite 310, Austin TX 78758. 512/873-4700. **Contact:** Human Resources. **Description:** Metrowerks is a developer of compiling software.

MICRO COMPUTER SYSTEMS, INC.
2300 Valley View Lane, Suite 800, Irving TX 75062. 972/659-1514. **Fax:** 972/659-1624. **Contact:** Human Resources. **World Wide Web address:** http://www.mcsdallas.com. **Description:** A developer of software. Products include Local Area Network (LAN) communication systems and configuration utilities for EISA computers. MCS specializes in effecting network systems interoperability from LANs to the Internet to the World Wide Web.

MICRO DATA BASE SYSTEMS INC.
P.O. Box 2438, West Lafayette IN 47906. 765/463-7200. **Contact:** Human Resources Department. **Description:** Produces and markets development software for PCs. Micro Data Base Systems produces database management systems for small computers and artificial intelligence systems that meet all business processing needs. Consulting and applications groups implement customized applications for a broad variety of clients worldwide using the firm's software tools. **NOTE:** The company is located at 1305 Cumberland Street, West Lafayette IN. **Educational backgrounds include:** Accounting; Business Administration; Computer Science; Marketing. **Benefits:** Dental Insurance; Life Insurance; Medical Insurance; Profit Sharing. **Corporate headquarters location:** This Location. **Other U.S. locations:** Shaumburg IL. **Operations at this facility include:** Administration; Computer Programming; Research and Development; Sales; Service.

MICRO DYNAMICS LTD.
8455 Colesville Road, Suite 820, Silver Spring MD 20910. 301/589-6300. **Contact:** Jim Tebay, Vice President, Marketing. **Description:** Develops software for document management and automation. **NOTE:** Interested jobseekers should address inquiries to the parent company: Jim Tebay, Vice President of Marketing, FormMaker, 2300 Windy Ridge Parkway, Suite 400N, Atlanta GA 30339. **Parent company:** FormMaker.

MICRO FIRMWARE, INC.
330 West Gray Street, Norman OK 73069-7111. 405/321-8333. **Fax:** 405/573-5535. **Contact:** Human Resources Department. **World Wide Web address:** http://www.firmware.com. **Description:** Micro Firmware is a developer and distributor of Phoenix ROM BIOS upgrade products.

MICRO HOUSE INTERNATIONAL

2477 North 55th Street, Suite 101, Boulder CO 80301. 303/443-3388. **Fax:** 800/488-2533. **Contact:** Human Resources. **E-mail address:** hr@microhouse.com. **World Wide Web address:** http://www.microhouse.com. **Description:** Micro House International develops technical software of an encyclopedic nature. Product line includes The Micro House Technical Library, DrivePro, EZ-Drive, and The Modem Technical Guide.

MICRO INTEGRATION CORPORATION

One Science Park, Frostburg MD 21532. 301/689-2811. **Contact:** Human Resources. **Description:** A software communications firm.

MICRO 2000, INC.

1100 East Broadway, 3rd Floor, Glendale CA 91205. 818/547-0125. **Fax:** 818/547-0397. **Contact:** Jacqui Mendelsohn, Vice President, Administration. **World Wide Web address:** http://www.micro2000.com. **Description:** Develops and markets computer diagnostic products for troubleshooting problems. Founded in 1990. **Common positions include:** Customer Service Representative; Sales Executive; Software Engineer; Technical Support Representative. **Educational backgrounds include:** Computer Science; Engineering. **Benefits:** Credit Union; Dental Insurance; Disability Coverage; Medical Insurance. **Corporate headquarters location:** This Location. **International locations:** Australia; Germany; Holland; United Kingdom. **Listed on:** Privately held. **Annual sales/revenues:** $5 - $10 million. **Number of employees at this location:** 45. **Number of projected hires for 1997 - 1998 at this location:** 20.

MICROCOM, INC.

500 River Ridge Drive, Norwood MA 02062. 617/551-1000. **Toll-free phone:** 800/822-8224. **Fax:** 617/551-1006. **Contact:** William Cashman, Human Resources Manager. **World Wide Web address:** http://www.microcom.com. **Description:** Microcom, Inc. designs, manufactures, and markets software and hardware remote access products for use with PCs and PC networks. The company's products enable users to access and communicate with online computer networks, including the Internet, America Online, CompuServe, and Prodigy; and corporate networks from remote locations. Microcom's products serve both central site network managers and individual remote users. Products designed for the central site include the High Density Management System, a dial-up access management system; and LANexpress, remote local area network access systems which include expressWATCH, a comprehensive remote network access management solution. Products designed for the individual remote user include high-performance V.34 PCMCIA, desktop, and other modems; Carbon Copy remote control/remote PC access software; LANexpress Remote client software, a remote node and remote control LAN access product; and ISDN terminal adapters. **Common positions include:** Computer Programmer; Software Engineer. **Educational backgrounds include:** Computer Science. **Benefits:** 401(k); Dental Insurance; Disability Coverage; Life Insurance; Medical Insurance; Stock Option; Tuition Assistance. **Special Programs:** Internships. **Corporate headquarters location:** This Location. **Operations at this facility include:** Administration; Research and Development; Sales; Service. **Listed on:** NASDAQ. **Number of employees at this location:** 260. **Number of employees nationwide:** 300.

MICROCRAFTS

17371 NE 67th Court, Suite 205, Redmond WA 98052. 206/250-0000. **Contact:** Human Resources. **Description:** MicroCrafts develops custom software around a client's needs.

MICROEDGE, INC.

P.O. Box 988, Apex NC 27502-0988. 919/303-7400. **Fax:** 919/303-8400. **Contact:** Human Resources. **E-mail address:** human_resources@slickedit.com. **World Wide Web address:** http://www.slickedit.com. **Description:** A developer of graphical programmers' editing software including Visual SlickEdit and SlickEdit text-mode version.

MICROFRAME INC.

21 Meridian Road, Edison NJ 08820. 908/494-4440. **Contact:** Human Resources. **Description:** Develops and markets software and hardware for computer security.

MICROFRONTIER, INC.
P.O. Box 71190, Des Moines IA 50325. 515/270-8109. **Fax:** 515/278-6828. **Contact:** Human Resources. **E-mail address:** mfi@microfrontier.com. **World Wide Web address:** http://www.microfrontier.com. **Description:** A developer of graphic arts software.

MICROGRAFX, INC.
1303 East Arapahoe Road, Richardson TX 75081. 972/234-1769. **Fax:** 972/234-6036. **Contact:** Human Resources. **World Wide Web address:** http://www.micrografx.com. **Description:** Micrografx, Inc. develops, markets, and supports a line of graphic application software products for IBM PCs and compatibles running under the Microsoft Windows operating environment. Its software is designed for both business and professional use and includes professional illustration, basic drawing and charting products, data-driven graphics, image editing, and reusable clip-art libraries. Micrografx, Inc. also offers systems software products designed to enhance the Windows and OS/2 operating environments. **Common positions include:** Account Representative; Applications Engineer; Computer Programmer; Sales Representative; Software Engineer; Systems Analyst. **Educational backgrounds include:** Business Administration; Computer Science; Engineering; Finance; Marketing. **Benefits:** 401(k); Dental Insurance; Disability Coverage; Employee Discounts; Life Insurance; Medical Insurance; Tuition Assistance. **Corporate headquarters location:** This Location. **International locations:** Australia; Italy; Japan; U.K. **Listed on:** NASDAQ. **Annual sales/revenues:** $5 - $10 million. **Number of employees at this location:** 200. **Number of employees worldwide:** 300.

MICROLEAGUE MULTIMEDIA, INC.
750 Dawson Drive, Newark DE 19713. 302/368-9990. **Contact:** Human Resources. **Description:** A developer of entertainment and educational software.

MICROLOG CORPORATION
20270 Goldenrod Lane, Germantown MD 20876. 301/428-3227. **Contact:** Human Resources. **Description:** Manufactures voice processing computer systems and the applicable software.

MICROMATH SCIENTIFIC SOFTWARE
P.O. Box 71550, Salt Lake City UT 84171. 801/943-0290. **Toll-free phone:** 800/942-6284. **Fax:** 801/943-0299. **Contact:** Human Resources. **World Wide Web address:** http://www.MicroMath. com. **Description:** A developer of software. **Corporate headquarters location:** This Location.

MICRONETICS DESIGN CORPORATION
1375 Piccard Drive, Suite 300, Rockville MD 20850. 301/258-2605. **Contact:** Human Resources. **Description:** Develops software for the M programming language. Companies use these products to develop their own applications.

MICROPROSE SOFTWARE INC.
180 Lakefront Drive, Hunt Valley MD 21030. 410/771-0440. **Contact:** Ms. Gayle Keidel, Human Resources. **Description:** A software developer that specializes in computer games.

MICRORIM INC.
15395 SE 30th Place, Bellevue WA 98007. 206/649-9500. **Fax:** 206/746-9438. **Contact:** Human Resources. **Description:** Develops database and Internet software.

MICROSOFT CORPORATION
One Microsoft Way, Suite 303, Redmond WA 98052-8303. 206/882-8080. **Contact:** Recruiting Department/NJB. **Description:** Designs, sells, and supports a product line of systems and applications microcomputer software for business, home, and professional use. Microsoft also produces related books and hardware products. The company has shown increasing interest in developing applications for interactive television. Software includes spreadsheet, desktop publishing, project management, graphics, word processing, and database applications as well as operating systems and programming languages. **Corporate headquarters location:** This Location. **Listed on:** NASDAQ.

Other U.S. locations:
- 10000 Torre Avenue, Cupertino CA 95014. 408/777-0314. This location is a research and development office.
- Three Park Plaza, Suite 1800, Irvine CA 92714. 714/263-0200. This location is a district sales office.
- 5335 Wisconsin Avenue NW, Suite 600, Washington DC 20015. 202/895-2000. This location is a sales office.

MICROSYSTEMS SOFTWARE, INC.

600 Worcester Road, Framingham MA 01702. 508/879-9000. **Toll-free phone:** 800/489-2001. **Fax:** 508/626-8515. **Contact:** Human Resources Department. **E-mail address:** jobs@microsys.com. **World Wide Web address:** http://www.microsys.com. **Description:** Microsystems Software is a developer of software and related hardware for the disabled. The HandiWARE product line offers screen magnifiers for the visually impaired, as well as on-screen keyboards and word prediction products. Cyber Patrol monitors and limits children's access to inappropriate Internet information and violent or sexually explicit Websites. Microsystems Software also develops CaLANdar, a software package that performs group scheduling. Foreign operations are conducted in the U.K. Founded in 1989. **Corporate headquarters location:** This Location. **Listed on:** Privately held.

MICROWARE SYSTEMS CORPORATION

1900 NW 114th Street, Des Moines IA 50325. 515/224-1929. **Contact:** Human Resources. **Description:** Microware Systems develops a real time operating system called OS-9.

MIDAS KAPITI

45 Broadway, New York NY 10006. 212/898-9500. **Contact:** Human Resources. **Description:** Develops software for the banking industry.

MINDSCAPE

88 Rowland Way, Novato CA 94945. 415/897-9900. **Fax:** 415/897-2747. **Contact:** Human Resources. **Description:** Develops, publishes, and licenses software for use with PCs and Nintendo. **Number of employees at this location:** 330.

MITCHELL INTERNATIONAL

9889 Willow Creek Road, San Diego CA 92131. 619/578-6550. **Fax:** 619/578-4752. **Recorded jobline:** 619/530-8915. **Contact:** Human Resources. **Description:** Mitchell is a provider of printed information and electronic software products for the automotive industry. Sales divisions are located nationwide. **Common positions include:** Accountant/Auditor; Adjuster; Administrative Services Manager; Automotive Mechanic; Budget Analyst; Computer Programmer; Construction and Building Inspector; Construction Contractor; Credit Manager; Customer Service Representative; Economist; Editor; Financial Analyst; Human Resources Manager; Human Service Worker; Insurance Agent/Broker; Librarian; Public Relations Specialist; Purchasing Agent/Manager; Quality Control Supervisor; Services Sales Representative; Software Engineer; Systems Analyst; Technical Writer/Editor. **Educational backgrounds include:** Accounting; Business Administration; Computer Science; Engineering; Finance; Marketing. **Benefits:** 401(k); Dental Insurance; Disability Coverage; Employee Discounts; Life Insurance; Medical Insurance; Pension Plan; Tuition Assistance. **Special Programs:** Internships. **Corporate headquarters location:** Stamford CT. **Other U.S. locations:** Chicago IL; Detroit MI; McLean VA; Milwaukee WI. **Subsidiaries include:** EH Boeckh (Milwaukee WI); NAG's (Detroit MI); Mitchell-Medical (VA). **Parent company:** Thomson Corporation. **Operations at this facility include:** Administration; Divisional Headquarters; Manufacturing; Regional Headquarters; Research and Development; Sales; Service. **Listed on:** Privately held. **Number of employees at this location:** 500. **Number of employees nationwide:** 730.

MOBIUS GROUP, INC.

68 T.W. Alexander Drive, Research Triangle Park NC 27709. 919/549-0444. **Contact:** Human Resources. **Description:** A developer of financial software.

MOBIUS MANAGEMENT SYSTEMS
One Ramada Plaza, New Rochelle NY 10801. 914/637-7200. **Contact:** Vicki Rolita, Human Resources. **Description:** Develops and sells business-related software products, including a report distribution program and an automated balance program.

MORSE WATCHMANS INC.
2 Morse Road, Oxford CT 06478. 203/264-4949. **Contact:** Human Resources. **Description:** Develops computer software security systems.

MOTOROLA EMTEK HEALTHCARE DIVISION
1501 West Fountainhead Parkway, Suite 190, Tempe AZ 85282. 602/902-2600. **Contact:** Human Resources. **Description:** Develops software used by health care providers.

MULTIDATA SYSTEMS INTERNATIONAL CORPORATION
9801 Manchester Road, St. Louis MO 63119. 314/968-6880. **Contact:** Human Resources. **Description:** Develops computer software for radiation treatment planning.

MUSITEK
410 Bryant Circle, Suite K, Ojai CA 93023. 805/646-8051. **Fax:** 805/646-8051. **Contact:** Human Resources. **World Wide Web address:** http://www.musitek.com. **Description:** A developer of music software. Product line includes MIDISCAN, a music-reading software that converts printed sheet music into multi-track MIDI files.

MUZE INC.
304 Hudson Street, New York NY 10013. 212/824-0300. **Contact:** Jeanne Petras, Director of Human Resources. **Description:** Develops software for touch-screen point-of-sales terminals that allows users access to a music database. **Common positions include:** Accountant/Auditor; Administrative Services Manager; Budget Analyst; Computer Programmer; Financial Analyst; Human Resources Manager; Purchasing Agent/Manager; Sales Executive; Sales Manager; Sales Representative; Software Engineer; Systems Analyst. **Educational backgrounds include:** Art/Design; Communications; Computer Science; Finance; Liberal Arts; Marketing. **Benefits:** 401(k); Medical Insurance. **Special Programs:** Internships. **Corporate headquarters location:** This Location. **Parent company:** MetroMedia. **Operations at this facility include:** Administration; Manufacturing; Research and Development; Sales; Service. **Listed on:** Privately held. **Number of employees at this location:** 120.

NDC AUTOMATION
3101 Latrobe Drive, Charlotte NC 28211-4849. 704/362-1115. **Contact:** Gail Hentz, Director of Human Resources. **Description:** NDC Automation acquires, develops, and markets hardware and software that are incorporated into automatic guided vehicle systems (AGVS), which are used by customers to transport materials between various locations within a manufacturing or distribution facility.

NTN COMMUNICATIONS, INC.
5966 Le Place Court, Suite 100, Carlsbad CA 92008. 760/438-7400. **Contact:** Human Resources. **Description:** A manufacturer of interactive video games including trivia and sports trivia games.

NARRATIVE COMMUNICATIONS CORPORATION
204 Second Avenue, Waltham MA 02154. 617/290-5300. **Contact:** Lauren Chatham, Office Manager. **World Wide Web address:** http://www.narrative.com. **Description:** A developer of software known as Enliven, which allows users to include sound, animation, and graphics on their Websites.

NATIONAL HEALTH ENHANCEMENT SYSTEMS INC.
3200 North Central Avenue, Suite 1700, Phoenix AZ 85012. 602/230-7575. **Contact:** Personnel Director. **Description:** National Health Enhancement Systems Inc. markets health check-up and evaluation systems, which include instrumentation and software for health care providers. **Common positions include:** Manufacturer's/Wholesaler's Sales Rep. **Educational backgrounds include:** Marketing. **Benefits:** 401(k); Disability Coverage; Medical Insurance.

NATIONAL SYSTEMS AND RESEARCH COMPANY

5475 Mark Dabbling Boulevard, Suite 200, Colorado Springs CO 80918. 719/590-8880. **Contact:** Human Resources Department. **Description:** Provides ADP support services, including software development and testing for commercial and government clients in the aerospace industry. **Common positions include:** Accountant/Auditor; Branch Manager; Computer Operator; Computer Programmer; Department Manager; Electrical/Electronics Engineer; Payroll Clerk; Receptionist; Secretary; Software Engineer; Systems Analyst; Typist/Word Processor. **Educational backgrounds include:** Business Administration; Computer Science; Engineering. **Benefits:** 401(k); Dental Insurance; Disability Coverage; Life Insurance; Medical Insurance; Savings Plan; Tuition Assistance. **Corporate headquarters location:** This Location. **Other U.S. locations:** Camarillo CA; Boulder CO; Golden CO; Loveland CO; Tulsa OK; Portland OR; Vancouver WA. **Subsidiaries include:** NSR Information, Inc. **Operations at this facility include:** Administration; Divisional Headquarters. **Listed on:** Privately held. **Number of employees at this location:** 95. **Number of employees nationwide:** 500.

NATIONS, INC.

788 Shrewsbury Avenue, Tinton Falls NJ 07724. 908/530-1818. **Contact:** Human Resources. **Description:** Designs and engineers software for the federal government.

NETFRAME SYSTEMS, INC.

1545 Barber Lane, Milpitas CA 95035. 408/474-1000. **Contact:** Human Resources. **Description:** A software developer.

NETMANAGE

10725 North DeAnza Boulevard, Cupertino CA 95014. 408/973-7171. **Contact:** Susan Lola, Human Resources. **Description:** A software company that develops products for internetworking applications, primarily for Microsoft Windows.

NETPHONIC COMMUNICATIONS INC.

1580 West El Camino Real, Mountain View CA 94040. 415/962-1111. **Contact:** Ken Rhie, President. **Description:** The developer of a software product called Web on Call, which is installed on a customer's server at the World Wide Web site. NetPhonic Communications' software enables users to access Web information via touch tone telephone, without the use of a modem.

NETPRO COMPUTING, INC.

7150 Camelback Road, Suite 100, Scottsdale AZ 85251. **Contact:** Toni Rucker, Human Resources. **E-mail address:** tonir@netpro.com. **World Wide Web address:** http://www.netpro.com. **Description:** A network software development firm.

NETWORK IMAGING CORPORATION

500 Huntmar Park Drive, Herndon VA 20170-5100. 703/318-8201. **Contact:** Human Resources. **Description:** Designs and markets open systems document imaging and optical disk storage software geared to businesses and government agencies that are looking to substantially reduce their enormous volume of paper documents, cut down on storage space, and improve work efficiency. The company's systems also accommodate blueprints, X-rays, and data in video and audio format. The company's software manages the storage of scanned data and allows users to quickly retrieve and access documents. Network Imaging systems can also serve as a key component for video- and audio-on-demand. **Corporate headquarters location:** This Location. **Listed on:** NASDAQ. **Number of employees nationwide:** 505.

NETWORK SPECIALISTS INC.

80 River Street, Suite 5B, Hoboken NJ 07030. 201/656-2121. **Fax:** 201/656-2727. **Contact:** Human Resources. **World Wide Web address:** http://www.nsisw.com. **Description:** A developer of network performance and fault-tolerant tools. Products are compatible with Novell NetWare, Microsoft Windows NT, and UNIX.

NEUMENON

44567 Pinetree Drive, Plymouth MI 48170. 313/207-8484. **Contact:** Diane Martz, Human Resources. **Description:** Designs inventory software for the automotive industry.

NEW TECHNOLOGY PARTNERS
40 South River Road, Building 44, Bedford NH 03110. 603/622-4400. **Fax:** 603/641-6934.
Contact: Human Resources. **E-mail address:** employment@ntp.com. **World Wide Web address:**
http://www.ntp.com. **Description:** A computer products and services firm offering software
products for Windows NT and UNIX add-ons, as well as technology consulting and training.

NEWTEK INC.
1200 SW Executive Drive, Topeka KS 66615. 913/228-8000. **Contact:** Human Resources.
Description: Designs and develops software used for animation and graphics.

NEXT SOFTWARE, INC.
900 Chesapeake Drive, Redwood City CA 94063. 415/366-0900. **Fax:** 415/780-4554. **Contact:**
Staffing Manager. **E-mail address:** resumes@next.com. **World Wide Web address:**
http://www.next.com. **Description:** A provider of custom development software for the World
Wide Web and Enterprise technology. **NOTE:** At the end of 1996, Apple purchased NeXT.
Interested jobseekers should call to find out more information. (At press time, all information was
current.) NeXT maintains an employment page at its Website. **Common positions include:**
Account Representative; Accountant/Auditor; Designer; District Manager; Marketing Manager;
Quality Assurance Engineer; Sales Manager; Software Engineer; Systems Engineer; Technical
Writer/Editor. **Benefits:** 401(k); Accident Insurance; Dental Insurance; Life Insurance; Medical
Insurance; STD/LTD Coverage; Vision Insurance. **Special Programs:** Internships. **Corporate
headquarters location:** This Location. **Number of employees nationwide:** 300.

NEXUS SOFTWARE, INC.
4020 Westchase Boulevard, Suite 220, Raleigh NC 27607. 919/233-1100. **Contact:** Human
Resources. **Description:** A manufacturer of software for the financial industry.

NOVA MICROSONICS
One Pearl Court, Allendale NJ 07401. 201/934-8338. **Fax:** 201/818-0273. **Contact:** Richard
Cialone, Staffing. **E-mail address:** rcialo@nms.com. **World Wide Web address:**
http://www.atl.com. **Description:** A software company that develops client/server image
management systems for use with medical ultrasound instrumentation. Founded in 1988. **Common
positions include:** Applications Engineer; Project Manager; Quality Control Supervisor; Software
Engineer. **Educational backgrounds include:** Computer Science; Engineering. **Benefits:** 401(k);
Dental Insurance; Disability Coverage; Life Insurance; Medical Insurance; Pension Plan; Profit
Sharing. **Corporate headquarters location:** Bothell WA. **Parent company:** Advanced
Technology Laboratories. **Annual sales/revenues:** $11 - $20 million. **Number of employees at
this location:** 85. **Number of employees nationwide:** 125.

NOVELL, INC.
1555 North Technology Way, Orem UT 84097. 801/222-6000. **Contact:** Human Resources.
World Wide Web address: http://www.novell.com. **Description:** Novell is singularly focused on
network software. The company sets and adheres to open standards, develops new tools and
systems, works in partnership with other companies, and makes computer networks easy to use and
manage. Novell has grown to become one of the largest software companies in the world by
building software that spans from operating systems, network infrastructure, and services, to
network applications and information access products. A $2 billion company, Novell has major
development sites in California, Utah, and New Jersey. Novell's leading products include NetWare
4, UnixWare 2, PerfectOffice 3, and GroupWise 4. Novell is built around four business units that
develop complementary system software components and applications used to deploy advanced
network-based information systems: the NetWare Systems Group, the UNIX Systems Group, the
Information Access and Management Group, and the Applications Group. **Corporate
headquarters location:** This Location. **Number of employees nationwide:** 7,900.
Other U.S. locations:
- 2180 Fortune Drive, Building 3, San Jose CA 95131. 408/434-2300. Recorded Jobline:
  800/624-4520. Activities at this location include marketing, software engineering,
  administration, and more.
- 5250 Seventy Seven Center Drive, Suite 350, Charlotte NC 28217. 704/527-0045. Contact:
  Gary Davis, Sales Coordinator. This location is a sales office.

NOW SOFTWARE, INC.

921 SW Washington Street, Suite 500, Portland OR 97205-2823. 503/274-2800. **Fax:** 503/274-6390. **Contact:** Human Resources. **E-mail address:** info@nowsoft.com. **World Wide Web address:** http://www.nowsoft.com. **Description:** Develops contact management software. **Corporate headquarters location:** This Location.

NU TECH SOFTWARE SOLUTIONS, INC.

11575 SW Pacific Highway, Suite 109, Tigard OR 97223-8671. 503/968-9035. **Fax:** 503/968-1877. **Contact:** Human Resources. **E-mail address:** info@nutech.com. **World Wide Web address:** http://www.nutech.com. **Description:** An Internet software solutions and services firm.

NUMEGA TECHNOLOGIES, INC.

9 Townsend West, Nashua NH 03063. 603/889-2386. **Toll-free phone:** 800/468-6342. **Fax:** 603/889-1135. **Contact:** Human Resources. **E-mail address:** jobs@numega.com. CompuServe: GO NUMEGA. **World Wide Web address:** http://www.numega.com. **Description:** A developer of software development tools including Bounds Checker, an advanced error detection tool; and SoftICE, an advanced debugger for Windows system code, device drivers, or complex applications.

NUMETRIX INC.

101 Merritt 7, Norwalk CT 06851. 203/847-3452. **Contact:** Human Resources. **Description:** Develops finite capacity software.

NYMA, INC.

7501 Greenway Center Drive, Suite 1200, Greenbelt MD 20770. 301/345-0832. **Contact:** Susan Nyce, Human Resources. **Description:** A computer services company specializing in custom software development and systems integration. The company operates in four business areas: Air Traffic Systems, Aerospace Systems, Information Systems, and Computer Products.

OMR SYSTEMS CORPORATION

101 Business Park Drive, Suite 220, Skillman NJ 08558. 609/497-2000. **Contact:** Human Resources. **Description:** Develops backroom trading software.

OBJECTIVITY, INC.

301-B East Evelyn Avenue, Mountain View CA 94041. 415/254-7100. **Contact:** Human Resources. **Description:** A manufacturer of computer database software.

OMTOOL

8 Industrial Way, Salem NH 03079. 603/898-8900. **Toll-free phone:** 800/886-7845. **Fax:** 603/890-6756. **Contact:** Human Resources. **E-mail address:** sales@omtool.com. **World Wide Web address:** http://www.omtool.com. **Description:** A developer of client/server software solutions serving the network fax, document management, and workgroup markets.

ONTARIO SYSTEMS CORPORATION

1150 Kilgore, Muncie IN 47305. 765/751-7000. **Contact:** Marty Lang, Human Resources. **Description:** Develops software for collection agencies and health care providers.

ONYX SOFTWARE CORPORATION

330 120th Avenue NE, Suite 210, Bellevue WA 98005. 206/451-8060. **Fax:** 206/451-8277. **Contact:** Recruiting. **E-mail address:** recruiting@onyxcorp.com. **World Wide Web address:** http://www.onyxcorp.com/. **Description:** Develops customer management software.

OPEN HORIZON INC.

601 Gateway Boulevard, Suite 800, South San Francisco CA 94080. 415/598-1200. **Contact:** Human Resources. **Description:** A developer of Java software.

OPEN MARKET, INC.

245 First Street, Cambridge MA 02142. 617/949-7000. **Fax:** 617/621-1703. **Contact:** Human Resources. **E-mail address:** sales@openmarket.com. **World Wide Web address:** http://www.openmarket.com. **Description:** Develops software that enables businesses to manage

secure access control to company information on the Internet. The company's products include OM-Transact, OM-SecureLink Executive, Open Market Secure Websaver, and WebReporter. Clients range from small to large corporations.

OPEN SOFTWARE ASSOCIATES, INC.
20 Trafalgar Square, Nashua NH 03063. 603/886-4330. **Fax:** 603/598-6877. **Contact:** Human Resources. **E-mail address:** sales@osa.com. **World Wide Web address:** http://www.osa.com. **Description:** A software firm specializing in Internet and intranet application development and deployment. Open Software Associates also offers professional computer services.

OPEN-CONNECT SYSTEMS INC.
2711 LBJ Freeway, Suite 800, Dallas TX 75234. 972/484-5200. **Contact:** Human Resources. **Description:** A developer of Java software.

ORACLE CORPORATION
500 Oracle Parkway, MS 20P2, Redwood Shores CA 94065. 415/506-7000. **Contact:** Recruiting. **Description:** A computer programming services company. **Common positions include:** Administrative Services Manager; Computer Programmer; Economist; Electrical/Electronics Engineer; Purchasing Agent/Manager; Software Engineer; Systems Analyst. **Educational backgrounds include:** Accounting; Business Administration; Computer Science; Engineering. **Benefits:** 401(k); Dental Insurance; Disability Coverage; Employee Discounts; Life Insurance; Medical Insurance; Profit Sharing; Savings Plan; Tuition Assistance. **Corporate headquarters location:** This Location. **Operations at this facility include:** Administration; Sales. **Number of employees nationwide:** 12,000.
Other U.S. locations:
- 3 Bethesda Metro Center, Suite 1400, Bethesda MD 20814. 301/657-7860. This location also offers training programs.
- 517 Route 1 South, Suite 4000, Iselin NJ 08830. 908/636-2000. This location is a sales office.
- 100 Summit Lake Drive, Valhalla NY 10595. 914/747-2900.

ORACLE CORPORATION
500 108th Avenue NE, Suite 1300, Bellevue WA 98004. 206/646-0200. **Contact:** Human Resources. **Description:** This location develops relational database software. **NOTE:** For commercial consultant positions, address resumes to Tammy Yeager. For sales positions, address resumes to Sheila Connolly. **Common positions include:** Administrative Services Manager; Computer Programmer; Economist; Electrical/Electronics Engineer; Purchasing Agent/Manager; Software Engineer; Systems Analyst. **Educational backgrounds include:** Accounting; Business Administration; Computer Science; Engineering. **Benefits:** 401(k); Dental Insurance; Disability Coverage; Employee Discounts; Life Insurance; Medical Insurance; Profit Sharing; Savings Plan; Tuition Assistance. **Corporate headquarters location:** Redwood Shores CA. **Other U.S. locations:** Nationwide. **Operations at this facility include:** Administration; Regional Headquarters; Sales. **Number of employees nationwide:** 12,000.

ORCA MEDICAL SYSTEMS
22118 20th Avenue SE, Suite 130, Bothell WA 98021. 206/489-2611. **Contact:** Human Resources. **Description:** ORCA Medical Systems develops automation software for hospital emergency rooms.

ORCAD
9300 SW Nimbus Avenue, Beaverton OR 97008. 503/671-9500. **Toll-free phone:** 800/671-9511. **Fax:** 503/671-9501. **BBS:** 503/671-9401. **Contact:** Human Resources. **E-mail address:** employment@orcad.com. **World Wide Web address:** http://www.orcad.com. **Description:** A developer of desktop electronic design automation (EDA) software. OrCAD's Windows-based software aids in the design of field-programmable gate arrays, complex programmable logic devices, and printed circuit boards. **Listed on:** NASDAQ. **Stock exchange symbol:** OCAD.

ORCOM SYSTEMS
1001 SW Disk Drive, Bend OR 97702. 541/389-0120. **Contact:** Human Resources. **Description:** Develops a variety of software packages for the utility industry.

OXFORD MOLECULAR
810 Glen Eagles Court, Suite 300, Towson MD 21286. 410/821-5980. **Contact:** Human Resources. **Description:** Develops software for the pharmaceutical industry.

PC DOCS INC.
124 Marriott Drive, Tallahassee FL 32301. 904/942-3627. **Contact:** Gary Pritchett, Human Resources. **Description:** Develops and supports computer software including PC Docs.

PDA, INC.
7701 College Boulevard, Overland Park KS 66210. 913/469-8700. **Contact:** Human Resources. **Description:** Designs computer software for the insurance industry. PDA also offers software consulting services.

PDC
111C Lindbergh Avenue, Livermore CA 94550. 510/449-6881. **Fax:** 510/449-6885. **Contact:** Allen Nickerson, Director Of Administration. **E-mail address:** anickers@pdc.com. **World Wide Web address:** http://www.pdc.com. **Description:** A software development company that integrates products and solutions into a large UNIX environment. **NOTE:** The company offers entry-level positions. **Common positions include:** Account Representative; Market Research Analyst; Project Manager; Sales Representative; Software Engineer; Systems Manager. **Educational backgrounds include:** Computer Science. **Benefits:** 401(k); Dental Insurance; Financial Planning; Life Insurance; Medical Insurance; Profit Sharing; Savings Plan. **Corporate headquarters location:** King of Prussia PA. **Other U.S. locations:** Chicago IL; Boston MA; New York NY; Dallas TX. **Listed on:** Privately held. **Annual sales/revenues:** $21 - $50 million. **Number of employees at this location:** 50. **Number of employees nationwide:** 150. **Number of projected hires for 1997 - 1998 at this location:** 25.

PKWARE, INC.
9025 North Deerwood Drive, Brown Deer WI 53223-2480. 414/354-8699. **Fax:** 414/354-8559. **Contact:** Human Resources. **E-mail address:** hr@pkware.com. **World Wide Web address:** http://www.pkware.com. **Description:** A developer of data compression software. Product line includes PKZIP.

PSI TECHNOLOGIES CORPORATION
1001 South 360, Building K, Suite 200, Austin TX 78746. 512/329-0081. **Contact:** Human Resources. **Description:** Develops retrieval and conversion software. **Corporate headquarters location:** This Location.

PACER INFO TECH INC.
2500 Maryland Road, Suite 400, Willow Grove PA 19090-1225. 215/657-7800. **Contact:** Human Resources. **Description:** Develops software and offers systems support for government contracts.

PALO ALTO SOFTWARE, INC.
144 East 14th Avenue, Eugene OR 97401. 541/683-6162. **Fax:** 541/683-6250. **Contact:** Human Resources. **E-mail address:** info@palo-alto.com. **World Wide Web address:** http://www.pasware.com. **Description:** Develops business plan and marketing software.

PARAGON PUBLISHING SYSTEMS
360-801 Route 101, Bedford NH 03110. 603/471-0077. **Contact:** Human Resources. **Description:** A firm providing software support for the publishing industry.

PARAMETRIC TECHNOLOGY CORPORATION
128 Technology Drive, Waltham MA 02154. **Fax:** 617/891/1069. **Contact:** Mike Cianciola, Employment Manager. **Description:** Parametric Technology Corporation designs and develops fully-integrated software products for mechnaical engineering and automated manufacturing based upon a parametric solids modeling system. The company has 105 offices in 23 countries. **Common positions include:** Applications Engineer; Computer Programmer; Educational Specialist; Mathematician; Sales Representative; Software Engineer; Systems Analyst. **Educational backgrounds include:** Computer Science; Engineering; Finance; Marketing. **Benefits:** 401(k);

Dental Insurance; ESOP; Medical Insurance; Vision Insurance. **Corporate headquarters location:** This Location. **Operations at this facility include:** Administration; Marketing; Research and Development; Sales. **Listed on:** NASDAQ. **Number of employees at this location:** 500. **Number of employees nationwide:** 1,600.
Other U.S. locations:
- 20800 Swenson Drive, Suite 250, Waukesha WI 53186. 414/798-9494. This location serves as a regional sales office.

PARASOFT CORPORATION
2031 South Myrtle Avenue, Monrovia CA 91016. 818/305-0041. **Toll-free phone:** 888/305-0041. **Fax:** 818/305-9048. **Contact:** Human Resources. **E-mail address:** jobs@parasoft.com. **World Wide Web address:** http://www.parasoft.com. **Description:** Develops various software using C and C++. **Corporate headquarters location:** This Location.

PARCPLACE SYSTEMS, INC.
999 East Arques Avenue, Sunnyvale CA 94086. 408/481-9090. **Contact:** Human Resources. **Description:** Produces and sells development tools software. These products help businesses create their own applications.

PARK CITY GROUP INC.
P.O. Box 5000, Park City UT 84060. 801/645-2041. **Contact:** Human Resources. **Description:** Manufactures back office operational software, from e-mail to inventory reports, for the retail industry. Park City Group has over 31 products.

PARSONS TECHNOLOGY CORPORATION
One Martha's Way, Hiawatha IA 52233. 319/395-9626. **Contact:** Human Resources. **Description:** Parsons Technology develops more than 80 software programs for a wide range of business applications and provides technical support for their software. **Corporate headquarters location:** This Location.

PASH & COMPANY
P.O. Box 830725, Birmingham AL 35283-0725. 205/324-0088. **Contact:** Human Resources. **Description:** Develops revenue enhancement software.

PATIENT FOCUS SYSTEMS
1140 Terex Road, Hudson OH 44236. **Toll-free phone:** 800/787-3477. **Contact:** Mike Haymond, Director of Product Development. **Description:** Develops medical software.

PAYSYS INTERNATIONAL
900 Winderley Place, Suite 200, Maitland FL 32751. 407/660-0343. **Contact:** Human Resources. **Description:** Develops credit card processing software for a variety of businesses.

PEACHTREE SOFTWARE
1505 Pavilion Place, Suite #C, Norcross GA 30093. 770/564-5800. **Contact:** Human Resources. **Description:** Develops, manufactures, and markets accounting software.

PENTAMATION ENTERPRISE INC.
225 Marketplace, Bethlehem PA 18018. 610/691-3616. **Contact:** Human Resources. **Description:** Develops software and computer systems for school districts and government facilities.

PEOPLESOFT, INC.
4440 Rosewood Drive, 2nd Floor, Pleasanton CA 94588-3031. 510/225-3000. **Toll-free phone:** 800/695-5293. **Contact:** Human Resources. **World Wide Web address:** http://www.peoplesoft.com. **Description:** Provides client/server applications and software solutions for businesses worldwide. PeopleSoft develops, markets, and supports a variety of enterprise solutions for accounting, materials management, distribution, manufacturing, and human resources, as well as offering industry-specific enterprise solutions to markets including financial services, health care, manufacturing, higher education, the public sector, and the federal government. PeopleSoft has been listed by *Fortune* magazine as one of America's fastest growing companies. In October

1996, PeopleSoft merged with Red Pepper Software, Inc., a provider of manufacturing and supply chain optimization solutions. **Corporate headquarters location:** This Location. **Listed on:** NASDAQ. **Stock exchange symbol:** PSFT. **Annual sales/revenues:** $300 million. **Number of employees worldwide:** 2,069.

PERSOFT, INC.

465 Science Drive, Madison WI 53711. 608/273-6000. **Toll-free phone:** 800/3685283. **Fax:** 608/273-8227. **Contact:** Human Resources. **E-mail address:** personnel@persoft.com. **World Wide Web address:** http://www.persoft.com. **Description:** A developer of communications software. Product line includes SmarTerm, PC-to-host connectivity software; and Intersect wireless bridges.

PERSONNEL DATA SYSTEMS INC.

670 Sentry Parkway, Blue Bell PA 19422. 610/828-4294. **Contact:** Human Resources Department. **Description:** Personnel Data Systems develops, markets, and supports software products for human resources departments.

PHILLIPS DESIGN

2726 Robert Avenue, Cincinnati OH 45211. 513/661-4494. **Contact:** Recruiting. **Description:** Offers contract computer programming services. **NOTE:** Phillips Design accepts resumes from experienced programmers only.

PHOENIX TECHNOLOGIES LTD.

846 University Avenue, Norwood MA 02062. 617/551-4000. **Contact:** Human Resources. **Description:** Phoenix Technologies designs, develops, and markets system software and end user software products and provides publishing services for personal computers and printers. The Peripherals Division is a designer, developer, and worldwide supplier of printer emulation software, page distribution languages, and controller hardware designs for the printing industry. Its PhoenixPage imaging software architecture enables printer manufacturers to offer products that are compatible with the PostScript language, the PCL printer language, and other imaging standards. Phoenix Technologies's PC Division works with leading vendors and standards committees to ensure that their products enable manufacturers to develop and deploy next-generation PCs quickly and cost-effectively. The company's Package Products Division is a single-source publisher of MS-DOS, Windows, and other software packages. More than 100 manufacturers worldwide rely on Phoenix Technologies to provide turnkey, cost-saving solutions for publishing industry standard software and documentation. **Corporate headquarters location:** This Location. **Number of employees at this location:** 330.

PHYCOM CORPORATION

3380 146th Place SE, Suite 414, Bellevue WA 98007. 206/644-1441. **Contact:** Human Resources Department. **Description:** Phycom Corporation develops case management software for the health care industry.

PHYSICIAN COMPUTER NETWORK

1200 The American Way, Morris Plains NJ 07950. 201/490-3100. **Contact:** Human Resources. **Description:** Develops software for medical practices.

PILOT SYSTEMS

325 North Corporate Drive, Suite 280, Milwaukee WI 53045. 414/792-0050. **Contact:** Human Resources Department. **Description:** Pilot Systems designs software packages for businesses. The company's products cover a wide range of topics.

PINDAR SYSTEMS, INC.

1661 North Swan Road, Suite 132, Tucson AZ 85712. 520/323-9500. **Fax:** 312/244-8161. **Contact:** Richard Hunt, Human Resources. **E-mail address:** pindar@pindarusa.com. **World Wide Web address:** http://www.primenet.com/~pindar. **Description:** A database technology management firm, manufacturer of The Catalog Management System.

PINNACLE TECHNOLOGY, INC.

P.O. Box 128, Kirklin IN 46050. 765/279-5157. **Fax:** 765/279-8039. **Contact:** Human Resources. **World Wide Web address:** http://www.pinnacletech.com. **Description:** A producer of desktop management and security software.

PITNEY BOWES SOFTWARE SYSTEMS

4343 Commerce Court, Suite 500, Lisle IL 60532-3618. 630/505-0572. **Contact:** Human Resources. **Description:** Develops computer software integration solutions.

PLANMAKER

1920 South Broadway, St. Louis MO 63104. 314/421-0670. **Fax:** 314/421-0668. **Contact:** Human Resources. **E-mail address:** planmaker@planmaker.com. **World Wide Web address:** http://www.planmaker.com. **Description:** A developer of business software.

PLATINUM SOFTWARE CORPORATION

195 Technology Drive, Irvine CA 92618. 714/453-4000. **Contact:** Human Resources. **Description:** Develops financial software for use in client/server environments. Platinum Software's products are geared towards small to mid-sized businesses. **Corporate headquarters location:** This Location. **Listed on:** NASDAQ.

PLATINUM TECHNOLOGY

1815 South Meyers Road, Oakbrook Terrace IL 60181. 630/620-5000. **Fax:** 630/691-0410. **Contact:** Deanna Jenca, Staffing Specialist. **Description:** PLATINUM TECHNOLOGY develops, markets, and supports an integrated line of system software and related educational programs for use with DB2, DB2/2, DB26000, Oracle, and Sybase. Products and services enable data center personnel of large and data-investigative organizations to operate their central databases more effectively. **Common positions include:** Accountant/Auditor; Computer Operator; Computer Programmer; Editor; Education Administrator; Financial Manager; Graphic Artist; Instructor/Trainer; Market Research Analyst; Marketing Manager; Public Relations Specialist; Receptionist; Secretary; Services Sales Representative; Software Developer; Technical Representative; Technical Writer/Editor; Telemarketer; Typist/Word Processor. **Educational backgrounds include:** Accounting; Business Administration; Computer Science; English; Finance; Journalism; Marketing. **Benefits:** 401(k); Dental Insurance; Disability Coverage; Flextime Plan; Health Club Discount; Life Insurance; Medical Insurance; Paid Vacation; Tuition Assistance. **Corporate headquarters location:** This Location. **Operations at this facility include:** Education; Marketing; Research and Development; Sales. **Listed on:** NASDAQ. **Number of employees at this location:** 320. **Number of employees nationwide:** 480.
Other U.S. locations:
- 400 Interstate North Parkway, Suite 900, Atlanta GA 30339. 770/850-1700. This location serves as a regional sales office.
- 404 Wyman Street, Suite 320, Waltham MA 02254. 617/891-6500. This location develops software.
- 620 West Germantown Pike, Plymouth Meeting PA 19462. 610/940-6020. Contact: Sam Porrecca, Hiring Manager. This location is a development lab.
- 1145 Herndon Parkway, Suite 200, Herndon VA 22070. 703/430-9247. This location develops and sells software.

POLICY MANAGEMENT SYSTEMS CORPORATION (PMSC)

12001 North Central Expressway, Suite 500, Dallas TX 75243. 972/778-7000. **Contact:** Barbara McGalloway, Personnel. **Description:** This is a sales location. Overall, Policy Management Systems Corporation develops and licenses standardized insurance software systems and provides automation and administrative support and information services to the worldwide insurance industry. The company also provides professional support services, which include implementation and integration assistance; consulting and education services; and information and outsourcing services. **Common positions include:** Computer Programmer; Insurance Agent/Broker; Systems Analyst. **Educational backgrounds include:** Computer Science. **Benefits:** 401(k); Dental Insurance; Disability Coverage; Employee Discounts; Life Insurance; Medical Insurance; Tuition Assistance. **Special Programs:** Internships. **Corporate headquarters location:** Columbia SC.

Operations at this facility include: Service. **Listed on:** New York Stock Exchange. **Number of employees nationwide:** 4,300.

POLICY MANAGEMENT SYSTEMS CORPORATION (PMSC)

P.O. Box 10, Columbia SC 29202. 803/735-4000. **Contact:** Recruiting. **Description:** Policy Management Systems Corporation develops and licenses standardized insurance software systems and provides automation and administrative support and information services to the worldwide insurance industry. The company also provides professional support services, which include implementation and integration assistance; consulting and education services; and information and outsourcing services. **Corporate headquarters location:** This Location. **Listed on:** New York Stock Exchange.

PORTA SYSTEMS CORPORATION

P.O. Box 11085, Charlotte NC 28220. 704/329-0700. **Contact:** Human Resources. **Description:** Develops software. Porta Systems Corporation supplies telephone companies worldwide with communication systems. This location designs, develops, and markets telecommunication software.

PORTLAND SOFTWARE

1000 SW Broadway, Suite 1850, Portland OR 97205. 503/220-2300. **Fax:** 503/220-8504. **Contact:** Simon King, Director of Development. **E-mail address:** simon@portsoft.com. **World Wide Web address:** http://www.portsoft.com. **Description:** A developer of Internet commerce tools through secure digital container technology for electronic software and data distribution. Portland Software offers a suite of encryption and electronic packaging applications. **Common positions include:** Software Engineer. **Corporate headquarters location:** This Location.

POSTALSOFT, INC.

4439 Norman Cooley Road, La Crosse WI 54601-8231. 608/788-8700. **Contact:** Human Resources. **Description:** Postalsoft develops software for mass mailings.

POWERQUEST CORPORATION

1083 North State Street, Orem UT 84057. 801/226-8977. **Fax:** 801/226-8941. **Contact:** Human Resources. **Description:** A software developer whose main product, Partition Magic, partitions a computer's hard disk without backing up or losing data. **Corporate headquarters location:** This Location.

PRE-ENGINEERING SOFTWARE

5800 One Perkins Place Drive, Suite 10D, Baton Rouge LA 70808. 504/769-3728. **Contact:** Human Resources. **Description:** The company writes software programs used to train civil engineers and other engineering professionals.

PRECISION COMPUTER SYSTEMS

4501 South Technology Drive, Sioux Falls SD 57106. 605/362-1260. **Contact:** Human Resources. **Description:** Develops, markets, and supports software applications for community banking and city governments.

PRECISION SYSTEMS INC.

11800 30th Court North, St. Petersburg FL 33716. 813/572-9300. **Contact:** Human Resources. **Description:** Develops software for voice enhancement services and voice dialing.

PREFERRED SYSTEMS, INC.

250 Captain Thomas Boulevard, West Haven CT 06516. 203/937-3000. **Toll-free phone:** 800/222-7638. **Fax:** 203/937-3032. **BBS:** 203/937-3019. **Contact:** Human Resources. **E-mail address:** sysop@prefsys.usa.com. **World Wide Web address:** http://www.prefsys.com. **Description:** A developer of database management software.

BYRON PREISS MULTIMEDIA COMPANY, INC.

24 West 25th Street, New York NY 10010. 212/989-6252. **Contact:** Human Resources Department. **Description:** Byron Preiss Multimedia Company, Inc. develops and publishes interactive multimedia software on CD-ROM and other multimedia formats for online delivery.

The company also co-publishes books and is developing its first online service. The Digital Bayhaus software features powerful search and retrieval methods, extensive cross-referencing of subjects, and complete indexing. The Crayon Multimedia software features children's software for home, school, and libraries. The Brooklyn Multimedia software features interactive games and entertainment software. The 21st Century Classics software includes classic literature for students and adults. The Arts and Commerce software features software for home, professional, and institutional graphic design, featuring leading artists, licenses, and designers.

PREMIA CORPORATION
1075 NW Murray Boulevard, Suite 268, Portland OR 97229. 503/641-6000. **Toll-free phone:** 888/4-PREMIA. **Contact:** Milo Wadlin, Human Resources. **E-mail address:** milow@premia.com. **World Wide Web address:** http://www.premia.com. **Description:** A developer and publisher of software for professional programmers.

PREMIER SOLUTIONS LTD.
325 Technology Drive, Malvern PA 19355. 610/251-6500. **Contact:** Human Resources. **Description:** Develops, markets, and supports software for the financial industry.

PRIMAVERA SYSTEMS INC.
2 Bala Plaza, Bala Cynwyd PA 19004. 610/667-8600. **Contact:** Eugene Eckenrode, Human Resources. **Description:** Develops and supports project management software.

PRINCETON FINANCIAL SYSTEMS INC.
600 College Road East, Princeton NJ 08540. 609/987-2400. **Fax:** 609/987-9320. **Contact:** Cara Verba, Human Resources. **Description:** Develops and supports financial software covering trading, mortgages, and other related topics.

PROFITKEY INTERNATIONAL, INC.
382 Main Street, Salem NH 03079. 603/898-9800. **Contact:** Patrick Patterson, Human Resources Director. **Description:** The company develops and markets business software and provides consulting and ancillary services. The company's products and services include transactional messaging middleware and distributed object technology, which facilitate communication among applications that reside on distributed and often incompatible hardware and software; industry-specific software applications the company has developed for manufacturers and for the retail petroleum and convenience store industry; and consulting services for enterprise messaging and for the manufacturing and financial services industries. Through the ASU consulting division, the company provides system integration consulting services to manufacturing businesses in California. **Corporate headquarters location:** This Location. **Subsidiaries include:** Bizware Computer Systems Inc.

PROGRAMMING RESOURCES COMPANY
875 Asylum Avenue, Hartford CT 06105. 860/728-1428. **Contact:** Human Resources. **Description:** Develops software. Programming Resources markets their products to the insurance industry. Currently the company is focusing on worker compensation applications.

PROGRESS SOFTWARE CORPORATION (PSC)
14 Oak Park, Bedford MA 01730. 617/280-4000. **Contact:** Human Resources. **Description:** Progress Software Corporation (PSC) supplies application development software to business and government industries worldwide. Advanced application development tools simplify and accelerate the creation, deployment, and maintenance of applications. The company's principal product is PROGRESS, a complete, integrated environment for developing and deploying mission-critical applications that are scalable, portable, and reconfigurable across a wide range of computing environments including client/server, host-based, and mixed. The product consists of the PROGRESS Application Development Environment and the PROGRESS DataServer Architecture, which includes the PROGRESS Relational Database Management System. Progress sells its products to organizations that develop and deploy major applications. Over half of the firm's worldwide revenues are derived from serving over 2,000 Application Partners who market PROGRESS-based applications. These organizations license PROGRESS to build and deliver sophisticated computerized information systems in diverse vertical markets such as manufacturing

resource planning, accounting, retail/point-of-sale, and stock trading. The balance of the firm's sales are to the MIS organizations of businesses and governments worldwide who license PROGRESS to develop and deploy a wide range of mission-critical applications. **Number of employees nationwide:** 629.

PROGRESSIVE NETWORKS
1111 3rd Avenue, Suite 500, Seattle WA 98101. 206/674-2700. **Fax:** 206/674-2699. **Contact:** Human Resources. **World Wide Web address:** http://www.realaudio.com. **Description:** Develops software that makes it possible for users to listen to audio over the Internet.

PROJECT SOFTWARE & DEVELOPMENT, INC.
20 University Road, Cambridge MA 02138. 617/661-1444. **Fax:** 617/661-1642. **Contact:** Human Resources. **World Wide Web address:** http://www.psdi.com. **Description:** Project Software & Development, Inc. develops, markets, and supports enterprisewide client/server applications software used to assist in maintaining and developing high-value capital assets facilities, systems, and production equipment. These products enable customers to reduce downtime, control maintenance expenses, cut spare parts inventories, improve purchasing efficiency, shorten product development cycles, and deploy productive assets and personnel more effectively. **Benefits:** Deferred Compensation; Profit Sharing; Stock Option. **Corporate headquarters location:** This Location. **Other U.S. locations:** Irvine CA; Denver CO; Watertown MA; Rockville MD; Dearborn MI; Rochelle Park NJ; Hauppauge NY; Irving TX; Bellevue WA. **Subsidiaries include:** Project Software Canada (Toronto); PSDI Australia Pty. Ltd. (Sydney); PSDI Benelux NV (The Netherlands); PSDI Deutschland GmbH (Munich); PSDI France SARL (Paris); and PSDI (UK) Ltd. (United Kingdom). **Listed on:** NASDAQ. **Number of employees nationwide:** 217. **Number of employees worldwide:** 383.

PROMODEL
1875 South Street, Orem UT 84097. 801/223-4600. **Contact:** Human Resources. **Description:** Develops and sells simulation software that is used in a wide range of industries including health care, manufacturing, and services.

PROVAR, INC.
2624 Lord Baltimore Drive, Suite D, Baltimore MD 21244. 410/265-8010. **Fax:** 410/265-8077. **Contact:** Human Resources. **Description:** Develops software, offers computer training, and provides systems integration services.

PURE ATRIA
20 Maguire Road, Lexington MA 02173. 617/676-2400. **Fax:** 617/676-2410. **Contact:** Human Resources. **E-mail address:** info@atria.com. **World Wide Web address:** http://www.atria.com. **Description:** Pure Atria develops, markets, and supports software that facilitates the management of complex software development, enhancement, and maintenance. Pure Atria's software configuration management products, ClearCase and ClearCase MultiSite, are used by professional software development teams to control the software development process, develop releases more efficiently, improve software quality, and shorten time to market. With global sales and support capabilities and an installed base of nearly 15,000 ClearCase licenses, Pure Atria is a leading provider of scalable, client/server SCM solutions. **Corporate headquarters location:** Sunnyvale CA. **Other U.S. locations:** Mountain View CA; Roseville CA; San Clemente CA; Orlando FL; Chicago IL; Iselin NJ; Dallas TX; McLean VA; Kirkland WA.

QAD INC.
6450 Via Real, Carpinteria CA 93013. 805/684-6614. **Fax:** 805/684-1890. **Contact:** Human Resources. **World Wide Web address:** http://www.qad.com. **Description:** This location houses administrative departments. Overall, QAD develops software including MFG/PRO, a software package designed to aid in supply and distribution management for large companies. MFG/PRO is produced in over twenty languages and QAD has offices worldwide. **Corporate headquarters location:** This Location.
Other U.S. locations:
- 695 Kenmoor Avenue SE, Suite 100, Grand Rapids MI 49546. 616/497-5457. This location focuses on developing software for the automotive industry.

- 10000 Midlantic Drive, Suite 200, Mount Laurel NJ 08054. 609/273-1717. This location serves as a technical support branch and regional sales office.

QUALITY SOFTWARE SYSTEMS INC.

200 Centennial Avenue, Piscataway NJ 08854. 908/885-1919. **Contact:** Human Resources. **Description:** Develops software to aid in warehouse management and development.

QUALITY SYSTEMS, INC.

17822 East 17th Street, Center Building 210, Tustin CA 92780. 714/731-7171. **Fax:** 714/731-9494. **Contact:** Raymond Mead, Manager, Human Resources. **E-mail address:** hr@qsii.com. **World Wide Web address:** http://www.aboveallqsii.com. **Description:** Quality Systems markets computerized information processing systems primarily to group dental and medical practices. The systems provide advanced computer-based automation in various aspects of group practice management, including the retention of patient information, treatment planning, appointment scheduling, billing, insurance claims processing, electronic insurance claims submission, allocation of income among group professionals, managed care reporting, word processing, and accounting. Founded in 1973. **NOTE:** The company offers entry-level positions. **Common positions include:** Account Manager; Account Representative; Applications Engineer; Internet Services Manager; Project Manager; Sales Representative; Software Engineer; Technical Writer/Editor. **Educational backgrounds include:** Computer Science; Economics; Marketing. **Benefits:** 401(k); Dental Insurance; Disability Coverage; Life Insurance; Medical Insurance. **Special Programs:** Internships. **Corporate headquarters location:** This Location. **Listed on:** NASDAQ. **Stock exchange symbol:** QSII. **Annual sales/revenues:** $11 - $20 million. **Number of employees at this location:** 160.

QUANTRA CORPORATION

2233 Lake Park Drive, Suite 400, Smyrna GA 30080. 770/431-0554. **Contact:** Human Resources. **Description:** Develops computer software for the mortgage and loan industry.

QUARK, INC.

1800 Grant Street, Denver CO 80203. 303/894-8888. **Contact:** Personnel Administrator. **Description:** A software engineering firm. The company's primary product is Quark Xpress, one of the leading products in desktop publishing.

QUARTERDECK CORPORATION

2401 Lemone Boulevard, Columbia MO 65201. 573/443-3282. **Contact:** Director of Human Resources. **Description:** Develops, markets, and supports software packages including multitasking, memory management products which enhance performance of DOS-based PC hardware and software, and several Internet products. **Stock exchange symbol:** Qdeck. **Number of employees worldwide:** 1,000.

QUARTERDECK SELECT

5770 Roosevelt Boulevard, Suite 400, Clearwater FL 34620. 813/523-9700. **Fax:** 813/523-2335. **Contact:** Human Resources. **Description:** Develops graphic, diagnostic, and Internet software. Quarterdeck Select also offers technical support.

QUEUE

338 Commerce Drive, Fairfield CT 06432. 203/335-0908. **Contact:** Human Resources. **Description:** Develops educational software. Queue's products range from early education software to software for much more advanced levels of education.

QUICKTURN DESIGN SYSTEMS, INC.

440 Clyde Street, Mountain View CA 94043. 415/967-3300. **Contact:** Human Resources. **Description:** A manufacturer of emulation systems, a design verification tool.

QUINTUS CORPORATION

47212 Mission Falls Court, Fremont CA 94539. **Toll-free phone:** 800/337-8941. **Contact:** Joyce Phillips, Human Resources. **Description:** Develops software known as Customer Q, a help desk application.

R&D SYSTEMS COMPANY
5225 North Academy Boulevard, Colorado Springs CO 80918. 719/590-8940. **Fax:** 719/548-1950. **Contact:** Pamela Green, Human Resource Specialist. **E-mail address:** pamelam@rdsystem.com. **Description:** Develops, sells, and supports accounting wholesale distribution software. **Common positions include:** Computer Programmer; Sales Representative. **Educational backgrounds include:** Accounting; Business Administration; Computer Science. **Benefits:** 401(k); Dental Insurance; Disability Coverage; Life Insurance; Medical Insurance; Profit Sharing; Telecommuting; Tuition Assistance. **Corporate headquarters location:** This Location. **Other U.S. locations:** Norcross AL; Dallas TX. **Listed on:** Privately held. **Annual sales/revenues:** $21 - $50 million. **Number of employees at this location:** 170. **Number of employees nationwide:** 250. **Number of projected hires for 1997 - 1998 at this location:** 45.

RIM SYSTEMS INC.
500 Technology Drive, Box 3094, Naperville IL 60566. 630/369-5300. **Contact:** Human Resources. **Description:** Develops software programs to help medical and dental insurance agencies manage their claim system.

RAIMA CORPORATION
1605 NW Sammamish Road, Issaquah WA 98027. 206/557-0200. **Contact:** Steve Smith, Owner. **Description:** Develops and markets database software.

RAINBOW TECHNOLOGIES
50 Technology Drive West, Irvine CA 92618. 714/454-2100. **Contact:** Human Resources. **Description:** Rainbow Technologies produces encryption keys for software developers to prevent piracy.

RATIONAL SOFTWARE
2800 San Tomas Expressway, Santa Clara CA 95051. 408/496-3600. **Contact:** Human Resources. **Description:** Develops object-oriented analysis software.

RAYTHEON E-SYSTEMS
1200 South Jupiter Road, Garland TX 75042-7111. 972/272-0515. **Toll-free phone:** 800/933-5359. **Contact:** Human Resources. **E-mail address:** employment@esi.org. **World Wide Web address:** http://www.raytheon.com. **Description:** This location develops high-technology software. Overall, Raytheon E-Systems designs, manufactures, and installs state-of-the-art communications and integrated command-and-control systems for military and industrial customers worldwide. The company is a complete systems engineering company offering technological innovation in the commercial and defense electronics industry.

REAL WORLD CORPORATION
P.O. Box 9516, Manchester NH 03108-9516. 603/641-0200. **Contact:** Human Resources. **World Wide Web address:** http://www.realworldcorp.com. **Description:** Develops software. Real World offers two computer accounting packages that can be used separately or together.

RED HAT SOFTWARE, INC.
3203 Yorktown Avenue, Suite 123, Durham NC 27713. 919/572-6500. **Fax:** 919/361-2711. **Contact:** Human Resources. **E-mail address:** marc@redhat.com. **World Wide Web address:** http://www.redhat.com. **Description:** Develops software for UNIX systems. Red Hat Software also offers technical support.

RELATIONAL TECHNOLOGY SYSTEMS
1601 Trapelo Road, Waltham MA 02154. 617/890-3726. **Fax:** 617/890-2953. **Contact:** Human Resources Department. **Description:** Relational Technology Systems develops field management systems for large organizations.

REMEDY CORPORATION
1505 Salado Drive, Mountain View CA 94043. 415/903-5200. **Contact:** Human Resources. **Description:** A designer and manufacturer of help desk software.

RESORT COMPUTER CORPORATION
12596 West Bayward, Suite 200, Lakewood CO 80228. 303/232-9417. **Contact:** Human Resources. **Description:** Develops software for the time-share resort industry.

RESTRAC
3 Allied Drive, Dedham MA 02026. 617/320-5600. **Contact:** Human Resources. **Description:** Manufactures and sells software that sorts and ranks resumes by criteria selected by the resume screener. Clients include Fidelity Management and Research, Hewlett-Packard, and Johnson & Johnson Services. **Corporate headquarters location:** This Location. **Listed on:** Privately held. **Number of employees at this location:** 125.

REYNOLDS & REYNOLDS
800 Concourse Parkway, Suite 100, Birmingham AL 35244. 205/444-5400. **Contact:** Human Resources. **Description:** Designs software for health management systems.

REYNOLDS & REYNOLDS MEDICAL SYSTEMS
9700 SW Nimbus Avenue, Beaverton OR 97008. 503/526-7700. **Contact:** Human Resources. **Description:** Develops hardware and software for the medical industry.

RISK MANAGEMENT TECHNOLOGIES
2150 Shattuck Avenue, Berkeley CA 94704. 510/548-7799. **Fax:** 510/548-7799. **Contact:** Carol Penskar, Chief Financial Officer. **World Wide Web address:** http://www.rmtech.com. **Description:** A developer of financial risk management and profitability software for financial institutions. Products include The RADAR System, an asset-liability management system designed to reduce interest rate risk exposure and maximize profits. **NOTE:** The company offers entry-level positions. **Common positions include:** Applications Engineer; Computer Programmer; Database Manager; Software Engineer. **Educational backgrounds include:** Computer Science; Finance. **Benefits:** 401(k); Dental Insurance; Disability Coverage; Job Sharing; Life Insurance; Medical Insurance; Telecommuting; Tuition Assistance. **Listed on:** Privately held. **Annual sales/revenues:** $5 - $10 million. **Number of employees at this location:** 35.

ROADSHOW INTERNATIONAL, INC.
8300 Greensboro Drive, Suite 400, McLean VA 22102. 703/790-8300. **Contact:** Human Resources. **Description:** Develops routing software for distribution companies.

ROCKWELL
2424 South 102nd Street, West Allis WI 53227. 414/321-8000. **Contact:** Human Resources Department. **Description:** A manufacturer of industrial computer software.

ROGUE WAVE SOFTWARE, INC.
850 SW 35th Street, Corvallis OR 97333. 541/754-3010. **Fax:** 541/757-6650. **Contact:** Human Resources. **E-mail address:** hr@roguewave.com. **World Wide Web address:** http://www.roguewave.com. **Description:** A provider of Java reusable cross-platform software parts and tools. **Corporate headquarters location:** This Location.

ROMTECH, INC.
2000 Cabot Boulevard, Suite 110, Langhorne PA 19047. 215/750-6606. **Fax:** 215/750-3722. **Contact:** Human Resources. **Description:** RomTech publishes and distributes CD-ROM software for the educational, home, and business markets. The company was founded in 1992 and is growing through a combination of acquisitions, new distribution agreements, and additional product development.

ROVACK
P.O. Box 858, Lake Elmo MN 55042. 612/779-9444. **Contact:** Greg Walker, Director of Operations. **Description:** A software developer for the health care community. **NOTE:** The physical address is 3549 Lake Elmo Avenue North, Lake Elmo MN 55042. **Common positions include:** Computer Programmer; Customer Service Representative; Services Sales Representative; Software Engineer; Systems Analyst. **Educational backgrounds include:** Accounting; Business Administration; Communications; Computer Science; Marketing. **Benefits:** Dental Insurance;

Disability Coverage; Employee Discounts; Life Insurance; Medical Insurance; Profit Sharing. **Corporate headquarters location:** This Location. **Operations at this facility include:** Administration; Divisional Headquarters; Manufacturing; Regional Headquarters; Research and Development; Sales; Service. **Listed on:** Privately held. **Number of employees at this location:** 31.

ROYALBLUE TECHNOLOGIES INC.
55 Broadway, One Exchange Plaza, 17th Floor, New York NY 10006. 212/267-5673. **Contact:** Human Resources. **Description:** Develops software for the NASDAQ stock market.

RYAN McFARLAND
8911 North Capital of Texas Highway, Austin TX 78759. 512/343-1010. **Toll-free phone:** 800/762-6265. **Fax:** 512/343-9487. **Contact:** Human Resources. **E-mail address:** rm_info@liant.com. **Description:** A provider of COBOL development tools and technologies for applications in client/server environments. **Parent company:** Liant Software Corp. (Framingham MA).

SAP AMERICA, INC.
950 Tower Lane, 12th Floor, Foster City CA 94404. 415/637-1655. **Contact:** Human Resources. **Description:** This location is a development facility. SAP develops software for client/server applications. **Corporate headquarters location:** Wayne PA.

SAP AMERICA, INC.
950 Winter Street, Suite 3800, Waltham MA 02154. **Toll-free phone:** 800/365-5727. **Toll-free fax:** 888/672-6726. **Contact:** Joseph Goss, College Relations Manager. **E-mail address:** joseph.goss@sap.ag.de. **World Wide Web address:** http://www.sap.com. **Description:** Develops, maintains, and implements software. SAP America is one of the largest independent software companies in the world. **NOTE:** Training programs are available. **Common positions include:** Consultant. **Educational backgrounds include:** Business Administration; Computer Science; Engineering. **Benefits:** 401(k); Dental Insurance; Disability Coverage; Life Insurance; Medical Insurance; Pension Plan; Profit Sharing; Tuition Assistance. **Corporate headquarters location:** Wayne PA. **Other U.S. locations:** Nationwide. **International locations:** Germany. **Parent company:** SAPAG. **Operations at this facility include:** Regional Headquarters. **Listed on:** DAX. **Annual sales/revenues:** More than $100 million. **Number of employees at this location:** 130. **Number of employees nationwide:** 2,000. **Number of employees worldwide:** 8,400. **Number of projected hires for 1997 - 1998 at this location:** 100.

SAP AMERICA, INC.
701 Lee Road, Wayne PA 19087. 610/521-4500. **Contact:** Human Resources. **Description:** Develops software. SAP America creates a variety of client/server computer software packages including financial programs, human resources, and materials management. **Coporate headquarters location:** This Location.

SAS INSTITUTE
SAS Campus Drive, Cary NC 27513-2414. 919/677-8000. **Fax:** 919/677-8123. **Contact:** Personnel. **World Wide Web address:** http://www.sas.com. **Description:** Develops software. The SAS Institute designs a variety of programs including warehouse management, statistics, inventory control, and others. **NOTE:** SAS asks that applicants call the jobline at ext. 15441 to get a position number before applying.

SCC
6285 Lookout Road, Boulder CO 80301. 303/581-5600. **Contact:** Human Resources. **Description:** Develops public safety computer software.

SCH TECHNOLOGIES
895 Central Avenue, 11th Floor, Cincinnati OH 45202. 513/579-0455. **Fax:** 513/579-1064. **Contact:** Lisa Sayers, Administrative Assistant. **Description:** SCH integrates, markets, and services horizontal software products to managers of medium to large open systems installations worldwide. SCH products address the needs of network management, distributed systems

management (which includes systems administration, data center management, and tape management), connectivity, and integrated office automation. SCH is a worldwide distributor of software to open systems users, as well as to Unisys, AT&T, and AT&T/NCR proprietary communities. SCH has chosen to concentrate its product mix in four specific areas: distributed systems management, network management, connectivity, and integrated office automation. The company also offers educational, consulting, and support services. **Common positions include:** Computer Programmer; Customer Service Representative; Marketing Specialist; Technical Representative; Technical Support Representative. **Educational backgrounds include:** Accounting; Art/Design; Business Administration; Communications; Liberal Arts; Marketing. **Benefits:** Disability Coverage; Life Insurance; Medical Insurance. **Corporate headquarters location:** This Location.

SCS/COMPUTE, INC.
2252 Welsch Industrial Court, St. Louis MO 63146. 314/997-7766. **Contact:** Etta Foster, Human Resources. **Description:** SCS/Compute, Inc. provides integrated software solutions designed to improve productivity for accounting and tax professionals. SCS/Compute's products focus on tax return calculation and financial statement preparation. SCS/Compute's tax return calculation products, Tax Machine and LMS/Tax, offer systems that integrate tax preparation, tax planning, depreciation, and client management into one software product. The products can be used in either single- or multi-user configurations, and Tax Machine is the only professional tax product which supports the UNIX operating system. The company's financial statement preparation product, Datawrite, is a client accounting general ledger and financial statement reporting software product designed to produce financial statements and specialized management reports. The Datawrite product family includes add-on modules for payroll processing, fixed asset accounting, business ration analysis, remote data entry, and file importation, with full integration to the SCS/Compute tax products. **Corporate headquarters location:** This Location.
Other U.S. locations:
- 3650 131st Avenue SE, Suite 400, Bellevue WA 98006. 206/643-2050. This location is the company's software development office.

SEA CORPORATION
20 Vernon Street, Norwood MA 02062. 617/762-9252. **Contact:** Human Resources Department. **Description:** Develops and sells custom-designed computer software.

SEA CORPORATION
221 3rd Street, Newport RI 02840. 401/847-2260. **Contact:** Human Resources. **Description:** Provides computer contract work for the government.

SEI
444 Market Street, 12th Floor, San Francisco CA 94111. 415/627-1900. **Contact:** Human Resources. **Description:** This location develops and sells investment accounting software to over 400 banking institutions. The company also offers consulting services from this office. SEI operates primarily in two business markets: trust and banking; and fund sponsor/investment advisory. The company provides investment and business solutions to firms in the investment business, who in turn serve their own investor clients. SEI provides direct investment solutions for $50 billion of investable capital and delivers systems and business solutions to organizations investing nearly $1 trillion. SEI is the largest provider of trust systems in the world, claiming over 40 percent of the U.S. trust market. **NOTE:** Please address resumes to SEI, Attn: Shelly Wolthusen, Human Resources Manager, 181 West Madison, 28th Floor, Chicago IL 60602. 312/460-2330. **Corporate headquarters location:** Wayne PA.

SOS COMPUTER SYSTEMS, INC.
720 East Timpanagos Parkway, Orem UT 84097. 801/222-0200. **Fax:** 801/222-0250. **Contact:** Human Resources. **Description:** Develops and sells operations software used by credit unions.

SPSS INC.
444 North Michigan Avenue, Chicago IL 60611. 312/329-2400. **Contact:** Willetta Hudson, Human Resources. **Description:** Develops, markets, and supports statistical software.

SQN INC.
P.O. Box 423, Rancocas NJ 08073. 609/261-5500. **Contact:** Human Resources. **Description:** Develops software for the banking industry.

STI
1225 Evans Road, Melbourne FL 32904. 407/723-3999. **Fax:** 407/984-3559. **Contact:** Corporate Recruiter, Human Resources. **E-mail address:** recruiter@sticomet.com. **World Wide Web address:** http://www.sticomet.com. **Description:** Develops software in a variety of environments for use in satellite technology.

SAFFIRE, INC.
776 East 930 South, American Fork UT 84003. 801/763-9003. **Contact:** Human Resources. **Description:** Saffire develops computer and video game software. Numerous products are licensed for the Super Nintendo and Sega Genesis game systems.

SAROS CORPORATION
10900 NE 8th Place, Suite 700, Bellevue WA 98004. 206/646-1066. **Contact:** Danna Lockerby, Recruiting. **Description:** Develops business networking software.

SAX SOFTWARE CORPORATION
950 Patterson Street, Eugene OR 97401. 541/344-2235. **Fax:** 541/344-2459. **Contact:** Human Resources. **E-mail address:** info@saxsoft.com. **World Wide Web address:** http://www.saxsoft.com. **Description:** A developer of Internet and Website software.

SCANTRON QUALITY COMPUTERS
20200 East 9 Mile Road, St. Clair Shores MI 48080. 810/774-7200. **Contact:** Human Resources. **Description:** Develops educational software and provides some hardware components.

SCIENTIFIC SOFTWARE-INTERCOMP, INC.
1801 California Street, Suite 295, Denver CO 80202. 303/292-1111. **Contact:** Ned Frazier, Vice President, Human Resources. **Description:** Scientific Software-Intercomp, Inc. provides high-technology software and solutions to professionals primarily in the oil and gas industries. More specifically, Scientific Software-Intercomp develops and markets proprietary computer software, furnishing management services in the petroleum and mining industries, and furnishing other electronic data processing services. **Corporate headquarters location:** This Location. **International locations:** Abu Dhabi; Cairo; Calgary; Kuala Lumpur; London; Paris; Quito.
Other U.S. locations:
• 10333 Richmond Avenue, Suite 1000, Houston TX 77042. 713/953-9272. This location develops software.

SCULPTURED SOFTWARE INC.
2144 South Highland Drive, Second Floor, Salt Lake City UT 84106. 801/486-2222. **Contact:** Human Resources. **Description:** Develops gaming software products for PCs and home entertainment systems.

SEAGATE SOFTWARE
600 East Diehl Road, Naperville IL 60563. 630/505-3300. **Contact:** Human Resources. **Description:** Develops and markets backup software.

SECURE DOCUMENT SYSTEMS
9485 Regency Square Boulevard, Jacksonville FL 32225. 904/725-2505. **Contact:** Human Resources. **Description:** Develops laser printing software.

SECURITIES INDUSTRY SOFTWARE CORPORATION
4725 Independence Street, Wheat Ridge CO 80033. 303/467-5050. **Contact:** Human Resources. **Description:** Develops financial software for brokerage houses and related companies.

SECURITY DYNAMICS TECHNOLOGIES
20 Crosby Drive, Bedford MA 01730. 617/687-7000. **Contact:** Human Resources. **Description:** Develops and markets security products and systems for computers. **Corporate headquarters location:** This Location.

SEER TECHNOLOGIES INC.
8000 Regency Parkway, Cary NC 27511. 919/380-5000. **Contact:** Human Resources. **Description:** Develops business applications.

SEMAPHORE, INC.
Three East 28th Street, 11th Floor, New York NY 10016. 212/545-7300. **Contact:** Human Resources. **Description:** Develops payroll software for accounting and engineering firms.

SERIF, INC.
One Chestnut Street, Suite 305, Nashua NH 03061. 603/889-8650. **Fax:** 603/889-1127. **Contact:** Human Resources. **E-mail address:** serif@serif.com. **World Wide Web address:** http://www.serif.com. **Description:** Serif, Inc. develops and supports software products for the desktop publishing and graphics markets.

7TH LEVEL, INC.
1110 East Collins Boulevard, Suite 122, Richardson TX 75081. 972/498-8100. **Contact:** Human Resources. **World Wide Web address:** http://www.7thlevel.com. **Description:** A developer and manufacturer of computer game software. **Corporate headquarters location:** This Location. **Listed on:** NASDAQ. **Stock exchange symbol:** SEVL.

SHARED SYSTEMS CORPORATION
15301 Dallas Parkway, Suite 600, Dallas TX 75248. 972/233-8356. **Contact:** Human Resources. **Description:** Shared Systems Corporation provides software and professional services to the financial, retail, and health care industries. **Corporate headquarters location:** Marlboro MA. **Parent company:** Stratus Computer, Inc. offers a broad range of computer platforms, application solutions, middleware, and professional services for critical online operations. Other Stratus subsidiaries include SoftCom Systems, Inc., a provider of data communications middleware and related professional services that bridge the gap between open distributed systems and legacy mainframe and mid-range systems used for online applications; and Isis Distributed Systems, Inc., a developer of advanced messaging middleware products that enable businesses to develop reliable, high-performance distributed computing applications involving networked desktop computers and shared systems.

SHERPA CORPORATION
611 River Oaks Parkway, San Jose CA 95134. 408/433-0455. **Toll-free phone:** 800/955-0455. **Fax:** 408/943-9507. **Contact:** Staffing. **E-mail address:** career@sherpa.com. **World Wide Web address:** http://www.sherpa.com. **Description:** Provides a product data management (PDM) software solution that integrates engineering and manufacturing environments. Sherpa Corporation markets PDM solution worldwide to *Fortune* 500 companies in the aerospace, automotive, consumer electronics, defense, electronics, medical products, and telecommunications industries. **Common positions include:** Accountant/Auditor; Administrative Assistant; Applications Engineer; Computer Programmer; Controller; Design Engineer; Human Resources Manager; Project Manager; Sales Engineer; Sales Manager; Software Engineer; Technical Support Representative; Technical Writer/Editor; Webmaster. **Educational backgrounds include:** Accounting; Computer Science; Engineering; Finance; Marketing. **Benefits:** 401(k); Dental Insurance; Disability Coverage; EAP; Life Insurance; Mass Transit Available; Medical Insurance; Section 125 Plan; Stock Option; Tuition Assistance; Vision Plan. **Special Programs:** Internships. **Corporate headquarters location:** This Location. **Other U.S. locations:** MI; NH. **International locations:** France; Germany; Italy; United Kingdom. **Operations at this facility include:** Divisional Headquarters. **Number of employees at this location:** 135. **Number of employees worldwide:** 200. **Number of projected hires for 1997 - 1998 at this location:** 40.

SHOWCASE CORPORATION
4131 Highway 52 North, Suite C-111, Rochester MN 55901. 507/288-5922. **Contact:** Human Resources. **World Wide Web address:** http://www.showcasecorp.com. **Description:** Develops software designed to access AS/400 systems. ShowCase's main product is STRATEGY.

SIERRA ON-LINE, INC.
3380 146th Place SE, Suite 200, Bellevue WA 98007. 206/649-9800. **Fax:** 206/641-7617. **Contact:** Recruiting - JL. **Description:** Sierra On-Line creates, produces, and distributes entertainment and educational software for the home, utilizing leading-edge interactive technologies. Sierra's products are designed for IBM-compatible and Macintosh systems. In 1992, Sierra acquired Bright Star Technology, an educational software publisher. In 1993, the company acquired Paris-based Coktel Vision, a leading European publisher of educational software. Sierra On-Line plans to release network-enabled educational products. **Corporate headquarters location:** This Location. **Other U.S. locations:** Oakland CA; Eugene OR. **Number of employees at this location:** 540.

SIERRA ON-LINE, INC.
P.O. Box 485, Coarsegold CA 93614-0485. 209/683-4468. **Contact:** Human Resources. **Description:** This is a research and development location. Overall, Sierra On-Line creates, produces, and distributes entertainment and educational software for the home, utilizing leading-edge interactive technologies. Sierra's products are designed for IBM compatible and Macintosh systems. Sierra continues to be a market leader in the development and publishing of entertainment software. In 1992, Sierra acquired Bright Star Technology, an educational software publisher. In 1993, they acquired Paris-based Coktel Vision, a leading European publisher of educational software. Sierra On-Line plans to release network-enabled educational products.

SIERRA ON-LINE, INC.
DYNAMICS DIVISION
1600 Mill Race Road, Eugene OR 97403. 541/343-0772. **Contact:** Human Resources. **Description:** Sierra On-Line designs, develops, and markets educational and entertainment software. This location, the Dynamics Division, sells and develops software and offers technical support. **NOTE:** For job openings at this location, resumes should be sent to the following address: Recruiting JL, Sierra On-Line Inc., 3380 146th Place SE, Suite 200, Bellevue WA 98007 or faxed to 206/641-7617.

SILVER PLATTER INFORMATION, INC.
100 River Ridge Drive, Norwood MA 02062. 617/769-2599. **Fax:** 617/769-8763. **Contact:** Carol Woolf, Director of Human Development. **E-mail address:** staffing@silverplatter.com. **Description:** A publisher and distributor of over 200 CD-ROM databases with software systems for data retrieval. **Common positions include:** Accountant/Auditor; Administrative Assistant; Chief Financial Officer; Controller; Customer Service Representative; Database Manager; Engineer; Human Resources Manager; Internet Services Manager; Librarian; Multimedia Designer; Public Relations Specialist; Sales Executive; Sales Manager; Sales Representative; Software Engineer; Systems Analyst; Systems Manager; Technical Writer/Editor. **Educational backgrounds include:** Computer Science; Engineering; Library Science; Marketing; Public Relations. **Benefits:** 401(k); Dental Insurance; Disability Coverage; Life Insurance; Medical Insurance; Tuition Assistance. **Corporate headquarters location:** This Location. **Other U.S. locations:** Pasadena CA; Newton MA. **International locations:** Hong Kong; London; Sydney. **Listed on:** Privately held. **Number of employees at this location:** 200. **Number of employees nationwide:** 220.

SILVON SOFTWARE INC.
900 Oakmont Lane, Suite 400, Westmont IL 60559. 630/655-3313. **Contact:** Human Resources. **Description:** Develops sales tracking software.

SIRSI
689 Discovery Drive, Huntsville AL 35806-2801. 205/922-9820. **Fax:** 205/922-9818. **Contact:** Human Resources. **E-mail address:** webmaster@sirsi.com. **World Wide Web address:** http://www.sirsi.com. **Description:** Develops software for library turnkey systems.

SITE/TECHNOLOGIES/INC.

2200 West Main Street, Suite 230B, Durham NC 27705. 919/416-3113. **Toll-free phone:** 800/722-0607. **Fax:** 919/286-1901. **Contact:** Human Resources. **World Wide Web address:** http://www.sitetech.com. **Description:** Site/technologies/inc. is a provider of quality control solutions for Web users. Products include SiteSweeper 1.0, which generates unified reports for large, complex sites located on multiple servers; and SiteMarks, which provides bookmark management for users of Netscape Navigator. **Corporate headquarters location:** This Location. **Listed on:** Privately held.

SKILLS BANK CORPORATION

7104 Ambassador Road, Park View Center One, Baltimore MD 21244. 410/818-5000. **Contact:** Human Resources. **Description:** A developer of educational software.

SKYWARD

5233 Coye Drive, Stevens Point WI 54481. 715/341-9406. **Contact:** Human Resources. **Description:** Develops school administration software.

SMITHWARE, INC.

2416 Hillsboro Road, Suite 201, Nashville TN 37212. 615/386-3100. **Fax:** 615/386-3135. **BBS:** 615/386-3295. **Contact:** Human Resources. **E-mail address:** info@smithware.com. CompuServe: 75470,546. **World Wide Web address:** http://www.smithware.com. **Description:** A developer, publisher, and distributor of Btrieve and Scalable SQL development tools for software developers. **Corporate headquarters location:** This Location.

SOFT ONE CORPORATION

1121 South Orem Boulevard, Orem UT 84058. 801/221-9400. **Contact:** Human Resources. **Description:** Produces the Interactive ClassAct Windows software training program on CD-ROM. The program can be used by all levels of computer users.

SOFTDESK, INC.

Seven Liberty Hill Road, Henniker NH 03242. 603/428-3199. **Fax:** 603/428-7901. **Contact:** Human Resources Department. **World Wide Web address:** http://www.softdesk.com. **Description:** Softdesk, Inc. is a leading supplier of architecture, engineering, and construction (AEC) application software for AutoCad and MicroSoft Windows. Softdesk Professional Products are targeted at designers and engineers and sold through a worldwide network of resellers and distributors. Softdesk Retail Products are targeted at non-professional home and office PC users and sold at the retail level. Softdesk applications cover the range of AEC disciplines, including civil/survey, building design and engineering, productivity, facilities management, imaging/scanning, AM/FM and utilities, and process and power plant design. In June 1995, Softdesk acquired IdeaGraphix, an Atlanta-based developer of AEC software for MicroStation.

SOFTLAB INC.

1000 Abernathy Road, Suite 1000, Atlanta GA 30328. 770/668-8811. **Fax:** 770/668-8812. **Contact:** Human Resources Manager. **E-mail address:** hr@softlabna.com. **Description:** Develops, markets, and supports software for banks and insurance companies. **Common positions include:** Quality Assurance Analyst; Software Developer; Training Specialist.

SOFTOUCH SYSTEMS, INC.

1300 South Meridian, Suite 600, Oklahoma City OK 73108. 405/947-8080. **Toll-free phone:** 800/944-3036. **Fax:** 405/947-8169. **Contact:** Mike Landreth, Human Resources. **E-mail address:** landy@softouch.com. **World Wide Web address:** http://www.softouch.com. **Description:** Develops custom software for the OS/2 operating system. **Corporate headquarters location:** This Location.

SOFTWARE AG

1000 Abernathy Road, Building 400, Suite 900, Atlanta GA 30328. 770/390-9258. **Contact:** Human Resources Department. **Description:** Software AG is the southeastern U.S. sales division of an international software company based in Darmstadt, Germany. Software AG sells computer software systems and provides professional services.

SOFTWARE AG OF NORTH AMERICA
1650 West 82nd Street, Suite 750, Bloomington MN 55431. 612/888-4404. **Contact:** Craig Larsen, Recruiting. **Description:** This location develops some software products but mainly serves as a regional sales office for Software AG.

SOFTWARE AG-REGIONAL OFFICES
5005 LBJ Freeway, Suite 700, Dallas TX 75244. 972/991-8900. **Contact:** Human Resources. **Description:** Distributes business applications for a German parent company. Software AG's Regional Offices supply businesses in the United States with customized applications for mainframe computers as well as offering technical support.

SOFTWARE ARTISTRY
9449 Priority Way West Drive, Indianapolis IN 46240. 317/843-1663. **Contact:** Human Resources Director. **Description:** Software Artistry develops, markets, and supports a family of internal and external customer support software. The company's primary product, Expert Advisor, is a complete problem management system that incorporates multiple diagnostic techniques to access a single solutions database to resolve customer problems. **Corporate headquarters location:** This Location. **Listed on:** NASDAQ.

SOFTWARE ENGINEERING OF AMERICA INC.
1230 Hempstead Turnpike, Franklin Square NY 11010. 516/328-7000. **Contact:** Human Resources. **Description:** Develops computer software for mainframes at large data centers. **NOTE:** Also at this location is Software Engineering of America's sister company, Advanced Productions, Inc.

SOFTWARE INTERPHASE, INC.
82 Cucumber Hill Road, Foster RI 02825. 401/397-2340. **Fax:** 401/397-6814. **Contact:** Human Resources. **E-mail address:** sinterphas@aol.com. **World Wide Web address:** http://www.sinterphase.com. **Description:** Provides engineering, programming, and documentation software solutions. **Corporate headquarters location:** This Location.

SOFTWARE PRODUCTIVITY SOLUTIONS
122 4th Avenue, Indiatlantic FL 32903. **Contact:** Human Resources. **Description:** Designs high-tech software for the government.

SOFTWARE PUBLISHING CORPORATION (SPC)
P.O. Box 11044, San Jose CA 95103. 408/537-3000. **Fax:** 408/537-3504. **Contact:** Human Resources. **E-mail address:** staffing@spco.com. **World Wide Web address:** http://www.spco.com. **Description:** A developer of visual communications IBM- and World Wide Web-compatible software. Founded in 1980. **NOTE:** The company maintains an employment page at its Website. The company's physical address is 111 North Market Street, San Jose CA. **Common positions include:** Software Engineer. **Benefits:** 401(k); Stock Purchase. **Listed on:** NASDAQ. **Stock exchange symbol:** SPCO. **Number of employees nationwide:** 70.

SOFTWARE SERVICES CORPORATION
650 Avis Drive, Suite 100, Ann Arbor MI 48108. 313/996-3636. **Contact:** Recruiting. **Description:** Develops custom-designed software. One major customer of Software Services Corporation is the Ford Motor Company.

SOFTWARE SOLUTIONS INC.
3425-B Corporate Way, Duluth GA 30136. 770/418-2000. **Contact:** Human Resources. **Description:** Develops software for distribution companies. Software Solutions also offers training and technical support.

SOFTWARE TECHNOLOGIES CORPORATION
P.O. Box 661090, Arcadia CA 91066. 818/445-7000. **Fax:** 818/445-5548. **Contact:** Human Resources. **E-mail address:** jobs-www@stc.com. **World Wide Web address:** http://www.stc.com. **Description:** A developer of software. **Corporate headquarters location:** This Location.

SOFTWARE 2000
25 Communication Way, Hyannis MA 02601. 508/778-2000. **Contact:** Human Resources. **Description:** A developer of business application software.

SOLID OAK SOFTWARE, INC.
P.O. Box 6826, Santa Barbara CA 93160. 805/967-9853. **Fax:** 805/967-1614. **Contact:** Human Resources. **World Wide Web address:** http://www.solidoak.com. **Description:** A developer of access-control software.

SOLOMON SOFTWARE
200 East Hardin Street, Findlay OH 45840. 419/424-0422. **Contact:** Human Resources. **Description:** Develops accounting software.

THE SOMBERS GROUP INC.
P.O. Box 5748, Newark DE 19714. 302/995-7187. **Contact:** Human Resources. **Description:** This location is the administrative office. Overall, The Sombers Group Inc. offers customized computer programming services for businesses.

SONIC FOUNDRY, INC.
100 South Baldwin Street, Suite 204, Madison WI 53703. 608/256-3133. **Toll-free phone:** 800/577-6642. **Contact:** Human Resources Department. **E-mail address:** CompuServe: 74774,1340 or GO SONIC. **World Wide Web address:** http://www.sfoundry.com. **Description:** A developer of digital audio editing software. Product line includes Sound Forge sound development tools.

SPECTRUM HOLOBYTE, INC.
2490 Mariner Square Loop, Alameda CA 94501. 510/522-3584. **Fax:** 510/522-3587. **Contact:** Human Resources. **Description:** Spectrum Holobyte, Inc. develops and publishes entertainment software for personal computers, including both IBM PCs and Macintoshes. The company also provides software for video game consoles including Sega and Nintendo and is creating software for next-generation game consoles. Spectrum Holobyte develops technologies including 3-D simulation, game artificial intelligence, and networked game-play, as well as next-generation console software. Over the past decade, the company has published more than 100 software titles for PCs and game consoles, with total sales surpassing 12 million units. Twenty-two of these titles have sold more than 100,000 units each. **Common positions include:** Accountant/Auditor; Attorney; Budget Analyst; Buyer; Computer Programmer; Customer Service Representative; Designer; Electrical/Electronics Engineer; Financial Analyst; Human Resources Manager; Public Relations Specialist; Purchasing Agent/Manager; Quality Control Supervisor; Systems Analyst; Technical Writer/Editor. **Educational backgrounds include:** Accounting; Art/Design; Business Administration; Communications; Computer Science; Engineering; Finance; Liberal Arts; Marketing; Mathematics. **Benefits:** 401(k); Dental Insurance; Disability Coverage; Employee Discounts; Life Insurance; Medical Insurance; Profit Sharing; Savings Plan; Tuition Assistance. **Corporate headquarters location:** This Location. **Other U.S. locations:** Hunt Valley MD; Seattle WA. **Subsidiaries include:** MicroProse U.S., Hunt Valley MD; MicroProse Europe, Chipping Sodbury, England. **Operations at this facility include:** Administration; Divisional Headquarters; Research and Development; Sales. **Listed on:** NASDAQ. **Number of employees at this location:** 200. **Number of employees nationwide:** 500.

SPECTRUM HUMAN RESOURCE SYSTEMS
1625 Broadway, Suite 2600, Denver CO 80202. 303/534-8813. **Contact:** Human Resources. **Description:** Develops computer software for human resources departments.

SPEECH SYSTEMS, INC.
2511 55th Street, Building C, Suite 200, Boulder CO 80301. 303/938-1110. **Fax:** 303/938-1874. **Contact:** Human Resources. **E-mail address:** employment@speechsys.com. **World Wide Web address:** http://www.speechsys.com. **Description:** A developer of advanced speech recognition technology, tools, and solutions. **Corporate headquarters location:** This Location.

SPIRE TECHNOLOGIES
P.O. Box 1970, Orem UT 84059. 801/226-3355. **Fax:** 801/224-3847. **Contact:** Human Resources. **Description:** Spire Technologies develops software and hardware, and provides technical support. **NOTE:** The company is located at 311 North State Street, Orem UT.

SPYGLASS, INC.
1240 East Diehl Road, Naperville IL 60563. 708/505-1010. **Fax:** 708/505-4944. **Contact:** Recruitment. **World Wide Web address:** http://www.spyglass.com. **Description:** This location is the corporate headquarters, and also houses the company's Technical Support Group and the Server Research & Development team. Spyglass is a developer of Internet filtering software that allows parents to block certain inappropriate material on the World Wide Web from young users. The primary product produced by the company is SurfWatch, which runs on Windows 95, Windows 3.1, and Macintosh operating systems. **Corporate headquarters location:** This Location.
Other U.S. locations:
- 111 Deerwood Road, Suite 100, San Ramon CA 94583. 510/855-3242. This location is a sales office.
- 3200 Farber Drive, Champaign IL 61821. 217/355-6000. This location houses the company's Client Technology Research & Development team.
- One Cambridge Center, Cambridge MA 02142. 617/679-4600. This location houses a sales office as well as the company's East Coast Research & Development team.

SPYGLASS, INC.
SURFWATCH SOFTWARE
175 South San Antonio Road, Suite 102, Los Altos CA 94022. 415/948-9500. **Fax:** 415/948-9577. **Contact:** Recruitment. **E-mail address:** jobs_la@spyglass.com. **World Wide Web address:** http://www.spyglass.com. **Description:** This location is the West Coast sales office as well as a research and development site. Spyglass is a developer of Internet filtering software that allows parents to block certain inappropriate material on the World Wide Web from young users. The primary product produced by the company is SurfWatch, which runs on Windows 95, Windows 3.1, and Macintosh operating systems. **Corporate headquarters location:** Naperville IL.

SRITEK INC.
8705 Freeway Drive, Macedonia OH 44056. 216/468-3380. **Contact:** Human Resources Department. **Description:** Sritek develops manufacturing and quality control software for businesses.

STAC, INC.
12636 High Bluff Drive, San Diego CA 92130. 619/794-4300. **Contact:** Human Resources. **Description:** A computer software developer.

STARPAK INTERNATIONAL
100 Garfield Street, Suite 400, Denver CO 80206. 303/399-2400. **Contact:** Human Resources. **Description:** A provider of computer software services.

STATE OF THE ART, INC.
56 Technology West Drive, Irvine CA 92718. 714/753-1222. **Contact:** Human Resources. **Description:** State of the Art, Inc. develops, markets, and supports high-end microcomputer accounting software which includes the M-A-S 90 Evolution/2 product line, consisting of 18 separate application modules that can be integrated into a comprehensive accounting system. **Number of employees at this location:** 200.

STATSOFT, INC.
2300 East 14th Street, Tulsa OK 74104. 918/749-1119. **Fax:** 918/749-2217. **Contact:** Human Resources Department. **E-mail address:** jobs@statsoft.com. **World Wide Web address:** http://www.statsoft.com. **Description:** Statsoft is a developer of statistics software. **Common positions include:** Statistician; Technician. **Educational backgrounds include:** Biology; Computer Science; Mathematics; Psychology; Statistics. **Corporate headquarters location:** This Location.

STERLING SOFTWARE, INC.
300 Crescent Court, Suite 1200, Dallas TX 75201-7832. 214/981-1000. **Contact:** Personnel Department. **World Wide Web address:** http://www.sterling.com. **Description:** A worldwide developer and supplier of software products and services. The company's business segments include systems management, which provides software products for large computer companies; applications management, which provides customers with software products for developing new client/server applications; and federal systems, which provides technical services to the federal government including the U.S. Department of Defense and NASA. The company's products are sold through offices in 17 countries and distributed worldwide. **Common positions include:** Accountant/Auditor; Blue-Collar Worker Supervisor; Buyer; Chemical Engineer; Chemist; Designer; Draftsperson; Electrical/Electronics Engineer; Human Resources Manager; Mechanical Engineer; Public Relations Specialist; Purchasing Agent/Manager; Quality Control Supervisor; Registered Nurse. **Educational backgrounds include:** Accounting; Business Administration; Chemistry; Computer Science; Engineering; Finance; Marketing; Physics. **Benefits:** 401(k); Dental Insurance; Disability Coverage; Life Insurance; Medical Insurance; Pension Plan; Profit Sharing; Tuition Assistance. **Corporate headquarters location:** This Location. **Operations at this facility include:** Administration; Manufacturing; Regional Headquarters; Sales. **Listed on:** New York Stock Exchange. **Number of employees nationwide:** 1,350.

Other U.S. locations:

- 11050 White Rock Road, Rancho Cordova CA 95670-6005. 909/889-2663. This location is a development facility for the company's Storage Management Division.
- 5900 Canogh Avenue, Woodland Hills CA 91367. 818/716-1616. This location acquires, develops, markets, and supports a broad range of computer software products and services.
- 3340 Peachtree Road NE, 11th Floor, Suite 1100, Atlanta GA 30326. 770/933-3778. This location develops software and services for client/server applications.
- 1406 Fort Crook Road South, Bellevue NE 68005. 402/291-8300. This location, as part of the Information Technology Division, develops ADP systems for the DOD.
- 25101 Chagrin Boulevard, Suite 260, Beachwood OH 44122. 216/464-4482. This location sells software to retail stores.
- 15301 Dallas Parkway, Suite 400, Dallas TX 75248. 972/788-2580. This location houses the Banking Systems Group.
- 1800 Alexander Bell Drive, Reston VA 20191. 703/264-8000. This location is the headquarters of the Systems Management Group and develops systems management software.

STERLING SOFTWARE, INC.
FEDERAL SYSTEMS GROUP
303 Twin Dolphin Drive, Suite 600, Redwood City CA 94065-1417. 415/802-7100. **Fax:** 415/969-3821. **Contact:** Andrea Germane, Recruiting Manager. **Description:** The Federal Systems Group of Sterling Software provides technical and professional services to the federal government under several multi-year contracts. As one of the largest contractors at the NASA/Ames Research Center, Sterling has been an integral partner in the progress of the center. **Common positions include:** Aerospace Engineer; Computer Programmer; Software Engineer. **Educational backgrounds include:** Computer Science; Engineering; Mathematics; Physics. **Benefits:** 401(k); Dental Insurance; Disability Coverage; Life Insurance; Medical Insurance; Tuition Assistance. **Listed on:** American Stock Exchange. **Number of employees at this location:** 700.

STINGRAY SOFTWARE
1201-F Raleigh Road, Suite 140, Chapel Hill NC 27514. 919/461-0672. **Toll-free phone:** 800/924-4223. **Fax:** 919/461-9811. **Contact:** Human Resources. **Description:** A developer of object-oriented software.

STORAGETEK TERIS
4 Westbrook Corporate Center, Suite 220, Westchester IL 60154. 708/409-9901. **Contact:** Mr. Kelvin Gouwens, Manager. **Description:** This location is a software support center for AS400, System 36, and networks.

STORAGETEK/NSC
2030 Main Street, Suite 1400, Irvine CA 92614. 714/622-5151. **Contact:** Human Resources. **Description:** An international software company. Storagetek/NSC designs, maintains, markets, and

services worldwide information retrieval subsystems for enterprise computer systems networks. **Corporate headquarters location:** This Location.
Other U.S. locations:
- 500 108th Avenue, Suite 460, Bellevue WA 98004. 206/454-5699. This location is a sales and customer service office.

STRANDWARE INC.
P.O. Box 634, Eau Claire WI 54702. **Fax:** 715/833-1995. **Contact:** Human Resources. **E-mail address:** info@strandware.com. **World Wide Web address:** http://www.strandware.com. **Description:** A developer and marketer of bar code labeling software.

STRUCTURAL DYNAMICS RESEARCH CORPORATION
2000 Eastman Drive, Milford OH 45150. 513/576-2400. **Fax:** 513/576-2922. **Contact:** Derek Jackson, Manager of Employment. **Description:** An international supplier of mechanical computer-aided engineering (MCAE) software and engineering services. Structural Dynamics Research has 55 offices. **Common positions include:** Aerospace Engineer; Civil Engineer; Computer Programmer; Mathematician; Mechanical Engineer; Software Engineer; Systems Analyst. **Educational backgrounds include:** Computer Science; Engineering; Mathematics. **Benefits:** 401(k); Dental Insurance; Disability Coverage; Employee Discounts; Life Insurance; Medical Insurance; Tuition Assistance. **Corporate headquarters location:** This Location. **Operations at this facility include:** Administration; Divisional Headquarters; Regional Headquarters; Research and Development; Sales; Service. **Listed on:** NASDAQ. **Number of employees at this location:** 1,000.
Other U.S. locations:
- 3945 Freedom Circle, Suite 560, Santa Clara CA 95054-1226. 408/980-0100.

SUMMIT MEDICAL SYSTEMS INC.
10900 Red Circle Drive, Suite 100, Minnetonka MN 55343. 612/939-2200. **Contact:** Director of Human Resources. **Description:** Develops and markets clinical database programs to the medical industry.

SUMMUS, INC.
950 Lake Murray Boulevard, Irmo SC 29063. 803/781-5674. **Fax:** 803/781-5679. **Contact:** Human Resources. **E-mail address:** wavelet@summus.com. **World Wide Web address:** http://www.summus.com. **Description:** A developer of compression technology software for digital communications.

SUNBURST COMMUNICATIONS INC.
101 Castleton Street, Pleasantville NY 10570. 914/747-3310. **Contact:** Miles Merwin, Human Resources. **Description:** Develops and markets educational software. Sunburst Communications's products focus on the K-12 levels.

SUNGARD BUSINESS SYSTEMS INC.
210 Automation Way, Birmingham AL 35210. 205/956-7500. **Contact:** Human Resources Manager. **Description:** A developer of employee benefit systems software.

SUNGARD DATA SYSTEMS, INC.
22134 Sherman Way, Canoga Park CA 91303. 818/884-5515. **Fax:** 818/346-5044. **Contact:** Human Resources. **Description:** This location develops and sells software. Overall, SunGard provides specialized computer services, mainly proprietary investment support systems for the financial services industry and disaster recovery services. SunGard Data Systems's investment accounting and portfolio systems maintain the books of record of all types of large investment portfolios including those managed by banks and mutual funds. The company's disaster recovery services include alternate-site backup, testing, and recovery services for IBM, DEC, Prime, Stratus, Tandem, and Unisys computer installations. The company's computer service unit provides remote-access IBM computer processing, direct marketing, and automated mailing services.

SUNGARD INSURANCE SYSTEMS
500 Northridge Road, Suite 400, Atlanta GA 30350. 770/587-6800. **Contact:** Human Resources Department. **Description:** This location sells and supports accounting and investment computer software. **NOTE:** All hiring is through the SunGard office in Minnesota. All applications should be sent to: SunGard, Attn: Human Resources, 601 Second Avenue South, Hopkins MN 55343. 612/936-8717. **Common positions include:** Computer Programmer; Systems Analyst. **Educational backgrounds include:** Computer Science. **Benefits:** Dental Insurance; Disability Coverage; Life Insurance; Medical Insurance; Savings Plan; Tuition Assistance. **Corporate headquarters location:** This Location. **Parent company:** SunGard Financial Systems. **Number of employees at this location:** 100.

SUNGARD INSURANCE SYSTEMS
3131 South Vaughn Way, Suite 650, Aurora CO 80014. 303/369-9700. **Contact:** Human Resources. **Description:** Develops financial software for insurance companies.

SUNGARD TRUST SYSTEMS
P.O. Box 240882, Charlotte NC 28224-0882. 704/527-6300. **Contact:** Jill Mikels, Human Resources Manager. **Description:** A developer of software for the banking industry.

SYBASE POWERSOFT CORPORATION
561 Virginia Road, Concord MA 01742. 508/287-1500. **Contact:** Human Resources. **Description:** Develops, markets, and supports application software tools for the emerging client/server market. **Number of employees at this location:** 106.

SYBASE, INC.
6475 Christie Avenue, Emeryville CA 94608. 510/922-3500. **Fax:** 510/922-8002. **Contact:** Human Resources. **Description:** Sybase, Inc. develops, markets, and supports a full line of relational database management software products and services for integrated, enterprisewide information management systems. **Corporate headquarters location:** This Location. **Number of employees at this location:** 2,350.
Other U.S. locations:
- 6550 Rock Spring Drive, Suite 800, Bethesda MD 20817. 301/896-1600. This location provides training services.
- 800 Bellevue Way NE, Suite 400, Bellevue WA 98004-4229. 206/451-7474.

SYKES ENTERPRISES INC.
777 North Fourth Street, Sterling CO 80751. 970/522-6638. **Contact:** Diane Brungardt, Manager of Human Resources. **Description:** A high-tech software company offering customer service support at this location. **NOTE:** Sykes seeks workers with DOS and Windows experience.

SYMANTEC CORPORATION
10201 Torre Avenue, Cupertino CA 95014-2132. 408/253-9600. **Contact:** Human Resources Staffing. **E-mail address:** jobs@symantec.com. **Description:** Symantec Corporation is a global organization that develops, manufactures, and markets software products for individuals and businesses around the world. Founded in 1982, the company is a vendor of utility software for stand alone and networked personal computers. In addition, it offers a wide range of project management products, productivity applications, and development languages and tools. With products available in both U.S. and international versions, Symantec maintains a strong presence. The company is organized into several product groups that are devoted to product marketing, engineering, technical support, quality assurance, and documentation. Finance, sales, and marketing are centralized at this location. **Common positions include:** Software Developer; Software Engineer. **Educational backgrounds include:** Accounting; Computer Science; Engineering; Marketing. **Benefits:** 401(k); Dental Insurance; Disability Coverage; Employee Discounts; Life Insurance; Medical Insurance; Profit Sharing; Tuition Assistance. **Special Programs:** Internships. **Corporate headquarters location:** This Location. **Listed on:** NASDAQ. **Number of employees nationwide:** 1,200.

SYMIX SYSTEMS, INC.
2800 Corporate Exchange Drive, Columbus OH 43231. 614/523-7000. **Contact:** Human Resources. **Description:** A developer and manufacturer of open, client/server software.

SYNCSORT
50 Tice Boulevard, CN18, Woodcliff Lake NJ 07675. 201/930-8200. **Contact:** Human Resources. **Description:** Develops operating sytems business software.

SYNERCOM TECHNOLOGIES
A DIVISION OF LOGICA, INC.
2500 City West Boulevard, Suite 1100, Houston TX 77042. 713/954-7000. **Contact:** Judy Alford, Human Resources. **Description:** Manufactures work management software systems for utility companies.

SYNOPSYS INC.
700 East Middlefield Road, Mountain View CA 94043. 415/962-5000. **Contact:** Human Resources. **Description:** Synopsys develops, markets, and supports high-level design automation (HLDA) software for designers of integrated circuits and electronic systems. Its products offer improved time to market, reduced development time, and reduced manufacturing costs when compared to earlier generations of electronic design automation (EDA) software. **Number of employees at this location:** 412.

SYNTELLECT INC.
15810 North 28th Avenue, Phoenix AZ 85023. 602/789-2800. **Contact:** Human Resources. **Description:** Develops software for telecommunications firms. Syntellect designs voice processing systems.

SYSTEM MANAGEMENT ARTS INC.
14 Mamaroneck Avenue, White Plains NY 10601. 914/948-6200. **Contact:** Department of Interest. **Description:** A developer of software used to manage large networks.

SYSTEM SOFTWARE ASSOCIATES
500 West Madison, Suite 3200, Chicago IL 60661. 312/641-2900. **Contact:** Norman Maskin, Corporate Recruiter. **Description:** System Software Associates develops, markets, and supports an integrated line of business application, computer-aided software engineering (CASE), and electronic data interchange (EDI) software, primarily for IBM minicomputers and workstations. **Number of employees nationwide:** 1,200.

SYSTEMS & COMPUTER TECHNOLOGY CORPORATION
Four Country View Road, Malvern PA 19355. 610/647-5930. **Fax:** 610/640-5162. **Contact:** Human Resources. **E-mail address:** dwiess@sctcorp.com. **World Wide Web address:** http://www.sctcorp.com. **Description:** A software development and services company, catering to the higher education, local government, utility, and manufacturing industries. The company operates through the following divisions: Information Resource Management (IRM); Software & Technology Services (STS); SCT Public Sector, Inc.; and SCT Utility Systems, Inc. **NOTE:** The company offers entry-level positions. **Common positions include:** Accountant/Auditor; Budget Analyst; Computer Programmer; Human Resources Specialist; Management Analyst/Consultant; Management Trainee; MIS Specialist; Systems Analyst; Technical Writer/Editor. **Educational backgrounds include:** Accounting; Communications; Computer Science; Economics; Liberal Arts; Marketing. **Benefits:** 401(k); Dental Insurance; Disability Coverage; Job Sharing; Life Insurance; Medical Insurance; Telecommuting; Tuition Assistance. **Corporate headquarters location:** This Location. **Listed on:** NASDAQ. **Annual sales/revenues:** More than $100 million.

SYSTEMS & PROGRAMMING SOLUTIONS
9325 North 107th Street, Milwaukee WI 53224. 414/362-8020. **Contact:** Human Resources. **Description:** A computer consulting and software programming firm. **Number of employees at this location:** 40.

SYSTEMS ALTERNATIVES
1705 Indian Wood Circle, Suite 100, Maumee OH 43537. 419/891-1100. **Contact:** Human Resources. **Description:** Develops software for the recycling industry.

SYSTEMS STRATEGIES INC.
2 Penn Plaza, Suite 650, New York NY 10121. 212/279-8400. **Contact:** Human Resources. **Description:** Develops connectivity and data integration software.

TAC SYSTEMS, INC.
1035 Putman Drive, Huntsville AL 35816. 205/721-1976. **Fax:** 205/721-0242. **Contact:** Human Resources. **World Wide Web address:** http://www.tacsys.com. **Description:** A software developer and provider of networking products.

TCAM SYSTEMS
7 Hanover Square, 8th Floor, New York NY 10004. 212/269-0200. **Contact:** Office Manager. **Description:** A software development firm. TCAM Systems produces software for the securities and brokerage industries.

TMS-SEQUOIA
P.O. Box 1358, Stillwater OK 74076. 405/377-0880. **Contact:** Human Resources. **Description:** Formerly known as Time Management Software, TMS-Sequoia is a developer of a wide range of software products.

TSI INTERNATIONAL
45 Danbury Road, P.O. Box 840, Wilton CT 06897. 203/761-8600. **Contact:** Ann Curry, Human Resources. **Description:** Develops electronic data interchange (EDI) software and markets its products to corporations.

TSW INTERNATIONAL
3301 Windy Ridge Parkway, Atlanta GA 30339. 770/952-8444. **Contact:** Allen Vaughn, Controller. **Description:** Develops software for client/server asset care applications used by manufacturing companies.

TACTICA CORPORATION
10300 SW, Portland OR 97223. 503/293-9585. **Contact:** Human Resources. **Description:** A developer of computer software.

TALX CORPORATION
1850 Borman Court, St. Louis MO 63146. 314/434-0046. **Contact:** Human Resources. **Description:** TALX Corporation develops and supports software for integrated voice communication.

TANGRAM ENTERPRISE SOLUTIONS, INC.
5511 Capital Center Drive, Suite 400, Raleigh NC 27606. 919/851-6000. **Contact:** Susan E. Barbee, Human Resources Manager. **E-mail address:** info@test.com. **Description:** Manufactures asset management software. Founded in 1982. **NOTE:** The company offers entry-level positions. **Common positions include:** Computer Programmer; Marketing Specialist; Sales Engineer; Sales Representative; Software Engineer; Systems Analyst; Technical Support Representative; Technical Writer/Editor. **Educational backgrounds include:** Computer Science; Engineering; Marketing. **Benefits:** 401(k); Dental Insurance; Disability Coverage; Life Insurance; Medical Insurance; Telecommuting. **Special Programs:** Internships. **Corporate headquarters location:** This Location. **Other U.S. locations:** Malvern PA. **Listed on:** NASDAQ. **Annual sales/revenues:** $11 - $20 million. **Number of employees nationwide:** 85. **Number of projected hires for 1997 - 1998 at this location:** 40.

TECHSMITH CORPORATION
P.O. Box 4758, East Lansing MI 48826. 517/333-2100. **Fax:** 517/333-1888. **Contact:** Human Resources. **E-mail address:** info@techsmith.com. **World Wide Web address:** http://www.techsmith.com. **Description:** Developers of SnagIt screen capture and print utilities;

NewsMonger, an Internet tool for monitoring Internet information; and Foray remote access products.

TECHTOOLS INC.
3-I Taggart Drive, Nashua NH 03060. 603/888-8400. **Contact:** Human Resources. **World Wide Web address:** http://www.techtools.com. **Description:** A developer of application development tools for MS Windows software products.

TELEDATA PRODUCTS INC.
2309 Renard Place SE, Suite 201, Albuquerque NM 87106. 505/243-2287. **Contact:** Samantha Lapin, President. **Description:** A computer products company that creates databases and software.

TENEX SYSTEMS INC.
3 Radner Corporate Center, Suite 160, Radnor PA 19087. 610/687-1160. **Contact:** Human Resources. **Description:** Designs software. TENEX Systems develops and supports software for schools and other educational facilities.

TETRAGENICS
130 North Main Street, Butte MT 59701. 406/782-9161. **Contact:** Human Resources. **Description:** Develops software and hardware. Tetragenics provides hydroelectric plants with monitoring and control software and hardware.

TEUBNER AND ASSOCIATES
P.O. Box 1994, Stillwater OK 74076. 405/624-2254. **Contact:** Human Resources. **Description:** A developer of communications software for use by large corporations.

3DO COMPANY
600 Galveston Drive, Building 5, Redwood City CA 94063. 415/261-3000. **Contact:** Human Resources. **Description:** A developer of video games.

TIMBERLINE SOFTWARE CORPORATION
P.O. Box 728, Beaverton OR 97075-0728. 503/626-6775. **Fax:** 503/641-7498. **Contact:** Susan Ross, Executive Secretary. **Description:** Timberline Software Corporation develops and markets computer software programs, primarily for the construction and property management industries. **NOTE:** The company is located at 9600 SW Nimbus Avenue, Beaverton OR. **Corporate headquarters location:** This Location.

TINGLEY SYSTEMS
31722 State Road 52, San Antonio FL 33576. 352/588-2250. **Contact:** Vivian C. Satz, Director of Human Resources. **Description:** A developer of software for the health care industry. Tingley Systems also operates as an Internet service provider.

TIVOLI SYSTEMS
9442 Capital of Texas Highway North, Suite 500, Austin TX 78759. 512/794-9070. **Contact:** Human Resources Director. **Description:** A developer of client-manager software for distributed environments.

TOPSPEED INC.
150 East Sample Road, Pompano Beach FL 33064. 954/785-4555. **Contact:** Human Resources. **Description:** Produces developmental software. Topspeed gears their software towards businesses with MIS departments that need assistance running their 16- or 32-bit applications. This company also offers technical support.

TOUCH N' GO SYSTEMS, INC.
406 G Street, Suite 210, Anchorage AK 99501. **Toll-free phone:** 800/691-6333. **Fax:** 907/274-9493. **Contact:** Human Resources. **E-mail address:** touchngo@touchngo.com. **World Wide Web address:** http://www.touchngo.com. **Description:** A developer of Touch N' Go, a personnel tracking product which allows managers to find out the status of employees. **Corporate headquarters location:** This Location.

TRADEWARE TECHNOLOGIES INC.
601 Corporate Circle, Golden CO 80401. 303/384-1400. **Contact:** Recruiting. **Description:** Develops maintenance software. **NOTE:** Tradeware Technologies is a small company that looks for qualified people with computer programming and engineering backgrounds.

TRANSARC CORPORATION
707 Grant Street, Gulf Tower 20th Floor, Pittsburgh PA 15219. 412/338-4400. **Contact:** Human Resources. **Description:** Develops AFS and DCE software.

TRAVELING SOFTWARE, INC.
18702 North Creek Parkway, Bothell WA 98011. 206/483-8088. **Fax:** 206/487-5677. **Contact:** Human Resources. **E-mail address:** hotjobs@travsoft.com. **World Wide Web address:** http://www.travsoft.com. **Description:** The manufacturer of Laplink, a remote file transfer product. **Common positions include:** Software Engineer. **Educational backgrounds include:** Computer Science. **Corporate headquarters location:** This Location.

TRI-COR INDUSTRIES, INC.
5 Eagle Center, Suite 8, O'Fallon IL 62269. 618/632-9804. **Contact:** Human Resources. **Description:** Develops software for Department of Defense applications and offers technical support.

TRILOGY DEVELOPMENT GROUP
6034 West Courtyard Drive, Suite 130, Austin TX 78730. 512/794-5900. **Contact:** Recruiting. **Description:** A developer of configuration software for a variety of industries, including automotive, utilities, insurance, shipping, and computers.

TRIMIN SYSTEMS INC.
3030 Center Point Drive, Suite 100, Roseville MN 55113. 612/636-7667. **Contact:** Human Resources. **Description:** Develops and markets manufacturing software.

TRIPOS INC.
1699 South Hanley Road, St. Louis MO 63144. 314/647-1099. **Contact:** Human Resources. **Description:** Develops science tool and analysis software for science applications.

TURBOPOWER SOFTWARE COMPANY
P.O. Box 49009, Colorado Springs CO 80949-9009. **Toll-free phone:** 800/333-4160. **Fax:** 719/260-7151. **Contact:** Human Resources. **E-mail address:** jobs@tpower.com. **World Wide Web address:** http://www.tpower.com. **Description:** A database technology and software development firm. Products include Memory Sleuth 1.0; Orpheus, for extending user interface design and database capabilities; Async Professional for Delphi, a serial communications library; B-Tree filer, a library of functions for network compatible databases; and Turbo Analyst, for organizing, tuning, and documenting Pascal source code.

20-21 INTERACTIVE
1815 South State Street, Suite 4500, Orem UT 84097. 801/225-8700. **Fax:** 801/225-1394. **Contact:** Human Resources. **Description:** Performs custom computer programming for multilevel marketing firms.

TYBRIN CORPORATION
1283 Eglin Parkway, Shalimar FL 32579. 904/651-1150. **Contact:** Human Resources. **Description:** Develops software for the Department of Defense.

U.S. SOFTWARE
14215 NW Science Park Drive, Portland OR 97229. 503/641-8446. **Fax:** 503/644-2413. **Contact:** Human Resources. **E-mail address:** info@ussw.com. **World Wide Web address:** http://www.ussw.com. **Description:** A developer of real-time development tools for embedded applications. **Corporate headquarters location:** This Location.

ULTIMUS
4915 Waters Edge Drive, Suite 135, Raleigh NC 27606. 919/233-7331. **Fax:** 919/233-7339. **Contact:** Human Resources. **E-mail address:** info@ultimus1.com. **World Wide Web address:** http://www.ultimus1.com. **Description:** Ultimus offers workflow automations through client/server Windows applications allowing users to design, simulate, implement, monitor, and measure workflow for various administrative business processes.

UNICOMP
1800 Sandy Plains Parkway, Suite 305, Marietta GA 30066. 770/424-3684. **Contact:** Human Resources. **Description:** Unicomp develops migration software that converts older computer programming systems into a UNIX-based system.

UNIDATA INC.
1099 18th Street, Suite 2200, Denver CO 80202. 303/294-0800. **Contact:** Human Resources. **Description:** Develops database software packages.

UNIFIED INFORMATION
13513 NE 126th Place, Unit B, Kirkland WA 98034. 206/820-8917. **Contact:** Human Resources. **Description:** Unified Information develops software that manages the scheduling of electricity for electric utility companies.

UNILINK SOFTWARE
P.O. Box 1630, 460 East Pearl Street, Jackson WY 83001. 307/733-5494. **Contact:** Brandon Elliot, Personnel. **Description:** Develops accounting software.

UNITED STATES DATA CORPORATION (USDATA)
2435 North Central Expressway, Richardson TX 75080-2722. 972/680-9700. **Fax:** 972/669-8318. **Contact:** Personnel Administrator. **Description:** Develops, markets, and supports application-enabler products that customers configure to implement a wide range of real-time monitoring, analysis, information management, and control solutions in worldwide industrial automation markets. USData also develops, markets, and supports integrated hardware, software, and systems solutions for automated identification and data collection applications that are sold to a broad base of customers throughout North America. The company also acts as a full-service distributor and value-added remarketer for manufacturers of bar code equipment. Technical centers, staffed with sales and support engineers, are located throughout the United States and Western Europe. **Common positions include:** Computer Programmer; Designer; Electrical/Electronics Engineer; Industrial Engineer; Manufacturer's/Wholesaler's Sales Rep.; Mathematician; Mechanical Engineer; Petroleum Engineer; Software Engineer; Systems Analyst; Technical Writer/Editor. **Educational backgrounds include:** Business Administration; Computer Science; Engineering; Marketing. **Benefits:** 401(k); Dental Insurance; Life Insurance; Medical Insurance; Tuition Assistance. **Special Programs:** Internships. **Corporate headquarters location:** This Location. **Other U.S. locations:** Atlanta GA; Chicago IL; Boston MA; Seattle WA. **Operations at this facility include:** Administration; Divisional Headquarters; Research and Development; Sales; Service. **Listed on:** Privately held. **Number of employees at this location:** 200. **Number of employees nationwide:** 300.

UNIVERSAL SYSTEMS, INC. (USI)
14585 Avion Parkway, Chantilly VA 20151. 703/222-2840. **Contact:** Human Resources. **Description:** A software development, systems integration, and documentation management firm. **Corporate headquarters location:** This Location.

UNIVERSAL SYSTEMS, INC. (USI)
1356 East 3300 South, Salt Lake City UT 84106. 801/484-9151. **Contact:** Human Resources. **Description:** A software development, systems integration, and documentation management firm. **Corporate headquarters location:** Chantilly VA.

VMARK SOFTWARE, INC.
50 Washington Street, Westborough MA 01581-1021. 508/366-3888. **Fax:** 508/389-8955. **Contact:** Mary Parker, Human Resources Manager. **E-mail address:** hr@vmark.com.

Description: Founded in 1984, VMARK Software designs, manufactures, sells, and supports a comprehensive suite of client/server products and services for the business solutions market. **Common positions include:** Computer Programmer; Financial Analyst; Software Engineer. **Educational backgrounds include:** Computer Science; Engineering; Marketing. **Benefits:** 401(k); Dental Insurance; Disability Coverage; Employee Discounts; Life Insurance; Medical Insurance; Tuition Assistance. **Corporate headquarters location:** This Location. **Other U.S. locations:** CA; CO; GA; IL; NC; NJ; TX; WA. **International locations:** Australia; Canada; France; Germany; Japan; Malaysia; Spain; South Africa; U.K. **Operations at this facility include:** Administration; Divisional Headquarters; Research and Development; Sales; Service. **Listed on:** NASDAQ. **Annual sales/revenues:** $51 - $100 million. **Number of employees at this location:** 230. **Number of employees nationwide:** 320. **Number of employees worldwide:** 470.

VNP SOFTWARE
One Kendall Square, Building 300, Cambridge MA 02139. 617/252-9220. **Contact:** Human Resources Department. **World Wide Web address:** http://www.vnp.com. **Description:** A provider of software development tools and consulting services. **NOTE:** VNP Software maintains an employment page at its Website.

VSI ENTERPRISES, INC.
5801 Goshen Springs Road, Norcross GA 30071. 404/242-7566. **Contact:** Human Resources. **Description:** Designs, produces, and markets tailored software for interactive group videoconferencing systems. Using standard telecommunication transmissions, VSI Enterprises's products allow many people at different geographic locations to see and hear one another on live television. The company's systems are designed with modular subsystems and open software to allow the systems to be reconfigured or expanded as needed. VSI Enterprises's clients include NYNEX, Sprint, and several federal agencies and government departments. **Corporate headquarters location:** This Location. **Subsidiaries include:** VSI; VSI Europe. **Listed on:** Boston Stock Exchange.

THE VANTIVE CORPORATION
2455 Augustine Drive, Santa Clara CA 95054. 408/982-5700. **Toll-free phone:** 800/582-6848. **Fax:** 408/982-5710. **Contact:** Human Resources. **E-mail address:** jobs@vantive.com. **World Wide Web address:** http://www.vantive.com. **Description:** A provider of integrated customer interaction software that allows businesses to enhance customer satisfaction by automating customer service, sales, marketing, field service, quality assurance, and help desk functions. Applications are based on a client/server architecture, for independent or enterprisewide customer information systems. **Common positions include:** Applications Engineer; Database Manager; Product Manager; Quality Assurance Engineer; Sales Engineer; Sales Representative; Software Engineer; Technical Writer/Editor; Telemarketer. **Corporate headquarters location:** This Location.

VARIATION SYSTEMS ANALYSIS, INC. (VSA)
300 Maple Park Boulevard, St. Clair Shores MI 48081. 810/774-2640. **Contact:** Human Resources. **Description:** A developer of dimensional management software for engineering applications.

VENTANA CORPORATION
1430 East Fort Lowell Road, Suite 301, Tucson AZ 85719. 520/325-8228. **Toll-free phone:** 800/368-6338. **Contact:** Lee Walker, Human Resources. **E-mail address:** walker@ventana.com. **World Wide Web address:** http://www.ventana.com. **Description:** A developer of group decision support software. Ventana's product line is called GroupSystems. **Common positions include:** Software Engineer. **Educational backgrounds include:** Computer Science; M.I.S. **Corporate headquarters location:** This Location.

VERIFONE, INC.
100 Kahelu Avenue, Mililani HI 96789-3909. 808/623-2911. **Contact:** Human Resources. **E-mail address:** rdjobs@verifone.com. **World Wide Web address:** http://www.verifone.com. **Description:** A research and development location for operating system software and hardware. **NOTE:** VeriFone maintains an employment page at its Website. **Common positions include:** Computer Programmer; Software Engineer; Test Engineer.

VERIFONE, INC.

800 El Camino Real, 4th Floor, Menlo Park CA 94025. 415/617-8000. **Fax:** 415/617-8019. **Contact:** Human Resources. **E-mail address:** icdjobs@verifone.com. **World Wide Web address:** http://www.verifone.com. **Description:** This location develops software for its Internet Commerce Division. **NOTE:** VeriFone maintains an employment page at its Website.

VERIFONE, INC.

Three Lagoon Drive, Redwood City CA 94065. 408/591-6500. **Contact:** Human Resources. **E-mail address:** misjobs@verifone.com. **World Wide Web address:** http://www.verifone.com. **Description:** This location develops client/server software. **NOTE:** VeriFone maintains an employment page at its Website.

VERILOG USA INC.

3010 LBJ Freeway, Suite 900, Dallas TX 75234. 214/241-6595. **Contact:** Human Resources. **Description:** A software engineering firm.

VERITAS SOFTWARE CORPORATION

1600 Plymouth Street, Mountain View CA 94043. 415/335-8000. **Fax:** 415/335-8050. **Contact:** Human Resources. **World Wide Web address:** http://www.veritas.com. **Description:** Veritas Software Corporation is a designer, developer, and marketer of enterprise data and storage management software including Volume Manager. Its products are designed to improve system performance and to reduce administration costs. **Corporate headquarters location:** This Location. **Listed on:** NASDAQ. **Stock exchange symbol:** VRTS. **Annual sales/revenues:** $24.1 million. **Number of employees nationwide:** 90.

VERITY INC.

894 Ross Drive, Sunnyvale CA 94089. 408/541-1500. **Contact:** Human Resources. **Description:** Develops and markets software tools and applications for searching, retrieving, and filtering information across the Internet.

VERMONT MICROSYSTEMS, INC.

11 Tigan Street, Winooski VT 05404. 802/655-2860. **Contact:** Human Resources. **Description:** A software developer. **Corporate headquarters location:** This Location.

VESON INC.

29 Broadway, New York NY 10006. 212/422-3000. **Contact:** Human Resources. **Description:** Develops a wide variety of computer software.

VIAGRAFIX
SOFTWARE DIVISION

One American Way, Pryor OK 74361. 918/825-7555. **Toll-free phone:** 800/233-3223. **Fax:** 918/825-6359. **BBS:** 918/825-4878. **Contact:** Human Resources. **E-mail address:** rogg@viagrafix.com. **CompuServe address:** DesignCad. **World Wide Web address:** http://www.viagrafix.com. **Description:** A developer of CAD software and provider of computer training services. **Corporate headquarters location:** This Location.

VIASOFT INC.

3033 North 44th Street, Phoenix AZ 85018. 602/952-0050. **Contact:** Human Resources. **Description:** Viasoft develops management and enterprise applications software. **Corporate headquarters location:** This location. **Listed on:** NASDAQ.
Other U.S. locations:
- 1231 Greenway Drive, Suite 380, Irving TX 75038. 214/550-1808. This location is a sales office.

VIEWLOGIC SYSTEMS, INC.

293 Boston Post Road West, Marlborough MA 01752-4615. 508/480-0881. **Fax:** 508/480-0882. **Contact:** Human Resources Department. **Description:** Viewlogic Systems, Inc. develops, markets, and supports software products that aid engineers in the design of advanced electronic products. Founded in 1984, Viewlogic has focused its attention on the specific needs of electronics firms,

providing them with a portfolio of state-of-the-art software products, available as stand alone or fully-integrated, and featuring open standards-based architectures that operate on every major computer platform. **Common positions include:** Accountant/Auditor; Electrical/Electronics Engineer; Software Engineer. **Educational backgrounds include:** Computer Science; Engineering. **Benefits:** Dental Insurance; Disability Coverage; Employee Discounts; Life Insurance; Medical Insurance; Pension Plan; Savings Plan; Tuition Assistance. **Special Programs:** Internships. **Corporate headquarters location:** This Location. **Other U.S. locations:** Camarillo CA; Fremont CA. **Operations at this facility include:** Regional Headquarters; Research and Development. **Number of employees at this location:** 200. **Number of employees nationwide:** 500.

VIEWPOINT DATALABS
625 South State Street, Orem UT 84058. 801/229-3000. **Fax:** 801/229-3300. **Contact:** Human Resources. **Description:** Viewpoint Datalabs creates 3-D computer models and performs custom 3-D computer graphics.

VIEWSOFT
250 West Center Street, Suite 300, Provo UT 84601. 801/377-0787. **Fax:** 801/377-0788. **Contact:** Human Resources. **Description:** Creates a variety of software development tools, mainly for Visual C++ developers.

VINCA CORPORATION
1815 South State Street, Suite 2000, Orem UT 84097. 801/223-3100. **Fax:** 801/223-3107. **Contact:** Human Resources. **Description:** Vinca created the Storage Access Networking architecture for computer networks and also supplies software products to provide its customers with increased data accessibility, business continuation, and storage management.

VIRTUS CORPORATION
114 MacKenan Drive, Suite 100, Cary NC 27511. 919/467-9700. **Fax:** 919/467-7929. **Contact:** Human Resources Manager. **E-mail address:** jobs@virtus.com. **World Wide Web address:** http://www.virtus.com. **Description:** A provider of virtual reality software for desktop computers featuring 3-D design and presentation graphics.

VISUAL COMPONENTS, INC.
15721 College Boulevard, Lenexa KS 66219. 913/599-6500. **Fax:** 913/599-6597. **Contact:** Human Resources. **E-mail address:** sally@visualcomp.com. **World Wide Web address:** http://www.visualcomp.com. **Description:** A manufacturer of object-based development tools. Product line includes ActiveX Controls, for client/server and Internet and intranet applications; Internet Tools, for platform independent Internet environments; and Code Libraries, which provides code that can be segmented and integrated into new and existing applications. **Corporate headquarters location:** This Location.

WALKER INTERACTIVE SYSTEMS
Marathon Plaza Three North, 303 Second Street, San Francisco CA 94107. 415/495-8811. **Fax:** 415/243-2843. **Contact:** Gregory Allen, Recruiter. **Description:** A developer of high-end financial applications software, primarily for *Fortune* 100 companies and government agencies. **Common positions include:** Computer Programmer; Consultant; Marketing Specialist; Sales Executive; Sales Manager; Software Engineer; Systems Analyst. **Educational backgrounds include:** Accounting; Engineering; Mathematics; Physics. **Benefits:** 401(k); Dental Insurance; Life Insurance; Medical Insurance; Pension Plan; Tuition Assistance. **Corporate headquarters location:** This Location. **Other U.S. locations:** Atlanta GA; Chicago IL; Boston MA. **International locations:** Australia; Singapore; United Kingdom. **Listed on:** NASDAQ. **Annual sales/revenues:** $51 - $100 million. **Number of employees at this location:** 280. **Number of projected hires for 1997 - 1998 at this location:** 30.

WALL DATA, INC.
11332 NE 122nd Way, Kirkland WA 98034. 206/814-9255. **Contact:** Human Resources. **Description:** Develops, markets, and supports Windows-based connectivity software and associated applications tools. Its software products provide PC users easy access to and use of

computer applications and data residing on multiple host mainframes and minicomputers in enterprisewide information systems networks. Wall Data's principal product line is the RUMBA family of PC-based connectivity software.

WANG LABORATORIES, INC.
600 Technology Park Drive, Billerica MA 01821. 508/967-5000. **Contact:** Judy Lyttle, Human Resources Director. **Description:** Wang is a leader in workflow, integrated imaging, document management, and related office software for client/server open systems; and a worldwide provider of integration and support services for office software and networks. Wang's Software Business is a provider of integrated client/server software that enables customers to streamline office information processes, cut costs, and respond more quickly to changes in market dynamics or customer needs. Principal markets for the Software Business are banking, insurance, government, and other enterprises with large-scale business processes. Wang's Solutions Integration Business provides extensive systems integration and other services to the U.S. government. Principal markets are federal agencies, the U.S. Department of Defense, and commercial not-for-profit hospitals and medical enterprises. The Customer Services Business provides maintenance and support services, network integration, installation, training, and other services to customers worldwide. With service delivery operations in more than 130 countries, Wang's Customer Service Business provides maintenance and support services for more than 3,500 products and 300 vendors. **Common positions include:** Accountant/Auditor; Attorney; Computer Programmer; Customer Service Representative; Electrical/Electronics Engineer; Financial Analyst; Human Resources Manager; Paralegal; Software Engineer; Systems Analyst; Technical Writer/Editor. **Educational backgrounds include:** Accounting; Communications; Computer Science; Finance. **Benefits:** Dental Insurance; Disability Coverage; Employee Discounts; Life Insurance; Medical Insurance; Savings Plan; Tuition Assistance. **Corporate headquarters location:** This Location. **Operations at this facility include:** Administration; Research and Development. **Listed on:** NASDAQ. **Other U.S. locations:**

- 160 Spear Street, Suite 330, San Francisco CA 94105-1543. 415/778-3950. This location is a service office.
- 600 Alexander Road, Building 2, 3rd Floor, Princeton NJ 08540. 609/951-4240. A customer service and sales location.

WANG SOFTWARE
622 Third Avenue, 30th Floor, New York NY 10017. 212/476-3000. **Contact:** Human Resources Department. **Description:** Develops and markets financial software.

WAYFARER COMMUNICATIONS, INC.
2041 Landings Drive, Mountain View CA 94043. 415/903-1720. **Fax:** 415/903-1730. **Contact:** Human Resources. **E-mail address:** info@wayfarer.com. **Description:** A software developer whose products allow the secure movement of various applications across the Internet and private intranets. **Corporate headquarters location:** This Location.

WELCOME SOFTWARE TECHNOLOGY CORPORATION
15995 North Barker Landing, Suite 275, Houston TX 77079. 713/558-0514. **Contact:** Human Resources Department. **Description:** Welcome Software Technology Corporation designs project management software.

WESTERN DATA SYSTEMS, INC.
26707 West Agoura Road, Calabasas CA 91302. 818/880-0800. **Contact:** Human Resources. **Description:** Western Data Systems develops defense systems software.

WESTGATE SOFTWARE INC.
4424 South 700 East, Suite 275, Salt Lake City UT 84107. 801/266-4484. **Contact:** Human Resources. **Description:** Develops software for use in film processing and the dry cleaning industry.

WESTRONICS
2575 South Highland Drive, Las Vegas NV 89109. 702/732-1414. **Contact:** Human Resources Department. **Description:** A manufacturer of video poker games.

WHITE PINE SOFTWARE
542 Amherst Street, Nashua NH 03063. **Fax:** 603/883-7920. **Contact:** Human Resources. **E-mail address:** hr@wpine.com. **World Wide Web address:** http://www.wpine.com. **Description:** A developer of computer-based information access and communications software offering cross-platform desktop videoconferencing, desktop-to-host connectivity, network access, and file transfer. **Corporate headquarters location:** This Location.

WILCO SYSTEMS
17 State Street, 22nd Floor, New York NY 10004. 212/269-3970. **Contact:** Human Resources Department. **Description:** Develops computer systems and software products for the financial and communications industries.

WIND RIVER SYSTEMS
1010 Atlantic Avenue, Alameda CA 94501. 510/748-4100. **Contact:** Human Resources. **Description:** A software engineering and development firm.

WINGRA TECHNOLOGIES, INC.
450 Science Drive, One West, Madison WI 53711. 608/238-4454. **Fax:** 608/238-8986. **Contact:** Personnel. **E-mail address:** personnel@wingra.com. **World Wide Web address:** http://www.wingra.com. **Description:** A developer of electronic messaging and network connectivity software.

WINNEBAGO SOFTWARE COMPANY
457 East South Street, Caledonia MN 55921. 507/724-5411. **Contact:** Nancy Hager, Vice President of Administration. **Description:** A developer of software for libraries. Winnebago Software Company's products are used internally by librarians as well as by library patrons.

WISMER MARTIN
12828 North Newport Highway, Mead WA 99021. 509/466-0396. **Contact:** Ron Berkshire, Human Resources. **Description:** Wismer Martin is the developer of SM*RT Practice and SM*RT Link medical practice management software. SM*RT Practice is used in thousands of providers' offices across the country. The upgrades that Wismer Martin have designed are based on input from physicians and their staffs. SM*RT Link allows electronic communication of health care information between geographically dispersed sites with differing information systems. It allows physicians, insurers, hospitals, medical laboratories, and other allied providers to electronically communicate confidential patient information in a secured manner. **Corporate headquarters location:** This Location.

WIZDOM CONTROLS
1300 Iroquois Avenue, Suite 125, Naperville IL 60565. 630/357-3311. **Contact:** Human Resources. **Description:** Wizdom Controls writes, produces, and distributes a variety of industrial automation software.

WOLFRAM RESEARCH, INC.
100 Trade Center Drive, Champaign IL 61820-7237. **Fax:** 217/398-0747. **Contact:** Personnel Department. **E-mail address:** resumes@wolfram.com. **World Wide Web address:** http://www.mathsource.wri.com. **Description:** A developer of mathematical software and services. Product line includes Mathematica software.

WONDERWARE, INC.
100 Technology Drive, Irvine CA 92718. 714/727-3200. **Contact:** Human Resources. **Description:** A developer of industrial applications software.

WORKGROUP SOLUTIONS, INC.
P.O. Box 460190, Aurora CO 80046-0190. **Fax:** 303/699-2793. **Contact:** Human Resources. **E-mail address:** info@wgs.com. **World Wide Web address:** http://www.wgs.com. **Description:** A developer and reseller of Web server software, Linux operating systems, Linux- and UNIX-related products and peripherals. Product line includes Linux Pro, WEBworx, and FlagShip.

WORKSTATION SOLUTIONS, INC.

One Overlook Drive, Amherst NH 03031-2800. **Contact:** Human Resources. **E-mail address:** info@worksta.com. **World Wide Web address:** http://www.worksta.com. **Description:** A designer and manufacturer of system administration software for computer networks. Workstation Solutions also provides support services.

XSOFT

3400 Hillview Avenue, Palo Alto CA 94304. 415/424-0111. **Contact:** Human Resources Department. **Description:** XSoft provides software designed to improve the way people create and use documents to capture, manage, and communicate ideas and information. The company develops and markets products that include document management and workflow solutions, office productivity packages, search and retrieval systems, and structured document publishing applications. XSoft also offers extensive product support, training, and consulting services. Its products and services are available in multiple operating systems, networks, and computer platforms. **Parent company:** Xerox Corporation develops, manufactures, markets, services, and finances information-processing products, including copiers, duplicators, scanners, electronic printing systems, word processing systems, personal computers, and computer peripherals. Xerox does business in over 120 countries.

XVT SOFTWARE, INC.

4900 Pearl East Circle, Suite 107, Boulder CO 80301. 303/443-4223. **Fax:** 303/443-0969. **Contact:** Manager of Human Resources. **Description:** XVT Software, Inc. is a leading provider of cross-platform graphical user interface (GUI) solutions for C and C++ developers. The company invented cross-platform development with the introduction of the XVT Portability Toolkit in 1988. Today, XVT offers powerful C and C++ development solutions that allow programmers to build a single application and then port to every major GUI without rewriting code. XVT development solutions include an interactive design tool for C, an application framework for C++ developers, and the XVT Portability Toolkit. In addition, XVT offers consulting and training and a Partners program to further assist customers. With more than 16,000 development licenses sold, XVT is one of the world's leading vendors of portable GUI development solutions. XVT Software is a private, employee-owned company. Revenue growth has exceeded 90 percent for five successive years and the company has been named to the *Inc.* 500 list of America's fastest-growing private companies. **Common positions include:** Computer Programmer; Customer Service Representative; Human Resources Manager; Public Relations Specialist; Quality Control Supervisor; Services Sales Representative; Software Engineer; Technical Writer/Editor. **Educational backgrounds include:** Accounting; Computer Science; Engineering; Finance; Marketing. **Benefits:** 401(k); Dental Insurance; Disability Coverage; Life Insurance; Medical Insurance; Tuition Assistance. **Corporate headquarters location:** This Location. **Operations at this facility include:** Administration; Customer Service; Marketing; Sales; Service; Support Services. **Number of employees at this location:** 125.

XACTWARE, INC.

1426 East 750 North, Orem UT 84097. 801/226-2251. **Fax:** 801/224-5218. **Contact:** Human Resources. **Description:** Manufactures software for the insurance and construction industries.

XCELLENET, INC.

Five Concourse Parkway, Suite 850, Atlanta GA 30328. 770/804-8100. **Fax:** 770/804-8102. **Contact:** Tom Darrow, Manager, Human Resources. **Description:** Founded in 1986, XcelleNet, Inc. develops and markets RemoteWare, a suite of client/server system development tools used to automate time-sensitive information delivery processes including order entry, inventory query, and price list and product catalog updates. Over 700 organizations worldwide have licensed RemoteWare to manage the exchange of information between central systems and over 213,000 remote or mobile field personnel, customers, suppliers, channels-of-distribution, or other corporate constituents. RemoteWare works by providing an architecture that incorporates a unique agent-server extension of the client/server model. This system was designed to provide open and flexible database access, API support, support for standard messaging interfaces, and compatibility with legacy systems. **Corporate headquarters location:** This Location.

XENERGY
3 Burlington Woods, Burlington MA 01803-4543. 617/273-5700. **Contact:** Human Resources Department. **Description:** A developer of software products used by utilities and energy companies.

XEROX IMAGING SYSTEMS INC.
9 Centennial Drive, Peabody MA 01960. 508/977-2000. **Contact:** Human Resources. **Description:** Develops scanning software.

XILINX, INC.
2100 Logic Drive, San Jose CA 95124. 408/879-5390. **Contact:** Human Resources Manager. **Description:** Xilinx, Inc. is the leading supplier of field programmable gate arrays (FPGAs) and related development system software used by electronic systems manufacturers for bringing complex products rapidly to the market. **Common positions include:** Accountant/Auditor; Computer Programmer; Customer Service Representative; Electrical/Electronics Engineer; Financial Analyst; Industrial Engineer; Software Engineer; Systems Analyst; Technical Writer/Editor. **Educational backgrounds include:** Accounting; Computer Science; Engineering. **Benefits:** 401(k); Dental Insurance; Disability Coverage; Life Insurance; Medical Insurance; Profit Sharing; Tuition Assistance. **Special Programs:** Internships. **Corporate headquarters location:** This Location. **Operations at this facility include:** Administration; Divisional Headquarters; Manufacturing; Regional Headquarters; Research and Development; Sales; Service. **Listed on:** NASDAQ. **Number of employees at this location:** 850. **Number of employees nationwide:** 900.

XYVISION, INC.
101 Edgewater Drive, Wakefield MA 01880. 617/245-4100. **Fax:** 617/246-5308. **Contact:** Michael Borin, Vice President of Human Resources. **Description:** Xyvision develops, markets, and supports software for document management, publishing, and prepress applications. The company combines its software with standard computer hardware, selected third-party software, and support services to create tightly integrated systems to improve productivity and strategic position. The company markets and supports its software worldwide. Xyvision offers two products, Parlance Publisher and Parlance Document Manager, as the core technologies in systems for document management and publishing applications (through Xyvision Publishing Systems). These systems are used to automate the production of reference books, journals, catalogs, directories, financial and legal materials, and technical manuals. Xyvision's color electronic prepress applications, through Contex Prepress Systems, are marketed primarily to commercial trade shops, printers, and prepress service organizations, as well as to consumer goods companies, advanced design firms, and packaging manufacturers. **Corporate headquarters location:** This Location.

Z-AXIS CORPORATION
7395 East Orchard Road, Suite A100, Greenwood Village CO 80111. 303/792-2400. **Contact:** Heidi O'Neil, Controller. **Description:** Z-Axis Corporation is a computer graphics firm. **Common positions include:** Animator; Commercial Artist; Computer Programmer; Electrical/Electronics Engineer; Operations/Production Manager; Services Sales Representative; Video Editor. **Educational backgrounds include:** Art/Design; Computer Science. **Benefits:** Dental Insurance; Disability Coverage; Life Insurance; Medical Insurance; Profit Sharing. **Corporate headquarters location:** This Location. **Operations at this facility include:** Administration; Production; Sales.

ZH COMPUTER, INC.
7600 Frances Avenue South, Suite 550, Minneapolis MN 55435-5939. 612/844-0915. **Fax:** 612/844-9025. **Contact:** Human Resources Department. **E-mail address:** scidev@zhcomp.com. CompuServe: 76500,1616. **World Wide Web address:** http://www.zhcomp.com. **Description:** A developer of software and products specializing in the application of advanced science and mathematics.

ZZ SOFT COMPANY
1256 South State Street, Suite 203, Orem UT 84058-8216. 801/221-9370. **Fax:** 801/221-9380. **Contact:** Human Resources. **Description:** Develops database software for publishing companies.

ZINC SOFTWARE, INC.
405 South 100 East, Pleasant Grove UT 84062. 801/785-8900. **Fax:** 801/785-8996. **Contact:** Human Resources. **Description:** Zinc Software makes computer development tools including application frameworks for C++ programming.

ZYDECOM
5220 Hollywood Avenue, Shreveport LA 71109. 318/636-5777. **Contact:** Human Resources. **Description:** A communications and communications software development firm.

ZYPCOM, INC.
2301 Industrial Parkway West, Hayward CA 94545-5029. 510/783-2501. **Fax:** 510/783-2414. **Contact:** Human Resources Manager. **E-mail address:** zypcom@tdl.com. **Description:** Develops, manufactures, and markets hardware and software products for data communications applications. **Common positions include:** Applications Engineer; Design Engineer; Electrical/Electronics Engineer; Sales Engineer; Sales Executive; Sales Manager; Software Engineer. **Educational backgrounds include:** Computer Science; Engineering; Marketing. **Benefits:** Dental Insurance; Disability Coverage; Employee Discounts; Life Insurance; Medical Insurance. **Special programs:** Apprenticeships; Internships. **Corporate headquarters location:** This Location. **Listed on:** Privately held. **Annual sales/revenues:** $11 - $20 million.

Note: Because addresses and telephone numbers of smaller companies can change rapidly, we recommend you call each company to verify the information below before inquiring about job opportunities. Mass mailings are not recommended.

Additional employers:

COMPUTER SOFTWARE, PROGRAMMING, AND SYSTEMS DESIGN

ABB Impell Corporation
5000 Executive Pkwy, San Ramon CA 94583-4210. 510/275-4400.

ACS Computer Services
525 Market Street, Suite 1400, San Francisco CA 94105-2722. 415/267-0300.

Activision
11601 Wilshire Blvd, Suite 300, Los Angeles CA 90025-1746. 310/473-9200.

Adept Technology Inc.
150 Rose Orchard Way, San Jose CA 95134-1358. 408/432-0888.

Adra Systems Inc.
2250 Lucien Way, Maitland FL 32751-7014. 407/667-3636.

Adtech Consultants Corp.
79 Stoneykill Rd, Wappingers Falls NY 12590-5458. 914/838-1990.

Advanced Control Technology
P.O. Box 1148, Albany OR 97321-3936. 541/967-8000.

Advanced Medical Management
2900 Wilcrest, Suite 110, Houston TX 77042. 713/358-5226.

Advantage Software Inc.
110 San Pedro Suite 700, San Antonio TX 78216-5508. 210/525-1421.

Air Routing International Corp.
2925 Briarpark Dr, Houston TX 77042-3729. 713/977-1020.

Akili Systems Group
3624 Oak Lawn Ave, Dallas TX 75219-4376. 214/526-5200.

All-Control Systems Inc.
905 Airport Rd, West Chester PA 19380-5983. 610/696-0244.

Allegro Development
3500 Maple, Dallas TX 75219-4284. 214/526-7071.

Allen Interactions
7900 W 78th St, Edina MN 55439-2523. 612/947-4055.

Alliant Techsystems Inc.
1050 NE Hostmark St, Poulsbo WA 98370-7538. 360/697-6600.

Anadac Inc.
2200 Clarendon Blvd, Suite 900, Arlington VA 22201-3331. 703/741-7000.

Analysis & Technology Inc.
137 Gaither Dr, Mount Laurel NJ 08054-1711. 609/778-9600.

Antrim Corp.
101 E Park Blvd, Plano TX 75074-5445. 214/422-1022.

Aonix
200 Wheeler Road, Burlington MA 01803-5152. 617/270-0030.

Applied Business Technology Corp.
361 Broadway, New York NY 10013-3903. 212/219-8945.

Applied Graphics Technologies
1112 S Wabash Ave, Chicago IL 60605-2309. 312/786-0600.

Arcon Corp.
260 Bear Hill Rd, Waltham MA 02154-1018. 617/890-3330.

ARIS Corp.
6720 Fort Dent Way, Suite 150, Tukwila WA 98188-2555. 206/433-2081.

Arrow Schweber Electronics
1180 Murphy Ave, San Jose CA
95131-2418. 408/441-9700.

Artesia Data Systems Inc.
12770 Merit Dr, Dallas TX
75251-1222. 972/788-0400.

Artwick Consulting
2004 Fox Drive, Champaign IL
61820-7332. 217/356-9796.

Asymetrix Corp.
110-110th Ave NE, Suite 700,
Bellevue WA 98004-5840.
206/462-0501.

**ATI Instruments North
America**
1001 Fourier Dr, Madison WI
53717-1924. 608/276-6300.

Atlanta Group Systems Inc.
3700 Crestwood Pkwy NW,
Duluth GA 30136-5583. 770/806-
8080.

Attachmate
1129 San Antonio Rd, Palo Alto
CA 94303-4310. 415/962-7100.

Auto Tell Services Inc.
600 Clark Ave, King Of Prussia
PA 19406-1433. 610/768-0200.

Autodesk Inc.
26200 Town Center Dr, Novi MI
48375. 810/347-9650.

Autodesk Retail Products
1725 220th Street, Suite C101,
Bothell WA 98021. 206/487-
2233.

Automated Solutions Corp.
1131 Rockingham Ln,
Richardson TX 75080-4326.
214/231-7361.

**Automatic Data Processing
(ADP)**
1700 Walt Whitman Road,
Melville NY 11747-3080.
516/694-7800.

**Automatic Data Processing
(ADP)**
7007 E Pleasant Valley Rd,
Cleveland OH 44131-5577.
216/447-1980.

**Automation Research Systems
Ltd.**
4480 King St, Alexandria VA
22302-1300. 703/820-9000.

Autosig Systems Inc.
4324 N Belt Line Rd, #C100,
Irving TX 75038-3534. 972/258-
8033.

Avant!
1208 East Arques Avenue,
Sunnyvale CA 94086. 408/369-
5400.

Avista Inc.
1 Insight Dr, Platteville WI
53818-3828. 608/348-8815.

Ball System Engineering
108 Olympia Drive, Suite 202,
Warner Robins GA 31093-3081.
912/922-4363.

Barrios Technology Inc.
1331 Gemini Street, Suite 300,
Houston TX 77058-2764.
713/480-1889.

Berkeley Systems
2095 Rose St, Berkeley CA
94709-1963. 510/540-5535.

Best Consulting
700 NE Multnomah St, Portland
OR 97232-4111. 503/236-5776.

Best Consulting
1100 E 6600 S Ste 200, Salt Lake
City UT 84121-2400. 801/266-
6138.

Best Consulting
2356 Gold Meadow Way Ste 160,
Gold River CA 95670. 916/638-
8085.

Best Programs Inc.
11413 Isaac Newton Sq S, Reston
VA 20190. 703/709-5200.

Biosym Molecular Simulations
9685 Scranton Rd, San Diego CA
92121-3752. 619/458-9990.

Bluebird Systems Inc.
5900 La Place Ct, Carlsbad CA
92008-8806. 760/438-2220.

BPS Reprographic Services
149 2nd St, San Francisco CA
94105-3714. 415/495-8700.

BR Data Service Co.
3301 Atlantic Ave, Brooklyn NY
11208-1914. 718/827-5220.

BTG Inc.
1450 Frazee Road, Suite 700, San
Diego CA 92108. 619/223-2208.

**Bull Worldwide Information
Systems**
1011 Oak St, Pittston PA 18640-
3771. 717/654-6040.

Business Control Systems Inc.
3939 Belt Line Road, Suite 450,
Dallas TX 75244-2219. 972/241-
8392.

Business Objects Inc.
2870 Zanker Road, San Jose CA
95134. 408/953-6000.

Cabletron Systems
16479 Dallas Pkwy, Suite 220,
Addison TX 75248-2643.
214/931-6222.

Cabletron Systems
174 Component Dr, San Jose CA
95131-1119. 408/383-1550.

Cadcentre Inc.
10700 Richmond Ave, Houston
TX 77042-4913. 713/977-1225.

CADD Consortium Inc.
2333 Globe Ave, Dallas TX
75228-4467. 214/320-1198.

Cadence Design Systems Inc.
5215 N O'Connor Blvd, Irving
TX 75039-3713. 972/869-0033.

Cair Systems Inc.
2100 Main Street SE, 4th Floor,
Irvine CA 92714-6237. 714/863-
1240.

Camax Systems Inc.
7851 Metro Pkwy, Minneapolis
MN 55425-1524. 612/854-5300.

Candle Corp.
12790 Merit Dr, Dallas TX
75251-1229. 972/991-0111.

Candle Corp.
9800 Richmond Ave, Suite 600,
Houston TX 77042-4527.
713/260-1700.

Cap Gemini America
10055 Red Run Blvd, Owings
Mills MD 21117-4892. 410/581-
5022.

Cap Gemini America
1001 4th Ave Plz # 3120, Seattle
WA 98154. 206/624-4600.

Cap Gemini America
5613 DTC Pkwy, Englewood CO
80111-3034. 303/220-1700.

Capitol Multimedia Inc.
200 Baker Avenue, Suite 300, Concord MA 01742. 508/287-5888.

Cbord Group Inc.
61 Brown Rd, Ithaca NY 14850-1247. 607/257-2410.

CDP Imaging Systems
6636 Cedar Ave S, Richfield MN 55423-2750. 612/861-0555.

Centrak
5500 Cenex Dr, Inver Grove MN 55077-1733. 612/552-5510.

Century Data Services Inc.
119 W 40th St, New York NY 10018-2500. 212/703-3500.

Cexec Inc.
8618 Westwood Center Drive, Vienna VA 22182-2222. 703/893-3220.

Cherry Bekaert and Holland
PO Box 27127, Richmond VA 23261-7127. 804/673-4224.

CIC Systems
165 University Ave, Westwood MA 02090-2320. 617/320-8300.

Cimarron
1830 Nasa Rd 1, Houston TX 77058-3502. 713/335-5800.

Cimtelligence Systems Inc.
1 Forbes Rd, Lexington MA 02173-7303. 617/861-1996.

Claritas/NPDC Inc.
11 W 42nd St, New York NY 10036-8002. 212/789-3580.

Clark Data Systems Microage
9777 W Gulf Bank Rd, Houston TX 77040-3132. 713/849-2828.

Code Com Inc.
406 W Horton St, Brenham TX 77833-2348. 409/830-0851.

Cogniseis Development Inc.
2401 Portsmouth St, Houston TX 77098-3903. 713/526-3273.

Cognos Corp.
67 S Bedford St, Burlington MA 01803-5152. 617/229-6600.

Columbia Ultimate Business Systems
14300 SE 1st St, Vancouver WA 98684-3500. 360/256-7358.

CompuCom
12600 SE 38th St, Bellevue WA 98006-5232. 206/644-9456.

Computer Associates International
1 Tech Dr, Andover MA 01810-2497. 508/685-1400.

Computer Associates International
1375 Corporate Center Pkwy, Santa Rosa CA 95407-5432. 707/528-6560.

Computer Associates International Inc.
Rte 206 & Orchard Rd, Princeton NJ 08543. 908/874-9000.

Computer Management Sciences Inc.
9 Hillside Ave, Waltham MA 02154-7552. 617/684-0004.

Computer Professionals
2101 W Commercial Blvd, Ft Lauderdale FL 33309-3071. 954/730-0730.

Computer Sciences Corporation
118 MacKenan Dr, Cary NC 27511-7921. 919/469-3325.

Computer Sciences Corporation
2600 Paramount Place, Fairborn OH 45324. 937/320-6300.

Computer Sciences Corporation
6100 Western Pl, Fort Worth TX 76107-4600. 817/777-4245.

Computer Sciences Corporation
PO Box 217, Clearfield UT 84015-0217. 801/777-4660.

Computer Systems Advisors
300 Tice Blvd, Woodcliff Lk NJ 07675-8405. 201/391-6500.

Computer Systems Development Inc.
242 Old New Brunswick Rd, Piscataway NJ 08854-3754. 908/562-0100.

Computer Task Group
5730 Oakbrook Parkway, Norcross GA 30093-1807. 770/263-3400.

Computer Technology Associates
41 Inverness Drive East, Englewood CO 80112-5412. 303/889-1200.

Computer-Aided Systems Inc.
30991 San Clemente St, Hayward CA 94544-7128. 510/429-2800.

Computerfocus Business Centers
Euclid Avenue Shopping Ctr, Bristol VA 24201. 540/466-4555.

Computize
8901 John W Carpenter Freeway, Dallas TX 75247-4523. 214/879-9900.

Compuware Corp.
3600 West 80th St, Bloomington MN 55431. 612/925-5900.

Compuware Corp.
10055 Miller Ave, Cupertino CA 95014-3468. 408/252-3801.

Compuware Corp.
15303 Dallas Pkwy, Suite 1150, Dallas TX 75248. 972/960-0960.

Comshare Inc.
4101 McEwen Rd, Suite 540, Dallas TX 75244-5130. 214/387-0171.

Comsys
4000 McEwen Rd S Ste 200, Dallas TX 75244-5014. 214/960-7053.

Comsys
4600 Marriott Dr Ste 510, Raleigh NC 27612. 919/787-1161.

Concentra
21 North Avenue, Burlington MA 01803-3301. 617/229-4600.

Consultants in Computer Software
8100 26th Ave South, Suite 160, Bloomington MN 55425-1304. 612/854-3600.

Control Systems
3939 Belt Line Road, Suite 450, Dallas TX 75244-2219. 972/484-2314.

Cooperative Computing Inc.
6207 Bee Caves Rd, Austin TX 78746-5146. 512/328-2300.

Corbel and Co.
PO Box 47720, Jacksonville FL 32247-8197. 904/399-5888.

CPS Systems Inc.
3400 Carlisle St, Dallas TX 75204-1264. 214/855-5277.

CRC Information Systems Inc.
435 Hudson St 7th Floor, New York NY 10014-3941. 212/620-5678.

Criterion Inc.
9425 N MacArthur Blvd, Irving TX 75063-4706. 214/401-2100.

CSC
4061 Powder Mill Rd, Calverton MD 20705-3149. 301/572-3211.

CSC Artemis
5251 Westheimer Rd, Suite 850, Houston TX 77056-5411. 713/626-1511.

CSC Consulting
300 Executive Dr, West Orange NJ 07052-3310. 201/243-0023.

CSC Continuum
9500 Arboretum Blvd, Austin TX 78759-6336. 512/345-5700.

CSC Healthcare Systems
415 McFarlan Rd, Kennett Square PA 19348. 610/444-1600.

CSC Healthcare Systems
34505 W 12 Mile Rd Ste 300, Farmington Hills MI 48331-3288. 810/553-0900.

CSC Index Inc.
5 Cambridge Ctr, Cambridge MA 02142-1407. 617/492-1500.

CSC Partners Inc.
29 Sawyer Rd, Waltham MA 02154-3427. 617/647-0116.

CSSI Computer Support Service Inc.
125 E John Carpenter Fwy Ste 270, Irving TX 75062-2238. 972/869-3966.

CTG Inc.
5300 NW 33rd Ave, Ft Lauderdale FL 33309-6377. 954/486-7105.

CTG Inc.
6000 Lombardo Ctr Ste 140, Seven Hills OH 44131-2579. 216/524-6441.

Cytrol Inc.
4620 W 77th St, Edina MN 55435-4924. 612/835-4884.

Dac Easy Inc.
17950 Preston Rd Ste 800, Dallas TX 75252-5691. 214/248-0305.

Dairyland Computer Consulting Co.
625 Lakeshore Dr S, Glenwood MN 56334-1549. 320/634-5331.

Dallas Systems Corp.
12740 Hillcrest Rd, Dallas TX 75230-2011. 214/233-3761.

Daly & Wolcott Inc.
141 James P Murphy Ind Rd, West Warwick RI 02893-2382. 401/823-8400.

Data Transformation Corp.
8300 Colesville Rd Ste 600, Silver Spring MD 20910-3243. 301/587-4580.

Datacard
P.O. Box 619, Simsbury CT 06070. 860/683-8981.

Datatrax Systems Corp.
650 S Taylor Ave, Louisville CO 80027-3032. 303/665-1030.

DataView Corporation
47 Pleasant St, Northampton MA 01060-3912. 413/586-4144.

Davis Thomas & Associates Inc. (DTA)
5353 Wayzata Blvd, Minneapolis MN 55416-1333. 612/591-6100.

DDC-I Inc.
5930 Lyndon B Johnson Fwy, Dallas TX 75240-6304. 214/458-9611.

Decision Systems Inc.
60 S 6th St, Minneapolis MN 55402-4423. 612/338-2585.

Delphi Information Systems
4572 S Hagadorn Rd, East Lansing MI 48823-5385. 517/337-9400.

Deltek Systems Inc.
2001 Gateway Pl, San Jose CA 95110-1013. 408/436-2136.

Digital Equipment Corp.
2702 International Ln, Madison WI 53704-3117. 608/246-3400.

Distribution Architects International Inc.
905 E Westchester Dr, Tempe AZ 85283-3000. 602/897-9576.

Document Sciences Corp.
6333 Greenwich Dr, San Diego CA 92122-5921. 619/625-2000.

Drew Technologies
2000 W Pioneer Pkwy, Peoria IL 61615-1835. 309/692-6161.

EA Systems Inc.
980 Atlantic Ave, Alameda CA 94501-1018. 510/748-4700.

EDG Inc.
1950 N Park Pl NW, Atlanta GA 30339-2044. 770/952-3030.

EDP Financial Services
257 Wyman Street, Waltham MA 02154-1226. 617/890-2400.

EDS Corp.
3215 Prospect Park Dr, Rancho Cordova CA 95670-6017. 916/636-4200.

EDS Corp.
13736 Riverport Dr, Maryland Heights MO 63043-4820. 314/344-5900.

EDS Corp. Seattle Regional Support Center
19351 8th Ave NE, Poulsbo WA 98370-8710. 360/697-9701.

EJR Computer Associates Inc.
5 Marine View Plz, Hoboken NJ 07030-5722. 201/795-3601.

EMC Corporation
12720 Hillcrest Rd Ste 802, Dallas TX 75230-2045. 972/233-5676.

EMC Corporation
488 Norristown Rd, Blue Bell PA 19422-2352. 610/834-7740.

EMD Associates Inc.
4245 Theurer Blvd, Winona MN 55987. 507/454-8254.

EMRC
1607 E Big Beaver Rd, Troy MI 48083-2066. 810/689-0077.

Emulex
14711 NE 29th Pl, Bellevue WA 98007-7666. 206/881-5773.

Engineering Geometry Systems
275 E South Temple, Salt Lake City UT 84111-1247. 801/575-6021.

Enterprise Solutions Ltd.
31416 Agoura Rd Ste 180, Westlake Village CA 91361. 818/597-8943.

ESRI
380 New York St, Redlands CA
92373-8118. 909/793-2853.

Executrain of Atlanta
1000 Abernathy Rd Bldg 400,
Atlanta GA 30328-5603.
770/396-9200.

Faac Inc.
825 Victors Way, Ann Arbor MI
48108-2738. 313/761-5836.

Fastech Integration Inc.
55 Old Bedford Rd, Lincoln MA
01773-1125. 617/259-3131.

FDSI Consulting
PO Box 3008, Bellevue WA
98009-3008. 206/637-6505.

Firesoft
128 Rogers St, Cambridge MA
02142-1024. 617/864-4010.

Fourth Shift Corporation
125 Cambridgepark Dr,
Cambridge MA 02140-2329.
617/492-2950.

Galileo International
5350 S Valentia Way, Englewood
CO 80112. 303/397-5000.

GE Capital IT Solutions
11333 Chimney Rock Rd,
Houston TX 77035-2901.
713/726-9822.

GE Co.
5420 Lyndon B Johnson Fwy,
Dallas TX 75240-6222. 214/788-
8200.

GE Information Services Inc.
8101 E Prentice Ave, Englewood
CO 80111-2926. 303/793-1300.

GEC-Marconi Avionics Inc.
2975 Northwoods Pkwy, Atlanta
GA 30366. 770/448-1947.

Genelco Inc.
5001 Lyndon B Johnson Fwy,
Dallas TX 75244-6120. 214/387-
5292.

Genesys Software Systems Inc.
5 Branch St, Methuen MA 01844-
1947. 508/685-5400.

Gensym Corp.
125 Cambridgepark Dr,
Cambridge MA 02140-2314.
617/547-2500.

Geodynamics Corp.
21171 S Western Ave, Torrance
CA 90501-1704. 310/320-2300.

GHG Corp.
1100 Hercules Ave, Houston TX
77058-2747. 713/488-8806.

Global Software Inc.
3200 Atlantic Ave, Raleigh NC
27604. 919/872-7800.

**Great Valley System
Consultants**
5 Great Valley Pkwy, Malvern
PA 19355-1426. 610/889-1319.

GTE Health Systems
2510 W Dunlap Ave, Phoenix AZ
85021-2723. 602/678-6000.

GTECH
4501 Osuna Road NE,
Albuquerque NM 87109-1210.
505/344-2400.

HB Maynard & Co. Inc.
400 Eight Parkway Ctr,
Pittsburgh PA 15220. 412/921-
2400.

HCL America Inc.
330 Potrero Ave, Sunnyvale CA
94086-4113. 408/733-0480.

HCM Inc.
3655 Torrance Blvd, Torrance
CA 90503-4813. 310/540-0050.

Health Data Sciences Corp.
268 W Hospitality Ln, San
Bernardino CA 92408-3241.
909/888-3282.

Herlick Data Systems
1524 E Central Ave, Redlands
CA 92374-4172. 909/798-0947.

**Hitachi Software Engineering
America Ltd.**
1111 Bayhill Dr, San Bruno CA
94066-3035. 415/615-9600.

Horizons Technology Inc.
3990 Ruffin Rd, San Diego CA
92123-1826. 619/292-8331.

Hughes Electro-Optical & Data
PO Box 902, El Segundo CA
90245-0902. 310/616-1375.

Hunter Associates Laboratory
11491 Sunset Hills Rd, Reston
VA 22090-5207. 703/471-6870.

IA Corp.
1900 Powell St, Emeryville CA
94608-1816. 510/450-7000.

IC Technologies Corp.
6400 Riverside Dr, Dublin OH
43017-5047. 614/798-1091.

Icon Resources Inc.
1050 North State, Suite 210,
Chicago IL 60610. 312/573-0142.

IDX
4901 Lyndon B Johnson Fwy,
Dallas TX 75244-6118. 972/458-
1060.

IIT Research Institute
4409 Forbes Blvd, Lanham MD
20706-4328. 301/459-3711.

Image Sciences Inc.
5910 N Central Expy, Dallas TX
75206-5140. 214/891-6500.

Imageware
5102 78th St, Lubbock TX
79424-3009. 806/798-2983.

IMI Systems Inc.
1500 Broadway, New York NY
10036-4015. 212/944-1555.

Indotronix International Inc.
7373 E Doubletree Ranch Rd,
Scottsdale AZ 85258-2035.
602/998-2112.

Indus Group
60 Spear St, San Francisco CA
94105-1506. 415/904-5000.

Info Group
46 Park St, Framingham MA
01702. 508/872-8383.

Infobases
305 N 500 W, Suite C, Provo UT
84601-2644. 801/375-2227.

**Information & Operation
Emporium**
44 Alpert Dr, Wappingers Falls
NY 12590-4619. 914/297-7700.

**Information Technology
Solutions Inc.**
255 South Boulevard East,
Petersburg VA 23805-2700.
804/733-2440.

Insci Corp.
200 Friberg Pkwy, Westborough
MA 01581-3911. 508/870-4000.

Integrated Custom Software
701 Hebron Avenue, Glastonbury
CT 06033. 860/657-3339.

**Integrated Microcomputer
Systems Inc.**
2 Research Pl, Rockville MD
20850-3295. 301/948-4790.

**Integrated Systems
Development Inc.**
400 136th Ave, Holland MI
49424-2923. 616/396-0880.

Interactive Inc.
5095 Murphy Canyon Rd, San
Diego CA 92123-4346. 619/560-
8525.

**Interactive Network
Technology Inc.**
5001 Lyndon B Johnson Fwy,
Dallas TX 75244-6120. 972/715-
2640.

Intercom Micro Systems
6418 Symposium Way, Clinton
MD 20735-3862. 301/856-2706.

**Intergraph Corp. Training
Center**
8111 Lyndon B Johnson Fwy,
Dallas TX 75251-1313. 214/497-
0861.

Interim Technologies
2777 N Stemmons Fwy, Dallas
TX 75207-2277. 214/631-0788.

Interim Technologies
630 3rd Ave, New York NY
10017-6705. 212/986-7600.

International Software Group
67 S Bedford St, Burlington MA
01803. 617/221-1450.

International Technegroup Inc.
5303 Dupont Cir, Milford OH
45150-2734. 513/576-3900.

Interplay Productions Inc.
16815 Von Karmen Ave, Irvine
CA 92606. 714/553-6655.

Intersystems Corp.
1 Memorial Dr, Cambridge MA
02142-1301. 617/621-0600.

IRI Software
200 5th Ave, Waltham MA
02154-8704. 617/890-1100.

ISI Systems Inc.
2 Tech Dr, Andover MA 01810-
2434. 508/682-5500.

Island Graphics Corp.
4000 Civic Center Dr, San Rafael
CA 94903-4171. 415/491-1000.

Ithaca Software Inc.
1301 Marina Village Pkwy,
Alameda CA 94501-1058.
510/523-5900.

ITIS
1 Park Ten Pl, Houston TX
77084-5062. 713/578-8800.

JG Van Dyke & Associates Inc.
141 National Business Pkwy,
Annapolis Junction MD 20701-
1026. 301/953-3600.

JG Van Dyke & Associates Inc.
6550 Rock Spring Dr, Bethesda
MD 20817. 301/897-8970.

JLI Inc.
4807 Lover's Ln, Dallas TX
75209-2822. 214/357-7470.

Johnson Intelligent Systems
1350 Nasa Rd 1, Houston TX
77058-3174. 713/333-4500.

Jon Goldman & Associates
2237 N Batavia St, Orange CA
92865. 714/283-5889.

Josten's Learning Corp.
8505 Freeport Pkwy, Irving TX
75063-2506. 972/929-4201.

**Kansas Geological Survey
Computer Service**
1930 Constant Ave, Lawrence KS
66047-3724. 913/864-4991.

Keane Inc.
6701 Center Dr W, Los Angeles
CA 90045-1568. 310/348-9300.

Keane Inc.
2525 Meridian Parkway, Durham
NC 27713-2261. 919/544-0891.

Kenan Systems Corp.
1 Main Street, Cambridge MA
02142-1517. 617/225-2200.

Key Four Inc.
5238 Royal Woods Parkway,
Tucker GA 30084-3079. 770/908-
4400.

Keynote Systems Inc.
500 Standard Life Building, 345
4th Avenue, Pittsburgh PA
15222. 412/261-0187.

Kirchman Corp.
PO Box 2269, Orlando FL 32802-
2269. 407/831-3001.

Knogo Corporation
350 Wireless Blvd, Hauppauge
NY 11788-3959. 516/232-2100.

Kodak Scientific Imaging
4 Science Park, New Haven CT
06511-1963. 203/786-5600.

KSI International
10338 Battleview Pkwy,
Manassas VA 20109. 703/330-
4500.

Lacerte Software Corp.
13155 Noel Road, Suite 2200,
Dallas TX 75240. 972/490-8500.

Lakeview Technology
401 16th Ave NW, Rochester
MN 55901-1853. 507/252-3440.

LB Consulting Inc.
44838 Huntingcross Dr, Novi MI
48375-3934. 810/349-5311.

LBMS Inc.
1800 West Loop South, 9th Floor,
Houston TX 77027-3211.
713/625-9300.

Lear Data Info-Services Inc.
5910 N Central Expy, Dallas TX
75206-5151. 214/360-9008.

**Lernout & Hauspie Speech
Products**
20 Mall Rd, Burlington MA
01803-4123. 617/238-0960.

Levi Ray and Shoup Inc.
2401 W Monroe St, Springfield
IL 62704-1439. 217/793-3800.

Lockheed Martin
4701 Forbes Blvd, Lanham MD
20706-4396. 301/306-8000.

Logicon, Inc.
4010 Sorrento Valley Blvd, San
Diego CA 92121-1405. 619/455-
1330.

Logicon, Inc.
2100 Washington Blvd, Arlington
VA 22204-5702. 703/486-3500.

Logistix
600 Charter Oak Ranch Rd,
Fountain CO 80817-4001.
719/382-7200.

Long Claim Services Inc.
4747 Lincoln Mall Dr, Matteson
IL 60443-3814. 708/747-4010.

Lotus Development Corp.
800 W El Camino Real, Mountain
View CA 94040-2567. 415/961-
8800.

Lotus Development Corp.
2010 Main Street, Suite 1100,
Irvine CA 92614. 714/261-2697.

Lotus Development Corp.
12750 Merit Drive, Suite 900,
Dallas TX 75251-1242. 214/490-
7706.

MacNeal-Schwendler
2975 Red Hill Ave, Costa Mesa
CA 92626-5923. 714/540-8900.

Macromedia
269 W Renner Rd, Richardson
TX 75080-1318. 972/680-2060.

Manager Software Products
131 Hartwell Ave, Lexington MA
02173-3126. 617/863-5800.

Mandex Inc.
12500 Fair Lakes Circle, Fairfax
VA 22033-3804. 703/227-0900.

Mantech Services Corporation
12015 Lee Jackson Memorial
Hwy, Fairfax VA 22033-3300.
703/218-6000.

Marketing Resources Plus
555 Twin Dolphin Dr, Redwood
City CA 94065-2102. 415/595-
1800.

Marketmax Inc.
100 Conifer Hill Dr, Suite 509,
Danvers MA 01923-1176.
508/777-0057.

Mathworks Inc.
24 Prime Park Way, Natick MA
01760-1528. 508/653-1415.

McHugh Freeman & Associates
20700 Swenson Dr, Waukesha
WI 53186-0904. 414/798-8600.

MCNC
PO Box 12889, Durham NC
27709-2889. 919/248-1800.

MDL Information Systems Inc.
14600 Catalina Street, San
Leandro CA 94577-6604.
510/895-1313.

Mechanical Dynamics Inc.
2301 Commonwealth Blvd, Ann
Arbor MI 48105-2945. 313/994-
3800.

Medical Systems Inc.
11520 N Central Expy, Dallas TX
75243-6605. 214/342-0789.

Mediqual Systems Inc.
1900 W Park Dr, Westborough
MA 01581-3942. 508/366-6365.

Medsamerica Inc.
12553 Gulf Freeway, Houston TX
77034. 800/456-6337.

Megacomm Associates Inc.
200 Jefferson Rd, Wilmington
MA 01887-1999. 508/658-8900.

MEI Technology Corporation
1050 Waltham St, Lexington MA
02173-8059. 617/862-3390.

Mentor Graphics Corp.
1001 Ridder Park Dr, San Jose
CA 95131-2314. 408/436-1500.

Mentor Graphics Corp.
400 Galleria Officentre,
Southfield MI 48034-8473.
810/356-7660.

**Mercer Management
Consulting**
33 Hayden Ave, Lexington MA
02173-7966. 617/861-7580.

Metro Information Services
7202 Glen Forest Dr, Richmond
VA 23226-3770. 804/285-7800.

Metters Industries Inc.
8200 Greensboro Dr, Suite 500,
Mc Lean VA 22102-3803.
703/821-3300.

Mettler Toledo Inc.
P.O. Box 71, Hightstown NJ
08520-1900. 609/448-3000.

Michael Baker Corp.
PO Box 12259, Pittsburgh PA
15231-0259. 412/269-6300.

Michael Baker Corp.
3601 Eisenhower Ave,
Alexandria VA 22304-6425.
703/960-8800.

Micro Majic Inc.
1955 S 1800 W, Woods Cross UT
84087-2314. 801/298-2001.

Micro Marketing
10455 Markison Rd, Dallas TX
75238-1651. 214/349-4600.

Micro One Inc.
8101 Ridgepoint Dr, Irving TX
75063-3165. 972/556-2223.

Micro-MRP Inc.
1065 E Hillsdale Blvd, Foster
City CA 94404-1615. 415/345-
6000.

MicroAge Computer Centers
707 W University Dr, College
Station TX 77840-1430. 409/846-
9727.

Microsoft Corporation
8300 Norman Center Dr,
Bloomington MN 55437-1027.
612/896-1001.

Microtec Research
2350 Mission College Blvd,
Santa Clara CA 95054-1532.
408/980-1300.

Microway Inc.
PO Box 79, Kingston MA 02364-
0079. 508/746-7341.

Milner Document Products
2845 Amwiler Rd, Atlanta GA
30360-2826. 770/263-5300.

Milsoft Integrated Solutions
4400 Buffalo Gap Rd, Abilene
TX 79606-2722. 915/695-1642.

Mincron SBC Corp.
333 N Sam Houston Pkwy E,
Houston TX 77060-2403.
713/999-7010.

MISI Co. Ltd.
811 Church Rd, Cherry Hill NJ
08002-1412. 609/661-0232.

Mitre Corp.
145 Wyckoff Rd, Eatontown NJ
07724-1878. 908/544-1414.

Modern Engineering
1700 Updike Court, Auburn Hills
MI 48321. 810/340-9080.

MRJ Inc.
200 Falls Corporate Ctr, West
Conshohocken PA 19428-2950.
610/940-4466.

Multimedia Abacus Corp.
9920 S La Cienega Blvd,
Inglewood CA 90301-4423.
310/645-0598.

Multisystems Inc.
10 Fawcett St, Cambridge MA
02138-1110. 617/864-5810.

Must Software International
101 Merritt 7, Norwalk CT
06851-1060. 203/845-5000.

NCI Information Systems Inc.
370 S 500 E, Clearfield UT
84015-4046. 801/776-1085.

Nettleship Group Inc.
2665 Main St, Santa Monica CA
90405-4054. 310/392-8585.

Nichols SELECT
1801 1st Ave South, Suite 400,
Birmingham AL 35233-1935.
205/320-6900.

Nimbus Information Systems
PO Box 7427, Charlottesville VA
22906-7427. 804/985-4300.

Northwest Payroll Services Inc.
5701 Kentucky Ave N,
Minneapolis MN 55428-3370.
612/536-9555.

Nova Technologies Inc.
311 E Vickery Blvd, Fort Worth
TX 76104-1352. 817/332-6682.

NYNEX DPI Co.
12946 Dairy Ashford Rd, Sugar
Land TX 77478. 713/240-9200.

O'Brien Computer Systems
201 Redwood Cir, Petaluma CA
94954-3849. 707/762-4053.

O'Brien's Computer Services
517 Sugar Hollow Rd, Piney Flats
TN 37686-3126. 423/323-2567.

Oasis Health Care Systems
100 Drakes Landing Rd,
Greenbrae CA 94904-3123.
415/925-0121.

Oberon Software Inc.
215 First St, Cambridge MA
02142-1102. 617/494-0990.

OGCI Software
1 Park Ten Pl, Houston TX
77084-5061. 713/579-2600.

On Technology Inc.
1 Cambridge Ctr, Cambridge MA
02142-1601. 617/374-1400.

Open Group
11 Cambridge Ctr, Cambridge
MA 02142-1405. 617/621-8700.

Open Systems Holdings Corp.
7626 Golden Triangle Dr, Eden
Prairie MN 55344-3732. 612/829-
0011.

Open Vision Technologies
7133 Koll Center Pkwy,
Pleasanton CA 94566-3101.
510/426-6400.

Opencon Systems
377 Hoes Ln, Piscataway NJ
08854-4138. 908/463-3131.

Origin Systems Inc.
5918 W Courtyard Dr, Austin TX
78730-5024. 512/434-4263.

Pacer Infotec
4099 SE International Way,
Portland OR 97222. 503/794-
1344.

Paragon Inc.
2318 Calle De Luna, Santa Clara
CA 95054-1003. 408/727-8824.

Paranet Inc.
1776 Yorktown St Ste 300,
Houston TX 77056-4114.
713/626-4800.

Paranet Inc.
14651 Dallas Pkwy, Dallas TX
75240-7476. 214/239-5544.

PDI
14755 27th Ave N, Plymouth MN
55447-4809. 612/553-9838.

Peerless Group Inc.
1212 E Arapaho Rd, Richardson
TX 75081-2459. 972/497-5500.

Perkins Consulting LLC
1211 SW 5th Ave Ste 1205,
Portland OR 97204-3713.
503/221-7590.

Petroleum Information Corp.
5333 Westheimer, Houston TX
77056. 713/840-8282.

Pilot Software Inc.
1 Canal Park, Cambridge MA
02141-2203. 617/374-9400.

Pixar
1001 W Cutting Blvd Ste 200,
Richmond CA 94804-2028.
510/236-4000.

Plexus Software Inc.
1310 Chesapeake Ter, Sunnyvale
CA 94089-1100. 408/747-1210.

Princeton Financial Systems
8 N Main St, Attleboro MA
02703-2282. 508/226-8504.

Process Software Corp.
959 Concord St, Framingham MA
01701-4682. 508/879-6994.

Professional Software Service
10635 N Ih 35, San Antonio TX
78233-6627. 210/653-0990.

Programart Corporation
124 Mount Auburn St, Cambridge
MA 02138-5758. 617/661-3020.

PSW Technologies
9050 Capital Tex Hwy N Ste 300,
Austin TX 78759. 512/343-6666.

Quality Systems Inc.
4000 Legato Rd, Fairfax VA
22033-4055. 703/352-9200.

Quarterdeck Office Systems
13160 Mindano, Marina del Rey
CA 90292. 310/309-3700.

Radix Systems Corp.
675 Massachusetts Ave,
Cambridge MA 02139-3309.
617/354-4015.

Raytheon Company
1847 W Main Rd, Portsmouth RI
02871-1037. 401/847-8000.

**Raytheon Systems Development
Co.**
353 James Record Rd SW,
Huntsville AL 35824-1515.
205/461-6100.

Recital Corp.
85 Constitution Ln, Danvers MA
01923-3630. 508/750-1066.

Republic Electronics Co.
49 Wireless Blvd, Hauppauge NY
11788. 516/273-9800.

Resumix Inc.
890 Ross Dr, Sunnyvale CA
94089. 408/744-3800.

Revelation Software Inc.
181 Harbor Dr, Stamford CT
06902-7474. 203/973-1000.

RLN Computer Solutions
1301 19th St, Ste 103, Plano TX
75074-5933. 214/424-2380.

RMS Technologies Inc.
4221 Forbes Blvd, Lanham MD
20706-4343. 301/925-9760.

RMS Technologies Inc.
2500 English Creek Ave, Egg
Harbor Township NJ 08234-
5562. 609/485-0615.

Rnet Computer Services
1400 Commerce Blvd, Sarasota
FL 34243-5023. 941/359-2111.

ROI Systems Inc.
435 Ford Rd, Minneapolis MN
55426-4913. 612/595-0500.

Ron Allen Consulting Service
835 Lakeview, Stansbury Park
UT 84074-9613. 801/882-3535.

SAI Multi-Interactive Systems
1701 N 8th St, McAllen TX
78501-2471. 210/631-4061.

SAI Software Consultants Inc.
5600 S Quebec St, Englewood
CO 80111-2207. 303/741-5452.

SAIC
3800 Watt Ave, Sacramento CA
95821-2672. 916/974-8800.

Sanchez Computer Associates
40 Valley Stream Pkwy, Malvern
PA 19355-1482. 610/296-8877.

Santa Cruz Operation
400 Encinal St # 1900, Santa
Cruz CA 95060-2115. 408/425-
7222.

Saztec International Inc.
43 Manning Rd, Billerica MA
01821. 508/262-9600.

**Science & Engineering
Associates Inc.**
PO Box 3722, Albuquerque NM
87190-3722. 505/884-2300.

**SCT Systems & Computer
Technology**
4100 Alpha Rd, Dallas TX
75244-4332. 972/383-7600.

SDC Computer Services
250 International Dr,
Williamsville NY 14231.
716/631-8433.

SE Warner Diskcopy Service
2225 Murray Holladay Rd, Salt
Lake City UT 84117-5382.
801/277-9444.

Seagate Software
134 Flanders Rd, Westborough
MA 01581-1017. 508/898-0100.

Secure Computing Corp.
2675 Long Lake Rd, Roseville
MN 55113-1117. 612/628-2700.

Seer Technologies Inc.
5 Penn Plz, New York NY
10001-1810. 212/643-6000.

Seer Technologies Inc.
1850 Parkway Pl, Marietta GA
30067-4439. 770/428-1773.

Sentinel Computer Services Inc.
39 S La Salle St, Chicago IL
60603-1603. 312/419-1490.

Servicesoft Corp.
50 Cabot St, Needham MA
02194-2819. 617/449-0049.

**Shaker Computer &
Management Service Inc.**
50 Century Hill Dr, Latham NY
12110-2116. 518/786-7200.

Shared Resource Management
3550 Lexington Ave N, St. Paul
MN 55126-8048. 612/486-0417.

**Sheshunoff Management
Service Inc.**
505 Barton Springs Rd, Austin
TX 78704-1248. 512/472-4000.

Simulation Sciences Inc.
601 Valencia Ave, Brea CA
92621-6346. 714/579-0412.

Softbridge Inc.
125 Cambridgepark Dr,
Cambridge MA 02140-2329.
617/576-2257.

Software Alliance Corp.
2150 Shattuck Ave Fl 11,
Berkeley CA 94704-1306.
510/548-7752.

Software Architects Inc.
122 W Carpenter Freeway, Irving
TX 75039. 972/791-0500.

Software of the Future
P.O. Box 12079, Arlington TX
76012. 214/264-2626.

Software Solutions Inc.
101 1st St, Los Altos CA 94022-
2750. 415/367-4994.

**Southfield Software & Systems
Co.**
40177 Sandpointe Way, Novi MI
48375-5341. 810/349-5258.

Sterling Commerce Inc.
PO Box 7160, Dublin OH 43017-
0760. 614/793-7000.

Sterling Resources Inc.
6 Forest Ave, Paramus NJ 07652-
5214. 201/843-6444.

Stonebridge Technologies
14800 Landmark Blvd, Suite 250,
Dallas TX 75240-8810. 972/404-
9755.

Strategic Decisions Group
2440 Sand Hill Rd, Menlo Park
CA 94025-6900. 415/854-9000.

Strategic Partners
5051 Andrew Dr, La Palma CA
90623-1404. 562/467-1512.

Strategic Simulations Inc.
675 Almanor Ave, Sunnyvale CA
94086-2901. 408/737-6800.

Strategic Software Systems Inc.
1508 Willow Lawn Dr, Richmond
VA 23230. 804/231-6633.

Stream Logic
1555 Adams Dr, Menlo Park CA
94025-1439. 415/325-4392.

STS Systems Inc.
350 Dunksferry Rd, Bensalem PA
19020-6543. 215/245-0710.

STW Inc.
212 E Franklin St, Grapevine TX
76051-5325. 817/329-1711.

Sulcus Computer Corporation
2150 East Highland, Phoenix AZ
85016. 602/808-3600.

**Summit Information
Systems/FIserv**
PO Box 3003, Corvallis OR
97339-3003. 541/758-5888.

SunGard
486 Potten Road, Waltham MA
02154. 617/890-2070.

SunGard Planning Solutions
1800 W Park Dr, Westborough
MA 01581-3912. 508/366-3225.

SunGard Shareholder Systems
951 Mariners Island Blvd, San
Mateo CA 94404-1558. 415/377-
3700.

Sunrise Support Services
1 Sierra Gate Plz, Roseville CA
95678-6609. 916/783-4200.

Sunsoft Inc.
6601 Center Dr W, Los Angeles
CA 90045-1582. 310/348-8649.

Support Systems Associates Inc.
1300 Veterans Hwy, Hauppauge
NY 11788-3063. 516/231-8998.

Surya Electronics Inc.
600 Windy Point Dr, Glendale
Heights IL 60139-3802. 630/858-
8000.

Switchview Inc.
801 E Campbell Rd, Richardson
TX 75081-6701. 972/918-9979.

Sybase Inc.
6 Concourse Pkwy NE, Atlanta
GA 30328-5351. 770/828-0500.

Sykes Enterprises Inc.
5215 N O'Connor Blvd, Irving
TX 75039-3738. 214/869-9062.

Sykes Enterprises Inc.
100 N Tampa St, Tampa FL
33602-5809. 813/274-1000.

Symantec
2500 Broadway, Santa Monica
CA 90404-3061. 310/453-4600.

Synetics Corp.
540 Edgewater Dr, Wakefield
MA 01880-6289. 617/245-9090.

Synon Inc.
1100 Larkspur Landing Cir,
Larkspur CA 94939-1827.
415/461-5000.

Synon Inc.
222 Las Colinas Blvd W, Irving
TX 75039-5427. 972/401-4010.

Synopsis
19500 NW Gibbs Road,
Beaverton OR 97006. 503/690-
6900.

Sysorex International Inc.
335 East Midefield Rd, Mountain
View CA 94043. 415/967-2200.

System Resources Corp.
128 Wheeler Rd, Burlington MA
01803-5170. 617/270-9228.

**Systems Application
Engineering Inc.**
3655 Westcenter Dr, Houston TX
77042-5297. 713/783-6020.

Systems Consultants
3122 Landover Dr, Carrollton TX
75007-3932. 972/492-1315.

Systems Integration & Analysis
8080 Dagget St, San Diego CA
92111-2326. 619/277-0700.

Taligent Inc.
10355 N De Anza Blvd,
Cupertino CA 95014-2027.
408/255-2525.

Tankle Business Services Inc.
280 Byberry Rd, Huntingdon
Valley PA 19006-4012. 215/947-
2990.

TBC
1400 East St, Pittsfield MA
01201-5329. 413/499-0188.

Tecnomatix Technologies Inc.
39810 Grand River Ave, Novi MI
48375-2101. 810/471-6140.

Teknekron Infoswitch Corp.
4425 Cambridge Rd, Fort Worth
TX 76155-2629. 817/267-3025.

Teknon Corp.
111 E Magnesium Rd, Spokane
WA 99208-5968. 509/467-6529.

Telos Consulting Services
1601 Trapelo Rd, Waltham MA
02154-7333. 617/890-5517.

Tesseract Corp.
475 Sansome St, San Francisco
CA 94111-3103. 415/981-1800.

Tiburon Inc.
11044 Research Blvd Ste B420,
Austin TX 78759. 512/345-8613.

Tidewater Consultants Inc.
160 Newtown Rd, Virginia Beach
VA 23462-2408. 757/497-8951.

Transition Systems Inc.
1 Boston Pl Fl 27, Boston MA
02108-4407. 617/723-4222.

Travel Technologies Group Inc.
5550 Lyndon B Johnson Fwy,
Dallas TX 75240-6263. 214/702-
1015.

Uniface Corp.
1320 Harbor Bay Pkwy, Alameda
CA 94502-6556. 510/748-6145.

Unisys Corporation
3199 Pilot Knob Rd, Eagan MN
55121-1328. 612/687-2200.

Unisys Corporation
1100 Corporate Dr, Farmington
NY 14425-9570. 716/924-0480.

**Unisys Corporation
Government Systems Group**
5151 Camino Ruiz, Camarillo CA
93012-8601. 805/987-6811.

Unisys Government Systems
8008 Westpark Dr, Mc Lean VA
22102-3106. 703/556-5000.

USCS International
11020 Sun Center Dr, Rancho
Cordova CA 95670-6114.
916/636-4500.

Vanstar Corp.
28 Centerpointe Dr Ste 110, La
Palma CA 90623-1054. 714/670-
6766.

Venturian Software
1300 Second St, Hopkins MN
55343. 612/931-2450.

Visual Integration Inc.
11770 Bernardo Plaza Ct, San
Diego CA 92128-2426. 619/592-
0286.

Volt Delta Resources Inc.
2401 N Glassell St, Orange CA
92865. 714/921-8000.

Whittaker Xyplex
2200 Lawson Ln, Santa Clara CA
95054-3310. 408/565-6000.

Work Management Solutions
119 Beach St, Boston MA 02111-
2511. 617/482-6677.

HARDWARE MANUFACTURERS

"There's no place to hide in a small company," warns the owner of one hardware assembly business. He advises people to take the initiative whenever possible and show a proprietary interest in the company they work for. This owner has been interested in computers since childhood, and he looks for people who show a similar zest for the industry.

Entry-level jobseekers may find that the desire to learn and succeed can compensate for a lack of formal education in computer assembly. In a small company, a candidate may also need excellent sales and interpersonal skills in addition to hardware knowledge.

ABL ELECTRONICS CORPORATION
10842 Beaver Dam Road, Hunt Valley MD 21030. 410/584-2700. **Contact:** Human Resources. **Description:** A manufacturer of computer cable.

ADPI
P.O. Box 499, Troy OH 45373. 937/339-2241. **Contact:** Human Resources Department. **Description:** Manufactures and sells strategic data logging and retrieval peripherals. ADPI produces data transferring units that log data from such technical equipment as seismographs. The company is currently developing a single board computer to interface with their data loggers.

ALI COMPUTERS
922 University City Boulevard, Blacksburg VA 24060. 540/552-6108. **Contact:** Human Resources Department. **Description:** Sells computer hardware, software, and peripherals. ALI Computers manufactures their own hardware as well as offering competitors' models. The company owns and operates a retail store that carries CPUs, monitors, printers, and other peripherals.

APS TECHNOLOGIES
6131 Deramus Avenue, Kansas City MO 64120. 816/483-1600. **Contact:** Human Resources Department. **Description:** Manufactures a variety of computer peripherals sold through the company's catalogs.

ASP COMPUTER PRODUCTS
285 North Wolfe Road, Sunnyvale CA 94086. 408/746-2965. **Contact:** Human Resources. **Description:** Provides businesses with various printer hardware. ASP Computer Products distributes printer sharing and printer server hardware for PCs to a variety of industries.

AST, INC.
7183 North Main Street, Clarkston MI 48346. 810/625-0191. **Contact:** Department of Interest. **Description:** A research and development firm specializing in the manufacture of hand-held data-gathering terminals. Founded in 1980.

AST RESEARCH, INC.
16215 Alton Parkway, Irvine CA 92618. **Contact:** Staffing Department. **Description:** AST Research, Inc. produces board-level enhancement products. The company also manufactures, distributes, and supports a line of personal computers under the Premium and Bravo brand names. The company conducts sales operations through 34 offices and subsidiaries worldwide and markets its products in 100 countries. **Common positions include:** Computer Programmer; Design Engineer; Hardware Engineer; Marketing Specialist; Project Engineer; Sales Representative; Software Engineer. **Corporate headquarters location:** This Location. **Listed on:** NASDAQ. **Number of employees nationwide:** 4,509.

ACCESS CORPORATION

4350 Glendale-Milford Road, Suite 250, Cincinnati OH 45242-3700. 513/786-8350. **Contact:** Liz Forg, Personnel Administrator. **Description:** Access Corporation produces document storage and retrieval systems. **Common positions include:** Computer Programmer; Systems Analyst. **Educational backgrounds include:** Computer Science. **Benefits:** 401(k); Dental Insurance; Disability Coverage; Life Insurance; Medical Insurance; Tuition Assistance. **Corporate headquarters location:** This Location. **Operations at this facility include:** Administration; Research and Development; Sales; Service.

ACER AMERICA CORPORATION

2641 Orchard Parkway, San Jose CA 95134. 408/433-4985. **Toll-free phone:** 800/SEE-ACER. **Fax:** 408/428-0947. **Contact:** Human Resources. **E-mail address:** careers@acer.com. Ftp address: ftp://ftp.acer.com/. **World Wide Web address:** http://www.acer.com/aac/. **Description:** This location is a manufacturing facility with an annual production capacity of over 600,000 systems, including notebooks. Acer America is one of the world's largest manufacturers of microcomputers and is one of the largest OEM suppliers. Acer also manufactures peripherals and components including monitors, notebooks, keyboards, CD-ROM drives, expansion cards, and multi-user and networked UNIX computer systems for commercial markets. Products include trademark Acer Aspire PCs; AcerNote color notebooks; AcerEntra and AcerPower desktop PCs; AcerAltos servers, Inter/intranet solutions, and software products; and AcerOpen components. **Common positions include:** Administrative Assistant; Computer Programmer; Contract/Grant Administrator; Network Administrator; Product Manager; Software Engineer; Systems Engineer; Technician. **Parent company:** Acer Group, Inc. (Taipei, Taiwan, R.O.C.). **Operations at this facility include:** Manufacturing. **Annual sales/revenues:** $1.4 billion. **Number of employees nationwide:** 1,300. **Number of employees worldwide:** 15,000.

ACMA COMPUTER

47988 Fremont Road, Fremont CA 94538. 510/623-1212. **Contact:** Human Resources. **Description:** A manufacturer of personal computers.

ADAGER CORPORATION

The Adager Way, Sun Valley ID 83353-3000. 208/726-9100. **Toll-free phone:** 800/533-7346. **Fax:** 208/726-8191. **Contact:** Human Resources. **E-mail address:** info@adager.com. **World Wide Web address:** http://www.adager.com. **Description:** A database utility firm. Adager's hardware is oriented to any type of HP3000 hardware, operating system, or IMAGE/SQL version.

ADVANCED LOGIC RESEARCH

9401 Jeronimo Way, Irvine CA 92618-1908. 714/581-6770. **Fax:** 714/581-9240. **Contact:** Irene Martinez, Personnel Manager. **Description:** A manufacturer and marketer of personal computers. **Common positions include:** Computer Engineer; Computer Programmer; Manufacturer's/Wholesaler's Sales Rep.; Marketing Specialist. **Number of employees nationwide:** 560.

ADVANCED TECHNOLOGICAL SOLUTIONS (ATS)

585 DeKalb Avenue, Brooklyn NY 11205. 718/780-2100. **Contact:** Brenda Frisby, Personnel Manager. **Description:** Advanced Technological Solutions (ATS) is a developer, manufacturer, and marketer of advanced information processing products, including computers and microelectronic technology, software, networking systems, and information technology-related services.

AMDAHL CORPORATION

85 Challenger Road, 3rd Floor, Ridgefield Park NJ 07660. 201/229-4400. **Contact:** Human Resources. **E-mail address:** jobs@amdahl.com. **Description:** This location is engaged in sales, service, and support. Amdahl Corporation designs, develops, manufactures, markets, and services large-scale, high-performance, general purpose computer systems including both hardware and software. Customers are primarily in large corporations, government agencies, and large universities with high-volume data processing requirements. Amdahl markets more than 470 different systems. **Corporate headquarters location:** Sunnyvale CA.

AMDAHL CORPORATION
770 The City Drive, Suite 4000, Orange CA 92868. 714/740-0440. **Contact:** Human Resources. **E-mail address:** jobs@amdahl.com. **Description:** This location is engaged in sales and service. Amdahl Corporation designs, develops, manufactures, markets, and services large-scale, high-performance, general purpose computer systems including both hardware and software. Customers are primarily in large corporations, government agencies, and large universities with high-volume data-processing requirements. Amdahl markets more than 470 different systems. **Corporate headquarters location:** Sunnyvale CA.

AMERICAN POWER CONVERSION (APC)
P.O. Box 278, West Kingston RI 02892. 401/789-5735. **Contact:** Human Resources. **World Wide Web address:** http://www.apcc.com. **Description:** Designs, develops, manufactures, and markets surge suppressers, uninterruptible power supplies (UPS), power conditioning equipment, and related software for computer and computer-related equipment, including local area networks (LANs), midrange computers, and engineering workstations. American Power Conversion also publishes a newsletter called *APC Currents* which provides news about the company and the company's products, as well as some of APC's customers. **NOTE:** The company is located at 132 Fairgrounds Road, West Kingston RI. **Common positions include:** Buyer; Computer Programmer; Credit Manager; Customer Service Representative; Economist; Electrical/Electronics Engineer; General Manager; Industrial Engineer; Industrial Production Manager; Materials Engineer; Mechanical Engineer; Operations/Production Manager; Public Relations Specialist; Purchasing Agent/Manager; Quality Control Supervisor; Software Engineer; Systems Analyst. **Educational backgrounds include:** Business Administration; Computer Science; Engineering. **Benefits:** Dental Insurance; Disability Coverage; ESOP; Incentive Plan; Life Insurance; Medical Insurance; Tuition Assistance. **Corporate headquarters location:** This Location. **International locations:** North Sydney, NSW/Australia (Asia Pacific headquarters); China; India; Galway, Ireland (European headquarters); Japan; Korea; Singapore. **Operations at this facility include:** Administration; Divisional Headquarters; Manufacturing; Sales; Service. **Listed on:** NASDAQ. **Stock exchange symbol:** APCC. **Number of employees at this location:** 1,400.

ANICOM, INC.
6133 North River Road, Suite 410, Rosemont IL 60018. 847/518-8700. **Contact:** Human Resources. **Description:** Engaged in the sale and distribution of multimedia wiring products. Anicom operates 33 locations throughout the U.S. **Corporate headquarters location:** This Location. **Listed on:** NASDAQ.

APPALACHIAN REGIONAL MANUFACTURING
1101 Lakeside Drive, Jackson KY 41339. 606/666-2433. **Contact:** Human Resources. **Description:** Manufactures computer keyboards.

APPLE COMPUTER, INC.
One Infinite Loop, MS 75-2CE, Cupertino CA 95014. **Fax:** 408/862-7080. **Recorded Jobline:** 408/974-9500. **Contact:** Employment Department. **Description:** Develops, manufactures, and markets personal computer systems. The company's desktop publishing and communications products are marketed internationally. **NOTE:** The jobline number in Sacramento is 916/314-2675. **Common positions include:** Computer Engineer; Computer Programmer; Marketing Specialist; Software Engineer. **Educational backgrounds include:** Business Administration; Computer Science; Finance. **Special Programs:** Internships. **Corporate headquarters location:** This Location. **Listed on:** NASDAQ.
Other U.S. locations:
- 2425 East Camelback Road, Suite 1100, Phoenix AZ 85016. 602/957-7144. This location is a sales office.
- 2911 Laguna Boulevard, Elk Grove CA 95758. 916/394-2600. This location manufactures several different CPUs.
- 2420 Ridgepoint Drive, Austin TX 78754. 512/919-2000. This location offers sales and technical support to companies and educational institutions.

AROMA COMPUTERS
331 East 1300 South, Orem UT 84058. 801/224-9551. **Contact:** Human Resources. **Description:** Aroma Computers uses different manufacturers' components to create custom-built computers for business and home use. The company also resells various software titles.

ARTEK COMPUTER SYSTEMS
47709 Fremont Boulevard, Fremont CA 94538. 510/490-8402. **Contact:** Human Resources. **Description:** Manufactures, develops, and markets customized software packages and computers. Artek Computer Systems uses parts from many different manufactures to create personalized systems for businesses. This company also designs software around a company's particular needs.

ARTIST GRAPHICS
900 Long Lake Road, St. Paul MN 55112. 612/631-7800. **Contact:** Human Resources. **Description:** Manufactures video graphics hardware. Artist Graphics supplies such large companies as BlueCross/BlueShield with video graphics cards for their computers.

ASPECT COMPUTER CORPORATION
201 Circle Drive North, Suite 102-103, Piscataway NJ 08854. 908/563-1304. **Contact:** Human Resources. **Description:** A manufacturer of computers.

ASPEN SYSTEMS, INC.
4026 Youngfield Street, Wheat Ridge CO 80033-3862. **Fax:** 303/431-7196. **Contact:** Human Resources. **E-mail address:** seanm@aspsys.com. **World Wide Web address:** http://www.aspsys.com. **Description:** A manufacturer of high-performance workstations and servers for the OEM, VAR, and retail industries.

ASTRO-MED, INC.
600 East Greenwich Avenue, West Warwick RI 02893. 401/828-4000. **Contact:** Director of Personnel. **Description:** Astro-Med is a supplier of specialty printer solutions. The company's specialty printers are total systems which display, monitor, analyze, and print data for aerospace, industrial, and medical applications. The machines, computer electronics, software, and consumables are all developed and manufactured by Astro-Med. Customers include such leading aircraft manufacturers as Boeing, Lockheed, McDonnell Douglas, and British Aerospace; automotive product manufacturers such as Chrysler, Ford, General Motors, Mercedes Benz, and Renault; telecommunications companies including AT&T, NYNEX, and MCI; electrical utility companies including Northeast Utilities, Hydro Quebec, and Florida Power & Light; steel companies including USX, Posco, and Bethlehem; such aluminum manufacturers as Alcoa and Reynolds; and paper manufacturers including International Paper, Kimberly-Clark, and Mead. Grass Instruments, a division of Astro-Med, is also at this location. **Number of employees at this location:** 250.

ATLANTIC DESIGN
5601 Wilkinson Boulevard, Charlotte NC 28208. 704/394-6341. **Contact:** Human Resources. **Description:** Engaged in the manufacture of keyboards and internal computer components.

AVALON COMPUTER SYSTEMS, INC.
30 West Sola Street, Santa Barbara CA 93101-2526. 805/965-9559. **Contact:** Human Resources Department. **Description:** A manufacturer of supercomputers.

AVNET COMPUTER
10H Centennial Drive, Peabody MA 01960. 508/532-2015. **Contact:** Human Resources. **Description:** This location produces computer hardware. **Parent company:** Avnet, Inc. (New York NY) operates throughout North America and Europe, and is the one of the nation's largest distributors of electronic components and computer products for industrial and military customers. The company also produces and distributes other electronic, electrical, and video communications products. Sutsidiaries include Hamilton Hallmark and Hallmark computers.

AVNET, INC.
60 South McKemy Avenue, Chandler AZ 85226. 602/961-6400. **Contact:** Sara Reinstein, Human Resources. **Description:** This location manufactures and distributes computers and computer-related products. Overall, Avnet, Inc. operates throughout North America and Europe as one of the nation's largest distributors of electronic components and computer products for industrial and military customers. The company also produces and distributes other electronic, electrical, and video communications products. **NOTE:** Jobseekers interested in management-level positions should address inquiries to: Avnet, Inc., Human Resources, 10950 Washington Boulevard, Culver City CA 90232.

AXONIX CORPORATION
844 South 200 East, Salt Lake City UT 84111. 801/521-9797. **Contact:** Personnel. **Description:** Axonix Corporation manufactures peripherals for laptop computers. Operations include technical support, customer service, and sales.

AYDIN CORPORATION
700 Dresher Road, Horsham PA 19044. 215/657-7510. **Contact:** Human Resources. **Description:** Manufactures computers.

BARRON CHASE SECURITIES
3340 Peachtree Road NE, Tower Place, Suite 1770, Atlanta GA 30326. 404/264-1693. **Contact:** Human Resources. **Description:** Designs and manufactures high-performance file servers and workstations which provide video-on-demand applications and services. The company's TRIUMPH products consist of a line of file servers for use with standard network systems. **Corporate headquarters location:** Boca Raton FL.

BEST POWER
P.O. Box 280, Necedah WI 54646. 608/565-7200. **Fax:** 608/565-2221. **Contact:** Human Resources. **E-mail address:** info@bestpower.com. **World Wide Web address:** http://www.bestpower.com. **Description:** A designer, manufacturer, and marketer of power protection products for PCs, LANs, WANs, and global networks. Best Power also provides support services.

BITWISE DESIGNS, INC.
Technology Center, Rotterdam Industrial Park, Schenectady NY 12306. 518/356-9741. **Contact:** Debbie Lauer, Human Resources. **World Wide Web address:** http://www.docstar.com. **Description:** Bitwise designs, manufactures, and distributes a diverse line of compatible personal computers through three different product lines: DeskStation, Super Portables, and BitBook. The DeskStation line includes desktop or workstation computers that can be operated as stand alone stations or as part of a computer network. The Super Portables, which are Bitwise's signature products, are a line of high-power specialized portable computers with desktop power and capability. The BitBook product line consists of notebook and laptop computers. **Subsidiaries include:** Electrograph; SST.

BOCA RESEARCH
Clint Moore Road, Boca Raton FL 33487. 561/241-8088. **Contact:** Human Resources. **Description:** A manufacturer of computer components and peripherals including network cards, video cards, and modems. **Corporate headquarters location:** This Location.

BREECE HILL TECHNOLOGIES
6287 Arapahoe Avenue, Boulder CO 80303. 303/449-2673. **Toll-free phone:** 800/941-0550. **Contact:** Human Resources. **World Wide Web address:** http://www.breecehill.com. **Description:** A developer and manufacturer of Digital Linear Tape (DLT) drive systems. **Corporate headquarters location:** This Location. **Number of employees at this location:** 100.

BULL HN WORLDWIDE INFORMATION SYSTEMS
1001 Pawtucket Boulevard, Lowell MA 01854. 508/275-3400. **Contact:** Human Resources. **Description:** Manufactures computers.

CI DESIGN
1695 West MacArthur, Costa Mesa CA 92626. 714/556-0888. **Contact:** Human Resources. **Description:** Designs and manufactures computer hardware and peripherals including storage and disk drive support.

CTI ELECTRONICS CORPORATION
110 Old South Avenue, Stratford CT 06497. 203/386-9779. **Contact:** Peter Mikan, President. **Description:** Manufactures computer peripherals including keyboards, trackballs, and joysticks.

CABLENET
501 West John Street, Matthews NC 28105. 704/847-1210. **Contact:** Human Resources. **Description:** Manufactures internal and external hardware to interface with computer printers, primarily for use with IBM mainframes.

CABLES TO GO
1501 Webster Street, Dayton OH 45404. 937/224-8646. **Contact:** Tim Gibson, Human Resources Administrator. **Description:** Manufactures computer cable.

CADEC SYSTEMS INC.
8 Perimeter Road East, Londonderry NH 03053. 603/668-1010. **Contact:** Human Resources. **Description:** A manufacturer of on-board computer systems for the trucking industry.

CALCOMP
2411 West La Palma Avenue, Anaheim CA 92801. 714/821-2000. **Contact:** Human Resources Department. **Description:** Designs, manufactures, sells, and services computer graphics hardware. **Common positions include:** Accountant/Auditor; Buyer; Computer Programmer; Credit Manager; Designer; Draftsperson; Electrical/Electronics Engineer; Financial Analyst; Industrial Engineer; Manufacturer's/Wholesaler's Sales Rep.; Marketing Specialist; Mechanical Engineer; Public Relations Specialist; Purchasing Agent/Manager; Quality Assurance Engineer; Safety Engineer; Software Engineer; Systems Analyst; Technical Writer/Editor; Transportation/Traffic Specialist. **Educational backgrounds include:** Accounting; Business Administration; Communications; Computer Science; Economics; Engineering; Finance; Liberal Arts; Marketing. **Benefits:** 401(k); Dental Insurance; Employee Discounts; Life Insurance; Medical Insurance; Pension Plan; Savings Plan; Tuition Assistance. **Corporate headquarters location:** This Location. **Parent company:** Lockheed Martin Corporation. **Operations at this facility include:** Administration; Divisional Headquarters; Manufacturing; Regional Headquarters; Research and Development; Sales; Service. **Listed on:** New York Stock Exchange. **Number of employees at this location:** 500. **Number of employees nationwide:** 800.

CALCOMP/DIGITIZER PRODUCTS DIVISION
14555 North 82nd Street, Scottsdale AZ 85260. 602/948-6540. **Contact:** Personnel Director. **Description:** Manufactures computer peripheral equipment, including digitizers. Nationally, Calcomp is a producer of computer graphic equipment. Customers include industrial and institutional firms using data processing equipment with a graphic output. **Corporate headquarters location:** Anaheim CA. **Parent company:** Sanders Association.

CAMBEX CORPORATION
360 2nd Avenue, Waltham MA 02154. 617/890-6000. **Fax:** 617/890-2899. **Contact:** Human Resources. **Description:** Cambex Corporation develops, manufactures, and markets a variety of direct access storage products that improve the performance of large and midsize IBM computers. These products include central and expanded memory, controller cache memory, disk array systems, disk and tape subsystems, and related software products. The products are used to enhance IBM System/3090, ES/9000, and RS/6000 computer systems. **Common positions include:** Branch Manager; Buyer; Computer Programmer; Customer Service Representative; Electrical/Electronics Engineer; Human Resources Manager; Manufacturer's/Wholesaler's Sales Rep.; Marketing Specialist; Operations/Production Manager; Quality Control Supervisor. **Educational backgrounds include:** Accounting; Business Administration; Communications; Computer Science; Engineering; Liberal Arts; Marketing. **Benefits:** Dental Insurance; Disability Coverage; Employee Discounts; Life Insurance; Medical Insurance; Profit Sharing; Savings Plan; Tuition

Assistance. **Corporate headquarters location:** This Location. **Other U.S. locations:** Scottsdale AZ; Thousand Oaks CA; Walnut Creek CA; Westport CT; Clearwater FL; Roswell GA; Schaumburg IL; Troy MI; Chesterfield MO; Charlotte NC; Clark NJ; Cincinnati OH; Blue Bell PA; Dallas TX; Reston VA. **Operations at this facility include:** Administration; Manufacturing; Research and Development; Sales; Service.

CHARLES RIVER DATA SYSTEMS
55 New York Avenue, Framingham MA 01701. 508/626-1000. **Contact:** Human Resources. **Description:** Charles River Data Systems develops and markets VME computers to manufacturers wishing to incorporate them into their proprietary systems.

CHASE RESEARCH INC.
545 Marriott Drive, Suite 100, Nashville TN 37214. 615/872-0770. **Contact:** Human Resources Department. **Description:** Chase Research is a manufacturer of computer hardware, including servers, I/O boards, print servers, and other data communications equipment.

CIRQUE CORPORATION
433 West Lawndale Drive, Salt Lake City UT 84115. 801/467-1100. **Contact:** Personnel. **Description:** Cirque Corporation manufactures the "glide point" which can be used instead of a mouse.

CITADEL COMPUTER CORPORATION
29 Armory Road, Milford NH 03055. 603/672-5500. **Contact:** Human Resources. **Description:** A manufacturer of durable computers for use in warehouses and on loading docks.

CITEL AMERICA, INC.
1111 Park Center Boulevard, Suite 340, Miami FL 33169. 305/621-0022. **Contact:** Human Resources Department. **Description:** Manufacturers surge protectors for computers.

CLEAR TO SEND ELECTRONICS
385 East Drive, Melbourne FL 32904. 407/728-0060. **Contact:** Human Resources. **Description:** Designs and manufactures data communications accessories, including modem eliminators.

COLAMCO INC.
975 Florida Central Parkway, Suite 1100, Longwood FL 32750. **Toll-free phone:** 800/327-2722. **Contact:** Angela Straiter, Office Manager. **Description:** Manufactures computer accessories including ribbons for printers and tape cartridges.

COLORADO MEMORY SYSTEMS, INC.
800 South Taft Avenue, Loveland CO 80537-6347. 970/669-8000. **Contact:** Human Resources. **Description:** Manufactures memory backup systems for computers. **Common positions include:** Computer Programmer; Electrical/Electronics Engineer; Financial Analyst; Hardware Engineer; Software Engineer. **Parent company:** Hewlett-Packard Company.

COMMAX TECHNOLOGIES, INC.
2031 Concourse Drive, San Jose CA 95131. 408/435-5000. **Toll-free phone:** 800/526-6629. **Fax:** 408/435-5020. **Contact:** Human Resources. **E-mail address:** sales@commax.com. **World Wide Web address:** http://www.commax.com. **Description:** A manufacturer of Pentium notebook computers. Product line includes V-star and V-star Plus notebook computers. **Corporate headquarters location:** This Location.

COMMUNICATIONS TECHNOLOGY CORPORATION (CTC)
9668 Highway 20 West, Madison AL 35758. 205/464-9211. **Contact:** Human Resources. **Description:** This location houses the corporate offices for CTC, a data communications testing equipment manufacturer. **Corporate headquarters location:** This Location.

COMPAQ COMPUTER CORPORATION
P.O. Box 692000, Mail Code 510106, Houston TX 77269-2000. 713/370-0670. **Contact:** Human

Resources Department. **E-mail address:** careerpaq@compaq.com. **World Wide Web address:** http://www.compaq.com. **Description:** Compaq Computer Corporation is a manufacturer of electronic computer systems and peripherals, including laptop and desktop personal computers, PC-based client servers, notebook computers, and tower systems. Compaq Computer Corporation sells its products through retail stores, warehouses, and resellers, as well as mail-order catalogs and telemarketers. The company conducts business in more than 90 countries. **Corporate headquarters location:** This Location.

COMPULINK
1205 Gandy Boulevard North, St. Petersburg FL 33702. 813/579-1500. **Contact:** Human Resources Department. **Description:** Compulink is a manufacturer of computer cable.

COMPUTER DEVICES INC.
34 Linnell Circle, Nutting Lake MA 01865. 508/663-4980. **Contact:** Personnel Manager. **Description:** Computer Devices Inc. designs, manufactures, sells, and rents portable and desktop thermal printing computer terminals and portable computers, serving the data communications needs of remote computer and minicomputer users. Customers of Computer Devices Inc. include users of time-shared computers, vendors of time-sharing services, and a wide range of individual users. **Common positions include:** Accountant/Auditor; Administrator; Branch Manager; Buyer; Computer Programmer; Credit Manager; Customer Service Representative; Department Manager; Draftsperson; Electrical/Electronics Engineer; Financial Analyst; Human Resources Manager; Industrial Engineer; Manufacturer's/Wholesaler's Sales Rep.; Mechanical Engineer; Operations/Production Manager; Purchasing Agent/Manager; Quality Control Supervisor; Systems Analyst. **Educational backgrounds include:** Accounting; Business Administration; Computer Science; Engineering; Finance; Marketing. **Benefits:** Dental Insurance; Disability Coverage; Life Insurance; Medical Insurance; Savings Plan; Tuition Assistance. **Corporate headquarters location:** This Location. **Operations at this facility include:** Administration; Manufacturing; Research and Development; Sales; Service.

COMPUTER PRODUCTS
Seven Elkins Street, South Boston MA 02127. 617/268-1170. **Fax:** 617/464-6679. **Contact:** Human Resources Department. **Description:** Computer Products is an international manufacturer of computers and computer-related products. **Corporate headquarters location:** This Location.

COMPUTER TECHNOLOGY CORPORATION
50 West Technecenter Drive, Milford OH 45150. 513/831-2340. **Contact:** Human Resources. **Description:** A manufacturer of computer hardware and software for factory floor operator interface.

COMPUTONE CORPORATION
1100 Northmeadow Parkway, Suite 150, Roswell GA 30076. 770/475-2725. **Contact:** Greg Alba, Manager of Human Resources. **Description:** Computone Corporation designs, manufactures, and markets hardware and software connectivity products for personal computers, servers, and workstations. The company's principal products are software-driven multi-port communications adapters that manage the flow of data between serial devices (e.g., terminals, printers, and modems) and the central processing unit of the host computer. The majority of Computone Corporation's multi-port systems are intelligent devices that incorporate on-board processors, memory chips, and related circuitry. These multi-port products are available for IBM personal computers and a number of industry-standard operating systems including UNIX, Xenix, IBM AIX, IBM OS/2, DOS, and Multi-user DOS. These products are offered under the trademarks IntelliPort, ValuePort, and IntelliCluster. Computone Corporation's multi-port communications server for Ethernet local area networks is IntelliServer. **Common positions include:** Buyer; Credit Manager; Customer Service Representative; Hardware Engineer; Manufacturer's/Wholesaler's Sales Rep.; Marketing Specialist; Operations/Production Manager; Quality Control Supervisor; Software Engineer; Systems Analyst; Technical Writer/Editor. **Benefits:** Disability Coverage; Life Insurance; Tuition Assistance. **Special Programs:** Internships. **Corporate headquarters location:** This Location. **Operations at this facility include:** Administration; Manufacturing; Research and Development; Sales; Service. **Number of employees at this location:** 70.

CONNECTICUT TECHNICAL CORPORATION
P.O. Box 4607, Stamford CT 06907. 203/964-1456. **Contact:** Human Resources Department. **Description:** Manufactures computer harnesses and cables.

CONTROL TECHNOLOGY, INC.
P.O. Box 59003, Knoxville TN 37950. 423/584-0440. **Contact:** Human Resources. **Description:** Manufactures back-up support units for programmable controllers.

CORNERSTONE IMAGING, INC.
1710 Fortune Drive, San Jose CA 95131. 408/435-8900. **Contact:** Human Resources. **Description:** Cornerstone Imaging, Inc. manufactures image display products for large-scale production document imaging processing systems. The company's products, which are based on proprietary silicon and software, improve the scanning and display of digitally stored images. **Corporate headquarters location:** This Location. **Number of employees at this location:** 98.

CRADEN PERIPHERALS
7860 Airport Highway, Pennsauken NJ 08109. 609/488-0700. **Contact:** Human Resources. **Description:** Manufactures and markets Craden printers.

CYBERCHRON CORPORATION
P.O. Box 160, US Route 9, Cold Spring NY 10516. 914/265-3700. **Contact:** Human Resources. **Description:** Manufactures ruggedized computers. Cyberchron produces computers that are made to withstand temperature and other environmental extremes. The military uses Cyberchron's products.

CYBERNETICS
111 Cybernetics Way, Yorktown VA 23693. 757/833-9000. **Contact:** Department of Interest. **Description:** Manufactures computer backup drives and other similar computer peripherals.

CYBERTECH INC.
935 Horsham Road, Horsham PA 19044. 215/957-6220. **Contact:** Human Resources. **Description:** Manufactures specialty printers. Cybertech designs and produces KIOFK printers that can be custom designed for a company's particular needs.

CYBEX COMPUTER PRODUCTS CORPORATION
4912 Research Drive, Huntsville AL 35805. 205/430-4000. **Toll-free phone:** 800/793-3758. **Fax:** 205/430-4030. **Contact:** Human Resources. **World Wide Web address:** http://www.cybex.com. **Description:** A developer of switching and transmission products for computer configurations. Products include keyboard/mouse/monitor switching systems, peripheral extension and expansion products, and phone-activated power control systems for PC, Macintosh, and Sun computer platforms.

DPC TECHNOLOGIES
207 Perry Parkway, Gaithersburg MD 20877. 301/330-9664. **Contact:** Human Resources. **Description:** A manufacturer of computer hardware and software.

DAISY DATA INC.
333 South Enola Drive, Enola PA 17025-2897. 717/732-8800. **Contact:** Human Resources. **Description:** Manufactures computers which are ruggedized for industrial environments.

DATA NET CORPORATION
10112 USA Today Way, Miramar FL 33025. 954/437-3535. **Contact:** Human Resources. **Description:** Manufactures data collection systems. Data Net provides companies with both the hardware and software for data collection.

DATA PAD CORPORATION
115 South State Street, Linden UT 84042. 801/225-0699. **Fax:** 801/785-0339. **Contact:** Human Resources. **Description:** Manufactures custom mouse pads. **Parent company:** Giraffics. **Number of employees at this location:** 90.

DATA RAY CORPORATION
P.O. Box 33637, Northglenn CO 80233. 303/451-1300. **Contact:** Human Resources. **Description:** A manufacturer of high-resolution CRT monitors.

DATA SET CABLE COMPANY
722 Danbury Road, Ridgefield CT 06877. 203/438-9684. **Contact:** Human Resources. **Description:** A manufacturer of custom computer cable assemblies.

DATA TECHNOLOGY CORPORATION (DTC)
1515 Centre Pointe Drive, Milpitas CA 95035. 408/942-4000. **Contact:** Page Frechette, Manager of Staffing. **Description:** Develops and manufactures computer peripherals, including printers, disk drives, terminals, controllers, and supplies. Products are marketed to both original equipment manufacturers and distributors. **Common positions include:** Computer Programmer; Editor; Electrical/Electronics Engineer; Systems Analyst; Technical Writer/Editor. **Educational backgrounds include:** Computer Science; Engineering. **Benefits:** Dental Insurance; Disability Coverage; Employee Discounts; Life Insurance; Medical Insurance; Profit Sharing; Savings Plan; Tuition Assistance; Vision Insurance. **Corporate headquarters location:** This Location. **Listed on:** New York Stock Exchange.

DATA TRANSLATION
100 Locke Drive, Marlborough MA 01752-1192. 508/481-3700. **Contact:** Human Resources. **Description:** Data Translation designs, develops, and manufactures high-performance digital media, data acquisition, and imaging products. The company's principal products are digital signal processing boards and software, which use personal computers to receive analog signals, convert them to digital form, and process the digital data. The company currently sells products through the following three business groups: digital media, data acquisition and imaging, and network distribution. One product, Media 100, enables video producers to produce broadcast quality videos on a Macintosh computer. Storing images and sounds on hard disk rather than videotape allows the user to access any scene or sound in real time. The company targets the corporate and institutional market, which includes a growing market of new users in addition to users of video production equipment.

DATACOM TECHNOLOGIES INC.
11001 31st Place West, Everett WA 98204. 206/355-0590. **Contact:** Human Resources. **Description:** Manufactures modems, test equipment, and other peripherals.

DATACUBE, INC.
300 Rosewood Drive, Danvers MA 01923. 508/777-4200. **Fax:** 508/777-3117. **Contact:** Pat Wonson, Human Resources Director. **E-mail address:** pat@datacube.com. **World Wide Web address:** http://www.datacube.com. **Description:** A manufacturer of board- and system-level hardware for image processing. **NOTE:** The company offers entry-level positions. **Common positions include:** Applications Engineer; Design Engineer; Sales Engineer; Software Engineer; Technical Writer/Editor. **Educational backgrounds include:** Engineering; Mathematics. **Benefits:** 401(k); Dental Insurance; Disability Coverage; Life Insurance; Medical Insurance; Tuition Assistance. **Corporate headquarters location:** This Location. **Listed on:** Privately held. **Annual sales/revenues:** $11 - $20 million. **Number of employees at this location:** 140. **Number of projected hires for 1997 - 1998 at this location:** 15.

DATALUX CORPORATION
155 Aviation Drive, Winchester VA 22602. 540/662-1500. **Contact:** Human Resources. **Description:** A manufacturer of Datalux brand computers.

DATAMETRICS CORPORATION
21135 Erwin Street, Woodland Hills CA 91367. 818/598-6200. **Contact:** Human Resources. **Description:** Founded in 1962, Datametrics Corporation designs, develops, and manufactures high-performance, high-reliability, high-speed color printers; high-resolution non-impact printer/plotters; ruggedized computers and workstations; and data-entry keyboards. The company serves a broad customer base that includes the U.S. Department of Defense, virtually all major U.S.

prime defense contractors, international defense companies, and selected industrial and commercial original equipment manufacturers. **Corporate headquarters location:** This Location.

DATAPRODUCTS CORPORATION

1757 Tapo Canyon Road, Simi Valley CA 93063. 805/578-4000. **Fax:** 805/578-4001. **Contact:** Personnel. **Description:** Develops, manufactures, and markets data handling and output equipment. Products include printers and digital communications equipment. **Common positions include:** Electrical/Electronics Engineer; Mechanical Engineer. **Educational backgrounds include:** Computer Science; Engineering; Marketing. **Benefits:** Dental Insurance; Disability Coverage; Employee Discounts; Medical Insurance; Pension Plan; Tuition Assistance. **Corporate headquarters location:** This Location. **Operations at this facility include:** Administration; Manufacturing; Research and Development; Sales; Service. **Listed on:** American Stock Exchange.

DATARAM CORPORATION

P.O. Box 7528, Princeton NJ 08543-7528. 609/799-0071. **Contact:** Dawn Craft, Human Resources Administrator. **Description:** Dataram develops memory and storage products that improve the performance of computer systems. The company offers a broad range of memory and storage products for workstations, servers, and minicomputers. Dataram primarily serves HP, DEC, Sun, and IBM users in such industries as manufacturing, finance, government, telecommunications, utilities, research, and education. **Listed on:** American Stock Exchange. **Stock exchange symbol:** DTM.

DATASOUTH COMPUTER CORPORATION

4216 Stuart Andrew Boulevard, Charlotte NC 28217. 704/523-8500. **Toll-free phone:** 800/476-2120. **Fax:** 704/525-6104. **Contact:** Personnel Manager. **Description:** Datasouth Computer Corporation designs, manufactures, and markets heavy-duty dot matrix and thermal printers used for high-volume print applications. The company's three printer product lines include Performax, an alternative to the line printer for high-speed report printing; XL series, for medium-volume printing applications; and Documas, which was the company's first narrow carriage printer. The company also manufactures a portable thermal printer, Freeliner, which is used primarily for printing one packing or shipping label at a time. The company sells its products through a network of nearly 100 distributors worldwide and direct to high-volume major accounts primarily in the transportation and travel, health care, and manufacturing and distribution industries. **Corporate headquarters location:** This Location. **Parent company:** Bull Run Corporation (Atlanta GA). **Listed on:** NASDAQ. **Number of employees nationwide:** 125.

DATAWARE DEVELOPMENT

1100 Abernathy Road NE, Suite 625, Atlanta GA 30328. 770/551-8118. **Contact:** Human Resources Department. **Description:** Dataware Development is a provider of optical disk storage for IBM computers. Optical disks, also called laser disks, are disks on which data is stored in the form of text, music, or pictures and then read by a laser which scans the surface.

DAYNA COMMUNICATIONS, INC.

849 West Levoy Drive, Salt Lake City UT 84123-2544. 801/269-7200. **Contact:** Human Resources. **Description:** Manufactures networking hardware for Macintosh and IBM computers.

DECISION ONE

P.O. Box 21069, Tulsa OK 74121. 918/627-1111. **Contact:** Human Resources. **Description:** Formerly Memorex-Telex, Decision One is an international supplier of plug-compatible computer equipment and accessories. Products include disk and tape storage devices, terminals, intelligent workstations and systems, controllers, printers, airline reservation systems, and a comprehensive range of computer supplies. The company operates in 27 countries around the world with distributor links in 50 other markets. **NOTE:** Please send resumes to Human Resources, 50 East Swedesford Road, Fraser PA 19355. **Corporate headquarters location:** Fraser PA.

DELL COMPUTER CORPORATION

One Dell Way, Round Rock TX 78682. 512/338-4400. **Contact:** Human Resources. **Description:** Dell Computer Corporation designs, develops, manufactures, markets, services, and supports high-performance personal computer systems and related equipment, including servers, workstations, notebooks, and desktop systems. The company also offers over 4,000 software packages and peripherals. Services include installation, support, and software integration. Dell Computer Corporation has computer production facilities in Texas and Ireland.

DESKSTATION TECHNOLOGY
13256 West 98th Street, Lenexa KS 66215. 913/599-1900. **Fax:** 913/599-4024. **Contact:** Human Resources. **World Wide Web address:** http://www.dti.com. **Description:** A designer and manufacturer of computer workstations based on RISC processors. Product line includes Raptor Reflex and Raptor 3 workstations.

DIAGRAPH CORPORATION
3401 Rider Trail South, Earth City MO 63045. 314/739-1221. **Contact:** Human Resources. **Description:** A manufacturer of ink-jet printers.

DIAMOND MULTIMEDIA SYSTEMS
312 SE Stonemill Drive, Suite 150, Vancouver WA 98684. 360/604-1400. **Contact:** Human Resources. **Description:** A manufacturer of modems.

DIGITAL EQUIPMENT CORPORATION
111 Powdermill Road, Maynard MA 01754-1499. 508/493-6488. **Contact:** U.S. Staffing and Planning Office. **Description:** Digital Equipment Corporation designs, manufactures, sells, and services computers and associated peripheral equipment, and related software and supplies. Applications include scientific research, computation, communications, education, data analysis, industrial control, time sharing, commercial data processing, graphic arts, word processing, health care, instrumentation, engineering, and simulation programs. **Common positions include:** Accountant/Auditor; Administrator; Computer Programmer; Credit Manager; Financial Analyst; Human Resources Manager; Manufacturer's/Wholesaler's Sales Rep.; Systems Analyst. **Educational backgrounds include:** Accounting; Business Administration; Computer Science; Finance; Marketing. **Benefits:** Dental Insurance; Disability Coverage; Employee Discounts; Life Insurance; Medical Insurance; Pension Plan; Savings Plan; Tuition Assistance. **Corporate headquarters location:** This Location. **Operations at this facility include:** Administration; Regional Headquarters; Sales; Service. **Listed on:** New York Stock Exchange. **Number of employees at this location:** 300. **Number of employees worldwide:** 77,800.
Other U.S. locations:
- 2700 North Central Avenue, Suite 700, Phoenix AZ 85004. 602/240-3400. This location is a sales and service office.
- 24 Executive Park, Irvine CA 92614. 714/261-4300. This location is a sales office.
- 44 Montgomery Street, Suite 250, San Francisco CA 94104. 415/477-6200. This location is a sales office.
- 2575 Augustine Drive, Santa Clara CA 95054. 408/727-0200. This facility produces mini-computers.
- 8085 South Chester Street, Englewood CO 80112-1478. 303/649-3000. This location is a sales office.
- 671 East River Park Lane, Suite 120, Boise ID 83706. 208/342-1015. This location is a marketing office.
- 6406 Ivy Lane, Greenbelt MD 20770. 301/459-7900. This location is a sales and service office.
- 721 Emerson Road, St. Louis MO 63141. 314/991-6400. Contact: Gloria Clayter, Personnel Director. This location is a sales office.
- One Digital Drive, Merrimack NH 03054. 603/884-5111. This facility produces peripheral equipment.
- 9 Northeastern Boulevard, Salem NH 03079. 603/894-3111. This location is a manufacturing facility.
- 5951 Jefferson Street NE, Albuquerque NM 87109. 505/343-7000. This location manufactures and services all types of mini-computers, including personal computers, laptops, and small VAX mainframe systems.
- One Liberty Plaza, 27th Floor, New York NY 10006. 212/978-1700. This location functions as a regional sales office as well as offering training services.

- 10101 Alliance Road, Cincinnati OH 45242. 513/984-7500. This location serves as a regional sales and service center for its customers.
- 7995 SW Mohawk Drive, Tualitan OR 97062-9195. 503/691-0400. This location is an administrative and service office.
- 1500 Ardmore Boulevard, Pittsburgh PA 15221. 412/244-7200. This location is a sales office.
- 30 Patewood Drive, Greenville SC 29607. 864/675-5000. This location is a sales and service office.
- 5310 Harvest Hill Road, Suite 200, Dallas TX 75230. 972/702-4000. This location is a sales and service office.
- 2500 Citywest Boulevard, Houston TX 77042. 713/953-3500. This location of Digital functions as a regional sales office.
- 6985 Union Park Center, Suite 400, Midvale UT 84047. 801/565-3000. This location is a sales and service office.

DIGITAL EQUIPMENT CORPORATION

305 Rockrimmon Boulevard South, Colorado Springs CO 80919. 719/548-2000. **Contact:** Nancy Casados, Employment Manager. **Description:** This facility produces peripheral equipment. Overall, Digital Equipment Corporation designs, manufactures, sells, and services computers, associated peripheral equipment, and related software and supplies. Applications include scientific research, computation, communications, education, data analysis, industrial control, time-sharing, commercial data processing, graphic arts, word processing, health care, instrumentation, engineering, and simulation programs. **Common positions include:** Systems Manager. **Educational backgrounds include:** Computer Science; Engineering; Mathematics. **Corporate headquarters location:** Maynard MA. **Operations at this facility include:** Sales. **Listed on:** New York Stock Exchange. **Number of employees nationwide:** 77,800.

DIGITAL PRODUCTS INC.

411 Waverley Oaks Road, Suite 227, Waltham MA 02154. 617/647-1234. **Contact:** Human Resources. **Description:** Digital Products manufactures peripheral sharing devices for the computer industry.

DIGITAL SYSTEMS CORPORATION

P.O. Box 158, Walkersville MD 21793-0158. 301/845-4141. **Fax:** 301/898-3331. **Contact:** Personnel Administrator. **World Wide Web address:** http://www.galaxysys.com. **Description:** Manufactures access control systems and security systems. The company also provides software support for the U.S. Navy. **Common positions include:** Administrative Assistant; Buyer; Computer Programmer; Design Engineer; Electrical/Electronics Engineer; General Manager; Internet Services Manager; Marketing Manager; MIS Specialist; Sales Engineer; Software Engineer; Systems Analyst; Systems Manager. **Educational backgrounds include:** Engineering; Marketing. **Benefits:** 401(k); Life Insurance; Medical Insurance. **Corporate headquarters location:** This Location. **Other U.S. locations:** Lexington Park MD. **Listed on:** Privately held. **Annual sales/revenues:** Less than $5 million. **Number of employees at this location:** 30.

DISC MANUFACTURING

4905 Moores Mill Road, Huntsville AL 35811. 205/859-9042. **Contact:** Human Resources. **Description:** Manufactures CD-ROM drives and computer CDs on which companies can store information.

DISC MANUFACTURING

1409 Faulk Road, Suite 102, Wilmington DE 19803. 302/479-2500. **Contact:** Human Resources. **Description:** This location is a sales and marketing office for CD-ROM drives and computer CDs manufactured at other locations.

DISPLAY TECHNOLOGIES

1001 North Francis Street, Carthage MO 64836. 417/358-0999. **Contact:** Ms. Jerry Reppond, Personnel Manager. **Description:** Assembles visual display monitors for computers.

DIVERSIFIED TECHNOLOGY, INC.
P.O. Box 748, Ridgeland MS 39158. 601/856-4121. **Contact:** Human Resources. **Description:** Manufactures computers that are used in a wide array of industrial applications.

DOMINO AMJET INC.
1290 Lakeside Drive, Gurnee IL 60031. 847/244-2501. **Contact:** Human Resources. **Description:** A manufacturer of inkjet and laser printers.

DURACOM COMPUTER SYSTEMS
2115 East Beltline Drive, Carrollton TX 75038. 972/416-7600. **Contact:** Human Resources. **Description:** Manufactures and sells computers. The company was founded in 1989.

EMC CORPORATION
P.O. Box 9103, Hopkinton MA 01748-9103. 508/435-1000. **Contact:** Human Resources. **Description:** EMC designs, manufactures, markets, and supports high-performance storage products. The company also provides related services for selected mainframe and midrange computer systems primarily manufactured by IBM and Unisys. EMC sells its products through a direct sales force, distributors, and original equipment manufacturers. **Corporate headquarters location:** This Location. **Listed on:** New York Stock Exchange. **Number of employees at this location:** 1,500.
Other U.S. locations:
- Nine Parkway Center, Suite 120, Pittsburgh PA 15220. 412/922-5222. This location is a sales office.

EPS TECHNOLOGIES, INC.
10069 Dakota Avenue, Box 278, Jefferson SD 57038. 605/966-5586. **Contact:** Human Resources. **Description:** A computer manufacturer and technical support provider.

ESE, INC.
P.O. Box 1107, Marshfield WI 54449. 715/387-4778. **Contact:** Human Resources. **Description:** A manufacturer of PCs. **NOTE:** The physical address is 3600 Downwind Drive, Marshfield WI.

ELECTRO STANDARDS LAB INC.
36 Western Industrial Drive, Cranston RI 02921. 401/943-1164. **Contact:** Human Resources. **Description:** A manufacturer of computer cables.

ELO-TOUCH SYSTEMS
105 Randolph Road, Oak Ridge TN 37830. 423/482-4100. **Contact:** Human Resources. **Description:** Designs and manufactures computer touch screen systems.

EMULEX CORPORATION
3535 Harbor Boulevard, Costa Mesa CA 92626. 714/662-5600. **Contact:** Sadie Herrera, Director of Human Resources. **Description:** Emulex Corporation specializes in intelligent interface technology for the computer industry. Emulex designs, manufactures, and markets data storage and network connectivity products, as well as advanced integrated circuits. These products are compatible with a variety of different systems.

ENCORE COMPUTER CORPORATION
6901 West Sunrise Boulevard, MS-111, Plantation FL 33313. 954/587-2900. **Contact:** Hope Lynch, Human Resources Director. **Description:** Specializes in minicomputers for aerospace, defense, simulation, energy, and information systems. **Common positions include:** Electrical/ Electronics Engineer; Software Engineer; Systems Analyst; Technical Writer/Editor. **Educational backgrounds include:** Computer Science; Engineering. **Benefits:** 401(k); Dental Insurance; Disability Coverage; Life Insurance; Medical Insurance; Savings Plan; Tuition Assistance. **Special Programs:** Internships. **Corporate headquarters location:** This Location. **Operations at this facility include:** Administration; Research and Development; Service. **Listed on:** NASDAQ. **Number of employees at this location:** 450. **Number of employees nationwide:** 720.
Other U.S. locations:
- 100 North Babcock Street, Melbourne FL 32935. 407/242-0337.

ENERTEC, INC.

811 West 5th Street, Lansdale PA 19446. 215/362-0966. **Fax:** 215/362-2404. **Contact:** Human Resources. **E-mail address:** info@enertec.com. **World Wide Web address:** http://www.enertec.com. **Description:** A manufacturer of SNIFFER, computerized continuous emissions monitoring (CEM) systems, used to test for state and EPA compliance for such pollutant emitting sources as incinerators, boilers, turbines, and co-generation plants.

EPSON PORTLAND INC.

3950 NW Aloclek Place, Hillsboro OR 97124-7199. 503/645-1118. **Toll-free phone:** 800/338-2349. **Recorded Jobline:** 503/645-1118x7002. **Contact:** Human Resources. **Description:** A manufacturer of computer printers. **Common positions include:** Production Worker. **Corporate headquarters location:** Torrance CA. **Parent company:** Epson Corporation. **Number of employees at this location:** 1,400.

EVEREX SYSTEMS INC.

5020 Brandin Court, Fremont CA 94538. 510/498-1111. **Fax:** 510/683-2099. **Contact:** Human Resources. **Description:** Manufactures computer peripheral equipment. **Benefits:** 401(k); Dental Insurance; Disability Coverage; Life Insurance; Medical Insurance. **Corporate headquarters location:** This Location. **Operations at this facility include:** Administration; Manufacturing; Sales; Service. **Listed on:** Privately held. **Number of employees at this location:** 100. **Number of employees nationwide:** 110.

EXABYTE CORPORATION

1685 38th Street, Boulder CO 80301. 303/442-4333. **Contact:** Human Resources. **Description:** Exabyte Corporation designs, manufactures, and markets cartridge tape subsystems for data storage applications. The company's products are used in a broad spectrum of computer systems based on 8mm helical scan, 4mm helical scan, and quarter-inch technologies. Its products are used in various computer systems ranging from personal computers to supercomputers. A large majority of its units are used with workstations, network file servers, and minicomputers. The capacity of these subsystems allows users to back up multiple disk drives, and in many cases, an entire computer system on a single tape cartridge.

EXTENDED SYSTEMS, INC.

P.O. Box 4937, Boise ID 83711. 208/322-7575. **Fax:** 208/327-5011. **Contact:** Stephen M. Shaffer, Human Resources Manager. **Description:** Designs, manufactures, and markets computer enhancement products, primarily printer sharing devices and network print servers. **Common positions include:** Accountant/Auditor; Computer Programmer; Electrical/Electronics Engineer; Financial Analyst; Human Resources Manager; Purchasing Agent/Manager; Software Engineer; Systems Analyst. **Educational backgrounds include:** Accounting; Business Administration; Computer Science; Engineering; Finance; Marketing. **Benefits:** 401(k); Dental Insurance; Disability Coverage; Life Insurance; Medical Insurance. **Special Programs:** Internships. **Corporate headquarters location:** This Location. **Other U.S. locations:** Bozeman MT. **Operations at this facility include:** Administration; Manufacturing; Research and Development; Sales; Service. **Listed on:** Privately held. **Number of employees at this location:** 150. **Number of employees nationwide:** 200.

FARGO ELECTRONICS, INC.

7901 Flying Cloud Drive, Eden Prairie MN 55344. 612/941-9470. **Fax:** 612/941-7836. **Contact:** Human Resources. **World Wide Web address:** http://www.fargo.com. **Description:** A developer and manufacturer of dye-sublimation and wax thermal transfer color page and card printers. **Corporate headquarters location:** This Location.

FASTCOMM COMMUNICATIONS CORPORATION

45472 Holiday Drive, Sterling VA 20166. 703/318-7750. **Contact:** Human Resources Department. **Description:** Fastcomm Communications specializes in the manufacture of modems.

FIELDWORKS, INC.

9961 Valley View Road, Eden Prairie MN 55344. 612/947-0856. **Contact:** Human Resources. **Description:** A manufacturer of ruggedized notebook computers.

FILETEK, INC.
9400 Key West Avenue, Rockville MD 20850. 301/251-0600. **Fax:** 301/251-1991. **Contact:** Debbie Mobley, Manager, Human Resources. **E-mail address:** djm@filetek.com. **World Wide Web address:** http://www.filetek.com. **Description:** Designs, develops, and manufactures mass storage systems. Founded in 1984. **Common positions include:** Accountant/Auditor; Financial Analyst; Sales Representative; Software Engineer. **Educational backgrounds include:** Computer Science; Engineering; Finance. **Benefits:** 401(k); Dental Insurance; Disability Coverage; Life Insurance; Mass Transit Available; Medical Insurance; Spending Account. **Corporate headquarters location:** This Location. **Listed on:** Privately held. **Number of employees at this location:** 100.

FORCE COMPUTERS, INC.
2001 Logic Drive, San Jose CA 95124. 408/369-6000. **Fax:** 408/371-7638. **Contact:** Sherry Maguire, Employment Representative. **World Wide Web address:** http://www. forcecomputers.com. **Description:** Manufactures embedded systems. The company supplies high-performance computer products to a broad range of telecommunication, industrial, and government customers worldwide. **Common positions include:** Buyer; Customer Service Representative; Electrical/Electronics Engineer; Financial Analyst; Hardware Engineer; Human Resources Manager; Manufacturing Engineer; MIS Specialist; Public Relations Specialist; Quality Control Supervisor; Sales Engineer; Software Engineer; Systems Manager; Technical Writer/Editor; Webmaster. **Educational backgrounds include:** Computer Science; Engineering; Marketing; Public Relations. **Benefits:** 401(k); Dental Insurance; Disability Coverage; Life Insurance; Medical Insurance; Tuition Assistance. **Corporate headquarters location:** This Location. **International locations:** Germany. **Operations at this facility include:** Divisional Headquarters; Regional Headquarters. **Listed on:** Privately held. **Number of employees at this location:** 110. **Number of employees nationwide:** 140. **Number of projected hires for 1997 - 1998 at this location:** 40.

FORMATION, INC.
121 Whittendale Drive, Moorestown NJ 08057. 609/234-5020. **Fax:** 609/234-8543. **Contact:** Kathy Cava, Manager of Human Resources. **Description:** Designs and manufactures communications products and real-time, high-performance storage and retrieval systems. The company's products are capable of integrating a number of inputs including video, audio, data/text, and radar, and can employ a variety of communications protocols. The company is currently supplying an open systems storage system using Redundant Array of Independent Disks (RAID) technology in connection with a major Internal Revenue Service contract. Formation also supplies plug compatible data storage systems for IBM AS/400 computers as well as data storage systems to open systems computer manufacturers and systems integrators. **Common positions include:** Computer Engineer; Electrical/Electronics Engineer; Mechanical Engineer; Software Engineer. **Educational backgrounds include:** Engineering. **Benefits:** Dental Insurance; Disability Coverage; Flextime Plan; Investment Plan; Life Insurance; Medical Insurance; Profit Sharing; Savings Plan; Tuition Assistance. **Corporate headquarters location:** This Location. **Operations at this facility include:** Administration; Manufacturing; Regional Headquarters; Research and Development; Sales; Service. **Number of employees at this location:** 150.

FOUNTAIN TECHNOLOGIES, INC.
50 Randolph Road, Somerset NJ 08873. 908/563-4800. **Fax:** 908/764-5687. **Contact:** Robin Kolodiy, Corporate Recruiter. **Description:** A manufacturer of IBM compatible computers. Founded in 1984. **NOTE:** The company offers entry-level positions. **Common positions include:** Account Representative; Accountant/Auditor; Computer Operator; Computer Programmer; Customer Service Representative; Database Manager; Electrical/Electronics Engineer; Graphic Artist; Industrial Engineer; Internet Services Manager; Marketing Specialist; Mechanical Engineer; MIS Specialist; Multimedia Designer; Online Content Specialist; Operations/Production Manager; Production Manager; Public Relations Specialist; Purchasing Agent/Manager; Quality Control Supervisor; Sales Executive; Sales Representative; Software Engineer; Systems Analyst; Technical Writer/Editor; Webmaster. **Educational backgrounds include:** Accounting; Art/Design; Business Administration; Communications; Computer Science; Engineering; Finance; Liberal Arts; Marketing; Public Relations. **Benefits:** Dental Insurance; Disability Coverage; Employee Discounts; Life Insurance; Medical Insurance. **Corporate headquarters location:** This Location.

Listed on: Privately held. **Annual sales/revenues:** More than $100 million. **Number of employees at this location:** 1,000.

FUJITSU COMPUTER PRODUCTS
2904 Orchard Parkway, San Jose CA 95134. 408/432-6333. **Contact:** Human Resources. **Description:** This location is a regional sales office. Fujitsu Computer Products manufactures printer drives and scanners.

FULLERTON COMPUTER INDUSTRIES
810 South Church Street, Winterville NC 28590. 919/355-3443. **Contact:** Human Resources. **Description:** A manufacturer of computer disk drives.

GATEWAY 2000
610 Gateway Drive, P.O. Box 2000, North Sioux City SD 57049. 605/232-2222. **Contact:** Human Resources. **Description:** A computer manufacturer whose business is generated primarily through mail order. **Common positions include:** Computer Support Technician; Customer Service Representative; Electrical/Electronics Engineer; Manufacturer's/Wholesaler's Sales Rep. **Educational backgrounds include:** Business Administration; Computer Science; Liberal Arts; Marketing; Technology. **Benefits:** Daycare Assistance; Dental Insurance; Disability Coverage; Employee Discounts; Life Insurance; Medical Insurance; Pension Plan; Profit Sharing; Tuition Assistance. **Corporate headquarters location:** This Location. **Other U.S. locations:** Sioux Falls SD. **Operations at this facility include:** Administration; Divisional Headquarters; Manufacturing; Regional Headquarters; Research and Development; Sales; Service. **Listed on:** NASDAQ. **Number of employees at this location:** 3,000. **Number of employees nationwide:** 3,800.

GEAC FASFAX
175 Ledge Street, Nashua NH 03060. 603/889-5152. **Contact:** Human Resources. **Description:** This location manufactures computer terminals.

GENICOM CORPORATION
One Solutions Way, Waynesboro VA 22980. 540/949-1000. **Contact:** Dick Gooch, Manager of Human Resources. **Description:** This location of GENICOM Corporation manufactures printers and relays. Overall, GENICOM is an international supplier of network system, service, and printer solutions. Through its Harris Adacom Network Service Solutions business unit, GENICOM provides integrated network system solutions including a broad portfolio of hardware and software products, technology consulting, and diagnostic and monitoring services. The company's Enterprising Service Solutions business unit provides logo and multivendor product field support and depot repair, express parts, and professional services. GENICOM's Impact and Laser Printing Solutions business unit designs, manufactures, and markets a wide range of computer printer technologies for general purpose applications. **Common positions include:** Accountant/Auditor; Buyer; Computer Programmer; Electrical/Electronics Engineer; Financial Analyst; Industrial Engineer; Marketing Specialist; Mechanical Engineer; Software Engineer. **Educational backgrounds include:** Accounting; Business Administration; Computer Science; Engineering; Finance; Physics. **Benefits:** 401(k); Dental Insurance; Disability Coverage; Life Insurance; Medical Insurance; Savings Plan; Tuition Assistance. **Corporate headquarters location:** Chantilly VA. **Other U.S. locations:** Nationwide. **Subsidiaries include:** GENICOM Pty. Ltd. (Australia); GENICOM Canada, Inc. (Canada); GENICOM S.A. (France); GENICOM GmbH (Germany); GENICOM SpA (Italy); GENICOM Limited (UK); and Harris Adacom Inc. (Canada). **Operations at this facility include:** Administration; Manufacturing; Research and Development; Service. **Listed on:** NASDAQ. **Number of employees nationwide:** 2,147.

GENROCO, INC.
255 Info Highway, Slinger WI 53086. 414/644-8700. **Fax:** 414/644-6667. **Contact:** Human Resources. **E-mail address:** emp@genroco.com. **World Wide Web address:** http://www.genroco.com. **Description:** A manufacturer of high-performance storage and network controllers for PCI- and SBus-based computer platforms. Products include the TURBOstor line of fast wide differential and Ultra SCSI, HiPPI networking interfaces, and digital video broadcast modules (DVB). **Common positions include:** Software Engineer. **Educational backgrounds include:** Computer Science.

HDS NETWORK SYSTEMS
400 Feheley Drive, King of Prussia PA 19406. 610/277-8300. **Contact:** Human Resources. **World Wide Web address:** http://www.hds.com. **Description:** Designs, manufactures, and markets a family of desktop computing devices, including multimedia capable X Window stations. The X Window System is an industry standard networking and communications protocol. X lets users interact with different computers by using a mouse, windows, familiar icons, and graphics. Since the X Window System is designed to operate across a network, computer users can access information from virtually any type of computer, running virtually any operating system. This allows computer users the flexibility to build environments consisting of different computers from a variety of manufacturers. HDS X stations compete with personal computers and workstations in the business environment.

HMT TECHNOLOGY
1055 Page Avenue, Fremont CA 94538. 510/490-3100. **Recorded Jobline:** 510/490-3000. **Contact:** Human Resources. **Description:** Manufactures hard disk drives. **Common positions include:** Accountant/Auditor; Computer Programmer; Designer; Electrical/Electronics Engineer; Industrial Engineer; Mechanical Engineer; Science Technologist; Software Engineer; Systems Analyst. **Educational backgrounds include:** Computer Science; Engineering. **Benefits:** 401(k); Dental Insurance; Disability Coverage; Employee Discounts; Life Insurance; Medical Insurance; Tuition Assistance. **Special Programs:** Internships. **Corporate headquarters location:** This Location. **Parent company:** Hitachi Metals. **Operations at this facility include:** Administration; Manufacturing; Research and Development; Sales; Service. **Number of employees at this location:** 500.

HMW ENTERPRISES INC.
207 North Franklin Street, Waynesboro PA 17268. 717/765-4690. **Contact:** Human Resources. **Description:** Manufactures industrial computers.

HAMMOND COMPUTER, INC.
NETWORK CENTER
70 East 3750 South, Salt Lake City UT 84115. 801/265-8500. **Contact:** Human Resources. **Description:** A manufacturer of computer hardware to house components.

HAYES MICROCOMPUTER PRODUCTS
5835 Peach Street Corner East, Atlanta GA 30092. 770/840-9200. **Contact:** Human Resources. **Description:** A manufacturer of modems.

HAZLOW ELECTRONICS INC.
49 Saint Bridgets Drive, Rochester NY 14605. 716/325-5323. **Contact:** Human Resources. **Description:** Manufactures cables for computers and other electronic devices.

HEURIKON CORPORATION
8310 Excelsior Drive, Madison WI 53717. 608/831-5500. **Contact:** Human Resources. **E-mail address:** hr@heurikon.com. **World Wide Web address:** http://www.heurikon.com. **Description:** A manufacturer of VMEbus CPUs and I/O boards and systems. **Common positions include:** Software Engineer.

HEWLETT-PACKARD COMPANY
EMPLOYMENT RESPONSE CENTER
Event #2498, Mail Stop 20-APP, 3000 Hanover Street, Palo Alto CA 94304-1181. 415/852-8473. **Fax:** 415/852-8138. **Contact:** Recruiter. **E-mail address:** hpresume@corp.hp.com. **World Wide Web address:** http://www.hp.com. **Description:** Hewlett-Packard designs and manufactures measurement and computation products and systems used in business, industry, engineering, science, health care, and education. Principal products are integrated instrument and computer systems (including hardware and software); computer systems and peripheral products; and medical electronic equipment and systems. **NOTE:** This location accepts collect calls. **Educational backgrounds include:** Accounting; Business Administration; Chemistry; Engineering; Finance; Marketing; Mathematics; Physics. **Benefits:** Dental Insurance; Disability Coverage; Employee Discounts; Life Insurance; Medical Insurance; Pension Plan; Profit Sharing;

Savings Plan; Tuition Assistance. **Special Programs:** Internships. **Corporate headquarters location:** This Location. **Operations at this facility include:** Administration; Manufacturing; Research and Development; Service. **Listed on:** New York Stock Exchange.
Other U.S. locations:

- 8000 Foothills Boulevard, Roseville CA 95747. 916/786-8000. This location produces business computer systems.
- 9606 Aero Drive, San Diego CA 92123. 619/279-3200. This location is a sales office.
- 16399 West Bernardo Drive, San Diego CA 92127. 619/487-4100. This location manufactures printers.
- 2850 Centerville Road, Wilmington DE 19808. 302/633-8000. This location manufactures analytical computers.
- 217 Southwest Boulevard East, Suite 102, Kokomo IN 46902. 765/455-3281. This location is a sales and service office.
- 100 Domain Drive, Exeter NH 03833. 603/772-1500. This location manufactures computers.
- 15115 SW Sequoia Parkway, Suite 100, Portland OR 97224-7156. 503/598-8000. Recorded Jobline: 541/754-0919. This location is a sales office.
- P.O. Box 8906, Vancouver WA 98668-8906. 360/254-8110. This location produces workstation printers including serial impact and serial inkjet printers.

HEWLETT-PACKARD COMPANY

11311 Chinden Boulevard, Boise ID 83714. **Contact:** Personnel Department. **World Wide Web address:** http://www.hp.com. **Description:** This location develops laser-jet printers, storage solution, and disk drive technologies. Overall, Hewlett-Packard designs and manufactures measurement and computation products used in business, industry, engineering, science, health care, and education. Principal products are integrated instrument and computer systems (including hardware and software), peripheral products, and medical electronic equipment systems. **NOTE:** Jobseekers should send resumes to: Employment Response Center, Event #2498, Hewlett-Packard Company, Mail Stop 20-APP, 3000 Hanover Street, Palo Alto CA 94304-1181. **Common positions include:** Computer Engineer; Electrical/Electronics Engineer; Financial Analyst; Human Resources Manager; Manufacturing Engineer; Marketing Specialist; Mechanical Engineer. **Educational backgrounds include:** Computer Science; Engineering; Human Resources; Marketing.

HEWLETT-PACKARD COMPANY

5301 Steven's Creek Boulevard, Santa Clara CA 95052. 408/246-4300. **Contact:** Staffing Department. **World Wide Web address:** http://www.hp.com. **Description:** This location contains several different divisions. The Test and Measurement Group designs and manufactures analytical and scientific computation products and systems used in business, engineering, science, health care, and education. The Integrated Circuits Business Division designs and markets ICs used in Hewlett-Packard and external customer products. The Computer Products Organization is responsible for the marketing, sales and support, and distribution of Hewlett-Packard PCs, low-end workstations, printers, calculators, palmtop computers, and related components. **NOTE:** Jobseekers should send resumes to: Employment Response Center, Event #2498, Hewlett-Packard Company, Mail Stop 20-APP, 3000 Hanover Street, Palo Alto CA 94304-1181. **Common positions include:** Computer Programmer; Customer Service Representative; Electrical/Electronics Engineer; Financial Analyst; Hardware Engineer; Marketing Specialist; Software Engineer. **Educational backgrounds include:** Computer Science; Engineering; Finance; Marketing. **Benefits:** Daycare Assistance; Disability Coverage; Employee Discounts; Life Insurance; Medical Insurance; Pension Plan; Profit Sharing; Savings Plan; Tuition Assistance. **Special Programs:** Internships. **Corporate headquarters location:** Palo Alto CA. **Operations at this facility include:** Administration; Manufacturing; Research and Development; Sales; Service. **Listed on:** New York Stock Exchange. **Number of employees nationwide:** 93,100.

HEWLETT-PACKARD CONVEX

3000 Waterview Parkway, Richardson TX 75080. 972/497-4000. **Contact:** Human Resources. **Description:** This location builds supercomputers. As a whole, Hewlett-Packard designs and manufactures measurement and computation products and systems used in business, industry, engineering, science, health care, and education. Principal products are integrated instrument and computer systems (including hardware and software); computer systems and peripheral products;

and medical electronic equipment and systems. **NOTE:** Jobseekers should send resumes to: Employment Response Center, Event #2498, Hewlett-Packard Company, Mail Stop 20-APP, 3000 Hanover Street, Palo Alto CA 94304-1181. **Common positions include:** Accountant/Auditor; Electrical/Electronics Engineer; Financial Analyst; Software Engineer; Technical Writer/Editor. **Educational backgrounds include:** Computer Science; Engineering. **Benefits:** 401(k); Dental Insurance; Disability Coverage; Life Insurance; Medical Insurance; Tuition Assistance. **Special Programs:** Internships. **Operations at this facility include:** Administration; Manufacturing; Research and Development; Sales; Service. **Listed on:** New York Stock Exchange. **Number of employees at this location:** 700.

HITACHI COMPUTER PRODUCTS
1800 East Imhoff Road, Norman OK 73071. 405/360-5500. **Contact:** Human Resources Department. **Description:** This location of Hitachi manufactures mainframe information storage computers.

HITACHI DATA SYSTEMS
951 East Bird, Richmond VA 23219. 804/644-7200. **Contact:** Human Resources. **Description:** Manufactures mainframe computers. This location serves as a sales and marketing office.

HITACHI DATA SYSTEMS
707 17th Street, Denver CO 80202. 303/291-1091. **Contact:** Human Resources. **Description:** Manufactures mainframe computers. This location serves as a sales and marketing office.

HITACHI DATA SYSTEMS
750 Central Expressway, Santa Clara CA 95050. 408/970-1000. **Contact:** Human Resources. **Description:** Manufactures mainframe computers. This location serves as a sales and marketing office.

HITACHI DIGITAL GRAPHICS INC.
815 Hermosa Drive, Sunnyvale CA 94086. 408/735-0577. **Contact:** Human Resources Department. **Description:** Manufactures graphics tablets with pressure sensitive pens for computers.

HOOLEON CORPORATION
411 South 6th Street, Cottonwood AZ 86326. 520/634-7515. **Contact:** Human Resources. **Description:** Customizes computer keyboards.

HOWTEK, INC.
21 Park Avenue, Hudson NH 03051. 603/882-5200. **Contact:** Connie Webster, Director of Human Resources. **Description:** Howtek, Inc. designs, engineers, and develops computer peripherals and systems for scanning and pre-press applications in the graphic arts, publishing, and commercial printing markets; contracts with third-party manufacturers to build company-developed products; and uses marketing and customer support resources to promote its products in a variety of distribution channels. Products include scanners and related hardware and software, and electronic prepress systems for newspaper, magazine, and commercial printing groups. Manufactured goods are sold throughout the United States, as well as Mexico, Central America, Canada, and the Far East. **Benefits:** Incentive Plan; Stock Option. **Corporate headquarters location:** This Location. **Subsidiaries include:** Color Fast, Inc., a printing and pre-press company. **Listed on:** American Stock Exchange. **Number of employees at this location:** 130.

HUSKY COMPUTERS, INC.
18167 US Highway 19 North, Suite 285, Clearwater FL 34624. 813/530-4141. **Contact:** Human Resources. **Description:** Husky Computers manufactures rugged laptop computers. This is a sales location.

I-O CORPORATION
2256 South 3600 West, Salt Lake City UT 84119. 801/973-6767. **Contact:** Human Resources. **Description:** Manufactures peripheral equipment for mainframes.

IBM CORPORATION

One Old Orchard Road, Armonk NY 10504. 914/765-1900. **Contact:** Central Employment. **Description:** International Business Machines (IBM) is a developer, manufacturer, and marketer of advanced information processing products, including computers and microelectronic technology, software, networking systems, and information technology-related services. The company strives to offer value worldwide, through its United States, Canada, Europe/Middle East/Africa, Latin America and Asia Pacific business units, by providing comprehensive and complete product choices. **NOTE:** Jobseekers should send a resume to IBM Staffing Services, Department 1DP, Building 051, P.O. Box 12195, Research Triangle Park NC 27709-2195. **Corporate headquarters location:** This Location.

Other U.S. locations:

- 9000 South Rita Road, Tucson AZ 85744. 520/799-1000. This facility develops data access and storage devices.
- 355 South Grand Avenue, Los Angeles CA 90071. 213/621-5100. This location functions as a regional marketing office.
- 425 Market Street, San Francisco CA 94105. 415/545-2000. This location functions as a regional sales office and offers technical support.
- 5600 Cottle Road, Building 78, San Jose CA 95193. 408/256-1600.
- 6300 Diagonal Highway, Boulder CO 80301. 303/924-6300. This location manufactures magnetic discs and tapes.
- 1301 K Street NW, Washington DC 20005. 202/515-4000. This location functions as a business support center.
- 1001 Jefferson Plaza, Wilmington DE 19801. 302/428-5059. This location is a marketing office.
- 1000 NW 51st Street, Boca Raton FL 33429. 561/640-6329.
- 3109 West Dr. Martin Luther King Boulevard, Tampa FL 33607. 813/872-2277. This location serves as a regional sales office.
- 3200 Windy Hill Road, Atlanta GA 30067. 404/238-7000. This is a sales office.
- One IBM Plaza, Chicago IL 60611. 312/245-7935.
- 6710 Rockledge Drive, Bethesda MD 20817. 301/803-6000.
- 4111 Highway 52 North, Rochester MN 55901-7829. 507/253-4011. This location is a manufacturing facility.
- 500 Maryville College Drive, St. Louis MO 63141. 314/469-4000. This location is a sales office.
- 8501 IBM Drive, Charlotte NC 28262-8563. 704/594-1000. This location manufactures, sells, and markets advanced information processing products, including computers and microelectronics technology, software, networking systems and information technology-related services.
- 6001 Indian School Road NE, Albuquerque NM 87110.
- 53 Knightsbridge Road, Piscataway NJ 08855. 908/885-3500. This location is a marketing office.
- 1701 North Street, Endicott NY 13760-0569. 607/755-9644.
- 590 Madison Avenue, New York NY 10022. 212/745-5500. This location is a sales office.
- 522 South Road, Poughkeepsie NY 12601-5400. 914/332-3900. This location is an administrative facility.
- 1133 Westchester Avenue, White Plains NY 10601. 914/641-5000.
- 11400 Burnett Road, Austin TX 78758. 512/823-0000. This location manufactures and develops facility electronic circuit cords, hardware, software, personal computers, advanced work stations circuit cards for IBM/1 and IBM/2 systems, and RISC/6000.
- 1605 LBJ Freeway, Dallas TX 75234. 972/280-4000. This location is a regional sales and marketing office.
- 1200 5th Avenue, Seattle WA 98101. 206/587-4400.
- 145 Summers Street, Charleston WV 25301. 304/347-7300.

IBM CORPORATION

P.O. Box 12195, Research Triangle Park NC 27709. 919/543-5221. **Toll-free fax:** 800/262-2494. **Recorded Jobline:** 800/964-4473. **Contact:** Hiring Manager. **World Wide Web address:** http://www.emp.ibm.com/carus.htm. **Description:** International Business Machines (IBM) is a developer, manufacturer, and marketer of advanced information processing products, including computers and microelectronics technology, software, networking systems and information technology-related services. The company strives to offer value through its United States, Canada, Europe/Middle East/Africa, Latin America and Asia Pacific business units, by providing comprehensive and complete product choices. **NOTE:** Send resumes to the Staffing Services Center, 1 DTA/051, P.O. Box 12195, Research Triangle Park NC 27709-2195. **Common positions include:** Chemical Engineer; Computer Operator; Computer Programmer; Data Entry Clerk; Electrical/Electronics Engineer; Manufacturing Engineer; Mechanical Engineer; Sales Representative; Secretary; Software Engineer; Systems Analyst; Technical Writer/Editor; Technician.

IOTECH, INC.
25971 Cannon Road, Cleveland OH 44146. 216/439-4091. **Fax:** 216/439-4093. **Contact:** Human Resources. **World Wide Web address:** http://www.iotech.com. **Description:** IOtech, Inc. develops, manufactures, and markets interfaces and data acquisition instruments used in a variety of applications. The company's hardware and software products are used primarily to support personal computers and engineering workstations. **Common positions include:** Applications Engineer; Design Engineer; Electrical/Electronics Engineer; Manufacturing Engineer; Mechanical Engineer; Software Engineer. **Educational backgrounds include:** Engineering. **Benefits:** 401(k); Dental Insurance; Disability Coverage; Life Insurance; Medical Insurance. **Corporate headquarters location:** This Location. **Listed on:** Privately held. **Number of employees at this location:** 90.

IPL SYSTEMS INC.
124 Acton Street, Maynard MA 01754. 508/461-1000. **Contact:** Controller. **Description:** The company produces plug-compatible products for IBM mid-range systems. Recent product introductions include add-in memory for the IBM System/36 and System/38, which the company markets to end users and distributors. IPL continues to be a manufacturer of IBM 43XX plug-compatible mainframes and mid-range, general purpose systems which it currently markets to OEM customers. **Common positions include:** Accountant/Auditor; Buyer; Computer Programmer; Electrical/Electronics Engineer; Human Resources Manager; Manufacturer's/Wholesaler's Sales Rep.; Marketing Specialist; Operations/Production Manager; Systems Analyst; Technical Writer/Editor. **Educational backgrounds include:** Business Administration; Engineering; Marketing. **Benefits:** Dental Insurance; Disability Coverage; Life Insurance; Medical Insurance; Savings Plan; Tuition Assistance. **Corporate headquarters location:** This Location. **Operations at this facility include:** Administration; Manufacturing; Research and Development; Sales; Service. **Listed on:** NASDAQ.

IIYAMA NORTH AMERICA INC.
650 Louis Drive, Suite 120, Warminster PA 18974. 215/957-6543. **Contact:** Human Resources. **Description:** Manufactures and sells color monitors.

IKEGAMI ELECTRONICS INC.
37 Brook Avenue, Maywood NJ 07607. 201/368-9171. **Contact:** Human Resources. **Description:** Manufactures and sells computer and broadcast monitors.

IMAGE GRAPHICS
917 Bridgeport Avenue, Shelton CT 06484. 203/926-0100. **Contact:** Human Resources. **Description:** Manufactures electronic data gathering equipment. Image Graphics also integrates document systems for businesses.

IMAGE SYSTEMS CORPORATION
11595 K-Tel Drive, Hopkins MN 55343. 612/935-1171. **Toll-free phone:** 800/462-4370. **Fax:** 612/935-1386. **Contact:** Laura Sorensen, Manager Of Human Resources. **Description:** Designs and manufactures high-resolution computer monitors. Applications include medical and document imaging and air traffic control. **Common positions include:** Account Manager; Design Engineer; Technical Writer/Editor. **Educational backgrounds include:** Business Administration; Engineering. **Benefits:** 401(k); Disability Coverage; Medical Insurance. **Corporate headquarters location:** This Location. **Annual sales/revenues:** $5 - $10 million. **Number of employees at this location:** 50.

IMAGITEX INC.
75 Northeastern Boulevard, Nashua NH 03062. 603/889-6600. **Contact:** Human Resources. **Description:** Manufactures scanners for computers.

IMATION CORPORATION
2100 15th Street North, Wahpeton ND 58075. 701/642-8711. **Contact:** Patty Zietlow, Human Resources Manager. **Description:** This location manufactures diskettes. Imation is a global technology solutions company, offering systems, products, and services for the handling, storage, transmission, and use of information. The company develops data storage products, medical imaging and photo products, printing and publishing systems, and customer support technologies and document imaging, and markets them under the trademark names Dry View laser imagers, Matchprint and Rainbow color proofing systems, Travan data cartridges, and LS-120 diskette technology. Imation is an independent spin-off of 3M's data storage and imaging businesses. **Stock exchange symbol:** IMN. **Annual sales/revenues:** $2.3 billion. **Number of employees worldwide:** 10,000.

IMATREX CORPORATION
325 Coosa Road, Boaz AL 35957. 205/593-0807. **Contact:** Human Resources. **Description:** Manufactures custom designed PCs and workstations.

INDUSTRIAL COMPUTER SOURCE
9950 Barnes Canyon Road, San Diego CA 92121-2720. 619/677-0877. **Contact:** Human Resources. **Description:** A manufacturer of ruggedized PC chassis and a reseller of computer hardware.

INFORMATION DIMENSIONS, INC.
6600 Frantz Road, P.O. Box 8007, Dublin OH 43016. 614/761-8083. **Fax:** 614/761-7290. **Contact:** Becky Bell, Human Resources Manager. **Description:** Develops document-management and text-retrieval systems.

INFOSAFE SYSTEMS, INC.
342 Madison Avenue, Suite 622, New York NY 10173. 212/867-7200. **Contact:** Alistair Weir, Human Resources Director. **Description:** Manufactures computer products that enable protection, retrieval, and monitoring of digital information use.

INTEGRAL PERIPHERALS
5775 Flatiron Parkway, Suite 100, Boulder CO 80301-5730. **Toll-free phone:** 800/333-8009. **Contact:** Human Resources Manager. **E-mail address:** ptate@integralnet.com. **World Wide Web address:** http://www.integralnet.com. **Description:** A developer and manufacturer of disk drives for mobile data storage applications, as well as PC card hard drives.

INTEGRATED CAD SERVICES INC.
1855 Williston Road, South Burlington VT 05403. 802/658-3272. **Contact:** Vice President. **Description:** Manufactures PCs. Integrated CAD Services also supports a sales and service team.

INTEL CORPORATION
2800 Center Drive, Dupont WA 98327. **Toll-free phone:** 800/628-8686. **Contact:** Human Resources. **E-mail address:** jobs@intel.com. **World Wide Web address:** http://www.intel.com. **Description:** This location manufactures PCs for original equipment manufacturers (OEM) who market the computers under their own brand names. Intel also plans to introduce light manufacturing and research and development at this site. Overall, Intel Corporation is a manufacturer of computer microprocessors and computer related parts. **NOTE:** Resumes should be sent to P.O. Box 1141, FM3-145, Folsom CA 95763. **Common positions include:** Engineer; Mechanical Engineer; Project Manager; Quality Assurance Engineer; Software Engineer; Systems Engineer. **Corporate headquarters location:** Santa Clara CA.

INTEL CORPORATION
15201 NW Greenbrier Parkway, Beaverton OR 97006. 503/677-7600. **Contact:** Human Resources. **E-mail address:** jobs@intel.com. **World Wide Web address:** http://www.intel.com. **Description:**

This location houses the design staff for the company's Enterprise Server Group. Intel is one of the largest semiconductor manufacturers in the world. Other operations include supercomputers; embedded control chips and flash memories; video technology software; multimedia hardware; personal computer enhancement products; and designing, making, and marketing microcomputer components, modules, and systems. Intel sells its products to original equipment manufacturers and other companies that incorporate them into their products. **Corporate headquarters location:** Santa Clara CA.

INTELLIGENT INSTRUMENTATION INC.
6550 South Bay Colony Drive, Suite 130, Tucson AZ 85706. 520/573-0887. **Contact:** Human Resources. **Description:** Manufactures data acquisition computer products and PC interface boards.

INTERACTIVE INC.
204 North Main Street, Humboldt SD 57035. 605/363-5117. **Contact:** Human Resources. **Description:** Manufactures sound exchange devices for computer video conferencing.

INTEREX
8447 East 35th Street North, Wichita KS 67226. 316/524-4747. **Contact:** Human Resources. **Description:** Manufactures computer accessories including cable, surge suppressers, and networking equipment.

INTERNATIONAL DATA PRODUCTS
20 Firstfield Road, Gaithersburg MD 20878. 301/590-8100. **Contact:** Human Resources. **Description:** A manufacturer of computers.

INTERVISION SYSTEMS, INC.
5237 Capital Boulevard, Raleigh NC 27604. 919/850-2511. **Fax:** 919/850-0562. **Contact:** Human Resources. **World Wide Web address:** http://www.intervisionsystems.com. **Description:** A manufacturer of wearable, belt-mounted computers featuring head-mounted VGAs and voice recognition for hands-free field operations used in emergency, firefighting, and military situations. **Corporate headquarters location:** This Location.

INTERVOICE, INC.
17811 Waterview Parkway, Dallas TX 75252. 972/454-8000. **Fax:** 972/907-1079. **Contact:** Human Resources. **Description:** InterVoice, Inc. develops, sells, and services interactive voice response systems that allow individuals to access a computer database using a telephone keypad, computer keyboard, or human voice. Applications are currently functioning in industries including insurance, banking, higher education, government, utilities, health care, retail distribution, transportation, and operator services. **Corporate headquarters location:** This Location.

IOLINE CORPORATION
12020 113th Avenue NE, Kirkland WA 98034. 206/821-2140. **Contact:** Human Resources. **Description:** Manufactures plotters for computers.

IOMEGA CORPORATION
1821 West Iomega Way, Roy UT 84067. 801/778-1000. **Contact:** Dan Henrie, Human Resources Director. **Description:** Iomega Corporation, established in 1980, creates information storage solutions that enhance the usefulness of personal computers and workstations in a growing variety of applications. Iomega's products help people manage their information storage needs. The company has three core products: Bernoulli, Ditto, and Zip. Bernoulli drives and disks provide reliable high-performance solutions for users who require unlimited removable storage. Bernoulli drives offer hard drive performance, while disks provide infinite capacity. Ditto tape drives and cartridges provide easy, dependable backup of hard drives. These Quarter-Inch Cartridge (QIC) compatible drives feature direct drive operation, quiet performance, and full-featured software, making backup easy and automatic. Zip drives are three drives in one, offering affordable expansion for hard drives, mobile storage with portable convenience, and easy backup information. The company has been public since 1983. **Corporate headquarters location:** This Location.

IRIS GRAPHICS INC.

Six Crosby Drive, Bedford MA 01730. 617/275-8777. **Fax:** 617/275-8590. **Contact:** Human Resources Manager. **Description:** Manufactures continuous ink-jet printers. The company's product line has expanded to serve such markets as desktop publishing, digital photography, fine art signage, 3-D design, packaging, and geotechnology. **Common positions include:** Chemist; Electrical/Electronics Engineer; Mechanical Engineer; Software Engineer; Technical Writer/Editor. **Educational backgrounds include:** Computer Science; Engineering; Marketing. **Benefits:** 401(k); Dental Insurance; Disability Coverage; Employee Discounts; Life Insurance; Medical Insurance; Profit Sharing; Tuition Assistance. **Corporate headquarters location:** This Location. **Other U.S. locations:** Nationwide. **Parent company:** Scitex Corp., Ltd. **Operations at this facility include:** Manufacturing; Research and Development. **Listed on:** Privately held. **Annual sales/revenues:** More than $100 million. **Number of employees at this location:** 280. **Number of projected hires for 1997 - 1998 at this location:** 50.

ITAC SYSTEMS, INC.

3121 Benton Street, Garland TX 75042. 972/494-3073. **Contact:** Human Resources Department. **Description:** Itac Systems manufactures the Mouse Track Trackball, a computer peripheral product. **Common positions include:** Accountant/Auditor; Adjuster; Collector; Credit Manager; Electrical/Electronics Engineer; Industrial Engineer; Investigator; Mechanical Engineer; Operations/Production Manager; Public Relations Specialist; Purchasing Agent/Manager; Quality Control Supervisor. **Educational backgrounds include:** Accounting; Business Administration; Engineering; Finance; Marketing. **Benefits:** Life Insurance; Medical Insurance. **Corporate headquarters location:** This Location. **Operations at this facility include:** Administration; Manufacturing; Research and Development; Sales; Service. **Listed on:** Privately held. **Number of employees at this location:** 25.

ITRONICS CORPORATION

P.O. Box 179, Spokane WA 99210-0179. 509/624-6600. **Contact:** Human Resources. **Description:** Manufactures laptop computers.

JBM ELECTRONICS COMPANY

4645 Laguardia Drive, St. Louis MO 63134. 314/426-7781. **Fax:** 314/426-0007. **Contact:** Human Resources. **Description:** Manufactures data communications equipment. JBM Electronics creates products that allow older units to interface with newer ones.

JCC CORPORATION

One Bridge Plaza, Suite 400, Fort Lee NJ 07024. 201/592-5023. **Contact:** Human Resources. **Description:** Manufactures and sells Internet access devices.

JVC AMERICA INC.

2 JVC Road, Tuscaloosa AL 35405. 205/556-7111. **Contact:** Human Resources. **Description:** This plant manufactures computer compact disks on which companies can store information.

K AND M ELECTRONICS, INC.

650 Danbury Road, Ridgefield CT 06877. **Fax:** 203/438-0321. **Contact:** Human Resources. **World Wide Web address:** http://www.web2.kme.com. **Description:** A manufacturer of cordless computer modems.

KYE INTERNATIONAL CORPORATION

2605 East Cedar Street, Ontario CA 91761. 909/923-3510. **Fax:** 909/923-1469. **Contact:** Human Resources Manager. **Description:** A manufacturer of computer hardware peripherals, including mice, scanners, CD-ROM drives, and networking products. **Common positions include:** Accountant/Auditor; Adjuster; Blue-Collar Worker Supervisor; Computer Programmer; Credit Manager; Customer Service Representative; Electrical/Electronics Engineer; Financial Analyst; General Manager; Industrial Engineer; Systems Analyst. **Benefits:** Dental Insurance; Disability Coverage; Medical Insurance. **Corporate headquarters location:** This Location. **Operations at this facility include:** Administration; Regional Headquarters; Sales; Service. **Listed on:** Privately held. **Number of employees at this location:** 20.

KANTEK INC.
15 Main Street, East Rockaway NY 11518. 516/593-3212. **Contact:** Personnel. **Description:** Manufactures glare reduction screens for computer monitors.

KAY COMPUTERS
722 Genevieve Street, Suite R, Solana Beach CA 92075. 619/481-0225. **Contact:** Human Resources. **Description:** A manufacturer of personal computers.

KENSINGTON MICROWARE, LTD.
2855 Campus Drive, San Mateo CA 94403. 415/572-2700. **Fax:** 415/572-9675. **Contact:** Human Resources. **E-mail address:** jobs@kensington.com. **World Wide Web address:** http://www.kensington.com. **Description:** A developer of computer accessories, peripherals, and software for the computer after-market. Products include mice and trackballs, surge suppressor systems, cable and lock security devices, and carrying cases. **Common positions include:** Buyer; Customer Service Representative; Human Resources Manager; Marketing Manager; Operations/Production Manager; Product Manager; Sales Manager; Systems Analyst; Technical Support Representative.

KEY TRONIC CORPORATION
P.O. Box 14687, Spokane WA 99214. 509/928-8000. **Fax:** 509/927-5248. **Contact:** Employment Administrator. **Description:** One of the world's largest independent manufacturers of computer keyboards and computer peripherals. **Common positions include:** Accountant/Auditor; Adjuster; Attorney; Blue-Collar Worker Supervisor; Budget Analyst; Buyer; Collector; Computer Programmer; Credit Manager; Customer Service Representative; Designer; Draftsperson; Economist; Electrical/Electronics Engineer; Electrician; Industrial Production Manager; Investigator; Mechanical Engineer; Operations Research Analyst; Public Relations Specialist; Purchasing Agent/Manager; Quality Control Supervisor; Services Sales Representative; Software Engineer; Systems Analyst; Transportation/Traffic Specialist; Travel Agent. **Educational backgrounds include:** Accounting; Business Administration; Computer Science; Engineering; Marketing. **Benefits:** 401(k); Bonus Award/Plan; Dental Insurance; Disability Coverage; Employee Discounts; Life Insurance; Medical Insurance; Profit Sharing; Tuition Assistance. **Corporate headquarters location:** This Location. **International locations:** Ireland; Mexico; Taiwan. **Operations at this facility include:** Administration; Manufacturing; Research and Development; Sales; Service. **Listed on:** NASDAQ. **Number of employees nationwide:** 850. **Number of employees worldwide:** 2,400.
Other U.S. locations:
* 4201 North Del Rey Boulevard, Las Cruces NM 88012. 505/382-8841. This location manufactures protective membranes for computer keyboards.

KINETIC SYSTEMS CORPORATION
900 North State Street, Lockport IL 60441. 815/838-0005. **Fax:** 815/838-0095. **Contact:** Human Resources. **World Wide Web address:** http://www.kscorp.com. **Description:** A manufacturer of high-performance data acquisition and control systems. Products include modules such as analog to digital converters; digital to analog converters; signal conditioners; timers; counters; buffer memories; computer interfaces; software packages and drivers; mainframes; and VXI slot-0 controllers. Founded in 1970. **Common positions include:** Designer; Software Engineer.

KINGSTON TECHNOLOGY
17600 Newhope, Fountain Valley CA 92708. **Fax:** 714/427-3555. **Contact:** Cyndi Christiansen, Human Resources Representative. **Description:** Manufactures memory upgrade, networking, and storage products. Founded in 1987. **NOTE:** The company offers entry-level positions. **Common positions include:** Account Manager; Accountant/Auditor; Administrative Assistant; Buyer; Computer Operator; Computer Programmer; Customer Service Representative; Design Engineer; Electrical/Electronics Engineer; Financial Analyst; Graphic Artist; Hardware Engineer; Human Resources Manager; Internet Services Manager; Manufacturing Engineer; Market Research Analyst; Marketing Specialist; Mechanical Engineer; MIS Specialist; Production Manager; Sales Representative; Software Engineer; Systems Analyst; Technical Writer/Editor. **Educational backgrounds include:** Accounting; Business Administration; Communications; Computer Science; Engineering; Finance; Marketing. **Benefits:** 401(k); Disability Coverage; Employee

Discounts; Life Insurance; Medical Insurance; Pension Plan. **Corporate headquarters location:** This Location. **International locations:** United Kingdom. **Listed on:** Privately held. **Annual sales/revenues:** More than $100 million. **Number of employees at this location:** 500.

KURZWEIL APPLIED INTELLIGENCE, INC.

411 Waverly Oaks Road, Waltham MA 02154. 617/893-5151. **Contact:** Human Resources. **Description:** The company develops, markets, and supports automated speech recognition systems used to create documents and interact with computers by voice. The company's speech recognition technology is speaker-independent, which means most users do not have to train the system on their voice to achieve satisfactory initial accuracy, and is speaker-adaptive, which means the system is able to adapt with use to the acoustic, phonetic, and linguistic patterns of individual users. The company's software technology is designed to run on 386, 486, or Pentium-based industry standard personal computers running MS-DOS and Windows. **Common positions include:** Software Engineer; Technical Writer/Editor. **Benefits:** 401(k); Dental Insurance; Life Insurance; Medical Insurance; Tuition Assistance. **Corporate headquarters location:** This Location. **Number of employees nationwide:** 100.

LEXMARK INTERNATIONAL, INC.

740 New Circle Road, Lexington KY 40511. 606/232-2379. **Contact:** Employment Manager. **Description:** Develops, manufactures, and markets PC printers, typewriters, keyboards, and related supplies. **Common positions include:** Accountant/Auditor; Chemical Engineer; Chemist; Computer Programmer; Electrical/Electronics Engineer; Industrial Engineer; Manufacturer's/Wholesaler's Sales Rep.; Marketing Specialist; Mechanical Engineer; Technical Writer/Editor. **Educational backgrounds include:** Accounting; Chemistry; Computer Science; Engineering; Finance; Marketing. **Benefits:** Dental Insurance; Employee Discounts; Medical Insurance; Savings Plan. **Corporate headquarters location:** Greenwich CT. **Other U.S. locations:** Boulder CO. **Operations at this facility include:** Manufacturing; Research and Development; Sales. **Number of employees at this location:** 4,000.

LINX DATA TERMINALS, INC.

625 Digital Drive, Suite 100, Plano TX 75075. 214/964-7090. **Contact:** Human Resources. **Description:** A manufacturer of data collection terminals.

LOCKHEED MARTIN TACTICAL DEFENSE SYSTEMS

3151 Zanker Road, San Jose CA 95134. 408/432-8000. **Fax:** 408/432-7961. **Contact:** Human Resources. **Description:** Lockheed Martin Tactical Defense Systems designs and builds 16-bit and 32-bit technical computing systems used in mil-spec environments. Applications include electronic warfare, signal intelligence, radar, sonar, and imaging where digital signal processing or general purpose computing is required. The company is also involved in systems engineering, software development tools, rugged computer systems, and integrated workstations of commercial architectures for proof-of-concept program phases. **Common positions include:** Accountant/Auditor; Buyer; Computer Programmer; Customer Service Representative; Human Resources Manager; Mechanical Engineer; Operations/Production Manager; Purchasing Agent/Manager; Quality Control Supervisor; Software Engineer; Systems Analyst. **Educational backgrounds include:** Accounting; Computer Science; Engineering; Finance; Marketing. **Benefits:** 401(k); Dental Insurance; Disability Coverage; Employee Discounts; Life Insurance; Medical Insurance; Savings Plan; Tuition Assistance. **Operations at this facility include:** Administration; Manufacturing; Research and Development; Sales; Service. **Number of employees at this location:** 50.
Other U.S. locations:
- 3655 Tampa Road, Oldsmar FL 34677. 813/855-5711. This location is a computer hardware manufacturing facility.
- 2751 Shepherd Road, St. Paul MN 55164-0525. 612/456-2222. This location designs defense computer systems.

LOGITECH, INC.

6505 Kaiser Drive, Fremont CA 94555. 510/795-8500. **Fax:** 510/713-5087. **Contact:** Human Resources Manager. **Description:** Designs, develops, manufactures, and markets computer hardware and software products. Logitech is a leading worldwide manufacturer of computer pointing devices (mice, trackballs, and joysticks) and imaging devices (scanners and cameras) for the PC, MAC, and other platforms. **Common positions include:** Accountant/Auditor; Buyer; Computer Programmer; Customer Service Representative; Electrical/Electronics Engineer; Financial Analyst; Manufacturer's/Wholesaler's Sales Rep.; Software Engineer; Systems Analyst; Technical Support Representative. **Educational backgrounds include:** Accounting; Computer Science; Engineering; Marketing. **Benefits:** 401(k); Dental Insurance; Disability Coverage; Employee Discounts; Life Insurance; Medical Insurance; Tuition Assistance. **Special Programs:** Internships. **Corporate headquarters location:** This Location. **Other U.S. locations:** Framingham MA; Dallas TX. **Parent company:** Logitech International S.A. **Operations at this facility include:** Administration; Research and Development; Sales. **Number of employees at this location:** 350.

LORTEC POWER SYSTEMS
10800 Middle Avenue, Building B, Elyria OH 44035-7822. 216/327-5050. **Contact:** Human Resources. **Description:** Manufactures Uninterrupted Power Systems (UPS) for computer backup.

LOWRY COMPUTER PRODUCTS
7100 Whitmore Lake Road, Brighton MI 48116. 810/229-7200. **Contact:** Human Resources. **Description:** A manufacturer of time accounting (punch card) and bar code data collection systems and software.

LOWRY COMPUTER PRODUCTS
700 Gale Drive, Suite 140, Campbell CA 95008. 408/980-5200. **Contact:** Human Resources. **Description:** Lowry manufactures bar code collection systems and the accompanying software.

LUCKY COMPUTER COMPANY
1701 North Greenville Avenue, Suite 602, Richardson TX 75081. 972/705-2600. **Contact:** Human Resources. **Description:** A computer assembly company.

MTI TECHNOLOGY CORPORATION
4905 East La Palma Avenue, Anaheim CA 92807. 714/970-0300. **Contact:** Kathy Nichols, Director of Human Resources. **Description:** MTI designs, manufactures, markets, and services high-performance storage solutions for the DEC, IBM, and open UNIX systems computing environments. These storage solutions integrate MTI's proprietary application and embedded software with its advanced servers and industry standard storage peripherals. This integration of software and server technologies, designated by the company as its Universal Storage Architecture, provides storage solutions designed to meet customers' information management requirements relating to high-performance, reliability, and data integrity. In addition, unlike storage solutions that are dedicated to a specific operating system, this architecture allows a customer to migrate the company's systems across multiple hosts, servers, and networking interfaces. The company's current products, many of which use its patented redundant arrays of independent disks (RAID), networking, fault tolerant, and other server technologies, include its StingRay, StingRAID, and FailSafe storage servers, NetBacker client/server application software, Infinity Automated Tape Library Series, and other systems and related application software.

MTL SYSTEMS
3481 Dayton Xenia Road, Beavercreek OH 45432. 937/426-3111. **Contact:** Human Resources. **Description:** Manufactures and sells computer monitors.

MAXI SWITCH INC.
2901 East Elvira Road, Tucson AZ 85706. 520/294-5450. **Contact:** Human Resources. **Description:** Manufactures computer keyboards.

MAXTOR CORPORATION
510 Cottonwood Drive, Milpitas CA 95135. 408/432-1700. **Contact:** Human Resources. **Description:** A large producer of hard disk drives and related electronic data storage equipment for computers, as well as related components for original equipment manufacturers. The company has production plants in Hong Kong and Singapore, 10 U.S. sales offices, and nine sales offices in

foreign nations. **Corporate headquarters location:** This Location. **Listed on:** NASDAQ. **Number of employees nationwide:** 8,000.
Other U.S. locations:
- 2190 Miller Drive, Longmont CO 80501. 303/678-2279. This location is a research and development facility.
- 1000 Abernathy Road, Building 400, Suite 1130, Atlanta GA 30328. 770/392-0584. This location is a regional sales office.

MAXVISION CORPORATION
2705 Artie Street, Suite 27, Huntsville AL 35805. 205/533-5800. **Fax:** 205/533-5801. **Contact:** Human Resources. **E-mail address:** personnel@maxvision.com. **World Wide Web address:** http://www.maxvision.com. **Description:** A designer and manufacturer of Pentium Pro and Alpha-based Windows NT workstations for CAD/CAM/CAE uses. MaxVision is a leading 3-D technology developer. **Corporate headquarters location:** This Location.

MEDIA LOGIC, INC.
P.O. Box 2258, Plainville MA 02762. 508/695-2006. **Fax:** 508/695-8593. **Contact:** Human Resources. **Description:** Media Logic, Inc. develops, manufactures, and sells certification and evaluation equipment for computer storage media, including floppy disks and magnetic tape. Certifiers are used by computer disk manufacturers to test each disk as it is manufactured and to sort into three industry-established categories. Evaluators are used by large consumers of disks to batch-test incoming disks. This equipment measures the recording performance and quality of digital information contained in flexible computer disks and magnetic tape cartridges. Media Logic also manufactures and sells industrial disk drives and heads through its subsidiary, Media Logic West located in California. **Benefits:** 401(k); Stock Option. **Corporate headquarters location:** This Location. **Subsidiaries include:** Media Logic Far East Limited (Hong Kong); Media Logic FSC, Inc.; and Cecil and Hume Associates, Inc. (dba Media Logic West). **Listed on:** American Stock Exchange. **Number of employees nationwide:** 70.

MEDIC COMPUTER SYSTEMS
8601 Six Forks Road, Suite 300, Raleigh NC 27615. 919/847-8102. **Contact:** Ms. Jan Guy, Human Resources Director. **Description:** At this location, Medic Computer Systems manufactures computer hardware and software systems for use in medical practices and hospitals.

MEDIC COMPUTER SYSTEMS
2020 North Loop West, Suite 140, Houston TX 77018. 713/688-3181. **Contact:** Human Resources. **Description:** This is a sales and engineering location. Overall, Medic Computer Systems manufactures computer hardware and software systems for use in medical practices and hospitals.

MEDIC COMPUTER SYSTEMS
Foster Plaza #6, 681 Andersen Drive, Pittsburgh PA 15220. 412/937-0690. **Contact:** Human Resources. **Description:** This a sales location. Overall, Medic Computer Systems manufactures computer hardware and software systems for use in medical practices and hospitals.

MEGABYTE COMPUTERS
941 Melbourne Road, Hurst TX 76053. 817/284-7793. **Contact:** Human Resources. **Description:** Builds generic computer systems to customer specifications.

MEIKO FEDERAL SYSTEMS
990 Monterey Boulevard, San Francisco CA 94127-2133. 415/334-6161. **Fax:** 415/334-2299. **Contact:** Human Resources. **E-mail address:** cs2@meiko.com. **World Wide Web address:** http://www.meiko.com. **Description:** A manufacturer of scalable systems constructed from commodity microprocessors. **Corporate headquarters location:** This Location.

METHEUS CORPORATION
1600 NW Compton Drive, Beaverton OR 97006. 503/690-1550. **Contact:** Sue Bublitz, Human Resources Administrator. **Description:** Designs, manufactures, and markets high-resolution graphics equipment for the high-technology market. **Common positions include:** Accountant/Auditor; Buyer; Customer Service Representative; Design Engineer; Electrical/

Electronics Engineer; Financial Analyst; Purchasing Agent/Manager; Software Developer. **Educational backgrounds include:** Accounting; Business Administration; Computer Science; Engineering; Technology. **Benefits:** 401(k); Dental Insurance; Disability Coverage; Life Insurance; Medical Insurance; Profit Sharing; Tuition Assistance. **Corporate headquarters location:** This Location. **Parent company:** STV (Singapore). **Operations at this facility include:** Administration; Manufacturing; Research and Development; Sales; Service. **Number of employees at this location:** 42. **Number of employees nationwide:** 47.

MICOM COMMUNICATIONS CORPORATION
4100 Los Angeles Avenue, Simi Valley CA 93063-3397. 805/583-8600. **Contact:** Dick Ballagh, Human Resources Manager. **Description:** A data communications equipment manufacturer, whose product line includes modems, multiplexors, data PABX's, and X.25 packet networks. **Common positions include:** Computer Programmer; Electrical/Electronics Engineer; Industrial Engineer; Manufacturer's/Wholesaler's Sales Rep.; Manufacturing Engineer; Software Engineer; Technical Writer/Editor. **Educational backgrounds include:** Business Administration; Communications; Computer Science; Engineering; Marketing; Mathematics. **Benefits:** Dental Insurance; Life Insurance; Medical Insurance; Profit Sharing; Tuition Assistance. **Corporate headquarters location:** This Location.

MICRO SOLUTIONS COMPUTER PRODUCTS INC.
132 West Lincoln Highway, De Kalb IL 60115. 815/756-3411. **Contact:** Human Resources. **Description:** Manufactures parallel printer port computer drives.

MICROGRAPHIC TECHNOLOGY CORPORATION
520 Logue Avenue, Mountain View CA 94043. 415/965-3700. **Contact:** Barry Cooper, Personnel. **Description:** Manufactures computer peripheral equipment. **Parent company:** Softnet.

MICRON ELECTRONICS
900 East Karcher Road, Nampa ID 83687. 208/893-3434. **Contact:** Human Resources. **Description:** Manufactures computer hardware. Micro Electronics also resells other companies' products.

MICROPOLIS CORPORATION
21211 Nordhoff Street, Chatsworth CA 91311. 818/709-3300. **Fax:** 818/718-5353. **Contact:** Human Resources. **Description:** Micropolis Corporation is a leading designer and manufacturer of SuperCapacity disk drives, information storage devices, and video systems. The company's customers are computer system manufacturers, distributors, value-added resellers, and system integrators. Principal products are disk drives and drive arrays with high-capacity, fast-access, high-reliability, and low-lifetime cost; and video servers. Micropolis's customers incorporate the company's disk drives into high-performance computer systems for single- and multi-user applications, including computer aided design and manufacturing systems, workstations, and local area network and video file servers. **Corporate headquarters location:** This Location.
Other U.S. locations:
- 19782 MacArthur Boulevard, Suite 320, Irvine CA 92715. 714/476-0411. This location is a field sales office.
- 100 Century Center Court, Suite 410, San Jose CA 95112-4512. 408/441-0333. This location is a regional sales office.
- One Stiles Road, Suite 303, Salem NH 03079. 603/898-3550. This location is a field sales office.

MICROSTAR COMPUTERS INC.
25 Kimberly Road, Building F, East Brunswick NJ 08816. 908/651-8686. **Contact:** Human Resources. **Description:** Manufactures laptop computers.

MICROTECH COMPUTERS, INC.
4824 Quail Crest Place, Lawrence KS 66049. 913/841-9513. **Contact:** Personnel Department. **Description:** Develops, manufactures, markets, installs, and services personal computers and related equipment. Primary customers are end users, retailers, corporations, and government agencies. **Common positions include:** Accountant/Auditor; Administrator; Advertising Clerk;

Buyer; Computer Programmer; Credit Manager; Customer Service Representative; Department Manager; Electrical/Electronics Engineer; Financial Analyst; Hardware Engineer; Human Resources Manager; Instructor/Trainer; Manufacturer's/Wholesaler's Sales Rep.; Marketing Specialist; Operations/Production Manager; Public Relations Specialist; Quality Control Supervisor; Software Engineer; Systems Analyst; Technical Writer/Editor. **Educational backgrounds include:** Accounting; Business Administration; Communications; Computer Science; Education; Engineering; Finance; Journalism; Liberal Arts; Marketing. **Benefits:** Dental Insurance; Disability Coverage; Life Insurance; Medical Insurance; Profit Sharing; Tuition Assistance. **Corporate headquarters location:** This Location. **Operations at this facility include:** Administration; Manufacturing; Research and Development; Sales; Service.

MICROTEST, INC.
4747 North 22nd Street, Phoenix AZ 85016. 602/952-6400. **Contact:** Human Resources. **Description:** Manufactures computer network and diagnostic products and electronics cables.

MICROTOUCH SYSTEMS, INC.
300 Griffin Park, Methuen MA 01844. 508/659-9000. **Fax:** 508/659-9100. **Contact:** Human Resources. **E-mail address:** touch@mts.mhs.compuserve.com. **Description:** MicroTouch Systems, Inc. is a manufacturer of touch-screen systems. The company's core product line is based on a patented analog capacitive touch technology which gives touch screens a combination of speed, accuracy, and durability. Products are used in a broad range of applications including point-of-sale terminals, self-service kiosks, gaming machines, industrial systems, ATMs, multimedia applications, and many other computer-based systems. MicroTouch also manufactures and markets TouchPen, a touch- and pen-sensitive digitizer used for pen-based and whiteboarding applications; TouchMate, a pressure-sensitive pad that makes any monitor placed on it touch-sensitive; and ThruGlass, a product that can sense a touch up to two inches of glass, allowing kiosks to be placed behind store windows for 24-hour access. The Factura Kiosk Division of MicroTouch specializes in the design and manufacture of custom kiosk housings. **Corporate headquarters location:** This Location.

MILTOPE CORPORATION
76 Pearl Street, Springfield VT 05156. 802/885-4100. **Contact:** Sandi D'Amore, Office Manager. **Description:** A manufacturer of disk drives.

MILTOPE GROUP, INC.
500 Richardson Road South, Hope Hull AL 36043. 334/284-8665. **Contact:** Ed Crowell, Vice President, Human Resources. **Description:** Miltope Group, Inc. is a holding company whose principal unit, wholly-owned Miltope Corporation, manufactures microcomputers and computer peripheral equipment for military and other applications that require reliable operation in severe land, sea, and airborne environments. Miltope Business Products, Inc., also wholly-owned, was formed in 1984 to produce commercial computer printer and document products.

MIPS TECHNOLOGIES INC.
2011 North Shoreline Boulevard, Mountain View CA 94043. 415/933-6477. **Contact:** Staffing. **Description:** Manufactures computers and related peripherals.

MODULAR COMPUTER SYSTEMS, INC.
1650 West McNab Road, Fort Lauderdale FL 33309. 954/974-1380. **Contact:** Phyllis Gordon, Human Resources Manager. **Description:** Manufactures computers designed for industrial automation, energy transportation, and communication systems.

MOST MANUFACTURING INC.
2180 Executive Circle, Colorado Springs CO 80906. 719/527-3400. **Contact:** Human Resources. **Description:** The company manufactures optical disk drives. **NOTE:** The company offers entry-level positions. **Common positions include:** Design Engineer; Electrical/Electronics Engineer. **Educational backgrounds include:** Engineering. **Benefits:** 401(k); Dental Insurance; Disability Coverage; Life Insurance; Medical Insurance; Tuition Assistance. **Corporate headquarters location:** Cypress CA. **Number of employees at this location:** 100.

MOTOROLA COMPUTER GROUP
2900 South Diablo Way, Tempe AZ 85282. 602/438-3080. **Contact:** Human Resources. **Description:** This location manufactures computer hardware, including Macintosh clones, UNIX servers, board- and platform-level systems, and more.

MOTOROLA COMPUTER GROUP
2100 East Elliot, Tempe AZ 85284. 602/438-3000. **Contact:** Human Resources. **Description:** This location manufactures computer hardware, including Macintosh clones, UNIX servers, board- and platform-level systems, and more.

MOTOROLA INFORMATION SYSTEMS GROUP
777 Passaic Avenue, Clifton NJ 07012. 201/470-9001. **Contact:** Human Resources. **Description:** Develops and sells modems and other networking equipment.

MOUNTAINGATE
A LOCKHEED MARTIN COMPANY
9393 Gateway Drive, Reno NV 89511. 702/851-9393. **Contact:** Department of Interest. **Description:** MountainGate is a manufacturer of data storage systems.

MULTI-TECH SYSTEMS INC.
2205 Woodale Drive, Mounds View MN 55112. 612/785-3500. **Contact:** Human Resources. **Description:** Manufactures modems and provides data communication services.

MULTIMAX, INC.
1441 McCormick Drive, Largo MD 20774. 301/925-8222. **Contact:** Human Resources. **Description:** A manufacturer of microcomputers and a provider of systems integration.

NAI TECHNOLOGIES SYSTEMS DIVISION
7125 Riverwood Drive, Columbia MD 21046. 410/312-5800. **Contact:** Human Resources. **Description:** Manufactures and sells computer hard drives and printers.

NBASE COMMUNICATIONS
12401 Middlebrook Road #160, Germantown MD 20874. 301/990-7100. **Contact:** Alan Brandt, Vice President. **Description:** NBase Communications is a manufacturer of computer hardware, primarily network connectivity products.

NCR CORPORATION
5512 East Morris Boulevard, Morristown TN 37813. 423/581-1620. **Contact:** Human Resources. **World Wide Web address:** http://www.ncr.com. **Description:** This location is a manufacturing facility. NCR Corporation is a worldwide provider of computer products and services. The company provides computer solutions to three targeted industries: retail, financial, and communication. NCR Computer Systems Group develops, manufactures, and markets computer systems. NCR Financial Systems Group is an industry leader in three target areas: financial delivery systems, relationship banking data warehousing solutions, and payments systems/item processing. NCR Retail Systems Group is a world leader in end-to-end retail solutions serving the food, general merchandise, and hospitality industries. NCR Worldwide Services provides data warehousing services solutions; end-to-end networking services; and designs, implements, and supports complex open systems environments. NCR Systemedia Group develops, produces, and markets a complete line of information products to satisfy customers' information technology needs including transaction processing media, auto identification media, business form communication products, managing documents and media, and a full line of integrated equipment solutions. NCR Corporation formerly operated as AT&T Global Information Solutions. **Annual sales/revenues:** More than $100 million. **Number of employees worldwide:** 38,000.
Other U.S. locations:
- 222 West Adams, Suite 1000, Chicago IL 60606. 312/781-3995. Contact: Ms. Beena John, Human Resource Consultant. This location is a sales office.
- 355 Fleet Street, Pittsburgh PA 15220. 412/922-4310. This location offers sales, service, and customer support.

NCR SYSTEMEDIA GROUP
9095 Washington Church Road, Miamisburg OH 45342-4428. 937/439-8200. **Contact:** Human Resources. **Description:** NCR Systemedia Group develops, produces, and markets a complete line of information technology products including transaction processing media, auto identification media, business form communication products, managing documents and media, and a full line of integrated equipment solutions. **Parent company:** NCR Corporation.

NCUBE
1825 NW 167th Place, Beaverton OR 97006. 503/629-5088. **Contact:** Human Resources. **Description:** nCUBE is a manufacturer of supercomputers.

NEC TECHNOLOGIES, INC.
1414 Massachusetts Avenue, Boxborough MA 01719. 508/264-8000. **Contact:** Dan McPhee, Director of Human Resources. **Description:** Manufactures, sells, and services small business computer systems, personal computers, laptop and notebook computers, monitors, multimedia, CD-ROMs, printers, and memory storage devices. **Common positions include:** Accountant/Auditor; Advertising Clerk; Attorney; Commercial Artist; Computer Programmer; Credit Manager; Customer Service Representative; Electrical/Electronics Engineer; Human Resources Manager; Marketing Specialist; Mechanical Engineer; Operations/Production Manager; Paralegal; Public Relations Specialist; Purchasing Agent/Manager; Quality Control Supervisor; Systems Analyst; Technical Writer/Editor. **Educational backgrounds include:** Accounting; Art/Design; Business Administration; Communications; Computer Science; Engineering; Finance; Liberal Arts; Marketing. **Benefits:** 401(k); Dental Insurance; Disability Coverage; Employee Discounts; Life Insurance; Medical Insurance; Pension Plan; Savings Plan; Tuition Assistance. **Corporate headquarters location:** This Location. **Parent company:** NEC Corporation (Tokyo, Japan). **Operations at this facility include:** Administration; Manufacturing; Research and Development; Sales; Service. **Listed on:** NASDAQ. **Number of employees nationwide:** 2,300.
Other U.S. locations:
- One NEC Drive, McDonough GA 30253. 770/957-6600. Contact: Wendy Watson, Human Resources Manager. This location is a manufacturing facility.

NATIONAL COMPUTER SYSTEMS INC.
4401 West 76th Street, Edina MN 55435. 612/830-7600. **Contact:** Betsy Shober, Human Resources Manager. **Description:** Manufactures scanners and collects data.

NATIONAL DATACOMPUTER, INC.
900 Middlesex Turnpike, Building Five, Billerica MA 01821. 508/663-7677. **Fax:** 508/667-1869. **Contact:** Human Resources. **Description:** National Datacomputer, Inc. designs, manufactures, and markets computerized systems used to automate the collection, processing, and communication of information related to product sales, distribution, and inventory control. The company's products and services include data communication networks, application-specific software, hand-held computers and related peripherals, and associated training and support services. The company's products facilitate rapid and accurate data collection, data processing, and two-way communication of information with a customer's host information system.

NATIONAL MICROCOMPUTERS INC.
5544 South Green Street, Murray UT 84123. 801/265-3700. **Contact:** Human Resources. **Description:** A microcomputer manufacturing facility.

NEMATRON CORPORATION
5840 Interface Drive, Ann Arbor MI 48103. 313/994-0501. **Contact:** Human Resources. **Description:** Manufactures rugged computers with touch-screen interfaces.

THE NETWORK CONNECTION
1324 Union Hill Road, Alpharetta GA 30201. 770/751-0889. **Fax:** 770/751-1884. **Contact:** Human Resources. **Description:** A manufacturer of digital video and file server technology as well as computer stations and software development. The Network Connection's product line includes the Cheetah series of performance servers, including the Cheetah Video On Demand Server. The

company serves the computer, telephony, avionics, and entertainment industries. **Corporate headquarters location:** This Location. **Listed on:** NASDAQ. **Stock exchange symbol:** TNCX.

NEWERTECH
TECHNOLOGY & PERFORMANCE DIVISION
4848 West Irving Street, Wichita KS 67209. 316/943-0222. **Fax:** 316/685-9368. **Contact:** Human Resources Department. **E-mail address:** info@newertech.com. **World Wide Web address:** http://www.newertech.com. **Description:** The Technology & Performance Division of Newertech manufactures docking bars, modems, PC cards, disk drives, and software.

NORAND CORPORATION
550 Second Street SE, Cedar Rapids IA 52401. 319/369-3100. **Fax:** 319/369-3791. **Contact:** James Harrington, Director of Human Resources. **Description:** Norand Corporation is a manufacturer and marketer of portable computerized data collection systems and hand-held radio frequency terminals used in a wide range of applications, including route accounting, inventory management, and warehouse data management. These systems are used by *Fortune* 500 companies and small businesses alike to improve accountability, productivity, and management control. **Common positions include:** Accountant/Auditor; Administrative Services Manager; Buyer; Computer Operator; Computer Programmer; Department Manager; Electrical/Electronics Engineer; Financial Analyst; General Manager; Human Resources Manager; Market Research Analyst; Mechanical Engineer; Software Engineer; Systems Analyst. **Educational backgrounds include:** Business Administration; Computer Science; Engineering; Finance; Marketing. **Benefits:** Dental Insurance; Disability Coverage; Life Insurance; Medical Insurance; Profit Sharing; Savings Plan; Stock Option; Tuition Assistance. **Corporate headquarters location:** This Location. **Operations at this facility include:** Administration; Manufacturing; Research and Development; Sales; Service. **Listed on:** NASDAQ. **Number of employees at this location:** 752. **Number of employees nationwide:** 835.

OCS INC. (OGLEVEE COMPUTER SYSTEMS)
150 Oglevee Lane, Connellsville PA 15425. 412/628-8360. **Contact:** Human Resources Department. **Description:** OCS Inc. (Oglevee Computer Systems) manufactures environmental control computer systems. **NOTE:** The posted phone number is for the parent company also. When calling, jobseekers should ask for Oglevee Computer Systems. **Parent company:** Oglevee Ltd.

OCTAGON SYSTEMS
6510 West 91st Avenue, Westminster CO 80030. 303/430-1500. **Contact:** Human Resources Department. **Description:** Octagon Systems is a manufacturer of embedded systems for industrial computers.

OKIDATA CORPORATION
532 Fellowship Road, Mount Laurel NJ 08054. 609/235-2600. **Contact:** Human Resources Department. **Description:** Okidata Corporation manufactures computer printers and fax machines.

OMNI TECH CORPORATION
N27 W23676 Paul Road, Pewaukee WI 53072. 414/523-3300. **Contact:** Human Resources Department. **Description:** Omni Tech Corporation manufactures PCs and offers custom networking services.

OPTELECOM, INC.
9300 Gaither Road, Gaithersburg MD 20877. 301/840-2121. **Contact:** Human Resources Department. **Description:** Optelecom, Inc. manufactures modems and interface cards for video, audio, and data transmission. Optelecom also performs coil windings.

OPTIMA TECHNOLOGY
17062 Murphy Avenue, Irvine CA 92614. 714/476-0515. **Contact:** Human Resources Department. **Description:** Optima Technology manufactures computer hard drives as well as some software products.

OUTPUT TECHNOLOGY CORPORATION
2310 North Fancher Road, Spokane WA 99212. 509/533-1257. **Contact:** D. Prindle, Personnel Administration. **E-mail address:** ddc@output.com. **World Wide Web address:** http://www.output.com. **Description:** Manufactures and distributes printers and related enhancement products for the computer industry. **Common positions include:** Account Representative; Administrative Assistant; Blue-Collar Worker Supervisor; Buyer; Chief Financial Officer; Clerical Supervisor; Controller; Customer Service Representative; Design Engineer; Draftsperson; Electrical/Electronics Engineer; Manufacturing Engineer; Market Research Analyst; Marketing Manager; Marketing Specialist; Mechanical Engineer; MIS Specialist; Production Manager; Project Manager; Purchasing Agent/Manager; Sales Manager; Sales Representative; Software Engineer; Technical Writer/Editor; Transportation/Traffic Specialist; Vice President of Marketing and Sales; Vice President of Operations. **Educational backgrounds include:** Accounting; Business Administration; Communications; Engineering; Finance; Marketing. **Benefits:** 401(k); Dental Insurance; Disability Coverage; Life Insurance; Medical Insurance; Tuition Assistance. **Corporate headquarters location:** This Location. **Listed on:** Privately held. **Number of employees at this location:** 130. **Number of projected hires for 1997 - 1998 at this location:** 6.

PC DESIGNS, INC.
2504 North Hemlock Circle, Broken Arrow OK 74012. 918/251-5550. **Contact:** Human Resources. **Description:** A manufacturer of personal computers.

PACKARD BELL NEC, INC.
One Packard Bell Way, Sacramento CA 95826. 916/388-0101. **Fax:** 916/388-5459. **Recorded Jobline:** 800/382-6444 or 916/379-6100 (California). **Contact:** Human Resources. **E-mail address:** 212-7051@mcimail.com. **World Wide Web address:** http://www.packardbell.com. **Description:** Packard Bell NEC is one of the world's largest designers, manufacturers, and marketers of a wide range of PC-compatible desktop and laptop computers. **NOTE:** Packard Bell NEC maintains an extensive employment opportunities page at their Website. Mail resumes to P.O. Box 299002, Attention: Human Resources Department 150-14, (insert appropriate job code), Sacramento CA 95829. **Common positions include:** Accountant/Auditor; Applications Engineer; Business Analyst; Computer Programmer; Contract/Grant Administrator; Database Manager; Design Engineer; Desktop Publishing Specialist; Editor; Employee Relations Director; Facilities Engineer; Financial Analyst; Graphic Artist; Industrial Engineer; Manufacturing Engineer; MIS Specialist; Process Engineer; Product Engineer; Quality Assurance Engineer; Safety Specialist; Sales Representative; Software Engineer; Systems Engineer; Technical Support Representative. **Corporate headquarters location:** This Location. **Other U.S. locations:** Westlake CA; Magna UT; Fife WA. **Subsidiaries include:** Zenith Data Systems Corp., 2150 East Lake Cook Road, Buffalo Grove IL 60089, 847/808-5000. **Operations at this facility include:** Manufacturing. **Annual sales/revenues:** $4 billion.

PACKARD BELL NEC, INC.
8285 West 3500 South, Magna UT 84044. **Fax:** 801/252-1934. **Recorded Jobline:** 800/382-6444. **Contact:** Human Resources. **E-mail address:** 212-7051@mcimail.com. **World Wide Web address:** http://www.packardbell.com. **Description:** This location is the company's international service and support facility. Packard Bell NEC is one of the world's largest designers, manufacturers, and marketers of a wide range of PC-compatible desktop and laptop computers. **NOTE:** Packard Bell NEC maintains an extensive employment opportunities page at their Website. **Common positions include:** Administrative Assistant; Customer Service Representative; Network Engineer; Sales Representative; Technical Support Representative. **Corporate headquarters location:** Sacramento CA. **Operations at this facility include:** Service; Support Services. **Annual sales/revenues:** $4 billion. **Number of employees at this location:** 1,200.

PARAVANT COMPUTER SYSTEMS INC.
780 South Apollo Boulevard, Atrium 1, Melbourne FL 32901. 407/727-3672. **Fax:** 407/725-0496. **Contact:** Human Resources. **Description:** Manufactures rugged, hand-held computer systems and applicable software for the military.

PERIPHERAL DYNAMICS INC.
5150 Campus Drive, Plymouth Meeting PA 19462. 610/825-7090. **Contact:** Human Resources.
Description: Manufactures scanners, optical readers, and other peripherals used in computer data entry.

PINNACLE MICRO, INC.
19 Technology Drive, Irvine CA 92618. 714/789-3000. **Toll-free phone:** 800/553-7070. **Fax:** 714/789-3157. **Contact:** Nancy Rosen, Human Resources Supervisor. **World Wide Web address:** http://www.pinnaclemicro.com. **Description:** An information storage technology firm. Pinnacle offers a CD-ROM information storage system with read/write and rewriteable capabilities. **Common positions include:** Account Manager; Account Representative; Administrative Assistant; Applications Engineer; MIS Specialist; Quality Control Supervisor; Sales Executive; Sales Manager; Sales Representative. **Educational backgrounds include:** Business Administration; Computer Science; Engineering. **Benefits:** 401(k); Dental Insurance; Disability Coverage; Life Insurance; Medical Insurance; Tuition Assistance. **Corporate headquarters location:** This Location. **Annual sales/revenues:** $51 - $100 million. **Number of employees at this location:** 100. **Number of employees worldwide:** 165.
Other U.S. locations:
- 4176 East Bijou, Colorado Springs CO 80910. 719/572-1200. This location is a manufacturing facility.
- 1330 Inverness Drive, Suite 450, Colorado Springs CO 80910. 719/591-4500. This location is a research and development facility.

PIXEL CRAFT INC.
3400 Arden Road, Hayward CA 94545. 510/562-2480. **Contact:** Human Resources. **Description:** Manufactures color scanners.

PIXELVISION
43 Nagog Park, Acton MA 01720. 508/264-9443. **Fax:** 508/264-9446. **Contact:** Human Resources. **Description:** A manufacturer of flat-paneled computer monitors.

PLASMON IDE
9625 West 76th Street, Suite 100, Eden Prairie MN 55344. 612/946-4100. **Contact:** Human Resources. **Description:** Manufactures optical and media storage products.

THE PORTLAND GROUP, INC. (PGI)
9150 SW Pioneer Court, Suite H, Wilsonville OR 97070. 503/682-2806. **Fax:** 503/682-2637. **Contact:** Human Resources. **E-mail address:** sales@pgroup.com. **World Wide Web address:** http://www.pgroup.com. **Description:** A developer of high-performance scalar and parallel compilers and software development tools. **Corporate headquarters location:** This Location.

PORTRAIT DISPLAYS, INC.
6665 Owens Drive, Pleasanton CA 94588. 510/227-2700. **Fax:** 510/227-2705. **Contact:** Human Resources. **World Wide Web address:** http://www.portrait.com. **Description:** A manufacturer of pivoting, portrait-capable computer monitors. Product line includes the Pivot 1700, featuring one of the largest vertical page displays on the market. **Listed on:** Privately held.

POWER COMPUTING CORPORATION
2555 North IH 35, Round Rock TX 78664. 512/244-8000. **Contact:** Human Resources. **Description:** Manufactures computers. Power Computing produces Macintosh clones.

PRACTICAL PERIPHERALS INC.
P.O. Box 5024, Thousand Oaks CA 91360. 805/497-4774. **Contact:** Human Resources. **Description:** Manufactures modems and related supplies and equipment.

PRECISION DIGITAL IMAGES CORPORATION
8520 154th Avenue NE, Redmond WA 98052. 206/882-0218. **Fax:** 206/867-9177. **Contact:** Human Resources. **Description:** Manufactures image processing and image collection devices for computers.

PRINTER RESOURCES INC.
1603 South Eastside Loop, Tucson AZ 85710. 520/721-2500. **Contact:** Human Resources. **Description:** Develops technological advances for computer printers.

PRINTRONIX INC.
17500 Cartwright Road, Irvine CA 92714. 714/863-1900. **Contact:** Julie Holder, Human Resources. **Description:** Designs, manufactures, and markets impact line printers and laser printers for use with minicomputers, microcomputers, and other computer systems. **Common positions include:** Accountant/Auditor; Buyer; Computer Programmer; Customer Service Representative; Draftsperson; Electrical/Electronics Engineer; Financial Analyst; Mechanical Engineer; Operations/Production Manager; Quality Control Supervisor; Technical Writer/Editor. **Educational backgrounds include:** Business Administration; Computer Science; Engineering. **Benefits:** Dental Insurance; Disability Coverage; Employee Discounts; Life Insurance; Medical Insurance; Profit Sharing; Savings Plan; Tuition Assistance. **Corporate headquarters location:** This Location. **Operations at this facility include:** Administration; Manufacturing; Research and Development; Sales; Service. **Listed on:** NASDAQ.

PROXIMA CORPORATION
9440 Carroll Park Drive, San Diego CA 92121-2298. 619/457-5500. **Contact:** Human Resources. **World Wide Web address:** http://www.prxm.com. **Description:** A multimedia projector company. Proxima manufactures computer peripheral desktop projectors, used to project images by accessing video signals through a PC's video port. **Listed on:** NASDAQ.

PUMA TECHNOLOGIES
One Tara Boulevard, Suite 210, Nashua NH 03062. 603/888-0666. **Fax:** 603/888-9817. **Contact:** Human Resources. **E-mail address:** info@ilink-corp.com. **World Wide Web address:** http://www.pumatech.com. **Description:** A manufacturer of networking computer products for the facilitation of transfer and translation of data between PCs and mobile devices. The company provides a range of support between platforms including palmtops, PDAs, and software applications.

PYRAMID TECHNOLOGY CORPORATION
3860 North First Street, San Jose CA 95134. 408/428-9000. **Fax:** 408/428-7110. **Contact:** Human Resources. **Description:** A computer services company that produces hardware for open systems.

QMS, INC.
999 Woodcock Road, Suite 104, Orlando FL 32803. 407/228-0380. **Contact:** Human Resources. **Description:** The sales office for a laser printer manufacturer. **Corporate headquarters location:** Mobile AL.

QMS, INC.
P.O. Box 81250, Mobile AL 36689-1250. 334/633-4300. **Contact:** Human Resources. **Description:** Manufactures laser printers. **Corporate headquarters location:** This Location.

QANTEL TECHNOLOGIES
48625 Warm Springs Boulevard, Fremont CA 94539. 510/659-8008. **Contact:** Human Resources. **Description:** Engaged in hardware manufacturing and software development.

QUANTUM CORPORATION
500 McCarthy Boulevard, Milpitas CA 95035. 408/894-4000. **Contact:** Human Resources. **Description:** Quantum designs, manufactures, and markets small hard disk drives used in desktop PCs, workstations, and notebook computers. The company's primary product line is the ProDrive series of high-performance, 3-1/2 inch hard disk drives. Matsushita-Kotobuki Electronics of Japan manufactures some of Quantum's drives. **Corporate headquarters location:** This Location. **Listed on:** NASDAQ. **Stock exchange symbol:** QNTM. **Annual sales/revenues:** $2 billion. **Number of employees nationwide:** 2,455.

QUATECH

662 Wolf Ledge, Akron OH 44312. 330/434-3154. **Toll-free phone:** 800/553-1170. **Fax:** 330/434-1409. **Contact:** Human Resources. **E-mail address:** sales@quatech.com. **World Wide Web address:** http://www.quatech.com. **Description:** A manufacturer of a line of communication, data acquisition, PCMCIA, and control products for IBM PC/XT, PC/AT, PS/2, and compatible systems.

RADIX CORPORATION

4855 Wiley Post Way, Salt Lake City UT 84116. 801/537-1717. **Contact:** Human Resources. **Description:** Manufactures hand-held computer systems used in meter readings and other applications.

RADYNE CORPORATION

5225 South 37th Street, Phoenix AZ 85040. 602/437-9620. **Contact:** Human Resources. **Description:** Manufactures satellite modems.

RARITAN COMPUTER, INC.

10-1 Ilene Court, Belle Mead NJ 08502. 908/874-4072. **Fax:** 908/874-5274. **Contact:** Human Resources. **E-mail address:** sales@raritan.com. **World Wide Web address:** http://www.raritan.com. **Description:** A designer and manufacturer of a line of products for sharing PCs and peripherals. Products include MasterConsole, a keyboard/video/mouse switch; CompuSwitch, a KVM switch allowing central control for up to four PCs; and Guardian, a virtual keyboard and mouse device which emulates keyboard and mouse signals.

READ-RITE CORPORATION

345 Los Coches Street, Milpitas CA 95035. 408/262-6700. **Contact:** Human Resources. **Description:** Read-Rite is an independent supplier of thin film magnetic recording heads for Winchester disk drives. The company entered the head stack assemblies (HSA) market with the late 1991 acquisition of Conner Peripherals and Maxtor's HSA operations. **Number of employees nationwide:** 2,500.

SCI SYSTEMS, INC.

5525 Astrozon Boulevard, Colorado Springs CO 80916. 719/380-5800. **Contact:** Gayla Pipkin, Human Resources Manager. **Description:** This location of SCI Systems, Inc. produces a wide range of assemblies for mass storage products including small tape back-up devices, large multiple-disk array systems, and high-capacity optical storage units. Computers using electronics produced at this location range from personal computers to high-performance workstations. The plant has been selected as the sole supplier of electronics for a major personal computer company's latest high-performance model. Worldwide, SCI Systems, Inc. designs, develops, manufactures, markets, distributes, and services electronic products for the computer, aerospace, defense, telecommunication, medical, and banking industries, as well as for the United States government. SCI is one of the world's largest electronics contract manufacturers and operates one of the largest surface mount technology (SMT) production capacities in the merchant market. Operational activities are conducted through a Commercial Division and a Government Division. The Commercial Division operates in five geographically organized business units: Eastern, Central, and Western Regions of North America; and European and Asian Regions. Each unit operates multiple plants which manufacture components, subassemblies, and finished products primarily for original equipment manufacturers. Design, engineering, purchasing, manufacturing, distribution, and support services are also offered, as well as several groups of proprietary products. The Governmental Division provides data management, instrumentation, communication, and computer subsystems to the U.S. government and its prime contractors and several foreign governments. **Corporate headquarters location:** Huntsville AL. **Other U.S. locations:** Arab AL; Lacey's Spring AL; San Jose CA; Watsonville CA; Graham NC; Hooksett NH; Rapid City SC. **International locations:** Canada; France; Ireland; Mexico; Scotland; Singapore; Thailand.

SCAN-GRAPHICS INC.

700 Abbott Drive, Broomall PA 19008. 610/328-1040. **Contact:** Human Resources. **Description:** Manufactures scanners and develops geo-spacial software.

SEAGATE TAPE

1650 Sunflower Avenue, Costa Mesa CA 92626. 714/641-0279. **Contact:** Human Resources. **Description:** This location manufactures tape drives.

SEAGATE TECHNOLOGY

4585 Scotts Valley Drive, Scotts Valley CA 95066. 408/438-6550. **Contact:** Corporate Employment. **Description:** Seagate Technology is a designer and manufacturer of data storage devices and related products including hard disk drives, tape drives, software, and systems for many different computer-related applications and operating systems. These products include 2.5 and 3.5 inch drives with memory storage capacity between 150 megabytes and one gigabyte. The company sells its products primarily through a sales force to OEMs (original equipment manufacturers) and through non-affiliated distributors. **Common positions include:** Accountant/Auditor; Administrator; Buyer; Chemist; Computer Programmer; Credit Manager; Customer Service Representative; Draftsperson; Electrical/Electronics Engineer; Financial Analyst; Human Resources Manager; Industrial Engineer; Marketing Specialist; Mechanical Engineer; Quality Control Supervisor; Systems Analyst; Technical Writer/Editor. **Educational backgrounds include:** Accounting; Business Administration; Chemistry; Computer Science; Engineering; Finance; Marketing. **Corporate headquarters location:** This Location. **Subsidiaries include:** Arcada Holdings, Inc., an information protection and storage management software company serving several operating systems; and Conners Storage Systems.
Other U.S. locations:
- 3081 Zanker Road, San Jose CA 95134. 408/456-4500. This location is mainly a tech support center with engineering and manufacturing facilities as well.
- 1830 Lefton Circle, Longmont CO 80501. 303/684-1500.
- 5000 Quorum Drive, Suite 485, Dallas TX 75240. 972/448-8050. This location is a sales and engineering office.

SEAGATE TECHNOLOGY

7801 Computer Avenue South, Bloomington MN 55435-5489. 612/844-8000. **Fax:** 612/844-7008. **Contact:** Mary E. Sybrant, Human Resources Specialist. **Description:** This location is a manufacturing facility. Overall, Seagate Technology is a designer and manufacturer of data storage devices and related products including hard disk drives, tape drives, software, and systems for many different computer-related applications and operating systems. These products include 2.5 and 3.5 inch drives with memory storage capacity between 150 megabytes and one gigabyte. The company sells its products primarily through a sales force to OEMs (original equipment manufacturers) and through non-affiliated distributors. **Common positions include:** Chemical Engineer; Mechanical Engineer; Software Engineer; Systems Analyst. **Educational backgrounds include:** Chemistry; Computer Science; Engineering; Physics. **Benefits:** 401(k); Dental Insurance; Disability Coverage; Employee Discounts; Life Insurance; Medical Insurance; Profit Sharing; Tuition Assistance. **Special programs:** Training programs. **Corporate headquarters location:** Scotts Valley CA. **Other U.S. locations:** CA; OK. **Annual sales/revenues:** More than $100 million. **Number of employees at this location:** 3,500. **Number of employees nationwide:** 89,000.

SELECTECH, LTD.

100 Allen Brook Lane, Williston VT 05495. 802/878-9600. **Contact:** John Catalano, President. **Description:** Manufactures infrared remote controls to aid consumers in computer-based presentations.

SHARP MANUFACTURING COMPANY OF AMERICA

Sharp Plaza Boulevard, Memphis TN 38193. 901/795-6510. **Contact:** Human Resources. **Description:** Manufactures computers and related equipment.

SHIP ANALYTICS INC.

183 Prov New London Turnpike, North Stonington CT 06359. 860/535-3092. **Contact:** Linda Gleason, Human Resources. **Description:** Designs and manufactures computer hardware and software that is incorporated into the company's ship simulators.

SIEMENS NIXDORF INFORMATION SYSTEMS, INC.
200 Wheeler Road, Burlington MA 01803. 617/273-0480. **Fax:** 617/221-0231. **Contact:** Human Resources/Employment Manager. **Description:** A manufacturer of computer systems, software, and peripherals. The company also offers consulting, planning, and implementation services. **NOTE:** The manufacturing facility in North Reading can also be contacted at this phone number. **Common positions include:** Customer Service Representative; Software Engineer; Systems Analyst. **Educational backgrounds include:** Business Administration; Communications; Computer Science; Marketing. **Benefits:** 401(k); Dental Insurance; Disability Coverage; Life Insurance; Medical Insurance; Tuition Assistance. **Corporate headquarters location:** This Location. **Operations at this facility include:** Administration; Sales; Service. **Listed on:** Privately held.
Other U.S. locations:
- 4701 Trousdale Drive, Suite 208, Nashville TN 37220-1320. 615/333-2610.

SILICON GRAPHICS/CRAY RESEARCH
SOFTWARE DEVELOPMENT
655 Lone Oak Drive, Eagan MN 55121. 612/452-6650. **Contact:** Carolyn Harrington, Personnel Director. **E-mail address:** crayjobs@cray.com. **World Wide Web address:** http://www.cray.com. **Description:** Silicon Graphics/Cray Research is an international producer of supercomputers and related peripherals, software, and operating systems. The company's machines are used in aircraft and construction simulation, weather forecasting, seismic analysis, data processing, nuclear and geophysical research, and automobile design. **Corporate headquarters location:** This Location.
Other U.S. locations:
- 1620 Olson Drive, Chippewa Falls WI 54729.

SKY COMPUTERS INC.
27 Industrial Avenue, Chelmsford MA 01824. 508/250-1920. **Contact:** Human Resources Department. **Description:** A manufacturer of computer equipment.

SMARTRONICS, INC.
P.O. Box 310, East Derry NH 03041. 603/437-1975. **Fax:** 603/434-5470. **Contact:** Human Resources. **E-mail address:** info@smartronics.com. **World Wide Web address:** http://www.smartronics.com. **Description:** A manufacturer of computer hardware such as programmable logic controllers, multi-drop networks, barcode entry terminals and controllers, and credit card readers. **NOTE:** The company's physical location is 37 Crystal Avenue, East Derry NH 03038.

SOFT VIEW COMPUTER PRODUCTS CORPORATION
2 Foxpoint Center, 6 Denny Road, Suite100, Wilmington DE 19809. 302/762-9229. **Contact:** Human Resources. **Description:** A manufacturer of a variety of computer-related products including keyboards, eyesaver screens, and ergonomic chairs.

SOFTEX, INC.
P.O. Box 7305, Columbia SC 29202. 803/739-9000. **Contact:** Human Resources. **Description:** A computer manufacturer.

SOURCE TECHNOLOGIES, INC.
628 Griffith Road, Charlotte NC 28217. 704/522-8500. **Contact:** Human Resources. **Description:** A manufacturer of computer printers.

SPECTRA INC.
P.O. Box 68-C, Hanover NH 03755. 603/643-4390. **Contact:** Human Resources. **Description:** Manufactures inkjet printer heads.

SPECTRA LOGIC
1700 North 55th Street, Boulder CO 80301. 303/449-7759. **Fax:** 303/939-8844. **Contact:** Human Resources. **E-mail address:** davej@spectralogic.com. **World Wide Web address:** http://www.spectralogic.com. **Description:** A manufacturer of backup hardware and software, and automated tape libraries. **Number of employees at this location:** 150.

SPECTRAGRAPHICS

9877 Waples Street, San Diego CA 92121. 619/450-0611. **Fax:** 619/450-0218. **Contact:** Deb Reinhard, Manager/Human Resources. **E-mail address:** akin@spectra.com. **World Wide Web address:** http://www.spectra.com. **Description:** The company designs and manufactures graphic workstations for the computer-aided design and manufacturing industries. **Common positions include:** Accountant/Auditor; Buyer; Computer Programmer; Software Engineer; Technical Writer/Editor. **Educational backgrounds include:** Computer Science. **Benefits:** 401(k); Dental Insurance; Disability Coverage; Life Insurance; Medical Insurance; Profit Sharing; Tuition Assistance. **Corporate headquarters location:** This Location. **Operations at this facility include:** Administration; Divisional Headquarters; Manufacturing; Research and Development. **Number of employees at this location:** 180. **Number of employees nationwide:** 230.

SPRENGNETHER INSTRUMENTS CORPORATION

4150 Laclede Avenue, St. Louis MO 63108. 314/535-1682. **Contact:** Human Resources. **Description:** Manufactures seismological and geophysical hand-held minicomputers mainly used by the Forest Service. Sprengnether Instruments also develops the applicable software for these systems.

STORAGE COMPUTER CORPORATION

11 Riverside Street, Nashua NH 03062-1373. 603/880-3005. **Fax:** 603/889-7232. **Contact:** Personnel Manager. **World Wide Web address:** http://www.storage.com. **Description:** Storage Computer Corporation designs, manufactures, and sells standards-based, high-performance, fault-tolerant storage solutions for use in client/server, online transaction processing, large database, multimedia, video-on-demand, and high-volume imaging applications. Worldwide distribution and services are provided through a comprehensive network of subsidiaries, joint ventures, resellers, integrators, and service providers in the United States, Latin America, Asia, Africa, and Europe. **Common positions include:** Accountant/Auditor; Buyer; Computer Programmer; Draftsperson; Electrical/Electronics Engineer; Mechanical Engineer; Quality Control Supervisor. **Educational backgrounds include:** Engineering. **Corporate headquarters location:** This Location. **Operations at this facility include:** Administration; Manufacturing; Research and Development; Sales; Service. **Listed on:** American Stock Exchange. **Number of employees at this location:** 60.

STORAGE SOLUTIONS INC.

550 West Avenue, Stamford CT 06902. 203/325-0035. **Contact:** Human Resources. **Description:** Manufactures computer storage systems.

STORAGE TECHNOLOGY CORPORATION

2270 South 88th Street, Louisville CO 80028. 303/673-5151. **Contact:** Human Resources. **World Wide Web address:** http://www.stortek.com. **Description:** Storage Technology Corporation supplies high-performance computer information storage and retrieval systems for mainframe and mid-frame computers and networks. Products include automated cartridge systems, random access subsystems, and fault-tolerant disk arrays. The company also distributes equipment; sells new peripherals, software, and hardware; and offers support services. **Corporate headquarters location:** This Location. **Operations at this facility include:** Administration; Divisional Headquarters; Manufacturing; Research and Development. **Listed on:** New York Stock Exchange. **Other U.S. locations:**

- 12200 Tech Road, Silver Spring MD 20904. 410/622-7866. This location is a sales and marketing office.
- 7600 Boone Avenue North, Minneapolis MN 55428. 612/339-6161. This location is a sales and technical support office.
- 349 South 200 East, Suite 540, Salt Lake City UT 84111. 801/322-3435. This location is an administrative facility.

STRATUS COMPUTER, INC.

2065 Hamilton Avenue, San Jose CA 95125. 408/559-5300. **Contact:** Human Resources. **E-mail address:** resumes@stratus.com. **World Wide Web address:** http://www.stratus.com. **Description:** This location is a development facility. Stratus offers a broad range of computer platforms, application solutions, middleware, and professional services for critical online operations. **Common positions include:** Software Engineer. **Corporate headquarters location:** Marlborough

MA. **Subsidiaries include:** Shared Systems Corporation, a provider of software and professional services to the financial services, retail, and health care industries; SoftCom Systems, Inc., a provider of data communications middleware and related professional services that bridge the gap between open distributed systems and legacy mainframe and mid-range systems used for online applications; and Isis Distributed Systems, Inc., a developer of advanced messaging middleware products that enable businesses to develop reliable, high-performance distributed computing applications involving networked desktop computers and shared systems.
Other U.S. locations:
• 4455 East Camelback Road, Suite 115A, Phoenix AZ 85018. 602/852-3000. This location has engineering, customer service, and sales personnel.
• 600 Embassy Row NE, Suite 460, Atlanta GA 30328. 770/392-1402. Offers sales and services for mainframe computer platforms for the Stratus computer.

SUN MICROSYSTEMS INC.
2550 Garcia Avenue, Mountain View CA 94043. 415/336-0470. **Contact:** Human Resources. **E-mail address:** staffing@sun.com. **World Wide Web address:** http://www.sun.com. **Description:** Sun Microsystems produces high-performance computer systems, workstations, servers, CPUs, peripherals, and operating system software. The company developed its own microprocessor called SPARC. Most products are sold to the engineering, scientific, technical, and commercial markets worldwide. **NOTE:** The company maintains an employment page at its Website. **Corporate headquarters location:** This Location. **Listed on:** NASDAQ.
Other U.S. locations:
• 7150 Campus Drive, Suite 150, Colorado Springs CO 80920. 719/528-4600. This location is a sales office.
• 2 Omni Way, Chelmsford MA 01824. 508/442-0000.
• 1700 Louisiana NE, Suite 300, Albuquerque NM 87110. 505/268-3350. This location is a sales office.
• 400 Atrium Drive, Somerset NJ 08873. 908/469-1000. This location is a sales office.
• 8705 SW Nimbus Avenue, Suite 300, Beaverton OR 97008. 503/627-0451. This location is a small sales office.

SYMBIOS LOGIC
3718 North Rock Road, Wichita KS 67226. 316/636-8000. **Contact:** Human Resources. **Description:** Formerly NCR Peripheral, this location of Symbios Logic manufactures and assembles computer peripheral equipment, including data transport and storage products.

SYMBOL TECHNOLOGIES INC.
340 Fischer Avenue, Costa Mesa CA 92626. 714/549-6000. **Contact:** Kathleen Deema, Human Resources. **Description:** Designs, manufactures, and sells various lines of portable and non-portable computers and systems for business information and bill collection applications. Clients include retail food stores, drug stores, and hardware stores. **NOTE:** Resumes should be sent to the Human Resources Department located at One Symbol Plaza, Holtsville NY 11742-1300. **Corporate headquarters location:** This Location. **Listed on:** American Stock Exchange.

SYQUEST TECHNOLOGY
47071 Bayside Parkway, Fremont CA 94538-6517. 510/226-4000. **Contact:** Human Resources. **Description:** Designs, develops, manufactures, and markets removable cartridge Winchester disk drives and data storage cartridges. **Number of employees nationwide:** 1,087.

SYQUEST TECHNOLOGY
3005 Center Green Drive, Suite 100, Boulder CO 80301. 303/938-2900. **Contact:** Human Resources. **Description:** Designs, develops, manufactures, and markets removable cartridge Winchester disk drives and data storage cartridges. **Number of employees nationwide:** 1,087.

SYSTEM CONNECTION
441 East Bay Boulevard, Provo UT 84606. 801/373-9800. **Contact:** Human Resources. **Description:** A manufacturer and distributor of computer cables and related equipment. **Number of employees at this location:** 100.

SYSTEM INDUSTRIES, INC./MTI

474 Portrero Avenue, Sunnyvale CA 94086. 408/730-1664. **Contact:** Heidi Hegel, Human Resources Manager. **Description:** System Industries, Inc./MTI manufactures, sells, and services disk and tape drive subsystems for DEC Computers. **Common positions include:** Accountant/Auditor; Administrator; Attorney; Blue-Collar Worker Supervisor; Buyer; Computer Programmer; Credit Manager; Customer Service Representative; Department Manager; Draftsperson; Electrical/Electronics Engineer; Financial Analyst; Human Resources Manager; Industrial Engineer; Marketing Specialist; Mechanical Engineer; Operations/Production Manager; Public Relations Specialist; Quality Control Supervisor; Services Sales Representative; Systems Analyst; Technical Writer/Editor; Transportation/Traffic Specialist. **Benefits:** Dental Insurance; Disability Coverage; Life Insurance; Medical Insurance; Pension Plan; Tuition Assistance. **Corporate headquarters location:** This Location. **Operations at this facility include:** Administration; Manufacturing; Research and Development; Service. **Listed on:** American Stock Exchange.

TV ONE MULTIMEDIA SOLUTIONS

1445 Jamike Drive, Suite 8, Erlanger KY 41018. 606/282-7303. **Fax:** 606/282-8225. **Contact:** Human Resources Department. **World Wide Web address:** http://www.tvone.com. **Description:** TV One offers a wide range of desktop video and presentation hardware and software products.

TALLY CORPORATION

5762 Peladeau Street, Emeryville CA 94608. 510/524-3950. **Fax:** 510/524-9954. **Contact:** Human Resources. **Description:** Designs, manufactures, and markets a family of color printers and presentation hardware and software products for the business and government marketplaces. These products work with PC, PS/2, and Apple Macintosh systems. The company's PostScript color printers includes both thermal wax transfer and dye-sublimation printers. The products all use a flexible RISC-based architecture that offers users a range of options and features. In 1992, the company introduced VideoShow HQ, its third-generation electronic presentation product. More recently, the company introduced VideoShow PRESENTER, its advanced hand-held remote control product for the PC Windows marketplace. In addition, the company's product line includes a number of other hardware and software presentation products and color printer supplies. Tally Corporation sells its products worldwide through a network of dealers and distributors in the U.S., Canada, Western Europe, Australia, and the Far East. The company's customers cover a broad range of industries and include many *Fortune* 500 companies. The company was founded in 1981 and has headquarters and manufacturing facilities in Berkeley CA and regional field personnel in the United States and Europe.

TALLY PRINTERS

P.O. Box 97018, Kent WA 98064-9718. 206/251-5500. **Contact:** Personnel Department. **Description:** Manufactures laser, serial, and line matrix printers.

TANDEM COMPUTERS INC.

901 Page Avenue, Fremont CA 94538. 510/354-4000. **Contact:** Human Resources. **Description:** Manufactures fault-tolerant computer systems.

TAPEDISK CORPORATION

200 Main Street, Suite 210, Menomonie WI 54751. 715/235-3388. **Fax:** 715/235-3818. **Contact:** Human Resources. **E-mail address:** tapedisk@tapedisk.com. **World Wide Web address:** http://www.tapedisk.com. **Description:** A manufacturer of tape-based secondary storage systems and crash guards.

TECMAR TECHNOLOGIES, INC.

1900 Pike Road, Suite E, Longmont CO 80501. 303/682-3700. **Toll-free phone:** 800/4BACKUP. **Fax:** 303/776-7706. **Contact:** Human Resources Department. **E-mail address:** hr@tecmar.com. **World Wide Web address:** http://www.tecmar.com. **Description:** A developer and manufacturer of tape storage technology including tape drives and tape backup products. Product line includes Wangtek, tape drives; WangDAT, 4mm digital audio tape drives; and ProLine, networking storage environments.

TEK INDUSTRIES, INC.
6 Progress Drive, Manchester CT 06040. 860/647-8738. **Contact:** Human Resources. **Description:** A manufacturer of computer-related products including touch-memory devices, smartcards, and telecommunications products.

TEKTRONIX, INC.
P.O. Box 500, M/S 55-545, Beaverton OR 97077-0001. 503/627-7111. **Contact:** Professional Staffing. **Description:** A large producer of electronic test and measurement, computer graphics, and communications equipment. Test and measurement products include oscilloscopes, logic analyzers, digitizers, and curve tracers. Communications equipment include vectorscopes, waveform monitors, signal generators, cable and fiber-optic testers, demodulators, and television routing and switching items. Computer graphics products include printers and terminals primarily for scientific and engineering uses. **Common positions include:** Broadcast Technician; Ceramics Engineer; Chemical Engineer; Chemist; Computer Programmer; Credit Manager; Customer Service Representative; Electrical/Electronics Engineer; General Manager; Human Resources Manager; Industrial Production Manager; Materials Engineer; Mechanical Engineer; Metallurgical Engineer; Operations/Production Manager; Public Relations Specialist; Purchasing Agent/Manager; Software Engineer; Systems Analyst; Technical Writer/Editor. **Educational backgrounds include:** Business Administration; Chemistry; Computer Science; Engineering; Finance. **Benefits:** 401(k); Dental Insurance; Disability Coverage; Employee Discounts; Life Insurance; Medical Insurance; Paid Vacation; Pension Plan; Profit Sharing; Sick Days; Tuition Assistance. **Special Programs:** Internships. **Corporate headquarters location:** This Location. **Operations at this facility include:** Administration; Divisional Headquarters; Manufacturing; Regional Headquarters; Research and Development; Sales; Service. **Listed on:** New York Stock Exchange. **Number of employees at this location:** 6,000.
Other U.S. locations:
• 3003 Bunker Hill Lane, Santa Clara CA 95052-8086. 408/496-0800. Contact: Sam Dunn, Human Resources. This location is a sales office.
• 6400 Enterprise Lane, Suite 200, Madison WI 53719. 608/274-8686. This location is a development and sales facility.

TEKTRONIX, INC.
P.O. Box 6026, Gaithersburg MD 20884. 301/948-7151. **Fax:** 301/921-0461. **Contact:** Human Resources. **Description:** Manufactures and markets electronic test and measurement equipment and a highly specialized line of computer graphics terminals and peripherals for the engineering and scientific market. **NOTE:** The physical address is 700 Professional Drive, Gaithersburg MD. **Common positions include:** Applications Engineer; Computer Programmer; Electrical/Electronics Engineer; Manufacturer's/Wholesaler's Sales Rep.; Sales Engineer; Systems Analyst. **Educational backgrounds include:** Business Administration; Computer Science; Engineering; Physics. **Benefits:** Daycare Assistance; Dental Insurance; Disability Coverage; Employee Discounts; Life Insurance; Medical Insurance; Pension Plan; Profit Sharing; Tuition Assistance. **Corporate headquarters location:** Beaverton OR. **Operations at this facility include:** Sales; Service. **Listed on:** New York Stock Exchange.

TELEBYTE TECHNOLOGY, INC.
270 Pulaski Road, Greenlawn NY 11740. 516/423-3232. **Fax:** 516/385-8184. **Contact:** Michael Breneisen, Controller. **Description:** Telebyte is a developer, manufacturer, and marketer of data communication and personal computer video products. The company's products are sold to OEMs, systems integrators, value-added resellers, and end users who develop communication systems based on serial data communication protocols and PC platforms. Telebyte's products provide solutions for connectivity in the areas of short haul modems, local area multiplexors, interface converters, fiber-optic communications devices, data communications test and monitoring equipment, switching devices, lightning and surge protection, and PC video applications. **Corporate headquarters location:** This Location.

TELEPAD CORPORATION
380 Herndon Parkway, Suite 1900, Herndon VA 20170. 703/834-9000. **Contact:** Patricia Askew, Human Resources Director. **Description:** TelePad designs, develops, and markets pen-based computing and mobile communications systems. By mid-1993, the company had developed

prototypes and completed its first production units of its TelePad SL, a portable, clipboard-sized pen computer weighing less than four and three-quarter pounds, including its battery. The company is now marketing the TelePad SL, related software and hardware products, and solutions to help businesses and government agencies manage, communicate, and process information. The company's products have been designed to make it easier to collect and transmit data by mobile, remote workers and integrate this data into the user's information system in order to create a field automation system.

TELOS CORPORATION
19886 Ashburn Road, Ashburn VA 20147. 703/724-3800. **Fax:** 703/318-1895. **Contact:** Human Resources Department. **Description:** This location of Telos Corporation manufactures PC-compatible computers and is also the company's corporate headquarters. Overall, Telos Corporation is a leader in providing complete computer information management solutions. The company produces hardware and software, and also offers integration, networking, and support services. **Common positions include:** Accountant/Auditor; Buyer; Draftsperson; Electrical/Electronics Engineer; Financial Analyst; Industrial Engineer; Mechanical Engineer; Operations/Production Manager; Quality Control Supervisor; Receptionist; Secretary; Software Engineer; Stock Clerk; Systems Analyst; Technical Writer/Editor. **Educational backgrounds include:** Accounting; Business Administration; Economics; Finance; Marketing. **Benefits:** Daycare Assistance; Dental Insurance; Disability Coverage; Life Insurance; Medical Insurance; Pension Plan; Tuition Assistance. **Corporate headquarters location:** This Location. **Operations at this facility include:** Administration; Manufacturing; Research and Development; Sales; Service. **Listed on:** NASDAQ. **Number of employees at this location:** 280. **Number of employees nationwide:** 2,100.

TELXON CORPORATION
P.O. Box 5582, Akron OH 44334-0582. 330/867-3700. **Contact:** Employee Services. **Description:** Manufactures hand-held microcomputer systems. **Corporate headquarters location:** This Location.

TELXON CORPORATION
3050 Citrus Circle, Suite 104, Walnut Creek CA 94596. 510/682-7044. **Contact:** Human Resources Department. **Description:** Telxon Corporation manufactures hand-held microcomputer systems.

TERA COMPUTER COMPANY
2815 Eastlake Avenue East, Seattle WA 98102. 206/325-0800. **Fax:** 206/325-2433. **Contact:** Human Resources Department. **E-mail address:** resume@tera.com. **World Wide Web address:** http://www.tera.com. **Description:** Tera Computer Company is a developer of supercomputer technologies. **Corporate headquarters location:** This Location.

TEXAS INSTRUMENTS, INC.
P.O. Box 6102, Temple TX 76502. 817/774-6001. **Contact:** Human Resources Department. **Description:** This location is a manufacturing facility for notebooks. Texas Instruments is one of the world's largest suppliers of semiconductor products. In 1954, the company commercialized the silicon transistor and, in 1958, created the first integrated circuit. TI's defense electronics business is a leading supplier of avionics, infrared, and weapons guidance systems to the U.S. Department of Defense and U.S. allies. The company is also a technology leader in high-performance notebook computers and model-based software development tools. TI sensors monitor and regulate pressure and temperature in products ranging from automobiles to air conditioning systems. **NOTE:** The company is located at 5701 Airport Road, Temple TX.

TEXAS MICROSYSTEMS, INC.
P.O. Box 42963, Houston TX 77242. 713/541-8200. **Toll-free phone:** 800/627-8700. **Contact:** Human Resources Department. **World Wide Web address:** http://www.texmicro.com. **Description:** Texas Microsystems is a provider of internetwork servers for the Novell marketplace. Texas Microsystems has an OEM agreement with Novell and manufactures industry standard servers, CPUs, chassis, systems, mobile systems, and monitors. Founded in 1976.

THOMAS-CONRAD CORPORATION
12301 Technology Boulevard, Austin TX 78727. 512/433-6000. **Fax:** 512/836-2840. **Contact:** Human Resources. **Description:** A computer peripheral manufacturer. Thomas-Conrad Corporation manufactures such products as monitors and printers.

TOSHIBA AMERICA INFORMATION SYSTEMS
P.O. Box 19724, Irvine CA 92713-9724. 714/583-3000. **Contact:** Human Resources. **Description:** Develops, markets, and supports computers, printers, faxes, and more.

TRANSITION NETWORKS
6475 City West Parkway, Minneapolis MN 55344. 612/941-7600. **Toll-free phone:** 800/526-9267. **Fax:** 612/941-2322. **Contact:** Human Resources. **E-mail address:** info@transition.com. **World Wide Web address:** http://www.transition.com. **Description:** A manufacturer of computer networking hardware. **Corporate headquarters location:** This Location.

TRI-PLEX SYSTEMS CORPORATION
9790-A Patuxent Woods Road, Columbia MD 21046. 410/290-7711. **Contact:** Human Resources. **Description:** A manufacturer of hardware for supercomputers.

TRIAD SYSTEMS CORPORATION
3055 Triad Drive, Livermore CA 94550. 510/449-0606. **Contact:** Human Resources. **Description:** Manufactures computers.

TRICORD SYSTEMS INC.
2800 NW Boulevard, Plymouth MN 55441. 612/557-9005. **Contact:** Human Resources. **Description:** A manufacturer of computer servers and related equipment.

TRIDENT MICROSYSTEMS
189 North Bernardo Drive, Mountain View CA 94043-5203. 415/691-9211. **Contact:** Rebecca Fernandez, Human Resources. **Description:** Designs, develops, and markets large-scale integrated circuit graphics products, including graphics controllers for IBM-compatible PCs; flat-panel controllers for notebook PCs; and video processing products for PCs. **Number of employees at this location:** 114.

UFO SYSTEMS, INC.
One Tobey Village Office Park, Pittsford NY 14534. 716/248-3372. **Contact:** Human Resources. **Description:** A manufacturer of computer hardware and provider of support services.

U.S. ROBOTICS, INC.
8100 North McCormick Boulevard, Skokie IL 60076. 847/982-5010. **Contact:** Human Resources. **World Wide Web address:** http://www.usr.com. **Description:** U.S. Robotics, Inc. is one of the largest independent manufacturers of modems; LAN and WAN access products and accessories; and computer communications software in the United States. **NOTE:** In February 1997, media reports indicated that 3Com planned a buyout of U.S. Robotics. **Common positions include:** Accountant/Auditor; Customer Service Representative; Electrical/Electronics Engineer; Technical Writer/Editor. **Number of employees nationwide:** 1,500.

UNISYS CORPORATION
13430 NW Freeway, Suite 500, Houston TX 77040. 713/744-2666. **Contact:** Human Resources Department. **Description:** This location manufactures and sells computers. Overall, Unisys Corporation is a provider of information services, technology, and software. The company employs 49,000 people in 100 countries. Unisys specializes in developing business critical solutions based on open information networks. The company's enabling software team creates a variety of software projects which facilitate the building of user applications and the management of distributed systems. The company's platforms group is responsible for UNIX operating systems running across a wide range of multiple processor server platforms, including all peripheral and communication drivers. The Unisys commercial parallel processing group develops a microkernel-based operating system, I/O device driver development, ATM hardware development, diagnostics, and system architects. The system management group is chartered with the overall management of

development programs for UNIX desktop and entry server products. **Corporate headquarters location:** Blue Bell PA.

UTICOR TECHNOLOGY, INC.
4140 Utica Ridge Road, Bettendorf IA 52722. 319/359-7501. **Contact:** Human Resources Department. **Description:** Uticor Technology, Inc. is a manufacturer of computer hardware and monitors.

VERBATIM CORPORATION
1200 W.T. Harris Boulevard, Charlotte NC 28262. 704/547-6500. **Contact:** Human Resources. **Description:** Verbatim manufactures computer disks, diskettes, and CDs.

VERTEX TECHNOLOGIES INC.
6671 Santa Barbara Road, Elkridge MD 21227. 410/796-0044. **Contact:** Dave Taylorson, Branch Manager. **Description:** Manufactures computer cables.

VIEWSONIC CORPORATION
20480 Business Parkway, Walnut CA 91789. 909/869-7976. **Contact:** Human Resources. **Description:** Viewsonic manufactures monitors for computers.

VISICOM LABORATORIES INC.
10052 Mesa Ridge Court, San Diego CA 92121. 619/457-2111. **Contact:** Human Resources. **Description:** A manufacturer of video graphics cards and software.

VISIONEER, INC.
34800 Campus Drive, Fremont CA 94555. 415/812-6400. **Contact:** Human Resources Director. **Description:** Manufactures office computer peripherals and software. Product line includes PaperPort, a document scanner which converts paper documents into monitor-displayed text; FormTyper, a software application which is compatible with PaperPort for the production of forms; and SCSI adapters.

VOCOLLECT INC.
701 Rodi Road, Pittsburgh PA 15235. 412/829-8145. **Contact:** Howard Eklund, Controller. **Description:** Vocollect Inc. is a manufacturer of voice activated computer-based data collection systems.

WEN TECHNOLOGY CORPORATION
103 Fairview Park Drive, Elmsford NY 10523. 914/347-7674. **Fax:** 914/347-4129. **Contact:** Sheree Wen, President. **Description:** A manufacturer of computer monitors and displays as well as other electronic products. **NOTE:** The company offers entry-level positions. **Common positions include:** Account Manager; Account Representative; Administrative Assistant; Applications Engineer; Buyer; Chief Financial Officer; Credit Manager; Customer Service Representative; Design Engineer; Designer; Electrical/Electronics Engineer; Industrial Engineer; Manufacturing Engineer; Mechanical Engineer; Multimedia Designer; Project Manager; Purchasing Agent/Manager; Quality Control Supervisor; Sales Engineer; Sales Executive; Sales Manager; Sales Representative; Secretary; Systems Manager. **Educational backgrounds include:** Accounting; Computer Science; Engineering; Finance; Marketing; Physics. **Benefits:** Medical Insurance. **Corporate headquarters location:** This Location. **Listed on:** Privately held. **Annual sales/revenues:** $51 - $100 million. **Number of employees at this location:** 40.

WPI TERMIFLEX
316 Daniel Webster Highway, Merrimack NH 03054. 603/424-3700. **Contact:** Patti Andersen, Human Resources Director. **Description:** Manufactures hand-held computers and terminals.

WABASH COMPUTER PRODUCTS
P.O. Box 470848, Tulsa OK 74147. 918/254-9704. **Contact:** Phil Odom, President. **Description:** Manufactures and sells its own brand of storage media.

WACOM TECHNOLOGY CORPORATION
501 SE Columbia Shores Boulevard, #300, Vancouver WA 98661. 360/750-8882. **Contact:** Human Resources. **Description:** Manufactures graphic tablets and graphic digitizing equipment for use with computers.

WANG FEDERAL
7900 Westpark Drive, McLean VA 22102. 703/827-3000. **Contact:** Human Resources. **Description:** A sales office for government contracts. Wang is a manufacturer of computer hardware.

WESTERN DIGITAL CORPORATION
DRIVE ENGINEERING DIVISION
5863 Rue Ferrari, San Jose CA 95138. 408/365-1190. **Contact:** Marianne Falk, Human Resources. **Description:** Western Digital primarily makes small-form factor rigid disk drives for PCs and file servers, and also integrates circuits and board products for graphics, storage control, and systems logic. Prior to March 1988, the company primarily made disk drive controllers, which are now integrated into disk drives. In March 1988, the company acquired Tandy Corporation's disk drive operations and began to make 3.5 inch magnetic disk drives for the IDE (intelligent drive electronics) interface.

WIN LABORATORIES
11090 Industrial Road, Manassas VA 20109-3958. 703/330-1426. **Contact:** Tina Shaffer, Human Resources. **Description:** A manufacturer of IBM-compatible computers. **Corporate headquarters location:** This Location.

WOLF COMPUTER SYSTEMS
7313 Ashcroft Drive, Suite 208, Houston TX 77081. 713/995-0279. **Contact:** Human Resources. **Description:** Wolf Computer Systems custom-builds computers, manufactures computer peripherals, develops anti-virus software, and installs computer networks.

WYSE TECHNOLOGY
3471 North 1st Street, San Jose CA 95134. 408/473-1200. **Contact:** Human Resources. **Description:** Manufactures computer terminals.

XANTE CORPORATION
4621 Springhill Avenue, Mobile AL 36608. 334/342-4840. **Contact:** Human Resources. **Description:** Manufactures laser printers.

XATA CORPORATION
500 East Travelers Trail, Burnsville MN 55337. 612/894-3680. **Contact:** Department of Interest. **Description:** Manufactures on-board computers for the trucking industry.

XIRCOM, INC.
2300 Corporate Center Drive, Thousand Oaks CA 91320. 805/376-9300. **Contact:** Human Resources. **Description:** A developer of modems and networking adapters.

XYCOM, INC.
750 North Maple Road, Saline MI 48176-1292. 313/429-4971. **Contact:** Joyce C. Girdis, Human Resources Director. **Description:** Develops, manufactures, and sells industrial microcomputers. Applications include the regulation and monitoring of continuous batch processes, and the control and monitoring of material handling equipment. **Common positions include:** Accountant/Auditor; Administrator; Applications Engineer; Buyer; Computer Programmer; Customer Service Representative; Department Manager; Draftsperson; Electrical/Electronics Engineer; Human Resources Manager; Industrial Engineer; Manufacturer's/Wholesaler's Sales Rep.; Marketing Specialist; Mechanical Engineer; Operations/Production Manager; Product Manager; Purchasing Agent/Manager; Software Engineer; Technical Writer/Editor; Technician; Test Engineer. **Educational backgrounds include:** Computer Science; Engineering; Marketing. **Benefits:** 401(k); Dental Insurance; Disability Coverage; Life Insurance; Medical Insurance; Pension Plan; Spending

Account; Tuition Assistance. **Corporate headquarters location:** This Location. **Operations at this facility include:** Administration; Manufacturing; Research and Development.

ZENDEX CORPORATION

6780 Sierra Court, Suite A, Dublin CA 94568. 510/828-3000. **Contact:** Human Resources. **Description:** Manufactures a variety of computers and computer boards, as well as PCs for use mainly in industrial applications.

ZENITH DATA SYSTEM DIRECT

3075 Research Drive, State College PA 16801. 814/238-1820. **Contact:** Human Resources Department. **Description:** A manufacturer of computers and computer software. Facilities here include production, warehouse, sales, repair, and service. **Corporate headquarters location:** Marlborough MA. **Number of employees at this location:** 100.

ZITEL CORPORATION

47211 Bayside Parkway, Fremont CA 94538. 510/440-9600. **Contact:** Human Resources. **Description:** Zitel designs, manufactures, and markets solid-state memory and storage peripherals for mainframe computers.

ZOLTRIX INC.

41786 Christy Street, Fremont CA 94538. 510/657-1188. **Contact:** Human Resources. **Description:** A manufacturer of fax modems as well as various multimedia products.

ZOOM TELEPHONICS INC.

207 South Street, Boston MA 02111. 617/423-1072. **Fax:** 617/423-3923. **Contact:** Martin Levin, Director of Human Resources. **E-mail address:** mlevin@zoomtel.com. **Description:** Zoom designs, produces, and markets internal, external, and PCMCIA modems and fax-modems that are sold internationally as packaged products and built into computers by major PC manufacturers. **Common positions include:** Accountant/Auditor; Design Engineer; Industrial Engineer; Mechanical Engineer; MIS Specialist; Software Engineer; Technical Writer/Editor. **Educational backgrounds include:** Business Administration; Engineering; Marketing. **Benefits:** 401(k); Dental Insurance; Life Insurance; Mass Transit Available; Medical Insurance; Tuition Assistance. **Corporate headquarters location:** This Location. **Operations at this facility include:** Administration; Manufacturing; Sales. **Listed on:** NASDAQ. **Annual sales/revenues:** $51 - $100 million. **Number of employees at this location:** 320.

ZYDACRON

7 Perimeter Road, Manchester NH 03103. 603/647-1000. **Fax:** 603/647-9470. **Contact:** Human Resources. **E-mail address:** info@zydacron.com. **World Wide Web address:** http://www.zydacron.com. **Description:** A designer and manufacturer of video codecs and communication products.

ZYKRONIX INC.

357 Inverness Drive South, Suite C, Englewood CO 80112. 303/799-4944. **Contact:** Human Resources. **Description:** Designs and manufactures PCs that are used mainly in industrial applications.

ZYXEL INC.

4920 East La Palma Avenue, Anaheim CA 92807. 714/693-0804. **Contact:** Human Resources. **Description:** A manufacturer of computer modems.

Note: Because addresses and telephone numbers of smaller companies can change rapidly, we recommend you call each company to verify the information below before inquiring about job opportunities. Mass mailings are not recommended.

Additional employers:

COMPUTER MANUFACTURERS

1-800-Computer Parts Inc.
2000 Derry Ashford Rd, Houston
TX 77099. 713/933-5858.

Action Instruments Inc.
8601 Aero Dr, San Diego CA
92123-1786. 619/279-5726.

Adra Systems Inc.
2 Executive Dr, Chelmsford MA
01824-2558. 508/937-3700.

Advanced Digital Systems Inc.
135 2nd Ave, Waltham MA
02154-1107. 617/431-2211.

Advanced Digital Technologies
334 E Lake Rd, Palm Harbor FL
34685-2427. 813/785-6399.

Advanced Modular Solutions
60 Codman Hill Rd, Boxboro MA
01719-1737. 508/266-9700.

Advanced Processing Labs Inc.
5871 Overlin Dr, San Diego CA
92121. 619/546-8626.

Affinitec Corp.
11737 Administration Dr, Saint
Louis MO 63146-3405. 314/569-
3450.

Altai Inc.
624 Six Flags Dr, Arlington TX
76011-6341. 817/649-1816.

Alvimer Inc.
7325 NW 31st St, Miami FL
33122-1240. 305/592-3921.

Amano Cincinnati Inc.
140 Harrison Avenue, Roseland
NJ 07068. 201/227-8256.

Amax Engineering Corp.
3288 Laurelview Ct, Fremont CA
94538-6535. 510/651-8886.

Amdahl Corporation
5000 Executive Pkwy, Suite 175,
San Ramon CA 94583-4210.
510/244-5200.

Amecom
5115 Calvert Rd, College Park
MD 20740-3898. 301/864-5600.

American Arium Inc.
14281 Chambers Rd, Tustin CA
92780. 714/731-1661.

American Magnetics Corp.
740 Watsoncenter Rd, Carson CA
90745. 213/775-8651.

American Research Corp.
602 Monterey Pass Rd, Monterey
Park CA 91754-2419. 213/265-
0835.

AMTX Inc.
5450 Campus Dr, Canandaigua
NY 14424-8207. 716/396-6800.

AnCor Co.
6130 Blue Circle Dr, Minnetonka
MN 55343-9109. 612/932-4001.

Applied Dynamics International
3800 Stone School Rd, Ann
Arbor MI 48108-2414. 313/973-
1300.

Aqua Measuring Instrument Co.
PO Box 369, La Verne CA
91750-0369. 909/392-5833.

Argo Instruments Inc.
188 Brooke Road, Winchester
VA 22603-5738. 540/665-0200.

ASA Computers
200 Trimble Road, Santa Clara
CA 95131. 408/232-5999.

Atalla Corp.
2304 Zanker Rd, San Jose CA
95131-1115. 408/435-8850.

Atlantic Logic Corp.
41 Canfield Rd, Cedar Grove NJ
07009-1201. 201/857-7878.

Atronics International Inc.
44700 Industrial Dr, Fremont CA
94538-6431. 510/656-8400.

Auto-Gas Systems Inc.
PO Box 6957, Abilene TX 79602.
915/676-3150.

Autocoach
1419 Upfield Dr, Carrollton TX
75006-6921. 214/553-1600.

Autologic Information International Inc.
1050 Rancho Conejo Blvd,
Thousand Oaks CA 91320-1717.
805/498-9611.

Automated Control Concepts
3535 State Route 66, Neptune NJ
07753-2622. 908/922-6611.

Award Software International
777 E Middlefield Rd, Mountain
View CA 94043-4023. 415/968-
3400.

Bailey Controls Division
400 Commons Way, Rockaway
NJ 07866-2044. 201/625-3005.

Barco Chromatics Inc.
2558 Mountain Industrial Blvd,
Tucker GA 30084-3810. 770/493-
7000.

Bas Micro Industries
6 Bendix, Irvine CA 92618.
714/457-8822.

BBN Manufacturing
50 Moulton St, Cambridge MA
02138-1119. 617/873-4000.

Beckman Instruments Inc.
90 Boroline Rd, Allendale NJ
07401-1613. 201/818-8900.

Bedford Control Systems Inc.
6 Executive Park Dr, North
Billerica MA 01862-1319.
508/667-2050.

Bek International
2804 NW 72nd Ave, Miami FL
33122-1310. 305/594-3756.

Bibco Inc.
326 E Main St, Benton Harbor
MI 49022-4414. 616/925-8445.

Bivar Inc.
4 Thomas, Irvine CA 92718-
2593. 714/951-8808.

Bonner Moore Associates Inc.
2727 Allen Pkwy, Houston TX
77019-2100. 713/522-6800.

Boyce Engineering
10555 Rockley Rd, Houston TX
77099-3581. 713/933-7210.

Bright Corp.
40 Shawmut Rd, Canton MA
02021-1409. 617/821-0320.

Business Automation Inc.
1572 N Main St, Orange CA
92667-3448. 714/998-6600.

Byk-Gardner USA, Inc.
Rivers Prk 2, 9104 Guilford Rd,
Columbia MD 21046. 301/483-
6500.

C-Net Technology Inc.
2199 Zanker Rd, San Jose CA
95131-2109. 408/954-8000.

Cambridge Management Corp.
16755 Von Karman Ave, Irvine
CA 92606. 714/261-8901.

**Canon Manufacturing, East
Coast Operations**
185 Monmouth Parkway, West
Long Branch NJ 07764-1019.
908/229-1100.

Cardinal Technologies Inc.
1827 Freedom Rd, Lancaster PA
17601-6759. 717/293-3000.

Cardkey Systems Inc.
1757 Tapo Canyon Rd, Simi
Valley CA 93063-3391. 805/522-
5555.

Centercore Inc.
110 Summit Dr, Exton PA 19341-
2838. 610/989-3980.

Central Data
1602 Newton Dr, Champaign IL
61821-1061. 217/359-8010.

Cerion Technologies
1401 Interstate Dr, Champaign IL
61821-1065. 217/359-3700.

Chaplet Systems USA Inc.
252 N Wolfe Rd, Sunnyvale CA
94086-4510. 408/732-7950.

Chyron Corp.
5 Hub Dr, Melville NY 11747-
3503. 516/845-2000.

Clearpoint Enterprises
25 Birch Street, Suite B41,
Milford MA 01757. 508/473-
6111.

Codar Technology Inc.
2405 Trade Center Ave,
Longmont CO 80503-7602.
303/776-0472.

Cognos Corp.
300 Lighting Way, Secaucus NJ
07094-3622. 201/601-0300.

**Columbus Instruments
International Corp.**
950 N Hague Ave, Columbus OH
43204-2121. 614/276-0861.

Command Data-Dallas
605 E Safari Pkwy, Grand Prairie
TX 75050-2327. 214/262-2692.

Compucard International
6501 NW 36th St, Miami FL
33166-6961. 305/871-9933.

Computer Language Research
2395 Midway Rd, Carrollton TX
75006-2521. 214/250-7000.

Computer Services Group Inc.
22 W 38th St, New York NY
10018-6204. 212/819-0122.

Computrac Inc.
222 Municipal Dr, Richardson
TX 75080-3583. 214/234-4241.

Computran Systems Corp.
100 1st St, Hackensack NJ
07601-2124. 201/489-7500.

Comtech Micro Systems Inc.
7950 Woodruff Ct, Springfield
VA 22151-2109. 703/321-5000.

Comtrade Inc.
1215 Bixby Dr, Hacienda Heights
CA 91745-1708. 818/961-6688.

Contrax Technologies Inc.
7509 Connelley Dr, Hanover MD
21076-1664. 410/760-6611.

Corporate Data Systems Corp.
Halsey Rd, Newton NJ 07860-
7052. 201/383-0754.

Cortron Inc.
1 Aegean Dr, Methuen MA
01844-1560. 508/975-5445.

Cray Research Inc.
3130 Crow Canyon Pl, San
Ramon CA 94583. 510/866-0600.

**Cray Research Inc. Business
Systems**
9480 Carroll Park Dr, San Diego
CA 92121-2256. 619/625-0500.

**Cray Research Inc. Super
Computing**
894 Ross Dr, Sunnyvale CA
94089-1443. 408/745-6466.

**Cray Research Inc.
Superserver Inc.**
8300 SW Creekside Pl, Beaverton
OR 97008-7101. 503/641-3151.

CTG Systems Inc.
PO Box 19676, Charlotte NC
28219-9676. 704/329-0233.

Cuplex Inc.
1500 Highway 66, Garland TX
75040-6727. 214/276-0333.

**Dainippon Screen Engineering
America Inc.**
3700 W Segerstrom Ave, Santa
Ana CA 92704-6410. 714/546-
9491.

Dapsco Inc.
3110 Kettering Blvd, Dayton OH
45439-1924. 937/294-5331.

Data Expert Corp.
1156 Sonora Ct, Sunnyvale CA
94086-5308. 408/737-8880.

Data General Corp.
1500 Rosecrans Ave, Manhattan
Beach CA 90266-3721. 310/643-
1200.

Data General Corp.
3835 N Freeway Blvd,
Sacramento CA 95834-1954.
916/923-5693.

Data General Corp.
Hwy 55 S, Apex NC 27502.
919/362-4800.

Data Trek Inc.
5838 Edison Pl, Carlsbad CA
92008-6596. 760/431-8400.

Datamax Corp.
4501 Parkway Commerce Blvd,
Orlando FL 32808-1089.
407/578-8007.

Decision Data Division II
4421 Stuart Andrew Blvd,
Charlotte NC 28217-1589.
704/527-7620.

**Delco Electronics Corp. Delco
Systems**
6767 Hollister Ave, Goleta CA
93117-3086. 805/961-5011.

Delfin Systems
3000 Patrick Henry Dr, Santa
Clara CA 95054-1839. 408/748-
1200.

Diab Data Inc.
323 Vintage Park Dr, Foster City
CA 94404-1136. 415/571-1700.

**Diamond Flower Electrical
Instruments Co.**
135 Main Ave, Sacramento CA
95838-2041. 916/568-1234.

Digital Equipment Corp.
701 E Gate Dr Ste 300, Mount
Laurel NJ 08054-3838. 609/273-
2000.

Digital Equipment Corp.
5471 Kearny Villa Rd, San Diego
CA 92123-1105. 619/292-1818.

Digital Equipment Corp.
100 Nagog Park, Acton MA
01720-3428. 508/264-7111.

Digital Equipment Corp.
9719 Lincoln Village Dr,
Sacramento CA 95827-3331.
916/362-2420.

Digital Equipment Corp.
1 Federal St, Springfield MA
01105-1160. 413/747-2800.

Digital Equipment Corp.
165 Dascomb Rd, Andover MA
01810-5886. 508/493-5111.

Digital Equipment Corp.
2809 Emerywood Pkwy,
Richmond VA 23294-3743.
804/756-1700.

Digital Sound Corp.
6307 Carpinteria Ave, Carpinteria
CA 93013-2901. 805/566-2000.

DP Technology Corp.
1150 Avenida Acaso, Camarillo
CA 93012-8719. 805/388-6000.

DTK Computer Inc. of Texas
10531 Wilcrest Dr, Houston TX
77099-2821. 713/568-6688.

Dyna Micro Inc.
48434 Milmont Dr, Fremont CA
94538-7326. 510/438-0233.

**Dynamic Analysis & Test
Associates**
2255 Faraday Ave, Carlsbad CA
92008-7209. 760/931-9511.

Dynamic Decisions Inc.
375 Raritan Center Pkwy, Edison
NJ 08837-3920. 908/225-8868.

Dynamic Technology Inc.
10901 Bren Rd E, Minnetonka
MN 55343-4410. 612/938-8280.

EIS Inc.
555 Herndon Pkwy, Herndon VA
22070-5226. 703/478-9808.

Electro Design Manufacturing
PO Box 2208, Decatur AL
35602-2208. 205/353-3855.

Electro-Radiation Inc.
39 Plymouth St, Fairfield NJ
07004-1615. 201/808-9033.

Electronic Designs Inc.
1 Research Dr, Westborough MA
01581-3922. 508/366-5151.

ELO Engineering Inc.
7770 Ranchers Rd NE,
Minneapolis MN 55432-2521.
612/571-2820.

Empac International Corp.
47490 Seabridge Dr, Fremont CA
94538-6548. 510/683-8800.

Energyline Systems Inc.
2065 Kittredge St, Berkeley CA
94704-1400. 510/644-8182.

Fidelis Group Inc.
94 Bridge St, Newton MA 02158-
1119. 617/964-9020.

First Equipment
4851 Keller Springs Rd, Dallas
TX 75248-5928. 214/380-2300.

First Interntional Computer
5020 Brandin Ct, Fremont CA
94538-3140. 510/252-7777.

Franklin Electronic Publishing
1 Franklin Plaza, Burlington NJ
08016. 609/386-2500.

Fujitsu Personal Systems Inc.
5200 Patrick Henry Dr, Santa
Clara CA 95054. 408/982-9500.

Fujitsu-ICL Systems Inc.
5429 Lyndon B Johnson Fwy,
Dallas TX 75240-2607. 214/716-
8300.

Gagetalker Corp.
11415 NE 128th St, Kirkland WA
98034-6332. 206/821-3200.

GBI Data & Sorting Systems
2030 Coolidge St, Hollywood FL
33020-2428. 954/920-0225.

Geac Computer
3707 W Cherry St, Tampa FL
33607-2528. 813/872-9990.

**Geac Control Transaction
Corp.**
130 Clinton Rd, Fairfield NJ
07004-2914. 201/575-9100.

Gemini Computers Inc.
PO Box 222417, Carmel CA
93922-2417. 408/373-8500.

General Digital Corp.
160 Chapel, Manchester CT
06040. 860/643-9260.

General Meters Corp.
1935 Dominion Way, Colorado
Springs CO 80918-8451.
719/522-9222.

Granite Communications Inc.
9 Townsend W, Nashua NH
03063-1217. 603/881-8666.

Graphics Microsystems Inc.
1284 Forgewood Ave, Sunnyvale
CA 94089-2215. 408/745-7745.

GRE America Inc.
425 Harbor Blvd, Belmont CA
94002-4048. 415/591-1400.

Greenleaf Distribution Inc.
4900 Patrick Henry Dr, Santa
Clara CA 95054-1822. 408/653-
0222.

GTX Corp.
2390 E Camelback Rd, Phoenix
AZ 85016-3452. 602/224-8700.

Haverly Systems Inc.
12 Hinchman Ave, Denville NJ
07834-2111. 201/627-1424.

HD Computer
988 Walsh Ave, Santa Clara CA
95050. 408/986-9898.

Hewlett-Packard Co.
104 Woodmere Rd, Folsom CA
95630-4705. 916/985-0900.

Hewlett-Packard Co.
690 E Middlefield Rd, Mountain
View CA 94043-4010. 415/968-
9200.

Hewlett-Packard Co.
15885 W Sprague Rd,
Strongsville OH 44136-1772.
216/243-7300.

Hewlett-Packard Co.
PO Box 2197, Colorado Springs
CO 80901-2197. 719/590-1900.

Hi-Q Computers
740 N Mary Ave, Sunnyvale CA
94086-2908. 408/245-5836.

Hitachi Data Systems Corp.
10277 Scripps Ranch Blvd, San
Diego CA 92131-1297. 619/537-
3000.

Hitachi Data Systems Corp.
5251 DTC Parkway #99,
Englewood CO 80111-2799.
303/843-6250.

Hitachi Data Systems Corp.
1705 Junction Ct Ste 200, San
Jose CA 95112-1023. 408/436-
2400.

**Honeywell Inc. Space &
Aviation Control**
PO Box 21111, Phoenix AZ
85036-1111. 602/436-2311.

Hughes Electronics Corp.
7200 Hughes Ter, Los Angeles
CA 90045-2400. 310/568-7200.

Hughes STX Corp. Systems
1577 Spring Hill Rd, Vienna VA
22182-2223. 703/827-6600.

Hunkar Laboratories Inc.
7007 Valley Ave, Cincinnati OH
45244-3031. 513/272-1010.

IBIS Corp.
11150 Sunset Hills Rd, Reston
VA 22090-5321. 703/478-0300.

IBM Corp.
30 Kraft Ave, Albany NY 12205-
5428. 518/459-6525.

IBM Corp.
1580 State Route 52, Hopewell
Junction NY 12533-6526.
914/892-7111.

IBM Corp.
8845 University Center Ln, San
Diego CA 92122-1084. 619/587-
5000.

IBM Corp.
2077 Gateway Pl, San Jose CA
95110-1016. 408/452-4800.

IBM Corp.
410 11th Ave SE, Olympia WA
98501-2371. 360/357-7726.

IBM Corp.
580 E Swedesford Rd, Wayne PA
19087-1608. 610/293-5400.

IBM Corp.
1111 Broadway, Oakland CA
94607-4036. 510/464-5000.

IBM Corp.
2005 Market St, Philadelphia PA
19103-7042. 215/851-3257.

IBM Corp.
2700 Gateway Oaks Dr,
Sacramento CA 95833-3501.
916/326-5000.

IBM Corp.
600 Anton Blvd, Costa Mesa CA
92626-7147. 714/438-5220.

IBM Corp.
PO Box 567, Riverside CA
92502-0567. 909/369-5300.

IBM Corp.
2535 Millikin Pkwy, Decatur IL
62526-2156. 217/875-7700.

IBM Corp.
4000 Executive Pkwy, San
Ramon CA 94583-4257. 510/277-
5000.

IBM Corp.
1333 N California Blvd, Walnut
Creek CA 94596-4534. 510/942-
2000.

IBM Corp.
14435 Cherry Lane Ct, Laurel
MD 20707-4991. 301/470-6010.

IBM Corp.
101 N Monroe St, Tallahassee FL
32301-1549. 904/599-4200.

IBM Corp. Marketing Division
2929 N Central Ave, Phoenix AZ
85012-2743. 602/217-2000.

ICD Inc.
1220 Rock St, Rockford IL
61101-1437. 815/968-2228.

Idac Inc.
20900 Swenson Dr, Waukesha
WI 53186-4050. 414/796-1060.

IMT Systems
7240 Brittmoore Rd, Houston TX
77041-3227. 713/937-2115.

Incredible Technologies Inc.
4010 Winnetka Ave, Rolling
Meadows IL 60008-1374.
847/870-7027.

Infax Inc.
2485 Lithonia Industrial Blvd,
Lithonia GA 30058-7642.
770/482-2755.

Infinite Graphics Inc.
4611 East Lake Street,
Minneapolis MN 55406-2305.
612/721-6283.

Integral Systems Inc.
5000 Philadelphia Way, Lanham
MD 20706-4417. 301/731-4233.

Integrated Business Solutions
5000 Executive Pkwy, San
Ramon CA 94583-4210. 510/275-
2500.

Integrated Design Engineering
1078 Headquarters Park, Fenton
MO 63026-1910. 314/343-0005.

Integrix Inc.
1200 Lawrence Dr, Newbury
Park CA 91320-1316. 805/375-
1055.

Intelligent Decisions
14121 Parke Long Ct, Chantilly
VA 22021-1647. 703/803-8070.

Intergraph Electronics
381 E Evelyn Ave, Mountain
View CA 94041-1530. 415/691-
9680.

Interlogic Industries
85 Marcus Dr, Melville NY
11747-4209. 516/420-8111.

International Parallel Machines
50 Conduit St, New Bedford MA
02745-6016. 508/990-2977.

Internet Inc.
1605 Western Ave, Chicago
Heights IL 60411-3149. 708/481-
7164.

ISX Corp.
4353 Park Terrace Dr, Westlake
Village CA 91361-4631.
818/706-2020.

Itron Inc.
2818 N Sullivan Rd, Spokane
WA 99216-1897. 509/924-9900.

Ivex Corp.
4355 International Blvd, Norcross
GA 30093-3018. 770/564-1148.

Jetta International Inc.
51 Stouts Ln, Monmouth Junction
NJ 08852-1916. 908/329-9651.

Julie Research Labs Inc.
508 W 26th St, New York NY
10001-5515. 212/633-6625.

Keydata International Inc.
111 Corporate Blvd, South
Plainfield NJ 07080-2409.
908/755-0350.

Kristel Corp.
555 Kirk Rd, Saint Charles IL
60174-3433. 630/443-1290.

Lab Volt Systems
PO Box 686, Farmingdale NJ
07727-0686. 908/938-2000.

LANCity Corp. of Bay Networks
200 Bulfinch Drive, Andover MA
01810-1428. 508/475-4050.

Landmark Graphics Corp.
15150 Memorial Dr, Houston TX
77079-4304. 713/560-1000.

Laserdata Inc.
300 Vesper Executive Park,
Tyngsboro MA 01879-2722.
508/649-4600.

Linksys Corp.
1401 Armstrong Ave, Irvine CA
92614. 714/261-1288.

Liuski International Inc.
6585 Crescent Dr, Norcross GA
30071-2901. 770/447-9454.

LJ Technical Systems Inc.
85 Corporate Dr, Holtsville NY
11742-2007. 516/234-2100.

Logic Covalent Systems Corp.
47436 Fremont Blvd, Fremont
CA 94538-6555. 510/683-3900.

Logic Process Corp.
10610 Metric Dr, Dallas TX
75243-5518. 214/340-5172.

Longshine Microsystems Inc.
10400 Pioneer Blvd, Santa Fe
Springs CA 90670-3734.
562/903-0899.

Lucent Technologies
3000 Skyline Dr, Mesquite TX
75149-1802. 214/284-2000.

MA Laboratories Inc.
1972 Concourse Dr, San Jose CA
95131-1719. 408/954-8188.

Macrolink Inc.
1500 N Kellogg Dr, Anaheim CA
92807-1930. 714/777-8800.

Manufacturing Technology Solutions Inc.
10220 Old Columbia Rd,
Columbia MD 21046-1754.
301/982-1135.

Maxpeed Inc.
1120 Chess Dr, Foster City CA
94404-1103. 415/345-5447.

Melard Technologies Inc.
28 Kaysal Ct, Armonk NY
10504-1344. 914/273-4488.

Mextell
159 Beeline Dr, Bensenville IL
60106-1601. 630/595-4146.

Micro Distribution
48879 Kato Rd, Fremont CA
94539-8070. 510/657-8077.

Microlab Inc.
6290 Edgewater Dr, Orlando FL
32810-4718. 407/297-1274.

Minarik Corp.
3720 E La Salle St, Phoenix AZ
85040-3976. 602/437-1400.

Minarik Electric Company
905 Thompson Ave, Glendale CA
91201-2011. 818/507-6500.

Mintronix Inc.
5251 Verdugo Way, Camarillo
CA 93012-8605. 805/482-1298.

MIS Inc.
45395 Northport Loop W,
Fremont CA 94538-6417.
510/226-9188.

Mitsubishi Electronics America
5665 Plaza Dr, Cypress CA
90630-5023. 714/220-2500.

Mitsubishi Electronics America
1050 E Arques Ave, Sunnyvale
CA 94086-4601. 408/730-5900.

Mizar Inc.
2410 Luna Rd, Carrollton TX
75006-6502. 214/277-4600.

Mnemonics Inc.
3900 Dow Rd, Melbourne FL
32934-9291. 407/254-7300.

Mobile Data Communcations Corp.
10850 N 24th Ave, Phoenix AZ
85029-4796. 602/678-3788.

Modular Automation Corp.
Rte 12 S Box 36, Greene NY
13778. 607/656-7101.

Modular Mining Systems Inc.
3289 E Hemisphere Loop, Tucson
AZ 85706-5028. 520/746-9127.

Monolith Corp.
5275 Capital Blvd, Raleigh NC
27604-2925. 919/878-1900.

Motion Analysis Corp.
3617 Westwind Blvd, Santa Rosa
CA 95403-1067. 707/579-6500.

Multi Media Communication Systems
8707 Skokie Blvd, Skokie IL
60077-2200. 847/673-8488.

Multi-Industry Tech Inc.
16717 Norwalk Blvd, Cerritos
CA 90703-1839. 562/921-6669.

NCR Corporation
17095 Via Del Campo, San Diego
CA 92127-1711. 619/485-1220.

NCR Corporation
965 Keynote Cir, Brooklyn
Heights OH 44131-1829.
216/459-0010.

Netpower Inc.
545 Oakmead Pkwy, Sunnyvale
CA 94086-4023. 408/522-9999.

New Dimensions Services Solutions
2440 Stanwell, Concord CA
94520. 510/356-5600.

Northwest Source Group Inc.
13975 Interurban Ave S, Tukwila
WA 98168-4721. 206/244-6205.

Novacor Inc.
1590 Oakland Rd, San Jose CA
95131-2445. 408/441-6500.

Numonics Corp.
101 Commerce Dr,
Montgomeryville PA 18936-
9628. 215/362-2766.

Octel Communications Corp.
455 Market St, San Francisco CA
94105-2440. 415/882-6000.

Omnidata International Inc.
PO Box 448, Logan UT 84323-
0448. 801/753-7760.

Onset Computer Corp.
PO Box 3450, Pocasset MA
02559-3450. 508/563-9000.

Optical Data Systems Inc.
1101 E Arapaho Road,
Richardson TX 75081-2336.
214/234-6400.

Optim Electronic Corp.
12401 Middlebrook Rd,
Germantown MD 20874-1525.
301/428-7200.

PagePoint Inc.
725 N Shoreline Blvd, Mountain View CA 94043-3208. 415/960-1200.

Pagg Corp.
425 Fortune Blvd, Milford MA 01757-1723. 508/478-8544.

Pailen-Johnson Associates Inc.
1370 Piccard Dr, Rockville MD 20850-4304. 301/948-5726.

Pao-Ku International Co. Ltd.
1057 Shore Rd, Naperville IL 60563-8758. 630/369-5199.

Paradigm Technology
71 Vista Montana, San Jose CA 95134-1507. 408/954-0500.

Parsytec Inc.
245 W Roosevelt Rd, West Chicago IL 60185-4804. 630/293-9500.

PC Craft Inc.
163 University Pkwy, Pomona CA 91768-4301. 909/869-6133.

PC Dynamics Inc.
10501 Fm 720, Frisco TX 75034-6211. 214/335-9841.

Person-System Integration Ltd.
2401 Huntington Ave, Alexandria VA 22303-1531. 703/960-5555.

Pinnacle Systems Inc.
280 N Bernardo, Mountain View CA 94043. 415/526-1600.

Pinpoint Solutions Inc.
501 E 1700 S, Salt Lake City UT 84105-2915. 801/466-6210.

PRB Associates Inc.
47 Airport View Dr, Hollywood MD 20636-9760. 301/373-2360.

Precise Connections Inc.
1114 Explorer St, Duncanville TX 75137-3012. 214/298-1040.

Protiva Computer
14800 McKinley Ave, Posen IL 60469-1525. 708/388-8200.

Radix II Inc.
6230 Oxon Hill Rd, Oxon Hill MD 20745-3035. 301/567-5200.

Raynet Electronics Co.
16810 Barker Springs Rd, Houston TX 77084-5037. 713/578-3802.

Recortec Inc.
1290 Lawrence Station Rd, Sunnyvale CA 94089-2220. 408/734-1290.

Research & Development Solutions
7921 Jones Branch Dr, McLean VA 22102-3306. 703/893-9533.

Rich Inc.
1400 Kensington Rd, Oakbrook IL 60521-2168. 630/574-7424.

Riso Inc.
300 Rosewood Dr, Danvers MA 01923-1389. 508/777-7377.

Rockwell Automation
8440 Darrow Rd, Twinsburg OH 44087-2310. 216/487-6000.

Samsung Electronics of America
105 Challenger Rd, Ridgefield Park NJ 07660-2113. 201/229-4000.

Scietex Digital Printing Inc.
3100 Research Blvd, Dayton OH 45420-4005. 937/259-3100.

SEEQ
47131 Bayside Pkwy, Fremont CA 94538-6517. 510/226-7400.

Sequel Inc.
2777 San Tomas Expressway, Santa Clara CA 95054. 408/987-1000.

SIIG
6078 Stewart Ave, Fremont CA 94538-3152. 510/657-8688.

Silicon Graphics Computer Systems
4600 S Ulster St, Denver CO 80237-2873. 303/796-0022.

Silicon Graphics Computer Systems
3000 Executive Pkwy, San Ramon CA 94583-2300. 510/277-1940.

Stallion Technologies Inc.
2880 Research Park Drive, Soquel CA 95073-2000. 408/761-9499.

Star Technologies Inc.
23162 La Cadena Dr, Laguna Hills CA 92653-1405. 714/768-6460.

Statesman Computers
PO Box 550, Charlotte Court House VA 23923-0550. 804/542-4933.

STB Systems Inc.
1651 N Glenville Dr, Richardson TX 75081-1956. 214/234-8750.

Stonehouse & Co.
4100 Spring Valley Rd, Dallas TX 75244-3629. 214/960-1566.

Sun Microsystems
2398 E Camelback Rd, Phoenix AZ 85016-9001. 602/224-9815.

Sun Microsystems
1920 Main St, Irvine CA 92714-7225. 714/833-1640.

Sun Moon Star
1941 Ringwood Ave, San Jose CA 95131-1721. 408/452-7811.

SuperCom Industries, Inc.
410 S Abbott Ave, Milpitas CA 95035-5257. 408/263-1600.

Symbolics Inc.
9000 Fullbright Ave, Chatsworth CA 91311-6125. 818/998-3600.

Symon Communications Inc.
10701 Corporate Dr, Stafford TX 77477-4013. 713/240-5555.

Syntellect, Inc.
1000 Holcomb Woods Pkwy, Roswell GA 30076-2575. 770/587-0700.

Sysorex Information Systems
485 Brooke Rd, Winchester VA 22603-5764. 540/662-1736.

System Integrators Inc.
1300 W National Dr, Sacramento CA 95834-1908. 916/929-9481.

T-R Associates Inc.
1 Export Ln, Archbald PA 18403-1957. 717/876-4067.

Tandem Computers Inc.
3760 Kilroy Airport Way, Long Beach CA 90806-2443. 562/595-7723.

Tandem Computers Inc.
2920 E Camelback Rd, Phoenix AZ 85016-4409. 602/224-5280.

Tandem Computers Inc.
2527 Camino Ramon, San Ramon CA 94583-4276. 510/275-6600.

Tandem Computers Inc.
10700 Parkridge Blvd, Reston
VA 22091-5429. 703/476-3252.

Tangent Computer
197 Airport Blvd, Burlingame CA
94010-2006. 415/342-9388.

Tatung Science & Technology
1840 McCarthy Blvd, Milpitas
CA 95035-7425. 408/383-0988.

Tektronix Inc.
40 Gill Lane, Woodbridge NJ
07095. 908/636-8616.

Telan Corp.
2088 Bergey Rd, Hatfield PA
19440-2317. 215/822-1234.

Teledyne Industries Inc.
19601 Nordhoff St, Northridge
CA 91324-2422. 818/886-2211.

Telesis Technologies Inc.
28181 River Dr, Circleville OH
43113-9726. 614/477-5000.

Texas Digital Systems Inc.
512 Fm 2818 Rd W, College
Station TX 77840. 409/693-9378.

Texas Instruments Inc.
5015 Campuswood Dr, East
Syracuse NY 13057-1222.
315/434-1600.

Texas Instruments Inc.
3030 Twin Dolphin Dr # 410,
Redwood City CA 94065.
415/594-0203.

Texas Instruments Inc.
Precision Automation
PO Box 242, Hunt Valley MD
21031-0242. 410/785-4800.

Tracor/AEL Industries Inc.
305 Richardson Rd, Lansdale PA
19446-1480. 215/822-2929.

Transend Corporation
225 Emerson St, Palo Alto CA
94301-1026. 415/324-5370.

U Tron Inc.
47448 Fremont Blvd, Fremont
CA 94538-6503. 510/656-3600.

UB Networks
200 Galleria Pkwy SE, Atlanta
GA 30339-5945. 770/955-7666.

UES Inc.
4401 Dayton Xenia Rd, Dayton
OH 45432-1805. 937/426-6900.

Ultracomp
400 N Lindburgh, St. Louis MO
63141. 314/567-0077.

US Assemblies Raleigh IMC
3303 Terminal Dr, Raleigh NC
27604-3889. 919/250-6700.

US Micro Express Inc.
1930 116th Ave NE, Bellevue
WA 98004-3044. 206/453-4046.

Valco Instruments Co. Inc.
PO Box 55603, Houston TX
77255-5603. 713/688-9345.

Validata Computer Corp.
428 S Perry St, Montgomery AL
36104-4274. 334/834-2324.

Valtronic USA Inc.
6168 Cochran Rd, Solon OH
44139-3306. 216/349-1239.

Vari-Tech Company
546 Leonard St NW, Grand
Rapids MI 49504-4203. 616/459-
7281.

Verbex Voice Systems
1090 King George Post Rd,
Edison NJ 08837-3701. 908/225-
5225.

Video Corporation of America
PO Box 1284, Weston CT 06883-
0284. 203/226-8800.

VMX Inc.
17217 Waterview Pkwy, Dallas
TX 75252-8004. 214/907-3000.

Xerox Corp.
250 N Halstead St, Pasadena CA
91107-3128. 818/351-2351.

YKE International Inc.
7616 Jamaica Ave, Jamaica NY
11421-1850. 718/296-0101.

Youngtron Inc.
63 E Broad St, Hatfield PA
19440-2464. 215/855-0438.

**COMPUTER PERIPHERAL
EQUIPMENT
MANUFACTURERS**

AAA Technology & Specialties
6219 Brittmoore Rd, Houston TX
77041-5114. 713/849-3366.

Able Communications Inc.
3629 W MacArthur Blvd, Ste
210, Santa Ana CA 92704.
714/979-7893.

ACC Technology Group
7 Whatney, Irvine CA 92718-
2806. 714/454-2441.

Access Beyond
1 Palmer Ter, Carlstadt NJ
07072-2714. 201/438-2400.

Accton Technology Corp.
1962 Zanker Rd, San Jose CA
95112-4216. 408/452-8900.

Adams-Smith Inc.
34 Tower St, Hudson MA 01749-
1742. 508/562-3801.

Adaptive Info Systems
26001 Pala, Mission Viejo CA
92691-2705. 714/587-9077.

Adax Inc.
614 Bancroft Way, Berkeley CA
94710-2224. 510/548-7047.

**Advanced Automation
Associates**
Eagle View Corp. Ctr 640 Rice
B, Exton PA 19341. 610/458-
8700.

Advanced Interlink Corp.
15181 Springdale St, Huntington
Beach CA 92649-1154. 714/894-
1675.

Advanced Matrix Technology
747 Calle Plano, Camarillo CA
93012-8556. 805/388-5799.

Advantest America Inc.
1100 Busch Pkwy, Buffalo Grove
IL 60089-4553. 847/634-2552.

Agency Management Service
PO Box 30001, College Station
TX 77842-3001. 409/693-6122.

Agile
875 Alfred Nobel Dr, Hercules
CA 94547-1800. 510/724-1600.

Aitech International
47971 Fremont Blvd, Fremont
CA 94538-6521. 510/226-8960.

Alacrity Systems Inc.
43 Newburgh Rd, Hackettstown
NJ 07840-3903. 908/813-2400.

Aladdin Knowledge Systems
270 Lexington Dr, Buffalo Grove
IL 60089-6930. 847/808-0300.

Alaris Inc.
47338 Fremont Blvd, Fremont
CA 94538-6501. 510/770-5700.

All Write Ribbon Inc.
3916 Bach Buxton Rd, Amelia
OH 45102-1014. 513/753-8300.

**Allied Telesyn International
Corp.**
19015 N Creek Pkwy, Bothell
WA 98011-8029. 206/487-8880.

Allsop Inc.
PO Box 23, Bellingham WA
98227-0023. 360/734-9090.

Allus Technology Corp.
PO Box 690907, Houston TX
77269-0907. 713/894-4455.

Almo Corp.
9815 Roosevelt Blvd,
Philadelphia PA 19114-1011.
215/698-4000.

Alpha Merics Corporation
4420 Shopping Ln, Simi Valley
CA 93063-3451. 805/520-3664.

Alpha Products Inc.
351 Irving Dr, Oxnard CA 93030-
5173. 805/981-8666.

Alpha Technologies Inc.
3767 Alpha Way, Bellingham
WA 98226-8302. 360/647-2360.

Alps Electric Inc.
7301 Orangewood Ave, Garden
Grove CA 92641-1411. 714/897-
1005.

Alta Research Corp.
1701 Clint Moore Rd, Boca
Raton FL 33487-2755. 954/428-
8535.

Alta Technology
9500 S 500 W, Sandy UT 84070-
6655. 801/562-1010.

Altek Corp.
12210 Plum Orchard Dr, Silver
Spring MD 20904-7802. 301/572-
2555.

American Advantech Corp.
750 E Arques Ave, Sunnyvale
CA 94086-3830. 408/245-6678.

American Covers Inc.
102 W 12200 S, Draper UT
84020-8983. 801/553-0600.

Amherst Merritt International
5565 Red Bird Center Dr, Dallas
TX 75237-1913. 214/339-0753.

Amherst Systems Inc.
30 Wilson Rd, Williamsville NY
14221-7082. 716/631-0610.

Amptron International Inc.
1028 Lawson St, Rowland
Heights CA 91748-1107.
818/912-5789.

Amtek Electronic Inc.
1150 N 5th St, San Jose CA
95112-4415. 408/971-8787.

Anacom General Corporation
1244 S Claudina St, Anaheim CA
92805-6232. 714/774-8080.

Anacomp Inc.
715 W Algonquin Rd, Arlington
Heights IL 60005-4415. 847/364-
6500.

Anagraph Inc.
3100 Pullman St, Costa Mesa CA
92626-4501. 714/540-2400.

Andataco
10140 Mesa Rim Rd, San Diego
CA 92121-2914. 619/453-9191.

Antex Electronics Corp.
16100 S Figueroa St, Gardena CA
90248-2617. 310/532-3092.

AOC International Ltd.
311 Sinclair Frontage Rd,
Milpitas CA 95035-5443.
408/956-1070.

Arco Electronics Inc.
2750 N 29th Ave, Hollywood FL
33020-1519. 305/925-2688.

Arcolectric Corp.
9001 Canoga Avenue, Canoga
Park CA 91304-1513. 818/700-
1933.

Area Electronics Systems Inc.
950 Fee Ana Street, Placentia CA
92670-6755. 714/993-0300.

Artecon
6305 El Camino Real, Carlsbad
CA 92009-1606. 760/931-5500.

Artisan Electronics Corp.
5 Eastmans Road, Parsippany NJ
07054-3776. 201/887-7100.

Assure Net Pathways
201 Ravendale Dr, Mountain
View CA 94043-5216. 415/964-
0707.

Atlantic Scientific Corp.
4300 Fortune Place, West
Melbourne FL 32904-1527.
407/725-8000.

ATS
1160 Ridder Park Drive, San Jose
CA 95131-2319. 408/441-7177.

Atto Technology Inc.
40 Hazelwood Drive, Amherst
NY 14228-2230. 716/691-1999.

Audio Digital Imaging Inc.
511 W Golf Rd, Arlington
Heights IL 60005-3904. 847/439-
1335.

Audiofax Inc.
2000 Powers Ferry Road,
Marietta GA 30067-9480.
770/933-7600.

Authorized Technical Services
477 Roland Way, Oakland CA
94621-2014. 510/638-4449.

Avo Biddle Instruments
510 Township Line Rd, Blue Bell
PA 19422-2701. 215/646-9200.

Azure Technologies Inc.
63 South St, Hopkinton MA
01748-2212. 508/435-3800.

**B&B Electronics
Manufacturing Co.**
707 Dayton Rd, Ottawa IL
61350-9545. 815/433-5100.

Barr Systems Inc.
4131 NW 28th Ln, Gainesville
FL 32606-6665. 352/371-3050.

Basic Measuring Instruments
3250 Jay St, Santa Clara CA
95054-3309. 408/970-3700.

Bass Inc.
2211 Arbor Blvd, Dayton OH
45439-1502. 937/293-5732.

BDT Products Inc.
17152 Armstrong Ave, Irvine CA
92714-5718. 714/660-1386.

Beall Technologies Inc.
200 Meadowlands Pkwy,
Secaucus NJ 07094-3616.
201/864-9433.

BEC Computer Corp.
22 Yearling Ct, Rockville MD
20850-3546. 301/294-2152.

BEI Sensors & Systems Co.
15771 Red Hill Ave, Tustin CA
92680-7303. 714/258-7500.

BEI Systron Donner Co.
2700 Systron Dr, Concord CA
94518-1355. 510/682-6161.

Belkin Components
1303 W Walnut Pkwy, Compton
CA 90220-5030. 310/898-1100.

Big Hand Productions Inc.
6305 N O'Connor, Ste 103, Irving
TX 75039. 972/506-9600.

Biscom Inc.
321 Billerica Rd, Chelmsford MA
01824-4100. 508/250-1800.

Blue Lance Inc.
1700 West Loop S, Houston TX
77027-3008. 713/680-1187.

Bomont Industries
4 Century Dr, Parsippany NJ
07054-4606. 201/984-3777.

BP Microsystems Inc.
1000 N Post Oak Rd, Houston
TX 77055-7237. 713/688-4600.

BRC Electronics
4301 Wiley Post Rd, Dallas TX
75244-2132. 214/385-3561.

BS Cable Co. Inc.
805 W 5th St, Lansdale PA
19446-2279. 215/361-6999.

Bus-Tech Inc.
129 Middlesex Tpke, Burlington
MA 01803-4404. 617/272-8200.

Buslogic Inc.
4151 Burton Dr, Santa Clara CA
95054-1564. 408/492-9090.

C-Tech Electronics Inc.
PO Box 2098, Tustin CA 92681-
2098. 714/573-4604.

C-Z Labs Inc.
135 N Rt 9 W, Congers NY
10920. 914/268-5056.

Cabletron Systems
5000 Birch St, Newport Beach
CA 92660-2131. 714/852-4126.

Cablexpress Corp.
500 East Brighton Avenue,
Syracuse NY 13210-4211.
315/476-3000.

Cactus Computer Systems Inc.
17 Industrial Rd, Fairfield NJ
07004-3017. 201/575-8810.

California Cartridge Co.
5009 Forni Dr, Concord CA
94520-8525. 510/825-8400.

**Cambridge Imaging
Technologies**
1430 Spring Hill Rd, McLean VA
22102-3000. 703/356-4600.

Camintonn Corporation
22 Morgan, Irvine CA 92718-
2022. 714/454-1500.

Canary Communications Inc.
1851 Zanker Rd, San Jose CA
95112-4213. 408/453-9201.

Canon USA Inc.
2051 Mission College Blvd,
Santa Clara CA 95054-1519.
408/982-5200.

Capital Design Consultants
26 Pawnee Ave, Mastic NY
11950-4516. 516/395-1106.

Carroll Touch
PO Box 1309, Round Rock TX
78680-1309. 512/244-3500.

Cigi International
1299 Orleans Dr, Sunnyvale CA
94089-1138. 408/752-2770.

**Communications Programs
Division**
1370 Main St, Waltham MA
02154-1622. 617/891-4700.

**Compaq Computer
Corporation**
61 Daggett Dr, San Jose CA
95134-2109. 408/383-9300.

Compex Inc.
4051 E La Palma Ave, Anaheim
CA 92807-1743. 714/630-7302.

Computer Enhancements
22641 Old Canal Rd, Yorba
Linda CA 92687-4601. 714/282-
8881.

Computer Optical Products Inc.
9305 Eton Ave, Chatsworth CA
91311-5810. 818/882-0424.

Computer Sports Systems
1100 Massachusetts Ave,
Cambridge MA 02138-5241.
617/492-6500.

Computer Support Corp.
15926 Midway Rd, Dallas TX
75244-2196. 214/661-8960.

**Computer Technology Link
Corp.**
9315 SW Nimbus Ave, Beaverton
OR 97008-7132. 503/646-3733.

Comstream
10180 Barnes Canyon Rd, San
Diego CA 92121-2724. 619/458-
1800.

Contec Microelectronics USA
2190 Bering Dr, San Jose CA
95131-2013. 408/434-6767.

Control Cable Inc.
7261 Ambassador Rd, Baltimore
MD 21244-2726. 410/298-4411.

Control Module Inc.
227 Brainard Rd, Enfield CT
06082. 860/745-2433.

Control Technology Co. Inc.
7608 N Hudson Ave, Oklahoma
City OK 73116-7798. 405/840-
3163.

Control Technology Co. Inc.
4116 29th St, Long Island City
NY 11101-3702. 718/361-2133.

Controlled Power Co.
1955 Stephenson Hwy, Troy MI
48083-2166. 810/528-3700.

Contronics Systems of Ohio Inc.
7100 N High St, Worthington OH
43085-2316. 614/841-0200.

Copar Corp.
5744 W 77th St, Burbank IL
60459-1305. 708/496-1859.

Corby Industries Inc.
1501 E Pennsylvania St,
Allentown PA 18103-1588.
610/433-1412.

Core International Inc.
6500 East Rogers Circle, Boca
Raton FL 33487-2699. 561/997-
6044.

Cornet Inc.
6800 Versar Center, Springfield
VA 22151-4177. 703/658-3400.

Corollary Inc.
2802 Kelvin Avenue, Irvine CA
92714-5826. 714/250-4040.

Corporate Systems Center
1294 Hammerwood Ave,
Sunnyvale CA 94089-2232.
408/737-7312.

Correct Type
67 Kent Ave, Brooklyn NY
11211-1926. 718/782-2601.

Costar Corp.
599 W Putnam Ave, Greenwich
CT 06830. 203/661-9700.

Covid Inc.
1711 W 17th St, Tempe AZ
85281-6229. 602/966-2221.

CSS Laboratories Inc.
1641 McGaw Ave, Irvine CA
92714-5631. 714/852-8161.

CTC Systems Inc.
415 Clyde Ave, Mountain View
CA 94043-2228. 415/966-1688.

CTC/Trigem Corp.
48400 Fremont Blvd, Fremont
CA 94538-6505. 510/770-8787.

CTX International Inc.
20530 Earlgate St, Walnut CA
91789-2909. 909/595-6146.

Curtis Young Corp.
1050 Taylors Ln, Cinnaminson
NJ 08077-2098. 609/665-6650.

Custom Cable Industries Inc.
3221 Cherry Palm Dr, Tampa FL
33619-8334. 813/623-2232.

Cybernet International
3811 Stinnaker Court, Fremont
CA 94538-7320. 510/623-3700.

Cyclades Corp.
41934 Christy St, Fremont CA
94538-3159. 510/770-9727.

Cygnet Systems Inc.
2560 Junction Ave, San Jose CA
95134-1902. 408/954-1800.

**Cymbolic Sciences
International**
26072 Merit Cir, Laguna Hills
CA 92653-7015. 714/582-3515.

Datacal Corp.
531 E Elliot Rd, Chandler AZ
85225-1118. 602/813-3100.

Dataforth Corp.
3331 E Hemisphere Loop, Tucson
AZ 85706-5011. 520/741-1404.

Dataserv Inc.
37562 Hills Tech Dr, Farmington
Hills MI 48331-5726. 810/489-
8400.

Dawn VME Products
47073 Warm Springs Blvd,
Fremont CA 94539-7454.
510/657-4444.

Dee One Systems
1558 Centre Pointe Dr, Milpitas
CA 95035-8011. 408/956-8286.

Dehaart Inc.
12 Wilmington Rd, Burlington
MA 01803-1729. 617/272-0794.

Deltec Electronics Corp.
2727 Kurtz St, San Diego CA
92110-3109. 619/291-4211.

Design Technology
11489 Woodside Ave, Santee CA
92071-4724. 619/448-2888.

Destiny Computer Inc.
3480 Investment Blvd, Hayward
CA 94545-3811. 510/783-2727.

Destiny Technology Corp.
3255 Scott Blvd, Santa Clara CA
95054-3013. 408/262-9400.

Devar Inc.
706 Bostwick Ave, Bridgeport
CT 06605-2327. 203/368-6751.

Diamond Flower Northeast Inc.
7 Elkins Rd, East Brunswick NJ
08816-2006. 201/390-2815.

Diamond Micro Solutions
1515 Aurora Dr, San Leandro CA
94577-3105. 510/351-4700.

Diaquest Inc.
1440 San Pablo Ave, Berkeley
CA 94702-1046. 510/526-7167.

Dicomed Inc.
12270 Nicollet Ave, Burnsville
MN 55337-1649. 612/895-3000.

Digicom Systems Inc.
188 Topaz St, Milpitas CA
95035-5429. 408/262-1277.

Digital Equipment Corp.
1 Digital Dr, Westminster MA
01473. 508/874-3111.

Digitech Industries Inc.
PO Box 2267, Danbury CT
06813-2267. 203/797-2676.

DIT-MCO International Corp.
5612 Brighton Ter, Kansas City
MO 64130-4530. 816/444-9700.

Diverse Logistics Inc.
2862 McGaw Ave, Irvine CA
92714-5836. 714/476-7171.

Dome Imaging Systems Inc.
400 5th Ave, Waltham MA
02154-8706. 617/895-1155.

Drawbase Software
222 3rd St, Cambridge MA
02142-1159. 617/868-6003.

Dymec Inc.
27 Katrina Rd, Chelmsford MA
01824. 508/256-0025.

Dynamic Decisions Inc.
53 Murray St, New York NY
10007-2201. 212/227-9888.

EIC Laboratories Inc.
111 Downey St, Norwood MA
02062-2612. 617/769-9450.

Electronic Solutions
6790 Flanders Dr, San Diego CA
92121-2902. 619/452-9333.

Electronic Specialists Inc.
171 S Main St, Natick MA
01760-4935. 508/655-1532.

Electroservice Laboratories
6085 Sikorsky St, Ventura CA
93003-7678. 805/644-2944.

Elexsys
1625 Plymouth St, Mountain
View CA 94043-1231. 415/903-
2400.

Elite Group Computer
45225 Northport Ct, Fremont CA
94538. 510/226-7333.

Elma Electronic Inc.
44350 S Grimmer Blvd, Fremont
CA 94538-6385. 510/656-3400.

Emcom Corp. of Delaware
840 Ave F, Plano TX 75074.
214/423-7183.

EME Corp.
1197 Baltimore Annapolis Blvd,
Arnold MD 21012-1808.
410/544-8563.

Engineered Data Products Inc.
2550 W Midway Blvd,
Broomfield CO 80020-7100.
303/465-2800.

Enhance Memory Products Inc.
18730 Oxnard Street Tarzana CA
91356-1454. 818/343-3066.

Ergotron Inc.
1181 Trapp Road, St. Paul MN
55121-1248. 612/452-8135.

Esca Corp.
11120 NE 33rd Place, Bellevue
WA 98004-1448. 206/822-6800.

Escod Industries
4709 Creekstone Drive, Durham
NC 27560. 919/941-5550.

ESI/EDP Systems Inc.
3200 Commonwealth Blvd,
Tallahassee FL 32303-3173.
904/575-0179.

ESS Technology Inc.
46107 Landing Parkway, Fremont
CA 94538-6407. 510/226-1088.

Etec Systems Inc.
26460 Corporate Avenue,
Hayward CA 94545-3914.
510/783-9210.

Eteq Microsystems Inc.
1900 McCarthy Blvd, Milpitas
CA 95035-7413. 408/432-8147.

Everest Electronic Equipment
2100 E Orangewood Ave,
Anaheim CA 92806-6193.
714/634-2200.

Extech Instruments Corp.
335 Bear Hill Road, Waltham
MA 02154-1020. 617/890-7440.

Falcon Systems Inc.
1417 North Market Boulevard,
Sacramento CA 95834-1936.
916/928-9255.

Farallon Computing Inc.
2470 Mariner Square Loop,
Alameda CA 94501-1010.
510/814-5100.

Faroudja Laboratories Inc.
750 Palomar Avenue, Sunnyvale
CA 94086-2914. 408/735-1492.

Fast Forward Video Inc.
18200 West McDurmott, Irvine
CA 92714-6710. 714/852-8404.

Faxstar Inc.
4001 Westerly Place, Newport
Beach CA 92660-2315. 714/724-
0806.

Focus Electronic Corp.
21078 Commerce Point Dr,
Walnut CA 91789-3051.
909/468-5533.

Focus Information Systems Inc.
48860 Milmont Dr, Fremont CA
94538-7369. 510/657-2845.

Follett Software Co.
1391 Corporate Dr, Mc Henry IL
60050-7040. 815/344-8700.

Franklin Datacom Inc.
733 Lakefield Rd, Westlake
Village CA 91361-2610.
805/373-8688.

Frontier Software Development
321 Billerica Rd, Chelmsford MA
01824-4100. 508/244-4000.

FTG Data Systems
8381 Katella Ave, Stanton CA
90680-3224. 714/995-3900.

Fujitsu Systems of America Inc.
11085 N Torrey Pines Rd, La
Jolla CA 92037-1007. 619/457-
9900.

Galaxy Networks Inc.
9330 De Soto Ave, Chatsworth
CA 91311-4926. 818/998-7851.

Gamma Link
1314 Chesapeake Terrace,
Sunnyvale CA 94089-1100.
408/744-1400.

General Micro Systems
PO Box 3689, Rancho
Cucamonga CA 91729-3689.
909/980-4863.

General Power Systems Inc.
17881 Cartwright Road, Irvine
CA 92614. 714/851-0800.

Georator Corp.
9617 Center Street, Manassas VA
22110-5521. 703/368-2101.

Geotest Inc.
18242 West McDurmott, Suite A,
Irvine CA 92614. 714/263-2222.

Grimes Company
1841 Enterprise Blvd, West
Sacramento CA 95691-3423.
714/671-3931.

GSI
17951 Sky Park Circle, Irvine CA
92714-6343. 714/261-7949.

H Co Computer Products Inc.
16812 Hale Ave, Irvine CA
92606. 714/833-3222.

H-R Industries Inc.
1302 E Collins Blvd, Richardson
TX 75081-2403. 214/301-6620.

Hach Associates Inc.
4994-B Indiana Ave, Winston-
Salem NC 27106. 910/744-7280.

Harris & Jeffries Inc.
888 Washington St, Dedham MA
02026-6021. 617/329-3200.

Hewlett-Packard Co.
5725 W Las Positas Blvd,
Pleasanton CA 94588-4084.
510/460-0282.

Hewlett-Packard Co.
1421 S Manhattan Ave, Fullerton
CA 92631-5221. 714/999-6700.

Hewlett-Packard Co.
1266 Kifer Rd, Sunnyvale CA
94086-5304. 408/746-5000.

Hewlett-Packard Co.
50 Fremont St, San Francisco CA
94105-2231. 415/882-6800.

Hi-Tech USA
1558 Centre Pointe Dr, Milpitas
CA 95035-8011. 408/262-8688.

HNC Software Inc.
5930 Cornerstone Ct W, San
Diego CA 92121-3728. 619/546-
8877.

HTI Voice Solutions Inc.
67 Forest St, Marlborough MA
01752-3088. 508/485-8400.

Hyperdata Technology Corp.
809 S Lemon Ave, Walnut CA
91789-2906. 909/468-2955.

I Bus Systems
9174 Sky Park Ct, San Diego CA
92123. 619/974-8400.

I/O Data Devices
2005 Hamilton Ave, San Jose CA
95125-5917. 408/377-7062.

Imagraph Corp.
11 Elizabeth Dr, Chelmsford MA
01824-4111. 508/256-4624.

Infotec Development Inc.
3611 S Harbor Blvd Ste 260,
Santa Ana CA 92704-6975.
714/549-0460.

Infotronic America Inc.
8834 N Capital Of Texas Hwy, Austin TX 78759-6396. 512/345-9646.

Integral Systems Inc.
2730 Shadelands Dr, Walnut Creek CA 94598-2515. 510/939-3900.

Integrated Network Corp.
757 Route 202-206, Bridgewater NJ 08807-2510. 908/218-1600.

Intellipower Inc.
15520 Rockfield Blvd, Irvine CA 92618. 714/587-0155.

Intellitools
55 Leveroni Ct, Novato CA 94949-5751. 415/382-5959.

Intercom Computer Systems
3182 Golansky Blvd, Woodbridge VA 22192-4221. 703/680-6999.

Interlink Computer Sciences
47370 Fremont Blvd, Fremont CA 94538-6566. 510/657-9800.

International Instrumentation
1200 Lawrence Dr, Thousand Oaks CA 91320. 805/376-9995.

International Power Machines
10451 Brockwood Rd, Dallas TX 75238-1641. 214/272-8000.

Invisible Software
1142 Chess Dr, Foster City CA 94404-1107. 415/570-5967.

IPC America Inc.
10300 Metric Blvd, Austin TX 78758-4937. 512/339-3500.

IPT
PO Box 12607, San Luis Obispo CA 93401. 805/541-3000.

ITG Inc.
2745 Hartland Rd, Falls Church VA 22043-3541. 703/698-8282.

J-Mark Computer Corp.
13111 Brooks Dr, Baldwin Park CA 91706-7901. 818/814-9472.

Jadtec Computer Group
1520 W Yale Ave, Orange CA 92687. 714/637-2900.

Jaton Corp.
556 S Milpitas Blvd, Milpitas CA 95035-5454. 408/942-9888.

JDS Microprocessing
22661 Lambert St, Lake Forest CA 92630-1612. 714/770-2263.

Jetfax Inc.
1376 Willow Rd, Menlo Park CA 94025-1516. 415/324-0600.

JMR Electronics Inc.
20400 Plummer St, Chatsworth CA 91311-5372. 818/993-4801.

Jones Futurex Inc.
3715 Atherton Rd, Rocklin CA 95765-3701. 916/632-3456.

Jovian Logic Corp.
47929 Fremont Blvd, Fremont CA 94538-6508. 510/651-4823.

JRL Systems Inc.
8305 W Highway 71, Austin TX 78735-8107. 512/288-6750.

Jupiter Technology Inc.
360 2nd Ave, Waltham MA 02154-3457. 617/894-9300.

Keithley Metrabyte
440 Myles Standish Blvd, Taunton MA 02780-7324. 508/880-3000.

Kelly Computer Systems Inc.
1060 L'Avenida St, Mountain View CA 94043-1422. 415/960-1010.

Kentek Information Systems
2945 Wilderness Pl, Boulder CO 80301-2255. 303/440-5500.

Keytec Inc.
1293 N Plano Rd, Richardson TX 75081-2424. 214/234-8617.

Kofax Image Products
3 Jenner, Irvine CA 92618. 714/727-1733.

L&K Micro Supply
711 Charcot Ave, San Jose CA 95131-2208. 408/954-0640.

L-Com Inc.
1755 Osgood St, North Andover MA 01845-1028. 508/682-6936.

Lan Art Corp.
145 Rosemary St, Needham MA 02194-3259. 617/444-1994.

Lancast/Casat Technology Inc.
12 Murphy Dr, Nashua NH 03062-1918. 603/880-1833.

Landsman Technology Inc.
7869 W Day Rd, Bainbridge Island WA 98110. 206/842-5480.

Lanex Corp.
10727 Tucker St, Beltsville MD 20705-2208. 301/595-9140.

LDDS Worldcom
1 Meadowlands Plz, East Rutherford NJ 07073-2100. 201/804-6400.

Leadman Electronic
2980 Gordon Ave, Santa Clara CA 95051-0710. 408/738-1751.

Leemah Datacom Security Corp.
6200 Paseo Padre Parkway, Freemont CA 94555. 510/786-0790.

Lemcom Systems Inc.
9033 N 24th Ave, Phoenix AZ 85021-2847. 602/944-1543.

Lexi Computer Systems Corp.
242 Neck Rd, Ward Hill MA 01835-8030. 508/521-1118.

Linkon Corp.
140 Sherman Street, Fairfield CT 06430. 212/753-2544.

Litton Industries Inc.
2500 N Airport Commerce Ave, Springfield MO 65803-9565. 417/862-0751.

Litton International Development Corp.
29851 Agoura Rd, Agoura Hills CA 91301. 818/991-9660.

Liuski International Inc.
10 Hub Dr, Melville NY 11747-3503. 516/454-8220.

Lockheed Commercial Electronic
MS 157 65 River Rd, Hudson NH 03051. 603/885-8550.

Logical Design Group Inc.
6301 Chapel Hill Rd, Raleigh NC 27607-5115. 919/851-1101.

Logicode Technology Inc.
1380 Flynn Rd, Camarillo CA 93012-8016. 805/388-9000.

Low Key Corp.
105 Maplewood Ave, Gloucester MA 01930-2731. 508/281-0448.

Lucas Deeco Products
31047 Genstar Rd, Hayward CA
94544-7831. 510/471-4700.

Madge Networks Inc.
2310 N 1st St, San Jose CA
95131-1011. 408/955-0700.

Mag Innovision
2801 S Yale St, Santa Ana CA
92704-5850. 714/751-2008.

Magnebit Corp.
4343 Viewridge Ave, San Diego
CA 92123-1619. 619/573-0727.

Magnecomp Corp.
471 Atlas St, Brea CA 92621-
3118. 909/693-9130.

Magnetec Corp.
7 Laser Ln, Wallingford CT
06492-1928. 203/949-9933.

Magnetic Technologies Corp.
770 Linden Ave, Rochester NY
14625-2764. 716/385-8711.

Magretech Inc.
7300 Hollister Ave, Goleta CA
93117-2806. 805/685-4551.

Maximum Strategy Inc.
801 Buckeye Ct, Milpitas CA
95035-7408. 408/383-1600.

MB Software
2225 E Randol Mill Rd Ste 305,
Arlington TX 76011-6306.
817/633-9400.

MBase Communications Inc.
16 Esquire Rd, North Billerica
MA 01862-2500. 508/671-9440.

MC Software Corporation
5032 Old Oak Ln, American Fork
UT 84003-9463. 801/756-9385.

**Measurex Management
Systems Division**
1280 Kemper Meadow Dr,
Cincinnati OH 45240-1632.
513/825-3931.

Media Logic ADL Inc.
1965 57th Ct N, Boulder CO
80301-2810. 303/939-9780.

Media Recovery Inc.
7929 Brookriver Dr, Dallas TX
75247-4900. 214/630-9625.

Media Recovery Inc.
12000 Crownpoint Dr, San
Antonio TX 78233-5360.
210/657-5071.

Mediashare Corp.
5927 Priestly Dr, Carlsbad CA
92008-8813. 760/931-7171.

Medplus Inc.
8600 Governors Hill Dr,
Cincinnati OH 45249-1388.
513/583-0500.

Meridian Data Inc.
5615 Scotts Valley Dr, Scotts
Valley CA 95066-3424. 408/438-
3100.

Merit Electronic Design
190 Rodeo Dr, Edgewood NY
11717-8317. 516/667-9699.

Metra Corporation
2205 Fortune Dr, San Jose CA
95131-1806. 408/432-1110.

Metricom Inc.
980 University Ave, Los Gatos
CA 95030-2319. 408/399-8200.

Micro Dynamics Corp.
7550 Market Place Dr, Eden
Prairie MN 55344. 612/941-8071.

Micro Equipment
2900 Jones Mill Road, Norcross
GA 30071-4606. 770/447-1726.

Micro General Corp.
14711 Bentley Circle, Tuston CA
27805-4615. 714/667-0557.

Micro Industries Corp.
8399 Green Meadows Dr N,
Westerville OH 43081-9486.
614/548-7878.

Micro Seven Inc.
1281 NE 25th Ave, Hillsboro OR
97124-5980. 503/693-6982.

Micro Systems Inc.
3447 Ocean View Blvd, Glendale
CA 91208-1508. 818/244-4600.

Micro-Integration Corp.
1 Science Park Way, Frostburg
MD 21532-2024. 301/689-0800.

Microland Electronic Corp.
1883 Ringwood Ave, San Jose
CA 95131-1721. 408/441-1688.

Micromint Inc.
4 Pk St, Vernon CT 06066.
860/871-6170.

Microsoft Corporation
2929 N Central Ave, Suite 1200,
Phoenix AZ 85012-2743.
602/266-0302.

Microspeed Inc.
2495 Industrial Parkway,
Hayward CA 94545. 510/259-
1270.

Microtek Lab Inc.
3715 Doolittle Drive, Redondo
Beach CA 90278-1226. 310/297-
5000.

Microwave Bypass Systems
25 Braintree Hill Park, Braintree
MA 02184-8702. 617/843-8260.

Miro Computer Products Inc.
955 Commercial St, Palo Alto CA
94303-4908. 415/855-0940.

Mitek Systems Inc.
10070 Carroll Canyon Rd, San
Diego CA 92131-1108. 619/635-
5900.

Mitsuba Corp.
1925 Wright Avenue, La Verne
CA 91750-5820. 909/392-2000.

Mitsumi Electronics Corp. Inc.
5808 West Campus Circle Drive,
Irving TX 75063-2655. 214/550-
7300.

Morley Co.
P.O. Box 629, Portsmouth NH
03801-4229. 603/436-5430.

Motorola Inc.
1701 Valley View Ln, Dallas TX
75234-9004. 214/888-2350.

Mouse Systems Corp.
47505 Seabridge Dr, Fremont CA
94538-6546. 510/656-1117.

Mutoh
3007 E Chambers St, Phoenix AZ
85040-3796. 602/276-5533.

NAI Technologies Inc.
2405 Trade Center Avenue,
Longmont CO 80503. 303/776-
5674.

National Computer Systems
1313 Lone Oak Rd, Eagan MN
55121-1334. 612/683-6000.

National Instrument Corp.
6504 Bridge Point Parkway,
Austin TX 78730-5039. 512/794-
0100.

Natural Microsystems
911 N Plum Grove Rd,
Schaumburg IL 60173-4751.
847/706-9700.

Neff Instrument Corp.
700 S Myrtle Ave, Monrovia CA
91016-3423. 818/357-2281.

Net Soft
31 Technology Drive, 2nd Floor,
Irvine CA 92618. 714/768-4013.

NetLink Inc.
1881 Worcester Rd, Framingham
MA 01701-5459. 508/879-6306.

Netspan Corp.
1411 E Campbell Rd, Suite 1000,
Richardson TX 75081-1971.
214/690-8844.

**Network Equipment
Technology Inc.**
800 Saginaw Dr, Redwood City
CA 94063-4740. 415/366-4400.

New Media Corporation
1 Technology Dr, Irvine CA
92718-2350. 714/453-0100.

Newgen Systems Corp.
3545 Cadillac Ave, Costa Mesa
CA 92626-1401. 714/641-8600.

Nex Test
2 Mid America Plz, Oakbrook
Terrace IL 60181. 630/574-3399.

Nikon Electronic Imaging
1300 Walt Whitman Rd, Melville
NY 11747-3012. 516/547-4355.

NMB Technologies Inc.
9730 Independence Ave,
Chatsworth CA 91311-4323.
818/341-3355.

Northern Computers Inc.
5007 S Howell Ave, Milwaukee
WI 53207-6157. 414/769-5980.

Nova Electric Inc.
100 School St, Bergenfield NJ
07621-2915. 201/385-0500.

Novadyne Computer Systems
1825 Monetary Ln, Carrollton TX
75006-7069. 214/389-5071.

O'Neil Products Development
8 Mason, Irvine CA 92718-2705.
714/458-1234.

OA Data Co.
3346 55th St, Woodside NY
11377-1937. 718/463-2115.

OAZ Communications Inc.
44920 Osgood Rd, Fremont CA
94539-6110. 510/226-0171.

Oce-Printing Systems
2530 East Cerritos Avenue,
Anaheim CA 92806-5627.
714/979-2240.

Oce-USA Inc.
5450 North Cumberland Avenue,
Chicago IL 60656. 630/351-2900.

Odetics Inc.
1585 S Manchester Ave,
Anaheim CA 92802-2907.
714/774-5000.

Olicom USA Inc.
900 E Park Blvd, Suite 250, Plano
TX 75074-5465. 214/423-7560.

Olivetti Supplies Inc.
137 4th St, Middletown PA
17057-5049. 717/944-5551.

Omnicomp Graphics Corp.
1734 W Sam Houston Pkwy N,
Houston TX 77043-2799.
713/464-2990.

**Omron Office Automation
Products Inc.**
3945 Freedom Cir, Santa Clara
CA 95054-1225. 408/727-1444.

Opcode Systems Inc.
3950 Fabian Way, Suite 100, Palo
Alto CA 94303-4605. 415/856-
3333.

Opticord Inc.
707 S Vermont St, Palatine IL
60067-7138. 847/705-1952.

Optivision Inc.
1480 Drew Ave, Davis CA
95616-4890. 916/757-4850.

Opus Systems Inc.
10050 Bubb Road, Cupertino CA
95014. 408/342-1060.

Orange Micro Inc.
1400 N Lakeview Ave, Anaheim
CA 92807-1896. 714/779-2772.

Orbit International Corp.
80 Cabot Ct, Hauppauge NY
11788-3729. 516/435-8300.

Orion Instruments Inc.
1376 Borregas Ave, Sunnyvale
CA 94089-1004. 408/747-0440.

Osicom Technologies
7402 Hollister Ave, Santa
Barbara CA 93117-2583.
805/968-4262.

OST Inc.
14225 Sullyfield Cir, Chantilly
VA 20151. 703/817-0400.

Pacific Computer Products Inc.
P.O. Box 8968, Fountain Valley
CA 92728. 714/549-7535.

Pam Pacific Associates Inc.
15314 East Valley Blvd, La
Puente CA 91746-3324. 818/333-
3009.

Parity Systems Inc.
110 Knowles Dr, Los Gatos CA
95030-1828. 408/378-1000.

Pathlight Technology & Ironics
767 Warren Rd, Ithaca NY
14850-1255. 607/266-4000.

PC Guardian
1133 Francisco Blvd E, San
Rafael CA 94901-5427. 415/459-
0190.

PCN
15 Crawford St, Needham MA
02194-2618. 617/433-3300.

PE Systems Inc.
500 Huntmar Park Dr, Herndon
VA 22070-5100. 703/691-3498.

Peak Technologies
21041 S Western Ave, Torrance
CA 90501-1711. 310/781-9222.

Pedcom Inc.
5500 Stewart Ave, Fremont CA
94538-3184. 510/490-3688.

Peerless Sytems
2381 Rosecrans Ave, El Segundo
CA 90245-4919. 310/536-0908.

Pentax Technologies Corp.
100 Technology Dr, Broomfield
CO 80021-3414. 303/460-1600.

Personal Data Systems Inc.
100 W Rincon Ave, Campbell
CA 95008-2818. 408/866-1126.

Petronic International Inc.
2300 Zanker Rd, San Jose CA
95131-1114. 408/943-1717.

Phoenix Inc.
525 Almanor Ave, Sunnyvale CA
94086-3512. 408/524-2345.

Pioneer New Media Technology
2265 E 220th St, Long Beach CA
90810-1643. 310/952-2111.

Pivot Point
600 W Cummings Park, Woburn
MA 01801-6369. 617/932-0932.

Plexcom Inc.
2255 Agate Ct, Simi Valley CA
93065-1841. 805/522-3333.

Plextor
4255 Burton Dr, Santa Clara CA
95054-1512. 408/980-1838.

Polywell Computers Inc.
1461 San Mateo Ave, South San
Francisco CA 94080-6505.
415/583-7222.

Powell Electronics Inc.
PO Box 8765, Philadelphia PA
19101-8765. 215/365-1900.

PQR Inc.
313 E Broad St, Palmyra NJ
08065-1607. 609/829-0707.

Practical Automation Inc.
45 Woodmont Rd, Milford CT
06460-2840. 203/882-5640.

Precision Connector Design Inc.
2 Technology Dr, Peabody MA
01960-7907. 508/532-8800.

Precision Handling Devices Inc.
617 Airport Rd, Fall River MA
02720-4707. 508/679-5282.

Presenta Technologies Corp.
12806 Schabarum Ave, Irwindale
CA 91706-6805. 818/960-0420.

Preston Scientific
1180 N Blue Gum St, Anaheim
CA 92806-2409. 714/632-3700.

Primax Electronics Inc. USA
521 Almanor Ave, Sunnyvale CA
94086-3512. 408/522-1200.

Printed Circuit Corp.
10 Micro Dr, Woburn MA 01801-
5702. 617/935-9570.

Printer Works
3481 Arden Rd, Hayward CA
94545-3929. 510/887-6116.

Printrak International Inc.
1250 N Tustin Ave, Anaheim CA
92807-1617. 714/666-2700.

Printware Inc.
1270 Eagan Industrial Rd, St.
Paul MN 55121-1231. 612/456-
1400.

Pro-Log Corporation
12 Upper Ragsdale Dr, Monterey
CA 93940-5730. 408/372-4593.

Procom Technology Inc.
2181 Dupont Dr, Irvine CA
92715-1301. 714/852-1000.

Proto Technology Corp.
55 Green St, Clinton MA 01510-
3002. 508/368-8588.

Proton Corp.
13855 Struikman Rd, Cerritos CA
90703-1031. 562/404-2222.

Proxim Inc.
295 Bernardo Ave, Mountain
View CA 94043-5205. 415/960-
1630.

PSINET Inc.
510 Huntmar Park Dr, Herndon
VA 22070-5100. 703/904-7187.

Psitech Inc.
18368 Bandilier Cir, Fountain
Valley CA 92708-7001. 714/964-
7818.

Pycon Inc.
3501 Leonard Ct, Santa Clara CA
95054-2043. 408/727-1213.

QC Optics Inc.
154 Middlesex Tpke, Burlington
MA 01803-4495. 617/272-4949.

QMS
41 E Daggett Dr, San Jose CA
95134. 408/432-0377.

Qualtec Data Products Inc.
47767 Warm Springs Blvd,
Fremont CA 94539-7470.
510/490-8911.

Quintar Company
370 Amapola Ave, Torrance CA
90501-1475. 310/320-5700.

Racal-Datacom
2900 Gordon Avenue, Santa
Clara CA 95051-0718. 408/774-
0346.

Racal-Guardata Inc.
480 Spring St # 900, Herndon VA
22070-5215. 703/471-0892.

Racore Computer Products Inc.
170 Knowles Drive, Suite 206,
Los Gatos CA 95030-1833.
408/374-8290.

Rad Network Devices Inc.
3505 Cadillac Ave, Costa Mesa
CA 92626-1431. 714/436-9700.

Radcom
900 Corporate Dr, Mahwah NJ
07430-2013. 201/529-2020.

Radius Inc.
215 Moffett Park Dr, Sunnyvale
CA 94089-1322. 408/541-6100.

Raxco Software Inc.
2440 Research Blvd, Rockville
MD 20850-3238. 301/258-2620.

Raytheon E-Systems
7700 Arlington Blvd, Falls
Church VA 22042-2902.
703/560-5000.

Real Applications Ltd.
21051 Warner Center Ln,
Woodland Hills CA 91367-6512.
818/226-6673.

Recoton Corporation
2950 Lake Emma Rd, Lake Mary
FL 32746-3705. 407/333-8900.

Recoton Corporation
145 E 57th St, New York NY
10022-2141. 212/644-0220.

Reliable Communications Inc.
PO Box 816, Angels Camp CA
95222-0816. 209/736-0421.

Retro-Fit Inc.
455 Fortune Blvd, Milford MA
01757-1723. 508/478-2222.

Road Warrior International
16580 Harbor Blvd, Huntington
Beach CA 92647-3614. 714/847-
1799.

Rose Electronics
PO Box 742571, Houston TX
77274-2571. 713/933-7673.

S3 Inc.
2831 Mission College Boulevard,
Santa Clara CA 95052. 408/980-
5400.

Sager Modern Computer Inc.
18005 Cortney Ct, Rowland
Heights CA 91748-1203.
818/964-8682.

SBE Inc.
4550 Norris Canyon Rd, San
Ramon CA 94583-1369. 510/355-
2000.

Scientific Research Management Corp.
1714 Ringwood Ave, San Jose CA 95131-1711. 408/437-1800.

Screenprint Dow Inc.
271 Ballardvale St, Wilmington MA 01887-1081. 617/935-6395.

Scriptel Holding Inc.
4153 Arlingate Plz, Columbus OH 43228-4115. 614/276-8402.

Seaport Imaging Inc.
1340 South DeAnza Blvd, San Jose CA 95129. 408/366-6400.

Secure Computing Inc.
2151 Salvio St, Concord CA 94520-2483. 510/827-5707.

Sejin America Inc.
2004 Martin Ave, Santa Clara CA 95050-2700. 408/980-7550.

Sequent Computer Systems Inc.
1430 Spring Hill Rd, Mc Lean VA 22102-3000. 703/442-9100.

Servcomp Inc.
10000 Old Katy Rd, Houston TX 77055-6035. 713/935-3600.

Sherwood Terminal
21056 Forbes St, Hayward CA 94545. 510/623-8900.

Silicon Star International Inc.
47889 Fremont Blvd, Fremont CA 94538-6506. 510/623-0500.

Simpact Inc.
9210 Sky Park Ct, San Diego CA 92123-4302. 619/565-1865.

Smart Games Interactive Inc.
2075 Case Pkwy S, Twinsburg OH 44087-2361. 216/963-0660.

SMK Electronics Corp.
625 Fee Ana Street, Placentia CA 92870. 714/996-0960.

Solectek Corporation
6370 Nancy Ridge Dr, Suite 109, San Diego CA 92121-3212. 619/450-1220.

Sonnet Technologies Inc.
18004 Sky Park Cir, Irvine CA 92714-6428. 714/261-2800.

Sony Corp. of America
3300 Zanker Rd, San Jose CA 95134-1940. 408/432-0190.

Sora Power
19060 S Dominguez Hills Dr, Rancho Dominguez CA 90220-6404. 310/884-5200.

Source Electronics Corp.
26 Clinton Dr, Hollis NH 03049-6577. 603/595-2906.

South Bay Circuits Inc.
210 Hillsdale Ave, San Jose CA 95136-1392. 408/978-8992.

South Hills Datacom
760 Beechnut Dr, Pittsburgh PA 15205-1804. 412/921-9000.

Spacetech IMC
100 Ft. John Street, Lowell MA 01852. 508/970-0330.

Spear Voice Systems Inc.
48501 Warm Springs Blvd, Fremont CA 94539-7750. 510/623-1800.

Specialix Inc.
745 Camden Ave, Campbell CA 95008-4146. 408/378-7919.

Specialty Products
2120 Denton Drive, Suite 104, Austin TX 78758. 512/832-8292.

Sprite Inc.
1120 Stewart Ct, Sunnyvale CA 94086-3918. 408/773-8888.

Stand Microsystems Corp.
6 Hughes, Irvine CA 92718-2021. 714/707-2400.

Starlight Networks Inc.
205 Ravendale Dr, Mountain View CA 94043-5216. 415/967-2774.

Stereographics Corp.
2171 Francisco Blvd E, San Rafael CA 94901. 415/459-4500.

Sterling Electronics Corp.
4201 Southwest Fwy, Houston TX 77027-7298. 713/627-9800.

Strawberry Tree Inc.
160 S Wolfe Rd, Sunnyvale CA 94086-6504. 408/736-8800.

Sudbury Systems Inc.
490 Boston Post Road, Sudbury MA 01776. 508/443-1100.

Sumitomo Electric USA Inc.
3235 Kifer Rd, Santa Clara CA 95051-0815. 408/737-8517.

Support Systems International Corp.
150 S 2nd St, Richmond CA 94804-2110. 510/234-9090.

Sutron Corp.
21300 Ridgetop Cir, Sterling VA 20166-6520. 703/406-2800.

Sutton Designs Inc.
215 N Cayuga St, Ithaca NY 14850-4323. 607/277-4301.

SVEC America Computer Corp.
1761 East Reynolds Avenue, Irvine CA 92614. 714/756-2233.

Symbol Technologies Inc.
2145 Hamilton Ave, San Jose CA 95125-5905. 408/369-2600.

Symbus Technology Inc.
950 Winter St, Waltham MA 02154-1226. 617/890-4100.

Sync Research Inc.
40 Parker, Irvine CA 92718-2013. 714/588-2070.

Systech Corp.
10505 Sorrento Valley Rd, San Diego CA 92121. 619/453-8970.

Systems Maintenance Service
537 Great Rd, Littleton MA 01460-1208. 508/952-4560.

Systems Research Labs Inc.
2800 Indian Ripple Rd, Dayton OH 45440-3696. 937/426-6000.

Tadpole Technology Inc.
12012 Technology Blvd, Austin TX 78727-6201. 512/219-2200.

Talaris Systems Inc.
PO Box 261580, San Diego CA 92196-1580. 619/587-0787.

Talking Technology Inc.
1125 Atlantic Ave, Alameda CA 94501-1145. 510/522-3800.

Tasking Company
333 Elm St, Dedham MA 02026-4530. 617/320-9400.

TCL Inc.
41829 Albrae St, Fremont CA 94538-3120. 510/657-3800.

TDK Corp. of America
1600 Feehanville Dr, Mount Prospect IL 60056-6024. 847/803-6100.

TDK Systems Development Center
136 New Mohawk Rd, Nevada City CA 95959-3262. 916/265-5395.

Teac America Inc.
7733 Telegraph Rd, Montebello CA 90640-6540. 213/726-0303.

Tec America Inc.
2710 Lakeview Ct, Fremont CA 94538-6534. 510/651-5333.

Tec America Inc.
4401 Bankers Cir, Atlanta GA 30360-2709. 770/449-3040.

Technetics Inc.
481 Cypress Lane, El Cajon CA 92020. 619/440-5121.

Technically Elite
6330 Saint Ignacio, San Jose CA 95119. 408/370-4300.

Technology 80 Inc.
658 Mendelssohn Ave N, Minneapolis MN 55427-4348. 612/542-9545.

Telcom Semiconductor
1300 Terra Bella Ave, Mountain View CA 94043. 415/968-9241.

Telenetics Corp.
26772 Vista Ter, Lake Forest CA 92630-8110. 714/455-4000.

Telepartner International
135 South Rd, Farmington CT 06032-2556. 860/674-2640.

Teleprocessing Products Inc.
4565 Industrial St, Simi Valley CA 93063-3464. 805/522-8147.

Telesensory Corporation
PO Box 7455, Mountain View CA 94039-7455. 415/960-0920.

Telog Instruments Inc.
830 Canning Pkwy, Victor NY 14564-8940. 716/742-3000.

Telxon Corp.
6333 Rothway St, Houston TX 77040-5040. 713/307-2500.

Tenkey Publishing Inc.
5422 Carrier Dr, Orlando FL 32819-8394. 407/351-0966.

Texas ISA Inc.
14825 Saint Mary's Ln, Houston TX 77079-2904. 713/493-9925.

Theos Software Corp.
1801 Oakland Blvd, Suite 315, Walnut Creek CA 94596-5041. 510/935-1118.

Titan Corporation
3033 Science Park Rd, San Diego CA 92121-1101. 619/552-9500.

TMC
631 S Milpitas Blvd, Milpitas CA 95035-5473. 408/262-1074.

Trace
1040 E Brokaw Rd, San Jose CA 95131-2309. 408/441-8040.

Transettlements Inc.
1745 Phoenix Blvd, Atlanta GA 30349-5534. 770/996-8109.

Transtech Parallel Systems Corp.
20 Thornwood Dr, Ithaca NY 14850-1265. 607/257-6502.

Trident Systems Inc.
10201 Lee Hwy, Suite 300, Fairfax VA 22030-2222. 703/273-1012.

Troy
2331 South Pullman St, Santa Ana CA 92705-5571. 714/250-3280.

TRW Inc.
1760 Glenn Curtiss St Bldg Dh, Carson CA 90746. 310/764-9464.

Twinhead Corp.
1537 Centre Pointe Dr, Milpitas CA 95035-8013. 408/945-0808.

Ultima International
3358 Gateway Blvd, Fremont CA 94538-6525. 510/659-1580.

Umax Technologies Inc.
3353 Gateway Blvd, Fremont CA 94538-6526. 510/651-8883.

Unicom Electric Inc.
11980 Telegraph Rd, Santa Fe Springs CA 90670-3797. 562/946-9650.

Unify Corporation
3927 Lennane Dr, Sacramento CA 95834-1922. 916/928-6400.

Vanguard R Squared Inc.
11211 E Arapahoe Rd, Englewood CO 80112-3819. 303/799-9292.

Vanstar Corp.
5964 W Las Positas Blvd, Pleasanton CA 94588-8540. 510/734-4000.

Vector Electronics Co.
12460 Gladstone Ave, Sylmar CA 91342-5374. 818/365-9661.

Verbatim Corporation
535 Independence Pkwy, Chesapeake VA 23320-5117. 757/547-5477.

Versyss Inc.
1 Hartfield Blvd, East Windsor CT 06088-9500. 860/627-5151.

Vertex Technologies
61 Executive Blvd, Farmingdale NY 11735-4710. 516/293-9880.

Vidar Systems Corp.
460 Springpark Pl, Herndon VA 22070-5237. 703/471-7070.

Visicom-Microlithics Corporation
17301 W Colfax Ave, Golden CO 80401-4800. 303/277-0271.

Voyetra Technologies
5 Odell Plz, Yonkers NY 10701-1406. 914/966-0600.

Western Telematic Inc.
5 Sterling, Irvine CA 92618. 714/586-9950.

Westrex International
25 Denby Rd, Boston MA 02134-1605. 617/254-1200.

Winchester Systems Inc.
400 W Cummings Park, Woburn MA 01801-6519. 617/933-8500.

Xecom
374 Turquoise St, Milpitas CA 95035-5431. 408/945-6640.

Xerox Color Graphics
5853 Rue Ferrari, San Jose CA 95138-1857. 408/225-2800.

Xtend Micro Products Inc.
2 Faraday, Irvine CA 92618. 714/699-1400.

Z-World Engineering
1724 Picasso Ave, Davis CA 95616-0547. 916/757-3737.

Znyx Corporation
48501 Warm Springs Blvd, Fremont CA 94539-7750. 510/249-0800.

COMPUTERS AND RELATED EQUIPMENT

Accel Technologies Inc.
6825 Flanders Dr, San Diego CA
92121-2986. 619/554-1000.

Accent
20 Trafalgar Sq, Nashua NH
03063-1985. 603/886-1570.

Adaptec Inc. Boulder Technology
6151 Lookout Rd, Boulder CO
80301-3359. 303/516-4000.

Advanced Digital Information Corp.
PO Box 97057, Redmond WA
98073-9757. 206/881-8004.

Akashic Memories Corporation
305 W Tasman Dr, San Jose CA
95134-1719. 408/944-9080.

Alps Electric Inc.
3553 N 1st St, San Jose CA
95134-1803. 408/432-6000.

American Teleprocessing Corp.
10681 Haddington Dr, Houston
TX 77043-3239. 713/973-1616.

Areal Technology Inc.
2146 Bering Dr, San Jose CA
95131-2013. 408/241-8290.

Avatar Systems Corp.
1455 McCarthy Blvd, Milpitas
CA 95035-7433. 408/321-0110.

Aviv Corp.
4 Fourth Ave, Burlington MA
01803-3304. 617/270-6900.

Aydin Electro-Fab Company
960 River Rd, Croydon PA
19021-7540. 215/788-0401.

Aztek Inc.
15 Marconi, Irvine CA 92718-
2701. 714/770-8406.

Bering Technology
1357 Dell Ave, Campbell CA
95008-6609. 408/364-6500.

Boundless Technologies
9430 Research Blvd, Austin TX
78759-6543. 512/346-2447.

Boundless Technologies
100 Marcus Blvd, Hauppauge NY
11788-3762. 516/342-7400.

Box Hill Systems Corp.
161 Ave Of The Amer, New
York NY 10013. 212/989-4455.

Braemar Inc.
11481 Rupp Dr, Burnsville MN
55337-1276. 612/890-5135.

CD Technology
762 San Aleso Ave, Sunnyvale
CA 94086-1445. 408/752-8500.

Cecorp
8 Chrysler, Irvine CA 92718-
2008. 714/583-0792.

Cirexx Corp.
3391 Keller St, Santa Clara CA
95054-2617. 408/988-3980.

CLM Systems Inc.
4805 West Laurel Street, Tampa
FL 33607. 813/286-8755.

Clone Software
4956 Ward Rd, Wheat Ridge CO
80033-3865. 303/423-0371.

Codonics Inc.
17991 Englewood Dr, Cleveland
OH 44130-3493. 216/243-1198.

Computer Products Center Inc.
21 Morgan Street, Irvine CA
92718-2005. 714/588-9800.

Computron Display Systems
1697 W Imperial Ct, Mount
Prospect IL 60056-5554.
847/952-8800.

Control Memory Factory
1450 Koll Cir, San Jose CA
95112-4612. 408/437-1122.

Cyber Touch
853 Lawrence Dr, Newbury Park
CA 91320-2232. 805/499-5000.

Data 1 Inc.
6416 Parkland Dr, Sarasota FL
34243-4060. 941/751-3336.

Deringer Manufacturing Company
1250 Townline Rd, Mundelein IL
60060-4448. 847/566-4100.

Diamond Multimedia Systems
7101 Super Dr, Albany OR
97321. 541/967-2400.

Digital Equipment Corp. Northeast Technical Center
334 South St, Shrewsbury MA
01545-4172. 508/841-3111.

Display Technologies Inc.
1071 Davis Rd, Elgin IL 60123-
1313. 847/931-2100.

DSI
1 Inverness Dr E, Englewood CO
80112-5519. 303/754-2000.

EBM Systems Inc.
7701 Greenbelt Rd, Greenbelt
MD 20770-2037. 301/441-8200.

Eesof Inc.
5601 Lindero Canyon Rd,
Westlake Village CA 91362-
4050. 818/879-6200.

Emeritus Technologies
2750 N Clovis Ave, Fresno CA
93727-7724. 209/292-8888.

Epson America Inc.
20770 Madrona Ave, Torrance
CA 90503-3777. 310/782-0770.

FED Corp.
1580 State Route 52, Hopewell
Junction NY 12533-6526.
914/892-1900.

Feith Systems and Software Inc.
425 Maryland Dr, Ft Washington
PA 19034-2501. 215/646-8000.

GBS Computer Systems
1035 N Meridian Rd,
Youngstown OH 44509-1016.
330/797-2700.

General Disk Corporation
4010 Moorpark Ave, San Jose
CA 95117-4101. 408/432-0505.

Genesco Technology Corporation
1620 S Sunkist St, Anaheim CA
92806-5811. 714/938-0348.

Georgens Industries Inc.
3346 Industrial Ct, San Diego CA
92121-1003. 619/481-8114.

Giga Trend Inc.
2234 Rutherford Rd, Carlsbad
CA 92008-8814. 760/931-9122.

Gigatek Memory Systems
1989 Palomar Oaks Way,
Carlsbad CA 92009-1307.
760/438-9010.

Graphon Corp.
150 Harrison Ave, Campbell CA
95008-1424. 408/370-4080.

Haushahn Systems & Engineering
5730 Eagle Drive, Grand Rapids MI 49512-4015. 616/285-3311.

Hewlett-Packard Co.
700 71st Ave, Greeley CO 80634-9776. 970/350-4000.

IEM Inc.
PO Box 1889, Fort Collins CO 80522-1889. 970/221-3005.

IGM Communications Inc.
4041 Home Rd, Bellingham WA 98226-9120. 360/733-4567.

Imation
8500 S Rita Rd, Tucson AZ 85747-9142. 520/574-8600.

Imnet Inc.
8601 Dunwoody Pl, Atlanta GA 30350-2550. 770/998-2200.

Industrial Technology Research
817 W Broad St, Bethlehem PA 18018-5223. 610/867-0101.

Informer Computer Systems
12833 Monarch St, Garden Grove CA 92641-3921. 714/891-1112.

Inmartech
1180 Miraloma Way, Sunnyvale CA 94086-4606. 408/253-6968.

Innovative Data Technology
15092 Avenue Of Science, San Diego CA 92128-3404. 619/676-1777.

Innovative Electronics Inc.
10110 USA Today Way, Hollywood FL 33025-3903. 954/432-0300.

Intecolor Corp.
2150 Boggs Rd, Duluth GA 30136-5890. 770/623-9145.

Integrated Software Design
171 Forbes Blvd, Mansfield MA 02048-1148. 508/339-4928.

ISI Inc.
1050 National Pkwy, Schaumburg IL 60173-4519. 847/490-1155.

Josten's Learning Corp.
2860 Old Rochester Rd, Springfield IL 62703-5632. 217/744-8700.

JTS Corporation
166 Baypointe Pkwy, San Jose CA 95134-1621. 408/468-1800.

Kastle Systems Inc.
1501 Wilson Blvd, Arlington VA 22209-2403. 703/524-7911.

Kobe Precision Inc.
31031 Huntwood Ave, Hayward CA 94544-7007. 510/487-3200.

Lockheed Martin
1 Ivybrook Blvd, Warminster PA 18974-1772. 215/443-7500.

Magnetic Recovery Technology Corp.
25431 Rye Canyon Rd, Valencia CA 91355-1206. 805/257-2262.

Maple Systems Inc.
1930 220th St SE, Bothell WA 98021-8471. 206/486-4477.

Megadata Corp.
35 Orville Dr, Bohemia NY 11716-2522. 516/589-6800.

Micro Design International Inc.
6985 University Blvd, Winter Park FL 32792-6715. 407/677-8333.

Micro Net Technology Inc.
80 Technology Dr, Irvine CA 92718-2301. 714/453-6000.

Micro Products Center Inc.
1840 County Line Rd, Huntingdon Valley PA 19006-1721. 215/355-4600.

Microcom Inc.
50001 Spring Valley Rd, Dallas TX 75244. 214/383-1383.

Micronet USA
201 San Antonio Cir, Mountain View CA 94040-1234. 415/948-6200.

Mitron Systems Corp.
9130-U Red Branch Road, Columbia MD 21045. 410/992-7700.

Mitsuba Southeast Inc.
2999 Pacific Dr, Norcross GA 30071-1872. 770/368-0532.

Most Inc.
11205 Knott Ave, Cypress CA 90630-5495. 714/898-9400.

Mountain Optech Inc.
4775 Walnut St, Boulder CO 80301-2579. 303/444-2851.

National Micrographics Systems Inc.
11941 Bournefield Way, Silver Spring MD 20904-7816. 301/622-4300.

National Micronetics Inc.
71 Smith Ave, Kingston NY 12401-3911. 914/338-0333.

Omron Systems Inc.
55 Commerce Dr, Schaumburg IL 60173-5302. 847/843-0515.

Ontrack Computer Systems
6321 Bury Dr, Eden Prairie MN 55346-1739. 612/937-1107.

OPT Industries Inc.
300 Red School Ln, Phillipsburg NJ 08865-2299. 908/454-2600.

Overland Data Inc.
8975 Balboa Ave, San Diego CA 92123-1507. 619/571-5555.

Ozo Diversified Automation
7450 E Jewell Ave, Suite A, Denver CO 80231-3200. 303/368-0401.

Pacific Business Systems Inc.
7 Hammond, Irvine CA 92718-1607. 714/768-8130.

Panasonic Industries Co.
1600 McCandless Dr, Milpitas CA 95035-8002. 408/945-5600.

Parallel Storage Solutions Co.
116 S Central Ave, Elmsford NY 10523-3503. 914/347-7044.

Parking Products Inc.
2517 Wyandotte Rd, Willow Grove PA 19090-1219. 215/657-7500.

PCs Technologies Inc.
4250 Wissahickon Ave, Philadelphia PA 19129-1234. 215/226-2220.

PF Micro
3598 Cadillac Ave, Costa Mesa CA 92626-1416. 714/549-4669.

Phase X Systems Inc.
19545 NW Von Neumann Dr, Beaverton OR 97006-6902. 503/531-2400.

Philips Laser Magnetic Storage
4425 Arrowswest Dr, Colorado
Springs CO 80907-3445.
719/593-7900.

Photonics Systems Inc.
6975 Wales Road, Northwood
OH 43619-1015. 419/666-0762.

Positran Inc.
800 East Main Street, Norristown
PA 19401-4104. 610/277-0500.

Pragmatech Inc.
101 Nicholson Lane, San Jose
CA 95134-1359. 408/943-1151.

Prima Computers
1933 O'Toole Avenue, Suite
A101, San Jose CA 95131.
408/432-1212.

Prima Storage Systems
3350 Scott Boulevard, Santa
Clara CA 95054-3108. 408/727-
2600.

**Proprietary Control Systems
Corporation**
18528 South Dominguez Hills
Drive, Rancho Dominguez CA
90220-6415. 310/638-0400.

PSI
150 Wright Brothers Drive, Salt
Lake City UT 84116-2885.
801/521-0366.

Qualstar Corp.
6709 Independence Avenue,
Canoga Park CA 91303. 818/882-
5822.

Reed Technology
20251 Century Boulevard,
Germantown MD 20874-1162.
301/428-3700.

**Roth Computer Register
Company**
1600 Saw Mill Run Blvd,
Pittsburgh PA 15210-3434.
412/884-5700.

Seagate Recording Media
47010 Kato Road, Fremont CA
94538-7332. 510/490-3222.

Sigma Designs Inc.
46501 Landing Parkway, Fremont
CA 94538-6421. 510/770-0100.

Storage Concepts Inc.
2652 McGaw Avenue, Irvine CA
92714-5840. 714/852-8511.

Storage Dimensions Inc.
1656 McCarthy Boulevard,
Milpitas CA 95035-7417.
408/954-0710.

Storagetek
1600 Riviera Avenue, Walnut
Creek CA 94596-3568. 510/934-
8000.

Storagetek
1990 West Camelback Road,
Phoenix AZ 85015-3464.
602/246-9899.

**Synektron Corp./A TDK Group
Company**
10500 SW Nimbus Avenue,
Portland OR 97223-4310.
503/684-3090.

Telemechanics Inc.
1636 5th Avenue, Bay Shore NY
11706-3425. 516/231-3833.

Televideo Systems Inc.
2345 Harris Way, San Jose CA
95131. 408/954-8333.

Todd Enterprises Inc.
65 East Beth Page Road,
Plainview NY 11803. 718/343-
1040.

Toshiba America
1251 Avenue of the Americas,
New York NY 10020-1104.
212/596-0600.

Transitional Technology Inc.
5401 East La Palma Avenue,
Anaheim CA 92807-2022.
714/693-1133.

Transmation Inc.
977 Mount Read Boulevard,
Rochester NY 14606-2895.
716/254-9000.

Turtle Mountain Corp.
380 Oak Grove Parkway, St. Paul
MN 55127-8506. 612/481-1427.

US Design Corporation
9075 Guilford Road, Columbia
MD 21046-2725. 410/381-3000.

Value Added Inc.
3230 Peachtree Corners Circle,
Norcross GA 30092-3655.
770/662-5800.

Wagers Business Systems Inc.
1955 University Avenue West, St.
Paul MN 55104-3427. 612/644-
3830.

Yaskawa Electric America Inc.
2942 MacArthur Blvd,
Northbrook IL 60062-2005.
847/291-2340.

Zenith Data Systems Corp.
2455 Horse Pen Road, Herndon
VA 20171. 703/713-3000.

COMPUTER COMPONENTS MANUFACTURERS

*K*evin *Flournoy of H.F. Henderson Industries, Inc. supervises operators responsible for manufacturing printed circuit boards, harnesses, battery packs, and other components. Kevin takes pride in the fact that products manufactured by his company are used by the Department of Defense for the security of our country.*

Chiachen Tsao, an engineer at AVM Technology, Inc., designs and builds sound cards for computers. At AVM, designers combine a knowledge of engineering with an understanding of musical concepts.

Both professionals agree that jobseekers should not enter the field of component design and manufacturing unless they are sure they are going to enjoy the work. Kevin and Chiachen list ambition, enthusiasm, and a willingness to learn new skills as important for success in this industry. People with an engineering background will have an advantage over others and will command a higher salary. A final recommendation to all jobseekers: Strive to show potential employers that you possess a superior level of dedication to the job and a desire to succeed in this industry.

ABB CEAG POWER SUPPLIES
One Pine Lake Parkway North, Palm Coast FL 32137-3608. 904/445-0311. **Contact:** Personnel. **Description:** A manufacturer of power supply units for computers.

AVM TECHNOLOGY, INC.
9774 South 700 East, Sandy UT 84070. 801/571-0967. **Contact:** Human Resources. **World Wide Web address:** http://www.avmtechnology.com. **Description:** A designer and producer of PC audio products. Product line includes Apex internal sound cards and Summit SST external MIDI/Digital recording modules.

ACCEL GRAPHICS INC.
1942 Zanker Road, San Jose CA 95112. 408/441-1556. **Contact:** Human Resources. **Description:** Manufactures and distributes accelerator boards. Accel Graphics provides end users and resellers with accelerator boards for graphic programs used in architectural and geographical applications.

ADAPTEC, INC.
691 South Milpitas Boulevard, Milpitas CA 95035. 408/945-8600. **Fax:** 408/262-2533. **Contact:** Human Resources. **World Wide Web address:** http://www.adaptec.com. **Description:** Adaptec is a supplier of high-performance intelligent subsystems and associated software. The company also produces VLSI circuits used to control the flow of data between a microcomputer's central processing unit (CPU) and its peripherals, including storage devices, laser printers, and network file servers. Products, mostly based on Small Computer System Interface (SCSI) Technology, include host adapters, proprietary VLSI circuits, and software-based development systems. The company's I/O solutions allow advanced computer applications, like multimedia, disk arrays, and high-speed/high resolution printing. The company's key products, called IOware, speed the transfer of data from one part of the computer system to another. **Number of employees nationwide:** 1,300.

ADAPTIVE SOLUTIONS, INC.
1400 NW Compton Drive, Suite 340, Beaverton OR 97006. 503/690-1236. **Fax:** 503/690-1249. **Contact:** Human Resources. **Description:** Adaptive Solutions, Inc. designs, develops, manufactures, and markets high-performance pattern recognition computing products that are marketed towards researchers and to original equipment manufacturers to be designed into end use applications systems. The company has been developing an end use product which is an accelerator

board to enhance the Adobe software application Photoshop. The company's initial target markets include pattern recognition applications including handwritten and printed character recognition, image processing, and neural network research. Other potential markets include speech processing and recognition, image compression and decompression, signal processing, process control, and forecasting. The company's products are based upon a parallel processor architecture and technology designed to be optimized for pattern recognition tasks. The company's products include system, board, and chip-level hardware, as well as software development tools and algorithms. **Corporate headquarters location:** This Location. **Number of employees at this location:** 50.

ALTERA CORPORATION

2610 Orchard Parkway, San Jose CA 95134-2020. 408/894-7000. **Fax:** 408/435-5065. **Contact:** Human Resources. **Description:** Designs, develops, and markets programmable logic integrated circuits and associated computer engineering development software and hardware. The company's semiconductor products, which are known as Erasable Programmable Logic Devices (EPLDs) and Flexible Logic Element Matrix (FLEM) devices, are standard logic chips that customers configure for specific end use applications using the company's proprietary software. The company's customers enjoy the benefits of low development costs, short lead times, and standard product inventories when compared to application specific integrated circuits (ASICs) and high-density and low-power consumption when compared to Transistor Transistor Logic (TTL). **Common positions include:** Accountant/Auditor; Automotive Engineer; Computer Programmer; Design Engineer; Manager of Information Systems; Mechanical Engineer; Research Technician; Sales Representative.

AMERICAN COMPUTER ASSEMBLY INC.

100 Chimney Point Drive, Ogdensburg NY 13669. 315/393-3575. **Contact:** Nicole Scott, Human Resources Representative. **Description:** A computer parts manufacturing company. American Computer Assembly Inc. is a subcontractor for printed circuit board assemblies.

AMERICAN MEGATRENDS

6145 Northbelt Parkway, Suite F, Norcross GA 30071-2972. 770/263-8181. **Contact:** Human Resources. **Description:** American Megatrends is a manufacturer of computer motherboards. **Corporate headquarters location:** This Location.

AMTELCO

4800 Curtin Drive, McFarland WI 53557. 608/838-4194. **Contact:** Tom Prough, Human Resources. **E-mail address:** irene@amtelcom.com. **World Wide Web address:** http://www.amtelcom.com. **Description:** A designer and manufacturer of PC switching boards and Multibus boards.

ANGIA COMMUNICATIONS

441 East Bay Boulevard, Provo UT 84606. 801/371-0540. **Fax:** 801/373-9847. **Contact:** Human Resources. **Description:** Angia Communications designs and manufactures a variety of computer cards, including those used in laptop computers and fax modems.

APEX PC SOLUTIONS, INC.

20031 142nd Avenue NE, Woodinville WA 98072. 206/402-9393. **Contact:** Mary Flynn. **Description:** Manufactures a variety of computer components, including concentrated switches. Apex PC Solutions also builds racks to hold peripherals.

BELL MICROPRODUCTS INC.

1941 Ringwood Avenue, San Jose CA 95131. 408/451-9400. **Fax:** 408/451-1600. **Contact:** Human Resources. **Description:** Founded in 1987, Bell Microproducts Inc. markets and distributes a select group of semiconductor and computer products to original equipment manufacturers and value-added resellers. Semiconductor products include memory, logic microprocessors, peripheral, and specialty components. Computer products include disk, tape and optical drives and subsystems, drive controllers, computers, and board-level products. The company also provides a variety of manufacturing and value-added services to its customers, including the supply of board-level products to customer specifications on a turnkey basis, as well as certain types of components and subsystem testing services, systems integration and disk drive formatting and testing, and the packaging of electronic component kits to customer specifications. **Corporate headquarters location:** This Location.

BRUMKO MAGNETICS
150 Binfield Street, Elkhorn NE 68022. 402/289-2400. **Contact:** Vicki Hamke, Human Resources Administrator. **Description:** Manufactures computer heads and components.

CSPI/SCANALYTICS DIVISION
40 Linnell Circle, Billerica MA 01821. 508/663-7598. **Toll-free phone:** 800/325-3110. **Fax:** 508/663-0150. **Contact:** Rose Doyon, Human Resources Manager. **E-mail address:** rdoyon@cspi.com. **World Wide Web address:** http://www.cspi.com. **Description:** CSPI designs, manufactures, and markets embedded processors which are small, low-power, special-purpose computers that enhance a system's ability to perform high-speed arithmetic. These processors are primarily used for defense, medical, industrial, and research applications. The company also develops and markets turnkey image analysis workstations targeted toward the biological sciences. Products include SuperCard DSP, which processes incoming signal streams in real time; products that isolate or reformat data; and Scanalytics products, including application-specific software packages for PC workstations. The responsibilities of the company's software engineers include the development of software involving automatic image analysis, advanced data visualization, and database development. **Common positions include:** Applications Engineer. **Educational backgrounds include:** Computer Science; Engineering; Mathematics; Physics. **Benefits:** 401(k); Dental Insurance; Disability Coverage; Financial Planning; Life Insurance; Medical Insurance; Tuition Assistance. **Corporate headquarters location:** This Location. **Other U.S. locations:** San Diego CA; MD. **Listed on:** NASDAQ. **Stock exchange symbol:** CSPI. **Annual sales/revenues:** $11 - $20 million. **Number of employees at this location:** 95. **Number of employees nationwide:** 105. **Number of projected hires for 1997 - 1998 at this location:** 10.

CALIDAD ELECTRONICS INC.
1920 SE Industrial Drive, Edinburg TX 78539. 210/381-0909. **Contact:** Human Resources. **Description:** The company assembles computer boards. **Corporate headquarters location:** Newark NY.

CAM GRAPHICS COMPANY INC.
15 Ranick Drive West, Amityville NY 11701. 516/842-3400. **Contact:** Human Resources. **Description:** Manufactures memory switches and touch-view screens. CAM Graphics Company supplies businesses and manufacturers with assorted memory-related devices and touch-view screens.

CENTENNIAL TECHNOLOGIES, INC.
37 Manning Road, Billerica MA 01821. 508/670-0646. **Contact:** Human Resources. **Description:** The company designs, manufactures, and sells personal computer cards used in portable computers and industrial applications, and font cartridges used in laser printers. The PC cards enhance the utility of portable computers and electronic equipment by adding memory, data/fax capabilities, and custom applications. The company's laser printer font cartridges broaden the capabilities of laser printers with applications in desktop publishing, word processing, and spreadsheet preparation. Customers include Digital Equipment Corporation, Texas Instruments, Inc., Fujitsu of America, Inc., and Xerox Corporation. **Corporate headquarters location:** This Location.

CHAMPION COMPUTER TECHNOLOGIES
749 Miner Road, Cleveland OH 44143. 216/646-2500. **Contact:** Human Resources Department. **Description:** A manufacturer of memory and PCMA products.

COLORGRAPHIC COMMUNICATIONS CORPORATION
P.O. Box 80448, Atlanta GA 30366. 770/455-3921. **Contact:** Human Resources Department. **Description:** A manufacturer of graphic adaptor boards that produce split screens on computers.

COMPUNETICS
3863 Rochester Road, Troy MI 48083. 810/524-6376. **Contact:** Human Resources. **Description:** A manufacturer of surface-mounted circuits.

COMSTAT CORPORATION
1720 Spectrum Drive, Lawrenceville GA 30243. 770/822-1962. **Contact:** Human Resources Department. **Description:** Manufactures a variety of electronic components for computers.

CREATIVE CONTROLLERS INC.
128 Kendrick Lane, Picayune MS 39466. 601/798-0577. **Contact:** Larry Upthegrove, General Manager. **Description:** Manufactures boards that interface with IBM computers to enable them to be compatible with all types of printers.

CYRIX CORPORATION
2703 North Central Expressway, Richardson TX 75080. 972/968-8388. **Fax:** 214/699-9857. **Contact:** Human Resources. **World Wide Web address:** http://www.cyrix.com. **Description:** Cyrix Corporation supplies high-performance microprocessors to the personal computer industry. The company designs, manufactures, and markets x86 software-compatible processors for the desktop and portable computer markets. Product line includes the Cyrix 5x86, a high-performance, low-power processor aimed at the mobile computer market; the Cyrix 6x86, a sixth-generation architecture, high-performance processor designed to be competitive with Intel's Pentium; and the P200, P166, and P155 desktop PCs. **Corporate headquarters location:** This Location. **Listed on:** NASDAQ. **Stock exchange symbol:** CYRX. **Annual sales/revenues:** $250 million. **Number of employees nationwide:** 320.

DATA DELAY DEVICES INC.
3 Mount Prospect Avenue, Clifton NJ 07013. 201/773-2299. **Contact:** Human Resources. **Description:** Manufactures delay wires.

DATA ENTRY PRODUCTS INC.
302 Third Street SE, Loveland CO 80537. 970/663-7337. **Contact:** Human Resources. **Description:** Manufactures keyboard components. Data Entry Products markets several keyboard components including snap dome switches.

DAYSTAR DIGITAL
5556 Atlanta Highway, Flowery Branch GA 30542. 770/967-2077. **Fax:** 770/967-3018. **Contact:** Human Resources. **World Wide Web address:** http://www.daystar.com. **Description:** DayStar Digital is a manufacturer of computer processor upgrade cards for Macintosh platform computers. Upgrade cards increase processing up to 11 times the original speed. The company's products are designed to provide optimum use of media/publishing applications. **Corporate headquarters location:** This Location.

DIGI INTERNATIONAL INC.
618 Grassmere Park Drive, Suite 6, Nashville TN 37211. 615/834-8000. **Contact:** Trish Holland, Human Resources. **Description:** This location, which formerly operated under the name Arnet Corporation prior to being bought by Digi International, manufactures expansion boards for computers. The company is a leading provider of data communications hardware and software that enable connectivity solutions for multi-user environments, remote access, and LAN connect markets. Digi International supports most major microcomputer and workstation architectures and most popular single- and multi-user systems, and provides cross-platform compatibility and software and technical support services. These products are produced under the DigiBoard, Arnet, Star Gate, and MiLAN trade names and marketed to a broad range of worldwide distributors, system integrators, value-added resellers (VARs), and OEMs. Digi acquired MiLAN Technology in 1994 and entered the LAN connectivity market. Products in this market reside directly on the LAN to improve its performance and increase its flexibility. **Corporate headquarters location:** Minnetonka MN.

DISCOPYLABS
48641 Milmont Drive, Fremont CA 94538. 510/651-5100. **Contact:** Human Resources. **Description:** DisCopyLabs assembles discs for software developers.

DISTRIBUTED PROCESSING TECHNOLOGY
140 Candace Drive, Maitland FL 32751. 407/830-5522. **Contact:** Human Resources. **Description:** Manufactures and distributes RAID and SCSI computerboards.

ESH, INC.
3020 South Park Drive, Tempe AZ 85282. 602/438-1112. **Fax:** 602/431-9633. **Contact:** Beth Moser, Human Resources Manager. **Description:** Manufactures computer boards. **Common positions include:** Blue-Collar Worker Supervisor; Electrical/Electronics Engineer; Operations/Production Manager; Quality Control Supervisor; Systems Analyst. **Educational backgrounds include:** Computer Science; Engineering; Marketing. **Benefits:** 401(k); Dental Insurance; Disability Coverage; Life Insurance; Medical Insurance; Tuition Assistance. **Special Programs:** Internships. **Corporate headquarters location:** This Location. **Operations at this facility include:** Administration; Manufacturing; Research and Development; Sales. **Listed on:** Privately held. **Number of employees at this location:** 95.

ENSONIQ CORPORATION
P.O. Box 3035, Malvern PA 19355-0735. 610/647-3930. **Fax:** 610/647-8908. **Contact:** Human Resources. **E-mail address:** hr@ensoniq.com. **World Wide Web address:** http://www.ensoniq.com. **Description:** A manufacturer of computer sound cards. **NOTE:** The physical address of the company is 155 Great Valley Parkway, Malvern PA. **Corporate headquarters location:** This Location.

EVERGREEN TECHNOLOGIES, INC.
915 SW 8th Street, Corvallis OR 97330-6211. 541/757-0934. **Contact:** Human Resources Department. **Description:** Manufactures upgrades for CPUs.

FIREPOWER, INC.
Menlo Park Design Center, 190 Independence Drive, Menlo Park CA 94025. **Fax:** 415/462-3051. **Contact:** Human Resources. **E-mail address:** lhorner@firepower.com. **World Wide Web address:** http://www.firepower.com. **Description:** A manufacturer of PowerPC microprocessors. **Common positions include:** Design Engineer; Engineer; Hardware Engineer; Materials Manager; Quality Assurance Engineer; Systems Engineer; Test Engineer; Tool Engineer. **Number of employees nationwide:** 46.

FUJITSU COMPUTER PRODUCTS OF AMERICA INC.
7300 NE Evergreen Parkway, Hillsboro OR 97124-5827. 503/681-7300x2445. **Recorded Jobline:** 503/693-2030. **Contact:** Human Resources Manager. **Description:** This location manufactures computer components including hard drives, laptop computers, and notebook computers. **Corporate headquarters location:** San Jose CA. **Parent company:** Fujitsu Corporation. **Number of employees at this location:** 430.

FUJITSU MICROELECTRONICS INC.
21015 SE Stark Street, Gresham OR 97030-2015. 503/669-6000. **Fax:** 503/669-6109. **Recorded Jobline:** 503/669-6075. **Contact:** Human Resources. **Description:** This location is a manufacturing facility of computers and computer components. **Common positions include:** Test Operator. **Corporate headquarters location:** San Jose CA. **Parent company:** Fujitsu Corporation. **Number of employees at this location:** 600.

FUJITSU MICROELECTRONICS INC.
3545 North 1st Street, San Jose CA 95134. 408/922-9000. **Contact:** Human Resources. **Description:** Adminstrative offices. Fujitsu Microelectronics manufactures computer components and computer products. **Corporate headquarters location:** This Location.

GADCO CORPORATION
12A Manor Parkway, Salem NH 03079. 603/898-8000. **Contact:** Human Resources. **Description:** Manufactures printed circuit boards.

GENERAL AUTOMATION INC.

17731 Mitchell North, Irvine CA 92614. 714/250-4800. **Contact:** Human Resources. **Description:** Develops, manufactures, and sells state-of-the-art computer and electronic components for worldwide distribution. **Corporate headquarters location:** This Location.

GENICOM CORPORATION

900 Clopper Road, Suite 110, Gaithersburg MD 20878. 301/258-5060. **Contact:** Human Resources. **Description:** This location of GENICOM Corporation develops control boards for printers. Overall, GENICOM is an international supplier of network system, service, and printer solutions. Through its Harris Adacom Network Service Solutions business unit, GENICOM provides integrated network system solutions including a broad portfolio of hardware and software products, technology consulting, and diagnostic and monitoring services. The company's Enterprising Service Solutions business unit provides logo and multivendor product field support and depot repair, express parts, and professional services. GENICOM's Impact and Laser Printing Solutions business unit designs, manufactures, and markets a wide range of computer printer technologies for general purpose applications. **Corporate headquarters location:** Chantilly VA. **Subsidiaries include:** GENICOM Pty. Ltd. (Australia); GENICOM Canada, Inc. (Canada); GENICOM S.A. (France); GENICOM GmbH (Germany); GENICOM SpA (Italy); GENICOM Limited (UK); and Harris Adacom Inc. (Canada). **Number of employees nationwide:** 2,147.

GLOBE MANUFACTURING SALES INC.

1159 US Route 22 East, Mountainside NJ 07092. 908/232-7301. **Fax:** 908/232-2590. **Contact:** Human Resources. **Description:** Manufactures computer brackets that hold the computer chip and other plastic parts.

HAUPPAUGE COMPUTER WORKS INC.

91 Cabot Court, Hauppauge NY 11788. 516/434-1600. **Contact:** Human Resources. **Description:** Manufactures PC circuit boards.

H.F. HENDERSON INDUSTRIES INC.

45 Fairfield Place, West Caldwell NJ 07006. 201/227-9250. **Contact:** Human Resources. **Description:** Manufactures printed circuit boards.

HERCULES COMPUTER TECHNOLOGY

3839 Spinnaker Court, Fremont CA 94538. 510/623-6030. **Contact:** Human Resources. **Description:** A manufacturer of window graphics accelerator cards.

HEWLETT-PACKARD COMPANY

1000 Northeast Circle Boulevard, Corvallis OR 97330. 541/757-2000. **Recorded Jobline:** 541/754-0919. **Contact:** Human Resources. **E-mail address:** resume@hp.com. **World Wide Web address:** http://www.jobs.hp.com. **Description:** This plant manufactures integrated circuit chips and other computer parts. Overall, Hewlett-Packard is engaged in the design and manufacturing of measurement and computation products and systems used in business, industry, engineering, science, health care, and education; principal products are integrated instrument and computer systems (including hardware and software), computer systems and peripheral products, and medical electronic equipment and systems. **NOTE:** Jobseekers should send resumes to: Employment Response Center, Event #2498, Hewlett-Packard Company, Mail Stop 20-APP, 3000 Hanover Street, Palo Alto CA 94304-1181. **Corporate headquarters location:** Palo Alto CA. **Number of employees at this location:** 5,000.

HOYA CORPORATION USA
MEMORY DIVISION

960 Rincon Circle, San Jose CA 95131. 408/435-5630. **Fax:** 408/435-5610. **Contact:** Human Resources. **E-mail address:** monica@mail.hoyausa.com. **Description:** Develops, produces, and markets glass thin-film rigid disks for use in hard drives for mobile computing applications. **NOTE:** The company offers entry-level positions. **Common positions include:** Machine Operator. **Benefits:** 401(k); Dental Insurance; Disability Coverage; Employee Discounts; Life Insurance; Medical Insurance; Savings Plan; Tuition Assistance. **Corporate headquarters location:** This Location. **Parent company:** Hoya Corporation (Japan). **Subsidiaries include:** Continuum; Probe

Tech. **Operations at this facility include:** Divisional Headquarters. **Listed on:** Privately held. **Number of employees at this location:** 185. **Number of employees worldwide:** 1,000. **Number of projected hires for 1997 - 1998 at this location:** 40.

HURLETRON INC.
1938 East Fairchild Street, Danville IL 61832. 217/446-6500. **Contact:** Human Resources. **Description:** Manufactures printed circuit boards.

HUTCHINSON TECHNOLOGY
40 West Highland Park, Hutchinson MN 55350. 320/587-1962. **Fax:** 320/587-1290. **Contact:** Randall Dostal, Staffing Specialist. **Description:** A manufacturer of precision spring components for disk drives. **Common positions include:** Chemical Engineer; Design Engineer; Electrical/Electronics Engineer; Industrial Engineer; Mechanical Engineer. **Educational backgrounds include:** Engineering. **Benefits:** 401(k); Dental Insurance; Disability Coverage; Life Insurance; Medical Insurance; Profit Sharing; Tuition Assistance. **Special Programs:** Internships. **Corporate headquarters location:** This Location. **Other U.S. locations:** Sioux Falls SD; Eau Claire WI. **Operations at this facility include:** Administration; Manufacturing; Research and Development; Sales. **Listed on:** NASDAQ.

I² TECHNOLOGY
101 Billerica Avenue, Building 6, North Billerica MA 01862. 508/670-3700. **Contact:** Human Resources. **Description:** A contract manufacturer of computer circuit boards.

INFORMATION TRANSFER INC.
100 Bloomfield Industrial Park, Bloomfield NY 14469. 716/657-7074. **Contact:** Human Resources. **Description:** Manufactures electronic data switches for computer systems.

INNOVEX
1313 5th Street South, Hopkins MN 55343. 612/938-4155. **Contact:** Human Resources. **Description:** Manufactures lead-wire assemblies for heads of computer hard drives. **Corporate headquarters location:** This Location.

INTEL CORPORATION
5200 NE Elam Young Parkway, Hillsboro OR 97124. 503/696-8080. **Toll-free phone:** 800/238-0486. **Recorded Jobline:** 503/696-2580. **Contact:** Human Resources. **Description:** This location is a manufacturing facility. Intel is a manufacturer of computer microprocessors and computer related parts. Intel operates six sites in the Washington County region of Oregon. **NOTE:** Resumes should be sent to the Staffing Department, SM3-145, Box 1141, Fulton CA 95763-1141. **Common positions include:** Administrative Assistant; Administrative Worker/Clerk; Engineer; Engineering Technician; Production Worker. **Corporate headquarters location:** Santa Clara CA. **Number of employees at this location:** 9,000.

INTERFACE ELECTRONICS INC.
4579 Abbotts Bridge Road, Suite 8, Duluth GA 30155. 770/623-1066. **Contact:** Human Resources. **Description:** Manufactures computer components and hardware including cables, motherboards, hard drives, and monitors.

KOMAG, INC.
275 South Hillview Drive, Milpitas CA 95035. 408/946-2300. **Contact:** Human Resources. **Description:** Komag develops, manufactures, and markets sputtered thin-film media for use in high-capacity, high-performance 5-1/4 inch, 3-1/2 inch, and smaller Winchester disk drive storage devices. Komag also makes thin-film recording heads. **Number of employees nationwide:** 3,100.

LSI LOGIC
1551 McCarthy Boulevard, Milpitas CA 95035. 408/433-8000. **Contact:** Personnel Manager. **Description:** LSI Logic is a designer, manufacturer, and marketer of specialized integrated circuits primarily in the desktop, personal computer, networking, and digital video applications markets. The company also provides computer-aided engineering design, technology services, and software

design tools to customers in the electronic data processing, telecommunications, and automation industries. **Listed on:** New York Stock Exchange.

M TECHNOLOGY, INC.
1931 Hartog Drive, San Jose CA 95131. **Contact:** Human Resources Department. **World Wide Web address:** http://www.mtiusa.com. **Description:** M Technology is a researcher, developer, and manufacturer of computer mother boards. The company markets its products through computer systems dealers, systems integrators, value-added resellers, and distributors. **Corporate headquarters location:** This Location.

MGV MANUFACTURING
451 Lanier Road, Madison AL 35758. 205/772-1100. **Contact:** Human Resources Department. **Description:** MGV Manufacturing manufactures computer memory modules.

MPM GOLDEN RAM
8 Whatney, Irvine CA 92618. 714/753-1200. **Contact:** Personnel Department. **Description:** MPM Golden RAM manufactures third party memory modules for computers in order to increase RAM.

MATRIX CORPORATION
1203 New Hope Road, Raleigh NC 27610. 919/231-8000. **Fax:** 919/231-8001. **Contact:** Human Resources Department. **E-mail address:** shels@pcgate.matrix.com or sheldonj@prodigy.com. **World Wide Web address:** http://www.matrix.com. **Description:** Matrix Corporation is a designer and manufacturer of computer boards for use in harsh environments.

MEGATEK CORPORATION
16868 Via Del Campo Court, San Diego CA 92127. 619/675-4000. **Fax:** 619/618-7910. **Recorded Jobline:** 619/550-5345. **Contact:** Kaley Mish, Director, Human Resources Department. **E-mail address:** jobs@megatek.com. **World Wide Web address:** http://www.megatek.com. **Description:** Megatek Corporation designs and manufactures graphic accelerator boards. **Common positions include:** Electrical/Electronics Engineer; Software Engineer. **Educational backgrounds include:** Computer Science; Engineering. **Benefits:** 401(k); Dental Insurance; Disability Coverage; Life Insurance; Medical Insurance; Tuition Assistance. **Corporate headquarters location:** Tokyo, Japan. **Parent company:** Nihon Unisys. **Operations at this facility include:** Administration; Manufacturing; Research and Development; Sales; Service. **Number of employees at this location:** 150.

MICRONICS COMPUTERS, INC.
45365 Northpart Loop West, Fremont CA 94538. 510/651-2300. **Contact:** Human Resources Department. **Description:** Micronics Computers, Inc. supplies high-performance system boards -- which contain the microprocessor and supporting memory components and logic circuitry that form the core of a personal computer -- for high-end IBM compatible PCs sold by original equipment manufacturers and value-added resellers. **Common positions include:** Accountant/Auditor; Buyer; Credit Clerk and Authorizer; Credit Manager; Customer Service Representative; Electrical/Electronics Engineer; Financial Manager; Human Resources Manager; Manufacturer's/Wholesaler's Sales Representative; Marketing Manager; Payroll Clerk; Purchasing Agent/Manager; Quality Control Supervisor; Receptionist; Software Engineer; Technical Representative. **Educational backgrounds include:** Accounting; Business Administration; Computer Science; Engineering; Finance; Marketing. **Benefits:** Dental Insurance; Disability Coverage; Employee Discounts; Life Insurance; Medical Insurance; Profit Sharing; Savings Plan. **Corporate headquarters location:** This Location. **Parent company:** Softnet (Lake Forest IL). **Operations at this facility include:** Administration; Manufacturing; Research and Development; Sales; Service. **Listed on:** NASDAQ. **Number of employees at this location:** 240.

MICROTECH INTERNATIONAL
158 Commerce Street, East Haven CT 06512. 203/468-6223. **Contact:** Mike Valenti, Director of Operations. **Description:** Manufactures memory and memory upgrades for computers.

MICROTEK INTERNATIONAL, INC.
3300 NW 211th Terrace, Hillsboro OR 97214. 503/645-7333. **Fax:** 503/629-8460. **Contact:** Human Resources Department. **E-mail address:** info@microtek.com. **World Wide Web address:** http://www.microtek.com. **Description:** Microtek International, Inc. is a manufacturer of microprocessor development tools, in-circuit emulators (ICE), and debuggers for embedded systems development.

MULTI MICRO SYSTEMS (MMS)
2124 Zanker Road, San Jose CA 95131. 408/437-1555. **Contact:** Human Resources. **Description:** A manufacturer of rack mount computer components including chassis, cards, backplane boards, and keyboards.

MYLEX CORPORATION
34551 Ardenwood Boulevard, Fremont CA 94555. 510/796-6100. **Contact:** Human Resources Department. **Description:** Mylex Corporation designs and manufactures disk array controllers, system boards, and network interface cards, as well as supporting proprietary software for a wide range of personal computers, workstations, and servers. A world leader in RAID (Redundant Array of Independent Disks) technology, Mylex Corporation believes it is the standard setter for RAID solutions and fully-integrated PCI systems. According to DISK/TREND report, the company has shipped more disk array controllers than all of its worldwide competitors combined. With the introductions of a MicroChannel controller, a dual-connector PCI controller, and an external SCSI controller, Mylex Corporation believes that it now offers a RAID solution for virtually every microcomputer-based platform in the industry today. **Corporate headquarters location:** This Location.

NETWORK CONTROLS INTERNATIONAL INC.
9 Woodlawn Green, Charlotte NC 28217. 704/527-4357. **Contact:** Human Resources. **Description:** Manufactures circuit boards to integrate computer systems in banks.

NUMBER NINE VISUAL TECHNOLOGY
18 Hartwell Avenue, Lexington MA 02173. 617/674-8649. **Fax:** 617/674-2919. **Contact:** Human Resources Director. **Description:** Number Nine is a manufacturer and supplier of high-performance visual technology solutions, including video/graphics accelerator subsystems, chips, and productivity-enhancing software. **Corporate headquarters location:** This Location.

PLX TECHNOLOGY, INC.
390 Potrero Avenue, Sunnyvale CA 94086. 408/774-9060. **Toll-free phone:** 800/759-3735. **Fax:** 408/774-2169. **Contact:** Bill Hart, Vice President Of Operations. **E-mail address:** bhart@plxtech.com. **World Wide Web address:** http://www.plxtech.com. **Description:** A manufacturer of computer chips. **NOTE:** The company offers entry-level positions. **Common positions include:** Applications Engineer; Design Engineer. **Educational backgrounds include:** Engineering. **Benefits:** 401(k); Dental Insurance; Disability Coverage; Medical Insurance. **Listed on:** Privately held. **Annual sales/revenues:** $11 - $20 million. **Number of employees at this location:** 25.

PNY ELECTRONICS
200 Anderson Avenue, Moonachie NJ 07074. 201/438-6300. **Toll-free phone:** 800/234-4597. **Fax:** 201/842-0584. **Contact:** Miriam Bailleman, Human Resources. **World Wide Web address:** http://www.pny.com. **Description:** A manufacturer and designer of computer memory. Founded in 1985. **NOTE:** The company offers entry-level positions. **Common positions include:** Account Manager; Account Representative; Design Engineer; Quality Control Supervisor; Sales and Marketing Representative. **Educational backgrounds include:** Accounting; Computer Science; Engineering. **Benefits:** 401(k); Dental Insurance; Disability Coverage; Life Insurance; Medical Insurance; Tuition Assistance. **Corporate headquarters location:** This Location. **Listed on:** Privately held. **Annual sales/revenues:** More than $100 million. **Number of employees at this location:** 250. **Number of employees nationwide:** 320. **Number of employees worldwide:** 420.

PROMISE TECHNOLOGY INC.
1460 Koll Circle, San Jose CA 95112. 408/452-0948. **Contact:** Human Resources. **Description:** Promise Technology manufactures high-performance hard drive controller cards for use in IBM computers.

RADISYS CORPORATION
5445 NE Dawson Creek Drive, Hillsboro OR 97124. 503/615-1100. **Fax:** 503/615-1150. **Contact:** Human Resources. **Description:** RadiSys Corporation is an independent designer and manufacturer of embedded computer technology used by OEMs in the manufacturing automation, telecommunications, medical devices, transportation, test and measurement, and retail automation industries. Unlike general purpose computers, embedded computers are incorporated into systems and equipment to perform a single or limited number of complex applications. RadiSys offers a broad range of embedded computer subsystems, board-level modules and chip-level products at varying levels of customization, from standard products to full custom solutions. RadiSys combines advanced design and manufacturing expertise with other value-added services to provide reduced time-to-market, high product quality, and stable product supply. RadiSys' customers include ABB, Allen-Bradley, Applied Materials, Ascom, FSI International, GE, Harris, Hewlett-Packard, Network Express, Philips, and Universal Instruments. **Corporate headquarters location:** This Location.

QDI COMPUTER
41456 Christy Street, Fremont CA 94538. 510/668-4933. **Contact:** Human Resources. **Description:** A manufacturer of computer components and motherboards. **Corporate headquarters location:** This Location.

QUANTUM PERIPHERALS COLORADO
1450 Infinite Drive, Louisville CO 80027. 303/604-4000. **Fax:** 303/604-5070. **Contact:** Human Resources Recruiter. **Description:** Quantum Peripherals Colorado is a leading developer of MR disk heads for the rigid disk drive industry. Quantum Peripherals Colorado has begun shipping its Phantom product line, which can operate at aerial densities ranging from 300 MB per square inch to 500 MB per square inch. The company's current technology is extendible to its Hornet product line, operating from 900 MB per square inch to 1,500 MB per square inch. Additionally, Quantum Peripherals Colorado is developing heads using giant MR technology. **NOTE:** This location formerly operated under the name Rocky Mountain Magnetics, Inc. (RMMI). **Common positions include:** Chemical Engineer; Materials Engineer. **Educational backgrounds include:** Chemistry; Engineering; Physics. **Benefits:** 401(k); Dental Insurance; Disability Coverage; Life Insurance; Medical Insurance; Tuition Assistance. **Special Programs:** Internships. **Corporate headquarters location:** Milpitas CA. **Other U.S. locations:** Shrewsbury MA. **Parent company:** Quantum Corp. **Operations at this facility include:** Manufacturing; Research and Development. **Number of employees at this location:** 550. **Number of employees nationwide:** 3,500.

QUICKLOGIC CORPORATION
2933 Bunker Hill Lane, Suite 100, Santa Clara CA 95054. 408/987-2000. **Contact:** Human Resources. **Description:** A manufacturer and distributor of field programmable logic units.

RACORE COMPUTER PRODUCTS INC.
2355 South 1070 West, Salt Lake City UT 84119. 801/973-9779. **Contact:** Human Resources. **Description:** Manufactures network boards.

RICHARD MANUFACTURING COMPANY
6250 Bury Drive, Eden Prairie MN 55346. 612/934-3000. **Contact:** Human Resources. **Description:** Manufactures components for computer disk drives.

SCJ ASSOCIATES INC.
60 Commerce Drive, Rochester NY 14623. 716/359-0600. **Contact:** Human Resources. **Description:** Assembles printed circuit boards. SCJ Associates also tests products and wraps cable.

SIMPLE TECHNOLOGY INC.
3001 Daimler Street, Santa Ana CA 92705. 714/476-1180. **Contact:** Human Resources. **Description:** Simple Technology designs memory upgrades and manufactures PC boards.

SMART MODULAR TECHNOLOGIES
4305 Cushing Parkway, Fremont CA 94538. 510/770-5777. **Contact:** Human Resources. **Description:** A manufacturer and distributor of computer components including Notebook, Desktop, and PC cards, fax modems, and memory modules.

SPECTRUM CONTROLS INC.
2700 Richards Road, Bellevue WA 98005. 206/746-9481. **Contact:** Human Resources. **Description:** Manufactures operator interfaces for industrial computers as well as other electronic parts.

STREAMLOGIC
2250 East Devon, Des Plaines IL 60018. 847/297-2280. **Contact:** Personnel. **Description:** Formerly Micropolis Corporation, Streamlogic is a leading manufacturer of RAID computer subsystems.

SUNDISK CORPORATION
140 Caspian Court, Sunnyvale CA 94089. 408/542-0500. **Contact:** Human Resources. **Description:** Manufacturer of computer components including memory cards.

SYNTRONIC INSTRUMENTS INC.
100 Industrial Road, Addison IL 60101. 630/543-6444. **Contact:** Human Resources. **Description:** Manufactures deflective yolks for CRTs.

SYSTEMSOFT CORPORATION
Two Vision Drive, Natick MA 01760. 508/651-0088. **Fax:** 508/651-8188. **Contact:** Human Resources. **Description:** SystemSoft Corporation is a supplier of PCMCIA (Personal Computer Memory Card International Association) and other system-level software to the rapidly growing market of mobile computers, comprised of laptops, notebooks, subnotebooks, and personal computing devices. PCMCIA is a published industry standard which enables personal computers and electronic devices to automatically recognize, install, and configure peripherals (including modems, flash memory, and network cards) incorporated in credit card-sized PCMCIA cards. System-level software provides both a connectivity layer, which facilitates the addition, configuration, and use of peripheral devices, and a hardware adaptation layer, which comprises the communication link between a computer operating system and hardware. Sales and support facilities are located in Santa Clara CA. **Benefits:** 401(k); Incentive Plan; Stock Option. **Corporate headquarters location:** This Location. **Listed on:** NASDAQ. **Annual sales/revenues:** Over $9 million. **Number of employees nationwide:** 76.
Other U.S. locations:
* 2350 Mission College Boulevard, Suite 450, Santa Clara CA 95054. 408/988-6756. Contact: J.S. Dix, Human Resources Administrator. This location develops firmware and code.

SYSTRAN CORPORATION
4126 Linden Avenue, Dayton OH 45432. 937/252-5601. **Contact:** Human Resources. **Description:** Manufactures computer circuit boards and other electronic equipment.

TL INDUSTRIES
2541 Tracy Road, Rossford OH 43619. 419/666-8144. **Contact:** Human Resources. **Description:** Manufactures computer chip boards.

TAMSCO MANUFACTURING
210 2nd Avenue East, Polson MT 59860. 406/883-3307. **Contact:** Human Resources. **Description:** A manufacturer of printed circuit boards and computer cables.

TANDON ASSOCIATES

2125-B Matera Road, Simi Valley CA 93065. 805/582-3200. **Contact:** Human Resources. **Description:** This location manufactures hard drives for computers.

TANDY WIRE AND CABLE COMPANY

3500 McCart Avenue, Fort Worth TX 76110. 817/921-2023. **Contact:** Marianne Wiestner, Human Resource Manager. **Description:** Manufactures wire and cable, including computer cable. **Common positions include:** Blue-Collar Worker Supervisor; Customer Service Representative; Draftsperson; Electrical/Electronics Engineer; Systems Analyst. **Educational backgrounds include:** Business Administration; Engineering. **Benefits:** Dental Insurance; Employee Discounts; Life Insurance; Medical Insurance; Profit Sharing; Savings Plan; Tuition Assistance. **Parent company:** Tandy Corporation. **Operations at this facility include:** Manufacturing. **Listed on:** New York Stock Exchange.

TANISYS TECHNOLOGY

12201 Tech Boulevard, Suite 130, Austin TX 78727. 512/258-3570. **Contact:** Human Resources. **Description:** A designer and manufacturer of memory modules and testers.

TECHNOLOGY WORKS

4030 West Braker Lane, Suite 500, Austin TX 78759. 512/794-8533. **Contact:** Human Resources Department. **Description:** Technology Works manufactures and sells computer memory.

TRANSERA CORPORATION

345 East 800 South, Orem UT 84097. 801/224-0088. **Contact:** Personnel. **Description:** TransEra Corporation operates through its three subsidiaries at this location. Improve Technologies manufactures CPU upgrade chips; Vantage Controls manufactures automotive, home, and commercial lighting systems; and HT Basic International manufactures software that utilizes a high-tech engineering language, and test and measurement hardware.

TRUEVISION INC.

7340 Shadeland Station, Indianapolis IN 46256-3919. 317/841-0332. **Contact:** Human Resources. **E-mail address:** jobs@truevision.com. **World Wide Web address:** http://www.truevision.com. **Description:** This location houses engineering, design, and technical support offices. Truevision is a developer of videographics PC digital video cards for desktop video production. **Corporate headquarters location:** Santa Clara CA. **Listed on:** NASDAQ. **Stock exchange symbol:** TRUV.

TRUEVISION INC.

2500 Walsh Avenue, Santa Clara CA 95051. 408/562-4200. **Fax:** 408/566-4123. **Contact:** Human Resources. **E-mail address:** jobs@truevision.com. **World Wide Web address:** http://www.truevision.com. **Description:** A developer of videographics PC digital video cards for desktop video production. **Corporate headquarters location:** This Location. **Listed on:** NASDAQ. **Stock exchange symbol:** TRUV.

VME MICROSYSTEMS INTERNATIONAL

12090 South Memorial Parkway, Huntsville AL 35803. 205/880-0444. **Contact:** Human Resources. **Description:** Manufactures I/O cards for computers.

VAC CORPORATION

4027 Will Rogers Parkway, Oklahoma City OK 73108. 405/943-9651. **Contact:** Human Resources. **Description:** Manufactures magnet assemblies for use in computer hard drives.

XCEED TECHNOLOGY

48700 Structural Drive, Chesterfield MI 48051. 810/598-8030. **Contact:** Shellie Simons, Human Resources. **Description:** Manufactures custom-designed computer cards for PCs, and color cards for Apple/Macintosh.

SEMICONDUCTOR MANUFACTURERS

"Make sure you enjoy what you are doing." Sound advice from Iain Mackie, Director of Engineering at Advanced Hardware Architecture. In addition to loving the work, Iain says jobseekers should be detail-oriented and possess the ability to learn new skills quickly.

When recruiting new candidates, companies look for graduates with either a bachelor's or master's degree and excellent grades in engineering and related courses. Prospects for jobseekers remain high with unemployment in this cyclical field hovering at about 1.5 percent.

Iain sees the lines between hardware and software design blurring even further than they already have. Industry professionals will need to understand the applications the chips they design are used for, in addition to having mastery of engineering principles. The technology to make better, faster chips will continue to grow at an incredible rate. The only thing limiting what people can do with these products, says Iain, is the human imagination.

ADVANCED HARDWARE ARCHITECTURE
2365 NE Hopkins Court, Pullman WA 99163. 509/334-1000. **Contact:** Human Resources. **Description:** The firm designs integrated circuits.

ADVANCED MICRO DEVICES, INC. (AMD)
One AMD Place, P.O. Box 3453, Sunnyvale CA 94088-3453. 408/732-2400. **Contact:** Human Resources. **World Wide Web address:** http://www.amd.com. **Description:** Designs, develops, manufactures, and markets complex monolithic integrated circuits for use by manufacturers of a broad range of electronic equipment and systems, primarily for instrument applications, and products involved in computation and communication. Advanced Micro Devices, Inc. operates worldwide. **Common positions include:** Computer Engineer; Electrical/Electronics Engineer; Materials Engineer. **Corporate headquarters location:** This Location. **International locations:** Worldwide. **Number of employees nationwide:** 11,550.
Other U.S. locations:
• 5204 East Ben White Boulevard, Mail Stop 556, Austin TX 78741. 512/385-8542.

AMERICAN COMPUTER DIGITAL COMPONENTS
440 Cloverleaf Drive, Baldwin Park CA 91706. 818/336-1388. **Contact:** Vincent Tseng, Personnel. **Description:** A semiconductor manufacturing firm. **Common positions include:** Mechanical Engineer; Sales Representative.

AMKOR ELECTRONICS, INC.
1345 Enterprise Drive, West Chester PA 19380-5959. 610/431-9600. **Contact:** Mr. Chris Roberts, Recruiting. **Description:** Distributes semiconductors. **Common positions include:** Computer Programmer; Customer Service Representative; Electrical/Electronics Engineer; Sales Representative; Systems Analyst.

AVEX ELECTRONICS, INC.
4807 Bradford Drive, Huntsville AL 35805. 205/722-6000. **Contact:** Betsy Mosgrove, Human Resources Department. **Description:** A manufacturer of semiconductors and related devices. **Common positions include:** Computer Programmer; Design Engineer; Hardware Engineer. **Parent company:** J.M. Huber Corporation.

BKC SEMICONDUCTORS INC.

Six Lake Street, Lawrence MA 01841-3011. 508/681-0392. **Contact:** Human Resources. **Description:** BKC Semiconductors manufactures a wide range of semiconductor diodes for signal switching, voltage conversion, rectification, and surge suppression of electrical units. The current product line includes silicon switching diodes, rectifiers, zener diodes, transient suppressers, Shottky diodes, germanium diodes, photo detector diodes, and outsourced products. BKC Semiconductors generally sells its products to the military and the commercial/industrial markets. The manufacturing process comprises six operations: wafer fabrication, assembly, testing, military processing, finishing operations, and quality assurance. All manufacturing operations are performed at this facility. **Benefits:** 401(k). **Corporate headquarters location:** This Location. **Subsidiaries include:** Souza Semiconductors, Inc.; BKC Photo Detector Division, Inc.; and Clearwater Enterprises of Massachusetts. **Listed on:** NASDAQ. **Number of employees nationwide:** 103.

CATALYST SEMICONDUCTOR, INC.

1250 Borregas Avenue, Sunnyvale CA 94089. 408/542-1000. **Contact:** Human Resources. **Description:** Founded in 1985, Catalyst Semiconductor is a supplier of nonvolatile semiconductors that provide design solutions for a broad range of applications including computers, wireless communications, networks, instrumentation, and automotive systems. Catalyst's family of nonvolatile devices includes FLASH, Serial and Parallel EEPROMs (electrical erasable read only memory), and NVRAMs. The company is dedicated to maintaining its position in EEPROM products while actively pursuing development of future generations of FLASH products. Catalyst's products are marketed through a direct sales force and a worldwide network of representatives and distributors. The company's employees work in technology development, engineering, marketing, and sales. **International locations:** Japan; Philippines; Singapore; Thailand. **Corporate headquarters location:** This Location.

CHERRY SEMICONDUCTOR CORPORATION

2000 South County Trail, East Greenwich RI 02818. 401/885-3600. **Contact:** Human Resources. **Description:** The semiconductor division principally manufactures linear integrated circuits for use in automotive, computer, and industrial control equipment. **Corporate headquarters location:** Waukegan IL. **Parent company:** The Cherry Corporation manufactures and distributes a wide range of industrial components to the computer, automotive, consumer, and commercial markets worldwide. The principal segments are electromechanical devices (snap-action, selector, and automobile special-use switches principally for use in automobiles, home appliances, office and industrial equipment, and vending machines); electronic assemblies and displays (keyboards, keyboard switches, gas discharge displays, and automotive electronics for use in a variety of products, including data entry terminals, automobiles, industrial and commercial control devices, business machines, and amusement products); and semiconductor devices.

CHIP EXPRESS

2323 Owen Street, Santa Clara CA 95054. 408/988-2445. **Contact:** Human Resources. **Description:** A manufacturer of semiconductor chips.

CHIPS AND TECHNOLOGIES, INC.

2950 Zanker Road, San Jose CA 95134. 408/434-0600. **Contact:** Human Resources. **Description:** Chips and Technologies manufactures advanced semiconductor devices for the worldwide personal computer industry. The company's products are found in a wide range of systems from compact portables to high-performance desktop computers. This location is engaged in research and development **Corporate headquarters location:** This Location.

COLLMER SEMICONDUCTOR, INC.

14368 Proton Road, Dallas TX 75244. 214/233-1589. **Contact:** Human Resources. **Description:** A manufacturer and distributor of semiconductors.

CRYSTAL SEMICONDUCTOR CORPORATION

P.O. Box 17847, Austin TX 78760. 512/445-7222. **Contact:** Tommy Morrow, Human Resources. **Description:** A semiconductor company that designs, markets, and tests computer chips for audio, digital, multimedia, and telecommunication products. **Corporate headquarters location:** This Location.

DALLAS SEMICONDUCTOR

4401 South Beltwood Parkway, Dallas TX 75244-3292. 972/371-4000. **Contact:** Human Resources. **Description:** Manufactures semiconductors. **Common positions include:** Electrical/Electronics Engineer; Mechanical Engineer; Software Engineer. **Educational backgrounds include:** Computer Science; Engineering; Physics. **Benefits:** 401(k); Dental Insurance; Disability Coverage; Life Insurance; Medical Insurance; Profit Sharing; Tuition Assistance. **Corporate headquarters location:** This Location. **Operations at this facility include:** Administration; Manufacturing; Research and Development; Sales; Service. **Listed on:** New York Stock Exchange. **Number of employees at this location:** 1,250.

DATA I/O CORPORATION

P.O. Box 97046, Redmond WA 98073-9746. 206/881-6444. **Fax:** 206/882-1043. **Recorded Jobline:** 206/867-6963. **Contact:** Employment Department. **Description:** Produces semiconductors and related devices. **NOTE:** The physical address of the company is 10525 Willows Road, Redmond WA. **Common positions include:** Electrical/Electronics Engineer; Software Engineer. **Educational backgrounds include:** Engineering. **Benefits:** 401(k); Dental Insurance; Disability Coverage; Life Insurance; Medical Insurance; Profit Sharing; Tuition Assistance. **Corporate headquarters location:** This Location. **Other U.S. locations:** Nationwide. **Number of employees at this location:** 285. **Number of employees nationwide:** 320.

DIGITAL EQUIPMENT CORPORATION

75 Reed Road, Hudson MA 01749. 508/568-4000. **Contact:** Human Resources. **Description:** This location designs, develops, and manufactures semiconductors. Overall, Digital Equipment Corporation designs, manufactures, sells, and services computers, associated peripheral equipment, and related software and supplies. Applications and programs include use in scientific research, computation, communications, education, data analysis, industrial control, time-sharing, commercial data processing, graphic arts, word processing, health care, instrumentation, engineering, and simulation. **Common positions include:** Electrical/Electronics Engineer; Process Engineer; Software Engineer. **Educational backgrounds include:** Computer Science; Engineering. **Benefits:** 401(k); Dental Insurance; Disability Coverage; Employee Discounts; Life Insurance; Medical Insurance; Pension Plan; Savings Plan; Tuition Assistance. **Corporate headquarters location:** Maynard MA. **Operations at this facility include:** Divisional Headquarters; Manufacturing; Research and Development. **Listed on:** New York Stock Exchange. **Number of employees at this location:** 2,000. **Number of employees nationwide:** 77,800.

DIONICS INC.

65 Rushmore Street, Westbury NY 11590. 516/997-7474. **Fax:** 516/997-7479. **Contact:** Human Resources. **Description:** Dionics Inc. designs, manufactures, and sells silicon semiconductor electronic products as individual discrete components, as multi-component integrated circuits, and as multi-component hybrid circuits.

HARRIS CORPORATION
SEMICONDUCTOR SECTOR

P.O. Box 883, Melbourne FL 32902. 407/724-7000. **Contact:** Denny Marini, Director of Human Resources. **Description:** This location manufactures semiconductors. With worldwide sales of more than $3 billion, Harris Corporation is focused on four major businesses: advanced electronic systems, semiconductors, electronic systems, and Lanier office systems. Electronic systems include advanced information processing and communication systems and software for defense applications, air traffic control, avionics, satellite communications, space exploration, mobile-radio networks, simulation, energy management, law enforcement, electronic systems testing, and newspaper composition. The semiconductor sector includes advanced analog, digital, and mixed-signal integrated circuits and discrete semiconductors for power, signal processing, data acquisitions, and logic applications. Markets include automotive systems, wireless communications, telecommunication line cards, video and imaging systems, industrial equipment, computer peripherals, and military and aerospace systems. The communications sector includes broadcast, radio communication, and telecommunications products and systems including transmitters and equipment for TV and radio; HF, VHF, and UHF radio communication equipment; microwave radios; digital telephone switches; telephone subscriber-loop equipment; and in-building paging equipment. Lanier Worldwide sells, services, and provides supplies for copying, facsimile, dictation, optical-based information management, continuous recording, and PC-based health care management systems through 1,600 locations in 80 countries. **Common positions include:** Accountant/Auditor; Administrator; Attorney; Blue-Collar Worker Supervisor; Computer Programmer; Customer Service Representative; Department Manager; Editor; Electrical/Electronics Engineer; Financial Analyst; General Manager; Human Resources Manager; Industrial Engineer; Manufacturer's/Wholesaler's Sales Rep.; Marketing Specialist; Mechanical Engineer; Metallurgical Engineer; Operations/Production Manager; Public Relations Specialist; Purchasing Agent/Manager; Quality Control Supervisor; Reporter; Systems Analyst. **Educational backgrounds include:** Accounting; Business Administration; Communications; Computer Science; Economics; Engineering; Finance; Marketing; Physics. **Benefits:** Dental Insurance; Disability Coverage; Employee Discounts; Life Insurance; Medical Insurance; Pension Plan; Profit Sharing; Savings Plan; Tuition Assistance. **Operations at this facility include:** Administration; Divisional Headquarters; Manufacturing; Research and Development; Sales; Service. **Listed on:** New York Stock Exchange.

HARRIS CORPORATION

330 Twin Dolphin Drive, River City CA 94065. 415/594-3000. **Contact:** Human Resources. **Description:** Manufactures semiconductors.

HARRIS SEMICONDUCTOR

125 Crestwood Road, Mountain Top PA 18707. 717/474-6761. **Contact:** Human Resources. **Description:** This location manufactures semiconductors. **Corporate headquarters location:** Melbourne FL. **Parent company:** With worldwide sales of more than $3 billion, Harris Corporation is focused on four major businesses: advanced electronic systems, semiconductors, electronic systems, and Lanier office systems. Electronic systems includes advanced information processing and communication systems and software for defense applications, air traffic control, avionics, satellite communications, space exploration, mobile-radio networks, simulation, energy management, law enforcement, electronic systems testing, and newspaper composition. The semiconductor sector is comprised of advanced analog, digital, and mixed-signal integrated circuits and discrete semiconductors for power, signal processing, data-acquisitions, and logic applications. Markets include automotive systems, wireless communications, telecommunication line cards, video and imaging systems, industrial equipment, computer peripherals, and military and aerospace systems. The communications sector includes broadcast, radio communication, and telecommunications products and systems, including transmitters and equipment for TV and radio; HF, VHF, and UHF radio communication equipment; microwave radios; digital telephone switches; telephone subscriber-loop equipment; and in-building paging equipment. Lanier Worldwide sells, services, and provides supplies for copying, facsimile, dictation, optical-based information management, continuous recording, and PC-based health care management systems through 1,600 locations in 80 countries. **Number of employees nationwide:** 28,000.

INTEL CORPORATION

2200 Mission College Boulevard, P.O. Box 58119, Santa Clara CA 95052-8119. 408/987-8080. **Recorded Jobline:** 800/637-7510. **Contact:** Staffing Department. **Description:** Intel is one of the largest semiconductor manufacturers in the world. Other operations include supercomputers; embedded control chips and flash memories; video technology software; multimedia hardware; personal computer enhancement products; and designing, making, and marketing microcomputer components, modules, and systems. Intel sells its products to original equipment manufacturers and other companies that incorporate them into their products. **NOTE:** Please send resumes to Staffing Department, MS FM4-145, P.O. Box 1141, Folsom CA 95763-1141. **Corporate headquarters location:** This Location. **Listed on:** NASDAQ. **Stock exchange symbol:** INTEL. **Other U.S. locations:**

- 5000 West Chandler Boulevard, Chandler AZ 85226-3699. 602/554-8080. This location is a manufacturing facility.
- 4100 Sara Road, Rio Rancho NM 87124. 505/893-7000. This location is a manufacturing facility.

- 734 East Utah Valley Drive, American Fork UT 84003. 801/763-2200. This location is a sales and support facility.

LSI COMPUTER SYSTEMS INC.
1235 Walt Whitman Road, Melville NY 11747. 516/271-0400. **Contact:** Human Resources. **Description:** A manufacturer of integrated circuits and microchips.

LANSDALE SEMICONDUCTOR INC.
2502 West Huntington Drive, Tempe AZ 85282. 602/438-0123. **Fax:** 602/438-0138. **Contact:** Cheryl Warianka, Human Resources. **E-mail address:** cheryl@lansdale.com. **World Wide Web address:** http://www.syspac.com/~lansdale. **Description:** A manufacturer of semiconductors. Lansdale manufactures products discontinued by such major manufacturers as Intel, Motorola, National/Fairchild, and Harris. Lansdale supports spare parts requirements by purchasing lines from the original manufacturer. **Corporate headquarters location:** This Location.

LATTICE SEMICONDUCTOR CORPORATION
5555 NE Moore Court, Hillsboro OR 97124. 503/681-0118. **Contact:** Mr. Terry Dols, Human Resources. **Description:** Lattice Semiconductor Corporation is a leader in the design, development, and marketing of high-speed CMOS Programmable Logic Devices (PLDs) in both low-density and high-density ranges. The world's fastest CMOS PLDs provide electronic engineers with quick-to-design and easy-to-configure components. Lattice products are sold worldwide through independent sales representatives and distributors, primarily to original equipment manufacturers of microcomputers, graphic systems, workstations, peripherals, telecommunications, military, and industrial controls. CMOS PLDs are assembled in 20 to 207 PIN standard packages and offered with various speed, power, and packaging options in commercial, industrial, and military temperature versions. **Corporate headquarters location:** This Location. **Number of employees at this location:** 398.

M/A-COM, INC.
1011 Pawtucket Boulevard, Lowell MA 01853. 508/442-4003. **Contact:** Jim Sullivan, Director of Human Resources. **Description:** Designs, develops, and manufactures microwave semiconductors and subsystems for commercial wireless, cellular, satellite, and automotive communications, and for government and defense applications. **Common positions include:** Accountant/Auditor; Buyer; Computer Programmer; Draftsperson; Financial Analyst; Microwave Engineer; Microwave Technician; Operations/Production Manager; Sales Engineer. **Educational backgrounds include:** Accounting; Business Administration; Engineering; Physics. **Benefits:** 401(k); Dental Insurance; Disability Coverage; Life Insurance; Medical Insurance; Tuition Assistance. **Corporate headquarters location:** Wakefield MA.

MATSUSHITA SEMICONDUCTOR CORPORATION
1111 39th Avenue SE, Puyallup WA 98374. 206/841-6066. **Fax:** 206/841-6611. **Contact:** Personnel. **E-mail address:** hr@masca.com. **Description:** This location is a manufacturer of integrated circuits. **NOTE:** The company offers entry-level positions. **Common positions include:** Chemical Engineer; Electrical/Electronics Engineer; Materials Engineer; Mechanical Engineer; Physicist/Astronomer. **Educational backgrounds include:** Engineering. **Benefits:** 401(k); Dental Insurance; Disability Coverage; Employee Discounts; Life Insurance; Medical Insurance; Pension Plan; Profit Sharing; Savings Plan; Tuition Assistance. **Corporate headquarters location:** This Location. **Parent company:** Matsushita Electronics Company (Japan) is a leading international manufacturer and marketer of consumer and industrial electronics under the trade names Panasonic, Technics, and Quasar. The company supplies DRAM and Microcontroller semiconductor devices and devices to electronics and equipment manufacturers worldwide. **Operations at this facility include:** Administration; Manufacturing. **Listed on:** Privately held. **Annual sales/revenues:** More than $100 million. **Number of employees at this location:** 400. **Number of projected hires for 1997 - 1998 at this location:** 40.

MICREL
1849 Fortune Drive, San Jose CA 95131. 408/944-0800. **Contact:** Human Resources. **Description:** A semiconductor manufacturer, Micrel designs and manufactures analog integrated circuits.

MICROCHIP TECHNOLOGY INC.

2355 West Chandler Boulevard, Chandler AZ 85224. 602/786-7200. **Fax:** 602/786-7790. **Recorded Jobline:** 602/786-7777. **Contact:** Mr. Randy Jacobs, Manager, Human Resources. **E-mail address:** resumes@ccmail.microchip.com. **World Wide Web address:** http://www.microchip2.com. **Description:** A semiconductor manufacturer. Microchip Technology specializes in 8-bit micro controllers, EPROMs, EEPROMs, and other non-volatile memory products. **Common positions include:** Accountant/Auditor; Applications Engineer; Budget Analyst; Buyer; Clerical Supervisor; Computer Programmer; Customer Service Representative; Designer; Electrical/Electronics Engineer; Financial Analyst; Human Resources Manager; Systems Analyst. **Educational backgrounds include:** Accounting; Business Administration; Chemistry; Engineering; Marketing. **Benefits:** 401(k); Dental Insurance; Disability Coverage; Life Insurance; Medical Insurance; Profit Sharing; Savings Plan; Tuition Assistance. **Special Programs:** Internships. **Corporate headquarters location:** This Location. **Operations at this facility include:** Administration; Manufacturing; Research and Development; Sales. **Listed on:** NASDAQ. **Other U.S. locations:**

- 2674 North First Street, Suite 215, San Jose CA 95134.

MICRON TECHNOLOGY

8000 South Federal Way, P.O. Box 6, Boise ID 83707-0006. 208/368-3700. **Fax:** 208/368-4641. **Recorded Jobline:** 208/368-4141. **Contact:** Personnel. **Description:** Micron Technology designs, manufactures, and markets semiconductor memory components, including Dynamic Random Access Memories (DRAMs), Static Random Access Memories (SRAMs), and board-level and system-level products and personal computers. **Common positions include:** Chemical Engineer; Computer Programmer; Electrical/Electronics Engineer; Software Engineer. **Educational backgrounds include:** Chemistry; Computer Science; Engineering. **Benefits:** 401(k); Dental Insurance; Disability Coverage; Life Insurance; Medical Insurance; Profit Sharing; Tuition Assistance. **Special Programs:** Internships. **Corporate headquarters location:** This Location. **Subsidiaries include:** Micron Europe Limited; Micron Semiconductor (Deutschland) GmbH; Micron Semiconductor Asia Pte. Ltd.; Micron Quantum Devices, Inc. **Operations at this facility include:** Administration; Manufacturing; Regional Headquarters; Research and Development; Sales. **Listed on:** New York Stock Exchange. **Number of employees at this location:** 6,300.

MICROSEMI CORPORATION

580 Pleasant Street, Watertown MA 02172. 617/926-0404. **Contact:** Personnel Manager. **Description:** This location is a manufacturing facility. Microsemi Corporation manufactures and markets semiconductors and similar products and provides related services principally for military, aerospace, medical, computer, telecommunications, and other markets. Major products include high-reliability silicon rectifiers and zener diodes, low-leakage and high-voltage diodes, temperature-compensated zener diodes, and a family of subminiature high-power transient suppresser diodes. These components are used throughout the electronics industry, with almost all electronic equipment employing zener diodes and rectifiers to control the direction of electrical flow, to regulate voltage, and to protect sensitive circuitry from line surges and transient volume spikes. **Common positions include:** Accountant/Auditor; Blue-Collar Worker Supervisor; Chemical Engineer; Customer Service Representative; Designer; Electrical/Electronics Engineer; Electrician; Industrial Production Manager; Mechanical Engineer; Operations/Production Manager; Quality Control Supervisor. **Educational backgrounds include:** Accounting; Business Administration; Chemistry; Engineering. **Benefits:** 401(k); Dental Insurance; Disability Coverage; Employee Discounts; Life Insurance; Medical Insurance; Pension Plan; Profit Sharing; Savings Plan; Tuition Assistance. **Corporate headquarters location:** Santa Ana CA. **Other U.S. locations:** AZ; CO. **Operations at this facility include:** Administration; Manufacturing; Sales; Service. **Listed on:** NASDAQ. **Number of employees at this location:** 315.

MICROSEMI CORPORATION

P.O. Box 1390, Scottsdale AZ 85252. 602/941-6300. **Contact:** Lorraine Greager, Director of Personnel. **Description:** This location houses wafer fabrication facilities. Microsemi Corporation manufactures and markets semiconductors and other products and provides related services principally for military, aerospace, medical, computer, telecommunications, and other markets. Major products include high-reliability silicon rectifiers and zener diodes, low-leakage and high-voltage diodes, temperature-compensated zener diodes, and a family of subminiature high-power transient suppressor diodes. These components are used throughout the electronics industry, with almost all electronic equipment employing zener diodes and rectifiers to control the direction of electrical flow, regulate voltage, and protect sensitive circuitry from line surges and transient volume spikes. **NOTE:** The company's physical address is 8700 East Thomas Road, Scottsdale AZ.

MICROSEMI CORPORATION/COLORADO

800 Hoyt Street, Broomfield CO 80020. 303/469-2161. **Contact:** Lynn Willis, Human Resources Manager. **Description:** This location manufactures rectifiers and diodes. Microsemi Corporation manufactures and markets semiconductors and other products and provides related services principally for military, aerospace, medical, computer, and telecommunications markets. Major products include high-reliability silicon rectifiers and zener diodes, low-leakage and high-voltage diodes, temperature-compensated zener diodes, and a family of subminiature high-power transient suppresser diodes. These components are used throughout the electronics industry, with almost all electronic equipment employing zener diodes and rectifiers to control the direction of electrical flow, regulate voltage, and protect sensitive circuitry from line surges and transient volume spikes. **Common positions include:** Accountant/Auditor; Blue-Collar Worker Supervisor; Buyer; Chemist; Draftsperson; Electrical/Electronics Engineer; Human Resources Manager; Industrial Production Manager; Manufacturer's/Wholesaler's Sales Rep.; Operations/Production Manager; Purchasing Agent/Manager; Quality Control Supervisor. **Educational backgrounds include:** Business Administration; Chemistry; Engineering; Marketing. **Benefits:** 401(k); Dental Insurance; Disability Coverage; Life Insurance; Medical Insurance; Profit Sharing; Tuition Assistance. **Corporate headquarters location:** Santa Ana CA. **Other U.S. locations:** Scottsdale AZ; Watertown MA. **Operations at this facility include:** Administration; Manufacturing; Research and Development; Sales. **Listed on:** NASDAQ. **Number of employees at this location:** 360. **Number of employees nationwide:** 1,200.

MOTOROLA, INC.

26635 West Agoura, Suite 201, Calabasas CA 91302. 818/878-6800. **Contact:** Human Resources. **Description:** Manufactures semiconductors. This location serves as a regional sales office.

MOTOROLA, INC.

2200 West Broadway, Mesa AZ 85202. 602/655-2011. **Recorded Jobline:** 602/303-6000. **Contact:** Employment Center. **Description:** Manufactures semiconductors at this location. **NOTE:** Please send resumes to the employment center at 1438 West Broadway, Suite B100, Tempe AZ 85282.

NEC ELECTRONICS, INC.

475 Ellis Street, San Jose CA 94039. 408/588-6000. **Contact:** Human Resources Department. **Description:** One of the largest semiconductor manufacturers in the world, NEC Electronics supplies a wide variety of micro-electronics components used in computers, telecommunications, data communications, instrumentation, automobiles, and consumer applications. **Common positions include:** Computer Programmer; Electrical/Electronics Engineer; Financial Analyst; Marketing Specialist; Technical Writer/Editor. **Educational backgrounds include:** Business Administration; Computer Science; Engineering; Marketing. **Benefits:** Dental Insurance; Disability Coverage; Life Insurance; Medical Insurance; Pension Plan; Tuition Assistance; Vision Insurance. **Corporate headquarters location:** This Location. **Parent company:** NEC Corporation. **Operations at this facility include:** Administration.

NEC ELECTRONICS, INC.

P.O. Box 619022, Roseville CA 95661-9022. 916/786-3900. **Contact:** Human Resources. **Description:** One of the largest semiconductor manufacturers in the world, NEC Electronics supplies a wide variety of micro-electronics components used in computers, telecommunications, data communications, instrumentation, automobiles, and consumer applications.

NATIONAL SEMICONDUCTOR CORPORATION

3333 West 9000 South, West Jordan UT 84088-8838. 801/562-7000. **Contact:** Employment Manager. **Description:** Produces semiconductors for electronics applications. Nationally, the company designs, develops, manufactures, and markets a variety of electronic products for use by consumers, industry, and government. Products include transistors, integrated circuits, LEDs, and electronic checkout systems for supermarkets. **Common positions include:** Chemical Engineer; Electrical/Electronics Engineer; Physicist/Astronomer. **Educational backgrounds include:** Chemistry; Engineering. **Benefits:** Dental Insurance; Disability Coverage; Life Insurance; Medical Insurance; Profit Sharing; Savings Plan; Tuition Assistance. **Corporate headquarters location:** Santa Clara CA. **Operations at this facility include:** Manufacturing; Research and Development. **Listed on:** New York Stock Exchange.

NATIONAL SEMICONDUCTOR CORPORATION

2900 Semiconductor Drive, MS 14-270, Santa Clara CA 95051. 408/721-5000. **Contact:** Staffing. **Description:** Designs, develops, and manufactures microprocessors, consumer products, integrated circuits, memory systems, computer products, telecommunication systems, and high-speed bipolar circuits. National Semiconductor operates seven plants in the U.S., the Pacific Rim, and Scotland. **Common positions include:** Computer Programmer; Department Manager; Electrical/Electronics Engineer; Financial Analyst; Human Resources Manager; Industrial Engineer; Mechanical Engineer; Physicist/Astronomer; Technical Writer/Editor. **Educational backgrounds include:** Business Administration; Computer Science; Engineering; Finance; Marketing; Mathematics; Physics. **Benefits:** Dental Insurance; Disability Coverage; Employee Discounts; Life Insurance; Medical Insurance; Pension Plan; Profit Sharing; Savings Plan; Tuition Assistance. **Special Programs:** Internships. **Corporate headquarters location:** This Location. **Operations at this facility include:** Administration; Manufacturing; Research and Development; Sales; Service. **Listed on:** New York Stock Exchange.

NATIONAL SEMICONDUCTOR CORPORATION
DATA MANAGEMENT DIVISION

MS 10-04, 333 Western Avenue, South Portland ME 04106. 207/775-8100. **Contact:** Human Resources. **Description:** This location serves as headquarters for the Data Management Division and designs and manufactures semiconductors for a variety of applications, including computers and communications technology. The company has invested $600 million in the South Portland site to build a state-of-the-art, Class I, 8-inch, semiconductor wafer-manufacturing facility. Production is scheduled to begin in mid-1997. Overall, National Semiconductor provides technological solutions to the communications, analog-intensive, and personal systems markets. National Semiconductor, with sales of $2.4 billion, is one of the largest U.S. semiconductor merchants, and one of the top 12 *Fortune* 500 IC manufacturers in the world. **Corporate headquarters location:** Santa Clara CA. **Number of employees at this location:** 1,300.

NOVELLUS SYSTEMS INC.

3970 North First Street, San Jose CA 95134. 408/943-9700. **Fax:** 408/943-3422. **Contact:** Professional Staffing. **E-mail address:** novellushr@aol.com. **Description:** A semiconductor capital equipment manufacturer. **NOTE:** The company offers entry-level positions. **Common positions include:** Draftsperson; Electrical/Electronics Engineer; Financial Analyst; Manufacturing Engineer; Mechanical Engineer; MIS Specialist; Sales Engineer; Software Engineer; Systems Analyst; Technical Writer/Editor. **Educational backgrounds include:** Chemistry; Engineering. **Benefits:** 401(k); Dental Insurance; Disability Coverage; Life Insurance; Medical Insurance; Profit Sharing; Stock Option; Tuition Assistance. **Corporate headquarters location:** This Location. **Other U.S. locations:** Nationwide. **International locations:** Japan; Korea; Taiwan. **Operations at this facility include:** Administration; Manufacturing; Research and Development; Sales; Service. **Listed on:** NASDAQ. **Annual sales/revenues:** More than $100 million. **Number of employees at this location:** 650. **Number of employees nationwide:** 800. **Number of employees worldwide:** 1,050. **Number of projected hires for 1997 - 1998 at this location:** 150.

PHILIPS SEMICONDUCTOR

9201 Pan American Freeway NE, Albuquerque NM 87113. 505/822-7000. **Fax:** 505/822-7494. **Contact:** Scott Sprague, Recruiter. **Description:** Manufactures and distributes integrated circuits for semiconductors. Its manufacturing facilities are located in California and New Mexico. The company is a division of Philips Electronics North America. **Common positions include:** Chemical Engineer; Design Engineer; Electrical/Electronics Engineer; Process Engineer; Product Engineer; Technician; Test Engineer. **Educational backgrounds include:** Computer Science; Engineering. **Benefits:** Dental Insurance; Disability Coverage; Employee Discounts; Life Insurance; Medical Insurance; Pension Plan; Savings Plan. **Corporate headquarters location:**

Sunnyvale CA. **Operations at this facility include:** Manufacturing. **Number of employees at this location:** 1,200. **Number of employees nationwide:** 2,500.

POWEREX, INC.
200 Hillis Street, Youngwood PA 15697. 412/925-7272. **Contact:** Human Resources Director. **Description:** A manufacturer of semiconductors.

QUALITY SEMICONDUCTOR, INC.
851 Martin Avenue, Santa Clara CA 95050. 408/450-8000. **Contact:** Human Resources. **Description:** A manufacturer of semiconductors.

RAMTRON CORPORATION
1850 Ramtron Drive, Colorado Springs CO 80921. 719/481-7000. **Contact:** Human Resources. **Description:** A manufacturer of semiconductors.

RAYTHEON SEMICONDUCTOR DIVISION
5580 Morehouse Drive, Suite 100, San Diego CA 92121. 619/457-1000. **Fax:** 619/455-6314. **Contact:** Julie Sanfilippo, Human Resources Representative. **Description:** Produces digital video processors for multimedia video applications including PC-to-TV, cable-to-TV, and professional video editing systems. Products also include a family of analog-to-digital converters for video signal processing and signal capture. **Common positions include:** Design Engineer; Electrical/Electronics Engineer. **Educational backgrounds include:** Engineering. **Benefits:** 401(k); Dental Insurance; Disability Coverage; Life Insurance; Medical Insurance; Pension Plan; Tuition Assistance. **Corporate headquarters location:** Lexington MA. **Parent company:** Raytheon, a diversified, international, technology-based company. Overseas facilities and offices are located in over 20 countries, principally in Europe, the Middle East, and the Pacific Rim. The company has four main business segments: electronics, major appliances, aircraft, and energy and environmental construction. **Operations at this facility include:** Administration; Research and Development; Sales. **Listed on:** New York Stock Exchange. **Number of employees at this location:** 60.

RAYTHEON SEMICONDUCTOR DIVISION
350 Ellis Street, Mountain View CA 94043. 415/968-9211. **Fax:** 415/966-7838. **Contact:** Human Resources. **Description:** This location is engaged in the manufacture, design, and development of semiconductor devices for commercial use. Products include standard linear and data conversion devices, Linear Arrays, LSI/VLSI gate arrays, high-density PROMS, and transistors. **Common positions include:** Designer; Electrical/Electronics Engineer; Manufacturer's/Wholesaler's Sales Rep.; Systems Analyst. **Educational backgrounds include:** Engineering. **Benefits:** Dental Insurance; Disability Coverage; Employee Discounts; Life Insurance; Medical Insurance; Pension Plan; Savings Plan; Tuition Assistance; Vision Insurance. **Corporate headquarters location:** Lexington MA. **Parent company:** Raytheon is a diversified, international, multi-industry technology-based company ranked among the 100 largest U.S. industrial corporations. Raytheon has 110 facilities in 28 states, plus the District of Columbia. Overseas facilities and representative offices are located in 26 countries, principally in Europe, the Middle East, and the Pacific Rim. The company has four business segments: Electronics; Major Appliances; Aircraft Products; and Energy and Environmental. **Operations at this facility include:** Administration; Manufacturing; Research and Development; Sales; Service. **Listed on:** New York Stock Exchange. **Number of employees at this location:** 600.

ROCKWELL SEMICONDUCTOR SYSTEMS
4311 Jamboree Road, P.O. Box C, Newport Beach CA 92658-8902. 714/221-4600. **Contact:** Personnel Director. **Description:** Manufactures modem chips and semiconductors. **Parent company:** Rockwell International provides products for the printing, military, automotive, and aerospace industries through its electronics, automotive, and graphics divisions. Products include military and commercial communication equipment, guidance systems, electronics, components for automobiles, and printing presses. A major client of Rockwell is the U.S. government. Rockwell provides the government with parts and services for bombers, as well as power systems for the space station, and is a major contractor for the Space Shuttle Orbiter program.

SGS-THOMSON MICROELECTRONICS

1310 Electronics Drive, Carrollton TX 75006. 972/466-6000. **Contact:** Human Resources Department. **Description:** This location manufactures microchips. SGS-THOMSON Microelectronics is a global independent semiconductor company that designs, develops, manufactures, and markets a broad range of semiconductor integrated circuits and discrete devices used in a variety of microelectronic applications. These applications include telecommunications systems, computer systems, consumer products, automotive products, and industrial automation and control systems. SGS-THOMSON Microelectronics was formed in June 1987 as a result of the combination of divisions from Thomson Semiconducteurs, Thomson-CSF, and SGS Microelettronica. Since its formation, SGS-THOMSON Microelectronics has significantly broadened and upgraded its range of products and technologies and has strengthened its manufacturing and distribution capabilities in Europe, North America, and the Asian Pacific region, while at the same time restructuring its operations to improve efficiency. **Listed on:** New York Stock Exchange. **Stock exchange symbol:** STM.

SGS-THOMSON MICROELECTRONICS

1000 East Bell Road, Phoenix AZ 85022. 602/867-6100. **Contact:** Marc Underwood, Director of Personnel. **Description:** This location manufactures microchips. SGS-THOMSON Microelectronics is a global independent semiconductor company that designs, develops, manufactures, and markets a broad range of semiconductor integrated circuits and discrete devices used in a variety of microelectronic applications. These applications include telecommunications systems, computer systems, consumer products, automotive products, and industrial automation and control systems. SGS-THOMSON Microelectronics was formed in June 1987 as a result of the combination of divisions from Thomson Semiconducteurs, the microelectronics business of the French state-controlled defense electronics company Thomson-CSF, and SGS Microelettronica, the microelectronics business owned by STET-Societa Finanziaria Telefonica p.a., the Italian state-controlled telephone company. Since its formation, SGS-THOMSON Microelectronics has significantly broadened and upgraded its range of products and technologies and has strengthened its manufacturing and distribution capabilities in Europe, North America, and the Asian Pacific region, while at the same time restructuring its operations to improve efficiency. **Corporate headquarters location:** Montgomeryville PA. **Listed on:** New York Stock Exchange. **Stock exchange symbol:** STM.

SAMSUNG AUSTIN SEMICONDUCTOR

8200 North Mopac Expressway, Suite 300, Austin TX 78759. 512/342-1904. **Contact:** Human Resources. **Description:** Manufactures semiconductors. **NOTE:** Parts of this plant were still under construction in early 1996. The company may be seeking new employees more actively in 1997 when the construction is closer to completion.

SEMATECH

2706 Montopolis Drive, Austin TX 78741. 512/356-3500. **Fax:** 512/356-3083. **Contact:** Personnel. **E-mail address:** staffing.hr@sematech.org. **World Wide Web address:** http://www.sematech.org. **Description:** A consortium of U.S. semiconductor manufacturers working with government and academia to sponsor and conduct research in semiconductor manufacturing technology for the U.S. semiconductor industry. SEMATECH develops advanced semiconductor manufacturing methods, materials, and equipment, and validates its development in a proving facility that simulates its members' production lines. Results are transferred to consortium members, including the Department of Defense, who use the devices for both military and commercial applications.

SEMICON COMPONENTS, INC.

10 North Avenue, Burlington MA 01803. 617/272-9015. **Contact:** Richard Allard, Personnel Director. **Description:** Designs, manufactures, and sells a variety of discrete semiconductors and high-density, double-sided, and multi-layered printed circuit boards. Semicon markets these products primarily to computer OEM's. **Corporate headquarters location:** This Location.

SIERRA SEMICONDUCTOR CORPORATION

2222 Qume Drive, San Jose CA 95131. 408/263-9300. **Fax:** 408/263-1272. **Contact:** Professional Staffing. **Description:** Sierra Semiconductor designs, manufactures, and sells semiconductors for the communications market. **Common positions include:** Accountant/Auditor; Electrical/Electronics Engineer; Financial Analyst; Food Scientist/Technologist; Science Technologist; Software Engineer; Systems Analyst. **Educational backgrounds include:** Computer Science; Engineering. **Benefits:** 401(k); Dental Insurance; Disability Coverage; Employee Discounts; Life Insurance; Medical Insurance; Stock Option; Vision Insurance. **Corporate headquarters location:** This Location. **Subsidiaries include:** PMC-Sierra and Prometheus. **Operations at this facility include:** Administration; Divisional Headquarters; Manufacturing; Research and Development; Sales. **Listed on:** NASDAQ. **Number of employees at this location:** 300. **Number of employees nationwide:** 325.

SILICONIX INC.
2201 Laurelwood Road, Santa Clara CA 95054. 408/988-8000. **Fax:** 408/970-3929. **Contact:** Supervisor, Employee Relations/Communications. **Description:** Manufactures semiconductor products. **Common positions include:** Accountant/Auditor; Buyer; Computer Programmer; Electrical/Electronics Engineer; Financial Analyst; Human Resources Manager; Industrial Engineer; Metallurgical Engineer; Systems Analyst. **Educational backgrounds include:** Accounting; Business Administration; Computer Science; Engineering; Marketing. **Benefits:** 401(k); Dental Insurance; Disability Coverage; Employee Discounts; Life Insurance; Medical Insurance; Pension Plan; Profit Sharing; Tuition Assistance. **Special Programs:** Internships. **Corporate headquarters location:** This Location. **Parent company:** AEG Capitol Corporation (Germany). **Operations at this facility include:** Administration; Manufacturing; Research and Development; Sales. **Listed on:** NASDAQ. **Number of employees at this location:** 800.

SIMTEK CORPORATION
1465 Kelly Johnson Boulevard, Colorado Springs CO 80920. 719/531-9444. **Fax:** 719/531-9481. **Contact:** Brian Stephens, Director, Quality Assurance. **Description:** Simtek Corporation develops, produces, and markets high-performance nonvolatile semiconductor memories. Unlike volatile memories, Simtek's products retain their information when power is removed. Simtek's products are targeted for use in such commercial electronic equipment as high-density hard disk drives, modems, smart utility meters, home and commercial security systems, portable telephones, instrumentation, and numerous military systems, including communications, radar, sonar, and smart weapons.

SOLITRON DEVICES, INC.
3301 Electronics Way, West Palm Beach FL 33407. 561/848-4311x255. **Fax:** 561/881-5652. **Contact:** Linda M. Petteruti, Human Resources Administrator. **Description:** A manufacturer of semiconductors. **Common positions include:** Accountant/Auditor; Advertising Clerk; Budget Analyst; Buyer; Chemical Engineer; Clerical Supervisor; Cost Estimator; Credit Clerk and Authorizer; Credit Manager; Customer Service Representative; Department Manager; Designer; Draftsperson; Electrical/Electronics Engineer; Electrician; Financial Manager; General Manager; Human Resources Manager; Industrial Engineer; Machinist; Manufacturer's/Wholesaler's Sales Rep.; Marketing Manager; Mechanical Engineer; Order Clerk; Payroll Clerk; Precision Assembler; Purchasing Agent/Manager; Quality Control Supervisor; Receptionist; Secretary; Stock Clerk; Systems Analyst; Tool and Die Maker. **Educational backgrounds include:** Accounting; Business Administration; Chemistry; Engineering; Finance; Marketing; Physics. **Benefits:** 401(k); Dental Insurance; Disability Coverage; Life Insurance; Medical Insurance; Pension Plan; Profit Sharing; Tuition Assistance. **Corporate headquarters location:** This Location. **Other U.S. locations:** Nationwide. **Operations at this facility include:** Administration; Divisional Headquarters; Manufacturing; Regional Headquarters; Research and Development; Sales. **Listed on:** NASDAQ. **Number of employees at this location:** 120.

SUBMICRON SYSTEMS CORPORATION
6330 Hedgewood Drive, Suite 150, Allentown PA 18106. 610/391-1804. **Contact:** Human Resources. **Description:** A manufacturer of semiconductors.

TECCOR ELECTRONICS INC.
1801 Hurd Drive, Irving TX 75038. 972/580-1515. **Contact:** Human Resources Department. **Description:** Produces electronic power controls and a wide variety of other electronic and related equipment. A second plant at this location produces semiconductor power devices, solid state

relays, and a variety of silicon chips and rectifiers. Markets for these products are manufacturers of lighting fixtures, appliances, heating and air conditioning equipment, and power hand tools which are distributed nationwide. **Common positions include:** Accountant/Auditor; Administrator; Blue-Collar Worker Supervisor; Buyer; Ceramics Engineer; Chemical Engineer; Chemist; Computer Programmer; Customer Service Representative; Department Manager; Draftsperson; Electrical/Electronics Engineer; Industrial Engineer; Purchasing Agent/Manager; Quality Control Supervisor; Sales Executive; Systems Analyst. **Educational backgrounds include:** Accounting; Business Administration; Chemistry; Engineering; Marketing. **Benefits:** Dental Insurance; Disability Coverage; Employee Discounts; Life Insurance; Medical Insurance; Pension Plan; Profit Sharing; Savings Plan; Tuition Assistance. **Corporate headquarters location:** This Location. **Listed on:** New York Stock Exchange.

TEXAS INSTRUMENTS, INC.
P.O. Box 655474, Dallas TX 75265. 972/995-2000. **Contact:** University Relations. **Description:** Designs, develops, and manufactures semiconductor memories, microprocessors, large-scale integrated circuits, and other semiconductor devices and materials; electronic calculators; minicomputers; microcomputers; electronic data terminals; clad metal systems; industrial and commercial electronic controls; opto-electronics, airborne, and ground-based radar systems; and electro-optics equipment. Plants and sales offices are located worldwide. **Common positions include:** Computer Programmer; Electrical/Electronics Engineer; Industrial Engineer; Mechanical Engineer; Systems Analyst. **Educational backgrounds include:** Computer Science; Engineering. **Benefits:** Dental Insurance; Disability Coverage; Life Insurance; Medical Insurance; Pension Plan; Profit Sharing; Savings Plan; Tuition Assistance. **Corporate headquarters location:** This Location. **Operations at this facility include:** Divisional Headquarters; Research and Development. **Listed on:** New York Stock Exchange. **Number of employees at this location:** 25,000. **Number of employees nationwide:** 56,000.
Other U.S. locations:
- 5613 DTC Parkway, Suite 500, Englewood CO 80111. 303/488-9300. Contact: Jo L. Fleming, Office Administrator. This location is a regional office serving three separate divisions: Semiconductors, Digital Systems/Sales, and Digital Systems/Service.
- P.O. Box 1443, Houston TX 77251-1443. 713/274-2000. This location manufactures semiconductors.

TOKYO ELECTRON AMERICA INC.
P.O. Box 17200, Austin TX 78760. 512/424-1000. **Contact:** Human Resources Department. **Description:** Manufactures semiconductors and related equipment. **NOTE:** The company is located at 2400 Grove Boulevard, Austin TX.

TRANSWITCH CORPORATION
8 Progress Drive, Shelton CT 06884. 203/929-8810. **Fax:** 203/926-9453. **Contact:** Human Resources. **E-mail address:** hr@txc.com. **World Wide Web address:** http://www.txc.com. **Description:** A supplier of high-speed semiconductor solutions for broadband network telecommunications and data communications applications. Products include very large scale integration (VLSI) devices for original equipment manufacturers in four markets: telephone networks, local area networks (LANs), wide area networks (WANs), and cable TV (CATV) systems. **Corporate headquarters location:** This Location. **Listed on:** NASDAQ. **Stock exchange symbol:** TXCC.

TRIQUINT SEMICONDUCTOR, INC.
3625A SW Murray Boulevard, Beaverton OR 97116. **Fax:** 503/644-3198. **Contact:** Alan Savage, Human Resources Administrator. **Description:** TriQuint Semiconductor designs, develops, manufactures, and markets a broad range of high-performance analog and mixed signal integrated circuits for the wireless communications, telecommunications, and computing markets. The company uses its proprietary gallium arsenide technology to enable its products to overcome the performance barriers of silicon devices in a variety of applications. Gallium arsenide integrated circuits have inherent physical properties that allow its electrons to move up to five times faster than those of silicon. These properties enable gallium arsenide integrated circuits to operate at much higher speeds than silicon devices or operate at the same high speeds with lower power consumption. The company sells its products worldwide to customers such as Alcatel, AT&T,

Digital Equipment, Hughes Aircraft, IBM, NEC, M/A COM, Motorola, Northern Telecom, Phillips, Rockwell, Siemens, and TRW. **Common positions include:** Designer; Electrical/Electronics Engineer; Financial Analyst; Product Engineer; Test Engineer; Test Operator. **Educational backgrounds include:** Business Administration; Engineering. **Benefits:** 401(k); Daycare Assistance; Dental Insurance; Disability Coverage; Life Insurance; Medical Insurance; Profit Sharing; Stock Option; Tuition Assistance. **Corporate headquarters location:** This Location. **Operations at this facility include:** Administration; Manufacturing; Sales. **Number of employees at this location:** 260.

TSENG LABS
Six Terry Drive, Newtown PA 18940-1831. 215/968-0502. **Contact:** Human Resources. **Description:** A developer of custom semiconductor devices and board-level enhancement products that expand the graphics capabilities and performance of IBM and IBM-compatible personal computers. **Number of employees at this location:** 60.

WESTERN DIGITAL CORPORATION
8105 Irvine Center Drive, Irvine CA 92718. 714/932-5000. **Contact:** Director of Compensation. **Description:** Designs, manufactures, and markets a line of proprietary semiconductor components and digital subsystems for use in the computer, computer peripheral, and communications markets. Western Digital markets over 150 products to a large base of original equipment manufacturers. **Corporate headquarters location:** This Location.

ZILOG INC.
2601 11th Avenue, North Extension, Nampa ID 83687. 208/466-4551. **Contact:** Kurt Williams, Director, Human Resources. **Description:** Manufactures semiconductors. **Common positions include:** Chemical Engineer; Electrical/Electronics Engineer; Industrial Engineer; Mechanical Engineer. **Educational backgrounds include:** Engineering. **Benefits:** 401(k); Dental Insurance; Disability Coverage; Life Insurance; Medical Insurance; Profit Sharing; Tuition Assistance. **Corporate headquarters location:** Campbell CA. **Operations at this facility include:** Administration; Manufacturing. **Listed on:** NASDAQ. **Number of employees at this location:** 500. **Number of employees nationwide:** 1,500.

Note: Because addresses and telephone numbers of smaller companies can change rapidly, we recommend you call each company to verify the information below before inquiring about job opportunities. Mass mailings are not recommended.

Additional employers:

SEMICONDUCTORS AND RELATED DEVICES

Aavid Engineering Inc.
PO Box 400, Laconia NH 03247-0400. 603/528-3400.

Advanced Cerametrics Inc.
245 N Main St, Lambertville NJ 08530-1416. 609/397-2900.

Advanced Power Technology
405 SW Columbia St, Bend OR 97702-1000. 541/382-8028.

Advanced Semiconductor Inc.
7525 Ethel Ave, North Hollywood CA 91605-1912. 818/982-1200.

Aegis Inc.
50 Welby Rd, New Bedford MA 02745-1100. 508/998-3141.

All American Semiconductor
16115 NW 52nd Ave, Hialeah FL 33014-6205. 305/621-8282.

Allegro Microsystems Inc.
115 NE Cutoff, Worcester MA 01606-1224. 508/853-5000.

AlliedSignal Micro Electronic
9140 Rt 108, Columbia MD 21045. 410/964-4000.

Alpha Industries Inc.
20 Sylvan Rd, Woburn MA 01801-1845. 617/935-5150.

American Power Devices Inc.
69 Bennett St, Lynn MA 01905-3003. 617/592-6090.

American Xtal Technology
4311 Solar Way, Fremont CA 94538. 510/833-0553.

Anadigics Inc.
35 Technology Dr, Warren NJ 07059-5148. 908/668-5000.

Analog Devices
1500 Spacepark Dr, Santa Clara CA 95052. 408/727-9222.

Analog Devices Inc.
Semiconductor Division
804 Woburn St, Wilmington MA 01887-3494. 617/935-5565.

Applied Micro Circuits Corp.
6195 Lusk Blvd, San Diego CA
92121-2793. 619/450-9333.

Aries Electronics Inc.
62A Trenton Ave, Frenchtown NJ
08825-1221. 908/996-6841.

Arlon Inc.
1100 Governor Lea Road, Bear
DE 19701. 302/834-2100.

Array Microsystems Inc.
1420 Quail Lake Loop, Colorado
Springs CO 80906-4694.
719/540-7900.

Astec Semiconductor
255 Sinclair Frontage Rd,
Milpitas CA 95035-5415.
408/263-8300.

ATR
1020 Commercial Street, San
Jose CA 95112. 408/277-0400.

Augat Inc. Interconnection
452 John Dietch Boulevard,
Attleboro Falls MA 02763.
508/477-9050.

Aura Vision Corp.
47865 Fremont Blvd, Fremont
CA 94538-6506. 510/252-6800.

Aurora Electronics Inc.
2030 Main Street, Suite 1120,
Irvine CA 92614. 714/660-1232.

**Austek Microsystems
Proprietary**
2903 Bunker Hill Ln, Santa Clara
CA 95054-1141. 408/988-8556.

**Automated Semiconductor
Equipment Corp.**
1777 Hamilton Ave, San Jose CA
95125-5430. 408/445-1422.

Automatic Control Systems Inc.
30 Waverly St, Taunton MA
02780-1460. 508/880-2400.

Automation Devices Inc.
7050 W Ridge Rd, Fairview PA
16415-2099. 814/474-5561.

Autronics Corp.
314 E Live Oak Ave, Arcadia CA
91006-5617. 818/445-5470.

AVG
363 Saint Paul Blvd, Carol
Stream IL 60188-1851. 630/510-
7171.

Axis
1290 Oakmead Pkwy, Sunnyvale
CA 94086-4036. 408/522-9599.

Bicron Electronics Company
50 Barlow St, Canaan CT 06018-
2035. 860/824-5125.

Bio-Rad Semiconductor
520 Clyde Ave, Mountain View
CA 94043-2212. 415/903-4880.

Bipolarics
108 Albright Way, Los Gatos CA
95030-1827. 408/379-4543.

**Bogue Electric Manufacturing
Co.**
100 Pennsylvania Ave, Paterson
NJ 07503-2510. 201/523-2200.

Bryant Electric Inc.
185 Plains Rd, Milford CT
06460-2473. 203/876-3600.

Cal-Tron Systems Inc.
857 Sandhill Ave, Carson CA
90746-1210. 310/516-1125.

California Micro Devices Corp.
215 Topaz St, Milpitas CA
95035-5430. 408/263-3214.

Calogic Corp.
237 Whitney Pl, Fremont CA
94539-7664. 510/656-2900.

**Cascade Design Automation
Corp.**
3650 131st Ave SE, Suite 650,
Bellevue WA 98006-1334.
206/643-0200.

Central Semiconductor Corp.
145 Adams Ave, Hauppauge NY
11788-3603. 516/435-1110.

CFM Technologies Inc.
1336 Enterprise Drive, West
Chester PA 19380-5959.
610/696-8300.

Cirrus Logic Inc.
350 Interlocken Boulevard,
Broomfield CO 80021-3426.
303/466-5228.

Citadel Technologies
2021 The Alameda, Suite 240,
San Jose CA 95126-1110.
408/241-1150.

Codi Semiconductor Inc.
161 Tices Lane, East Brunswick
NJ 08816-2040. 908/390-2828.

Comlinear Corporation
4800 Wheaton Dr, Fort Collins
CO 80525-9483. 970/226-0500.

Compensated Devices Inc.
166 Tremont St, Melrose MA
02176-2296. 617/665-1071.

Component Intertechnologies
2426 Perry Hwy, Hadley PA
16130-0096. 412/253-3161.

Composite Modules
1 Mill St, Attleboro MA 02703-
2903. 508/226-0420.

Composite Technical Ceramics
25 Lark Industrial Pkwy,
Greenville RI 02828-3024.
401/949-2000.

Concept Systems Design Inc.
2800 Bayview Drive, Fremont
CA 94538-7328. 510/226-8155.

Consolidated Industries Inc.
4015 Pulaski Pike NW,
Huntsville AL 35810-1901.
205/859-6890.

Contacts Metals and Weld Inc.
PO Box 2266, Indianapolis IN
46206-2266. 317/634-8884.

CP Clare Corp.
78 Cherry Hill Drive, Beverly
MA 01915. 617/246-4000.

Crimson Semiconductor Inc.
460 W 34th St, New York NY
10001-2320. 212/947-8585.

Cryco Quartz Inc.
8107 Altoga Dr, Austin TX
78724-2613. 512/926-8931.

**Crystal Specialty International
Electronics**
2853 Janitell Rd, Colorado
Springs CO 80906-4104.
719/540-0990.

Crystal Systems Inc.
27 Congress St, Salem MA
01970-5575. 508/745-0088.

Crystallume
125 Constitution Dr, Menlo Park
CA 94025-1126. 415/324-9681.

Crystallume
3506 Bassett St, Santa Clara CA
95054-2704. 408/653-1700.

Crystaloid Electronics Co.
5282 Hudson Dr, Hudson OH
44236-3738. 216/655-2429.

CTS Microelectronics
1201 Cumberland Ave, West
Lafayette IN 47906-1317.
765/463-2565.

Cubic Memory Inc.
27 Janis Way, Scotts Valley CA
95066-3506. 408/438-1887.

Cypress Semiconductor Inc.
2401 E 86th St, Minneapolis MN
55425-2704. 612/851-5200.

**Data Instruments Advanced
Silicon**
1253 Reamwood Ave, Sunnyvale
CA 94089-2226. 408/744-0452.

Delta Design Inc.
5785 Kearny Villa Rd, San Diego
CA 92123-1111. 619/292-5000.

Diamond Images
120-C Albright Way, Los Gatos
CA 95030-1802. 408/866-8460.

Die Mate Corp.
3945 25th Ave, Schiller Park IL
60176-2117. 847/671-3535.

Diodes Inc.
3050 E Hillcrest Dr, Suite 200,
Westlake Village CA 91362-
3154. 805/446-4800.

Disco Hi-Tec America
7909 SW Cirrus Dr, Beaverton
OR 97008-5971. 503/644-0323.

DSP Group Inc.
3120 Scott Boulevard, Santa
Clara CA 95054. 408/986-4300.

DuPont Photomasks Inc.
555 Union Blvd, Allentown PA
18103-1229. 610/434-3111.

DuPont Photomasks Inc.
592 South Rd, Poughkeepsie NY
12601-5501. 914/463-5300.

DuPont Photomasks Inc.
2920 Coronado Dr, Santa Clara
CA 95054-3203. 408/492-1900.

DuPont Photomasks Inc.
2235 Zanker Rd, San Jose CA
95131-1120. 408/435-8335.

Ecliptek Corporation
3545-B Cadillac Ave, Costa Mesa
CA 92626-1401. 714/433-1200.

Edal Industries Inc.
51 Commerce St, East Haven CT
06512-3851. 203/467-2591.

8x8
2445 Mission College Blvd,
Santa Clara CA 95054-1214.
408/727-1885.

El Dorado Technical Service
2340 Harris Way, San Jose CA
95037. 408/943-0140.

Elantec Semiconductor Inc.
1996 Tarob Ct, Milpitas CA
95035-6824. 408/945-1323.

Electro Ceramic Industries
75 Kennedy St, Hackensack NJ
07601-5262. 201/342-2630.

Electro-Films Inc.
111 Gilbane St, Warwick RI
02886-6901. 401/738-9150.

Electronic Devices Inc.
21 Grey Oaks Ave, Yonkers NY
10710-3205. 914/965-4400.

Electronic Space Systems Corp.
Old Powder Mill Rd, Concord
MA 01742. 508/369-7200.

Emcore Corp.
394 Elizabeth Ave, Somerset NJ
08873-5118. 908/271-9090.

EMF Systems Inc.
121 N Science Park Rd, State
College PA 16803-2211.
814/237-5738.

**Empak Inc. Electronics
Division**
4405 Arrowswest Dr, Colorado
Springs CO 80907-3445.
719/528-2600.

Energy Sciences Inc.
42 Industrial Way, Wilmington
MA 01887-4605. 508/694-9000.

Epitaxx Inc.
7 Graphics Dr, West Trenton NJ
08628-1547. 609/538-1800.

Epitaxy Inc.
PO Box 4778, Santa Clara CA
95056-4778. 408/988-2161.

ERG Materials & Aerospace
900 Stanford Ave, Oakland CA
94608-2320. 510/658-9785.

ESI
150 Dan Rd, Canton MA 02021-
2820. 617/821-1222.

Exel Microelectronics Inc.
PO Box 49038, San Jose CA
95161-9038. 408/432-0500.

Expertech Inc.
5100 Scotts Valley Dr, Scotts
Valley CA 95066-3552. 408/439-
9300.

Exsil Inc.
6541 Via Del Oro, San Jose CA
95119-1207. 408/629-3142.

F&S Alloys and Minerals Corp.
605 3rd Ave, New York NY
10158-0180. 212/490-1356.

Fair Child Technologies
47613 Warm Springs Blvd,
Fremont CA 94539-7470.
510/659-8370.

Far 'C'ing Interprises
18675 Adams Ct, Morgan Hill
CA 95037-2805. 408/776-0001.

Film Microelectronics Inc.
530 Turnpike St, North Andover
MA 01845-5812. 508/975-3385.

Firing Circuits Inc.
PO Box 2007, Norwalk CT
06852-2007. 203/846-1633.

Fluoroware Inc.
102 North Jonathan Boulevard,
Chaska MN 55318-2341.
612/448-3131.

Frequency Electric Inc.
55 Charles Lindbergh Blvd,
Uniondale NY 11553-3633.
516/794-4500.

**Frontier Semiconductor
Measurement Inc.**
1631 North 1st Street, San Jose
CA 95112-4516. 408/452-8898.

Fuel Cell Manufacturing Corp.
318 Industrial Ln, Torrington CT
06790-7704. 860/496-1111.

Fujitsu Microelectronics Inc.
2603 Main St, Irvine CA 92714-
6232. 714/724-8777.

**Fusion Semiconductor Systems
Corp.**
770 Lucerne Drive, Sunnyvale
CA 94086-3844. 408/774-6260.

Gaiser Tool Co.
4544 McGrath Street, Ventura
CA 93003-5704. 805/644-5583.

GE Co.
205 Great Valley Parkway,
Malvern PA 19355-1308.
610/251-7000.

GE Co. Carolina Products Plant
900 N George St, Goldsboro NC 27530-2432. 919/731-5100.

GEC Plessey Semiconductors
1500 Green Hills Rd, Scotts Valley CA 95066-4922. 408/438-2900.

General Ceramics Tekform Products
2770 E Coronado St, Anaheim CA 92806-2401. 714/630-2340.

General Instrument Corp.
11801 Miriam Dr, El Paso TX 79936-7431. 915/592-5700.

General Instrument Corp.
181 W Madison St, Chicago IL 60602-4510. 312/541-5000.

General Instrument Corp.
172 Spruce St, Westbury NY 11590-3220. 516/333-8400.

General MicroCircuits Corp.
1133 N Main St, Mooresville NC 28115-2359. 704/663-5975.

General MicroCircuits Corp.
780 Boston Rd, Billerica MA 01821-5925. 508/663-9101.

Genesis
3066 Scott Blvd, Santa Clara CA 95054-3325. 408/986-1636.

Gentron Corporation
7345 E Acoma Dr, Scottsdale AZ 85260-3124. 602/443-1288.

Genus Inc.
4 Tucker Dr, Newburyport MA 01950-1924. 508/463-1500.

Germanium Power Devices Corp.
York St, Andover MA 01810. 508/475-5982.

GMT Microelectronic
950 Rittenhouse Road, Norristown PA 19403-5239. 610/666-7950.

Goldstar Electron Inc.
3003 N 1st St, San Jose CA 95134-2004. 408/432-1331.

H-Square Co.
1289 Reamwood Ave, Sunnyvale CA 94089-2234. 408/734-2543.

Hitachi America Ltd.
500 Park Blvd, Itasca IL 60143-1289. 630/773-4864.

Hitachi Semiconductor America
6431 Longhorn Dr, Irving TX 75063-2738. 214/580-0088.

Hittite Microwave Corp.
21 Cabot Rd, Woburn MA 01801-1003. 617/933-7267.

Honeywell SSEC
12001 Highway 55, Plymouth MN 55441-4744. 612/954-2300.

Hughes Microelectronics Circuits
500 Superior Ave, Newport Beach CA 92663-3627. 714/759-2411.

Hutson Industries Inc.
1000 Hutson Cir, Frisco TX 75034. 214/377-2402.

IBM Corp.
1580 State Route 52, Hopewell Junction NY 12533-6531. 914/894-2121.

IC Designs
12020 113th Avenue NE, Kirkland WA 98034-6920. 206/821-9202.

IC Works Inc.
3725 N 1st St, San Jose CA 95134-1708. 408/922-0202.

ICT Inc.
2123 Ringwood Ave, San Jose CA 95131-1725. 408/434-0678.

IDT
1566 Moffett St, Salinas CA 93905-3342. 408/424-7726.

II-VI Inc.
375 Saxonburg Blvd, Saxonburg PA 16056-9430. 412/352-4455.

Inspex Inc.
47 Manning Rd, Billerica MA 01821-3925. 508/667-5500.

Integrated Circuit Systems Inc.
2435 Boulevard Of Generals, Norristown PA 19403-3661. 610/630-5300.

Integrated Silicon Solutions Inc.
2231 Lawson Ln, Santa Clara CA 95054. 408/733-4774.

Integrated Solutions Inc.
836 North Street, Tewksbury MA 01876-1253. 508/640-1400.

Integrated Wave Technologies
4042 Clipper Ct, Fremont CA 94538-6540. 510/353-0260.

International Electronic Research Corp.
135 West Magnolia Blvd, Burbank CA 91502-1722. 818/842-7277.

International Leadframe
3320 Victor Ct, Santa Clara CA 95054-2316. 408/980-0782.

International Microcircuits Inc.
525 Los Coches St, Milpitas CA 95035-5423. 408/263-6300.

International Rectifier Corp.
233 Kansas St, El Segundo CA 90245-4316. 310/322-3331.

Interpoint Corp.
10301 Willows Rd NE, Redmond WA 98052-2529. 206/882-3100.

Invax Technologies
233 Weddell Rd, Sunnyvale CA 94089. 408/745-6240.

IPEC Clean
1922 Avenida Del Oro, Oceanside CA 92056-5898. 760/758-0994.

Irvine Sensors Corp.
3001 Red Hill Ave, Costa Mesa CA 92626-4529. 714/549-8211.

Ixys Corp.
3540 Bassett St, Santa Clara CA 95054-2704. 408/982-0700.

JMS Southeast Inc.
105 Temperature Lane, Statesville NC 28677-9639. 704/873-1835.

Johnson Matthey Inc.
10080 Willow Creek Rd, San Diego CA 92131-1623. 619/566-9510.

Kearney-National Inc.
108 Corporate Park Drive, White Plains NY 10604. 914/694-6700.

LAM Research Inc.
16 Jonspin Road, Wilmington MA 01887-1093. 508/657-3933.

Lanxide Corp.
PO Box 6077, Newark DE
19714-6077. 302/456-6200.

Lawrence Semiconductor Labs
2300 W Huntington Dr, Tempe
AZ 85282-3130. 602/438-2300.

Leach International
6900 Orangethorpe Ave, Buena
Park CA 90620-1351. 714/739-
1150.

Linfinity Microelectronics Inc.
11861 Western Ave, Garden
Grove CA 92641-2119. 714/898-
8121.

Lite On
720 S Hillview Drive, Milpitas
CA 95035-5455. 408/946-4873.

Litton Industries Inc.
2203 West Walnut Street,
Garland TX 75042. 214/487-
4100.

**Litton Industries Inc. Solid
State Division**
3251 Olcott St, Santa Clara CA
95054-3006. 408/988-1331.

Lockheed Martin
Ridge Hill, Yonkers NY 10710.
914/968-2500.

Logic Devices Inc.
628 E Evelyn Ave, Sunnyvale CA
94086-6400. 408/737-3300.

Lumex Opto/Components Inc.
290 E Helen Rd, Palatine IL
60067-6955. 847/359-2790.

M and W Systems
2346 Tripaldi Way, Hayward CA
94545-5021. 510/887-7008.

M/A-Com PHI
1742 Crenshaw Blvd, Torrance
CA 90501-3385. 310/320-6160.

Marco International
5603 Arapahoe Ave, Boulder CO
80303-1332. 303/449-9191.

Marlow Industries
10451 Vista Park Rd, Dallas TX
75238-1645. 214/340-4900.

Material Research Corp.
2350 Mission College Blvd,
Santa Clara CA 95054-1535.
408/748-0902.

Matrix Integrated Systems
4050 Lakeside Dr, Richmond CA
94806-1936. 510/222-2727.

Maxim Integrated Products
120 San Gabriel Dr, Sunnyvale
CA 94086-5150. 408/737-7600.

**Melcor Materials Electronic
Products Corp.**
1040 Spruce St, Trenton NJ
08648-4548. 609/393-4178.

MEMC
2400 Walsh Ave, Santa Clara CA
95051-1303. 408/988-2600.

Metelics Corp.
975 Stewart Dr, Sunnyvale CA
94086-3913. 408/737-8181.

Metron Technology
770 Lucerne Dr, Sunnyvale CA
94086-3844. 408/481-1010.

MGI Electronics
1203 W Geneva Dr, Tempe AZ
85282-3432. 602/967-8011.

Micon Inc.
8540 Mosley Rd, Houston TX
77075-1116. 713/947-9470.

Micro Glass Inc.
6200 E Molloy Rd, East Syracuse
NY 13057-1022. 315/437-7571.

Micro Linear Corp.
2092 Concourse Dr, San Jose CA
95131-1861. 408/433-5200.

Microchip Technology Inc.
1200 S 52nd St, Tempe AZ
85281-6922. 602/804-3400.

Micropac Industries Inc.
905 E Walnut St, Garland TX
75040-6611. 214/272-3571.

Microphase Corp.
587 Connecticut Ave, Norwalk
CT 06854-1711. 203/866-8000.

Microsemi Corporation
P.O. Box 26890, Santa Ana CA
92799-6890. 714/979-8220.

Microwave Technology Inc.
4268 Solar Way, Fremont CA
94538-6388. 510/651-6700.

Minco Technology Labs Inc.
1805 Rutherford Ln, Austin TX
78754-5112. 512/834-2022.

Mitsubishi Semiconductor
3 Diamond Ln, Durham NC
27704-9410. 919/479-3333.

Mosaic Semiconductor Inc.
7420 Carroll Rd, San Diego CA
92121-2334. 619/271-4565.

Mosel Vitelic Corp.
3910 N 1st St, San Jose CA
95134-1501. 408/433-6000.

Motorola Semiconductor Inc.
2733 Albright Rd, Kokomo IN
46902-3996. 765/457-6634.

Motorola Semiconductor Inc.
1155 Business Center Drive,
Horsham PA 19044-2319.
215/957-4100.

Motorola Semiconductor Inc.
100 Passaic Ave, Fairfield NJ
07004-3508. 201/808-2400.

Mouser Electronics
958 North Main, Mansfield TX
76063-4829. 817/483-4422.

MTS Micro Electronics
3361 E Miraloma Ave, Anaheim
CA 92806-1931. 714/577-3990.

MX-Com Inc.
4800 Bethania Station Rd,
Winston-Salem NC 27105-1200.
910/744-5050.

National Hybrid Inc.
2200 Smithtown Ave,
Ronkonkoma NY 11779-7329.
516/981-2400.

National Semiconductor Corp.
1111 W Bardin Rd, Arlington TX
76017-5999. 817/468-6300.

Nchip Inc.
1971 N Capitol Ave, San Jose CA
95132-1057. 408/945-9991.

**New Jersey Semiconductor
Products**
20 Stern Ave, Springfield NJ
07081-2905. 201/376-2922.

NIC Components Corp.
6000 New Horizons Blvd,
Amityville NY 11701-1146.
516/226-7500.

Nippert Co.
801 Pittsburgh Dr, Delaware OH
43015-3812. 614/363-1981.

NTE Electronics Inc.
44 Farrand St, Bloomfield NJ
07003-2516. 201/748-5089.

Oki Semiconductor
11155 SW Leveton Dr, Tualatin
OR 97062-8094. 503/692-9100.

Ontrak Systems Inc.
77 W Montague Expy, Milpitas
CA 95035-6264. 408/262-5200.

Optical Associates Inc.
1425 McCandless Dr, Milpitas
CA 95035-8023. 408/263-4944.

Optical Specialties Inc.
4281 Technology Dr, Fremont
CA 94538-6339. 510/490-6400.

Orbit Semiconductor Inc.
169 W Java Dr, Sunnyvale CA
94089-1016. 408/744-1800.

**Performance Semiconductor
Corp.**
630 E Weddell Dr, Sunnyvale CA
94089-1751. 408/734-8200.

Philips Semiconductors
PO Box 3409, Sunnyvale CA
94088-3409. 408/991-2000.

Phoenix Semiconductors Inc.
1706 W 10th Pl, Tempe AZ
85281-5212. 602/968-6389.

Photronics Inc.
1913 Tarob Ct, Milpitas CA
95035-6825. 408/262-8800.

Piconics Inc.
26 Cummings Rd, Tyngsboro
MA 01879-1498. 508/649-7501.

PM Electro Tech Inc.
3350 Scott Blvd, Santa Clara CA
95054-3108. 408/727-5501.

PMC-Sierra Inc.
9400 SW Gemini Dr, Beaverton
OR 97008-7105. 503/520-1800.

**Polishing Corp. of America
(PCA)**
430 Martin Ave, Santa Clara CA
95050-2911. 408/988-6000.

Poly-Flex Circuits
28 Kenney Dr, Cranston RI
02920-4486. 401/463-3180.

Polyfet RF Devices
1110 Avenida Acaso, Camarillo
CA 93012-8725. 805/484-4210.

Power Spectra Inc.
919 Hermosa Ct, Sunnyvale CA
94086-4103. 408/737-7977.

PPC Products Corp.
7516 Central Industrial Dr,
Riviera Beach FL 33404-3499.
561/848-9606.

Prometrix Corp.
2500 Augustine Dr, Santa Clara
CA 95054-3001. 408/970-9500.

Prosource Industries
1040 Avenue M, Grand Prairie
TX 75050-1916. 214/660-1400.

QT Optoelectronics
610 N Mary Ave, Sunnyvale CA
94086-2906. 408/720-1440.

Recticon Enterprises
114 S Washington St, Pottstown
PA 19464-5978. 610/327-4220.

Renard Manufacturing Co. Inc.
3305 NW 79th Ave, Miami FL
33122-1068. 305/592-1500.

Rippey Corp.
5000 Hillsdale Circle, El Dorado
Hills CA 95762. 916/939-4332.

Rockwell International Corp.
2427 W Hillcrest Dr, Newbury
Park CA 91320-2202. 805/498-
6715.

RSM Sensitron Inc.
221 W Industry Ct, Deer Park NY
11729-4605. 516/586-7600.

RTI Export Inc. Barbados
1260 Red Fox Rd, Arden Hills
MN 55112-6944. 612/636-9770.

SDL Inc.
80 Rose Orchard Way, San Jose
CA 95134-1356. 408/943-9411.

Sechan Electronics Inc.
525 Furnace Hills Pike, Lititz PA
17543-8954. 717/627-4141.

Sela USA Inc.
790 Lucerne Dr, Sunnyvale CA
94086-3838. 408/988-5151.

Semi Conductors Inc.
3680 Investment Ln, Riviera
Beach FL 33404-1762. 561/842-
0305.

Semiconductor Circuits Inc.
49 Range Rd, Windham NH
03087-2019. 603/893-2330.

**Semiconductor Packaging
Material Co. Inc.**
431 Fayette Ave, Mamaroneck
NY 10543-2211. 914/698-5353.

Semiware Inc.
1234 Yard Ct, San Jose CA
95133-1048. 408/297-7900.

Semtech Inc.
1111 Comstock St, Santa Clara
CA 95054-3407. 408/727-6562.

Sensarray Corp.
3410 Garrett Dr, Santa Clara CA
95054-2803. 408/727-4656.

Sensotron Inc.
5881 Engineer Dr, Huntington
Beach CA 92649-1127. 714/898-
5618.

**SGS-THOMSON
Microelectronics**
16350 W Bernardo Dr, San Diego
CA 92127-1802. 619/485-8900.

**SGS-THOMSON
Microelectronics**
141 Commerce Dr,
Montgomeryville PA 18936-
9641. 215/361-6400.

Shinkawa USA Inc.
2620 Augustine Dr, Santa Clara
CA 95054-2903. 408/727-2601.

Siemens Components Inc.
10950 N Tantau Ave, Cupertino
CA 95014-0716. 408/777-4500.

Siemens Electrochemical
200 Richland Creek Dr, Princeton
IN 47671-0001. 812/386-1000.

Signal Processing Technology
4755 Forge Rd, Colorado Springs
CO 80907-3519. 719/528-2300.

Silicon Magic Corp.
4500 Great America Parkway,
Santa Clara CA 95054. 408/969-
3000.

Silicon Systems Inc.
14351 Myford Rd, Tustin CA
92680-7039. 714/731-7110.

Silicon Transistor Corp.
27 Katrina Rd, Chelmsford MA
01824-2864. 508/256-3321.

Silicon Valley Group Inc.
2240 Ringwood Ave, San Jose
CA 95131-1716. 408/434-0500.

Sipex Corporation
22 Linnell Cir, Billerica MA
01821-3901. 508/667-8700.

Solid State Devices Inc.
14849 Firestone Blvd, La Mirada
CA 90638-6017. 714/670-7734.

Solid State Measurements Inc.
110 Technology Dr, Pittsburgh
PA 15275-1026. 412/787-0620.

Solid State Testing Inc.
56 Middlesex Tpke, Burlington
MA 01803-4969. 617/272-0972.

Solitec Wafer Processing Co.
685 River Oaks Pkwy, San Jose
CA 95134-1907. 408/955-9939.

Space Power Electronics Inc.
305 Jeffry Ln, Glen Gardner NJ
08826-3221. 908/537-2184.

ST Research Corp.
8419 Terminal Rd H, Newington
VA 22122. 703/550-7000.

Stec Instruments Inc.
1080 E Duane Ave, Sunnyvale
CA 94086-2628. 408/730-8795.

Strasbaugh Inc.
825 Buckley Rd, San Luis Obispo
CA 93401-8130. 805/541-6424.

Sumitomo Sitix Silicon Inc.
537 Grandin Rd, Maineville OH
45039-9762. 513/583-2600.

Sun Disk Corporation
3270 Jay St, Santa Clara CA
95054-3309. 408/542-0500.

Sun Power Corporation
430 Indio Way, Sunnyvale CA
94086-4202. 408/991-0900.

Surtech Labs
580 Pleasant St, Watertown MA
02172-2408. 617/924-9280.

Synergy Semiconductor Corp.
3250 Scott Boulevard, Santa
Clara CA 95054. 408/730-1313.

Talon Technology Corp.
1819 Firman Dr, Richardson TX
75081-1842. 214/680-9913.

Tech Etch Inc.
45 Aldrin Rd, Plymouth MA
02360-4803. 508/747-0300.

Technical Components Co. Inc.
6633 99th St, Chicago Ridge IL
60415-1296. 708/424-1900.

Technics Inc.
7060 Koll Center Pkwy,
Pleasanton CA 94566. 510/417-1500.

Tedea Huntleigh Celesco
PO Box 7964, Canoga Park CA
91309-7964. 818/884-6860.

Tegal Corp.
2201 S McDowell Blvd,
Petaluma CA 94954. 707/763-5600.

Tektron Micro Electronics Inc.
7483 Candlewood Rd, Hanover
MD 21076-3142. 410/850-4200.

Telegenix Inc.
PO Box 5550, Cherry Hill NJ
08034-0511. 609/424-5220.

Tensleep Design Inc.
3809 S 2nd St, Austin TX 78704-7058. 512/447-5558.

Texas Instruments Inc.
PO Box 60448, Midland TX
79711-0448. 915/561-6500.

Texas Instruments Inc.
399 Thornall St, Edison NJ
08837-2236. 908/906-0033.

Texas Optoelectronics Inc.
714 Shepherd Dr, Garland TX
75042-6833. 214/487-0085.

Thermco Systems
1465 N Batavia St, Orange CA
92667-3504. 714/639-2340.

Toko America Inc.
1250 Feehanville Dr, Mount
Prospect IL 60056-6023.
847/635-3200.

Tosch SMD Inc.
3600 Gantz Rd, Grove City OH
43123. 614/875-7912.

Toshiba America
1060 Rincon Cir, San Jose CA
95131-1325. 408/456-8900.

Total Technologies
2110 S Anne St, Santa Ana CA
92704-4409. 714/241-0406.

Transducer Systems Inc.
4000 Bridge St, Drexel Hill PA
19026-2711. 610/284-2508.

Transistor Devices Inc.
85 Horsehill Rd, Cedar Knolls NJ
07927-2097. 201/267-1900.

Trazar Co.
3250 Keller St, Santa Clara CA
95054-2613. 408/970-9501.

TSI Microelectronics Corp.
5 Southside Rd, Danvers MA
01923-1408. 508/774-8722.

Tylan General Inc.
9577 Chesapeake Dr, San Diego
CA 92123-1304. 619/571-1222.

**Ultra Clean Technology
Systems**
150 Independence Dr, Menlo
Park CA 94025-1136. 415/323-4100.

Ultrapointe Corp.
163 Baypointe Pkwy, San Jose
CA 95134-1622. 408/894-7080.

**United States Dynamics
Corporation**
425 Bayview Avenue, Amityville
NY 11701-2622. 516/842-5600.

**United Technologies
Microelectronics Center**
1575 Garden of the Gods Rd,
Colorado Springs CO 80907-3415. 719/594-8000.

Unitek
47333 Warm Springs Blvd,
Fremont CA 94539-7462.
510/623-8544.

Unity Micro Electronics Inc.
47456 Fremont Blvd, Fremont
CA 94538-6503. 510/661-2700.

Universal Semiconductor Inc.
1925 Zanker Rd, San Jose CA
95112-4292. 408/436-1906.

Vadem
1960 Zanker Rd, San Jose CA
95112-4216. 408/467-2100.

Veeder-Root Company
125 Powder Forest Dr, Simsbury
CT 06070. 860/651-2700.

Versitron
27 McCullough Dr, New Castle
DE 19720-6611. 302/323-8600.

Viking Components
11 Columbia, Aliso Viejo CA
92656-1460. 714/643-7255.

Virginia Semiconductor Inc.
1501 Powhatan St,
Fredericksburg VA 22401-4647.
540/373-2900.

Vitesse Semiconductor Corp.
741 Calle Plano, Camarillo CA
93012-8543. 805/388-3700.

Vitrus
881 Main Street, Pawtucket RI
02860-4930. 401/724-9350.

Vixel Corp.
325 Interlocken Pkwy,
Broomfield CO 80021-3484.
303/460-0700.

VLSI Technology Inc.
1109 McKay Drive, San Jose CA
95131-1797. 408/434-3000.

Voltage Multipliers Inc.
8711 W Roosevelt Avenue,
Visalia CA 93291-9458. 209/651-
1402.

Weltronic/Technitron Corp.
150 East Saint Charles Road,
Carol Stream IL 60188-2075.
630/462-8250.

Westar Rep Co.
26500 Agoura Road, Calabasas
CA 91302-1952. 818/880-0594.

Williams Advanced Material
2978 Main Street, Buffalo NY
14214-1099. 716/837-1000.

Xetel
2525 Brockton Drive Austin TX
78758-4463. 512/834-2266.

Xicor Inc.
1511 Buckeye Drive, Milpitas
CA 95035-7431. 408/432-8888.

Zilog Inc.
210 East Hacienda Avenue,
Campbell CA 95008-6600.
408/370-8000.

Zoran Corporation
2041 Mission College Boulevard,
Santa Clara CA 95054-1518.
408/986-1314.

NETWORKING AND SYSTEMS SERVICES

*R*ay *Johansen, a Technical Support Engineer at Cabletron Systems, emphasizes that jobseekers need to seize the initiative, display excellent communication skills, and possess the desire to learn new technologies in order to thrive in the field of computer networking. Like other aspects of the computer industry, networking technology constantly evolves, requiring a support engineer to stay educated on the latest changes. Travel to customer sites is required and the hectic pace is not for everyone.*

The U.S. Air Force provided Ray with solid computer training and he turned that knowledge into a career. Ray recommends that jobseekers may also want to look into specialized courses at a technical school or community college to get hands-on experience. Within the next five years, Ray sees home-based computer networks accounting for a growing percentage of the demand for the services of support engineers.

ADC KENTROX
14375 NW Science Park Drive, Portland OR 97229. 503/643-1681. **Contact:** Human Resources. **E-mail address:** info@kentrox.com. **World Wide Web address:** http://www.kentrox.com. **Description:** A manufacturer of standards-based, high-speed digital network access and service internetworking products for connectivity of local area networks (LANs) and customer premises equipment over wide area networks (WANs).

ACCESS BEYOND
1300 Quince Orchard Boulevard, Gaithersburg MD 20878. 301/417-0552. **Contact:** Human Resources. **Description:** Designs, develops, manufactures, and markets data communications networking systems and specialized electronic instrumentation equipment. The company was formerly known as Penril Datacomm.

ACE*COMM
704 Quince Orchard Road, Gaithersburg MD 20878. 301/721-3000. **Contact:** Human Resources Department. **Description:** A developer of information systems for the telecommunications industry.

ADAC HEALTHCARE INFORMATION SYSTEMS
5 Greenway Plaza, Suite 1900, Houston TX 77046. 713/960-1907. **Fax:** 713/960-0164. **Contact:** Lourdes Welsh, Recruiter. **E-mail address:** lwelsh@hcis.adaclabs.com. **World Wide Web address:** http://www.adaclabs.com. **Description:** Formerly Community Health Computing Corporation, Adac Healthcare Information Systems designs, develops, and supports hospital information systems, for both clinical and business applications including UNIX, Oracle, and Client-Server Operating Systems Environment. **Common positions include:** Applications Engineer; Market Research Analyst; Marketing Specialist; Technical Writer/Editor. **Educational backgrounds include:** Accounting; Biology; Business Administration; Chemistry; Computer Science; Finance; Mathematics. **Benefits:** 401(k); Dental Insurance; Disability Coverage; Life Insurance; Medical Insurance; Tuition Assistance. **Corporate headquarters location:** Milpitas CA. **Parent company:** ADAC Labs. **Listed on:** NASDAQ. **Number of employees at this location:** 750. **Number of projected hires for 1997 - 1998 at this location:** 60.

ADVANCED ENGINEERING & RESOURCE ASSOCIATES, INC. (AERA)
1755 Jefferson Davis Highway, Suite 800, Arlington VA 22202. 703/412-7188. **Fax:** 703/412-7198. **Contact:** Human Resources. **E-mail address:** interactive@aera.com. **Description:** An engineering services firm offering expertise in propulsion systems, satellite systems, advanced

logistics, and mechanical systems. AERA also provides multimedia and network services including LANs and WANs; and interactive multimedia including CD-ROM, 3D and 2D animation, digital video and audio, and hypertext and graphics.

AEROCOMM
13228 West 99th Street, Lenexa KS 66215. 913/492-2320. **Contact:** Human Resources. **Description:** Manufactures wireless printer sharing products.

AFFINITY
16 Portland Road, Highland NJ 07732. 908/872-2240. **Contact:** Human Resources Department. **Description:** Affinity provides software and systems integration services.

ALLSTAR SYSTEMS, INC.
6401 Southwest Freeway, Houston TX 77074. 713/795-2302. **Fax:** 713/795-2049. **Contact:** Jill Banta, Human Resources. **E-mail address:** jbanta@allstar.com. **World Wide Web address:** http://www.allstar.com. **Description:** Sells computers and offers networking services. Founded in 1981. **Common positions include:** Account Representative; Accountant/Auditor; Buyer; Computer Programmer; Computer Support Technician; Customer Service Representative; Purchasing Agent/Manager; Sales Executive; Sales Representative. **Educational backgrounds include:** Accounting; Computer Science. **Benefits:** 401(k); Dental Insurance; Medical Insurance; Profit Sharing. **Corporate headquarters location:** This Location. **Other U.S. locations:** Dallas TX. **Listed on:** Privately held. **Annual sales/revenues:** More than $100 million. **Number of employees at this location:** 200. **Number of employees nationwide:** 300. **Number of projected hires for 1997 - 1998 at this location:** 20.

ALLTEL INFORMATION SERVICES
4001 North Rodney Parham Road, Little Rock AR 72212. 501/220-5100. **Contact:** Human Resources. **Description:** Offers networking services and technical support to businesses.

AMDAHL CORPORATION
1250 East Arques Avenue, Sunnyvale CA 94088. 408/746-6000. **Contact:** Lisa Carter, Staffing Consultant. **E-mail address:** jobs@amdahl.com. **Description:** Designs, develops, manufactures, markets, and services large-scale, high-performance, general purpose computer systems, including both hardware and software. Customers are primarily in large corporations, government agencies, and large universities with high-volume data-processing requirements. Amdahl markets more than 470 different systems. **Common positions include:** Accountant/Auditor; Computer Programmer; Customer Service Representative; Financial Analyst; Human Resources Manager; Mechanical Engineer; Software Engineer; Systems Analyst; Technical Writer/Editor; Transportation/Traffic Specialist. **Educational backgrounds include:** Accounting; Business Administration; Communications; Computer Science; Engineering; Finance; Liberal Arts; Marketing. **Benefits:** 401(k); Daycare Assistance; Dental Insurance; Disability Coverage; Employee Discounts; Life Insurance; Medical Insurance; Pension Plan; Profit Sharing; Savings Plan; Tuition Assistance. **Special Programs:** Internships. **Corporate headquarters location:** This Location. **Other U.S. locations:** IL; NY; TX. **Operations at this facility include:** Administration; Manufacturing; Regional Headquarters; Research and Development; Sales; Service. **Listed on:** NASDAQ. **Number of employees at this location:** 3,500.

AMDAHL CORPORATION
2111 East Highland Avenue, Suite 375, Phoenix AZ 85016. 602/955-0770. **Contact:** Human Resources Department. **Description:** This location is a customer service and sales office. Amdahl Corporation designs, manufactures, develops, markets, and services large-scale, high-performance, general purpose computer systems, including both hardware and software. Customers are primarily large corporations, government agencies, and large universities with high-volume data processing requirements. The company markets more than 470 different systems. Amdahl provides consulting and services along with applications development for a broad range of client/server environments. Amdahl is a leader in IT solutions. **Common positions include:** Accountant/Auditor; Budget Analyst; Electrical/Electronics Engineer; Financial Analyst; Management Analyst/Consultant; Management Trainee; Public Relations Specialist; Services Sales Representative; Software Engineer; Systems Analyst; Technical Writer/Editor. **Benefits:** 401(k); Daycare Assistance; Dental

Insurance; Disability Coverage; Employee Discounts; Life Insurance; Medical Insurance; Profit Sharing; Savings Plan; Tuition Assistance. **Corporate headquarters location:** Sunnyvale CA. **Listed on:** NASDAQ. **Number of employees nationwide:** 6,000.

AMERICABLE INC.
7450 Flying Cloud Drive, Eden Prairie MN 55344. 612/942-3800. **Contact:** Human Resources. **Description:** Supplies businesses with networking hardware and software.

AMERICAN MANAGEMENT SYSTEMS INC.
12601 Fair Lakes Circle, Fairfax VA 22033. 703/227-6000. **Contact:** Human Resources. **Description:** This location is engaged in aspects of operations. American Management Systems helps large organizations solve complex management problems by applying information technology and systems engineering. Industries and markets served include financial service institutions; insurance companies; federal agencies; state and local government; colleges and universities; telecommunications firms; health care providers; and energy companies. **NOTE:** Interested jobseekers should contact Human Resources at 4050 Legato Road, Fairfax VA 22033. **Common positions include:** Computer Engineer; Computer Programmer; Computer Scientist; Manager of Information Systems; Software Engineer. **Number of employees nationwide:** 4,250.

AMERICAN MANAGEMENT SYSTEMS INC.
4050 Legato Road, Fairfax VA 22033. 703/267-8000. **Contact:** Judy Blair, Vice President, Human Resources. **Description:** American Management Systems helps large organizations solve complex management problems by applying information technology and systems engineering. Industries and markets served include financial service institutions; insurance companies; federal agencies; state and local government; colleges and universities; telecommunications firms; health care providers; and energy companies. **Corporate headquarters location:** This Location. **Number of employees nationwide:** 4,250.

AMERICAN SYSTEMS CORPORATION (ASC)
14200 Park Meadow Drive, Chantilly VA 22021. 703/968-5006. **Toll-free phone:** 800/733-2721. **Fax:** 703/968-5151. **Contact:** John Sweeney, Recruiter. **Description:** Provides information technology services to the Department of Defense and other government agencies, the intelligence community, and commercial clients. **NOTE:** The company offers entry-level positions. **Common positions include:** Accountant/Auditor; Administrative Assistant; Computer Animator; Computer Programmer; Database Manager; Electrical/Electronics Engineer; Financial Analyst; Internet Services Manager; Mechanical Engineer; MIS Specialist; Network Administrator; Online Content Specialist; Software Engineer; Systems Analyst; Systems Engineer; Systems Manager; Telecommunications Manager. **Educational backgrounds include:** Accounting; Computer Science; Engineering; Physics. **Benefits:** 401(k); Dental Insurance; Disability Coverage; ESOP; Life Insurance; Medical Insurance; Savings Plan. **Special Programs:** Internships. **Other U.S. locations:** Arlington VA; Norfolk VA. **Subsidiaries include:** American Communications Corporation. **Listed on:** Privately held. **Annual sales/revenues:** $51 - $100 million. **Number of employees at this location:** 250. **Number of employees nationwide:** 800. **Number of projected hires for 1997 - 1998 at this location:** 150.

AMERITECH LIBRARY SERVICES
400 Dynix Drive, Provo UT 84604. 801/223-5200. **Fax:** 801/223-5202. **Contact:** Human Resources. **Description:** Ameritech Library Services installs automated computer networks at libraries.

ANSTEC, INC.
1410 Spring Hill Road, McLean VA 22102. 703/848-7200. **Contact:** Staffing Manager. **Description:** Anstec, Inc. is a provider of computer systems integration and design services. **Common positions include:** Computer Engineer; Computer Programmer; Customer Service Representative; Production Coordinator.

APCON, INC.
17938 SW Boones Ferry Road, Portland OR 97224. 503/639-6700. **Fax:** 503/639-6740. **Contact:** Human Resources. **E-mail address:** hr@apcon.com. **World Wide Web address:**

http://www.apcon.com. **Description:** A designer and manufacturer of network and SCSI connectivity products.

THE APEX GROUP
7151 Columbia Gateway Drive, Columbia MD 21046. 410/290-1606. **Contact:** Human Resources. **Description:** A firm offering computer network integration and systems services.

APPLIED SYSTEMS, INC.
200 Applied Parkway, University Park IL 60466. 708/534-5575. **Contact:** Director of Human Resources. **Description:** Provides computer systems integration and design services. **Common positions include:** Computer Programmer; Software Engineer; Systems Specialist.

APPRISE CORPORATION
310 Waterview Boulevard, Parsippany NJ 07054. 201/316-4000. **Contact:** Lynn Roesch, Human Resources. **Description:** This location of Apprise Corporation offers information systems services. **Number of employees nationwide:** 5,600.

ARTISOFT, INC.
2202 North Forbes Boulevard, Tucson AZ 85745. 520/670-7100. **Fax:** 520/670-7101. **Recorded Jobline:** 520/670-4201. **Contact:** Human Resources. **Description:** Artisoft, Inc. is one of the world's leading providers of networking solutions for small and growing businesses. The company offers the award-winning LANtastic network operating system as well as network management, backup, and multiplatform connectivity systems. **Common positions include:** Civil Engineer; Software Engineer. **Educational backgrounds include:** Computer Science; Engineering. **Benefits:** 401(k); Dental Insurance; Disability Coverage; Employee Discounts; Life Insurance; Medical Insurance; Savings Plan; Tuition Assistance. **Corporate headquarters location:** This Location. **Operations at this facility include:** Administration; Manufacturing; Research and Development; Sales; Service. **Listed on:** NASDAQ. **Number of employees at this location:** 400. **Number of employees nationwide:** 420.

ASANTE TECHNOLOGIES
821 Fox Lane, San Jose CA 95131. 408/435-8401. **Fax:** 408/894-9150. **Contact:** Human Resources Director. **E-mail address:** hr@asante.com. **World Wide Web address:** http://www.asante.com. **Description:** Designs, develops, and manufactures Ethernet and Fast Ethernet networking systems. **Common positions include:** Design Engineer; Electrical/Electronics Engineer; Manufacturing Engineer; Software Engineer. **Educational backgrounds include:** Computer Science; Engineering. **Benefits:** 401(k); Dental Insurance; Disability Coverage; Life Insurance; Medical Insurance; Telecommuting; Tuition Assistance. **Operations at this facility include:** Administration; Research and Development; Sales; Service. **Listed on:** NASDAQ. **Annual sales/revenues:** $51 - $100 million. **Number of employees at this location:** 170. **Number of employees nationwide:** 200.

ASTEA INTERNATIONAL INC.
455 Business Center Drive, Horsham PA 19044. 215/682-2500. **Contact:** Human Resources. **Description:** Develops, markets, and supports a variety of applications for client/server and host-based environments that permit organizations of various sizes across a wide range of industries to automate and integrate field service and customer support functions. Astea also offers a full range of consulting services, training, and customer support. **NOTE:** Entry-level positions are offered. **Common positions include:** Computer Programmer; Customer Service Representative; Software Engineer; Technical Writer/Editor. **Educational backgrounds include:** Business Administration; Computer Science; Mathematics. **Benefits:** 401(k); Dental Insurance; Disability Coverage; Life Insurance; Medical Insurance; Pension Plan; Profit Sharing. **Corporate headquarters location:** This Location. **International locations:** Australia; France; Germany; Israel; the Netherlands; New Zealand; United Kingdom. **Operations at this facility include:** Administration; Divisional Headquarters; Sales; Service. **Listed on:** NASDAQ. **Annual sales/revenues:** $21 - $50 million. **Number of employees at this location:** 200. **Number of employees worldwide:** 450.
Other U.S. locations:
- 100 South Ellsworth Avenue, 9th Floor, San Mateo CA 94401. 415/696-3260. This location is a customer service center.

- 55 Middlesex Turnpike, Bedford MA 01730. 617/275-5440. This location is a field service office.

ATLANTIC NETWORK SYSTEMS
8205 Brown Leigh, Raleigh NC 27612. 919/786-4388. **Contact:** Human Resources. **Description:** A manufacturer of computer networking products.

ATLIS SYSTEMS INC.
8728 Colesville Road, Suite 300, Silver Spring MD 20910. 301/578-4200. **Contact:** Human Resources. **Description:** Provides various computer-related services, including document coding and systems integration.

ATTACHMATE CORPORATION
1000 Alderman Drive, Alpharetta GA 30202. 770/442-4000. **Fax:** 770/442-4369. **Recorded Jobline:** 770/442-4010. **Contact:** Human Resources Department. **Description:** Attachmate Corporation designs, manufactures, and markets PC to mainframe data communications products worldwide. Products include IRMA, Crosstalk, Crossfax, Remote LAN node, Quickapp, DCA/Microsoft Select, and many others. **Common positions include:** Computer Programmer; Economist; Financial Analyst; Marketing Specialist; Production Manager; Secretary; Software Engineer; Technical Writer/Editor. **Educational backgrounds include:** Business Administration; Computer Science; Engineering; Finance; Marketing. **Benefits:** Dental Insurance; Disability Coverage; Incentive Plan; Life Insurance; Medical Insurance; Profit Sharing; Tuition Assistance. **Special Programs:** Internships. **Corporate headquarters location:** This Location. **Operations at this facility include:** Administration; Research and Development; Sales; Service. **Listed on:** Privately held. **Number of employees at this location:** 550. **Number of employees nationwide:** 1,000.
Other U.S. locations:
- 3617 131st Avenue SE, Bellevue WA 98006. 206/644-4010.

AUSPEX SYSTEMS, INC.
5200 Great America Parkway, Santa Clara CA 95054-1139. 408/492-0900. **Contact:** Leesa Gidaro, Senior Recruiter. **E-mail address:** employment@auspex.com. **World Wide Web address:** http://www.auspex.com. **Description:** Auspex Systems, Inc. is a leading provider of high-performance network data servers that are optimized to move large amounts of data quickly and reliably from central information repositories to users' workstations. The Auspex network data server product line offers a cost-effective solution for client/server environments that require high reliability, a scalable architecture, fast response times, and large storage capacities up to 900 gigabytes per server. Founded in 1987. **Common positions include:** Budget Analyst; Credit Manager; Manufacturing Engineer; Marketing Specialist; Mechanical Engineer; Sales Representative; Software Engineer. **Educational backgrounds include:** Computer Science. **Corporate headquarters location:** This Location. **Operations at this facility include:** Administration; Manufacturing; Research and Development; Sales; Service. **Listed on:** NASDAQ. **Stock exchange symbol:** ASPX. **Annual sales/revenues:** More than $100 million. **Number of employees worldwide:** 500.

G. AUSTIN COMPUTER NETWORKING
2040 Loop 336 West, Suite 125, Conroe TX 77304. 409/539-6361. **Contact:** Human Resources. **Description:** A provider of networking services.

AUTOMATED SYSTEMS DESIGN (ASD)
645 Hembree Parkway, Suite D, Roswell GA 30076. 770/740-2300. **Contact:** Human Resources. **Description:** Installs cabling for computer networks.

AXON NETWORKS, INC.
199 Wells Avenue, Newton MA 02159. 617/630-9600. **Fax:** 617/630-9604. **Contact:** Human Resources Department. **Description:** AXON Networks manages LANs and WANs through the development and manufacture of products that integrate hardware and software. **Corporate headquarters location:** This Location. **Parent company:** 3Com Corporation (Santa Clara CA).

BDM FEDERAL, INC.

1501 BDM Way, McLean VA 22102. 703/848-5000. **Fax:** 703/848-5006. **Contact:** Sue Krieger, Manager Of Corporate Employment. **E-mail address:** hrmclean@bdm.com. **World Wide Web address:** http://www.bdm.com. **Description:** BDM International, Inc. is a global provider of information technology and other technology solutions, systems, and support for governmental and commercial clients. The largest BDM company, BDM Federal, Inc., provides advanced information technology and other technology services and systems to the federal government. BDM Technologies, Inc. brings highly-focused information technology expertise and solutions to industry, commercial enterprises, and state and local governments. The companies of BDM operate in the U.S., Europe, the Middle East, and many other locations worldwide. **NOTE:** The company offers entry-level positions. **Common positions include:** Accountant/Auditor; Administrative Assistant; Applications Engineer; Biomedical Engineer; Budget Analyst; Chemical Engineer; Chemist; Computer Operator; Computer Programmer; Cost Estimator; Database Manager; Design Engineer; Environmental Engineer; Financial Analyst; Geologist/Geophysicist; Graphic Artist; Internet Services Manager; Management Analyst/Consultant; Operations/Production Manager; Secretary; Software Engineer; Statistician; Systems Analyst; Technical Writer/Editor; Telecommunications Manager; Typist/Word Processor. **Educational backgrounds include:** Business Administration; Communications; Computer Science; Engineering; Mathematics; Physics. **Benefits:** 401(k); Daycare Assistance; Dental Insurance; Disability Coverage; Employee Discounts; Life Insurance; Medical Insurance; Savings Plan; Stock Option; Tuition Assistance. **Special Programs:** Internships. **Corporate headquarters location:** This Location. **Operations at this facility include:** Administration; Divisional Headquarters; Regional Headquarters; Research and Development; Service. **Listed on:** NASDAQ. **Annual sales/revenues:** More than $100 million. **Number of employees at this location:** 300. **Number of employees worldwide:** 8,000. **Number of projected hires for 1997 - 1998 at this location:** 600.
Other U.S. locations:
- 1250 Academy Park Loop, Suite 220, Colorado Springs CO 80910. 719/380-8959. Contact: Robert Hohlstein, Manager.

BDM INTERNATIONAL INC.

950 Explorer Boulevard, Huntsville AL 35806. 205/922-5000. **Contact:** Human Resources. **Description:** Provides computer systems support, develops software, and aids the government in computer simulations. BDM International is a diversified, growth-oriented, professional, and technical services firm whose operating subsidiaries provide research and development and contract support to clients in the following industries: national defense, communications, energy, transportation, environmental, and other areas. More than 85 percent of BDM's revenue and earnings are derived from tests, experiments, designs, analyses, research, and other services intended to strengthen the defense of the United States and allied nations. BDM International, Inc. is a global provider of information technology and other technology solutions, systems, and support for governmental and commercial clients. In addition to BDM Federal, BDM International is also the parent of BDM Technologies, Inc., which brings highly-focused information technology expertise and solutions to industry, commercial enterprises, and state and local governments. The companies of BDM operate in the U.S., Europe, and the Middle East.

BFG SYSTEMS

3502 Lake Linda Drive, Suite 420, Orlando FL 32817. 407/382-3499. **Contact:** Human Resources. **Description:** Offers systems integration solutions. BFG Systems supplies *Fortune* 500 companies with networking and systems integration hardware and software.

BTG, INC.

1945 Old Gallows Road, Vienna VA 22182. 703/556-6518. **Fax:** 703/761-3245. **Contact:** Karen Wall, Director of Human Resources. **World Wide Web address:** http://www.btg.com. **Description:** An information technology company providing complete solutions to specific system and product needs of the United States government, its agencies and departments, and other customers, combining systems development, integration, and engineering with the value-added reselling of hardware, software, and integrated systems. Founded in 1982. **Common positions include:** Software Engineer. **Corporate headquarters location:** This Location. **Other U.S. locations:** San Diego CA; Annapolis Junction MD; San Antonio TX. **Subsidiaries include:** BDS, Inc. (a value-added reseller of computer software and hardware and a provider of integrated information systems) located in McLean VA and Brussels, Belgium; Advanced Computer Technology, Inc. (a value-added reseller of software and hardware and a seller of high-performance computers) with two locations in Arlington VA; and Delta Research Corporation (a software developer and systems integrator focusing on project planning, cost controls, and environmental engineering, primarily for the government) which has facilities in Arlington VA, Niceville FL, Colorado Springs CO, and Santa Maria CA. **Listed on:** NASDAQ. **Number of employees worldwide:** 650.

BANYAN SYSTEMS INC.
120 Flanders Road, Westborough MA 01581. 508/898-1000. **Contact:** Human Resources. **Description:** Founded in 1983, Banyan is a software manufacturer, specializing in networking software products, including the VINES network operating system. The company's products are internationally marketed to all business segments. **Corporate headquarters location:** This Location. **Number of employees nationwide:** 670.

BAY NETWORKS COMMUNICATIONS
P.O. Box 58185, Santa Clara CA 95052. 408/988-2400. **Contact:** Human Resources. **Description:** Bay Networks designs, manufactures, markets, and supports diverse, high-performance local area network (LAN) systems. Its systems distribute management and processing power through intelligent hubs to provide efficient, reliable delivery of data across LANs. **Corporate headquarters location:** This Location. **Listed on:** NASDAQ. **Stock exchange symbol:** BNET. **Annual sales/revenues:** $400 million. **Number of employees nationwide:** 1,250.
Other U.S. locations:
- Eight Federal Street, Billerica MA 01821-3559. 508/670-8888.

BAY STATE COMPUTER GROUP
52 Roland Street, Boston MA 02129. 617/623-3100. **Toll-free phone:** 800/831-1113. **Fax:** 617/666-5966. **Contact:** Human Resources. **World Wide Web address:** http://www.bscg.com. **Description:** A reseller and systems integrator. Bay State's products include UNIX, NT, and Windows and Netware operating systems. Services include LAN and WAN development and implementation, client/server network design, and database management tools. **Common positions include:** Account Manager; Account Representative; Accountant/Auditor; Administrative Assistant; Buyer; Controller; Credit Manager; Customer Service Representative; Human Resources Manager; Internet Services Manager; Operations/Production Manager; Project Manager; Purchasing Agent/Manager; Sales Engineer; Sales Representative; Systems Analyst. **Educational backgrounds include:** Accounting; Business Administration; Computer Science; Engineering; Finance; Liberal Arts; Marketing. **Benefits:** 401(k); Dental Insurance; Disability Coverage; Employee Discounts; Life Insurance; Medical Insurance; Tuition Assistance. **Corporate headquarters location:** This Location. **Other U.S. locations:** Edison NJ. **Parent company:** U.S. Office Products. **Listed on:** NASDAQ. **Stock exchange symbol:** OFIS. **Annual sales/revenues:** More than $100 million. **Number of employees at this location:** 100. **Number of employees nationwide:** 15,000. **Number of projected hires for 1997 - 1998 at this location:** 50.

BETA TECH INC.
P.O. Box 426, Roseville MI 48066. 810/772-4612. **Contact:** Human Resources. **Description:** Offers technical support for computer networking systems.

BLACK BOX CORPORATION
P.O. Box 12800, Pittsburgh PA 15241. 412/746-5500. **Contact:** Human Resources. **Description:** A manufacturer of communications, networking, and other computer products.

BROCK INTERNATIONAL, INC.
2859 Paces Ferry Road, Suite 1000, Atlanta GA 30342. 770/431-1200. **Fax:** 770/431-1201. **Recorded Jobline:** 770/431-1238. **Contact:** Corporate Recruiter. **World Wide Web address:** http://www.broc.com. **Description:** Brock International, Inc., founded in 1984, develops, markets, and supports software systems that automate the integrated sales, marketing, and customer service functions of business organizations in a wide range of industries. These sales performance solutions provide end-users with a closed-loop business solution. Brock has operations in 26 countries with over 1,000 clients and 40,000 worldwide users as well as key strategic business partnerships with

Microsoft, Oracle, Sybase, Informix, Hewlett-Packard, IBM, Sun MicroSystems, DEC, and XcelleNet. **Common positions include:** Computer Programmer; MIS Specialist; Software Engineer; Systems Analyst; Technical Writer/Editor. **Educational backgrounds include:** Computer Science. **Benefits:** 401(k); Dental Insurance; Disability Coverage; Life Insurance; Medical Insurance; Profit Sharing. **Corporate headquarters location:** This Location. **Operations at this facility include:** Administration; Divisional Headquarters; Regional Headquarters; Research and Development; Sales; Service. **Listed on:** NASDAQ. **Stock exchange symbol:** BROC. **Annual sales/revenues:** $21 - $50 million. **Number of employees at this location:** 230.

BRODART
BRODART AUTOMATION
500 Arch Street, Williamsport PA 17705. 717/326-2461. **Contact:** Human Resources. **Description:** Brodart fulfills the supply needs of libraries. Brodart Automation, a division of Brodart, provides libraries with computers and software needed in automating its services. The company manufactures, sells, and distributes its own products.

BROOKTROUT TECHNOLOGY, INC.
410 First Avenue, Needham MA 02194-2317. 617/449-4100. **Fax:** 617/449-9009. **Contact:** Human Resources. **Description:** Brooktrout Technology, Inc. designs, manufactures, and markets software, hardware, and systems solutions for electronic messaging applications in telecommunications and networking environments worldwide. These products help direct the flow of electronic information efficiently, and address the growing use of faxes in business communications. Products are sold through a direct sales force to original equipment manufacturers, value-added resellers, and systems integrators. Brooktrout currently holds two U.S. patents, one for fax-on-demand document retrieval technology, and the other for the use of Direct Inward Dialing telephone service with computer-based facsimile messaging systems. **Benefits:** Stock Option. **Corporate headquarters location:** This Location. **Subsidiaries include:** Brooktrout Networks Group, Inc. (Texas); Brooktrout Securities Corporation; Brooktrout Technology Europe, Ltd. **Listed on:** NASDAQ. **Number of employees nationwide:** 60.

BULL HN WORLDWIDE INFORMATION SYSTEMS
300 Concord Road, Billerica MA 01821-4199. 508/294-6000. **Contact:** Gwen Bates, Director of Human Resources. **Description:** Bull HN Information Systems, Inc. is the U.S.-based component of Groupe Bull, a French information provider. Bull is one of the largest information systems companies with a presence in more than 100 countries. A major systems and technologies integrator with a comprehensive range of solutions, services, and support capabilities, Bull offers an evolutionary strategy, the Distributed Computing Model, which allows users to integrate multivendor systems in a flexible, open environment. **Common positions include:** Accountant/Auditor; Computer Programmer; Data Processor; Financial Analyst; Marketing Specialist; Sales Executive. **Corporate headquarters location:** This Location. **Parent company:** Bull Data Systems, Inc.
Other U.S. locations:
- 13430 North Black Canyon Highway, Phoenix AZ 85029. 602/862-8000.
- 1001 Pawtucket Boulevard, Lowell MA 01854. 508/275-3400. This location is a manufacturing facility.

BYTEX CORPORATION
Four Technology Drive, Westborough MA 01581. 508/366-8000. **Contact:** Human Resources Manager. **Description:** A datacommunications and internetworking company providing manufacturing, sales, and service of an intelligent switching hub used in both local and wide area computer networks. **Common positions include:** Electrical/Electronics Engineer; Software Engineer; Technical Writer/Editor. **Educational backgrounds include:** Computer Science; Engineering; Marketing. **Benefits:** 401(k); Dental Insurance; Disability Coverage; Life Insurance; Medical Insurance; Tuition Assistance. **Corporate headquarters location:** Minneapolis MN. **Parent company:** Network Systems Corporation. **Operations at this facility include:** Engineering and Design; Manufacturing; Marketing; Service. **Listed on:** NASDAQ. **Number of employees at this location:** 200. **Number of employees nationwide:** 1,300.

CD CONNECTION
6550A Best Friend Road, Norcross GA 30092. 770/446-1332. **Contact:** Larry Gee, Product Manager. **Description:** Manufactures CD-ROM networking hardware and software.

CGI SYSTEMS, INC.
AN IBM COMPANY
300 Berwyn Park, Suite 100, Berwyn PA 19312. 610/695-8100. **Fax:** 610/695-8585. **Contact:** Jim Lauckner, Manager/Recruiter. **Description:** A software services company. CGI Systems is a worldwide organization spanning 12 countries. The company was acquired by IBM in June 1993. Services include full life cycle systems development, repository-based development with PACBASE (an SAA and A/D cycle supported CASE product), LAN support, and additional on-site support services and training in emerging technologies. **NOTE:** Consulting positions are available for individuals with at least two years application development experience in IBM Mainframe, IBM/DEC Minicomputer or IBM Compatible Microcomputer environments, or LAN/WAN Consultants in Novell, Banyon, or UNIX environments. **Common positions include:** Computer Programmer; Systems Analyst. **Benefits:** 401(k); Dental Insurance; Disability Coverage; Employee Discounts; Health Club Discount; Life Insurance; Medical Insurance; Profit Sharing; Savings Plan; Tuition Assistance. **Corporate headquarters location:** Malvern PA. **Other U.S. locations:** Washington DC; FL; Chicago IL; Boston MA; New York NY. **Parent company:** IBM Corporation. **Operations at this facility include:** Divisional Headquarters; Regional Headquarters. **Number of employees at this location:** 100. **Number of employees nationwide:** 450.

CABLETRON NETWORK EXPRESS
305 East Eisenhower Parkway, Ann Arbor MI 48108. 313/761-5005. **Contact:** Human Resources. **Description:** This location develops remote access products for networks. As a whole, Cabletron develops, manufactures, markets, installs, and supports a wide range of standards-based local area network (LAN) connectivity products, including network management software, high-speed adapter cards, smart hubs, and other network interconnection equipment. **Corporate headquarters location:** Rochester NH. **Listed on:** New York Stock Exchange.

CABLETRON SYSTEMS, INC.
P.O. Box 5005, Rochester NH 03866-5005. 603/332-9400. **Fax:** 603/332-8007. **Contact:** Recruiter. **Description:** Cabletron Systems, Inc. develops, manufactures, markets, installs, and supports a wide range of standards-based local area network (LAN) connectivity products, including network management software, high-speed adapter cards, smart hubs, and other network interconnection equipment. **NOTE:** The company is located at 35 Industrial Way, Rochester NH. **Common positions include:** Accountant/Auditor; Customer Service Representative; Electrical/Electronics Engineer; Mechanical Engineer; Software Engineer; Technical Writer/Editor. **Special Programs:** Internships. **Corporate headquarters location:** This Location. **Other U.S. locations:** Nationwide. **Operations at this facility include:** Administration; Sales. **Listed on:** New York Stock Exchange. **Number of employees at this location:** 2,000.

CAM DATA SYSTEMS, INC.
17520 Newhope Street, Fountain Valley CA 92708. 714/241-9241. **Contact:** Human Resources. **Description:** CAM Data Systems, Inc., founded in 1983, designs, develops, markets, and services inventory management, point-of-sale, order entry, and accounting systems for small- to medium-sized retailers and wholesalers. CAM's software generally operates on IBM and IBM-compatible computers in a variety of different operating environments. These currently include DOS, Novell, UNIX, PCMOS, and Citrix. CAM provides its products and services on a direct basis only. CAM also offers after-sale services. These services include phone support seven days a week, nationwide hardware and software service, database conversions, and regular program enhancements. CAM Data has installed nearly 4,000 systems in North America. **Corporate headquarters location:** This Location. **Number of employees at this location:** 116.

CAMCO ENTERPRISES
37 Highland Avenue, Washington PA 15301. 412/222-6001. **Contact:** Human Resources. **Description:** Manufactures and sells computer systems and provides networking services and Website design.

CEDALION EDUCATION

4020 Stirrup Creek Drive, Durham NC 27703. 919/361-1944. **Contact:** Human Resources. **Description:** Develops information technology systems and software.

CEDAR CLIFFS SYSTEMS CORPORATION

6601 Springfield Center Drive, Springfield VA 22150. 703/922-7717. **Contact:** Human Resources. **Description:** A computer systems integration firm.

THE CENTECH GROUP

4200 Wilson Boulevard, Suite 700, Arlington VA 22203. 703/525-6852. **Contact:** Human Resources. **Description:** A government contractor that provides computer network support.

CENTURY SYSTEMS

8033 University Boulevard, Suite C, Des Moines IA 50325. 515/223-8088. **Contact:** Human Resources. **Description:** A computer networking firm.

CHANEY SYSTEMS, INC.

5100 South Calhoun Road, New Berlin WI 53151. 414/679-3908. **Fax:** 414/679-3715. **Contact:** Penny Wells, Recruiter. **E-mail address:** wells@chaney.net. **World Wide Web address:** http://www.chaney.net. **Description:** A systems integration and consulting firm providing information technology solutions, design, implementation, support, and training to *Fortune* 500 companies. Founded in 1984. **NOTE:** The company offers training programs. **Common positions include:** Applications Engineer; Computer Programmer; Database Manager; Information Systems Consultant; Internet Services Manager; Management Analyst/Consultant; MIS Specialist; Network Engineer; Project Manager; Software Engineer; Systems Analyst; Systems Manager; Webmaster. **Educational backgrounds include:** Business Administration; Computer Science. **Benefits:** 401(k); Dental Insurance; Disability Coverage; Employee Discounts; Life Insurance; Medical Insurance. **Corporate headquarters location:** This Location. **Other U.S. locations:** Milwaukee WI; Madison WI. **Listed on:** Privately held. **Number of employees at this location:** 70. **Number of projected hires for 1997 - 1998 at this location:** 30 - 50.

CISCO SYSTEMS, INC.

170 West Tasman, San Jose CA 95134. 415/326-1941. **Contact:** Human Resources. **Description:** Cisco Systems, Inc. develops, manufactures, markets, and supports high-performance internetworking systems that enable customers to build large-scale integrated computer networks. Its products connect and manage communications among local and wide area networks that employ a variety of protocols, media interfaces, network topologies, and cable systems. **Number of employees at this location:** 882.

CITATION COMPUTER SYSTEMS, INC.

424 South Woods Mill Road, Suite 200, Chesterfield MO 63017. 314/579-7900. **Contact:** Ms. Pat Morris, Human Resources Director. **Description:** Citation Computer Systems, Inc. provides networked microcomputer-based integrated information systems for health care institutions. The company designs, develops, markets, and services cost-effective proprietary health care information systems for laboratories, financial/administrative departments, and order communications/results reporting areas. These systems are marketed to hospitals, group practices, clinics, reference laboratories, and nursing homes. **Common positions include:** Accountant/Auditor; Computer Programmer; Customer Service Representative; Payroll Clerk; Receptionist; Sales Representative; Secretary; Technical Representative. **Educational backgrounds include:** Computer Science; Marketing; Medical Technology. **Benefits:** 401(k); Dental Insurance; Disability Coverage; Life Insurance; Medical Insurance; Tuition Assistance. **Other U.S. locations:** Richardson TX; Madison WI. **Operations at this facility include:** Administration; Divisional Headquarters; Research and Development; Sales; Service. **Listed on:** NASDAQ. **Number of employees at this location:** 110. **Number of employees nationwide:** 185.

CODENOLL TECHNOLOGY CORPORATION

200 Corporate Boulevard South, Yonkers NY 10701. 914/965-6300. **Contact:** Joseph Ferraro, Human Resources. **Description:** Codenoll designs, develops, manufactures, and markets high-performance computer networking products primarily based on fiber-optic and related software,

and electronic semiconductor and opto-electronic technology. The company's two product technology areas are CodeNet fiber-optic Local Area Network hardware and software products for use with personal computers, workstations, minicomputers, mainframes, communications servers, and Gateways; and CodeLink point-to-point fiber-optic transmission equipment for commercial and military data communications and telecommunications, and component products including fiber-optic transmitters, receivers, LEDs, and detectors.

COLE-LAYER-TRUMBLE COMPANY
3199 Klepinger Road, Dayton OH 45406. 937/276-5261. **Contact:** Becky McCoy, Human Resources Department. **Description:** Cole-Layer-Trumble creates computer systems for state and local governments. **Common positions include:** Real Estate Appraiser. **Benefits:** Dental Insurance; Disability Coverage; Life Insurance; Medical Insurance; Pension Plan; Savings Plan; Tuition Assistance. **Corporate headquarters location:** This Location. **Operations at this facility include:** Administration; Regional Headquarters; Research and Development; Sales; Service. **Listed on:** Privately held. **Number of employees at this location:** 100. **Number of employees nationwide:** 600.

COLUMBIA SERVICES GROUP INC.
2751 Prosperity Avenue, Suite 600, Fairfax VA 22031. 703/528-8100. **Contact:** Michelle Anderson, Vice President of Human Resources. **Description:** Assists businesses with the operation of their computer systems.

COMPAQ DALLAS
8404 Esters Boulevard, Irving TX 75063. 972/929-1700. **Contact:** Human Resources. **Description:** Formerly known as Networth, Compaq Dallas manufactures Fast Ethernet networking hardware.

COMPATIBLE SYSTEMS
P.O. Box 17220, Boulder CO 80308-7220. 303/444-9532. **Contact:** Human Resources. **E-mail address:** jobs@compatible.com. **World Wide Web address:** http://www.compatible.com. **Description:** Compatible Systems designs and manufactures a line of interconnectivity products for Internet access, Wide Area Networks (WANs), and intranetworks.

COMPRO SYSTEMS, INC.
7798 Jessup Road, Route 175, Jessup MD 20794. 301/596-1770. **Contact:** Human Resources. **Description:** A systems integration firm.

COMPUCARE COMPANY
12110 Sunset Hills Road, Suite 600, Reston VA 20190-3224. 703/709-2300. **Fax:** 703/709-2490. **Contact:** Human Resources Specialist. **Description:** A developer of both hardware and software information systems for hospitals and other health care facilities. **Common positions include:** Computer Programmer; Software Engineer; Systems Analyst. **Educational backgrounds include:** Business Administration; Computer Science. **Benefits:** 401(k); Dental Insurance; Disability Coverage; Life Insurance; Medical Insurance; Tuition Assistance. **Corporate headquarters location:** This Location. **Other U.S. locations:** Fountain Valley CA; Bloomington MN; Plano TX. **Subsidiaries include:** Antrim Corporation; Health Systems Integration, Inc. **Operations at this facility include:** Administration; Research and Development; Sales; Service. **Listed on:** Privately held. **Number of employees at this location:** 150. **Number of employees nationwide:** 800.

COMPUCOM INC.
10100 North Central Expressway, Dallas TX 75231. 214/265-3600. **Contact:** Human Resources. **Description:** CompuCom is a leading PC services integration company providing product procurement, advanced configuration, network integration, and support services. CompuCom serves the needs of large businesses through more than 40 sales and service locations nationwide. **Common positions include:** Accountant/Auditor; Buyer; Computer Operator; Computer Programmer; Customer Service Representative; Software Engineer; Systems Analyst. **Educational backgrounds include:** Business Administration; Computer Science. **Corporate headquarters location:** This Location. **Operations at this facility include:** Administration; Divisional

Headquarters; Regional Headquarters; Sales; Service. **Number of employees at this location:** 800. **Number of employees nationwide:** 1,600.
Other U.S. locations:
* 100 Eagle Rock Avenue, East Hanover NJ 07936. 201/887-1000.

COMPUSERVE INC.

222 West Las Colinas Boulevard, Suite 1710, Irving TX 75039. 972/869-6300. **Contact:** Human Resources. **Description:** The Network Services Division of CompuServe provides domestic and international public data network services. Services include fixed-X.25 and fixed-QLLC asynchronous dial access, frame relay, point-of-sale access, IBM enhanced protocol services, and hybrid public/private network solutions. This location of CompuServe is strictly a sales office for the network services division. This division does not represent the CompuServe Information Service. There are approximately 25 other CompuServe branches across the U.S. **NOTE:** Send resumes to Human Resources, 5000 Arlington Center Boulevard, Columbus OH 43220. **Common positions include:** Account Manager; Manufacturer's/Wholesaler's Sales Rep. **Educational backgrounds include:** Business Administration; Communications; Computer Science; Marketing; MBA. **Benefits:** 401(k); Dental Insurance; Disability Coverage; Life Insurance; Medical Insurance; Profit Sharing; Tuition Assistance. **Corporate headquarters location:** Columbus OH. **Parent company:** H&R Block. **Listed on:** New York Stock Exchange.

COMPUTER & CONTROLS SOLUTIONS (CCSI)

1510 Stone Ridge Drive, Stone Mountain GA 30083. 770/491-1131. **Contact:** Human Resources. **Description:** A high-tech systems integration firm.

COMPUTER AID, INC.

1209 Hausman Road, Allentown PA 18104. 610/395-5120. **Contact:** Kathy Bartholomew, Human Resources. **Description:** Installs computer systems for businesses.

COMPUTER AID, INC.

2345 Rice Street, Suite 112, Roseville MN 55113. 612/490-5700. **Contact:** Bob Reinert, Recruiter. **Description:** Offers systems support on a contract basis.

COMPUTER BASED SYSTEMS INC.

2750 Prosperity Avenue, Suite 300, Fairfax VA 22031. 703/849-8080. **Contact:** Human Resources. **Description:** Offers networking and systems integration services to the government.

COMPUTER CABLE CONNECTION INC.

2810 Harland Drive, Bellevue NE 68005. 402/731-9528. **Contact:** Human Resources. **Description:** Computer Cable Connection installs office networking systems and provides various custom cabling services.

COMPUTER HORIZONS CORPORATION

747 Third Avenue, 15th Floor, New York NY 10017. 212/371-9600. **Fax:** 212/980-4676. **Contact:** Recruiting Department. **Description:** Computer Horizons is a full-service technology solutions company offering contract staffing, outsourcing, re-engineering, migration, and downsizing and network management. The company has a worldwide network of 33 offices. **Common positions include:** Computer Programmer; Management Analyst/Consultant; Software Engineer; Systems Analyst. **Educational backgrounds include:** Computer Science. **Benefits:** 401(k); Dental Insurance; Disability Coverage; Life Insurance; Medical Insurance; Tuition Assistance. **Corporate headquarters location:** Chicago IL. **Subsidiaries include:** Horizons Consulting, Inc.; Unified Systems Solutions, Inc.; Strategic Outsourcing Services, Inc.; Birla Horizons International. **Operations at this facility include:** Administration; Sales. **Listed on:** NASDAQ. **Annual sales/revenues:** More than $100 million. **Number of employees worldwide:** 3,000.

COMPUTER HORIZONS CORPORATION

49 Old Bloomfield Avenue, Mountain Lakes NJ 07046. 201/402-7400. **Contact:** Human Resources. **Description:** Computer Horizons is a full-service technology solutions company offering contract staffing, outsourcing, re-engineering, migration, and downsizing and network management. **Corporate headquarters location:** Chicago IL. **Other U.S. locations:** Nationwide.

Subsidiaries include: Horizons Consulting, Inc.; Unified Systems Solutions, Inc.; Strategic Outsourcing Services, Inc.; and Birla Horizons International Ltd. **Listed on:** NASDAQ. **Number of employees nationwide:** 1,500. **Number of employees worldwide:** 3,000.

COMPUTER HORIZONS CORPORATION
6400 Shafer Court, Rosemont IL 60018. 847/698-6800. **Toll-free phone:** 800/877-2421. **Fax:** 847/698-6823. **Contact:** Staffing Manager. **Description:** Computer Horizons is a full-service technology solutions company offering contract staffing, outsourcing, re-engineering, migration and downsizing, and network management. Founded in 1969. **Common positions include:** Computer Programmer; MIS Specialist; Software Engineer; Systems Analyst. **Corporate headquarters location:** Chicago IL. **Listed on:** NASDAQ. **Annual sales/revenues:** More than $100 million. **Number of employees at this location:** 225. **Number of employees worldwide:** 3,000. **Number of projected hires for 1997 - 1998 at this location:** 500.

COMPUTER HORIZONS CORPORATION
6450 Rockside Woods Boulevard South, Suite 270, Independence OH 44131. 216/524-8816. **Fax:** 216/524-9015. **Contact:** Recruiting. **Description:** Computer Horizons is a full-service technology solutions company offering contract staffing, outsourcing, re-engineering, migration and downsizing, and network management. Founded in 1969.

COMPUTER IDENTICS ID MATRIX
Five Shawmut Road, Canton MA 02021. 617/821-0830. **Fax:** 617/828-8942. **Contact:** Patricia Green, Human Resources. **Description:** Computer Identics ID Matrix provides data collection integrators with complete solutions, including scanning components, networking, software tools, and support services. Customers are primarily involved in materials handling and factory automation environments, and include systems integrators, original equipment manufacturers, value-added resellers, and end users performing in-house systems integration. Products and related services fall into four major categories: omnidirectional scanning systems, intelligent fixed position line scanners, data collection terminals, and networking products. Computer Identics ID Matrix's foreign subsidiaries are located in Belgium, Canada, England, France, and Germany. **Benefits:** 401(k). **Corporate headquarters location:** This Location.
Other U.S. locations:
- 1641 North First Street, Suite 225, San Jose CA 95112. 408/437-2460. This location is a sales office.
- 1080 Holcomb Bridge Road, Roswell GA 30076. 770/640-0737. This location is a field office.
- 1821 Walden Office Square, Suite 356, Schaumburg IL 60173. 847/397-7967. This location is a sales office.
- 6311 North O'Connor Road, Irving TX 75039. 972/869-7684. This location is a sales office.

COMPUTER METHODS CORPORATION
13740 Merriman Road, Livonia MI 48150. 313/522-5187. **Contact:** Human Resources. **World Wide Web address:** http://www.cmethods.com/. **Description:** Offers a variety of computer systems services. Computer Methods Corporation's staff includes software engineers with a knowledge of C, C++, and other programming languages; business consultants with skills in COBOL and client/server environments; and various individuals with Microsoft certification.

COMPUTER NETWORK TECHNOLOGY CORPORATION
6500 Wedgewood Road, Maple Grove MN 55311-3640. 612/550-8000. **Contact:** Tom Morris, Human Resources. **Description:** Computer Network Technology Corporation designs, manufactures, markets, and supports Channel Networking products that enable the high-speed transmission of information among local and geographically dispersed computing systems, primarily IBM and IBM-compatible, as well as related peripheral devices. **Common positions include:** Accountant/Auditor; Budget Analyst; Buyer; Electrical/Electronics Engineer; Financial Analyst; Software Engineer; Technical Writer/Editor. **Educational backgrounds include:** Business Administration; Computer Science; Liberal Arts. **Benefits:** 401(k); Daycare Assistance; Dental Insurance; Disability Coverage; Employee Discounts; Life Insurance; Medical Insurance; Profit Sharing; Tuition Assistance. **Corporate headquarters location:** This Location. **Operations**

at this facility include: Administration; Manufacturing; Research and Development. **Number of employees at this location:** 300. **Number of employees nationwide:** 400.

COMPUTER PROFESSIONALS, INC. (CPI)
River Hill Conference Center, 54 Marina Road, Lake Wylie SC 29710. 803/831-9111. **Contact:** Human Resources. **Description:** Computer Professionals is a provider of information systems, as well as consulting and integration services. **Number of employees at this location:** 350.

COMPUTER SCIENCES CORPORATION
2100 East Grand Avenue, El Segundo CA 90245. 310/615-0311. **Contact:** Scott Sharpe, Vice President of Corporate Personnel. **Description:** Computer Sciences Corporation primarily services the U.S. government. The four sectors of the company include the Systems Group division, the Consulting division, the Industry Services Group, and the CSC divisions. The Systems Group division designs, engineers, and integrates computer-based systems and communications systems, providing all the hardware, software, training, and related elements necessary to operate a system. The Consulting division includes consulting and technical services in the development of computer and communication systems to non-federal organizations. The Industry Services Group provides service to health care, insurance, and financial services, as well as providing large-scale claim processing and other insurance-related services. CSC Health Care markets business systems and services to the managed health care industry, clinics, and physicians. CSC Enterprises provides consumer credit reports and account management services to credit grantors. **Common positions include:** Business Analyst; Computer Scientist; Economist; Engineer; Geologist/Geophysicist; Mathematician; Operations Research Analyst; Physicist/Astronomer; Statistician. **Corporate headquarters location:** This Location. **Listed on:** New York Stock Exchange. **Number of employees nationwide:** 20,000.
Other U.S. locations:
- 711 Daily Drive, Camarillo CA 93010. 805/987-9641. This location is a government contractor.
- 4045 Hancock Street, San Diego CA 92110. 619/225-8401. **Contact:** Ms. Jo Jefferds, Manager of Human Resources. This location is engaged in research and development for the U.S. Navy.
- 1250 Academy Park Loop, Suite 240, Colorado Springs CO 80910. 719/596-7500. **Contact:** Patty Johnson, Human Resources. This location works under government contracts.
- 7459 Candlewood Road, Hanover MD 21076. 410/684-3500. **Contact:** Mary Ellen Huffman, Human Resources.
- 15245 Shady Grove Road, Rockville MD 20850. 301/921-3000.
- 3170 Fairview Park Drive, Falls Church VA 22042. 703/876-1000. **Contact:** Jack Farrell, Director. This location serves as the headquarters for the Federal Division.

COMSO, INC.
6303 Ivy Lane, Suite 300, Greenbelt MD 20770. 301/345-0046. **Contact:** Human Resources. **Description:** A computer systems integration firm.

COMWARE INTERNATIONAL
P.O. Box 410, Solvang CA 93464. 805/686-1262. **Contact:** David Retz, President. **Description:** Manufactures networking equipment for remote dial-in uses between computers.

CONCURRENT COMPUTER CORPORATION
2101 West Cypress Creek Road, Fort Lauderdale FL 33309. 954/973-5300. **Fax:** 954/973-5301. **Contact:** Human Resources. **Description:** Concurrent Computer Corporation provides secure computer systems, software, technical support, and other services to companies in various industries including academic, aerospace/defense, CAD engineering, and scientific. **Common positions include:** Software Engineer. **Educational backgrounds include:** Computer Science; Engineering; Marketing. **Special Programs:** Internships. **Operations at this facility include:** Administration; Divisional Headquarters; Manufacturing; Research and Development; Sales; Service. **Listed on:** American Stock Exchange. **Number of employees at this location:** 300. **Number of employees nationwide:** 425.

CONCURRENT COMPUTER CORPORATION
13555 SE 38th Street, Suite 260, Bellevue WA 98006. 206/641-9830. **Contact:** Human Resources. **Description:** This location is a sales office. Overall, Concurrent Computer Corporation provides secure computer systems, software, technical support, and other services to companies in various industries including academic, aerospace/defense, CAD engineering, and scientific.

CONCURRENT COMPUTER CORPORATION
3151 Airway Avenue, Costa Mesa CA 92626. 714/641-4741. **Contact:** Human Resources. **Description:** This location provides computer services. Overall, Concurrent Computer Corporation provides secure computer systems, software, technical support, and other services to companies in various industries including academic, aerospace/defense, CAD engineering, and scientific.

CONNECTRONIX CORPORATION
2121 South 3600 West, Salt Lake City UT 84119. 801/975-7477. **Contact:** Personnel. **Description:** Manufactures networking products.

CONTINENTAL RESOURCES, INC.
175 Middlesex Turnpike, Bedford MA 01730. 617/275-0850. **Toll-free phone:** 800/937-4688. **Fax:** 617/533-0212. **Contact:** Janet McPherson, Human Resources/Recruiter. **Description:** Continental Resources is a national organization with eight regional centers. The company configures, integrates, sells, services, and supports computer systems and electronic test equipment. The company offers rental and lease programs as well as PCs, networks, peripherals, software, and computer supplies. **NOTE:** The company offers entry-level positions. **Common positions include:** Account Representative; Administrative Assistant; Buyer; Computer Programmer; Customer Service Representative; Human Resources Manager; Sales Engineer; Sales Executive; Sales Representative; Software Engineer; Systems Analyst. **Educational backgrounds include:** Accounting; Business Administration; Computer Science; Marketing. **Benefits:** 401(k); Dental Insurance; Disability Coverage; Employee Discounts; Life Insurance; Medical Insurance; Savings Plan; Tuition Assistance. **Special Programs:** Internships. **Corporate headquarters location:** This Location. **Other U.S. locations:** Milpitas CA; Torrance CA; Orlando FL; Bensenville IL; Beltsville MD; Mount Laurel NJ; Somerset NJ; New York NY. **Subsidiaries include:** Wall Industries, a maker of AC/DC power sources and DC/DC converters; and Continental Leasing, a lease financing company. **Operations at this facility include:** Administration; Divisional Headquarters; Manufacturing; Sales; Service. **Annual sales/revenues:** More than $100 million. **Number of employees at this location:** 120. **Number of employees nationwide:** 275.

CONTROL DATA SYSTEMS, INC.
4201 Lexington Avenue North, Arden Hills MN 55126. 612/482-2100. **Fax:** 612/482-2791. **Contact:** Human Resources. **E-mail address:** info@cdc.com. **Description:** Control Data Systems provides a range of services that include design consulting, program management, application development, network integration, and life cycle support. Control Data Systems creates custom solutions depending on a client's needs. An independent company since 1992, Control Data is developing two network infrastructure solutions: message integration (designing and building messaging solutions for large corporations and government institutions that want widely dispersed users and applications to exchange information) and product data management (the company's consultants use the workflow, product configuration, and imaging capabilities of the Metaphase concurrent engineering software to help manufacturers track design and tooling data from initial sketches on a CAD terminal to the final checkout on the assembly line floor). **Common positions include:** Computer Programmer; Industrial Engineer; Services Sales Representative; Software Engineer; Systems Analyst. **Educational backgrounds include:** Computer Science; Engineering; Mathematics. **Benefits:** 401(k); Dental Insurance; Disability Coverage; Employee Discounts; Life Insurance; Medical Insurance; Profit Sharing; Savings Plan; Tuition Assistance. **Special Programs:** Internships. **Corporate headquarters location:** This Location. **Other U.S. locations:** Nationwide. **Operations at this facility include:** Administration; Divisional Headquarters; Regional Headquarters; Sales; Service. **Listed on:** NASDAQ. **Number of employees at this location:** 700. **Number of employees nationwide:** 2,800.

CONTROL DATA SYSTEMS, INC.

1306 Orleans Drive, Sunnyvale CA 94089. 408/541-4200. **Contact:** Human Resources. **E-mail address:** info@cdc.com. **Description:** Control Data Systems provides a range of services that include design consulting, program management, application development, network integration, and life cycle support. Control Data Systems creates custom solutions depending on a client's needs. An independent company since 1992, Control Data is developing two network infrastructure solutions: message integration (designing and building messaging solutions for large corporations and government institutions that want widely dispersed users and applications to exchange information) and product data management (the company's consultants use the workflow, product configuration, and imaging capabilities of the Metaphase concurrent engineering software to help manufacturers track design and tooling data from initial sketches on a CAD terminal to the final checkout on the assembly line floor). **NOTE:** Interested jobseekers should send resumes to 4201 Lexington Avenue North, Arden Hills MN 55126. **Common positions include:** Computer Programmer; Technical Writer/Editor. **Educational backgrounds include:** Computer Science; Mathematics. **Benefits:** Dental Insurance; Disability Coverage; Employee Discounts; Life Insurance; Medical Insurance; Pension Plan; Profit Sharing; Savings Plan; Tuition Assistance. **Special Programs:** Internships. **Corporate headquarters location:** Minneapolis MN. **Operations at this facility include:** Divisional Headquarters; Research and Development; Sales. **Listed on:** New York Stock Exchange.

CONTROL SYSTEMS INTERNATIONAL INC.

4210 Shawnee Mission Parkway, Suite 200A, Fairway KS 66205-2513. 913/432-4442. **Contact:** Human Resources. **Description:** Designs integrated automation systems.

COOK MANUFACTURING CORPORATION

P.O. Box 1737, Duncan OK 73534. 405/252-2885. **Contact:** Human Resources. **Description:** A manufacturer of computer systems to run CN and CNC machines.

CORDANT INC.

11400 Commerce Park Drive, Reston VA 20191. 703/758-7000. **Contact:** Ken Christians, Employment Manager. **Description:** Offers systems integration solutions.

CORE TECHNOLOGY CORPORATION

7435 Westshire Drive, Lansing MI 48917. 517/627-1521. **Fax:** 517/627-8944. **Contact:** Human Resources. **E-mail address:** admin@ctc-core.com. **World Wide Web address:** http://www.ctc-core.com. **Description:** A provider of connectivity solutions for the Unisys environment.

CREATIVE COMPUTER APPLICATIONS, INC.

26115-A Mureau Road, Calabasas CA 91302. 818/880-6700. **Contact:** Human Resources. **Description:** Creative Computer Applications, Inc. designs and manufactures computer-based clinical information systems and products that automate the acquisition and management of clinical data for the health care industry. The company sells its products and systems primarily to hospitals, clinics, reference laboratories, veterinaries, other health care institutions, and original equipment manufacturers. The company also generates revenue through service contracts with customers which provide technical support and repair services for specified periods of time. **Corporate headquarters location:** This Location.

CROSSCOM CORPORATION

450 Donald Lynch Boulevard, Marlborough MA 01752. 508/481-4060. **Fax:** 508/229-5597. **Contact:** Human Resources Manager. **Description:** Develops, manufactures, markets, and supports advanced internetworking products. **Benefits:** 401(k); Dental Insurance; Disability Coverage; Life Insurance; Medical Insurance. **Corporate headquarters location:** This Location. **Other U.S. locations:** CA; CO; FL; GA; IL; KS; NC; NY; PA; TX; VA; WA; WI. **Operations at this facility include:** Administration; Manufacturing; Research and Development; Sales; Service. **Listed on:** NASDAQ. **Number of employees at this location:** 165. **Number of employees nationwide:** 215.

CUBIX CORPORATION

2800 Lockheed Way, Carson City NV 89706. 702/888-1000. **Fax:** 702/888-1002. **Contact:** Joanne L. Lanigir, Personnel Manager. **World Wide Web address:** http://www.cubix.com. **Description:**

A designer and manufacturer of computer networking products. The company's products are optimized for use as mission-critical communication, specialty, and file servers. **Common positions include:** Customer Service Representative; Electrical/Electronics Engineer; Manufacturing Engineer; Mechanical Engineer; MIS Specialist; Sales Engineer; Sales Representative; Secretary; Software Engineer. **Educational backgrounds include:** Computer Science. **Benefits:** 401(k); Dental Insurance; Disability Coverage; Life Insurance; Medical Insurance; Tuition Assistance. **Corporate headquarters location:** This Location. **Other U.S. locations:** Tampa FL; Keego Harbor MI. **Operations at this facility include:** Manufacturing. **Annual sales/revenues:** $51 - $100 million. **Number of employees at this location:** 140. **Number of projected hires for 1997 - 1998 at this location:** 20.

CYBER DIGITAL, INC.
400 Oser Avenue, Suite 1650, Hauppauge NY 11788-3641. 516/231-1200. **Contact:** Personnel. **Description:** Cyber Digital designs, develops, manufactures, and markets digital switching and networking systems that enable simultaneous communication of voice and data to a large number of users. The company's systems are based on its proprietary software technology which permits the modemless transmission of data between a variety of incompatible and dissimilar end user equipment, including computers, printers, workstations, and data terminals, over standard telephone lines. **Common positions include:** Design Engineer; Electrical/Electronics Engineer; Purchasing Agent/Manager; Telecommunications Manager. **Educational backgrounds include:** Business Administration; Communications; Engineering; Marketing. **Benefits:** Disability Coverage; Medical Insurance. **Corporate headquarters location:** This Location. **Operations at this facility include:** Administration; Manufacturing; Research and Development; Sales. **Stock exchange symbol:** CYBD. **Annual sales/revenues:** Less than $5 million.

D-LINK SYSTEMS
5 Music, Irvine CA 92618. 714/455-1688. **Contact:** Human Resources. **Description:** A manufacturer of networking hardware.

DCE CORPORATION
5 Hillandale Avenue, Stamford CT 06902. 203/358-3940. **Contact:** Eric Sheperd, President. **Description:** Manufactures fax servers that automate faxing capabilities for multiple users simultaneously through their computers. DCE Corporation's system works with all types of networks and midrange computers. **Number of employees at this location:** 50.

DALLAS DIGITAL CORPORATION
624 Krona, Suite 160, Plano TX 75074. 972/424-2800. **Contact:** Human Resources. **Description:** Offers networking solutions. Dallas Digital provides businesses with networking computer systems hardware and software.

DATA COMMUNICATION FOR BUSINESS INC.
807 Pioneer Street, Champaign IL 61820. 217/352-3207. **Contact:** Human Resources. **Description:** A manufacturer of data communications equipment for wide area computer applications.

DATA GENERAL CORPORATION
4400 Computer Drive, Westborough MA 01580. 508/366-8911. **Contact:** Vice President. **Description:** Designs, manufactures, and markets general purpose computer systems and related products and services, including peripheral equipment, software services, training, and maintenance. Data General markets directly to end users and OEMs and offers six product families whose applications include industrial manufacturing for controlling discrete assembly line operations, monitoring continuous production processes, testing, production planning, inventory management, and environmental surveillance. The company's products are also used in business data systems. **Corporate headquarters location:** This Location.
Other U.S. locations:
- 8480 East Orchard Road, Suite 3300, Englewood CO 80111. 303/486-7600. This location is a sales office.
- 4170 Ashford Dunwoody Road, Suite 300, Atlanta GA 30319. 404/705-2500. This location is a sales office.

- Route 9, Cofflin Drive, Southborough MA 01772. 508/480-7000. This location is a manufacturing facility.
- 62 T.W. Alexander Drive, Research Triangle Park NC 27709. 919/248-5970. This location performs research and development, and provides some customer support.
- Park 80 West Plaza 1, Battle Brook NJ 07663. 201/587-8700. This location is a sales office.
- 7927 Jones Branch Drive, Suite 200, McLean VA 22102-3377. 703/827-9600. This location is a sales office.

DATA GENERAL CORPORATION
2603 Main Street, Suite 400, Irvine CA 92714-6232. 714/250-6006. **Fax:** 714/724-3507. **Contact:** Human Resources. **Description:** This facility is responsible for sales and support. Overall, Data General Corporation designs, manufactures, and markets general purpose computer systems and related products and services, including peripheral equipment, software services, training, and maintenance. The company markets directly to end-users and original equipment manufacturers. The company offers six product families whose applications include industrial manufacturing for controlling discrete assembly line operations, monitoring continuous production processes, testing, production planning, inventory management, and environmental surveillance. Products are also used in business data systems. **Common positions include:** Project Manager; Software Developer; Software Engineer. **Educational backgrounds include:** Business Administration; Computer Science. **Benefits:** Dental Insurance; Disability Coverage; Employee Discounts; Life Insurance; Medical Insurance; Pension Plan; Savings Plan; Stock Option; Tuition Assistance. **Corporate headquarters location:** Westborough MA. **Operations at this facility include:** Regional Headquarters; Sales; Service. **Listed on:** New York Stock Exchange. **Number of employees at this location:** 100.

DATA RESEARCH ASSOCIATES (DRA)
1276 North Warson Road, P.O. Box 8495, St. Louis MO 63132-1806. 314/432-1100. **Contact:** Human Resources. **Description:** Data Research Associates (DRA), founded in 1975, is an automation systems integrator for libraries and other information providers. DRA's products and services improve efficiency in traditional library operations and give libraries networking capabilities. DRA's automation systems and networking services are adaptable for use in libraries of various sizes. DRA's customer base includes more than 500 system installations representing over 1,700 individual libraries in the United States, Australia, Canada, Europe, the Pacific Rim, and South America.

DATA RETRIEVAL CORPORATION
11801 West Silver Spring Drive, Suite 200, Milwaukee WI 53225. 414/536-1960. **Contact:** Human Resources. **Description:** Provides networking solutions including hardware and consulting services.

DATATECH INDUSTRIES INC.
23 Madison Road, Fairfield NJ 07004. 201/808-4000. **Contact:** Human Resources. **Description:** Datatech Industries specializes in installing mainframes and networking hardware for businesses. This company will also help service workstations.

DAXUS CORPORATION
915 Penn Avenue, Pittsburgh PA 15222. 412/577-4100. **Contact:** Human Resources. **Description:** Provides computer systems integration services. **Corporate headquarters location:** This Location.

DELTA COMPUTEC, INC. (DCI)
900 Huyler Street, Teterboro NJ 07608. 201/440-8585. **Contact:** Human Resources. **Description:** Provides businesses with integrated networking solutions.

DIALOGIC CORPORATION
1515 Route 10, Parsippany NJ 07050. 201/993-3000. **Contact:** Steve Wentzell, Human Resources Director. **Description:** Dialogic is a leader in open computer technology positioned in two industries: computing and telecommunications. Dialogic's products are compatible and scalable across systems of all levels of size and complexity, from one port to several thousand, on a customer's premises or in a central office environment.

DIGI INTERNATIONAL INC.
11001 Bren Road East, Minnetonka MN 55343. 612/943-9020. **Contact:** Human Resources. **Description:** Provides data communications hardware and software that enable connectivity solutions for multi-user environments, remote access, and LAN connect markets. These products support most major microcomputer and workstation architectures and most popular single- and multi-user systems. Digi International provides cross-platform compatibility and software and technical support services. Digi International's products are produced under the DigiBoard, Arnet, Star Gate, and MiLAN trade names and marketed to a broad range of worldwide distributors, system integrators, value-added resellers (VARs), and OEMs. The company acquired MiLAN Technology in 1994 thus entering the LAN connectivity market. Products in this market reside directly on the LAN to improve its performance and increase its flexibility. **Corporate headquarters location:** This Location.

DIGICON CORPORATION
6903 Rockledge Drive, Suite 600, Bethesda MD 20817-1818. 301/564-6400. **Contact:** Human Resources. **Description:** A computer information systems integration firm.

DIGITAL LINK CORPORATION
217 Humboldt Court, Sunnyvale CA 94089. 408/745-6200. **Fax:** 408/745-6250. **Contact:** Human Resources. **Description:** Digital Link develops, manufactures, and markets high-speed digital access products for the WAN (wide area networks) marketplace. The company provides DSU/CSU (Digital Service Units/Channel Service Units), inverse multiplexer, SMDS (Switched Multimegabit Data Service), ATM, enhanced T1/E1, and Prelude access products. The multiplexer takes low-speed inputs and concentrates them onto one high-speed T1 line; the inverse multiplexer reverses the process. T1/E1 ports for digitized voice or video enable efficient line utilization. The company markets its products primarily through a direct sales force, value-added resellers, and original equipment manufacturers. In addition to six sales offices in the United States, the company has two international offices in the United Kingdom and Germany. **Corporate headquarters location:** This Location.

R.K. DIXON COMPANY
5111 Tremont Avenue, Suite C, Davenport IA 52807. 319/391-8200. **Contact:** Human Resources. **Description:** R.K. Dixon sells and services copiers, fax machines, and filing systems, and installs computer networks for offices.

DYNATECH COMMUNICATIONS INC.
12650 Darby Brook Court, Woodbridge VA 22192. 703/550-0011. **Contact:** Human Resources. **Description:** Manufactures local and wide area networking systems.

ECCS, INC.
One Sheila Drive, Eatontown NJ 07724. 908/747-6995. **Contact:** Human Resources. **Description:** ECCS designs and configures computer systems that include hardware, software, networking, and communications products. ECCS's mass storage enhancement products include RAID (Redundant Array of Independent Disks) products and technology; external disk, optical, and tape systems; internal disk and tape storage devices; and RAM. The company also provides related technical services. **Corporate headquarters location:** This Location.

ECI SYSTEMS & ENGINEERING
596 Lynnhaven Parkway, Virginia Beach VA 23452. 757/498-5000. **Contact:** Human Resources. **Description:** Builds computer systems under contract for the U.S. government.

ECI TELEMATICS INTERNATIONAL, INC.
1201 West Cyprus Creek Road, Fort Lauderdale FL 33309. 954/772-3070. **Contact:** Human Resources Department. **Description:** ECI Telematics International provides network applications for voice and data. **Common positions include:** Electrical/Electronics Engineer; Systems Analyst. **Educational backgrounds include:** Computer Science; Engineering. **Benefits:** 401(k); Dental Insurance; Disability Coverage; Life Insurance; Medical Insurance; Tuition Assistance. **Corporate headquarters location:** This Location. **Other U.S. locations:** Calabasas CA. **Operations at this**

facility include: Administration; Manufacturing; Research and Development. **Number of employees at this location:** 220.

EDS (ELECTRONIC DATA SYSTEMS CORPORATION)
5400 Legacy Drive, Plano TX 75024. 214/605-6000. **Contact:** Staffing Department. **Description:** Provides consulting, systems development, systems integration, and systems management for large-scale and industry-specific applications. **NOTE:** When corresponding with the Staffing Department, letters should be addressed to Staffing, Loc H4-GB-35. **Corporate headquarters location:** This Location. **Parent company:** General Motors Corporation. **Number of employees nationwide:** 75,000.
Other U.S. locations:
* 9201 North 25th Avenue, Suite 200, Phoenix AZ 85021. 602/997-7391.
* 180 Blue Ravine Road, Suite 1A, Folsom CA 95630. 916/351-8938.
* 6950 West Jefferson Avenue, Lakewood CO 80235. 303/763-3700. Contact: Steve Molitor, Director of Personnel.
* 248 Chapman Road, Newark DE 19702. 302/454-7622.
* 4800 Six Forks Road, Six Forks Center, Raleigh NC 27609. 919/783-8000.
* 13600 EDS Drive, Herndon VA 20171. 703/742-2000.

EDS (ELECTRONIC DATA SYSTEMS CORPORATION)
25 Hanover Road, Florham Park NJ 07932. 201/301-7502. **Contact:** Kathy Chan, Recruiter. **World Wide Web address:** http://www.eds.com. **Description:** EDS provides information technology services including system interaction development, management, and consulting. **Common positions include:** Account Manager; Administrative Assistant; Computer Operator; Computer Programmer; Database Manager; Human Resources Manager; MIS Specialist; Online Content Specialist; Project Manager; Software Engineer; Systems Analyst; Technical Writer/Editor; Telecommunications Manager. **Educational backgrounds include:** Computer Science. **Benefits:** 401(k); Dental Insurance; Disability Coverage; Employee Discounts; Life Insurance; Medical Insurance; Pension Plan; Savings Plan; Telecommuting. **Corporate headquarters location:** Plano TX. **Other U.S. locations:** Nationwide. **Listed on:** London Stock Exchange; New York Stock Exchange. **Number of employees worldwide:** 95,000.

EDS/UNIGRAPHICS
10824 Hope Street, Cypress CA 90630. 714/952-0311. **Contact:** Recruiter. **Description:** Provides integrated hardware, software, and network solutions to *Fortune* 500 companies. EDS/Unigraphics focuses primarily on international corporations in the service, wholesale, distribution, and transportation industries. **Common positions include:** Computer Programmer; Designer; Human Resources Manager; Mechanical Engineer; Software Engineer; Systems Analyst. **Educational backgrounds include:** Computer Science; Mathematics; Physics. **Benefits:** 401(k); Dental Insurance; Disability Coverage; Employee Discounts; Life Insurance; Medical Insurance; Pension Plan; Tuition Assistance. **Special Programs:** Internships. **Corporate headquarters location:** Plano TX. **Subsidiaries include:** EDS Electronics and Data Systems. **Parent company:** General Motors. **Operations at this facility include:** Regional Headquarters; Sales. **Listed on:** NASDAQ; New York Stock Exchange. **Number of employees at this location:** 550.

E.G. SOFTWARE, INC.
621 Southwest Morrison, Suite 1025, Portland OR 97205. 503/294-7025. **Fax:** 503/294-7130. **Contact:** Human Resources. **E-mail address:** resume@egsoftware.com. **World Wide Web address:** http://www.egsoftware.com. **Description:** e.g. Software is a developer and publisher of software products for the Internet and for Novell networks.

EMF
60 Foundry Street, Keene NH 03431. 603/352-8400. **Contact:** Human Resources. **Description:** Offers third party computer support and maintenance.

EMS (ELECTRONIC MAIL SYSTEMS)
P.O. Box 5233, Manchester NH 03108. 603/624-0585. **Contact:** Irvin Corliss, Director of Operations. **Description:** Creates networks that allow corporations to send and translate data to different company locations. **Parent company:** ITN (International Teleputer Network).

EDGE SYSTEMS, INC.
3010-A Woodcreek Drive, Downers Grove IL 60515. 630/810-9669. **Contact:** Human Resources. **Description:** A systems integrator.

ENTERPRISING SERVICE SOLUTIONS
P.O. Box 809022, Dallas TX 75380. 214/386-2000. **Contact:** Human Resources Department. **Description:** The company offers networking and systems integration services. Enterprising Service Solutions provides networking and installation services, help desk support, and integration design.

ENTEX INFORMATION SERVICES
6 International Drive, Rye Brook NY 10573. 914/935-3600. **Contact:** Human Resources. **Description:** Entex Information Services provides systems integration services. The company also resells software.

ENTEX INFORMATION SERVICES
2100 South York Road, Oak Brook IL 60521. 630/368-3300. **Contact:** Human Resources. **Description:** Entex Information Services provides systems integration services. The company also resells software.

ESSENTIAL COMMUNICATIONS, INC.
4374 Alexander Boulevard NE, Suite T, Albuquerque NM 87107. 505/344-0080. **Toll-free phone:** 800/278-7897. **Fax:** 505/344-0408. **Contact:** John Gibbon, Human Resources Department. **E-mail address:** jgibbon@esscom.com. **World Wide Web address:** http://www.esscom.com. **Description:** Essential Communications, Inc. is a provider of gigabit networking products for servers and workstations. **Corporate headquarters location:** This Location.

FTP SOFTWARE, INC.
100 Brickstone Square, Fifth Floor, Andover MA 01810. 508/685-4000. **Toll-free phone:** 800/282-4FTP. **Fax:** 508/659-6557. **Contact:** Human Resources Department. **Description:** FTP Software provides open systems products and services that help companies and institutions interconnect their personal computers with virtually every major computer and network.

FAIR, ISAAC AND COMPANY
120 North Redwood Drive, San Rafael CA 94903. 415/472-2211. **Contact:** Human Resources Department. **Description:** Fair, Isaac and Company develops data management systems and services for the consumer credit, personal lines insurance, and direct marketing industries. The company utilizes database enhancement software, predictive modeling, adaptive control, and systems automation to aid customers in decision-making. Established in 1956, Fair, Isaac pioneered the credit risk scoring technologies now employed by most major U.S. consumer credit grantors. Its rule-based decision management systems, originally developed to screen consumer credit applicants, are now routinely employed in all phases of the credit account cycle: direct mail solicitation (i.e. credit cards and lines of credit), application processing, card reissuance, online credit authorization, and collection. **Common positions include:** Account Representative; Actuary; Client/Server Specialist; Computer Graphics Specialist; Computer Programmer; Credit Manager; Customer Service Representative; Economist; Financial Analyst; Human Resources Manager; Management Analyst/Consultant; Marketing Specialist; Mathematician; Project Manager; Quality Control Supervisor; Recruiter; Software Engineer; Statistician; Systems Analyst; Technical Writer/Editor. **Educational backgrounds include:** Computer Science; Mathematics; Operations; Statistics. **Benefits:** 401(k); Dental Insurance; Disability Coverage; Life Insurance; Medical Insurance; Pension Plan; Profit Sharing; Savings Plan; Tuition Assistance. **Corporate headquarters location:** This Location. **Subsidiaries include:** Dynamark (Minneapolis MN). **Operations at this facility include:** Divisional Headquarters; Research and Development; Sales. **Listed on:** NASDAQ. **Number of employees at this location:** 500. **Number of employees nationwide:** 700.
Other U.S. locations:
* 10 Corporate Circle, Suite 330, New Castle DE 19720. 302/324-8015. This location is a sales office.

FILENET CORPORATION
3565 Harbor Boulevard, Costa Mesa CA 92626. 714/966-3400. **Contact:** Ms. Dana Flynn, Senior Human Resources Representative. **Description:** FileNet develops and markets document-imaging and work flow management solutions for companies in various industries. FileNet has subsidiaries in six European countries. **Common positions include:** Accountant/Auditor; Computer Programmer; Customer Service Representative; Financial Analyst; Human Resources Manager; Management Analyst/Consultant; Software Developer; Systems Analyst; Teacher/Professor; Technical Writer/Editor. **Educational backgrounds include:** Business Administration; Computer Science; Engineering; Marketing. **Benefits:** 401(k); Dental Insurance; Disability Coverage; Employee Discounts; ESOP; Life Insurance; Medical Insurance; Profit Sharing; Savings Plan; Tuition Assistance. **Special Programs:** Internships. **Corporate headquarters location:** This Location. **Operations at this facility include:** Administration; Manufacturing; Research and Development; Sales; Service. **Listed on:** NASDAQ. **Number of employees at this location:** 560. **Number of employees nationwide:** 1,000.
Other U.S. locations:
- 100 First Plaza, Suite 1600, San Francisco CA 94105-2636. 415/957-2700. This location is a sales office.

FORE SYSTEMS
174 Thorn Hill Road, Warrendale PA 15086. 412/772-6600. **Contact:** Human Resources. **World Wide Web address:** http://www.fore.com. **Description:** Manufactures LAN and WAN switches. Fore Systems also develops internetworking software and network management software.

FUTRON CORPORATION
7315 Wisconsin Avenue, Suite 900, Bethesda MD 20814. 301/907-7100. **Contact:** Human Resources. **Description:** Offers networking and other computer services to the government.

G.E. INFORMATION SERVICES
401 North Washington Street, Rockville MD 20850. 301/340-4000. **Contact:** Human Resources. **Description:** Supplies companies with computers, software, networking services, and other related business productivity solutions.

GRC INTERNATIONAL, INC.
1900 Gallows Road, Vienna VA 22182. 703/506-5000. **Fax:** 703/448-6890. **Contact:** Linda Thompson, Director of Human Resources. **Description:** An international provider of knowledge-based professional services and technology-based product solutions to government and commercial customers. GRC International's activities encompass sophisticated telecommunications products, network systems analysis, and network software development operations for the commercial market. The company creates large-scale decision-support systems and software engineering environments; applies operations research and mathematical modeling to business and management systems; and implements advanced database technology. GRC International also provides studies and analysis capabilities for policy development and planning; modeling and simulation of hardware and software used in real-time testing of sensor, weapon, and battlefield management command, control, and communication systems; and testing and evaluation. **Corporate headquarters location:** This Location. **Listed on:** New York Stock Exchange. **Number of employees worldwide:** 1,300.
Other U.S. locations:
- 300 North Continental Boulevard, Suite 550, El Segundo CA 90245. 310/640-1661.
- 2221 Camino del Rio South, Suite 305, San Diego CA 92108. 619/681-1890. This location specializes in computer systems engineering.
- 31255 Cedar Valley Drive, Suite 300, Westlake Village CA 91362. 818/991-3033. This location is engaged in technical research.
- 1980 North Atlantic Avenue, Suite 1030, Cocoa Beach FL 32931. 407/784-4030. This location houses the Precision Technologies Division.
- 7200 North 9th Avenue, Suite 1, Pensacola FL 32504. 904/477-6378.
- One 11th Avenue, Suite D-2, Shalimar FL 32579. 904/651-3373.
- 6100 Bandera Road, Suite 507, San Antonio TX 78238. 210/520-7878. This location is involved in technical research.
- 2121 Eisenhower Avenue, Suite 200, Alexandria VA 22314. 703/548-8168.

- Crystal Gateway Two Suite 200, 1225 Jefferson Davis Highway, Arlington VA 22202. 703/414-5000.
- 14120 Parke Long Court, Suite 101, Chantilly VA 20151. 703/803-1304. This location performs systems integration.

GRC INTERNATIONAL, INC.

5383 Hollister Avenue, Santa Barbara CA 93111. 805/964-7724. **Fax:** 805/964-9834. **Contact:** Kristi Waldron, Technical Recruiter. **E-mail address:** wjobs@grci.com. **World Wide Web address:** http://www.grci.com. **Description:** An international provider of knowledge-based professional services and technology-based product solutions to government and commercial customers. GRC International's activities encompass sophisticated telecommunications products, network systems analysis, and network software development operations for the commercial market. The company creates large-scale decision-support systems and software engineering environments; applies operations research and mathematical modeling to business and management systems; and implements advanced database technology. GRC International also provides studies and analysis capabilities for policy development and planning; modeling and simulation of hardware and software used in real-time testing of sensor, weapon, and battlefield management command, control, and communication systems; and testing and evaluation. **Common positions include:** Computer Programmer; Database Manager; Design Engineer; Marketing Manager; Marketing Specialist; Operations/Production Manager; Sales Engineer; Software Engineer; Systems Analyst; Systems Manager; Telecommunications Manager. **Educational backgrounds include:** Business Administration; Computer Science; Engineering; Mathematics; Physics. **Benefits:** 401(k); Dental Insurance; Disability Coverage; Employee Discounts; Medical Insurance; Pension Plan; Profit Sharing; Tuition Assistance. **Special Programs:** Internships. **Corporate headquarters location:** Vienna VA. **Other U.S. locations:** Nationwide. **Operations at this facility include:** Regional Headquarters. **Listed on:** New York Stock Exchange. **Stock exchange symbol:** GRH. **Number of employees at this location:** 120.

GTI CORPORATION

9715 Business Park Avenue, San Diego CA 92131. 619/537-2500. **Contact:** Human Resources. **Description:** GTI Corporation is a multinational manufacturer and marketer of networking and network access products. Through GTI-Promptus, it is an emerging technology leader in on-demand management for worldwide networking. Through GTI-Valor Electronics, it is a leading supplier of magnetics-based components for signal processing and power-transfer functions in networking products. **Corporate headquarters location:** This Location.

GENERAL DATACOMM, INC.

1579 Straits Turnpike, Middlebury CT 06762-1299. 203/574-1118. **Fax:** 203/598-7944. **Contact:** Louise Palowski, Manager of Corporate Human Resources. **E-mail address:** hr@gdc.com. **World Wide Web address:** http://www.gdc.com. **Description:** General DataComm is a provider of business solutions for enterprise and telecommunications networks based on Asynchronous Transfer Mode (ATM) products and services. The company operates in three areas: ATM products, Internetworking products, and network access products. The company designs, assembles, markets, installs, and maintains products and services that enable telecommunications common carriers, corporations, and governments to build, upgrade, and manage their global telecommunication networks. General DataComm's networks transmit information via telephone lines, microwave, satellites, fiber optics, and other media between computers and terminals or information-processing systems. Founded in 1969. **Common positions include:** Accountant/Auditor; Applications Engineer; Computer Programmer; Customer Service Representative; Human Resources Manager; Marketing Manager; Public Relations Specialist; Registered Nurse; Technical Writer/Editor. **Benefits:** Dental Insurance; Disability Coverage; Life Insurance; Medical Insurance; Pension Plan; Profit Sharing; Tuition Assistance. **Corporate headquarters location:** This Location. **Listed on:** New York Stock Exchange. **Stock exchange symbol:** GDC. **Annual sales/revenues:** $221.2 million. **Number of employees nationwide:** 1,800.
Other U.S. locations:
- 2135 South Cherry Street, Suite 250, Denver CO 80222. 303/782-3600. This location is a sales office.
- 6520 Powers Ferry Road, Suite 370, Atlanta GA 30339. 770/955-0682. This location is a regional sales office.

- 30800 Telegraph Road, Suite 1930, Bingham Farms MI 48025. 810/540-4110. Contact: Steve Doherty, District Sales Manager. This location serves as a regional sales and service office.
- 42 Broadway, Suite 1627, New York NY 10004. 212/248-7220. Contact: Jerry Iris, District Sales Manager. This location serves as a regional sales and service office.
- 6450 Rockside Woods Boulevard, Suite 100, Independence OH 44131. 216/642-1190.

GENERAL KINETICS INC.
14130-C Sullyfield Circle, Chantilly VA 22021-1615. 703/802-9300. **Fax:** 703/818-3706. **Contact:** Human Resources. **Description:** A leading designer, producer, and marketer of secure digital facsimile equipment and secure local area networks. General Kinetics sells its products to the U.S. and foreign governments, through its Secure Communications Division (formerly Cryptek, Inc.) which is also at this location. Through its Electronic Enclosure Division (Johnstown PA and Orlando FL), General Kinetics also designs and manufactures precision enclosures for electronic systems, principally for sale to the U.S. Department of Defense and U.S. Navy. Food Technology Corporation (also at this location), a subsidiary of the company, manufactures and distributes food testing equipment and machine vision equipment. **Benefits:** ESOP. **Corporate headquarters location:** This Location. **Listed on:** American Stock Exchange. **Number of employees nationwide:** 120.

GENICOM CORPORATION
14800 Conference Center Drive, Suite 400, Chantilly VA 20151. 703/802-9200. **Fax:** 703/802-9039. **Contact:** Human Resources. **Description:** An international supplier of network system, service, and printer solutions. Through its Harris Adacom Network Service Solutions business unit, GENICOM Corporation provides integrated network system solutions including a broad portfolio of hardware and software products, technology consulting, and diagnostic and monitoring services. The company's Enterprising Service Solutions business unit provides logo and multivendor product field support and depot repair, express parts, and professional services. GENICOM's Impact and Laser Printing Solutions business unit designs, manufactures, and markets a wide range of computer printer technologies for general purpose applications. **Common positions include:** Accountant/Auditor; Services Sales Representative. **Educational backgrounds include:** Business Administration; Computer Science; Marketing. **Benefits:** 401(k); Dental Insurance; Disability Coverage; Life Insurance; Medical Insurance. **Corporate headquarters location:** This Location. **Other U.S. locations:** Nationwide. **Subsidiaries include:** GENICOM Pty. Ltd. (Australia); GENICOM Canada, Inc. (Canada); GENICOM S.A. (France); GENICOM GmbH (Germany); GENICOM SpA (Italy); GENICOM Limited (UK); and Harris Adacom Inc. (Canada). **Operations at this facility include:** Administration; Divisional Headquarters; Sales. **Listed on:** NASDAQ. **Number of employees at this location:** 30. **Number of employees nationwide:** 2,150.

GLOBAL TURNKEY SYSTEMS, INC.
20 Waterview Boulevard, Third Floor, Parsippany NJ 07054. 201/331-1010. **Fax:** 201/331-0042. **Contact:** Vice President of Finance and Administration. **Description:** Offers midrange computer systems and software services for book and magazine publishers. **Common positions include:** Accountant/Auditor; Computer Programmer; Systems Analyst; Technical Writer/Editor. **Educational backgrounds include:** Accounting; Business Administration; Computer Science; Finance; Marketing. **Benefits:** 401(k); Dental Insurance; Disability Coverage; Life Insurance; Medical Insurance; Tuition Assistance. **Corporate headquarters location:** This Location. **Other U.S. locations:** Denver CO. **Operations at this facility include:** Sales; Service. **Listed on:** Privately held. **Number of employees at this location:** 60.

HBO & COMPANY
301 Perimeter Center North, Atlanta GA 30346. 770/393-6000. **Contact:** Christine Rumsey, Vice President of Human Resources. **World Wide Web address:** http://www.hboc.com. **Description:** HBO & Company (HBOC) is an information solutions company that provides information systems and technology to health care enterprises including hospitals, integrated delivery networks, and managed care organizations. HBOC's primary products are Pathways 2000, a family of client/server-based applications that allow the integration and uniting of health care providers; STAR, Series, and HealthQuest transaction systems; TRENDSTAR decision support system; and QUANTUM enterprise information system. The company also offers outsourcing services that include strategic information systems planning, data center operations, receivables management,

business office administration, and major system conversions. **Corporate headquarters location:** This Location. **Subsidiaries include:** HBO & Company Canada Ltd.; HBO & Company (UK) Limited. **Listed on:** NASDAQ. **Number of employees nationwide:** 1,900.
Other U.S. locations:

- 587 East State Road 434, Longwood FL 32750. 407/831-8444. This location offers sales and technical support.
- 303 Perimeter Center North, Atlanta GA 30346. 404/395-4200.
- 10101 Claude Freeman Drive, Charlotte NC 28262. 704/549-7000. This location offers sales and technical support.
- One Penn Center West, Suite 120, Pittsburgh PA 15276. 412/787-7780. This location is involved in technical support and sales.

HBO & COMPANY/CYCARE BUSINESS GROUP
700 Locust Street, Suite 500, P.O. Box 1278, Dubuque IA 52004-1278. 319/557-3360. **Fax:** 319/557-3951. **Recorded Jobline:** 319/557-3995. **Contact:** Tracy Gallery, Human Resources Coordinator. **E-mail address:** tgallery@dbq.cycare.com. **World Wide Web address:** http://www.cycare.com. **Description:** HBO & Company/CyCare Business Group is a leading provider of management information systems, related support services, and electronic data interchange services for medical group practices, faculty practice plans, and medical enterprises. **Common positions include:** Computer Programmer; Customer Service Representative; Systems Analyst; Technical Writer/Editor. **Educational backgrounds include:** Business Administration; Communications; Computer Science; English; Journalism; Liberal Arts. **Benefits:** 401(k); Dental Insurance; Disability Coverage; Life Insurance; Medical Insurance; Tuition Assistance. **Corporate headquarters location:** Scottsdale AZ. **Other U.S. locations:** San Diego CA; Atlanta GA; Chicago IL; Minneapolis MN; Bedminster NJ; Dallas TX. **Operations at this facility include:** Administration; Research and Development; Service. **Listed on:** New York Stock Exchange. **Number of employees at this location:** 325. **Number of employees nationwide:** 470.

JACK HENRY AND ASSOCIATES, INC.
663 West Highway 60, Monett MO 65708. 417/235-6652. **Fax:** 417/235-8406. **Contact:** Human Resources. **Description:** Jack Henry and Associates, Inc. provides integrated computer systems for in-house data processing to banks and other financial institutions. The company also provides data conversion, software installation, and software customization for the implementation of its systems, as well as customer maintenance and support services. **Number of employees nationwide:** 180.

HEWLETT-PACKARD COMPANY
COMMERCIAL SYSTEMS DIVISION
19111 Pruneridge Avenue, Cupertino CA 95014. 408/725-8900. **Contact:** Personnel Department. **Description:** This location produces business computer systems for interactive, online data processing, database management, and distributed data processing. Hewlett-Packard designs and manufactures measurement and computation products and systems used in business, industry, engineering, science, health care, and education. Principal products are integrated instrument and computer systems (including hardware and software); computer systems and peripheral products; and medical electronic equipment and systems. **NOTE:** Jobseekers should send resumes to: Employment Response Center, Event #2498, Hewlett-Packard Company, Mail Stop 20-APP, 3000 Hanover Street, Palo Alto CA 94304-1181. **Common positions include:** Computer Programmer; Electrical/Electronics Engineer; Financial Analyst; Marketing Specialist. **Educational backgrounds include:** Computer Science; Engineering; Finance; Marketing. **Benefits:** Credit Union; Dental Insurance; Disability Coverage; Employee Discounts; Life Insurance; Medical Insurance; Pension Plan; Profit Sharing; Savings Plan; Tuition Assistance. **Corporate headquarters location:** Palo Alto CA. **Operations at this facility include:** Administration; Divisional Headquarters; Manufacturing; Research and Development; Service. **Listed on:** American Stock Exchange; New York Stock Exchange; Tokyo Stock Exchange.

HEWLETT-PACKARD COMPANY
WORK STATION DIVISION
300 Apollo Drive, Chelmsford MA 01824. 508/256-6600. **Contact:** Human Resources. **Description:** This location manufactures computer systems. Overall, Hewlett-Packard is a designer and manufacturer of measurement and computation products and systems used in business,

industry, engineering, science, health care, and education. Principal products are integrated instrument and computer systems (including hardware and software); computer systems and peripheral products; and medical electronic equipment and systems. **NOTE:** Jobseekers should send resumes to: Employment Response Center, Event #2498, Hewlett-Packard Company, Mail Stop 20-APP, 3000 Hanover Street, Palo Alto CA 94304-1181.

I-NET, INC.
6700 Rockledge Drive, Suite 100, Bethesda MD 20817. 301/214-0900. **Contact:** Human Resources. **Description:** Provides network integration services.

IPC INFORMATION SYSTEMS
88 Pine Street, New York NY 10005. 212/825-9060. **Contact:** Jody Tracey, Manager of Human Resources. **Description:** A specialized, state-of-the-art telecommunications company, providing network communications solutions for the financial industry. Through its Information Transport Systems (ITS) business, IDC Information Systems provides its customers with voice, data, and video solutions through the design, integration, implementation, and support of local and wide area networks. ITS solutions incorporate the latest technology and are supported by a team of experienced, qualified systems engineers. In addition to its 12 domestic locations, the company supports its customers through a network of distributors in 24 foreign countries throughout Asia, Europe, and North America. **Corporate headquarters location:** This Location. **Number of employees nationwide:** 775.

IPC INFORMATION SYSTEMS
One Station Place, Third Floor South, Stamford CT 06902. 203/326-7000. **Fax:** 203/326-7086. **Contact:** Jody Tracey, Manager of Human Resources. **Description:** This location is involved in systems engineering. IPC Information Systems is a specialized, state-of-the-art telecommunications company, providing network communications solutions for the financial industry. Through its Information Transport Systems (ITS) business, the company provides its customers with voice, data, and video solutions through the design, integration, implementation, and support of local and wide area networks. ITS solutions incorporate the latest technology and are supported by a team of experienced, qualified systems engineers. In addition to its 12 domestic locations, the company supports its customers through a network of distributors in 24 foreign countries throughout Asia, Europe, and North America. **NOTE:** Interested jobseekers should address inquiries to Jody Tracey, Manager of Human Resources, 88 Pine Street, 14th & 15th Floors, Wall Street Plaza, New York NY 10005. **Common positions include:** Electrical/Electronics Engineer; General Manager; Software Engineer. **Benefits:** 401(k); Dental Insurance; Disability Coverage; Life Insurance; Medical Insurance; Profit Sharing; Tuition Assistance. **Corporate headquarters location:** New York NY. **Listed on:** NASDAQ. **Number of employees at this location:** 70. **Number of employees nationwide:** 775.

IDEA
Seven Oak Park, Bedford MA 01730. 617/275-2800. **Fax:** 617/533-0500. **Contact:** Human Resources. **Description:** Provides networking and communication solutions. **Common positions include:** Accountant/Auditor; Computer Programmer; Software Engineer; Systems Analyst; Technical Writer/Editor. **Educational backgrounds include:** Computer Science; Engineering. **Benefits:** 401(k); Disability Coverage; Life Insurance; Medical Insurance; Profit Sharing; Tuition Assistance. **Corporate headquarters location:** This Location. **Operations at this facility include:** Administration; Research and Development; Sales. **Number of employees at this location:** 150. **Number of employees nationwide:** 700.

IMATION CORPORATION
One Imation Place, Oakdale MN 55128-3414. 612/704-4000. **Toll-free phone:** 888/466-3456. **Fax:** 800/537-4675. **Contact:** Human Resources. **E-mail address:** info@imation.com. **World Wide Web address:** http://www.imation.com. **Description:** A global technology solutions company offering systems, products, and services for the handling, storage, transmission, and use of information. Businesses include data storage products, medical imaging and photo products, printing and publishing systems, and customer support technologies and document imaging. Imation is an independent spin-off of 3M's data storage and imaging businesses. **Corporate headquarters location:** This Location. **Listed on:** Chicago Stock Exchange; New York Stock

Exchange. **Stock exchange symbol:** IMN. **Annual sales/revenues:** $2.3 billion. **Number of employees worldwide:** 10,000.

INFORMATION MANAGEMENT CONSULTANTS, INC. (IMC)
7915 Westpark Drive, McLean VA 22102. 703/893-3100. **Fax:** 703/734-1751. **Contact:** Recruiter. **E-mail address:** personnel@imcinc.com. **World Wide Web address:** http://www.imcinc.com. **Description:** IMC provides a wide range of information systems planning, development, and implementation services to public and private sector clients in the United States and overseas. IMC has over 300 employees with in-depth experience in information systems planning, systems development, networking and telecommunications, and electronic imaging technology. **NOTE:** Entry-level positions and training programs are offered. **Common positions include:** Computer Programmer; Internet Services Manager; Management Analyst/Consultant; Stationary Engineer; Systems Analyst; Technical Writer/Editor. **Educational backgrounds include:** Business Administration; Computer Science; Engineering. **Benefits:** 401(k); Dental Insurance; Disability Coverage; Life Insurance; Medical Insurance. **Operations at this facility include:** Administration; Regional Headquarters; Sales. **Listed on:** Privately held. **Annual sales/revenues:** $21 - $50 million. **Number of employees at this location:** 300.

INFORMATION SYSTEMS & NETWORK CORPORATION
10411 Motor City Drive, Bethesda MD 20817. 301/469-0400. **Contact:** Human Resources. **Description:** A provider of computer services to the U.S. government.

INTEGRATED NETWORK SERVICES (INS)
1325 Northmeadow Parkway, Suite 110, Atlanta GA 30076-3896. 770/751-0350. **Contact:** Human Resources. **Description:** Integrated Network Services performs computer network installations for hospitals. **Corporate headquarters location:** This Location.

INTEGRATED SYSTEMS ANALYSTS, INC.
2800 Shirlington Road, Suite 1100, Arlington VA 22206. 703/824-0700. **Contact:** Human Resources. **Description:** A computer systems integration services firm.

INTERGRAPH CORPORATION
Mail Stop IW2000, Huntsville AL 35894-0003. 205/730-2000. **Contact:** Director of Personnel. **Description:** Intergraph develops, manufactures, markets, and maintains interactive computer graphics systems that support the creation, analysis, display, output, and maintenance of virtually every type of design, drawing, map, or other graphic representation. The company's hardware products include workstations, servers, and peripherals. Software products include operating systems, database management applications, and over 1,200 graphics software programs for the CADD/CAM, engineering, design, and manufacturing industries. Customers include companies in the utilities, transportation, building, process, vehicle design, electronics, manufacturing, and publishing industries. **Listed on:** NASDAQ. **Number of employees at this location:** 9,500.

INTERLINK COMPUTER SCIENCES
2443 Warrenville Road, Suite 600, Lisle IL 60532. 708/245-5151. **Contact:** Human Resources. **Description:** Provides interconnectivity solutions for IBM mainframes.

INTERNET COMMUNICATIONS CORPORATION
7100 East Belleview Avenue, Suite 201, Greenwood Village CO 80111. 303/770-7600. **Fax:** 303/770-2706. **Contact:** Human Resources. **Description:** Specializes in the design, support, and management of local area and wide area networks. Internet Communications Corporation also provides Internet access to corporate customers. The company's diverse client base includes health care and insurance companies, finance and securities firms, educational and government organizations, and retail and manufacturing concerns. Since the company's founding in 1986, Internet Communications has maintained an average annual growth rate of over 50 percent.

INTERPHASE CORPORATION
13800 Senlac, Dallas TX 75234. 214/654-5000. **Toll-free phone:** 800/777-3722. **Fax:** 214/654-5500. **Contact:** Sherri Rhodes, Human Resources Manager. **E-mail address:** resumes@iphase.com. **World Wide Web address:** http://www.iphase.com. **Description:**

Interphase Corporation is a developer, manufacturer, and marketer of networking and mass storage controllers, as well as stand alone networking devices for computer systems. Historically, the product line consisted of mass storage controllers. However, in recent years, development and marketing efforts have focused more heavily on networking products, which now comprise over 50 percent of the company's revenues. Many of the networking products are sold to original equipment manufacturers and to value-added resellers, systems integrators, and large end users. **Common positions include:** Design Engineer; Electrical/Electronics Engineer; Software Engineer. **Educational backgrounds include:** Computer Science; Engineering. **Benefits:** 401(k); Dental Insurance; Disability Coverage; Life Insurance; Medical Insurance; Pension Plan; Tuition Assistance; Vision Insurance. **Corporate headquarters location:** This Location. **Operations at this facility include:** Administration; Manufacturing; Research and Development; Sales. **Listed on:** NASDAQ. **Annual sales/revenues:** Less than $5 million. **Number of employees at this location:** 200.

JASTECH INC.
1375 East 9th, Cleveland OH 44114. 216/566-7600. **Fax:** 216/566-0165. **Contact:** Human Resources. **Description:** Services computer information systems. Jastech also offers consulting support in mainframe and client/server environments.

JOSTENS LEARNING CORPORATION
7878 North 16th Street, Suite 100, Phoenix AZ 85020. 602/678-7272. **Contact:** Ken Ross, Director of Compliance and Human Resources. **Description:** Designs computer systems. **NOTE:** Send resumes to Ken Ross, 9920 Pacific Heights Boulevard, Suite 500, San Diego CA 92121.

KASPIA SYSTEMS
8625 SW Cascade Avenue, Suite 602, Beaverton OR 97008. 503/644-1800. **Fax:** 503/520-0600. **Contact:** Human Resources. **E-mail address:** info@kaspia.com. **World Wide Web address:** http://www.kaspia.com. **Description:** A developer of fully-automated network monitoring and reporting systems.

KESTREL ASSOCIATES
5203 Leesburg Pike, Suite 800, Falls Church VA 22041. 703/379-2261. **Contact:** Human Resources. **Description:** A computer systems integration firm.

KLEINSCHMIDT INC.
450 Lake Cook Road, Deerfield IL 60015. 847/845-1000. **Contact:** Personnel. **Description:** Offers third party computer networking services.

LANSHARK SYSTEMS, INC.
784 Morrison Road, Columbus OH 43230. 614/751-1111. **Fax:** 614/751-1112. **Contact:** Scott Sharkey, President/CEO. **Description:** LANshark Systems, Inc. is a developer of LAN-based enterprise messaging software products. Products include mail and mail-enabled applications, network user applications, and network administration tools and utilities. LANshark is a leading third-party software developer in the Banyan VINES market and is diversifying its product line to include products for MAPI-1, OS/2, NT, Win'95, ENS for NT, HP/UX, Sun, and Macintosh. **Common positions include:** Computer Programmer; Customer Service Representative; Management Trainee; Services Sales Representative; Software Engineer; Systems Analyst. **Educational backgrounds include:** Business Administration; Computer Science; Engineering; Marketing. **Benefits:** 401(k); Casual Dress; Dental Insurance; Disability Coverage; Life Insurance; Medical Insurance; Pension Plan; Profit Sharing; Tuition Assistance. **Special Programs:** Internships. **Corporate headquarters location:** This Location. **Operations at this facility include:** Administration; Divisional Headquarters; Manufacturing; Regional Headquarters; Research and Development; Sales; Service. **Listed on:** Privately held. **Number of employees at this location:** 15. **Number of employees nationwide:** 20.

LANSYSTEMS
300 Park Avenue South, 15th Floor, New York NY 10010. 212/995-7700. **Contact:** Human Resources. **Description:** A systems integrator and software development company. LANSystems develops and markets a wide range of business-related software products.

LANTRONIX
15353 Barranca Parkway, Irvine CA 92618. 714/453-3990. **Fax:** 714/453-3995. **Contact:** Human Resources Department. **E-mail address:** sales@lantronix.com. **World Wide Web address:** http://www.lantronix.com. **Description:** Lantronix is a developer of network connectivity products. The company's product line includes Ethernet and Fast Ethernet for workgroup applications. **Corporate headquarters location:** This Location. **Annual sales/revenues:** $41 million.

LIANT SOFTWARE CORPORATION
959 Concord Street, Framingham MA 01701. 508/872-8700. **Fax:** 508/872-9102. **Contact:** Arnold Salvatore, Human Resources Department. **E-mail address:** abs@liant.com. **World Wide Web address:** http://www.liant.com. **Description:** Liant Software Corporation is a developer of network-based programming and software development tools that enhance client/server systems and architectures. **Educational backgrounds include:** Business Administration; Computer Science. **Corporate headquarters location:** This Location. **Listed on:** Privately held. **Number of employees at this location:** 100.

LOGIC WORKS
111 Campus Drive, Princeton NJ 08540. 609/514-1177. **Contact:** Human Resources. **Description:** A computer networking, client/server database firm.

LOGTEC
2900 Presidential Drive, Fairborn OH 45324. 937/429-2928. **Contact:** Personnel. **Description:** Offers computer systems support for the government. **NOTE:** Retired military professionals with a computer background have an excellent chance with this company.

MRJ INC.
10560 Arrowhead Drive, Fairfax VA 22030. 703/385-0700. **Contact:** Elaine White, Director of Human Resources. **Description:** A technology solutions government contractor.

MRK TECHNOLOGY LTD.
3 Summit Park Drive, Suite 300, Independence OH 44131. 216/520-4300. **Contact:** Human Resources. **Description:** A computer integration firm.

MTS-BOSTON INFORMATION SERVICES, INC.
62 Massachusetts Avenue, Lexington MA 02173. 617/863-1414. **Contact:** Manager, Client Services. **Description:** MTS-Boston is an information services company serving the high-technology industries through database technology. **Common positions include:** Computer Programmer; Electrical/Electronics Engineer; MIS Specialist; Network Engineer; Software Engineer; Systems Analyst; Technical Writer/Editor. **Educational backgrounds include:** Computer Science. **Corporate headquarters location:** This Location. **Other U.S. locations:** Washington DC. **Operations at this facility include:** Regional Headquarters. **Listed on:** Privately held.

MACRO COMPUTER PRODUCTS
2523 Products Court, Rochester Hills MI 48309. 810/853-5353. **Contact:** Human Resources. **Description:** A systems integrator offering both hardware and software solutions. Macro Computer Products assembles tape and disk storage subsystems for mainframe and open systems. The company also develops computer imaging systems.

MANAGEMENT INFORMATION SYSTEMS GROUP, INC.
P.O. Box 13966, Research Triangle Park NC 27709-3966. 919/549-8700. **Contact:** Human Resources Department. **Description:** Offers EDI services that help process business documents over a network.

McDATA CORPORATION
310 Interlocken Parkway, Broomfield CO 80021. 303/460-9200. **Contact:** Human Resources Department. **Description:** McData Corporation provides businesses with high-band networking solutions.

MEDIA LINK TECHNOLOGIES
18 West Mercer Road, Suite 300, Seattle WA 98119. 206/284-9770. **Contact:** Human Resources. **Description:** A multimedia networking company providing LAN services and communications management systems solutions.

MERCURY COMPUTER SYSTEMS
199 Riverneck Road, Chelmsford MA 01824. 508/256-1300. **Contact:** Human Resources. **Description:** A technology company specializing in high-performance computer solutions. Specifically, Mercury Computer Systems develops hardware and software for medical imaging and digital video editing.

MICRODYNE CORPORATION
3601 Eisenhower Avenue, Suite 300, Alexandria VA 22304. 703/739-0500. **Contact:** Heide Steiner, Director of Human Resources. **Description:** Designs, manufactures, markets, and supports a broad line of data communications hardware products that enable local area network and remote network access communications. Microdyne is a leading supplier of LAN adapter cards which provide the essential connection between computers -- including personal computers, file servers, minicomputers, and mainframe computers -- and the network. The company's products support all established computer bus structures, wiring systems, and network topologies, including Ethernet, Fast Ethernet, and Token Ring. Microdyne is divided into three businesses: networking products, which support inter-computer network communications; aerospace telemetry, which manufactures receivers used in a variety of applications involving missiles, aircraft, satellites, and other space vehicles; and manufacturer support services. **Corporate headquarters location:** This Location. **Other U.S. locations:** Carson CA; San Jose CA; Ocala FL; Indianapolis IN. **International locations:** Berkeshire, England; Munich, Germany. **Listed on:** NASDAQ. **Number of employees worldwide:** 470.

MICROS SYSTEMS, INC.
12000 Baltimore Avenue, Beltsville MD 20705. 301/210-8041. **Fax:** 301/210-3727. **Contact:** Lisa Goldman, Human Resources. **Description:** Micros Systems, Inc. develops point-of-sale systems and property management systems for the hospitality industry. **Common positions include:** Accountant/Auditor; Blue-Collar Worker Supervisor; Buyer; Clerical Supervisor; Computer Programmer; Customer Service Representative; Draftsperson; Software Engineer; Systems Analyst; Technical Writer/Editor. **Educational backgrounds include:** Computer Science; Engineering. **Benefits:** 401(k); Dental Insurance; Disability Coverage; Life Insurance; Medical Insurance; Tuition Assistance. **Other U.S. locations:** Nationwide. **Operations at this facility include:** Administration; Divisional Headquarters; Manufacturing; Research and Development; Sales; Service. **Listed on:** NASDAQ. **Number of employees at this location:** 450. **Number of employees nationwide:** 650.

MICROTEK SYSTEMS INC.
5343 North 118th Court, Milwaukee WI 53225. 414/466-2224. **Contact:** Human Resources. **Description:** An integrator of networking systems.

MICROTEST ENTERPRISE GROUP
22 Cotton Road, Nashua NH 03063. 603/880-0300. **Fax:** 603/880-7229. **Contact:** Human Resources. **World Wide Web address:** http://www.logicraft.com. **Description:** A developer of enterprise CD-ROM networking systems. **Corporate headquarters location:** This Location.

MNEMONIC SYSTEMS
2001 L Street NW, Suite 1000, Washington DC 20036. 202/785-4530. **Contact:** Human Resources. **Description:** A computer systems integration firm.

MOTOROLA COMPUTER GROUP
COMMERCIAL SYSTEMS DIVISION
1150 Kifer Road, Suite 100, Sunnyvale CA 94086. 408/749-0510. **Contact:** Human Resources. **Description:** Markets and services multi-function computer systems for business applications. **NOTE:** Please send resumes to Human Resources, 1438 West Broadway, Tempe AZ 85282. 602/303-6050. **Common positions include:** Computer Programmer; Software Engineer.

Educational backgrounds include: Accounting; Business Administration; Communications; Computer Science; Engineering; Finance; Marketing. **Benefits:** Dental Insurance; Disability Coverage; Employee Discounts; Life Insurance; Medical Insurance; Profit Sharing; Savings Plan; Tuition Assistance. **Corporate headquarters location:** Schaumburg IL. **Parent company:** Motorola, Inc. **Operations at this facility include:** Divisional Headquarters. **Listed on:** New York Stock Exchange.

MOTOROLA INFORMATION SYSTEMS GROUP
20 Cabot Boulevard, Mansfield MA 02048. 508/261-4725. **Fax:** 508/337-7172. **Contact:** Personnel Department. **Description:** A supplier of network products and integrated network solutions. The company operates in 52 countries through its 11 international subsidiaries and established distributor organization. Motorola Information Systems Group offers network management systems, T1/E1 switches, Frame Relay products, X.25 PADs and switches, multiplexers, DSU/CSUs, modems, LAN internetworking devices, and professional services. **Common positions include:** Electrical/Electronics Engineer; Marketing Manager; Services Sales Representative; Software Engineer; Technical Writer/Editor. **Educational backgrounds include:** Computer Science; Engineering; Marketing. **Benefits:** Dental Insurance; Disability Coverage; EAP; Employee Discounts; Life Insurance; Medical Insurance; Pension Plan; Profit Sharing; Savings Plan; Tuition Assistance. **Corporate headquarters location:** Schaumburg IL. **Parent company:** Motorola, Inc. **Operations at this facility include:** Administration; Divisional Headquarters; Manufacturing; Research and Development; Sales; Service. **Listed on:** New York Stock Exchange. **Number of employees at this location:** 2,400. **Number of employees nationwide:** 3,400.

NCI INFORMATION SYSTEMS, INC.
8260 Greensboro Drive, Suite 400, McLean VA 22102. 703/903-0325. **Fax:** 703/903-9750. **Contact:** Human Resources. **Description:** Provides telecommunications, information systems technology, imaging systems and data management, and facilities management services.

NCR CORPORATION
1700 South Patterson Boulevard, Dayton OH 45479-0001. 937/445-5000. **Contact:** Jeanne Bauer, Human Resources. **World Wide Web address:** http://www.ncr.com. **Description:** NCR Corporation is a worldwide provider of computer products and services. The company provides computer solutions to three targeted industries: retail, financial, and communication. NCR Computer Systems Group develops, manufactures, and markets computer systems. NCR Financial Systems Group is an industry leader in three target areas: financial delivery systems, relationship banking data warehousing solutions, and payments systems/item processing. NCR Retail Systems Group is a world leader in end-to-end retail solutions serving the food, general merchandise, and hospitality industries. NCR Worldwide Services provides data warehousing services solutions; end-to-end networking services; and designs, implements, and supports complex open systems environments. NCR Systemedia Group develops, produces, and markets a complete line of information products to satisfy customers' information technology needs including transaction processing media, auto identification media, business form communication products, managing documents and media, and a full line of integrated equipment solutions. NCR Corporation formerly operated as AT&T Global Information Solutions. **Educational backgrounds include:** Business Administration; Computer Science. **Benefits:** 401(k); Dental Insurance; Disability Coverage; Life Insurance; Medical Insurance; Pension Plan; Profit Sharing; Savings Plan; Tuition Assistance. **Special Programs:** Internships. **Corporate headquarters location:** This Location. **Operations at this facility include:** Divisional Headquarters. **Annual sales/revenues:** More than $100 million. **Number of employees nationwide:** 19,000. **Number of employees worldwide:** 38,000. **Other U.S. locations:**

- 2323 East Magnolia, Suite 110, Phoenix AZ 85034. 602/306-1859. Contact: Sherry Daly, Field Manager. This location is a manufacturing and sales office.
- 151 National Drive, Glastonbury CT 06033. 860/659-5300. This location is a remote support center for data processing, as well as an item image center.
- Two Oak Way, Berkeley Heights NJ 07922. 908/790-2500. This location is involved in the sale of computer products and scanning equipment, as well as software consulting and engineering.

NCR CORPORATION
100 North Sepulveda Boulevard, El Segundo CA 90245. 310/524-6500. **Contact:** Human Resources. **Description:** NCR Corporation is a worldwide provider of computer products and services. The company provides computer solutions to three targeted industries: retail, financial, and communication. NCR Computer Systems Group develops, manufactures, and markets computer systems. NCR Financial Systems Group is an industry leader in three target areas: financial delivery systems, relationship banking data warehousing solutions, and payments systems/item processing. NCR Retail Systems Group is a world leader in end-to-end retail solutions serving the food, general merchandise, and hospitality industries. NCR Worldwide Services provides data warehousing services solutions; end-to-end networking services; and designs, implements, and supports complex open systems environments. NCR Systemedia Group develops, produces, and markets a complete line of information products to satisfy customers' information technology needs including transaction processing media, auto identification media, business form communication products, managing documents and media, and a full line of integrated equipment solutions. NCR Corporation formerly operated as AT&T Global Information Solutions. **Common positions include:** Computer Programmer; Software Engineer; Systems Analyst; Technical Writer/Editor. **Educational backgrounds include:** Computer Science. **Benefits:** 401(k); Dental Insurance; Disability Coverage; Employee Discounts; Life Insurance; Medical Insurance; Pension Plan; Stock Option. **Special Programs:** Internships. **Corporate headquarters location:** Dayton OH. **Other U.S. locations:** Nationwide. **Operations at this facility include:** Research and Development. **Annual sales/revenues:** More than $100 million. **Number of employees at this location:** 750. **Number of employees worldwide:** 38,000.

NCR CORPORATION
450 East John Carpenter Freeway, Irving TX 75062. 972/650-2710. **Contact:** Joy Maffeo, Human Resources Consultant. **E-mail address:** joy.maffeo@dallastx.ncr.com. **World Wide Web address:** http://www.ncr.com. **Description:** This is a sales and service location. NCR Corporation is a worldwide provider of computer products and services. The company provides computer solutions to three targeted industries: retail, financial, and communication. NCR Computer Systems Group develops, manufactures, and markets computer systems. NCR Financial Systems Group is an industry leader in three target areas: financial delivery systems, relationship banking data warehousing solutions, and payments systems/item processing. NCR Retail Systems Group is a world leader in end-to-end retail solutions serving the food, general merchandise, and hospitality industries. NCR Worldwide Services provides data warehousing services solutions; end-to-end networking services; and designs, implements, and supports complex open systems environments. NCR Systemedia Group develops, produces, and markets a complete line of information products to satisfy customers' information technology needs including transaction processing media, auto identification media, business form communication products, managing documents and media, and a full line of integrated equipment solutions. NCR Corporation formerly operated as AT&T Global Information Solutions. **Educational backgrounds include:** Business Administration; Computer Science. **Benefits:** 401(k); Dental Insurance; Disability Coverage; Life Insurance; Medical Insurance; Pension Plan; Savings Plan; Tuition Assistance. **Corporate headquarters location:** Dayton OH. **Other U.S. locations:** Nationwide. **Operations at this facility include:** Administration; Sales; Service. **Listed on:** New York Stock Exchange. **Annual sales/revenues:** More than $100 million. **Number of employees nationwide:** 19,000. **Number of employees worldwide:** 38,000.

NCR CORPORATION
3325 Platt Springs Road, West Columbia SC 29170. 803/796-9740. **Contact:** Human Resources. **Description:** NCR Corporation is a worldwide provider of computer products and services. The company provides computer solutions to three targeted industries: retail, financial, and communication. NCR Computer Systems Group develops, manufactures, and markets computer systems. NCR Financial Systems Group is an industry leader in three target areas: financial delivery systems, relationship banking data warehousing solutions, and payments systems/item processing. NCR Retail Systems Group is a world leader in end-to-end retail solutions serving the food, general merchandise, and hospitality industries. NCR Worldwide Services provides data warehousing services solutions; end-to-end networking services; and designs, implements, and supports complex open systems environments. NCR Systemedia Group develops, produces, and markets a complete line of information products to satisfy customers' information technology needs

including transaction processing media, auto identification media, business form communication products, managing documents and media, and a full line of integrated equipment solutions. NCR Corporation formerly operated as AT&T Global Information Solutions. **Common positions include:** Computer Programmer; Customer Service Representative; Electrical/Electronics Engineer; Financial Analyst; General Manager; Human Resources Manager; Industrial Engineer; Management Analyst/Consultant; Management Trainee; Quality Control Supervisor; Software Engineer; Systems Analyst; Technical Writer/Editor. **Educational backgrounds include:** Business Administration; Computer Science; Electronics; Engineering; Finance; Marketing; Mathematics. **Benefits:** 401(k); Dental Insurance; Disability Coverage; Employee Discounts; Life Insurance; Medical Insurance; Pension Plan; Salary Continuation; Tuition Assistance; Wellness Program. **Special Programs:** Internships. **Corporate headquarters location:** Dayton OH. **Other U.S. locations:** Nationwide. **Parent company:** AT&T. **Operations at this facility include:** Administration; Divisional Headquarters; Manufacturing; Research and Development; Service. **Annual sales/revenues:** More than $100 million. **Number of employees at this location:** 1,300. **Number of employees nationwide:** 19,000. **Number of employees worldwide:** 38,000.

NRAD
53560 Hull Street, San Diego CA 92152. 619/553-7812. **Contact:** Personnel. **Description:** The U.S. Navy's research and development laboratories. Areas of research include satellite communications, information management systems, sensor systems, systems integration, environmental remediation, and software development.

NATIONAL TECH TEAM, INC.
22000 Garrison Avenue, Dearborn MI 48124. 313/277-2277. **Contact:** Human Resources. **Description:** National Team Tech provides network design, programming, and training.

NAVIGATION TECHNOLOGIES
740 East Arques Avenue, Sunnyvale CA 94086. 408/737-3200. **Toll-free phone:** 800/888-7222. **Fax:** 408/737-4122. **Contact:** Human Resources. **Description:** A developer of digital databases for in-vehicle navigation systems. **NOTE:** The company offers entry-level positions. **Common positions include:** Accountant/Auditor; Geographer; Market Research Analyst; Marketing Manager; Marketing Specialist; Sales Executive; Sales Representative. **Educational backgrounds include:** Geography. **Benefits:** 401(k); Dental Insurance; Disability Coverage; Life Insurance; Medical Insurance. **Corporate headquarters location:** This Location. **Other U.S. locations:** Fargo ND. **International locations:** Canada; Europe; Japan. **Listed on:** Privately held. **Annual sales/revenues:** Less than $5 million. **Number of employees at this location:** 150. **Number of employees nationwide:** 400. **Number of employees worldwide:** 800.

NETIS TECHNOLOGY, INC.
1606 Centre Pointe Drive, Milpitas CA 95035. 408/263-0368. **Fax:** 408/263-4624. **Contact:** Human Resources. **World Wide Web address:** http://www.netistech.com. **Description:** Provides systems integration and networking services. **Common positions include:** Administrative Assistant; Hardware Engineer; Sales Executive; Sales Manager; Software Engineer. **Educational backgrounds include:** Computer Science; Marketing. **Benefits:** 401(k); Dental Insurance; Employee Discounts; Medical Insurance; Profit Sharing. **Listed on:** Privately held. **Annual sales/revenues:** $11 - $20 million. **Number of employees at this location:** 20.

NETWORK COMPUTING DEVICES, INC.
350 North Bernardo Avenue, Mountain View CA 94043. 415/694-0650. **Toll-free phone:** 800/866-4080. **Fax:** 415/961-7711. **Contact:** Human Resources Department. **E-mail address:** resumes@ncd.com. **World Wide Web address:** http://www.ncd.com. **Description:** This location develops and manufactures hardware and software which enable computers to be networked. Network Computing Devices, Inc. provides desktop information access solutions for network computing environments. The company is a leading worldwide supplier of X Window System terminals and PC-X server software products which integrate Microsoft Windows- and DOS-based PCs into X/UNIX networks. NCD also supplies the Z-Mail family of cross platform electronic-mail and messaging software for open systems environments, as well as Mariner, an Internet access and navigation software tool that provides a unified interface to all Internet resources. **Common positions include:** Administrative Assistant; Buyer; Marketing Manager; Sales Representative;

Software Engineer; Systems Engineer; Technical Support Representative; Webmaster. **Benefits:** 401(k); Credit Union; Dental Insurance; Disability Coverage; EAP; Flexible Benefits; Life Insurance; Long-Term Care; Medical Insurance; Sabbatical Leave; Stock Purchase; Tuition Assistance; Vision Insurance. **Corporate headquarters location:** This Location. **Listed on:** NASDAQ. **Stock exchange symbol:** NCDI.
Other U.S. locations:
- 10900 NE 8th Street, Suite 900, Bellevue WA 98004. 206/635-0849. This location is a sales office.

NETWORK COMPUTING DEVICES, INC.
9590 SW Gemini Drive, Beaverton OR 97005. 503/641-2200. **Toll-free phone:** 800/866-4080. **Fax:** 415/691-1312. **Contact:** Human Resources. **E-mail address:** resumes@ncd.com. **World Wide Web address:** http://www.ncd.com. **Description:** Network Computing Devices, Inc. provides desktop information access solutions for network computing environments. The company is a leading worldwide supplier of X Window System terminals and PC-X server software products which integrate Microsoft Windows- and DOS-based PCs into X/UNIX networks. NCD also supplies the Z-Mail family of cross platform electronic-mail and messaging software for open systems environments, as well as Mariner, an Internet access and navigation software tool that provides a unified interface to all Internet resources. **Common positions include:** Buyer; Marketing Manager; Sales Representative; Software Engineer; Technical Support Representative; Webmaster. **Benefits:** 401(k); Credit Union; Dental Insurance; Disability Coverage; EAP; Flexible Benefits; Life Insurance; Long-Term Care; Medical Insurance; Sabbatical Leave; Stock Option; Tuition Assistance; Vision Insurance. **Corporate headquarters location:** Mountain View CA. **Listed on:** NASDAQ. **Stock exchange symbol:** NCDI.

NETWORK GENERAL CORPORATION
4200 Bohannon Drive, Menlo Park CA 94025. 415/473-2000. **Fax:** 415/321-0854. **Contact:** Human Resources. **Description:** Network General Corporation designs, manufactures, markets, and supports software-based analysis and monitoring tools primarily for managing enterprisewide computer networks. The company's tool line consists of software and network interface cards used with portable PC-compatible computers to monitor and analyze individual local area networks (LANs) or wide area networks (WANs) segments.

NETWORK INFOSERVE, INC.
P.O. Box 172068, Tampa FL 33672-2068. 813/229-1178. **Contact:** Human Resources. **Description:** A software integration services firm.

NETWORK PERIPHERALS, INC.
1371 McCarthy Boulevard, Milpitas CA 95035. 408/321-7300. **Contact:** Human Resources. **Description:** A computer network products firm, specializing in LANs. **Listed on:** NASDAQ.

NETWORK SIX INC.
475 Kilvert Street, Warwick RI 02886. 401/732-9000. **Contact:** Human Resources. **Description:** Network Six Inc. is a provider of systems integration and consulting services to state government human services agencies. The company provides project management, systems design, software development, hardware procurement and installation, training, and data conversion services.

NETWORK SYSTEMS CORPORATION
7600 Boone Avenue North, Minneapolis MN 55428. 612/424-4888. **Contact:** Dave Gilbertson, Human Resources Manager. **Description:** A data telecommunications firm, specializing in the local networking of mainframe computers and subsystems. **Number of employees at this location:** 1,000.

NETWORKS, INC.
3265 Meridian Parkway, Suite 104, Fort Lauderdale FL 33331. 954/389-3880. **Contact:** Karen Anghelescu, Administrative Assistant. **Description:** Designs and manufactures custom computer systems for clients in a variety of industries.

NEWCOME ELECTRONIC SYSTEMS

9005 Antares Avenue, Columbus OH 43240. 614/848-5688. **Fax:** 614/848-9921. **Contact:** Kelly Fox, Director of Human Resources. **Description:** Designs and installs voice and data distribution systems, primarily to support the communications needs of commercial enterprises in local area networking environments. Voice and data networking projects represent approximately 80 percent of annual sales. In addition, Newcome Electronic Systems also maintains a strong presence in the audio/video marketplace, designing and installing custom systems in both commercial and residential applications. Founded in 1978. **NOTE:** The company offers entry-level positions, training programs, and apprenticeships. **Common positions include:** Computer Animator; Computer Operator; Computer Programmer; Cost Estimator; Draftsperson; Electrician; General Manager; Human Resources Manager; Marketing Manager; Marketing Specialist; MIS Specialist; Operations/Production Manager; Purchasing Agent/Manager; Sales Engineer; Sales Executive; Sales Manager; Sales Representative; Secretary; Systems Analyst; Telecommunications Manager; Typist/Word Processor; Vice President; Video Production Coordinator. **Educational backgrounds include:** Accounting; Business Administration; Communications; Computer Science; Engineering; Liberal Arts. **Benefits:** 401(k); Disability Coverage; Employee Discounts; Flexible Schedule; Life Insurance; Profit Sharing. **Special Programs:** Internships. **Corporate headquarters location:** This Location. **Listed on:** Privately held. **Annual sales/revenues:** $11 - $20 million. **Number of employees at this location:** 50.

NORTHROP GRUMMAN CORPORATION
DATA SYSTEMS & SERVICES DIVISION

South Oyster Bay Road, Bethpage NY 11758. 516/575-0574. **Contact:** Human Resources. **Description:** This division of Northrop Grumman is one of the leading designers and integrators of large-scale information systems for the federal government, with specialties in image processing, high-performance computing, command and control, logistics, and other information needs. The company is also involved in computer hardware and software support, engineering operations, and maintenance services. **Corporate headquarters location:** Los Angeles CA.

OLIVETTI NORTH AMERICA, INC.

22425 East Appleway Avenue, Liberty Lake WA 99019-9534. 509/927-5600. **Contact:** Human Resources. **Description:** Provides branch automation solutions, systems, and services to the financial industry, and is a major vendor for third-party maintenance. **Common positions include:** Accountant/Auditor; Budget Analyst; Buyer; Computer Programmer; Dispatcher; Electronics Technician; Financial Manager; General Manager; Human Resources Manager; Inspector/ Tester/Grader; Market Research Analyst; Marketing Manager; Secretary; Software Engineer; Systems Analyst; Technical Writer/Editor. **Educational backgrounds include:** Accounting; Business Administration; Computer Science; Marketing. **Benefits:** Dental Insurance; Disability Coverage; Life Insurance; Medical Insurance; Pension Plan; Savings Plan; Tuition Assistance. **Corporate headquarters location:** This Location. **Parent company:** Olivetti. **Operations at this facility include:** Administration; Sales; Service. **Listed on:** Privately held. **Number of employees at this location:** 500. **Number of employees nationwide:** 1,100.
Other U.S. locations:
• One Vandegraaff Drive, Suite 202, Burlington MA 01803. 617/272-2010.

OMNIDATA COMMUNICATIONS

906 North Main, Suite 3, Wichita KS 67203. 316/264-5068. **Fax:** 316/264-7031. **Contact:** Human Resources. **World Wide Web address:** http://www.omnidata.com. **Description:** A supplier of a wide range of data communication and computer products and services including modems, multiplexers, T-1, ISDN, integrated network management systems, and local area networks (LAN).

ORACLE GOVERNMENT

196 Van Buren Street, Herndon VA 22070. 703/904-8200. **Contact:** Human Resources. **Description:** A computer systems integration company providing solutions and services to federal and commercial clients. **NOTE:** Please send resumes to Human Resources, 3 Bethesda Metro Center, Suite 1400, Bethesda MD 20814. 301/657-7860.

OREGON DIGITAL COMPUTER PRODUCTS INC.
887 NW Grant Avenue, Corvallis OR 97330. 541/753-5554. **Contact:** Human Resources. **Description:** Offers a variety of network services. Oregon Digital Computer Products installs, maintains, and supports business computer networking systems.

THE ORKAND CORPORATION
7799 Leesburg Pike, Suite 700 North, Falls Church VA 22043-2499. 703/610-4200. **Contact:** Peter Nasou, Recruiting. **Description:** Designs information-based computer systems under government contracts and for private industry.

PRC, INC.
1500 PRC Drive, McLean VA 22102-5050. 703/556-1000. **Fax:** 703/556-2269. **Contact:** Human Resources. **Description:** PRC is a global provider of scientific and technology-based systems and services to government and commercial clients. The company provides computer systems integration, systems engineering, software development, environmental engineering, and consulting services to clients worldwide. **Common positions include:** Accountant/Auditor; Aerospace Engineer; Attorney; Budget Analyst; Chemical Engineer; Computer Operator; Computer Programmer; Editor; Electrical/Electronics Engineer; Employment Interviewer; Financial Analyst; Graphic Artist; Industrial Engineer; Mechanical Engineer; Meteorologist; Nuclear Engineer; Receptionist; Secretary; Software Engineer; Systems Analyst; Technical Support Representative; Typist/Word Processor. **Educational backgrounds include:** Accounting; Computer Science; Economics; Engineering; Finance; Geology; Liberal Arts. **Benefits:** 401(k); Dental Insurance; Disability Coverage; Employee Discounts; Life Insurance; Medical Insurance; Pension Plan; Savings Plan; Tuition Assistance. **Corporate headquarters location:** This Location. **Parent company:** Black & Decker. **Operations at this facility include:** Administration; Regional Headquarters. **Listed on:** New York Stock Exchange. **Number of employees at this location:** 3,500. **Number of employees nationwide:** 7,000.

PATTON ELECTRONICS COMPANY
7622 Rickenbacker Drive, Gaithersburg MD 20879. 301/975-1000. **Contact:** Human Resources. **Description:** Manufactures data communications equipment.

PERIFITECH INC.
1265 Ridge Road, Hinckley OH 44233. 216/278-2070. **Contact:** Human Resources. **Description:** Provides businesses with network servers, workstations, and storage devices. Perifitech also offers support services for their products.

PERLE SYSTEMS INC.
630 Oakmont Lane, Westmont IL 60559. 630/789-3171. **Toll-free phone:** 800/337-3753. **Contact:** Human Resources. **Description:** Manufactures remote controllers for network computing for businesses.

PETRO VEND INC.
6900 Santa Fe Drive, Hodgkins IL 60525. 708/485-4200. **Contact:** Human Resources. **Description:** Develops computer-based fuel management systems for oil companies.

PHAMIS INC.
1001 4th Avenue Plaza, Suite 1500, Seattle WA 98154. 206/622-9558. **Fax:** 206/623-3950. **Contact:** Human Resources. **Description:** PHAMIS, Inc. develops, markets, installs, and services enterprisewide, patient-centered health care information systems for use by large and medium-sized health care providers. The company operates solely in the industry segment of health care information systems. The PHAMIS-LASTWORD system is an integrated hardware and software solution that constructs an online lifetime medical record that is accessible simultaneously throughout the health care delivery enterprise. The LASTWORD system collects, stores, and organizes patient-centered health care information as a single relational database that enables immediate online access to current and prior episodes of patient care. The online medical record assists health care providers to better monitor and manage the cost and quality of patient care and to support enterprisewide decision making in finance and administration. **Corporate headquarters location:** This Location. **Number of employees at this location:** 226.

PHASENET SYSTEMS
19545 NW Von Neumann Drive, Beaverton OR 97006. 503/531-2480. **Fax:** 503/531-2401. **Contact:** Human Resources. **E-mail address:** karen@phasenet.com. **World Wide Web address:** http://www.phasenet.com. **Description:** A provider of network computing solutions and Internet systems.

THE PINNACLE GROUP
P.O. Box 630, Athens AL 35612. 205/233-5363. **Contact:** Human Resources. **Description:** The Pinnacle Group specializes in computer customization implementation. The company installs and supports computer software networks for other companies.

POLYCHROME AMERICAS
11100 West 82nd Street, Lenexa KS 66214. 913/492-3322. **Contact:** Bill Palafox, Human Resources. **Description:** This location of Polychrome develops and supports integrated systems for the graphic arts industry.

POMEROY COMPUTER RESOURCES
1020 Petersburg Road, Hebron KY 41048. 606/282-7111. **Contact:** Kim Garner, Human Resources Manager. **Description:** Pomeroy Computer Resources sells, installs, and services microcomputers and microcomputer equipment primarily for business, professional, educational, and government customers. The company also derives revenue from customer support services, including network analysis and design, systems configuration, custom installation, training, maintenance, and repair. The company was formed in 1992, and merged several months later with eight related businesses, five of which owned and operated franchises of ComputerLand Corporation in Ohio and Kentucky. **Common positions include:** Accountant/Auditor; Computer Programmer; Purchasing Agent/Manager; Services Sales Representative; Software Engineer; Systems Analyst. **Educational backgrounds include:** Accounting; Business Administration; Computer Science; Marketing. **Benefits:** 401(k); Life Insurance; Medical Insurance; Profit Sharing. **Corporate headquarters location:** This Location. **Other U.S. locations:** Knoxville KY; Lexington KY; Cincinnati OH; Louisville OH; Kingsport TN; Nashville TN. **Subsidiaries include:** Xenas Multimedia. **Operations at this facility include:** Administration; Divisional Headquarters; Sales; Service. **Listed on:** NASDAQ. **Number of employees at this location:** 120. **Number of employees nationwide:** 450.

PRAGMATICS, INC.
8301 Greensboro Drive, Suite 225, McLean VA 22102. 703/761-4033. **Contact:** Human Resources. **Description:** A computer systems integration firm.

PROGRESSIVE BUSINESS SOLUTIONS
5907 Breckenridge Parkway, Tampa FL 33610. 813/621-1117. **Contact:** Human Resources. **Description:** A computer systems integration firm.

RGI INC.
5203 Leesburg Pike, Suite 1300, Falls Church VA 22041. 703/820-4900. **Contact:** Human Resources. **Description:** Offers computer hardware and software support services to the government on a contractual basis.

RACAL-DATACOM
P.O. Box 407044, Fort Lauderdale FL 33340-7044. 954/846-5250. **Fax:** 954/846-5025. **Contact:** Human Resources. **Description:** Racal-Datacom Inc. manufactures data communications equipment including WANs, LANs, and access products. The company also offers related services including project management, installation, consultation, network integration, maintenance, disaster recovery, and training. **Common positions include:** Electrical/Electronics Engineer; Marketing Manager; Software Engineer. **Educational backgrounds include:** Accounting; Business Administration; Communications; Computer Science; Engineering; Finance; Marketing. **Benefits:** 401(k); Dental Insurance; Disability Coverage; Employee Discounts; Life Insurance; Medical Insurance; Tuition Assistance. **Other U.S. locations:** Nationwide. **International locations:** Worldwide. **Parent company:** Racal Corporation. **Operations at this facility include:**

Administration; Divisional Headquarters; Manufacturing; Research and Development; Sales; Service. **Number of employees at this location:** 900. **Number of employees nationwide:** 2,500.

RACAL-DATACOM
1100 Jupiter Road, Suite 190, Plano TX 75074. 972/509-4700. **Contact:** Personnel. **Description:** This location is a sales and technical service office. Racal-Datacom Inc. manufactures data communications equipment including WANs, LANs, and access products. The company also offers related services including project management, installation, consultation, network integration, maintenance, disaster recovery, and training. **NOTE:** Professional hiring is done primarily through the Florida office (8600 NW 41st Street, Miami FL 33166). **Corporate headquarters location:** Sunrise FL.

RACAL-DATACOM
60 Codman Hill Road, Boxborough MA 01719. 508/263-9929. **Contact:** Human Resources. **Description:** Racal-Datacom designs and supplies Ethernet and Token Ring Network Interface Cards, hubs, and media products for Local Area Network (LAN) connectivity in multivendor environments. **Common positions include:** Accountant/Auditor; Buyer; Electrical/Electronics Engineer; Marketing Specialist; Software Engineer. **Educational backgrounds include:** Business Administration; Engineering; Marketing. **Corporate headquarters location:** Sunrise FL. **Parent company:** Racal Corporation. **Operations at this facility include:** Administration; Research and Development; Sales.

REYNOLDS & REYNOLDS
Executive Plaza IV, 11350 McCormick Road, Hunt Valley MD 21031. 410/771-9211. **Contact:** Robin Poe, Human Resources Representative. **Description:** This location is a sales office and provides technical support services. Overall, Reynolds & Reynolds develops computer systems for the automotive and health care industries. The company also supplies these industries with business forms. **Operations at this facility include:** Manufacturing; Sales; Service.

REYNOLDS & REYNOLDS
2388 Schuetz Road, Suite C-80, St. Louis MO 63146. 314/569-3345. **Contact:** Chris Jones, Manager. **Description:** This location serves as a regional sales office. Overall, Reynolds & Reynolds develops computer systems for the automotive and health care industries. The company also supplies these industries with business forms.

RISC MANAGEMENT
320 North Michigan Avenue, Suite 704, Chicago IL 60601. 312/332-9494. **Contact:** Recruiting. **Description:** Offers businesses systems integration services. Risc Management sells integration software and provides technical support.

SFA, INC.
1401 McCormick Drive, Largo MD 20774. 301/925-9400. **Fax:** 301/925-8568. **Contact:** Lisa Broome, Director, Human Resources. **Description:** SFA, Inc. is a diversified international supplier of products and services aimed at helping clients capitalize on leading edge systems and technologies. SFA conducts advanced research studies; designs and develops state-of-the-art prototypes; and produces custom hardware and software systems for defense, communications, and other commercial applications. **Common positions include:** Accountant/Auditor; Aerospace Engineer; Biological Scientist; Biomedical Engineer; Buyer; Chemist; Electrical/Electronics Engineer; Financial Analyst; Materials Engineer; Mechanical Engineer; Metallurgical Engineer; Physicist/Astronomer; Purchasing Agent/Manager; Software Engineer; Systems Analyst; Technical Writer/Editor. **Educational backgrounds include:** Accounting; Biology; Communications; Computer Science; Engineering; Marketing; Mathematics; Physics. **Benefits:** 401(k); Dental Insurance; Disability Coverage; Life Insurance; Medical Insurance; Tuition Assistance. **Corporate headquarters location:** This Location. **Other U.S. locations:** Washington DC; Columbia MD; Frederick MD; Landover MD; Lexington Park MD. **Subsidiaries include:** SFA DataComm, Inc.; SFA SACOM. **Operations at this facility include:** Administration; Manufacturing; Research and Development; Sales; Service. **Listed on:** Privately held. **Number of employees at this location:** 300.

STMS (SOLUTIONS THAT MAKE SENSE)
44880 Falcon Place, Sterling VA 20166. 703/318-7867. **Contact:** Human Resources. **Description:** A computer network technologies firm. This location is a sales office and provides technical support.

S2, INC.
600 Embassy Row, Suite 300, Atlanta GA 30328. 770/551-2551. **Contact:** Marjorie Reeves, Human Resources. **Description:** S2, Inc. is a provider of data communications middleware and related professional services that bridge the gap between open distributed systems and legacy mainframe and midrange systems used for online applications. **Corporate headquarters location:** Marlboro MA. **Parent company:** Stratus Computer, Inc. offers a broad range of computer platforms, application solutions, middleware, and professional services for critical online operations. Other Stratus subsidiaries include Shared Systems Corporation, a provider of software and professional services to the financial services, retail, and health care industries; and Isis Distributed Systems, Inc., a developer of advanced messaging middleware products that enable businesses to develop reliable, high performance distributed computing applications involving networked desktop computers and shared systems.

SABREDATA
1321 Rutherford Lane, #200, Austin TX 78753. 512/835-1152. **Contact:** Barbara Taylor, Human Resources. **Description:** Founded in 1988, this computer and network integration company is a provider of local area networks (LANs), as well as a reseller of computer hardware in conjunction with networking projects.

SAFEGUARD SCIENTIFICS, INC.
435 Devon Park Drive, Wayne PA 19087. 610/293-0600. **Fax:** 610/293-0601. **Contact:** Personnel. **Description:** A strategic information systems holding company that has operations in three industry segments: Information Technology, Metal Finishing, and Commercial Real Estate. Over 95 percent of Safeguard Scientifics's sales are in the Information Technology segment, which consists of the delivery of personal computer services, including procurement and configuration of personal computers, application software and related products, network integration, and technical support; and the design, development, and sale of strategic business applications systems. Safeguard Scientifics provides its subsidiaries with active strategic management, operating guidance, and financing. **Corporate headquarters location:** This Location. **Subsidiaries include:** CompuCom Systems, Inc.; Tangram Enterprise Solutions, Inc.; Premier Solutions Ltd. **Listed on:** New York Stock Exchange.

SCAN-OPTICS, INC.
169 Progress Drive, Manchester CT 06040. 860/649-5713. **Fax:** 860/649-7283. **Contact:** Human Resources Department. **Description:** Scan-Optics designs and manufactures information processing systems used for imaging, data capture, document processing, and information management. Scan-Optics's systems make it possible to process very large volumes of paper using such features as high-speed paper movement, optical character recognition, intelligent character recognition, high-speed image capture, image processing, and image storage and retrieval systems. Scan-Optics's system encompasses hardware, software, and integration technologies for complete solutions. Typical applications for Scan-Optics systems include the processing of credit card sales drafts, mail order forms, federal and state tax forms, health care forms, automobile registrations, shareholder proxies, and payroll time cards. Scan-Optics markets to subscription and catalog fulfillment companies, manufacturers, government tax and employment agencies, financial institutions, health care organizations, and other business entities. **Corporate headquarters location:** East Hartford CT. **Other U.S. locations:** Irvine CA.

SCHOONMAKER & ASSOCIATES, INC.
1046 New Hampshire Street, Lawrence KS 66044. 913/843-3941. **Contact:** Human Resources. **Description:** Sets up and supports office computer networks.

SEQUENT COMPUTER SYSTEMS, INC.
15450 SW Koll Parkway, Beaverton OR 97006-6063. 503/626-5700. **Contact:** Staffing Manager. **Description:** Sequent Computer Systems is a leading provider of enterprise architecture and open

solutions for large corporations. The company's Enterprise Division provides architectural consulting, professional services, and industry-leading servers that enable customers to migrate successfully from host-based, proprietary computer environments to open, client/server architectures. Working closely with leading hardware and software partners, the Enterprise Division helps customers design and implement enterprisewide architectures and systems for online transaction processing, decision support, and workgroup computing. Sequent's Platform Division develops -- solely and in partnership with other companies -- high-performance, Intel-based, symmetric multiprocessing servers that run on the UNIX and Microsoft Windows NT operating systems and support enterprisewide applications and information services. The Platform Division's products are sold by the Enterprise Division and through a variety of indirect channels, including OEM resellers, value-added resellers, and distributors. Since the company was founded in 1983, Sequent has installed more than 5,500 large scale systems in 52 countries worldwide. **Corporate headquarters location:** This Location. **Number of employees nationwide:** 1,600.
Other U.S. locations:

- 11100 Santa Monica Boulevard, Suite 200, Los Angeles CA 90025. 310/444-7115. This location is a sales office.
- One Market Spear Street Tower, Suite 700, San Francisco CA 94105. 408/970-3640. This location is a sales office which provides customer service and technical support.
- 4100 East Mississippi Avenue, Suite 1800, Denver CO 80222. 303/758-4931. This location is a sales office.
- 1050 Crowne Pointe Parkway, Suite 1800, Atlanta GA 30338. 770/913-3855. This location is a sales office.
- 6250 River Road, Suite 7050, Rosemont IL 60018. 847/318-0050. This location is a field sales office.
- 25 Mall Road, Sixth Floor, Burlington MA 01803. 617/229-8881. This location is a sales office.
- 26999 Central Park Boulevard, Suite 200, Southfield MI 48076. 810/355-4050. This location is a sales office.
- 70 South Wood Avenue, First Floor, Iselin NJ 08830. 908/549-8500. This location is a sales office.
- 757 Third Avenue, Third Floor, New York NY 10017. 212/750-9300. This location is a sales office.
- 15303 Dallas Parkway, Suite 1350, Dallas TX 75248. 972/661-1900. This location is sales office.
- 11 Greenway Plaza, Suite 1404, Houston TX 77086. 713/871-3191. This location is a sales office.

SHARED MEDICAL SYSTEMS CORPORATION
51 Valley Stream Parkway, Malvern PA 19355. 610/219-3359. **Fax:** 610/219-8266. **Contact:** Human Resources. **Description:** Shared Medical Systems is a leading provider of health care information systems and service solutions to hospitals, multi-entity health care corporations, integrated health networks, physician groups, and other health care providers in North America and Europe. The company also provides a full complement of solutions for the newly emerging community health information networks, which include payers and employers as well as providers. Shared Medical Systems offers a comprehensive line of health care information systems, including clinical, financial, administrative, ambulatory, and decision support systems, for both the public and private health care sectors. These systems are offered on computers operating at the customer site, at the SMS Information Services Center, or as part of a distributed network. The company also provides a portfolio of professional services including systems installation, support, and education. In addition, the company provides specialized consulting services for the design and integration of software and networks, facilities management, information systems planning, and systems-related process re-engineering. **Common positions include:** Accountant/Auditor; Computer Programmer; Human Resources Manager; Licensed Practical Nurse; Medical Records Technician; Pharmacist; Radiologic Technologist; Registered Nurse; Software Engineer; Systems Analyst; Technical Writer/Editor. **Educational backgrounds include:** Accounting; Business Administration; Communications; Computer Science; Finance; Health Care; Marketing; Mathematics. **Benefits:** 401(k); Dental Insurance; Disability Coverage; Employee Discounts; Life Insurance; Medical Insurance; Savings Plan; Tuition Assistance. **Corporate headquarters location:** This Location. **Other U.S. locations:** Los Angeles CA; Oakland CA; San Francisco CA; Santa Barbara CA;

Denver CO; Washington DC; Miami FL; Atlanta GA; Chicago IL; Indianapolis IN; New Orleans LA; Boston MA; Ann Arbor MI; Kansas City MO; St. Louis MO; Charlotte NC; Somerset NJ; Cleveland OH; Philadelphia PA; Pittsburgh PA; Nashville TN; Dallas TX; Seattle WA. **Operations at this facility include:** Administration; Regional Headquarters; Research and Development; Sales; Service. **Listed on:** NASDAQ. **Number of employees at this location:** 2,500. **Number of employees nationwide:** 4,000.

SHARED MEDICAL SYSTEMS CORPORATION
2201 Broadway #605, Oakland CA 94612. 510/444-3434. **Fax:** 510/465-4136. **Contact:** Janice Griffen, Recruiter. **World Wide Web address:** http://www.smed.com. **Description:** This location is a technical support office. Overall, Shared Medical Systems is a leading provider of health care information systems and service solutions to hospitals, multi-entity health care corporations, integrated health networks, physician groups, and other health care providers in North America and Europe. The company also provides a full complement of solutions for the newly emerging community health information networks, which include payers and employers as well as providers. Shared Medical Systems offers a comprehensive line of health care information systems, including clinical, financial, administrative, ambulatory, and decision support systems, for both the public and private health care sectors. These systems are offered on computers operating at the customer site, at the SMS Information Services Center, or as part of a distributed network. The company also provides a portfolio of professional services including systems installation, support, and education. In addition, the company provides specialized consulting services for the design and integration of software and networks, facilities management, information systems planning, and systems-related process re-engineering. **Common positions include:** Computer Programmer; Database Manager; MIS Specialist; Systems Analyst; Systems Manager; Technical Writer/Editor. **Educational backgrounds include:** Business Administration; Computer Science; Health Care. **Benefits:** 401(k); Dental Insurance; Disability Coverage; Life Insurance; Mass Transit Available; Medical Insurance; Pension Plan; Profit Sharing. **Corporate headquarters location:** Malvern PA. **Other U.S. locations:** Nationwide. **Stock exchange symbol:** SHRMED. **Annual sales/revenues:** $5 - $10 million. **Number of employees at this location:** 150. **Number of employees worldwide:** 4,500. **Number of projected hires for 1997-1998 at this location:** 20.

SHERIKON INC.
14500 Avion Parkway, Suite 200, Chantilly VA 20151-1108. 703/803-7000. **Contact:** Human Resources. **Description:** A government contractor that offers computer support services.

SHIVA CORPORATION
28 Crosby Drive, Bedford MA 01730. 617/270-8300. **Fax:** 617/270-8585. **Contact:** Human Resources. **E-mail address:** hr@shiva.com. **Description:** Shiva Corporation, incorporated in 1985, initially specialized in the development of networking products for AppleTalk LANs. In 1991, the company began its transition toward the sale of remote access products with the introduction of NetModem/E, its first multiplatform, multiprotocol remote access product. In August 1995, Shiva acquired Spider Systems, a leader in ISDN technology. **Benefits:** 401(k); Computer Loans; Credit Union; Dental Insurance; Disability Coverage; EAP; ESOP; Fitness Program; Life Insurance; Medical Insurance; Spending Account; Tuition Assistance. **Corporate headquarters location:** This Location. **Listed on:** NASDAQ. **Stock exchange symbol:** SHVA.

SIGNAL CORPORATION
3040 Williams Drive, Suite 200, Fairfax VA 22031. 703/205-0500. **Fax:** 703/205-0560. **Contact:** Human Resources. **Description:** A technology firm engaged in computer network engineering, software design and development, office automation, telecommunications systems integration, image processing and data compression, and computer systems security. **Listed on:** Privately held.

SILICON GRAPHICS
2011 North Shoreline Boulevard, Mountain View CA 94043-1389. 415/960-1980. **Contact:** Human Resources. **Description:** Silicon Graphics makes a family of workstation and server systems that are used by engineers, scientists, and other creative professionals to develop, analyze, and simulate complex, three-dimensional objects. Systems include Onyx, IRIS, Indigo, and Challenge, and all use RISC processors in their circuitry. In 1992, the company acquired MIPS Computer Systems. **NOTE:** Send resumes to Human Resources Department, P.O. Box 7311,

Mountain View CA 94039-7311. **Corporate headquarters location:** This Location. **Number of employees nationwide:** 3,750.
Other U.S. locations:
- 6501 America's Parkway NE, Suite 565, Albuquerque NM 87110. 505/884-1400. This location is a district office offering sales and support services.

SOFTECH, INC.

460 Totten Pond Road, Waltham MA 02154-1960. 617/890-6900. **Contact:** Human Resources. **Description:** A computer systems integration company that provides a wide array of computing solutions, both off-the-shelf and custom developed, to solve complex business solutions. The company is organized into three business units: The Products Group, which is composed of 13 branch offices within the U.S. that market hardware, off-the-shelf software, and service offerings to commercial customers; The Systems Integration Services Group, which is composed of two offices that market high-end service offerings throughout the U.S.; and The Massively Parallel Software Division, which is a group that was formed to capitalize on technology previously released by the company. The Products Group is organized under a wholly-owned subsidiary of Information Decisions, Inc. **NOTE:** All interested applicants should send resumes to: SofTech, Inc., Human Resources, 3260 Eagle Park Drive NE, Grand Rapids MI 49505. 616/957-2330. **Corporate headquarters location:** This Location.

SOURCE DIGITAL SYSTEMS

1919 Gallows Road, Vienna VA 22182. 703/821-6800. **Contact:** Human Resources. **Description:** A systems integrator that also resells value-added computer products.

SPUR PRODUCTS CORPORATION

9288 West Emerald Street, Boise ID 83704. 208/377-0001. **Contact:** Human Resources. **Description:** Spur Products builds and installs interfaces that enable office computer networks to operate more smoothly.

STANDARD MICROSYSTEMS CORPORATION

80 Arkay Drive, Hauppauge NY 11788. 516/273-3100. **Fax:** 516/435-0373. **Contact:** Ms. Joey Halpin, Human Resources. **Description:** Standard Microsystems is a supplier of products used to connect personal computers over local area networks. The Desktop Networks Business Unit designs, produces, and markets network interface cards (adapters). The Enterprise Networks Business Unit designs, produces, and markets wiring hubs, LAN switches, and supporting software. All of these products allow PC users to communicate over LANs. The Component Products Division designs, produces, and markets large-scale integrated circuits. Standard Microsystems's PC input/output circuits incorporate multiple functions on individual chips. These functions include floppy disk control; serial, parallel, and game port control; hard disk interface; keyboard and mouse control; real-time clock and support for high-speed transmission; and plug-and-play protocols. **Corporate headquarters location:** This Location. **Number of employees at this location:** 650.

STANDARD TECHNOLOGY INC.

6116 Executive Boulevard, Suite 700, Rockville MD 20852. 301/770-2800. **Contact:** Human Resources. **Description:** Provides businesses with information technologies solutions and engineering and technical support.

STRATEGIC TECHNOLOGIES

301 Gregson Drive, Cary NC 27511. 919/481-9797. **Contact:** Gail Davis, Human Resources Director. **Description:** A company that sells and integrates corporate information systems. **Listed on:** Privately held.

SULCUS COMPUTER CORPORATION

Sulcus Center, 41 North Main Street, Greensburg PA 15601. 412/836-2000. **Contact:** Human Resources. **Description:** Develops, manufactures, markets, and installs microcomputer systems designed to automate the creation, handling, storage, and retrieval of information and documents. The company designs its systems primarily for the hospitality and real estate industries and the legal profession. The company's sales practices are currently systems oriented, rather than individual sales of hardware or software, toward the vertical marketing of its integrated products. Systems include a network of hardware, software, and cabling as well as stand alone systems for full-service training, maintenance, and support. The company has installed systems throughout North and South America, Europe, Africa, Asia and Australia.

SUNGARD DATA SYSTEMS, INC.
1285 Drummers Lane, Wayne PA 19087. 610/341-8700. **Contact:** Karen Bilinski, Human Resources Director. **Description:** Provides specialized computer services, mainly proprietary investment support systems for the financial services industry and disaster recovery services. SunGard Data Systems's investment accounting and portfolio systems maintain the books of record of all types of large investment portfolios including those managed by banks and mutual funds. The company's disaster recovery services include alternate-site backup, testing, and recovery services for IBM, DEC, Prime, Stratus, Tandem, and Unisys computer installations. The company's computer service unit provides remote-access IBM computer processing, direct marketing, and automated mailing services. **Corporate headquarters location:** This Location. **Number of employees at this location:** 2,100.

SYMPLEX COMMUNICATIONS CORPORATION
5 Research Drive, Ann Arbor MI 48103. 313/995-1555. **Contact:** Human Resources. **E-mail address:** webmaster@symplex.com. **World Wide Web address:** http://www.symplex.com. **Description:** A manufacturer of internetworking devices for the LAN-to-WAN market, as well as frame relay and leased line applications. Product line includes the DirectRoute WAN ISDN router and Datamizer IV, for WAN connectivity.

SYSOREX INFORMATION SYSTEMS
3975 Fair Ridge Drive, Fairfax VA 22033. 703/273-9200. **Contact:** Human Resources. **Description:** A systems integration firm.

SYSTEMHOUSE
5850 T.G. Lee Boulevard, Suite 510, Orlando FL 32822. 407/888-1800. **Contact:** Sandra McCaskill, Recruiter. **Description:** Systemhouse offers systems integration computer services, combining hardware, software, and network communications for large companies. The company also offers telecommunications, client/server architecture, rapid applications development, UNIX-based applications development, and large-scale network management. In addition, Systemhouse supports voice and data technologies worldwide; provides a blend of technical support and professional services to assist in the development, testing, and deployment of next generation software and hardware products for the communications industry; and offers expertise in computer interface technology and software development. **Common positions include:** Accountant/Auditor; Computer Programmer; Electrical/Electronics Engineer; Systems Analyst; Technical Writer/Editor. **Educational backgrounds include:** Accounting; Computer Science; Engineering; Marketing; Mathematics. **Benefits:** Dental Insurance; Disability Coverage; Life Insurance; Medical Insurance; Pension Plan; Savings Plan; Tuition Assistance. **Special Programs:** Internships. **Corporate headquarters location:** Dallas TX. **Other U.S. locations:** Los Angeles CA; Boulder CO; Washington DC; Chicago IL; New York NY. **Operations at this facility include:** Service. **Listed on:** New York Stock Exchange. **Number of employees at this location:** 150. **Number of employees nationwide:** 2,000.

SYSTEMS RESEARCH & APPLICATION
4300 Fair Lakes Court, South Building Suite 500, Fairfax VA 22033. 703/558-4700. **Fax:** 703/227-8268. **Contact:** Human Resources. **Description:** Assists companies with systems integration, networking, and software development.

T&B COMPUTING, INC.
P.O. Box 302, Ann Arbor MI 48106. 313/930-3800. **Contact:** Kristine Dreffs, Human Resources. **Description:** Provides systems and services for the publishing and accounting industries. **Common positions include:** Accountant/Auditor; Administrator; Computer Programmer; Services Sales Representative; Systems Analyst. **Educational backgrounds include:** Business Administration; Engineering. **Benefits:** Dental Insurance; Disability Coverage; Life Insurance; Medical Insurance;

Tuition Assistance. **Corporate headquarters location:** This Location. **Operations at this facility include:** Administration; Manufacturing; Research and Development; Sales; Service.

TDS COMPUTING SERVICES, INC. (TDS/CS)

8401 Greenway Boulevard, Suite 300, Middleton WI 53562-0980. 608/845-4600. **Contact:** Pam Lokken, Human Resources. **Description:** An information systems company, providing systems development, integration, and operations services to Telephone and Data Systems, Inc. (TDS) and non-affiliated companies. By delivering advanced information system solutions in such key areas as billing, network operations, customer contact, financial/human resource management, and office automation, TDS/CS helps create and sustain competitive advantages for clients. **Corporate headquarters location:** Chicago IL. **Parent company:** TDS (Telephone and Data Systems, Inc.) provides high-quality telecommunications services through local telephone companies, consolidated cellular markets, managed non-consolidated cellular markets, and paging operations centers, to telephone, cellular telephone, and radio paging customers across the United States. TDS's strategic business units are TDS Telecommunications Corporation, which provides local telephone and access service to rural and suburban areas across the nation and pursues an active program of acquiring operating telephone companies; United States Cellular Corporation, which manages and invests in cellular systems throughout the nation; and American Paging, Inc., which operates paging and voice mail systems, offering a wide variety of service packages including basic local paging service, statewide, regional, and nationwide paging, and voice mail with fax options. TDS's other associated service companies include American Communications Consultants, Inc., an engineering and management consulting company, which provides the TDS family of companies and a strong base of non-affiliated clients with innovative, leading-edge solutions in many telecommunications technologies; and Suttle Press, Inc., TDS's commercial printing subsidiary, which provides high-quality flat color and full color printed products to a growing base of commercial customers. **Number of employees nationwide:** 4,100.

TANDEM COMPUTERS INC.

10435 North Tantau Avenue, Cupertino CA 95014-2599. 408/725-6000. **Contact:** Human Resources. **Description:** Tandem Computers is a designer and producer of networks and fault-tolerant computer systems for the online transaction processing markets. The company also manufactures disk storage devices. Fault-tolerance in the computer market is obtained by backup and component redundancy, and is used in the communications, banking, manufacturing, distribution, and transportation industries. Through subsidiary UB Networks, the company provides service, integration, and software for networks. **Corporate headquarters location:** This Location. **Listed on:** New York Stock Exchange.
Other U.S. locations:

* 444 Market Street, Suite 1450, San Francisco CA 94111. 415/398-9440. This location is a sales office.
* 6400 South Fiddler's Green Circle, Suite 1775, Englewood CO 80111. 303/779-6766. This location is a district sales office.
* 1575 DeLucchi Lane, Suite 115, Reno NV 89502-6578. 702/825-9595. **Contact:** Ted Lungren, Office Manager. This location is a service office.

TECHMATICS

Fair Lakes Building 2, 12450 Fair Lakes Circle Suite 800, Fairfax VA 22033. 703/802-8300. **Contact:** Human Resources. **Description:** Develops and supports information systems technology.

TECHNICAL AND MANAGEMENT SERVICES CORPORATION (TAMSCO)

4041 Powder Mill Road, Suite 500, Calverton MD 20705. 301/595-0710. **Contact:** Human Resources Department. **Description:** Offers products and services oriented to ADP and telecommunications system development, manufacturing, and integration. These products and services include requirements definition, system engineering, systems design, telecommunications network design, software development, electronics and telecommunications equipment, hardware development and manufacturing, systems integration, and implementation. **Common positions include:** Aerospace Engineer; Aircraft Mechanic/Engine Specialist; Budget Analyst; Computer Programmer; Designer; Draftsperson; Electrical/Electronics Engineer; Financial Analyst; Management Analyst/Consultant; Mechanical Engineer; Software Engineer; Systems Analyst; Technical Writer/Editor. **Educational backgrounds include:** Business Administration; Computer

Science; Engineering. **Benefits:** 401(k); Dental Insurance; Disability Coverage; Life Insurance; Medical Insurance; Tuition Assistance. **Corporate headquarters location:** This Location. **Number of employees at this location:** 40. **Number of employees nationwide:** 500.

TELECOMMUNICATIONS SYSTEMS, INC.
275 West Street, Suite 400, Annapolis MD 21401. 410/263-7616. **Contact:** Human Resources. **Description:** Provides systems engineering and integration.

3COM CORPORATION
5400 Bayfront Plaza, Santa Clara CA 95052-8145. 408/764-5000. **Contact:** Professional Staffing. **E-mail address:** 3com_resumes@3mail.3com.com. **Description:** 3Com is a billion-dollar *Fortune* 500 company delivering global data networking solutions to organizations around the world. The company designs, manufactures, markets, and supports a broad range of ISO 9000-compliant global data networking solutions including router, hubs, remote access servers, switches, and adapters for Ethernet, Token Ring, and high-speed networks. These products enable computers to communicate at high speeds and share resources including printers, disk drives, modems, and minicomputers. **NOTE:** In February 1997, media reports indicated that 3Com planned a buyout of U.S. Robotics. **Common positions include:** Accountant/Auditor; Administrator; Buyer; Customer Service Representative; Electrical/Electronics Engineer; Financial Analyst; Human Resources Manager; Marketing Specialist; Operations/Production Manager; Sales Representative; Software Engineer; Systems Analyst. **Educational backgrounds include:** Accounting; Business Administration; Computer Science; Engineering; Finance; Marketing. **Benefits:** 401(k); Computer Loans; Dental Insurance; Disability Coverage; Employee Discounts; Life Insurance; Medical Insurance; Pension Plan; Profit Sharing; Savings Plan; Stock Option; Tuition Assistance. **Special Programs:** Internships. **Corporate headquarters location:** This Location. **Other U.S. locations:** North Billerica MA. **Operations at this facility include:** Administration; Divisional Headquarters; Manufacturing; Research and Development; Sales; Service. **Listed on:** NASDAQ. **Number of employees nationwide:** 2,800.
Other U.S. locations:
- 2390 East Camelback Road, Suite 300, Phoenix AZ 85016. 602/553-1038. This location is a sales and technical support office.
- 118 Turnpike Road, Southborough MA 01772. 508/460-8900. At this location, the company assembles global networking products.

3M HEALTH INFORMATION SYSTEMS
575 West Murray Boulevard, Murray UT 84123-4611. 801/265-4400. **Fax:** 801/263-8474. **Contact:** Human Resources Manager. **Description:** 3M Health Information Systems is a leading supplier of clinically-based computer systems and software to more than 3,000 hospitals, health networks and enterprises, managed care organizations, outpatient facilities, and medical group practices. The company provides the health care market with clinical information systems that are widely known for their innovative capabilities in the areas of data integration, total quality management, and expert system technology. **NOTE:** This company offers entry-level positions. **Common positions include:** Computer Programmer; Customer Service Representative; Medical Records Technician; Registered Nurse; Software Engineer; Systems Analyst. **Educational backgrounds include:** Business Administration; Computer Science; Health Care; Marketing. **Benefits:** 401(k); Dental Insurance; Disability Coverage; Employee Discounts; Life Insurance; Medical Insurance; Pension Plan; Profit Sharing; Savings Plan; Tuition Assistance. **Special Programs:** Internships. **Corporate headquarters location:** This Location. **Other U.S. locations:** Wallingford CT; Silver Spring MD. **Parent company:** 3M (St. Paul MN). **Operations at this facility include:** Administration; Divisional Headquarters; Research and Development; Sales; Service. **Listed on:** Amsterdam Stock Exchange; Chicago Stock Exchange; Frankfurt Stock Exchange; New York Stock Exchange; Pacific Exchange; Paris Stock Exchange; Swiss Stock Exchange; Tokyo Stock Exchange. **Stock exchange symbol:** MMM. **Number of employees at this location:** 300. **Number of employees nationwide:** 400.

TRESP ASSOCIATES, INC.
4900 Seminary Road, Suite 700, Alexandria VA 22311. 703/845-9400. **Contact:** Gary Frenn, Human Resources Director. **Description:** TRESP Associates, Inc. is a diversified, high-technology company specializing in the application of information systems technologies in solving complex

technical problems for government and commercial clients. Areas of specialization include network engineering and LAN technology, including support and network administration; applied information technology, including research and analysis, information engineering, maintenance, administration, and program support; logistics, including inventory and distribution, logistic systems management, and maintenance operations; and training. TRESP Associates's major projects include research, MIS, and administrative support for the U.S. Department of Transportation's Federal Highway Administration; integrated management and technical support for the Defense Information Systems Agency's Center for Information Management's Office of Technical Integration; facilities management for the Office of the Secretary of Defense; logistics support for the U.S. Army Strategic Logistics Agency; conference and meeting planning for the U.S. Department of Transportation Urban Mass Transit Administration; LAN/office automation for the U.S. Army; network engineering/LAN technology for the U.S. Department of Justice's National Institute of Corrections; and program coordination and evaluation for the NASA Office of Education.

TRI-COR INDUSTRIES, INC.
8181 Professional Place, Suite 201, Landover MD 20785. 301/731-6140. **Contact:** Human Resources. **Description:** This location is an administrative office. In general, Tri-Cor Industries provides computer services under contract for the federal government. Responsibilities include networking, software development, operations, security, and more.

TUCKER AND ASSOCIATES INC.
616 Girod Street, New Orleans LA 70130. 504/522-4627. **Contact:** Human Resources. **Description:** Provides technical support for companies that are changing their computer networks or setting up a new network. The firm also provides management consulting.

TURNER SYSTEMS COMPANY (TSC)
582 Forest Ridge Drive, Youngstown OH 44512. 330/629-8670. **Contact:** Robert Turner, Owner. **Description:** Provides software and hardware integration services.

U.B. NETWORKS, INC.
3990 Freedom Circle, Santa Clara CA 95054. 408/496-0111. **Fax:** 408/970-7353. **Contact:** Staffing Specialists. **E-mail address:** jpeters@ub.com or dfinucan@ub.com. **Description:** Designs computer networks. U.B. Networks was formerly Ungermann-Bass, Inc. **Common positions include:** Accountant/Auditor; Buyer; Customer Service Representative; Electrical/Electronics Engineer; Financial Analyst; Human Resources Manager; Public Relations Specialist; Software Engineer; Systems Analyst; Technical Writer/Editor. **Educational backgrounds include:** Computer Science; Engineering; Finance. **Benefits:** 401(k); Daycare Assistance; Dental Insurance; Disability Coverage; Employee Discounts; Life Insurance; Medical Insurance; Tuition Assistance. **Corporate headquarters location:** This Location. **Parent company:** Tandem Computers. **Operations at this facility include:** Administration; Divisional Headquarters; Research and Development; Sales; Service. **Listed on:** Privately held. **Number of employees at this location:** 450. **Number of employees nationwide:** 750.
Other U.S. locations:
* 312 Plum Street, Suite 1475, Cincinnati OH 45202. 513/684-4000.

U.S. COMPUTER GROUP
4 Dubon Court, Farmingdale NY 11735. 516/753-6080. **Contact:** Human Resources. **Description:** U.S. Computer Group installs and supports computer systems, and manufactures computers.

U.S. CONNECT MILWAUKEE
MICRO INFORMATION SERVICES
11425 West Lake Park Drive, Mequon WI 53224. 414/577-6600. **Contact:** Human Resources. **Description:** Provides computer networking services. **Number of employees at this location:** 40.

UNISYS CORPORATION
P.O. Box 500, Blue Bell PA 19424. 215/986-4011. **Contact:** Recruiting and Staffing Department. **World Wide Web address:** http://www.unisys.com. **Description:** Unisys Corporation is a provider of information services, technology, and software. Unisys specializes in developing

business critical solutions based on open information networks. The company's enabling software team creates a variety of software projects which facilitate the building of user applications and the management of distributed systems. The company's platforms group is responsible for UNIX Operating Systems running across a wide range of multiple processor server platforms, including all peripheral and communication drivers. The Unisys commercial parallel processing team develops microkernel-based operating systems, I/O device drivers, ATM hardware, diagnostics, and system architectures. The system management group is chartered with the overall management of development programs for UNIX desktop and entry server products. **Corporate headquarters location:** This Location. **Number of employees worldwide:** 49,000.

Other U.S. locations:

- 2525 East Camelback Road, Suite 1100, Phoenix AZ 85016. 602/224-4200. This location houses a sales force and an engineering staff.
- 3101 Pegasus Drive, Bakersfield CA 93308. 805/391-3800. This location is a service center.
- 1000 Marina Boulevard, Brisbane CA 94005. 415/875-4400.
- 2049 Century Park East, Suite 310, Los Angeles CA 90067. 310/208-1511.
- 25725 Jeronimo Road, Mission Viejo CA 92691. 714/380-5000. This location is an engineering office.
- 10850 Via Frontera, San Diego CA 92127. 619/451-3000. This location is manufacturing facility.
- 2700 North First Street, San Jose CA 95134-2028. 408/434-2848. This location is a manufacturing facility.
- 6025 South Quebec Street, Suite 200, Englewood CO 80111. 303/694-8000. This location offers repair, marketing, and technical support services.
- 7000 West Palmetto Park Road, Suite 201, Boca Raton FL 33433. 561/750-5800. This location is a regional headquarters.
- 4151 Ashford Dunwoody Road NE, Atlanta GA 30319. 404/851-3000. This location maintains a staff of information services specialists and global customer support representatives.
- 333 Butterfield Road, One Unisys Center, Lombard IL 60148. 630/969-5550. This location is a sales office.
- 28 Atlantic Place, South Portland ME 04106. 207/773-8119.
- 41100 Plymouth Road, Plymouth MI 48170-1856. 313/451-4000. This location supports an engineering, assembly, and marketing staff.
- P.O. Box 64942, St. Paul MN 55164. 612/635-7777. This location is a sales, manufacturing, and software development location.
- Two Oak Way, Berkeley Heights NJ 07922. 908/771-5000. This location is an administrative facility.
- 2476 Swedesford Road, Paoli PA 19301. 610/993-7400. This location is a software engineering facility.
- 105 Westpark Drive, Suite 410, Brentwood TN 37027. 615/371-7800. This location is a sales and repair services office.
- 12010 Sunrise Valley Drive, Reston VA 22091. 703/620-7000. Contact: Bob Birch, Human Resources. This location offers various computer services to the federal government.

UNISYS CORPORATION
322 North 2200 West, Mail Stop D07, Salt Lake City UT 84116-2979. 801/594-5000. **Contact:** Summer Dawson, Human Resources. **Description:** The Salt Lake City location is involved in scalable commercial transaction processing systems based on distributed computing technologies. Overall, Unisys Corporation is a provider of information services, technology, and software. Unisys specializes in developing business critical solutions based on open information networks. The company also provides networking and communication strategies, definition, engineering, and business development for non-host communications and networking products. Unisys is also engaged in maximizing investment in parallel processing for both hardware and software development, as well as ICV support. The company's enabling software team creates a variety of software projects which facilitate the building of user applications and the management of distributed systems. The company's platforms group is responsible for UNIX Operating Systems running across a wide range of multiple processor server platforms, including all peripheral and communication drivers. The Unisys commercial parallel processing team develops microkernel-based operating systems, I/O device drivers, ATM hardware, diagnostics, and system architectures.

The system management group is chartered with the overall management of development programs for UNIX desktop and entry server products. **Corporate headquarters location:** Blue Bell PA. **Number of employees worldwide:** 49,000.

UNITECH

12450 Fair Lakes Circle, Suite 625, Fairfax VA 22033. 703/502-9600. **Contact:** Human Resources. **Description:** Offers a wide variety of computer services including network design, network management, LAN/WAN implementation, and more.

USER TECHNOLOGY ASSOCIATES

2711 Jefferson Davis Highway, Arlington VA 22554. 703/418-6846. **Contact:** Human Resources. **Description:** A provider of information management technology and systems integration.

VAIL RESEARCH

5285 Shawnee Road, Suite 210, Alexandria VA 22312. 703/642-0901. **Contact:** Human Resources. **Description:** A government contractor providing computer systems integration and engineering services.

VALINOR INC.

Seven Perimeter Road, Manchester NH 03103. 603/668-1776. **Contact:** Human Resources Department. **Description:** A computer systems company involved in the sale of Microsoft products as well as Microsoft training and networking.

VANSTAR

3005 South 48th Street, Tempe AZ 85282. 602/431-6700. **Contact:** Human Resources. **Description:** Vanstar sells and repairs computers as well as installing hardware and networking systems.

VEDA INC.

2001 North Beauregard Street, Suite 1200, Alexandria VA 22311. 703/575-3100. **Contact:** Personnel. **E-mail address:** personnel@veda.com. **World Wide Web address:** http://www.veda.com. **Description:** Provides computer systems support. Veda personnel possess skills covering a wide range of systems integration topics. The company assists the military with software engineering, systems integration, and project development. **Corporate headquarters location:** This Location.
Other U.S. locations:
- 197 Eglin Parkway, Suite 201, Fort Walton Beach FL 32548-4468. 904/243-6644. This location offers computer systems support to nearby Airforce bases.
- 22309 Exploration Drive, Lexington Park MD 20653. 301/863-4302. This location offers computer support to the Naval Air Warfare Center.
- P.O. Box 896, Dahlgren VA 22448. 540/663-3400. This location offers computer support to the Naval Surface Warfare Center.
- 2101 Executive Drive, Tower Box 58, Hampton VA 23666. 757/825-0973. This location offers computer systems support to the Joint Warfighting Center.

VITRO CORPORATION

1601 Research Boulevard, Rockville MD 20850. 301/738-4000. **Contact:** Human Resources Department. **Description:** Vitro Corporation provides computer integrated systems design services to a broad range of clients.

VOLT INFORMATION SCIENCES, INC.

1221 Avenue of the Americas, 47th Floor, New York NY 10020. 212/704-2400. **Contact:** Human Resources. **Description:** Volt Information Sciences installs and supports computer systems and is involved in electronic manufacturing.

WCS COMPANY

P.O. Box 248, Newbury Park CA 91319-0248. 805/498-7621. **Contact:** Human Resources. **Description:** WCS installs, maintains, and supports computer networks.

WYBRITE INC.
3839 Washington Avenue North, Minneapolis MN 55412. 612/588-7501. **Contact:** Ms. Bev Rust, Office Manager. **Description:** Custom-designs, installs, and maintains networking systems, primarily for corporate customers. Wybrite also has a retail division operates under the name PC Solutions.

X INSIDE INC.
1801 Broadway, Suite 1710, Denver CO 80202. 303/298-7478. **Toll-free phone:** 800/946-7433. **Fax:** 303/298-1406. **Contact:** Human Resources. **E-mail address:** jobs@xinside.com. **World Wide Web address:** http://www.xinside.com. **Description:** A provider of networked graphics technology. Products include X Windows display servers; Motif window manager and development; OpenGL development; and custom development.

XDB SYSTEMS, INC.
9861 Broken Land Parkway, Columbia MD 21046. 410/312-9300. **Contact:** Human Resources. **Description:** A client/server database provider.

XYPLEX INC.
330 Codman Hill Road, Boxborough MA 01719-1739. 508/264-9900. **Contact:** Dennis Fitzgerald, Vice President Human Resources. **Description:** Designs, manufactures, markets, and supports high-performance data communications network systems. **Number of employees nationwide:** 300.

XYQUAD INC.
2921 South Brentwood Boulevard, St. Louis MO 63144. 314/961-5995. **Contact:** Human Resources. **Description:** A computer integration company that sells, installs, and maintains computer networks. **Corporate headquarters location:** This Location.
Other U.S. locations:
• 221 East Lake Street, Addison IL 60101. 630/530-8884.

COMPUTER CONSULTANTS

*M*ike *Maziarz, a Staff Consultant with CSC, found his job through* The Boston JobBank. *While the reference book streamlined his efforts, Mike landed the position by displaying his desire to succeed in the field. His recommendation to jobseekers is to fine tune your oral and written communication skills, display the ability to learn new technologies, and possess a general desire to overcome challenges. Those candidates with computer science backgrounds will command an advantage over other jobseekers only if they can apply their knowledge to real life situations and be ready to learn new skills.*

Mike sees the field remaining tight, with several larger corporations retaining a majority of the consulting business. Smaller firms can remain competitive on a local scale if they can outbid a larger firm on a particular project. But no matter the size of one's company, a consultant's daily routine constantly changes and requires a person to remain flexible and ambitious. Mike gives one final piece of advice to jobseekers and consultants alike: In this field of organized chaos, there's no room to sit back and get comfortable.

ACI AUTOMATED CONCEPTS, INC.
90 Woodbridge Center Drive, Woodbridge NJ 07095. 908/602-0200. **Contact:** Human Resources. **Description:** A software consulting firm for systems integration.

AIM INC.
4403 Forbes Boulevard, Lanham MD 20706. 301/794-8200. **Contact:** Lori Poindexter, Recruiting. **Description:** Provides computer consulting services.

AMS COURSEWARE DEVELOPERS
300 Chapel Road, Manchester CT 06040. 860/646-3264. **Contact:** Human Resources. **Description:** Develops computer training courses. AMS Courseware Developers creates custom designed training courses for specific computer applications used by business clientele.

ABACUS TECHNOLOGY CORPORATION
5454 Wisconsin Avenue, Suite 1100, Chevy Chase MD 20815. 301/907-8500. **Contact:** Human Resources. **Description:** An information technology consulting firm.

ADVANCED RESOURCE TECHNOLOGIES, INC. (ARTI)
4900 Seminary Road, Suite 1200, Alexandria VA 22311. 703/998-4200. **Contact:** Human Resources. **Description:** Offers a wide variety of computer services including network engineering, information management, systems analysis, and more. Advanced Resource Technologies provides most of its services to the government but is currently branching out into the commercial sector. **Listed on:** Privately held.

AJILON
210 West Pennsylvania Avenue, Suite 650, Towson MD 21204-4532. 410/821-0435. **Contact:** Melanie Porter, Vice President, Human Resources. **Description:** Formerly Adia Information Technology, Ajilon is a provider of computer consulting services. **Common positions include:** Information Systems Consultant; Software Engineer. **Educational backgrounds include:** Data Processing.
Other U.S. locations:
- 5333 North 7th Street, Suite A201, Phoenix AZ 85014. 602/230-1008.
- One Tower Bridge, 100 Front Street, West Conshohocken PA 19428. 610/834-8290.

ALLDATA CORPORATION
9412 Big Horn Boulevard, Elk Grove CA 95758-1100. 916/684-5200. **Contact:** Department of Interest. **Description:** Provides computer consulting services. **Common positions include:** Accountant/Auditor; Electrical/Electronics Engineer; Financial Analyst; Mechanical Engineer; Software Engineer. **Number of employees at this location:** 300.

AMERICAN TECHNICAL SERVICES
5575 Technology Center Drive, Suite 218, Colorado Springs CO 80919. 719/522-1616. **Toll-free phone:** 800/636-4287. **Fax:** 719/522-1619. **Contact:** Manager. **Description:** A computer consulting firm specializing in information technology, automated and communications systems, and technical engineering services. **Common positions include:** Computer Programmer; Systems Analyst. **Benefits:** 401(k); Dental Insurance; Medical Insurance.

ANALYSTS INTERNATIONAL CORPORATION (AiC)
7615 Metro Boulevard, Minneapolis MN 55439-3050. **Toll-free phone:** 800/776-3553. **Fax:** 612/897-4551. **Contact:** Senior Staffing Assistant. **Description:** AiC is an international computer consulting firm. The company assists clients in developing systems in a variety of industries using different programming languages and software. This involves systems analysis, design, and development. **NOTE:** A minimum of one to two years of programming experience is required. **Common positions include:** Computer Programmer; Management Analyst/Consultant; Software Engineer; Systems Analyst; Technical Writer/Editor. **Educational backgrounds include:** Business Administration; Computer Science; Engineering; Mathematics. **Benefits:** 401(k); Dental Insurance; Disability Coverage; Employee Discounts; Life Insurance; Medical Insurance; Pension Plan; Profit Sharing; Savings Plan; Tuition Assistance. **Corporate headquarters location:** This Location. **Operations at this facility include:** Administration; Sales; Service. **Listed on:** NASDAQ. **Number of employees at this location:** 400. **Number of employees nationwide:** 2,700. **Number of employees worldwide:** 3,700.
Other U.S. locations:
- 11024 North 28th Drive, Suite 240, Phoenix AZ 85029. 602/789-7200.
- 1850 Gateway Boulevard, Suite 120, Concord CA 94520-3277. 510/687-5522.
- 7800 East Union Avenue, Suite 600, Denver CO 80237-2755. 303/721-6200. Contact: Kimberly Shocole, Manager of Human Resources.
- 621 NW 53rd Street, Suite 140, Boca Raton FL 33487-8211. 561/241-5912. Contact: Blanca Posada, Human Resources Representative.
- 600 North Westshore Boulevard, Suite 304, Tampa FL 33609-1145. 813/281-0458.
- 2365 Harrodsburg Road, Suite B450, Lexington KY 40504-3335. 606/223-0001.
- 5750 Castle Creek Parkway, Suite 259, Indianapolis IN 46250. 317/842-1100.
- 3000 Town Center, Suite 570, Southfield MI 48075-1297. 810/353-7230.
- 3101 Broadway, Suite 204, Kansas City MO 64111-1858. 816/531-5050.
- 600 Emerson Road, Suite 200, St. Louis MO 63141-6708. 314/997-1746.
- 2700 Gateway Center Boulevard, Suite 600, Morrisville NC 27560-9121. 919/460-6141.
- 6910 Pacific Street, Suite 204, Omaha NE 68106-1045. 402/558-2800.
- 111 Wood Avenue South, Iselin NJ 08830. 201/535-9844.
- One Penn Plaza, Suite 1910, New York NY 10119-0002. 212/465-1660.
- 471 East Broad Street, Suite 2001, Columbus OH 43215-3861. 614/224-6790.
- 1033 La Posada Drive, Suite 300, Austin TX 78752-3824. 512/206-2700. Contact: Jan Collins, Human Resources Manager.
- 3030 LBJ Freeway LB 52, Suite 820, Dallas TX 75234. 972/243-2001.
- 1415 North Loop West, Suite 300, Houston TX 77008. 713/869-3420.

ANALYSTS INTERNATIONAL CORPORATION (AiC)
1101 Perimeter Drive, Suite 500, Schaumburg IL 60173-5060. 847/619-4673. **Fax:** 847/605-9489. **Contact:** Sandra Lamar, Recruiter. **E-mail address:** slamar@analysts.com. **Description:** AiC is an international computer consulting firm. The company assists clients in developing systems in a variety of industries using different programming languages and software. This involves systems analysis, design, and development. **Common positions include:** Applications Engineer; Computer Programmer; Software Engineer; Systems Analyst; Systems Manager. **Educational backgrounds include:** Computer Science; Engineering. **Benefits:** 401(k); Dental Insurance; Disability Coverage; Employee Discounts; Life Insurance; Medical Insurance; Pension Plan; Profit Sharing; Tuition

Assistance. **Corporate headquarters location:** Minneapolis MN. **Listed on:** NASDAQ. **Annual sales/revenues:** More than $100 million. **Number of employees at this location:** 300.

ANALYSTS INTERNATIONAL CORPORATION (AiC)
Corporate Plaza One, Suite 350, 6450 Rockside Wood Boulevard South, Cleveland OH 44131-2220. 216/524-8990. **Fax:** 216/524-9535. **Contact:** Human Resources. **E-mail address:** cleveland@analysts.com. **World Wide Web address:** http://www.analysts.com. **Description:** AiC is an international computer consulting firm. The company assists clients in developing systems in a variety of industries using different programming languages and software. This involves systems analysis, design, and development. **Common positions include:** Computer Programmer; Electrical/Electronics Engineer; Systems Analyst. **Educational backgrounds include:** Computer Science. **Benefits:** 401(k); Dental Insurance; Disability Coverage; Life Insurance; Medical Insurance; Tuition Assistance. **Corporate headquarters location:** Minneapolis MN. **Listed on:** NASDAQ. **Stock exchange symbol:** ANALY. **Annual sales/revenues:** more than $100 million. **Number of employees at this location:** 100. **Number of employees nationwide:** 3,000.

ANALYSTS INTERNATIONAL CORPORATION (AiC)
16 West Main Street, Suite 200, Rochester NY 14614-1601. 716/325-6640. **Fax:** 716/325-6273. **Contact:** Recruiting Department. **E-mail address:** recruit@servtech.com. **World Wide Web address:** http://www.analysts.com. **Description:** AiC is an international computer consulting firm. The company assists clients in developing systems in a variety of industries using different programming languages and software. This involves systems analysis, design, and development. **Common positions include:** Computer Programmer; Software Engineer; Systems Analyst. **Educational backgrounds include:** Computer Science; Engineering; Mathematics. **Benefits:** 401(k); Dental Insurance; Disability Coverage; Life Insurance; Medical Insurance; Profit Sharing; Tuition Assistance. **Corporate headquarters location:** Minneapolis MN. **Operations at this facility include:** Service. **Listed on:** NASDAQ. **Number of employees worldwide:** 3,700.

ANALYSTS INTERNATIONAL CORPORATION (AiC)
10655 NE 4th Street, Suite 804, Bellevue WA 98004-5022. 206/454-2500. **Contact:** Human Resources. **Description:** AiC is an international computer consulting firm. The company assists clients in developing systems in a variety of industries using different programming languages and software. This involves systems analysis, design, and development. **Common positions include:** Software Engineer. **Corporate headquarters location:** Minneapolis MN. **Other U.S. locations:** 40 nationwide. **Listed on:** NASDAQ. **Number of employees at this location:** 80. **Number of employees worldwide:** 3,700.

ANALYSTS INTERNATIONAL CORPORATION (AiC)
SOUTHERN REGION/BRANCH OFFICE
1100 Johnson Ferry Road, Suite 850, Atlanta GA 30342-1746. 404/256-5190. **Contact:** Vicki Brown, Recruiter. **Description:** AiC is an information systems consulting firm. AiC provides analytical and programming services, including consulting, systems analysis, design, programming, instruction, and technical writing. **Common positions include:** Computer Programmer; Software Engineer; Technical Writer/Editor. **Educational backgrounds include:** Computer Science; Engineering; Mathematics. **Benefits:** 401(k); Dental Insurance; Disability Coverage; Life Insurance; Medical Insurance; Profit Sharing; Tuition Assistance. **Corporate headquarters location:** Minneapolis MN. **Operations at this facility include:** Regional Headquarters. **Listed on:** NASDAQ. **Annual sales/revenues:** More than $100 million. **Number of employees at this location:** 100. **Number of employees worldwide:** 3,700.

ANDERSEN CONSULTING
425 Walnut Street, Cincinnati OH 45202. 513/651-2444. **Contact:** Human Resources. **Description:** Offers technology consulting services. Andersen Consulting helps businesses apply their computer technology more effectively by combing general business knowledge with information systems skills. **Parent company:** Arthur Andersen & Company.

APPLIED SCIENCE ASSOCIATES INC.
292 Three Degree Road, Butler PA 16001. 412/284-7300. **Contact:** Human Resources. **Description:** Offers computer and multimedia training programs.

ASIX
777 108th Avenue NE, #1830, Bellevue WA 98004. 206/635-0709. **Contact:** Human Resources. **Description:** ASIX is a computer consulting firm.

ATLANTIC DATA SERVICES
One Battlemart Park, Quincy MA 02169. 617/770-3333. **Contact:** Human Resources. **Description:** Provides computer consulting services. Atlantic Data Services offers contract computer consulting on a variety of team oriented computer projects to the banking industry.

BEST INC.
12910 Totem Lake Boulevard, Suite 270, Kirkland WA 98034. 206/814-8104. **Contact:** Human Resources. **Description:** Offers computer consulting services. Best Inc. provides businesses with support for mainframes, personal computers, systems configurations, information systems, and software testing.

THE BLACKSTONE GROUP
155 108th Avenue NE, Suite 503, Bellevue WA 98004. 206/646-7800. **Contact:** Human Resources. **Description:** Provides software and information systems consulting services.

BUSINESS INFORMATION TECHNOLOGY, INC.
1800 Sutter Street, Suite 770, Concord CA 94520. 510/671-0595. **Fax:** 510/671-2525. **Contact:** Trish Prokop, Employment Administrator. **E-mail address:** trish.prokop@internetmcl.com. **World Wide Web address:** http://www.bitcorp.com/bitcorp. **Description:** A leading software consulting firm serving clients in a variety of industries. The firm implements HRMS, financial, distribution, manufacturing, and student systems. **Common positions include:** Applications Engineer; Computer Programmer; Consultant; Sales Representative; Software Engineer; Systems Analyst. **Educational backgrounds include:** Accounting; Computer Science; Finance. **Benefits:** 401(k); Dental Insurance; Disability Coverage; Life Insurance; Medical Insurance; Profit Sharing. **Corporate headquarters location:** This Location. **Other U.S. locations:** Chicago IL; Philadelphia PA; Dallas TX. **Parent company:** Ciber. **Listed on:** NASDAQ. **Number of employees at this location:** 220. **Number of projected hires for 1997 - 1998 at this location:** 30.

BUSINESS SOLUTIONS INTERNATIONAL, INC.
150 Town Acres Lane, Roselle IL 60172-1521. 630/307-3660. **Fax:** 630/307-3662. **Contact:** Rebecca Larson, Human Resources Representative. **Description:** An information systems consulting firm. Business Solutions International, Inc. assists public- and private-sector managers in improving the overall performance of their organizations. **Common positions include:** Computer Programmer; Management Analyst/Consultant; Software Engineer; Systems Analyst. **Educational backgrounds include:** Computer Science; Engineering; Liberal Arts. **Benefits:** Dental Insurance; Life Insurance; Medical Insurance. **Operations at this facility include:** Administration; Sales.

CSC CONSULTING
255 East 5th Street, 27th Floor, Cincinnati OH 45202. 513/381-4440. **Contact:** Human Resources. **Description:** Offers consulting services for systems integration and any applicable software. **Parent company:** Computer Sciences Corporation primarily services the U.S. government. The four sectors of the company include the Systems Group division, the Consulting division, the Industry Services Group, and the CSC divisions. The Systems Group division designs, engineers, and integrates computer-based systems and communications systems, providing all the hardware, software, training, and related elements necessary to operate a system. The Consulting division includes consulting and technical services in the development of computer and communication systems to non-federal organizations. The Industry Services Group provides service to health care, insurance, and financial services, as well as providing large-scale claim processing and other insurance-related services. CSC Health Care markets business systems and services to the managed health care industry, clinics, and physicians. CSC Enterprises provides consumer credit reports and account management services to credit grantors.
Other U.S. locations:
- Two Ravinia Drive, Suite 1150, Atlanta GA 30346. 770/677-3200.
- 5885 Landerbrook Drive, Suite 300, Cleveland OH 44124-4031. 216/449-3600.

CSC PARTNERS
One Newton Executive Park, Newton Lower Falls MA 02162-1435. 617/332-3900. **Contact:** Ms. Pat Belheen, Recruiting Manager. **Description:** A consulting company specializing in systems integration, systems design, and applications development for the commercial and private sectors. **Common positions include:** Computer Programmer; Software Engineer; Systems Analyst. **Educational backgrounds include:** Computer Science. **Benefits:** 401(k); Dental Insurance; Disability Coverage; Life Insurance; Medical Insurance; Profit Sharing; Tuition Assistance. **Corporate headquarters location:** Waltham MA. **Parent company:** Computer Sciences Corporation primarily services the U.S. government. The four sectors of the company include the Systems Group division, the Consulting division, the Industry Services Group, and the CSC divisions. The Systems Group division designs, engineers, and integrates computer-based systems and communications systems, providing all the hardware, software, training, and related elements necessary to operate a system. The Consulting division includes consulting and technical services in the development of computer and communication systems to non-federal organizations. The Industry Services Group provides service to health care, insurance, and financial services, as well as providing large-scale claim processing and other insurance-related services. CSC Health Care markets business systems and services to the managed health care industry, clinics, and physicians. CSC Enterprises provides consumer credit reports and account management services to credit grantors. **Listed on:** New York Stock Exchange. **Number of employees at this location:** 200. **Number of employees nationwide:** 1,600.

CTG (COMPUTER TASK GROUP, INC.)
370 East South Temple, #250, Salt Lake City UT 84111. 801/363-0800. **Contact:** Human Resources. **Description:** A computer consulting firm that performs programming and networking services for corporate clients.

CTG (COMPUTER TASK GROUP, INC.)
28411 Northwestern Highway, Suite 200, Southfield MI 48034. 810/746-6090. **Contact:** Human Resources. **Description:** Offers consulting services to corporate clients for their computer networking needs.

CAMBRIDGE COMPUTER ASSOCIATES
90 Sherman Street, Cambridge MA 02140. 617/868-1111. **Contact:** Human Resources. **Description:** A computer consulting firm.

CAP GEMINI AMERICA
25800 Science Park Drive, Suite 180, Cleveland OH 44122. 216/464-8616. **Contact:** Ms. Stacy Cook, Recruiting Manager. **Description:** A provider of information technology consulting services. The services Cap Gemini America performs include systems integration; application design, development, and documentation; systems conversions and migrations; and information technology consulting. The company delivers these services as large-scale total responsibility projects or extensions of the clients' staff to a wide variety of corporations in all industries. Fifty branch offices are located nationwide. **Common positions include:** Computer Programmer; Electrical/Electronics Engineer; Project Manager; Systems Analyst; Technical Writer/Editor. **Educational backgrounds include:** Computer Science; Engineering; M.I.S.; Mathematics. **Benefits:** Medical Insurance. **Number of employees nationwide:** 3,000.
Other U.S. locations:
- 111 Wood Avenue South, Iselin NJ 08830. 908/906-0400.
- 1114 Avenue of the Americas, 29th Floor, New York NY 10036. 212/944-6464. Contact: Laurie Jadick, Recruiter.
- 890 SW First Avenue, Portland OR 97201-5824. 503/295-1909. Contact: Greg Lueck, Human Resources Manager.
- 4321 West College Avenue, Appleton WI 54914. 414/730-3856.

CENTURY TECHNOLOGIES INC.
8405 Colesville Road, Suite 400, Silver Spring MD 20910. 301/585-4800. **Contact:** Human Resources. **Description:** Offers consulting services to software developers.

CIBER, INC.
5251 DTC Parkway, Suite 1400, Greenwood Village CO 80111. 303/220-0100. **Contact:** Human Resources. **Description:** Ciber, Inc. is a nationwide provider of information technology consultation services, including computer applications software and software implementation services. **Corporate headquarters location:** This Location.
Other U.S. locations:
- 1915 Highway 52 North, Suite 105, Rochester MN 55901. 507/280-9267.
- 12312 Olive Boulevard, Suite 175, St. Louis MO 63141. 314/434-7900. Contact: Amy Crane, Recruiter.

COMMAND TECHNOLOGIES CORPORATION
P.O. Box 670, Warrenton VA 27186. 540/349-8623. **Contact:** Human Resources. **Description:** A computer systems engineering firm offering a wide variety of computer services including technical assistance, network engineering, and more. **Corporate headquarters location:** This Location.

COMPLETE BUSINESS SOLUTIONS, INC. (CBSI)
32605 West Twelve Mile, Suite 250, Farmington Hills MI 48334. 810/488-2088. **Fax:** 810/488-2089. **Contact:** Human Resources. **Description:** CBSI is an international systems integration consulting firm specializing in strategic systems development, systems integration, software application development, projects, contracts, and business process/system re-engineering. Founded in 1985, the company's performance has enabled it to grow at a compound growth rate of 73 percent. **Common positions include:** Computer Programmer; Systems Analyst. **Educational backgrounds include:** Computer Science. **Benefits:** 401(k); Dental Insurance; Disability Coverage; Medical Insurance; Tuition Assistance. **Corporate headquarters location:** This Location. **Other U.S. locations:** CA; IL; NJ. **Operations at this facility include:** Administration; Divisional Headquarters; Sales; Service. **Number of employees nationwide:** 750.

COMPUTER AID, INC.
901 Market Street, 13th Floor, Wilmington DE 19801. 302/888-5500. **Contact:** Human Resources. **Description:** A computer consulting company.

COMPUTER CONSULTING GROUP
84 Villa Road, Greenville SC 29615. 864/232-7940. **Contact:** Human Resources. **Description:** A computer consultancy.

COMPUTER DATA SYSTEMS INC.
2000 Century Parkway, Suite 650, Atlanta GA 30345. 404/320-8990. **Contact:** Human Resources. **Description:** Computer Data Systems Inc. is a professional services firm which provides computer consulting and data processing for both federal and commercial clients at sites throughout the United States. **Corporate headquarters location:** Rockville MD.

COMPUTER HORIZONS CORPORATION
6312 South Fiddler's Green Circle, Suite 230E, Englewood CO 80111. 303/220-9444. **Contact:** Mark Seleck, Recruiting. **Description:** Offers a variety of consulting and support services.

COMPUTER MANAGEMENT SCIENCES, INC.
1051 Perimeter Drive, Suite 705, Schaumburg IL 60173. 847/619-2966. **Toll-free phone:** 800/725-2674. **Fax:** 847/619-2988. **Contact:** Al Calalang, Senior Corporate Recruiter. **E-mail address:** ecalalang@aol.com. **World Wide Web address:** http://www.cmsx.com. **Description:** A provider of information technology consulting and custom software development services. **Common positions include:** Computer Programmer; Systems Analyst. **Educational backgrounds include:** Computer Science. **Benefits:** 401(k); Dental Insurance; Disability Coverage; ESOP; Life Insurance; Medical Insurance; Profit Sharing; Spending Account; Tuition Assistance. **Corporate headquarters location:** Jacksonville FL. **Other U.S. locations:** Hartford CT; Atlanta GA; Boston MA; Charlotte NC; Cleveland OH; Greenville SC. **Operations at this facility include:** Regional Headquarters. **Listed on:** NASDAQ. **Stock exchange symbol:** CMSX. **Annual sales/revenues:** $21 - $50 million. **Number of employees at this location:** 20. **Number of employees nationwide:** 500.

COMPUTER MANAGEMENT SCIENCES, INC.
8133 Bay Meadows Way, Jacksonville FL 32256-8128. 904/737-8955. **Toll-free phone:** 800/725-2674. **Contact:** Human Resources Department. **Description:** A provider of information technology consulting and custom software development services. **Corporate headquarters location:** This Location. **Stock exchange symbol:** CMSX. **Annual sales/revenues:** $21 - $50 million.

COMPUTER PEOPLE INC.
707 SW Washington Street, Suite 510, Portland OR 97205. 503/224-6070. **Contact:** Mike Stout, Recruiter. **Description:** Computer People Inc. is a computer consulting firm that provides programming assistance on a permanent or temporary basis to clients in the business sector. **Corporate headquarters location:** El Segundo CA. **Number of employees at this location:** 100. **Other U.S. locations:**
- 200 Galleria Parkway NW, Suite 450, Atlanta GA 30339. 770/951-1772.
- 1601 5th Avenue, Suite 540, Seattle WA 98101. 206/628-0950.

COMPUTER PROFESSIONALS, INC. (CPI)
112 Tryon Street, Charlotte NC 28284. 704/334-8303. **Contact:** Human Resources. **Description:** Offers on-site computer consulting services in mainframe and client/server environments.

COMPUTER RESOURCE MANAGEMENT INC.
950 Herndon Parkway, Suite 360, Herndon VA 20170. 703/435-7613. **Contact:** Human Resources. **Description:** Offers computer consulting and network services.

COMPUWARE CORPORATION
7261 Engle Road, Suite 400, Middleburg Heights OH 44130. 216/234-8446. **Contact:** George Palton, Human Resources. **Description:** Offers computer consulting services.

COMSYS TECHNICAL SERVICES INC.
P.O. Box 7947, Gaithersburg MD 20898. 301/921-3600. **Contact:** Human Resources. **Description:** A provider of contract programming and computer and software consulting services.

CONNECT COMPUTER COMPANY
7101 Metro Boulevard, Minneapolis MN 55439. 612/944-0181. **Contact:** Human Resources. **Description:** Offers computer consulting services to a variety of businesses.

THE CONSTELL GROUP
216 Stelton Road, Suite E-1, Piscataway NJ 08854. 908/968-9310. **Toll-free phone:** 800/736-9310. **Fax:** 800/270-4946. **Contact:** Stuart Grader, Recruiting Manager. **E-mail address:** constell@soho.ios.com. **World Wide Web address:** http://www.constell.com. **Description:** Provides consulting for client/server development, mainframe and legacy systems, industry-specific analysis, application-specific analysis, and network development. Founded in 1984. **Common positions include:** Account Representative; Applications Engineer; Computer Operator; Computer Programmer; Database Manager; Internet Services Manager; Project Manager; Sales Executive; Sales Manager; Sales Representative; Secretary; Systems Analyst; Systems Manager. **Educational backgrounds include:** Computer Science. **Benefits:** 401(k); Dental Insurance; Disability Coverage; Life Insurance; Medical Insurance; Tuition Assistance. **Corporate headquarters location:** Elmwood Park NJ. **Other U.S. locations:** Boston MA; New York NY. **Listed on:** Privately held. **Annual sales/revenues:** $11 - $20 million. **Number of employees at this location:** 80. **Number of employees nationwide:** 140. **Number of projected hires for 1997 - 1998 at this location:** 50.

CORNELL TECHNICAL SERVICES
1921 Gallows Road, Suite 880, Vienna VA 22182. 703/734-8599. **Fax:** 703/734-2949. **Contact:** David Lindsay, Director of Recruiting. **World Wide Web address:** http://www.cornelltech.com. **Description:** A provider of computer consulting services. Founded in 1991. **NOTE:** The company offers entry-level positions. **Common positions include:** Computer Programmer; Systems Analyst; Technical Writer/Editor. **Educational backgrounds include:** Computer Science; Engineering. **Benefits:** 401(k); Dental Insurance; Disability Coverage; Financial Planning; Life Insurance; Medical Insurance; Tuition Assistance. **Corporate headquarters location:** This Location. **Other**

U.S. locations: Atlanta GA; Richmond VA. **Annual sales/revenues:** $11 - $20 million. **Number of employees nationwide:** 200. **Number of projected hires for 1997 - 1998 at this location:** 100.

CORPORATE INTELLIGENCE, INC. (CII)
7200 Falls of the News, Suite 202, Raleigh NC 27615. 919/676-8300. **Contact:** Human Resources. **Description:** Offers software consulting services.

COST MANAGEMENT SYSTEMS CORPORATION
301 Maple Avenue West, White Oak Tower Building, Vienna VA 22180-4300. 703/938-2600. **Contact:** Human Resources. **Description:** Offers a variety of computer services to the Department of Defense and commercial clients. Services include networking, systems integration, and more.

DDR, INC.
8111 LBJ Freeway, Suite 1155, Dallas TX 75251. 972/783-9981. **Contact:** Shanna Shults, Account Executive. **E-mail address:** ddrdal@gte.net. **Description:** DDR is a technical recruiting firm. The company provides consultants specializing in planning, design, development, and implementation of computer-based business systems. **Common positions include:** Computer Animator; Computer Operator; Computer Programmer; Database Manager; Financial Analyst; Hardware Engineer; Operations/Production Manager; Project Manager; Software Engineer; Systems Analyst; Technical Writer/Editor. **Educational backgrounds include:** Computer Science; Economics; Engineering; Finance; Mathematics; Physics. **Benefits:** Dental Insurance; Life Insurance; Medical Insurance. **Corporate headquarters location:** This Location. **Annual sales/revenues:** $5 - $10 million. **Number of employees nationwide:** 130.

DATA AID, INC.
Two Chase Corporate Drive, Suite 105, Birmingham AL 35244. **Toll-free phone:** 800/987-8878. **Fax:** 205/987-1014. **Contact:** Sandy Caldwell, College Recruiter. **Description:** A full-service computer consulting firm. **NOTE:** Entry-level positions are offered. **Common positions include:** Computer Programmer; Management Analyst/Consultant; MIS Specialist; Software Engineer; Systems Analyst; Technical Writer/Editor; Telecommunications Manager. **Educational backgrounds include:** Computer Science; Engineering. **Benefits:** 401(k); Dental Insurance; Disability Coverage; Life Insurance; Medical Insurance. **Corporate headquarters location:** This Location. **Other U.S. locations:** Montgomery AL; Tampa FL; Atlanta GA; St. Louis MO. **Parent company:** Corestaff. **Operations at this facility include:** Administration; Sales. **Listed on:** NASDAQ. **Number of employees at this location:** 300. **Number of employees nationwide:** 2,500.

DATA BASE CONSULTANTS INC.
4835 LBJ Freeway, Suite 900, Dallas TX 75244. 972/392-0955. **Contact:** Human Resources. **Description:** Offers consulting services for businesses with Oracle databases.

DATA SYSTEMS & SOFTWARE INC.
200 Route 17, Mahwah NJ 07430. 201/529-2026. **Contact:** Human Resources. **Description:** Data Systems & Software is a leading provider of consulting and development services for computer software and systems to high-technology companies in Israel and the United States, principally in the area of state-of-the-art, embedded real-time systems.

DATAQUEST COMPUTER
3990 Westerly Place, Suite 100, Newport Beach CA 92660. 714/476-9117. **Contact:** Human Resources. **Description:** A computer consulting firm.

DAUGHERTY SYSTEMS
One City Place, Suite 240, St. Louis MO 63141. 314/432-8200. **Fax:** 314/432-8217. **Contact:** Clyde J. Donovan, Manager, Human Resources. **E-mail address:** donovan@daugherty.com. **World Wide Web address:** http://www.daughtery.com. **Description:** A computer consulting firm. **NOTE:** The company offers training programs. **Common positions include:** Computer Programmer; Database Manager; Information Systems Consultant; MIS Specialist; Software Engineer; Systems Analyst; Systems Manager; Webmaster. **Educational backgrounds include:** Computer Science; Engineering; Mathematics. **Benefits:** 401(k); Dental Insurance; Disability

Coverage; Life Insurance; Medical Insurance; Tuition Assistance. **Corporate headquarters location:** This Location. **Other U.S. locations:** Atlanta GA; Chicago IL; Baltimore MD; Dallas TX. **Listed on:** Privately held. **Annual sales/revenues:** $21 - $50 million. **Number of employees at this location:** 190. **Number of employees nationwide:** 300. **Number of projected hires for 1997 - 1998 at this location:** 60.

DECISION CONSULTANTS INC.
28411 Northwestern Highway, Suite 750, Southfield MI 48034. 810/352-8650. **Contact:** Human Resources. **Description:** Offers computer software support for a variety of applications.

DELTAM SYSTEMS, INC.
1400 Fashion Island Boulevard #303, San Mateo CA 94404. 415/571-0555. **Fax:** 415/571-1469. **Contact:** Recruiter. **E-mail address:** hudspeth@deltamsys.com. **Description:** A computer consulting contracting firm. Founded in 1984. **Common positions include:** Computer Programmer; Information Systems Consultant; Systems Analyst. **Benefits:** 401(k); Dental Insurance; Mass Transit Available; Medical Insurance. **Annual sales/revenues:** $11 - $20 million. **Number of employees at this location:** 125. **Number of projected hires for 1997 - 1998 at this location:** 200 consultants.

DENNEY CONSULTING
4629 South Woodlake Lane SE, Conyers GA 30208-4622. 770/922-8585. **Contact:** Human Resources. **Description:** Denney Consulting is a computer consulting firm. The company provides services to companies operating with AS-400 or RPG programming languages and to individuals and companies using PC platforms.

DIGITAL SYSTEMS RESEARCH, INC.
4301 North Fairfax Drive, Arlington VA 22203. 703/522-6067. **Contact:** Human Resources. **Description:** A consulting company offering networking and systems integration, and other services.

ECOM ELITE COMPUTER CONSULTANTS
10333 NW Freeway, Suite 414, Houston TX 77092. 713/686-9740. **Contact:** Human Resources. **Description:** A computer consulting firm.

EXECUTRAIN OF FLORIDA
One Urban Center, 4830 West Kennedy Boulevard, Suite 700, Tampa FL 33609. 813/288-2000. **Contact:** Human Resources. **Description:** A firm which trains businesses in the use of computer software.

EXECUTRAIN OF TEXAS
12201 Merit Drive, Suite 350, Dallas TX 75251. 972/387-1212. **Contact:** Human Resources. **Description:** Trains employees from client companies on virtually every kind of software available today.

EXPERTEC, INC.
340 North Main Street, Suite 304, Plymouth MI 48170. 313/451-2272. **Contact:** Human Resources Manager. **Description:** Expertec, Inc. is a computer consulting firm. The company also develops software.

FUSION SYSTEMS GROUP LTD.
#1 Walsh Court, New York NY 10005. 212/285-8001. **Contact:** Human Resources. **Description:** Offers software consulting services to businesses.

GRC INTERNATIONAL, INC.
621 Lynnhaven Parkway, Suite 360, Virginia Beach VA 23451. 757/463-0683. **Fax:** 757/463-2996. **Contact:** Human Resources. **Description:** This location provides computer training for a broad customer base. GRC International is an international provider of knowledge-based professional services and technology-based product solutions to government and commercial customers. GRC International's activities encompass sophisticated telecommunications products,

network systems analysis, and network software development operations for the commercial market. The company creates large-scale decision-support systems and software engineering environments; applies operations research and mathematical modeling to business and management systems; and implements advanced database technology. GRC International also provides studies and analysis capabilities for policy development and planning; modeling and simulation of hardware and software used in real-time testing of sensor, weapon, and battlefield management command, control, and communication systems; and testing and evaluation. **NOTE:** Please direct all employment inquiries to Linda Thompson, Director of Human Resources, GRC International, Inc., 1900 Gallows Road, Vienna VA 22182. **Corporate headquarters location:** Vienna VA. **Other U.S. locations:** Nationwide. **Listed on:** New York Stock Exchange. **Number of employees worldwide:** 1,300.

GREENBRIER & RUSSEL INC.
1450 East American Lane, Suite 1700, Schaumburg IL 60173. 847/706-4000. **Contact:** Sherry Harmon, Recruiter. **Description:** A computer consulting company. Greenbrier & Russel also provides software training services.

HRH
545 Eighth Avenue, Suite 401, New York NY 10018. 212/868-1126. **Contact:** Human Resources. **Description:** Macintosh computer consultants. **NOTE:** Positions are available for people with experience with networking and training.

IMI SYSTEMS
Four Kings Highway East, Haddonfield NJ 08033. 609/795-5000. **Fax:** 609/795-9850. **Contact:** Recruiting Manager. **Description:** An information technology consulting firm. IMI Systems provides computer consulting personnel to other corporations.

INFORMATION HORIZON
20 Waterview Boulevard, 4th Floor, Parsippany NJ 07054. 201/402-9364. **Contact:** Human Resources. **Description:** Offers computer consulting services specializing in financial applications.

INFORMATION TECHNOLOGY SERVICES INC.
2310 130th Avenue, Suite B201, Bellevue WA 98005. 206/869-4040. **Contact:** Human Resources. **Description:** Information Technology Services provides consulting, networking, programming, and Internet services for companies in various industries.

INTEGRATED BUSINESS SOLUTIONS
1050 17th Street NW, Suite 600, Washington DC 20036. 202/467-6899. **Contact:** Ann Moon, Principal. **Description:** A computer consulting firm.

INTERIM TECHNOLOGY
823 Commerce Drive, Oak Brook IL 60521-1919. 630/574-3030. **Contact:** Staffing Specialist. **Description:** A worldwide information systems consulting firm with 60 offices on four continents. Services include software development, project management, quality/productivity metrics, testing, strategic consulting, and client/server development using object-oriented development techniques on distributed systems. **Common positions include:** Accountant/Auditor; Administrator; Advertising Clerk; Branch Manager; Computer Programmer; Customer Service Representative; Department Manager; Electrical/Electronics Engineer; Human Resources Manager; Instructor/Trainer; Marketing Specialist; Quality Control Supervisor; Services Sales Representative; Statistician; Systems Analyst; Technical Writer/Editor. **Educational backgrounds include:** Communications; Computer Science; Mathematics. **Benefits:** Dental Insurance; Life Insurance; Medical Insurance; Savings Plan; Tuition Assistance. **Special Programs:** Internships. **Corporate headquarters location:** This Location. **Operations at this facility include:** Administration; Research and Development; Sales; Service. **Listed on:** Australian Stock Exchange. **Number of employees at this location:** 2,500.
Other U.S. locations:
*   4600 South Ulster Street, Suite 630, Denver CO 80237-2869. 303/694-1500.
*   10 Parsonage Road, Suite 212, Edison NJ 08837. 908/494-9333.
*   9 Polito Avenue, 9th Floor, Lyndhurst NJ 07071. 201/392-0800.

INTERIM TECHNOLOGY
324 East Wisconsin Avenue, Suite 530, Milwaukee WI 53202. 414/765-0550. **Contact:** Craig Rondou, Staffing Specialist. **Description:** A worldwide information systems consulting firm with 60 offices on four continents. Services include software development project management, quality/productivity metrics, testing, strategic consulting, and client/server development using object-oriented development techniques on distributed systems. **Common positions include:** Computer Programmer; Internet Services Manager; Management Analyst/Consultant; MIS Specialist; Software Engineer; Systems Analyst; Technical Writer/Editor. **Educational backgrounds include:** Business Administration; Computer Science; Mathematics. **Benefits:** 401(k); Disability Coverage; Life Insurance; Medical Insurance; Pension Plan; Savings Plan; Tuition Assistance. **Corporate headquarters location:** Oak Brook IL. **Other U.S. locations:** Nationwide. **Parent company:** Interim. **Operations at this facility include:** Administration; Divisional Headquarters; Sales. **Listed on:** NASDAQ. **Annual sales/revenues:** More than $100 million. **Number of employees at this location:** 50. **Number of employees nationwide:** 400,000.

INTERVISE CONSULTANTS, INC.
6001 Montrose Road, Suite 1000, Rockville MD 20852. 301/984-6800. **Contact:** Human Resources. **Description:** A computer systems engineering consulting firm.

JRA INFORMATION SERVICES, INC.
4100 McEwen, Suite 230, Dallas TX 75244. 972/702-8900. **Contact:** Human Resources. **Description:** Provides computer consulting and technical consulting services.

KEANE, INC.
Cali Center, 100 Walnut Street, Clark NJ 07066. 908/396-4321. **Contact:** Human Resources. **Description:** Offers computer consulting services.

KEANE, INC.
485 Devon Park Drive, Suite 119, Wayne PA 19087. 610/687-5590. **Contact:** Human Resources. **Description:** This location offers businesses a variety of computer consulting services. Keane also develops, markets, and manages software for its clients and will assist in project management.

KEANE, INC.
400 North Executive Drive, Brookfield WI 53005. 414/797-4980. **Contact:** Human Resources. **Description:** Offers computer consulting services.

KEYSTONE COMPUTER ASSOCIATES
1055 Virginia Drive, Fort Washington PA 19034. 215/643-3800. **Contact:** Human Resources. **Description:** A computer consulting firm.

LEGAL DECISIONS SYSTEMS INC. (LDSi)
2021 East Hennepin Avenue, LL-30, Minneapolis MN 55413. 612/378-1108. **Contact:** Jerry Snyder, President. **Description:** Provides computer consulting services. LDSi offers support for networks and applications.

LUCENT TECHNOLOGIES SERVICES COMPANY, INC.
7042 Alamo Downs Parkway, Suite 300, San Antonio TX 78238. 210/520-7701. **Fax:** 210/520-8547. **Contact:** Kathy Brown, Human Resources. **Description:** This location provides computer training, primarily for UNIX applications. Overall, Lucent Technologies manufactures data communications products, computer products, switching and transmission equipment, and components. **Common positions include:** Computer Programmer; Software Engineer; Systems Analyst; Technical Writer/Editor; Telecommunications Manager; Transportation/Traffic Specialist; Typist/Word Processor.

MCSI TECHNOLOGIES, INC.
8401 Colesville Road, Suite 305, Silver Spring MD 20910. 301/495-4444. **Contact:** Human Resources. **Description:** A computer design consulting firm.

MS TECH INC.
Forest Hills Plaza #218, Pittsburgh PA 15221. 412/824-2767. **Contact:** Human Resources. **Description:** Provides computer consulting services.

MACFADDEN & ASSOCIATES
1320 Fenwick Lane, Suite 600, Silver Spring MD 20910. 301/588-5900. **Contact:** Human Resources. **Description:** A computer consulting firm.

MASTECH SYSTEMS CORPORATION
1004 McKee Road, Oakdale PA 15071. 412/787-2100. **Contact:** Human Resources. **Description:** Offers computer consulting services.

MICROCOMPUTER CONSULTANT GROUP
1440 Canal Street, Suite 1703, New Orleans LA 70112. 504/522-2581. **Contact:** Human Resources. **Description:** A computer consulting firm.

OMNI RESOURCES
131 West Wilson Street, Suite 1003, Madison WI 53703. 608/284-2040. **Contact:** Shelley Udell, Account Manager. **Description:** A computer consulting firm.

OMNI RESOURCES
2670 South Ashland Avenue, Green Bay WI 54304. 414/499-8232. **Contact:** Therese Baier, Resource Consulting Manager. **Description:** A computer consulting firm.

ON-LINE RESOURCES INC.
2170 West State Road 434, Suite 220, Longwood FL 32779. 407/869-7724. **Fax:** 407/869-8346. **Contact:** Steve Zappia, Recruiting. **Description:** Offers contract computer consulting services.

PC ETC. INC.
462 7th Avenue, 4th Floor, New York NY 10018. 212/736-5870. **Contact:** Michael Grele, Human Resources Director. **Description:** PC Etc. develops and offers instructor-led and computer-based personal computer training programs, and provides consulting services, primarily to large business and public sector organizations. The company's instructor-led training programs include a wide range of introductory and advanced classes in operating systems, including MS/DOS, Microsoft Windows, OS/2, and Apple Macintosh System 7.0; word processing; spreadsheets; databases; communications; executive overviews; integrated software packages; computer graphics; and desktop publishing. The company's computer-based training programs include offerings on Lotus Notes, CC Mail, Microsoft Office, and Lotus Smartsuite. The consulting division provides computer personnel, on a temporary basis, to a client for special projects. PC Etc. operates nine training facilities, including four in New York City, one in Northern California, three in Ontario, Canada, and one in Metropark NJ. **Corporate headquarters location:** This Location.

PKS INFORMATION SERVICES, INC.
11707 Miracle Hills Drive, Omaha NE 68154. 402/498-8250. **Contact:** Human Resources. **Description:** Provides computer outsourcing services.

PSI INTERNATIONAL INC.
10306 Eaton Place, Suite 400, Fairfax VA 22030. 703/352-8700. **Contact:** Human Resources. **Description:** Offers computer consulting services and develops software.

PARAGON COMPUTER PROS INC.
20 Commerce Drive, Suite 226, Cranford NJ 07016. 908/709-6767. **Contact:** Human Resources. **Description:** Offers computer consulting services for a variety of businesses.

PENCOM SYSTEMS INC.
40 Fulton Street, New York NY 10038. 212/513-7777. **Fax:** 212/227-1854. **Contact:** Cliff Kahan, Recruiting. **Description:** Provides computer consulting services. Pencom Systems helps businesses with open systems management and software consulting.

PERCEPTICS CORPORATION
725 Pellissippi Parkway, Knoxville TN 37932-3350. 423/966-9200. **Contact:** Jim Disney, Human Resources Director. **Description:** Provides consulting, design, development, and fabrication of systems for image processing, pattern recognition, and computer vision.

PERSONAL PC CONSULTANTS
11026 Prairie Hills Drive, Omaha NE 68154. 402/393-4548. **Contact:** Human Resources. **Description:** A computer consulting firm.

THE PRESIDIO CORPORATION
5100-J Philadelphia Way, Lanham MD 20706. 301/459-2200. **Contact:** Human Resources. **Description:** Provides information management services.

PRINCETON INFORMATION
120 Wood Avenue South, Suite 404, Iselin NJ 08830. 908/906-5660. **Contact:** Human Resources. **Description:** Offers computer consulting services.

QUALITY SOFTWARE ENGINEERING, INC.
1065 Executive Parkway Drive, Suite 305, St. Louis MO 63141. 314/579-9898. **Fax:** 314/579-9853. **Contact:** Lisa Scherer, Technical Staffing Specialist. **E-mail address:** qse@qse.com. **World Wide Web address:** http://www.qse.com/qse. **Description:** An information technology consulting and services firm. The Information Systems Consulting Division provides services to professionals including application system development and maintenance, database administration, system installation and management, network installation and management, technical writing, training, quality assurance, project management, and management consulting. The Information Technologies Division provides services for clients who outsource those functions, including hardware and software evaluation and selection, installation, system integration, project development services, training, and end user support. **Common positions include:** Applications Engineer; Computer Operator; Computer Programmer; Database Manager; Internet Services Manager; MIS Specialist; Project Manager; Software Engineer; Systems Manager; Technical Writer/Editor; Webmaster. **Educational backgrounds include:** Business Administration; Computer Science; Engineering. **Benefits:** 401(k); Dental Insurance; Disability Coverage; Life Insurance; Medical Insurance; Tuition Assistance. **Corporate headquarters location:** This Location.

RCG
411 West 40th Street, New York NY 10018. 212/642-6000. **Contact:** Lori-Ann Lee, Human Resources. **Description:** A computer consulting firm.

RWD TECHNOLOGIES, INC.
10480 Little Pawtuxet Parkway, Suite 1200, Columbia MD 21044. 410/730-4377. **Contact:** Human Resources. **Description:** A computer engineering consulting firm.

REOHR GROUP INC.
P.O. Box 80240, Valley Forge PA 19484-0240. 610/768-7150. **Contact:** Technical Recruiter. **Description:** Offers computer consulting services to a variety of businesses.

SAI SOFTWARE CONSULTANTS
2313 Timber Shadows, Suite 200, Kingwood TX 77345. 713/358-1858. **Contact:** Human Resources. **Description:** Provides various computer and software consulting services.

SCB COMPUTERS
1365 West Brierbrook Road, Germantown TN 38138. 901/754-6577. **Contact:** Debbie Perdzock, Vice President, Recruiting. **Description:** A computer consulting firm with clients nationwide.

SETA CORPORATION
6862 Elm Street, McLean VA 22101. 703/821-8178. **Contact:** Human Resources. **Description:** A government contractor providing computer services.

SILVERLINE INDUSTRIES INC.
53 Knights Road, Piscataway NJ 08854. 908/248-3363. **Contact:** Human Resources. **Description:** Offers computer consulting services. **NOTE:** Silverline also hires people with programming knowledge.

SIRCO ASSOCIATES INC.
901 Wilshire Drive, Suite 540, Troy MI 48099. 810/362-2200. **Contact:** Human Resources. **Description:** Offers computer consulting services.

SMITH MICRO SOFTWARE, INC.
2955 Valmont, Suite 230, Boulder CO 80301. 303/444-5115. **Contact:** Human Resources. **Description:** A consulting firm specializing in computers and computer applications in business and industry. **NOTE:** Human Resources may be contacted at the corporate office at the following address: 51 Columbia, Suite 200, Aliso Viejo CA 92656. **Corporate headquarters location:** Aliso Viejo CA.

SOFTWARE ARCHITECTS
Three Westbrook Corporate Center, Suite 400, Westchester IL 60154. 708/531-0011. **Contact:** Mary Olson, Manager of Staffing & Placement. **Description:** A computer consulting firm for *Fortune* 1000 companies.

SOFTWARE SYNERGY INC. (SSI)
10000 Allisonville Road, Fishers IN 46038. 317/849-4444. **Contact:** Human Resources. **Description:** A computer consulting firm.

STAFFWARE INC.
1235 North Loop West, Suite 1100, Houston TX 77008. 713/880-0232. **Contact:** Human Resources. **Description:** Provides computer consulting services.

STATISTICA
30 West Gude Drive, Suite 300, Rockville MD 20850. 301/424-1911. **Contact:** Carol Dantzig, Human Resources. **Description:** Provides professional and computer systems engineering consulting services.

STERLING INFORMATION GROUP
1717 West 6th Street, Suite 340, Austin TX 78703. 512/344-1000. **Contact:** Human Resources. **Description:** A software consulting and development company.

STRATAGEM INC.
10850 West Park Place, Suite 250, Milwaukee WI 53224. 414/359-4747. **Contact:** Human Resources. **Description:** Offers computer consulting services for information systems.

SYSTEMS & PROGRAMMING CONSULTANTS
100 Galleria Parkway, Suite 270, Atlanta GA 30339. 770/612-4999. **Toll-free phone:** 800/226-3112. **Fax:** 770/612-4995. **Contact:** Kathy Kallinay, Personnel Consultant. **E-mail address:** spc@atl.mindspring.com. **World Wide Web address:** http://www.spc1.com. **Description:** A provider of information management solutions for a wide variety of companies, from small, specialized firms to *Fortune* 500 organizations. **Common positions include:** Computer Programmer; Database Manager; MIS Specialist; Software Engineer; Systems Analyst; Technical Writer/Editor. **Educational backgrounds include:** Computer Science. **Benefits:** 401(k); Disability Coverage; Life Insurance; Medical Insurance; Pension Plan; Profit Sharing. **Corporate headquarters location:** Charlotte NC. **Other U.S. locations:** Tampa FL; Raleigh NC; Nashville TN; Dallas TX. **Listed on:** Privately held. **Annual sales/revenues:** $21 - $50 million. **Number of employees at this location:** 50. **Number of employees nationwide:** 450. **Number of projected hires for 1997 - 1998 at this location:** 50.

SYSTEMS NETWORK
3001 North Rock Point Drive East, Suite 340, Tampa FL 33607. 813/288-0808. **Contact:** Human Resources. **Description:** A computer consulting firm.

SYSTEMS SOLUTIONS INC.
2108 East Thomas Road, Suite 300, Phoenix AZ 85016. 602/955-5566. **Contact:** Human Resources. **Description:** A full-service computer group. Systems Solutions offers a wide range of support including Internet access, consulting services, software development, and more.

SYTEL, INC.
6430 Rockledge Drive, Suite 400, Bethesda MD 20817. 301/530-1000. **Contact:** Human Resources. **Description:** An information technology firm performing a variety of computer services for the government.

TSG TECHNICAL SERVICES
2970 Clairmont Road NE, Suite 1020, Atlanta GA 30329-1634. 404/633-0646. **Contact:** Human Resources. **Description:** TSG Technical Services is a project-based information technology consulting firm. Clients include AT&T and Walt Disney Inc.

T.S.R. INC.
400 Oser Avenue, Hauppauge NY 11767. 516/231-0333. **Contact:** Recruiter. **Description:** Provides computer consulting services.

TASC INC.
2555 University Boulevard, Fairborn OH 45324. 937/426-1040. **Contact:** Human Resources. **Description:** Offers computer consulting services to businesses and the government.

TECHNICAL SOFTWARE SOLUTIONS, INC.
901 Sandy Spring Road, Suite 505, Laurel MD 20707. **Toll-free phone:** 888/369-0040. **Fax:** 301/369-0045. **Contact:** Steven Haversack, Chief of Operations. **E-mail address:** MD@solutions1.com. **World Wide Web address:** http://www.solutions1.com. **Description:** A consulting firm specializing in the areas of software engineering; programming and design; hardware and software valuation, selection, and training; and network integration. **Common positions include:** Computer Programmer; MIS Manager; Software Engineer; Systems Analyst; Technical Writer/Editor. **Educational backgrounds include:** Communications; Computer Science; Engineering. **Benefits:** 401(k); Dental Insurance; Life Insurance; Medical Insurance. **Corporate headquarters location:** This Location. **Listed on:** Privately held. **Annual sales/revenues:** $5 - $10 million. **Number of employees at this location:** 50. **Number of employees nationwide:** 150.

TECHNISOURCE INC.
1901 West Cypress Creek Road, Ft. Lauderdale FL 33309. 954/493-8601. **Contact:** Recruiter. **Description:** Offers a variety of computer support services including network consulting and software development outsourcing.

TECHNIUM, INC.
8745 West Higgins Road, Suite 480, Chicago IL 60631. 773/380-0555. **Fax:** 773/380-0568. **Contact:** Jason Hanold, People and Learning. **E-mail address:** jhanold@technium.com. **World Wide Web address:** http://www.technium.com. **Description:** Technium provides computer consulting services focusing on client/server technologies. The company's client base represents a variety of industries, from consumer products and health care to financial services and software. Technium provides a full range of services to deploy client/server applications, including architecture planning, application analysis, visualization, and design; graphical user interface development using Visual C++, Visual Basic, PowerBuilder, and Delphi; object-oriented development with C, C++, and Smalltalk; relational database development in SQL Server, Microsoft Access, Oracle, and Sybase; and decision support systems development using OLAP and Data Warehousing technologies. Technium invests substantially in its consultants' development, with ongoing education keeping consultants abreast of the latest advances in technologies. **NOTE:** The company offers entry-level positions. **Common positions include:** Computer Programmer; Internet Services Manager; MIS Specialist; Software Engineer; Systems Analyst. **Educational backgrounds include:** Computer Science; Information Systems; Marketing. **Benefits:** 401(k); Dental Insurance; Disability Coverage; Employee Discounts; Life Insurance; Mass Transit Available; Medical Insurance; Profit Sharing; Savings Plan; Tuition Assistance. **Corporate headquarters location:** This Location. **Other U.S. locations:** Dallas TX; Milwaukee WI.

Operations at this facility include: Administration; Divisional Headquarters; Regional Headquarters; Research and Development; Sales; Service. **Listed on:** Privately held. **Annual sales/revenues:** $5 - $10 million. **Number of employees at this location:** 60. **Number of employees nationwide:** 75.

TECHNOLOGY GROUP
100 North Tampa Boulevard, Suite 1935, Tampa FL 33602. 813/229-7006. **Contact:** Human Resources. **Description:** A computer consulting firm.

3X CORPORATION
750 Lakeview Plaza Boulevard, Worthington OH 43085. 614/433-9406. **Contact:** Human Resources. **Description:** A computer consulting firm.

TRECOM BUSINESS SYSTEMS, INC.
333 Thornall Road, Edison NJ 08837-2246. 908/549-4100. **Fax:** 908/549-2375. **Contact:** Recruiting. **World Wide Web address:** http://www.trecom.com. **Description:** Provides computer consulting services. Trecom offers its clients a variety of services including outsourcing solutions, consulting, and systems integration. Trecom expects to conduct over $180 million in sales nationwide. **Corporate headquarters location:** This Location.
Other U.S. locations:
- 3021 Lorna Road, Suite 100, Birmingham AL 35216. 205/978-9424.
- 185 Plains Road, Milford CT 06460. 203/882-0700. Contact: Frank Feroleto, Recruiting.
- 5110 Eisenhower Boulevard, Suite 105, Tampa FL 33634. 813/888-7400.
- 1979 Lakeside Parkway, Suite 250, Tucker GA 30084. 770/621-0300. Contact: Cathy Turk, Recruiting.
- 45 Broadway, 15th Floor, New York NY 10006. 212/809-6600. Contact: Ira Glass, Recruiting.
- 13355 Noel Road, Suite 500, Dallas TX 75240. 972/774-4450.

VIP SYSTEMS
One West Pennsylvania Avenue, Suite 700, Towson MD 21204. 410/832-8300. **Contact:** Human Resources. **Description:** Offers computer consulting services.

VIAGRAFIX
TRAINING DIVISION
5 South Vann Street, Pryor OK 74361. 918/825-6700. **Toll-free phone:** 800/842-4723. **Fax:** 918/825-6744. **Toll-free fax:** 800/842-3294. **Contact:** Rick Ogg, Human Resources Manager. **E-mail address:** rogg@viagrafix.com. **World Wide Web address:** http://www.viagrafix.com. **Description:** A developer of CAD software and a provider of computer training services. **Corporate headquarters location:** This Location.

WESSON, TAYLOR, WELLS & ASSOCIATES
2300 Yorkmount Road, Suite 240, Charlotte NC 28217. 704/357-0895. **Contact:** Human Resources. **Description:** A software consulting firm.

PAUL YAO COMPANY
1075 Bellevue Way NE, Suite 300, Seattle WA 98004-4276. 206/747-1355. **Contact:** Human Resources. **Description:** Paul Yao Company trains software engineers in current computer technology.

ZENACOMP
North Laurel Park Drive, Suite 351, Livonia MI 48152. 313/464-3700. **Contact:** Human Resources. **Description:** A computer consulting firm.

INTERNET PROVIDERS

Young. Evolving. Exciting. This is how Web Producer Ken Brooks describes the Internet Services field. Ken's advice to jobseekers is to "know the languages." The more programming languages and computer skills a person knows, the higher the pay, and the more job opportunities will become available. Education helps, but is certainly not necessary to navigate the Net. The Internet is something you can learn on your own.

Ken sees this field becoming more interactive and recommends keeping up on the latest Internet languages. An interested jobseeker can find books on HTML, Java Script, and C++, or sign up for a course at a local college. Daily responsibilities may include updating the company's Website with new, fresh information, and exploring marketing options. Assuring repeat visits to your Website is essential, and this can be accomplished by proactive site management. The number of companies offering Internet Services may decrease due to competition, but the demand for Internet professionals should remain high.

AUTOMATED GRAPHICS
4590 Graphics Drive, White Plains MD 20695. 301/843-7185. **Contact:** Human Resources. **Description:** Offers Internet services. Automated Graphics will host Websites or help a business access the Internet.

BITWISE INTERNET TECHNOLOGIES INC.
22 Drydock Avenue, Boston MA 02210. 617/261-4700. **Contact:** Human Resources. **Description:** Provides Internet access. Bitwise Internet Technologies provides individuals and businesses with access to the Internet including corporate dial up and World Wide Web home page services.

BRIGADOON INC.
P.O. Box 53168, Bellevue WA 98015-3168. 206/562-1960. **Contact:** Human Resources. **Description:** Provides Internet services for business and personal use.

CONNECTNET
6370 Lusk Boulevard, San Diego CA 92121. 619/450-0254. **Contact:** Human Resources. **Description:** Provides access to the Internet and supports World Wide Web home pages.

DELTACOMM ONLINE
507 Airport Boulevard, Suite 109, Morrisville NC 27560. 919/467-1143. **Fax:** 919/460-4531. **Contact:** Human Resources. **World Wide Web address:** http://www.delta.com. **Description:** A regional Internet service provider.

DIMENSION X
181 Fremont Street, Suite 120, San Francisco CA 94105. 415/243-0900. **Contact:** Allison Crawford, Office Manager. **Description:** Sets up Websites for various customers. Dimension X also develops 2-D and 3-D animation tools.

DVORAK INTERNET DEVELOPMENT, INC.
P.O. Box 1524, Broomfield CO 80038. **Contact:** Human Resources. **E-mail address:** info@dvorak.com. **World Wide Web address:** http://www.dvorak.com. **Description:** A provider of automated Internet systems including Websprite, an information gathering system.

E.CENTRAL
1640 Logan, Denver CO 80203. 303/830-0123. **Contact:** Ted Pinkowitz, President. **World Wide Web address:** http://www.ecentral.com. **Description:** E.Central provides access to the Internet for companies and individuals, as well as e-mail accounts and home pages on the World Wide Web.

EXCITE INC.
1091 North Shoreline Boulevard, Mountain View CA 94043. 415/943-1200. **Fax:** 415/943-1299. **Contact:** Human Resources. **E-mail address:** resumes@excite.com. **World Wide Web address:** http://www.excite.com. **Description:** Offers World Wide Web services and products with its proprietary concept searching technology. In addition to Web navigation services, Excite features site reviews, editorial columns, news, and city.net, a regional information page. **NOTE:** Excite maintains an employment page at its Website. **Common positions include:** Account Manager; Account Representative; Administrative Assistant; Advertising Manager; Client/Server Specialist; Computer Programmer; Internet Specialist; Public Relations Manager; Sales Manager; Sales Representative; Systems Engineer; Systems Specialist. **Benefits:** Medical Insurance; Stock Option. **Corporate headquarters location:** This Location. **Listed on:** NASDAQ. **Stock exchange symbol:** XCIT. **Annual sales/revenues:** $8.2 million.

I/PRO
785 Market Street, 13th Floor, San Francisco CA 94103. 415/975-5800. **Fax:** 415/975-5818. **Contact:** Human Resources. **E-mail address:** info@ipro.com. **World Wide Web address:** http://www.ipro.com. **Description:** A provider of services and software for the measurement and analysis of Website usage. Products allow Web operators to analyze site usage while advertisers are able to determine optimal site selection. **Corporate headquarters location:** This Location.

INFOSEEK CORPORATION
2620 Augustine Drive, Suite 250, Santa Clara CA 95054. 408/567-2700. **Toll-free phone:** 800/781-4636. **Fax:** 408/986-1889. **Contact:** Human Resources. **E-mail address:** jobs@infoseek.com. **World Wide Web address:** http://www.infoseek.com. **Description:** An Internet search service. Founded in 1994. **NOTE:** Infoseek maintains an employment page at its Website. **Common positions include:** Associate Editor; Computer Programmer; Internet Specialist; Marketing Manager; Product Manager; Production Engineer; Software Engineer. **Corporate headquarters location:** This Location. **Listed on:** NASDAQ. **Stock exchange symbol:** SEEK.

INKTOMI CORPORATION
2168 Shattuck, Berkeley CA 94704. 510/883-7300. **Contact:** Human Resources. **Description:** Creator and operator of HotBot, a search engine on the World Wide Web.

INTERNATIONAL DISCOUNT TELECOMMUNICATIONS CORPORATION (IDT)
294 State Street, Hackensack NJ 07601. 201/928-1000. **Contact:** Jonathan Rand, Manager of Human Resources. **Description:** An Internet service provider. **NOTE:** Interested jobseekers should address inquiries to the attention of Jonathan Rand, Manager of Human Resources, IDT Corporation, 190 Main Street, Hackensack NJ 07601.

LYCOS, INC.
293 Boston Post Road, Marlboro MA 01752. 508/229-0717. **Fax:** 508/229-2866. **Contact:** Human Resources. **E-mail address:** webmaster@lycos.com. **World Wide Web address:** http://www.lycos.com. **Description:** An Internet services provider which finds, indexes, and filters information on the Internet and World Wide Web. Founded in 1995. **Corporate headquarters location:** This Location. **Listed on:** NASDAQ. **Stock exchange symbol:** LCOS. **Annual sales/revenues:** $4.4 million.

MAGNET INTERACTIVE STUDIOS
3255 Grace Street NW, Washington DC 20007. 202/625-1111. **Contact:** Dina White, Director of Human Resources. **Description:** Creates sites on the World Wide Web for other companies. Magnet Interactive Studios also produces multimedia materials including CD-ROMs.

NETCOM ON-LINE COMMUNICATION SERVICES, INC.

2 North Second Street, Plaza A, San Jose CA 95113. 408/881-1815. **Toll-free phone:** 800/638-2661. **Fax:** 408/345-2669. **Contact:** Human Resources. **E-mail address:** nc0200@corp.netcom.com. **World Wide Web address:** http://www.netcom.com. **Description:** NETCOM is an international Internet service provider. **NOTE:** NETCOM maintains an extensive employment page at its Website. **Common positions include:** Computer Programmer; Engineer; Software Engineer. **Corporate headquarters location:** This Location. **Listed on:** NASDAQ. **Stock exchange symbol:** NETC. **Annual sales/revenues:** $100 million.

NET.GENESIS CORPORATION

68 Rogers Street, Cambridge MA 02142-1119. 617/577-9800. **Fax:** 617/577-9850. **Contact:** Human Resources. **E-mail address:** sales@netgen.com. **World Wide Web address:** http://www.netgen.com. **Description:** A provider of solutions for improving Internet service performance and usage. **Corporate headquarters location:** This Location.

NETPROFIT

1455 South State Street, Suite A, Orem UT 84058. 801/224-1811. **Fax:** 801/224-2332. **Contact:** Human Resources. **World Wide Web address:** http://www.netprofile.com. **Description:** Provides high-end Internet Website development including database integration, Java programming, and CGI. **Common positions include:** Account Manager; Accountant/Auditor; Applications Engineer; Computer Animator; Computer Operator; Computer Programmer; Electrical/Electronics Engineer; Sales Representative; Secretary; Webmaster. **Educational backgrounds include:** Art/Design; Computer Science; Engineering. **Benefits:** Dental Insurance; Employee Discounts; Medical Insurance; Tuition Assistance. **Special Programs:** Internships. **Corporate headquarters location:** This Location. **Annual sales/revenues:** Less than $5 million. **Number of employees at this location:** 15.

NETSCAPE COMMUNICATIONS CORPORATION

501 East Middlefield Road, Mountain View CA 94043. 415/254-1900. **Fax:** 415/428-4072. **Contact:** Human Resources. **World Wide Web address:** http://home.netscape.com. **Description:** An Internet provider and developer of software including Constellation. **NOTE:** Netscape offers a voluminous employment opportunities page and company information at their Website. It includes current job openings and e-mail addresses. Each department has its own e-mail address. **Common positions include:** Client/Server Specialist; Computer Engineer; Computer Programmer; Sales Representative; Software Engineer; Technical Writer/Editor. **Corporate headquarters location:** This Location. **Other U.S. locations:** Manhattan Beach CA; Washington DC; Atlanta GA; Chicago IL; New York NY; Dallas TX. **Listed on:** NASDAQ. **Stock exchange symbol:** NSCP. **Annual sales/revenues:** $231 million.

NETWORK PUBLISHING

One East Center Street, Suite 300, Provo UT 84601. 801/377-9399. **Fax:** 801/377-9390. **Contact:** Human Resources. **Description:** An Internet site developer that also offers consulting and maintenance services to companies that have Internet sites.

NETWORK SOLUTIONS INC. (NSI)

505 Huntmar Park Drive, Herndon VA 22070. 703/742-0400. **Fax:** 703/742-4823. **Contact:** Ivan Yopp, Staffing Manager. **E-mail address:** ivany@netsol.com. **Description:** Provides top level domain registration services for the Internet. Founded in 1979. **Common positions include:** Computer Programmer; Systems Manager. **Parent company:** Science Applications International Corp. (McLean VA).

SRT INTERNET SERVICES

24 2nd Avenue SE, Minot ND 58701. 701/858-7873. **Contact:** Human Resources. **Description:** SRT Internet Services sets up home pages on the Internet for companies.

STARWAVE CORPORATION

13810 SE Eastgate Way, Bellevue WA 98005. 206/957-2000. **Fax:** 206/957-2009. **Contact:** Human Resources. **E-mail address:** hr@starwave.com. **World Wide Web address:** http://www.starwave.com. **Description:** A provider of interactive Internet services. Services includes sports; entertainment, including film, music, and popular culture; and family interests. Founded in 1993 by Paul Allen, co-founder of Microsoft. **Common positions include:** Advertising Manager; Editor; Editorial Assistant; Graphic Designer; Marketing Specialist; Production Manager; Project Manager; Software Engineer; Writer. **Special Programs:** Internships.

SUPERNET INC.

999 18th Street, Suite 2640, Denver CO 80202. 303/296-8202. **Contact:** Department of Interest. **Description:** One of the largest Internet access providers in Colorado. SuperNet also creates home pages on the World Wide Web for customers.

TIAC (THE INTERNET ACCESS COMPANY)

100 Sylvan Road, Suite G600, Woburn MA 01801. 617/276-7200. **Fax:** 617/275-2224. **Contact:** Linda Vaughn, Human Resources Director. **World Wide Web address:** http://www.tiac.net. **Description:** TIAC (The Internet Access Company) provides access to the Internet for customers across the Northeast. The company's software enables users to browse the World Wide Web, send and receive e-mail, and enjoy a wide variety of other online services. The company provides 24-hour, 7-day customer service. TIAC also produces a monthly magazine for customers. **Corporate headquarters location:** This Location.

UUNET TECHNOLOGIES, INC.

3060 Williams Drive, Fairfax VA 22031-4648. 703/206-5600. **Toll-free phone:** 800/488-6383. **Fax:** 703/206-5927. **Contact:** Human Resources. **E-mail address:** jobs@uu.net. **Description:** A commercial Internet service provider. UUNET Technologies also provides a variety of Internet access options, applications, and consulting services. **NOTE:** In August 1996, the company announced plans to merge with MFS Communications Company, Inc. **Corporate headquarters location:** This Location.

WORLDS INC.

605 Market Street, 14th Floor, San Francisco CA 94105. **Fax:** 415/547-1631. **Contact:** Human Resources. **E-mail address:** hr@worlds.net. **World Wide Web address:** http://www.worlds.net. **Description:** A 3-D Internet technology firm offering a Web browser, authoring tools, and server software. **Common positions include:** Communications Specialist. **Corporate headquarters location:** This Location.

YAHOO! CORPORATION

3400 Central Expressway, Suite 201, Santa Clara CA 95051. 408/731-3300. **Fax:** 408/731-3301. **Contact:** Human Resources. **E-mail address:** hr@yahoo.com. **World Wide Web address:** http://www.yahoo.com. **Description:** An Internet search engine. **NOTE:** Yahoo! maintains an extensive employment page at its Website. The company does not accept phone calls, contact by fax, e-mail, or mail. **Common positions include:** Administrative Assistant; Advertising Manager; Associate Editor; Managing Editor; Marketing Manager; Product Manager; Program Manager; Project Manager; Sales Manager; Software Engineer. **Corporate headquarters location:** This Location.

ONLINE SERVICES

What *are my goals? What do I want to accomplish? Is this company right for me? Mark Stavish, Vice President of Human Resources at America Online, says that if you know the answers to these questions, you have an edge over other jobseekers.*

When hiring new people, Mark looks for candidates that possess strong technical skills, a keen business acumen, and an ability to self-motivate. Online companies require the services of database administrators, software developers, and other creative individuals. Experienced professionals have an advantage, but first time jobseekers should by no means despair, especially if they have the qualities employers look for.

Online services have begun to change peoples' daily routines, and Mark expects that trend to continue. Competition for jobs will remain keen, although there should be a growing number of opportunities for computer professionals in this field into the next century.

ACCURATE DATA ONLINE CORPORATION
1250 Grumman Place, Suite A, Titusville FL 32780. 407/268-2622. **Contact:** Human Resources. **Description:** Conducts online data processing for credit unions.

ADHESIVE MEDIA, INC.
101 West Sixth Street, Suite 210, Austin TX 78701. 512/478-9900. **Fax:** 512/478-9934. **Contact:** Human Resources. **E-mail address:** adhesive-media@eden.com. **World Wide Web address:** http://www.eden.com. **Description:** Adhesive Media specializes in traditional and interactive information services, software development, consulting services, and operates the Eden Matrix Online service.

AIRLINE TARIFF PUBLISHING COMPANY
Washington Dulles International Airport, P.O. Box 17415, Washington DC 20041. 703/492-2320. **Contact:** Human Resources. **Description:** Offers online services. Airline Tariff Publishing provides airlines with online information concerning fares and regulations.

AMERICA ONLINE INC. (AOL)
8619 Westwood Center Drive, Vienna VA 22182-2285. 703/448-8700. **Contact:** Vice President of Human Resources. **Description:** America Online (AOL) is a global leader in the market for interactive services. AOL offers its more than 3.5 million members a wide variety of services via personal computer and other intelligent devices. These include electronic mail, conferencing, software, computing support, interactive magazines and newspapers, and online classes, as well as access to the Internet. AOL is also a provider of data network services, new media and interactive marketing services, and multimedia and CD-ROM production services. The company has alliances with Time Warner, Capital Cities/ABC, Knight-Ridder, Tribune, Hachette, IBM, and American Express. AOL has plans to expand its operations to Europe. **Benefits:** 401(k); ESOP; Incentive Plan. **Corporate headquarters location:** This Location. **Listed on:** American Stock Exchange; Chicago Stock Exchange; NASDAQ; Pacific Exchange. **Number of employees at this location:** 530. **Number of employees nationwide:** 2,500.

CHEMICAL ABSTRACTS SERVICE
P.O. Box 3012, Columbus OH 43210-0012. **Contact:** Human Resources. **Description:** Designs, develops, operates, and markets chemical information services, including CAS Online and Chemical Abstracts. **Common positions include:** Chemist. **Benefits:** 401(k); Dental Insurance;

Disability Coverage; Employee Discounts; Life Insurance; Medical Insurance; Pension Plan; Savings Plan; Tuition Assistance. **Corporate headquarters location:** Washington DC. **Parent company:** American Chemical Society. **Operations at this facility include:** Administration; Divisional Headquarters; Research and Development; Sales. **Listed on:** Privately held. **Number of employees at this location:** 1,200. **Number of employees nationwide:** 2,000.

CITYSEARCH
790 East Colorado Boulevard, Suite 200, Pasadena CA 91101. 818/405-0050. **Fax:** 818/405-9929. **Contact:** Sharon Smith, Director of Human Resources. **World Wide Web address:** http://www.citysearch.com. **Description:** CitySearch provides online information services to the city of Pasadena. The company's World Wide Web site includes various community activities as well as places to visit within the city.

COMPUSERVE INC.
5000 Arlington Center Boulevard, Columbus OH 43220. 614/457-8600. **Contact:** Judy Reinhard, Human Resources Director. **Description:** CompuServe Inc. provides business information and network communication services to large corporations, government agencies, and individual customers worldwide. **Common positions include:** Account Manager; Computer Programmer; Customer Service Representative; Field Engineer; Marketing Specialist. **Educational backgrounds include:** Business Administration; Communications; Computer Science; Marketing. **Benefits:** 401(k); Dental Insurance; Disability Coverage; Employee Discounts; Life Insurance; Medical Insurance; Savings Plan; Tuition Assistance. **Special Programs:** Internships. **Corporate headquarters location:** This Location. **Parent company:** H&R Block. **Operations at this facility include:** Administration; Research and Development; Service. **Number of employees nationwide:** 2,200.

CONNECT INC.
515 Ellis Street, Mountain View CA 94043. 415/254-4000. **Contact:** Human Resources Department. **Description:** The operator of an online service.

DATEQ INFORMATION NETWORK, INC.
5555 Triangle Parkway, Suite 400, Norcross GA 30092. 770/446-8282. **Contact:** Human Resources. **Description:** DATEQ Information Network, Inc. provides information to the automobile insurance and car rental industries through a centralized information network. The network enables the company to access a variety of information databases, retrieve and merge data in response to a specific customer inquiry, convert data into customer formats, and accept inquiries and transmit information to customers through delivery systems utilizing software developed by the company. The delivery systems enable customers to establish mainframe or personal computer links to the company's centralized information network. The company also provides a telephone service center through which phone-in requests from customers for information may be processed. DATEQ's main product is the motor vehicle report, a historical report of an individual's driving record available from the Department of Motor Vehicles. Other products include a database of licensed drivers designed to reveal undisclosed drivers within a household, a vehicle registration database designed to verify actual ownership of vehicles, and an automobile insurance claims database designed to supply information on a driver's prior claims history. **Corporate headquarters location:** This Location. **Number of employees at this location:** 55.

DECISIVE QUEST, INC.
735 North Plano Road, Richardson TX 75081. 972/480-9070. **Fax:** 972/480-0348. **Contact:** Human Resources. **World Wide Web address:** http://www.questmatch.com. **Description:** Provides online employment services through customer use of its company-developed database software called QuestMatch. Clients include both jobseekers and firms with open positions. Founded in 1993.

DELPHI INTERNET SERVICES CORPORATION
1030 Massachusetts Avenue, Cambridge MA 02138. 617/491-3342. **Fax:** 617/441-4902. **Contact:** Recruiting. **E-mail address:** career@delphi.com. **World Wide Web address:** http://www.delphi.com. **Description:** An online services firm offering news, entertainment, games, and chat among others. **NOTE:** Delphi maintains an employment opportunities page at its Website.

Common positions include: Computer Programmer. **Corporate headquarters location:** This Location.

DOW JONES TELERATE INC.

Harborside Financial Center, 600 Plaza 2, Jersey City NJ 07311. 201/938-4000. **Contact:** Andrew David, Assistant Manager, Staffing. **Description:** Operates an online financial information network. Dow Jones Telerate provides financial data on securities, stocks, commodities, and other Dow Jones newswires to bank trading rooms worldwide.

EARTHWEB

Three Park Avenue, 38th Floor, New York NY 10016. 212/725-6550. **Contact:** Clark Silva, Human Resources Manager. **Description:** An Internet media technology firm that develops online services for communities as well as chat software based on Java applications.

FUTURESOURCE

955 Parkview Boulevard, Lombard IL 60148. 630/620-8444. **Contact:** Gail Jensen, Human Resources Manager. **Description:** An online, real-time, financial news service provider.

INFORMATION AMERICA INC.

600 West Peachtree Street, Suite 1200, Atlanta GA 30308. 404/892-1800. **Contact:** Human Resources. **Description:** An online information company that provides access to public and court records for clients including lawyers, bankers, investigators, and those involved with commercial transactions and business litigation. Information America maintains over 10 offices nationwide and has information from over 20 states. **Corporate headquarters location:** This Location. **Number of employees nationwide:** 200.

JUNO ONLINE SERVICES, L.P.

120 West 45th Street, 39th Floor, New York NY 10036. 212/403-8800. **Contact:** Human Resources. **Description:** Sets up e-mail accounts on the Internet for subscribers.

KNIGHT-RIDDER, INC.
DIALOG INFORMATION SERVICES INC.

2440 El Camino Real, Mountain View CA 94040. **Contact:** Human Resources. **Description:** An international online information vendor. Offering online access to more than 300 databases containing over 80 million records, the service is the largest of its kind in the world. Dialog online information retrieval serves the business, scientific, technology, medical, education, and medical communities in more than 100 countries. Knight-Ridder, a major newspaper publishing company, owns 28 dailies in 15 states, and three non-dailies in suburban areas. The company also produces publications such as *Myrtle Beach's Golf, CubaNews* newsletter in Miami and *Northland Outdoors* in Grand Forks. The larger papers include the *Miami Herald, Philadelphia Inquirer, Philadelphia Daily News, Detroit Free Press,* and *San Jose Mercury News.* Knight-Ridder also has interests in the information distribution market through Business Information Services, with subsidiaries Knight-Ridder Information, Inc., Knight-Ridder Financial, and Technimetrics. Knight-Ridder Financial provides real-time financial news and pricing information through primary products MoneyCenter, Digital Datafeed, ProfitCenter, and TradeCenter. Knight-Ridder also has interests in cable television and other businesses. TKR Cable, a 50-50 joint venture with Liberty Media Corporation, serves 344,000 basic subscribers in New Jersey and New York and manages Kentucky systems with 277,000 subscribers. Through TKR Cable Partners, Knight-Ridder owns a 15 percent share of TCI/TKR L.P. cable systems with 867,000 subscribers in five states. Other interests include partial ownership of the Seattle Times Company, two paper mills, a newspaper advertising sales company, and SCI Holdings. **Common positions include:** Accountant/Auditor; Computer Programmer; Customer Service Representative; Financial Analyst; Librarian; Services Sales Representative; Software Engineer; Technical Writer/Editor. **Educational backgrounds include:** Accounting; Biology; Business Administration; Chemistry; Communications; Computer Science; Finance; Library Science; Marketing; Mathematics. **Benefits:** Credit Union; Dental Insurance; Life Insurance; Medical Insurance; Pension Plan; Profit Sharing; Savings Plan; Tuition Assistance. **Corporate headquarters location:** This Location. **Operations at this facility include:** Administration; Research and Development; Sales; Service.

LEGI-SLATE
777 North Capital Street, Washington DC 20002. 202/898-2300. **Contact:** Human Resources Department. **Description:** Offers online services specializing in tracking state and federal legislation.

LEXIS-NEXIS
9393 Springboro Pike, Miamisburg OH 45342. 937/865-6800. **Fax:** 937/865-7476. **Contact:** Staffing Department. **Description:** Lexis-Nexis is an online, full text database service including legal, news, business, and general information. The service enables lawyers, business professionals, and government agencies to electronically research thousands of resources from their own computers.

MOORE DATA MANAGEMENT SERVICES
2117 West River Road North, Minneapolis MN 55411. 612/588-7200. **Contact:** Human Resources. **Description:** A primary provider of online computer services to the real estate industry. This location also houses Moore Graphics Services, which provides commercial printing and electronic, on-demand, database publishing to *Fortune* 500 companies. **Corporate headquarters location:** This Location. **International locations:** Worldwide. **Parent company:** Moore Corporation. **Operations at this facility include:** Manufacturing. **Listed on:** New York Stock Exchange. **Number of employees at this location:** 100. **Number of employees nationwide:** 1,200.

NCR CORPORATION
135 Pennsylvania Avenue, Framingham MA 01701. 617/244-3550. **Contact:** Human Resources. **Description:** This location provides online services. Overall, NCR Corporation is a worldwide provider of computer products and services. The company provides computer solutions to three targeted industries: retail, financial, and communication. NCR Computer Systems Group develops, manufactures, and markets computer systems. NCR Financial Systems Group is an industry leader in three target areas: financial delivery systems, relationship banking data warehousing solutions, and payments systems/item processing. NCR Retail Systems Group is a world leader in end-to-end retail solutions serving the food, general merchandise, and hospitality industries. NCR Worldwide Services provides data warehousing services solutions; end-to-end networking services; and designs, implements, and supports complex open systems environments. NCR Systemedia Group develops, produces, and markets a complete line of information products to satisfy customers' information technology needs including transaction processing media, auto identification media, business form communication products, managing documents and media, and a full line of integrated equipment solutions. NCR Corporation formerly operated as AT&T Global Information Solutions. **Annual sales/revenues:** more than $100 million. **Number of employees worldwide:** 38,000.

NETS INC.
25 First Street, Cambridge MA 02141. 617/252-5000. **Contact:** Human Resources. **Description:** Formerly AT&T New Media Services, Nets Inc. is an online service company featuring content ranging from business newspapers and magazines to travel services and research tools. **Number of employees at this location:** 300.

NEWSNET
945 East Haverford Road, Bryn Mawr PA 19010. 610/527-8030. **Contact:** Human Resources. **Description:** Maintains an online database of publications and newswires.

PRODIGY, INC.
445 Hamilton Avenue, White Plains NY 10601. 212/253-7000. **Fax:** 914/448-3467. **Contact:** Human Resources. **E-mail address:** nicole@prodigy.com. **World Wide Web address:** http://www.prodigy.com. **Description:** Provides online services to subscribers. Services include a Web browser, chat rooms, e-mail, personal Web pages, live news, investment tools, educational tools, medical information, games, music, and bulletin boards. **Corporate headquarters location:** This Location. **Subsidiaries include:** Africa Online; GES; Prodigy Services Corp.; Prodigy Ventures, Inc.

QUOTRON SYSTEMS, INC.

200 North Sepulveda, Suite 500, El Segundo CA 90245. **Toll-free phone:** 800/426-4770. **Contact:** Human Resources. **Description:** Operates an extensive international data communications network for delivering such services as electronic stock quotation systems and business and economic news. Sales offices are located in 10 states across the country. **Corporate headquarters location:** This Location.
Other U.S. locations:
- 1700 Broadway, 31st Floor, New York NY 10019.

STRATUS COMPUTER, INC.

55 Fairbanks Boulevard, MS M22-PER, Marlborough MA 01752. 508/460-2000. **Fax:** 508/480-0243. **Contact:** Lorri Estrada, Human Resources. **E-mail address:** resumes@stratus.com. **World Wide Web address:** http://www.stratus.com. **Description:** Stratus offers a broad range of computer platforms, application solutions, middleware, and professional services for critical online operations. **Corporate headquarters location:** This Location. **Subsidiaries include:** Shared Systems Corporation, a provider of software and professional services to the financial services, retail, and health care industries; SoftCom Systems, Inc., a provider of data communications middleware and related professional services that bridge the gap between open distributed systems and legacy mainframe and midrange systems used for online applications; and Isis Distributed Systems, Inc., a developer of advanced messaging middleware products that enable businesses to develop reliable, high-performance distributed computing applications involving networked desktop computers and shared systems. Isis is also located at this address.

TRACK DATA

56 Pine Street, Seventh Floor, New York NY 10005. 212/248-9090. **Contact:** Human Resources Department. **Description:** Formerly known as Global Market Information, Track Data electronically provides trading information, news, and third-party database services on stocks, bonds, commodities, and other securities through its Dial/Data and Track OnLine services. The company's InfoVest services combine an online database with software that enables tailored analyses and reports on publicly traded companies. The company's AIQ Systems division produces expert systems software for individual and professional investors.

WIRELESS FINANCIAL SERVICES INC.

7 West 54th Street, New York NY 10019. 212/445-9374. **Contact:** Human Resources. **Description:** Offers online financial services.

Visit our exciting job and career site at http://www.careercity.com

MISCELLANEOUS COMPUTER PRODUCTS AND SERVICES

*I*n his role as Video Producer, Jon Tang uses computers to create interesting multimedia CD-ROMs. Requirements for the job include shooting video footage, recording audio tracks, adding backgrounds to the clips, and compressing the work using a computer. His advice to jobseekers in this field is to get hands-on experience. One way to do this is through an internship. These opportunities introduce jobseekers to people in the field and often establish valuable contacts, and, in Jon's case, a permanent position.

If you're still in college, Jon recommends specializing in computer science. Otherwise, a jobseeker can learn how to use the necessary software packages from technical classes or by reading the manuals that accompany different products. It's extremely helpful to get some type of exposure, as many employers like to screen samples of a candidate's work.

In the future, as computer and video technology advances, better quality CD-ROM products will emerge. Prices of equipment will continue to fall, allowing more companies to invest in in-house video production. This may create a rise in competition among multimedia CD-ROM production companies and certainly in the demand for video producers.

ACCRAM
2901 West Clarendon Avenue, Phoenix AZ 85017. 602/264-0288. **Contact:** Department of Interest. **Description:** Offers services including repairs, sales, systems integration, and cabling.

ACTEL CORPORATION
955 East Arques Avenue, Sunnyvale CA 94086-4521. 408/739-1010. **Contact:** Michelle Begun, Vice President of Human Resources. **Description:** Manufactures integrated circuits and develops software. Actel is a world leader in high-performance field programmable gate arrays. Actel is also the home of anti-fuse technology. **Common positions include:** Computer-Aided Designer; Design Engineer; Manufacturing Engineer; Marketing Specialist; Software Engineer; Systems Analyst; Test Engineer. **Number of employees at this location:** 200.

ADVANCED LASER GRAPHICS
1101 30th Avenue NW, Washington DC 20007. 202/342-2100. **Contact:** Human Resources. **Description:** A computer firm whose services include technical support and electronic publishing.

ALPHA MICROSYSTEMS
2722 South Fairview Street, Santa Ana CA 92704. 714/957-8500. **Fax:** 714/641-7678. **Contact:** Human Resources Department. **World Wide Web address:** http://www.alphamicro.com. **Description:** Alpha Microsystems is an information technology products and services provider concentrating on vertical market niches, served directly through both value-added resellers and distributors. The company also provides consulting, networking, software support, and maintenance services, with approximately 50 locations in North America and the United Kingdom. **Corporate headquarters location:** This Location.

ALPHANET SOLUTIONS
7 Ridgedale Avenue, Cedar Knolls NJ 07927. 201/267-0088. **Contact:** Human Resources. **Description:** Provides *Fortune* 500 companies with computer supplies. Alphanet Solutions also offers computer training courses and other support services.

APPLIED TECHNOLOGY
2615 Camino Del Rio South, Suite 400, San Diego CA 92108. 619/297-9101. **Contact:** Human Resources. **Description:** Performs computer-related work under government contracts.

ARCATA ASSOCIATES, INC.
4220 Arcata Way, North Las Vegas NV 89030. 702/399-9966. **Contact:** Lynn Clayton, Human Resources Director. **Description:** Provides computer engineering services and technical support under contract by the U.S. government.

ARRAY MICROSYSTEMS, INC.
987 University Avenue, Los Gatos CA 95030. 408/399-1505. **Fax:** 408/399-1506. **Contact:** Human Resources. **E-mail address:** jobs@array.com. **World Wide Web address:** http://www.array.com. **Description:** A multimedia technology firm that develops video e-mail products.

BANCTEC
4201 Taggart Creek Road, Suite 110, Charlotte NC 28208. 704/392-1424. **Contact:** Service Manager. **Description:** This location houses the service division. Overall, BancTec designs, manufactures, and sells image-enable processing computer systems throughout the world. **Number of employees nationwide:** 500.

BANTA ISG
2600 North Main Street, Spanish Fork UT 84660-9596. 801/798-0800. **Contact:** Gordon Lindstrom, Human Resources Director. **Description:** Provides complete software project management -- software documentation printing, magnetic disk and CD-ROM replication, electronic document indexing and retrieval software, CD-ROM mastering, software packaging services, complete turnkey literature management, electronic order fulfillment, 1-800 order services, and customer product inventory management. **Parent company:** Banta Corporation is a technology and market leader in printing and digital imaging. The corporation serves publishers of educational and general books, special-interest magazines, consumer and business catalogs, and direct marketing materials. In addition to printing and digital imaging, Banta Corporation offers multimedia and software packages, interactive media, point-of-purchase materials, and single-use products. Banta Corporation operates through the following groups: Banta Book Group; Banta Catalog Group; Banta Digital Group; Banta Direct Marketing Group; Banta Information Services Group; Banta Publications Group; Signs, Displays, Labels & Stamps; and Single-Use Products.

BROWN DISC PRODUCTS COMPANY, INC.
1120-B Elkton Drive, Colorado Springs CO 80907-3568. 719/593-1015. **Contact:** Eva Rider, Human Resources. **Description:** Brown Disc Products Company, Inc. is a manufacturer and supplier of magnetic media for desktop computers. Brown Disc's major line of business is software duplicating and turnkey packaging and fulfillment. Customers include American Medical Association, MCI, Nationwide Insurance, Texas Instruments, Hewlett-Packard, and Coors. The company also continues to be a nationwide supplier to Kinko's. **Number of employees at this location:** 220.

C-PHONE CORPORATION
6714 Netherlands Drive, Wilmington NC 28405. 910/395-6100. **Contact:** Human Resources. **Description:** A provider of video conferencing technology.

CACI, INC.
1100 North Glebe Road, Arlington VA 22201. 703/841-7800. **Contact:** Carol Harvey, Human Resources Representative. **Description:** An international high-technology and professional services corporation. CACI is a leader in advanced information systems, systems engineering, logistics sciences, proprietary analytical software products, market analysis consulting services, and information products and systems. **Common positions include:** Accountant/Auditor; Computer Programmer; Department Manager; Electrical/Electronics Engineer; Financial Analyst; Marketing Specialist; Purchasing Agent/Manager; Quality Control Supervisor; Systems Analyst; Technical Writer/Editor. **Educational backgrounds include:** Accounting; Business Administration; Computer Science; Engineering; Finance; Marketing; Mathematics. **Benefits:**

Dental Insurance; Disability Coverage; Life Insurance; Medical Insurance; Pension Plan; Tuition Assistance. **Corporate headquarters location:** This Location. **Operations at this facility include:** Administration. **Number of employees nationwide:** 1,500.

CTG (COMPUTER TASK GROUP, INC.)
800 Delaware Avenue, Buffalo NY 14209. 716/882-8000. **Fax:** 716/887-7246. **Recorded Jobline:** 800/992-5350x1234. **Contact:** Louis Boyle, Corporate Director, Sourcing Operations. **Description:** CTG is an integrated information technology services company which develops and delivers workable solutions. CTG provides services in two interrelated areas: Professional Software Services and Information Technology Consulting. In Professional Services, CTG provides information technology skills. With these services, customers are looking for IT skills on a temporary basis. In Information Services, CTG provides IT solutions. The company focuses on three major offerings, including Business Consulting, Development and Integration, and Managed Support, which provide planning, analysis, implementation, and support services. CTG serves companies in most industries through a wide network of offices in North America and Europe. **Common positions include:** Computer Programmer; Management Analyst/Consultant; Recruiter; Software Engineer; Systems Analyst. **Educational backgrounds include:** Business Administration; Computer Science; Mathematics. **Benefits:** 401(k); Dental Insurance; Disability Coverage; Employee Discounts; Life Insurance; Medical Insurance; Pension Plan; Savings Plan; Tuition Assistance. **Corporate headquarters location:** This Location. **Other U.S. locations:** Nationwide. **Listed on:** Amsterdam Stock Exchange; New York Stock Exchange. **Number of employees nationwide:** 4,500.

COMMUNICATION INTELLIGENCE CORPORATION (CIC)
275 Shoreline Drive, Suite 520, Redwood Shores CA 94065. 415/802-7888. **Fax:** 415/802-7777. **Contact:** Human Resources. **Description:** Communication Intelligence Corporation develops, markets, and licenses handwriting recognition and related technologies for the emerging pen-based computer market. The company has created a natural input recognition system which allows a computer to recognize hand-printed character input. **Corporate headquarters location:** This Location.

COMNET CORPORATION
4200 Parliament Place, Suite 600, Lanham MD 20706. 301/918-0400. **Contact:** Trent L. Lutz, Human Resources Manager. **Description:** Provides a broad range of computer services. Comnet Corporation's principal services are facilities management (systems tailored for use by a single customer with specific needs), and remote data processing services (systems tailored for use by a variety of customers). **Common positions include:** Accountant/Auditor; Administrative Worker/Clerk; Attorney; Budget Analyst; Financial Analyst; Human Resources Manager; Purchasing Agent/Manager. **Educational backgrounds include:** Accounting; Business Administration; Economics; Finance; Liberal Arts. **Benefits:** 401(k); Dental Insurance; Disability Coverage; Life Insurance; Medical Insurance; Tuition Assistance. **Corporate headquarters location:** This Location. **Operations at this facility include:** Administration. **Listed on:** NASDAQ. **Annual sales/revenues:** Less than $5 million. **Number of employees at this location:** 35.

COMPDISK
6228 West Oakton, Morton Grove IL 60053. 847/965-8404. **Contact:** Human Resources. **Description:** Duplicates floppy disks. Compdisk provides businesses with large quantities of floppy disks.

COMPUTER BUSINESS SERVICES, INC. (CBSI)
508 East 6th Street, Sheridan IN 46069. 317/758-4415. **Contact:** Human Resources. **Description:** CBSI manufactures computer hardware and software, and also develops training materials.

COMPUTING DEVICES INTERNATIONAL
8100 34th Street, Bloomington MN 55425. 612/853-4636. **Contact:** Human Resources Department. **Description:** An electronic information services company. Computing Devices International designs and manufactures software as well as hardware. The company is also military

contracted and designs computer chips for aviation applications. **Corporate headquarters location:** This Location.
Other U.S. locations:
- 40 Lake Bellevue Drive, Suite 100, Bellevue WA 98005-2480. 206/453-7658.

COMTREX SYSTEMS CORPORATION

102 Executive Drive, Suite 102, Moorestown NJ 08057. 609/778-0090. **Fax:** 609/778-9322. **Contact:** Lisa Mudrick, Human Resources. **Description:** Comtrex specializes in point-of-sale solutions for the food service and hospitality industries. The company designs, develops, assembles, and markets electronic terminals and computer software which provide retailers with transaction processing, in-store controls, and management information. Comtrex's products are sophisticated terminals which combine traditional cash register functions with the control and data gathering capabilities of a computerized system. The company develops and licenses the use of software programs which provide enhanced reporting capabilities for its terminals systems and facilitate local and remote polling of information transfer between computers and the company's terminal systems. **Corporate headquarters location:** This Location.

COMVERSE TECHNOLOGY INC.

170 Crossways Park Drive, Woodbury NY 11797-2048. 516/677-7200. **Contact:** Ms. Teri Caperna, Personnel Manager. **Description:** Comverse Technology Inc. designs, develops, manufactures, markets, and supports special purpose computer and telecommunications systems for multimedia communications and information processing applications. The company's systems are used in a broad range of applications by fixed and wireless telephone network operators, government agencies, financial institutions, and other public and commercial organizations worldwide. The company has developed two main product lines: the TRILOGUE family of telephone-accessed, multimedia messaging, and information processing systems; and the AUDIODISK family of multiple channel, multimedia digital recording systems.

CORPORATE DISK COMPANY

1226 Michael Drive, Wood Dale IL 60191. 630/616-0700. **Contact:** Human Resources. **Description:** Provides businesses with computer disks. Corporate Disk duplicates and sells a variety of computer disks.

DATAFLEX

2145 Calumet Street, Clearwater FL 34625. 813/562-2336. **Contact:** Dave Castel, Human Resources Director. **Description:** A provider of personal computers, repair services, network design, consulting, education, and supplies. **Corporate headquarters location:** This Location. **Number of employees at this location:** 92.

DELEX SYSTEMS INC.

1953 Gallows Road, Suite 700, Vienna VA 22182. 703/734-8300. **Fax:** 703/893-5338. **Contact:** Director of Human Resources. **Description:** Offers operationally-oriented systems engineering and analyses. Delex Systems provides computer-based training systems and intelligence analysis to the defense community. **Common positions include:** Accountant/Auditor; Aerospace Engineer; Computer Programmer; Economist; Electrical/Electronics Engineer; Financial Analyst; Human Resources Manager; Purchasing Agent/Manager; Software Engineer; Systems Analyst; Technical Writer/Editor. **Educational backgrounds include:** Computer Science; Engineering; Mathematics; Physics. **Benefits:** 401(k); Dental Insurance; Disability Coverage; Life Insurance; Medical Insurance; Tuition Assistance. **Corporate headquarters location:** This Location. **Other U.S. locations:** CA; OH. **Operations at this facility include:** Administration; Sales. **Listed on:** Privately held. **Number of employees at this location:** 140. **Number of employees nationwide:** 250.

DIAMOND MULTIMEDIA SYSTEMS

2880 Junction Avenue, San Jose CA 95134. 408/325-7000. **Fax:** 408/325-7070. **Contact:** Human Resources. **Description:** A computer graphics company.

DIEBOLD
1133 Corporate Drive, Farmington NY 14425-9570. 716/924-7121. **Contact:** Human Resources. **Description:** Formerly known as Griffin Technology, Diebold designs, manufactures, and markets microcomputer systems and identification cards. The company's prime market is colleges and universities throughout the United States.

FRONTIER ENGINEERING, INC.
P.O. Box 1023, Stillwater OK 74076. 405/624-1769. **Contact:** Human Resources. **Description:** A computer hardware and software engineering firm.

GB TECH INC.
2200 Space Park Drive, Suite 400, Houston TX 77058. 713/333-3703. **Contact:** Human Resources. **Description:** Performs aerospace and computer-related engineering. GB Tech contracts with several large organizations, including McDonnell Douglas and NASA. **Corporate headquarters location:** This Location.

HADRON INC.
4900 Seminary Road, Alexandria VA 22311. 703/824-0400. **Contact:** Human Resources. **Description:** Offers computer engineering services.

HALIFAX CORPORATION
5250 Cherokee Avenue, Alexandria VA 22312. 703/750-2202. **Fax:** 703/658-2999. **Contact:** Personnel Department. **Description:** Halifax Corporation is an electronics and facilities support services company, serving U.S. government agencies, systems integrators, financial institutions, and the educational community worldwide since 1967. The company installs, operates, maintains, and supports computer systems and equipment including LANs, workstations, and communications systems. **Common positions include:** Accountant/Auditor; Aerospace Engineer; Buyer; Clerical Supervisor; Computer Programmer; Customer Service Representative; Electrical/Electronics Engineer; Electrician; Payroll Clerk; Purchasing Agent/Manager; Quality Control Supervisor; Receptionist; Secretary; Services Sales Representative; Software Engineer; Stationary Engineer; Systems Analyst. **Educational backgrounds include:** Accounting; Business Administration; Computer Science; Marketing. **Benefits:** 401(k); Dental Insurance; Disability Coverage; Employee Discounts; Life Insurance; Medical Insurance; Profit Sharing; Tuition Assistance. **Corporate headquarters location:** This Location. **Other U.S. locations:** Nationwide. **Operations at this facility include:** Administration; Regional Headquarters; Sales; Service. **Listed on:** American Stock Exchange. **Number of employees at this location:** 70. **Number of employees nationwide:** 490.

HEWLETT-PACKARD COMPANY
3404 East Harmony Road, Fort Collins CO 80525. 970/229-3800. **Contact:** Director of Personnel. **Description:** This location is home to some of Hewlett-Packard's leading-edge technologies including Graphics Software and Hardware Labs; Engineering Systems Lab; a RISC VLSI Design Lab; Hewlett-Packard Open View, Hewlett-Packard's industry-leading software for managing customers' networks and systems; and Manufacturing Test Division, a market leader in providing the electronic manufacturing marketplace with test system solutions. **NOTE:** Jobseekers should send resumes to: Employment Response Center, Event #2498, Hewlett-Packard Company, Mail Stop 20-APP, 3000 Hanover Street, Palo Alto CA 94304-1181. **Common positions include:** Computer Engineer; Electrical/Electronics Engineer; Mechanical Engineer; Software Engineer. **Educational backgrounds include:** Computer Science; Engineering; Mathematics; Physics. **Benefits:** 401(k); Dental Insurance; Disability Coverage; Employee Discounts; Life Insurance; Medical Insurance; Pension Plan; Profit Sharing; Savings Plan; Tuition Assistance. **Special Programs:** Internships. **Corporate headquarters location:** Palo Alto CA. **Other U.S. locations:** Nationwide. **Operations at this facility include:** Administration; Divisional Headquarters; Manufacturing; Regional Headquarters; Research and Development. **Listed on:** New York Stock Exchange. **Number of employees at this location:** 5,000.

INFONATIONAL
210 West 520 North, Holland Square, Orem UT 84057. 801/224-7676. **Contact:** Human Resources. **Description:** Infonational uses high-end scanning devices to scan information onto computer disks in order to build databases for its customers.

INFORMATION RESOURCES, INC.
150 North Clinton Street, Chicago IL 60661. 312/726-1221. **Fax:** 312/726-5304. **Contact:** Associate Director of Recruitment. **Description:** Information Resources, Inc. develops and maintains computerized proprietary databases, decision support software, and analytical models to assist clients, primarily in the consumer packaged goods industry, in testing and evaluating their marketing plans for new products, media advertising, price, and sales promotions. Total store data is selected from nearly 2,700 retail outlets in 75 markets and numerous other communities across the U.S., and from nearly 500 drug stores and 250 mass-merchandise stores. **Common positions include:** Accountant/Auditor; Computer Programmer; Human Resources Manager; Market Research Analyst; Marketing Manager; Secretary; Systems Analyst. **Educational backgrounds include:** Business Administration; Computer Science; Economics; Finance; Liberal Arts; Marketing; Mathematics. **Benefits:** 401(k); Daycare Assistance; Dental Insurance; Disability Coverage; Life Insurance; Medical Insurance; Tuition Assistance. **Corporate headquarters location:** This Location. **Other U.S. locations:** Los Angeles CA; San Francisco CA; Darien CT; Waltham MA; Fairfield NJ; Cincinnati OH. **Operations at this facility include:** Administration; Divisional Headquarters; Regional Headquarters. **Listed on:** NASDAQ. **Number of employees at this location:** 1,400. **Number of employees nationwide:** 5,800.

INFORMATION SYSTEMS AND SERVICES, INC. (ISSI)
8405 Coldesville Road, Silver Springs MD 20910. 301/588-3800. **Contact:** Human Resources. **Description:** Information Systems and Services is an enterprise software, networking, and services firm.

INFORMATION TECHNOLOGY SOLUTIONS
Two Eaton Street, Suite 908, Hampton VA 23669. 757/723-3544. **Contact:** Human Resources Department. **Description:** Information Technology Solutions is a government contractor providing information technology services and manufacturing computer hardware and software. **Listed on:** Privately held.

INNODATA CORPORATION
95 Rockwell Place, Brooklyn NY 11217. 718/855-0044. **Fax:** 718/522-9235. **Contact:** Human Resources Department. **Description:** Innodata Corporation is a worldwide electronic publishing company specializing in data conversion for CD-ROM, print, and online database publishers. The company also offers medical transcription services to health care providers through its Statline division.

INSTITUTE FOR SCIENTIFIC INFORMATION
3501 Market Street, Philadelphia PA 19104. 215/386-0100. **Contact:** Human Resources. **Description:** Maintains one of the world's largest scientific databases. Institute for Scientific Information supplies researchers and professors with needed information on print, diskette, CD-ROMs, and magnetic tape. This company also offers online services and technical support.

JC COMPUTER SERVICES, INC.
4705 Eisenhower Avenue, Alexandria VA 22304. 703/461-0860. **Fax:** 703/370-4017. **Contact:** Gale M. Sampson, Vice President of Government Operations. **E-mail address:** gales@jccs.com. **World Wide Web address:** http://www.jccs.com. **Description:** A provider of computer and network services including maintenance and repair, network design and installation, multivendor hardware and software sales, security and risk analysis, HTML design, Internet services, training, and information technology consulting. Founded in 1985. **NOTE:** The company offers entry-level positions. **Common positions include:** Administrative Assistant; Applications Engineer; Computer Programmer; Customer Service Representative; Database Manager; Design Engineer; Electrical/Electronics Engineer; Internet Services Manager; MIS Specialist; Multimedia Designer; Online Content Specialist; Software Engineer; Systems Analyst; Technical Writer/Editor; Telecommunications Manager. **Educational backgrounds include:** Business Administration;

Communications; Computer Science; Engineering; Liberal Arts. **Benefits:** Dental Insurance; Employee Discounts; Life Insurance; Medical Insurance; Tuition Assistance. **Corporate headquarters location:** This Location. **Listed on:** Privately held. **Annual sales/revenues:** Less than $5 million. **Number of employees at this location:** 50.

KAO INFOSYSTEMS COMPANY
26200 SW 95th, Suite 302, Wilsonville OR 97070-9241. 503/682-7611. **Contact:** Personnel. **Description:** A software duplication company. **NOTE:** This company hires through temporary agencies. Check with the company to find which agency they use. Staffing fluctuates widely according to the number of orders to be filled.

KRONOS INC.
400 Fifth Avenue, Waltham MA 02154. 617/890-3232. **Contact:** Human Resources Department. **Description:** Kronos Inc. designs, develops, and markets fully-integrated software and intelligent data collection terminals that enhance productivity in the workplace. The company has 24 locations in the U.S., one in the U.K., and two in Canada, as well as dealers worldwide. **Common positions include:** Electrical/Electronics Engineer; Software Engineer; Technical Writer/Editor. **Educational backgrounds include:** Computer Science; Engineering; Mathematics. **Benefits:** 401(k); Dental Insurance; Disability Coverage; Employee Discounts; Life Insurance; Medical Insurance; Tuition Assistance. **Special Programs:** Internships. **Corporate headquarters location:** This Location. **Operations at this facility include:** Administration; Divisional Headquarters; Research and Development; Sales; Service. **Number of employees at this location:** 900.

LIANT SOFTWARE SERVICES
8711 Burnet Road, Building C, Austin TX 78757. 512/371-7028. **Fax:** 512/371-7609. **Contact:** Human Resources. **World Wide Web address:** http://www.liant.com. **Description:** A provider of software development and turnkey services which include packaging production, media duplication, and software distribution. **Parent company:** Liant Software Corporation (Framingham MA).

LIEBERT CORPORATION
EMERSON ELECTRIC
1050 Dearborn Drive, P.O. Box 29186, Columbus OH 43229. 614/888-0246. **Contact:** Mr. Terry E. Martin, Selection and Development. **Description:** A company specializing in support systems for computer rooms and related applications.

LUCAS DIGITAL LTD.
P.O. Box 2459, San Rafael CA 94912. 415/258-2200. **Recorded Jobline:** 415/258-2100. **Contact:** Human Resources Department. **E-mail address:** hrdept@ldlhr.com. **World Wide Web address:** http://www.ldlhr.com. **Description:** Lucas Digital Ltd., a digital effects company engaged in film productions, includes Industrial Light & Magic, a visual effects company; and Skywalker Sound, a state-of-the-art audio facility. **Common positions include:** Database Manager; Network Administrator; Software Engineer; Systems Specialist; Technician; Video Maintenance Engineer. **Corporate headquarters location:** This Location.

MAXIMA CORPORATION
4200 Parliament Place, Lanham MD 20706-1849. 301/459-2000. **Fax:** 301/459-2003. **Contact:** Denise Bailey-King, Director of Personnel. **Description:** Maxima Corporation is a computer facilities management firm.

MERLIN SOFTWARE SERVICES
1420 Presidential Drive, Richardson TX 75081. 214/235-9551. **Contact:** Human Resources Department. **Description:** Merlin Software Services is a CD-ROM duplication firm.

MICROELECTRONICS & COMPUTER TECHNOLOGY
3500 West Balcones Center Drive, Austin TX 78759. 512/343-3446. **Fax:** 512/338-3447. **Contact:** Human Resources Department. **E-mail address:** employment@mcc.com. **World Wide Web address:** http://www.mcc.com. **Description:** A research and development company specializing in

computer technology. **Common positions include:** Buyer; Computer Programmer; Controller; Design Engineer; Electrical/Electronics Engineer; Hardware Engineer; Human Resources Manager; Internet Services Manager; Marketing Manager; MIS Specialist; Project Manager; Software Engineer; Systems Analyst; Webmaster. **Educational backgrounds include:** Business Administration; Computer Science; Engineering; Finance; Marketing; Mathematics; Physics. **Benefits:** Dental Insurance; Disability Coverage; Employee Discounts; Life Insurance; Medical Insurance; Profit Sharing; Savings Plan; Telecommuting; Tuition Assistance. **Corporate headquarters location:** This Location. **Other U.S. locations:** Santa Clara CA. **Annual sales/revenues:** $21 - $50 million. **Number of employees at this location:** 130.

MINNESOTA SUPER COMPUTER CENTER INC.

1200 Washington Avenue South, Minneapolis MN 55415. 612/337-0200. **Contact:** Personnel Department. **Description:** A computer simulation center affiliated with the University of Minnesota.

THE WALDAC GROUP

5050 West Lemon Street, Tampa FL 33609. 813/286-8086. **Contact:** Human Resources. **Description:** The Waldac Group sells hardware and software, offers computer training classes, and provides networking support.

WAVE TECHNOLOGIES INTERNATIONAL

10845 Olive Boulevard, Suite 250, St. Louis MO 63141. 314/995-5767. **Contact:** Human Resources. **Description:** Offers helpdesk support, courseware products, and computer training programs.

Note: Because addresses and telephone numbers of smaller companies can change rapidly, we recommend you call each company to verify the information below before inquiring about job opportunities. Mass mailings are not recommended.

Additional employers:

COMPUTER PROCESSING AND DATA PREPARATION SERVICES

ADP Automatic Data Processing
8701 Mylander Ln, Towson MD 21286-2102. 410/821-0600.

ADP Automatic Data Processing
2010 Crow Canyon Pl, San Ramon CA 94583. 510/866-1100.

ADP Automatic Data Processing
13141 Northwest Fwy, Houston TX 77040-6307. 713/939-4600.

ADP Automatic Data Processing
12610 Park Plaza Dr, Cerritos CA 90703-8558. 562/924-4999.

ADP Automatic Data Processing
9310 Tech Center Dr Ste 170, Sacramento CA 95826-2564. 209/944-5862.

ADP Automatic Data Processing
1949 E Sunshine St 1-201, Springfield MO 65804-1601. 417/887-5884.

ADP Automatic Data Processing
1950 Hassell Rd, Hoffman Estates IL 60195-2300. 847/490-1987.

ADP Automatic Data Processing
5680 New Northside Dr NW, Atlanta GA 30328-4612. 770/955-3600.

ADP Automatic Data Processing
8100 Cedar Avenue South, Bloomington MN 55425-1802. 612/854-1700.

ADP Automatic Data Processing
500 West 7th Street, Cincinnati OH 45203-1594. 513/852-5200.

ADP Automatic Data Processing
10155 SE Sunnyside Rd, Clackamas OR 97015-9765. 503/654-6800.

ADP Automatic Data Processing
33 W State St, Binghamton NY 13901-2331. 607/724-4667.

ADP Automatic Data Processing
1 ADP Plz, Milford CT 06460-3061. 203/783-3000.

Advanced Automation Associates
43 Manning Rd, Billerica MA 01821-3925. 508/262-9600.

AIM
219 E Main St, Mechanicsburg PA 17055-6541. 717/766-3461.

Americom Direct Marketing
1065 Bristol Road, Mountainside NJ 07092. 908/355-2300.

Anacomp Inc.
2520 Pilot Knob Rd, Mendota
Heights MN 55120-1137.
612/683-1000.

Applications Research Corp.
250 W Street Rd, Warminster PA
18974-3207. 215/674-4525.

ASAP Inc.
3000 France Ave S, Minneapolis
MN 55416-4223. 612/926-7589.

**Automated Concepts
Consulting**
8770 W Bryn Mawr Ave,
Chicago IL 60631-3515. 312/380-
4200.

B&G Processing Systems Inc.
PO Box 306, Fraser MI 48026.
810/247-0695.

Berkley Information Services
10 Roundwind Rd, Luverne MN
56156. 507/283-9195.

Bisys Inc.
2091 Springdale Rd, Cherry Hill
NJ 08003-4005. 609/424-0150.

Bisys Inc.
575 University Ave, Norwood
MA 02062-2636. 617/551-8884.

Bisys Inc.
11 E Greenway Plz Ste 300,
Houston TX 77046-1102.
713/622-8911.

Boris Systems Inc.
4660 S Hagadorn Rd, East
Lansing MI 48823-5353.
517/332-7702.

Botal Associates Inc.
7 Dey St, New York NY 10007-
3201. 212/227-7370.

Broad Data Systems Inc.
2710 Yorktowne Blvd, Brick NJ
08723-7966. 908/255-6260.

Business Records Corp.
1431 Tallevast Rd, Sarasota FL
34243-5035. 941/351-4981.

Buypass Corporation
360 Interstate North Pkwy,
Atlanta GA 30339-2204.
770/953-2664.

Cap Gemini America
100 Washington Ave S,
Minneapolis MN 55401-2150.
612/375-9881.

Cap Gemini America
4445 Lake Forest Dr, Cincinnati
OH 45242-3751. 513/563-6622.

Care Computer Systems Inc.
636 120th Ave NE, Bellevue WA
98005-3039. 206/451-8272.

CHMC
1437 Gordon St, Allentown PA
18102-5627. 610/439-1717.

Comdisco Technical Services
800 Albion Ave, Schaumburg IL
60193-4523. 847/985-8660.

Commercial Data Processing
4 Sperry Rd, Fairfield NJ 07004-
2005. 201/882-1660.

**Computer Associates
International**
2600 Maitland Center Pkwy,
Maitland FL 32751-7221.
407/661-3900.

Computer Bank Inc.
1612 Shadywood Ln, Mount
Pleasant TX 75455-5637.
903/572-4336.

Computer Data Systems Inc.
5015 Bradford Drive NW,
Huntsville AL 35805-1906.
205/830-1400.

Computer Intelligence Corp.
3344 N Torrey Pines Ct, La Jolla
CA 92037-1024. 619/450-1667.

Computer Support Centres
1920 Thoreau Dr N, Schaumburg
IL 60173-4151. 847/397-8000.

Computer Task Group
30 N Union St, Rochester NY
14607-1345. 716/325-4220.

Computoservice Inc.
1315 Stadium Road, Mankato
MN 56001-5355. 507/625-1691.

Comvestrix
1100 Valley Brook Ave,
Lyndhurst NJ 07071-3608.
201/935-8300.

Data Conversions
2501 Hilldale Boulevard,
Arlington TX 76016-1964.
817/429-2255.

Data Entry Company Inc.
424 Arch Street, Oakland MD
21550-1915. 301/334-1234.

Data Entry Services
23750 Elmira, Redford MI
48239-1405. 313/533-2299.

Data Exchange Center Inc.
9041 N Deerwood Dr, Milwaukee
WI 53223-2437. 414/355-5906.

Data Plus Computer Service
150 Main St, Pawtucket RI
02860-4118. 401/724-2220.

Data Shop Inc.
1230 Plainfield Ave, Janesville
WI 53545-0434. 608/752-2580.

Datatronic Systems Corp.
PO Box 44128, Panorama City
CA 91412-0128. 818/988-5290.

Deltanet Inc.
100 1st St, San Francisco CA
94105-2637. 415/995-8700.

Digital Image
36524 Grand River Ave,
Farmingtn Hills MI 48335-2868.
810/477-5600.

Digital Imaging & Technology
1151 N Magnolia Ave, Anaheim
CA 92801-2639. 714/229-9311.

**Disabled Programmers Inc.
(DPI)**
151 Martinvale Ln, San Jose CA
95119-1319. 408/629-3700.

**Diversified International
Sciences**
9901 Business Pkwy, Lanham
MD 20706-1840. 301/731-9070.

Domestic Data Solutions Inc.
51 Rose Ln, Medford NY 11763-
1328. 516/698-4919.

Dyncorp
2727 Hamner Ave, Norco CA
91760-1995. 909/735-3300.

Ecta Corp.
321 Norristown Rd, Ambler PA
19002-2755. 215/540-0250.

EDS Electronic Data Systems
2525 Kell Blvd, Wichita Falls TX
76308-1049. 817/322-4978.

EDS Electronic Data Systems
2600 Technology Dr, Orlando FL
32804-8094. 407/297-0870.

Erdman Anthony & Associates
2165 Brighton Henrietta Town L,
Rochester NY 14623-2703.
716/325-1866.

Etc. Data Processing Services
1 Jeanne Dr, Newburgh NY
12550-1702. 914/564-6000.

Filmet
7436 Washington Avenue,
Pittsburgh PA 15219. 412/351-
3510.

Financial Services Inc.
21 Harristown Rd, Glen Rock NJ
07452-3314. 201/652-6000.

Financial Technologies
14300 Sullyfield Cir, Chantilly
VA 20151. 703/631-4400.

First Data Corp.
414-F Gallimore Dairy Rd,
Greensboro NC 27409-9725.
910/665-3300.

First Health Services Corp.
4300 Cox Road, Richmond VA
23060. 804/965-7400.

First Image Management Co.
117 W Commerce St, San
Antonio TX 78205-2405.
210/226-8211.

FIserv of Mendota Heights
1333 Northland Dr, Mendota
Heights MN 55120-1141.
612/683-4600.

FIserv Seattle Inc.
15375 SE 30th Pl, Bellevue WA
98007-6500. 206/562-8700.

FIserv Tampa Inc.
5802 Benjamin Center Dr, Tampa
FL 33634-5204. 813/885-5800.

**Florida Informanagement
Service**
P.O. Box 1547, Orlando FL
32802. 407/841-1712.

GE Co.
20 Waterview Blvd, Parsippany
NJ 07054-1229. 201/299-2000.

Gelco Information Networks
10700 Prairie Lakes Dr, Eden
Prairie MN 55344-3886. 612/947-
1501.

General Technical Services Inc.
10100 Old Columbia Road,
Columbia MD 21046-2623.
301/621-8222.

Grimbac
5365 Hill-23 Drive, Flint MI
48507-3906. 801/239-5553.

GTE Data Services Inc.
7901 Freeport Blvd, Sacramento
CA 95832-9701. 916/665-3200.

Hand Graphic Systems Inc.
323 West 8th Street, Suite 203,
Kansas City MO 64105. 816/474-
4263.

Health Management Systems
401 Park Ave S, New York NY
10016-8808. 212/685-4545.

Helix
310 S Racine Ave, Chicago IL
60607-2841. 312/421-6000.

IBM Corp.
140 E Town St, Columbus OH
43215-5125. 614/225-2500.

ICC Services Inc.
8734 W Chester Pike, Upper
Darby PA 19082-2618. 610/853-
3040.

IMI Systems Inc.
14180 Dallas Pkwy, Suite 450,
Dallas TX 75240-4370. 972/788-
2311.

IMI Systems Inc.
1 Gatehall Dr, Parsippany NJ
07054-4514. 201/292-9200.

Information Retrieval Methods
1525 N Stemmons Fwy, Dallas
TX 75207-3410. 972/242-1480.

**Information Systems
Consulting Inc.**
1100 Main St, Kansas City MO
64105-2105. 816/842-3000.

Interactive Marketing Systems
11 W 42nd St, New York NY
10036-8002. 212/789-3600.

International Business Service
8150 Leesburg Pike, Vienna VA
22182-2714. 703/892-0660.

Intuition Systems Inc.
11860 31st Ct N, St Petersburg
FL 33713. 813/573-3511.

**ISI Information Service
International**
100 International Dr, Mount
Olive NJ 07828-1383. 201/691-
3500.

ITS Corporation
P.O. Box 1148, Ventura CA
93002-1148. 805/641-0093.

Ivey Seright
424 8th Ave N, Seattle WA
98109-4705. 206/623-8113.

JCI Data Processing Inc.
200 Route 130 S, Cinnaminson
NJ 08077-2892. 609/786-2600.

JTS Computer Services Inc.
30 Corporate Woods, Suite 350,
Rochester NY 14623-1454.
716/273-6100.

Kaiser
25 N Via Monte, Walnut Creek
CA 94598-2510. 510/926-0156.

Kale Design
50 Oreland Mill Rd, Suite 200,
Oreland PA 19075-1305.
215/884-5400.

Keane Inc.
200 Galleria Pkwy SE, Atlanta
GA 30339-5944. 770/850-7270.

LJ Gonzer Associates
1225 Raymond Blvd, Newark NJ
07102-2919. 201/624-5600.

M&I-New England
1525 Washington St, Braintree
MA 02184-7599. 617/849-1600.

**Management Applied
Programming**
3415 S Sepulveda Blvd, Los
Angeles CA 90034-6060.
310/397-7220.

MCC Corp.
535 Mountain Ave, New
Providnce NJ 07974-2000.
908/582-9500.

Med Quist Inc.
5 Greentree Center, Suite 311,
Marlton NJ 08053. 609/596-8877.

Medical Records Corp.
3637 Green Rd, Cleveland OH
44122-5717. 216/464-2244.

Metro Information Services
14651 Dallas Pkwy, Suite 132,
Dallas TX 75240-7476. 972/490-
4782.

Micropublication Systems Inc.
2500 Belshore Avenue, Carson
CA 90746-3516. 310/763-7575.

Midwest Business Systems Inc.
2035 28th St SE, Grand Rapids
MI 49508-1594. 616/452-1755.

National Business Systems Inc.
2905 W Service Rd, Eagan MN
55121-1244. 612/688-0202.

NCR Corporation
2031 Old Trenton Rd, Cranbury
NJ 08512-1417. 609/443-1800.

Nova Information Systems Inc.
5 Concourse Pkwy NE, Atlanta
GA 30328-6101. 770/392-9106.

O'Neil & Associates
425 N Findlay St, Dayton OH
45404-2203. 937/461-1852.

OAO Corporation
201 Main St, Fort Worth TX
76102-3121. 817/870-9841.

On-Line Financial Services Inc.
900 Commerce Dr, Oak Brook IL
60521-1967. 630/571-7900.

Package Fulfillment Center Inc.
1401 Lakeland Ave, Bohemia NY
11716-3316. 516/567-7000.

Palarco Inc.
996 Old Eagle School Rd, Wayne
PA 19087-1806. 610/687-3410.

Payroll
205 Regency Executive Park Dr,
Charlotte NC 28217-3989.
704/523-8434.

Pergament Graphic Systems
38 E 30th St, New York NY
10016-7316. 212/213-8310.

Perot Systems Corporation
12377 Merit Dr, Dallas TX
75251-2224. 214/383-5600.

**Philips Communications &
Processing Service**
6606 Lyndon B Johnson Fwy,
Dallas TX 75240-6533. 214/861-
1700.

**Pinkerton Computer
Consultants**
1900 N Beauregard St,
Alexandria VA 22311-1716.
703/820-5571.

**Pinkerton Computer
Consultants**
4 Interplex Dr Ste 111, Trevose
PA 19053-6963. 215/639-9535.

Pixel Works
600 Cutlass Dr, Novato CA
94947-4785. 415/892-0925.

Pointer's International Corp.
8224 White Settlement Rd, Fort
Worth TX 76108-1603. 817/246-
4931.

Premier Operating Systems Inc.
27 W Main St, Freeport IL
61032-4215. 815/233-3653.

Presort Services Inc.
3584 Roger B Chaffee Mem Dr
SE, Wyoming MI 49548-2328.
616/247-1177.

Pro Data Concepts
588 Route 70, Brick NJ 08723-
4014. 908/920-8890.

Professional Color Service Inc.
909 Hennepin Ave, Minneapolis
MN 55403-1806. 612/673-8900.

Professional Mail Services Inc.
5608 Spring Ct, Raleigh NC
27604-2966. 919/876-9651.

Purvis Systems Inc.
7001 Brush Hollow Rd, Westbury
NY 11590-1743. 516/997-5800.

QC Data Inc.
777 Grant St, Denver CO 80203-
3501. 303/837-1444.

Quadax Inc.
4079 Executive Pkwy,
Westerville OH 43081-3859.
614/882-1200.

Quantum Systems & Software
300 Arboretum Pl, Richmond VA
23236-3465. 804/320-4800.

QuestPoint
397 N Sam Houston Pkwy E,
Houston TX 77060-2402.
713/448-2100.

QuestPoint
8827 Staples Mill Rd, Richmond
VA 23228-2014. 804/264-7300.

Rapid Design Service Inc.
4424 Interpoint Blvd, Dayton OH
45424-5709. 937/236-8602.

RAR Graphics
1395 NW 17th Ave, Delray
Beach FL 33445-2551. 561/272-
5850.

Renkim Corp.
13333 Allen Rd, Southgate MI
48195-2294. 313/374-8300.

Response Data Corp.
1050 Wall St W, Lyndhurst NJ
07071-3615. 201/460-9000.

Rexnord Data Systems
1817 Leslie Ave, Alexandria VA
22301-1223. 703/683-8410.

Richard M. Walsh Associates
530 Main St, Dickson City PA
18519-1524. 717/383-4510.

Ring Computer Enterprises
7777 Fay Ave, La Jolla CA
92037-4326. 619/455-5745.

Risk Sciences Group Inc.
5620 Glenridge Dr NE, Atlanta
GA 30342-1334. 404/256-0830.

Robert F. White/ADP
209 W Jackson Blvd, Chicago IL
60606-6907. 312/322-9600.

**RPI Virtual Entertainment
Group**
PO Box 14607, San Francisco CA
94114-0607. 415/777-3226.

Sabre Group
6801 Governors Lake Pkwy,
Norcross GA 30071-1139.
770/246-5939.

Sales Technologies Inc.
1300 Morris Dr, Wayne PA
19087-5559. 610/408-2800.

**San Diego Data Processing
Corp.**
5975 Santa Fe St, San Diego CA
92109-1696. 619/490-0600.

Satellite Image Systems
990 Atherton Dr, Murray UT
84123-3401. 801/262-1792.

Scene Science
PO Box 25329, Rochester NY
14625-0329. 716/264-1470.

Science Systems & Applications
5900 Princess Garden Pkwy,
Lanham MD 20706-2925.
301/731-9300.

**Security Pacific Information
Service Corp.**
10174 Old Grove Rd, San Diego
CA 92131-1649. 619/530-9214.

Sharkbyte
433 Las Colinas Blvd E, Irving
TX 75039-5581. 214/444-8855.

SIAC
2 Metrotech Ctr, Brooklyn NY
11201-3838. 212/383-4871.

Sierra Systems Consultants Inc.
400 Las Colinas Blvd E, Irving
TX 75039-5579. 214/263-3373.

Skyline Digital Images
12345 Portland Ave, Burnsville
MN 55337-1529. 612/895-6373.

SMS Inc.
4251 Plymouth Rd, Ann Arbor
MI 48105-3638. 313/994-8300.

Southeastern Computer Service
112 North Madison, Douglas GA
31533-4616. 912/384-7175.

Standard Data Corp.
440 9th Ave, New York NY
10001-1620. 212/564-4433.

Sun Microsystems
12 E Greenway Plz, Houston TX
77046-1203. 713/622-0072.

SunGard Financial Systems
601 2nd Ave S, Hopkins MN
55343-7779. 612/935-3300.

Systems Analysis Inc.
3610 156th Ave SE, Bellevue
WA 98006-1729. 206/641-3100.

Tandem Computers Inc.
Bldg 1306 Concourse Drive,
Linthicum Heights MD 21090-
1000. 410/859-8800.

TDEC
7200 Wisconsin Ave, Suite 905,
Bethesda MD 20814-4811.
301/718-0703.

Technical Industries Inc.
25631 Little Mack Ave, St Clair
Shores MI 48081. 810/777-0160.

Tekgraf Inc.
6721 Portwest Dr, Houston TX
77024-8019. 713/868-9330.

Thomcomp Inc.
555 E North Ln, Conshohocken
PA 19428-2246. 610/834-1120.

Tisco
P.O. Box 280, San Dimas CA
91773-2541. 909/592-2679.

Tobin & Associates Inc.
675 Panorama Trail, Suite 2,
Rochester NY 14625-2406.
716/586-2103.

Trade Service Systems
1777 Sentry Pkwy W, Blue Bell
PA 19422-2213. 215/542-8300.

Type A-Scan Inc.
200 Varick Street, New York NY
10014. 212/367-8406.

Ultradata Corp.
5020 Franklin Dr, Pleasanton CA
94588-3354. 510/463-8356.

Unco Data Systems
6030 Culligan Way, Minnetonka
MN 55345-5917. 612/935-4466.

USLife Systems Corporation
6363 Forest Park Road, Dallas
TX 75235-5400. 214/637-1179.

Working Knowledge Inc.
738 Elgin Rd, Newtown Square
PA 19073-3203. 610/359-9080.

**MISC. COMPUTER
RELATED SERVICES**

Adistra Corp.
101 Union St, Plymouth MI
48170-1692. 313/425-2600.

Advanced Computer
315 Bollay Dr, Goleta CA 93117-
2994. 805/961-0200.

Altair Corp.
350 Barclay Blvd, Lincolnshire
IL 60069-3606. 847/634-9540.

Alternative Resources Corp.
13760 Noel Rd, Suite 250, Dallas
TX 75240-4362. 972/934-0505.

Alternative Resources Corp.
100 Tri State International,
Lincolnshire IL 60069-4435.
847/317-1000.

ARI Network Services Inc.
330 E Kilbourn Ave, Milwaukee
WI 53202-6636. 414/278-7676.

ASC Solutions Inc.
3740 N Josey Ln, Carrollton TX
75007-2472. 972/492-0569.

Ashman Smith & Company
10100 Brecksville Rd, Cleveland
OH 44141-3206. 216/526-5300.

Ball Corporation
5580 Morehouse Dr, San Diego
CA 92121. 619/457-5550.

Berger & Co.
1350 17th St, Denver CO 80202-
1508. 303/571-4557.

Bindco
1089 Mills Way, Redwood City
CA 94063-3119. 415/363-2200.

Buckley-Dement LP
612 S Clinton Street, Chicago IL
60607-4306. 312/427-3862.

CACI International Inc.
3100 Presidential Drive, Fairborn
OH 45324-2039. 937/429-2771.

CE Services
2895 113th Street, Grand Prairie
TX 75050-6481. 214/988-4888.

CGI Systems
500 5th Avenue, New York NY
10110-0002. 212/575-6280.

Check Free Corp.
25 Crossroads Drive, Owings
Mills MD 21117-5450. 410/581-
2900.

Chubb Institute
8 Sylvan Way, Parsippany NJ
07054-3801. 201/285-9700.

CMSI Inc.
1900 Main St, Irvine CA 92714-
7200. 714/863-3011.

Compex Corporation
5500 Cherokee Ave, Suite 500,
Alexandria VA 22312-2321.
703/642-5910.

Computer Horizons Corp.
521 Plymouth Rd, Plymouth
Meeting PA 19462-1638.
610/825-5565.

Computer Horizons Corp.
3710 Corporex Park Drive,
Tampa FL 33619-1160. 813/626-
6366.

Computer People Inc.
4400 Campbell's Run Rd,
Pittsburgh PA 15205. 412/562-
0233.

Computer Recovery Service
15475 N Greenway Hayden
Loop, Scottsdale AZ 85260-1614.
602/443-2588.

Crusoe Communications Inc.
387 Passaic Avenue, Fairfield NJ
07004-2014. 201/882-1022.

CSC Consulting
1160 West Swedesford Rd,
Berwyn PA 19312. 610/251-
0660.

Data Systems Analysts Inc.
10400 Eaton Pl, Fairfax VA
22030-2208. 703/591-3704.

Digital Consulting & Software
1 Sugar Creek Ctr Blvd, Sugar
Land TX 77478-3435. 713/242-
9181.

Digital Consulting Inc.
204 Andover St, Andover MA
01810-5697. 508/470-3870.

Donovan Data Systems Inc.
115 W 18th St, New York NY
10011-4113. 212/633-8100.

Evolving Systems Inc.
6892 South Yosemite St,
Englewood CO 80155. 303/689-
1000.

FIServ
885 S Village Oaks Dr, Covina
CA 91724-3697. 818/339-9011.

Gensym Corporation
85 W Algonquin Rd, Arlington
Heights IL 60005-4422. 847/593-
7820.

IBM Corp.
3 Bedford Farms, Bedford NH
03110-6522. 603/472-4100.

IBM Corp.
301 Congress Ave, Austin TX
78701-4041. 512/473-8000.

IBM Corp.
100 E Pratt St, Baltimore MD
21202-1009. 410/332-2200.

IBM Corp.
1745 Jefferson Davis Hwy,
Arlington VA 22202-3402.
703/769-2000.

IBM Corp.
PO Box 137, Pleasantville NJ
08232-0137. 609/484-2400.

ICT Inc.
18310 Montgomery Village Ave,
Gaithersburg MD 20879-3551.
301/948-0200.

Information Access
362 Lakeside Dr, Foster City CA
94404-1146. 415/378-5000.

Information Spectrum Inc.
1040 Kings Hwy North, Cherry
Hill NJ 08034-1908. 609/667-
6161.

Interactive Business Systems
2625 Butterfield Rd Ste 114W,
Oak Brook IL 60521-1237.
630/571-9100.

Internet Engineering
155 2nd St, Cambridge MA
02141-2125. 617/441-5050.

Inturnet Inc.
811 Alpha Dr, Richardson TX
75081-2862. 214/783-0066.

KCMS Inc.
3401 East West Hwy, Hyattsville
MD 20782-1915. 301/853-6601.

Keane Inc.
4501 Erskine Rd Ste 120,
Cincinnati OH 45242-4727.
513/794-1400.

Lasernet
11216 Waples Mill Rd, Fairfax
VA 22030-6099. 703/591-4232.

Lemans Group Ltd.
455 S Gulph Rd, King Of Prussia
PA 19406-3114. 610/337-8686.

Logic Systems Inc.
158 W 76th St, New York NY
10023-8457. 212/501-0396.

Management Assistance Corp.
of America
8600 Boeing Dr, El Paso TX
79925-1226. 915/772-4975.

Matrix Computer Consulting
65 State Rt 4, River Edge NJ
07661-1949. 201/488-4181.

MCRB
9171 Oso Ave, Chatsworth CA
91311-6049. 818/407-4300.

Media Connection New York
443 Park Ave S, New York NY
10016-7322. 212/686-3845.

Metro Information Services
Relections II, Virginia Beach VA
23450. 757/486-1900.

MicroAge Computer Centers
757 Springdale Dr, Exton PA
19341-2836. 610/524-6665.

Moore Communications
5767 Uplander Way, Culver City
CA 90230-6605. 310/670-3605.

N2K
435 Devon Park Dr, Wayne PA
19087-1935. 610/293-4700.

New Horizons Computer
Learning Center
1231 E Dyer Rd Ste 140, Santa
Ana CA 92705-5606. 714/556-
1220.

New Resources Corp.
3315 E Algonquin Rd Ste 500,
Rolling Meadows IL 60008.
847/797-5800.

OCLC (On-Line Computer
Library Center)
6565 Frantz Rd, Dublin OH
43017-5308. 614/764-6000.

Optimum Services & Systems
4500 Forbes Rd, Lanham MD
20706-4372. 301/459-9100.

Ovid Technologies Inc.
333 7th Ave, New York NY
10001-5004. 212/563-3006.

Policy Management Systems
8300 Central Park, Waco TX
76712-6529. 817/299-4000.

Quebecor Integrated Media
4918 20th St E, Tacoma WA
98424-1922. 206/922-9393.

Shaw Data Services
122 E 42nd St, New York NY
10168-0082. 212/682-8877.

SoftWorld Services
18939 120th Ave NE Ste 111,
Bothell WA 98011. 206/483-
8989.

Solution Technologies Inc.
702 Lisburn Rd, Camp Hill PA
17011. 717/763-5611.

Stoner Associates Inc.
PO Box 86, Carlisle PA 17013-
0086. 717/243-1900.

Structured Logic Co. Inc.
330 7th Ave, New York NY
10001-5010. 212/947-7510.

SunGard
504 Totten Pond Rd, Waltham
MA 02154-1904. 617/466-9800.

SunGard Capital Markets
560 Lexington Ave, New York
NY 10022-6828. 212/371-1116.

Tangent International
Computer Consultants
30 Broad St, New York NY
10004-2304. 212/809-8200.

Tel Tech Corp. Consulting Division
39 Broadway, New York NY 10006-3003. 212/514-5440.

Unisys Corporation
4700 Rockside Rd, Independence OH 44131-2155. 216/573-2840.

Vista Environmental Info Inc.
5060 Shoreham Pl, San Diego CA 92122-5904. 619/450-6100.

COMPUTER REPAIR SERVICES/RENTAL AND LEASING

*T*he *daily routine of a PC Technician changes constantly. As the cavalry of the computer world, technicians are called in whenever something goes wrong. Sometimes it's as simple as plugging the computer into the wall or removing a disk that's stuck in the hard drive. Other times, a technician must determine the specific reason for why a printer won't function properly.*

Traveling to client sites is one major requirement of the job. One professional in the industry said he often travels over 200 miles a day and visits anywhere from five to ten customers. He enjoys dealing with different people on a daily basis and overcoming a variety of problems. Sometimes the job can be slightly frustrating because time constraints don't always allow an on-site technician to discover the root of a problem.

As processors become faster and perform more tasks, PC Technicians will have to study the latest technologies to stay competitive. Repair professionals will also need to know software basics, to distinguish between hardware and software problems. The demand for computer repair professionals should continue to rise.

ABL DATA SYSTEMS
32 East 14 Mile Road, Madison Heights MI 48071. 810/588-1083. **Contact:** Human Resources. **Description:** Services computer hardware for individuals and business clients. ABL Data Systems repairs CPUs, monitors, printers, and other peripherals for IBM, IBM clones, and some Apple units.

ALPHA MICROSYSTEMS
7001 Peachtree Industrial Boulevard, Franklin Office Park, Suite 402, Norcross GA 30092. 770/447-0303. **Contact:** Human Resources. **Description:** This location is a repair and rebuild facility under various maintenance agreements. Alpha Microsystems is an information technology products and services provider concentrating on vertical market niches, served through value-added resellers and distributors. The company also provides consulting, networking, software support, and maintenance services, with approximately 50 locations in North America and the United Kingdom. **Corporate headquarters location:** Santa Ana CA.

ALPHA MICROSYSTEMS
1438 Elmhurst Road, Elk Grove Village IL 60007. 847/593-6661. **Contact:** Human Resources. **Description:** This location is a technical support center, providing repair services for PCs and printers. Alpha Microsystems is an information technology products and services provider concentrating on vertical market niches, served directly through both value-added resellers and distributors. The company also provides consulting, networking, software support, and maintenance services, with approximately 50 locations in North America and the United Kingdom. **Corporate headquarters location:** Santa Ana CA.

ALTEC LANSING MULTIMEDIA
Route 6 & Route 209, Milford PA 18337. 717/296-4434. **Contact:** Human Resources. **Description:** Repairs and replaces speakers. Altec Lansing Multimedia repairs and replaces speakers for both stereo systems and computers including IBM, IBM compatible, and Macintosh.

AMERICAN COMPUTER SOLUTIONS
5854 Highland Ridge, Cincinnati OH 45232. 513/242-5600. **Contact:** Human Resources. **Description:** Repairs hardware and sells both hardware and software. American Computer Solutions repairs virtually any type of hard-drive including both IBM and Macintosh. This company also sells a wide variety of software to businesses.

APPLIED SOLUTIONS
501 Greene Street, Augusta GA 30901-1460. 706/722-8649. **Contact:** Human Resources. **Description:** Applied Solutions provides support to businesses through purchasing and maintaining computer systems.

ARIX CORPORATION
1024 Morse Avenue, Sunnyvale CA 94089. 408/541-8700. **Contact:** Human Resources. **Description:** A computer refurbishing and retail firm. Arix also offers systems support.

ARKANSAS COMPUTER SERVICE, INC. (ACS)
8023 Chicot Road, Little Rock AR 72209. 501/568-0286. **Contact:** Human Resources. **Description:** ACS is involved in computer repair, computer sales, and networking assistance.

BRS COMPUTER SERVICE
3401 East McDowell Road, Phoenix AZ 85008. 602/277-3282. **Contact:** Judy Welch, Service Manager. **Description:** Sells and services computer hardware.

CHMC
COMPUTER HARDWARE MAINTENANCE COMPANY
2010 Cabot Boulevard West, Langhorne PA 19047. 215/752-2221. **Contact:** Human Resources Department. **Description:** Sells and repairs computer hardware.

CNS INC.
100 Ford Road, Denville NJ 07834. 201/625-4056. **Fax:** 201/625-9489. **Contact:** Human Resources Department. **Description:** A full-service computer hardware repair center.

CARLISLE IMAGING
1617 Southwood Drive, Nashua NH 03063. 603/595-2522. **Contact:** Nancy Boisvert, Human Resources. **Description:** Repairs Canon and IBM bubble-jet printers.

CHESS INC.
410 Raritan Way, Denver CO 80204. 303/573-5133. **Contact:** Human Resources Department. **Description:** Specializes in the repair and sale of computer equipment and printers.

CIRCUIT TEST, INC.
14601 McCormick Drive, Tampa FL 33626. 813/855-6685. **Contact:** Human Resources Department. **Description:** Circuit Test specializes in the repair of computer monitors.

COLUMBIA COMPUTER MAINTENANCE
8600 SE Stark Street, Portland OR 97216. 503/252-7566. **Contact:** Donna Uleen, General Manager. **Description:** Provides repair services for many different types of computer hardware.

COMPUTER MAINTENANCE CENTER
4615 Hawkins Street NE, Albuquerque NM 87109. 505/345-8800. **Contact:** Owner. **Description:** Computer Maintenance Center specializes in computer repair.

COMPUTER SUPPORT GROUP
4430 East Miami River Road, Cleveland OH 45002. 513/481-4636. **Contact:** Robert Oherron. **Description:** Engaged in the sale and service of computers to corporations.

COMPUTER SUPPORT OF SIOUX CITY
4200 War Eagle Drive, Sioux City IA 51109. 712/277-0095. **Contact:** Human Resources. **Description:** Services computer parts.

COMPUTER SYSTEMS REPAIR
200 Meadowlands Parkway, Secaucus NJ 07094. 201/617-7711. **Contact:** Ms. Dana Hobbs, Human Resources. **Description:** A computer service and repair company.

CONNECTING POINT COMPUTER CENTER
4G COMPUTERS
1515 Wyoming, Missoula MT 59801. 406/728-5454. **Fax:** 406/721-2744. **Contact:** John Oetinger, Owner. **E-mail address:** cp4g@montana.com. **World Wide Web address:** http://www.montana. com/cp4g/cp4g.htm. **Description:** Engaged in computer sales and service. **Educational backgrounds include:** Computer Science. **Benefits:** Dental Insurance; Employee Discounts; Medical Insurance. **Corporate headquarters location:** This Location. **Annual sales/revenues:** Less than $5 million.

DCT SYSTEMS GROUP
6610 Bay Circle, Suite A, Norcross GA 30071. 770/729-8181. **Contact:** Human Resources. **Description:** Services and resells all types of computers.

DATA FLOW INC.
215 Bobbie Street, Bossier City LA 71112. 318/746-2131. **Contact:** Human Resources. **Description:** Data Flow offers computer services and sales.

DATA STORAGE TECH
14752 Sinclair Circle, Tustin CA 92680. 714/730-0722. **Contact:** Human Resources. **Description:** Installs and repairs personal computers.

DATA SUPPORT SYSTEMS
7318 Impala Drive, Richmond VA 23228. 804/261-6963. **Contact:** Human Resources. **Description:** Repairs virtually every type of personal computer and monitor.

DATAFIX
1203 Nettleton Circle, Jonesboro AR 72401. 501/972-5330. **Contact:** Steve Taylor, Manager. **Description:** Provides repair services for most types of computers and printers.

DATAPRODUCTS CORPORATION
528 Route 13 South, Milford NH 03055. 603/673-9100. **Contact:** Human Resources. **Description:** This location is a repair center. Dataproducts Corporation develops, manufactures, and markets data handling and output equipment. Products include printers and digital communications equipment.

DATASERV COMPUTER MAINTENANCE INC.
19011 Lake Drive East, Chanhassen MN 55317. 612/906-6000. **Contact:** Human Resources. **Description:** Provides businesses with computer maintenance services.

DECISION DATA SERVICE, INC.
400 Horsham Road, Suite D, Horsham PA 19044. 215/957-9500. **Contact:** Human Resources. **Description:** Decision Data Service, Inc. has become a leading independent supplier of maintenance services and supplies for the IBM midrange market. Decision Data serves 27,500 customers, providing on-site maintenance and depot repair service for systems, PCs, printers, terminals, controllers, tape drives, DASD drives, and networks. The company also offers system installs, removals, upgrades, and technical support. More than half of the company's employees are customer service engineers who are trained on IBM's System 34, 36 and 38; AS/400; RS/6000; Series/1; and peripherals, as well as on systems and peripherals from QANTEL, Compaq, Hewlett-Packard, Genicom, Wang, Seagate, Printronix, Dataproducts, Magna, Fujitsu, Decision Datas/IIS, IDEA, and Acknowledge. **Common positions include:** Accountant/Auditor; Adjuster; Computer Programmer; Contract/Grant Administrator; Customer Service Representative; Electrical/ Electronics Engineer; Systems Analyst. **Educational backgrounds include:** Accounting; Business Administration; Computer Science; Engineering; Finance; Marketing. **Benefits:** 401(k); Dental Insurance; Disability Coverage; Employee Discounts; Life Insurance; Medical Insurance; Pension Plan; Tuition Assistance. **Corporate headquarters location:** This Location. **Other U.S. locations:**

Phoenix AZ; Fremont CA. **Operations at this facility include:** Administration; Divisional Headquarters; Regional Headquarters; Research and Development; Sales; Service. **Number of employees at this location:** 300. **Number of employees nationwide:** 1,500.

DECISION ONE
N92 W14612 Anthony Avenue, Menomonee Falls WI 53051. 414/255-4634. **Contact:** Human Resources Department. **Description:** Decision One is a computer repair company.

DECISION ONE
2323 Industrial Parkway West, Hayward CA 94545. 510/732-3000. **Contact:** Human Resources. **Description:** Decision One is a computer repair company.

ECONOCOM USA
965 Ridge Lake Boulevard, Memphis TN 38120. 901/685-0021. **Contact:** Michael Drake, President. **Description:** A lessor of computer hardware to commercial customers.

ENTEX INFORMATION SERVICES
12100 East Iliff Avenue, Suite 200, Aurora CO 80014. 303/745-9600. **Contact:** Human Resources. **Description:** Sells and services microcomputer systems and products, related peripherals, and software. Entex provides training classes for software products as well as technical and support services. **NOTE:** Jobseekers should apply in person and must fill out an application. **Number of employees at this location:** 203.

FEDERAL COMPUTER CORPORATION
2745 Hartland Road, Falls Church VA 22043. 703/698-7711. **Contact:** Human Resources Department. **Description:** Provides computer maintenance services.

GTE LEASING CORPORATION
201 North Franklin Street, Suite 1800, Tampa FL 33602. 813/229-6000. **Contact:** Human Resources. **Description:** Leases computers and telephone equipment.

GEAC COMPUTER
1415 Combermere, Troy MI 48083. 810/585-3895. **Contact:** Human Resources. **Description:** This location is a service office. Geac Computer is a hardware manufacturer.

GENERAL DIAGNOSTICS
50 West Liberty Street, Reno NV 89501. 702/324-3343. **Contact:** Carol Olsen, Manager & Corporate Administrator. **Description:** Repairs computer hardware as well as electronics products.

INTERNATIONAL ASSEMBLY SPECIALISTS
945 East Ohio Street, Tucson AZ 85714. 520/294-0898. **Contact:** Mr. Dale Marthaler, General Manager. **Description:** Refurbishes IBM products.

KENNSCO INC.
14700 28th Avenue, Plymouth MN 55447. 612/559-5100. **Contact:** Dan Anderson, Manager. **Description:** Sells and services computer hardware.

KEY COMPUTER SERVICES OF CHELSEA
37 West 17th Street, Suite 2E, New York NY 10011. 212/206-8060. **Contact:** Personnel. **Description:** A computer rental company.

L.M.S. TECHNICAL SERVICES INC.
21 Grand Avenue, Farmingdale NY 11735. 516/694-2034. **Fax:** 516/694-2315. **Contact:** Human Resources. **Description:** Offers computer maintenance services.

LOGICAL SOLUTIONS, INC.
One Pope Road, Windham ME 04062. 207/892-7536. **Contact:** Human Resources. **Description:** A computer repair and sales office.

LOGISTICS MANAGEMENT
6269 Shelby Drive, Memphis TN 38141. 901/541-2200. **Contact:** Charlene Myler, Human Resources Director. **Description:** Provides computer repair services as well as inventory management and storage of computers, primarily for Canon, Apple, and NEC models.

MAGNETIC DATA INC.
6754 Shady Oak Road, Eden Prairie MN 55344. 612/941-0453. **Contact:** Human Resources. **Description:** Refurbishes computer parts.

MAINTECH
39 Patterson Avenue, Wallington NJ 07057. 201/614-1700. **Contact:** Human Resources. **Description:** Provides on-site computer maintenance services.

MICRO TECHNOLOGY CONCEPTS INC.
258 Johnson Avenue, Brooklyn NY 11206. 718/456-9100. **Contact:** Human Resources. **Description:** Assembles and repairs computers.

MICROAGE COMPUTER CENTERS, INC.
225 North Bluff, Suite 12, St. George UT 84770. 801/628-4423. **Contact:** Human Resources. **Description:** This location specializes in computer repairs. MicroAge Computer Centers is a business unit of MicroAge that services franchised resellers. MicroAge is a *Fortune* 500 service company providing information technology products and services to corporations worldwide, and institutions and governmental agencies throughout the country. Other business units of MicroAge include MicroAge Technologies, marketing to value-added resellers; MicroAge Infosystems Services, coordinating and servicing large-account marketing efforts in conjunction with franchised resellers; MicroAge Product Services, providing distribution, logistics, technical, and outsourcing services; and MicroAge Channel Services, providing purchasing and marketing services for resellers and vendors. **Corporate headquarters location:** Tempe AZ.

NCE COMPUTER GROUP
P.O. Box 0122, Whitehouse NJ 08888-0122. 908/534-3888. **Contact:** Human Resources. **Description:** Provides technical support and computer repair services for corporations.
Other U.S. locations:
- 9717 Pacific Heights Boulevard, San Diego CA 92121. 619/452-7974.

NATIONAL COMPUTER INC.
4683 Whipple Avenue NW, Canton OH 44718. 330/649-9922. **Contact:** Human Resources. **Description:** Assembles and repairs computer systems.

NOVA TECHNOLOGY SERVICES
8580B Dorsey Run Road, Jessup MD 20794. 301/470-0008. **Contact:** Personnel. **Description:** Provides computer maintenance services, including fiberoptic network cabling. Nova Technology Services also sells hardware and is a Microsoft solutions provider.

OCE PRINTING SYSTEMS USA
5600 Broken Sound Boulevard NW, Boca Raton FL 33487. 561/997-3100. **Contact:** Human Resources. **Description:** Services computer printers and copiers.

PINNACLE DATA SYSTEMS, INC.
2155 Dublin Road, Columbus OH 43228. 614/487-1150. **Contact:** Human Resources. **Description:** A builder of custom computers and provider of computer repair services.

PROGRESSIVE COMPUTER SYSTEMS, INC.
100 Europa Drive, Suite 100, Chapel Hill NC 27514. 919/929-3080. **Contact:** Human Resources. **Description:** A computer firm providing hardware assembly, sales, and service.

SARCOM COMPUTER RENTALS
600F Lakeview Plaza Boulevard, Worthington OH 43085. **Toll-free phone:** 800/589-2052. **Contact:** Tom Hayes, Manager. **Description:** A computer rental company.

SERV-A-COMP INC.
1813 South Market Street, Chattanooga TN 37408. 423/265-8010. **Contact:** Carol Glantz, Personnel. **Description:** Provides repair services for computers, printers, and monitors. Serv-A-Comp also sells refurbished hardware.

SERVITECH INC.
1509 Brook Drive, Downers Grove IL 60515. 630/620-8750. **Contact:** Personnel. **Description:** Offers computer repair services.

STORES AUTOMATED SYSTEMS, INC. (SASI)
311 Sinclair Street, Bristol PA 19007. 215/785-4321. **Fax:** 215/785-5329. **Contact:** Debbie Kenderdine, Human Resources Manager. **E-mail address:** hr@sasipos.com. **World Wide Web address:** http://www.sasipos.com. **Description:** Sells and services computer-based point-of-sale systems. **Common positions include:** Account Representative; Computer Programmer; Project Manager; Systems Analyst. **Educational backgrounds include:** Computer Science; Marketing. **Benefits:** 401(k); Dental Insurance; Disability Coverage; Life Insurance; Medical Insurance; Pension Plan; Tuition Assistance. **Corporate headquarters location:** This Location. **Other U.S. locations:** CA; IL; MO; OH; TX. **Listed on:** Privately held. **Annual sales/revenues:** $21 - $50 million. **Number of employees at this location:** 150. **Number of employees nationwide:** 250. **Number of projected hires for 1997 - 1998 at this location:** 15.

SUNRISE RESOURCES, INC.
5500 Wayzata Boulevard, Golden Valley MN 55416. 612/593-1904. **Contact:** Human Resources. **Description:** Leases computers.

SYSTEMS & SERVICE PROS, INC.
2030 Upland Way, Philadelphia PA 19151. **Fax:** 215/878-0425. **Contact:** Greg Fecca, Vice President of Operations. **Description:** Provides computer maintenance and two-way radio sales and service. Customers of Systems & Service Pros include mostly large, *Fortune* 500 companies. The company performs on-site or on-call technical services. **NOTE:** The company offers entry-level positions as well as training programs. **Common positions include:** Administrative Services Manager; Buyer; Computer Support Technician; Services Sales Representative; Software Engineer. **Educational backgrounds include:** Business Administration; Communications; Computer Science. **Benefits:** 401(k); Dental Insurance; Disability Coverage; Life Insurance; Mass Transit Available; Medical Insurance; Tuition Assistance. **Special Programs:** Internships. **Corporate headquarters location:** This Location. **Operations at this facility include:** Administration; Sales; Service. **Annual sales/revenues:** $5 - $10 million. **Number of employees at this location:** 30. **Number of employees nationwide:** 60.

SYSTEMS MAINTENANCE INC.
3785 Presidential Parkway, Suite 101, Atlanta GA 30340. 770/451-8008. **Contact:** Human Resources. **Description:** Repairs various computer systems.

TCBC/COMPUTER MECHANICS
5 Hanover Square, New York NY 10004. 212/363-5935. **Contact:** Administration. **Description:** Sells and repairs computers for other corporations.

TELOS SYSTEMS SOLUTIONS
6767 Old Madison Pike NW, Huntsville AL 35806. 205/922-8000. **Contact:** Maria Heflin, Human Resources. **Description:** Offers hardware maintenance services to the federal government.

UNIVERSAL DATA CONSULTANTS, INC. (UDC)
6630 Bay Circle, Norcross GA 30071. 770/446-6733. **Contact:** Human Resources. **Description:** Engaged in computer repair and sales.

VIRTUAL TECHNOLOGY
2950 Waterview Drive, Rochester Hills MI 48309. 810/853-6000. **Contact:** Margie Diskin, Human Resources Manager. **Description:** Sells, leases, repairs, and services various types of computer hardware and software by such companies as IBM and Hewlett-Packard.

WIZARD COMPUTER SERVICES, INC.
6908 Engle Road, Suite J, Middleburg Heights OH 44130. 216/891-0060. **Contact:** Human Resources. **Description:** A computer hardware servicing business.

Note: Because addresses and telephone numbers of smaller companies can change rapidly, we recommend you call each company to verify the information below before inquiring about job opportunities. Mass mailings are not recommended.

Additional employers:

COMPUTER MAINTENANCE AND REPAIR

ABL Data Systems
24390 W 10th Mile Rd, Southfield MI 48034. 810/358-0500.

Alpha Microsystems
13003 Southwest Fwy, Stafford TX 77477-4114. 713/240-5005.

Alpha Microsystems
1600 N Interstate 35, Carrollton TX 75006-8606. 214/323-1662.

Alternative Solutions
235 Andover St, Andover MA 01810-5630. 508/475-6255.

Automated Wagering
401 Hackensack Ave, Hackensack NJ 07601-6411. 201/488-3300.

Budget Electronics
4803 Torrance Blvd, Torrance CA 90503-4111. 310/214-1883.

C&L Terminals
11616 E Montgomery Dr, Spokane WA 99206-6607. 509/891-8845.

Cambria Corp.
5090 Central Hwy, Pennsauken NJ 08109-4637. 609/665-3600.

Carlton Engineering
14 Ballardvale Rd, Andover MA 01810-4819. 508/475-4388.

Circuit Test
4601 Cromwell, Suite 2, Memphis TN 38118. 901/795-5300.

Colorado Business Computer Systems Inc.
6000 E Evans Ave, Denver CO 80222-5406. 303/759-4589.

Computer Action Inc.
4801 Little Rd, Arlington TX 76017-1079. 817/572-7962.

Computer Nurse
2101 Webster St, Oakland CA 94612-3027. 510/645-1520.

Computex USA
11702 Imperial Hwy, Norwalk CA 90650-2818. 562/929-3685.

Connecting Point
2236 Washington Blvd, Ogden UT 84401-1410. 801/621-5244.

CRH Systems
116 W California St, Floydada TX 79235-2703. 806/983-2445.

DCI
900 Hayler Street, Teterboro NJ 07608. 800/477-8586.

Digital Equipment Corp.
7402 Westshire St, Lansing MI 48917-8687. 517/627-8315.

Digital Servicenter
8301 Professional Pl, Landover MD 20785-2237. 301/306-2200.

Distributed Logic Corp.
2652 McGaw Ave, Irvine CA 92714-5840. 714/937-5700.

Entex Information Services Inc.
2030 North Loop W, Houston TX 77018-8126. 713/686-0303.

ESSC
4 D Crosby Dr, Bedford MA 01730-1402. 617/275-2777.

Express Computer Repair
12008 Bustleton Ave, Philadelphia PA 19116-2108. 215/698-2102.

General Analytics Corporation
7918 Jones Branch Dr, Mc Lean VA 22102-3307. 703/847-4660.

General Diagnostics Inc.
3600 Parkway Ln, Hilliard OH 43026-1281. 614/777-9434.

Grant Computer
2779 Clairmont Rd NE, Atlanta GA 30329-2605. 404/248-0486.

Grumman Systems Support Corp.
10 Kingsbridge Rd, Fairfield NJ 07004-2104. 201/882-6869.

HSI Business Center
1901 Beaser Ave, Ashland WI 54806-3604. 715/682-8830.

Huntprint Sales & Service
3700 West Silverado Drive, Taylorsville UT 84118. 801/967-2618.

Information Decisions Inc.
4909 Waters Edge Dr, Raleigh NC 27606-2462. 919/859-2555.

Ken's Office Systems Inc.
503 S Saginaw Rd, Midland MI 48640-4584. 517/631-0133.

Laser Impact
10435 Burnet Rd, Austin TX 78758-4450. 512/832-9151.

Mansfield Typewriter Co.
1150 National Pkwy, Mansfield OH 44906-1979. 419/529-2444.

Michigan Business Systems Inc.
24399 Telegraph Rd, Southfield MI 48034-3033. 810/356-2300.

Micro Exchange Corp.
20 Hanes Dr, Wayne NJ 07470-4722. 201/872-1200.

MicroAge Solutions Inc.
3905 Annapolis Ln N, Plymouth MN 55447-5473. 612/551-1100.

NE Systems & Peripherals Inc.
4475 S Clinton Ave, South Plainfield NJ 07080-1200. 908/757-1226.

Oakland Microcentre
1808 Franklin St, Oakland CA
94612-3410. 510/836-0670.

Ohio Business Machines
1728 Saint Clair Ave NE,
Cleveland OH 44114-2008.
216/579-1345.

Peachwood Technology Co.
2516 Glade St, Muskegon MI
49444-1318. 616/733-8314.

Peak Technologies Group Inc.
4181 Centurion Way, Dallas TX
75244-2312. 972/661-2225.

Pentax
3709 Alliance Dr, Greensboro
NC 27407-2059. 910/632-0404.

Pre-Owned Electronics Inc.
125 Middlesex Turnpike, Bedford
MA 01730-1429. 617/275-4600.

Precision Business Systems
5112 S College Ave, Fort Collins
CO 80525-3865. 970/223-6444.

**Priolo Communications &
Cable Inc.**
309 Bradley Ave, Staten Island
NY 10314-5135. 718/370-9700.

**Professional Software
Engineering Inc.**
11838 Rock Landing Rd,
Newport News VA 23606-4232.
757/873-1100.

Pulau Electronics Corp.
12443 Research Pkwy, Orlando
FL 32826-3243. 407/380-9191.

Pyramid Business Systems Inc.
1626 Martin Luther King Jr Way,
Oakland CA 94612-1329.
510/832-1137.

Rodan Inc.
301 10th St NW, Conover NC
28613-2419. 704/464-2816.

Sentinel Technologies
20900 Swenson Dr, Waukesha
WI 53186-4050. 414/797-9140.

**SMS Systems Maintenance
Service Inc.**
28 Travis St, Boston MA 02134-
1253. 617/787-7784.

Strong Computers
1670 Washington Blvd, Ogden
UT 84404-5751. 801/393-3421.

Sybell Computers
6633 Hillcroft St, Houston TX
77081-4816. 713/981-5663.

**Technical Computer Support
Service**
6197 Cornerstone Ct E, San
Diego CA 92121-3710. 619/455-
5300.

Technical Support Services
6830 Broadway, Denver CO
80221-2851. 303/430-0933.

Technical Support Services
81 Croton Ave, Ossining NY
10562-4206. 914/762-5910.

Telos Field Engineering
7 Railroad Ave, Bedford MA
01730-2145. 617/275-4771.

Turbotronics
718 Black Horse Pike,
Turnersville NJ 08012-1053.
609/228-9200.

United Technologists
401 E Manchester Blvd,
Inglewood CA 90301-1356.
310/674-9711.

US Microplus Inc.
701 N Central Expy, Richardson
TX 75080-5319. 972/437-6845.

Vanstar Corp.
94 N Main St, Wharton NJ
07885-1607. 201/442-4900.

Vanstar Corp.
600 N Bell Ave, Carnegie PA
15106-4304. 412/429-2090.

Var Support Inc.
1563 Como Ave, St. Paul MN
55108-2546. 612/645-8542.

Wilson Group
24565 Hallwood Ct, Farmingtn
Hills MI 48335-1667. 616/459-
2230.

**COMPUTER RENTAL AND
LEASING**

ACS Equipment Corporation
10696 Haddington Dr, Houston
TX 77043-3247. 713/827-0993.

Bay Resources Inc.
4605 Albema Cloud Dr, Orlando
FL 32811-7374. 407/423-4140.

Centron DPL Company Inc.
6455 City West Pkwy, Eden
Prairie MN 55344-3246. 612/829-
2800.

Computerental Corp.
4270 Henninger Ct, Chantilly VA
22021-2931. 703/968-7956.

Electro Rent Corporation
3333 Earhart Dr, Carrollton TX
75006-5056. 972/437-3383.

O/E Automation Inc.
3290 W Big Beaver Rd, Troy MI
48084-2906. 810/643-2035.

COMPUTER DISTRIBUTORS, RESELLERS, AND WHOLESALERS

*R*ising *to the top of this field will certainly bring its rewards. Jamie Borst, a reseller at Continental Resources, says a top professional can easily command a salary of six figures. One way to enter the industry is to acquire a position at a distributor. These companies tend to be larger and hire more people. Since distributors deal with both resellers and manufacturers, a shrewd, hardworking individual can develop the necessary contacts to move up in the field.*

Companies look for goal-oriented individuals with the ability to learn new technologies quickly. A computer science background helps, but the display of good communication skills and the desire to succeed may prove to be more important. Solid sales skills and a keen business acumen will help as well.

Reselling includes more than just passing a computer on to another owner. A typical day might include contacting manufacturers, cultivating client accounts, configuring systems, or arranging the outsourcing of engineers. As technology continues to advance, so will the needs of the customers, and that means a good reseller will have to keep abreast of the current technology to stay above the competition.

ADI SYSTEMS
2115 Ringwood Avenue, San Jose CA 95131. 408/944-0100. **Fax:** 408/954-7747. **Contact:** Jenny Traenkle, Human Resources Manager. **Description:** Distributes computer monitors.

ABACUS COMPUTERS INC.
498 Church Street, Salem OR 97301. 503/362-8921. **Contact:** Ron O'Seanecy, Personnel Department. **Description:** A computer reseller.

ACCESS GRAPHICS TECHNOLOGY
1426 Pearl Street, Boulder CO 80302. 303/938-9333. **Contact:** Human Resources. **Description:** A computer wholesaler. **Corporate headquarters location:** This Location.

ACOM COMPUTER INC.
2850 East 29th Street, Long Beach CA 90806. 562/424-7899. **Contact:** Human Resources. **Description:** Resells laser printers. Acom Computer Inc. also distributes software to various businesses including Quickcheck for Windows.

ACTION DATA DISTRIBUTORS INC.
25 Walpole Park South, Walpole MA 02081. 508/660-1010. **Contact:** Human Resources. **Description:** Distributes computer hard drives and monitors to resellers. Action Data Distributors Inc. provides resellers with Acer products and Action's own line.

ADVANCED COMPUTER LINK INC.
2155 Bering Drive, San Jose CA 95131. 408/432-0270. **Contact:** Human Resources. **Description:** Distributes memory cards and CPUs. Advanced Computer Link Inc. supplies individuals, resellers, and retailers with both IBM and Macintosh memory upgrades and hard drives.

ADVANCED COMPUTER RESOURCES
5086 Holt Boulevard, Montclair CA 91736. 909/482-4678. **Contact:** Human Resources. **Description:** Distributes a variety of hardware and peripherals. Advanced Computer Resources provides users with memory and motherboard upgrades, CD-ROM drives, CPUs, and other IBM and clone devices.

ADVANCED TECH COMPUTERS
1230 East Washington Street, Colton CA 92324. 909/370-1035. **Contact:** Human Resources. **Description:** Supplies businesses and individuals with computer hardware. Sells and repairs IBM and IBM compatible CPUs, monitors, and peripherals including modems.

AEGIS ASSOCIATES, INC.
98 Galen Street, Watertown MA 02172. **Contact:** Jack Jandreau, Human Resources. **Description:** A computer reseller that builds their own computers, as well as rents them. Aegis Associates is one of *Inc.*'s 500 Fastest-Growing Private Companies.

AHEARN & SOPER
540 North Commercial Street, Manchester NH 03101. 603/647-2700. **Contact:** Cathy Thomas, Sales Operations Manager. **Description:** Distributes various types of printers and scanners for consumer use. **Corporate headquarters location:** Canada.

ALABAMA CAD/CAM
4900 Corporate Drive, Suite B, Huntsville AL 35805. 205/837-8372. **Contact:** Human Resources. **Description:** A sales office that sells a broad range of computer hardware and software.

ALLIED COMPUTER GROUP
600 West Virginia Street, Milwaukee WI 53208. 414/437-3888. **Contact:** Human Resources Department. **Description:** A computer wholesaler that also provides systems integration services.

ALPHANUMERIC SYSTEMS
3700 Barrett Drive, Raleigh NC 27609. 919/781-7575. **Contact:** Human Resources. **Description:** A computer products reseller and networking firm.

ALTECON DATA COMMUNICATIONS INC.
1333 Strad Avenue, North Tonawanea NY 14120. 716/693-2121. **Contact:** Human Resources. **Description:** Distributes networking hardware. Altecon Data Communications supplies various businesses with the hardware and other peripherals to link computer systems together.

AMBASSADOR BUSINESS SOLUTIONS
425 North Martingale Road, Schaumburg IL 60173. 847/706-3400. **Contact:** Human Resources. **Description:** Provides hardware to businesses and individuals. Ambassador Business Solutions, a subsidiary of Canon, sells Canon brand name laptops, printers, and work stations. This company also supplies Hewlett-Packard with printer engines. **Parent company:** Canon USA.

AMERICA II ELECTRONICS, INC.
2600 118th Avenue, St. Petersburg FL 33716. 813/573-0900. **Contact:** Human Resources. **Description:** A distributor of computer chips and other components.

AMERICAN COMPUTER HARDWARE CORPORATION
2205 South Wright Street, Santa Ana CA 92705. 714/549-2688. **Contact:** Human Resources. **Description:** A distributor of computer printers and enclosures for several manufacturers.

AMERIQUEST TECHNOLOGIES, INC.
6100 Hollywood Boulevard, Hollywood FL 33024. 954/967-2397. **Contact:** Human Resources. **Description:** AmeriQuest Technologies, Inc. is one of the world's largest computer products distributors and a provider of hardware solutions. **Subsidiaries include:** CDS Distribution, Inc. and Robec, Inc. are distributors that market and sell hardware products for the personal computer market. AmeriQuest/Kenfil Inc. is a distributor that markets and sells software products for the personal computer market. CMS Enhancements, Inc. is a supplier of hard disk drive subsystems for

IBM compatible and other leading personal and business computers, including Apple, Compaq, and others. AmeriQuest currently markets more than 2,000 products to original equipment manufacturers, value-added resellers, and dealers throughout the United States and in many foreign countries, including national and regional distributors and large reseller computer chains.

AMERIQUEST TECHNOLOGY SYSTEM GROUP
425 Privet Road, Horsham PA 19044. 215/675-9300. **Contact:** Harold Streets, Human Resources. **Description:** A wholesaler of computer equipment.

AMPRO COMPUTERS INC.
990 Almanor Avenue, Sunnyvale CA 94086. 408/522-2100. **Contact:** Human Resources. **Description:** Distributes embedded computer systems. Ampro Computers supplies companies in such fields as the medical industry, the military, and transportation with various embedded computer systems.

ANZAC COMPUTER EQUIPMENT CORPORATION
31057 Genstar Road, Hayward CA 94544. 510/475-4600. **Contact:** Human Resources. **Description:** Sells printers and corresponding interfaces. Anzac Computer Equipment Corporation supplies businesses using IBM's System AS400 with printers.

ARC TECHNOLOGIES GROUP INC.
322 Fifth Avenue, Suite 400, Pittsburgh PA 15222. 412/281-4427. **Contact:** Human Resources. **Description:** Resells computer hardware and software to businesses.

AURO COMPUTER SYSTEMS
6820 Lauffer Road, Columbus OH 43231. 614/523-1733. **Contact:** Human Resources. **Description:** Sells computer hardware. Auro Computer Systems sells IBM and IBM compatible hardware including CPUs, monitors, and peripherals to businesses or individuals. This company also supports a service department.

AURORA ELECTRONICS
9477 Waples Street, Suite 150, San Diego CA 92121. 619/552-1213. **Fax:** 619/546-8053. **Contact:** Human Resources. **Description:** Provides materials management and specialized distribution services to the computer industry. **Common positions include:** Accountant/Auditor; Buyer; Computer Programmer; Credit Manager; Customer Service Representative; Electrical/Electronics Engineer; Financial Analyst; General Manager; Human Resources Manager; Industrial Engineer; Mechanical Engineer; Systems Analyst. **Educational backgrounds include:** Accounting; Engineering; Finance. **Benefits:** 401(k); Daycare Assistance; Dental Insurance; Disability Coverage; Employee Discounts; Life Insurance; Medical Insurance; Profit Sharing; Stock Option; Tuition Assistance. **Special Programs:** Internships. **Corporate headquarters location:** Irvine CA. **Other U.S. locations:** Irvine CA; Sacramento CA; Marina Del Rey CA; Memphis TN. **Operations at this facility include:** Administration; Divisional Headquarters; Manufacturing; Sales; Service. **Listed on:** American Stock Exchange. **Number of employees at this location:** 150. **Number of employees nationwide:** 560.

AURORA ELECTRONICS
4755 Alla Road, Marina Del Rey CA 90292. 310/827-0999. **Fax:** 310/301-6089. **Contact:** Renee Dubronski, Human Resources Administrator. **Description:** Provides materials management and specialized distribution services to the computer industry. The company serves original equipment manufacturers, service organizations, and resellers worldwide. **Common positions include:** Account Representative; Buyer; Computer Programmer; Purchasing Agent/Manager; Sales Manager; Sales Representative; Systems Analyst. **Educational backgrounds include:** Business Administration; Liberal Arts; Marketing. **Benefits:** 401(k); Dental Insurance; Disability Coverage; Employee Discounts; Financial Planning; Life Insurance; Medical Insurance; Tuition Assistance. **Corporate headquarters location:** Irvine CA. **Other U.S. locations:** San Diego CA; San Jose CA; Sacramento CA. **International locations:** The Netherlands. **Listed on:** American Stock Exchange. **Annual sales/revenues:** $51 - $100 million. **Number of employees at this location:** 150. **Number of employees worldwide:** 400.

BCM ADVANCED RESEARCH

11100 Metric Boulevard, Austin TX 78758. 512/833-9888. **Contact:** Human Resources. **Description:** This branch of BCM Advanced Research offers PC sales and technical services. **Corporate headquarters location:** Irvine CA.

BARRISTER INFORMATION SYSTEMS CORPORATION

465 Main Street, Seventh Floor, Buffalo NY 14203-1788. 716/845-5010. **Fax:** 716/845-5033. **Contact:** Donna Lonca, Human Resources Manager. **Description:** A distributor of systems and software to the legal market. The company's primary goal is to provide the benefits of automation to lawyers so they can improve productivity and manage their practices more effectively. Today, the company has hundreds of law firm clients in 40 states who depend on the company as a supplier of minicomputers, personal computers, local area networks, software, systems integration services, software support, and equipment maintenance. The company has locations in 20 other states, in addition to New York. **Common positions include:** Accountant/Auditor; Attorney; Buyer; Computer Programmer; Financial Analyst; Human Resources Manager; Public Relations Specialist; Purchasing Agent/Manager; Services Sales Representative; Software Engineer; Systems Analyst. **Educational backgrounds include:** Accounting; Business Administration; Communications; Computer Science; Engineering; Finance; Marketing. **Benefits:** 401(k); Disability Coverage; Life Insurance; Medical Insurance; Tuition Assistance. **Corporate headquarters location:** This Location. **Operations at this facility include:** Administration; Manufacturing; Research and Development; Sales; Service. **Listed on:** American Stock Exchange. **Number of employees at this location:** 100. **Number of employees nationwide:** 200.

BENEFIT COMPUTER SYSTEMS

14510 NE 20th Street, Bellevue WA 98007. 206/641-9145. **Contact:** Human Resources. **Description:** Resells CAD software.

BETHCO INC.

2700 Northeast Expressway, Suite B450, Atlanta GA 30345. 404/636-7330. **Contact:** Human Resources. **Description:** Resells computer equipment to businesses.

BUSINESS EQUIPMENT & SUPPLY COMPANY

P.O. Box 986, Columbus MS 39701. 601/328-6860. **Contact:** Human Resources. **Description:** Supplies businesses with office equipment. Business Equipment & Supply stocks hardware and can special order software.

CALDWELL-SPARTIN

1640 Powers Ferry Road, Building 15, Marietta GA 30067. 770/955-1528. **Contact:** Donna Robertson, Human Resources Director. **Description:** A computer hardware and software distributor that also provides temporary and permanent placement services. **Common positions include:** Blue-Collar Worker Supervisor; Branch Manager; Chemist; Civil Engineer; Computer Programmer; Customer Service Representative; Draftsperson; Electrical/Electronics Engineer; Industrial Engineer; Mechanical Engineer; Operations/Production Manager; Services Sales Representative; Systems Analyst; Technical Writer/Editor. **Educational backgrounds include:** Art/Design; Business Administration; Computer Science; Engineering; Marketing. **Benefits:** Dental Insurance; Life Insurance; Medical Insurance; Tuition Assistance. **Corporate headquarters location:** This Location. **Operations at this facility include:** Administration; Sales; Service.

CALL BUSINESS SYSTEMS, INC.

455 East 400 South, Suite 102, Salt Lake City UT 84111. 801/364-7007. **Fax:** 801/364-7099. **Contact:** Stephen J. McEntire, Director of Operations. **World Wide Web address:** http://www.cis.com. **Description:** Engaged in the sale of software and hardware MRP packages to manufacturers. **Common positions include:** Computer Programmer; Management Analyst/Consultant; Systems Analyst. **Educational backgrounds include:** Computer Science. **Benefits:** 401(k); Dental Insurance; Disability Coverage; Life Insurance; Medical Insurance; Pension Plan; Profit Sharing. **Corporate headquarters location:** This Location. **Listed on:** Privately held. **Annual sales/revenues:** $5 - $10 million. **Number of employees at this location:** 20.

CARRERA COMPUTERS, INC.
23181 Verdugo Drive, Laguna Hills CA 92653. 714/707-5051. **Contact:** Human Resources. **Description:** A reseller of Digital Equipment computers.

CHANGE CORPORATION ·
2835 Pan American Freeway NE, Suite C, Albuquerque NM 87107-1606. 505/343-8888. **Contact:** Human Resources. **Description:** Change Corporation is a computer reseller and also provides computer repair services and networking services.

CLARION OFFICE SUPPLIES INC.
101 East Main Street, Little Falls NJ 07424. 201/785-8383. **Contact:** Human Resources. **Description:** Distributes a wide variety of office supplies including computer hardware. Clarion Office Supplies provides individuals and businesses with most major brands of CPUs, monitors, and more. This company makes available a 1,000-page catalog.

COLOURCOMP CORPORATION
P.O. Box 10520, Phoenix AZ 85064. 602/870-8898. **Contact:** Human Resources. **Description:** A computer reseller.

COMARK, INC.
444 Scott Drive, Bloomingdale IL 60108. 630/351-9700. **Contact:** Christine Schuver, Human Resources Manager. **Description:** A distributor of computer media and peripheral equipment. **Common positions include:** Accountant/Auditor; Buyer; Computer Programmer; Credit Manager; Customer Service Representative; Manufacturer's/Wholesaler's Sales Rep.; Operations/Production Manager. **Educational backgrounds include:** Marketing. **Benefits:** Daycare Assistance; Dental Insurance; Employee Discounts; Life Insurance; Medical Insurance; Savings Plan. **Corporate headquarters location:** This Location. **Number of employees at this location:** 200.

COMPUTER AMERICA
1925 Francisco Boulevard, San Rafael CA 94901. 415/257-1010. **Contact:** Human Resources. **Description:** A computer products reseller.

COMPUTER DECISIONS INTERNATIONAL INC.
22260 Haggerty, Suite 300, Northville MI 48167. 810/347-4600. **Contact:** Human Resources. **Description:** Resells computer hardware and software.

COMPUTER MARKETPLACE, INC.
1490 Railroad Street, Corona CA 91720. 909/735-2102. **Fax:** 909/735-5717. **Contact:** Human Resources. **Description:** Computer Marketplace is engaged in the national wholesale distribution of new and used computer equipment to dealers, computer maintenance companies, leasing companies, equipment brokers, and end users. A majority of this equipment is manufactured by IBM and includes computer peripheral equipment, upgrades, and parts, particularly for AS/400, System/36, PS/2, and RISC System/6000 computer systems. Computer Marketplace also sells new computer equipment produced by IBM, Sun Microsystems, Compaq, and Motorola, among other manufacturers. **Corporate headquarters location:** This Location.

COMPUTER RECYCLERS
1005 State Street, Orem UT 84057. 801/226-1892. **Fax:** 801/226-2129. **Contact:** Human Resources. **Description:** Computer Recyclers sells new and used computers, computer components, and software.

COMPUTER SYSTEMS INTEGRATORS INC.
P.O. Box 798, Wappinger Falls NY 12590. 914/298-7625. **Contact:** Human Resources. **Description:** A reseller of educational software. Computer Systems Integrators provides schools with educational hardware and software for kindergarten through twelfth grade.

COMPUTERSOURCE
3814 Williams Boulevard, Kenner LA 70065. 504/443-4100. **Contact:** Human Resources. **Description:** A computer reseller that also services computers.

CONNECTING POINT COMPUTER CENTER
529 North Monroe Street, Green Bay WI 54301. 414/435-2335. **Contact:** Human Resources Department. **Description:** A reseller of computers and camcorders. This location is also a camcorder connecting point.

CRANEL INC.
8999 Gemini Parkway, Columbus OH 43240. 614/431-8000. **Contact:** Human Resources. **Description:** A distributor of a wide range of computer hardware.

CREATIVE COMPUTERS
23710 El Toro Road, #F-1, Lake Forest CA 92630. 714/859-3300. **Fax:** 714/859-8057. **Contact:** Ed Matibag, General Manager. **E-mail address:** edmat@et.cc-inc.com. **World Wide Web address:** http://www.macmall.com. **Description:** A reseller of computers and computer-related products. **Corporate headquarters location:** This Location.

DCI INC.
3915 Thorn Hill Drive, Lilibum GA 30247. 770/593-2025. **Contact:** Recruiting Department. **Description:** DCI Inc. is a computer resale company that also repairs printers, monitors, and central processing units.

DFI
2131 NW 79th Avenue, Miami FL 33122. 305/477-1988. **Contact:** Personnel. **Description:** A wholesaler of computers.

DP CONNECTIONS, INC.
348 Crompton Street, Charlotte NC 28241. 704/588-7500. **Contact:** Human Resources. **Description:** A computer reseller and acquisition specialist.

DTC COMPUTER SUPPLIES
P.O. Box 2834, Rancho Cucamonga CA 91729-2834. 909/466-7680. **Contact:** Human Resources. **Description:** A distributor of computer supplies. DTC also manufactures magnetic computer tape.

DTK COMPUTERS
770 Epperson Drive, City of Industry CA 91748. 818/810-8880. **Contact:** Human Resources. **Description:** A computer wholesale firm.

DTK COMPUTERS
8400 NW 17th Street, Miami FL 33126. 305/477-7440. **Contact:** Human Resources. **Description:** A computer wholesale firm.

DALY COMPUTERS
7-1 Metropolitan Court, Gaithersburg MD 20878. 301/670-0381. **Contact:** Human Resources. **Description:** A computer reseller.

DARTEK COMPUTER SUPPLY
175 Ambassador Drive, Naperville IL 60540. 630/941-1000. **Contact:** Human Resources. **Description:** This location is a distribution center. Overall, Dartek manufactures a wide variety of computer supplies including hardware, software, and computer office equipment.

DASH, INC.
8226 Nieman Road, Lenexa KS 66214. 913/681-3001. **Contact:** Human Resources. **Description:** A computer distributor, reseller, and provider of configuration services.

DATA STORAGE MARKETING
5718 Central Avenue, Boulder CO 80301. 303/442-4747. **Contact:** Human Resources. **Description:** A distributor of computers and peripherals. **Corporate headquarters location:** This Location.

DATA SYSTEMS HARDWARE INC.
22455 Davis Drive, Suite 110, Sterling VA 20164. 703/450-1700. **Contact:** Human Resources. **Description:** Sells almost every brand of high-end printer. Data Systems also carries printer supplies.

DATA SYSTEMS NETWORK CORPORATION
34705 West 12 Mile Road, Farmington Hills MI 48331. 810/489-7117. **Contact:** Human Resources. **Description:** A computer reseller.

DATAC
401 Whitney Avenue, Suite 406, Gretna LA 70056. 504/368-2097. **Contact:** Human Resources. **Description:** Sells computer hardware and software including the Integraph brand.

DATACO DEREX INC.
546 NW 77th Street, Boca Raton FL 33487. 561/994-6993. **Contact:** Human Resources. **Description:** Engaged in the sale and service of computer printers.

DATAEASE INTERNATIONAL INC.
7 Cambridge Drive, Trumbull CT 06611. 203/374-8000. **Contact:** Human Resources. **Description:** Distributes relational database software.

DICKENS DATA SYSTEMS INC.
1175 North Meadow Parkway, Suite 150, Roswell GA 30076. 770/475-8860. **Contact:** Human Resources Department. **Description:** A supplier of computers, peripherals, and software to the multi-user microcomputer industry. Dickens Data Systems also provides integration services.

DIGITAL SOLUTIONS INC. (DSI)
1101 13th Avenue, Altoona PA 16601. 814/944-0405. **Contact:** Human Resources. **Description:** A reseller of computer hardware. Digital Solutions also writes their own correctional software.

ECC CORPORATION
430 West Main Street, Trappe PA 19426. 610/489-2227. **Contact:** Human Resources. **Description:** A computer reseller.

ELCOM INTERNATIONAL
400 Blue Hill Drive, Westwood MA 02090. 617/551-3380. **Contact:** Human Resources. **Description:** A computer reseller.

EMTEC
300 Ben Hamby Drive, Greenville SC 29615. 864/987-7333. **Contact:** Carl Krezdorn, Manager. **Description:** A computer reseller, dealing primarily with corporations. Emtec also has a full-service department.

ENTEX INFORMATION SERVICES
725 Canton Street, Norwood MA 02062. 617/255-1551. **Contact:** Human Resources. **Description:** Resells computers and offers support services for their products.

ENTEX INFORMATION SERVICES
4281 Olympic Boulevard, Erlanger KY 41018. 606/371-9995. **Contact:** Human Resources. **Description:** This location of Entex Information Services distributes computer hardware.

ENTRE COMPUTER CENTER
13400 Bishop's Lane, Suite 270, Brookfield WI 53005. 414/821-1060. **Contact:** Human Resources. **Description:** Engaged in computer sales and service, as well as offering computer systems training.

EVANS SYSTEMS, INC. (ESI)
P.O. Box 2480, Bay City TX 77404-2480. 409/245-2424. **Fax:** 409/244-5070. **Contact:** Human Resources. **Description:** ESI is a vertically integrated company whose business segments include

Distributor Information Systems Corporation, which provides information systems and software for distributors and convenience store owners and operators, Way Energy, which distributes wholesale and retail refined petroleum products and lubricants and owns and operates convenience stores in southern Texas and Louisiana; Chem-Way Systems, Inc., which produces, packages, and markets automotive aftermarket chemical products in 23 states; and EDCO Environmental Systems, Inc. which provides environmental remediation services and new installations of underground storage tanks. **Annual sales/revenues:** $100 million.

EXECUTIVE IMAGING SYSTEMS INC.
P.O. Box 2380, Cherry Hill NJ 08034. 609/424-5898. **Contact:** Human Resources. **Description:** Resells computers, printers, and peripherals.

FEDERAL DATA CORPORATION
4800 Hampden Lane, Suite 1100, Bethesda MD 20814. 301/986-0800. **Contact:** EEO. **Description:** A wholesaler of computers. Federal Data Corporation's primary customers are government contractors. **Common positions include:** Computer Programmer; Software Engineer; Systems Analyst. **Educational backgrounds include:** Business Administration; Computer Science. **Benefits:** Medical Insurance; Savings Plan. **Corporate headquarters location:** This Location. **Number of employees at this location:** 145.

FISHER/UNITECH INC.
1550 West Maple Road, Troy MI 48084. 810/816-1500. **Contact:** Human Resources. **Description:** Resells high-end computers.

FUJITSU COMPUTER PRODUCTS
1741 Junction Avenue, Santa Clara CA 95112. 408/436-5599. **Contact:** Human Resources. **Description:** Distributes disk drives and printers.

GBC GLOBELLE
100 GBC Court, West Berlin NJ 08091. 609/767-2500. **Contact:** Human Resources. **Description:** Sells microcomputer systems, peripherals, and support products. GBC Globelle also provides systems support, services, and network systems training. **Number of employees at this location:** 130.

GE CAPITAL IT SOLUTIONS
700 Canal Street, Stamford CT 06902. 203/357-1464. **Contact:** Human Resources Department. **Description:** GE Capital IT Solutions, formerly AmeriData Technologies, Inc., is a nationwide reseller of computer products and services to commercial, governmental, and educational users. The company's products and services include procurement, value-added systems, systems integration, networking services, support, maintenance, facilities management and outsourcing, software and business consulting services, and rental services. GE Capital IT Solutions currently markets its computer products and business services through its offices in approximately 70 cities throughout the country. **Benefits:** 401(k). **Corporate headquarters location:** This Location. **Parent company:** G.E. Capital. **Listed on:** New York Stock Exchange. **Number of employees nationwide:** 2,100.
Other U.S. locations:
- 500 Fairway Drive, Suite 108, Deerfield Beach FL 33441. 954/421-3484.
- 6410 Atlantic Boulevard, Suite 130, Norcross GA 30071. 770/734-2700.
- 220 Girard Street, Gaithersburg MD 20884. 301/258-2965. Contact: Dorothy Carbo, Human Resources.
- 10200 51st Avenue North, Minneapolis MN 55442. 612/557-2500. Contact: Jean Green, Human Resources Manager.
- 8514 Eager Road, Suite G, St. Louis MO 63144. 314/963-5500.
- 3920 Park Avenue, Edison NJ 08820. 908/321-1100.
- 4474 Sigma Road, Dallas TX 75244. 972/776-1900.

GTSI (GOVERNMENT TECHNOLOGY SERVICES, INC.)
4100 Lafayette Center Drive, Chantilly VA 20151. 703/631-3333. **Fax:** 703/222-5240. **Contact:** Human Resources Department. **Description:** Government Technology Services, Inc. specializes in

providing government customers with a broad selection of brand name computer hardware, software, and peripheral products. GTSI sells directly to all departments and agencies of the federal government, many state and local governments, and indirectly to the government market through hundreds of system integrators and prime contractors. GTSI's selection of desktop, mobile, and engineering workstation computing products continues to expand along with its ability to provide technical expertise in systems integration and network configuration to an ever-expanding government market. Recent acquisitions include Falcon Microsystems, Inc. **Common positions include:** Accountant/Auditor; Budget Analyst; Customer Service Representative; Financial Analyst; Software Engineer; Systems Analyst. **Educational backgrounds include:** Business Administration; Marketing. **Benefits:** 401(k); Dental Insurance; Disability Coverage; Life Insurance; Medical Insurance. **Special Programs:** Internships. **Corporate headquarters location:** This Location. **Operations at this facility include:** Administration; Sales; Service. **Listed on:** NASDAQ. **Number of employees at this location:** 430.

GATES/FA DISTRIBUTING INC.
39 Pelham Ridge Drive, Greenville SC 29615. 864/234-0736. **Contact:** Human Resources Director. **Description:** Gates/FA Distributing Inc. distributes microcomputers, networking software, and computer peripheral equipment including monitors, hard-disk drives, modems, and other related equipment. The company also packages computer systems, offers systems integration services, and provides technical support services. **Common positions include:** Buyer; Computer Operator; Computer Programmer; Credit Clerk and Authorizer; Credit Manager; Customer Service Representative; Human Resources Manager; Marketing Manager; Payroll Clerk; Services Sales Representative. **Educational backgrounds include:** Accounting; Business Administration; Computer Science; Marketing. **Benefits:** 401(k); Credit Union; Dental Insurance; Disability Coverage; Life Insurance; Medical Insurance. **Corporate headquarters location:** This Location. **Operations at this facility include:** Administration; Regional Headquarters; Sales. **Number of employees at this location:** 300. **Number of employees nationwide:** 400.

GEAC COMPUTER
320 Nevada Street, Newtonville MA 02160. 617/965-6310. **Contact:** Human Resources. **Description:** This location distributes products manufactured by the company. Geac Computer develops and markets business applications software in the areas of human resources, materials, management, manufacturing, health care, and higher education. Among the company's products is the SmartStream series of financial software.

GLOBAL COMPUTER SUPPLIES COMPANY
22 Harbor Park Drive, Port Washington NY 11050. 516/625-4300. **Contact:** Human Resources. **Description:** Sells computers and related supplies.

GRAHAM MICROAGE COMPUTER CENTER
133 South Pennsylvania Street, Indianapolis IN 46204. 317/634-8202. **Contact:** Mark Graham, General Manager. **Description:** Sells and services computers for business and educational customers.

HEMISPHERE CORPORATION
2815 East 3300 South, Salt Lake City UT 84109. 801/466-8899. **Contact:** Personnel. **Description:** Hemisphere Corporation distributes accounting software used by contractors.

HIBBITT KARLSSON & SORENSEN
1080 Main Street, Pawtucket RI 02860. 401/727-4200. **Contact:** Human Resources. **Description:** Sells Abacus software.

ICS COMPUTER SYSTEMS CORPORATION
2301 North Central Expressway, Suite 158, Plano TX 75075. 214/644-2371. **Contact:** Human Resources. **Description:** A computer reseller.

ICAS COMPUTER SYSTEMS, INC.
44 North Morris Street, Dover NJ 07801. 201/366-1900. **Contact:** Human Resources. **Description:** A computer reseller.

IMAGE MICRO SYSTEMS, INC.
4929 Wilshire Boulevard, Suite 104, Los Angeles CA 90010. 310/815-1000. **Contact:** Human Resources. **Description:** A reseller of computer hardware and software.

IMPRINTER CORPORATION
432 South 22nd Street, Heath OH 43056. 614/522-2158. **Contact:** Human Resources. **Description:** A distributor of computer hardware.

INACOM CORPORATION
10810 Farnam Drive, Omaha NE 68154. 402/392-3900. **Contact:** Human Resources. **Description:** A wholesaler of computer products.

INACOM INFORMATION SYSTEMS
8601 Dunwoody Place, Suite 538, Atlanta GA 30350. 770/643-8533. **Contact:** Human Resources Department. **Description:** A microcomputer reseller also offering sales advice, technical and software support, customized training, and networking specialists. Primary customers are mid-sized to large corporate accounts. **NOTE:** All hiring is done through the Nebraska office. All applications should be sent to Human Resources, Inacom Information Systems, 10810 Farnam Drive, Omaha NE 68154. 800/843-2762. **Common positions include:** Customer Service Representative; Services Sales Representative; Technician. **Educational backgrounds include:** Business Administration; Computer Science. **Benefits:** Dental Insurance; Disability Coverage; Employee Discounts; Life Insurance; Medical Insurance; Savings Plan. **Corporate headquarters location:** Omaha NE. **Operations at this facility include:** Administration; Regional Headquarters; Sales; Service.
Other U.S. locations:
- 1009 Lenox Drive, Building 4 East Suite 105, Lawrenceville NJ 08648. 609/896-2927.

INACOMP COMPUTER CENTERS
23400 Michigan Avenue, Suite 160, Dearborn MI 48124. 313/274-0090. **Contact:** Human Resources. **Description:** Resells most major brands of hardware.

INACOMP COMPUTER CENTERS
800 Kirts Boulevard, Suite 100, Troy MI 48084. 810/244-0520. **Contact:** Kim Stolarski, Recruiting. **Description:** Resells most major brands of hardware.

INDELIBLE BLUE, INC.
3209 Gresham Lake Road, Suite 135, Raleigh NC 27615. 919/878-9700. **Contact:** Human Resources. **Description:** A reseller of software.

INFO SYSTEMS OF NORTH CAROLINA INC.
7500 East Independence Boulevard, Charlotte NC 78227. 704/535-7180. **Contact:** Human Resources. **Description:** A diverse computer company that offers a variety of services. Info Systems of North Carolina writes programs, resells computer hardware and software, and offers technical support.

INFORMATION DECISIONS INC. (IDI)
3260 Eagle Park Drive NE, Grand Rapids MI 49505. 616/957-2330. **Contact:** Serena Norris, Human Resources Director. **Description:** A computer reseller.

INFORMATION INTEGRATION, INC.
901 Russell Avenue, Gaithersburg MD 20879. 301/921-0246. **Contact:** Human Resources. **Description:** Resells computer systems and develops software.

INGRAM MICRO
1600 East Saint Andrew Place, Santa Ana CA 92705. 714/566-1000. **Contact:** Human Resources Department. **Description:** Ingram Micro is a computer distributor and reseller of hardware and software to large corporations. **Corporate headquarters location:** This Location.
Other U.S. locations:
- 696 Park North Boulevard, Clarkston GA 30021-1963. 404/296-6487.

INSIGHT ELECTRONICS, INC.
9980 Huennekens Street, San Diego CA 92121. **Fax:** 619/677-3171. **Contact:** Human Resources/Computers. **Description:** A distributor of computers and semiconductors.

INTELISYS INC.
12015 Lee Jackson Highway, Fairfax VA 22033. 703/385-0053. **Contact:** Human Resources. **Description:** Engaged in the resale of computer hardware.

INTELLIGENT ELECTRONICS, INC.
411 Eagleview Boulevard, Exton PA 19341. 610/458-5500. **Contact:** Human Resources. **Description:** Intelligent Electronics, Inc. is a leading wholesale distributor of office productivity solutions, including personal computers and related equipment. The company sells primarily to small- and medium-sized businesses through a network that consists of franchised and affiliated customers. **Corporate headquarters location:** This Location. **Number of employees at this location:** 3,000.

JBA INTERNATIONAL INC.
161 Gaither Drive, Mount Laurel NJ 08054. 609/231-9400. **Contact:** Human Resources. **Description:** Sells computer hardware and software.

LINK COMPUTER CORPORATION
317 East Pleasant Valley Boulevard, Altoona PA 16602. 814/946-3085. **Contact:** Human Resources. **Description:** A reseller of computers.

MANKATO BUSINESS PRODUCTS
1715 Commerce Drive, North Mankato MN 56003. 507/625-7440. **Contact:** Human Resources. **Description:** Resells computers, faxes, and other office equipment.

McBRIDE & ASSOCIATES
5555 McLeod Road NE, Albuquerque NM 87109. 505/837-7500. **Contact:** Human Resources. **Description:** A computer resale business.

MERISEL INC.
200 North Continental Boulevard, El Segundo CA 90245. 310/615-3080. **Contact:** Human Resources. **Description:** Distributes microcomputer hardware and software products to over 65,000 computer resellers worldwide, including Apple, AST Research, Compaq, Hewlett-Packard, IBM, Intel, Lotus, Microsoft, Novell, Okidata, and Sun. Merisel currently maintains 21 distribution centers that serve the U.S., Canada, Europe, Latin America, and Australia, and operates an export sales group that serves the rest of the world. **Corporate headquarters location:** This Location. **Number of employees nationwide:** 3,072.
Other U.S. locations:
- 4100 Westpark Drive SW, Atlanta GA 30336-2635. 404/344-1400.

MICRO PROFESSIONALS INC.
14800 South McKinley, Posen IL 60469. 708/396-8500. **Contact:** Human Resources. **Description:** Sells computers.

MICROAGE COMPUTER CENTERS, INC.
2400 South MicroAge Way, MS 5, Tempe AZ 85282. 602/968-3168. **Contact:** Human Resources. **Description:** MicroAge Computer Centers is a business unit of MicroAge, a *Fortune* 500 service company, that provides information technology products and services to corporations worldwide, and institutions and governmental agencies throughout the country. MicroAge's other business units include MicroAge Technologies, marketing to value-added resellers; MicroAge Infosystems Services, coordinating and servicing large-account marketing efforts in conjunction with franchised resellers; MicroAge Product Services, providing distribution, logistics, technical, and outsourcing services; and MicroAge Channel Services, providing purchasing and marketing services for resellers and vendors. *Computerworld* magazine conducted a survey of PC support services to large endusers who rated MicroAge first among its reseller peers and second among vendors. MicroAge received the Datamation QUEST award recognizing project management capabilities in connection

with a major customer's enterprisewide technology project. **Corporate headquarters location:** This Location.

MICROMAX DISTRIBUTION
509A North Military Way, Norfolk VA 23502. 757/466-9018. **Contact:** Paul Spektor, President. **Description:** A distributor of computers and related peripherals.

MISCO
One Misco Plaza, Holmdel NJ 07733. 908/264-1000. **Contact:** Human Resources. **Description:** Resells computer supplies and peripherals including faxes and modems.

NEW MMI CORPORATION
2400 Reach Road, Williamsport PA 17701. 717/327-9575. **Contact:** Human Resources. **Description:** Sells computers and offers service and repair support.

NISSEI SANGYO AMERICA (NSA) - HITACHI
100 Lowder Brook Drive, Suite 2400, Westwood MA 02090. 617/461-8300. **Fax:** 617/461-8664. **Contact:** Human Resources. **World Wide Web address:** http://www.nas-hitachi.com. **Description:** As a member of Hitachi's group of companies, this location imports and distributes computer monitors. **Common positions include:** Administrative Assistant; Administrative Services Manager; Assistant Manager; Budget Analyst; General Manager; Marketing Manager; Marketing Specialist; Operations/Production Manager; Sales Executive; Sales Manager; Sales Representative. **Educational backgrounds include:** Accounting; Business Administration; Computer Science. **Benefits:** 401(k); Dental Insurance; Disability Coverage; Employee Discounts; Life Insurance; Medical Insurance; Pension Plan; Profit Sharing; Tuition Assistance. **Corporate headquarters location:** Rolling Meadows IL. **Other U.S. locations:** Nationwide. **Parent company:** Nissei Sangyo Company Ltd. (Tokyo, Japan). **Operations at this facility include:** Divisional Headquarters. **Number of employees at this location:** 25. **Number of employees nationwide:** 200.

NU DATA INC.
1950 Swarthmore Avenue, Lakewood NJ 08701. 908/842-5757. **Contact:** Human Resources. **Description:** Distributes workstation hardware and computer cables.

NU DESIGN ENGINEERING AMERICA INC.
5198 West 76th Street, Edina MN 55439. 612/832-0420. **Contact:** Human Resources. **Description:** Sells computers, scanners, monitors, and other desktop publishing-related hardware.

ONE SOURCE MICRO PRODUCTS, INC.
3305 Lathrop Street, South Bend IN 46628. 219/288-7455. **Contact:** Human Resources. **Description:** A computer hardware wholesaler.

PC PROFESSIONAL INC.
1615 Webster Street, Oakland CA 94612. 510/465-5700. **Contact:** Human Resources. **Description:** A value-added reseller of various types of computer hardware to corporate customers and consumers.

PREMIO COMPUTER
2186 NW 89th Place, Miami FL 33172. 305/471-0199. **Contact:** Human Resources. **Description:** A computer reseller.

PRINTEK INC.
1517 Townline Road, Benton Harbor MI 49022. 616/925-3200. **Contact:** Human Resources. **Description:** Sells computer printers and ribbons.

PROPHET 21 INC.
19 West College Avenue, Yardley PA 19067. 215/493-8900. **Contact:** Human Resources. **Description:** Supplies computer systems, including hardware and software, to the distribution industry.

PYGMY COMPUTER SYSTEMS INC.
13501 SW 128th Street, Suite 204, Miami FL 33186. 305/253-1212. **Contact:** Human Resources. **World Wide Web address:** http://www.pygmy.com. **Description:** Sells pocket computers and associated software.

QUILL CORPORATION
100 Schelter Road, Lincolnshire IL 60069-3621. 847/634-4850. **Contact:** Human Resources. **Description:** Resells computers and other office supplies.

SAMPO CORPORATION OF AMERICA
5550 Peachtree Industrial Boulevard, Norcross GA 30071. 770/449-6220. **Fax:** 770/447-1109. **Contact:** Personnel Manager. **Description:** Sampo Corporation of America distributes computer monitors. **Common positions include:** Credit Manager; Customer Service Representative; Electrical/Electronics Engineer; Operations/Production Manager; Quality Control Supervisor. **Benefits:** 401(k); Dental Insurance; Disability Coverage; Employee Discounts; Life Insurance; Medical Insurance; Savings Plan; Tuition Assistance. **Other U.S. locations:** Norcross GA. **Operations at this facility include:** Administration; Sales; Service. **Number of employees at this location:** 50.

SKYWALKER COMMUNICATIONS
1000 South Service Road West, O'Fallon MO 63366. 314/272-7045. **Contact:** Human Resources. **Description:** A distributor of satellite and computer products.

SOFTBANK SERVICES GROUP
699 Hertel Avenue, Buffalo NY 14207. 716/871-6400. **Fax:** 716/871-6404. **Contact:** Human Resources Department. **Description:** Sells and resells computer support services to hardware manufacturers.

SOFTWARE SPECTRUM INC.
3140 Merritt Drive, Garland TX 75041. 972/840-6600. **Fax:** 972/864-7878. **Contact:** Human Resources. **Description:** A reseller of microcomputer software and services to businesses and government agencies. Software Spectrum also offers technical support and volume software license services. **Common positions include:** Accountant/Auditor; Advertising Clerk; Blue-Collar Worker Supervisor; Buyer; Collector; Computer Programmer; Credit Manager; Customer Service Representative; Electrical/Electronics Engineer; Human Resources Manager; Operations/Production Manager; Public Relations Specialist; Purchasing Agent/Manager; Quality Control Supervisor; Services Sales Representative; Systems Analyst; Technical Writer/Editor. **Educational backgrounds include:** Accounting; Art/Design; Business Administration; Communications; Computer Science; Economics; Engineering; Finance; Liberal Arts; Marketing. **Benefits:** 401(k); Dental Insurance; Disability Coverage; Employee Discounts; Life Insurance; Medical Insurance; Profit Sharing. **Corporate headquarters location:** This Location. **Other locations:** Worldwide. **Subsidiaries include:** Spectrum Integrated Services. **Operations at this facility include:** Administration; Sales; Service. **Listed on:** NASDAQ. **Number of employees at this location:** 500. **Number of employees nationwide:** 575.

SOFTWORKS COMPUTER
10202 North Enterprise Drive, Mequon WI 53092. 414/242-8800. **Contact:** Human Resources. **Description:** A builder and reseller of computers.

SOUTHERN ELECTRONICS DISTRIBUTORS
4916 North Royal Atlanta Drive, Tucker GA 30085. 770/491-8962. **Toll-free phone:** 800/284-2782. **Contact:** Human Resources Department. **Description:** A wholesaler of computer systems and cellular phones. **Corporate headquarters location:** This Location.

STREAM INTERNATIONAL
105 Rosemont Road, Westwood MA 02090. 617/751-1000. **Contact:** Human Resources. **Description:** Resells computer software. Stream can order almost any software title from most of the major developers. Stream also offers support services.

STREAM INTERNATIONAL

275 Dan Road, Canton MA 02021. 617/575-6800. **Contact:** Human Resources. **Description:** Resells computer software. Stream can order almost any software title from most of the major developers. This location is a regional office offering software support.

SUPERCOM INDUSTRIES INC.

298 Fernwood Avenue, Edison NJ 08837. 908/417-1900. **Contact:** Human Resources. **Description:** Distributes computers.

SUPPLY TECH INC.

1000 Campus Drive, Ann Arbor MI 48104-6700. 313/998-4000. **Contact:** Ann Swain, Human Resources. **Description:** A distributor of EDI software.

SYLVEST MANAGEMENT SYSTEMS CORPORATION

10001 Derek Wood Lane, Suite 225, Lanham MD 20706. 301/459-2700. **Contact:** Human Resources. **World Wide Web address:** http://www.sylvest.com. **Description:** A computer products reseller and systems integration firm.

TAITRON COMPONENTS

25202 Anza Drive, Santa Clarita CA 91355. 805/257-6060. **Contact:** Human Resources. **Description:** Distributors of electronic components, including transistors, diodes, and semiconductors.

TECH DATA CORPORATION

5350 Tech Data Drive, Vay Vista Complex, Clearwater FL 34620. 813/539-7429. **Contact:** Human Resources. **Description:** A distributor of microcomputer-related hardware and software products to value-added resellers and computer retailers throughout the United States, Canada, Europe, Latin America, and the Caribbean. Tech Data Corporation purchases its products directly from manufacturers and publishers in large quantities, maintains an inventory of more than 25,000 products, and sells to an active base of over 50,000 customers. The company offers manufacturers of microcomputer hardware and publishers of software the ability to reach low-volume customers on a cost-efficient basis. Tech Data Corporation provides its customers with products in networking, mass storage, peripherals, software, and systems from more than 600 manufacturers and publishers including Adobe, Aldus, Apple, Borland, Corel, Hewlett-Packard, IBM, Intel, Lotus, Microsoft, and Novell. The company also maintains a staff of technical advisers who assist customers by telephone. **Common positions include:** Accountant/Auditor; Budget Analyst; Buyer; Clerical Supervisor; Computer Operator; Computer Programmer; Credit Manager; Customer Service Representative; Department Manager; Economist; Editor; Education Administrator; Employment Interviewer; Financial Analyst; General Manager; Human Resources Manager; Management Trainee; Manufacturer's/Wholesaler's Sales Rep.; Marketing Specialist; Operations/Production Manager; Purchasing Agent/Manager; Quality Control Supervisor; Services Sales Representative; Software Engineer; Systems Analyst; Technical Writer/Editor; Wholesale and Retail Buyer. **Educational backgrounds include:** Accounting; Business Administration; Computer Science; Finance; Marketing. **Benefits:** 401(k); Daycare Assistance; Dental Insurance; Disability Coverage; Employee Discounts; Life Insurance; Medical Insurance; Savings Plan; Tuition Assistance. **Corporate headquarters location:** This Location. **Operations at this facility include:** Administration; Divisional Headquarters; Regional Headquarters; Sales; Service. **Listed on:** NASDAQ. **Number of employees at this location:** 300. **Number of employees nationwide:** 1,500.

TECHNOLOGY SQUARED

5198 West 76th Street, Edina MN 55439. 612/832-5622. **Contact:** Human Resources. **Description:** A computer distributor conducting most of its business through catalog orders.

TIGER DIRECT

8700 West Flagler Street, 4th Floor, Miami FL 33174. 305/229-1119. **Contact:** Mary Caparro, Human Resources Manager. **Description:** A reseller of computer hardware and software.

TRI-STAR COMPUTER
2424 West 14th Street, Tempe AZ 85281. 602/731-4926. **Contact:** Human Resources.
Description: A computer reseller.

UINTA BUSINESS SYSTEMS INC.
2250 South Main Street, Salt Lake City UT 84115. 801/486-5041. **Contact:** Human Resources.
Description: Sells computer hardware and peripherals.

UNIVERSAL COMPUTER SYSTEMS, INC. (UCS)
FORD DEALER COMPUTER SYSTEMS, INC. (FDCS)
6700 Hollister, Houston TX 77040. 713/718-1800. **Toll-free phone:** 800/883-3031. **Contact:**
Recruiting Department. **Description:** UCS and FDCS sell turnkey computer systems to automobile
dealerships worldwide. These systems operate on powerful IBM mainframe and Texas
Instruments/Hewlett-Packard computers located at the dealership sites with the necessary terminals
and printers. **Common positions include:** Accountant/Auditor; Buyer; Clerical Supervisor;
Computer Operator; Computer Programmer; Customer Service Representative; Dispatcher;
Employment Interviewer; Manufacturer's/Wholesaler's Sales Rep.; Order Clerk; Paralegal; Payroll
Clerk; Photographic Process Worker; Postal Clerk/Mail Carrier; Precision Assembler; Prepress
Worker; Printing Press Operator; Purchasing Agent/Manager; Quality Control Supervisor;
Receptionist; Secretary; Services Sales Representative; Stock Clerk; Systems Analyst; Translator;
Travel Agent; Typist/Word Processor. **Educational backgrounds include:** Accounting; Business
Administration; Communications; Economics; Finance; Liberal Arts; Marketing. **Benefits:** Dental
Insurance; Disability Coverage; Free Parking; Health Club Discount; Life Insurance; Medical
Insurance; Paid Vacation; Savings Plan. **Corporate headquarters location:** This Location. **Other
U.S. locations:** Southfield MI. **Operations at this facility include:** Administration;
Manufacturing; Research and Development; Sales; Service. **Listed on:** Privately held. **Number of
employees nationwide:** 1,350.

UNLIMITED OFFICE PRODUCTS
2344 Flatbush Avenue, Brooklyn NY 11234. 718/252-6500. **Fax:** 718/252-8585. **Contact:** Joe
Rivera, Administrative Aid. **Description:** Distributes computers and other office equipment.

WESTERN MICRO TECHNOLOGY, INC.
254 East Hacienda Avenue, Campbell CA 95008. 408/725-1660. **Contact:** Human Resources.
Description: A distributor of computers, hardware, software, and other computer-related
equipment.

XL CONNECT
8044 Montgomery Road, Suite 601, Cincinnati OH 45236. 513/792-4500. **Contact:** Human
Resources Department. **Description:** XL Connect is one of the 10 largest computer resellers in the
United States in terms of sales. The company sells, installs, and services microcomputers, UNIX
workstations, turnkey local and wide area network systems, computer software, and peripheral
products for business, professional, educational, and governmental customers. XL Connect also
offers a wide range of sophisticated customer support and consulting services. The company
currently operates 23 branch sales offices in 15 states. The company does not maintain any retail
facilities.

*Note: Because addresses and telephone numbers of smaller companies can change
rapidly, we recommend you call each company to verify the information below before
inquiring about job opportunities. Mass mailings are not recommended.*

Additional employers:

**COMPUTER EQUIPMENT
AND SOFTWARE
WHOLESALE**

ACC Microelectonics Corp.
14205 Burnet, Austin TX 78728-
6527. 512/310-7315.

Althon Micro Inc.
280 Paseo Tesora, Walnut CA
91789-2723. 909/594-3128.

American Power Conversion
16775 Addison Rd, Dallas TX
75248-6922. 800/800-4272.

Artex International Corp.
3111 Route 38, Mount Laurel NJ
08054-9754. 609/234-2388.

Avnet CMG
3011 S 52nd St, Tempe AZ
85282-3209. 602/414-6500.

Azerty Inc.
13 Centre Dr, Orchard Park NY
14127-4103. 716/662-0200.

Campbell Services Inc.
21700 Northwestern Hwy,
Southfield MI 48075-4906.
810/559-5955.

Cardexpert Technology
47881 Fremont Blvd, Fremont
CA 94538-6506. 510/252-1118.

CDW Computer Centers Inc.
1020 E Lake Cook Rd, Buffalo
Grove IL 60089. 847/465-6000.

Centerline Software Inc.
10 Fawcett St, Cambridge MA
02138-1110. 617/498-3000.

Centron
618 Shoemaker Rd, King Of
Prussia PA 19406-4203. 610/337-
8240.

Cimlinc Inc.
1222 Hamilton Pkwy, Itasca IL
60143-1160. 630/250-0090.

Compucom Systems Inc.
6 N Broad St, Woodbury NJ
08096-4635. 609/848-2300.

Computer Parts and Service
10205 51st Ave N, Plymouth MN
55442-3206. 612/553-1514.

Computer Sales International
PO Box 16264, Saint Louis MO
63105-0964. 314/997-7010.

Computer Warehouse
574 Dorchester Ave, Boston MA
02127-2722. 617/269-0299.

D&H Distributing Co.
P.O. Box 5967, Harrisburg PA
17110-2511. 717/236-8001.

Daewoo International America
85 Challenger Rd, Ridgefield
Park NJ 07660-2104. 201/229-
4500.

Daisytek Inc.
500 N Central Expressway, Plano
TX 75074-6772. 972/881-4700.

Danka
12 Edison Pl, Springfield NJ
07081-1310. 201/376-0055.

Data General Corp.
5710 LBJ Freeway, Dallas TX
75240. 972/702-1400.

Digital Equipment Corp.
3201 Enterprise Pkwy,
Beachwood OH 44122-7320.
216/831-6000.

Digital Equipment Corp.
20 Corporate Pl S, Piscataway NJ
08854-6144. 908/562-4000.

Digital Inc.
2113 Wells Branch Pkwy, Austin
TX 78728-6970. 512/990-3400.

Digital Products Corp.
800 NW 33rd St, Pompano Beach
FL 33064-2046. 305/783-9600.

Encore Computer
8111 LBJ Freeway, Suite 300,
Dallas TX 75251-1325. 214/238-
9548.

ETI
66 Bovet Rd, San Mateo CA
94402-3127. 415/345-9100.

Execu-Flow Systems Inc.
1 Ethel Rd, Edison NJ 08817-
2838. 908/287-9191.

Flextronics
1299 Commerce Dr, Richardson
TX 75081-2406. 214/680-6500.

Glide-Write
109 Bonaventura Dr, San Jose
CA 95134-2106. 408/383-9009.

Global Ultimace Systems Inc.
4 North St, Waldwick NJ 07463-
1842. 201/445-5050.

Hewlett-Packard Co.
2000 West Loop S, Houston TX
77027-3513. 713/439-5300.

Hewlett-Packard Co.
3701 Koppers St, Baltimore MD
21227-1024. 410/644-5800.

Hewlett-Packard Co.
538 Rod Hollow Rd, Melville NY
11747. 516/753-0555.

**Hewlett-Packard Co. Sales &
Support**
250 N Patrick Blvd, Brookfield
WI 53045-5826. 414/792-8800.

IBM Corp.
1100 Berkshire Blvd,
Wyomissing PA 19610-1221.
610/320-7200.

IBM Corp.
800 N Frederick Ave,
Gaithersburg MD 20879-3326.
301/240-0111.

IBM Corp.
600 Parsippany Rd, Parsippany
NJ 07054-3715. 201/386-5600.

IBM Corp. System and Supplies
431 Ridgeroot, Dayton NJ 08810.
908/329-7000.

Icon Office Solutions
1702 West 3rd Street, Tempe AZ
85281-7235. 602/929-7800.

I3S
1330 Riverband, Dallas TX
75247. 214/631-1031.

JDS Micro Devices
1850 South 10th St, San Jose CA
95112. 408/494-1400.

Josten's Learning Corp.
9920 Pacific Heights Blvd, San
Diego CA 92121-4330. 619/587-
0087.

Kao Infosystems Company
40 Grissom Rd, Plymouth MA
02360-4841. 508/747-5520.

Micro Distribution
892 Main Street, Wilmington MA
01887-4409. 508/658-7100.

Micro Focus Inc.
2465 E Bayshore Rd, Palo Alto
CA 94303-3297. 415/856-4161.

Micromedex Inc.
6200 S Syracuse Way,
Englewood CO 80111-4740.
303/486-6400.

Microscript
99 Rosewood Dr, Danvers MA
01923-1300. 508/777-5202.

Microworx Direct Inc.
793 South Goodman St,
Rochester NY 14620-2516.
716/271-0050.

Navarre Corp.
7400 49th Ave N, New Hope MN
55428-4258. 612/535-8333.

NCR Corporation
805 Central Ave, Cincinnati OH
45202-1947. 513/852-7200.

New Concepts Corporation
2341 S Friebus Ave, Tucson AZ
85713-4250. 520/323-6645.

Novaquest Information Systems
19951 Mariner Ave, Suite 100,
Torrance CA 90503-1653.
714/549-0159.

Okidata
2470 Windy Hill Rd, Marietta
GA 30067-8613. 770/933-5352.

Pantex Computer Inc.
10301 Harwin Dr, Houston TX
77036-1501. 713/988-1688.

PC Parts Express
1221 Champion Cir, Carrollton
TX 75006-6854. 972/406-8583.

PC Service Source
2350 Valley View Ln, Dallas TX
75234-5754. 214/406-8583.

PCs Compleat Inc.
495 Technology Ctr W,
Marlborough MA 01752.
508/480-8500.

Rave Computer
36960 Metro Ct, Sterling Heights
MI 48312-1014. 810/939-8230.

Reed Technology
One Progress Dr, Horsham PA
19044. 215/641-6000.

Richard Young Products
508 S Military Trl, Deerfield
Beach FL 33442-3011. 954/426-
8100.

Ross Systems Inc.
555 Twin Dolphin Dr, Redwood
City CA 94065-2102. 415/593-
2500.

Software Group Inc.
1120 Jupiter Rd Ste 100, Plano
TX 75074-7040. 214/424-1579.

Software House International
2 Riverview Dr, Somerset NJ
08873-1150. 908/805-9160.

Solutions IQ
1260 116th Avenue NE, Bellevue
WA 98004. 206/451-2727.

Sovran
2535 Pilot Knob Rd, St. Paul MN
55120-1120. 612/884-7301.

**Storage Technology
Corporation**
952 Echo Ln Ste 370, Houston
TX 77024-2757. 713/464-3022.

**Storage Technology
Corporation**
100 Matsonford Rd, Bldg 5, Suite
110, Radnor PA 19087. 610/989-
9050.

Sunquest Information Systems
4801 E Broadway Blvd, Tucson
AZ 85711-3609. 520/570-2000.

Tandberg Data Inc.
2685-A Park Center Dr, Simi
Valley CA 93065-6211. 805/579-
1000.

Thinking Machines Corp.
14 Crosby Dr, Bedford MA
01730-1402. 617/276-0400.

Transnet Corp.
45 Columbia Rd, Somerville NJ
08876-3576. 908/253-0500.

Vanstar Corp.
7676 Hillmont St, Houston TX
77040-6400. 713/939-3500.

NOTE: *While every effort is made to keep the addresses and phone numbers of these companies up-to-date, employment services often move or change hands and are therefore more difficult to track. Please notify the publisher if you find any discrepancies.*

EXECUTIVE SEARCH FIRMS OF ALABAMA

CLARK PERSONNEL SERVICE OF MOBILE, INC.
4315 Downtowner, Loop North, Mobile AL 36609. 334/342-5511. **Fax:** 334/343-5588. **Contact:** Donna Clark, President. **Description:** An executive search firm operating on a contingency basis. Company pays fee. **Specializes in the areas of:** Accounting/Auditing; Computer Science/Software; Engineering; Industrial; Manufacturing; Personnel/Labor Relations; Sales and Marketing; Transportation. **Positions commonly filled include:** Accountant/Auditor; Administrative Services Manager; Aerospace Engineer; Agricultural Engineer; Architect; Bank Officer/Manager; Biological Scientist; Biomedical Engineer; Blue-Collar Worker Supervisor; Branch Manager; Chemical Engineer; Chemist; Civil Engineer; Clerical Supervisor; Computer Programmer; Credit Manager; Draftsperson; Electrical/Electronics Engineer; Human Resources Manager; Industrial Engineer; Manufacturer's/Wholesaler's Sales Rep.; Mechanical Engineer; Metallurgical Engineer; Mining Engineer; Nuclear Engineer; Operations/Production Manager; Petroleum Engineer; Purchasing Agent/Manager; Quality Control Supervisor; Software Engineer; Stationary Engineer; Structural Engineer; Systems Analyst. **Corporate headquarters location:** This Location. **Average salary range of placements:** $30,000 - $50,000. **Number of placements per year:** 200 - 499.

COMPUSEARCH OF BIRMINGHAM
MANAGEMENT RECRUITERS INTERNATIONAL
P.O. Box 381626, Birmingham AL 35238-1626. 205/871-3550. **Contact:** Office Manager. **Description:** An executive search firm. **Specializes in the areas of:** Accounting/Auditing; Administration/MIS/EDP; Advertising; Architecture/Construction/Real Estate; Banking; Chemical; Communications; Computer Hardware/Software; Design; Electrical; Engineering; Food Industry; General Management; Health/Medical; Insurance; Legal; Manufacturing; Operations Management; Personnel/Labor Relations; Printing/Publishing; Procurement; Retail; Sales and Marketing; Technical and Scientific; Textiles; Transportation.

COMPUSEARCH OF MOBILE
MANAGEMENT RECRUITERS INTERNATIONAL
3263 Demetropolis Road, Suite 6C, Mobile AL 36603. 334/602-0104. **Contact:** R.C. Brock, Manager. **Description:** An executive search firm. **Specializes in the areas of:** Accounting/Auditing; Administration/MIS/EDP; Advertising; Architecture/Construction/Real Estate; Banking; Chemical; Communications; Computer Hardware/Software; Design; Electrical; Engineering; Food Industry; General Management; Health/Medical; Insurance; Legal; Manufacturing; Operations Management; Personnel/Labor Relations; Printing/Publishing; Procurement; Retail; Sales and Marketing; Technical and Scientific; Textiles; Transportation.

DUNHILL OF SOUTH BIRMINGHAM
2738 18th Street South, Birmingham AL 35209. 205/877-4580. **Toll-free phone:** 800/548-0116. **Fax:** 205/877-4590. **Contact:** Peggy Clarke, President. **Description:** An executive search firm operating on a contingency basis. The firm also provides temporary and contract services. Company pays fee. **Specializes in the areas of:** Accounting/Auditing; Administration/MIS/EDP; Banking; Computer Science/Software; Finance; Manufacturing; Personnel/Labor Relations; Secretarial; Technical and Scientific. **Positions commonly filled include:** Accountant/Auditor; Administrative Services Manager; Bank Officer/Manager; Budget Analyst; Computer Programmer; Credit Manager; Financial Analyst; Health Services Manager; Human Resources Manager; Market Research Analyst; MIS Specialist; Operations/Production Manager; Purchasing Agent/Manager; Quality Control Supervisor; Software Engineer; Systems Analyst; Telecommunications Manager. **Benefits available to temporary workers:** Bonuses; Insurance; Paid Holidays; Vacation Pay.

Corporate headquarters location: Woodbury NY. **Other U.S. locations:** Nationwide. **Number of placements per year:** 50 - 99.

F-O-R-T-U-N-E PERSONNEL CONSULTANTS
3311 Bob Wallace Avenue SW, Huntsville AL 35805. 205/534-7282. **Contact:** President. **Description:** An executive search firm operating on a contingency basis. Company pays fee. **Specializes in the areas of:** Accounting/Auditing; Administration/MIS/EDP; Computer Science/Software; Engineering; Finance; General Management; Industrial; Manufacturing; Personnel/Labor Relations. **Positions commonly filled include:** Accountant/Auditor; Buyer; Ceramics Engineer; Computer Programmer; Design Engineer; Designer; Electrical/Electronics Engineer; Environmental Engineer; Financial Analyst; General Manager; Human Resources Specialist; Industrial Engineer; Industrial Production Manager; Materials Engineer; Mechanical Engineer; Metallurgical Engineer; MIS Specialist; Pharmacist; Quality Control Supervisor; Telecommunications Manager; Transportation/Traffic Specialist. **Average salary range of placements:** More than $50,000. **Number of placements per year:** 100 - 199.

NATIONAL LABOR LINE
805 Oakwood Avenue NW, Huntsville AL 35811-1629. 205/535-9541. **Fax:** 205/535-0744. **Contact:** Carol Rouse, Owner. **Description:** An executive search firm that also handles contract services and career/outplacement counseling. The firm specializes in technical positions, particularly computer and engineering professionals. Company pays fee. **Specializes in the areas of:** Administration/MIS/EDP; Computer Science/Software; Engineering; Manufacturing; Sales and Marketing. **Corporate headquarters location:** This Location. **Average salary range of placements:** More than $50,000. **Number of placements per year:** 1 - 49.

PERSONNEL INC.
P.O. Box 1413, Huntsville AL 35807-0413. 205/536-4431. **Fax:** 205/539-0583. **Contact:** Bill Breen, Owner/Manager. **Description:** An executive search firm specializing in engineering, computer applications, senior management, human resources, and manufacturing materials management. The firm also offers career/outplacement counseling. **Specializes in the areas of:** Administration/MIS/EDP; Computer Science/Software; Engineering; Manufacturing; Personnel/Labor Relations; Technical and Scientific. **Positions commonly filled include:** Accountant/Auditor; Aerospace Engineer; Buyer; Computer Programmer; Design Engineer; Electrical/Electronics Engineer; Environmental Engineer; Industrial Engineer; Industrial Production Manager; Mechanical Engineer; MIS Specialist; Operations/Production Manager; Purchasing Agent/Manager; Quality Control Supervisor; Software Engineer; Systems Analyst; Technical Writer/Editor; Telecommunications Manager. **Average salary range of placements:** $30,000 - $50,000. **Number of placements per year:** 100 - 199.

SALES CONSULTANTS/MANAGEMENT RECRUITERS OF BIRMINGHAM
P.O. Box 381626, Birmingham AL 35238-1626. 205/871-1128. **Fax:** 205/871-1148. **Contact:** Cleve A. Park, Owner. **Description:** This company provides executive search and recruitment, interim placement, outplacement, compatibility assessment, video-conferencing, and other staffing services. Company pays fee. **Specializes in the areas of:** Accounting/Auditing; Administration/MIS/EDP; Banking; Computer Science/Software; Engineering; Finance; Food Industry; Health/Medical; Industrial; Insurance; Manufacturing; Personnel/Labor Relations; Printing/Publishing; Sales and Marketing. **Positions commonly filled include:** Accountant/ Auditor; Actuary; Adjuster; Branch Manager; Claim Representative; Computer Programmer; Design Engineer; Electrical/Electronics Engineer; General Manager; Health Services Manager; Industrial Engineer; Manufacturer's/Wholesaler's Sales Rep.; Mechanical Engineer; MIS Specialist; Quality Control Supervisor; Systems Analyst. **Corporate headquarters location:** This Location. **Number of placements per year:** 50 - 99.

SNELLING PERSONNEL SERVICES
400 Vestavia Parkway, Suite 221, Vestavia Hills AL 35216-3750. 205/822-7878. **Fax:** 205/979-7663. **Contact:** Roy Bosko, Manager. **World Wide Web address:** http://www.snellingsearch.com. **Description:** A full-service personnel agency and executive search firm. Company pays fee. **Specializes in the areas of:** Computer Science/Software; Engineering; Personnel/Labor Relations; Sales and Marketing. **Positions commonly filled include:** Administrative Services Manager;

Aerospace Engineer; Blue-Collar Worker Supervisor; Chemical Engineer; Chemist; Computer Programmer; Design Engineer; Electrical/Electronics Engineer; Human Resources Manager; MIS Specialist; Telecommunications Manager. **Benefits available to temporary workers:** Medical Insurance; Vacation Days. **Corporate headquarters location:** Dallas TX. **International locations:** Worldwide. **Average salary range of placements:** $30,000 - $50,000. **Number of placements per year:** 200 - 499.

SNELLING PERSONNEL SERVICES
400 14th Street SE, Decatur AL 35601. **Contact:** Recruiter. **Description:** An executive search firm, operating on a contingency basis. Company pays fee. **Specializes in the areas of:** Computer Science/Software; Engineering; Health/Medical; Sales and Marketing. **Positions commonly filled include:** Civil Engineer; Computer Programmer; Design Engineer; Electrical/Electronics Engineer; Industrial Engineer; Manufacturer's/Wholesaler's Sales Rep.; Mechanical Engineer; MIS Specialist; Physical Therapist; Services Sales Representative; Software Engineer; Structural Engineer. **Corporate headquarters location:** Dallas TX. **Other U.S. locations:** Nationwide. **Average salary range of placements:** $30,000 - $50,000. **Number of placements per year:** 1 - 49.

PERMANENT EMPLOYMENT AGENCIES OF ALABAMA

A-1 EMPLOYMENT SERVICE
1015 Montlamar Drive, Suite 130, Mobile AL 36609-1713. 334/343-9702. **Fax:** 334/343-9706. **Contact:** Dorothy Robney, Owner/Manager. **Description:** An employment agency. Company pays fee. **Specializes in the areas of:** Accounting/Auditing; Administration/MIS/EDP; Advertising; Banking; Computer Hardware/Software; Finance; Food Industry; General Management; Health/Medical; Industrial; Insurance; Legal; Manufacturing; Nonprofit; Personnel/Labor Relations; Printing/Publishing; Retail; Sales and Marketing; Secretarial; Transportation. **Positions commonly filled include:** Accountant/Auditor; Administrative Assistant; Bookkeeper; Buyer; Chemical Engineer; Chemist; Claim Representative; Clerk; Computer Operator; Computer Programmer; Credit Manager; Customer Service Representative; Data Entry Clerk; Draftsperson; Driver; Editor; EDP Specialist; Electrical/Electronics Engineer; Factory Worker; Hotel Manager; Industrial Designer; Industrial Engineer; Legal Secretary; Light Industrial Worker; Management Trainee; Manufacturing Engineer; Marketing Specialist; Mechanical Engineer; Medical Secretary; Nurse; Operations/Production Manager; Public Relations Specialist; Purchasing Agent/Manager; Quality Control Supervisor; Receptionist; Sales Representative; Secretary; Software Engineer; Systems Analyst; Technical Writer/Editor; Technician; Typist/Word Processor. **Number of placements per year:** 200 - 499.

AEROTEK, INC.
4920 Corporate Drive, Suite G, Huntsville AL 35805. 205/830-1995. **Toll-free phone:** 800/377-3144. **Fax:** 205/830-2332. **Contact:** Human Resources. **E-mail address:** aerotek. huntsville@internetmci.com. **Description:** An employment agency with divisions specializing in engineering, telecommunications, laboratory support, and computer applications. **Specializes in the areas of:** Computer Science/Software; Engineering; Industrial; Manufacturing. **Positions commonly filled include:** Aerospace Engineer; Aircraft Mechanic/Engine Specialist; Chemical Engineer; Civil Engineer; Computer Programmer; Cost Estimator; Design Engineer; Designer; Draftsperson; Electrical/Electronics Engineer; Environmental Engineer; Industrial Engineer; Industrial Production Manager; Mechanical Engineer; Metallurgical Engineer; MIS Specialist; Nuclear Engineer; Quality Control Supervisor; Software Engineer; Structural Engineer; Systems Analyst; Technical Writer/Editor. **Corporate headquarters location:** Baltimore MD. **Other U.S. locations:** Nationwide. **Average salary range of placements:** $30,000 - $50,000. **Number of placements per year:** 100 - 199.

PERFORM STAFFING SERVICE
3107 Independence Drive, Birmingham AL 35209. 205/870-8170. **Contact:** Jerry Sulzby, Vice President/General Manager. **Description:** An employment agency. Company pays fee. **Specializes in the areas of:** Accounting/Auditing; Banking; Computer Hardware/Software. **Positions commonly filled include:** Bookkeeper; Clerk; Data Entry Clerk; EDP Specialist; Factory Worker; Light Industrial Worker; Nurse; Receptionist; Secretary; Stenographer; Typist/Word Processor. **Number of placements per year:** 50 - 99.

J.L. SMALL ASSOCIATES
P.O. Box 157, Birmingham AL 35201-0157. 205/252-9933. **Fax:** 205/254-1126. **Contact:** Jim Small, Owner. **Description:** An employment agency that focuses on executive and management recruiting. The company operates on a contingency basis. Company pays fee. **Specializes in the areas of:** Accounting/Auditing; Computer Science/Software; Engineering; Finance; General Management; Health/Medical; Industrial; Manufacturing; Personnel/Labor Relations. **Positions commonly filled include:** Accountant/Auditor; Chemical Engineer; Chemist; Civil Engineer; Computer Programmer; Credit Manager; Design Engineer; Electrical/Electronics Engineer; Environmental Engineer; Financial Analyst; General Manager; Human Resources Manager; Industrial Engineer; Industrial Production Manager; Mechanical Engineer; Metallurgical Engineer; MIS Specialist; Purchasing Agent/Manager; Quality Control Supervisor; Structural Engineer; Systems Analyst. **Number of placements per year:** 1 - 49.

SNELLING PERSONNEL SERVICES
1813 University Drive, Huntsville AL 35801. 205/533-1410. **Fax:** 205/534-6691. **Contact:** George Barnes, Owner/Manager. **Description:** An employment agency. Company pays fee. **Specializes in the areas of:** Computer Science/Software; Engineering; Health/Medical; Sales and Marketing. **Positions commonly filled include:** Accountant/Auditor; Aerospace Engineer; Civil Engineer; Computer Programmer; Customer Service Representative; Electrical/Electronics Engineer; Industrial Engineer; Manufacturer's/Wholesaler's Sales Rep.; Mechanical Engineer; Nuclear Medicine Technologist; Occupational Therapist; Pharmacist; Physical Therapist; Physicist/Astronomer; Registered Nurse; Software Engineer; Speech-Language Pathologist; Systems Analyst. **Number of placements per year:** 100 - 199.

VIP PERSONNEL, INC.
478 Palisades Boulevard, Birmingham AL 35209. 205/879-8889. **Fax:** 205/879-8919. **Contact:** Bonnie Wainwright, Owner. **Description:** An employment agency. Company pays fee. **Specializes in the areas of:** Accounting/Auditing; Administration/MIS/EDP; Advertising; Banking; Broadcasting; Computer Science/Software; Fashion; Finance; General Management; Health/Medical; Legal; Manufacturing; Nonprofit; Personnel/Labor Relations; Retail; Sales and Marketing; Secretarial. **Positions commonly filled include:** Accountant/Auditor; Adjuster; Administrative Services Manager; Advertising Clerk; Bank Officer/Manager; Branch Manager; Brokerage Clerk; Budget Analyst; Buyer; Claim Representative; Clerical Supervisor; Computer Programmer; Counselor; Credit Manager; Customer Service Representative; Dental Assistant/Dental Hygienist; Education Administrator; Financial Analyst; General Manager; Health Services Manager; Hotel Manager; Human Resources Specialist; Industrial Production Manager; Management Trainee; Manufacturer's/Wholesaler's Sales Rep.; Mathematician; Medical Records Technician; Paralegal; Public Relations Specialist; Quality Control Supervisor; Radio/TV Announcer/Broadcaster; Systems Analyst; Travel Agent; Typist/Word Processor; Underwriter/Assistant Underwriter. **Number of placements per year:** 1000+.

EXECUTIVE SEARCH FIRMS OF ARIZONA

COMPUTECH CORPORATION
4375 North 75th Street, Scottsdale AZ 85251. 602/947-7534. **Fax:** 602/947-7537. **Contact:** Bob Dirickson, President. **Description:** An executive search firm. Company pays fee. **Specializes in the**

areas of: Computer Science/Software. **Positions commonly filled include:** Computer Programmer; Consultant; Software Engineer; Systems Analyst; Technical Writer/Editor. **Number of placements per year:** 100 - 199.

ELECTRONIC POWER SOURCE

1507 West Loughlin Drive, Chandler AZ 85224. 602/821-1946. **Fax:** 602/786-3057. **Contact:** Garry Moore, Owner. **Description:** A nationwide executive search firm which specializes in analog design, power supplies, ballasts, magnetics, and other high-tech industries. Company pays fee. **Specializes in the areas of:** Computer Science/Software; Engineering; Industrial; Manufacturing. **Positions commonly filled include:** Aerospace Engineer; Computer Programmer; Designer; Electrical/Electronics Engineer; Industrial Engineer; Mechanical Engineer; Systems Analyst. **Number of placements per year:** 1 - 49.

MANAGEMENT RECRUITERS INTERNATIONAL (MRI) TUCSON-FOOTHILLS

6262 North Swan Road, Suite 100, Tucson AZ 85718-3600. 520/529-6818. **Fax:** 520/529-6877. **Contact:** Ms. Lorian E. Roethlein, Manager. **Description:** An executive search firm. **Specializes in the areas of:** Accounting/Auditing; Computer Science/Software; Engineering; Manufacturing. **Positions commonly filled include:** Accountant/Auditor; Computer Programmer; Design Engineer; Electrical/Electronics Engineer; General Manager; Human Resources Manager; Industrial Engineer; MIS Specialist; Software Engineer; Systems Analyst. **Average salary range of placements:** More than $50,000. **Number of placements per year:** 1 - 49.

McEVOY & JOHNSON ASSOCIATES, INC.

8337 East San Rosendo Drive, Scottsdale AZ 85258. 602/998-3522. **Fax:** 602/948-9791. **Contact:** Donna Johnson, President. **E-mail address:** mjassoc@primenet.com. **Description:** An executive search firm. Company pays fee. **Specializes in the areas of:** Computer Science/Software. **Positions commonly filled include:** Computer Programmer; Software Engineer; Systems Analyst. **Average salary range of placements:** More than $50,000. **Number of placements per year:** 50 - 99.

MOSQUEDA & MIJAS

10 Sherwood Lane, Flagstaff AZ 86001-5205. 520/779-3013. **Contact:** Jerry Mosqueda, President. **E-mail address:** gmosqueda@aol.com. **Description:** An executive search firm. **NOTE:** Appointments are required for application. Company pays fee. **Specializes in the areas of:** Aerospace; Automotive; Computer Graphics; Electronics; Finance; Information Technology; Telecommunications. **Positions commonly filled include:** Accountant/ Auditor; Administrator; Computer Programmer; Operations/Production Manager; Procurement Specialist. **Number of placements per year:** 1 - 49.

PROFESSIONAL SEARCH

7434 East Stetson Drive, Scottsdale AZ 85251. 602/994-4400. **Contact:** Julie Beauvais, Office Administrator. **Description:** An executive search firm. Company pays fee. **Specializes in the areas of:** Computer Science/Software. **Positions commonly filled include:** Computer Programmer; Systems Analyst. **Number of placements per year:** 100 - 199.

SOFTWARE TRANSITIONS

4400 North Scottsdale Road, Suite 9-277, Scottsdale AZ 85251. 602/946-5591. **Contact:** Tom Kernan, Human Resources. **Description:** An executive search firm. Company pays fee. **Specializes in the areas of:** Computer Hardware/Software; Engineering; Technical and Scientific. **Number of placements per year:** 1 - 49.

TSS CONSULTING, LTD.

2425 East Camelback, Suite 375, Phoenix AZ 85016. 602/955-7000. **Toll-free phone:** 800/489-2425. **Fax:** 602/957-3948. **Contact:** George A. Armes, Senior Consultant. **E-mail address:** mcdjoh@aol.com. **Description:** An executive search firm that places workers primarily in high-tech industries including telecommunications, aerospace, electronics, petrochemicals, computer systems, semiconductors and microelectronics, and information technology. Company pays fee. **Specializes in the areas of:** Computer Science/Software; Engineering. **Positions commonly filled include:** Electrical/Electronics Engineer; MIS Manager; Software Engineer; Telecommunications Manager. **Number of placements per year:** 1 - 49.

UNIVERSAL SEARCH
1944 West Thunderbird Road, Suite 5, Phoenix AZ 85023. 602/863-0037. **Fax:** 602/942-2008. **Contact:** Max R. Shunk, President. **Description:** An executive search firm. Founded in 1979. Company pays fee. **Specializes in the areas of:** Computer Science/Software; Engineering; General Management; Health/Medical; Industrial; Manufacturing; Technical and Scientific. **Positions commonly filled include:** Buyer; Chemical Engineer; Computer Programmer; Electrical/Electronics Engineer; Human Resources Specialist; Industrial Engineer; Industrial Production Manager; Management Analyst/Consultant; Mechanical Engineer; MIS Specialist; Operations/Production Manager; Purchasing Agent/Manager; Quality Control Supervisor; Software Engineer; Telecommunications Manager. **Average salary range of placements:** More than $50,000. **Number of placements per year:** 1 - 49.

PERMANENT EMPLOYMENT AGENCIES OF ARIZONA

ERICKSON AND ASSOCIATES, INC.
HC 32 Box 362A, Prescott AZ 86303. 520/776-0045. **Fax:** 520/776-0823. **Contact:** Elvin Erickson, President. **Description:** An employment agency. **Specializes in the areas of:** Administration/MIS/EDP; Computer Hardware/Software; Technical and Scientific. **Positions commonly filled include:** Aerospace Engineer; Computer Programmer; EDP Specialist; Electrical/Electronics Engineer; Mechanical Engineer; Physicist/Astronomer; Systems Analyst; Technical Writer/Editor.

GENERAL EMPLOYMENT ENTERPRISES, INC.
100 West Clarendon Avenue, Suite 2240, Phoenix AZ 85013. 602/265-7800. **Fax:** 602/265-1779. **Contact:** Don Cuppy, Manager. **Description:** An employment agency. Company pays fee. **Specializes in the areas of:** Computer Science/Software; Engineering; Manufacturing; Technical and Scientific. **Positions commonly filled include:** Aerospace Engineer; Biochemist; Biomedical Engineer; Chemical Engineer; Chemist; Civil Engineer; Computer Programmer; Designer; Draftsperson; Electrical/Electronics Engineer; Geologist/Geophysicist; Industrial Engineer; Internet Services Manager; Mechanical Engineer; Metallurgical Engineer; MIS Specialist; Operations/Production Manager; Quality Control Supervisor; Software Engineer; Structural Engineer; Systems Analyst; Technical Writer/Editor; Telecommunications Manager. **Corporate headquarters location:** Oak Brook IL. **Other U.S. locations:** Nationwide. **Average salary range of placements:** $30,000 - $50,000. **Number of placements per year:** 100 - 199.

PRIORITY STAFFING, INC.
4600 South Mill Avenue, Suite 275, Tempe AZ 85282. 602/491-2191. **Fax:** 602/491-5702. **Contact:** Lamonte Thomas, Executive Vice President. **Description:** An employment agency providing temporary and permanent placements on a contingency basis. Founded in 1993. **Specializes in the areas of:** Accounting/Auditing; Administration/MIS/EDP; Banking; Computer Science/Software; Engineering; Finance; General Management; Health/Medical; Industrial; Insurance; Legal; Manufacturing; Personnel/Labor Relations; Printing/Publishing; Sales and Marketing; Secretarial; Technical and Scientific. **Positions commonly filled include:** Accountant/Auditor; Administrative Services Manager; Advertising Clerk; Bank Officer/Manager; Blue-Collar Worker Supervisor; Branch Manager; Chemical Engineer; Claim Representative; Clerical Supervisor; Computer Programmer; Customer Service Representative; Design Engineer; Designer; Education Administrator; Financial Analyst; General Manager; Hotel Manager; Human Resources Specialist; Human Service Worker; Industrial Engineer; Industrial Production Manager; Insurance Agent/Broker; Internet Services Manager; Management Analyst/Consultant; Manufacturer's/Wholesaler's Sales Rep.; Mechanical Engineer; MIS Specialist; Multimedia Designer; Operations/Production Manager; Paralegal; Public Relations Specialist; Quality Control

Supervisor; Services Sales Representative; Social Worker; Surveyor; Systems Analyst; Technical Writer/Editor; Typist/Word Processor.

TECH/AID OF ARIZONA
1438 West Broadway, Suite B-225, Tempe AZ 85282. 602/894-6161. **Contact:** Manager. **Description:** An employment agency. Company pays fee. **Specializes in the areas of:** Architecture/Construction/Real Estate; Cable TV; Computer Hardware/Software; Construction; Engineering; Manufacturing; Technical and Scientific. **Positions commonly filled include:** Aerospace Engineer; Architect; Buyer; Ceramics Engineer; Draftsperson; Electrical/Electronics Engineer; Estimator; Factory Worker; Industrial Engineer; Mechanical Engineer; Metallurgical Engineer; Mining Engineer; Operations/Production Manager; Petroleum Engineer; Purchasing Agent/Manager; Quality Control Supervisor; Technical Writer/Editor; Technician. **Number of placements per year:** 1000+.

VOLT TECHNICAL SERVICES
3020 East Camelback Road, Suite 365, Phoenix AZ 85016. 602/955-7750. **Contact:** Regional Manager. **Description:** An employment agency. Company pays fee. **Specializes in the areas of:** Accounting/Auditing; Art/Design; Chemical; Computer Science/Software; Engineering; Personnel/Labor Relations. **Positions commonly filled include:** Accountant/Auditor; Aerospace Engineer; Buyer; Chemical Engineer; Chemist; Civil Engineer; Commercial Artist; Computer Programmer; Draftsperson; Electrical/Electronics Engineer; Human Resources Manager; Industrial Engineer; Mechanical Engineer; Metallurgical Engineer; Physicist/Astronomer; Purchasing Agent/Manager; Quality Control Supervisor; Statistician; Systems Analyst; Technical Writer/Editor.

EXECUTIVE SEARCH FIRMS OF ARKANSAS

EXECUTIVE RECRUITERS OUTPLACEMENT CONSULTANTS
P.O. Box 21810, Little Rock AR 72221-1810. 501/224-7000. **Fax:** 501/224-8534. **Contact:** Greg Downs, Vice President. **Description:** An executive search firm. **Specializes in the areas of:** Accounting/Auditing; Bookkeeping; Computer Programming; Data Processing; Data Security; Engineering; Finance; Sales and Marketing; Software Engineering.

MANAGEMENT RECRUITERS OF LITTLE ROCK
Redding Building, Suite 314, 1701 Centerview Drive, Little Rock AR 72211-4313. 501/224-0801. **Fax:** 501/224-0798. **Contact:** Noel K. Hall, Managing Partner. **E-mail address:** litrock!ltr@mrnet.com. **Description:** An executive search firm operating on a contingency basis. Company pays fee. **Specializes in the areas of:** Accounting/Auditing; Administration/MIS/EDP; Advertising; Architecture/Construction/Real Estate; Banking; Chemical; Communications; Computer Hardware/Software; Design; Electrical; Engineering; Food Industry; General Management; Health/Medical; Insurance; Legal; Manufacturing; Operations Management; Personnel/Labor Relations; Printing/Publishing; Procurement; Retail; Sales and Marketing; Technical and Scientific; Textiles; Transportation. **Positions commonly filled include:** Accountant/Auditor; Agricultural Engineer; Computer Programmer; Design Engineer; Electrical/Electronics Engineer; Environmental Engineer; Industrial Engineer; Industrial Production Manager; Mechanical Engineer; MIS Specialist; Physical Therapist; Physician; Quality Control Supervisor; Registered Nurse; Systems Analyst. **Corporate headquarters location:** Cleveland OH. **Average salary range of placements:** More than $50,000. **Number of placements per year:** 50 - 99.

SPENCER & ASSOCIATES, INC.
212 North 34th Street, Rogers AR 72756. 501/631-1300. **Fax:** 501/631-1551. **Contact:** James Spencer, Manager. **Description:** An executive search firm operating on a contingency basis. Founded in 1988. Company pays fee. **Specializes in the areas of:** Banking; Computer

Science/Software. **Positions commonly filled include:** Computer Programmer; Systems Analyst. **Average salary range of placements:** $30,000 - $50,000. **Number of placements per year:** 1 - 49.

TURNAGE EMPLOYMENT SERVICE GROUP
1225 Breckenridge Drive, Suite 206, Little Rock AR 72205. 501/224-6870. **Fax:** 501/224-5709. **Contact:** Linda Dicus, Office Manager. **Description:** An executive search firm operating on a contingency basis. Company pays fee. **Specializes in the areas of:** Administration/MIS/EDP; Computer Science/Software; Engineering; General Management; Health/Medical; Industrial; Insurance; Legal; Manufacturing; Personnel/Labor Relations; Sales and Marketing; Secretarial; Technical and Scientific. **Positions commonly filled include:** Accountant/Auditor; Administrative Services Manager; Bank Officer/Manager; Blue-Collar Worker Supervisor; Branch Manager; Buyer; Chemical Engineer; Clerical Supervisor; Computer Programmer; Cost Estimator; Credit Manager; Dental Assistant/Dental Hygienist; Design Engineer; Draftsperson; Electrical/Electronics Engineer; Environmental Engineer; Financial Analyst; General Manager; Health Services Manager; Hotel Manager; Human Resources Specialist; Industrial Production Manager; Licensed Practical Nurse; Management Analyst/Consultant; Management Trainee; Metallurgical Engineer; MIS Specialist; Operations/Production Manager; Paralegal; Property and Real Estate Manager; Public Relations Specialist; Purchasing Agent/Manager; Quality Control Supervisor; Registered Nurse; Restaurant/Food Service Manager; Services Sales Representative; Software Engineer; Systems Analyst; Technical Writer/Editor; Travel Agent; Typist/Word Processor; Underwriter/Assistant Underwriter. **Corporate headquarters location:** This Location. **Other area locations:** Conway AR; Helena AR; Russeldale AR. **Average salary range of placements:** $20,000 - $29,999. **Number of placements per year:** 200 - 499.

PERMANENT EMPLOYMENT AGENCIES OF ARKANSAS

PERFORMANCE STAFFING SERVICE
5 Office Park Drive, Suite 103, Little Rock AR 72211. 501/228-6335. **Contact:** Vickie Siebenmorgen, Manager/Senior Recruiter. **Description:** An employment agency. Performance Staffing is a full-service agency that also provides temporary and contract services and outplacement counseling. The company focuses on accounting, data processing, and technical placement. Founded in 1989. Company pays fee. **Specializes in the areas of:** Accounting/Auditing; Advertising; Computer Science/Software; Engineering; Finance; Industrial; Manufacturing. **Positions commonly filled include:** Accountant/Auditor; Administrative Services Manager; Budget Analyst; Chemical Engineer; Computer Programmer; Credit Manager; Design Engineer; Draftsperson; Economist; Electrical/Electronics Engineer; Financial Analyst; Internet Services Manager; Mathematician; Mechanical Engineer; Software Engineer; Structural Engineer; Systems Analyst. **Average salary range of placements:** $30,000 - $50,000. **Number of placements per year:** 200 - 499.

EXECUTIVE SEARCH FIRMS OF CALIFORNIA

ACCESS TECHNOLOGY
600 Anton Boulevard, Suite 920, Costa Mesa CA 92626. 714/850-1000. **Fax:** 714/850-1916. **Contact:** R.J. Nadel, President. **Description:** An executive search firm. **Specializes in the areas of:** Computer Science/Software; Engineering; Manufacturing; Technical and Scientific. **Positions**

commonly filled include: Electrical/Electronics Engineer; Software Engineer; Technical Writer/Editor. **Number of placements per year:** 1 - 49.

ADVANCED TECHNOLOGY CONSULTANTS, INC. (ATC)
536 Weddell Drive, Suite 7, Sunnyvale CA 94089. 408/734-0635. **Fax:** 408/734-5833. **Contact:** Reza Vakili, Professional Staffing Manager. **E-mail address:** atci@ix.netcom.com. **Description:** An executive search firm focusing on all areas of the software industry from development to marketing and support. Industry concentrations include software architecture, design and development, quality assurance, technical marketing, product marketing management, customer support, technical writing, and network administration. Company pays fee. **Specializes in the areas of:** Computer Science/Software; Engineering; Sales and Marketing. **Positions commonly filled include:** Computer Programmer; Design Engineer; Internet Services Manager; Software Engineer; Strategic Relations Manager; Systems Analyst; Technical Writer/Editor. **Corporate headquarters location:** This Location. **Average salary range of placements:** More than $50,000. **Number of placements per year:** 50 - 99.

AFFORDABLE EXECUTIVE RECRUITERS
5627 Sepulveda Boulevard, Suite 207, Van Nuys CA 91411. 818/782-8554. **Contact:** Fred Gerson, President. **Description:** An executive search firm. **Specializes in the areas of:** Accounting/Auditing; Banking; Computer Science/Software; Engineering; Finance; General Management; Sales and Marketing. **Positions commonly filled include:** Accountant/Auditor; Bank Officer/Manager; Computer Programmer; Human Resources Manager; Internet Services Manager; Software Engineer; Systems Analyst; Typist/Word Processor. **Number of placements per year:** 1 - 49.

ALLIED SEARCH INC.
2030 Union Street, Suite 206, San Francisco CA 94123. 415/921-2200. **Contact:** Don May, Manager. **Description:** An executive search firm. Company pays fee. **Specializes in the areas of:** Accounting/Auditing; Administration/MIS/EDP; Banking; Computer Science/Software; Finance; General Management; Health/Medical; Legal; Personnel/Labor Relations; Retail. **Average salary range of placements:** More than $50,000. **Number of placements per year:** 50 - 99.

ALLIED SEARCH INC.
8530 Wilshire Boulevard, Suite 404, Beverly Hills CA 90211. 213/680-4040. **Contact:** Don May, Manager. **Description:** An executive search firm. Company pays fee. **Specializes in the areas of:** Accounting/Auditing; Administration/MIS/EDP; Banking; Computer Science/Software; Finance; General Management; Health/Medical; Legal; Personnel/Labor Relations; Retail. **Positions commonly filled include:** Accountant/Auditor; Actuary; Attorney; Bank Officer/Manager; Budget Analyst; Buyer; Computer Programmer; Financial Analyst; Hotel Manager; Human Resources Manager; MIS Specialist; Software Engineer; Systems Analyst. **Average salary range of placements:** More than $50,000. **Number of placements per year:** 50 - 99.

ANKENBRANDT GROUP
4685 MacArthur Court, Suite 480, Newport Beach CA 92660. 714/955-1455. **Fax:** 714/955-2029. **Contact:** Office Manager. **E-mail address:** ankgrp@ix.netcom.com. **Description:** An executive search firm. Company pays fee. **Specializes in the areas of:** Accounting/Auditing; Administration/MIS/EDP; Computer Science/Software; Sales and Marketing; Technical and Scientific. **Positions commonly filled include:** Accountant/Auditor; Internet Services Manager; Management Analyst/Consultant; Marketing Specialist; Sales Representative; Systems Analyst. **Number of placements per year:** 100 - 199.

ASSOCIATED SOFTWARE CONSULTANTS, INC.
1509 North Sepulveda Boulevard, Manhattan Beach CA 90266-5109. 310/545-5646. **Fax:** 310/546-3433. **Contact:** Marshall Biggs, President. **Description:** An executive search firm. Company pays fee. **Specializes in the areas of:** Administration/MIS/EDP; Computer Hardware/Software. **Positions commonly filled include:** Computer Operator; Computer Programmer; Data Entry Clerk; EDP Specialist; MIS Specialist; Software Engineer; Systems Analyst; Technical Writer/Editor. **Number of placements per year:** 50 - 99.

BARNES & ASSOCIATES
1101 Dove Street, Suite 238, Newport Beach CA 92660. 714/253-6750. **Fax:** 714/253-6753. **Contact:** Meredith Schwarz, Managing Partner. **E-mail address:** msbarnes@ix.netcom.com. **Description:** An executive search firm. Company pays fee. **Specializes in the areas of:** Computer Hardware/Software; Sales and Marketing. **Positions commonly filled include:** Computer Programmer; Internet Services Manager; Manufacturer's/Wholesaler's Sales Rep.; Software Engineer; Strategic Relations Manager; Systems Analyst. **Average salary range of placements:** . More than $50,000. **Number of placements per year:** 50 - 99.

ROBERT BEECH WEST INC.
383 South Palm Canyon Drive, Palm Springs CA 92262. 760/864-1380. **Fax:** 760/864-1382. **Contact:** Robert Beech, Director. **Description:** An executive search firm which places sales and marketing professionals with four or more years experience within the computer software industry. Company pays fee. **Specializes in the areas of:** Computer Science/Software; Sales and Marketing. **Positions commonly filled include:** Customer Service Representative; Manufacturer's/Wholesaler's Sales Rep.; Sales Representative; Software Engineer. **Average salary range of placements:** More than $50,000. **Number of placements per year:** 50 - 99.

BRYSON MYERS COMPANY
2083 Old Middlefield Way, Suite 806, Mountain View CA 94043. 415/964-7600. **Contact:** Chris Myers, Partner. **Description:** An executive search firm. Company pays fee. **Specializes in the areas of:** Computer Hardware/Software; Engineering; Health/Medical; Technical and Scientific. **Positions commonly filled include:** Biomedical Engineer; Ceramics Engineer; Computer Programmer; Electrical/Electronics Engineer; Industrial Designer; Industrial Engineer; Manufacturing Engineer; MIS Specialist; Software Engineer; Technical Writer/Editor. **Number of placements per year:** 50 - 99.

BULLIS & COMPANY, INC.
120 Quintara Street, San Francisco CA 94116-1359. 415/753-6140. **Fax:** 415/753-6653. **Contact:** R.J. Bullis, President. **Description:** An executive search firm. Company pays fee. **Specializes in the areas of:** Architecture/Construction/Real Estate; Computer Science/Software; Engineering; Finance; General Management; Manufacturing. **Average salary range of placements:** More than $50,000. **Number of placements per year:** 1 - 49.

CN ASSOCIATES
4040 Civic Center Drive, Suite 200, San Raphael CA 94903. 415/883-1114. **Fax:** 415/883-3321. **Contact:** Charles Nicolosi, Principal. **E-mail address:** chanic@ix.netcom.com. **Description:** An executive search firm. Company pays fee. **Specializes in the areas of:** Administration/MIS/EDP; Computer Science/Software; Engineering; Sales and Marketing; Technical and Scientific. **Positions commonly filled include:** Branch Manager; Computer Programmer; Internet Services Manager; Management Analyst/Consultant; MIS Specialist; Multimedia Designer; Radio/TV Announcer/Broadcaster; Software Engineer; Systems Analyst; Telecommunications Manager. **Average salary range of placements:** More than $50,000. **Number of placements per year:** 1 - 49.

CALIFORNIA SEARCH AGENCY, INC.
111 Pacifica, Suite 250, Irvine CA 92718. 714/453-8424. **Fax:** 714/453-9121. **Contact:** Don Crane, President. **World Wide Web address:** http://www.jobagency.com. **Description:** An executive search firm specializing in executive management, engineering, manufacturing, marketing, sales, and technical placement in state-of-the-art research and development and service companies. Company pays fee. **Specializes in the areas of:** Computer Science/Software; Engineering; Food Industry; General Management; Industrial; Manufacturing; Personnel/Labor Relations; Sales and Marketing; Technical and Scientific. **Positions commonly filled include:** Aerospace Engineer; Agricultural Scientist; Biomedical Engineer; Buyer; Chemical Engineer; Civil Engineer; Computer Programmer; Cost Estimator; Designer; Draftsperson; Electrical/Electronics Engineer; General Manager; Geologist/Geophysicist; Industrial Engineer; Mechanical Engineer; Metallurgical Engineer; Mining Engineer; Nuclear Engineer; Petroleum Engineer; Purchasing Agent/Manager; Quality Control Supervisor; Science Technologist; Software Engineer; Stationary

Engineer; Structural Engineer; Systems Analyst; Technical Writer/Editor. **Average salary range of placements:** More than $50,000. **Number of placements per year:** 50 - 99.

COAST TO COAST EXECUTIVE SEARCH
4040 Civic Center Drive, Suite 200, San Rafael CA 94903-1900. 415/492-2870. **Fax:** 415/491-4710. **Contact:** Alan Horiwitz, President. **E-mail address:** coastalan@aol.com. **Description:** An executive search firm specializing in sales, marketing, research and development, and technical positions in the consumer products, medical, and computer hardware/software industries. Company pays fee. **Specializes in the areas of:** Banking; Computer Science/Software; Food Industry; Health/Medical; Sales and Marketing; Technical and Scientific. **Positions commonly filled include:** Food Scientist/Technologist. **Corporate headquarters location:** This Location. **Average salary range of placements:** More than $50,000. **Number of placements per year:** 1 - 49.

COMPUTER NETWORK RESOURCES INC.
28231 Tinajo, Mission Viejo CA 92692. 714/951-5929. **Fax:** 714/951-6013. **Contact:** Ken Miller, President. **Description:** An executive search firm specializing in information technology for insurance and financial applications. Company pays fee. **Specializes in the areas of:** Computer Science/Software; Insurance; Sales and Marketing. **Positions commonly filled include:** Systems Analyst; Technical Writer/Editor. **Number of placements per year:** 1 - 49.

CORPORATE DYNAMIX
602 Santa Monica Boulevard, Santa Monica CA 90401-2502. 310/260-1390. **Fax:** 310/458-2177. **Contact:** David Sterenfeld, Principal. **E-mail address:** corpdyn@aol.com. **Description:** An executive search firm focusing on the computer hardware/software marketplace. The firm places sales, management, and marketing candidates. **Specializes in the areas of:** Computer Science/Software; Sales and Marketing. **Positions commonly filled include:** Sales and Marketing Manager; Software Engineer; Systems Analyst. **Corporate headquarters location:** This Location. **Average salary range of placements:** More than $50,000. **Number of placements per year:** 100 - 199.

CORPORATE SEARCH
6457 Edgemoor Way, San Jose CA 95129. 408/996-3000. **Contact:** Hal Wilson, President. **Description:** A technical recruiting firm. Company pays fee. **Specializes in the areas of:** Computer Science/Software; Design; Engineering; Technical and Scientific. **Positions commonly filled include:** Aerospace Engineer; Biomedical Engineer; Chemical Engineer; Computer Programmer; Electrical/Electronics Engineer; Financial Analyst; Industrial Engineer; Management Analyst/Consultant; Mechanical Engineer; Metallurgical Engineer; Nuclear Engineer; Systems Analyst; Technical Writer/Editor; Travel Agent. **Number of placements per year:** 1 - 49.

DALEY CONSULTING & SEARCH
1866 Clayton Road, Suite 211, Concord CA 94520. 510/798-3866. **Fax:** 510/798-4475. **Contact:** Mike Daley, Owner. **E-mail address:** mdaley@dpsearch.com. **World Wide Web address:** http://www.dpsearch.com. **Description:** Daley Consulting & Search is an executive recruiting and search firm that specializes in the placement of data processing and information systems professionals in the San Francisco Bay area and in Sacramento. Founded in 1985. Company pays fee. **Specializes in the areas of:** Administration/MIS/EDP; Computer Science/Software; Information Systems. **Positions commonly filled include:** Computer Programmer; Data Entry Clerk; MIS Specialist; Multimedia Designer; Software Engineer; Systems Analyst; Technical Writer/Editor; Telecommunications Manager. **Corporate headquarters location:** This Location. **Other U.S. locations:** Sacramento CA. **Average salary range of placements:** More than $50,000. **Number of placements per year:** 1 - 49.

DATA CENTER PERSONNEL
24007 Ventura Boulevard, Suite 240, Calabasas CA 91302. 818/225-2830. **Fax:** 818/225-2840. **Contact:** Jim Auld, President. **E-mail address:** datacenter@earthlink.net. **Description:** An executive search firm. Company pays fee. **Specializes in the areas of:** Administration/MIS/EDP; Computer Science/Software; Information Systems. **Positions commonly filled include:** Computer Programmer; Software Engineer; Systems Analyst; Telecommunications Manager. **Average salary range of placements:** More than $50,000. **Number of placements per year:** 50 - 99.

DRUMMER PERSONNEL, INC.
2121 South El Camino, Suite 515B, San Mateo CA 94403. 415/345-7691. **Fax:** 415/345-8496. **Contact:** Geri Geller, Consultant. **Description:** An executive search and employment agency. Company pays fee. **Specializes in the areas of:** Administration/MIS/EDP; Computer Science/Software; Sales and Marketing; Secretarial. **Positions commonly filled include:** Customer Service Representative; Management Trainee; Manufacturer's/ Wholesaler's Sales Rep. **Corporate headquarters location:** This Location. **Average salary range of placements:** $30,000 - $50,000. **Number of placements per year:** 50 - 99.

DUNHILL PROFESSIONAL SEARCH OF SAN FRANCISCO
220 Montgomery Street, Box 2909, San Francisco CA 94104. 415/956-3700. **Contact:** George Curtiss, President. **Description:** An executive search firm. **Specializes in the areas of:** Banking; Computer Hardware/Software; Electronics; Health/Medical; Paper; Sales and Marketing.

DUNHILL PROFESSIONAL SEARCH OF SAN JOSE
1190 Saratoga Avenue, Suite 210, San Jose CA 95129. 408/236-3262. **Fax:** 408/296-7866. **Contact:** Kevin A. P. Keifer, President. **Description:** An executive search firm. Company pays fee. **Specializes in the areas of:** Computer Science/Software; Engineering; Industrial; Manufacturing; Sales and Marketing; Secretarial; Technical and Scientific. **Positions commonly filled include:** Chemical Engineer; Computer Programmer; Customer Service Representative; Electrical/Electronics Engineer; Manufacturer's/Wholesaler's Sales Rep.; Mechanical Engineer; Operations/Production Manager; Quality Control Supervisor; Software Engineer. **Number of placements per year:** 1 - 49.

DYNAMIC SYNERGY CORPORATION
2730 Wilshire Boulevard, Suite 550, Santa Monica CA 90403-4747. 310/573-7300. **Contact:** Personnel. **Description:** An executive search firm. **Specializes in the areas of:** Computer Science/Software; Sales and Marketing.

EAGLE SEARCH ASSOCIATES
336 Bon Air Center, #295, Greenbrae CA 94904. 415/398-6066. **Fax:** 415/398-0858. **Contact:** Mark Gideon, Executive Director. **Description:** An executive search firm. Company pays fee. **Specializes in the areas of:** Computer Science/Software; Sales and Marketing. **Positions commonly filled include:** Systems Analyst. **Number of placements per year:** 1 - 49.

ETHOS CONSULTING, INC.
100 Pine Street, Suite 750, San Francisco CA 94111-5208. 415/397-2211. **Fax:** 415/397-0856. **Contact:** Conrad E. Prusak, President. **E-mail address:** conrad@ethos-net.com. **Description:** A senior-level executive search firm. Company pays fee. **Specializes in the areas of:** Banking; Computer Programming; Finance; Food Industry; General Management; Health/Medical; Retail; Sales and Marketing; Transportation. **Positions commonly filled include:** Bank Officer/Manager; General Manager; Management Analyst/Consultant. **Corporate headquarters location:** This Location. **Average salary range of placements:** More than $50,000. **Number of placements per year:** 1 - 49.

EXCEL TECHNICAL DATA PROCESSING SEARCH
30100 Town Center Drive, Suite 0-129, Laguna Niguel CA 92677. 714/240-0438. **Fax:** 714/240-0408. **Contact:** Bob Langieri, Director. **E-mail address:** langieri@msn.com. **Description:** An executive search firm. Excel Technical places all levels of program analysts, systems analysts, operations, technical support, and communications specialists. This location is also involved in placing contract programmers on temporary assignments. Company pays fee. **Specializes in the areas of:** Administration/MIS/EDP; Computer Science/Software. **Positions commonly filled include:** Computer Operator; Computer Programmer; EDP Specialist; MIS Specialist; Software Engineer; Systems Analyst. **Corporate headquarters location:** This Location. **Average salary range of placements:** More than $50,000. **Number of placements per year:** 50 - 99.

EXECUTIVE DIRECTION
369 Pine Street, San Francisco CA 94104-3302. 415/394-5500. **Fax:** 415/956-5186. **Contact:** Fred Naderi, President. **E-mail address:** edi@netcom.com. **World Wide Web address:**

http://www.edir.com. **Description:** An executive search firm. **Specializes in the areas of:** Computer Hardware/Software; Engineering. **Positions commonly filled include:** Computer Programmer; General Manager; Management Analyst/Consultant; MIS Manager; Multimedia Designer; Science Technologist; Software Engineer; Systems Analyst. **Average salary range of placements:** More than $50,000. **Number of placements per year:** 1 - 49.

EXECUTIVE SEARCH CONSULTANTS
2108 Appaloosa Circle, Petaluma CA 94954-4643. 707/763-0100. **Fax:** 707/765-6983. **Contact:** Peg Iversen, Owner. **E-mail address:** esc@crl.com. **Description:** An executive search firm operating on both retainer and contingency bases. Company pays fee. **Specializes in the areas of:** Computer Science/Software; Engineering; Technical and Scientific. **Positions commonly filled include:** Computer Programmer; Electrical/Electronics Engineer; MIS Specialist; Multimedia Designer; Software Engineer; Systems Analyst; Telecommunications Manager. **Corporate headquarters location:** This Location. **Average salary range of placements:** More than $50,000. **Number of placements per year:** 1 - 49.

FISHER & ASSOCIATES
1063 Lenor Way, San Jose CA 95128. 408/554-0156. **Contact:** Gary Fisher, Owner. **Description:** An executive search firm focusing on high-tech industries. Company pays fee. **Specializes in the areas of:** Administration/MIS/EDP; Computer Science/Software; Engineering; General Management; Sales and Marketing. **Positions commonly filled include:** General Manager; Software Engineer. **Number of placements per year:** 1 - 49.

THE GOODMAN GROUP
P.O. Box 150960, San Rafael CA 94915. 415/455-1500. **Contact:** Recruitment. **Description:** An executive search firm. Company pays fee. **Specializes in the areas of:** Computer Hardware/Software; Engineering; Health/Medical; Insurance; MIS/EDP; Sales and Marketing. **Positions commonly filled include:** Aerospace Engineer; Biomedical Engineer; Civil Engineer; Computer Programmer; EDP Specialist; Electrical/Electronics Engineer; General Manager; Management Analyst/Consultant; Physician; Sales Representative; Systems Analyst. **Number of placements per year:** 50 - 99.

GORELICK & ASSOCIATES
3013 Daimler, Santa Ana CA 92705. 714/660-5050. **Fax:** 714/660-1705. **Contact:** Michael Gorelick, Vice President. **Description:** An executive search firm. **Specializes in the areas of:** Computer Science/Software; Consumer Package Goods; Food Industry; General Management; Sales and Marketing; Telecommunications. **Positions commonly filled include:** Branch Manager; Management Analyst/Consultant; Manufacturer's/Wholesaler's Sales Rep. **Number of placements per year:** 50 - 99.

GRANT & ASSOCIATES
220 Montgomery Street, Suite 303, San Francisco CA 94104-3402. 415/986-1500. **Fax:** 415/986-1630. **Contact:** Susan Grant, President. **E-mail address:** grantsgg@aol.com. **Description:** A generalist retained executive search firm. **Specializes in the areas of:** Computer Science/Software; Engineering; General Management; Manufacturing; Sales and Marketing; Technical and Scientific. **Positions commonly filled include:** Aerospace Engineer; Biochemist; Biomedical Engineer; Branch Manager; Chemical Engineer; Electrical/Electronics Engineer; General Manager; Human Resources Manager; Internet Services Manager; Management Analyst/Consultant; Manufacturer's/Wholesaler's Sales Rep.; Mechanical Engineer; MIS Specialist; Multimedia Designer; Operations/Production Manager; Science Technologist; Services Sales Representative; Software Engineer; Strategic Relations Manager; Systems Analyst; Telecommunications Manager. **Corporate headquarters location:** This Location. **Average salary range of placements:** More than $50,000. **Number of placements per year:** 1 - 49.

GRID SYSTEMS
10951 Hillhaven Avenue, Tujunga CA 91042-1417. 818/352-8000. **Contact:** President. **E-mail address:** grid@compuserve.com. **Description:** An executive search firm. Founded in 1980. Company pays fee. **Specializes in the areas of:** Biology; Computer Science/Software;

Engineering; Health/Medical; Technical and Scientific. **Positions commonly filled include:** Aerospace Engineer; Biochemist; Biological Scientist; Biomedical Engineer; Chemical Engineer; Chemist; Computer Programmer; Design Engineer; Designer; Electrical/Electronics Engineer; General Manager; Industrial Engineer; Industrial Production Manager; Internet Services Manager; Mathematician; Mechanical Engineer; Metallurgical Engineer; MIS Specialist; Multimedia Designer; Software Engineer; Systems Analyst; Technical Writer/Editor; Telecommunications Manager. **Corporate headquarters location:** This Location. **Average salary range of placements:** More than $50,000. **Number of placements per year:** 1 - 49.

HIRE GROUND

502 Flynn Avenue, Redwood City CA 94063-2927. 415/369-8300. **Fax:** 415/369-4992. **Contact:** Kim Kawar, Owner. **E-mail address:** kkawar@interserve.com. **Description:** An executive and technical search firm for high-tech engineering companies. Company pays fee. **Specializes in the areas of:** Computer Science/Software. **Positions commonly filled include:** Design Engineer; Mining Engineer; MIS Specialist; Multimedia Designer; Software Engineer. **Corporate headquarters location:** This Location. **Average salary range of placements:** More than $50,000.

IMPACT, INC.

1000 Fremont Avenue, Suite S, Los Altos CA 94024. 415/941-9400. **Fax:** 415/917-1424. **Contact:** Mary Voss, President. **Description:** A retainer executive search firm. Company pays fee. **Specializes in the areas of:** Computer Science/Software; Engineering; Personnel/Labor Relations. **Positions commonly filled include:** Computer Programmer; Design Engineer; Designer; Electrical/Electronics Engineer; Human Resources Manager; Industrial Engineer; Mathematician; Mechanical Engineer; MIS Specialist; Multimedia Designer; Quality Control Supervisor; Software Engineer; Systems Analyst; Telecommunications Manager. **Benefits available to temporary workers:** 401(k). **Corporate headquarters location:** This Location. **Average salary range of placements:** More than $50,000. **Number of placements per year:** 1000+.

INTELLEX PERSONNEL SERVICES

28047 Dorothy Drive, Suite 200, Agoura Hills CA 91301. 818/865-0099. **Fax:** 818/865-9153. **Contact:** Gloria Esposito, Account Representative/Recruiter. **Description:** An executive search firm. **Specializes in the areas of:** Computer Science/Software. **Positions commonly filled include:** Computer Programmer; Software Engineer; Systems Analyst; Technical Writer/Editor. **Number of placements per year:** 1 - 49.

INTELLISOURCE COMPUTER CAREERS, LTD.

27921 Via Estancia, San Juan Capistrano CA 92675. 714/496-7927. **Fax:** 714/496-0822. **Contact:** Mark Rinovato, President. **E-mail address:** careers@intellisource.com. **World Wide Web address:** http://www.intellisource.com. **Description:** An executive search firm which specializes in the recruitment and placement of high-technology professionals with software engineering and information technology expertise. IntelliSource has successfully staffed start-ups, rapidly growing mid-sized firms, and *Fortune* 500 companies. Company pays fee. **Specializes in the areas of:** Computer Science/Software; Engineering. **Positions commonly filled include:** Computer Programmer; Electrical/Electronics Engineer; Software Engineer; Systems Analyst. **Corporate headquarters location:** This Location. **Average salary range of placements:** More than $50,000. **Number of placements per year:** 1 - 49.

KABL ABILITY NETWORK

1727 State Street, Santa Barbara CA 93101. 805/563-2398. **Contact:** Brad Naegle, President. **Description:** Executive search consultants for firms seeking permanent or temporary executives in managerial and technical positions. Company pays fee. **Specializes in the areas of:** Administration/MIS/EDP; Computer Science/Software; Executives; General Management; Nonprofit; Personnel/Labor Relations. **Positions commonly filled include:** Computer Programmer; Electrical/Electronics Engineer; General Manager; Human Resources Manager; Industrial Engineer; Industrial Production Manager; Management Analyst/Consultant; Software Engineer; Systems Analyst. **Corporate headquarters location:** This Location. **Average salary range of placements:** More than $50,000. **Number of placements per year:** 1 - 49.

KLENIN GROUP
32107 Lindero Canyon Road, Suite 218, Westlake Village CA 91361. 818/597-3434. **Fax:** 818/597-3438. **Contact:** Larry Klenin, President. **Description:** An executive search firm which specializes in high-tech sales and management. **Specializes in the areas of:** Computer Science/Software; Sales and Marketing. **Positions commonly filled include:** Systems Analyst. **Average salary range of placements:** More than $50,000. **Number of placements per year:** 50 - 99.

JOHN KUROSKY & ASSOCIATES
Three Corporate Park Drive, Suite 210, Irvine CA 92714. 714/851-6370. **Fax:** 714/851-8465. **Contact:** William Mason, Executive Vice President. **E-mail address:** jka@ix.netcom.com. **Description:** An executive search firm. John Kurosky & Associates places executives, managers, and technical specialists within high-tech companies. Company pays fee. **Specializes in the areas of:** Accounting/Auditing; Administration/MIS/EDP; Biology; Computer Science/Software; Engineering; Finance; Food Industry; General Management; Health/Medical; Industrial; Manufacturing; Personnel/Labor Relations; Sales and Marketing; Technical and Scientific; Transportation. **Positions commonly filled include:** Aerospace Engineer; Agricultural Engineer; Biological Scientist; Biomedical Engineer; Chemical Engineer; Civil Engineer; Electrical/Electronics Engineer; Industrial Production Manager; Operations/Production Manager; Purchasing Agent/Manager; Quality Control Supervisor; Systems Analyst. **Average salary range of placements:** More than $50,000. **Number of placements per year:** 200 - 499.

LARKIN ASSOCIATES
604 Santa Monica Boulevard, Santa Monica CA 90401. 310/260-0080. **Fax:** 310/260-0090. **Contact:** Managing Director. **Description:** An executive search firm that places personnel involved with interactive multimedia and software publishing. Company pays fee. **Specializes in the areas of:** Art/Design; Computer Science/Software; Sales and Marketing; Technical and Scientific. **Positions commonly filled include:** Computer Programmer; Multimedia Designer. **Average salary range of placements:** More than $50,000. **Number of placements per year:** 1 - 49.

LARSEN, ZILLIACUS & ASSOCIATES
601 West 5th Street, Suite 710, Los Angeles CA 90071-2004. **Contact:** Sandy Hogan, Director of Research. **E-mail address:** sandy@directnet.com. **World Wide Web address:** http://www. directnet.com~sandy. **Description:** An executive search firm. Company pays fee. **Specializes in the areas of:** Accounting/Auditing; Administration/MIS/EDP; Banking; Computer Science/Software; Engineering; Finance; Manufacturing; Nonprofit; Personnel/Labor Relations. **Positions commonly filled include:** Accountant/Auditor; Bank Officer/Manager; General Manager; Human Resources Manager; MIS Manager; Software Engineer. **Average salary of placements:** More than $50,000. **Number of placements per year:** 1 - 49.

LIFTER & ASSOCIATES
10918 Lurline Avenue, Chatsworth CA 91311. 818/998-0283. **Fax:** 818/341-7979. **Contact:** Barbara and Jay Lifter, Office Managers. **E-mail address:** lifterasso@aol.com. **Description:** An executive search firm. Company pays fee. **Specializes in the areas of:** Administration/MIS/EDP; Banking; Computer Science/Software; Consulting; Finance; Information Technology; Insurance; Manufacturing. **Positions commonly filled include:** Computer Programmer; Management Analyst/Consultant; MIS Manager; Multimedia Designer; Software Engineer; Systems Analyst. **Average salary range of placements:** More than $50,000. **Number of placements per year:** 1 - 49.

MALIBU GROUP
30423 Canwood Street, Suite 131, Agoura Hills CA 91301. 818/889-9125. **Fax:** 818/889-2054. **Contact:** Bob Woodall, Owner/Manager. **Description:** An executive search firm. Company pays fee. **Specializes in the areas of:** Computer Science/Software; Engineering; Fashion; Finance; Food Industry; Health/Medical; Industrial; Insurance; Manufacturing; Printing/Publishing; Retail; Sales and Marketing; Technical and Scientific; Transportation. **Positions commonly filled include:** Biomedical Engineer; Buyer; Ceramics Engineer; Chemical Engineer; Chemist; EDP Specialist;

Hotel Manager; Manufacturing Engineer; Marketing Specialist; MIS Specialist; Sales Representative; Technical Representative. **Number of placements per year:** 50 - 99.

MANAGEMENT RECRUITERS
16027 Ventura Boulevard, Suite 320, Encino CA 91436. 818/906-3155. **Fax:** 818/906-0642. **Contact:** Loren Kaun, Manager. **E-mail address:** mri-la@internetmci.com. **World Wide Web address:** http://www.resource.net/mri-la. **Description:** An executive search firm. Company pays fee. **Specializes in the areas of:** Accounting/Auditing; Administration/MIS/EDP; Advertising; Architecture/Construction/Real Estate; Banking; Chemical; Communications; Computer Hardware/Software; Construction; Electrical; Engineering; Finance; Food Industry; General Management; Health/Medical; Insurance; Legal; Manufacturing; Operations Management; Personnel/Labor Relations; Pharmaceutical; Printing/Publishing; Procurement; Retail; Sales and Marketing; Technical and Scientific; Textiles; Transportation. **Positions commonly filled include:** Accountant/Auditor; Administrative Services Manager; Bank Officer/Manager; Biochemist; Biological Scientist; Biomedical Engineer; Computer Programmer; Construction and Building Inspector; Cost Estimator; Engineer; Financial Analyst; General Manager; Health Services Manager; Internet Services Manager; MIS Specialist; Occupational Therapist; Operations/Production Manager; Recreational Therapist; Registered Nurse; Respiratory Therapist; Securities Sales Representative; Software Engineer; Systems Analyst; Telecommunications Manager. **Average salary range of placements:** More than $50,000. **Number of placements per year:** 100 - 199.

MANAGEMENT RECRUITERS OF BURLINGAME
111 Anza Boulevard, Suite 109, Burlingame CA 94010. 415/548-4800. **Fax:** 415/548-4805. **Contact:** Don Hirschbein, President. **Description:** An executive search firm which also operates as a temporary agency. Company pays fee. **Specializes in the areas of:** Accounting/Auditing; Administration/MIS/EDP; Advertising; Architecture/Construction/Real Estate; Banking; Communications; Computer Hardware/Software; Construction; Electrical; Engineering; Finance; Food Industry; General Management; Health/Medical; Insurance; Legal; Manufacturing; Operations Management; Personnel/Labor Relations; Pharmaceutical; Printing/Publishing; Procurement; Retail; Sales and Marketing; Technical and Scientific; Textiles; Transportation. **Number of placements per year:** 100 - 199.

MANAGEMENT RECRUITERS OF OAKLAND
SALES CONSULTANTS
480 Roland Way, Suite 103, Oakland CA 94621. 510/635-7901. **Toll-free phone:** 800/581-7901. **Fax:** 510/562-7237. **Contact:** Tom Thrower, General Manager. **E-mail address:** mrscrecruit@internetmci.com. **World Wide Web address:** http://www.mrinet.com. **Description:** An executive search firm. The firm provides both temporary and permanent staffing. Company pays fee. **Specializes in the areas of:** Administration/MIS/EDP; Computer Science/Software; Engineering; Finance; Food Industry; General Management; Health/Medical; Industrial; Manufacturing; Sales and Marketing; Technical and Scientific. **Positions commonly filled include:** Administrative Services Manager; Biochemist; Biological Scientist; Biomedical Engineer; Branch Manager; Chemical Engineer; Chemist; Computer Programmer; Construction and Building Inspector; Construction Contractor; Credit Manager; Design Engineer; Dietician/Nutritionist; EEG Technologist; EKG Technician; Electrical/Electronics Engineer; Environmental Engineer; Food Scientist/Technologist; General Manager; Health Services Manager; Hotel Manager; Human Resources Manager; Industrial Engineer; Industrial Production Manager; Internet Services Manager; Management Analyst/Consultant; Management Trainee; Manufacturer's/Wholesaler's Sales Rep.; Market Research Analyst; Mechanical Engineer; Metallurgical Engineer; Mining Engineer; MIS Specialist; Multimedia Designer; Nuclear Engineer; Nuclear Medicine Technologist; Occupational Therapist; Operations/Production Manager; Petroleum Engineer; Pharmacist; Physical Therapist; Physician; Psychologist; Public Relations Specialist; Purchasing Agent/Manager; Quality Control Supervisor; Radiologic Technologist; Recreational Therapist; Respiratory Therapist; Restaurant/Food Service Manager; Science Technologist; Services Sales Representative; Software Engineer; Speech-Language Pathologist; Stationary Engineer; Strategic Relations Manager; Surgical Technician; Systems Analyst; Telecommunications Manager. **Corporate headquarters location:** Cleveland OH. **Other U.S. locations:** Nationwide. **Average salary range of placements:** $30,000 - $50,000. **Number of placements per year:** 500 - 999.

MANAGEMENT RECRUITERS OF ORANGE
One City Boulevard West, Suite 710, Orange CA 92668. 714/978-0500. **Fax:** 714/978-8064. **Contact:** John Lewis, General Manager. **Description:** An executive search firm. Company pays fee. **Specializes in the areas of:** Administration/MIS/EDP; Computer Science/Software; Health/Medical; Printing/Publishing; Sales and Marketing. **Positions commonly filled include:** Computer Programmer; EDP Specialist; Health Services Manager; Insurance Agent/Broker; Nurse; Physician; Software Engineer; Surgical Technician; Systems Analyst; Telecommunications Manager; Underwriter/Assistant Underwriter. **Number of placements per year:** 200 - 499.

MANAGEMENT RECRUITERS OF PLEASANTON
4125 Mohr Avenue, Suite M, Pleasanton CA 94566-4740. 510/462-8579. **Fax:** 510/462-0208. **Contact:** Mike Machi, President. **E-mail address:** mriptown@aol.com. **Description:** An executive search firm. Company pays fee. **Specializes in the areas of:** Accounting/Auditing; Administration/ MIS/EDP; Architecture/Construction/Real Estate; Banking; Communications; Computer Hardware/Software; Construction; Electrical; Engineering; Finance; Food Industry; General Management; Health/Medical; Personnel/Labor Relations; Pharmaceutical; Printing/ Publishing; Procurement; Retail; Sales and Marketing; Technical and Scientific; Textiles; Transportation. **Average salary range of placements:** More than $50,000. **Number of placements per year:** 50 - 99.

MANAGEMENT RECRUITERS OF SAN FRANCISCO
591 Redwood Highway, Suite 2225, Mill Valley CA 94941. 415/981-5950. **Fax:** 415/383-1426. **Contact:** Eric Wheel, Manager. **E-mail address:** ericmri@well.com. **World Wide Web address:** http://www.mrimvca.com. **Description:** An executive search firm. Company pays fee. **Specializes in the areas of:** Administration/MIS/EDP; Advertising; Banking; Computer Hardware/Software; Construction; Engineering; Legal; Manufacturing; Operations Management; Technical and Scientific. **Number of placements per year:** 200 - 499.

MANAGEMENT SOLUTIONS
99 Almaden Boulevard, Suite 600, San Jose CA 95113. 408/292-6600. **Fax:** 408/298-5714. **Contact:** Rich Williams, President. **World Wide Web address:** http://www.mgmtsolutions.com. **Description:** An executive search firm operating on a retainer and contingency basis. The firm also acts as a temporary agency and provides contract services. Company pays fee. **Specializes in the areas of:** Accounting/Auditing; Computer Science/Software; Engineering; Finance; Manufacturing; Personnel/Labor Relations. **Positions commonly filled include:** Accountant/ Auditor; Budget Analyst; Computer Programmer; Credit Manager; Design Engineer; Draftsperson; Electrical/Electronics Engineer; Environmental Engineer; Financial Analyst; Human Resources Manager; Industrial Engineer; Internet Services Manager; Mechanical Engineer; Metallurgical Engineer; MIS Specialist; Multimedia Designer; Operations/Production Manager; Quality Control Supervisor; Systems Analyst; Technical Writer/Editor; Telecommunications Manager. **Corporate headquarters location:** This Location. **Other U.S. locations:** Walnut Creek CA; Portland OR. **Number of placements per year:** 1000+.

MASTER SEARCH
P.O. Box 9070, Santa Rosa CA 95405. 707/538-4000. **Contact:** Gregory Masters, President. **Description:** An executive search firm. Company pays fee. **Specializes in the areas of:** Computer Science/Software; Engineering; Food Industry; General Management; Industrial; Manufacturing; Technical and Scientific; Transportation. **Positions commonly filled include:** Biological Scientist; Biomedical Engineer; Chemical Engineer; Electrical/Electronics Engineer; Food Scientist/ Technologist; General Manager; Manufacturing Engineer; Mechanical Engineer; Operations/ Production Manager; Quality Control Supervisor; Science Technologist; Software Engineer; Transportation/Traffic Specialist. **Number of placements per year:** 1 - 49.

MATA & ASSOCIATES
180 Harbor Drive, Suite 208, Sausalito CA 94965. 415/332-2893. **Fax:** 415/332-3916. **Contact:** Dick Mata, Owner. **Description:** An executive search firm for high-tech companies specializing in data processing, computer engineering, software engineering, networking, client/server development, technical support, and telecommunications. Company pays fee. **Specializes in the areas of:** Computer Science/Software. **Positions commonly filled include:** Computer

Programmer; Software Engineer; Systems Analyst; Technical Writer/Editor; Telecommunications Manager. **Corporate headquarters location:** This Location. **Average salary range of placements:** More than $50,000. **Number of placements per year:** 1 - 49.

McCOY LIMITED
22624 Lyons Avenue, Suite 109, Newhall CA 91321-2761. 408/734-8647. **Toll-free phone:** 800/829-6269. **Fax:** 408/734-8441. **Contact:** Joel Burris, Technical Services Manager. **E-mail address:** joel@smartlink.net. **World Wide Web address:** http://www.mccoy.com. **Description:** An executive search firm specializing in the computer industry. Company pays fee. **Specializes in the areas of:** Administration/MIS/EDP; Computer Science/Software; Engineering; Sales and Marketing. **Positions commonly filled include:** Computer Programmer; Design Engineer; Designer; Multimedia Designer; Quality Control Supervisor; Software Engineer; Systems Analyst; Technical Writer/Editor. **Corporate headquarters location:** This Location. **Average salary range of placements:** More than $50,000. **Number of placements per year:** 100 - 199.

JAMES MOORE & ASSOCIATES
90 New Montgomery, Suite 412, San Francisco CA 94105. 415/392-3933. **Fax:** 415/896-0931. **Contact:** Ann Mitchell, Director of Research. **Description:** An executive search firm which focuses on client/server networking, GUI, and object-oriented technology. **Specializes in the areas of:** Computer Hardware/Software. **Positions commonly filled include:** Computer Programmer; Internet Services Manager; Software Engineer; Systems Analyst. **Average salary range of placements:** More than $50,000.

NATIONAL CAREER CHOICES
1300 Santa Barbara Street, Santa Barbara CA 93101. 805/965-0511. **Fax:** 805/966-9857. **Contact:** Gary Kravetz, CEO. **E-mail address:** ncc@west.net. **Description:** National Career Choices is an executive recruitment firm specializing in biotech, medical personnel, computer engineering and software, sales professionals, and management. The firm is a part of Santa Barbara Placement, a temporary and permanent placement agency. The firm also encompasses Applied Micro Solutions, Inc., which handles employee leasing. Company pays fee. **Specializes in the areas of:** Administration/MIS/EDP; Computer Science/Software; Engineering; Finance; General Management; Health/Medical; Personnel/Labor Relations; Technical and Scientific. **Positions commonly filled include:** Accountant/Auditor; Actuary; Administrative Services Manager; Advertising Clerk; Aerospace Engineer; Biological Scientist; Biomedical Engineer; Chemical Engineer; Computer Programmer; Electrical/Electronics Engineer; Financial Analyst; General Manager; Manufacturer's/Wholesaler's Sales Rep.; Mechanical Engineer; MIS Specialist; Nuclear Engineer; Nuclear Medicine Technologist; Quality Control Supervisor; Science Technologist; Software Engineer; Strategic Relations Manager; Systems Analyst; Technical Writer/Editor; Telecommunications Manager. **Other U.S. locations:** Ventura CA. **Average salary range of placements:** More than $50,000. **Number of placements per year:** 50 - 99.

NATIONAL RESOURCES
41661 Enterprise Circle North, Suite 117, Temecula CA 92590. 909/694-5577. **Fax:** 909/699-3855. **Contact:** Gene Jenkins, Owner. **Description:** An executive search firm. Company pays fee. **Specializes in the areas of:** Computer Science/Software. **Positions commonly filled include:** Computer Programmer; Software Engineer; Systems Analyst. **Number of placements per year:** 100 - 199.

NATIONAL SEARCH ASSOCIATES
2035 Corte Del Nogal, Suite 100, Carlsbad CA 92009. 760/431-1115. **Fax:** 760/431-0660. **Contact:** Philip Peluso, President. **Description:** An executive search firm. **Specializes in the areas of:** Biology; Computer Science/Software. **Positions commonly filled include:** Biological Scientist; Biomedical Engineer; Chemist; Software Engineer.

NATIONS STAFFING SOLUTIONS/ASSOCIATES RESOURCE INTERNATIONAL
18952 MacArthur Boulevard, Suite 100, Irvine CA 92715. **Toll-free phone:** 888/252-2520. **Toll-free fax:** 888/252-2550. **Recorded jobline:** 714/222-9833. **Contact:** Rita Harris, Marketing Manager. **Description:** An executive search firm that also places people on a temporary or permanent basis in various fields including mortgage banking, financial services, information

systems, credit and collection, administrative, technical training, and sales. Company pays fee. **Specializes in the areas of:** Administration/MIS/EDP; Banking; Computer Science/Software; Credit and Collection; Finance; Insurance. **Positions commonly filled include:** Accountant/ Auditor; Bank Officer/Manager; Branch Manager; Clerical Supervisor; Computer Programmer; Credit Manager; Customer Service Representative; Management Trainee; MIS Specialist; Securities Sales Representative; Software Engineer; Systems Analyst; Technical Writer/Editor; Typist/Word Processor; Underwriter/Assistant Underwriter. **Corporate headquarters location:** This Location. **Number of placements per year:** 500 - 999.

NEW VENTURE DEVELOPMENT, INC.
596 Canyon Vista Drive, Thousand Oaks CA 91320. 805/498-8506. **Fax:** 805/498-2735. **Contact:** David R. Du Ket, President. **Description:** An executive search firm. **Specializes in the areas of:** Accounting/Auditing; Administration/MIS/EDP; Computer Science/Software; Engineering; Manufacturing; Sales and Marketing; Telecommunications. **Positions commonly filled include:** Biological Scientist; Computer Programmer; Electrical/Electronics Engineer; Mechanical Engineer; Software Engineer; Structural Engineer; Systems Analyst; Telecommunications Manager. **Number of placements per year:** 50 - 99.

OMNISEARCH, INC.
1875 South Grant Street, Suite 310-D, San Mateo CA 94402. 415/574-6090. **Fax:** 415/574-4109. **Contact:** David Scardifield, Principal. **Description:** An executive search firm specializing in high-technology industry placements such as software/hardware engineering, technical support, systems engineering, and sales positions. Company pays fee. **Specializes in the areas of:** Computer Science/Software. **Positions commonly filled include:** Computer Programmer; Electrical/ Electronics Engineer; Software Engineer; Systems Analyst; Telecommunications Manager. **Corporate headquarters location:** This Location. **Other U.S. locations:** Nationwide. **Average salary range of placements:** More than $50,000. **Number of placements per year:** 50 - 99.

ONLINE PROFESSIONAL SEARCH
1030 Trellis Lane, Alameda CA 94502-7053. 510/769-7111. **Contact:** Anita Ho, President. **Description:** An executive search firm that recruits and places candidates with a minimum of five years experience in data processing. Online Professional Search specializes in the placement of programmers/analysts, Unix administrators, Visual Basic programmers, Oracle programmers, and CICS/MVS/IMS systems programmers. **Specializes in the areas of:** Administration/MIS/EDP; Banking; Computer Science/Software. **Positions commonly filled include:** Computer Programmer; MIS Specialist; Systems Analyst. **Corporate headquarters location:** This Location. **Average salary range of placements:** More than $50,000. **Number of placements per year:** 1 - 49.

PC PERSONNEL
226 Airport Parkway, Suite 625, San Jose CA 95110. 408/452-0500. **Fax:** 408/452-0584. **Contact:** Karl Beckstrand, Technical Recruiter. **E-mail address:** netjobs@aol.com. **World Wide Web address:** http://www.pcpersonnel.com. **Description:** An agency specializing in computer network experts. Regular and contract work is available for systems analysts, network administrators, technical support, network engineers, and IS managers. Company pays fee. **Specializes in the areas of:** Computer Science/Software. **Positions commonly filled include:** Internet Services Manager; MIS Specialist; Systems Analyst; Telecommunications Manager. **Benefits available to temporary workers:** Childcare; Medical Insurance; Paid Holiday; Paid Vacation. **Corporate headquarters location:** San Francisco CA. **Average salary range of placements:** $30,000 - $50,000. **Number of placements per year:** 1 - 49.

PEDEN & ASSOCIATES
2000 Broadway Street, Redwood City CA 94063. 415/367-1181. **Fax:** 415/367-7525. **Contact:** Ann Peden, President. **E-mail address:** apeden@pedenassoc.com. **Description:** An executive search firm that places software developers in start-up companies located in the San Francisco Bay Area. Founded in 1980. **NOTE:** A minimum of two years experience in the industry is required; also needed is experience with various software. Company pays fee. **Specializes in the areas of:** Computer Science/Software. **Positions commonly filled include:** Electrical/Electronics Engineer;

Software Engineer. **Corporate headquarters location:** This Location. **Average salary range of placements:** More than $50,000. **Number of placements per year:** 1 - 49.

PROFILES EXECUTIVE SEARCH
970 West 190th Street, Suite 600, Torrance CA 90502. 310/523-3400. **Fax:** 310/523-3401. **Contact:** Bari Kaplan, Executive Recruiter. **Description:** An executive search firm. Company pays fee. **Specializes in the areas of:** Computer Science/Software. **Positions commonly filled include:** Software Engineer. **Number of placements per year:** 1 - 49.

PROSEARCH & ASSOCIATES
4500 Campus, Suite 130, Newport Beach CA 92660. 714/452-0630. **Fax:** 714/951-2602. **Contact:** Sharon Dodson, Office Manager. **Description:** An executive search firm that also does temporary placement on retainer and contingency bases. **Specializes in the areas of:** Accounting/Auditing; Administration/MIS/EDP; Computer Science/Software; Personnel/Labor Relations; Sales and Marketing; Secretarial; Technical and Scientific. **Positions commonly filled include:** Accountant/Auditor; Administrative Services Manager; Bank Officer/Manager; Branch Manager; Clinical Lab Technician; Computer Programmer; MIS Specialist; Registered Nurse; Software Engineer; Systems Analyst; Technical Writer/Editor. **Average salary range of placements:** $30,000 - $50,000.

PROTOCOL SEARCH & SELECTION
650 Hampshire Road, Westlake Village CA 91361. 805/371-0069. **Fax:** 805/371-0048. **Contact:** Chris Salcido, Branch Manager. **Description:** An executive search firm. Company pays fee. **Specializes in the areas of:** Computer Hardware/Software. **Positions commonly filled include:** Computer Programmer; MIS Specialist; Systems Analyst. **Average salary range of placements:** More than $50,000. **Number of placements per year:** 100 - 199.

RJ ASSOCIATES
23730 Canzonet Street, Woodland Hills CA 91367. 818/715-7121. **Fax:** 818/715-9438. **Contact:** Judith Fischer, President. **Description:** RJ Associates provides executive search services for middle/upper financial, accounting, and information technology positions. Company pays fee. **Specializes in the areas of:** Accounting/Auditing; Computer Science/Software; Finance; General Management. **Positions commonly filled include:** Accountant/Auditor; Budget Analyst; Financial Analyst; Management Analyst/Consultant; Systems Analyst. **Average salary range of placements:** More than $50,000. **Number of placements per year:** 50 - 99.

RML GROUP
23195 La Cadena Drive, Suite 102, Laguna Hills CA 92653. 714/458-7787. **Fax:** 714/707-2755. **Contact:** Manager. **Description:** An executive search firm. Founded in 1980. Company pays fee. **Specializes in the areas of:** Computer Hardware/Software; Engineering. **Positions commonly filled include:** Computer Programmer; Marketing Specialist; Software Engineer; Systems Analyst. **Number of placements per year:** 1 - 49.

RAPHAEL RECRUITMENT
4655 Cherryvale Avenue, Suite 100, Soquel CA 95073. 408/464-2760. **Fax:** 408/479-9046. **Contact:** Kent Halpern, President. **E-mail address:** raphael@cruzio.com. **World Wide Web address:** http://www.cruzio.com. **Description:** An executive search firm representing client companies in the Bay Area. Company pays fee. **Specializes in the areas of:** Computer Science/Software; Engineering; Manufacturing. **Positions commonly filled include:** Computer Programmer; Design Engineer; Electrical/Electronics Engineer; Mechanical Engineer; Multimedia Designer; Software Engineer. **Corporate headquarters location:** This Location. **Average salary range of placements:** More than $50,000. **Number of placements per year:** 1 - 49.

RAYCOR GROUP, INC.
1874 South Pacific Coast Highway, Suite 180, Redondo Beach CA 90277. 310/791-5090. **Fax:** 310/791-5089. **Contact:** Halle Soy Bel, Owner. **Description:** An executive search firm. Company pays fee. **Specializes in the areas of:** Accounting/Auditing; Computer Science/Software. **Positions commonly filled include:** Accountant/Auditor; Computer Programmer; Systems Analyst. **Number of placements per year:** 1 - 49.

SAI (SLOAN ASSOCIATES INC.)
2855 Mitchell Drive, Suite 117, Walnut Creek CA 94598. 510/932-3000. **Fax:** 510/932-3857. **Contact:** Steve Sloan, President. **Description:** An executive search firm. Company pays fee. **Specializes in the areas of:** Computer Science/Software. **Positions commonly filled include:** Branch Manager; Systems Analyst. **Number of placements per year:** 50 - 99.

SMC GROUP
26772 Vista Terrace, Lake Forest CA 92630-8110. 714/855-4545. **Contact:** Shala Shashani, President. **E-mail address:** hshm59a@prodigy.com. **Description:** An executive search firm specializing in the LAN/WAN industry. Company pays fee. **Specializes in the areas of:** Computer Hardware/Software. **Positions commonly filled include:** MIS Manager; Software Engineer; Telecommunications Manager. **Number of placements per year:** 50 - 99.

S.R. & ASSOCIATES
110 Newport Center Drive, Suite 202, Newport Beach CA 92660. 714/756-3271. **Fax:** 714/640-7268. **Contact:** Steve Ross, President. **Description:** An executive search firm that also develops and markets software applications for recruiting industries. Company pays fee. **Specializes in the areas of:** Accounting/Auditing; Computer Science/Software; Sales and Marketing. **Positions commonly filled include:** Accountant/Auditor; Computer Programmer; Credit Manager; Human Resources Manager; Management Analyst/Consultant; Manufacturer's/Wholesaler's Sales Rep.; Services Sales Representative; Systems Analyst. **Corporate headquarters location:** This Location. **Average salary range of placements:** More than $50,000. **Number of placements per year:** 50 - 99.

SALES CONSULTANTS OF SACRAMENTO
4320 Auburn Boulevard, Suite 2100, Sacramento CA 95841. 916/344-3737. **Fax:** 916/481-7099. **Contact:** Ron Whitney, Manager. **Description:** An executive search firm, operating on both retainer and contingency bases. Founded in 1978. **Specializes in the areas of:** Computer Science/Software; Sales and Marketing; Technical and Scientific. **Average salary range of placements:** More than $50,000. **Number of placements per year:** 50 - 99.

SANFORD ROSE ASSOCIATES
580 Broadway, Suite 226, Laguna Beach CA 92651. 714/497-5728. **Fax:** 714/497-4086. **Contact:** Bob Dudley, President. **E-mail address:** caorange@aol.com. **Description:** An executive search firm specializing in information services. Company pays fee. **Specializes in the areas of:** Computer Science/Software. **Positions commonly filled include:** Computer Programmer; Electrical/Electronics Engineer; MIS Specialist; Purchasing Agent/Manager; Software Engineer; Systems Analyst. **Average salary range of placements:** More than $50,000. **Number of placements per year:** 1 - 49.

SARVER & CARRUTH ASSOCIATES
P.O. Box 1967, Buellton CA 93427-1967. 805/686-4425. **Fax:** 805/686-5941. **Contact:** Catherine Sarver, Principal. **E-mail address:** csarver@ix.netcom.com. **Description:** An executive search firm specializing in the semiconductor and computer head and disk industries. Company pays fee. **Specializes in the areas of:** Computer Hardware/Software; Engineering; Technical and Scientific. **Positions commonly filled include:** Chemical Engineer; Electrical/Electronics Engineer; Industrial Engineer; Mechanical Engineer; Metallurgical Engineer; Software Engineer. **Number of placements per year:** 1 - 49.

SHARP PERSONNEL & SEARCH
1665 East 4th Street, Suite 204, Santa Ana CA 92701. 714/667-6909. **Fax:** 714/667-2916. **Contact:** Manager. **Description:** An executive search firm that also provides temporary placement. Company pays fee. **Specializes in the areas of:** Accounting/Auditing; Banking; Computer Science/Software; Finance; Health/Medical; Insurance; Personnel/Labor Relations; Secretarial. **Positions commonly filled include:** Accountant/Auditor; Bank Officer/Manager; Financial Analyst; Human Resources Manager; Software Engineer; Systems Analyst; Underwriter/Assistant Underwriter. **Benefits available to temporary workers:** Credit Union; Dental Insurance; Paid Holidays; Paid Vacation. **Average salary range of placements:** $30,000 - $50,000. **Number of placements per year:** 100 - 199.

SHULMAN-ALEXANDER INC.

3528 Sacramento Street, Suite 103, San Francisco CA 94118-1847. 415/931-7096. **Fax:** 415/206-1525. **Contact:** Principal. **Description:** An executive search firm. Company pays fee. **Specializes in the areas of:** Advertising; Computer Hardware/Software; Marketing; Public Relations. **Positions commonly filled include:** Internet Services Manager; Multimedia Designer; Public Relations Specialist; Strategic Relations Manager; Technical Writer/Editor. **Number of placements per year:** 1 - 49.

SPECTRAWEST

39899 Balentine Drive, Suite 218, Newark CA 94560. 510/490-4500. **Fax:** 510/490-6877. **Contact:** Fred Arredondo, Director. **Description:** An executive search firm which conducts searches for hardware design and software engineers. Clients are predominantly located in the San Francisco Bay Area. Company pays fee. **Specializes in the areas of:** Computer Science/Software; Engineering. **Positions commonly filled include:** Chemical Engineer; Electrical/Electronics Engineer; Metallurgical Engineer; Software Engineer. **Corporate headquarters location:** This Location. **Average salary range of placements:** More than $50,000. **Number of placements per year:** 1 - 49.

SPLAINE & ASSOCIATES INC.

15951 Los Gatos Boulevard, Los Gatos CA 95032-3488. 408/354-3664. **Fax:** 408/356-6329. **Contact:** Charles Splaine, President. **E-mail address:** cs@exec-search.com. **World Wide Web address:** http://www.exec-search.com. **Description:** An executive search firm. Company pays fee. **Specializes in the areas of:** Computer Hardware/Software; Data Communications; Electronics. **Average salary range of placements:** More than $50,000. **Number of placements per year:** 1 - 49.

ADELE STEINMETZ

711 Colorado Avenue, Suite 4, Palo Alto CA 94303. 415/321-3723. **Fax:** 415/321-3703. **Contact:** Adele Steinmetz, Owner. **E-mail address:** astnmetz@ix.netcom. **Description:** An executive search firm. **Specializes in the areas of:** Computer Hardware/Software; Engineering; Manufacturing; Sales and Marketing; Technical and Scientific. **Positions commonly filled include:** Chemical Engineer; Chemist; Computer Programmer; Design Engineer; Designer; Draftsperson; Electrical/Electronics Engineer; Environmental Engineer; Financial Analyst; General Manager; Industrial Engineer; Mathematician; MIS Manager; Operations/Production Manager; Science Technologist; Software Engineer; Systems Analyst; Telecommunications Manager. **Number of placements per year:** 1 - 49.

FRED STUART PERSONNEL SERVICES

5855 East Naples Plaza, Suite 310, Long Beach CA 90803-5078. 562/439-0921. **Contact:** Fred Stuart, Owner. **Description:** An executive search firm. Company pays fee. **Specializes in the areas of:** Accounting/Auditing; Banking; Computer Hardware/Software; Finance. **Positions commonly filled include:** Computer Operator; Computer Programmer; EDP Specialist; MIS Specialist; Software Engineer; Systems Analyst.

SYSTEMS RESEARCH GROUP

162 South Rancho Santa Fe Road, Suite B80, Encinitas CA 92024. 760/436-1575. **Fax:** 760/634-3614. **Contact:** Stephen Gebletz, President. **Description:** An executive search firm specializing in high-technology industries with a concentration on CAD/CAM/CAE, rapid prototyping, product data management, and computer graphics. Company pays fee. **Specializes in the areas of:** Administration/MIS/EDP; Computer Science/Software; Engineering; Sales and Marketing; Technical and Scientific. **Positions commonly filled include:** Architect; Biochemist; Biological Scientist; Biomedical Engineer; Chemical Engineer; Civil Engineer; Computer Programmer; Design Engineer; Designer; Draftsperson; Electrical/Electronics Engineer; Geographer; Mechanical Engineer; Science Technologist; Software Engineer; Strategic Relations Manager; Systems Analyst. **Average salary range of placements:** More than $50,000. **Number of placements per year:** 50 - 99.

TECHKNOWLEDGE
7700 Irvine Center Drive, Suite 245, Irvine CA 92718. 714/453-1533. **Fax:** 714/456-8323. **Contact:** John Cowallin, CEO/Recruiter. **Description:** TechKnowledge specializes in placing information technology professionals. Company pays fee. **Specializes in the areas of:** Computer Science/Software; General Management. **Positions commonly filled include:** Computer Programmer; MIS Specialist; Software Engineer; Systems Analyst; Telecommunications Manager. **Average salary range of placements:** More than $50,000. **Number of placements per year:** 100 - 199.

TOD INC.
690 Market Street, San Francisco CA 94104. 415/392-0700. **Fax:** 415/392-0752. **Contact:** Pat Hardy, Operations Manager. **E-mail address:** ronfrede@netcom.com. **Description:** An executive search firm. Company pays fee. **Specializes in the areas of:** Accounting/Auditing; Banking; Computer Science/Software; Finance; Insurance; Legal; Manufacturing; Personnel/Labor Relations; Sales and Marketing; Secretarial. **Positions commonly filled include:** Accountant/ Auditor; Administrative Services Manager; Bank Officer/Manager; Buyer; Claim Representative; Clerical Supervisor; Clinical Lab Technician; Computer Programmer; Counselor; Credit Manager; Customer Service Representative; Draftsperson; Financial Analyst; Human Resources Specialist; Internet Services Manager; Market Research Analyst; Medical Records Technician; MIS Specialist; Multimedia Designer; Paralegal; Purchasing Agent; Quality Control Supervisor; Securities Sales Representative; Services Sales Representative; Software Engineer; Statistician; Systems Analyst; Technical Writer/Editor; Video Production Coordinator. **Average salary range of placements:** $30,000 - $50,000. **Number of placements per year:** 1000+.

TRIPLE-J SERVICES
1508 18th Street, Suite 302, Bakersfield CA 93389. 805/321-0695. **Fax:** 805/321-0882. **Contact:** Jack Jones, Manager. **E-mail address:** quack@bak2.lightspeed.net. **Description:** An executive search firm. Company pays fee. **Specializes in the areas of:** Computer Science/Software; Engineering; Food Industry; Oil and Gas; Technical and Scientific. **Positions commonly filled include:** Agricultural Engineer; Chemical Engineer; Civil Engineer; Computer Programmer; Cost Estimator; Design Engineer; Designer; Draftsperson; Electrical/Electronics Engineer; Environmental Engineer; Industrial Engineer; Mechanical Engineer; MIS Specialist; Petroleum Engineer; Software Engineer; Structural Engineer; Systems Analyst. **Average salary range of placements:** More than $50,000. **Number of placements per year:** 50 - 99.

WALKER & TORRENTE
P.O. Box 707, Belvedere-Tiburon CA 94920-0707. 415/435-9178. **Fax:** 415/435-9144. **Contact:** William T. Walker, Partner. **Description:** A retainer and contingency search firm. Company pays fee. **Specializes in the areas of:** Accounting/Auditing; Administration/MIS/EDP; Banking; Computer Science/Software; Economics; Finance; Personnel/Labor Relations. **Positions commonly filled include:** Accountant/Auditor; Bank Officer/Manager; Budget Analyst; Computer Programmer; Economist; Financial Analyst; General Manager; Human Resources Manager; Industrial Engineer; Management Analyst/Consultant; Mathematician; MIS Specialist; Systems Analyst. **Average salary range of placements:** More than $50,000. **Number of placements per year:** 1 - 49.

WENDELL ASSOCIATES
P.O. Box 7376, San Jose CA 95150-7376. 408/725-1345. **Contact:** Oliver Peeler, Owner. **Description:** An executive search firm. Company pays fee. **Specializes in the areas of:** Computer Hardware/Software; Health/Medical; Sales and Marketing. **Positions commonly filled include:** Biomedical Engineer; Computer Programmer; Electrical/Electronics Engineer; Mechanical Engineer; Software Engineer.

WESTERN TECHNICAL RESOURCES
451 Los Gatos Boulevard, Suite 102, Los Gatos CA 95032. 408/258-8533. **Toll-free phone:** 800/600-5351. **Fax:** 408/358-8535. **Contact:** Bruce West, Principal. **Description:** Provides consulting and contract labor for most technical disciplines. Company pays fee. **Specializes in the areas of:** Computer Science/Software; Engineering; Health/Medical; Technical and Scientific. **Positions commonly filled include:** Aerospace Engineer; Biochemist; Biological Scientist;

Biomedical Engineer; Chemical Engineer; Chemist; Clinical Lab Technician; Computer Programmer; Design Engineer; Designer; Draftsperson; Electrical/Electronics Engineer; Environmental Engineer; Industrial Engineer; Mathematician; Mechanical Engineer; MIS Specialist; Nuclear Engineer; Petroleum Engineer; Software Engineer; Structural Engineer; Systems Analyst; Technical Writer/Editor. **Benefits available to temporary workers:** Frequent Flyer Miles; Health Insurance; Paid Holidays; Vacation. **Corporate headquarters location:** This Location. **Average salary range of placements:** More than $50,000. **Number of placements per year:** 1 - 49.

PERMANENT EMPLOYMENT AGENCIES OF CALIFORNIA

A PERMANENT SUCCESS EMPLOYMENT SERVICES
12658 Washington Boulevard, Suite 104, Los Angeles CA 90066. 310/305-7376. **Fax:** 310/306-2929. **Contact:** Darrell W. Gurney, Owner. **Description:** An employment agency focusing on permanent placement of career professionals in sales/marketing, accounting/finance, human resource, and MIS/computer fields. The company also places secretarial and administrative support. Company pays fee. **Specializes in the areas of:** Accounting/Auditing; Administration/ MIS/EDP; Computer Science/Software; Finance; Personnel/Labor Relations; Sales and Marketing; Secretarial. **Positions commonly filled include:** Accountant/Auditor; Administrative Services Manager; Budget Analyst; Claim Representative; Clerical Supervisor; Computer Programmer; Credit Manager; Customer Service Representative; Financial Analyst; Human Resources Specialist; Manufacturer's/Wholesaler's Sales Rep.; MIS Specialist; Multimedia Designer; Public Relations Specialist; Services Sales Representative; Systems Analyst; Technical Writer/Editor; Telecommunications Manager; Typist/Word Processor. **Average salary range of placements:** $30,000 - $50,000. **Number of placements per year:** 50 - 99.

ANSWERS UNLIMITED
8640 West Third Street, Los Angeles CA 90048. 310/276-7711. **Fax:** 310/276-6156. **Contact:** Christine Lieber, President. **Description:** An employment agency which also provides temporary placement. Company pays fee. **Specializes in the areas of:** Accounting/Auditing; Administration/ MIS/EDP; Advertising; Art/Design; Computer Science/Software; Finance; General Management; Insurance; Legal; Manufacturing; Nonprofit; Personnel/Labor Relations; Printing/ Publishing; Secretarial. **Number of placements per year:** 200 - 499.

ARROWSTAFF SERVICES
2010 North First Street, Suite 300, San Jose CA 95131-2037. 408/437-8989. **Fax:** 408/437-0624. **Contact:** Recruiter. **Description:** A professional management and staffing organization focusing on placement of engineering, programming, and technical personnel. Company pays fee. **Specializes in the areas of:** Administration/MIS/EDP; Computer Science/Software; Engineering; Industrial; Manufacturing. **Positions commonly filled include:** Buyer; Chemical Engineer; Chemist; Computer Programmer; Design Engineer; Designer; Draftsperson; Electrical/Electronics Engineer; Internet Services Manager; Landscape Architect; Mechanical Engineer; MIS Specialist; Multimedia Designer; Purchasing Agent/Manager; Quality Control Supervisor; Software Engineer; Strategic Relations Manager; Systems Analyst; Technical Writer/Editor; Telecommunications Manager. **Number of placements per year:** 1000+.

BLAINE & ASSOCIATES
10100 Santa Monica Boulevard, Suite 470, Los Angeles CA 90067-4008. 310/785-0560. **Fax:** 310/785-9670. **Contact:** Carrie Policella, President. **Description:** An employment agency. Company pays fee. **Specializes in the areas of:** Accounting/Auditing; Administration/MIS/EDP; Advertising; Computer Science/Software; Finance; Legal; Personnel/Labor Relations; Secretarial.

Positions commonly filled include: Administrative Services Manager; Advertising Clerk; Clerical Supervisor; Customer Service Representative; Human Resources Specialist; Paralegal; Typist/ Word Processor. **Corporate headquarters location:** This Location. **Other U.S. locations:** Phoenix AZ. **Average salary range of placements:** $30,000 - $50,000. **Number of placements per year:** 100 - 199.

CT ENGINEERING

P.O. Box 1062, El Segundo CA 90245. 310/643-8333. **Contact:** Manager. **Description:** An employment agency. The agency's primary focus is on placements in engineering industries. **Specializes in the areas of:** Accounting/Auditing; Clerical; Computer Science/Software; Engineering; Secretarial; Technical and Scientific. **Positions commonly filled include:** Administrative Assistant; Aerospace Engineer; Clerk; Computer Programmer; Draftsperson; Industrial Engineer; Mechanical Engineer; Physicist/Astronomer; Purchasing Agent/Manager; Receptionist; Secretary; Statistician; Stenographer; Systems Analyst; Technical Writer/Editor; Technician; Typist/Word Processor.

CALIFORNIA JOB CONNECTION

4041 MacArthur Boulevard, Cerritos CA 90703. 562/809-7785. **Toll-free phone:** 800/645-2971. **Fax:** 562/403-3427. **Contact:** Hazel Xavier, Branch Manager. **Description:** An employment agency. Company pays fee. **Specializes in the areas of:** Banking; Computer Science/Software; Industrial; Manufacturing; Personnel/Labor Relations; Sales and Marketing; Secretarial. **Positions commonly filled include:** Accountant/Auditor; Administrative Services Manager; Advertising Clerk; Blue-Collar Worker Supervisor; Branch Manager; Buyer; Civil Engineer; Clerical Supervisor; Computer Programmer; Customer Service Representative; General Manager; Human Resources Specialist; Management Trainee; Public Relations Specialist; Purchasing Agent/ Manager; Quality Control Supervisor; Technical Writer/Editor; Telecommunications Manager; Typist/Word Processor. **Number of placements per year:** 100 - 199.

CAREER VISIONS

9640-B Mission Gorge Road, Suite 235, Santee CA 92071. 619/448-7082. **Contact:** Recruiter. **Description:** An employment agency. Company pays fee. **Specializes in the areas of:** Accounting/Auditing; Administration/MIS/EDP; Advertising; Architecture/Construction/Real Estate; Art/Design; Banking; Computer Science/Software; Engineering; Finance; Food Industry; General Management; Health/Medical; Industrial; Insurance; Legal; Manufacturing; Nonprofit; Personnel/Labor Relations; Sales and Marketing; Secretarial; Technical and Scientific; Transportation. **Positions commonly filled include:** Accountant/Auditor; Administrative Services Manager; Advertising Clerk; Aircraft Mechanic/Engine Specialist; Architect; Attorney; Automotive Mechanic; Bank Officer/Manager; Biomedical Engineer; Blue-Collar Worker Supervisor; Branch Manager; Buyer; Chemical Engineer; Civil Engineer; Claim Representative; Clerical Supervisor; Clinical Lab Technician; Computer Programmer; Counselor; Credit Manager; Customer Service Representative; Designer; Draftsperson; Electrical/Electronics Engineer; Financial Analyst; General Manager; Health Services Manager; Hotel Manager; Human Resources Manager; Human Service Worker; Industrial Engineer; Industrial Production Manager; Insurance Agent/Broker; Landscape/Grounds Maintenance; Librarian; Library Technician; Management Analyst/Consultant; Management Trainee; Manufacturer's/Wholesaler's Sales Rep.; Mechanical Engineer; Medical Records Technician; Operations/Production Manager; Paralegal; Property and Real Estate Manager; Public Relations Specialist; Purchasing Agent/Manager; Quality Control Supervisor; Restaurant/Food Service Manager; Securities Sales Representative; Services Sales Representative; Software Engineer; Structural Engineer; Surveyor; Systems Analyst; Technical Writer/Editor; Underwriter/Assistant Underwriter; Wholesale and Retail Buyer. **Number of placements per year:** 200 - 499.

THE COMPUTER RESOURCES GROUP, INC.

275 Battery Street, Suite 800, San Francisco CA 94111. 415/398-3535. **Contact:** Jackie Autry, Vice President. **Description:** An employment agency. **Specializes in the areas of:** Computer Hardware/Software; MIS/EDP. **Positions commonly filled include:** Computer Programmer; EDP Specialist; MIS Specialist; Systems Analyst. **Number of placements per year:** 1000+.

DATA CAREERS PERSONNEL SERVICES, INC.
3320 Fourth Avenue, San Diego CA 92103. 619/291-9994. **Fax:** 619/291-9835. **Contact:** Bob Shelton, President. **Description:** Data Careers Personnel Services, Inc. provides temporary, contract, and permanent placement for the computer industry. Company pays fee. **Specializes in the areas of:** Accounting/Auditing; Computer Hardware/Software; Engineering; Finance; MIS/EDP; Software Engineering; Temporary Assignments. **Positions commonly filled include:** Accountant/Auditor; Bookkeeper; Computer Engineer; Computer Operator; Computer Programmer; Customer Service Representative; Data Entry Clerk; EDP Specialist; Financial Analyst; Software Engineer; Systems Analyst; Typist/Word Processor. **Average salary range of placements:** $30,000 - $50,000. **Number of placements per year:** 50 - 99.

DATA SYSTEMS SEARCH CONSULTANTS
1756 Lacassie Avenue, Suite 202, Walnut Creek CA 94596. 510/256-0635. **Contact:** John R. Martinez, President. **Description:** An employment agency. Company pays fee. **Specializes in the areas of:** Computer Hardware/Software. **Positions commonly filled include:** Computer Operator; Computer Programmer; EDP Specialist; MIS Specialist; Operations/Production Manager; Systems Analyst.

EASTRIDGE INFOTECH
2355 Northside Drive, Suite 180, San Diego CA 92108. 619/260-2109. **Fax:** 619/280-0843. **Contact:** Richard L. Olson Jr., Recruiter. **E-mail address:** eastridge@sisna.com. **Description:** An employment agency which also provides temporary placement. **Specializes in the areas of:** Computer Science/Software; Technical and Scientific. **Positions commonly filled include:** Computer Programmer; Draftsperson; Internet Services Manager; Software Engineer; Systems Analyst. **Number of placements per year:** 1 - 49.

EMPLOYMENT DEVELOPMENT DEPARTMENT
1325 Pine Street, Redding CA 96001. 916/225-2191. **Fax:** 916/241-4074. **Recorded jobline:** 916/225-2284. **Contact:** Jim Saims, Employment Program Representative. **Description:** The state of California's program for unemployed professionals. The agency offers career/outplacement counseling. **Specializes in the areas of:** Accounting/Auditing; Administration/MIS/EDP; Advertising; Banking; Biology; Computer Science/Software; Education; Engineering; Food Industry; General Management; Health/Medical; Industrial; Insurance; Personnel/Labor Relations; Retail; Sales and Marketing; Secretarial; Technical and Scientific.

FASTEK TECHNICAL SERVICES
4479 Stoneridge Drive, Pleasanton CA 94588. 510/462-1050. **Fax:** 510/462-1139. **Contact:** Technical Recruiter. **Description:** An employment agency. **Specializes in the areas of:** Accounting/Auditing; Computer Science/Software; Engineering; Technical and Scientific. **Positions commonly filled include:** Accountant/Auditor; Aerospace Engineer; Agricultural Engineer; Agricultural Scientist; Architect; Biological Scientist; Biomedical Engineer; Broadcast Technician; Buyer; Chemical Engineer; Chemist; Civil Engineer; Clinical Lab Technician; Computer Programmer; Cost Estimator; Designer; Draftsperson; Editor; Electrical/Electronics Engineer; Electrician; Geographer; Geologist/Geophysicist; Human Resources Manager; Industrial Engineer; Landscape Architect; Mathematician; Mechanical Engineer; Metallurgical Engineer; Mining Engineer; Petroleum Engineer; Physicist/Astronomer; Purchasing Agent/Manager; Quality Control Supervisor; Science Technologist; Software Engineer; Stationary Engineer; Structural Engineer; Systems Analyst; Technical Writer/Editor; Transportation/Traffic Specialist. **Number of placements per year:** 500 - 999.

GPL ENGINEERING
3031 Tisch Way, Suite 810, San Jose CA 95128. 408/243-1077. **Fax:** 408/241-2652. **Contact:** Guy Leo, Owner. **Description:** An employment agency. Company pays fee. **Specializes in the areas of:** Computer Science/Software; Electrical; Engineering. **Positions commonly filled include:** Electrical/Electronics Engineer; Software Engineer. **Number of placements per year:** 1 - 49.

GENERAL EMPLOYMENT
21515 Hawthorne Boulevard, #665, Torrance CA 90503. 310/540-9151. **Fax:** 310/316-2095. **Contact:** Karen Weaver, Agency Manager. **Description:** An employment agency which specializes

in placement within the information technology industry, including programmer, software engineer, and network support technician positions. Company pays fee. **Specializes in the areas of:** Computer Science/Software. **Positions commonly filled include:** Internet Services Manager; MIS Specialist; Multimedia Designer; Software Engineer; Systems Analyst; Technical Writer/ Editor; Telecommunications Manager. **Corporate headquarters location:** Oakbrook Terrace IL. **Other U.S. locations:** Nationwide. **Number of placements per year:** 100 - 199.

GOULD PERSONNEL SERVICES
850 Colorado Boulevard, Suite 104, Los Angeles CA 90041. 213/256-5800. **Fax:** 213/255-0414. **Contact:** Warren Gould, President. **Description:** Gould Personnel Services is an employer-retained firm that handles permanent positions only. **Specializes in the areas of:** Accounting/Auditing; Advertising; Architecture/Construction/Real Estate; Banking; Computer Science/Software; Economics; Finance; General Management; Health/Medical; Insurance; Legal; Manufacturing; Nonprofit; Personnel/Labor Relations; Printing/Publishing; Sales and Marketing; Secretarial. **Positions commonly filled include:** Accountant/Auditor; Adjuster; Administrative Services Manager; Bank Officer/Manager; Branch Manager; Buyer; Chiropractor; Claim Representative; Clerical Supervisor; Collector; Computer Programmer; Counselor; Credit Manager; Customer Service Representative; Draftsperson; EKG Technician; Investigator; Licensed Practical Nurse; Management Trainee; Physical Therapist; Registered Nurse. **Average salary range of placements:** $20,000 - $29,999. **Number of placements per year:** 100 - 199.

GREAT 400 GROUP INTERNATIONAL
500 East Carson Street, Suite 105, Carson CA 90745. 310/518-9627. **Fax:** 310/522-0103. **Contact:** Jules Howard, Manager/Senior Recruiter. **Description:** An employment agency which specializes in placing engineering, computer, and telecommunications professionals. Company pays fee. **Specializes in the areas of:** Computer Hardware/Software; Engineering; Technical and Scientific. **Positions commonly filled include:** Accountant/Auditor; Administrative Assistant; Aerospace Engineer; Ceramics Engineer; Chemical Engineer; Chemist; Civil Engineer; Computer Operator; Computer Programmer; EDP Specialist; Electrical/Electronics Engineer; Industrial Engineer; Legal Secretary; Manufacturing Engineer; Mechanical Engineer; Metallurgical Engineer; MIS Specialist; Operations/Production Manager; Quality Control Supervisor; Secretary; Software Engineer; Systems Analyst; Technical Writer/Editor; Technician; Typist/Word Processor. **Average salary range of placements:** More than $50,000. **Number of placements per year:** 1 - 49.

ROBERT HALF INTERNATIONAL, INC.
10877 Wilshire Boulevard, Suite 1605, Los Angeles CA 90024. 310/286-6800. **Contact:** Recruiter. **Description:** An employment agency. Company pays fee. **Specializes in the areas of:** Accounting/Auditing; Banking; Computer Hardware/Software; Finance; MIS/EDP. **Positions commonly filled include:** Accountant/Auditor; Bank Officer/Manager; Bookkeeper; Computer Programmer; Credit Manager; Data Entry Clerk; EDP Specialist; Financial Analyst; Human Resources Manager; Systems Analyst. **Number of placements per year:** 500 - 999.

JUSTUS PERSONNEL SERVICES
10680 West Pico Boulevard, Suite 210, Los Angeles CA 90064. 310/204-6711. **Fax:** 310/204-2941. **Contact:** Janet Justus, Owner. **Description:** An employment agency. Company pays fee. **Specializes in the areas of:** Computer Operations; Entertainment; Finance; Government; Health/Medical; Legal; Manufacturing; Nonprofit; Real Estate. **Positions commonly filled include:** Accountant/Auditor; Administrative Assistant; Bookkeeper; Clerk; Customer Service Representative; Legal Secretary; Sales and Marketing Manager; Secretary; Systems Analyst; Typist/Word Processor. **Number of placements per year:** 200 - 499.

MS DATA PERSONNEL SERVICE CORPORATION
151 Kalmus Drive, Suite H9A, Costa Mesa CA 92626. 714/540-4430. **Contact:** Personnel Manager. **Description:** An employment agency. **Specializes in the areas of:** Computer Hardware/Software; MIS/EDP; Technical and Scientific; Temporary Assignments. **Positions commonly filled include:** Computer Operator; Computer Programmer; Data Entry Clerk; EDP Specialist; MIS Specialist; Receptionist; Systems Analyst; Typist/Word Processor.

RICHARD MARIES AGENCY
200 West Pondera Avenue, Lancaster CA 93534. 805/942-0466. **Contact:** Office Manager. **Description:** An employment agency. Company pays fee. **Specializes in the areas of:** Accounting/Auditing; Clerical; Computer Hardware/Software; Health/Medical; Secretarial. **Positions commonly filled include:** Accountant/Auditor; Administrative Assistant; Aerospace Engineer; Bookkeeper; Civil Engineer; Claim Representative; Clerk; Customer Service Representative; Data Entry Clerk; Draftsperson; General Manager; Hotel Manager; Insurance Agent/Broker; Legal Secretary; Medical Secretary; Nurse; Purchasing Agent/Manager; Receptionist; Secretary; Typist/Word Processor. **Number of placements per year:** 50 - 99.

MULTAX SYSTEMS, INC.
505 North Sepulveda Boulevard, Suite 7, Manhattan Beach CA 90266-6743. 310/379-8398. **Fax:** 310/379-1142. **Contact:** Randy Heinesh, General Manager. **Description:** An employment agency. Company pays fee. **Specializes in the areas of:** Computer Science/Software; Engineering. **Positions commonly filled include:** Aerospace Engineer; Computer Programmer; Designer; Draftsperson; Software Engineer; Structural Engineer; Systems Analyst. **Number of placements per year:** 200 - 499.

PRO STAFF PERSONNEL SERVICES
990 West 190th Street, Torrance CA 90502. 310/353-2411. **Fax:** 310/353-2416. **Contact:** Crystal Flood, Staffing Supervisor. **Description:** Pro Staff Personnel Services specializes in temporary and permanent placement. **Specializes in the areas of:** Accounting/Auditing; Administration/MIS/EDP; Computer Science/Software; Industrial; Personnel/Labor Relations. **Positions commonly filled include:** Accountant/Auditor; Administrative Services Manager; Bank Officer/Manager; Blue-Collar Worker Supervisor; Branch Manager; Claim Representative; Computer Programmer; Customer Service Representative; Environmental Engineer; Financial Analyst; Human Resources Manager; Internet Services Manager; MIS Specialist; Paralegal; Purchasing Agent/Manager; Services Sales Representative. **Number of placements per year:** 1000+.

RENIOR STAFFING SERVICES INC.
3710 Grand Avenue, Oakland CA 94610. 510/836-2220. **Fax:** 510/836-0321. **Contact:** Greg Dencker, Vice President. **Description:** An employment agency. **Specializes in the areas of:** Banking; Computer Science/Software; Personnel/Labor Relations; Real Estate. **Positions commonly filled include:** Accountant/Auditor; Administrative Services Manager; Architect; Attorney; Computer Programmer; Construction and Building Inspector; Cost Estimator; Customer Service Representative; Design Engineer; Electrical/Electronics Engineer; Environmental Engineer; General Manager; Human Resources Specialist; Human Service Worker; Management Analyst/Consultant; Mining Engineer; Property and Real Estate Manager; Real Estate Agent; Services Sales Representative; Software Engineer. **Corporate headquarters location:** This Location. **Average salary range of placements:** $20,000 - $29,999. **Number of placements per year:** 1000+.

SANTA BARBARA PLACEMENT
1300 Santa Barbara Street, #B, Santa Barbara CA 93101-2017. **Contact:** Gary Kravetz, CEO. **Description:** An employment agency operating on both a retainer and contingency basis. Santa Barbara Placement offers permanent and temporary placements, and also provides career/outplacement counseling. **Specializes in the areas of:** Accounting/Auditing; Banking; Computer Science/Software; Engineering; General Management; Health/Medical; Legal; Manufacturing; Nonprofit; Personnel/Labor Relations; Secretarial. **Positions commonly filled include:** Accountant/Auditor; Architect; Chemical Engineer; Computer Programmer; General Manager; Human Resources Specialist; Management Trainee; Systems Analyst; Typist/Word Processor. **Average salary range of placements:** $20,000 - $29,999. **Number of placements per year:** 500 - 999.

SEARCH WEST OF ENCINO
2049 Century Park East, Suite 650, Los Angeles CA 90067. 310/284-8888. **Contact:** General Manager. **Description:** An employment agency. Company pays fee. **Specializes in the areas of:** Accounting/Auditing; Advertising; Banking; Computer Hardware/Software; Engineering; Finance; Food Industry; Health/Medical; Insurance; MIS/EDP; Real Estate; Sales and Marketing; Technical

and Scientific. **Positions commonly filled include:** Accountant/Auditor; Administrative Assistant; Aerospace Engineer; Bank Officer/Manager; Biomedical Engineer; Bookkeeper; Chemical Engineer; Civil Engineer; Computer Operator; Computer Programmer; Credit Manager; Customer Service Representative; Draftsperson; Economist; EDP Specialist; Electrical/Electronics Engineer; Financial Analyst; Hotel Manager; Human Resources Manager; Industrial Designer; Industrial Engineer; Insurance Agent/Broker; Marketing Specialist; Nurse; Operations/Production Manager; Purchasing Agent/Manager; Quality Control Supervisor; Sales Representative; Systems Analyst; Technician; Underwriter/Assistant Underwriter. **Number of placements per year:** 1000+.

SEARCH WEST OF ONTARIO

2151 Convention Center Way, Suite 121-B, Ontario CA 91764. 909/937-0100. **Contact:** Nate Reddicks, General Manager. **Description:** An employment agency. Company pays fee. **Specializes in the areas of:** Accounting/Auditing; Architecture/Construction/Real Estate; Banking; Computer Hardware/Software; Construction; Engineering; Finance; Food Industry; Health/Medical; Insurance; Manufacturing; Personnel/Labor Relations; Printing/Publishing; Sales and Marketing; Technical and Scientific. **Positions commonly filled include:** Accountant/Auditor; Actuary; Aerospace Engineer; Agricultural Engineer; Architect; Attorney; Bank Officer/Manager; Biological Scientist; Biomedical Engineer; Buyer; Ceramics Engineer; Chemical Engineer; Chemist; Civil Engineer; Computer Programmer; Credit Manager; Dietician/Nutritionist; Draftsperson; Economist; EDP Specialist; Electrical/Electronics Engineer; Financial Analyst; Food Scientist/Technologist; General Manager; Hotel Manager; Human Resources Manager; Industrial Designer; Marketing Specialist; Mechanical Engineer; Metallurgical Engineer; Mining Engineer; MIS Specialist; Petroleum Engineer; Physicist/Astronomer; Purchasing Agent/Manager; Quality Control Supervisor; Sales Representative; Statistician; Systems Analyst; Technical Writer/Editor; Underwriter/Assistant Underwriter. **Number of placements per year:** 1000+.

SHARF, WOODWARD & ASSOCIATES

5900 Sepulveda Boulevard, #104, Sherman Oaks CA 91411. 818/989-2200. **Fax:** 818/781-5554. **Contact:** Office Manager. **Description:** Sharf, Woodward & Associates is a technical computer contract and permanent placement firm. **Specializes in the areas of:** Computer Science/Software. **Positions commonly filled include:** Computer Programmer; MIS Specialist; Software Engineer; Systems Analyst. **Average salary range of placements:** More than $50,000. **Number of placements per year:** 200 - 499.

SNELLING PERSONNEL SERVICES

26229 Eden Landing Road, Suite 3, Hayward CA 94545. 510/538-3750. **Contact:** Gloria S. Capron, CEO. **Description:** An employment agency. Company pays fee. **Specializes in the areas of:** Accounting/Auditing; Advertising; Banking; Clerical; Computer Hardware/Software; Construction; Food Industry; Legal; Manufacturing; Printing/Publishing; Real Estate; Secretarial; Technical and Scientific. **Positions commonly filled include:** Accountant/Auditor; Administrative Assistant; Bookkeeper; Credit Manager; Customer Service Representative; Data Entry Clerk; Legal Secretary; Receptionist; Secretary; Typist/Word Processor. **Number of placements per year:** 200 - 499.

SOURCE ENGINEERING

1290 Oakmead Parkway, Suite 318, Sunnyvale CA 94086. 408/738-8440. **Fax:** 408/730-1042. **Contact:** Managing Director. **Description:** An employment agency. Company pays fee. **Specializes in the areas of:** Computer Science/Software; Engineering; Sales and Marketing; Technical and Scientific. **Positions commonly filled include:** Branch Manager; Computer Programmer; Customer Service Representative; Electrical/Electronics Engineer; General Manager; Management Analyst/Consultant; Software Engineer; Systems Analyst. **Number of placements per year:** 50 - 99.

SUNDAY & ASSOCIATES, INC.

P.O. Box 847, Petaluma CA 94953. 510/644-0440. **Contact:** Michael Sunday, Owner. **Description:** An employment agency. **Specializes in the areas of:** Computer Hardware/Software; Technical and Scientific. **Positions commonly filled include:** Computer Programmer; Software Engineer; Systems Analyst; Technical Writer/Editor.

SYSTEM ONE

3021 Citrus Circle, Suite 230, Walnut Creek CA 94598. 510/932-8801. **Fax:** 510/932-3651. **Contact:** Dave Doyle, Owner. **E-mail address:** system1search@ccnet.com. **World Wide Web address:** http://www.ccnet.com~system. **Description:** An employment agency. Company pays fee. **Specializes in the areas of:** Clerical; Computer Science/Software; Engineering; Food Industry; Manufacturing; Technical and Scientific. **Positions commonly filled include:** Bookkeeper; Chemical Engineer; Electrical/Electronics Engineer; Industrial Engineer; Legal Secretary; Manufacturing Engineer; Mechanical Engineer; Secretary; Typist/Word Processor. **Number of placements per year:** 1 - 49.

SYSTEMATICS AGENCY

606 Wilshire Boulevard, Suite 202, Santa Monica CA 90401. 310/395-4991. **Fax:** 310/395-2254. **Contact:** Peter J. Locke, Owner. **Description:** An employment agency. Company pays fee. **Specializes in the areas of:** Administration/MIS/EDP; Computer Science/Software. **Positions commonly filled include:** Computer Programmer; MIS Specialist; Software Engineer; Systems Analyst; Telecommunications Manager. **Average salary range of placements:** $30,000 - $50,000. **Number of placements per year:** 100 - 199.

SYSTEMS CAREERS

211 Sutter Street, Suite 607, San Francisco CA 94108. 415/434-4770. **Contact:** Wayne Sarchett, Principal. **Description:** An employment agency. Company pays fee. **Specializes in the areas of:** Banking; Computer Hardware/Software; Food Industry; Manufacturing; MIS/EDP; Sales and Marketing; Transportation. **Positions commonly filled include:** Computer Programmer; EDP Specialist; Industrial Engineer; Management Analyst/Consultant; Software Engineer. **Number of placements per year:** 1 - 49.

TAD RESOURCES INTERNATIONAL

3450 East Spring Street, #103, Long Beach CA 90806. 562/426-0446. **Fax:** 562/426-7876. **Contact:** Timothy O. Lane, Recruiter. **Description:** An employment agency. Company pays fee. **Specializes in the areas of:** Accounting/Auditing; Administration/MIS/EDP; Computer Science/Software; Engineering; Insurance; Legal; Manufacturing; Secretarial; Technical and Scientific. **Positions commonly filled include:** Accountant/Auditor; Buyer; Customer Service Representative; Designer; Draftsperson; Electrical/Electronics Engineer; Environmental Engineer; Insurance Agent/Broker; Management Analyst/Consultant; Mechanical Engineer; Medical Records Technician; Mining Engineer; MIS Specialist; Operations/Production Manager; Paralegal; Petroleum Engineer; Quality Control Supervisor; Software Engineer; Structural Engineer; Technical Writer/Editor; Typist/Word Processor. **Benefits available to temporary workers:** 401(k); Dental Insurance; Medical Insurance; Vision Plan. **Corporate headquarters location:** Cambridge MA. **Other U.S. locations:** Nationwide. **International locations:** Worldwide. **Average salary range of placements:** $30,000 - $50,000. **Number of placements per year:** 500 - 999.

TECH SEARCH

2015 Bridgeway, Suite 301, Sausalito CA 94965. 415/332-1282. **Fax:** 415/332-1285. **Contact:** Roger M. King, President. **Description:** An employment agency. Company pays fee. **Specializes in the areas of:** Computer Science/Software. **Positions commonly filled include:** Computer Programmer; Software Engineer; Systems Analyst; Technical Writer/Editor. **Number of placements per year:** 200 - 499.

TECHNICAL DIRECTIONS, INC. (TDI)

8880 Rio San Diego Drive, Suite 925, San Diego CA 92108. 619/297-5611. **Contact:** Office Manager. **Description:** An employment agency. Company pays fee. **Specializes in the areas of:** Banking; Computer Hardware/Software; Finance; Manufacturing; MIS/EDP; Technical and Scientific; Temporary Assignments. **Positions commonly filled include:** Computer Programmer; EDP Specialist; Systems Analyst; Technical Writer/Editor. **Number of placements per year:** 50 - 99.

TRATTNER NETWORK

729 Sequoia Valley Road, Mill Valley CA 94941-2637. 415/380-1199. **Fax:** 415/389-1189. **Contact:** Jim Trattner, President. **E-mail address:** jtrattner@trattnet.com. **World Wide Web**

address: http://www.trattnet.com. **Description:** An employment agency. Trattner Network focuses on high-tech placements. Founded in 1975. **Specializes in the areas of:** Computer Science/Software; Engineering. **Positions commonly filled include:** Computer Programmer; Internet Services Manager; MIS Specialist; Software Engineer; Systems Analyst. **Average salary range of placements:** More than $50,000. **Number of placements per year:** 100 - 199.

ZEIGER TECHNICAL CAREERS, INC.
20969 Ventura Boulevard, Suite 217, Woodlands Hills CA 91364. 818/999-9394. **Fax:** 818/999-5036. **Contact:** Stephen A. Zeiger, President/CEO. **Description:** An employment agency. Company pays fee. **Specializes in the areas of:** Computer Science/Software; Engineering; General Management; Manufacturing; Personnel/Labor Relations. **Positions commonly filled include:** Biomedical Engineer; Chemical Engineer; Computer Programmer; Design Engineer; Designer; Electrical/Electronics Engineer; General Manager; Human Resources Manager; Industrial Engineer; Industrial Production Manager; Management Analyst/Consultant; Mechanical Engineer; Metallurgical Engineer; MIS Specialist; Nuclear Engineer; Operations/Production Manager; Physicist/Astronomer; Software Engineer; Statistician; Systems Analyst. **Average salary range of placements:** More than $50,000. **Number of placements per year:** 100 - 199.

EXECUTIVE SEARCH FIRMS OF COLORADO

THE BRIDGE
P.O. Box 740297, Arvada CO 80006-0297. 303/422-1900. **Fax:** 303/422-0016. **Contact:** Alex B. Wilcox, Principal. **Description:** A diversified executive search and management consulting firm offering professional search/consulting services. Company pays fee. **Specializes in the areas of:** Accounting/Auditing; Administration/MIS/EDP; Chemical; Computer Science/Software; Engineering; Environmental; Manufacturing; Personnel/Labor Relations; Sales and Marketing. **Positions commonly filled include:** Accountant/Auditor; Biological Scientist; Chemist; Computer Programmer; Engineer; Forester/Conservation Scientist; General Manager; Geologist/Geophysicist; Human Resources Manager; Purchasing Agent/Manager; Quality Control Supervisor; Technical Writer/Editor. **Corporate headquarters location:** This Location. **Average salary range of placements:** More than $50,000. **Number of placements per year:** 1 - 49.

CAREER FORUM, INC.
4350 Wadsworth Boulevard, Suite 300, Wheat Ridge CO 80033. 303/425-8721. **Fax:** 303/425-0535. **Contact:** Stan Grebe, President. **Description:** An executive search firm. Company pays fee. **Specializes in the areas of:** Accounting/Auditing; Administration/MIS/EDP; Advertising; Architecture/Construction/Real Estate; Banking; Computer Science/Software; Engineering; Finance; General Management; Industrial; Insurance; Manufacturing; Printing/Publishing; Retail; Sales and Marketing; Secretarial; Technical and Scientific; Transportation. **Positions commonly filled include:** Accountant/Auditor; Administrative Services Manager; Aerospace Engineer; Agricultural Engineer; Bank Officer/Manager; Biomedical Engineer; Branch Manager; Broadcast Technician; Buyer; Chemical Engineer; Civil Engineer; Claim Representative; Clerical Supervisor; Computer Programmer; Cost Estimator; Credit Manager; Customer Service Representative; Electrical/Electronics Engineer; General Manager; Industrial Engineer; Management Trainee; Manufacturer's/Wholesaler's Sales Rep.; Mechanical Engineer; Metallurgical Engineer; Mining Engineer; Nuclear Engineer; Operations/Production Manager; Quality Control Supervisor; Services Sales Representative; Software Engineer; Structural Engineer; Systems Analyst; Underwriter/Assistant Underwriter. **Number of placements per year:** 200 - 499.

CAREER MARKETING ASSOCIATES
7100 East Belleview Avenue, Suite 102, Greenwood Village CO 80111. 303/779-8890. **Fax:** 303/779-8139. **Contact:** Terry Leyden CPC, General Manager. **World Wide Web address:** http://www.cmagroup.com. **Description:** An executive search firm offering both retainer and contingency services. Company pays fee. **Specializes in the areas of:** Administration/MIS/EDP;

Biology; Computer Science/Software; Engineering; Health/Medical; Manufacturing. **Positions commonly filled include:** Biological Scientist; Biomedical Engineer; Computer Programmer; Design Engineer; Electrical/Electronics Engineer; Internet Services Manager; Mechanical Engineer; MIS Specialist; Multimedia Designer; Occupational Therapist; Physical Therapist; Software Engineer; Systems Analyst; Toxicologist. **Benefits available to temporary workers:** Incentive Plan; Medical Insurance; Paid Vacations. **Corporate headquarters location:** This Location. **Average salary range of placements:** More than $50,000. **Number of placements per year:** 100 - 199.

CARLSON, BENTLEY ASSOCIATES
3889 Promontory Court, Boulder CO 80304. 303/443-6500. **Contact:** Donald E. Miller, President. **Description:** An executive search firm. Company pays fee. **Specializes in the areas of:** Computer Science/Software. **Positions commonly filled include:** Computer Programmer; Systems Analyst. **Number of placements per year:** 1 - 49.

CASEY SERVICES, INC.
6143 South Willow Drive, Suite 102, Greenwood Village CO 80111. 303/721-9211. **Fax:** 303/721-9508. **Recorded jobline:** 303/721-9211. **Contact:** Pat Melin, Placement Manager. **Description:** An executive search firm operating on a contingency basis. The firm also provides temporary placements. The company focuses on placing accounting, finance, and information systems personnel. Company pays fee. **Specializes in the areas of:** Accounting/Auditing; Administration/MIS/EDP; Banking; Computer Science/Software; Finance; General Management; Health/Medical; Sales and Marketing; Secretarial. **Positions commonly filled include:** Accountant/Auditor; Bank Officer/Manager; Budget Analyst; Clerical Supervisor; Computer Programmer; Cost Estimator; Credit Manager; Customer Service Representative; General Manager; MIS Specialist; Operations/Production Manager; Quality Control Supervisor; Services Sales Representative; Software Engineer; Systems Analyst; Telecommunications Manager. **Benefits available to temporary workers:** Dental Insurance; Medical Insurance. **Corporate headquarters location:** Denver CO. **Other U.S. locations:** Chicago IL. **Number of placements per year:** 200 - 499.

DANIELS & PATTERSON CORPORATE SEARCH, INC.
1732 Marion Street, Denver CO 80218. 303/830-1230. **Fax:** 303/832-6162. **Contact:** Ruby Chavez Patterson, President. **Description:** A permanent and contract human resources firm providing EDP, MIS, IS, telecommunications, clerical/office support, and contract personnel. Company pays fee. **Specializes in the areas of:** Computer Hardware/Software; Computer Science/Software. **Positions commonly filled include:** Administrative Assistant; Computer Programmer; Data Analyst; Data Entry Clerk; MIS Specialist; Secretary; Software Engineer; Systems Analyst. **Corporate headquarters location:** This Location. **Average salary range of placements:** $30,000 - $50,000. **Number of placements per year:** 1 - 49.

DUNHILL PERSONNEL OF BOULDER
P.O. Box 488, Niwot CO 80544. 303/444-5531. **Contact:** Francis Boruff, Owner. **Description:** An executive search firm. **Specializes in the areas of:** Computer Hardware/Software; Engineering; Technical and Scientific.

DUNHILL PERSONNEL OF DENVER SOUTH
6909 South Holly Circle, Suite 305, Englewood CO 80112. 303/721-0525. **Fax:** 303/721-0747. **Contact:** Sandra Funt, Vice President. **E-mail address:** dunhillden@aol.com. **Description:** An executive search firm specializing in all areas of health care including clinical, nursing, administration, and allied health, as well as in data processing and information systems. Company pays fee. **Specializes in the areas of:** Computer Science/Software; Health/Medical. **Positions commonly filled include:** Clinical Lab Technician; Computer Programmer; Dietician/Nutritionist; Health Services Manager; Medical Records Technician; MIS Specialist; Nuclear Medicine Technologist; Occupational Therapist; Pharmacist; Physical Therapist; Physician; Psychologist; Radiologic Technologist; Registered Nurse; Respiratory Therapist; Social Worker; Speech-Language Pathologist; Systems Analyst. **Average salary range of placements:** $30,000 - $50,000. **Number of placements per year:** 1 - 49.

GEOSEARCH, INC.

5585 Erindale Drive, Suite 104, Colorado Springs CO 80918-6966. 719/260-7087. **Fax:** 719/260-7389. **Contact:** Richard Serby, President. **E-mail address:** map@usa.net. **World Wide Web address:** http://www.geosearch.com. **Description:** A retainer and contingency executive search firm. **Specializes in the areas of:** Computer Science/Software; Engineering; Technical and Scientific. **Positions commonly filled include:** Civil Engineer; Computer Programmer; Environmental Engineer; Forester/Conservation Scientist; Geographer; Landscape Architect; Market Research Analyst; Software Engineer; Surveyor; Systems Analyst; Technical Writer/Editor; Urban/Regional Planner. **Average salary range of placements:** $30,000 - $50,000. **Number of placements per year:** 200 - 499.

HALLMARK PERSONNEL SYSTEMS, INC.

6825 East Tennessee Avenue, Suite 637, Denver CO 80224. 303/388-6190. **Fax:** 303/355-3760. **Contact:** Joe Sweeney, CPC, General Manager. **Description:** An executive search firm. Company pays fee. **Specializes in the areas of:** Administration/MIS/EDP; Computer Science/Software; Engineering; General Management; Industrial; Personnel/Labor Relations; Sales and Marketing; Software Development. **Positions commonly filled include:** Aerospace Engineer; Biochemist; Biological Scientist; Biomedical Engineer; Computer Programmer; Engineer; Geologist/ Geophysicist; Industrial Production Manager; Quality Control Supervisor; Science Technologist; Structural Engineer; Systems Analyst; Technical Writer/Editor; Telecommunications Manager. **Corporate headquarters location:** This Location. **Other U.S. locations:** Nationwide. **Average salary range of placements:** More than $50,000. **Number of placements per year:** 1 - 49.

HEALTHCARE RECRUITERS OF THE ROCKIES, INC.

6860 South Yosemite Court, Suite 200, Englewood CO 80112. 303/779-8570. **Fax:** 303/779-7974. **Contact:** Richard Moore, President. **Description:** An executive search firm for the health care industry. Professionals are placed in sales, marketing, field service, management, and engineering positions. The firm specializes in medical device and health care information systems. Company pays fee. **Specializes in the areas of:** Computer Science/Software; Health/Medical; Sales and Marketing; Technical and Scientific. **Positions commonly filled include:** Biochemist; Biological Scientist; Biomedical Engineer; Chemical Engineer; Chemist; Clinical Lab Technician; Design Engineer; Health Services Manager; Industrial Engineer; Industrial Production Manager; Manufacturer's/Wholesaler's Sales Rep.; Mechanical Engineer; MIS Specialist; Occupational Therapist; Pharmacist; Physical Therapist; Physician; Registered Nurse; Services Sales Representative; Software Engineer; Systems Analyst. **Corporate headquarters location:** Dallas TX. **Average salary range of placements:** $30,000 - $50,000. **Number of placements per year:** 50 - 99.

I.J. & ASSOCIATES, INC.

2525 South Wadsworth Boulevard, Suite 106, Lakewood CO 80227. 303/984-2585. **Fax:** 303/984-2589. **Contact:** Ila Larson, Owner. **E-mail address:** ilarson793@aol.com. **World Wide Web address:** http://www.ijassoc.com. **Description:** An executive search firm for the computer industry. **Specializes in the areas of:** Computer Science/Software. **Positions commonly filled include:** Computer Programmer; Systems Analyst.

PINNACLE SOURCE

9250 East Costilla Avenue, Englewood CO 80112-3643. 303/792-5300. **Fax:** 303/792-5301. **Contact:** Jordan Greenberg, President. **Description:** A retainer and contingency search firm. Company pays fee. **Specializes in the areas of:** Computer Science/Software; Sales and Marketing; Technical and Scientific. **Positions commonly filled include:** Branch Manager; Computer Programmer; MIS Specialist; Services Sales Representative; Software Engineer; Systems Analyst. **Corporate headquarters location:** This Location. **Average salary range of placements:** More than $50,000. **Number of placements per year:** 50 - 99.

PROFESSIONAL SEARCH AND PLACEMENT

4901 East Dry Creek Road, Littleton CO 80122-4010. 303/779-8004. **Contact:** John Turner, President. **Description:** An executive search firm specializing in computer technology. The firm recruits nationally and offers permanent and contract assignments. Founded in 1971. **Specializes in** the areas of: Computer Science/Software. **Positions commonly filled include:** Computer Programmer; Software Engineer; Systems Analyst; Technical Writer/Editor. **Benefits available to temporary workers:** Medical Insurance. **Corporate headquarters location:** This Location. **Average salary range of placements:** More than $50,000. **Number of placements per year:** 200 - 499.

SNELLING PERSONNEL SERVICES
2480 West 26th Avenue, Suite 140B, Denver CO 80211-5304. 303/964-8200. **Fax:** 303/964-9312. **Contact:** F. Daryl Gatewood, President. **Description:** An executive search firm. Snelling also places temporary personnel. Company pays fee. **Specializes in the areas of:** Accounting/Auditing; Administration/MIS/EDP; Computer Science/Software; Engineering; Finance; Personnel/Labor Relations; Sales and Marketing; Secretarial; Technical and Scientific. **Positions commonly filled include:** Accountant/Auditor; Administrative Services Manager; Aerospace Engineer; Bank Officer/Manager; Clerical Supervisor; Computer Programmer; Electrical/Electronics Engineer; Financial Analyst; Human Resources Manager; Public Relations Specialist; Securities Sales Representative; Services Sales Representative; Systems Analyst; Typist/Word Processor. **Benefits available to temporary workers:** 401(k) Plan; Medical Insurance; Paid Holidays; Paid Vacations. **Corporate headquarters location:** Dallas TX. **International locations:** Worldwide. **Average salary range of placements:** $30,000 - $50,000. **Number of placements per year:** 100 - 199.

STAR PERSONNEL SERVICE
2851 South Parker Road, Aurora CO 80014-2736. 303/695-1161. **Fax:** 303/695-1058. **Contact:** Paul Staffieri, Owner. **Description:** An executive search firm which also operates as a temporary and permanent employment agency. Company pays fee. **Specializes in the areas of:** Accounting/Auditing; Computer Science/Software; Engineering; Finance; General Management; Industrial; Manufacturing; Personnel/Labor Relations; Retail; Sales and Marketing. **Positions commonly filled include:** Accountant/Auditor; Computer Programmer; Credit Manager; Draftsperson; General Manager; Health Services Manager; Management Trainee; Mechanical Engineer; Property and Real Estate Manager; Purchasing Agent/Manager; Quality Control Supervisor; Software Engineer; Structural Engineer; Underwriter/Assistant Underwriter. **Average salary range of placements:** $30,000 - $50,000. **Number of placements per year:** 50 - 99.

TRIAD CONSULTANTS
8101 East Prentice Avenue, Suite 610, Englewood CO 80111. 303/220-8516. **Fax:** 303/220-5265. **Contact:** Ronald Burgy, Partner. **E-mail address:** ronburgy@tci-colorado.com. **World Wide Web address:** http://www.tci-colorado.com. **Description:** An executive search firm specializing in the placement of software developers, DBAs, project managers, CIOs, and management consultants, as well as sales and technical support personnel. Company pays fee. **Specializes in the areas of:** Administration/MIS/EDP; Computer Science/Software; Sales and Marketing. **Positions commonly filled include:** Computer Programmer; MIS Specialist; Software Engineer; Systems Analyst. **Corporate headquarters location:** This Location. **Average salary range of placements:** More than $50,000. **Number of placements per year:** 100 - 199.

J.Q. TURNER & ASSOCIATES, INC.
200 South Wilcox Street, Unit 217, Castle Rock CO 80104-1913. 303/671-0800. **Fax:** 303/688-0188. **Contact:** Jim Turner, President. **E-mail address:** inquest@aol.com. **World Wide Web address:** http://www.members.aol.com/inquest/www/jqt.htm. **Description:** An executive search firm focusing on the nationwide recruitment of technical professionals for clients in biomedical, industrial, and scientific instrumentation; computers and computer peripherals; robotics; telecommunications; and pyrotechnics. Company pays fee. **Specializes in the areas of:** Computer Hardware/Software; Computer Operations; Computer Science/Software; Engineering; Industrial; Manufacturing; Technical and Scientific; Telecommunications. **Positions commonly filled include:** Biomedical Engineer; Chemical Engineer; Electrical/Electronics Engineer; Industrial Engineer; Industrial Production Manager; Manufacturing Engineer; Manufacturing Manager; Mechanical Engineer; Optical Engineer; Quality Assurance Engineer; Quality Control Supervisor; Software Engineer. **Corporate headquarters location:** This Location. **Average salary range of placements:** More than $50,000. **Number of placements per year:** 1 - 49.

WELZIG, LOWE & ASSOCIATES
761 West Birch Court, Louisville CO 80027. 303/666-4195. **Contact:** Frank Welzig, President. **Description:** A retainer and contingency executive recruitment firm serving a wide range of industries and disciplines. **Specializes in the areas of:** Computer Science/Software; Engineering; Oil and Gas; Technical and Scientific. **Positions commonly filled include:** Aerospace Engineer; Computer Programmer; Design Engineer; Electrical/Electronics Engineer; MIS Specialist; Software Engineer; Systems Analyst.

PERMANENT EMPLOYMENT AGENCIES OF COLORADO

ELEVENTH HOUR STAFFING SERVICES
8586 East Arapahoe Road, Englewood CO 80112. 303/220-8892. **Fax:** 303/770-8243. **Contact:** Mary A. Smith, Manager. **Description:** An employment agency. The company also provides temporary placement, career/outplacement counseling, and contract services. Company pays fee. **Specializes in the areas of:** Accounting/Auditing; Administration/MIS/EDP; Advertising; Banking; Computer Science/Software; Finance; Food Industry; General Management; Health/Medical; Insurance; Legal; Manufacturing; Retail; Sales and Marketing; Secretarial; Technical and Scientific; Transportation. **Positions commonly filled include:** Accountant/Auditor; Administrative Services Manager; Advertising Clerk; Branch Manager; Brokerage Clerk; Budget Analyst; Buyer; Computer Programmer; Cost Estimator; Counselor; Credit Manager; Customer Service Representative; Health Services Manager; Human Resources Specialist; Insurance Agent/Broker; Librarian; Management Analyst/Consultant; Management Trainee; Paralegal; Purchasing Agent/Manager; Quality Control Supervisor; Securities Sales Representative; Services Sales Representative; Software Engineer; Systems Analyst; Technical Writer/Editor; Telecommunications Manager; Typist/Word Processor. **Corporate headquarters location:** This Location. **Other U.S. locations:** Fullerton CA; Ontario CA; Tustin CA; Kansas City KS; Springfield MO. **Number of placements per year:** 500 - 999.

ENSCICON CORPORATION
One Broadway, #201, Denver CO 80203. 303/722-0979. **Fax:** 303/722-1030. **Contact:** Lori Kochevar, Recruiter. **Description:** An employment agency. Enscicon supplies permanent placement and consulting services in computer science and engineering fields. Company pays fee. **Specializes in the areas of:** Administration/MIS/EDP; Computer Science/Software; Engineering; Industrial; Manufacturing; Technical and Scientific. **Positions commonly filled include:** Aerospace Engineer; Architect; Biomedical Engineer; Chemical Engineer; Chemist; Civil Engineer; Computer Programmer; Design Engineer; Draftsperson; Electrical/Electronics Engineer; Environmental Engineer; Industrial Engineer; Mechanical Engineer; Mining Engineer; MIS Specialist; Nuclear Engineer; Petroleum Engineer; Science Technologist; Software Engineer; Structural Engineer; Technical Writer/Editor. **Benefits available to temporary workers:** 401(k) Plan; Dental Insurance; Medical Insurance; Paid Holidays; Paid Vacations. **Number of placements per year:** 200 - 499.

SOS STAFFING SERVICES
501 South Cherry Street, Suite 400, Denver CO 80222. 303/329-3776. **Contact:** Ms. Sandra Houck, Office Manager. **Description:** SOS Staffing Services is an employment agency. **Specializes in the areas of:** Accounting/Auditing; Banking; Clerical; Computer Hardware/Software; Engineering; Finance; Health/Medical; Insurance; Legal; Personnel/Labor Relations; Printing/Publishing; Real Estate; Sales and Marketing; Technical and Scientific. **Number of placements per year:** 200 - 499.

EXECUTIVE SEARCH FIRMS OF CONNECTICUT

CHARTER PERSONNEL SERVICES
P.O. Box 1070, 52 Federal Road, Danbury CT 06813-1070. 203/744-6440. **Fax:** 203/748-2122.
Contact: James T. Fornabaio, President. **Description:** An executive search firm which also
provides temporary placements. Company pays fee. **Specializes in the areas of:** Accounting/
Auditing; Administration/MIS/EDP; Biology; Computer Science/Software; Engineering; General
Management; Health/Medical; Industrial; Legal; Manufacturing; Secretarial; Technical and
Scientific. **Positions commonly filled include:** Accountant/Auditor; Administrative Services
Manager; Biological Scientist; Biomedical Engineer; Chemical Engineer; Chemist; Electrical/
Electronics Engineer; Industrial Engineer; Mechanical Engineer; Paralegal; Pharmacist; Physician;
Quality Control Supervisor; Software Engineer; Statistician; Typist/Word Processor; Veterinarian.
Benefits available to temporary workers: Medical Insurance. **Number of placements per year:**
200 - 499.

CHAVES & ASSOCIATES
Seven Whitney Street Extension, Westport CT 06880. 203/222-2222. **Fax:** 203/222-2223.
Contact: Victor Chaves, CPC, President. **Description:** An executive search firm which also
operates as an employment agency. Company pays fee. **Specializes in the areas of:** Computer
Science/Software. **Positions commonly filled include:** Computer Programmer; Internet Services
Manager; MIS Specialist; Multimedia Designer; Software Engineer; Systems Analyst. **Average
salary range of placements:** More than $50,000.

CORPORATE STAFFING SOLUTIONS
98 Mill Plain Road, Danbury CT 06811. 203/744-6020. **Contact:** Office Manager. **Description:**
An executive search firm which also operates as a temporary agency. Company pays fee.
Specializes in the areas of: Administration/MIS/EDP; Computer Science/Software; Electronics;
General Management; Manufacturing; Personnel/Labor Relations; Sales and Marketing; Technical
and Scientific. **Number of placements per year:** 1000+.

DATA TRENDS INC.
4133 Whitney Avenue, Hamden CT 06518-1432. 203/287-8485. **Fax:** 203/248-8138. **Contact:**
George Flohr, Manager. **E-mail address:** dti2@ix.netcom.com. **Description:** An executive search
firm operating on a contingency basis. Data Trends specializes in placing information systems
professionals in Connecticut, suburban New York, and western Massachusetts. Company pays fee.
Specializes in the areas of: Administration/MIS/EDP; Computer Science/Software. **Positions
commonly filled include:** Computer Programmer; Internet Services Manager; MIS Specialist;
Software Engineer; Systems Analyst. **Corporate headquarters location:** This Location. **Average
salary range of placements:** $30,000 - $50,000. **Number of placements per year:** 100 - 199.

DEVELOPMENT SYSTEMS, INC.
The Highland Building, 402 Highland Avenue, Suite 6, Cheshire CT 06410. 203/272-1117.
Contact: Arnold C. Bernstein, President. **Description:** An executive search firm. Company pays
fee. **Specializes in the areas of:** Administration/MIS/EDP; Computer Hardware/Software;
Personnel/Labor Relations; Retail. **Number of placements per year:** 1 - 49.

DIVERSITY RECRUITING SERVICES INC.
621 Farmington Avenue, Hartford CT 06105. 860/231-7786. **Fax:** 860/231-9730. **Contact:**
Douglas Dorsey, President. **Description:** An executive search firm. Company pays fee. **Specializes
in the areas of:** Accounting/Auditing; Computer Science/Software; Engineering; Finance; Food
Industry; General Management; Sales and Marketing. **Positions commonly filled include:**
Accountant/Auditor; Budget Analyst; Chemical Engineer; Civil Engineer; Financial Analyst;
Industrial Engineer; Management Trainee; Mechanical Engineer; MIS Specialist; Restaurant/Food

Service Manager; Securities Sales Representative; Software Engineer; Systems Analyst; Telecommunications Manager; Underwriter/Assistant Underwriter. **Average salary range of placements:** More than $50,000. **Number of placements per year:** 1 - 49.

EXECUTIVE REGISTER INC.
34 Mill Plain Road, Danbury CT 06811. 203/743-5542. **Contact:** J. Scott Williams, President. **Description:** An employment agency which also offers executive recruiting services. Company pays fee. **Specializes in the areas of:** Accounting/Auditing; Computer Science/Software; Engineering; Information Systems. **Positions commonly filled include:** Accountant/Auditor; Chemical Engineer; Computer Programmer; Electrical/Electronics Engineer; Financial Analyst; Mechanical Engineer; MIS Specialist; Multimedia Designer; Software Engineer; Systems Analyst. **Number of placements per year:** 50 - 99.

STANLEY HERZ & CO.
300 Broad Street, Suite 502, Stamford CT 06901. 203/358-9500. **Fax:** 203/357-9848. **Contact:** Stanley Herz, Personnel. **Description:** An executive search firm. Company pays fee. **Specializes in the areas of:** Computer Science/Software; Finance; Technical and Scientific. **Positions commonly filled include:** Financial Manager; MIS Manager. **Number of placements per year:** 1 - 49.

HUNTINGTON GROUP
6527 Main Street, Trumbull CT 06611. 203/261-1166. **Fax:** 203/452-9153. **Contact:** Marsha Nelson, Associate. **Description:** An executive search firm which also operates as an employment agency. Company pays fee. **Specializes in the areas of:** Computer Science/Software; Sales and Marketing. **Positions commonly filled include:** Account Representative; Quality Assurance Engineer; Sales Manager; Systems Analyst.

LUTZ ASSOCIATES
9 Stephen Street, Manchester CT 06040. 860/647-9338. **Fax:** 860/647-7918. **Contact:** Al Lutz, Owner. **Description:** An executive search firm. Founded in 1987. Company pays fee. **Specializes in the areas of:** Automation/Robotics; Computer Science/Software; Engineering; Industrial; Manufacturing; Technical and Scientific. **Positions commonly filled include:** Aerospace Engineer; Biomedical Engineer; Chemical Engineer; Chemist; Computer Programmer; Design Engineer; Designer; Engineer; MIS Specialist; Quality Control Supervisor; Science Technologist; Software Engineer; Systems Analyst; Technical Writer/Editor; Telecommunications Manager. **Average salary range of placements:** More than $50,000. **Number of placements per year:** 1 - 49.

MANAGEMENT RECRUITERS INTERNATIONAL
57 Danbury Road, Wilton CT 06897-4439. 203/834-1111. **Contact:** Robert Schmidt, Manager. **Description:** An executive search firm. **Specializes in the areas of:** Accounting/Auditing; Administration/MIS/EDP; Advertising; Architecture/Construction/Real Estate; Banking; Communications; Computer Science/Software; Construction; Electrical; Engineering; Finance; Food Industry; General Management; Health/Medical; Insurance; Legal; Manufacturing; MIS/EDP; Operations Management; Personnel/Labor Relations; Printing/Publishing; Procurement; Retail; Sales and Marketing; Technical and Scientific; Textiles; Transportation.

MANAGEMENT RECRUITERS INTERNATIONAL
1700 Post Road, Suite 5E, Fairfield CT 06430. 203/255-2299. **Contact:** Rush Oster, Manager. **Description:** An executive search firm. Company pays fee. **Specializes in the areas of:** Accounting/Auditing; Administration/MIS/EDP; Advertising; Architecture/Construction/Real Estate; Banking; Chemical; Communications; Computer Science/Software; Construction; Electrical; Engineering; Food Industry; General Management; Health/Medical; Insurance; Legal; Manufacturing; Operations Management; Personnel/Labor Relations; Pharmaceutical; Printing/Publishing; Retail; Sales and Marketing; Technical and Scientific; Textiles; Transportation. **Number of placements per year:** 50 - 99.

MAXWELL-MARCUS STAFFING CONSULTANTS
266 Broad Street, Milford CT 06460-0591. 203/874-5424. **Fax:** 203/874-5571. **Contact:** Dan Regan, President. **E-mail address:** maxwell@interserv.com. **World Wide Web address:** http://www.espan.com/spot/maxmarc/. **Description:** An executive search firm. Company pays fee.

Specializes in the areas of: Administration/MIS/EDP; Computer Science/Software; Engineering; Finance; General Management; International Executives; Manufacturing; Sales and Marketing. Positions commonly filled include: Biological Scientist; Chemical Engineer; Chemist; Electrical/ Electronics Engineer; General Manager; Management Analyst/Consultant; Management Trainee; Mechanical Engineer; Metallurgical Engineer; Operations/Production Manager; Software Engineer. Average salary range of placements: More than $50,000. Number of placements per year: 50 - 99.

CAROL A. McINNIS & ASSOCIATES

203 Broad Street, Suite 6, Milford CT 06460. 203/876-7110. Fax: 203/783-1230. Contact: Chris McInnis, Research Manager. Description: An executive search firm. Company pays fee. Specializes in the areas of: Biology; Computer Science/Software; Health/Medical; Medical Technology; Technical and Scientific. Positions commonly filled include: Biochemist; Biological Scientist; Clinical Lab Technician; Computer Programmer; Licensed Practical Nurse; MIS Specialist; Nuclear Medicine Technologist; Pharmacist; Physical Therapist; Registered Nurse; Respiratory Therapist; Software Engineer; Statistician; Systems Analyst; Technical Writer/Editor; Telecommunications Manager. Average salary range of placements: More than $50,000. Number of placements per year: 50 - 99.

McKNIGHT PERSONNEL SERVICES

P.O. Box 1172, Stamford CT 06904. 203/357-1891. Fax: 203/323-5612. Contact: Richard F. McKnight, President. Description: An employment agency and executive search firm affiliated with Manpower Inc. of Stamford. Company pays fee. Specializes in the areas of: Accounting/ Auditing; Clerical; Finance; Personnel/Labor Relations; Sales and Marketing; Secretarial; Software Engineering. Positions commonly filled include: Accountant/Auditor; Administrative Assistant; Bank Officer/Manager; Bookkeeper; Buyer; Clerk; EDP Specialist; Electrical/Electronics Engineer; Financial Analyst; General Manager; Legal Secretary; Marketing Specialist; Receptionist; Sales Representative; Secretary; Software Engineer; Typist/Word Processor. Average salary range of placements: More than $50,000. Number of placements per year: 50 - 99.

PRH MANAGEMENT, INC.

2777 Summer Street, Stamford CT 06905. 203/327-3900. Fax: 203/327-6324. Contact: Peter R. Hendelman, President. Description: An executive search firm. Company pays fee. Specializes in the areas of: Accounting/Auditing; Advertising; Bookkeeping; Computer Science/Software; Engineering; Finance; General Management; Sales and Marketing; Technical and Scientific. Positions commonly filled include: Accountant/Auditor; Branch Manager; Software Engineer; Telecommunications Manager. Average salary range of placements: More than $50,000.

EDWARD J. POSPESIL & COMPANY

44 Long Hill Road, Guilford CT 06437-1870. 203/458-6566. Fax: 203/458-6564. Contact: Ed Pospesil, Owner/Principal. E-mail address: ejpco@aol.com. Description: An executive search firm. Company pays fee. Specializes in the areas of: Administration/MIS/EDP; Computer Science/Software; Technical and Scientific. Positions commonly filled include: Computer Programmer; Internet Services Manager; Management Analyst/Consultant; MIS Specialist; Software Engineer; Systems Analyst; Technical Writer/Editor; Telecommunications Manager. Number of placements per year: 1 - 49.

SIGER & ASSOCIATES

966 Westover Road, Stamford CT 06902. 203/348-0976. Fax: 203/348-0698. Contact: Ray Milo, President. E-mail address: rmilo711@aol.com. Description: An executive search firm. Company pays fee. Specializes in the areas of: Banking; Computer Science/Software; Consulting; Finance; Health/Medical. Average salary range of placements: More than $50,000. Number of placements per year: 1 - 49.

SOURCE SERVICES CORPORATION

Enterprise Corporate Tower, One Corporate Drive, Suite 215, Shelton CT 06484-6209. 203/944-9001. Fax: 203/926-1414. Contact: Page Soltvedt, Product Sales Manager. Description: An executive search firm. Company pays fee. Specializes in the areas of: Accounting/Auditing; Banking; Computer Science/Software; Engineering; Finance. Positions commonly filled include:

Accountant/Auditor; Actuary; Bank Officer/Manager; Budget Analyst; Computer Programmer; Credit Manager; Design Engineer; Financial Analyst; Human Resources Manager; Internet Services Manager; MIS Specialist; Multimedia Designer; Operations/Production Manager; Software Engineer; Statistician; Systems Analyst; Technical Writer/Editor; Telecommunications Manager. **Average salary range of placements:** $30,000 - $50,000.

WALLACE ASSOCIATES
49 Leavenworth Street, Suite 200, P.O. Box 11294, Waterbury CT 06724. 203/575-1311. **Fax:** 203/879-2407. **Contact:** Greg Gordon, Principal. **Description:** An executive search firm. Company pays fee. **Specializes in the areas of:** Computer Hardware/Software; Engineering; General Management; Health/Medical; Manufacturing; Packaging; Physician Executive; Technical and Scientific. **Positions commonly filled include:** Biological Scientist; Biomedical Engineer; Chemical Engineer; Chemist; Computer Programmer; EDP Specialist; Electrical/Electronics Engineer; Industrial Designer; Industrial Engineer; Manufacturing Engineer; Mechanical Engineer; Metallurgical Engineer; MIS Specialist; Operations Research Analyst; Quality Control Supervisor; Software Engineer. **Number of placements per year:** 1 - 49.

PERMANENT EMPLOYMENT AGENCIES OF CONNECTICUT

A&A PERSONNEL SERVICES
91 Bainton Road, West Hartford CT 06117-2816. 860/549-5262. **Contact:** Lewis Schweitzer, Manager. **Description:** An employment agency. Company pays fee. **Specializes in the areas of:** Accounting/Auditing; Banking; Computer Science/Software; Engineering; Finance; Insurance; Manufacturing; MIS/EDP; Sales and Marketing; Technical and Scientific. **Positions commonly filled include:** Accountant/Auditor; Actuary; Aerospace Engineer; Attorney; Bank Officer/Manager; Bookkeeper; Civil Engineer; Claim Representative; Computer Programmer; Credit Manager; Customer Service Representative; EDP Specialist; Electrical/Electronics Engineer; Financial Analyst; Human Resources Manager; Industrial Designer; Industrial Engineer; Mechanical Engineer; Purchasing Agent/Manager; Sales Representative; Statistician; Systems Analyst; Technician; Underwriter/Assistant Underwriter. **Number of placements per year:** 50 - 99.

BAILEY ASSOCIATES
1208 Main Street, Branford CT 06405. 203/488-2504. **Contact:** Mary Briem, Senior Technical Consultant. **Description:** An employment agency. Company pays fee. **Specializes in the areas of:** Computer Science/Software. **Positions commonly filled include:** Computer Programmer; Customer Service Representative; Electrical/Electronics Engineer; MIS Specialist; Multimedia Designer; Software Engineer; Systems Analyst. **Average salary range of placements:** More than $50,000. **Number of placements per year:** 200 - 499.

DATA PROS
340 Broad Street, Suite 201, Windsor CT 06095-3030. 860/688-0020. **Fax:** 860/683-8903. **Contact:** Len Collyer, President. **Description:** An employment agency. Company pays fee. **Specializes in the areas of:** Computer Science/Software; Engineering; Insurance. **Positions commonly filled include:** Computer Programmer; Electrician; Emergency Medical Technician; Engineer; Systems Analyst. **Number of placements per year:** 100 - 199.

FAIRFIELD EMPLOYMENT AGENCY
410 Asylum Street, Hartford CT 06103. 203/293-2722. **Contact:** Rubye Banks, Assistant Director. **Description:** An employment agency. Company pays fee. **Specializes in the areas of:** Accounting/Auditing; Administration/MIS/EDP; Architecture/Construction/Real Estate; Banking;

Computer Science/Software; Engineering; Finance; General Management; Industrial; Manufacturing; Personnel/Labor Relations; Retail; Sales and Marketing; Secretarial. **Positions commonly filled include:** Accountant/Auditor; Actuary; Administrative Services Manager; Advertising Clerk; Architect; Blue-Collar Worker Supervisor; Branch Manager; Budget Analyst; Chemical Engineer; Civil Engineer; Claim Representative; Computer Programmer; Construction and Building Inspector; Cost Estimator; Credit Manager; Customer Service Representative; Design Engineer; Draftsperson; EEG Technologist; Electrical/Electronics Engineer; Environmental Engineer; General Manager; Health Services Manager; Human Resources Specialist; Industrial Engineer; Librarian; Management Trainee; Mechanical Engineer; MIS Specialist; Operations/ Production Manager; Paralegal; Quality Control Supervisor; Radio/TV Announcer/Broadcaster; Registered Nurse; Services Sales Representative; Software Engineer; Structural Engineer; Systems Analyst; Technical Writer/Editor; Typist/Word Processor. **Corporate headquarters location:** This Location. **Average salary range of placements:** More than $50,000. **Number of placements per year:** 50 - 99.

HALLMARK TOTALTECH INC.
1090 Elm Street, Rocky Hill CT 06067. 860/529-7500. **Toll-free phone:** 800/876-4255. **Fax:** 860/529-9800. **Contact:** President. **Description:** An employment agency. Company pays fee. **Specializes in the areas of:** Administration/MIS/EDP; Computer Science/Software; Engineering; Personnel/Labor Relations; Technical and Scientific. **Positions commonly filled include:** Aircraft Mechanic/Engine Specialist; Blue-Collar Worker Supervisor; Buyer; Clinical Lab Technician; Computer Programmer; Construction and Building Inspector; Cost Estimator; Customer Service Representative; Design Engineer; Designer; Draftsperson; Electrical/Electronics Engineer; Environmental Engineer; Geologist/Geophysicist; Industrial Engineer; Industrial Production Manager; Internet Services Manager; Landscape Architect; Management Analyst/Consultant; Mechanical Engineer; MIS Specialist; Multimedia Designer; Operations/Production Manager; Property and Real Estate Manager; Quality Control Supervisor; Science Technologist; Services Sales Representative; Software Engineer; Structural Engineer; Systems Analyst; Technical Support Representative; Telecommunications Manager; Typist/Word Processor. **Corporate headquarters location:** This Location. **Average salary range of placements:** $30,000 - $50,000. **Number of placements per year:** 200 - 499.

J.G. HOOD ASSOCIATES
599 Riverside Avenue, Westport CT 06880. 203/226-1126. **Contact:** Joyce Hood, Owner/President. **Description:** An employment agency. Company pays fee. **Specializes in the areas of:** Administration/MIS/EDP; Computer Science/Software; Engineering; Finance; Industrial; Manufacturing; Personnel/Labor Relations; Sales and Marketing. **Positions commonly filled include:** Accountant/Auditor; Aerospace Engineer; Biomedical Engineer; Chemical Engineer; Chemist; Computer Programmer; Cost Estimator; Designer; Electrical/Electronics Engineer; Human Resources Manager; Industrial Engineer; Industrial Production Manager; Quality Control Supervisor; Science Technologist; Software Engineer; Structural Engineer; Systems Analyst. **Number of placements per year:** 1 - 49.

W.R. LAWRY, INC.
P.O. Box 832, Simsbury CT 06070. 860/651-0281. **Fax:** 860/651-8324. **Contact:** Bill Lawry, Owner. **Description:** An employment agency. **NOTE:** The agency is located at 6 Wilcox Street, Simsbury CT. Company pays fee. **Specializes in the areas of:** Computer Science/Software; Engineering; General Management; Technical and Scientific. **Positions commonly filled include:** Biological Scientist; Engineer; General Manager. **Number of placements per year:** 1 - 49.

R.J. PASCALE & COMPANY
500 Summer Street, Stamford CT 06901. 203/358-8155. **Fax:** 203/969-3990. **Contact:** Ron Pascale, President. **Description:** An employment agency. Company pays fee. **Specializes in the areas of:** Accounting/Auditing; Administration/MIS/EDP; Computer Science/Software; Finance. **Positions commonly filled include:** Accountant/Auditor; Budget Analyst; Financial Analyst; Management Analyst/Consultant; Software Engineer; Systems Analyst. **Number of placements per year:** 50 - 99.

Q.S.I.
P.O Box 5606, Hamden CT 06578. 203/287-8900. **Fax:** 203/248-0911. **Contact:** Kim Feuer, Technical Recruiter. **Description:** Q.S.I. is an employment agency specializing in MIS technical staffing. Company pays fee. **Specializes in the areas of:** Accounting/Auditing; Computer Science/Software; Finance; Legal. **Positions commonly filled include:** Accountant/Auditor; Attorney; Computer Programmer; Financial Analyst; MIS Specialist; Systems Analyst; Telecommunications Manager. **Average salary range of placements:** More than $50,000. **Number of placements per year:** 1 - 49.

RESOURCE ASSOCIATES
730 Hopmeadow Street, Simsbury CT 06070. 860/651-4918. **Fax:** 860/651-3137. **Contact:** Eric Grossman, President. **Description:** Resource Associates is an employment agency. **Specializes in the areas of:** Accounting/Auditing; Computer Science/Software; Finance. **Positions commonly filled include:** Accountant/Auditor; Computer Programmer; Financial Analyst; Software Engineer; Systems Analyst.

SUPER SYSTEMS, INC.
345 North Main Street, Suite 321, West Hartford CT 06117-2508. 860/523-4246. **Fax:** 860/233-6943. **Contact:** Mary Ann Salas, CPC, President. **Description:** An employment agency. **Specializes in the areas of:** Computer Science/Software; Data Processing; Engineering; Health/Medical; Technical and Scientific. **Positions commonly filled include:** Aerospace Engineer; Agricultural Engineer; Biomedical Engineer; Chemical Engineer; Civil Engineer; Computer Programmer; Electrical/Electronics Engineer; Industrial Engineer; Licensed Practical Nurse; Management Analyst/Consultant; Materials Engineer; Mechanical Engineer; Mining Engineer; Nuclear Engineer; Occupational Therapist; Petroleum Engineer; Physical Therapist; Respiratory Therapist; Software Engineer; Stationary Engineer; Structural Engineer; Systems Analyst. **Number of placements per year:** 50 - 99.

TEMTEK TECHNICAL PERSONNEL
266 Broad Street, Milford CT 06460. 203/877-8634. **Fax:** 203/877-6394. **Contact:** Bill Jacowleff, Recruiter. **E-mail address:** temtek@aol.com. **Description:** Temtek Technical Personnel is an employment agency that also offers executive search services. Company pays fee. **Specializes in the areas of:** Biology; Computer Science/Software; Engineering; Food Industry; Manufacturing; Technical and Scientific. **Positions commonly filled include:** Architect; Biological Scientist; Broadcast Technician; Chemist; Computer Programmer; Designer; Draftsperson; Electrician; Engineer; Food Scientist/Technologist; Geologist/Geophysicist; Library Technician; Quality Control Supervisor; Science Technologist; Systems Analyst; Technical Writer/Editor. **Average salary range of placements:** More than $50,000.

WORKFORCE ONE
235 Interstate Lane, Waterbury CT 06705. 203/759-2180. **Fax:** 203/759-2182. **Contact:** Donald Rulli, Operations Manager. **Description:** Workforce One is an employment agency. **Specializes in the areas of:** Accounting/Auditing; Computer Science/Software; Electronics; Finance; Manufacturing; Personnel/Labor Relations; Secretarial; Technical and Scientific. **Positions commonly filled include:** Accountant/Auditor; Administrative Services Manager; Aerospace Engineer; Aircraft Mechanic/Engine Specialist; Bank Officer/Manager; Blue-Collar Worker Supervisor; Branch Manager; Budget Analyst; Buyer; Chemical Engineer; Chemist; Civil Engineer; Clerical Supervisor; Clinical Lab Technician; Computer Programmer; Construction and Building Inspector; Construction Contractor; Cost Estimator; Credit Manager; Customer Service Representative; Design Engineer; Designer; Draftsperson; Electrical/Electronics Engineer; Environmental Engineer; Financial Analyst; General Manager; Human Resources Specialist; Industrial Engineer; Industrial Production Manager; Management Analyst/Consultant; Mechanical Engineer; Medical Records Technician; MIS Specialist; Purchasing Agent/Manager; Quality Control Supervisor; Software Engineer; Stationary Engineer; Structural Engineer; Systems Analyst; Technical Writer/Editor; Typist/Word Processor. **Average salary range of placements:** $30,000 - $50,000. **Number of placements per year:** 1 - 49.

EXECUTIVE SEARCH FIRMS OF DELAWARE

J.B. GRONER EXECUTIVE SEARCH INC.
P.O. Box 101, Claymont DE 19703. 302/792-9228. Fax: 302/792-9446. **Contact:** James Groner, President. **Description:** A small professional placement firm primarily handling permanent placements as well as some contract placement in the computer and engineering fields. J.B. Groner Executive Search Inc. operates on retainer and contingency bases. Company pays fee. **Specializes in the areas of:** Accounting/Auditing; Administration/MIS/EDP; Banking; Computer Science/ Software; Economics; Engineering; Finance; General Management; Health/Medical; Industrial; Insurance; Legal; Manufacturing; Nonprofit; Personnel/Labor Relations; Sales and Marketing; Technical and Scientific. **Positions commonly filled include:** Accountant/Auditor; Administrative Services Manager; Aerospace Engineer; Architect; Attorney; Biochemist; Biological Scientist; Biomedical Engineer; Branch Manager; Budget Analyst; Buyer; Chemical Engineer; Chemist; Civil Engineer; Claim Representative; Computer Programmer; Construction Manager; Counselor; Credit Manager; Customer Service Representative; Design Engineer; Economist; Editor; Education Administrator; Electrical/Electronics Engineer; Environmental Engineer; Financial Analyst; General Manager; Geologist/Geophysicist; Health Services Manager; Hotel Manager; Human Resources Manager; Industrial Engineer; Industrial Production Manager; Insurance Agent/Broker; Internet Services Manager; Landscape Architect; Librarian; Licensed Practical Nurse; Management Analyst/Consultant; Management Trainee; Manufacturer's/Wholesaler's Sales Rep.; Market Research Analyst; Mathematician; Mechanical Engineer; Medical Records Technician; Metallurgical Engineer; Mining Engineer; MIS Specialist; Multimedia Designer; Nuclear Engineer; Nuclear Medicine Technologist; Operations/Production Manager; Petroleum Engineer; Pharmacist; Physician; Physicist/Astronomer; Property and Real Estate Manager; Psychologist; Public Relations Specialist; Purchasing Agent/Manager; Quality Control Supervisor; Restaurant/Food Service Manager; Science Technologist; Securities Sales Representative; Services Sales Representative; Social Worker; Sociologist; Software Engineer; Speech-Language Pathologist; Stationary Engineer; Statistician; Strategic Relations Manager; Structural Engineer; Surgical Technician; Surveyor; Systems Analyst; Teacher/Professor; Technical Writer/Editor; Telecommunications Manager; Underwriter/Assistant Underwriter; Urban/Regional Planner. **Corporate headquarters location:** This Location. **Average salary range of placements:** $30,000 - $50,000. **Number of placements per year:** 1 - 49.

PERMANENT EMPLOYMENT AGENCIES OF DELAWARE

SNELLING PERSONNEL SERVICES
3617-A Silverside Road, Wilmington DE 19810. 302/478-6060. **Contact:** General Manager. **Description:** Snelling Personnel Services is an employment agency. Company pays fee. **Specializes in the areas of:** Accounting/Auditing; Banking; Clerical; Computer Hardware/Software; Engineering; Fashion; Finance; Food Industry; Personnel/Labor Relations; Sales and Marketing; Technical and Scientific; Transportation. **Positions commonly filled include:** Accountant/Auditor; Administrative Assistant; Administrative Worker/Clerk; Bank Officer/Manager; Biomedical Engineer; Bookkeeper; Buyer; Ceramics Engineer; Civil Engineer; Claim Representative; Clerk; Computer Operator; Computer Programmer; Credit Manager; Customer Service Representative; Data Entry Clerk; Editor; EDP Specialist; Electrical/Electronics Engineer; Financial Analyst; Food Scientist/Technologist; General Manager; Hotel Manager; Human Resources Manager; Industrial Engineer; Insurance Agent/Broker; Legal Secretary; Marketing Specialist; Mechanical Engineer; Medical Secretary; Metallurgical Engineer; Petroleum Engineer; Purchasing Agent/Manager; Receptionist; Reporter; Sales Representative; Secretary;

Stenographer; Systems Analyst; Typist/Word Processor. **Number of placements per year:** 100 - 199.

PERMANENT EMPLOYMENT AGENCIES OF WASHINGTON DC

C ASSOCIATES
P.O. Box 15420, 1619 G Street SE, Washington DC 20003-3132. 202/544-0821. **Fax:** 202/547-8357. **Contact:** John Capozzi, Jr., President. **Description:** An employment agency. **Specializes in the areas of:** Computer Science/Software.

PORTFOLIO
1730 K Street NW, Suite 1350, Washington DC 20006. 202/293-5700. **Contact:** Human Resources. **Description:** A temporary and permanent employment agency. **Specializes in the areas of:** Art/Design; Computer Graphics; Multimedia; Technical Writing.

EXECUTIVE SEARCH FIRMS OF FLORIDA

ACTIVE PROFESSIONALS
2572 Atlantic Boulevard, Suite One, Jacksonville FL 32207. 904/396-7148. **Fax:** 904/396-6321. **Contact:** Faye Rustin, Owner. **Description:** An executive search firm which also operates as an employment agency. Company pays fee. **Specializes in the areas of:** Accounting/Auditing; Architecture/Construction/Real Estate; Computer Hardware/Software; Engineering; Finance; General Management; Industrial; Insurance; Manufacturing; Personnel/Labor Relations; Sales and Marketing; Secretarial. **Number of placements per year:** 100 - 199.

ASH & ASSOCIATES EXECUTIVE SEARCH
4778 West Commercial Boulevard, Fort Lauderdale FL 33319. 305/735-3355. **Fax:** 305/735-0069. **Contact:** Janis Ash, President. **Description:** An executive search firm. Company pays fee. **Specializes in the areas of:** Accounting/Auditing; Administration/MIS/EDP; Computer Science/Software; Engineering; Insurance; Technical and Scientific. **Positions commonly filled include:** Accountant/Auditor; Attorney; Biochemist; Biological Scientist; Biomedical Engineer; Chemist; Computer Programmer; Environmental Engineer; Human Resources Manager; Insurance Agent/Broker; Internet Services Manager; Mechanical Engineer; Metallurgical Engineer; MIS Specialist; Physical Therapist; Securities Sales Representative; Software Engineer; Structural Engineer; Systems Analyst; Technical Writer/Editor; Telecommunications Manager; Transportation/Traffic Specialist; Underwriter/Assistant Underwriter. **Corporate headquarters location:** This Location. **Average salary range of placements:** More than $50,000. **Number of placements per year:** 50 - 99.

BRYAN & ASSOCIATES
WORKNET, ETC.
1279 Kingsley Avenue, Suite 116, Orange Park FL 32073-4604. 904/264-7549. **Fax:** 904/264-7912. **Contact:** Bryan Cornwall, President. **Description:** An executive search firm. **Specializes in the areas of:** Accounting/Auditing; Administration/MIS/EDP; Advertising; Banking; Broadcasting; Computer Science/Software; Economics; Finance; Food Industry; General Management; Health/Medical; Industrial; Insurance; Legal; Manufacturing; Personnel/Labor Relations; Printing/Publishing; Retail; Sales and Marketing; Secretarial; Technical and Scientific; Transportation. **Positions commonly filled include:** Accountant/Auditor; Adjuster; Administrative

Services Manager; Advertising Clerk; Aircraft Mechanic/Engine Specialist; Architect; Attorney; Bank Officer/Manager; Biochemist; Blue-Collar Worker Supervisor; Branch Manager; Broadcast Technician; Brokerage Clerk; Budget Analyst; Buyer; Chemist; Claim Representative; Clerical Supervisor; Clinical Lab Technician; Computer Programmer; Construction and Building Inspector; Construction Contractor; Cost Estimator; Counselor; Credit Manager; Customer Service Representative; Designer; Draftsperson; Economist; Editor; Education Administrator; Electrician; Environmental Engineer; Financial Analyst; General Manager; Health Services Manager; Hotel Manager; Human Resources Manager; Human Service Worker; Industrial Engineer; Insurance Agent/Broker; Librarian; Management Analyst/Consultant; Management Trainee; Manufacturer's/ Wholesaler's Sales Rep.; Market Research Analyst; Medical Records Technician; Multimedia Designer; Occupational Therapist; Operations/Production Manager; Paralegal; Public Relations Specialist; Quality Control Supervisor; Radio/TV Announcer/Broadcaster; Recreational Therapist; Reporter; Restaurant/Food Service Manager; Securities Sales Representative; Services Sales Representative; Software Engineer; Stationary Engineer; Strategic Relations Manager; Systems Analyst; Technical Writer/Editor; Telecommunications Manager; Transportation/Traffic Specialist; Typist/Word Processor; Underwriter/Assistant Underwriter; Video Production Coordinator. **Benefits available to temporary workers:** Medical Insurance; Paid Holidays; Paid Vacation. **Corporate headquarters location:** This Location. **Average salary range of placements:** $30,000 - $50,000. **Number of placements per year:** 200 - 499.

COLLI ASSOCIATES OF TAMPA

P.O. Box 2865, Tampa FL 33601. 813/681-2145. **Fax:** 813/661-5217. **Contact:** Carolyn Colli, Manager. **Description:** An executive search firm specializing in all fields of manufacturing and engineering. Company pays fee. **Specializes in the areas of:** Computer Science/Software; Engineering; Industrial; Manufacturing. **Positions commonly filled include:** Aerospace Engineer; Biomedical Engineer; Buyer; Chemical Engineer; Designer; Electrical/Electronics Engineer; Human Resources Manager; Industrial Engineer; Industrial Production Manager; Mechanical Engineer; Metallurgical Engineer; Nuclear Engineer; Operations/Production Manager; Purchasing Agent/Manager; Quality Control Supervisor; Software Engineer; Stationary Engineer; Structural Engineer. **Corporate headquarters location:** This Location. **Average salary range of placements:** $30,000 - $50,000. **Number of placements per year:** 1 - 49.

CRITERION EXECUTIVE SEARCH

5420 Bay Center Drive, Suite 101, Tampa FL 33609-3469. 813/286-2000. **Fax:** 813/287-1660. **Contact:** Richard James, President. **Description:** An executive search firm. Company pays fee. **Specializes in the areas of:** Administration/MIS/EDP; Computer Science/Software; Engineering; Insurance; Legal; Manufacturing; Technical and Scientific. **Positions commonly filled include:** Accountant/Auditor; Adjuster; Administrative Services Manager; Aerospace Engineer; Agricultural Engineer; Biological Scientist; Biomedical Engineer; Branch Manager; Chemical Engineer; Chemist; Civil Engineer; Electrical/Electronics Engineer; Human Resources Manager; Industrial Engineer; Industrial Production Manager; Insurance Agent/Broker; Manufacturer's/Wholesaler's Sales Rep.; Mechanical Engineer; Metallurgical Engineer; Mining Engineer; Nuclear Engineer; Nuclear Medicine Technologist; Operations/Production Manager; Petroleum Engineer; Purchasing Agent/Manager; Radiologic Technologist; Software Engineer; Stationary Engineer; Structural Engineer; Underwriter/Assistant Underwriter. **Number of placements per year:** 100 - 199.

GALLIN ASSOCIATES

2454 North McMullen Booth Road, Clearwater FL 34619. 813/724-8303. **Fax:** 813/724-8503. **Contact:** Office Manager. **Description:** An executive search firm. **Specializes in the areas of:** Chemical; Computer Science/Software; Engineering; MIS/EDP; Personnel/Labor Relations. **Positions commonly filled include:** Chemical Engineer; Chemist; Computer Programmer; Design Engineer; Environmental Engineer; Human Resources Manager; Internet Services Manager; MIS Specialist; Systems Analyst. **Average salary range of placements:** More than $50,000. **Number of placements per year:** 1 - 49.

KEYS EMPLOYMENT AGENCY

P.O. Box 1973, Big Pine Key FL 33043. 305/872-9692. **Fax:** 305/872-9692. **Contact:** Donna Glenn, President. **E-mail address:** dg305@aol.com. **Description:** An executive search firm. Company pays fee. **Specializes in the areas of:** Computer Science/Software. **Positions commonly**

filled include: Accountant/Auditor; Adjuster; Administrative Services Manager; Advertising Clerk; Architect; Attorney; Bank Officer/Manager; Biological Scientist; Branch Manager; Broadcast Technician; Budget Analyst; Civil Engineer; Clerical Supervisor; Clinical Lab Technician; Computer Programmer; Construction and Building Inspector; Construction Contractor; Cost Estimator; Counselor; Credit Manager; Customer Service Representative; Dental Assistant/Dental Hygienist; Dental Lab Technician; Designer; Dietician/Nutritionist; Draftsperson; Education Administrator; Electrical/Electronics Engineer; Electrician; Emergency Medical Technician; Financial Analyst; General Manager; Health Services Manager; Human Resources Manager; Human Service Worker; Landscape Architect; Librarian; Library Technician; Licensed Practical Nurse; Management Analyst/Consultant; Management Trainee; Mathematician; Medical Records Technician; Occupational Therapist; Operations/Production Manager; Paralegal; Pharmacist; Physical Therapist; Preschool Worker; Property and Real Estate Manager; Psychologist; Public Relations Specialist; Purchasing Agent/Manager; Quality Control Supervisor; Radio/TV Announcer/Broadcaster; Registered Nurse; Reporter; Respiratory Therapist; Restaurant/ Food Service Manager; Securities Sales Representative; Services Sales Representative; Social Worker; Sociologist; Structural Engineer; Surveyor; Systems Analyst; Teacher/Professor; Technical Writer/Editor; Transportation/Traffic Specialist; Travel Agent; Urban/Regional Planner. **Average salary range of placements:** More than $50,000. **Number of placements per year:** 1 - 49.

KOERNER GROUP
9900 West Sample Road, Suite 300, Coral Springs FL 33065. 954/755-6676. **Contact:** Office Manager. **E-mail address:** dskgrp@aol.com. **Description:** Koerner Group is an executive search firm. Company pays fee. **Specializes in the areas of:** Accounting/Auditing; Banking; Computer Science/Software; Engineering; Finance; General Management; Industrial; Manufacturing; Personnel/Labor Relations. **Positions commonly filled include:** Accountant/Auditor; Attorney; Biological Scientist; Biomedical Engineer; Budget Analyst; Chemical Engineer; Chemist; Civil Engineer; Computer Programmer; Credit Manager; Electrical/Electronics Engineer; Financial Analyst; Hotel Manager; Human Resources Manager; Industrial Engineer; Industrial Production Manager; Mechanical Engineer; Public Relations Specialist; Purchasing Agent/Manager; Quality Control Supervisor; Registered Nurse; Software Engineer; Statistician; Systems Analyst; Technical Writer/Editor. **Number of placements per year:** 50 - 99.

R.H. LARSEN & ASSOCIATES, INC.
1401 East Broward Boulevard, Suite 101, Fort Lauderdale FL 33301-2118. 954/763-9000. **Fax:** 954/463-9318. **Contact:** Human Resources. **E-mail address:** emcdon5310@aol.com. **Description:** R.H. Larsen & Associates is a retained executive search firm. **Specializes in the areas of:** Accounting/Auditing; Administration/MIS/EDP; Architecture/Construction/Real Estate; Banking; Computer Science/Software; Engineering; Finance; Food Industry; General Management; Health/Medical; Manufacturing; Personnel/Labor Relations; Printing/Publishing; Sales and Marketing. **Average salary range of placements:** More than $50,000. **Number of placements per year:** 1 - 49.

MANAGEMENT RECRUITERS OF INDIALANTIC
134 5th Avenue, Suite 208, Indialantic FL 32903-3170. 407/951-7644. **Fax:** 407/951-4235. **Contact:** Lawrence Cinco, General Manager. **Description:** Management Recruiters of Indialantic is a contingency search firm. Company pays fee. **Specializes in the areas of:** Computer Science/Software; Engineering; Manufacturing. **Positions commonly filled include:** Agricultural Engineer; Computer Programmer; Design Engineer; Designer; Electrical/Electronics Engineer; Industrial Engineer; Mechanical Engineer; MIS Specialist; Systems Analyst. **Other U.S. locations:** Nationwide. **Average salary range of placements:** More than $50,000. **Number of placements per year:** 50 - 99.

MANAGEMENT RECRUITERS OF MIAMI
815 NW 57th Avenue, Suite 110, Miami FL 33126. 305/264-4212. **Fax:** 305/264-4251. **Contact:** Del Diaz, President. **Description:** Management Recruiters of Miami is an executive search firm. Company pays fee. **Specializes in the areas of:** Accounting/Auditing; Administration/MIS/EDP; Architecture/ Construction/Real Estate; Computer Science/Software; Engineering; Finance; Health/Medical; Industrial; Manufacturing; Personnel/Labor Relations; Sales and Marketing.

Positions commonly filled include: Accountant/Auditor; Administrative Services Manager; Aerospace Engineer; Agricultural Engineer; Biological Scientist; Biomedical Engineer; Branch Manager; Buyer; Ceramics Engineer; Chemical Engineer; Chemist; Civil Engineer; Clinical Lab Technician; Computer Programmer; Construction Contractor; Cost Estimator; Customer Service Representative; Designer; EEG Technologist; EKG Technician; Electrical/Electronics Engineer; Financial Analyst; General Manager; Geologist/Geophysicist; Health Services Manager; Human Resources Manager; Industrial Engineer; Industrial Production Manager; Licensed Practical Nurse; Manufacturer's/ Wholesaler's Sales Rep.; Materials Engineer; Mechanical Engineer; Medical Records Technician; Metallurgical Engineer; Mining Engineer; Nuclear Engineer; Nuclear Medicine Technologist; Occupational Therapist; Petroleum Engineer; Physical Therapist; Property and Real Estate Manager; Public Relations Specialist; Purchasing Agent/Manager; Quality Control Supervisor; Radiologic Technologist; Recreational Therapist; Registered Nurse; Respiratory Therapist; Science Technologist; Services Sales Representative; Software Engineer; Stationary Engineer; Structural Engineer; Surveyor; Systems Analyst; Technical Writer/Editor; Wholesale and Retail Buyer. **Number of placements per year:** 50 - 99.

MANAGEMENT RECRUITERS OF ST. PETERSBURG
9500 Koger Boulevard, Suite 203, St. Petersburg FL 33702. 813/577-2116. **Contact:** Manager. **Description:** Management Recruiters of St. Petersburg is an executive search firm. **Specializes in the areas of:** Accounting/Auditing; Administration/MIS/EDP; Advertising; Architecture/Construction/Real Estate; Banking; Chemical; Communications; Computer Hardware/Software; Construction; Electrical; Engineering; Finance; Food Industry; General Management; Health/Medical; Industrial; Insurance; Legal; Manufacturing; Operations Management; Personnel/Labor Relations; Pharmaceutical; Printing/Publishing; Procurement; Sales and Marketing; Technical and Scientific; Textiles; Transportation.

MANAGEMENT RECRUITERS OF TALLAHASSEE
1406 Hays Street, Suite 7, Tallahassee FL 32301. 904/656-8444. **Contact:** Kitte Carter, Manager. **Description:** An executive search firm. **Specializes in the areas of:** Accounting/Auditing; Administration/MIS/EDP; Advertising; Architecture/Construction/Real Estate; Banking; Chemical; Communications; Computer Hardware/Software; Construction; Design; Electrical; Engineering; Finance; Food Industry; General Management; Health/Medical; Industrial; Insurance; Legal; Manufacturing; Operations Management; Personnel/Labor Relations; Pharmaceutical; Printing/ Publishing; Procurement; Retail; Sales and Marketing; Technical and Scientific; Textiles; Transportation.

MANAGEMENT RECRUITERS OF TAMPA
500 North Westshore Boulevard, Suite 540, Tampa FL 33609. 813/281-2353. **Contact:** Ron Cottick, Manager. **Description:** An executive search firm. **Specializes in the areas of:** Accounting/Auditing; Administration/MIS/EDP; Advertising; Architecture/Construction/Real Estate; Banking; Chemical; Communications; Computer Hardware/Software; Construction; Design; Electrical; Engineering; Finance; Food Industry; General Management; Health/Medical; Industrial; Insurance; Legal; Manufacturing; Operations Management; Personnel/Labor Relations; Printing/Publishing; Procurement; Retail; Sales and Marketing; Technical and Scientific; Textiles; Transportation.

MANAGEMENT SEARCH INTERNATIONAL, INC.
2170 West State Road 434, Suite 454, Longwood FL 32779. 407/865-7979. **Fax:** 407/865-7670. **Contact:** John Edward Clark, Managing Partner. **Description:** An executive search firm. Company pays fee. **Specializes in the areas of:** Administration/MIS/EDP; Art/Design; Computer Science/Software; Engineering; Insurance; Printing/Publishing; Sales and Marketing; Telecommunications. **Positions commonly filled include:** Computer Programmer; Electrical/Electronics Engineer; General Manager; Health Services Manager; Industrial Engineer; Insurance Agent/Broker; Management Analyst/Consultant; Mechanical Engineer; Metallurgical Engineer; Mining Engineer; MIS Manager; Operations/Production Manager; Petroleum Engineer; Software Engineer; Telecommunications Manager; Underwriter/Assistant Underwriter. **Average salary range of placements:** $30,000 - $50,000. **Number of placements per year:** 1 - 49.

MANKUTA GALLAGHER & ASSOCIATES INC.

8333 West McNab Road, Tamarac FL 33321. 954/720-7357. **Toll-free phone:** 800/797-4276. **Fax:** 954/720-5813. **Contact:** Managing Partner. **E-mail address:** stanley@satelnet.org. **Description:** An executive search firm. Company pays fee. **Specializes in the areas of:** Administration/MIS/EDP; Biology; Computer Science/Software; Engineering; Manufacturing. **Positions commonly filled include:** Biochemist; Biological Scientist; Biomedical Engineer; Chemical Engineer; Chemist; Civil Engineer; Computer Programmer; Design Engineer; Electrical/Electronics Engineer; Industrial Engineer; Internet Services Manager; Management Analyst/Consultant; Mechanical Engineer; MIS Specialist; Occupational Therapist; Physical Therapist; Physician; Quality Control Supervisor; Registered Nurse; Science Technologist; Software Engineer; Speech-Language Pathologist; Systems Analyst. **Average salary range of placements:** $50,000. **Number of placements per year:** 50 - 99.

NORRELL TECHNICAL SERVICES

1801 Sarno Road, Suite 1, Melbourne FL 32951. 407/259-8619. **Toll-free phone:** 800/689-8367. **Fax:** 407/255-1930. **Contact:** David Keldsen, Division Manager. **E-mail address:** dave@norrelltech.com. **World Wide Web address:** http://www.norrelltech.com. **Description:** Norrell Technical Services is an executive search firm that offers short-term, long-term, managed, and contract-to-hire staffing opportunities. Company pays fee. **Specializes in the areas of:** Computer Science/Software; Engineering; Industrial; Manufacturing; Personnel/Labor Relations; Technical and Scientific. **Positions commonly filled include:** Aerospace Engineer; Buyer; Chemical Engineer; Chemist; Civil Engineer; Computer Programmer; Cost Estimator; Design Engineer; Designer; Draftsperson; Electrical/Electronics Engineer; Food Scientist/Technologist; Human Resources Manager; Industrial Engineer; Industrial Production Manager; Internet Services Manager; Mechanical Engineer; MIS Specialist; Multimedia Designer; Software Engineer; Systems Analyst; Technical Writer/Editor. **Benefits available to temporary workers:** 401(k); Dental Insurance; Disability; Life Insurance; Medical Insurance. **Corporate headquarters location:** Atlanta GA. **International locations:** Worldwide. **Average salary range of placements:** More than $50,000.

THE PERSONNEL INSTITUTE

725 West Lorraine Drive, Deltona FL 32725-8606. 407/860-2084. **Fax:** 407/860-1231. **Contact:** Dr. William E. Stuart, President. **Description:** The Personnel Institute is an executive search firm. Company pays fee. **Specializes in the areas of:** Computer Hardware/Software; Engineering; Manufacturing; MIS/EDP; Technical and Scientific. **Positions commonly filled include:** Aerospace Engineer; Bank Officer/Manager; Biological Scientist; Biomedical Engineer; Computer Programmer; EDP Specialist; Industrial Designer; Industrial Engineer; Marketing Specialist; Mechanical Engineer; MIS Specialist; Operations/Production Manager; Physicist/Astronomer; Quality Control Supervisor; Systems Analyst. **Number of placements per year:** 200 - 499.

PRO-FINDERS INC.

6170 Central Avenue, Suite B, St. Petersburg FL 33707-1523. 813/345-6138. **Contact:** Walt Shepard, President. **Description:** An executive search firm. Company pays fee. **Specializes in the areas of:** Accounting/Auditing; Administration/MIS/EDP; Computer Science/Software; Engineering; General Management; Industrial; Manufacturing; Personnel/Labor Relations; Retail; Technical and Scientific. **Average salary range of placements:** $30,000 - $50,000. **Number of placements per year:** 100 - 199.

JACK RICHMAN & ASSOCIATES

P.O. Box 25412, Tamarac FL 33320. 954/940-0721. **Fax:** 954/389-9572. **Contact:** Jack Richman, President. **E-mail address:** jrafl@aol.com. **World Wide Web address:** http://www.netmar.com/~edp/jra.html. **Description:** An executive search firm specializing in the computer industry. Company pays fee. **Specializes in the areas of:** Administration/MIS/EDP; Computer Science/Software; Engineering. **Positions commonly filled include:** Computer Programmer; Systems Analyst; Technical Writer/Editor. **Corporate headquarters location:** This Location. **Other U.S. locations:** Nationwide. **Average salary range of placements:** More than $50,000. **Number of placements per year:** 1 - 49.

ROMAC INTERNATIONAL
500 West Cypress Creek Road, Suite 200, Fort Lauderdale FL 33309. 954/928-0800. **Fax:** 954/771-7649. **Contact:** Manager. **Description:** An executive search firm. **Specializes in the areas of:** Accounting/Auditing; Computer Hardware/Software; Government. **Positions commonly filled include:** Accountant/Auditor; Computer Programmer; EDP Specialist; MIS Specialist; Systems Analyst.

SALES CONSULTANTS OF FORT LAUDERDALE
100 West Cypress Creek Road, Suite 965, Fort Lauderdale FL 33309. 954/772-5100. **Contact:** Jeff Taylor, Manager. **Description:** An executive search firm. **Specializes in the areas of:** Accounting/Auditing; Administration/MIS/EDP; Advertising; Architecture/ Construction/Real Estate; Banking; Chemical; Communications; Computer Hardware/ Software; Construction; Design; Electrical; Engineering; Finance; Food Industry; General Management; Health/Medical; Industrial; Insurance; Legal; Manufacturing; Operations Management; Personnel/Labor Relations; Printing/Publishing; Procurement; Retail; Sales and Marketing; Technical and Scientific; Textiles; Transportation.

SALES CONSULTANTS OF JACKSONVILLE
9471 Baymeadows Road, Suite 204, Jacksonville FL 32256. 904/737-5770. **Contact:** General Manager. **Description:** An executive search firm. **Specializes in the areas of:** Accounting/ Auditing; Administration/MIS/EDP; Advertising; Architecture/Construction/Real Estate; Banking; Chemical; Communications; Computer Hardware/Software; Construction; Design; Electrical; Engineering; Finance; Food Industry; General Management; Health/Medical; Industrial; Insurance; Legal; Manufacturing; Operations Management; Personnel/Labor Relations; Pharmaceutical; Printing/Publishing; Procurement; Retail; Sales and Marketing; Technical and Scientific; Textiles; Transportation.

SALES CONSULTANTS OF SARASOTA
1343 Main Street, Suite 600, Sarasota FL 34236. 941/365-5151. **Contact:** Rose L. Castellano, Managing Partner. **Description:** An executive search firm. **Specializes in the areas of:** Accounting/Auditing; Administration/MIS/EDP; Advertising; Architecture/ Construction/Real Estate; Banking; Chemical; Communications; Computer Hardware/Software; Construction; Data Communications; Electrical; Engineering; Finance; Food Industry; General Management; Health/ Medical; Industrial; Insurance; Legal; Manufacturing; Operations Management; Personnel/Labor Relations; Printing/Publishing; Procurement; Retail; Sales and Marketing; Technical and Scientific; Textiles; Transportation.

SNELLING SEARCH
9500 Koger Boulevard, St. Petersburg FL 33702. 813/577-9111. **Toll-free phone:** 800/484-5552. **Fax:** 813/577-1071. **Contact:** James Conwell, Manager. **Description:** An executive search firm. **Specializes in the areas of:** Computer Science/Software; Engineering; Manufacturing; Printing/Publishing. **Positions commonly filled include:** Aerospace Engineer; Agricultural Engineer; Aircraft Mechanic/Engine Specialist; Chemical Engineer; Civil Engineer; Computer Programmer; Design Engineer; Designer; Electrical/Electronics Engineer; Industrial Engineer; Industrial Production Manager; Mechanical Engineer; Nuclear Engineer; Software Engineer; Structural Engineer; Systems Analyst. **Average salary range of placements:** $30,000 - $50,000. **Number of placements per year:** 200 - 499.

THE STEWART SEARCH GROUP, INC.
P.O. Box 2588, Ponte Vedra Beach FL 32082. 904/285-6622. **Fax:** 904/285-0076. **Contact:** James H. Stewart, President. **Description:** An executive search firm. **NOTE:** The Stewart Search Group's physical address is 201 ATP Tour Boulevard, Suite 130, Ponte Vedra Beach FL. **Specializes in the areas of:** Biology; Chemical; Computer Science/Software; Engineering; Finance; Marketing. **Positions commonly filled include:** Computer Programmer; Financial Manager; Health Care Administrator; Research Assistant. **Number of placements per year:** 1 - 49.

SYSTEM ONE TECHNICAL, INC.
4902 Eisenhower Boulevard, Suite 370, Tampa FL 33634. 813/249-1757. **Toll-free phone:** 800/741-8324. **Fax:** 813/880-9141. **Contact:** Recruiter. **E-mail address:** sotcorp@ibm.net.

Description: A full-service executive search firm specializing in information technology, telecommunications, engineering, and other technical disciplines. Company pays fee. **Specializes in the areas of:** Administration/MIS/EDP; Computer Science/Software; Engineering; General Management; Manufacturing; Technical and Scientific; Telecommunications. **Benefits available to temporary workers:** 401(k); Insurance Plan; Paid Vacation. **Corporate headquarters location:** This Location. **Other U.S. locations:** Orlando FL; Atlanta GA; Baltimore MD; Charlotte NC; Raleigh NC; Valley Forge PA; Dallas TX. **Average salary range of placements:** $20,000 - $29,999. **Number of placements per year:** 100 - 199.

TEMP FORCE/DIVACK MANAGEMENT
1111 North Westshore Boulevard, Suite 102, Tampa FL 33607. 813/286-2222. **Fax:** 813/289-0465. **Contact:** Susan Divack, President. **Description:** An executive search firm. Company pays fee. **Specializes in the areas of:** Accounting/Auditing; Administration/MIS/EDP; Computer Hardware/Software; General Management; Legal; Manufacturing; Personnel/Labor Relations; Secretarial; Technical and Scientific. **Positions commonly filled include:** Accountant/Auditor; Administrative Assistant; Bookkeeper; Clerk; Computer Operator; Computer Programmer; Customer Service Representative; Data Entry Clerk; EDP Specialist; Factory Worker; Legal Secretary; Light Industrial Worker; Marketing Specialist; Medical Secretary; Operations/ Production Manager; Quality Control Supervisor; Receptionist; Secretary; Systems Analyst; Typist/Word Processor.

PERMANENT EMPLOYMENT AGENCIES OF FLORIDA

ALPHA PERSONNEL
ALPHA TEMPS
10707 66th Street, Suite B, Pinellas Park FL 34666-2336. 813/544-5627. **Contact:** Director. **Description:** An employment agency which provides both permanent and temporary placement. Company pays fee. **Specializes in the areas of:** Accounting/Auditing; Banking; Computer Science/Software; Legal; Manufacturing; Nonprofit; Personnel/Labor Relations; Secretarial. **Number of placements per year:** 200 - 499.

AVAILABILITY, INC.
5340 West Kennedy Boulevard, Suite 100, Tampa FL 33609. 813/286-8800. **Fax:** 813/286-0574. **Contact:** Ed Hart, President. **Description:** An employment agency which provides temporary and permanent placements as well as contract services. **Specializes in the areas of:** Accounting/ Auditing; Administration/MIS/EDP; Banking; Clerical; Computer Science/Software; Engineering; Finance; General Management; Health/Medical; Legal; Personnel/Labor Relations; Printing/ Publishing; Secretarial. **Positions commonly filled include:** Accountant/Auditor; Administrative Services Manager; Advertising Clerk; Bank Officer/Manager; Branch Manager; Brokerage Clerk; Claim Representative; Clerical Supervisor; Computer Programmer; Cost Estimator; Credit Manager; Customer Service Representative; Dental Assistant/Dental Hygienist; Dentist; Editor; Education Administrator; Emergency Medical Technician; Financial Analyst; General Manager; Health Services Manager; Human Resources Specialist; Human Service Worker; Internet Services Manager; Management Analyst/Consultant; Management Trainee; Medical Records Technician; Occupational Therapist; Operations/Production Manager; Paralegal; Public Relations Specialist; Recreational Therapist; Registered Nurse; Respiratory Therapist; Software Engineer; Systems Analyst; Technical Writer/Editor; Telecommunications Manager; Typist/Word Processor; Underwriter/Assistant Underwriter.

BELMONT TRAINING & EMPLOYMENT
17800 NW 27th Avenue, Opa Locka FL 33056. 954/628-3838. **Fax:** 954/628-2331. **Contact:** Steven L. Davis, Center Coordinator. **Description:** An employment agency. **Specializes in the**

areas of: Administration/MIS/EDP; Computer Science/Software; Education; Food Industry; General Management; Retail; Sales and Marketing; Secretarial. **Positions commonly filled include:** Accountant/Auditor; Administrative Services Manager; Advertising Clerk; Automotive Mechanic; Blue-Collar Worker Supervisor; Clerical Supervisor; Computer Programmer; Construction Contractor; Customer Service Representative; EKG Technician; Financial Analyst; Human Resources Specialist; Management Trainee; Market Research Analyst; Paralegal; Registered Nurse; Restaurant/Food Service Manager; Teacher/Professor; Typist/Word Processor. **Number of placements per year:** 200 - 499.

CAPITAL DATA, INC.
P.O. Box 2244, Palm Harbor FL 34682-2244. 813/784-4100. **Toll-free phone:** 800/771-4100. **Toll-free fax:** 800/787-4172. **Contact:** Jack Logan, Manager. **E-mail address:** as400cdo@aol.com. **Description:** A full-service employment agency which offers permanent, temporary, and contract placements; executive search; and career/outplacement counseling in the information systems industries. Company pays fee. **Specializes in the areas of:** Computer Science/Software; Technical and Scientific. **Positions commonly filled include:** Computer Programmer; Software Engineer; Systems Analyst; Technical Writer/Editor; Telecommunications Manager. **Average salary range of placements:** More than $50,000. **Number of placements per year:** 100 - 199.

CAREER PLANNERS, INC.
5730 Corporate Way, Suite 100, West Palm Beach FL 33407. 561/683-8785. **Fax:** 561/683-4047. **Contact:** Deborah M. Finley, President. **Description:** A full-service employment agency offering temporary and permanent placements. Company pays fee. **Specializes in the areas of:** Accounting/ Auditing; Administration/MIS/EDP; Architecture/Construction/ Real Estate; Banking; Computer Science/Software; Engineering; Finance; Legal; Manufacturing; Personnel/Labor Relations; Sales and Marketing; Secretarial. **Positions commonly filled include:** Accountant/ Auditor; Administrative Services Manager; Bank Officer/Manager; Branch Manager; Budget Analyst; Buyer; Civil Engineer; Computer Programmer; Cost Estimator; Credit Manager; Design Engineer; Draftsperson; Environmental Engineer; Financial Analyst; Management Trainee; Market Research Analyst; Mechanical Engineer; MIS Specialist; Operations/Production Manager; Paralegal; Petroleum Engineer; Property and Real Estate Manager; Purchasing Agent/Manager; Quality Control Supervisor; Securities Sales Representative; Services Sales Representative; Systems Analyst; Technical Writer/Editor; Telecommunications Manager; Typist/Word Processor; Underwriter/Assistant Underwriter. **Average salary range of placements:** $30,000 - $50,000. **Number of placements per year:** 200 - 499.

DECISION CONSULTANTS
13535 Southern Sound Drive, Suite 220, Clearwater FL 34622. 813/573-2626. **Contact:** Manager. **Description:** An employment agency. **Specializes in the areas of:** Administration/MIS/EDP; Computer Hardware/Software.

EN-DATA CORPORATION
P.O. Box 2949, Sanford FL 32772-2949. 407/323-0033. **Fax:** 407/323-0685. **Contact:** Maria Kingston, Operations Manager. **Description:** En-Data Corporation is an employment agency that provides desktop publishing placement, contract programming services, and information systems consulting. Company pays fee. **Specializes in the areas of:** Computer Science/Software; Data Processing; Information Systems. **Positions commonly filled include:** Computer Programmer; MIS Specialist; Multimedia Designer; Software Engineer; Systems Analyst; Technical Writer/Editor. **Average salary range of placements:** More than $50,000. **Number of placements per year:** 50 - 99.

EXECUTIVE DIRECTIONS INC.
450 North Park Road, Suite 302, Hollywood FL 33021. 954/962-9444. **Fax:** 954/963-4333. **Contact:** Bob Silverman, Director. **E-mail address:** edifla@newpower.com. **Description:** An employment agency. Company pays fee. **Specializes in the areas of:** Administration/MIS/EDP; Computer Science/Software. **Positions commonly filled include:** Computer Programmer; Management Analyst/Consultant; Software Engineer; Systems Analyst; Telecommunications

Manager. **Average salary range of placements:** $30,000 - $50,000. **Number of placements per year:** 100 - 199.

OLSTEN STAFFING SERVICES

801 Brickell Avenue, Miami FL 33131-2951. **Contact:** Recruiter. **Description:** Olsten Staffing Services is an employment agency. **Specializes in the areas of:** Accounting/Auditing; Banking; Broadcasting; Computer Science/Software; Finance; Health/Medical; Insurance; Legal; Personnel/Labor Relations. **Positions commonly filled include:** Accountant/Auditor; Clerical Supervisor; Computer Programmer; Customer Service Representative; Paralegal; Preschool Worker; Systems Analyst; Typist/Word Processor. **Number of placements per year:** 200 - 499.

PERSONNEL ONE, INC.

7890 West Flagler Street, Miami FL 33144. 305/662-2500. **Fax:** 305/662-6700. **Contact:** Office Manager. **Description:** An employment agency. Company pays fee. **Specializes in the areas of:** Accounting/Auditing; Banking; Computer Science/Software; Engineering; General Management; Legal; Personnel/Labor Relations; Sales and Marketing; Secretarial; Technical and Scientific. **Positions commonly filled include:** Accountant/Auditor; Bank Officer/Manager; Biological Scientist; Branch Manager; Chemist; Clerical Supervisor; Clinical Lab Technician; Computer Programmer; Construction and Building Inspector; Construction Contractor; Counselor; Credit Manager; Engineer; Financial Analyst; Human Resources Manager; Human Service Worker; Manufacturer's/Wholesaler's Sales Rep.; Paralegal; Property and Real Estate Manager; Public Relations Specialist; Services Sales Representative; Systems Analyst. **Number of placements per year:** 200 - 499.

AARON STEWART PERSONNEL INC.

201 SE 15th Terrace, Deerfield Beach FL 33441-4428. 954/480-9902. **Contact:** Human Resources. **Description:** An employment agency that also operates as a temporary and executive search firm. **Specializes in the areas of:** Accounting/Auditing; Banking; Computer Science/Software; Finance; Insurance; Legal; Nonprofit; Personnel/Labor Relations; Printing/Publishing; Secretarial. **Number of placements per year:** 200 - 499.

U.S. PERSONNEL/STAFFING SOLUTIONS

13535 Feather Sound Drive, Clearwater FL 34622-2259. 813/571-3111. **Fax:** 813/572-7588. **Recorded jobline:** 813/469-8106. **Contact:** Corry Tyle, Marketing Manager. **Description:** An employment agency which provides supplemental staffing and direct placement to companies. Company pays fee. **Specializes in the areas of:** Administration/MIS/EDP; Computer Science/Software; Industrial; Secretarial. **Positions commonly filled include:** Claim Representative; Clerical Supervisor; Customer Service Representative; Typist/Word Processor. **Number of placements per year:** 1000+.

EXECUTIVE SEARCH FIRMS OF GEORGIA

A.D. & ASSOCIATES

5589 Woodsong Drive, Suite 104, Dunwoody GA 30338. 770/393-0021. **Contact:** A. Dwight Hawksworth, President. **Description:** A.D. & Associates is an executive search firm. Company pays fee. **Specializes in the areas of:** Accounting/Auditing; Administration/MIS/EDP; Computer Hardware/Software; Finance; General Management; Personnel/Labor Relations; Sales and Marketing; Technical and Scientific; Telecommunications. **Positions commonly filled include:** Accountant/Auditor; Chief Financial Officer; Credit Manager; Electrical/Electronics Engineer; Marketing Specialist; Sales Manager; Sales Representative; Systems Analyst. **Number of placements per year:** 1 - 49.

AE ASSOCIATES
6351 M Jonesboro Road, Morrow GA 30260. 770/968-8002. **Contact:** Victor Cannon, Vice President. **Description:** An executive search firm. **Specializes in the areas of:** Accounting/Auditing; Computer Science/Software; Engineering; Personnel/Labor Relations. **Positions commonly filled include:** Accountant/Auditor; Buyer; Computer Programmer; Engineer; Financial Analyst; Human Resources Manager; Telecommunications Manager. **Average salary range of placements:** $30,000 - $50,000. **Number of placements per year:** 1 - 49.

ARJAY & ASSOCIATES
2386 Clower Street, Snellville GA 30278-6134. 770/979-3799. **Fax:** 770/985-7696. **Contact:** David L. Hubert, Director of Operations. **Description:** An executive search firm. Company pays fee. **Specializes in the areas of:** Computer Science/Software; Engineering; Industrial; Manufacturing. **Positions commonly filled include:** Computer Programmer; EKG Technician; Industrial Engineer; Mechanical Engineer; Metallurgical Engineer; Mining Engineer; Nuclear Engineer; Quality Control Supervisor; Software Engineer; Systems Analyst. **Average salary range of placements:** More than $50,000. **Number of placements per year:** 1 - 49.

ASHLEY-NOLAN INTERNATIONAL, INC.
2000 Powers Ferry Center, Suite 2-3, Marietta GA 30067. 770/956-8010. **Fax:** 770/956-1551. **Contact:** Pamela Johns, President. **E-mail address:** ashnol@atlanta.com. **Description:** An executive search firm. Company pays fee. **Specializes in the areas of:** Accounting/Auditing; Administration/MIS/EDP; Banking; Computer Science/Software; Engineering; Finance; Manufacturing; Personnel/Labor Relations; Sales and Marketing; Technical and Scientific. **Positions commonly filled include:** Accountant/Auditor; Administrative Services Manager; Budget Analyst; Buyer; Chemical Engineer; Civil Engineer; Construction and Building Inspector; Construction Contractor; Design Engineer; Electrical/Electronics Engineer; Financial Analyst; Human Resources Manager; Industrial Engineer; Mechanical Engineer; Metallurgical Engineer; MIS Specialist; Purchasing Agent/Manager; Software Engineer; Strategic Relations Manager; Systems Analyst; Telecommunications Manager. **Average salary range of placements:** More than $50,000. **Number of placements per year:** 1 - 49.

THE BOWERS GROUP, INC.
1850 Parkway Place, Suite 420, Marietta GA 30067-4439. 770/421-1019. **Fax:** 770/509-1095. **Contact:** Brenda Bowers, President. **Description:** An executive search firm specializing in the placement of computer software and hardware professionals. Company pays fee. **Specializes in the areas of:** Computer Science/Software. **Positions commonly filled include:** Computer Programmer; MIS Specialist; Systems Analyst. **Corporate headquarters location:** This Location. **Average salary range of placements:** More than $50,000. **Number of placements per year:** 1 - 49.

BRADSHAW & ASSOCIATES
1850 Parkway Place, Suite 420, Marietta GA 30067. 770/426-5600. **Fax:** 770/427-1727. **Contact:** Rod Bradshaw, President. **Description:** Bradshaw & Associates specializes in the recruitment and placement of entry- to upper-level professional positions in a wide range of fields on a contingency or retained basis. Company pays fee. **Specializes in the areas of:** Accounting/Auditing; Biotechnology; Chemical; Computer Hardware/Software; Data Processing; Finance. **Positions commonly filled include:** Accountant/Auditor; Aerospace Engineer; Agricultural Engineer; Architect; Bank Officer/Manager; Biochemist; Biological Scientist; Budget Analyst; Chemical Engineer; Chemist; Civil Engineer; Computer Programmer; Construction and Building Inspector; Construction Contractor; Design Engineer; Electrical/Electronics Engineer; Environmental Engineer; Financial Analyst; General Manager; Geographer; Geologist/Geophysicist; Health Services Manager; Human Resources Manager; Industrial Engineer; Internet Services Manager; Management Analyst/Consultant; Management Trainee; Manufacturer's/Wholesaler's Sales Rep.; Mathematician; Mechanical Engineer; Metallurgical Engineer; Mining Engineer; MIS Specialist; Nuclear Engineer; Operations/Production Manager; Petroleum Engineer; Pharmacist; Physical Therapist; Physician; Public Relations Specialist; Quality Control Supervisor; Registered Nurse; Respiratory Therapist; Securities Sales Representative; Services Sales Representative; Software Engineer; Statistician; Strategic Relations Manager; Structural Engineer; Surgical Technician;

Systems Analyst; Telecommunications Manager; Transportation/Traffic Specialist. **Corporate headquarters location:** This Location.

COMPREHENSIVE SEARCH

316 South Lewis Street, La Grange GA 30240-3144. 706/884-3232. **Fax:** 706/884-4106. **Contact:** Merritt Shelton, Researcher/Database Manager. **E-mail address:** compsrch@wp-lag.mindspring.com. **Description:** An executive search firm which specializes in the building products and interior furnishings industries. Company pays fee. **Specializes in the areas of:** Administration/MIS/EDP; Advertising; Architecture/Construction/Real Estate; Art/Design; Computer Science/Software; Industrial; Manufacturing; Retail; Sales and Marketing. **Positions commonly filled include:** Administrative Services Manager; Architect; Computer Programmer; Construction and Building Inspector; Construction Contractor; Design Engineer; Designer; Draftsperson; Industrial Engineer; Industrial Production Manager; Internet Services Manager; Management Trainee; Manufacturer's/Wholesaler's Sales Rep.; MIS Specialist; Services Sales Representative; Software Engineer; Systems Analyst. **Corporate headquarters location:** This Location. **Other U.S. locations:** Atlanta GA; Brooklyn NY; New York NY. **Average salary range of placements:** More than $50,000. **Number of placements per year:** 50 - 99.

CORPORATE SEARCH CONSULTANTS

47 Perimeter Center East, Suite 260, Atlanta GA 30346-2001. 770/399-6205. **Fax:** 770/399-6416. **Contact:** Harriet Rothberg, President. **E-mail address:** barry@rothberg.com. **World Wide Web address:** http://www.rothberg.com. **Description:** A nationwide executive search firm. Company pays fee. **Specializes in the areas of:** Accounting/Auditing; Administration/MIS/EDP; Architecture/Construction/Real Estate; Computer Science/Software; Engineering; Finance; General Management; Health/Medical; Industrial; Manufacturing; Personnel/Labor Relations; Sales and Marketing; Secretarial; Technical and Scientific. **Positions commonly filled include:** Accountant/Auditor; Administrative Services Manager; Architect; Blue-Collar Worker Supervisor; Branch Manager; Budget Analyst; Buyer; Civil Engineer; Clerical Supervisor; Clinical Lab Technician; Computer Programmer; Construction and Building Inspector; Construction Contractor; Cost Estimator; Credit Manager; Customer Service Representative; Design Engineer; Designer; Draftsperson; Electrical/Electronics Engineer; Environmental Engineer; Financial Analyst; General Manager; Health Services Manager; Landscape Architect; Librarian; Management Analyst/Consultant; Management Trainee; Manufacturer's/Wholesaler's Sales Rep.; Market Research Analyst; Mechanical Engineer; Medical Records Technician; Metallurgical Engineer; MIS Specialist; Multimedia Designer; Occupational Therapist; Operations/Production Manager; Physical Therapist; Physician; Physician Assistant; Purchasing Agent/Manager; Quality Control Supervisor; Respiratory Therapist; Services Sales Representative; Software Engineer; Speech-Language Pathologist; Structural Engineer; Systems Analyst; Technical Writer/Editor; Telecommunications Manager; Transportation/Traffic Specialist; Typist/Word Processor. **Corporate headquarters location:** This Location. **Average salary range of placements:** $30,000 - $50,000. **Number of placements per year:** 1000+.

DUNHILL PROFESSIONAL SEARCH

3340 Peachtree Road NE, Suite 2570, Atlanta GA 30326. 404/261-3751. **Contact:** Recruiter. **Description:** An executive search firm. **Specializes in the areas of:** Banking; Computer Hardware/Software; Health/Medical; Sales and Marketing.

EXECUTIVE FORCE, INC.

2271 Winding Way, Suite 210, Tucker GA 30084. 770/939-6123. **Fax:** 770/939-0484. **Contact:** Jim Sayers, President. **Description:** A search firm that specializes in the data processing industry, with an emphasis on sales and management for the software and services industry. Company pays fee. **Specializes in the areas of:** Banking; Computer Science/Software; Sales and Marketing. **Positions commonly filled include:** Bank Officer/Manager; Branch Manager; General Manager. **Average salary range of placements:** More than $50,000. **Number of placements per year:** 50 - 99.

EXPRESS PERSONNEL SERVICES

914 West Tyler Street, Griffin GA 30223. 770/227-9103. **Fax:** 770/227-1139. **Contact:** Phillip Purser, General Manager. **E-mail address:** xpress1134@aol.com. **World Wide Web address:**

http://www.monster.com/expresspersonnel. **Description:** A full-service human resources outsourcing firm. Company pays fee. **Specializes in the areas of:** Accounting/Auditing; Administration/MIS/EDP; Banking; Computer Science/Software; Engineering; Finance; General Management; Industrial; Manufacturing; Personnel/Labor Relations; Sales and Marketing; Technical and Scientific. **Positions commonly filled include:** Chemical Engineer; Clerical Supervisor; Computer Programmer; Credit Manager; Customer Service Representative; Electrical/Electronics Engineer; Electrician; Human Resources Manager; Industrial Engineer; Industrial Production Manager; Mechanical Engineer; MIS Specialist; Operations/Production Manager; Systems Analyst; Typist/Word Processor. **Corporate headquarters location:** This Location. **Number of placements per year:** 1000+.

FOX-MORRIS ASSOCIATES
9000 Central Park West, Suite 150, Atlanta GA 30328. 770/393-0933. **Fax:** 770/399-4499. **Contact:** Ty Smith, Branch Manager. **Description:** An executive search firm. **Specializes in the areas of:** Accounting/Auditing; Computer Hardware/Software; Engineering; Finance; Manufacturing; MIS/EDP; Personnel/Labor Relations; Sales and Marketing. **Number of placements per year:** 1000+.

HOWIE AND ASSOCIATES, INC.
875 Old Roswell Road, Suite F100, Roswell GA 30076. 770/998-0099. **Fax:** 770/993-7406. **Contact:** Ellen Howie-Brown, President. **E-mail address:** howieinc@ik.netcom.com. **Description:** An executive search firm operating on a contingency basis. Company pays fee. **Specializes in the areas of:** Computer Science/Software; Data Processing. **Positions commonly filled include:** Computer Programmer; Internet Services Manager; Software Engineer; Systems Analyst; Technical Writer/Editor; Telecommunications Manager. **Corporate headquarters location:** This Location. **Average salary range of placements:** More than $50,000. **Number of placements per year:** 100 - 199.

JES SEARCH FIRM, INC.
3475 Lenox Road, Suite 970, Atlanta GA 30326. 404/262-7222. **Fax:** 404/266-3533. **Contact:** Brenda Evers, President. **Description:** An executive search firm. Company pays fee. **Specializes in the areas of:** Computer Science/Software. **Positions commonly filled include:** Software Engineer. **Number of placements per year:** 50 - 99.

LEADER INSTITUTE, INC.
340 Interstate North Parkway, Suite 250, Atlanta GA 30339. 770/984-2700. **Fax:** 770/984-2990. **Contact:** Rick Zabor, President. **Description:** An executive search firm. Company pays fee. **Specializes in the areas of:** Administration/MIS/EDP; Computer Science/Software; Engineering; Manufacturing. **Positions commonly filled include:** Computer Programmer; Electrical/Electronics Engineer; Industrial Engineer; Mechanical Engineer; Purchasing Agent/Manager; Software Engineer; Systems Analyst. **Number of placements per year:** 50 - 99.

BOB MADDOX ASSOCIATES
3134 West Roxboro Road NE, Atlanta GA 30324-2542. 404/231-0558. **Fax:** 404/231-1074. **Contact:** Robert E. Maddox, President. **Description:** An executive search firm operating on both retainer and contingency bases. Company pays fee. **Specializes in the areas of:** Computer Science/Software; Industrial; Manufacturing; Nonprofit; Sales and Marketing. **Positions commonly filled include:** Branch Manager; General Manager; Manufacturer's/Wholesaler's Sales Rep.; Services Sales Representative. **Corporate headquarters location:** This Location. **Other U.S. locations:** Charlotte NC. **Average salary range of placements:** More than $50,000. **Number of placements per year:** 50 - 99.

MANAGEMENT RECRUITERS
3700 Crestwood Parkway NW, Suite 320, Duluth GA 30136-5585. 770/925-2266. **Fax:** 770/325-1090. **Contact:** David Riggs, Manager. **E-mail address:** mri.atlanta@internetmci.com. **Description:** An executive search firm. Company pays fee. **Specializes in the areas of:** Biology; Computer Science/Software; Finance; Sales and Marketing; Technical and Scientific. **Positions commonly filled include:** Biochemist; Biological Scientist; Biomedical Engineer; Chemical Engineer; Credit Manager; Environmental Engineer; Financial Analyst; Market Research Analyst;

Multimedia Designer; Operations/Production Manager; Science Technologist; Securities Sales Representative; Software Engineer. **International locations:** Worldwide. **Average salary range of placements:** More than $50,000. **Number of placements per year:** 200 - 499.

MANAGEMENT RECRUITERS INTERNATIONAL/SALES CONSULTANTS
P.O. Box 72527, Marietta GA 30007-2527. 770/643-9990. **Fax:** 770/509-9014. **Contact:** Larry Dougherty, Office Manager. **E-mail address:** mri-cobb@atl.mindspring.com. **Description:** An executive search firm that provides video conferencing, interexecutive placement, and outplacement services. Company pays fee. **Specializes in the areas of:** Computer Science/ Software; Engineering; General Management; Logistics; Manufacturing; Technical and Scientific; Transportation. **Positions commonly filled include:** Branch Manager; Budget Analyst; Computer Programmer; General Manager; Industrial Engineer; Management Analyst/Consultant; Manufacturer's/Wholesaler's Sales Rep.; Marketing/Public Relations Manager; MIS Specialist; Operations/Production Manager; Software Engineer; Systems Analyst; Telecommunications Manager; Transportation/Traffic Specialist. **Corporate headquarters location:** Cleveland OH. **Average salary range of placements:** More than $50,000. **Number of placements per year:** 100 - 199.

MANAGEMENT RECRUITERS OF ATLANTA
5901 Peachtree-Dunwoody Road NE, Suite 370C, Atlanta GA 30328. 770/394-1300. **Contact:** General Manager. **Description:** An executive search firm. **Specializes in the areas of:** Accounting/ Auditing; Administration/MIS/EDP; Advertising; Architecture/Construction/Real Estate; Banking; Chemical; Communications; Computer Hardware/Software; Construction; Electrical; Engineering; Finance; Food Industry; General Management; Health/Medical; Insurance; Legal; Manufacturing; Operations Management; Personnel/Labor Relations; Pharmaceutical; Printing/Publishing; Procurement; Retail; Sales and Marketing; Technical and Scientific; Textiles; Transportation.

MANAGEMENT RECRUITERS OF ATLANTA
30 Woodstock Street, Roswell GA 30076. 404/998-1555. **Contact:** Art Katz, Manager. **Description:** An executive search firm. **Specializes in the areas of:** Accounting/Auditing; Administration/MIS/EDP; Advertising; Architecture/Construction/Real Estate; Banking; Chemical; Communications; Computer Hardware/Software; Construction; Electrical; Engineering; Finance; Food Industry; General Management; Health/Medical; Insurance; Legal; Manufacturing; Operations Management; Personnel/Labor Relations; Pharmaceutical; Printing/Publishing; Procurement; Retail; Sales and Marketing; Technical and Scientific; Textiles; Transportation.

MANAGEMENT RECRUITERS OF ATLANTA (SOUTH)
406 Lime Creek Drive, Suite B, Peachtree City GA 30269. 770/486-0603. **Fax:** 770/631-7684. **Contact:** Ron Wise/Sandra Wise, Owners. **Description:** An executive search firm. **Specializes in the areas of:** Accounting/Auditing; Administration/MIS/EDP; Advertising; Architecture/ Construction/Real Estate; Banking; Chemical; Communications; Computer Hardware/Software; Electrical; Engineering; Finance; Food Industry; General Management; Health/Medical; Insurance; Legal; Manufacturing; Operations Management; Personnel/Labor Relations; Pharmaceutical; Printing/Publishing; Procurement; Sales and Marketing; Technical and Scientific; Textiles; Transportation. **Positions commonly filled include:** Branch Manager; Chemical Engineer; Engineer; Mathematician. **Average salary range of placements:** More than $50,000. **Number of placements per year:** 50 - 99.

MANAGEMENT RECRUITERS OF COLUMBUS
233 12th Street, Suite 818A, Columbus GA 31901-2449. 706/571-9611. **Fax:** 706/571-3288. **Contact:** Recruiter. **Description:** An executive search firm that specializes in information systems technology. Company pays fee. **Specializes in the areas of:** Computer Science/Software. **Positions commonly filled include:** Computer Programmer; MIS Specialist; Systems Analyst.

MANAGEMENT RECRUITERS OF MARIETTA
274 North Marietta Parkway, Suite C, Marietta GA 30060. 770/423-1443. **Fax:** 770/423-1303. **Contact:** James K. Kirby, Managing Principal. **E-mail address:** jkirby@mindspring.com. **Description:** An executive search firm specializing in information technology, management consulting, manufacturing, and logistics distribution. Company pays fee. **Specializes in the areas**

of: Computer Science/Software; Engineering; Fashion; General Management; Industrial; Logistics; Manufacturing; Personnel/Labor Relations; Sales and Marketing; Transportation. **Positions commonly filled include:** Accountant/Auditor; Aerospace Engineer; Biomedical Engineer; Chemical Engineer; Computer Programmer; Design Engineer; Electrical/Electronics Engineer; General Manager; Human Resources Manager; Industrial Engineer; Management Analyst/ Consultant; Mechanical Engineer; MIS Specialist; Operations/Production Manager; Purchasing Agent/Manager; Quality Control Supervisor; Software Engineer; Systems Analyst; Transportation/ Traffic Specialist. **Corporate headquarters location:** This Location. **Average salary range of placements:** More than $50,000. **Number of placements per year:** 50 - 99.

NEIS
141 Village Parkway, Building 5, Marietta GA 30067. 770/952-0081. **Fax:** 770/952-0218. **Contact:** Mike Mauldin, President. **Description:** An executive search firm providing tailored recruitment programs to organizations ranging from information-only to complete retained search assignments. NEIS's primary areas of focus are technology, health care, sales, marketing, and services. Company pays fee. **Specializes in the areas of:** Administration/MIS/EDP; Computer Science/Software; Engineering; Finance; General Management; Health/Medical; Personnel/Labor Relations; Printing/Publishing; Retail; Sales and Marketing; Transportation. **Positions commonly filled include:** Accountant/Auditor; Administrative Services Manager; Aerospace Engineer; Agricultural Engineer; Branch Manager; Chemical Engineer; Civil Engineer; Clinical Lab Technician; Computer Programmer; Construction Contractor; Credit Manager; Customer Service Representative; Design Engineer; Designer; Dietician/Nutritionist; EEG Technologist; EKG Technician; Electrical/Electronics Engineer; Emergency Medical Technician; Environmental Engineer; Financial Analyst; General Manager; Health Services Manager; Hotel Manager; Human Resources Manager; Industrial Engineer; Internet Services Manager; Licensed Practical Nurse; Management Analyst/Consultant; Management Trainee; Manufacturer's/Wholesaler's Sales Rep.; Market Research Analyst; Mechanical Engineer; Medical Records Technician; Metallurgical Engineer; Mining Engineer; MIS Specialist; Multimedia Designer; Nuclear Engineer; Nuclear Medicine Technologist; Occupational Therapist; Operations/Production Manager; Paralegal; Petroleum Engineer; Pharmacist; Physical Therapist; Physician; Property and Real Estate Manager; Psychologist; Purchasing Agent/Manager; Radiologic Technologist; Recreational Therapist; Registered Nurse; Respiratory Therapist; Restaurant/ Food Service Manager; Social Worker; Sociologist; Software Engineer; Structural Engineer; Surgical Technician; Systems Analyst; Technical Writer/Editor; Telecommunications Manager; Typist/Word Processor. **Corporate headquarters location:** This Location.

PERSONNEL OPPORTUNITIES INC.
5064 Roswell Road, Suite D-301, Atlanta GA 30342. 404/252-9484. **Fax:** 404/252-9821. **Contact:** President. **Description:** An executive search firm. Founded in 1966. **Specializes in the areas of:** Accounting/Auditing; Computer Hardware/Software; Engineering; Health/Medical; Insurance; Legal; Manufacturing; Sales and Marketing. **Positions commonly filled include:** Accountant/Auditor; Chemical Engineer; Chemist; Claim Representative; Computer Programmer; Manufacturing Engineer; Mechanical Engineer; Nurse; Sales Representative; Software Engineer; Systems Analyst. **Number of placements per year:** 100 - 199.

PINNACLE CONSULTING GROUP
6621 Bay Circle, Suite 180, Norcross GA 30071. 770/447-3770. **Contact:** Debra Fogarty, Director of Recruiting. **Description:** An executive search firm. Company pays fee. **Specializes in the areas of:** Administration/MIS/EDP; Computer Science/Software; General Management; Printing/Publishing; Retail; Sales and Marketing. **Positions commonly filled include:** Computer Programmer; General Manager; Services Sales Representative; Software Engineer; Systems Analyst; Wholesale and Retail Buyer. **Number of placements per year:** 1 - 49.

SANFORD ROSE ASSOCIATES
3525 Holcomb Bridge Road, Suite 2B, Norcross GA 30092. 770/449-7200. **Fax:** 770/449-0516. **Contact:** Mr. Don Patrick, President. **Description:** An executive search firm. Company pays fee. **Specializes in the areas of:** Computer Science/Software; Engineering; General Management; Sales and Marketing; Technical and Scientific; Wireless Communications. **Positions commonly filled**

include: Branch Manager; Electrical/Electronics Engineer; General Manager; Science Technologist; Software Engineer. **Number of placements per year:** 1 - 49.

SAVANNAH MANAGEMENT RECRUITERS

P.O. Box 22548, Savannah GA 31403. 912/232-0132. **Fax:** 912/232-0136. **Contact:** Managing Partner. **Description:** An executive search firm. Company pays fee. **Specializes in the areas of:** Computer Hardware/Software; Engineering; General Management; Industrial; Manufacturing; Technical and Scientific; Transportation. **Positions commonly filled include:** Biochemist; Biomedical Engineer; Chemical Engineer; Chemist; Civil Engineer; Electrical/Electronics Engineer; Industrial Engineer; Manufacturing Engineer; Mechanical Engineer; Operations/ Production Manager; Purchasing Agent/Manager; Quality Control Supervisor; Software Engineer. **Number of placements per year:** 1 - 49.

SOLUTION SOURCE, INC.

3761 Venture Drive, Duluth GA 30136-5528. 770/418-1051. **Fax:** 770/476-2871. **Contact:** Brian Whitfield, Vice President. **Description:** An executive search firm that provides full-service consulting and project outsourcing to many of Atlanta's largest companies. Company pays fee. **Specializes in the areas of:** Computer Science/Software. **Positions commonly filled include:** Computer Programmer; MIS Specialist; Systems Analyst; Technical Writer/Editor. **Benefits available to temporary workers:** Medical Insurance. **Corporate headquarters location:** This Location. **Average salary range of placements:** More than $50,000. **Number of placements per year:** 50 - 99.

TAURION CORPORATION

P.O. Box 956716, Duluth GA 30136. 770/449-7155. **Fax:** 770/449-6421. **Contact:** John Puchis, Vice President. **Description:** An executive search firm. Company pays fee. **Specializes in the areas of:** Computer Science/Software; Data Processing. **Positions commonly filled include:** Computer Programmer; Software Engineer; Systems Analyst. **Number of placements per year:** 1 - 49.

TOAR CONSULTANTS INC.

1176 Grimes Bridge Road, Suite 200, Roswell GA 30075-3934. 770/993-7663. **Fax:** 770/998-5853. **Contact:** Human Resources. **E-mail address:** 71521.413@compuserve.com. **Description:** An executive search firm. Company pays fee. **Specializes in the areas of:** Accounting/Auditing; Administration/MIS/EDP; Banking; Computer Science/Software; Engineering; Finance; General Management; Insurance; Manufacturing; Printing/Publishing. **Positions commonly filled include:** Accountant/Auditor; Bank Officer/Manager; Chemical Engineer; Chemist; Civil Engineer; Computer Programmer; Design Engineer; Electrician; Engineer; MIS Manager; Operations/ Production Manager; Quality Control Supervisor; Systems Analyst. **Average salary range of placements:** More than $50,000. **Number of placements per year:** 50 - 99.

PERMANENT EMPLOYMENT AGENCIES OF GEORGIA

BUSINESS PROFESSIONAL GROUP, INC.

3490 Piedmont Road, Suite 212, Atlanta GA 30305. 404/262-2577. **Fax:** 404/262-3463. **Contact:** Michelle Abel, Vice President of Operations. **Description:** An employment agency. **Specializes in the areas of:** Accounting/Auditing; Administration/MIS/EDP; Advertising; Computer Science/ Software; Engineering; Finance; General Management; Industrial; Insurance; Manufacturing; Personnel/Labor Relations; Sales and Marketing; Technical and Scientific; Transportation. **Positions commonly filled include:** Accountant/Auditor; Buyer; Chemist; Claim Representative;

Electrical/Electronics Engineer; Financial Analyst; Industrial Engineer; Management Trainee; Mechanical Engineer; Metallurgical Engineer; Software Engineer; Technical Writer/Editor; Underwriter/Assistant Underwriter. **Number of placements per year:** 100 - 199.

MSI INTERNATIONAL CORPORATE
245 Peachtree Center Avenue NE, Suite 2500, Atlanta GA 30303. 404/659-5236. **Fax:** 404/659-7139. **Contact:** Eric J. Lindberg, President. **Description:** An employment agency. **Specializes in the areas of:** Accounting/Auditing; Banking; Computer Hardware/Software; Data Processing; Finance.

OLSTEN TECHNICAL SERVICES
6065 Roswell Road, Suite 220, Atlanta GA 30328. 404/255-4222. **Fax:** 404/843-9618. **Contact:** Tim Long or Deb Veal, Professional Technical Consultants. **Description:** Olsten Technical Services is one of America's largest staffing services. The agency provides flexible scheduling for numerous temporary and full-time regular hire opportunities. Company pays fee. **Specializes in the areas of:** Computer Science/Software; Engineering; Technical and Scientific. **Positions commonly filled include:** Aerospace Engineer; Architect; Chemical Engineer; Civil Engineer; Computer Programmer; Design Engineer; Draftsperson; Electrical/Electronics Engineer; Environmental Engineer; Industrial Engineer; Mechanical Engineer; Metallurgical Engineer; MIS Specialist; Software Engineer; Structural Engineer; Systems Analyst; Technical Writer/Editor; Telecommunications Manager. **Corporate headquarters location:** Melville NY. **International locations:** Worldwide. **Average salary range of placements:** $30,000 - $50,000. **Number of placements per year:** 1000+.

PREMIER STAFFING
6 Piedmont Circle NE, Suite 204, Atlanta GA 30305-1515. 404/237-1113. **Fax:** 404/237-0954. **Contact:** Melanie Robertson, Operations Manager. **Description:** An employment agency. Founded in 1970. Company pays fee. **Specializes in the areas of:** Computer Science/Software; Insurance; Secretarial. **Positions commonly filled include:** Accountant/Auditor; Claim Representative; Clerical Supervisor; Typist/Word Processor. **Benefits available to temporary workers:** 401(k); Medical Insurance; Paid Vacation; Paid Holidays. **Number of placements per year:** 500 - 999.

RANDSTAD STAFFING SERVICES
196 Alps Road, Suite 1, Athens GA 30606-4068. 706/548-9590. **Contact:** Marcelle Lacy, Manager. **Description:** An employment agency. Founded in 1960. Company pays fee. **Specializes in the areas of:** Administration/MIS/EDP; Banking; Computer Hardware/Software; Industrial; Manufacturing; Sales and Marketing. **Positions commonly filled include:** Blue-Collar Worker Supervisor; Clerical Supervisor; Computer Programmer; Customer Service Representative; Industrial Production Manager; Quality Control Supervisor; Services Sales Representative; Surveyor; Typist/Word Processor. **Benefits available to temporary workers:** 401(k); Dental Insurance; Medical Insurance. **Corporate headquarters location:** Atlanta GA. **Other U.S. locations:** Nationwide. **Average salary range of placements:** Less than $20,000. **Number of placements per year:** 1000+.

RANDSTAD STAFFING SERVICES
6681 Roswell Road NE, Atlanta GA 30328. 404/250-1008. **Fax:** 404/252-0605. **Contact:** Lisa Brantley, Branch Manager. **Description:** A temporary and permanent employment agency. Company pays fee. **Specializes in the areas of:** Accounting/Auditing; Computer Science/Software; Industrial; Manufacturing; Personnel/Labor Relations; Technical and Scientific. **Positions commonly filled include:** Accountant/Auditor; Administrative Services Manager; Advertising Clerk; Blue-Collar Worker Supervisor; Branch Manager; Claim Representative; Clerical Supervisor; Credit Manager; Customer Service Representative; Draftsperson; Financial Analyst; General Manager; Health Services Manager; Hotel Manager; Human Resources Specialist; Industrial Engineer; Internet Services Manager; Management Trainee; Market Research Analyst; Mechanical Engineer; MIS Specialist; Quality Control Supervisor; Software Engineer; Statistician; Telecommunications Manager; Typist/Word Processor. **Benefits available to temporary workers:** 401(k); Medical Insurance; Paid Holidays; Paid Vacation. **Average salary range of placements:** $30,000 - $50,000. **Number of placements per year:** 1000+.

RANDSTAD STAFFING SERVICES

2015 South Park Place, Atlanta GA 30339. 770/937-7000. **Contact:** Eric Vonk, CEO. **Description:** An employment agency. Company pays fee. **Specializes in the areas of:** Accounting/Auditing; Advertising; Banking; Computer Science/Software; Education; General Management; Industrial; Legal; Personnel/Labor Relations; Printing/Publishing; Retail; Sales and Marketing; Secretarial; Technical and Scientific; Transportation. **Positions commonly filled include:** Accountant/Auditor; Administrative Services Manager; Advertising Clerk; Brokerage Clerk; Budget Analyst; Buyer; Claim Representative; Computer Programmer; Cost Estimator; Counselor; Credit Manager; Customer Service Representative; Electrician; Financial Analyst; Human Resources Specialist; Industrial Engineer; Paralegal; Teacher/Professor; Travel Agent; Typist/Word Processor; Underwriter/Assistant Underwriter. **Benefits available to temporary workers:** 401(k); Credit Union; Dental Insurance; Disability Coverage; Life Insurance; Medical Insurance; Paid Holidays; Paid Vacation; Vision Insurance. **Corporate headquarters location:** Atlanta GA. **Other U.S. locations:** Nationwide. **Average salary range of placements:** Less than $20,000. **Number of placements per year:** 100 - 199.

SNELLING PERSONNEL SERVICES

1337 Canton Road, Suite D-3, Marietta GA 30066-6054. 770/423-1177. **Fax:** 770/423-0558. **Contact:** President. **E-mail address:** snelling@cyberatl.net. **World Wide Web address:** http://www.cyberatl.net. **Description:** An employment agency, specializing in sales and information systems. Founded in 1986. Company pays fee. **Specializes in the areas of:** Computer Science/Software; Engineering; Health/Medical; Sales and Marketing; Secretarial. **Positions commonly filled include:** Chemical Engineer; Clerical Supervisor; Computer Programmer; Credit Manager; Customer Service Representative; Environmental Engineer; Manufacturer's/Wholesaler's Sales Rep.; Mechanical Engineer; Physical Therapist; Services Sales Representative; Software Engineer. **Corporate headquarters location:** Dallas TX. **Other U.S. locations:** Nationwide. **Average salary range of placements:** $30,000 - $50,000. **Number of placements per year:** 100 - 199.

SOUTHERN EMPLOYMENT SERVICE

1233 54th Street, Columbus GA 31904. 706/327-6533. **Fax:** 706/323-7920. **Contact:** Human Resources. **Description:** An employment agency. **Specializes in the areas of:** Accounting/Auditing; Administration/MIS/EDP; Banking; Computer Hardware/Software; General Management; Industrial; Manufacturing; Personnel/Labor Relations; Sales and Marketing; Secretarial. **Positions commonly filled include:** Accountant/Auditor; Administrative Assistant; Bookkeeper; Civil Engineer; Clerk; Computer Operator; Computer Programmer; Credit Manager; Customer Service Representative; Data Entry Clerk; Manufacturing Engineer; Marketing Specialist; Mechanical Engineer; MIS Specialist; Operations/Production Manager; Quality Control Supervisor; Receptionist; Sales Manager; Secretary; Systems Analyst; Typist/Word Processor. **Number of placements per year:** 100 - 199.

TYLER STAFFING SERVICES

750 Hammond Drive, Building 4, Atlanta GA 30328. 404/250-9525. **Fax:** 404/851-1754. **Contact:** Roy Abernathy, President. **Description:** An employment agency that also offers contract services. Founded in 1979. Company pays fee. **Specializes in the areas of:** Computer Science/Software; Engineering; Insurance; Manufacturing; Printing/Publishing; Secretarial; Technical and Scientific. **Positions commonly filled include:** Adjuster; Claim Representative; Customer Service Representative; Design Engineer; Designer; Draftsperson; Electrical/Electronics Engineer; Mechanical Engineer; Occupational Therapist; Physical Therapist; Purchasing Agent/Manager; Quality Control Supervisor; Software Engineer; Typist/Word Processor; Underwriter/Assistant Underwriter. **Number of placements per year:** 50 - 99.

WPPS SOFTWARE STAFFING

230 Peachtree Street NW, Suite 1514, Atlanta GA 30303. 404/588-9100. **Fax:** 404/588-1395. **Contact:** Sharon Simpson, Customer Service Manager. **Description:** An employment agency. Founded in 1978. **Specializes in the areas of:** Accounting/Auditing; Administration/MIS/EDP; Advertising; Architecture/Construction/Real Estate; Banking; Computer Science/Software;

Finance; Legal; Nonprofit; Personnel/Labor Relations; Printing/Publishing; Sales and Marketing; Secretarial. **Positions commonly filled include:** Accountant/Auditor; Administrative Services Manager; Clerical Supervisor; Computer Programmer; Customer Service Representative; Designer; Editor; Human Resources Specialist; Human Service Worker; Paralegal; Software Engineer; Technical Writer/Editor; Typist/Word Processor. **Benefits available to temporary workers:** 401(k); Bonus Plan; Medical Insurance; Paid Holidays; Paid Vacation. **Corporate headquarters location:** Charlotte NC. **Other US locations:** Nationwide. **Average salary range of placements:** $20,000 - $29,999. **Number of placements per year:** 1 - 49.

EXECUTIVE SEARCH FIRMS OF HAWAII

DUNHILL PROFESSIONAL SEARCH OF HAWAII
Davies Pacific Center, 841 Bishop Street, Suite 420, Honolulu HI 96813. 808/524-2550. **Fax:** 808/533-2196. **Contact:** Nadine Stollenmaier, President. **E-mail address:** jobsrus@aloha.net. **World Wide Web address:** http://www.dunhillstaff.com. **Description:** An executive search firm. Company pays fee. **Specializes in the areas of:** Accounting/Auditing; Administration/MIS/EDP; Architecture/Construction/Real Estate; Banking; Computer Science/Software; Engineering; Finance; Food Industry; General Management; Industrial; Manufacturing; Sales and Marketing; Technical and Scientific. **Positions commonly filled include:** Accountant/Auditor; Bank Officer/Manager; Computer Operator; Computer Programmer; Data Entry Clerk; Engineer; Insurance Agent/Broker; Secretary. **Average salary range of placements:** $30,000 - $50,000. **Number of placements per year:** 1 - 49.

SALES CONSULTANTS OF HONOLULU
33 King Street, Honolulu HI 96813. 808/533-3282. **Contact:** Don Bishop, Recruiter. **Description:** An executive search firm. **Specializes in the areas of:** Computer Science/Software; Printing/ Publishing; Sales and Marketing; Transportation. **Positions commonly filled include:** Computer Programmer; Hotel Manager; Industrial Engineer; Software Engineer. **Other U.S. locations:** Nationwide. **Average salary range of placements:** $30,000 - $50,000. **Number of placements per year:** 50 - 99.

PERMANENT EMPLOYMENT AGENCIES OF IDAHO

VOLT SERVICES GROUP
8100 West Emerald Drive, Suite 120, Boise ID 83704. 208/375-9947. **Contact:** Recruiting. **Description:** An employment agency offering placements in clerical postions and technical positions for computer professionals. **Specializes in the areas of:** Accounting/Auditing; Clerical; Computer Programming; Engineering.

WESTERN STAFF SERVICES
1111 South Orchard Street, Suite 157, Boise ID 83705. 208/345-3828. **Contact:** Davida Corn, Office Manager. **Description:** An employment agency that places people in a wide variety of industries including the computer field. **Specializes in the areas of:** Clerical; Computer Hardware/Software; Industrial.

EXECUTIVE SEARCH FIRMS OF ILLINOIS

THE ABILITY GROUP
1011 East State Street, Rockford IL 61104. 815/964-0119. **Fax:** 815/964-9965. **Contact:** J. Lowell Hawkinson, Chairman. **Description:** An executive search firm and employment agency. Company pays fee. **Specializes in the areas of:** Accounting/Auditing; Administration/MIS/EDP; Computer Science/Software; Engineering; Industrial; Manufacturing; Personnel/Labor Relations; Technical and Scientific. **Positions commonly filled include:** Accountant/Auditor; Aerospace Engineer; Aircraft Mechanic/Engine Specialist; Automotive Mechanic; Biomedical Engineer; Blue-Collar Worker Supervisor; Buyer; Chemical Engineer; Civil Engineer; Clerical Supervisor; Computer Programmer; Customer Service Representative; Design Engineer; Designer; Draftsperson; Electrical/Electronics Engineer; Electrician; Environmental Engineer; Human Resources Manager; Industrial Engineer; Management Trainee; Mathematician; Mechanical Engineer; Metallurgical Engineer; MIS Specialist; Purchasing Agent/Manager; Software Engineer; Stationary Engineer; Statistician; Structural Engineer; Systems Analyst; Technical Writer/Editor. **Benefits available to temporary workers:** Insurance; Paid Holidays. **Corporate headquarters location:** This Location. **Other U.S. locations:** Oregon IL; Beloit WI. **Average salary range of placements:** $30,000 - $50,000. **Number of placements per year:** 100 - 199.

ACCORD INC.
1543 Westchester Boulevard, Westchester IL 60154. 708/345-7900. **Contact:** Chester Dombrowski, Manager. **Description:** An executive search firm. Company pays fee. **Specializes in the areas of:** Computer Science/Software; Engineering; Manufacturing; MIS/EDP; Technical and Scientific. **Positions commonly filled include:** Computer Programmer; Customer Service Representative; Draftsperson; EDP Specialist; Electrical/Electronics Engineer; General Manager; Industrial Engineer; Industrial Production Manager; Mechanical Engineer; MIS Specialist; Quality Control Supervisor; Sales Representative; Software Engineer; Technical Writer/Editor; Technician. **Number of placements per year:** 50 - 99.

AMERICAN ENGINEERING COMPANY
5151 North Harlem Avenue, Suite 312, Chicago IL 60656-3610. 773/775-0030. **Fax:** 773/775-0028. **Contact:** Anthony Davero, President. **Description:** A retainer and contingency search firm. Company pays fee. **Specializes in the areas of:** Accounting/Auditing; Computer Science/Software; Engineering; Finance; Manufacturing; Sales and Marketing; Secretarial; Technical and Scientific. **Positions commonly filled include:** Accountant/Auditor; Administrative Services Manager; Attorney; Blue-Collar Worker Supervisor; Budget Analyst; Buyer; Clerical Supervisor; Computer Programmer; Cost Estimator; Counselor; Credit Manager; Customer Service Representative; Design Engineer; Designer; Draftsperson; Electrical/Electronics Engineer; Financial Analyst; Human Resources Manager; Industrial Engineer; Industrial Production Manager; Insurance Agent/Broker; Internet Services Manager; Management Analyst/Consultant; Manufacturer's/Wholesaler's Sales Rep.; Market Research Analyst; Mechanical Engineer; Metallurgical Engineer; MIS Specialist; Operations/Production Manager; Paralegal; Public Relations Specialist; Purchasing Agent/Manager; Quality Control Supervisor; Software Engineer; Stationary Engineer; Systems Analyst; Technical Writer/Editor; Telecommunications Manager; Typist/Word Processor; Underwriter/Assistant Underwriter. **Corporate headquarters location:** This Location. **Average salary range of placements:** More than $50,000. **Number of placements per year:** 200 - 499.

B.D.G. SOFTWARE NETWORK
1785 Woodhaven Drive, Crystal Lake IL 60014-1941. 815/477-2334. **Fax:** 815/477-2265. **Contact:** Barry D. Gruner, President. **Description:** An executive search firm operating on both retainer and contingency bases. The firm specializes in the placement of MIS and technical support professionals. Company pays fee. **Specializes in the areas of:** Computer Science/Software. **Positions commonly filled include:** Computer Programmer; Management Analyst/Consultant; Market Research Analyst; MIS Specialist; Software Engineer; Systems Analyst; Telecommunications Manager. **Corporate headquarters location:** This Location. **Other U.S. locations:** Nationwide. **Average salary range of placements:** More than $50,000. **Number of placements per year:** 1 - 49.

BRITANNIA
160 East Chicago Street, Elgin IL 60120-5524. 847/697-4600. **Fax:** 847/697-4608. **Contact:** Ellen Gann, Division Coordinator. **Description:** A contingency search firm and employment agency. Company pays fee. **Specializes in the areas of:** Computer Science/Software; Engineering; Industrial; Manufacturing; Sales and Marketing; Secretarial; Technical and Scientific. **Positions commonly filled include:** Architect; Biological Scientist; Civil Engineer; Computer Programmer; Customer Service Representative; Design Engineer; Draftsperson; Electrical/Electronics Engineer; Industrial Engineer; Industrial Production Manager; Mechanical Engineer; Operations/Production Manager; Purchasing Agent/Manager; Quality Control Supervisor; Software Engineer; Systems Analyst; Technical Writer/Editor; Telecommunications Manager. **Average salary range of placements:** $30,000 - $50,000. **Number of placements per year:** 50 - 99.

BUSINESS SYSTEMS OF AMERICA, INC.
150 North Wacker Drive, Suite 2970, Chicago IL 60606. 312/849-9222. **Fax:** 312/849-9260. **Contact:** Lou Costabile, Recruiter. **World Wide Web address:** http://www.bussyam.com. **Description:** An executive search firm providing consultants in the areas of LAN and PC technical support. The firm also provides temporary services. Company pays fee. **Specializes in the areas of:** Computer Science/Software. **Positions commonly filled include:** Electrical/Electronics Engineer; Software Engineer; Systems Analyst; Technical Writer/Editor. **Corporate headquarters location:** This Location. **Average salary range of placements:** $30,000 - $50,000. **Number of placements per year:** 50 - 99.

C-U EMPLOYMENT SERVICE
510 West Park Avenue, Apartment E, Champaign IL 61820-3971. 217/351-7105. **Fax:** 217/351-7333. **Contact:** Mr. Burke, Owner. **Description:** A contingency search firm and employment agency. Company pays fee. **Specializes in the areas of:** Computer Science/Software; Engineering; Industrial; Manufacturing; Sales and Marketing. **Positions commonly filled include:** Accountant/Auditor; Aerospace Engineer; Bank Officer/Manager; Branch Manager; Buyer; Civil Engineer; Clerical Supervisor; Computer Programmer; Credit Manager; Customer Service Representative; Design Engineer; Draftsperson; Electrical/Electronics Engineer; Environmental Engineer; Financial Consultant; General Manager; Human Resources Manager; Industrial Engineer; Industrial Production Manager; Management Trainee; Mechanical Engineer; Metallurgical Engineer; MIS Specialist; Operations/Production Manager; Psychologist; Purchasing Agent/Manager; Quality Control Supervisor; Restaurant/Food Service Manager; Services Sales Representative; Software Engineer; Statistician; Structural Engineer; Systems Analyst; Telecommunications Manager; Typist/Word Processor. **Corporate headquarters location:** This Location. **Number of placements per year:** 1 - 49.

C.E.S. ASSOCIATES
112 South Grant Street, Hinsdale IL 60521. 630/654-2596. **Fax:** 630/654-2713. **Contact:** Dr. James F. Baker, President. **Description:** An executive search firm. Company pays fee. **Specializes in the areas of:** Accounting/Auditing; Administration/MIS/EDP; Architecture/Construction/Real Estate; Banking; Computer Science/Software; Education; Engineering; Finance; General Management; Health/Medical; Insurance; Manufacturing; Personnel/Labor Relations; Sales and Marketing. **Positions commonly filled include:** Accountant/Auditor; Bank Officer/Manager; Chemical Engineer; Chemist; Computer Programmer; Counselor; Financial Analyst; Health Services Manager; Hotel Manager; Human Resources Manager; Nuclear Medicine Technologist; Occupational Therapist; Operations/Production Manager; Pharmacist; Physical Therapist; Physician; Property and Real Estate Manager; Psychologist; Public Relations Specialist; Purchasing Agent/Manager; Radiologic Technologist; Real Estate Agent; Recreational Therapist; Registered Nurse; Respiratory Therapist; Social Worker; Speech-Language Pathologist; Systems Analyst; Teacher/Professor. **Number of placements per year:** 1 - 49.

C.R.T., INC.
218 South Main Street, Wauconda IL 60084. 847/816-0610. **Fax:** 847/487-0195. **Contact:** William Mellor, Executive Recruiter. **Description:** An executive search firm. Company pays fee. **Specializes in the areas of:** Computer Science/Software. **Positions commonly filled include:** Computer Programmer; Operations/Production Manager; Software Engineer; Systems Analyst; Technical Writer/Editor. **Number of placements per year:** 1 - 49.

CHICAGO FINANCIAL SEARCH

125 South Wacker Drive, Suite 300, Chicago IL 60606. 312/207-0400. **Contact:** Mike Kelly, Manager. **Description:** An executive search firm. Company pays fee. **Specializes in the areas of:** Accounting/Auditing; Banking; Brokerage; Computer Science/Software; Finance. **Positions commonly filled include:** Accountant/Auditor; Computer Operator; Computer Programmer; MIS Specialist. **Number of placements per year:** 100 - 199.

COMPUPRO

1117 South Milwaukee Avenue, Suite B9, Libertyville IL 60048. 847/549-8603. **Fax:** 847/549-7429. **Contact:** Doug Baniqued, Human Resources. **Description:** An executive search firm. Company pays fee. **Specializes in the areas of:** Administration/MIS/EDP; Computer Science/Software. **Positions commonly filled include:** Computer Programmer; Financial Analyst; Management Analyst/Consultant; Software Engineer; Systems Analyst; Technical Writer/Editor. **Number of placements per year:** 1 - 49.

COMPUTER FUTURES EXCHANGE, INC.

1111 Westgate, Oak Park IL 60301. 708/445-8000. **Fax:** 708/445-1498. **Contact:** Corey D. Gimbel, President. **Description:** An executive search firm. Company pays fee. **Specializes in the areas of:** Administration/MIS/EDP; Banking; Computer Science/Software; Economics; Finance. **Positions commonly filled include:** Computer Programmer; Economist; Mathematician; Securities Sales Representative; Software Engineer; Systems Analyst. **Number of placements per year:** 50 - 99.

CORPORATE CONSULTANTS

480 Central Avenue, Northfield IL 60093-3016. 847/446-5627. **Fax:** 847/446-3536. **Contact:** Allen Arends, President. **Description:** Corporate Consultants is a recruitment firm specializing in data/telecommunications and computer-related industries. Company pays fee. **Specializes in the areas of:** Computer Science/Software; Sales and Marketing. **Positions commonly filled include:** Computer Programmer; Customer Service Representative; Manufacturer's/Wholesaler's Sales Rep.; Market Research Analyst; MIS Specialist; Multimedia Designer; Services Sales Representative; Software Engineer; Strategic Relations Manager; Systems Analyst; Technical Writer/Editor; Telecommunications Manager. **Corporate headquarters location:** This Location. **Other U.S. locations:** Nationwide. **Average salary range of placements:** More than $50,000. **Number of placements per year:** 1 - 49.

DATA CAREER CENTER, INC.

225 North Michigan Avenue, Suite 930, Chicago IL 60601. 312/565-1060. **Fax:** 312/565-0246. **Contact:** Larry Chaplik, President. **Description:** An executive search firm. Company pays fee. **Specializes in the areas of:** Computer Science/Software. **Positions commonly filled include:** Computer Programmer; Management Analyst/Consultant; Software Engineer; Systems Analyst. **Number of placements per year:** 1 - 49.

DATAQUEST INC.

7105 Virginia Road, Suite 2B, Crystal Lake IL 60014-7985. 815/356-7500. **Fax:** 815/477-2359. **Contact:** Robert Vincent, Managing Consultant. **E-mail address:** dtaqst@aol.com. **Description:** Dataquest is an executive search firm operating on a contingency basis and specializing in placing information systems professionals. Company pays fee. **Specializes in the areas of:** Computer Science/ Software. **Positions commonly filled include:** Computer Programmer; Systems Analyst. **Corporate headquarters location:** This Location. **Average salary range of placements:** $30,000 to more than $50,000. **Number of placements per year:** 50 - 99.

DUNHILL PROFESSIONAL SEARCH

5005 Newport Drive, Suite 201, Rolling Meadows IL 60008. 847/398-3400. **Contact:** Russ Kunke, Consultant. **Description:** An executive search firm. Company pays fee. **Specializes in the areas of:** Administration/MIS/EDP; Computer Science/Software; Engineering; Sales and Marketing. **Positions commonly filled include:** Branch Manager; Computer Programmer; Electrical/Electronics Engineer; Software Engineer; Systems Analyst. **Number of placements per year:** 1 - 49.

DYNAMIC SEARCH SYSTEMS, INC.
3800 North Wilke Road, Suite 485, Arlington Heights IL 60004. 847/259-3444. **Fax:** 847/259-3480. **Contact:** Michael J. Brindise, President. **Description:** An employment services firm, founded in 1983, that specializes in MIS professional staffing. Company pays fee. **Specializes in the areas of:** Administration/MIS/EDP; Computer Science/Software. **Positions commonly filled include:** Computer Programmer; Internet Services Manager; Management Analyst/Consultant; MIS Specialist; Software Engineer. **Average salary range of placements:** More than $50,000. **Number of placements per year:** 200 - 499.

EXECUTIVE CONCEPTS INC.
1000 East Woodfield Road, Schaumburg IL 60173-4728. 847/605-8300. **Fax:** 847/605-8089. **Contact:** Tom Werle, President. **Description:** An executive search firm operating on both retainer and contingency bases that specializes in data processing. Company pays fee. **Specializes in the areas of:** Computer Science/Software. **Positions commonly filled include:** Computer Programmer; Systems Analyst. **Corporate headquarters location:** This Location. **Average salary range of placements:** $30,000 - $50,000. **Number of placements per year:** 50 - 99.

EXECUTIVE SEARCH INTERNATIONAL
4300 North Brandywine Drive, Suite 104, Peoria IL 61614-5550. 309/685-6273. **Contact:** Robert Vaughan, President. **Description:** An executive search firm. Company pays fee. **Specializes in the areas of:** Computer Science/Software; Engineering; General Management; Industrial; Manufacturing; Sales and Marketing; Technical and Scientific. **Positions commonly filled include:** Biomedical Engineer; Branch Manager; Chemical Engineer; Civil Engineer; Computer Programmer; Electrical/Electronics Engineer; General Manager; Human Resources Manager; Industrial Engineer; Industrial Production Manager; Manufacturer's/Wholesaler's Sales Rep.; Mechanical Engineer; MIS Specialist; Nuclear Medicine Technologist; Software Engineer; Systems Analyst; Telecommunications Manager. **Corporate headquarters location:** This Location. **Other U.S. locations:** Nationwide. **Average salary range of placements:** More than $50,000. **Number of placements per year:** 50 - 99.

EXECUTIVE SEARCH NETWORK
7 1/2 East Miner Street, Arlington Heights IL 60004. 847/394-1805. **Fax:** 847/394-1841. **Contact:** Mike Tollefson, President. **Description:** An executive search firm. Company pays fee. **Specializes in the areas of:** Computer Hardware/Software; Engineering; Technical and Scientific. **Positions commonly filled include:** Aerospace Engineer; Buyer; Ceramics Engineer; Chemical Engineer; Computer Programmer; Electrical/Electronics Engineer; Industrial Engineer; Manufacturing Engineer; Marketing Specialist; Mechanical Engineer; Metallurgical Engineer; Purchasing Agent/Manager; Quality Control Supervisor; Software Engineer; Technical Representative; Technical Writer/Editor. **Number of placements per year:** 1 - 49.

GENERAL EMPLOYMENT ENTERPRISES, INC.
280 Shuman Boulevard, Suite 185, Naperville IL 60563. 630/983-1233. **Fax:** 630/983-2993. **Contact:** Colin Reed, Manager. **Description:** An executive search firm specializing in recruitment and placement, both permanent and contract, in information systems and technologies, accounting/finance, and engineering. Company pays fee. **Specializes in the areas of:** Accounting/Auditing; Administration/MIS/EDP; Computer Science/Software; Engineering; Technical and Scientific. **Positions commonly filled include:** Accountant/Auditor; Budget Analyst; Chemical Engineer; Computer Programmer; Credit Manager; Customer Service Representative; Design Engineer; Designer; Draftsperson; Electrical/Electronics Engineer; Financial Analyst; Industrial Engineer; Industrial Production Manager; Mechanical Engineer; Metallurgical Engineer; MIS Specialist; Multimedia Designer; Purchasing Agent/Manager; Quality Control Supervisor; Software Engineer; Structural Engineer; Systems Analyst; Technical Writer/Editor; Telecommunications Manager. **Corporate headquarters location:** Oakbrook Terrace IL. **Other U.S. locations:** Nationwide. **Average salary range of placements:** More than $50,000. **Number of placements per year:** 1 - 49.

GIRMAN GROUP
40 Shuman Boulevard, Suite 160, Naperville IL 60563. 630/260-0040. **Fax:** 630/260-0094.

Contact: Jerrilyn Girman, Partner. **Description:** An executive search firm. **Specializes in the areas of:** Administration/MIS/EDP; Computer Hardware/Software; Manufacturing; Sales and Marketing; Technical and Scientific. **Positions commonly filled include:** Computer Programmer; Consultant; Sales Manager; Sales Representative; Software Engineer; Systems Analyst. **Number of placements per year:** 50 - 99.

DAVID GOMEZ & ASSOCIATES

20 North Clark Street, Suite 3535, Chicago IL 60602-5002. 312/346-5525. **Fax:** 312/346-1438. **Contact:** Consultant. **E-mail address:** 102542.2313@compuserve.com. **Description:** A generalist executive search firm serving all areas of business. Company pays fee. **Specializes in the areas of:** Accounting/Auditing; Administration/MIS/EDP; Advertising; Art/Design; Banking; Computer Science/Software; Finance; General Management; Insurance; Manufacturing; Personnel/Labor Relations; Printing/Publishing; Sales and Marketing; Secretarial. **Positions commonly filled include:** Accountant/Auditor; Administrative Services Manager; Advertising Clerk; Bank Officer/Manager; Branch Manager; Budget Analyst; Computer Programmer; Credit Manager; Customer Service Representative; Design Engineer; Designer; Financial Analyst; General Manager; Human Resources Manager; Librarian; Management Analyst/Consultant; Market Research Analyst; MIS Specialist; Operations/Production Manager; Software Engineer; Systems Analyst; Telecommunications Manager; Typist/Word Processor; Underwriter/Assistant Underwriter. **Corporate headquarters location:** This Location. **Average salary range of placements:** More than $50,000. **Number of placements per year:** 200 - 499.

HILLON & REDELL LTD.

P.O. Box 25067, Chicago IL 60625. 773/271-4449. **Fax:** 773/271-9689. **Contact:** John T. Redell, Jr., Managing Partner. **Description:** An executive search firm. Company pays fee. **Specializes in the areas of:** Administration/MIS/EDP; Computer Science/Software. **Positions commonly filled include:** Computer Programmer; Software Engineer; Systems Analyst; Technical Writer/Editor. **Number of placements per year:** 50 - 99.

THE HUNTER RESOURCE GROUP, INC.

One North LaSalle, 24th Floor, Chicago IL 60602. 312/201-0302. **Fax:** 312/201-0402. **Contact:** Frank Scarpelli, President. **E-mail address:** fshunter@aol.com. **Description:** An executive search firm specializing in technology, human resources, accounting, and finance. The Hunter Resource Group works with mid-sized and *Fortune* 500 firms throughout North America, Europe, and Asia. Company pays fee. **Specializes in the areas of:** Administration/MIS/EDP; Computer Science/Software; General Management; Personnel/Labor Relations. **Positions commonly filled include:** Accountant/Auditor; Computer Programmer; Financial Analyst; General Manager; Human Resources Manager; Internet Services Manager; MIS Specialist; Multimedia Designer; Software Engineer; Systems Analyst. **Corporate headquarters location:** This Location. **Average salary range of placements:** More than $50,000.

INTERVIEWING CONSULTANTS INC.
ICI TEMPORARIES

19 South LaSalle Street, Suite 900, Chicago IL 60603. 312/263-1710. **Contact:** Ron 'Gia' Giambarberee, Owner/Manager. **Description:** This location functions as an executive search firm, a permanent employment agency, and a temporary employment agency. Company pays fee. **Specializes in the areas of:** Accounting/Auditing; Banking; Clerical; Computer Science/Software; Engineering; General Management; Insurance; Office Support; Sales and Marketing; Secretarial; Temporary Assignments. **Positions commonly filled include:** Accountant/Auditor; Adjuster; Administrative Services Manager; Advertising Clerk; Brokerage Clerk; Budget Analyst; Buyer; Claim Representative; Clerical Supervisor; Computer Programmer; Cost Estimator; Customer Service Representative; Electrical/Electronics Engineer; Financial Analyst; General Manager; Human Resources Manager; Management Analyst/Consultant; Management Trainee; Manufacturer's/Wholesaler's Sales Rep.; Mechanical Engineer; Medical Records Technician; Paralegal; Property and Real Estate Manager; Securities Sales Representative; Services Sales Representative; Software Engineer; Systems Analyst; Underwriter/Assistant Underwriter. **Number of placements per year:** 200 - 499.

JOHNSON PERSONNEL COMPANY
861 North Madison Street, Rockford IL 61107. 815/964-0840. **Fax:** 815/964-0855. **Contact:** Darrell Johnson, Owner. **Description:** An executive search firm. Company pays fee. **Specializes in the areas of:** Accounting/Auditing; Computer Science/Software; Engineering; Manufacturing; Personnel/Labor Relations. **Positions commonly filled include:** Accountant/Auditor; Buyer; Ceramics Engineer; Computer Programmer; Electrical/Electronics Engineer; Financial Analyst; Human Resources Manager; Industrial Engineer; Materials Engineer; Mechanical Engineer; Metallurgical Engineer; Operations/Production Manager; Quality Control Supervisor; Software Engineer; Statistician; Systems Analyst. **Number of placements per year:** 1 - 49.

JERRY L. JUNG CO.
140 Iowa Avenue, Belleville IL 62220-3940. 618/277-8881. **Fax:** 618/277-8386. **Contact:** Jerry Jung, Principal. **Description:** An executive search firm. **Specializes in the areas of:** Computer Science/Software; Engineering; Manufacturing. **Positions commonly filled include:** Aerospace Engineer; Agricultural Engineer; Chemical Engineer; Computer Programmer; Design Engineer; Designer; Draftsperson; Electrical/Electronics Engineer; Environmental Engineer; General Manager; Industrial Engineer; Industrial Production Manager; Mechanical Engineer; Metallurgical Engineer; Operations/Production Manager; Quality Control Supervisor; Software Engineer. **Corporate headquarters location:** This Location. **Other U.S. locations:** Nationwide. **Average salary range of placements:** More than $50,000. **Number of placements per year:** 1 - 49.

SAMUEL F. KROLL & ASSOCIATES
1804 North Naper Boulevard, Naperville IL 60563-8800. 630/505-5825. **Fax:** 630/505-5826. **Contact:** Samuel F. Kroll, Principal. **E-mail address:** 73361.2153@compuserve.com. **Description:** An executive search firm that offers permanent opportunities for information systems professionals ranging from hands-on technical specialists to directors of MIS. Company pays fee. **Specializes in the areas of:** Administration/MIS/EDP; Computer Science/Software. **Positions commonly filled include:** Computer Programmer; MIS Specialist; Systems Analyst; Telecommunications Manager. **Corporate headquarters location:** This Location. **Average salary range of placements:** More than $50,000. **Number of placements per year:** 1 - 49.

ARLENE LEFF & ASSOCIATES
203 North LaSalle Street, Suite 2100, Chicago IL 60601. 312/558-1350. **Fax:** 312/558-1350. **Contact:** Arlene Leff, Owner. **Description:** An executive search firm. The company has two divisions: Support Level and Executive Search. Company pays fee. **Specializes in the areas of:** Accounting/Auditing; Administration/MIS/EDP; Advertising; Computer Science/Software; Engineering; Fashion; Finance; General Management; Health/Medical; Industrial; Insurance; Printing/Publishing; Retail; Sales and Marketing; Secretarial. **Positions commonly filled include:** Accountant/Auditor; Administrative Services Manager; Advertising Clerk; Bank Officer/Manager; Blue-Collar Worker Supervisor; Branch Manager; Buyer; Chemical Engineer; Computer Programmer; Construction Contractor; Cost Estimator; Credit Manager; Design Engineer; Electrical/Electronics Engineer; General Manager; Human Resources Manager; Industrial Engineer; Management Trainee; MIS Specialist; Purchasing Agent/Manager; Quality Control Supervisor; Services Sales Representative; Software Engineer; Systems Analyst; Technical Writer/Editor; Typist/Word Processor. **Number of placements per year:** 50 - 99.

LYNCO MANAGEMENT PERSONNEL
P.O. Box 343, Geneva IL 60134-0343. 630/801-1600. **Contact:** Lynn Buehler, General Manager. **Description:** A contingency search firm. Company pays fee. **Specializes in the areas of:** Accounting/Auditing; Administration/MIS/EDP; Banking; Computer Science/Software; Engineering; Finance; Food Industry; General Management; Industrial; Manufacturing; Personnel/Labor Relations; Printing/Publishing; Sales and Marketing; Technical and Scientific. **Positions commonly filled include:** Accountant/Auditor; Administrative Services Manager; Agricultural Engineer; Bank Officer/Manager; Biomedical Engineer; Blue-Collar Worker Supervisor; Branch Manager; Budget Analyst; Buyer; Chemical Engineer; Chemist; Civil Engineer; Computer Programmer; Data Analyst; Design Engineer; Education Administrator; Electrical/Electronics Engineer; Environmental Engineer; Financial Analyst; General Manager; Hotel Manager; Human Resources Manager; Industrial Engineer; Industrial Production Manager; Insurance Agent/Broker; Internet Services Manager; Management Analyst/Consultant;

Management Trainee; Manufacturer's/Wholesaler's Sales Rep.; Mechanical Engineer; Metallurgical Engineer; MIS Specialist; Nuclear Engineer; Operations/Production Manager; Paralegal; Petroleum Engineer; Public Relations Specialist; Purchasing Agent/Manager; Quality Control Supervisor; Restaurant/Food Service Manager; Services Sales Representative; Software Engineer; Stationary Engineer; Strategic Relations Manager; Systems Analyst; Technical Writer/Editor; Transportation/ Traffic Specialist; Video Production Coordinator. **Corporate headquarters location:** This Location. **Other U.S. locations:** Aurora IL. **Average salary range of placements:** $30,000 - $50,000. **Number of placements per year:** 100 - 199.

MACRO RESOURCES
47 West Polk Street, Suite 204, Chicago IL 60605-2000. **Fax:** 312/663-0883. **Contact:** Frank Roti, President. **Description:** A retainer and contingency search firm. Company pays fee. **Specializes in the areas of:** Banking; Computer Science/Software; Engineering; Finance. **Positions commonly filled include:** Electrical/Electronics Engineer; Mathematician; MIS Specialist; Software Engineer. **Corporate headquarters location:** This Location. **Average salary range of placements:** More than $50,000. **Number of placements per year:** 1 - 49.

MANAGEMENT RECRUITERS INTERNATIONAL
211 Landmark Drive, Suite E1, Normal IL 61761. 309/452-1844. **Fax:** 309/452-0403. **Contact:** M. Allan Snedden, Owner/General Manager. **Description:** An executive search firm. **Specializes in the areas of:** Accounting/Auditing; Administration/MIS/EDP; Computer Programming; Engineering; Finance; Health/Medical; Manufacturing; Personnel/Labor Relations. **Positions commonly filled include:** Accountant/Auditor; Computer Programmer; EDP Specialist; Human Resources Manager; Industrial Engineer; Manufacturing Engineer; MIS Specialist; Systems Analyst. **Number of placements per year:** 50 - 99.

MANAGEMENT RECRUITERS OF DES PLAINES, INC.
1400 East Touhy Avenue, Suite 160, Des Plaines IL 60018. 847/297-7102. **Fax:** 847/297-8477. **Contact:** Manager. **Description:** An executive search firm. **Specializes in the areas of:** Accounting/Auditing; Administration/MIS/EDP; Advertising; Architecture/ Construction/Real Estate; Banking; Chemical; Communications; Computer Science/Software; Design; Electrical; Engineering; Finance; Food Industry; General Management; Health/Medical; Industrial; Insurance; Legal; Manufacturing; Operations Management; Personnel/Labor Relations; Pharmaceutical; Printing/Publishing; Procurement; Retail; Sales and Marketing; Technical and Scientific; Textiles; Transportation.

MANAGEMENT RECRUITERS OF ROCKFORD, INC.
1463 South Bell School Road, Suite 3, Rockford IL 61108. 815/399-1942. **Contact:** D. Michael Carter, Manager. **Description:** An executive search firm. **Specializes in the areas of:** Accounting/Auditing; Administration/MIS/EDP; Advertising; Architecture/Construction/Real Estate; Banking; Chemical; Communications; Computer Science/Software; Design; Electrical; Engineering; Film Production; Food Industry; General Management; Health/Medical; Industrial; Insurance; Legal; Manufacturing; Operations Management; Personnel/Labor Relations; Pharmaceutical; Printing/Publishing; Procurement; Retail; Sales and Marketing; Technical and Scientific; Textiles; Transportation.

MANAGEMENT RECRUITERS OF SPRINGFIELD
124 East Laurel Street, Suite B, Springfield IL 62704-3946. 217/544-2051. **Fax:** 217/544-2055. **Contact:** Mark Cobb, General Manager. **E-mail address:** sprfield!mac@mrinet.com. **World Wide Web address:** http://www.mrinet.com. **Description:** A retainer and contingency executive search firm. Company pays fee. **Specializes in the areas of:** Architecture/Construction/Real Estate; Biotechnology; Computer Science/Software; Engineering; Manufacturing; Pharmaceutical; Technical and Scientific. **Positions commonly filled include:** Biochemist; Biological Scientist; Biomedical Engineer; Chemical Engineer; Chemist; Civil Engineer; Clinical Lab Technician; Construction Contractor; Cost Estimator; Electrical/Electronics Engineer; Environmental Engineer; Mechanical Engineer; Operations/Production Manager; Quality Control Supervisor; Software Engineer. **Corporate headquarters location:** Cleveland OH. **Other U.S. locations:** Nationwide. **Average salary range of placements:** More than $50,000. **Number of placements per year:** 1 - 49.

PAUL MAY & ASSOCIATES (PMA)

730 North Franklin, Suite 612, Chicago IL 60610-3526. 312/649-8400. **Fax:** 312/649-8999. **Contact:** Paul May, President. **E-mail address:** PMA4jobs@aol.com. **Description:** An executive search firm operating on a contingency basis that specializes in recruiting a wide range of information systems professionals. **Specializes in the areas of:** Computer Science/Software; Sales and Marketing. **Positions commonly filled include:** Computer Programmer; Internet Services Manager; Management Trainee; Systems Analyst. **Corporate headquarters location:** This Location. **Average salary range of placements:** $30,000 - $50,000. **Number of placements per year:** 50 - 99.

M.W. McDONALD & ASSOCIATES, INC.

P.O. Box 541, Barrington IL 60010-0541. 630/238-0980. **Fax:** 630/238-0984. **Contact:** Pamela Niedermeier, Director of Research and Operations. **E-mail address:** mwminc@ais.net. **Description:** An executive search firm operating on a retainer basis. **Specializes in the areas of:** Computer Science/Software; Engineering; Sales and Marketing; Technical and Scientific. **Positions commonly filled include:** Computer Programmer; Design Engineer; Designer; Draftsperson; Electrical/Electronics Engineer; Industrial Engineer; Mechanical Engineer; Metallurgical Engineer; Multimedia Designer; Systems Analyst. **Corporate headquarters location:** This Location. **Other U.S. locations:** San Francisco CA. **Average salary range of placements:** More than $50,000. **Number of placements per year:** 50 - 99.

JUAN MENEFEE & ASSOCIATES

503 South Oak Park Avenue, Oak Park IL 60304. 708/848-7722. **Fax:** 708/848-6008. **Contact:** Juan F. Menefee, President. **Description:** An executive search firm. Company pays fee. **Specializes in the areas of:** Accounting/Auditing; Computer Science/Software; Engineering; Food Industry; General Management; Health/Medical; Insurance; Legal; Manufacturing; Personnel/Labor Relations; Sales and Marketing; Secretarial. **Positions commonly filled include:** Accountant/Auditor; Administrative Services Manager; Attorney; Bank Officer/Manager; Branch Manager; Chemist; Computer Programmer; Electrical/Electronics Engineer; Health Services Manager; Human Resources Manager; Industrial Engineer; Management Analyst/Consultant; Management Trainee; Mechanical Engineer; Physician; Public Relations Specialist; Registered Nurse; Software Engineer; Travel Agent. **Number of placements per year:** 50 - 99.

NU-WAY SEARCH

P.O. Box 494, Lake Zurich IL 60047-0494. **Contact:** Steve Riess, Counselor. **Description:** A contingency search firm. Company pays fee. **Specializes in the areas of:** Computer Science/Software. **Positions commonly filled include:** Computer Programmer; Internet Services Manager; Software Engineer; Statistician; Systems Analyst; Technical Writer/Editor; Telecommunications Manager. **Corporate headquarters location:** This Location. **Number of placements per year:** 50 - 99.

OFFICEMATES5 OF WHEELING

1400 East Lake Cook Road, Suite 115, Buffalo Grove IL 60089. 847/459-6160. **Contact:** Human Resources. **Description:** An executive search firm. **Specializes in the areas of:** Accounting/Auditing; Administration/MIS/EDP; Advertising; Architecture/Construction/Real Estate; Banking; Chemical; Communications; Computer Science/Software; Electrical; Engineering; Finance; Food Industry; General Management; Health/Medical; Legal; Manufacturing; Operations Management; Pharmaceutical; Printing/Publishing; Procurement; Retail; Sales and Marketing; Technical and Scientific; Textiles; Transportation.

OMEGA TECHNICAL CORPORATION

15 Spinning Wheel Road, Suite 120, Hinsdale IL 60521. 630/986-8116. **Fax:** 630/986-0036. **Contact:** Recruiter. **Description:** An executive search firm. Company pays fee. **Specializes in the areas of:** Administration/MIS/EDP; Computer Science/Software; Engineering; Industrial; Manufacturing; Personnel/Labor Relations; Technical and Scientific. **Positions commonly filled include:** Aerospace Engineer; Agricultural Engineer; Biomedical Engineer; Chemical Engineer; Civil Engineer; Computer Programmer; Designer; Draftsperson; Electrical/Electronics Engineer; Electrician; Industrial Engineer; Industrial Production Manager; Mechanical Engineer; Metallurgical Engineer; Nuclear Engineer; Purchasing Agent/Manager; Quality Control

Supervisor; Software Engineer; Structural Engineer; Systems Analyst; Technical Writer/Editor. **Number of placements per year:** 100 - 199.

PROFESSIONAL PLACEMENT SERVICES
1314 West Northwest Highway, Suite 112, Arlington Heights IL 60004. 847/253-5300. **Contact:** Patrick M. Latimer, President. **Description:** An executive search firm specializing in the data processing field. Company pays fee. **Specializes in the areas of:** Administration/MIS/EDP; Computer Science/Software. **Positions commonly filled include:** Computer Programmer; MIS Specialist; Software Engineer; Systems Analyst; Technical Writer/Editor. **Corporate headquarters location:** This Location. **Other U.S. locations:** Nationwide. **Average salary range of placements:** $30,000 - $50,000. **Number of placements per year:** 1 - 49.

PROFESSIONAL SEARCH CENTER LIMITED
1515 East Woodfield Road, Suite 830, Schaumburg IL 60173-6046. 847/330-3250. **Fax:** 847/330-3255. **Contact:** Jerry Hirschel, CPC, President. **E-mail address:** hdhunter1@aol.com. **Description:** An executive search firm operating on both retainer and contingency bases that specializes in placing information systems professionals. **Specializes in the areas of:** Computer Science/Software. **Positions commonly filled include:** Computer Programmer; Systems Analyst.

QUANTUM PROFESSIONAL SEARCH
QUANTUM STAFFING SERVICES
100 West 22nd Street, Suite 115, Lombard IL 60148. 630/916-7300. **Fax:** 630/916-8338. **Contact:** Patrick Brady, President. **E-mail address:** patbrady1@aol.com. **Description:** An executive search firm. Company pays fee. **Specializes in the areas of:** Computer Science/Software; General Management; Industrial; Manufacturing; Sales and Marketing; Secretarial. **Positions commonly filled include:** Branch Manager; Computer Programmer; Customer Service Representative; General Manager; Industrial Engineer; Industrial Production Manager; Manufacturer's/Wholesaler's Sales Rep.; Operations/Production Manager; Purchasing Agent/Manager; Services Sales Representative; Software Engineer; Systems Analyst; Technical Writer/Editor. **Corporate headquarters location:** This Location. **Average salary range of placements:** More than $50,000. **Number of placements per year:** 100 - 199.

RESPONSIVE SEARCH INC.
999 Oakmont Plaza, Suite 535, Westmont IL 60559-5517. 630/789-1300. **Fax:** 630/789-1338. **Contact:** Keith D. Hansel, President. **Description:** An executive search firm operating on both retainer and contingency bases that specializes in the computer industry. **Specializes in the areas of:** Computer Science/Software. **Positions commonly filled include:** Computer Programmer; Systems Analyst. **Corporate headquarters location:** This Location. **Other U.S. locations:** Nationwide. **Average salary range of placements:** More than $50,000. **Number of placements per year:** 50 - 99.

SHS, INC.
205 West Wacker Drive, Suite 600, Chicago IL 60606. 312/419-0370. **Fax:** 312/419-8953. **Contact:** Ric Pantaleo, Manager. **Description:** An executive search firm. Company pays fee. **Specializes in the areas of:** Administration/MIS/EDP; Advertising; Art/Design; Broadcasting; Computer Science/Software; Food Industry; General Management; Health/Medical; Industrial; Manufacturing; Printing/Publishing; Sales and Marketing. **Positions commonly filled include:** Administrative Services Manager; Blue-Collar Worker Supervisor; Branch Manager; Computer Programmer; Editor; General Manager; Health Services Manager; Industrial Production Manager; Insurance Agent/Broker; Management Trainee; Manufacturer's/Wholesaler's Sales Rep.; Operations/Production Manager; Science Technologist; Services Sales Representative; Software Engineer; Systems Analyst; Technical Writer/Editor. **Corporate headquarters location:** This Location. **Average salary range of placements:** More than $50,000.

SALES CONSULTANTS
MANAGEMENT RECRUITERS INTERNATIONAL, INC.
3400 Dundee Road, Suite 340, Northbrook IL 60062. 847/509-9000. **Contact:** Dan Cook, Manager. **Description:** An executive search firm. **Specializes in the areas of:** Accounting/Auditing; Administration/MIS/EDP; Advertising; Architecture/Construction/Real Estate; Banking;

Chemical; Communications; Computer Science/Software; Design; Electrical; Engineering; Finance; Food Industry; General Management; Health/Medical; Industrial; Insurance; Legal; Manufacturing; Operations Management; Personnel/Labor Relations; Pharmaceutical; Printing/ Publishing; Procurement; Retail; Sales and Marketing; Technical and Scientific; Textiles; Transportation.

SALES CONSULTANTS OF CHICAGO
6420 West 127th Street, Suite 209, Palos Heights IL 60463. 708/371-9677. **Fax:** 708/371-9832. **Contact:** Jack White, President. **Description:** An executive search firm. Company pays fee. **Specializes in the areas of:** Administration/MIS/EDP; Banking; Computer Hardware/Software; Engineering; Finance; Food Industry; General Management; Health/Medical; Industrial; Manufacturing; Sales and Marketing; Technical and Scientific. **Positions commonly filled include:** Biomedical Engineer; Chemical Engineer; Chemist; Civil Engineer; EDP Specialist; Electrical/Electronics Engineer; Industrial Designer; Industrial Engineer; Manufacturing Engineer; Marketing Specialist; Mechanical Engineer; Metallurgical Engineer; Operations/Production Manager; Quality Control Supervisor. **Number of placements per year:** 1 - 49.

SALES CONSULTANTS OF OAK BROOK
1415 West 22nd Street, Suite 725, Oak Brook IL 60521. 630/990-8233. **Fax:** 630/990-2973. **Contact:** Gary Miller, Manager. **Description:** An executive search firm. **Specializes in the areas of:** Accounting/Auditing; Administration/MIS/EDP; Advertising; Architecture/Construction/Real Estate; Banking; Chemical; Communications; Computer Science/Software; Design; Finance; Food Industry; General Management; Health/Medical; Industrial; Insurance; Legal; Manufacturing; Operations Management; Personnel/Labor Relations; Pharmaceutical; Printing/Publishing; Procurement; Regulatory Affairs; Sales and Marketing; Technical and Scientific; Textiles; Transportation.

SEARCH DYNAMICS INC.
9420 West Foster Avenue, Suite 200, Chicago IL 60656-1006. 773/992-3900. **Contact:** George Apostle, President. **Description:** A retainer and contingency search firm. **Specializes in the areas of:** Computer Science/Software; Engineering; Manufacturing; Technical and Scientific. **Positions commonly filled include:** Design Engineer; Designer; Draftsperson; Industrial Engineer; Industrial Production Manager; Metallurgical Engineer; Quality Control Supervisor; Software Engineer. **Number of placements per year:** 50 - 99.

SEVCOR INTERNATIONAL, INC.
One Pierce Place, Suite 400E, Itasca IL 60143. 630/250-3088. **Fax:** 630/250-3089. **Contact:** J. Randy Severinsen, President. **Description:** An executive search firm. Company pays fee. **Specializes in the areas of:** Computer Science/Software; Insurance; Technical and Scientific. **Positions commonly filled include:** Actuary; Computer Programmer; Systems Analyst. **Number of placements per year:** 100 - 199.

SMITH SCOTT & ASSOCIATES
P.O. Box 941, Lake Forest IL 60045. 847/295-9517. **Fax:** 847/295-9534. **Contact:** Gary J. Smith, Managing Partner. **E-mail address:** gjsmith@interaccess.com. **Description:** An executive search firm. Company pays fee. **Specializes in the areas of:** Computer Science/Software; Information Technology; Personnel/Labor Relations. **Positions commonly filled include:** Human Resources Manager; Systems Analyst. **Average salary range of placements:** More than $50,000. **Number of placements per year:** 1 - 49.

STONE ENTERPRISES, LTD.
645 North Michigan Avenue, Suite 800, Chicago IL 60611. 312/404-9300. **Fax:** 312/404-9388. **Contact:** Susan L. Stone, President. **Description:** An executive search firm. **Specializes in the areas of:** Accounting/Auditing; Computer Hardware/Software; Distribution; Engineering; Manufacturing; Sales Promotion; Telecommunications.

SYNERGISTICS ASSOCIATES LTD.
400 North State Street, Suite 400, Chicago IL 60610. 312/467-5450. **Fax:** 312/822-0246. **Contact:** Alvin Borenstine, President. **Description:** A retainer executive search firm specializing in chief

information officers and computer technology professionals. Company pays fee. **Specializes in the areas of:** Computer Science/Software. **Average salary range of placements:** More than $50,000. **Number of placements per year:** 1 - 49.

SYSTEMS ONE, LTD.

1100 East Woodfield Road, Schaumburg IL 60173-5116. 847/619-9300. **Fax:** 847/619-0071. **Contact:** Ed Hiloy, Manager. **Description:** An executive search firm operating on both retainer and contingency bases that encourages clients to pursue information technology-based strategic corporate goals. Company pays fee. **Specializes in the areas of:** Computer Science/Software. **Positions commonly filled include:** Computer Programmer; Management Analyst/Consultant; MIS Specialist; Software Engineer; Systems Analyst; Telecommunications Manager. **Corporate headquarters location:** This Location. **Average salary range of placements:** More than $50,000.

TSC MANAGEMENT SERVICES

P.O. Box 384, Barrington IL 60011. 847/381-0167. **Fax:** 847/381-2169. **Contact:** Robert Stanton, President. **Description:** An executive search firm. **E-mail address:** tscmgtserv@aol.com. Company pays fee. **Specializes in the areas of:** Computer Science/Software; Engineering; Industrial; Interactive Entertainment; Manufacturing; Technical and Scientific. **Positions commonly filled include:** Computer Programmer; Designer; Electrical/Electronics Engineer; Industrial Engineer; Mechanical Engineer; Quality Control Supervisor; Software Engineer; Statistician; Systems Analyst; Technical Writer/Editor. **Corporate headquarters location:** This Location. **International locations:** Worldwide. **Average salary range of placements:** More than $50,000. **Number of placements per year:** 50 - 99.

ROY TALMAN & ASSOCIATES

203 North Wabash, Suite 1120, Chicago IL 60601-2413. 312/630-0130. **Toll-free phone:** 800/279-6135. **Fax:** 312/630-0135. **Contact:** Ilya Talman, President. **E-mail address:** talman@mcs.com. **World Wide Web address:** http://www.roy-talman.com. **Description:** An executive search firm specializing in finding information technology professionals for companies utilizing or developing cutting edge computer systems. Company pays fee. **Specializes in the areas of:** Administration/ MIS/EDP; Banking; Computer Science/Software; Engineering; Finance. **Positions commonly filled include:** Computer Programmer; EDP Specialist; Financial Analyst; Mathematician; MIS Specialist; Software Engineer; Systems Analyst. **Corporate headquarters location:** This Location. **Average salary range of placements:** $30,000 - $50,000. **Number of placements per year:** 100 - 199.

TECHNICAL RECRUITING CONSULTANTS

215 North Arlington Heights Road, Suite 102, Arlington Heights IL 60004. 847/394-1101. **Contact:** Dick Latimer, President. **Description:** An executive search firm. Company pays fee. **Specializes in the areas of:** Administration/MIS/EDP; Computer Science/Software; Engineering; Manufacturing. **Positions commonly filled include:** Civil Engineer; Computer Programmer; EDP Specialist; Electrical/Electronics Engineer; Industrial Engineer; Manufacturing Engineer; Mechanical Engineer; MIS Specialist; Software Engineer; Systems Analyst. **Number of placements per year:** 50 - 99.

MARK WILCOX ASSOCIATES, LTD.

799 Roosevelt Road, Building #3, Suite 108, Glen Ellyn IL 60137. 630/790-4300. **Fax:** 630/790-4495. **Contact:** Mark A. Wilcox, President. **Description:** A full-service human resources firm. The firm provides clients with a full range of retained, project-level, contract, and contingency search services. Mark Wilcox Associates also provides on-site recruiting and human resources support services. Company pays fee. **Specializes in the areas of:** Administration/MIS/EDP; Computer Science/Software; Engineering; Industrial; Manufacturing; Personnel/Labor Relations; Sales and Marketing; Technical and Scientific. **Positions commonly filled include:** Accountant/Auditor; Aerospace Engineer; Agricultural Engineer; Buyer; Chemical Engineer; Chemist; Civil Engineer; Computer Programmer; Designer; Electrical/Electronics Engineer; General Manager; Health Services Manager; Human Resources Manager; Industrial Engineer; Industrial Production Manager; Mechanical Engineer; Metallurgical Engineer; Operations/Production Manager; Petroleum Engineer; Public Relations Specialist; Purchasing Agent/Manager; Quality Control Supervisor; Software Engineer; Structural Engineer; Systems Analyst; Technical Writer/Editor.

Benefits available to temporary workers: Dental Insurance; Medical Insurance; Paid Holidays; Paid Vacation. **Corporate headquarters location:** This Location. **Average salary range of placements:** More than $50,000. **Number of placements per year:** 100 - 199.

WILSON-DOUGLAS-JORDAN

70 West Madison Street, Chicago IL 60602-4205. 312/782-0286. **Contact:** John Wilson, President. **Description:** An executive search firm operating on a retainer basis that specializes in the information technology industry. Wilson-Douglas-Jordan's clients include several *Fortune* 100 companies and Big Six accounting firms. Company pays fee. **Specializes in the areas of:** Computer Science/Software. **Positions commonly filled include:** Computer Programmer; Management Analyst/Consultant; MIS Specialist; Software Engineer; Systems Analyst; Telecommunications Manager. **Corporate headquarters location:** This Location. **Average salary range of placements:** More than $40,000. **Number of placements per year:** 50 - 99.

XAGAS & ASSOCIATES

701 East State Street, Geneva IL 60134. 630/232-7044. **Fax:** 630/232-7154. **Contact:** Steve Xagas, President/Founder. **Description:** An executive search firm. Company pays fee. **Specializes in the areas of:** Automation/Robotics; Automotive; Computer Science/Software; Engineering; Food Industry; Industrial; Manufacturing; Plastics; Technical and Scientific. **Positions commonly filled include:** Electrical/Electronics Engineer; General Manager; Mechanical Engineer; Metallurgical Engineer; Operations/Production Manager; Quality Control Supervisor; Radio/TV Announcer/ Broadcaster; Software Engineer; Statistician. **Number of placements per year:** 50 - 99.

PERMANENT EMPLOYMENT AGENCIES OF ILLINOIS

ABLEST STAFFING

611 East State Street, Geneva IL 60134. 630/232-1883. **Fax:** 630/232-1904. **Contact:** Carolyn Melka, Office Manager. **Description:** An employment agency. **Specializes in the areas of:** Computer Science/Software; Health/Medical; Legal; Manufacturing.

AVAILABILITY, INC.

P.O. Box 562, Alton IL 62002. 618/467-6449. **Contact:** Lee J. Hamel, Manager. **Description:** An employment agency. **Specializes in the areas of:** Accounting/Auditing; Administration/MIS/EDP; Banking; Clerical; Computer Hardware/Software; Engineering; Finance; Sales and Marketing; Technical and Scientific.

BELL PERSONNEL INC.

5368 West 95th Street, Oak Lawn IL 60453. 708/636-3151. **Fax:** 708/636-9315. **Contact:** Human Resources. **Description:** An employment agency. Company pays fee. **Specializes in the areas of:** Accounting/Auditing; Administration/MIS/EDP; Advertising; Banking; Computer Science/ Software; Finance; Insurance; Legal; Sales and Marketing; Secretarial. **Positions commonly filled include:** Accountant/Auditor; Advertising Clerk; Buyer; Clerical Supervisor; Cost Estimator; Credit Manager; Customer Service Representative; General Manager; Human Resources Specialist; Management Trainee; Paralegal; Typist/Word Processor. **Average salary range of placements:** $30,000 - $50,000.

CAREER MANAGEMENT ASSOCIATES

P.O. Box 5863, Rockford IL 61125. 815/229-7815. **Contact:** Darlene Furst, President. **Description:** An employment agency. **Specializes in the areas of:** Accounting/Auditing; Banking; Clerical; Computer Science/Software; Engineering; Finance; Health/Medical; Insurance; Manufacturing; MIS/EDP; Secretarial.

CEMCO SYSTEMS
2015 Spring Road, Suite 250, Oak Brook IL 60521. 630/573-5050. **Contact:** David Gordan, General Manager. **Description:** An employment agency. **Specializes in the areas of:** Computer Science/Software; MIS/EDP.

MICHAEL DAVID ASSOCIATES, INC.
180 North Michigan Avenue, Suite 1016, Chicago IL 60601. 312/236-4460. **Fax:** 312/236-5401. **Contact:** Director of Administrative Services. **Description:** An employment agency. Company pays fee. **Specializes in the areas of:** Accounting/Auditing; Clerical; Computer Science/Software; Engineering; Finance; Food Industry; Health/Medical; Insurance; Legal; Manufacturing; Personnel/Labor Relations; Sales and Marketing; Technical and Scientific. **Number of placements per year:** 200 - 499.

DUNHILL OF CHICAGO
68 East Wacker Place, 12th Floor, Chicago IL 60601. 312/346-0933. **Contact:** George Baker, Owner/President. **Description:** An employment agency. **Specializes in the areas of:** Accounting/Auditing; Banking; Computer Science/Software; Finance; Sales and Marketing.

GENERAL EMPLOYMENT ENTERPRISES, INC.
1101 Perimeter Drive, Suite 735, Schaumburg IL 60173. 847/240-1233. **Fax:** 847/240-1671. **Contact:** Frank Anichini, Manager. **Description:** An employment agency and contract services firm founded in 1893. Company pays fee. **Specializes in the areas of:** Accounting/Auditing; Administration/MIS/EDP; Banking; Computer Hardware/Software. **Positions commonly filled include:** Accountant/Auditor; Bank Officer/Manager; Computer Programmer; Customer Service Representative; Financial Analyst; Internet Services Manager; MIS Specialist; Software Engineer; Systems Analyst; Telecommunications Manager. **Corporate headquarters location:** Oakbrook Terrace IL. **Other U.S. locations:** Nationwide. **Average salary range of placements:** $30,000 - $50,000. **Number of placements per year:** 100 - 199.

MULLINS & ASSOCIATES, INC.
520 South Northwest Highway, Barrington IL 60010. 847/382-1800. **Fax:** 847/382-1329. **Contact:** Terri Mullins, Vice President. **Description:** An employment agency. **Specializes in the areas of:** Computer Science/Software; Engineering. **Positions commonly filled include:** Chemical Engineer; Computer Programmer; Design Engineer; Designer; Draftsperson; Industrial Engineer; MIS Specialist; Software Engineer; Systems Analyst. **Corporate headquarters location:** This Location. **Average salary range of placements:** More than $50,000. **Number of placements per year:** 200 - 499.

THE MURPHY GROUP
1211 West 22nd Street, Suite 221, Oak Brook IL 60521. 630/574-2840. **Contact:** William A. Murphy II, Vice President/General Manager. **Description:** An employment agency. **Specializes in the areas of:** Accounting/Auditing; Advertising; Banking; Clerical; Computer Science/Software; Engineering; Finance; Food Industry; Insurance; Manufacturing; MIS/EDP; Printing/Publishing; Sales and Marketing; Secretarial; Technical and Scientific.

NETWORK RESOURCE GROUP INC.
920 South Spring Street, Springfield IL 62704. **Toll-free phone:** 800/519-1000. **Toll-free fax:** 800/519-2425. **Contact:** Martine Davis, Recruiter. **World Wide Web address:** http://www.nrgjobs.com. **Description:** An employment agency. Company pays fee. **Specializes in the areas of:** Computer Science/Software. **Positions commonly filled include:** Computer Programmer; Internet Services Manager; MIS Specialist; Software Engineer; Systems Analyst; Telecommunications Manager. **Average salary range of placements:** $30,000 - $50,000. **Number of placements per year:** 1 - 49.

OLSTEN INFORMATION TECHNOLOGY STAFFING
16 West Ontario, Chicago IL 60610. 312/661-0490. **Fax:** 312/661-0491. **Contact:** Stephanie Bethea, Recruiting Coordinator. **E-mail address:** olsten@xnet.com. **World Wide Web address:** http://www.olsten-chicago.com. **Description:** An employment agency. Company pays fee. **Specializes in the areas of:** Accounting/Auditing; Banking; Computer Science/Software;

Industrial; Insurance; Legal; Sales and Marketing. **Positions commonly filled include:** Computer Programmer; MIS Specialist; Multimedia Designer; Software Engineer; Systems Analyst; Technical Writer/Editor. **Corporate headquarters location:** This Location. **Other U.S. locations:** Nationwide. **Number of placements per year:** 1000+.

OMNI ONE
2200 East Devon Avenue, Suite 246, Des Plaines IL 60018. 847/299-1233. **Contact:** Steve Leibovitz, Agency Manager. **Description:** An employment agency. **Specializes in the areas of:** Computer Science/Software; Engineering; MIS/EDP; Technical and Scientific.

OPPORTUNITY PERSONNEL SERVICE
200 West Adams Street, Suite 1702, Chicago IL 60606. 312/704-9898. **Contact:** Gwen Hudson, President. **Description:** An employment agency. **Specializes in the areas of:** Accounting/Auditing; Clerical; Computer Science/Software; Finance; Insurance; Legal; Manufacturing; Personnel/Labor Relations; Sales and Marketing.

PS INC.
141 West Jackson Boulevard, Suite 2188, Chicago IL 60604. 312/922-3222. **Fax:** 312/922-2281. **Contact:** Mary Parker, President. **Description:** An employment agency that provides both permanent and temporary placement. Company pays fee. **Specializes in the areas of:** Accounting/Auditing; Administration/MIS/EDP; Computer Science/Software; Finance; Personnel/Labor Relations; Sales and Marketing; Secretarial. **Positions commonly filled include:** Accountant/Auditor; Administrative Services Manager; Advertising Clerk; Computer Programmer; Customer Service Representative; Human Resources Specialist; Internet Services Manager; Market Research Analyst; MIS Specialist; Purchasing Agent/Manager; Software Engineer; Systems Analyst; Telecommunications Manager; Typist/Word Processor. **Number of placements per year:** 100 - 199.

PERSONNEL CONNECTION INC.
960 Clocktower Drive, Suite E, Springfield IL 62704. 217/787-9022. **Fax:** 217/787-9231. **Contact:** Carla J. Oller, President. **Description:** An employment agency. Company pays fee. **Specializes in the areas of:** Accounting/Auditing; Administration/MIS/EDP; Banking; Computer Science/Software; General Management; Insurance; Personnel/Labor Relations; Sales and Marketing; Secretarial. **Positions commonly filled include:** Accountant/Auditor; Branch Manager; Buyer; Claim Representative; Clerical Supervisor; Computer Programmer; Credit Manager; Customer Service Representative; Human Resources Manager; Industrial Engineer; Management Trainee; Mechanical Engineer; Medical Records Technician; Purchasing Agent/Manager; Systems Analyst; Underwriter/Assistant Underwriter; Wholesale and Retail Buyer.

WALSH & COMPANY
731 South Durkin Drive, Springfield IL 62704. 217/793-0200. **Fax:** 217/793-9078. **Contact:** Jack Walsh, Owner. **Description:** An employment agency. Company pays fee. **Specializes in the areas of:** Computer Science/Software; Sales and Marketing; Secretarial. **Positions commonly filled include:** Computer Programmer; Manufacturer's/Wholesaler's Sales Rep.; MIS Specialist; Software Engineer; Systems Analyst. **Average salary range of placements:** $30,000 - $50,000. **Number of placements per year:** 50 - 99.

WEBB EMPLOYMENT SERVICE
318 Park Avenue, Rockford IL 61101. 815/963-0644. **Fax:** 815/963-0649. **Contact:** Deborah Webb, Owner. **Description:** An employment agency. Company pays fee. **Specializes in the areas of:** Accounting/Auditing; Administration/MIS/EDP; Architecture/Construction/Real Estate; Computer Science/Software; General Management; Industrial; Manufacturing; Sales and Marketing; Technical and Scientific. **Positions commonly filled include:** Accountant/Auditor; Aerospace Engineer; Architect; Bank Officer/Manager; Blue-Collar Worker Supervisor; Broadcast Technician; Budget Analyst; Buyer; Civil Engineer; Computer Programmer; Construction and Building Inspector; Construction Manager; Cost Estimator; Dental Assistant/Dental Hygienist; Design Engineer; Designer; Draftsperson; Electrical/Electronics Engineer; Electrician; General Manager; Industrial Engineer; Industrial Production Manager; Management Trainee; Manufacturer's/Wholesaler's Sales Rep.; Mechanical Engineer; Metallurgical Engineer; MIS

Specialist; Multimedia Designer; Purchasing Agent/Manager; Quality Control Supervisor; Services Sales Representative; Software Engineer; Stationary Engineer; Structural Engineer; Surveyor; Systems Analyst; Technical Writer/Editor. **Corporate headquarters location:** This Location. **Average salary range of placements:** $30,000 - $50,000. **Number of placements per year:** 1 - 49.

WILLS & COMPANY, INC.
222 East Wisconsin, Suite 100, Lake Forest IL 60045. 847/735-1622. **Fax:** 847/735-1633. **Contact:** Don Wills, President. **Description:** An employment agency. **Specializes in the areas of:** Computer Hardware/Software; Data Processing.

WORLD EMPLOYMENT SERVICE
1213 Dundee Road, Buffalo Grove IL 60089. 847/870-0900. **Fax:** 847/870-0906. **Contact:** Helga E. Jones, Manager. **Description:** An employment agency. Company pays fee. **Specializes in the areas of:** Accounting/Auditing; Administration/MIS/EDP; Advertising; Banking; Computer Science/Software; Engineering; Finance; Food Industry; General Management; Health/Medical; Industrial; Insurance; Legal; Manufacturing; Personnel/Labor Relations; Retail; Sales and Marketing; Technical and Scientific. **Positions commonly filled include:** Accountant/Auditor; Adjuster; Administrative Services Manager; Advertising Clerk; Architect; Bank Officer/Manager; Brokerage Clerk; Budget Analyst; Buyer; Chemical Engineer; Claim Representative; Clerical Supervisor; Computer Programmer; Counselor; Credit Manager; Customer Service Representative; Design Engineer; Draftsperson; Electrical/Electronics Engineer; Financial Analyst; Health Services Manager; Hotel Manager; Human Resources Specialist; Industrial Engineer; Industrial Production Manager; Insurance Agent/Broker; Internet Services Manager; Management Analyst/Consultant; Management Trainee; Manufacturer's/Wholesaler's Sales Rep.; Market Research Analyst; Mechanical Engineer; Medical Records Technician; Metallurgical Engineer; MIS Specialist; Operations/Production Manager; Paralegal; Property and Real Estate Manager; Public Relations Specialist; Purchasing Agent/Manager; Restaurant/Food Service Manager; Services Sales Representative; Software Engineer; Statistician; Strategic Relations Manager; Systems Analyst; Technical Writer/Editor; Telecommunications Manager; Transportation/Traffic Specialist; Travel Agent; Typist/Word Processor; Underwriter/Assistant Underwriter. **Corporate headquarters location:** This Location. **Average salary range of placements:** $20,000 - $29,999. **Number of placements per year:** 100 - 199.

EXECUTIVE SEARCH FIRMS OF INDIANA

THE CONSULTING FORUM, INC.
9200 Keystone Crossing, Suite 700, Indianapolis IN 46240. 317/580-4800. **Fax:** 317/580-4801. **Contact:** Don Kellner, President. **E-mail address:** tcf@iquest.net. **Description:** An executive search firm operating on both retainer and contingency bases. The firm specializes in the recruitment of information systems professionals, software engineers, and computer specialists. Company pays fee. **Specializes in the areas of:** Administration/MIS/EDP; Computer Science/Software. **Positions commonly filled include:** Computer Programmer; MIS Specialist; Software Engineer; Systems Analyst; Technical Writer/Editor. **Corporate headquarters location:** This Location. **Average salary range of placements:** $30,000 - $50,000. **Number of placements per year:** 100 - 199.

EXECUSEARCH
105 East Jefferson Boulevard, Suite 800, South Bend IN 46601-1811. 219/233-9353. **Contact:** Manager. **Description:** An executive search firm. **Specializes in the areas of:** Accounting/Auditing; Administration/MIS/EDP; Advertising; Architecture/Construction/Real Estate; Banking; Chemical; Communications; Computer Hardware/Software; Construction; Electrical; Engineering; Finance; Food Industry; General Management; Health/Medical; Insurance;

Legal; Manufacturing; Personnel/Labor Relations; Pharmaceutical; Printing/Publishing; Procurement; Retail; Sales and Marketing; Technical and Scientific; Textiles; Transportation.

GREAT LAKES SEARCH
6327 University Commons, South Bend IN 46635. 219/271-1360. **Fax:** 219/271-8775. **Contact:** Pamela Olmstead, Manager. **E-mail address:** pamo@skycnet.net. **World Wide Web address:** http://www.fwi.com/. **Description:** An executive search firm. Great Lakes Search provides direct placement services for Midwest-based clients. **Specializes in the areas of:** Accounting/Auditing; Computer Science/Software; Finance; Manufacturing; Personnel/Labor Relations. **Positions commonly filled include:** Computer Programmer; Human Resources Manager; MIS Specialist; Operations/Production Manager; Purchasing Agent/Manager; Quality Control Supervisor; Systems Analyst. **Corporate headquarters location:** Fort Wayne IN. **Average salary range of placements:** More than $50,000. **Number of placements per year:** 1 - 49.

THE MALLARD GROUP
3322 Oak Borough, Fort Wayne IN 46804. 219/436-3970. **Fax:** 219/436-7012. **Contact:** Robert Hoffman, Director. **Description:** Founded in 1988, this executive search firm operates on a contingency basis. The Mallard Group deals with domestic and international recruitment of technical professionals. Company pays fee. **Specializes in the areas of:** Computer Science/Software; Engineering; Manufacturing; Sales and Marketing. **Positions commonly filled include:** Design Engineer; Electrical/Electronics Engineer; Human Resources Manager; Industrial Engineer; Mechanical Engineer; Metallurgical Engineer; Purchasing Agent/Manager; Software Engineer. **Corporate headquarters location:** This Location. **Average salary range of placements:** $30,000 - $50,000. **Number of placements per year:** 1 - 49.

MANAGEMENT RECRUITERS INC.
15209 Herriman Boulevard, Noblesville IN 46060. 317/773-4323. **Fax:** 317/773-9744. **Contact:** H. Peter Isenberg, President. **Description:** An executive search firm. Company pays fee. **Specializes in the areas of:** Administration/MIS/EDP; Computer Science/Software; Engineering; General Management; Industrial; Manufacturing; Sales and Marketing. **Positions commonly filled include:** Computer Programmer; Designer; Draftsperson; Electrical/Electronics Engineer; Food Scientist/Technologist; Industrial Engineer; Industrial Production Manager; Manufacturer's/ Wholesaler's Sales Rep.; Mechanical Engineer; Metallurgical Engineer; Pharmacist; Software Engineer; Systems Analyst. **Corporate headquarters location:** Cleveland OH. **Other U.S. locations:** Nationwide. **Average salary range of placements:** More than $50,000. **Number of placements per year:** 1 - 49.

MANAGEMENT RECRUITERS OF EVANSVILLE
Riverside 1 Building, Suite 209, 101 Court Street, Evansville IN 47708. 812/464-9155. **Fax:** 812/422-6718. **Contact:** Manager. **Description:** An executive search firm. **Specializes in the areas of:** Accounting/Auditing; Administration/MIS/EDP; Advertising; Architecture/Construction/ Real Estate; Banking; Chemical; Communications; Computer Hardware/ Software; Construction; Electrical; Engineering; Finance; Food Industry; General Management; Health/Medical; Insurance; Legal; Manufacturing; Personnel/Labor Relations; Pharmaceutical; Printing/Publishing; Procurement; Retail; Sales and Marketing; Technical and Scientific; Textiles; Transportation.

MANAGEMENT RECRUITERS OF INDIANAPOLIS
3905 Vincennes Road, Suite 301, Indianapolis IN 46268. 317/228-3300. **Contact:** Manager. **Description:** An executive search firm. **Specializes in the areas of:** Accounting/Auditing; Administration/MIS/EDP; Advertising; Architecture/Construction/Real Estate; Banking; Chemical; Communications; Computer Hardware/Software; Construction; Electrical; Engineering; Finance; Food Industry; General Management; Health/Medical; Insurance; Legal; Manufacturing; Operations Management; Personnel/Labor Relations; Pharmaceutical; Printing/Publishing; Procurement; Retail; Sales and Marketing; Technical and Scientific; Textiles; Transportation.

MANAGEMENT RECRUITERS OF RICHMOND
STAFFING SOLUTIONS OF RICHMOND
2519 East Main Street, Suite 101, Richmond IN 47374-5864. 765/935-3356. **Contact:** Mr. Rande Martin, Manager. **Description:** An executive search firm. **Specializes in the areas of:**

Accounting/Auditing; Administration/MIS/EDP; Advertising; Architecture/Construction/ Real Estate; Banking; Chemical; Communications; Computer Hardware/Software; Construction; Electrical; Engineering; Finance; Food Industry; General Management; Health/Medical; Insurance; Legal; Manufacturing; Operations Management; Personnel/Labor Relations; Pharmaceutical; Printing/Publishing; Procurement; Retail; Sales and Marketing; Technical and Scientific; Textiles; Transportation.

MANAGEMENT SERVICES
P.O. Box 830, Middlebury IN 46540-0830. 219/825-3909. **Fax:** 219/825-7115. **Contact:** Office Manager. **Description:** A full-service professional employment agency specializing in administrative, technical, and executive management placement throughout the Midwest. Founded in 1976, Management Services also acts as an executive search firm and operates on a contingency basis. Company pays fee. **Specializes in the areas of:** Computer Science/Software; Engineering; Manufacturing; Personnel/Labor Relations; Sales and Marketing. **Positions commonly filled include:** Accountant/Auditor; Buyer; Chemical Engineer; Computer Programmer; Electrical/Electronics Engineer; Financial Analyst; General Manager; Human Resources Manager; Industrial Engineer; Industrial Production Manager; Landscape Architect; Mechanical Engineer; Metallurgical Engineer; MIS Specialist; Operations/Production Manager; Purchasing Agent/ Manager; Quality Control Supervisor; Software Engineer; Systems Analyst; Transportation/Traffic Specialist. **Corporate headquarters location:** This Location. **Other U.S. locations:** Nationwide. **Average salary range of placements:** $30,000 - $50,000. **Number of placements per year:** 1 - 49.

NATIONAL CORPORATE CONSULTANTS, INC.
ADVANTAGE SERVICES, INC.
409 East Cook Road, Suite 200, Fort Wayne IN 46825. 219/493-4506. **Contact:** Manager. **Description:** An executive search firm. **Specializes in the areas of:** Accounting/Auditing; Administration/MIS/EDP; Banking; Chemical; Computer Hardware/Software; Engineering; Finance; Food Industry; Manufacturing; Pharmaceutical; Technical and Scientific; Transportation. **Number of placements per year:** 50 - 99.

OAKWOOD INTERNATIONAL INC.
3935 Lincoln Way East, Suite A, Mishawaka IN 45644. 219/255-9861. **Fax:** 219/257-8914. **Contact:** Scott Null, President. **E-mail address:** 72633.1765@compuserve.com. **World Wide Web address:** http://www.quest.net/interact/oakwood. **Description:** A contingency search firm. Company pays fee. **Specializes in the areas of:** Administration/MIS/EDP; Computer Science/Software; Engineering; General Management; Manufacturing; Sales and Marketing; Technical and Scientific. **Positions commonly filled include:** Aerospace Engineer; Biomedical Engineer; Branch Manager; Chemical Engineer; Chemist; Civil Engineer; Computer Programmer; Construction Contractor; Design Engineer; Designer; Draftsperson; Electrical/Electronics Engineer; Environmental Engineer; Industrial Engineer; Internet Services Manager; Manufacturer's/Wholesaler's Sales Rep.; Mechanical Engineer; Metallurgical Engineer; MIS Specialist; Multimedia Designer; Quality Control Supervisor; Software Engineer; Structural Engineer; Systems Analyst; Technical Writer/Editor; Telecommunications Manager. **Corporate headquarters location:** This Location. **Average salary range of placements:** More than $50,000. **Number of placements per year:** 1 - 49.

OFFICEMATES5 OF INDIANAPOLIS
1099 North Meridian Street, Landmark Building, Suite 640, Indianapolis IN 46204. 317/237-2787. **Contact:** Manager. **Description:** An executive search firm. **Specializes in the areas of:** Accounting/Auditing; Administration/MIS/EDP; Advertising; Architecture/ Construction/Real Estate; Banking; Chemical; Communications; Computer Hardware/ Software; Construction; Electrical; Engineering; Finance; Food Industry; General Management; Health/Medical; Insurance; Legal; Manufacturing; Personnel/Labor Relations; Pharmaceutical; Printing/Publishing; Procurement; Retail; Sales and Marketing; Technical and Scientific; Textiles; Transportation.

OFFICEMATES5 OF INDIANAPOLIS (NORTH)
8888 Keystone Crossing Boulevard, Suite 1420, Indianapolis IN 46240. 317/843-2512. **Contact:**

Manager. **Description:** Officemates5 is an executive search firm. **Specializes in the areas of:** Accounting/Auditing; Administration/MIS/EDP; Advertising; Architecture/Construction/Real Estate; Banking; Chemical; Communications; Computer Hardware/Software; Construction; Electrical; Engineering; Finance; Food Industry; General Management; Health/Medical; Legal; Manufacturing; Personnel/Labor Relations; Pharmaceutical; Printing/Publishing; Procurement; Retail; Sales and Marketing; Technical and Scientific; Textiles; Transportation. **Other U.S. locations:** Nationwide.

QUIRING ASSOCIATES INC.

7321 Shadeland Station Way, Suite 150, Indianapolis IN 46256-3935. 317/841-0102. **Fax:** 317/577-8240. **Contact:** Patti Quiring, CPC, Human Resources. **World Wide Web address:** http://www.iquest.net/quiring. **Description:** An executive search firm. Company pays fee. **Specializes in the areas of:** Accounting/Auditing; Administration/MIS/EDP; Banking; Computer Science/Software; Engineering; Finance; Health/Medical; Industrial; Insurance; Manufacturing; Nonprofit; Personnel/Labor Relations; Sales and Marketing; Technical and Scientific. **Positions commonly filled include:** Accountant/Auditor; Bank Officer/Manager; Computer Programmer; Computer Scientist; Engineer; Financial Consultant; Manufacturing Engineer; Scientist. **Number of placements per year:** 1 - 49.

UNIQUE, INC.

8765 North Guion Road, Suite A, Indianapolis IN 46268. 317/875-8281. **Fax:** 317/875-3127. **Contact:** Jennifer Flora, President. **Description:** An executive search firm specializing in sales, data processing, telecommunications, office support, and high-technology. Company pays fee. **Specializes in the areas of:** Computer Science/Software; Food Industry; General Management; Legal; Personnel/Labor Relations; Printing/Publishing; Sales and Marketing; Secretarial. **Positions commonly filled include:** Branch Manager; Clerical Supervisor; Computer Programmer; Customer Service Representative; General Manager; Hotel Manager; Human Resources Manager; Management Trainee; Paralegal; Restaurant/Food Service Manager; Services Sales Representative; Software Engineer; Systems Analyst; Telecommunications Manager; Typist/Word Processor. **Corporate headquarters location:** This Location. **Other U.S. locations:** Nationwide. **Number of placements per year:** 500 - 999.

PERMANENT EMPLOYMENT AGENCIES OF INDIANA

ACT-1 STAFFING, INC.

8555 River Road, Suite 185, Indianapolis IN 46240. 317/574-3400. **Fax:** 317/587-7730. **Contact:** Human Resources. **Description:** An employment agency. Company pays fee. **Specializes in the areas of:** Accounting/Auditing; Banking; Computer Science/Software; Finance; General Management; Industrial; Insurance; Legal; Manufacturing; Nonprofit; Personnel/Labor Relations; Sales and Marketing; Secretarial; Technical and Scientific. **Positions commonly filled include:** Accountant/Auditor; Administrative Services Manager; Bank Officer/Manager; Biochemist; Brokerage Clerk; Chemical Engineer; Chemist; Civil Engineer; Clerical Supervisor; Clinical Lab Technician; Computer Programmer; Customer Service Representative; Editor; Electrician; General Manager; Human Resources Specialist; Management Trainee; Paralegal; Services Sales Representative; Typist/Word Processor; Underwriter/Assistant Underwriter. **Other U.S. locations:** OH. **Number of placements per year:** 1000+.

ALPHA RAE PERSONNEL, INC.

127 West Berry Street, Suite 200, Fort Wayne IN 46802. 219/426-8227. **Fax:** 219/426-1152. **Contact:** Rae Pearson, President. **Description:** An employment agency. **Specializes in the areas of:** Computer Science/Software; Data Processing; Engineering; Legal; Sales and Marketing; Software Engineering.

BILL CALDWELL EMPLOYMENT SERVICE

123 Main Street, Suite 307, Evansville IN 47708. 812/423-8006. **Fax:** 812/423-8008. **Contact:** Carmen M. Caldwell, Manager. **Description:** A full-service employment agency. **Specializes in the areas of:** Accounting/Auditing; Banking; Computer Science/Software; Engineering; Finance; General Management; Health/Medical; Industrial; Manufacturing; Retail; Sales and Marketing. **Positions commonly filled include:** Accountant/Auditor; Actuary; Adjuster; Administrative Services Manager; Advertising Clerk; Agricultural Engineer; Architect; Attorney; Automotive Mechanic; Bank Officer/Manager; Biochemist; Biological Scientist; Biomedical Engineer; Blue-Collar Worker Supervisor; Branch Manager; Broadcast Technician; Brokerage Clerk; Budget Analyst; Buyer; Chemical Engineer; Chemist; Civil Engineer; Claim Representative; Clerical Supervisor; Clinical Lab Technician; Computer Programmer; Construction Contractor; Cost Estimator; Customer Service Representative; Design Engineer; Designer; Draftsperson; Economist; Editor; EEG Technologist; Electrical/Electronics Engineer; Electrician; Financial Analyst; Food Scientist/Technologist; General Manager; Health Services Manager; Human Resources Specialist; Industrial Engineer; Internet Services Manager; Librarian; Management Analyst/Consultant; Market Research Analyst; Materials Engineer; Mechanical Engineer; Metallurgical Engineer; Mining Engineer; Operations/Production Manager; Pharmacist; Physical Therapist; Public Relations Specialist; Quality Control Supervisor; Radio/TV Announcer/Broadcaster; Radiologic Technologist; Real Estate Agent; Recreational Therapist; Restaurant/Food Service Manager; Services Sales Representative; Sociologist; Software Engineer; Statistician; Structural Engineer; Systems Analyst; Teacher/Professor; Telecommunications Manager; Typist/Word Processor; Underwriter/Assistant Underwriter.

CAREER CONSULTANTS, INC.

107 North Pennsylvania, Indianapolis IN 46204. 317/639-5601. **Contact:** Manager. **Description:** An employment agency. Company pays fee. **Specializes in the areas of:** Accounting/Auditing; Computer Science/Software; Engineering; Food Industry; Industrial; Information Systems; Manufacturing; Technical and Scientific. **Positions commonly filled include:** Accountant/Auditor; Computer Programmer; Electrical/Electronics Engineer; Financial Analyst; Human Resources Manager; Industrial Engineer; Industrial Production Manager; Mechanical Engineer; Metallurgical Engineer; Purchasing Agent/Manager; Quality Control Supervisor; Software Engineer; Statistician; Systems Analyst; Technical Writer/Editor. **Number of placements per year:** 50 - 99.

CORPORATE STAFFING RESOURCES

100 East Wayne Street, Suite 100, South Bend IN 46601. 219/233-8209. **Fax:** 219/280-2661. **Recorded jobline:** 219/237-9675. **Contact:** Charmaine Woolridge, Placement Coordinator. **Description:** Founded in 1987, this is a full-service placement firm with four divisions: temporary clerical and industrial, technical/contract, professional, and outplacement services. The technical division provides highly trained and experienced employees including engineers, CAD drafters, computer programmers, supervisors, lab technicians, accountants, electronics technicians, mechanical technicians, and other technical positions. **Specializes in the areas of:** Accounting/ Auditing; Administration/MIS/EDP; Computer Science/Software; Engineering; Industrial; Manufacturing; Technical and Scientific. **Positions commonly filled include:** Accountant/Auditor; Administrative Services Manager; Architect; Biological Scientist; Buyer; Chemical Engineer; Chemist; Civil Engineer; Computer Programmer; Cost Estimator; Credit Manager; Customer Service Representative; Design Engineer; Designer; Draftsperson; Electrical/Electronics Engineer; Electrician; Environmental Engineer; Financial Analyst; Human Resources Manager; Industrial Engineer; Industrial Production Manager; Mechanical Engineer; Metallurgical Engineer; MIS Specialist; Operations/Production Manager; Purchasing Agent/Manager; Quality Control Supervisor; Software Engineer; Structural Engineer; Systems Analyst; Technical Writer/Editor. **Benefits available to temporary workers:** 401(k); Medical Insurance; Paid Holidays; Paid Vacation. **Corporate headquarters location:** This Location. **Other U.S. locations:** IN; OH; MI; MO. **Average salary range of placements:** $30,000 - $50,000. **Number of placements per year:** 1000+.

EMPLOYMENT RECRUITERS INC.

P.O. Box 1624, Elkhart IN 46515-1624. 219/262-2654. **Fax:** 219/262-0095. **Contact:** Suzanne Pedler, President. **Description:** Founded in 1982, Employment Recruiters Inc. places professionals with manufacturers throughout the Midwest. Company pays fee. **Specializes in the areas of:**

Computer Science/Software; Engineering; General Management; Industrial; Manufacturing; Sales and Marketing. **Positions commonly filled include:** Chemical Engineer; Chemist; Computer Programmer; Credit Manager; Customer Service Manager; Customer Service Representative; Design Engineer; Designer; Draftsperson; Electrical/Electronics Engineer; Environmental Engineer; Food Scientist/Technologist; General Manager; Industrial Engineer; Industrial Production Manager; Manufacturer's/Wholesaler's Sales Rep.; Market Research Analyst; Mathematician; Metallurgical Engineer; Mining Engineer; MIS Specialist; Operations/Production Manager; Purchasing Agent/Manager; Quality Control Supervisor; Safety Engineer; Science Technologist; Software Engineer; Stationary Engineer; Statistician; Strategic Relations Manager; Structural Engineer; Systems Analyst; Technical Writer/Editor. **Corporate headquarters location:** Elkhart IN. **Average salary range of placements:** $30,000 - $50,000. **Number of placements per year:** 1 - 49.

MAYS & ASSOCIATES INC.
941 East 86th Street, Suite 109, Indianapolis IN 46240. 317/253-9999. **Fax:** 317/253-2749. **Contact:** Roger R. Mays, President. **Description:** An employment agency. Company pays fee. **Specializes in the areas of:** Computer Science/Software; Engineering; Manufacturing. **Positions commonly filled include:** Aerospace Engineer; Biomedical Engineer; Chemical Engineer; Computer Programmer; Designer; Electrical/Electronics Engineer; General Manager; Human Resources Manager; Industrial Engineer; Mechanical Engineer; Metallurgical Engineer; Purchasing Agent/Manager; Quality Control Supervisor; Software Engineer; Structural Engineer; Systems Analyst. **Number of placements per year:** 50 - 99.

THE REGISTRY INC.
600 King Cole Building, 7 North Meridian Street, Indianapolis IN 46204-3033. 317/634-1200. **Fax:** 317/263-3845. **Contact:** Director of Operations. **Description:** An employment agency. **Specializes in the areas of:** Accounting/Auditing; Administration/MIS/EDP; Bookkeeping; Clerical; Computer Science/Software; Health/Medical; Legal; Real Estate.

SNELLING & SNELLING
1000 East 80th Place, Merrillville IN 46410-5644. 219/769-2922. **Fax:** 219/755-0557. **Contact:** Cheri K. Elser, General Manager. **Description:** An employment agency providing permanent placement as privately owned franchisee of Snelling & Snelling, Inc. of Dallas TX. Founded in 1985. Company pays fee. **Specializes in the areas of:** Computer Science/Software; Food Industry; Sales and Marketing; Secretarial. **Positions commonly filled include:** Bookkeeper; Restaurant/Food Service Manager; Secretary; Services Sales Representative. **Average salary range of placements:** $20,000 - $29,999. **Number of placements per year:** 100 - 199.

EXECUTIVE SEARCH FIRMS OF IOWA

ATKINSON SEARCH & PLACEMENT INC.
508 North Second Street, Fairfield IA 52556. 515/472-3666. **Fax:** 515/472-7270. **Contact:** Arthur Atkinson, President. **Description:** An executive search firm. Company pays fee. **Specializes in the areas of:** Computer Science/Software. **Positions commonly filled include:** Engineer; Marketing Specialist; Sales Representative; Software Engineer; Technical Writer/Editor. **Number of placements per year:** 1 - 49.

BURTON PLACEMENT SERVICES
1651 Lincoln Way, Clinton IA 52732. 319/243-6791. **Fax:** 319/243-6895. **Contact:** Heather Bush, Staffing Consultant. **Description:** An executive search firm that provides placement for computer, human resource, management, and technical professionals. Burton also places industrial and clerical professionals in temporary assignments. **Specializes in the areas of:** Accounting/Auditing; Administration/MIS/EDP; Banking; Computer Science/Software; Engineering; Finance; General Management; Industrial; Legal; Manufacturing; Personnel/Labor Relations; Printing/Publishing;

Retail; Sales and Marketing; Secretarial; Technical and Scientific. **Positions commonly filled include:** Accountant/Auditor; Administrative Services Manager; Advertising Clerk; Aircraft Mechanic/Engine Specialist; Bank Officer/Manager; Claim Representative; Clerical Supervisor; Computer Programmer; Counselor; Credit Manager; General Manager; Health Services Manager; Health Services Worker; Hotel Manager; Human Resources Manager; Industrial Engineer; Management Trainee; Medical Records Technician; Public Relations Specialist; Purchasing Agent/Manager; Quality Control Supervisor; Systems Analyst; Telecommunications Manager; Underwriter/Assistant Underwriter. **Corporate headquarters location:** Sterling IL. **Other U.S. locations:** Dixon IL; Oregon IL; Rochelle IL; Sycamore IL.

BYRNES & RUPKEY INC.

3356 Kimball Avenue, Waterloo IA 50702. 319/234-6201. **Fax:** 319/234-6360. **Contact:** Lois Rupkey, Executive Vice President. **Description:** Byrnes & Rupkey is a career placement and executive search firm. **Specializes in the areas of:** Administration/MIS/EDP; Architecture/Construction/Real Estate; Banking; Computer Science/Software; Engineering; Finance; General Management; Health/ Medical; Industrial; Insurance; Manufacturing; Personnel/Labor Relations; Retail; Sales and Marketing; Secretarial; Technical and Scientific. **Positions commonly filled include:** Civil Engineer; Computer Programmer; Draftsperson; Electrical/Electronics Engineer; Human Resources Manager; Manufacturer's/Wholesaler's Sales Rep.; Mechanical Engineer; Occupational Therapist; Petroleum Engineer; Physical Therapist; Physician; Restaurant/Food Service Manager; Software Engineer. **Corporate headquarters location:** This Location. **Other U.S. locations:** Warwick RI.

EXECUTIVE SEARCH ASSOCIATES

300 Pioneer Bank Building, Sioux City IA 51101. 712/277-8103. **Contact:** Mary Kay Bohan, Office Manager. **Description:** Executive Search Associates is an executive search and recruiting firm. Company pays fee. **Specializes in the areas of:** Accounting/Auditing; Administration/MIS/EDP; Computer Science/Software; Engineering; Finance; Food Industry; General Management; Health/Medical; Insurance; Legal; Manufacturing; Personnel/Labor Relations; Retail; Sales and Marketing; Secretarial; Telecommunications. **Positions commonly filled include:** Accountant/Auditor; Bank Officer/Manager; Branch Manager; Brokerage Clerk; Budget Analyst; Civil Engineer; Claim Representative; Clerical Supervisor; Computer Programmer; Cost Estimator; Counselor; Credit Manager; Customer Service Representative; Electrical/Electronics Engineer; Financial Analyst; General Manager; Health Services Manager; Hotel Manager; Human Resources Manager; Industrial Engineer; Industrial Production Manager; Insurance Agent/Broker; Internet Services Manager; Librarian; Licensed Practical Nurse; Management Trainee; Manufacturer's/Wholesaler's Sales Rep.; Market Research Analyst; Metallurgical Engineer; MIS Specialist; Multimedia Designer; Occupational Therapist; Operations/Production Manager; Paralegal; Pharmacist; Physical Therapist; Property and Real Estate Manager; Public Relations Specialist; Purchasing Agent/Manager; Quality Control Supervisor; Restaurant/Food Service Manager; Securities Sales Representative; Services Sales Representative; Software Engineer; Systems Analyst; Telecommunications Manager; Travel Agent; Typist/Word Processor; Underwriter/Assistant Underwriter. **Corporate headquarters location:** This Location. **Average salary range of placements:** $20,000 - $29,999. **Number of placements per year:** 100 - 199.

MANAGEMENT RECRUITERS

1312 Fourth Street SW, Suite 102, Mason City IA 50401. 515/424-1680. **Fax:** 515/424-6868. **Contact:** Cheryl Plagge, President. **E-mail address:** cplagge@willowtree.com. **Description:** An executive search firm which handles permanent and interim assignments nationwide with a focus on information systems positions. Company pays fee. **Specializes in the areas of:** Administration/ MIS/EDP; Computer Science/Software; Food Industry; Insurance; Manufacturing; Technical and Scientific. **Positions commonly filled include:** Computer Programmer; Management Analyst/Consultant; MIS Specialist; Software Engineer; Systems Analyst; Telecommunications Manager. **Corporate headquarters location:** Cleveland OH. **Other U.S. locations:** Nationwide. **Average salary range of placements:** More than $50,000. **Number of placements per year:** 1 - 49.

MANAGEMENT RECRUITERS
150 First Avenue NE, Suite 400, Cedar Rapids IA 52401. 319/366-8441. **Fax:** 319/366-1103.
Contact: Office Manager. **Description:** Management Recruiters is an executive search firm.
Specializes in the areas of: Accounting/Auditing; Administration/MIS/EDP; Advertising;
Architecture/Construction/Real Estate; Banking; Chemical; Communications; Computer
Hardware/Software; Design; Electrical; Engineering; Food Industry; General Management;
Health/Medical; Insurance; Legal; Manufacturing; Operations Management; Personnel/Labor
Relations; Printing/Publishing; Procurement; Retail; Sales and Marketing; Technical and
Scientific; Textiles; Transportation.

MANAGEMENT RECRUITERS
806 5th Street, Suite 209, Coralville IA 52241. 319/354-4320. **Contact:** John Sims, Manager.
Description: Management Recruiters is an executive search firm. **Specializes in the areas of:**
Accounting/Auditing; Administration/MIS/EDP; Advertising; Architecture/Construction/Real
Estate; Banking; Chemical; Communications; Computer Hardware/Software; Design; Electrical;
Engineering; Food Industry; General Management; Health/Medical; Insurance; Legal;
Manufacturing; Operations Management; Personnel/Labor Relations; Printing/Publishing;
Procurement; Retail; Sales and Marketing; Technical and Scientific; Textiles; Transportation.

MANAGEMENT RECRUITERS OF DUBUQUE
700 Locust Street, Suite 707, Dubuque IA 52001-6824. 319/583-1554. **Fax:** 319/583-0414.
Contact: Michael Pratt, General Manager. **E-mail address:** rsgh40a@prodigy.com. **Description:**
Management Recruiters of Dubuque is an executive/management search and recruiting firm.
Company pays fee. **Specializes in the areas of:** Banking; Computer Science/Software; Sales and
Marketing. **Positions commonly filled include:** Bank Officer/Manager; Computer Programmer;
MIS Specialist; Software Engineer; Systems Analyst; Telecommunications Manager. **Corporate
headquarters location:** Cleveland OH. **Other U.S. locations:** Nationwide. **Average salary range
of placements:** $30,000 - $50,000. **Number of placements per year:** 1 - 49.

MANAGEMENT RECRUITERS QUAD CITIES, INC.
OFFICEMATES5 QUAD CITIES, INC./COMPUSEARCH OF QUAD CITIES
Alpine Centre South, Penthouse, 2435 Kimberly Road, Bettendorf IA 52722. 319/359-3503.
Contact: Jerry Herrmann, Manager. **Description:** An executive search firm. Three different
divisions of the same company operate out of this location. **Specializes in the areas of:**
Accounting/Auditing; Administration/MIS/EDP; Advertising; Architecture/ Construction/Real
Estate; Banking; Communications; Computer Hardware/Software; Design; Electrical; Engineering;
Food Industry; General Management; Health/Medical; Insurance; Legal; Manufacturing;
Operations Management; Personnel/Labor Relations; Printing/Publishing; Procurement; Retail;
Sales and Marketing; Technical and Scientific; Textiles; Transportation.

MID-STATES TECHNICAL STAFFING
5309 Victoria Avenue, Davenport IA 52807-2989. 319/359-7042. **Fax:** 319/359-6331. **Contact:**
Connie Whitcomb, Sales Administrator. **Description:** A contingency search firm and temporary
agency. Company pays fee. **Specializes in the areas of:** Architecture/Construction/Real Estate;
Art/Design; Computer Science/Software; Engineering; General Management; Industrial;
Manufacturing; Technical and Scientific. **Positions commonly filled include:** Aerospace Engineer;
Agricultural Engineer; Architect; Chemical Engineer; Chemist; Civil Engineer; Computer
Programmer; Construction Contractor; Cost Estimator; Design Engineer; Designer; Draftsperson;
Electrical/Electronics Engineer; Electrician; Environmental Engineer; Food Scientist/Technologist;
Geographer; Geologist/Geophysicist; Industrial Engineer; Industrial Production Manager;
Landscape/Grounds Maintenance; Management Analyst/Consultant; Mathematician; Mechanical
Engineer; Metallurgical Engineer; Mining Engineer; Multimedia Designer; Nuclear Engineer;
Operations/Production Manager; Petroleum Engineer; Purchasing Agent/Manager; Quality Control
Supervisor; Software Engineer; Stationary Engineer; Statistician; Structural Engineer; Surveyor;
Systems Analyst; Technical Writer/Editor. **Corporate headquarters location:** This Location.
Other U.S. locations: Chicago IL; Louisville KY; Minneapolis MN. **Average salary range of
placements:** $30,000 - $50,000. **Number of placements per year:** 200 - 499.

PERMANENT EMPLOYMENT AGENCIES OF IOWA

BREI & ASSOCIATES, INC.
2598 28th Avenue, Marion IA 52302. 319/377-9196. **Fax:** 319/377-9219. **Contact:** Randy Brei, President. **E-mail address:** rbrei@netins.net. **World Wide Web address:** http://www.netins.net/showcase/rdbrei/. **Description:** Brei & Associates, Inc. is an employment agency that focuses on placement of electronics engineers and software developers. Areas covered by the company include software design, hardware design, and systems engineering. Company pays fee. **Specializes in the areas of:** Computer Science/Software; Engineering. **Positions commonly filled at this location include:** Computer Programmer; Design Engineer; Electrical/Electronics Engineer; Mechanical Engineer; Software Engineer; Systems Analyst; Technical Writer/Editor. **Average salary range of placements:** More than $50,000. **Number of placements per year:** 1 - 49.

CAMBRIDGE STAFFING SOLUTIONS
100 First Avenue Northeast, Suite 109, Cedar Rapids IA 52401-1109. 319/366-7771. **Contact:** Office Manager. **Description:** Cambridge Staffing Solutions is an employment agency. Company pays fee. **Specializes in the areas of:** Accounting/Auditing; Clerical; Computer Science/Software; Engineering; Health/Medical; Manufacturing; Sales and Marketing; Secretarial. **Positions commonly filled at this location include:** Accountant/Auditor; Aerospace Engineer; Architect; Bank Officer/Manager; Blue-Collar Worker Supervisor; Clerical Supervisor; Computer Programmer; Design Engineer; Human Resources Manager; Industrial Designer; Industrial Engineer; Mechanical Engineer; MIS Specialist; Physical Therapist; Physician; Purchasing Agent/Manager; Quality Control Supervisor; Respiratory Therapist; Sales Representative; Secretary; Typist/Word Processor. **Number of placements per year:** 100 - 199.

EXECUTIVE RESOURCES
3816 Ingersoll Avenue, Des Moines IA 50312. 515/287-6880. **Fax:** 515/255-9445. **Contact:** Ms. Gerry Mullane, President. **Description:** Executive Resources is an employment agency. **Specializes in the areas of:** Computer Hardware/Software; Data Processing; Insurance; Sales and Marketing.

KEY EMPLOYMENT SERVICES
1001 Office Park Road, Suite 320, West Des Moines IA 50265. 515/224-0446. **Fax:** 515/224-6599. **Contact:** Don Jayne, CPC, Owner. **Description:** Key Employment Services is a permanent employment agency. **Specializes in the areas of:** Accounting/Auditing; Data Processing; Engineering; Finance; Insurance; Personnel/Labor Relations; Sales and Marketing; Software Engineering.

McGLADREY SEARCH GROUP
400 Locust, Suite 640, Des Moines IA 50309. 515/281-9200. **Fax:** 515/284-1545. **Contact:** Thomas Hamilton, Manager. **Description:** McGladrey Search Group is a permanent employment agency. **Specializes in the areas of:** Accounting/Auditing; Banking; Computer Hardware/Software; Data Processing; Engineering; Finance; Health/Medical; Manufacturing.

NATE VIALL & ASSOCIATES
Post Office Box 12238, Des Moines IA 50312. 515/274-1729. **Fax:** 515/274-5646. **Contact:** Nate Viall, President. **Description:** Nate Viall & Associates is a permanent employment agency. Company pays fee. **Specializes in the areas of:** Administration/MIS/EDP; Computer Science/Software. **Positions commonly filled at this location include:** Computer Programmer; Systems Analyst.

EXECUTIVE SEARCH FIRMS OF KANSAS

COMPUSEARCH OF OVERLAND PARK
9401 Indian Creek Parkway, Suite 920, Overland Park KS 66210. 913/661-9400. **Fax:** 913/661-9030. **Contact:** Office Manager. **Description:** Compusearch of Overland Park is an executive search firm. **Specializes in the areas of:** Accounting/Auditing; Administration/MIS/EDP; Advertising; Architecture/Construction/Real Estate; Banking; Chemical; Communications; Computer Hardware/Software; Design; Electrical; Engineering; Food Industry; General Management; Health/Medical; Insurance; Legal; Manufacturing; Operations Management; Personnel/Labor Relations; Printing/Publishing; Procurement; Retail; Sales and Marketing; Technical and Scientific; Textiles; Transportation.

DUNHILL PERSONNEL
3706 SW Topeka Boulevard, Suite 202, Topeka KS 66609-1239. 913/267-2773. **Contact:** Bob Washatka, President/Owner. **Description:** An executive search firm. **Specializes in the areas of:** Banking; Computer Hardware/Software; Health/Medical; Office Support; Rubber.

MANAGEMENT RECRUITERS OF OVERLAND PARK
9401 Indian Creek Parkway, Suite 920, Overland Park KS 66210. 913/661-9300. **Fax:** 913/661-9030. **Contact:** Office Manager. **Description:** An executive search firm. **Specializes in the areas of:** Accounting/Auditing; Administration/ MIS/EDP; Advertising; Architecture/Construction/Real Estate; Banking; Chemical; Communications; Computer Hardware/Software; Design; Electrical; Engineering; Food Industry; General Management; Health/Medical; Insurance; Legal; Manufacturing; Operations Management; Personnel/Labor Relations; Printing/Publishing; Procurement; Retail; Sales and Marketing; Technical and Scientific; Textiles; Transportation.

MANAGEMENT RECRUITERS OF TOPEKA
3400 SW VanBuren, Topeka KS 66611. 913/267-5430. **Contact:** Kirk Hawkins, Manager. **Description:** A contingency executive search firm. **Specializes in the areas of:** Accounting/Auditing; Administration/MIS/EDP; Advertising; Architecture/Construction/Real Estate; Banking; Chemical; Communications; Computer Hardware/Software; Design; Electrical; Engineering; Food Industry; General Management; Health/Medical; Insurance; Legal; Manufacturing; Operations Management; Personnel/Labor Relations; Printing/Publishing; Procurement; Retail; Sales and Marketing; Technical and Scientific; Textiles; Transportation. **Corporate headquarters location:** Cleveland OH. **Other U.S. locations:** Nationwide. **Average salary range of placements:** More than $50,000.

OFFICEMATES5 OF OVERLAND PARK
SALES CONSULTANTS OF OVERLAND PARK
9401 Indian Creek Parkway, Suite 920, Overland Park KS 66210. 913/661-9111. **Fax:** 913/661-9030. **Contact:** Office Manager. **Description:** An executive search firm. **Specializes in the areas of:** Accounting/Auditing; Administration/MIS/EDP; Advertising; Architecture/Construction/Real Estate; Banking; Chemical; Communications; Computer Hardware/Software; Design; Electrical; Engineering; Food Industry; General Management; Health/Medical; Insurance; Legal; Manufacturing; Operations Management; Personnel/Labor Relations; Printing/Publishing; Procurement; Retail; Sales and Marketing; Technical and Scientific; Textiles; Transportation.

TEMTECH
10000 West 75th Street, Suite 118, Shawnee Mission KS 66204-2241. 913/831-1821. **Fax:** 913/831-1834. **Contact:** Carlene White, President. **Description:** A contract engineering company providing technical, engineering, and manufacturing executives on both temporary and permanent assignments. **Specializes in the areas of:** Architecture/Construction/Real Estate; Art/Design; Computer Science/Software; Engineering; Food Industry; General Management; Health/Medical; Manufacturing; Personnel/Labor Relations; Sales and Marketing; Technical and Scientific. **Positions commonly filled include:** Accountant/Auditor; Administrative Services Manager; Aerospace Engineer; Agricultural Engineer; Architect; Biomedical Engineer; Branch Manager; Buyer; Chemical Engineer; Chemist; Civil Engineer; Clerical Supervisor; Customer Service

Representative; Dental Assistant/Dental Hygienist; Design Engineer; Designer; Draftsperson; EEG Technologist; EKG Technician; Electrical/Electronics Engineer; Emergency Medical Technician; Environmental Engineer; Financial Analyst; General Manager; Geologist/Geophysicist; Human Resources Manager; Industrial Engineer; Industrial Production Manager; Landscape Architect; Licensed Practical Nurse; Mechanical Engineer; Medical Records Technician; Metallurgical Engineer; Mining Engineer; MIS Specialist; Occupational Therapist; Operations/Production Manager; Paralegal; Pharmacist; Physical Therapist; Purchasing Agent/Manager; Quality Control Supervisor; Registered Nurse; Services Sales Representative; Software Engineer; Structural Engineer; Systems Analyst; Technical Writer/Editor; Telecommunications Manager; Transportation/Traffic Specialist; Typist/Word Processor. **Benefits available to temporary workers:** Holiday; Vacation. **Corporate headquarters location:** This Location. **Other U.S. locations:** Nationwide. **Average salary range of placements:** More than $50,000. **Number of placements per year:** 100 - 199.

PERMANENT EMPLOYMENT AGENCIES OF KANSAS

BUSINESS SPECIALISTS
105 South Broadway Street, Suite 200, Wichita KS 67202-4217. 316/267-7375. **Fax:** 316/267-1085. **Contact:** Bim Heineman, Vice President. **Description:** An employment agency. Company pays fee. **Specializes in the areas of:** Accounting/Auditing; Administration/MIS/EDP; Advertising; Architecture/Construction/Real Estate; Banking; Computer Science/Software; Engineering; Finance; Food Industry; General Management; Health/Medical; Industrial; Insurance; Legal; Manufacturing; Nonprofit; Personnel/Labor Relations; Printing/Publishing; Retail; Sales and Marketing; Secretarial; Technical and Scientific. **Corporate headquarters location:** This Location. **Other U.S. locations:** Denver CO; San Francisco CA. **Number of placements per year:** 1000+.

DP CAREER ASSOCIATES
6405 Metcalf, Suite 425, Shawnee Mission KS 66202. 913/236-8288. **Contact:** Office Manager. **Description:** An employment agency. Company pays fee. **Specializes in the areas of:** Computer Hardware/Software; MIS/EDP; Technical and Scientific. **Positions commonly filled include:** Computer Operator; Computer Programmer; Customer Service Representative; Data Entry Clerk; EDP Specialist; Marketing Specialist; MIS Specialist; Systems Analyst; Technical Writer/Editor; Technician; Typist/Word Processor. **Number of placements per year:** 100 - 199.

DUNHILL OF WICHITA
317 South Hydraulic, Wichita KS 67211. 316/265-9541. **Contact:** Harold Wood, President. **Description:** An employment agency. **Specializes in the areas of:** Accounting/Auditing; Banking; Computer Hardware/Software; Engineering; Manufacturing; Technical and Scientific. **Positions commonly filled include:** Accountant/Auditor; Bank Officer/Manager; Computer Operator; Computer Programmer; General Manager; Industrial Designer; Industrial Engineer; Purchasing Agent/Manager; Technician.

MORGAN HUNTER CORPORATE SEARCH
6800 College Boulevard, Suite 550, Overland Park KS 66211. 913/491-3434. **Contact:** Jerry Hellebusch, President/Owner. **Description:** An employment agency. Company pays fee. **Specializes in the areas of:** Accounting/Auditing; Administration/MIS/EDP; Computer Science/Software; Finance; Insurance; Secretarial. **Positions commonly filled include:** Accountant/Auditor; Administrative Services Manager; Budget Analyst; Claim Representative; Clerical Supervisor; Computer Programmer; Credit Manager; Customer Service Representative; Financial Analyst; Systems Analyst; Underwriter/Assistant Underwriter. **Number of placements per year:** 200 - 499.

EXECUTIVE SEARCH FIRMS OF KENTUCKY

FIRST CHOICE SERVICES
1711 Ashley Circle, Suite 6, Bowling Green KY 42104. 502/782-9152. **Contact:** Bob Toth, President. **Description:** First Choice Services is an executive search firm. **Specializes in the areas of:** Accounting/Auditing; Administration/MIS/EDP; Advertising; Architecture/Construction/Real Estate; Banking; Chemical; Communications; Computer Hardware/Software; Design; Electrical; Engineering; Food Industry; General Management; Health/Medical; Insurance; Legal; Manufacturing; Operations Management; Personnel/Labor Relations; Printing/Publishing; Procurement; Real Estate; Retail; Sales and Marketing; Technical and Scientific; Textiles; Transportation.

KAREN MARSHALL ASSOCIATES
6304 Deep Creek Drive, Prospect KY 40059. 502/228-0800. **Fax:** 502/228-0663. **Contact:** Karen Marshall, Systems Recruiter. **Description:** Karen Marshall Associates is a contingency executive search firm. Company pays fee. **Specializes in the areas of:** Administration/MIS/EDP; Computer Science/Software. **Positions commonly filled include:** Computer Programmer; Internet Services Manager; MIS Specialist; Software Engineer; Statistician; Systems Analyst; Telecommunications Manager. **Corporate headquarters location:** Louisville KY. **Other U.S. locations:** Nationwide. **Average salary range of placements:** $30,000 - $50,000. **Number of placements per year:** 1 - 49.

SALES CONSULTANTS OF LOUISVILLE
4360 Brownsboro Road, Suite 240, Louisville KY 40207. 502/897-2902. **Contact:** Manager. **Description:** An executive search firm. **Specializes in the areas of:** Accounting/Auditing; Administration/MIS/EDP; Advertising; Architecture/Construction/Real Estate; Banking; Chemical; Communications; Computer Hardware/Software; Design; Electrical; Engineering; Food Industry; General Management; Health/Medical; Insurance; Legal; Manufacturing; Operations Management; Personnel/Labor Relations; Printing/Publishing; Procurement; Retail; Sales and Marketing; Technical and Scientific; Textiles; Transportation.

SNELLING PERSONNEL SERVICES
4010 DuPont Circle, Suite 419, Louisville KY 40207. 502/895-9494. **Contact:** Steve Steinmetz, President. **Description:** An executive search firm. **Specializes in the areas of:** Administration/MIS/EDP; Computer Hardware/Software; Engineering; Manufacturing; Sales and Marketing; Technical and Scientific.

PERMANENT EMPLOYMENT AGENCIES OF KENTUCKY

C.M. MANAGEMENT SERVICES
698 Perimeter Drive, Suite 200, Lexington KY 40517. 606/266-5000. **Contact:** Office Manager. **Description:** An employment agency. Company pays fee. **Specializes in the areas of:** Accounting/Auditing; Banking; Clerical; Computer Hardware/Software; Engineering; Food Industry; Manufacturing; Sales and Marketing; Secretarial; Technical and Scientific. **Positions commonly filled include:** Accountant/Auditor; Administrative Assistant; Bank Officer/Manager; Biological Scientist; Bookkeeper; Computer Programmer; Customer Service Representative; Draftsperson; EDP Specialist; Electrical/Electronics Engineer; Factory Worker; Financial Analyst; Human Resources Manager; Industrial Engineer; Legal Secretary; Light Industrial Worker; Marketing Specialist; Mechanical Engineer; Metallurgical Engineer; Mining Engineer;

Receptionist; Sales Representative; Secretary; Stenographer; Systems Analyst; Typist/Word Processor. **Number of placements per year:** 50 - 99.

COMPUTER CAREER CONSULTANTS
1800 Meidinger Tower, Louisville KY 40202. 502/589-3100. **Fax:** 502/589-3107. **Contact:** Recruiter. **Description:** An employment agency. Company pays fee. **Specializes in the areas of:** Computer Hardware/Software; MIS/EDP. **Positions commonly filled include:** Computer Programmer; EDP Specialist; MIS Specialist; Sales Representative; Systems Analyst.

HOUCK CAREER CONSULTANTS (HCC)
7404 Old Coach Road, Suite 110, Crestwood KY 40014-9787. 502/241-2882. **Fax:** 502/241-6411. **Contact:** Recruiter. **Description:** An employment agency. Company pays fee. **Specializes in the areas of:** Computer Hardware/Software. **Positions commonly filled include:** Computer Programmer; EDP Specialist; MIS Specialist; Software Engineer; Systems Analyst. **Number of placements per year:** 1 - 49.

MANPOWER TEMPORARY SERVICES
600 Perimeter Drive, Lexington KY 40517-4120. 606/268-1331. **Contact:** Maggie Coats, Human Resources. **Description:** An employment agency. **Specializes in the areas of:** Accounting/ Auditing; Administration/MIS/EDP; Computer Science/Software; Industrial; Secretarial; Technical and Scientific. **Positions commonly filled include:** Accountant/Auditor; Chemist; Claim Representative; Clinical Lab Technician; Computer Programmer; Draftsperson; Electrical/ Electronics Engineer; Human Resources Manager; Industrial Engineer; Industrial Production Manager; MIS Specialist; Operations/Production Manager; Quality Control Supervisor; Software Engineer; Systems Analyst; Typist/Word Processor. **Benefits available to temporary workers:** 401(k); Life Insurance; Medical Insurance; Paid Vacation. **Corporate headquarters location:** Milwaukee WI. **Number of placements per year:** 1000+.

EXECUTIVE SEARCH FIRMS OF LOUISIANA

CLERTECH GROUP, INC.
7029 Canal Boulevard, New Orleans LA 70124. 504/486-9733. **Fax:** 504/482-1475. **Contact:** George Dorko, Director of Technical Recruiting. **Description:** A retainer and contingency search firm. Company pays fee. **Specializes in the areas of:** Computer Science/Software; Economics; Electrical; Engineering; Technical and Scientific. **Positions commonly filled include:** Computer Programmer; Cost Estimator; Design Engineer; Economist; Electrical/Electronics Engineer; Mechanical Engineer; MIS Specialist; Nuclear Engineer; Operations/Production Manager; Software Engineer; Structural Engineer; Systems Analyst; Telecommunications Manager. **Corporate headquarters location:** This Location. **Other U.S. locations:** Nationwide. **Average salary range of placements:** More than $50,000. **Number of placements per year:** 1 - 49.

MANAGEMENT RECRUITERS OF BATON ROUGE
P.O. Box 3553, Baton Rouge LA 70821-3553. 504/383-1234. **Contact:** Cecilia Franklin, Manager. **Description:** An executive search firm. **Specializes in the areas of:** Accounting/Auditing; Administration/MIS/EDP; Advertising; Architecture/Construction/Real Estate; Banking; Chemical; Communications; Computer Hardware/Software; Design; Electrical; Engineering; Food Industry; General Management; Health/Medical; Insurance; Legal; Manufacturing; Operations Management; Personnel/Labor Relations; Printing/Publishing; Procurement; Retail; Sales and Marketing; Technical and Scientific; Textiles; Transportation.

MANAGEMENT RECRUITERS OF BATON ROUGE
SALES CONSULTANTS OF BATON ROUGE
5551 Corporate Boulevard, Lamar Building, Suite 2-H, Baton Rouge LA 70808-2512. 504/928-2212. **Fax:** 504/928-1109. **Contact:** Gregg Fell, Manager. **Description:** An executive search firm.

Specializes in the areas of: Accounting/Auditing; Administration/MIS/EDP; Advertising; Architecture/Construction/Real Estate; Banking; Chemical; Communications; Computer Hardware/ Software; Design; Electrical; Engineering; Food Industry; General Management; Health/Medical; Insurance; Legal; Manufacturing; Operations Management; Personnel/Labor Relations; Printing/ Publishing; Procurement; Retail; Sales and Marketing; Technical and Scientific; Textiles; Transportation.

MANAGEMENT RECRUITERS OF METAIRIE
COMPUSEARCH/SALES CONSULTANTS
P.O. Box 6605, Metairie LA 70009. 504/831-7333. **Contact:** Office Manager. **Description:** An executive search firm. **Specializes in the areas of:** Accounting/Auditing; Administration/MIS/EDP; Advertising; Architecture/Construction/Real Estate; Banking; Chemical; Communications; Computer Hardware/Software; Design; Electrical; Engineering; Food Industry; General Management; Health/Medical; Insurance; Legal; Manufacturing; Operations Management; Personnel/Labor Relations; Printing/Publishing; Procurement; Retail; Sales and Marketing; Technical and Scientific; Textiles; Transportation.

SNELLING PERSONNEL SERVICES
1500 Louisville Avenue, Monroe LA 71201. 318/387-6090. **Fax:** 318/361-6097. **Contact:** Wayne Williamson, President. **Description:** An executive search firm. Company pays fee. **Specializes in the areas of:** Accounting/Auditing; Administration/MIS/EDP; Banking; Computer Science/ Software; Engineering; Health/Medical; Manufacturing; Personnel/Labor Relations; Sales and Marketing. **Positions commonly filled include:** Accountant/Auditor; Bank Officer/Manager; Biological Scientist; Chemical Engineer; Chemist; Civil Engineer; Claim Representative; Computer Programmer; Construction Contractor; Draftsperson; Electrical/Electronics Engineer; Human Resources Manager; Industrial Engineer; Management Trainee; Mechanical Engineer; Occupational Therapist; Paralegal; Pharmacist; Physical Therapist; Physician; Quality Control Supervisor; Registered Nurse; Respiratory Therapist; Restaurant/Food Service Manager; Systems Analyst. **Number of placements per year:** 200 - 499.

SNELLING PERSONNEL SERVICES
7809 Jefferson Highway, Suite C-1, Baton Rouge LA 70809. 504/927-0096. **Contact:** Office Manager. **Description:** An executive search firm. **Specializes in the areas of:** Accounting/ Auditing; Banking; Computer Hardware/Software; Engineering; Sales and Marketing. **Number of placements per year:** 50 - 99.

TALLEY & ASSOCIATES, INC.
1105 West Prien Lake Road, Suite D, Lake Charles LA 70601. 318/474-JOBS. **Fax:** 318/474-0885. **Contact:** Burt Bollotte, Manager. **Description:** A full-service search firm. **Specializes in the areas of:** Accounting/Auditing; Administration/MIS/EDP; Advertising; Architecture/Construction/Real Estate; Banking; Biology; Computer Science/Software; Engineering; Fashion; Finance; Food Industry; General Management; Health/Medical; Legal; Manufacturing; Personnel/Labor Relations; Printing/Publishing; Retail; Sales and Marketing; Secretarial; Technical and Scientific; Transportation. **Positions commonly filled include:** Accountant/Auditor; Adjuster; Administrative Services Manager; Advertising Clerk; Architect; Attorney; Bank Officer/Manager; Biochemist; Blue-Collar Worker Supervisor; Branch Manager; Broadcast Technician; Buyer; Chemical Engineer; Chemist; Civil Engineer; Claim Representative; Clerical Supervisor; Clinical Lab Technician; Computer Programmer; Construction Contractor; Cost Estimator; Credit Manager; Customer Service Representative; Dental Assistant/Dental Hygienist; Dentist; Design Engineer; Dietician/Nutritionist; Draftsperson; Electrical/Electronics Engineer; Environmental Engineer; Financial Analyst; General Manager; Geologist/Geophysicist; Health Services Manager; Hotel Manager; Human Resources Manager; Human Service Worker; Industrial Engineer; Industrial Production Manager; Insurance Agent/Broker; Landscape Architect; Licensed Practical Nurse; Management Trainee; Manufacturer's/Wholesaler's Sales Rep.; Mechanical Engineer; Medical Records Technician; MIS Specialist; Occupational Therapist; Operations/Production Manager; Paralegal; Petroleum Engineer; Pharmacist; Physical Therapist; Physician; Property and Real Estate Manager; Psychologist; Public Relations Specialist; Purchasing Agent/Manager; Quality Control Supervisor; Radio/TV Announcer/Broadcaster; Radiologic Technologist; Registered Nurse; Restaurant/Food Service Manager; Securities Sales Representative; Social Worker;

Sociologist; Systems Analyst; Technical Writer/Editor; Telecommunications Manager; Transportation/Traffic Specialist; Travel Agent; Typist/Word Processor; Video Maintenance Engineer.

EXECUTIVE SEARCH FIRMS OF MAINE

COMPUSOURCE
56 Industrial Park Road, Suite 6, Saco ME 04072. 207/284-1188. **Contact:** Lincoln Page, President. **Description:** A permanent placement executive search firm. **Specializes in the areas of:** Computer Hardware/Software; Computer Operations; Computer Programming.

SELECTECH
605 U.S. Route One, Scarborough ME 04074. 207/883-6561. **Toll-free phone:** 800/434-6561. **Fax:** 207/883-2964. **Contact:** Chris Griffith, President. **E-mail address:** chrisg@selectech.com. **World Wide Web address:** http://www.selectech.com. **Description:** An executive search firm for high-tech fields. SelecTech focuses on computer science, information systems, key management, and sales/marketing. **Specializes in the areas of:** Administration/MIS/EDP; Computer Science/ Software; Sales and Marketing; Technical and Scientific. **Positions commonly filled include:** Computer Programmer; MIS Specialist; Multimedia Designer; Software Engineer; Systems Analyst; Technical Writer/Editor; Telecommunications Manager. **Corporate headquarters location:** This Location. **Average salary range of placements:** More than $50,000. **Number of placements per year:** 1 - 49.

EXECUTIVE SEARCH FIRMS OF MARYLAND

BEAR TREES CONSULTING INC.
815B Rockville Pike, Suite 250, Rockville MD 20852-1214. 301/217-5976. **Fax:** 301/309-8769. **Contact:** Brad Lewis, President. **E-mail address:** blewis5950@aol.com. **Description:** An executive search firm operating on both a retainer and contingency basis that specializes in the placement of technical and high-technology sales professionals. Company pays fee. **Specializes in the areas of:** Computer Science/Software; Sales and Marketing; Technical and Scientific. **Positions commonly filled include:** Computer Programmer; MIS Specialist; Multimedia Designer; Software Engineer; Systems Analyst; Technical Writer/Editor; Telecommunications Manager. **Corporate headquarters location:** This Location. **Average salary range of placements:** More than $50,000. **Number of placements per year:** 100 - 199.

COMPUTER MANAGEMENT INC.
809 Glen Eagle Court, Suite 205, Towson MD 21286. 410/583-0050. **Fax:** 410/494-9410. **Contact:** Janet Miller, Recruiter. **Description:** An executive search firm that focuses on placement of technical professionals. Company pays fee. **Specializes in the areas of:** Computer Operations; Information Systems. **Positions commonly filled include:** Computer Programmer; Network Engineer; Software Engineer; Systems Analyst. **Number of placements per year:** 1 - 49.

COMTEX, INC.
12024 Blackberry Terrace, North Potomac MD 20878. 301/340-6963. **Fax:** 301/340-7008. **Contact:** Lynn Argain, President. **Description:** An executive search firm operating on a contingency basis placing experienced sales and pre-sales systems engineers. Company pays fee. **Specializes in the areas of:** Computer Hardware/Software; Network Administration; Technical and Scientific. **Positions commonly filled include:** Design Engineer; Software Engineer. **Corporate**

headquarters location: Rockville MD. **Other U.S. locations:** Nationwide. **Average salary range of placements:** More than $50,000. **Number of placements per year:** 1 - 49.

EMPLOYER EMPLOYEE EXCHANGE, INC.
260 Gateway Drive, Bel Air MD 21014. 410/821-1900. **Toll-free phone:** 800/821-1902. **Fax:** 410/821-1904. **Contact:** Mike Worrell, Technical Recruiter. **E-mail address:** EEE11@aol.com. **Description:** An executive search firm. Company pays fee. **Specializes in the areas of:** Administration/MIS/EDP; Computer Science/Software; Engineering; Sales and Marketing; Secretarial; Technical and Scientific. **Positions commonly filled include:** Accountant/Auditor; Chemical Engineer; Chemist; Civil Engineer; Clerical Supervisor; Computer Programmer; Customer Service Representative; Draftsperson; Electrical/ Electronics Engineer; Industrial Engineer; Management Trainee; Manufacturer's/Wholesaler's Sales Rep.; MIS Specialist; Services Sales Representative; Software Engineer; Structural Engineer; Systems Analyst; Technical Writer/Editor; Telecommunications Manager; Typist/Word Processor. **Benefits available to temporary workers:** Bonuses; Medical Insurance; Paid Holidays; Paid Vacation. **Corporate headquarters location:** Towson MD. **Average salary range of placements:** $30,000 - $50,000. **Number of placements per year:** 100 - 199.

A.G. FISHKIN AND ASSOCIATES, INC.
P.O. Box 34413, Bethesda MD 20827. 301/770-4944. **Contact:** Anita Fishkin, President. **Description:** An executive search firm. Company pays fee. **Specializes in the areas of:** Administration/MIS/EDP; Computer Hardware/Software; Technical and Scientific; Telecommunications. **Positions commonly filled include:** Computer Programmer; Electrical/ Electronics Engineer; Management Analyst/Consultant; Sales Representative; Systems Analyst.

FUTURES, INC.
8600 LaSalle Road, Suite 615, Towson MD 21286. 410/337-2001. **Contact:** Daniel Otakie, CPC, President. **Description:** An executive search firm. Company pays fee. **Specializes in the areas of:** Accounting/Auditing; Administration/MIS/EDP; Architecture/Construction/ Real Estate; Banking; Computer Hardware/Software; Engineering; Fashion; Finance; Food Industry; General Management; Health/Medical; Insurance; Legal; Manufacturing; Personnel/Labor Relations; Retail; Sales and Marketing; Technical and Scientific; Transportation. **Positions commonly filled include:** Accountant/Auditor; Administrative Assistant; Aerospace Engineer; Architect; Biological Scientist; Biomedical Engineer; Bookkeeper; Buyer; Ceramics Engineer; Chemical Engineer; Chemist; Civil Engineer; Claim Representative; Computer Programmer; Customer Service Representative; Draftsperson; EDP Specialist; Electrical/Electronics Engineer; Hotel Manager; Industrial Designer; Industrial Engineer; Legal Secretary; Manufacturing Engineer; Mechanical Engineer; Medical Secretary; Metallurgical Engineer; MIS Specialist; Nurse; Purchasing Agent/ Manager; Quality Control Supervisor; Sales Representative; Software Engineer; Systems Analyst; Technical Writer/Editor; Typist/Word Processor.

L.S. GROSS & ASSOCIATES
28 Allegheny Avenue, Baltimore MD 21204. 410/821-9351. **Fax:** 410/583-1901. **Contact:** Linda Gross, President. **Description:** A retainer and contingency search firm. **Specializes in the areas of:** Administration/MIS/EDP; Advertising; Biology; Computer Science/Software; Engineering; Finance; Food Industry; General Management; Health/Medical; Industrial; Legal; Manufacturing; Personnel/Labor Relations; Printing/Publishing. **Positions commonly filled include:** Administrative Services Manager; Attorney; Biochemist; Biological Scientist; Biomedical Engineer; Buyer; Chemical Engineer; Computer Programmer; Economist; Electrical/Electronics Engineer; Environmental Engineer; Food Scientist/Technologist; Health Services Manager; Human Resources Manager; Industrial Engineer; Internet Services Manager; Management Analyst/ Consultant; MIS Specialist; Paralegal; Physical Therapist; Physician; Purchasing Agent/Manager; Registered Nurse; Software Engineer; Structural Engineer; Surgical Technician; Systems Analyst. **Corporate headquarters location:** This Location. **Average salary range of placements:** More than $50,000. **Number of placements per year:** 1 - 49.

JDG ASSOCIATES LIMITED
1700 Research Boulevard, Rockville MD 20850. 301/340-2210. **Fax:** 301/762-3117. **Contact:** Joseph DeGioia, President. **E-mail address:** degioia@jdgsearch.com. **Description:** JDG

Associates Limited searches and recruits specialists and serves the fields of information technology, engineering, management consulting, quantitative science, and association management. Of its clients, 35 percent are placed on a retainer basis and 65 percent are placed on a contingency basis. Company pays fee. **Specializes in the areas of:** Administration/MIS/EDP; Computer Science/ Software; Economics; Engineering; Nonprofit. **Positions commonly filled include:** Budget Analyst; Computer Programmer; Economist; Electrical/Electronics Engineer; Financial Analyst; Industrial Engineer; Management Analyst/Consultant; Mathematician; MIS Specialist; Multimedia Designer; Science Technologist; Software Engineer; Statistician; Systems Analyst; Telecommunications Manager; Transportation/Traffic Specialist. **Corporate headquarters location:** This Location. **Average salary range of placements:** More than $50,000. **Number of placements per year:** 200 - 499.

E.G. JONES ASSOCIATES, LTD.

1505 York Road, Lutherville Titonium MD 21093-5628. 410/337-4925. **Toll-free phone:** 800/944-6676. **Fax:** 410/337-5057. **Contact:** Edward G. Jones, President. **Description:** A professional recruitment firm specializing in chemical engineering, packaging, real estate, information technology, and presidents for emerging firms. **Specializes in the areas of:** Administration/MIS/EDP; Computer Science/Software; Engineering. **Positions commonly filled include:** Chemical Engineer; Computer Programmer; MIS Specialist; Property and Real Estate Manager; Software Engineer; Systems Analyst. **Average salary range of placements:** More than $50,000. **Number of placements per year:** 50 - 99.

MANAGEMENT RECRUITERS OF ANNAPOLIS

2083 West Street, Suite 5A, Annapolis MD 21401. 410/841-6600. **Contact:** John Czajkowski, Manager. **Description:** An executive search firm. **Specializes in the areas of:** Accounting/Auditing; Administration/MIS/EDP; Advertising; Architecture/Construction/ Real Estate; Banking; Chemical; Communications; Computer Hardware/Software; Design; Electrical; Engineering; Finance; Food Industry; General Management; Health/Medical; Insurance; Legal; Manufacturing; Operations Management; Personnel/Labor Relations; Pharmaceutical; Printing/ Publishing; Procurement; Retail; Sales and Marketing; Technical and Scientific; Textiles; Transportation.

MANAGEMENT RECRUITERS OF BALTIMORE
SALES CONSULTANTS OF BALTIMORE

9515 Deereco Road, Suite 900, Timonium MD 21093. 410/252-6616. **Fax:** 410/252-7076. **Contact:** Ken Davis, General Manager. **Description:** An executive search firm. Company pays fee. **Specializes in the areas of:** Accounting/Auditing; Administration/MIS/EDP; Advertising; Architecture/Construction/Real Estate; Banking; Chemical; Communications; Computer Hardware/ Software; Design; Electrical; Engineering; Finance; Food Industry; General Management; Health/Medical; Insurance; Legal; Manufacturing; Operations Management; Personnel/Labor Relations; Pharmaceutical; Printing/Publishing; Procurement; Retail; Sales and Marketing; Technical and Scientific; Textiles; Transportation. **Positions commonly filled include:** Accountant/Auditor; Aerospace Engineer; Bank Officer/Manager; Biological Scientist; Biomedical Engineer; Chemical Engineer; Civil Engineer; Claim Representative; Clerical Supervisor; Computer Programmer; Credit Manager; Customer Service Representative; Electrical/Electronics Engineer; Financial Analyst; Industrial Engineer; Industrial Production Manager; Insurance Agent/ Broker; Mechanical Engineer; Operations/Production Manager; Paralegal; Quality Control Supervisor; Securities Sales Representative; Systems Analyst. **Number of placements per year:** 200 - 499.

MANAGEMENT RECRUITERS OF FREDERICK
OFFICEMATES5 OF BALTIMORE

201 Thomas Johnson Drive, Suite 202, Frederick MD 21702. 301/663-0600. **Fax:** 301/663-0454. **Contact:** Ms. Pat Webb, Owner/Manager. **Description:** An executive search firm. **Specializes in the areas of:** Accounting/Auditing; Administration/MIS/EDP; Advertising; Architecture/ Construction/Real Estate; Banking; Chemical; Communications; Computer Hardware/ Software; Design; Electrical; Engineering; Finance; Food Industry; General Management; Health/Medical; Insurance; Legal; Manufacturing; Operations Management; Personnel/Labor Relations;

Pharmaceutical; Printing/Publishing; Procurement; Retail; Sales and Marketing; Technical and Scientific; Textiles; Transportation.

MANAGEMENT RECRUITERS OF WASHINGTON, D.C.
1100 Wayne Avenue, Suite 1080, Silver Spring MD 20910. 301/589-5400. **Fax:** 301/589-3033. **Contact:** Manager. **Description:** An executive search firm. **Specializes in the areas of:** Accounting/Auditing; Administration/MIS/EDP; Advertising; Architecture/ Construction/Real Estate; Banking; Chemical; Communications; Computer Hardware/Software; Design; Electrical; Engineering; Finance; Food Industry; General Management; Health/Medical; Insurance; Legal; Manufacturing; Operations Management; Personnel/Labor Relations; Pharmaceutical; Printing/ Publishing; Procurement; Retail; Sales and Marketing; Technical and Scientific; Textiles; Transportation.

PROFESSIONAL PERSONNEL SERVICES
1420 East Joppa Road, Towson MD 21286. 410/823-5630. **Fax:** 410/821-9423. **Contact:** Neal Fisher, President. **Description:** An executive search firm focusing on information systems industries. Company pays fee. **Specializes in the areas of:** Computer Science/Software. **Positions commonly filled include:** Computer Programmer; Systems Analyst. **Number of placements per year:** 1 - 49.

SALES CONSULTANTS OF BALTIMORE CITY
575 South Charles Street, Suite 401, Baltimore MD 21201. 410/727-5750. **Fax:** 410/727-1253. **Contact:** Steven Braun, President. **Description:** An executive search firm. Company pays fee. **Specializes in the areas of:** Advertising; Computer Hardware/Software; Fashion; Food Industry; General Management; Health/Medical; Industrial; Insurance; Manufacturing; Personnel/Labor Relations; Printing/Publishing; Retail; Secretarial; Technical and Scientific; Transportation. **Positions commonly filled include:** Administrative Assistant; Bookkeeper; Buyer; Claim Representative; Credit Manager; Customer Service Representative; Data Entry Clerk; EDP Specialist; Electrical/Electronics Engineer; Legal Secretary; Marketing Specialist; Medical Secretary; MIS Specialist; Nurse; Public Relations Specialist; Purchasing Agent/Manager; Receptionist; Recruiter; Sales Representative; Secretary; Software Engineer; Systems Analyst; Typist/Word Processor. **Number of placements per year:** 100 - 199.

SALES CONSULTANTS OF COLUMBIA
10320 Little Patuxent Parkway, Suite 511, Columbia MD 21044. 410/992-4900. **Fax:** 410/992-4905. **Contact:** David Rubin, General Manager. **Description:** An executive search firm. Company pays fee. **Specializes in the areas of:** Computer Hardware/Software; Data Communications; Engineering; Sales and Marketing; Telecommunications. **Positions commonly filled include:** Sales Representative; Telecommunications Analyst. **Number of placements per year:** 1 - 49.

SALES CONSULTANTS OF LUTHERVILLE
9515 Deereco Road, Suite 801, Timonium MD 21093. 410/252-6616. **Fax:** 410/252-7076. **Contact:** Ken Davis, President. **Description:** An executive search firm operating on a contingency basis. Company pays fee. **Specializes in the areas of:** Administration/MIS/EDP; Banking; Computer Science/Software; Engineering; Finance; Food Industry; Industrial; Insurance; Manufacturing; Sales and Marketing; Secretarial. **Positions commonly filled include:** Administrative Services Manager; Chemical Engineer; Clerical Supervisor; Computer Programmer; Customer Service Representative; Design Engineer; Electrical/Electronics Engineer; Industrial Engineer; Insurance Agent/Broker; Management Analyst/Consultant; Manufacturer's/Wholesaler's Sales Rep.; Mechanical Engineer; MIS Specialist; Operations/Production Manager; Securities Sales Representative; Services Sales Representative; Software Engineer; Systems Analyst; Typist/Word Processor; Underwriter/Assistant Underwriter. **Corporate headquarters location:** Cleveland OH. **Other U.S. locations:** Nationwide. **Average salary range of placements:** More than $50,000. **Number of placements per year:** 200 - 499.

SALES CONSULTANTS OF PRINCE GEORGES COUNTY
7515 Annapolis Road, Suite 304, Hyattsville MD 20784. 301/731-4201. **Contact:** Tom Hummel,

Manager. **Description:** This location of Sales Consultants is an executive search firm. **Specializes in the areas of:** Accounting/Auditing; Administration/MIS/EDP; Advertising; Architecture/ Construction/Real Estate; Banking; Chemical; Communications; Computer Hardware/Software; Design; Electrical; Engineering; Finance; Food Industry; General Management; Health/Medical; Insurance; Legal; Manufacturing; Operations Management; Personnel/Labor Relations; Pharmaceutical; Printing/Publishing; Procurement; Retail; Sales and Marketing; Technical and Scientific; Textiles; Transportation.

SEEK INTERNATIONAL, INC.
15 Stablemere Court, Baltimore MD 21209-1062. 410/653-9680. **Fax:** 410/653-9682. **Contact:** Heather Finley, Account Executive. **E-mail address:** seekint@clark.net. **Description:** Seek International is an executive search firm specializing in the high-tech computer market. The company works with businesses of various sizes. **Specializes in the areas of:** Administration/ MIS/EDP; Computer Science/Software; Engineering; Sales and Marketing. **Positions commonly filled include:** Computer Programmer; Internet Services Manager; MIS Specialist; Sales Representative; Software Engineer; Systems Analyst; Telecommunications Manager. **Corporate headquarters location:** This Location. **Average salary range of placements:** More than $50,000. **Number of placements per year:** 200 - 499.

SUDINA SEARCH
375 West Padonia Road, Timonium MD 21093-2100. 410/252-6900. **Fax:** 410/252-8033. **Contact:** Chuck Sudina, President. **E-mail address:** sudina@access.digex.net. **World Wide Web address:** http://www.sudinasearch.com. **Description:** Sudina Search is one of the largest executive search firms in the Mid-Atlantic region specializing in information systems, accounting/finance, and health care. Company pays fee. **Specializes in the areas of:** Accounting/Auditing; Computer Science/Software; Finance; Health/Medical. **Positions commonly filled include:** Accountant/Auditor; Computer Programmer; Financial Analyst; Health Services Manager; MIS Specialist; Multimedia Designer; Software Engineer; Systems Analyst; Telecommunications Manager. **Corporate headquarters location:** Baltimore MD. **Other U.S. locations:** Nationwide. **Average salary range of placements:** More than $50,000. **Number of placements per year:** 200 - 499.

VEY MARK ASSOCIATES, INC.
8753 Ruppert Court, Ellicott City MD 21043-5451. 410/992-8422. **Fax:** 410/992-8934. **Contact:** Harvey M. Weisberg, President. **Description:** An executive search firm operating on both retainer and contingency bases. Company pays fee. **Specializes in the areas of:** Computer Science/Software; Engineering. **Positions commonly filled include:** Computer Programmer; Electrical/ Electronics Engineer; Mathematician; Software Engineer; Systems Analyst. **Corporate headquarters location:** This Location. **Average salary range of placements:** More than $50,000. **Number of placements per year:** 1 - 49.

WINSTON SEARCH, INC.
16 Greenmeadow Drive, Suite 305, Timonium MD 21093. 410/560-1111. **Fax:** 410/560-0112. **Contact:** Tom Winston, President. **Description:** Winston Search is a retained search company offering career and outplacement counseling. Company pays fee. **Specializes in the areas of:** Accounting/Auditing; Administration/MIS/EDP; Advertising; Computer Science/Software; Engineering; Finance; General Management; Industrial; Manufacturing; Personnel/Labor Relations; Sales and Marketing; Secretarial. **Positions commonly filled include:** Accountant/Auditor; Aerospace Engineer; Biomedical Engineer; Branch Manager; Chemical Engineer; Chemist; Civil Engineer; Design Engineer; Designer; Electrical/Electronics Engineer; Financial Analyst; General Manager; Human Resources Manager; Industrial Engineer; Industrial Production Manager; Management Analyst/ Consultant; Manufacturer's/Wholesaler's Sales Rep.; Market Research Analyst; Mechanical Engineer; MIS Specialist; Nuclear Engineer; Operations/Production Manager; Software Engineer; Systems Analyst; Telecommunications Manager. **Corporate headquarters location:** This Location. **Average salary range of placements:** More than $50,000. **Number of placements per year:** 100 - 199.

PERMANENT EMPLOYMENT AGENCIES OF MARYLAND

ADMINISTRATIVE PERSONNEL
1112 Wayne Avenue, Silver Spring MD 20910. 301/565-3900. **Fax:** 301/588-9044. **Contact:** Robert L. McDermott, Owner. **Description:** An employment agency. Company pays fee. **Specializes in the areas of:** Accounting/Auditing; Administration/MIS/EDP; Banking; Clerical; Computer Hardware/Software; Finance; Mortgage; Sales and Marketing. **Positions commonly filled include:** Accountant/Auditor; Administrative Assistant; Administrative Worker/Clerk; Bank Officer/Manager; Bookkeeper; Clerk; Computer Operator; Computer Programmer; Credit Manager; Customer Service Representative; Data Entry Clerk; EDP Specialist; Legal Secretary; Medical Secretary; Receptionist; Sales Executive; Sales Representative; Secretary; Stenographer; Typist/Word Processor. **Other U.S. locations:** Washington DC; VA. **Number of placements per year:** 1000+.

CAREER BYTE, INC.
8604 Second Avenue, Suite 160, Silver Spring MD 20910. 301/587-5626. **Fax:** 301/587-0323. **Contact:** Carl Hollenbach, President. **Description:** An employment agency. Company pays fee. **Specializes in the areas of:** Computer Science/Software. **Positions commonly filled include:** Computer Programmer; MIS Specialist; Software Engineer; Systems Analyst; Telecommunications Manager. **Number of placements per year:** 50 - 99.

J.R. ASSOCIATES
152 Rollins Avenue, Suite 200, Rockville MD 20852. 301/984-8885. **Contact:** Daniel Keller, President. **Description:** An employment agency. Company pays fee. **Specializes in the areas of:** Administration/MIS/EDP; Computer Hardware/Software; Engineering; Sales and Marketing; Technical and Scientific. **Positions commonly filled include:** Computer Programmer; Data Analyst; EDP Specialist; Financial Analyst; Marketing Specialist; MIS Specialist; Sales Engineer; Sales Representative; Systems Analyst; Telecommunications Analyst. **Number of placements per year:** 50 - 99.

MANAGEMENT RECRUITERS INTERNATIONAL
SALES CONSULTANTS
1395 Piccard Drive, Suite 330, Rockville MD 20850. 301/417-9100. **Toll-free phone:** 800/875-9630. **Fax:** 301/417-9620. **Contact:** Veronica Gordon, Office Manager. **Description:** An executive search firm. Company pays fee. **Specializes in the areas of:** Computer Science/Software; Telecommunications. **Positions commonly filled include:** Telecommunications Manager. **Corporate headquarters location:** Cleveland OH. **Other U.S. locations:** Nationwide. **International locations:** Worldwide. **Average salary range of placements:** More than $50,000. **Number of placements per year:** 200 - 499.

OPPORTUNITY SEARCH INC.
P.O. Box 751, Olney MD 20830. 301/924-4741. **Fax:** 301/924-1318. **Contact:** Marc Tappis, President. **Description:** An employment agency. Company pays fee. **Specializes in the areas of:** Computer Science/Software; Engineering. **Positions commonly filled include:** Computer Programmer; Electrical/Electronics Engineer; Software Engineer; Systems Analyst. **Number of placements per year:** 1 - 49.

QUEST SYSTEMS, INC.
4701 Sangamore Road, Suite 260N, Bethesda MD 20816. 301/229-4200. **Fax:** 301/229-0965. **Contact:** Tom Carter, Manager. **E-mail address:** questsyst@aol.com. **World Wide Web address:** http://www.questsyst.com. **Description:** An employment agency. Company pays fee. **Specializes in the areas of:** Administration/MIS/EDP; Computer Science/Software. **Positions commonly filled include:** Computer Programmer; MIS Specialist; Software Engineer; Systems Analyst.

Corporate headquarters location: This Location. **Other U.S. locations:** Atlanta GA; Baltimore MD; Philadelphia PA. **Average salary range of placements:** $30,000 - $50,000. **Number of placements per year:** 1000+.

QUEST SYSTEMS, INC.

11350 McCormick Road, Executive Plaza One, Suite 408, Hunt Valley MD 21031. 410/771-6600. **Fax:** 410/771-1907. **Contact:** Barry Bollinger, Manager. **E-mail address:** questsyst@aol.com. **World Wide Web address:** http://www.questsyst.com. **Description:** An employment agency specializing in computer technologies and information systems. Company pays fee. **Specializes in the areas of:** Administration/MIS/EDP; Computer Science/Software. **Positions commonly filled include:** Computer Programmer; MIS Specialist; Software Engineer; Systems Analyst. **Corporate headquarters location:** Bethesda MD. **Other U.S. locations:** Atlanta GA; Baltimore MD; Philadelphia PA. **Average salary range of placements:** $30,000 - $50,000. **Number of placements per year:** 1000+.

TECHNICAL PROFESSIONAL SEARCH GROUP

1305 Warwick Drive, Lutherville MD 21093. 410/296-4944. **Fax:** 410/321-4834. **Contact:** Dan Jones, Senior Consultant. **Description:** An employment agency. **Specializes in the areas of:** Communications; Computer Hardware/Software.

TECHNICAL TALENT LOCATORS, LTD.

8850 Stanford Boulevard, Suite 1850, Columbia MD 21045-4753. 410/995-6051. **Fax:** 410/995-6281. **Contact:** Stephen Horn, President. **Description:** An employment agency specializing in the placement of software engineers, systems engineers, and associated technical personnel. Company pays fee. **Specializes in the areas of:** Computer Science/Software; Engineering; Technical and Scientific. **Positions commonly filled include:** Computer Programmer; Design Engineer; Software Engineer; Systems Analyst. **Corporate headquarters location:** This Location. **Average salary range of placements:** More than $50,000. **Number of placements per year:** 50 - 99.

TRI-SERV INC.

22 West Padonia Road, Suite C-353, Timonium MD 21093. 410/561-1740. **Fax:** 410/252-7417. **Contact:** Walter J. Braczynski, President. **Description:** Provides permanent placement for all levels of technical personnel. Company pays fee. **Specializes in the areas of:** Computer Hardware/Software; Engineering; Manufacturing; Technical and Scientific. **Positions commonly filled include:** Aerospace Engineer; Ceramics Engineer; Chemical Engineer; Chemist; Civil Engineer; Computer Operator; Computer Programmer; Draftsperson; Electrical/Electronics Engineer; Industrial Engineer; Mechanical Engineer; Operations/Production Manager; Quality Control Supervisor; Systems Analyst; Technical Writer/Editor; Technician. **Corporate headquarters location:** This Location. **Average salary range of placements:** $30,000 - $50,000. **Number of placements per year:** 200 - 499.

EXECUTIVE SEARCH FIRMS OF MASSACHUSETTS

ABBOTT'S OF BOSTON, INC.

P.O. Box 588, Waltham MA 02254. 617/332-1100. **Fax:** 617/890-0714. **Contact:** Ivan Samuels, President. **Description:** An executive search firm operating on a contingency basis. Company pays fee. **Specializes in the areas of:** Computer Science/Software; Engineering; Health/Medical; Industrial; Manufacturing; Personnel/Labor Relations; Technical and Scientific. **Positions commonly filled include:** Biological Scientist; Biomedical Engineer; Computer Programmer; Design Engineer; Electrical/Electronics Engineer; Health Services Manager; Human Resources Manager; Internet Services Manager; Mechanical Engineer; MIS Specialist; Multimedia Designer; Software Engineer; Statistician; Systems Analyst. **Corporate headquarters location:** Newton

MA. **Average salary range of placements:** More than $50,000. **Number of placements per year:** 1 - 49.

ARCHITECTS

2 Electronics Avenue, Danvers MA 01923. 508/777-8500. **Fax:** 508/774-5620. **Contact:** Bob Jones, Principal. **Description:** An executive search firm operating on both retainer and contingency bases. The firm specializes in the information systems and software development industries. Company pays fee. **Specializes in the areas of:** Computer Science/Software; Engineering. **Positions commonly filled include:** Computer Programmer; Internet Services Manager; MIS Specialist; Multimedia Designer; Science Technologist; Software Engineer; Systems Analyst. **Corporate headquarters location:** This Location. **Average salary range of placements:** More than $50,000. **Number of placements per year:** 100 - 199.

CARTER/MACKAY EXECUTIVE

111 Speen Street, Framingham MA 01701. 508/626-2240. **Fax:** 508/879-2327. **Contact:** Human Resources. **Description:** An executive search firm operating on a contingency basis. The firm specializes in the placement of sales representatives, sales managers, and marketing and implementation consultants in the medical, pharmaceutical, biotechnology, software, telecommunications, and data communications fields. Company pays fee. **Specializes in the areas of:** Computer Science/Software; Sales and Marketing. **Positions commonly filled include:** Registered Nurse. **Corporate headquarters location:** This Location. **Other U.S. locations:** Nationwide. **Average salary range of placements:** $30,000 - $50,000.

CLEAR POINT CONSULTANTS

3 Centennial Drive, Peabody MA 01960. 508/750-4000. **Fax:** 508/774-1135. **Contact:** Cynthia Thornton, Vice President. **E-mail address:** carol@clearpnt.com. **World Wide Web address:** http://www.clearpnt.com. **Description:** An executive search firm operating on a contingency basis. The firm specializes in the placement of information design and delivery professionals. Company pays fee. **Specializes in the areas of:** Computer Science/Software; Printing/Publishing; Technical and Scientific. **Positions commonly filled include:** Editor; Multimedia Designer; Technical Writer/Editor. **Corporate headquarters location:** This Location. **Average salary range of placements:** More than $50,000. **Number of placements per year:** 100 - 199.

THE CLIENT SERVER GROUP, INC.

10 West Central Street, Natick MA 01760. 508/653-2900. **Fax:** 508/653-2977. **Contact:** John Plugis, President. **Description:** An executive search firm which operates on a contingency basis. Positions are primarily in the information management and software development areas of the financial, consulting, and software development industries. **NOTE:** A minimum of three years of commercial, post-academic experience is required for most positions. **Specializes in the areas of:** Banking; Computer Science/Software; Insurance. **Positions commonly filled include:** Computer Programmer; Internet Services Manager; Quality Assurance Engineer; Software Engineer; Systems Analyst; Telecommunications Manager. **Corporate headquarters location:** This Location. **Average salary range of placements:** More than $50,000. **Number of placements per year:** 1 - 49.

COMPUTER SECURITY PLACEMENT

P.O. Box 204-B, Northborough MA 01532. 508/393-7803. **Fax:** 508/393-6802. **Contact:** Cameron Carey, President. **Description:** An executive placement firm operating on a contingency basis. The firm specializes in placing high-level professionals in the fields of data security and disaster recovery planning. Company pays fee. **Specializes in the areas of:** Banking; Computer Science/ Software; Finance; Food Industry; Insurance. **Positions commonly filled include:** Systems Analyst. **Corporate headquarters location:** This Location. **Other U.S. locations:** Nationwide. **Average salary range of placements:** More than $50,000. **Number of placements per year:** 1 - 49.

DiFRONZO ASSOCIATES

2 Electronics Avenue, Danvers MA 01923. 508/774-5100. **Fax:** 508/774-5620. **Contact:** Alfred M. DiFronzo, CPC, President. **Description:** An executive search firm operating on a contingency basis. The firm specializes in the placement of information systems professionals. Areas of

placements' technical expertise include IBM AS/400, IBM/MVS, PC's, HP3000, RPG 400, COBOL, DB2, CICS, IMS, PowerBuilder, Visual Basic, Access, Oracle, Sybase, Foxpro, and Powerhouse. Company pays fee. **Specializes in the areas of:** Administration/MIS/EDP; Computer Science/Software; Engineering. **Positions commonly filled include:** Computer Programmer; MIS Specialist; Software Engineer; Systems Analyst. **Corporate headquarters location:** This Location. **Average salary range of placements:** $30,000 - $50,000. **Number of placements per year:** 50 - 99.

EASTWOOD PERSONNEL ASSOCIATES, INC.

P.O. Box 462, Franklin MA 02038. 508/528-8111. **Fax:** 508/528-1221. **Contact:** Rick Hohenberger, President. **Description:** An executive search firm operating on a contingency basis. The firm specializes in financial and computer software placement. Company pays fee. **Specializes in the areas of:** Banking; Computer Science/Software. **Positions commonly filled include:** Bank Officer/Manager; Computer Programmer; Financial Analyst; MIS Specialist; Software Engineer; Systems Analyst. **Corporate headquarters location:** This Location. **Other U.S. locations:** Nationwide. **Number of placements per year:** 1 - 49.

FORD & FORD

105 Chestnut Street, Needham MA 02192. 617/449-8200. **Fax:** 617/444-7335. **Contact:** Bernard Ford, Principal. **Description:** An executive search firm operating on both retainer and contingency bases. Company pays fee. **Specializes in the areas of:** Accounting/Auditing; Administration/ MIS/EDP; Advertising; Art/Design; Computer Science/Software; Fashion; Personnel/Labor Relations; Printing/Publishing; Retail; Sales and Marketing. **Positions commonly filled include:** Branch Manager; Environmental Engineer; General Manager; Human Resources Manager; Internet Services Manager; Market Research Analyst; MIS Specialist; Multimedia Designer; Public Relations Specialist; Technical Writer/Editor. **Corporate headquarters location:** This Location.

HM ASSOCIATES

2 Electronics Avenue, Danvers MA 01923. 508/762-7474. **Fax:** 508/739-9071. **Contact:** Hugh MacKenzie, Principal. **Description:** An executive search firm operating on a contingency basis. Company pays fee. **Specializes in the areas of:** Computer Science/Software; Engineering; Technical and Scientific. **Positions commonly filled include:** Aerospace Engineer; Biomedical Engineer; Computer Programmer; Internet Services Manager; Mechanical Engineer; MIS Specialist; Software Engineer; Systems Analyst. **Corporate headquarters location:** This Location. **Other U.S. locations:** Nationwide. **Average salary range of placements:** More than $50,000. **Number of placements per year:** 1 - 49.

HARBOR CONSULTING CORPORATION

10 State Street, Woburn MA 01801. 617/938-6886. **Fax:** 617/932-0421. **Contact:** Mike Moore, President. **Description:** An executive search firm operating on a retainer basis. The firm specializes in the computer software and electronics industries. Company pays fee. **Specializes in the areas of:** Computer Science/Software; Engineering. **Positions commonly filled include:** Computer Programmer; Internet Services Manager; Software Engineer; Systems Analyst; Telecommunications Manager. **Corporate headquarters location:** This Location. **Average salary range of placements:** More than $50,000. **Number of placements per year:** 50 - 99.

HARVEST PERSONNEL

65 James Street, Worcester MA 01603. 508/792-4545. **Contact:** Sue Soper, Internet Coordinator. **E-mail address:** resumes@harvestpersonnel.com. **World Wide Web address:** http://www. harvestpersonnel.com. **Description:** An executive search firm operating on a contingency basis. **Specializes in the areas of:** Accounting/Auditing; Administration/MIS/EDP; Banking; Biology; Computer Science/Software; Engineering; Finance; General Management; Health/Medical; Insurance; Legal; Secretarial; Technical and Scientific. **Positions commonly filled include:** Accountant/Auditor; Aerospace Engineer; Attorney; Bank Officer/Manager; Biological Scientist; Biomedical Engineer; Budget Analyst; Buyer; Claim Representative; Clinical Lab Technician; Computer Programmer; Dental Assistant/Dental Hygienist; Dentist; Design Engineer; Designer; Dietician/Nutritionist; Draftsperson; EEG Technologist; EKG Technician; Electrical/Electronics Engineer; Environmental Engineer; Financial Analyst; General Manager; Health Services Manager; Human Resources Manager; Industrial Engineer; Industrial Production Manager; Internet

Services Manager; Licensed Practical Nurse; Mechanical Engineer; Medical Records Technician; Metallurgical Engineer; MIS Specialist; Multimedia Designer; Nuclear Medicine Technologist; Occupational Therapist; Operations/Production Manager; Paralegal; Pharmacist; Physical Therapist; Purchasing Agent/Manager; Quality Control Supervisor; Registered Nurse; Respiratory Therapist; Science Technologist; Services Sales Representative; Software Engineer; Structural Engineer; Surgical Technician; Technical Writer/Editor; Typist/Word Processor; Underwriter/ Assistant Underwriter. **Corporate headquarters location:** Ware MA. **Other U.S. locations:** Hartford CT; Marlborough MA. **Average salary range of placements:** $30,000 - $50,000. **Number of placements per year:** 1000+.

JUDGE TECHNICAL SERVICES

200 Foxboro Boulevard, Suite 700, Foxboro MA 02035. **Toll-free phone:** 800/765-5874. **Fax:** 508/698-2122. **Contact:** Thomas Anderson, Vice President. **E-mail address:** careers@tiac.net. **World Wide Web address:** http://www.brainiac.com/tsnetwork. **Description:** An executive search firm that specializes in the placement of engineers, information systems professionals, software engineers, computer-aided designers and drafters, and technical support professionals. Company pays fee. **Specializes in the areas of:** Administration/MIS/EDP; Biology; Computer Science/ Software; Engineering; Industrial; Legal; Personnel/Labor Relations; Technical and Scientific. **Positions commonly filled include:** Aerospace Engineer; Biological Scientist; Buyer; Chemical Engineer; Civil Engineer; Computer Programmer; Cost Estimator; Design Engineer; Designer; Draftsperson; Editor; Human Resources Manager; Industrial Engineer; Internet Services Manager; MIS Specialist; Operations/Production Manager; Quality Control Supervisor; Software Engineer; Structural Engineer; Systems Analyst. **Corporate headquarters location:** Bala Cynwyd PA. **Other U.S. locations:** Nationwide. **Average salary range of placements:** $30,000 - $50,000. **Number of placements per year:** 1000+.

THE KOTEEN ASSOCIATES

70 Walnut Street, Wellesley MA 02181. 617/239-0011. **Fax:** 617/965-5886. **Contact:** Anne Koteen, President. **E-mail address:** recruit@koteenassoc.com. **Description:** An executive search firm. Company pays fee. **Specializes in the areas of:** Computer Science/Software. **Positions commonly filled include:** Computer Programmer; Internet Services Manager; Management Analyst/Consultant; MIS Specialist; Software Engineer; Systems Analyst; Telecommunications Manager. **Average salary range of placements:** More than $50,000. **Number of placements per year:** 50 - 99.

LAKE AND MANNING

5 Evergreen Lane, Hopedale MA 01747. 508/473-6955. **Fax:** 508/473-6956. **Contact:** Audrey Lake, Partner. **E-mail address:** lakemannin@aol.com. **Description:** An executive search firm. Company pays fee. **Specializes in the areas of:** Advertising; Computer Hardware/Software; Engineering; General Management; Printing/Publishing; Sales and Marketing; Technical and Scientific. **Positions commonly filled include:** Commercial Artist; Computer Programmer; Editor; Electrical/Electronics Engineer; Industrial Designer; Marketing Specialist; Operations/Production Manager; Public Relations Specialist; Reporter; Sales Representative; Software Engineer; Systems Analyst; Technical Illustrator; Technical Writer/Editor; Typist/Word Processor.

THE LITTLETON GROUP

136 Main Street, Acton MA 01720. 508/263-7221. **Fax:** 508/263-7740. **Contact:** Carl Tomforde, President. **Description:** An executive search firm operating on a contingency basis. The Littleton Group recruits and places high-technology professionals in various disciplines, primarily in the New England area. Company pays fee. **Specializes in the areas of:** Accounting/Auditing; Computer Hardware/Software; Engineering; General Management; Manufacturing; Personnel/ Labor Relations; Sales and Marketing; Technical and Scientific. **Positions commonly filled include:** Accountant/Auditor; Buyer; Computer Programmer; Draftsperson; Electrical/Electronics Engineer; Industrial Engineer; Manufacturing Engineer; Marketing Specialist; Mechanical Engineer; Operations/Production Manager; Purchasing Agent/Manager; Quality Control Supervisor; Sales Representative; Software Engineer. **Corporate headquarters location:** This Location. **Average salary range of placements:** More than $50,000. **Number of placements per year:** 1 - 49.

LOGIX INC.
1601 Trapelo Road, Waltham MA 02154. 617/890-0500. **Fax:** 617/890-3535. **Contact:** David M. Zell, President/CEO. **Description:** An executive search firm. Company pays fee. **Specializes in the areas of:** Administration/MIS/EDP; Biology; Computer Science/Software; Engineering; Technical and Scientific. **Positions commonly filled include:** Biological Scientist; Biomedical Engineer; Chemist; Computer Programmer; Electrical/Electronics Engineer; Software Engineer; Systems Analyst. **Number of placements per year:** 200 - 499.

MANAGEMENT RECRUITERS INTERNATIONAL
607 Boylston Street, Suite 700, Boston MA 02116. 617/262-5050. **Contact:** Jack Nehiley, General Manager. **Description:** An executive search firm. **Specializes in the areas of:** Accounting/ Auditing; Administration/MIS/EDP; Advertising; Architecture/Construction/Real Estate; Banking; Communications; Computer Hardware/Software; Construction; Electrical; Engineering; Finance; Food Industry; General Management; Health/Medical; Industrial; Insurance; Manufacturing; MIS/EDP; Operations Management; Personnel/Labor Relations; Pharmaceutical; Printing/ Publishing; Procurement; Retail; Sales and Marketing; Technical and Scientific; Transportation.

MANAGEMENT RECRUITERS INTERNATIONAL OF BRAINTREE
639 Granite Street, Braintree MA 02184. 617/848-1666. **Contact:** Steve Morse, Manager. **Description:** An executive search firm. **Specializes in the areas of:** Accounting/Auditing; Administration/MIS/EDP; Advertising; Architecture/Construction/Real Estate; Banking; Communications; Computer Hardware/Software; Construction; Electrical; Engineering; Finance; Food Industry; General Management; Health/Medical; Insurance; Legal; Manufacturing; Operations Management; Personnel/Labor Relations; Pharmaceutical; Printing/Publishing; Procurement; Retail; Sales and Marketing; Technical and Scientific; Textiles; Transportation.

MANAGEMENT RECRUITERS INTERNATIONAL OF SPRINGFIELD
1500 Main Street, Suite 1822, Springfield MA 01115. 413/781-1550. **Contact:** Barbara Suman, Administrative Manager. **Description:** An executive search firm. **Specializes in the areas of:** Accounting/Auditing; Administration/MIS/EDP; Advertising; Architecture/Construction/Real Estate; Banking; Communications; Computer Hardware/Software; Electrical; Engineering; Finance; Food Industry; General Management; Health/Medical; Personnel/Labor Relations; Pharmaceutical; Printing/Publishing; Procurement; Sales and Marketing; Technical and Scientific; Textiles; Transportation.

MANAGEMENT RECRUITERS INTERNATIONAL OF WESTBOROUGH
Westborough Office Park, 2000 West Park Drive, Westborough MA 01581. 508/366-9900. **Contact:** Irene Garrity, Manager. **Description:** An executive search firm. **Specializes in the areas of:** Accounting/Auditing; Administration/MIS/EDP; Advertising; Architecture/Construction/Real Estate; Banking; Communications; Computer Hardware/Software; Construction; Electrical; Engineering; Finance; Food Industry; General Management; Health/Medical; Insurance; Legal; Personnel/Labor Relations; Pharmaceutical; Procurement; Retail; Sales and Marketing; Technical and Scientific; Textiles; Transportation.

F.L. MANNIX & COMPANY, INC.
10 Village Road, Weston MA 02193. 617/894-9660. **Contact:** F.L. Mannix, President. **Description:** Founded in 1959, F.L. Mannix & Company is a retainer executive search and consulting firm that places professionals in the high-technology, broadcast, and various other industries. **Specializes in the areas of:** Administration/MIS/EDP; Broadcasting; Computer Science/Software; Engineering. **Positions commonly filled include:** Computer Programmer; Electrical/Electronics Engineer; Internet Services Manager; MIS Specialist; Systems Analyst; Telecommunications Manager. **Corporate headquarters location:** This Location. **Other U.S. locations:** Nationwide. **Average salary range of placements:** $30,000 - $50,000.

McDEVITT ASSOCIATES
90 Madison Street, Worcester MA 01608-2030. 508/752-5226. **Fax:** 508/752-9860. **Contact:** Larry McDevitt, Executive Recruiter. **E-mail address:** mcdassoc@ultranet.com. **Description:** An executive search firm. The company is online nationwide with over 650 smaller agencies via Nationwide Interchange Service, Top Echelon, and Staffing Interchange. McDevitt Associates also

places professionals in long-term contracting assignments. Company pays fee. **Specializes in the areas of:** Accounting/Auditing; Administration/MIS/EDP; Computer Science/Software; Engineering; Finance; General Management; Insurance; Manufacturing; Personnel/Labor Relations; Sales and Marketing. **Positions commonly filled include:** Accountant/Auditor; Claim Representative; Computer Programmer; Customer Service Representative; Design Engineer; Draftsperson; Human Resources Manager; Internet Services Manager; MIS Specialist; Quality Control Supervisor; Systems Analyst; Telecommunications Manager; Underwriter/Assistant Underwriter. **Corporate headquarters location:** This Location. **Average salary range of placements:** $30,000 - $50,000. **Number of placements per year:** 50 - 99.

MORENCY ASSOCIATES

2 Electronics Avenue, Danvers MA 01923-1009. 508/750-4460. **Fax:** 508/750-4465. **Contact:** Marcia Morency, President. **Description:** An executive search firm. Company pays fee. **Specializes in the areas of:** Accounting/Auditing; Administration/MIS/EDP; Computer Hardware/Software; Finance; Printing/Publishing; Sales and Marketing; Technical and Scientific. **Positions commonly filled include:** Accountant/Auditor; Administrative Assistant; Buyer; Claim Representative; Computer Programmer; Contract/Grant Administrator; EDP Specialist; Marketing Specialist; MIS Specialist; Sales Representative; Software Engineer. **Number of placements per year:** 1 - 49.

MORGAN & ASSOCIATES

P.O. Box 379, Granby MA 01033. 413/467-9156. **Fax:** 413/467-3003. **Contact:** Diane R. Morgan, Owner. **E-mail address:** morgan.assoc@the-spa.com. **Description:** An executive search firm operating on both retainer and contingency bases. Founded in 1981. Company pays fee. **Specializes in the areas of:** Computer Science/Software; Engineering; Manufacturing. **Positions commonly filled include:** Aerospace Engineer; Computer Programmer; Design Engineer; Electrical/Electronics Engineer; Engineer; Mechanical Engineer; Metallurgical Engineer; Operations/Production Manager; Physical Therapist; Software Engineer; Systems Analyst. **Corporate headquarters location:** This Location. **Average salary range of placements:** $30,000 - $50,000. **Number of placements per year:** 50 - 99.

NEW BOSTON SELECT GROUP INC.

7 Wheeling Avenue, Woburn MA 01801. 617/939-1500. **Fax:** 617/939-1695. **Contact:** C. Scott Bevins, Vice President/Business Development. **World Wide Web address:** http://www.newboston.com. **Description:** A consortium of recruitment firms. Company pays fee. **Specializes in the areas of:** Accounting/Auditing; Banking; Computer Hardware/Software; Engineering; Finance; MIS/EDP. **Positions commonly filled include:** Accountant/Auditor; Computer Programmer; EDP Specialist; Electrical/Electronics Engineer; Financial Analyst; Systems Analyst. **Benefits available to temporary workers:** 401(k); Medical Insurance. **Average salary range of placements:** More than $50,000. **Number of placements per year:** 1000+.

NEW DIMENSIONS IN TECHNOLOGY, INC.

74 Atlantic Avenue, Suite 101, Marblehead MA 01945. 617/639-0866. **Fax:** 617/639-0863. **Contact:** Beverly A. Kahn, President. **E-mail address:** resumes@ndt.com. **World Wide Web address:** http://www.ndt.com. **Description:** An executive search firm operating on both retainer and contingency bases. The firm specializes in director- and officer-level searches, as well as recruitment of professionals for advanced development, information technology, consumer electronics, online services, and consulting. Company pays fee. **Specializes in the areas of:** Banking; Computer Hardware/Software; Consulting; Engineering; General Management; Information Technology; Insurance; Start-up Organizations. **Positions commonly filled include:** Computer Programmer; Consultant; Electrical/Electronics Engineer; Financial Analyst; Information Systems Consultant; Management Analyst/Consultant; Software Engineer; Systems Analyst. **Corporate headquarters location:** This Location. **Average salary range of placements:** More than $50,000.

NEW ENGLAND SEARCH, INC.

P.O. Box 1248, Mill Street, Webster MA 01570. 508/943-3000. **Fax:** 508/943-9958. **Contact:** T. Michael Owens, Director. **Description:** An executive search firm operating on both retainer and contingency bases. Company pays fee. **Specializes in the areas of:** Accounting/Auditing;

Administration/MIS/EDP; Computer Science/Software; Engineering; Finance; Manufacturing; Personnel/Labor Relations. **Positions commonly filled include:** Accountant/Auditor; Administrative Services Manager; Computer Programmer; Design Engineer; Electrical/Electronics Engineer; Financial Analyst; General Manager; Human Resources Manager; Industrial Engineer; Industrial Production Manager; Mechanical Engineer; MIS Specialist; Quality Control Supervisor; Software Engineer; Systems Analyst; Telecommunications Manager. **Corporate headquarters location:** This Location. **Average salary range of placements:** $30,000 - $50,000. **Number of placements per year:** 50 - 99.

PALADIN PERSONNEL CONSULTANTS
3 Bridle Path, Sherborn MA 01770. 508/651-1909. **Fax:** 508/651-1908. **Contact:** Andrew Sideman, Principal. **Description:** An executive search firm. **Specializes in the areas of:** Computer Science/Software; Engineering; Sales and Marketing; Technical and Scientific. **Positions commonly filled include:** Computer Programmer; Electrical/Electronics Engineer; Systems Analyst; Technical Writer/Editor.

NORMAN POWERS ASSOCIATES
P.O. Box 3221, Framingham MA 01701. 508/877-2025. **Contact:** Human Resources. **Description:** An executive search firm operating on both retainer and contingency bases. The firm specializes in the placement of professionals in the commercial and military electronics industry, from high-tech research and development companies to manufacturers of technical products such as networking and communications equipment. Company pays fee. **Specializes in the areas of:** Computer Science/Software; Engineering; Manufacturing. **Positions commonly filled include:** Computer Programmer; Design Engineer; Electrical/Electronics Engineer; Mechanical Engineer; Software Engineer; Systems Analyst.

PROGRESSIVE SEARCH ASSOCIATES, INC. (PSA)
465 Auburn Street, Auburndale MA 02166. 617/244-1250. **Contact:** David J. Abrams, President. **E-mail address:** djab@ix.netcom.com. **Description:** An executive search firm operating on both retainer and contingency bases. Company pays fee. **Specializes in the areas of:** Computer Science/Software; Sales and Marketing; Technical and Scientific. **Positions commonly filled include:** Internet Services Manager; MIS Specialist; Software Engineer; Technical Writer/Editor; Telecommunications Manager. **Corporate headquarters location:** This Location. **Average salary range of placements:** More than $50,000. **Number of placements per year:** 50 - 99.

PROSEARCH
591 North Avenue, Wakefield MA 01880. 617/245-2224. **Fax:** 617/224-4252. **Contact:** Mr. Patrick Cresse, Senior Technical Recruiter. **Description:** An executive search firm which operates on a contingency basis. The firm specializes in professional MIS placement in the financial services and software industries. **Specializes in the areas of:** Administration/MIS/EDP; Computer Science/Software. **Positions commonly filled include:** Computer Programmer; Internet Services Manager; MIS Specialist; Systems Analyst. **Corporate headquarters location:** This Location. **Average salary range of placements:** Less than $20,000. **Number of placements per year:** 1 - 49.

J.E. RANTA ASSOCIATES
112 Washington Street, Marblehead MA 01945. 617/639-0788. **Fax:** 617/631-9828. **Contact:** Ed Ranta, Owner/President. **Description:** An executive search firm offering retained and contingency placement services for industry positions in financial services. J.E. Ranta Associates is affiliated with several other search firms throughout the U.S. Company pays fee. **Specializes in the areas of:** Administration/MIS/EDP; Banking; Computer Science/Software; Technical and Scientific. **Positions commonly filled include:** Computer Programmer; Internet Services Manager; Management Analyst/Consultant; MIS Specialist; Systems Analyst; Telecommunications Manager. **Corporate headquarters location:** This Location. **Average salary range of placements:** More than $50,000. **Number of placements per year:** 1 - 49.

RESOURCES OBJECTIVES INC.
185 Devonshire Street, Boston MA 02110. 617/338-5500. **Fax:** 617/338-5558. **Contact:** Human Resources. **Description:** An executive search firm operating on a retainer basis. Company pays fee.

Specializes in the areas of: Banking; Computer Science/Software; Legal; Sales and Marketing. **Positions commonly filled include:** Accountant/Auditor; Attorney; Bank Officer/Manager; Branch Manager; Physician; Software Engineer; Technical Writer/Editor. **Corporate headquarters location:** This Location. **Other U.S. locations:** Nationwide. **Number of placements per year:** 50 - 99.

LOUIS RUDZINSKY ASSOCIATES, INC.
394 Lowell Street, Suite 17, Lexington MA 02173. 617/862-6727. **Contact:** Howard Rudzinsky, Vice President. **Description:** An executive search firm. **Specializes in the areas of:** Computer Hardware/Software; Engineering; Sales and Marketing; Technical and Scientific. **Number of placements per year:** 50 - 99.

SALES CONSULTANTS INTERNATIONAL
P.O. Box 420, Sagamore Beach MA 02562. 508/888-8704. **Contact:** Ed Cahan, Manager. **Description:** An executive search firm. **Specializes in the areas of:** Accounting/Auditing; Administration/MIS/EDP; Advertising; Architecture/Construction/Real Estate; Banking; Communications; Computer Hardware/Software; Construction; Electrical; Engineering; Finance; Food Industry; General Management; Health/Medical; Operations Management; Personnel/Labor Relations; Pharmaceutical; Printing/Publishing; Procurement; Retail; Sales and Marketing; Technical and Scientific; Textiles; Transportation.

SALES CONSULTANTS OF PLYMOUTH COUNTY
567 Pleasant Street, Suite 8, Brockton MA 02401. 508/587-2030. **Fax:** 508/587-9261. **Contact:** Milt Feinson, President. **Description:** An executive search firm. **Specializes in the areas of:** Accounting/Auditing; Administration/MIS/EDP; Advertising; Architecture/Construction/Real Estate; Banking; Communications; Computer Hardware/Software; Electrical; Engineering; Finance; Food Industry; General Management; Health/Medical; Insurance; Legal; Manufacturing; Operations Management; Personnel/Labor Relations; Pharmaceutical; Printing/Publishing; Procurement; Retail; Sales and Marketing; Technical and Scientific; Textiles; Transportation. **Corporate headquarters location:** Cleveland OH. **Other U.S. locations:** Nationwide. **Average salary range of placements:** $30,000 - $50,000.

SALES CONSULTANTS OF WELLESLEY
888 Worcester Street, Suite 95, Wellesley MA 02181. 617/235-7700. **Fax:** 617/237-7207. **Contact:** Arthur J. Durante, General Manager. **Description:** An executive search firm. **Specializes in the areas of:** Accounting/Auditing; Administration/MIS/EDP; Advertising; Architecture/Construction/Real Estate; Banking; Communications; Computer Hardware/Software; Construction; Electrical; Engineering; Finance; Food Industry; General Management; Health/Medical; Insurance; Legal; Manufacturing; Operations Management; Personnel/Labor Relations; Pharmaceutical; Printing/Publishing; Procurement; Retail; Sales and Marketing; Technical and Scientific; Textiles; Transportation.

SELECTED EXECUTIVES INC.
76 Winn Street, Woburn MA 01801. 617/933-1500. **Fax:** 617/933-4145. **Contact:** Mr. Lee R. Sanborn, Jr., Manager. **Description:** An executive search firm operating on both retainer and contingency bases. The firm specializes in placement for equal opportunity employers. Company pays fee. **Specializes in the areas of:** Accounting/Auditing; Computer Science/Software; Engineering; Finance; Manufacturing; Personnel/Labor Relations; Sales and Marketing. **Positions commonly filled include:** Accountant/Auditor; Actuary; Administrative Services Manager; Aerospace Engineer; Aircraft Mechanic/Engine Specialist; Architect; Attorney; Automotive Mechanic; Bank Officer/Manager; Biological Scientist; Budget Analyst; Buyer; Chemical Engineer; Chemist; Civil Engineer; Claim Representative; Clinical Lab Technician; Computer Programmer; Counselor; Credit Manager; Customer Service Representative; Design Engineer; Electrical/Electronics Engineer; Environmental Engineer; Financial Analyst; General Manager; Health Services Manager; Human Resources Manager; Industrial Engineer; Management Analyst/Consultant; Management Trainee; Manufacturer's/Wholesaler's Sales Rep.; Market Research Analyst; Mechanical Engineer; Metallurgical Engineer; MIS Specialist; Nuclear Engineer; Operations/Production Manager; Public Relations Specialist; Purchasing Agent/Manager; Quality Control Supervisor; Registered Nurse; Restaurant/Food Service Manager; Securities Sales

Representative; Services Sales Representative; Software Engineer; Stationary Engineer; Statistician; Strategic Relations Manager; Systems Analyst; Technical Writer/Editor; Telecommunications Manager; Underwriter/Assistant Underwriter. **Corporate headquarters location:** This Location. **Average salary range of placements:** More than $50,000. **Number of placements per year:** 200 - 499.

L.A. SILVER ASSOCIATES
463 Worcester Road, Framingham MA 01701. 508/879-2603. **Contact:** Mr. Lee Silver, President. **Description:** An executive search firm. Company pays fee. **Specializes in the areas of:** Accounting/Auditing; Computer Hardware/Software; Distribution; Engineering; Finance; Food Industry; International Executives; Manufacturing; Personnel/Labor Relations; Real Estate; Retail. **Number of placements per year:** 100 - 199.

STRAUBE ASSOCIATES
855 Turnpike Street, North Andover MA 01845. 508/687-1993. **Fax:** 508/687-1886. **Contact:** Stanley H. Straube, President. **Description:** Straube Associates is an executive search firm. Company pays fee. **Specializes in the areas of:** Banking; Computer Science/Software; Engineering; General Management; Health/Medical; Manufacturing; Personnel/Labor Relations; Sales and Marketing; Technical and Scientific. **Number of placements per year:** 1 - 49.

S.B. WEBSTER & ASSOCIATES, INC.
P.O. Box 1007, Duxbury MA 02331. 617/934-6603. **Fax:** 617/934-9181. **Contact:** William L. Webster, President. **Description:** An executive search firm operating on a retainer basis. Company pays fee. **Specializes in the areas of:** Administration/MIS/EDP; Computer Science/Software; General Management; Manufacturing. **Positions commonly filled include:** Management Trainee; MIS Specialist; Operations/Production Manager; Software Engineer; Telecommunications Manager. **Corporate headquarters location:** This Location. **Other U.S. locations:** Nationwide. **Average salary range of placements:** More than $50,000. **Number of placements per year:** 1 - 49.

PERMANENT EMPLOYMENT AGENCIES OF MASSACHUSETTS

ABILITY SEARCH OF NEW ENGLAND
P.O. Box 883, Framingham MA 01701. 508/872-2060. **Fax:** 508/872-2345. **Contact:** Jerry Vengrow, President. **Description:** An employment agency. Company pays fee. **Specializes in the areas of:** Accounting/Auditing; Administration/MIS/EDP; Biology; Computer Science/Software; Engineering; Finance; General Management; Industrial; Legal; Manufacturing; Personnel/Labor Relations; Printing/Publishing; Sales and Marketing; Secretarial; Technical and Scientific. **Positions commonly filled include:** Accountant/Auditor; Aerospace Engineer; Agricultural Engineer; Architect; Attorney; Biological Scientist; Biomedical Engineer; Blue-Collar Worker Supervisor; Budget Analyst; Buyer; Chemical Engineer; Chemist; Civil Engineer; Clerical Supervisor; Clinical Lab Technician; Computer Programmer; Credit Manager; Customer Service Representative; Dental Assistant/Dental Hygienist; Dental Lab Technician; Designer; Draftsperson; Economist; EEG Technologist; EKG Technician; Electrical/Electronics Engineer; Electrician; Financial Analyst; Geologist/Geophysicist; Human Resources Manager; Industrial Engineer; Industrial Production Manager; Management Analyst/Consultant; Mathematician; Mechanical Engineer; Medical Records Technician; Metallurgical Engineer; Nuclear Engineer; Nuclear Medicine Technologist; Operations/Production Manager; Paralegal; Physician; Physicist/ Astronomer; Purchasing Agent/Manager; Quality Control Supervisor; Radiologic Technologist; Science Technologist; Software Engineer; Statistician; Structural Engineer; Systems Analyst; Technical Writer/Editor; Transportation/Traffic Specialist. **Number of placements per year:** 50 - 99.

ADDITIONAL TECHNICAL SUPPORT SERVICES

P.O. Box 9018, 1466 Main Street, Waltham MA 02254. 617/893-5600. **Contact:** Office Manager. **Description:** An employment agency. Company pays fee. **Specializes in the areas of:** Computer Hardware/Software; Engineering; Manufacturing; MIS/EDP; Technical and Scientific. **Positions commonly filled include:** Aerospace Engineer; Civil Engineer; Computer Programmer; Draftsperson; Electrical/Electronics Engineer; Industrial Engineer; Mechanical Engineer; MIS Specialist; Purchasing Agent/Manager; Quality Control Supervisor; Sales Representative; Secretary; Systems Analyst; Technical Writer/Editor; Technician. **Number of placements per year:** 500 - 999.

BRENTWOOD PERSONNEL

1408 Providence Highway, Suite 322, Norwood MA 02062. 617/848-6330. **Contact:** Jeffery Kurtz, President. **Description:** An employment agency. **Specializes in the areas of:** Computer Hardware/Software; Engineering; MIS/EDP. **Positions commonly filled include:** Aerospace Engineer; Computer Programmer; Electrical/Electronics Engineer; Physicist/Astronomer. **Number of placements per year:** 50 - 99.

C.R.W. & ASSOCIATES

P.O. Box 66126, Newton MA 02166. 617/928-0408. **Contact:** Charles R. Watt, Managing Director. **Description:** An employment agency. Company pays fee. **Specializes in the areas of:** Administration/MIS/EDP; Computer Science/Software. **Positions commonly filled include:** Computer Programmer; Software Engineer; Systems Analyst. **Number of placements per year:** 100 - 199.

CENTOR PERSONNEL

185 New Boston Street, Woburn MA 01801. 617/935-2955. **Fax:** 617/935-2954. **Contact:** Paulette Centor, President. **World Wide Web address:** http://www1.usa1.com/~pcentor. **Description:** Centor Personnel is an employment agency. Company pays fee. **Specializes in the areas of:** Computer Science/Software; Sales and Marketing. **Positions commonly filled include:** Internet Services Manager; Market Research Analyst; MIS Specialist; Services Sales Representative. **Average salary range of placements:** More than $50,000. **Number of placements per year:** 1 - 49.

CLEARY CONSULTANTS INC.

21 Merchants Row, Boston MA 02109. 617/367-7189. **Fax:** 617/367-3202. **Contact:** Mary Cleary, President. **Description:** An employment agency. Company pays fee. **Specializes in the areas of:** Accounting/Auditing; Administration/MIS/EDP; Advertising; Banking; Computer Science/Software; Finance; Health/Medical; Legal; Personnel/Labor Relations; Sales and Marketing; Secretarial. **Positions commonly filled include:** Accountant/Auditor; Actuary; Administrative Services Manager; Advertising Clerk; Attorney; Bank Officer/Manager; Branch Manager; Brokerage Clerk; Clerical Supervisor; Customer Service Representative; Financial Analyst; Health Services Manager; Human Resources Manager; Licensed Practical Nurse; Medical Records Technician; Paralegal; Pharmacist; Physician; Quality Control Supervisor; Registered Nurse; Services Sales Representative; Software Engineer; Systems Analyst; Travel Agent. **Number of placements per year:** 200 - 499.

ROBERT H. DAVIDSON ASSOCIATES, INC.

1410 Providence Highway, Norwood MA 02062. 617/769-8350. **Fax:** 617/769-8391. **Contact:** Robert H. Davidson, President. **Description:** Robert H. Davidson Associates is an employment agency that provides search, recruiting, and consulting services to businesses in the computer, automotive, environmental, communications, financial, and electronics industries. Company pays fee. **Specializes in the areas of:** Accounting/Auditing; Computer Science/Software; Engineering; Health/Medical; Manufacturing; Sales and Marketing. **Positions commonly filled include:** Biological Scientist; Budget Analyst; Buyer; Chemist; Computer Programmer; Electrical/Electronics Engineer; Industrial Engineer; Mechanical Engineer; Metallurgical Engineer; Purchasing Agent/Manager; Software Engineer; Technical Writer/Editor. **Number of placements per year:** 1 - 49.

EF TECHNICAL RESOURCES
1218 Boylston Street, Newton MA 02164-1017. 617/969-7850. **Fax:** 617/969-9429. **Contact:** Elli Freihofer, Owner. **Description:** An employment agency. Company pays fee. **Specializes in the areas of:** Administration/MIS/EDP; Computer Science/ Software; Technical and Scientific. **Positions commonly filled include:** Computer Programmer; MIS Specialist; Software Engineer; Systems Analyst; Telecommunications Manager. **Number of placements per year:** 1 - 49.

HUMAN RESOURCE CONSULTANTS
1252 Elm Street, West Springfield MA 01089. 413/737-7563. **Fax:** 413/731-9897. **Contact:** John M. Turner, President. **Description:** An employment agency. Company pays fee. **Specializes in the areas of:** Biology; Computer Science/Software; Engineering; Health/Medical. **Positions commonly filled include:** Biomedical Engineer; Computer Programmer; Industrial Engineer; Mechanical Engineer; Physical Therapist; Respiratory Therapist; Software Engineer; Systems Analyst. **Number of placements per year:** 50 - 99.

I.T. RESOURCES
3 Wallis Court, Lexington MA 02173. 617/863-2661. **Fax:** 617/863-2686. **Contact:** Ken Loomis, President. **World Wide Web address:** http://www.ourworld.compuserve.com/homepages/kloomis. **Description:** An employment agency. Company pays fee. **Specializes in the areas of:** Administration/MIS/EDP; Computer Science/Software. **Positions commonly filled include:** Computer Programmer; MIS Specialist; Software Engineer; Systems Analyst; Telecommunications Manager. **Average salary range of placements:** More than $50,000. **Number of placements per year:** 50 - 99.

INTERACTIVE SOFTWARE PLACEMENT, INC. (ISPI)
465 Auburn Street, Newton MA 02166. 617/244-1250. **Fax:** 617/965-7998. **Contact:** Sean Leary, Human Resources. **Description:** An employment agency. **Specializes in the areas of:** Software Development; Software Engineering; Software Quality Assurance.

WILLIAM JAMES ASSOCIATES
800 Turnpike Street, North Andover MA 01845-6156. 508/685-0700. **Fax:** 508/685-7113. **Contact:** Bill Josephson, President. **Description:** An employment agency. **Specializes in the areas of:** Computer Science/Software. **Positions commonly filled include:** Computer Programmer; Software Engineer; Systems Analyst.

LANE EMPLOYMENT SERVICE, INC.
370 Main Street, Suite 820, Worcester MA 01608. 508/757-5678. **Contact:** Richard Lane, CPC/President. **Description:** An employment agency. **Specializes in the areas of:** Accounting/ Auditing; Banking; Clerical; Computer Hardware/Software; Engineering; Finance; Insurance; Manufacturing; MIS/EDP; Secretarial. **Positions commonly filled include:** Accountant/Auditor; Actuary; Administrative Assistant; Bank Officer/Manager; Biological Scientist; Bookkeeper; Claim Representative; Clerk; Computer Programmer; Credit Manager; Data Entry Clerk; EDP Specialist; Electrical/Electronics Engineer; Financial Analyst; General Manager; Human Resources Manager; Industrial Designer; Industrial Engineer; Legal Secretary; Marketing Specialist; Mechanical Engineer; Medical Secretary; Metallurgical Engineer; Purchasing Agent/Manager; Receptionist; Secretary; Stenographer; Systems Analyst; Typist/Word Processor.

JOHN LEONARD PERSONNEL ASSOCIATES
70 Walnut Street, Wellesley MA 02181. 617/431-5948. **Contact:** Lisa McPhee, Manager. **Description:** An employment agency. Company pays fee. **Specializes in the areas of:** Advertising; Architecture/Construction/Real Estate; Banking; Biotechnology; Computer Hardware/Software; Engineering; Finance; Health/Medical; Legal; Personnel/Labor Relations; Secretarial. **Positions commonly filled include:** Administrative Assistant; Bookkeeper; Clerk; Computer Operator; Customer Service Representative; Data Entry Clerk; Legal Secretary; Medical Secretary; Typist/Word Processor.

PLACEMENTS PLUS
210 Winter Street, Suite 204, Weymouth MA 02188. 617/331-0111. **Contact:** Sharon Sinnott, President. **Description:** An employment agency. Company pays fee. **Specializes in the areas of:**

Clerical; Computer Hardware/Software; Secretarial. **Positions commonly filled include:** Administrative Assistant; Bookkeeper; Clerk; Data Entry Clerk; Legal Secretary; Medical Secretary; Receptionist; Secretary; Stenographer; Typist/Word Processor. **Number of placements per year:** 50 - 99.

E.S. RANDO ASSOCIATES, INC.
P.O. Box 654, Wilmington MA 01887. 508/657-4730. **Fax:** 508/658-4650. **Contact:** Ed Rando, President. **Description:** An employment agency. **Specializes in the areas of:** Administration/ MIS/EDP; Computer Hardware/Software; Information Systems; Software Engineering. **Positions commonly filled include:** Computer Programmer; EDP Specialist; MIS Specialist; Software Engineer; Systems Analyst. **Number of placements per year:** 50 - 99.

REARDON ASSOCIATES, INC.
27 Cambridge Street, Burlington MA 01803. 617/272-2750. **Fax:** 617/229-6814. **Contact:** Michelle Kraemer, Division Manager. **Description:** An employment agency focusing on human resources, manufacturing, operations, finance, and MIS personnel. Company pays fee. **Specializes in the areas of:** Accounting/Auditing; Computer Science/Software; Manufacturing; Personnel/ Labor Relations; Secretarial. **Positions commonly filled include:** Accountant/Auditor; Administrative Assistant; Bookkeeper; Clerk; Customer Service Representative; Human Resources Specialist; MIS Specialist; Operations/Production Manager; Purchasing Agent/Manager; Receptionist; Secretary; Typist/Word Processor. **Benefits available to temporary workers:** Medical Insurance; Paid Holidays; Paid Vacation. **Corporate headquarters location:** Dedham MA. **Average salary range of placements:** $30,000 - $50,000. **Number of placements per year:** 100 - 199.

GEORGE D. SANDEL ASSOCIATES
P.O. Box 588, Waltham MA 02254. 617/890-0713. **Fax:** 617/890-0714. **Contact:** Ivan Samuels, President. **Description:** An employment agency. Company pays fee. **Specializes in the areas of:** Administration/MIS/EDP; Computer Science/Software; Engineering; Health/Medical; Industrial; Manufacturing; Personnel/Labor Relations; Sales and Marketing; Technical and Scientific. **Positions commonly filled include:** Aerospace Engineer; Biological Scientist; Biomedical Engineer; Computer Programmer; Electrical/Electronics Engineer; Health Services Manager; Human Resources Manager; Industrial Engineer; Mechanical Engineer; Physical Therapist; Physicist/Astronomer; Registered Nurse; Software Engineer; Systems Analyst. **Number of placements per year:** 1 - 49.

SNELLING PERSONNEL SERVICES
3 Courthouse Lane, Chelmsford MA 01824. 508/970-3434. **Fax:** 508/970-3637. **Contact:** Human Resources. **Description:** Snelling Personnel Services provides permanent, temporary, and temp-to-hire placement. Company pays fee. **Specializes in the areas of:** Accounting/Auditing; Administration/MIS/EDP; Advertising; Computer Science/Software; Manufacturing; Personnel/ Labor Relations; Printing/Publishing; Sales and Marketing; Secretarial. **Positions commonly filled include:** Accountant/Auditor; Adjuster; Administrative Services Manager; Blue-Collar Worker Supervisor; Buyer; Chemist; Claim Representative; Clerical Supervisor; Clinical Lab Technician; Computer Programmer; Cost Estimator; Counselor; Credit Manager; Customer Service Representative; Designer; Draftsperson; Electrician; Human Resources Manager; Insurance Agent/Broker; Management Trainee; Manufacturer's/Wholesaler's Sales Rep.; Paralegal; Property and Real Estate Manager; Public Relations Specialist; Purchasing Agent/Manager; Quality Control Supervisor; Services Sales Representative; Software Engineer; Systems Analyst. **Corporate headquarters location:** Dallas TX. **Number of placements per year:** 200 - 499.

SOURCE EDP
155 Federal Street, Suite 410, Boston MA 02110. 617/482-8211. **Fax:** 617/482-9084. **Contact:** Steve McMahan, Branch Manager. **Description:** An employment agency. Company pays fee. **Specializes in the areas of:** Computer Hardware/Software; MIS/EDP; Sales and Marketing. **Positions commonly filled include:** Computer Operator; Computer Programmer; EDP Specialist; MIS Specialist; Sales Representative; Systems Analyst; Technical Writer/Editor. **Number of placements per year:** 1000+.

TSI/TIMELY SOLUTIONS, INC.

170 Main Street, Tewksbury MA 01876-1765. 508/640-0548. **Fax:** 508/640-1739. **Contact:** Robert Boc, Sales Vice President. **Description:** An employment agency that offers both contract and permanent software staffing services. Areas of focus include information systems, software engineering, quality assurance, and technical writing. Company pays fee. **Specializes in the areas of:** Administration/MIS/EDP; Computer Science/Software. **Positions commonly filled include:** Computer Programmer; MIS Specialist; Systems Analyst; Technical Writer/Editor. **Average salary range of placements:** More than $50,000. **Number of placements per year:** 100 - 199.

TECH/AID

P.O. Box 670, Waltham MA 02254. 617/891-0800. **Contact:** Office Manager. **Description:** An employment agency. Company pays fee. **NOTE:** The physical address is 295 Weston Street, Waltham MA. **Specializes in the areas of:** Architecture/Construction/Real Estate; Cable TV; Computer Hardware/Software; Construction; Engineering; Manufacturing; Technical and Scientific. **Positions commonly filled include:** Aerospace Engineer; Architect; Buyer; Ceramics Engineer; Chemical Engineer; Civil Engineer; Draftsperson; Electrical/Electronics Engineer; Estimator; Industrial Designer; Industrial Engineer; Mechanical Engineer; Metallurgical Engineer; Mining Engineer; Operations/Production Manager; Petroleum Engineer; Purchasing Agent/Manager; Quality Control Supervisor. **Number of placements per year:** 1000+.

TECH/AID

400 Grove Street, Worcester MA 01605. 508/792-6255. **Contact:** Office Manager. **Description:** An employment agency. Company pays fee. **Specializes in the areas of:** Architecture/Construction/Real Estate; Cable TV; Computer Hardware/Software; Construction; Engineering; Manufacturing; Technical and Scientific. **Positions commonly filled include:** Aerospace Engineer; Architect; Buyer; Ceramics Engineer; Chemical Engineer; Civil Engineer; Draftsperson; Electrical/Electronics Engineer; Estimator; Factory Worker; Industrial Designer; Industrial Engineer; Mechanical Engineer; Metallurgical Engineer; Mining Engineer; Operations/Production Manager; Petroleum Engineer; Purchasing Agent/Manager; Quality Control Supervisor; Technical Writer/Editor; Technician. **Number of placements per year:** 1000+.

TECHNICAL PERSONNEL SERVICES INC.

867 Turnpike Street, Suite 213, North Andover MA 01845. 508/794-3347. **Fax:** 508/683-6450. **Contact:** Paul Donatio, General Manager. **Description:** An employment agency. **Specializes in the areas of:** Computer Science/Software; Engineering; Manufacturing; Technical and Scientific. **Positions commonly filled include:** Aerospace Engineer; Agricultural Engineer; Biomedical Engineer; Chemical Engineer; Civil Engineer; Computer Programmer; Designer; Draftsperson; Electrical/Electronics Engineer; Industrial Engineer; Mechanical Engineer; Metallurgical Engineer; Nuclear Engineer; Structural Engineer. **Number of placements per year:** 100 - 199.

TEMPXPRESS/THE FREEMAN GROUP

20 Park Plaza, Boston MA 02116-4303. 617/451-5105. **Fax:** 617/451-5109. **Contact:** John Dempsey, Senior Recruiter. **Description:** An employment agency. Company pays fee. **Specializes in the areas of:** Accounting/Auditing; Administration/MIS/EDP; Banking; Computer Science/Software; Engineering; Finance; General Management; Health/Medical; Legal; Personnel/ Labor Relations; Sales and Marketing; Secretarial; Technical and Scientific. **Positions commonly filled include:** Accountant/Auditor; Aerospace Engineer; Bank Officer/Manager; Blue-Collar Worker Supervisor; Budget Analyst; Computer Programmer; Customer Service Representative; Design Engineer; Electrical/Electronics Engineer; Financial Analyst; Health Services Manager; Human Resources Specialist; Human Service Worker; Internet Services Manager; Medical Records Technician; Multimedia Designer; Nuclear Engineer; Operations/ Production Manager; Paralegal; Software Engineer; Stationary Engineer; Structural Engineer; Systems Analyst; Technical Writer/Editor; Typist/Word Processor; Underwriter/ Assistant Underwriter. **Benefits available to temporary workers:** Health Insurance; Holiday/Vacation Pay. **Corporate headquarters location:** This Location. **Other U.S. locations:** Braintree MA. **Number of placements per year:** 1000+.

TESMER ALLEN ASSOCIATES

P.O. Box 1491, Westborough MA 01581. 508/366-1160. **Fax:** 508/366-6419. **Contact:** Bernard Tesmer, President. **Description:** An employment agency focusing on recruitment and placement of engineering and MIS personnel. Founded in 1970. Company pays fee. **Specializes in the areas of:** Administration/MIS/EDP; Computer Hardware/Software; Engineering. **Positions commonly filled include:** Ceramics Engineer; Chemical Engineer; Chemist; Computer Programmer; EDP Specialist; Electrical/Electronics Engineer; Environmental Engineer; Industrial Engineer; Manufacturing Engineer; Mechanical Engineer; Metallurgical Engineer; MIS Specialist; Multimedia Designer; Software Engineer; Systems Analyst. **Number of placements per year:** 50 - 99.

TOTAL TECHNICAL SERVICES, INC.
167 Pleasant Street, Attleboro MA 02703. 508/226-3880. **Toll-free phone:** 800/342-2364. **Fax:** 508/226-4363. **Contact:** Christopher H. Bernard, Branch Manager. **Description:** An employment agency focusing on technical placements. Company pays fee. **Specializes in the areas of:** Administration/MIS/EDP; Computer Science/Software; Engineering; Industrial; Manufacturing; Personnel/Labor Relations; Secretarial; Technical and Scientific. **Positions commonly filled include:** Administrative Services Manager; Bank Officer/Manager; Blue-Collar Worker Supervisor; Buyer; Chemical Engineer; Civil Engineer; Clerical Supervisor; Computer Programmer; Credit Manager; Design Engineer; Designer; Draftsperson; Electrical/Electronics Engineer; Electrician; Environmental Engineer; Human Resources Specialist; Industrial Engineer; Mechanical Engineer; MIS Specialist; Multimedia Designer; Nuclear Engineer; Paralegal; Petroleum Engineer; Purchasing Agent/Manager; Quality Control Supervisor; Software Engineer; Statistician; Structural Engineer; Systems Analyst; Technical Writer/Editor; Typist/Word Processor; Urban/Regional Planner. **Benefits available to temporary workers:** Medical Insurance; Paid Vacation. **Corporate headquarters location:** Waltham MA. **Other U.S. locations:** Atlanta GA; Boston MA; Dallas TX. **Average salary range of placements:** $20,000 - $29,999. **Number of placements per year:** 200 - 499.

H.L. YOH COMPANY
99 South Bedford Street, Suite 204, Burlington MA 01803. 617/273-5151. **Contact:** Carol Clark, New England Branch Manager. **Description:** H.L. Yoh places both temporary and permanent workers. **Specializes in the areas of:** Architecture/Construction/Real Estate; Computer Hardware/ Software; Engineering; Manufacturing; MIS/EDP; Personnel/Labor Relations; Technical and Scientific. **Positions commonly filled include:** Aerospace Engineer; Architect; Buyer; Chemical Engineer; Chemist; Civil Engineer; Commercial Artist; Computer Operator; Computer Programmer; Data Entry Clerk; Draftsperson; Driver; Editor; EDP Specialist; Electrical/Electronics Engineer; Human Resources Manager; Industrial Designer; Industrial Engineer; Manufacturing Engineer; Mechanical Engineer; Metallurgical Engineer; MIS Specialist; Operations/Production Manager; Purchasing Agent/Manager; Quality Control Supervisor; Reporter; Software Engineer; Systems Analyst; Technical Illustrator; Technical Writer/Editor; Technician; Typist/Word Processor. **Number of placements per year:** 200 - 499.

EXECUTIVE SEARCH FIRMS OF MICHIGAN

AJM PROFESSIONAL SERVICES
803 West Big Beaver Road, Suite 357, Troy MI 48084-4734. 810/244-2222. **Fax:** 810/244-2233. **Contact:** Jeffrey Jones, Principal. **E-mail address:** ajmps@aol.com. **Description:** An executive search firm specializing in information systems positions. Company pays fee. **Specializes in the areas of:** Administration/MIS/EDP; Computer Science/Software. **Positions commonly filled include:** Computer Programmer; Management Analyst/Consultant; MIS Specialist; Software Engineer; Systems Analyst. **Corporate headquarters location:** This Location. **Average salary range of placements:** More than $50,000. **Number of placements per year:** 100 - 199.

ACTION HUMAN RESOURCES, INC.
600 Renaissance Center, Detroit MI 48243. 313/446-6961. **Fax:** 810/234-0797. **Contact:** Valerie June, Executive Search Coordinator. **E-mail address:** action1@tir.com. **Description:** A minority-owned recruiting firm specializing in the recruitment of diversity candidates for mid- to upper-level

positions. Company pays fee. **Specializes in the areas of:** Accounting/Auditing; Administration/ MIS/EDP; Banking; Computer Science/Software; Engineering; Finance; General Management; Manufacturing; Personnel/Labor Relations; Telecommunications. **Positions commonly filled include:** Accountant/Auditor; Budget Analyst; Computer Programmer; Electrical/Electronics Engineer; Financial Analyst; Human Resources Manager; Industrial Engineer; Management Analyst/Consultant; Mechanical Engineer; MIS Specialist; Operations/Production Manager; Public Relations Specialist; Purchasing Agent/Manager; Software Engineer; Strategic Relations Manager; Systems Analyst; Telecommunications Manager; Transportation/Traffic Specialist. **Corporate headquarters location:** Flint MI. **Average salary range of placements:** More than $50,000. **Number of placements per year:** 1 - 49.

AMERICAN COMPUTER SERVICE
29777 Telegraph Road, Southfield MI 48034-1303. 810/857-1500. **Fax:** 810/857-4244. **Contact:** Dan Corp, President. **Description:** A contingency search firm. Company pays fee. **Specializes in the areas of:** Computer Science/Software; Sales and Marketing. **Positions commonly filled include:** Sales and Marketing Representative; Software Engineer. **Corporate headquarters location:** This Location. **Other U.S. locations:** Nationwide. **Average salary range of placements:** More than $50,000. **Number of placements per year:** 50 - 99.

THE AUCON COMPANY
3749 High Gate Road, Muskegon MI 49441. 616/798-4883. **Fax:** 616/798-4087. **Contact:** Raymond B. Audo, President. **Description:** An executive search firm. Company pays fee. **Specializes in the areas of:** Accounting/Auditing; Administration/MIS/EDP; Computer Science/Software; Engineering; General Management; Industrial; Manufacturing; Technical and Scientific. **Average salary range of placements:** More than $50,000. **Number of placements per year:** 1 - 49.

COLLINS & ASSOCIATES
10188 West H Avenue, Kalamazoo MI 49009-8506. 616/372-4300. **Fax:** 616/372-3921. **Contact:** Phil Collins, Principal. **E-mail address:** pcollins@serv01.net-link.net. **World Wide Web address:** http://www.net-link.net/collins. **Description:** An executive search firm. Company pays fee. **Specializes in the areas of:** Computer Science/Software. **Positions commonly filled include:** Computer Programmer; Internet Services Manager; MIS Specialist; Software Engineer; Systems Analyst; Telecommunications Manager. **Average salary range of placements:** $30,000 - $50,000. **Number of placements per year:** 1 - 49.

DUNHILL PROFESSIONAL SEARCH
7901 Sprinkle Road, Portage MI 49002. 616/324-3990. **Fax:** 616/324-3590. **Contact:** Ken Killman, Owner. **Description:** An executive search firm. Company pays fee. **Specializes in the areas of:** Administration/MIS/EDP; Computer Science/Software; Engineering; Health/Medical; Manufacturing. **Positions commonly filled include:** Accountant/Auditor; Chemical Engineer; Computer Programmer; Design Engineer; Electrical/Electronics Engineer; Financial Analyst; Industrial Engineer; Mechanical Engineer; Metallurgical Engineer; MIS Specialist; Occupational Therapist; Pharmacist; Physical Therapist; Software Engineer; Systems Analyst. **Corporate headquarters location:** This Location. **Average salary range of placements:** $30,000 - $50,000. **Number of placements per year:** 50 - 99.

EXECUTECH, INC.
2009 Hogback Road, Suite 9, Ann Arbor MI 48105-9732. 313/483-8454. **Fax:** 313/483-0740. **Contact:** Donald Frederick, CEO. **E-mail address:** extinc@aol.com. **Description:** A retainer and contingency search firm. Company pays fee. **Specializes in the areas of:** Computer Science/Software. **Positions commonly filled include:** Computer Programmer; Internet Services Manager; Metallurgical Engineer; MIS Specialist; Multimedia Designer; Quality Control Supervisor; Systems Analyst; Telecommunications Manager. **Average salary range of placements:** More than $50,000. **Number of placements per year:** 1 - 49.

EXECUTIVE RECRUITERS INTERNATIONAL
1150 Griswold, Suite 3000, Detroit MI 48226-1900. 313/961-6200. **Fax:** 313/963-1826. **Contact:** Kathleen A. Sinclair, President. **Description:** An executive search firm. Company pays fee.

Specializes in the areas of: Administration/MIS/EDP; Architecture/Construction/Real Estate; Automotive; Computer Science/Software; Engineering; Environmental; General Management; Industrial; Manufacturing; Personnel/Labor Relations; Printing/Publishing; Sales and Marketing; Technical and Scientific; Transportation. **Positions commonly filled include:** Accountant/ Auditor; Administrative Services Manager; Architect; Automotive Mechanic; Biological Scientist; Biomedical Engineer; Blue-Collar Worker Supervisor; Buyer; Chemical Engineer; Chemist; Civil Engineer; Computer Programmer; Construction and Building Inspector; Construction Contractor; Cost Estimator; Customer Service Representative; Designer; Draftsperson; Economist; Electrical/ Electronics Engineer; Electrician; Environmental Engineer; General Manager; Geologist/ Geophysicist; Human Resources Manager; Industrial Engineer; Industrial Production Manager; Landscape Architect; Management Analyst/Consultant; Management Trainee; Manufacturer's/ Wholesaler's Sales Rep.; Mechanical Engineer; Metallurgical Engineer; Mining Engineer; Nuclear Engineer; Operations/Production Manager; Petroleum Engineer; Public Relations Specialist; Purchasing Agent/Manager; Quality Control Supervisor; Services Sales Representative; Software Engineer; Stationary Engineer; Structural Engineer; Surveyor; Systems Analyst; Technical Writer/ Editor; Transportation/Traffic Specialist; Urban/Regional Planner; Water Transportation Specialist. **Corporate headquarters location:** This Location. **International locations:** Worldwide. **Number of placements per year:** 1 - 49.

GRS
934 West Fulton, Grand Rapids MI 49504. 616/242-7700. **Contact:** Bill Fischer, Senior Consultant. **Description:** A retainer and contingency search firm. Company pays fee. **Specializes in the areas of:** Accounting/Auditing; Administration/MIS/EDP; Computer Science/Software; Engineering; Finance; General Management; Manufacturing; Personnel/Labor Relations; Technical and Scientific. **Positions commonly filled include:** Blue-Collar Worker Supervisor; Branch Manager; Buyer; Chemical Engineer; Claim Representative; Computer Programmer; Design Engineer; Designer; Draftsperson; Electrical/Electronics Engineer; Financial Analyst; General Manager; Human Resources Manager; Industrial Engineer; Industrial Production Manager; Mechanical Engineer; Metallurgical Engineer; MIS Specialist; Purchasing Agent/Manager; Quality Control Supervisor; Software Engineer; Systems Analyst. **Corporate headquarters location:** This Location. **Average salary range of placements:** $30,000 - $50,000. **Number of placements per year:** 1 - 49.

ROBERT HALF INTERNATIONAL
201 West Big Beaver Road, Suite 310, Troy MI 48084. 810/524-3100. **Fax:** 810/524-3115. **Contact:** George Corser, Division Director. **Description:** An executive search firm. Robert Half International is one of the world's largest accounting, financial, banking, and information technology staffers. Company pays fee. **Specializes in the areas of:** Accounting/Auditing; Administration/MIS/EDP; Banking; Computer Science/Software; Finance. **Positions commonly filled include:** Accountant/Auditor; Bank Officer/Manager; Budget Analyst; Computer Programmer; Cost Estimator; Credit Manager; Financial Analyst; Internet Services Manager; Management Analyst/Consultant; MIS Specialist; Software Engineer; Systems Analyst; Technical Writer/Editor. **Corporate headquarters location:** Menlo Park CA. **International locations:** Worldwide. **Average salary range of placements:** $30,000 - $50,000. **Number of placements per year:** 500 - 999.

IBA SEARCH CONSULTANTS
8300 Thornapple River Drive, Caledonia MI 49316. 616/891-2160. **Fax:** 616/891-1180. **Contact:** Jim Lakatos, President. **E-mail address:** ibahunter@wingsbbs.com. **World Wide Web address:** http://www.web.wingsbbs.com/ibaresearch. **Description:** An executive search firm providing placement on a contingency basis. Company pays fee. **Specializes in the areas of:** Administration/ MIS/EDP; Computer Science/Software; Economics; Engineering; Manufacturing; Technical and Scientific. **Positions commonly filled include:** Aerospace Engineer; Agricultural Engineer; Biological Scientist; Biomedical Engineer; Chemical Engineer; Chemist; Civil Engineer; Computer Programmer; Designer; Draftsperson; Electrical/Electronics Engineer; General Manager; Geologist/ Geophysicist; Industrial Engineer; Industrial Production Manager; Manufacturer's/Wholesaler's Sales Rep.; Mechanical Engineer; Metallurgical Engineer; Mining Engineer; Nuclear Engineer; Operations/Production Manager; Petroleum Engineer; Quality Control Supervisor; Science Technologist; Software Engineer; Stationary Engineer; Structural Engineer; Surveyor; Systems

Analyst; Technical Writer/Editor; Telecommunications Manager. **Corporate headquarters location:** This Location. **Other U.S. locations:** Nationwide. **Average salary range of placements:** $30,000 - $50,000. **Number of placements per year:** 1 - 49.

MRI HUMAN RESOURCE SOLUTIONS
550 Stephenson Highway, Suite 407, Troy MI 48083-1152. 810/585-4200. **Contact:** Ed Moeller, Manager. **Description:** An executive search firm. **Specializes in the areas of:** Accounting/ Auditing; Administration/MIS/EDP; Advertising; Architecture/Construction/Real Estate; Banking; Chemical; Communications; Computer Hardware/Software; Design; Electrical; Engineering; Food Industry; General Management; Health/Medical; Insurance; Legal; Manufacturing; Operations Management; Personnel/Labor Relations; Printing/Publishing; Procurement; Retail; Sales and Marketing; Technical and Scientific; Textiles; Transportation.

MANAGEMENT RECRUITERS OF BATTLE CREEK
500 Country Pine Lane, Suite 1, Battle Creek MI 49015. 616/979-3939. **Fax:** 616/979-8899. **Contact:** Tony Richardson, Manager. **Description:** An executive search firm. Management Recruiters is the largest network of search firms in the world. Company pays fee. **Specializes in the areas of:** Accounting/Auditing; Administration/MIS/EDP; Advertising; Computer Science/ Software; Engineering; Finance; Health/Medical; Industrial; Manufacturing. **Positions commonly filled include:** Buyer; Chemical Engineer; Computer Programmer; Cost Estimator; Design Engineer; Designer; Financial Analyst; Health Services Manager; Industrial Engineer; Industrial Production Manager; Mechanical Engineer; Medical Records Technician; Metallurgical Engineer; MIS Specialist; Operations/Production Manager; Quality Control Supervisor; Software Engineer; Systems Analyst. **Corporate headquarters location:** Cleveland OH. **Other U.S. locations:** Nationwide. **Average salary range of placements:** $30,000 - $50,000. **Number of placements per year:** 100 - 199.

MANAGEMENT RECRUITERS OF BINGHAM FARMS
Bingham Office Park, 30300 Telegraph Road, Suite 285, Bingham Farms MI 48025. 810/647-2828. **Contact:** Fred Bawulski, General Manager. **Description:** An executive search firm. **Specializes in the areas of:** Accounting/Auditing; Administration/MIS/EDP; Advertising; Architecture/Construction/Real Estate; Banking; Chemical; Communications; Computer Hardware/ Software; Design; Electrical; Engineering; Food Industry; General Management; Health/Medical; Insurance; Legal; Manufacturing; Operations Management; Personnel/Labor Relations; Printing/ Publishing; Procurement; Retail; Sales and Marketing; Technical and Scientific; Textiles; Transportation.

MANAGEMENT RECRUITERS OF DEARBORN
Parklane Towers West, Suite 1224, Three Parklane Boulevard, Dearborn MI 48126-2591. 313/336-6650. **Contact:** Elaine Kozlowski, Manager. **Description:** An executive search firm. **Specializes in the areas of:** Accounting/Auditing; Administration/MIS/EDP; Advertising; Architecture/ Construction/Real Estate; Banking; Chemical; Communications; Computer Hardware/Software; Design; Electrical; Engineering; Food Industry; General Management; Health/Medical; Insurance; Legal; Manufacturing; Operations Management; Personnel/Labor Relations; Printing/Publishing; Procurement; Retail; Sales and Marketing; Technical and Scientific; Textiles; Transportation.

MANAGEMENT RECRUITERS OF FLINT
G5524 South Saginaw Road, Flint MI 48507. 810/695-0120. **Contact:** David Reed, Manager. **Description:** A technical search and recruitment firm. Company pays fee. **Specializes in the areas of:** Accounting/Auditing; Administration/MIS/EDP; Advertising; Architecture/Construction/Real Estate; Banking; Chemical; Communications; Computer Hardware/Software; Design; Electrical; Engineering; Food Industry; General Management; Health/Medical; Insurance; Legal; Manufacturing; Operations Management; Personnel/Labor Relations; Printing/Publishing; Procurement; Retail; Sales and Marketing; Technical and Scientific; Textiles; Transportation. **Positions commonly filled include:** Blue-Collar Worker Supervisor; Computer Programmer; Cost Estimator; Design Engineer; Designer; Electrical/Electronics Engineer; Electrician; Industrial Engineer; Industrial Production Manager; Mechanical Engineer; Software Engineer; Systems Analyst. **Corporate headquarters location:** This Location. **Average salary range of placements:** $30,000 - $50,000. **Number of placements per year:** 1 - 49.

MANAGEMENT RECRUITERS OF GRAND RAPIDS

146 Monroe Center, Suite 1126, Grand Rapids MI 49503. 616/336-8484. **Contact:** Manager. **Description:** An executive search firm. **Specializes in the areas of:** Accounting/Auditing; Administration/MIS/EDP; Advertising; Architecture/Construction/Real Estate; Banking; Chemical; Communications; Computer Hardware/Software; Design; Electrical; Engineering; Food Industry; General Management; Health/Medical; Insurance; Legal; Manufacturing; Operations Management; Personnel/Labor Relations; Printing/Publishing; Procurement; Retail; Sales and Marketing; Technical and Scientific; Textiles; Transportation.

MANAGEMENT RECRUITERS OF KALAMAZOO

4021 West Main Street, Suite 200, Kalamazoo MI 49006-2746. 616/381-1153. **Fax:** 616/381-8031. **Contact:** Manager. **Description:** An executive search firm. **Specializes in the areas of:** Accounting/Auditing; Administration/MIS/EDP; Advertising; Architecture/Construction/Real Estate; Banking; Chemical; Communications; Computer Hardware/Software; Design; Electrical; Engineering; Food Industry; General Management; Health/Medical; Insurance; Legal; Manufacturing; Operations Management; Personnel/Labor Relations; Printing/Publishing; Procurement; Retail; Sales and Marketing; Technical and Scientific; Textiles; Transportation.

MANAGEMENT RECRUITERS OF LANSING

2491 Cedar Park Drive, Holt MI 48842-2184. 517/694-1153. **Contact:** John Peterson, Manager. **Description:** An executive search firm. **Specializes in the areas of:** Accounting/Auditing; Administration/MIS/EDP; Advertising; Architecture/Construction/Real Estate; Banking; Chemical; Communications; Computer Hardware/Software; Design; Electrical; Engineering; Food Industry; General Management; Health/Medical; Insurance; Legal; Manufacturing; Operations Management; Personnel/Labor Relations; Printing/Publishing; Procurement; Retail; Sales and Marketing; Technical and Scientific; Textiles; Transportation.

MANAGEMENT RECRUITERS OF LIVONIA

37677 Professional Center Drive, Suite 100-C, Livonia MI 48154-1138. 313/953-9590. **Fax:** 313/953-0566. **Contact:** Don Eden, President. **Description:** A contingency search firm providing opportunities in the areas of engineering, health care, insurance, sales, and other professional, technical, and managerial positions. **Specializes in the areas of:** Accounting/Auditing; Administration/MIS/EDP; Computer Science/Software; Engineering; Finance; Food Industry; General Management; Health/Medical; Industrial; Sales and Marketing; Technical and Scientific. **Positions commonly filled include:** Accountant/Auditor; Administrative Services Manager; Branch Manager; Budget Analyst; Buyer; Chemical Engineer; Chemist; Computer Programmer; Cost Estimator; Credit Manager; Customer Service Representative; Designer; Economist; Electrical/Electronics Engineer; Electrician; Financial Analyst; General Manager; Health Services Manager; Human Resources Manager; Industrial Engineer; Industrial Production Manager; Manufacturer's/Wholesaler's Sales Rep.; Mechanical Engineer; Metallurgical Engineer; Occupational Therapist; Physical Therapist; Purchasing Agent/Manager; Quality Control Supervisor; Registered Nurse; Respiratory Therapist; Services Sales Representative; Software Engineer; Systems Analyst; Transportation/Traffic Specialist; Wholesale and Retail Buyer. **Corporate headquarters location:** This Location. **Average salary range of placements:** $30,000 - $50,000. **Number of placements per year:** 1 - 49.

MANAGEMENT RECRUITERS OF MUSKEGON

3145 Henry Street, Suite 203, Muskegon MI 49441. 616/755-6486. **Contact:** John R. Mitchell, Jr., Manager. **Description:** An executive search firm. **Specializes in the areas of:** Accounting/ Auditing; Administration/MIS/EDP; Advertising; Architecture/Construction/ Real Estate; Banking; Chemical; Communications; Computer Hardware/Software; Design; Electrical; Engineering; Food Industry; General Management; Health/Medical; Insurance; Legal; Manufacturing; Operations Management; Personnel/Labor Relations; Printing/Publishing; Procurement; Retail; Sales and Marketing; Technical and Scientific; Textiles; Transportation.

MANAGEMENT RECRUITERS OF ROCHESTER

2530 South Rochester Road, Rochester Hills MI 48307. 810/299-1900. **Contact:** Mark Angott, General Manager. **Description:** An executive search firm. **Specializes in the areas of:** Accounting/Auditing; Administration/MIS/EDP; Advertising; Architecture/Construction/Real

Estate; Banking; Chemical; Communications; Computer Hardware/ Software; Design; Electrical; Engineering; Food Industry; General Management; Health/ Medical; Insurance; Legal; Manufacturing; Operations Management; Personnel/Labor Relations; Printing/Publishing; Procurement; Retail; Sales and Marketing; Technical and Scientific; Textiles; Transportation.

METROTECH
28500 Southfield Road, Lathrup Village MI 48076. 810/557-8700. **Fax:** 810/557-8507. **Contact:** Michael Callaway, President. **Description:** An executive search firm. Company pays fee. **Specializes in the areas of:** Computer Science/Software; Engineering; Health/Medical; Industrial; Secretarial; Transportation. **Positions commonly filled include:** Computer Programmer; Design Engineer; EEG Technologist; EKG Technician; Electrical/Electronics Engineer; Industrial Engineer; Industrial Production Manager; Licensed Practical Nurse; Mechanical Engineer; MIS Specialist; Occupational Therapist; Operations/Production Manager; Physical Therapist; Quality Control Supervisor; Registered Nurse; Software Engineer; Systems Analyst; Technical Writer/ Editor; Typist/Word Processor. **Corporate headquarters location:** This Location. **Number of placements per year:** 1000+.

MICHIGAN SEARCH PLUS
25882 Orchard Lake Road, Suite 207, Farmington Hills MI 48336. 810/471-6110. **Contact:** Christy Greeneisen, President. **World Wide Web address:** http://www.michsrchpl@gnn.com. **Description:** A personnel search firm. Company pays fee. **Specializes in the areas of:** Administration/MIS/EDP; Computer Science/Software; Engineering; Finance; General Management; Industrial; Manufacturing; Personnel/Labor Relations; Sales and Marketing; Secretarial. **Positions commonly filled include:** Automotive Engineer; Biomedical Engineer; Blue-Collar Worker Supervisor; Buyer; Chemical Engineer; Chemist; Computer Programmer; Cost Estimator; Design Engineer; Designer; Draftsperson; Electrical/Electronics Engineer; Electrician; Environmental Engineer; Financial Analyst; Food Scientist/Technologist; General Manager; Human Resources Manager; Industrial Engineer; Industrial Production Manager; Manufacturer's/ Wholesaler's Sales Rep.; Mechanical Engineer; Metallurgical Engineer; MIS Specialist; Operations/Production Manager; Paralegal; Purchasing Agent/Manager; Quality Control Supervisor; Science Technologist; Software Engineer; Systems Analyst; Transportation/Traffic Specialist. **Corporate headquarters location:** This Location. **Number of placements per year:** 50 - 99.

OFFICE STAFFING RECRUITING
1550 East Beltline, Suite 300, Grand Rapids MI 49344. 616/949-2525. **Fax:** 616/949-2982. **Contact:** Pam McMaster, Team Manager. **Description:** An executive search firm with clients in Lansing, Grand Rapids, and the lake shore of Michigan. **Specializes in the areas of:** Accounting/ Auditing; Administration/MIS/EDP; Banking; Computer Science/Software; Engineering; Finance; General Management; Industrial; Insurance; Manufacturing; Personnel/ Labor Relations; Sales and Marketing; Secretarial. **Positions commonly filled include:** Accountant/Auditor; Administrative Services Manager; Branch Manager; Brokerage Clerk; Buyer; Claim Representative; Clerical Supervisor; Computer Programmer; Design Engineer; General Manager; Human Resources Specialist; Human Service Worker; Industrial Production Manager; Insurance Agent/Broker; Manufacturer's/Wholesaler's Sales Rep.; Mechanical Engineer; Operations/Production Manager; Purchasing Agent/Manager; Quality Control Supervisor; Securities Sales Representative; Software Engineer; Systems Analyst; Telecommunications Manager; Underwriter/Assistant Underwriter. **Average salary range of placements:** $30,000 - $50,000. **Number of placements per year:** 50 - 99.

OPEN PAGE SEARCH SERVICES
1354 Ardmoor Avenue, Ann Arbor MI 48103. 313/761-3556. **Fax:** 313/761-1554. **Contact:** Frederick Page, Principal. **Description:** A recruiting and placement firm placing professionals with client companies and consulting firms. Company pays fee. **Specializes in the areas of:** Administration/MIS/EDP; Computer Science/Software; Insurance. **Positions commonly filled include:** Actuary; Computer Programmer; Financial Analyst; Management Analyst/Consultant; Mathematician; MIS Specialist; Securities Sales Representative; Software Engineer; Systems Analyst; Underwriter/Assistant Underwriter. **Corporate headquarters location:** This Location. **Number of placements per year:** 1 - 49.

OPPORTUNITY KNOCKING, INC.
9677 Peer Road, South Lyon MI 48178. 810/437-3700. **Fax:** 810/437-0939. **Contact:** Calvin Weaver, Executive Placement Manager. **E-mail address:** blckbltcal@aol.com. **Description:** An executive search firm specializing in computer personnel. This firm is managed by Certified Personnel Consultants. Company pays fee. **Specializes in the areas of:** Computer Science/Software. **Positions commonly filled include:** Computer Programmer; Internet Services Manager; MIS Specialist; Software Engineer; Systems Analyst. **Corporate headquarters location:** This Location. **Average salary range of placements:** $30,000 - $50,000. **Number of placements per year:** 1 - 49.

PROFESSIONAL PERSONNEL CONSULTANTS INTERNATIONAL
28200 Orchard Lake Road, Farmington Hills MI 48334. 810/737-1750. **Fax:** 810/737-5866. **Contact:** Dan Mistura, President. **Description:** A management and executive search firm. Company pays fee. **Specializes in the areas of:** Accounting/Auditing; Administration/MIS/EDP; Computer Science/Software; Engineering; Finance; Health/Medical; Manufacturing; Personnel/ Labor Relations; Sales and Marketing; Secretarial; Technical and Scientific. **Positions commonly filled include:** Accountant/Auditor; Budget Analyst; Buyer; Chemist; Computer Programmer; Credit Manager; Designer; Electrical/Electronics Engineer; Financial Analyst; General Manager; Human Resources Manager; Industrial Engineer; Mechanical Engineer; Metallurgical Engineer; Nuclear Engineer; Occupational Therapist; Pharmacist; Physical Therapist; Physician; Purchasing Agent/Manager; Quality Control Supervisor; Software Engineer; Structural Engineer; Systems Analyst. **Corporate headquarters location:** This Location. **Other U.S. locations:** Nationwide. **Average salary range of placements:** More than $50,000. **Number of placements per year:** 200 - 499.

THE RESOURCE GROUP
12935 West Bayshore Drive, Traverse City MI 49684. 616/941-5063. **Fax:** 616/941-5079. **Contact:** Jack Moorlay, Vice President. **Description:** An executive search firm. Company pays fee. **Specializes in the areas of:** Computer Science/Software; Engineering. **Positions commonly filled include:** Designer; Electrical/Electronics Engineer; Mechanical Engineer; Software Engineer. **Number of placements per year:** 50 - 99.

SALES CONSULTANTS OF AUBURN HILLS
2701 University Drive, Suite 205, Auburn Hills MI 48326. 810/373-7177. **Fax:** 810/373-7759. **Contact:** Boe Embrey, Owner/President. **Description:** An executive search firm. Company pays fee. **Specializes in the areas of:** Computer Science/Software; Engineering; Health/Medical; Industrial; Manufacturing; Sales and Marketing. **Positions commonly filled include:** Biomedical Engineer; Branch Manager; Chemical Engineer; Civil Engineer; Electrical/Electronics Engineer; Manufacturer's/Wholesaler's Sales Rep.; Mechanical Engineer; Mining Engineer; Software Engineer. **Number of placements per year:** 50 - 99.

SALES CONSULTANTS OF FARMINGTON HILLS
30445 Northwestern Highway, Suite 360, Farmington Hills MI 48334. 810/626-6600. **Fax:** 810/626-7542. **Contact:** Mark Gilbert, Manager. **Description:** An executive search firm. **Specializes in the areas of:** Accounting/Auditing; Administration/MIS/EDP; Advertising; Architecture/Construction/Real Estate; Banking; Chemical; Communications; Computer Hardware/ Software; Design; Electrical; Engineering; Food Industry; General Management; Health/Medical; Insurance; Legal; Manufacturing; Operations Management; Personnel/Labor Relations; Printing/ Publishing; Procurement; Retail; Sales and Marketing; Technical and Scientific; Textiles; Transportation.

SHARROW & ASSOCIATES INC.
24735 Van Dyke Avenue, Center Line MI 48015. 810/759-6910. **Toll-free phone:** 800/344-5032. **Fax:** 810/759-6914. **Contact:** Douglas Sharrow, President. **E-mail address:** david@access.digex. net. **World Wide Web address:** http://www.access.digex.net/~david/saph.html. **Description:** An executive search firm. Company pays fee. **Specializes in the areas of:** Architecture/ Construction/Real Estate; Computer Science/Software; Engineering; Health/ Medical; Insurance; Legal. **Positions commonly filled include:** Chemical Engineer; Chemist; Computer Programmer; Construction and Building Inspector; Construction Contractor; Electrical/Electronics Engineer;

Industrial Engineer; Mechanical Engineer; Occupational Therapist; Patent Examiner; Physical Therapist; Property and Real Estate Manager; Systems Analyst. **Corporate headquarters location:** This Location. **Other U.S. locations:** Florence KY. **Average salary range of placements:** More than $50,000. **Number of placements per year:** 50 - 99.

SNELLING & SNELLING

30100 Telegraph Road, Suite 474, Bingham Farms MI 48025-4518. 810/644-4600. **Fax:** 810/644-4739. **Contact:** Jacqueline Dombrowski, President. **Description:** A contingency search firm also offering temporary assignments. Company pays fee. **Specializes in the areas of:** Computer Science/Software; Engineering; Secretarial; Technical and Scientific. **Positions commonly filled include:** Brokerage Clerk; Chemical Engineer; Chemist; Computer Programmer; Customer Service Representative; Design Engineer; Designer; Electrical/Electronics Engineer; Mechanical Engineer; Metallurgical Engineer; Physician; Quality Control Supervisor; Software Engineer; Systems Analyst; Typist/Word Processor. **Benefits available to temporary workers:** 401(k). **Corporate headquarters location:** This Location. **Other U.S. locations:** Royal Oak MI. **Average salary range of placements:** More than $50,000. **Number of placements per year:** 100 - 199.

THOMAS & ASSOCIATES OF MICHIGAN

P.O. Box 366, Union Pier MI 49129. 616/469-5760. **Fax:** 616/469-5774. **Contact:** Thomas J. Zonka, President. **Description:** An executive search firm linked to a national computerized network with over 650 professional firms represented by 2,000 recruiters. Company pays fee. **Specializes in the areas of:** Computer Science/Software; Engineering; General Management; Industrial; Manufacturing; Personnel/Labor Relations; Sales and Marketing; Transportation. **Positions commonly filled include:** Buyer; Chemical Engineer; Computer Programmer; Designer; Electrical/Electronics Engineer; General Manager; Human Resources Manager; Industrial Engineer; Industrial Production Manager; Manufacturer's/Wholesaler's Sales Rep.; Mechanical Engineer; Metallurgical Engineer; Operations/Production Manager; Purchasing Agent/Manager; Quality Control Supervisor; Software Engineer; Structural Engineer; Systems Analyst; Transportation/Traffic Specialist. **Corporate headquarters location:** This Location. **Other U.S. locations:** Nationwide. **Number of placements per year:** 200 - 499.

WING TIPS AND PUMPS, INC.

P.O. Box 99580, Troy MI 48099-9580. 810/641-0980. **Fax:** 810/641-0895. **Contact:** Mr. Verba Edwards, CEO. **Description:** A retainer and contingency search firm. Company pays fee. **Specializes in the areas of:** Accounting/Auditing; Banking; Computer Science/Software; Engineering; Finance; Health/Medical; Insurance; Legal; Manufacturing; Personnel/Labor Relations; Sales and Marketing; Technical and Scientific. **Positions commonly filled include:** Accountant/Auditor; Aerospace Engineer; Attorney; Automotive Mechanic; Buyer; Chemical Engineer; Computer Programmer; Cost Estimator; Design Engineer; Designer; Electrical/Electronics Engineer; Electrician; Environmental Engineer; General Manager; Health Services Manager; Hotel Manager; Human Resources Manager; Industrial Engineer; Insurance Agent/Broker; Licensed Practical Nurse; Mechanical Engineer; Metallurgical Engineer; MIS Specialist; Physician; Quality Control Supervisor; Registered Nurse; Software Engineer; Systems Analyst; Underwriter/Assistant Underwriter. **Corporate headquarters location:** This Location. **Other U.S. locations:** Nationwide. **Average salary range of placements:** More than $50,000. **Number of placements per year:** 1 - 49.

PERMANENT EMPLOYMENT AGENCIES OF MICHIGAN

BARMAN STAFFING SOLUTIONS

2976 Ivanrest, Grandville MI 49418. 616/531-4122. **Contact:** Paul Barman, President. **Description:** An employment agency. Company pays fee. **Specializes in the areas of:**

Accounting/Auditing; Architecture/Construction/Real Estate; Computer Science/Software; Design; Engineering; Food Industry; Health/Medical; Industrial; Manufacturing; MIS/EDP. **Positions commonly filled include:** Accountant/Auditor; Aerospace Engineer; Agricultural Engineer; Architect; Ceramics Engineer; Electrical/Electronics Engineer; Financial Analyst; General Manager; Human Resources Manager; Industrial Designer; Industrial Engineer; Interior Designer; Mechanical Engineer; Metallurgical Engineer; Purchasing Agent/Manager. **Number of placements per year:** 1 - 49.

CAREER QUEST, INC.
610 East Grand River Avenue, Lansing MI 48906. 517/485-3330. **Fax:** 517/485-8821. **Contact:** Tonya R. Fountain, Placement Coordinator. **Description:** An employment agency. Company pays fee. **Specializes in the areas of:** Accounting/Auditing; Administration/MIS/EDP; Art/Design; Banking; Computer Science/ Software; Economics; Finance; General Management; Insurance; Manufacturing; Sales and Marketing; Secretarial. **Positions commonly filled include:** Accountant/Auditor; Administrative Services Manager; Bank Officer/Manager; Blue-Collar Worker Supervisor; Branch Manager; Brokerage Clerk; Budget Analyst; Buyer; Claim Representative; Clerical Supervisor; Clinical Lab Technician; Computer Programmer; Customer Service Representative; Design Engineer; Designer; Draftsperson; Electrical/Electronics Engineer; Electrician; Financial Analyst; General Manager; Health Services Manager; Human Resources Specialist; Insurance Agent/Broker; Management Analyst/Consultant; Management Trainee; MIS Specialist; Property and Real Estate Manager; Public Relations Specialist; Quality Control Supervisor; Real Estate Agent; Services Sales Representative; Software Engineer; Systems Analyst; Typist/Word Processor. **Number of placements per year:** 50 - 99.

CLARK & HARTMAN PROFESSIONAL SEARCH INC.
315 North Main Street, Ann Arbor MI 48104. 313/996-3100. **Contact:** Lewis Clark, President. **Description:** An employment agency. **Specializes in the areas of:** Computer Hardware/Software. **Positions commonly filled include:** Computer Operator; Computer Programmer; Data Entry Clerk; Systems Analyst.

CORPORATE STAFFING RESOURCES
815 Main Street, Joseph MI 49085. 616/983-5803. **Fax:** 616/983-5827. **Recorded jobline:** 219/233-8209. **Contact:** Helen CKE Preston, Branch Manager. **Description:** An employment agency. Company pays fee. **Specializes in the areas of:** Accounting/Auditing; Administration/ MIS/EDP; Computer Science/Software; Engineering; General Management; Manufacturing; Personnel/Labor Relations; Sales and Marketing; Technical and Scientific. **Positions commonly filled include:** Accountant/Auditor; Administrative Services Manager; Advertising Clerk; Aerospace Engineer; Agricultural Engineer; Aircraft Mechanic/Engine Specialist; Biochemist; Biological Scientist; Biomedical Engineer; Blue-Collar Worker Supervisor; Chemical Engineer; Chemist; Civil Engineer; Clerical Supervisor; Computer Programmer; Cost Estimator; Credit Manager; Design Engineer; Designer; Draftsperson; Electrical/Electronics Engineer; Environmental Engineer; Financial Analyst; General Manager; Health Services Manager; Human Resources Specialist; Industrial Engineer; Industrial Production Manager; Internet Services Manager; Management Analyst/Consultant; Market Research Analyst; MIS Specialist; Multimedia Designer; Operations/Production Manager; Science Technologist; Software Engineer; Statistician; Structural Engineer; Systems Analyst; Technical Writer/Editor; Telecommunications Manager; Typist/Word Processor. **Number of placements per year:** 500 - 999.

EXECUTECH RESOURCE CONSULTANTS
5700 Crooks Road, Suite 105, Troy MI 48098. 810/828-3000. **Fax:** 810/828-3333. **Contact:** Jeff Bagnasco, Office Manager. **Description:** An employment agency. **Specializes in the areas of:** Automotive; Computer Programming; Environmental; Office Support; Plastics; Sales and Marketing. **Positions commonly filled include:** Administrative Worker/Clerk; Automotive Engineer; Computer Programmer; Environmental Engineer; Manufacturing Engineer; Plastics Engineer.

JOE L. GILES AND ASSOCIATES, INC.
18105 Parkside Street, Suite 14, Detroit MI 48221. 313/864-0022. **Fax:** 313/864-8351. **Contact:** Joe L. Giles, Owner/President. **Description:** An employment agency. Company pays fee.

Specializes in the areas of: Computer Hardware/Software; Engineering; MIS/EDP. **Positions commonly filled include:** EDP Specialist; Electrical/Electronics Engineer; Mechanical Engineer; MIS Specialist; Systems Analyst. **Number of placements per year:** 50 - 99.

WILLIAM HOWARD AGENCY

38701 Seven Mile Road, Suite 445, Livonia MI 48152. 313/464-6777. **Contact:** Christina Fortucci, Administrator. **Description:** An employment agency that also offers career counseling services. **Specializes in the areas of:** Accounting/Auditing; Advertising; Architecture/Construction/Real Estate; Banking; Computer Hardware/Software; Construction; Design; Education; Engineering; Food Industry; Health/Medical; Insurance; Legal; Manufacturing; MIS/EDP; Nonprofit; Printing/Publishing; Sales and Marketing; Secretarial; Technical and Scientific; Transportation. **Positions commonly filled include:** Accountant/Auditor; Actuary; Administrative Assistant; Advertising Account Executive; Aerospace Engineer; Agricultural Engineer; Architect; Attorney; Bank Officer/Manager; Biological Scientist; Biomedical Engineer; Bookkeeper; Ceramics Engineer; Civil Engineer; Claim Representative; Clerk; Commercial Artist; Computer Programmer; Credit Manager; Customer Service Representative; Data Entry Clerk; Dietician/Nutritionist; Draftsperson; Economist; EDP Specialist; Electrical/Electronics Engineer; Financial Analyst; General Manager; Hotel Manager; Human Resources Manager; Industrial Designer; Industrial Engineer; Insurance Agent/Broker; Interior Designer; Legal Secretary; Management Analyst/Consultant; Marketing Specialist; Mechanical Engineer; Medical Secretary; Metallurgical Engineer; Mining Engineer; Nurse; Petroleum Engineer; Physicist/Astronomer; Public Relations Specialist; Purchasing Agent/Manager; Receptionist; Secretary; Statistician; Stenographer; Systems Analyst; Technical Writer/Editor; Technician; Typist/Word Processor; Underwriter/Assistant Underwriter. **Number of placements per year:** 100 - 199.

KEY PERSONNEL

570 East 16th Street, Holland MI 49424. 616/396-7575. **Fax:** 616/396-3327. **Contact:** Manager. **Description:** An employment agency. Company pays fee. **Specializes in the areas of:** Accounting/Auditing; Administration/MIS/EDP; Art/Design; Computer Science/Software; Engineering; Industrial; Manufacturing; Personnel/Labor Relations. **Positions commonly filled include:** Accountant/Auditor; Administrative Services Manager; Civil Engineer; Designer; Draftsperson; Electrical/Electronics Engineer; Human Resources Manager; Industrial Engineer; Mechanical Engineer; Purchasing Agent/Manager; Quality Control Supervisor; Software Engineer. **Number of placements per year:** 500 - 999.

LUDOT PERSONNEL

P.O. Box 208, Southfield MI 48037. 810/353-9720. **Contact:** Michael Morton, Vice President. **Description:** An employment agency. Company pays fee. **Specializes in the areas of:** Automotive; Computer Hardware/Software; Engineering; Manufacturing; MIS/EDP. **Positions commonly filled include:** Accountant/Auditor; Aerospace Engineer; Economist; Electrical/Electronics Engineer; Industrial Engineer; Mechanical Engineer; Metallurgical Engineer; MIS Specialist; Quality Control Supervisor; Systems Analyst. **Number of placements per year:** 50 - 99.

MICHIGAN EMPLOYMENT SECURITY COMMISSION

P.O. Box 356, Escanaba MI 49829. 906/786-6841. **Contact:** Manager. **Description:** An employment agency. **NOTE:** The physical address is 2827 North Lincoln Road, Escanaba MI. **Specializes in the areas of:** Accounting/Auditing; Advertising; Architecture/Construction/Real Estate; Banking; Computer Hardware/Software; Design; Education; Engineering; Food Industry; Health/Medical; Insurance; Legal; Manufacturing; MIS/EDP; Nonprofit; Printing/Publishing; Sales and Marketing; Secretarial; Technical and Scientific; Transportation. **Positions commonly filled include:** Accountant/Auditor; Actuary; Administrative Assistant; Advertising Account Executive; Aerospace Engineer; Agricultural Engineer; Architect; Attorney; Bank Officer/Manager; Biological Scientist; Biomedical Engineer; Bookkeeper; Ceramics Engineer; Civil Engineer; Claim Representative; Clerk; Commercial Artist; Computer Programmer; Construction Trade Worker; Credit Manager; Customer Service Representative; Data Entry Clerk; Dietician/Nutritionist; Draftsperson; Driver; Economist; EDP Specialist; Electrical/Electronics Engineer; Factory Worker; Financial Analyst; General Manager; Hotel Manager; Human Resources Manager; Industrial Designer; Legal Secretary; Light Industrial Worker; Management Analyst/Consultant; Marketing

Specialist; Mechanical Engineer; Medical Secretary; Metallurgical Engineer; Mining Engineer; Nurse; Petroleum Engineer; Physicist/Astronomer; Public Relations Specialist; Purchasing Agent/ Manager; Receptionist; Reporter; Sales Representative; Secretary; Statistician; Stenographer; Systems Analyst; Technical Writer/Editor; Technician; Typist/Word Processor; Underwriter/ Assistant Underwriter. **Number of placements per year:** 1000+.

NATIONWIDE CAREER NETWORK
5445 Corporate Drive, Suite 160, Troy MI 48098-2683. **Contact:** Recruiter. **Description:** An employment agency. Company pays fee. **Specializes in the areas of:** Computer Science/Software; Engineering; Health/Medical. **Positions commonly filled include:** Biomedical Engineer; Chemical Engineer; Civil Engineer; Computer Programmer; Design Engineer; Environmental Engineer; Health Services Manager; Industrial Engineer; Industrial Production Manager; Mechanical Engineer; Medical Records Technician; MIS Specialist; Multimedia Designer; Physical Therapist; Physician; Registered Nurse; Software Engineer; Structural Engineer; Surgical Technician; Systems Analyst; Telecommunications Manager; Typist/Word Processor. **Average salary range of placements:** $30,000 - $50,000. **Number of placements per year:** 200 - 499.

PREFERRED EMPLOYMENT PLANNING
1479 West Bristol Road, Flint MI 48507-5523. 810/233-7200. **Fax:** 810/233-3095. **Contact:** Dan LePard, Manager. **Description:** An employment agency. **Specializes in the areas of:** Administration/MIS/EDP; Computer Science/Software; Engineering; Finance; Food Industry; General Management; Industrial; Manufacturing; Personnel/Labor Relations; Sales and Marketing; Secretarial; Technical and Scientific. **Positions commonly filled include:** Accountant/ Auditor; Administrative Services Manager; Advertising Clerk; Blue-Collar Worker Supervisor; Branch Manager; Ceramics Engineer; Chemical Engineer; Clerical Supervisor; Computer Programmer; Counselor; Credit Manager; Customer Service Representative; Design Engineer; Electrical/ Electronics Engineer; Environmental Engineer; General Manager; Hotel Manager; Human Resources Specialist; Human Service Worker; Industrial Engineer; Industrial Production Manager; Management Trainee; Manufacturer's/Wholesaler's Sales Rep.; Materials Engineer; Mechanical Engineer; Metallurgical Engineer; MIS Specialist; Operations/Production Manager; Pharmacist; Property and Real Estate Manager; Purchasing Agent/Manager; Quality Control Supervisor; Radio/TV Announcer/Broadcaster; Reporter; Restaurant/Food Service Manager; Securities Sales Representative; Services Sales Representative; Social Worker; Software Engineer; Speech-Language Pathologist; Systems Analyst; Technical Writer/Editor; Typist/Word Processor. **Average salary range of placements:** $20,000 - $29,999. **Number of placements per year:** 100 - 199.

RESOURCE TECHNOLOGIES CORPORATION
431 Stephenson Highway, Troy MI 48083. 810/585-4750. **Contact:** Manager. **Description:** An employment agency. **Specializes in the areas of:** Computer Science/Software; Design; Engineering.

SALES EXECUTIVES INC.
755 West Big Beaver Road, Suite 2107, Troy MI 48084. 810/362-1900. **Fax:** 810/362-0253. **Contact:** Mr. Dale Statson, President. **Description:** An employment agency. Company pays fee. **Specializes in the areas of:** Chemical; Computer Hardware/Software; Finance; General Management; Health/Medical; Plastics; Sales and Marketing. **Positions commonly filled include:** Marketing Specialist; Sales Manager; Sales Representative. **Number of placements per year:** 200 - 499.

SOURCE EDP
2000 Town Center, Suite 850, Southfield MI 48075. 810/352-6520. **Fax:** 810/352-7514. **Contact:** Brad Foster, Director. **Description:** An employment agency. Company pays fee. **Specializes in the areas of:** Administration/MIS/EDP; Computer Science/Software. **Positions commonly filled include:** Computer Programmer; Software Engineer; Systems Analyst; Technical Writer/Editor. **Number of placements per year:** 100 - 199.

SOURCE SERVICES
161 Ottawa NW, Suite 409-D, Grand Rapids MI 49503. 616/459-3600. **Fax:** 616/459-3670. **Contact:** Director. **Description:** Founded in 1962, this employment agency specializes in the

accounting, finance, and computer industries. Company pays fee. **Specializes in the areas of:** Accounting/Auditing; Computer Science/Software; Finance. **Positions commonly filled include:** Accountant/Auditor; Budget Analyst; Computer Programmer; Financial Analyst; MIS Specialist; Software Engineer; Systems Analyst. **Benefits available to temporary workers:** Medical Insurance. **Corporate headquarters location:** Dallas TX. **Average salary range of placements:** $30,000 - $50,000. **Number of placements per year:** 200 - 499.

WISE PERSONNEL SERVICES, INC.
200 Admiral Avenue, Kalamazoo MI 49002-3503. 616/323-2300. **Toll-free phone:** 800/842-2136. **Fax:** 616/323-8588. **Contact:** Margaret Ham, Account Executive/Recruiter. **Description:** An employment agency. Company pays fee. **Specializes in the areas of:** Accounting/Auditing; Administration/MIS/EDP; Computer Science/Software; Engineering; Sales and Marketing; Technical and Scientific. **Positions commonly filled include:** Accountant/Auditor; Administrative Services Manager; Architect; Blue-Collar Worker Supervisor; Branch Manager; Buyer; Chemical Engineer; Clerical Supervisor; Computer Programmer; Customer Service Representative; Design Engineer; Designer; Draftsperson; Electrical/Electronics Engineer; Electrician; Geologist/Geophysicist; Human Resources Manager; Industrial Engineer; Industrial Production Manager; Internet Services Manager; Mechanical Engineer; Quality Control Supervisor; Services Sales Representative; Software Engineer; Surveyor; Systems Analyst. **Benefits available to temporary workers:** Dental Insurance; Medical Insurance; Paid Holidays; Paid Vacation. **Number of placements per year:** 50 - 99.

EXECUTIVE SEARCH FIRMS OF MINNESOTA

COMPUTER EMPLOYMENT
5151 Edina Industrial Boulevard, Edina MN 55439-3013. **Fax:** 612/831-4684. **Contact:** Marty Koepp, Owner. **Description:** An executive search firm operating on a contingency basis. Company pays fee. **Specializes in the areas of:** Administration/MIS/EDP; Computer Science/Software. **Positions commonly filled include:** Computer Programmer; Management Analyst/Consultant; MIS Specialist; Software Engineer; Systems Analyst; Telecommunications Manager. **Corporate headquarters location:** This Location. **Average salary range of placements:** $30,000 - $50,000. **Number of placements per year:** 50 - 99.

EHS & ASSOCIATES, INC.
1516 West Lake Street, Suite 102, Minneapolis MN 55408. 612/824-3993. **Fax:** 612/824-4843. **Contact:** David Elhard, Vice President. **Description:** An executive search firm. Company pays fee. **Specializes in the areas of:** Administration/MIS/EDP; Art/Design; Computer Science/Software; Food Industry; Health/Medical. **Positions commonly filled include:** Computer Programmer; Dental Assistant/Dental Hygienist; Dentist; Hotel Manager; Pharmacist; Physical Therapist; Physician; Purchasing Agent/Manager; Registered Nurse; Restaurant/Food Service Manager; Software Engineer; Systems Analyst. **Number of placements per year:** 200 - 499.

ESP SYSTEMS PROFESSIONALS
701 4th Avenue South, Suite 1800, Minneapolis MN 55415-1819. 612/337-3000. **Fax:** 612/337-9199. **Contact:** Robert R. Hildreth, President. **Description:** An executive search firm. ESP Systems Professionals focuses on placing information systems professionals, from programmers to senior managers. Company pays fee. **Specializes in the areas of:** Computer Science/Software. **Positions commonly filled include:** Computer Programmer; Internet Services Manager; MIS Specialist; Software Engineer; Systems Analyst. **Average salary range of placements:** $30,000 - $50,000. **Number of placements per year:** 200 - 499.

GATEWAY SEARCH
15500 Wayzata Boulevard, Suite 262, Wayzata MN 55391-1438. 612/473-3137. **Fax:** 612/473-3276. **Contact:** James Bortolussi, President. **Description:** An executive search firm operating on a

contingency basis that specializes in placing client/server professionals. Company pays fee. **Specializes in the areas of:** Computer Science/Software. **Positions commonly filled include:** Computer Programmer; Software Engineer; Systems Analyst. **Corporate headquarters location:** This Location. **Average salary range of placements:** More than $50,000. **Number of placements per year:** 1 - 49.

T.H. HUNTER, INC.

526 Nicollet Mall, Suite 310, Minneapolis MN 55402. 612/339-0530. **Fax:** 612/338-4757. **Contact:** Martin Conroy, Executive Recruiter. **Description:** An executive search firm. Company pays fee. **Specializes in the areas of:** Accounting/Auditing; Administration/MIS/EDP; Advertising; Banking; Biology; Computer Science/Software; Economics; Engineering; Finance; Food Industry; Health/Medical; Insurance; Legal; Manufacturing; Nonprofit; Personnel/Labor Relations; Printing/Publishing; Retail; Sales and Marketing; Technical and Scientific. **Positions commonly filled include:** Accountant/Auditor; Actuary; Attorney; Bank Officer/Manager; Biomedical Engineer; Branch Manager; Budget Analyst; Buyer; Computer Programmer; Credit Manager; Economist; Electrical/Electronics Engineer; Financial Analyst; General Manager; Health Services Manager; Human Resources Manager; Internet Services Manager; Licensed Practical Nurse; Management Analyst/Consultant; Manufacturer's/Wholesaler's Sales Rep.; Mechanical Engineer; Mortgage Banker; Occupational Therapist; Physical Therapist; Physician; Quality Control Supervisor; Registered Nurse; Science Technologist; Software Engineer; Speech-Language Pathologist; Systems Analyst; Technical Writer/Editor; Telecommunications Manager; Urban/ Regional Planner. **Average salary range of placements:** More than $50,000. **Number of placements per year:** 50 - 99.

MANAGEMENT RECRUITERS OF MINNEAPOLIS
SALES CONSULTANTS OF MINNEAPOLIS

7550 France Avenue South, Suite 180, Edina MN 55435. 612/835-4466. **Contact:** Manager. **Description:** An executive search firm operating on a contingency basis. **Specializes in the areas of:** Accounting/Auditing; Administration/MIS/EDP; Advertising; Architecture/Construction/Real Estate; Banking; Chemical; Communications; Computer Hardware/Software; Construction; Design; Electrical; Engineering; Finance; Food Industry; General Management; Health/Medical; Industrial; Insurance; Legal; Manufacturing; Operations Management; Personnel/Labor Relations; Pharmaceutical; Printing/Publishing; Procurement; Retail; Sales and Marketing; Technical and Scientific; Textiles; Transportation.

MANAGEMENT RECRUITERS OF ROCHESTER

1903 South Broadway, Rochester MN 55904. 507/282-2400. **Contact:** Nona Vierkant, Office Manager. **Description:** An executive search firm operating on a contingency basis. Company pays fee. **Specializes in the areas of:** Computer Science/Software. **Positions commonly filled include:** Computer Programmer; Systems Analyst. **Average salary range of placements:** $30,000 - $50,000. **Number of placements per year:** 1 - 49.

NER (NATIONAL ENGINEERING RESOURCES), INC.

6200 Shingle Creek Parkway, Suite 160, Brooklyn Center MN 55430. 612/561-7610. **Toll-free phone:** 800/665-7610. **Fax:** 612/561-7675. **Contact:** Michelle Smith, Technical Recruiter. **E-mail address:** nerinc@sprynet.com. **World Wide Web address:** http://www.occ.com/ner/. **Description:** An executive search firm focusing on the placement of engineering, technical, and scientific personnel. **Specializes in the areas of:** Administration/MIS/EDP; Clerical; Computer Hardware/Software; Engineering; Industrial; Medical Technology; Oil and Gas; Petrochemical; Printing/Publishing; Technical and Scientific. **Positions commonly filled include:** Administrative Services Manager; Agricultural Engineer; Aircraft Mechanic/Engine Specialist; Ceramics Engineer; Computer Programmer; Materials Engineer; Medical Assistant; Metallurgical Engineer; Nuclear Engineer; Quality Control Supervisor; Software Engineer; Structural Engineer; Systems Analyst. **Number of placements per year:** 1 - 49.

NORTH AMERICAN RECRUITERS

1660 South Highway 100, Suite 519, St. Louis Park MN 55416. 612/591-1951. **Toll-free phone:** 800/886-7598. **Fax:** 612/591-5850. **Contact:** David Knutson, President. **Description:** An executive search firm operating on a retainer basis. Company pays fee. **Specializes in the areas of:**

Accounting/Auditing; Administration/MIS/EDP; Architecture/Construction/Real Estate; Computer Science/Software; Engineering; Finance; Food Industry; General Management; Health/Medical; Industrial; Manufacturing; Nonprofit; Printing/Publishing; Sales and Marketing; Technical and Scientific. **Positions commonly filled include:** Accountant/Auditor; Biomedical Engineer; Chemical Engineer; Civil Engineer; Computer Programmer; Construction Contractor; Customer Service Representative; Electrical/Electronics Engineer; Financial Analyst; General Manager; Industrial Engineer; Industrial Production Manager; Mechanical Engineer; Operations/Production Manager; Purchasing Agent/Manager; Services Sales Representative; Software Engineer; Structural Engineer; Systems Analyst; Telecommunications Manager. **Corporate headquarters location:** This Location. **Average salary range of placements:** More than $50,000. **Number of placements per year:** 100 - 199.

NORTHLAND RECRUITING INC.

10801 Wayzata Boulevard, Suite 325, Minnetonka MN 55305. 612/541-1060. **Fax:** 612/595-9878. **Contact:** David Gavin, President. **Description:** An executive search firm operating on a contingency basis. **Corporate headquarters location:** This Location. **Specializes in the areas of:** Computer Science/Software; Engineering. **Number of placements per year:** 100 - 199.

PROGRAMMING ALTERNATIVES OF MINNESOTA, INC.

6750 France Avenue South, Suite 144, Edina MN 55435. 612/922-7879. **Fax:** 612/922-3726. **Contact:** Kenneth Rosaro, Divisional Manager. **Description:** An executive search firm that focuses on information technology staffing and consulting. Company pays fee. **Specializes in the areas of:** Biotechnology; Computer Science/Software; Data Processing; Engineering; Industrial; Manufacturing; Technical and Scientific. **Positions commonly filled include:** Aerospace Engineer; Biological Scientist; Biomedical Engineer; Chemical Engineer; Computer Programmer; Design Engineer; Electrical/Electronics Engineer; Industrial Designer; Mechanical Engineer; Metallurgical Engineer; MIS Specialist; Operations/Production Manager; Physicist/Astronomer; Purchasing Agent/Manager; Quality Control Supervisor; Software Engineer; Systems Analyst; Technical Writer/Editor; Telecommunications Manager. **Corporate headquarters location:** Minneapolis MN. **Average salary range of placements:** More than $50,000. **Number of placements per year:** 50 - 99.

QUANTUM INFORMATION SYSTEMS PROFESSIONALS

6600 City West Parkway, Suite 310, Eden Prairie MN 55344. 612/829-5950. **Fax:** 612/829-5988. **Contact:** Doug Berg, President/Recruiter. **Description:** An executive search firm. Company pays fee. **Specializes in the areas of:** Computer Science/Software. **Positions commonly filled include:** Computer Programmer; Systems Analyst. **Number of placements per year:** 1 - 49.

REGENCY RECRUITERS, INC.

7101 York Avenue South, Suite 248, Edina MN 55435-4450. 612/921-3377. **Contact:** David Tetzloff, President. **Description:** An executive search firm focusing on engineering positions, including electrical and mechanical design, software, and manufacturing. Company pays fee. **Specializes in the areas of:** Computer Science/Software; Engineering; Industrial; Manufacturing; Technical and Scientific. **Positions commonly filled include:** Aerospace Engineer; Biomedical Engineer; Chemical Engineer; Civil Engineer; Computer Programmer; Draftsperson; Electrical/Electronics Engineer; Mechanical Engineer; Metallurgical Engineer; Quality Control Supervisor; Software Engineer; Structural Engineer; Technical Writer/Editor. **Average salary range of placements:** $30,000 - $50,000.

SEARCH SPECIALISTS

2655 North Shore Drive, Wayzata MN 55391. 612/449-8990. **Fax:** 612/449-0369. **Contact:** Craig Lindell, President. **E-mail address:** clindell@sprynet.com. **World Wide Web address:** http://www.cities~online.com/search. **Description:** An executive search, recruitment, and placement firm, focusing on such industries as architectural, engineering, data processing, software development, and sales and marketing. Company pays fee. **Specializes in the areas of:** Architecture/Construction/Real Estate; Computer Science/Software; Engineering; Sales and Marketing. **Positions commonly filled include:** Architect; Civil Engineer; Design Engineer; Draftsperson; Electrical/Electronics Engineer; Landscape Architect; Mechanical Engineer; MIS

Specialist; Software Engineer; Structural Engineer. **Average salary range of placements:** $20,000 - $29,999. **Number of placements per year:** 1 - 49.

STAFF CONNECTION, INC.

1000 Shelard Parkway, Suite 101, St. Louis Park MN 55426-4917. 612/545-2228. **Fax:** 612/545-3699. **Contact:** Craig Lyon, Secretary/Treasurer. **E-mail address:** sci@mm.com. **Description:** An executive search firm operating on a contingency basis that specializes in placing computer professionals. **Specializes in the areas of:** Computer Science/Software; Engineering; Technical and Scientific. **Positions commonly filled include:** Computer Programmer; Internet Services Manager; MIS Specialist; Software Engineer; Systems Analyst. **Corporate headquarters location:** Minneapolis MN. **Other U.S. locations:** Phoenix AZ; Las Vegas NV. **Average salary range of placements:** $30,000 - $50,000.

SYSTEMS SEARCH, INC.

P.O. Box 85, Champlin MN 55316. 612/323-9690. **Contact:** Mike Fitzpatrick, President. **E-mail address:** 74352.3305@compuserve.com. **Description:** An executive search firm of computer professionals that recruits and places other computer professionals including systems analysts, consultants, engineers, administrators, and programmers. Company pays fee. **Specializes in the areas of:** Administration/MIS/EDP; Computer Science/Software. **Positions commonly filled include:** Computer Programmer; Internet Services Manager; Management Analyst/Consultant; MIS Specialist; Multimedia Designer; Software Engineer; Systems Analyst; Telecommunications Manager. **Average salary range of placements:** $30,000 - $50,000. **Number of placements per year:** 1 - 49.

TOTAL SEARCH

1590 60th Avenue NE, Fridley MN 55432-5826. 612/571-0247. **Contact:** Tom Harrington, President. **Description:** An executive search firm operating on a contingency basis that specializes in placing food management, retail management, and computer science professionals. Company pays fee. **Specializes in the areas of:** Computer Science/Software; Food Industry; Printing/Publishing; Retail. **Positions commonly filled include:** Computer Programmer; Restaurant/Food Service Manager; Systems Analyst. **Corporate headquarters location:** This Location. **Number of placements per year:** 50 - 99.

PERMANENT EMPLOYMENT AGENCIES OF MINNESOTA

ALTERNATIVE STAFFING, INC.

8120 Penn Avenue South, Suite 570, Bloomington MN 55431-1326. 612/888-6077. **Contact:** Kim Howard, President. **Description:** An employment agency. Company pays fee. **Specializes in the areas of:** Accounting/Auditing; Clerical; Computer Hardware/Software; Legal; Manufacturing; Printing/Publishing; Sales and Marketing; Secretarial. **Positions commonly filled include:** Accountant/Auditor; Administrative Assistant; Bookkeeper; Claim Representative; Clerk; Computer Programmer; Credit Manager; Customer Service Representative; Data Entry Clerk; Draftsperson; Factory Worker; Financial Analyst; Human Resources Manager; Legal Secretary; Light Industrial Worker; Marketing Specialist; Medical Secretary; Purchasing Agent/Manager; Receptionist; Sales Representative; Secretary; Stenographer; Typist/Word Processor. **Number of placements per year:** 1000+.

BARTZ ROGERS & PARTNERS

6465 Wayzata Drive, Suite 777, Minneapolis MN 55426. 612/936-0657. **Fax:** 612/936-0142. **Contact:** Douglas Bartz, Partner. **Description:** An employment agency. Company pays fee. **Specializes in the areas of:** Computer Science/Software. **Positions commonly filled include:** Computer Programmer; Systems Analyst. **Number of placements per year:** 100 - 199.

MANPOWER TECHNICAL SERVICES

3601 Minnesota Drive, Suite 450, Saint Paul MN 55101-1716. 612/820-0365. **Fax:** 612/820-0350. **Contact:** Carol Cotter, Recruiter. **Description:** An employment agency. Company pays fee. **Specializes in the areas of:** Administration/MIS/EDP; Computer Science/Software; Engineering; Finance; Industrial; Manufacturing; Personnel/Labor Relations; Printing/Publishing; Technical and Scientific. **Positions commonly filled include:** Accountant/Auditor; Administrative Services Manager; Aerospace Engineer; Aircraft Mechanic/Engine Specialist; Bank Officer/Manager; Biochemist; Biological Scientist; Biomedical Engineer; Blue-Collar Worker Supervisor; Branch Manager; Broadcast Technician; Chemical Engineer; Computer Programmer; Credit Manager; Customer Service Representative; Design Engineer; Designer; Draftsperson; Editor; Electrical/ Electronics Engineer; Environmental Engineer; Financial Analyst; Food Scientist/Technologist; General Manager; Geologist/Geophysicist; Human Resources Specialist; Industrial Engineer; Industrial Production Manager; Insurance Agent/Broker; Internet Services Manager; Library Technician; Management Analyst/Consultant; Management Trainee; Manufacturer's/Wholesaler's Sales Rep.; Market Research Analyst; Materials Engineer; Mathematician; Mechanical Engineer; Metallurgical Engineer; MIS Specialist; Multimedia Designer; Nuclear Engineer; Operations/ Production Manager; Paralegal; Physicist/Astronomer; Purchasing Agent/Manager; Quality Control Supervisor; Science Technologist; Securities Sales Representative; Services Sales Representative; Software Engineer; Structural Engineer; Systems Analyst; Technical Writer/Editor; Telecommunications Manager; Typist/Word Processor; Underwriter/Assistant Underwriter; Video Production Coordinator. **Benefits available to temporary workers:** 401(k); Medical Insurance. **Corporate headquarters location:** Milwaukee WI. **Other U.S. locations:** Nationwide. **Average salary range of placements:** More than $50,000. **Number of placements per year:** 500 - 999.

TECHNICAL RESOURCES, INC.

7460 Market Place Drive, Eden Prairie MN 55344-3634. 612/941-9441. **Toll-free phone:** 800/298-5627. **Fax:** 612/941-9440. **Contact:** Alice E. Riggs, Account Executive. **E-mail address:** rsg@skypoint.com. **Description:** An employment agency that specializes in placing technical professionals in temporary and permanent positions. Company pays fee. **Specializes in the areas of:** Computer Science/Software; Engineering; Food Industry; Industrial; Manufacturing; Technical and Scientific. **Positions commonly filled include:** Aerospace Engineer; Architect; Buyer; Ceramics Engineer; Chemical Engineer; Chemist; Civil Engineer; Computer Programmer; Design Engineer; Designer; Draftsperson; Editor; Electrical/Electronics Engineer; Industrial Engineer; Industrial Production Manager; Materials Engineer; Mechanical Engineer; Metallurgical Engineer; MIS Specialist; Operations/Production Manager; Software Engineer; Statistician; Structural Engineer; Systems Analyst; Technical Writer/Editor. **Benefits available to temporary workers:** 401(k); Medical; Vacation. **Corporate headquarters location:** This Location. **Average salary range of placements:** $30,000 - $50,000. **Number of placements per year:** 100 - 199.

EXECUTIVE SEARCH FIRMS OF MISSISSIPPI

DUNHILL PROFESSIONAL SEARCH OF JACKSON

13 North Town Drive, Suite 220, Jackson MS 39211. 601/956-1060. **Contact:** Manager. **Description:** An executive search firm. **Specializes in the areas of:** Accounting/Auditing; Computer Hardware/Software; Engineering.

IMPACT PERSONNEL SERVICES

212 Haddon Circle, Brandon MS 39042-8046. 601/992-1591. **Fax:** 601/992-5037. **Contact:** Jan Prystupa, Human Resources. **Description:** A service-oriented executive search firm offering nationwide and worldwide placement. Company pays fee. **Specializes in the areas of:** Administration/MIS/EDP; Computer Science/Software; Engineering; General Management; Industrial; Manufacturing; Personnel/Labor Relations. **Positions commonly filled include:** Chemical Engineer; Civil Engineer; Computer Programmer; Design Engineer; Designer; Draftsperson; Electrical/Electronics Engineer; General Manager; Human Resources Manager;

Industrial Engineer; Industrial Production Manager; Internet Services Manager; Mechanical Engineer; Metallurgical Engineer; MIS Specialist; Petroleum Engineer; Physical Therapist; Physician; Quality Control Supervisor; Structural Engineer; Systems Analyst; Technical Writer/ Editor; Telecommunications Manager. **Number of placements per year:** 50 - 99.

PERMANENT EMPLOYMENT AGENCIES OF MISSISSIPPI

ANDRUS ASSOCIATES INC.
dba SERVICE SPECIALISTS LTD.
500 Greymont Avenue, Jackson MS 39202-3446. 601/948-8980. **Fax:** 601/948-8983. **Contact:** Elva Giddings, President. **Description:** An employment agency that also provides career counseling. Company pays fee. **Specializes in the areas of:** Accounting/Auditing; Administration/MIS/EDP; Banking; Computer Science/Software; Engineering; Finance; Industrial; Manufacturing; Personnel/ Labor Relations; Sales and Marketing; Secretarial. **Positions commonly filled include:** Accountant/Auditor; Adjuster; Administrative Services Manager; Blue-Collar Worker Supervisor; Buyer; Chemical Engineer; Computer Programmer; Design Engineer; Draftsperson; Environmental Engineer; Financial Analyst; Human Resources Specialist; Industrial Engineer; Industrial Production Manager; Management Trainee; Mechanical Engineer; Purchasing Agent/Manager; Services Sales Representative; Typist/Word Processor. **Average salary range of placements:** $30,000 - $50,000. **Number of placements per year:** 100 - 199.

COATS & COATS PERSONNEL
P.O. Box 1009, Meridian MS 39302. 601/693-2991. **Fax:** 601/693-9983. **Contact:** Tom Coats, Owner. **Description:** An employment agency. Company pays fee. **Specializes in the areas of:** Accounting/Auditing; Aerospace; Computer Science/Software; Engineering; Food Industry; General Management; Industrial; Manufacturing; Personnel/Labor Relations; Retail; Sales and Marketing; Secretarial; Technical and Scientific. **Positions commonly filled include:** Accountant/ Auditor; Aerospace Engineer; Blue-Collar Worker Supervisor; Branch Manager; Chemical Engineer; Civil Engineer; Computer Programmer; Counselor; Credit Manager; Design Engineer; Draftsperson; Electrical/Electronics Engineer; Environmental Engineer; General Manager; Human Resources Specialist; Industrial Engineer; Industrial Production Manager; Internet Services Manager; Management Trainee; Manufacturer's/Wholesaler's Sales Rep.; Mechanical Engineer; Medical Records Technician; MIS Specialist; Paralegal; Registered Nurse; Restaurant/Food Service Manager; Software Engineer; Speech-Language Pathologist; Typist/Word Processor. **Average salary range of placements:** $20,000 - $29,999. **Number of placements per year:** 500 - 999.

OPPORTUNITIES UNLIMITED
P.O. Box 1518, Pascagoula MS 39568. 601/762-8068. **Contact:** Manager. **Description:** An employment agency. **NOTE:** The company's physical address is 3903 Market Street, Pascagoula MS. **Specializes in the areas of:** Accounting/Auditing; Clerical; Computer Hardware/Software; Engineering; Finance; Manufacturing; Technical and Scientific. **Positions commonly filled include:** Aerospace Engineer; Buyer; Chemical Engineer; Civil Engineer; Computer Programmer; EDP Specialist; Electrical/Electronics Engineer; Financial Analyst; Human Resources Manager; Industrial Engineer; Legal Secretary; Mechanical Engineer; Metallurgical Engineer; MIS Specialist; Petroleum Engineer; Physicist/Astronomer; Quality Control Supervisor; Receptionist; Secretary; Systems Analyst; Technical Writer/Editor; Typist/Word Processor. **Number of placements per year:** 50 - 99.

TATUM PERSONNEL INC.
293 Highland Village, Jackson MS 39211. 601/362-3135. **Fax:** 601/981-5995. **Contact:** Jim Tatum, President. **Description:** An employment agency offering permanent and temporary

placements. **Specializes in the areas of:** Administration/MIS/EDP; Computer Science/Software; Engineering; Insurance; Manufacturing; Sales and Marketing; Secretarial. **Positions commonly filled include:** Accountant/Auditor; Bank Officer/Manager; Buyer; Chemical Engineer; Civil Engineer; Claim Representative; Clerical Supervisor; Computer Programmer; Cost Estimator; Credit Manager; Customer Service Representative; Draftsperson; Electrical/Electronics Engineer; Human Resources Manager; Industrial Engineer; Industrial Production Manager; Management Trainee; Manufacturer's/Wholesaler's Sales Rep.; Mechanical Engineer; Operations/Production Manager; Paralegal; Purchasing Agent/Manager; Quality Control Supervisor; Restaurant/Food Service Manager; Securities Sales Representative; Services Sales Representative; Software Engineer; Structural Engineer; Systems Analyst; Underwriter/Assistant Underwriter.

EXECUTIVE SEARCH FIRMS OF MISSOURI

ADVANCED CAREERS OF KANSAS CITY, INC.
6528 Raytown Road, Kansas City MO 64133. 816/358-3553. **Fax:** 816/358-3566. **Contact:** Hal Willis, Vice President. **Description:** An executive search firm operating on a contingency basis. Company pays fee. **Specializes in the areas of:** Accounting/Auditing; Computer Science/Software; Engineering; Finance; Food Industry; General Management; Health/Medical; Manufacturing; Personnel/Labor Relations. **Positions commonly filled include:** Accountant/Auditor; Chemical Engineer; Civil Engineer; Electrical/Electronics Engineer; Manufacturing Engineer; Mechanical Engineer; Sales Representative. **Average salary range of placements:** $30,000 - $50,000. **Number of placements per year:** 1 - 49.

BRADFORD & GALT CONSULTING SERVICES
12400 Olive Boulevard, Suite 430, St. Louis MO 63141. 314/434-9200. **Fax:** 314/434-9266. **Contact:** Staff Manager. **Description:** An executive search firm. Company pays fee. **Specializes in the areas of:** Computer Science/Software. **Positions commonly filled include:** Computer Programmer; Systems Analyst. **Number of placements per year:** 50 - 99.

BURNS EMPLOYMENT SERVICE, INC.
9229 Ward Parkway, Kansas City MO 64114-3335. 816/361-6444. **Fax:** 816/361-7747. **Contact:** Kevin Burns, President. **Description:** An executive search firm. Company pays fee. **Specializes in the areas of:** Accounting/Auditing; Advertising; Architecture/Construction/Real Estate; Banking; Biology; Computer Science/Software; Education; Engineering; Finance; Food Industry; General Management; Industrial; Insurance; Manufacturing; Sales and Marketing; Technical and Scientific. **Positions commonly filled include:** Accountant/Auditor; Administrative Services Manager; Architect; Attorney; Bank Officer/Manager; Biochemist; Biological Scientist; Biomedical Engineer; Branch Manager; Budget Analyst; Buyer; Chemical Engineer; Chemist; Civil Engineer; Claim Representative; Clerical Supervisor; Computer Programmer; Cost Estimator; Credit Manager; Customer Service Representative; Design Engineer; Designer; Draftsperson; Economist; Education Administrator; Electrical/Electronics Engineer; Environmental Engineer; Financial Analyst; Food Scientist/Technologist; General Manager; Geologist/Geophysicist; Industrial Engineer; Industrial Production Manager; Management Trainee; Market Research Analyst; Materials Engineer; Mechanical Engineer; Metallurgical Engineer; MIS Specialist; Operations/Production Manager; Petroleum Engineer; Public Relations Specialist; Purchasing Agent/Manager; Quality Control Supervisor; Science Technologist; Services Sales Representative; Software Engineer; Structural Engineer; Surveyor; Systems Analyst; Technical Writer/Editor; Telecommunications Manager; Transportation/Traffic Specialist; Underwriter/Assistant Underwriter. **Average salary range of placements:** More than $50,000. **Number of placements per year:** 500 - 999.

DECK & DECKER EMPLOYMENT SERVICE
1900 North Providence Road, Suite 319, Columbia MO 65202-3710. 573/449-0876. **Fax:** 573/449-0878. **Recorded jobline:** 573/449-0890. **Contact:** Jack W. Rogers, MA Ed., Owner. **E-mail**

address: deckerjobs@aol.com. **Description:** An executive search firm that also offers contract services and career counseling. Company pays fee. **Specializes in the areas of:** Accounting/Auditing; Computer Science/Software; General Management; Industrial; Insurance; Manufacturing; Personnel/Labor Relations; Printing/Publishing; Retail; Sales and Marketing; Secretarial; Technical and Scientific. **Positions commonly filled include:** Accountant/Auditor; Adjuster; Administrative Services Manager; Advertising Clerk; Bank Officer/Manager; Blue-Collar Worker Supervisor; Branch Manager; Broadcast Technician; Brokerage Clerk; Budget Analyst; Buyer; Claim Representative; Clerical Supervisor; Clinical Lab Technician; Computer Programmer; Construction and Building Inspector; Construction Contractor; Cost Estimator; Counselor; Credit Manager; Customer Service Representative; Dental Assistant/Dental Hygienist; Draftsperson; Editor; Electrical/Electronics Engineer; Electrician; Engineer; Financial Analyst; Food Scientist/Technologist; General Manager; Health Services Manager; Hotel Manager; Human Resources Manager; Human Service Worker; Industrial Production Manager; Insurance Agent/Broker; Internet Services Manager; Landscape Architect; Licensed Practical Nurse; Management Analyst/Consultant; Management Trainee; Manufacturer's/Wholesaler's Sales Rep.; Market Research Analyst; Medical Records Technician; MIS Specialist; Occupational Therapist; Property and Real Estate Manager; Public Relations Specialist; Purchasing Agent/Manager; Quality Control Supervisor; Radio/TV Announcer/Broadcaster; Real Estate Agent; Reporter; Restaurant/Food Service Manager; Securities Sales Representative; Services Sales Representative; Social Worker; Software Engineer; Surveyor; Systems Analyst; Technical Writer/Editor; Telecommunications Manager; Transportation/Traffic Specialist; Underwriter/Assistant Underwriter. **Benefits available to temporary workers:** Medical Insurance. **Average salary range of placements:** $20,000 - $29,999. **Number of placements per year:** 100 - 199.

GRUEN & ASSOCIATES, INC.
9270 Olive Boulevard, St. Louis MO 63134. 314/567-1478. **Fax:** 314/567-0567. **Contact:** Brian Gruen, Senior Account Executive. **E-mail address:** bgruen@aol.com. **Description:** An executive search firm that specializes in placements in the information systems field. The company provides both retainer and contingency searches. Company pays fee. **Specializes in the areas of:** Computer Science/Software; Engineering. **Positions commonly filled include:** Computer Programmer; MIS Specialist; Software Engineer; Systems Analyst. **Corporate headquarters location:** Medora IL. **Other U.S. locations:** Nationwide. **Number of placements per year:** 50 - 99.

JRL EXECUTIVE RECRUITERS
2700 Rockcreek Parkway, Suite 303, North Kansas City MO 64117. 816/471-4022. **Fax:** 816/471-8634. **Contact:** Larry Eason, President. **Description:** An executive search firm that specializes in placing technical personnel and senior-level management in various industries including automotive, manufacturing, utilities, plastics, food processing, medical, pharmaceutical, electronics, and information technology. Company pays fee. **Specializes in the areas of:** Computer Science/Software; Engineering; Food Industry; General Management; Industrial; Manufacturing; Personnel/Labor Relations; Printing/Publishing; Technical and Scientific; Transportation. **Positions commonly filled include:** Biomedical Engineer; Chemical Engineer; Chemist; Civil Engineer; Construction Manager; Cost Estimator; Design Engineer; Designer; Environmental Engineer; Food Scientist/Technologist; General Manager; Human Resources Specialist; Industrial Engineer; Industrial Production Manager; Landscape Architect; Mechanical Engineer; Metallurgical Engineer; Mining Engineer; Nuclear Engineer; Operations/Production Manager; Petroleum Engineer; Property and Real Estate Manager; Quality Control Supervisor; Software Engineer; Stationary Engineer; Structural Engineer; Surveyor; Technical Writer/Editor; Telecommunications Manager. **Corporate headquarters location:** This Location. **Average salary range of placements:** More than $50,000. **Number of placements per year:** 50 - 99.

LLOYD, MARTIN & ASSOCIATES
2258 Schuetz Road, St. Louis MO 63146. 314/991-8500. **Fax:** 314/991-8055. **Contact:** Fred Lloyd, Owner. **Description:** An executive search firm operating on a contingency basis. The firm specializes in the placement of information systems professionals. Company pays fee. **Specializes in the areas of:** Administration/MIS/EDP; Computer Science/Software. **Positions commonly filled include:** Computer Programmer; Internet Services Manager; Management Analyst/Consultant; MIS Specialist; Multimedia Designer; Software Engineer; Systems Analyst; Technical Writer/Editor; Telecommunications Manager. **Corporate headquarters location:** This Location.

Average salary range of placements: More than $50,000. **Number of placements per year:** 50 - 99.

MANAGEMENT RECRUITERS OF KANSAS CITY
712 Broadway, Suite 500, Kansas City MO 64105. 816/221-2377. **Contact:** Steve Orr, Manager. **Description:** An executive search firm. **Specializes in the areas of:** Accounting/Auditing; Administration/MIS/EDP; Advertising; Architecture/Construction/Real Estate; Banking; Chemical; Communications; Computer Hardware/Software; Design; Electrical; Engineering; Finance; Food Industry; General Management; Health/Medical; Industrial; Insurance; Legal; Manufacturing; Operations Management; Personnel/Labor Relations; Pharmaceutical; Printing/Publishing; Procurement; Retail; Sales and Marketing; Technical and Scientific; Textiles; Transportation.

MANAGEMENT RECRUITERS OF SPRINGFIELD
1807 East Edgewood, Suite B, Springfield MO 65804. 417/882-6220. **Contact:** Manager. **Description:** An executive search firm. **Specializes in the areas of:** Accounting/Auditing; Administration/MIS/EDP; Advertising; Architecture/Construction/Real Estate; Banking; Chemical; Communications; Computer Hardware/Software; Design; Electrical; Engineering; Finance; Food Industry; General Management; Health/Medical; Industrial; Insurance; Legal; Manufacturing; Operations Management; Personnel/Labor Relations; Pharmaceutical; Printing/Publishing; Procurement; Retail; Sales and Marketing; Technical and Scientific; Textiles; Transportation.

MANAGEMENT RECRUITERS OF ST. LOUIS (AIRPORT)
3301 Rider Trail South, Suite 100, St. Louis MO 63045. 314/344-0959. **Contact:** Recruiter. **Description:** An executive search firm. Company pays fee. **Specializes in the areas of:** Accounting/Auditing; Administration/MIS/EDP; Advertising; Architecture/Construction/Real Estate; Banking; Chemical; Communications; Computer Hardware/Software; Design; Electrical; Engineering; Finance; Food Industry; General Management; Health/Medical; Industrial; Insurance; Legal; Manufacturing; Operations Management; Personnel/Labor Relations; Pharmaceutical; Printing/Publishing; Procurement; Retail; Sales and Marketing; Technical and Scientific; Textiles; Transportation.

MANAGEMENT RECRUITERS OF ST. LOUIS (CLAYTON)
11701 Borman Drive, Suite 250, St. Louis MO 63146. 314/991-4355. **Contact:** Phil Bertsch, Manager. **Description:** An executive search firm. **Specializes in the areas of:** Accounting/Auditing; Administration/MIS/EDP; Advertising; Architecture/Construction/Real Estate; Banking; Chemical; Communications; Computer Hardware/Software; Design; Electrical; Engineering; Finance; Food Industry; General Management; Health/Medical; Industrial; Insurance; Legal; Manufacturing; Operations Management; Personnel/Labor Relations; Pharmaceutical; Printing/Publishing; Procurement; Retail; Sales and Marketing; Technical and Scientific; Textiles; Transportation.

MANAGEMENT RECRUITERS OF ST. LOUIS (WEST COUNTY)
200 Fabricator Drive, Fenton MO 63026. 314/349-4455. **Contact:** J. Edward Travis, General Manager. **Description:** An executive search firm. **Specializes in the areas of:** Accounting/Auditing; Administration/MIS/EDP; Advertising; Architecture/Construction/Real Estate; Banking; Chemical; Communications; Computer Hardware/Software; Design; Electrical; Engineering; Finance; Food Industry; General Management; Health/Medical; Industrial; Insurance; Legal; Manufacturing; Operations Management; Personnel/Labor Relations; Pharmaceutical; Printing/Publishing; Procurement; Sales and Marketing; Technical and Scientific; Textiles; Transportation.

OFFICEMATES5 OF ST. LOUIS (DOWNTOWN)
211 North Broadway, Suite 2360, St. Louis MO 63102. 314/241-5866. **Contact:** Carol Zagarri, General Manager. **Description:** An executive search firm. **Specializes in the areas of:** Accounting/Auditing; Administration/MIS/EDP; Advertising; Architecture/Construction/Real Estate; Banking; Chemical; Communications; Computer Hardware/Software; Design; Electrical; Engineering; Finance; Food Industry; General Management; Health/Medical; Industrial; Insurance; Legal; Manufacturing; Operations Management; Personnel/Labor Relations; Printing/Publishing; Procurement; Sales and Marketing; Technical and Scientific; Textiles; Transportation.

RAICHE & ASSOCIATES, INC.
11021 Natural Bridge Road, Bridgeton MO 63044-2317. 314/895-4554. **Contact:** Human Resources. **Description:** An executive search firm. Company pays fee. **Specializes in the areas of:** Computer Hardware/Software. **Positions commonly filled include:** Computer Programmer; General Manager; Systems Analyst. **Number of placements per year:** 1 - 49.

THE RIVER BEND GROUP
36 Four Seasons Shopping Center, Suite 315, Chesterfield MO 63017. 314/579-9729. **Fax:** 314/469-8592. **Contact:** John M. Sroka, Owner. **E-mail address:** jsroka@aol.com. **Description:** An executive search firm that specializes in placing technical professionals. Company pays fee. **Specializes in the areas of:** Computer Science/Software; Technical and Scientific. **Positions commonly filled include:** Computer Programmer; Internet Services Manager; MIS Specialist; Software Engineer; Systems Analyst. **Corporate headquarters location:** This Location. **Average salary range of placements:** More than $50,000. **Number of placements per year:** 50 - 99.

TECHNICAL & MEDICAL RESOURCES, INC.
708 West R.D. Mize Road, Suite E, Blue Springs MO 64015. 816/224-1938. **Toll-free phone:** 800/211-2633. **Fax:** 816/224-1936. **Contact:** Sheri Scott, CPC, Recruiter. **Description:** An executive search firm and employment agency, operating on a retainer and contingency basis. Founded in 1991. Company pays fee. **Specializes in the areas of:** Administration/MIS/EDP; Computer Science/Software; Health/Medical. **Positions commonly filled include:** Administrative Services Manager; Dental Lab Technician; Dentist; Dietician/Nutritionist; Emergency Medical Technician; Health Services Manager; Licensed Practical Nurse; Management Analyst/Consultant; Medical Records Technician; Occupational Therapist; Pharmacist; Physical Therapist; Physician; Registered Nurse; Respiratory Therapist; Social Worker; Speech-Language Pathologist; Strategic Relations Manager; Surgical Technician; Systems Analyst. **Corporate headquarters location:** This Location. **Average salary range of placements:** More than $50,000. **Number of placements per year:** 1 - 49.

TECHNICAL RESOURCES INTERNATIONAL
968 Chestnut Ridge Road, Ballwin MO 63021. 314/861-2059. **Toll-free phone:** 800/549-2059. **Fax:** 314/861-2269. **Contact:** Tony Montane, President. **Description:** An executive search firm. The firm also serves as a temporary agency for existing clients. Company pays fee. **Specializes in the areas of:** Computer Science/Software; Engineering; MIS. **Positions commonly filled include:** Computer Programmer; Design Engineer; Electrical/Electronics Engineer; Industrial Engineer; Mechanical Engineer; MIS Specialist; Software Engineer; Structural Engineer; Systems Analyst. **Benefits available to temporary workers:** Medical Insurance; Paid Vacation. **Corporate headquarters location:** This Location. **Average salary range of placements:** $20,000 - $50,000. **Number of placements per year:** 1 - 49.

PERMANENT EMPLOYMENT AGENCIES OF MISSOURI

HDB INC.
180 South Weidman Road, Suite 213, St. Louis MO 63021. 314/391-7799. **Fax:** 314/391-1224. **Contact:** Kathryn Davis-Wolfe, President. **E-mail address:** kwolfe@hdbinc.com. **Description:** An employment agency. Company pays fee. **Specializes in the areas of:** Administration/MIS/EDP; Computer Science/Software; Consulting. **Positions commonly filled include:** Computer Programmer; Systems Analyst. **Number of placements per year:** 50 - 99.

JACKSON EMPLOYMENT AGENCY
3450 Prospect Avenue, Kansas City MO 64128. 816/921-0181. **Contact:** Charles H. Jackson Sr., President. **Description:** An employment agency. Founded in 1969. **Specializes in the areas of:**

Biology; Computer Science/Software; Engineering; Finance; Food Industry; Personnel/Labor Relations. **Positions commonly filled include:** Accountant/Auditor; Aerospace Engineer; Agricultural Engineer; Agricultural Scientist; Architect; Attorney; Biological Scientist; Biomedical Engineer; Chemical Engineer; Chemist; Civil Engineer; Computer Programmer; Cost Estimator; Credit Manager; Electrical/Electronics Engineer; Electrician; Financial Analyst; Food Scientist/ Technologist; Human Resources Manager; Industrial Engineer; Mathematician; Mechanical Engineer; Metallurgical Engineer; Meteorologist; Mining Engineer; Nuclear Engineer; Petroleum Engineer; Purchasing Agent/Manager; Software Engineer; Stationary Engineer; Statistician; Structural Engineer; Systems Analyst; Telecommunications Manager; Transportation/Traffic Specialist. **Number of placements per year:** 50 - 99.

KENDALL & DAVIS COMPANY

11325 Concord Village Avenue, St. Louis MO 63123. 314/843-8838. **Toll-free phone:** 800/950-1551. **Contact:** Administrative Assistant. **Description:** An employment agency. **Specializes in the areas of:** Biotechnology; Computer Science/Software; Health/Medical.

KENNISON & ASSOCIATES INC.

3101 Broadway, Suite 280, Kansas City MO 64111. 816/753-4401. **Fax:** 816/753-3430. **Contact:** Gary S. Fawkes, Managing Partner. **Description:** An employment agency. Company pays fee. **Specializes in the areas of:** Computer Science/Software; Health/Medical. **Positions commonly filled include:** Computer Programmer; Management Analyst/Consultant; Occupational Therapist; Physical Therapist; Recreational Therapist; Respiratory Therapist; Speech-Language Pathologist; Systems Analyst; Technical Writer/Editor.

SNELLING PERSONNEL

16100 Chesterfield Parkway South, Suite 285, Chesterfield MO 63017. 314/532-1004. **Contact:** A.H. Harter, Jr., President. **Description:** An employment agency. Company pays fee. **Specializes in the areas of:** Accounting/Auditing; Administration/MIS/EDP; Advertising; Banking; Computer Hardware/Software; Engineering; Finance; Food Industry; General Management; Industrial; Insurance; Legal; Manufacturing; Retail; Sales and Marketing; Secretarial; Technical and Scientific. **Positions commonly filled include:** Accountant/Auditor; Administrative Assistant; Aerospace Engineer; Architect; Attorney; Biological Scientist; Biomedical Engineer; Bookkeeper; Buyer; Ceramics Engineer; Chemical Engineer; Chemist; Civil Engineer; Claim Representative; Clerk; Computer Operator; Computer Programmer; Credit Manager; Customer Service Representative; Data Entry Clerk; Draftsperson; EDP Specialist; Electrical/Electronics Engineer; Hotel Manager; Industrial Designer; Industrial Engineer; Legal Secretary; Manufacturing Engineer; Marketing Specialist; Mechanical Engineer; Medical Secretary; Metallurgical Engineer; MIS Specialist; Operations/Production Manager; Public Relations Specialist; Purchasing Agent/ Manager; Quality Control Supervisor; Receptionist; Recruiter; Sales Representative; Secretary; Software Engineer; Systems Analyst; Systems Engineer; Technical Illustrator; Technical Writer/ Editor; Technician; Typist/Word Processor; Underwriter/Assistant Underwriter. **Number of placements per year:** 500 - 999.

EXECUTIVE SEARCH FIRMS OF NEBRASKA

CHOICE ENTERPRISES

6675-A Sorenson Parkway, Omaha NE 68152. 402/571-8140. **Fax:** 402/571-5027. **Contact:** Jack L. Choice, Owner. **Description:** An executive search firm. Choice Enterprises operates through three divisions. Careers by Choice handles executive search services in data processing, engineering, and finance and accounting. CBC Temporaries is a temporary agency handling placements in the clerical, data entry, and word processing fields. CBC Construction Services offers general and electrical contracting services. **Specializes in the areas of:** Accounting/ Auditing; Administration/MIS/EDP; Computer Science/Software; Engineering; Finance;

Health/Medical; Manufacturing; Secretarial. **Average salary range of placements:** $30,000 -
$50,000. **Number of placements per year:** 1 - 49.

COMPUSEARCH OF LINCOLN
210 Gateway, Suite 434, Lincoln NE 68505-2438. 402/467-5549. **Fax:** 402/467-1150. **Contact:**
Bill Elam, Manager. **Description:** An executive search firm. **Specializes in the areas of:**
Accounting/Auditing; Administration/MIS/EDP; Advertising; Architecture/Construction/Real
Estate; Banking; Chemical; Communications; Computer Hardware/Software; Design; Electrical;
Engineering; Food Industry; General Management; Health/Medical; Insurance; Legal;
Manufacturing; Operations Management; Personnel/Labor Relations; Printing/Publishing;
Procurement; Retail; Sales and Marketing; Technical and Scientific; Textiles; Transportation.

COMPUSEARCH OF OMAHA
7171 Mercy Road, Suite 252, Omaha NE 68106. 402/397-8115. **Fax:** 402/397-6322. **Contact:** Les
Zanotti, Manager. **Description:** An executive search firm. **Specializes in the areas of:**
Accounting/Auditing; Administration/MIS/EDP; Advertising; Banking; Chemical;
Communications; Computer Hardware/Software; Design; Electrical; Engineering; Food Industry;
General Management; Health/Medical; Insurance; Legal; Manufacturing; Operations Management;
Personnel/Labor Relations; Printing/Publishing; Procurement; Retail; Sales and Marketing;
Technical and Scientific; Textiles; Transportation.

CORPORATE RECRUITERS, LTD.
202 South 71st Street, Omaha NE 68132. 402/393-5515. **Fax:** 402/393-4212. **Contact:** Linda
Malerbi, Vice President. **Description:** An executive search firm. Company pays fee. **Specializes in
the areas of:** Accounting/Auditing; Administration/MIS/EDP; Banking; Computer Science/
Software; Food Industry; General Management; Health/Medical; Industrial; Legal; Manufacturing;
Personnel/Labor Relations; Sales and Marketing; Secretarial; Transportation. **Number of
placements per year:** 50 - 99.

EXPRESS PERSONNEL
12119 Pacific Street, Omaha NE 68154. 402/333-5353. **Contact:** Larry Humberstone, Owner.
Description: An executive search firm that also operates as a temporary and employment agency.
Company pays fee. **Specializes in the areas of:** Administration/MIS/EDP; Computer
Science/Software; Engineering; Finance; Food Industry; General Management; Industrial;
Insurance; Manufacturing; Personnel/Labor Relations; Sales and Marketing; Secretarial; Technical
and Scientific; Transportation. **Positions commonly filled include:** Administrative Services
Manager; Agricultural Engineer; Attorney; Bank Officer/Manager; Branch Manager; Buyer;
Computer Programmer; Credit Manager; Dietician/Nutritionist; Electrical/Electronics Engineer;
Environmental Engineer; Financial Analyst; General Manager; Human Resources Specialist;
Management Trainee; MIS Specialist; Multimedia Designer; Operations/Production Manager;
Public Relations Specialist; Purchasing Agent/Manager; Quality Control Supervisor; Restaurant/
Food Service Manager; Software Engineer; Systems Analyst; Telecommunications Manager;
Transportation/Traffic Specialist; Travel Agent; Underwriter/Assistant Underwriter; Urban/
Regional Planner; Video Production Coordinator. **Average salary range of placements:** $30,000 -
$50,000. **Number of placements per year:** 100 - 199.

MANAGEMENT RECRUITERS OF LINCOLN
COMPUSEARCH OF LINCOLN
210 Gateway, Suite 434, Lincoln NE 68505-2438. 402/467-5534. **Contact:** Manager. **Description:**
An executive search firm. **Specializes in the areas of:** Accounting/Auditing; Administration/
MIS/EDP; Advertising; Banking; Chemical; Communications; Computer Hardware/Software;
Design; Electrical; Engineering; Food Industry; General Management; Health/Medical; Insurance;
Legal; Manufacturing; Operations Management; Personnel/Labor Relations; Printing/Publishing;
Procurement; Retail; Sales and Marketing; Technical and Scientific; Textiles; Transportation.

MANAGEMENT RECRUITERS OF OMAHA
OFFICEMATES5 OF OMAHA
7171 West Mercy Road, Suite 252, Omaha NE 68106. 402/397-8320. **Fax:** 402/397-6322.

Contact: Les Zanotti, Manager. **Description:** An executive search firm. **Specializes in the areas of:** Accounting/Auditing; Administration/MIS/EDP; Advertising; Architecture/Construction/Real Estate; Banking; Chemical; Communications; Computer Hardware/Software; Design; Electrical; Engineering; Food Industry; General Management; Health/Medical; Insurance; Legal; Manufacturing; Operations Management; Personnel/Labor Relations; Printing/Publishing; Procurement; Retail; Sales and Marketing; Technical and Scientific; Textiles; Transportation.

NOLL HUMAN RESOURCE SERVICES
2120 South 72nd Street, Suite 900, Omaha NE 68124. 402/391-7736. **Toll-free phone:** 800/536-7600. **Fax:** 402/391-6748. **Contact:** Recruiter. **Description:** An executive search firm. **Specializes in the areas of:** Administration/MIS/EDP; Banking; Computer Science/Software; General Management; Health/Medical; Secretarial; Transportation. **Positions commonly filled include:** Actuary; Bank Officer/Manager; Clerical Supervisor; Computer Programmer; Financial Analyst; General Manager; Health Services Manager; Licensed Practical Nurse; Management Analyst/Consultant; Occupational Therapist; Operations/Production Manager; Paralegal; Physical Therapist; Physician; Radiologic Technologist; Registered Nurse; Respiratory Therapist; Surgical Technician; Transportation/Traffic Specialist. **Other U.S. locations:** Dallas TX. **Number of placements per year:** 200 - 499.

PROFESSIONS
501 Olson Drive, Suite 2, Papillion NE 68046-5752. 402/331-6440. **Fax:** 402/331-8826. **Contact:** David Hawkins, President. **E-mail address:** jobsarus@neonramp.com. **World Wide Web address:** http://www.jobsarus.com. **Description:** An executive search firm that also operates as an employment agency. Company pays fee. **Specializes in the areas of:** Accounting/Auditing; Banking; Biology; Computer Hardware/Software; Engineering; Finance; Food Industry; Health/Medical; Insurance; Manufacturing; Personnel/Labor Relations; Sales and Marketing; Technical and Scientific; Transportation. **Average salary range of placements:** $30,000 - $50,000. **Number of placements per year:** 50 - 99.

RECRUITERS INTERNATIONAL, INC.
11330 Q Street, Omaha NE 68131. 402/339-9839. **Contact:** Kenneth H. Mertins, President. **Description:** An executive search firm operating on a retained and contingency basis. The firm also operates as an employment agency. Company pays fee. **Specializes in the areas of:** Accounting/Auditing; Administration/MIS/EDP; Advertising; Architecture/Construction/Real Estate; Banking; Computer Science/Software; Engineering; Finance; Food Industry; General Management; Health/Medical; Industrial; Insurance; Legal; Manufacturing; Personnel/Labor Relations; Printing/Publishing; Retail; Sales and Marketing; Technical and Scientific; Transportation. **Other U.S. locations:** Nationwide.

THE REGENCY GROUP, LTD.
EXECUTIVE SEARCH AND RECRUITING
256 North 115th Street, Suite 1, Omaha NE 68154-2521. 402/334-7255. **Fax:** 402/334-7148. **Contact:** Dan J. Barrow, CPC, General Manager. **E-mail address:** regencygrp@aol.com. **World Wide Web address:** http://www.gonix.com/regency/. **Description:** An executive search firm. Company pays fee. **Specializes in the areas of:** Administration/MIS/EDP; Computer Science/Software; Telecommunications. **Positions commonly filled include:** Computer Programmer; Design Engineer; Human Resources Manager; MIS Specialist; Multimedia Designer; Operations/Production Manager; Software Engineer; Systems Analyst; Technical Writer/Editor; Telecommunications Manager. **Average salary range of placements:** $30,000 - $50,000. **Number of placements per year:** 50 - 99.

RELIABLE NATIONAL PERSONNEL CONSULTANTS
Atrium Plaza, 11318 Davenport Street, Omaha NE 68154. 402/330-2814. **Fax:** 402/330-8164. **Contact:** Harlan Rohmberg, CPC, President. **Description:** A national recruiting firm with 300 affiliated offices. Company pays fee. **Specializes in the areas of:** Architecture/Construction/Real Estate; Computer Hardware/Software; Executives; Finance. **Positions commonly filled include:** Accountant/Auditor; Administrator; Architect; Data Processor; Engineer; Management. **Number of placements per year:** 50 - 99.

PERMANENT EMPLOYMENT AGENCIES OF NEBRASKA

EGGERS COMPANY
11272 Elm Street, Omaha NE 68144. 402/333-3480. **Fax:** 402/333-9759. **Contact:** James W. Eggers, CPC, President. **Description:** An employment agency. **Specializes in the areas of:** Accounting/Auditing; Banking; Computer Science/Software; Insurance; Manufacturing; Retail; Sales and Marketing.

HANSEN AGRI-PLACEMENT
P.O. Box 1172, Grand Island NE 68802. 308/382-7350. **Contact:** Jack Hansen, Owner. **Description:** Hansen Agri-Placement is an employment agency. Company pays fee. **Specializes in the areas of:** Accounting/Auditing; Agri-Business; Banking; Computer Hardware/Software; Engineering; Manufacturing; MIS/EDP; Sales and Marketing. **Positions commonly filled include:** Accountant/Auditor; Actuary; Advertising Clerk; Agricultural Engineer; Architect; Attorney; Bank Officer/Manager; Bookkeeper; Buyer; Chemist; Civil Engineer; Claim Representative; Clerk; Commercial Artist; Computer Operator; Computer Programmer; Credit Manager; Customer Service Representative; Data Entry Clerk; Draftsperson; EDP Specialist; Electrical/Electronics Engineer; Financial Analyst; General Manager; Hotel Manager; Human Resources Manager; Industrial Engineer; Insurance Agent/Broker; Legal Secretary; Manufacturing Engineer; Marketing Specialist; Mechanical Engineer; Medical Secretary; MIS Specialist; Operations/Production Manager; Purchasing Agent/Manager; Quality Control Supervisor; Receptionist; Sales Representative; Secretary; Stenographer; Systems Analyst; Technician; Typist/Word Processor. **Number of placements per year:** 1000+.

PROFESSIONAL RECRUITERS INC.
P.O. Box 24227, Omaha NE 68124. 402/397-2885. **Fax:** 402/397-7357. **Contact:** Wayne L. Smith, President. **Description:** An employment agency. Company pays fee. **NOTE:** The company's physical address is 7253 Grover Street, Omaha NE. **Specializes in the areas of:** Computer Science/Software; Engineering; Manufacturing; Personnel/Labor Relations. **Positions commonly filled include:** Agricultural Engineer; Ceramics Engineer; Chemical Engineer; Computer Programmer; Electrical/Electronics Engineer; Human Resources Manager; Industrial Engineer; Industrial Production Manager; Materials Engineer; Mechanical Engineer; Metallurgical Engineer; Purchasing Agent/Manager; Quality Control Supervisor; Software Engineer; Statistician; Systems Analyst. **Number of placements per year:** 200 - 499.

EXECUTIVE SEARCH FIRMS OF NEVADA

ACUMEN PERSONNEL
2909 West Charleston Boulevard, Las Vegas NV 89102. 702/877-6775. **Fax:** 702/878-9297. **Contact:** Lynn Murray, Owner. **Description:** Fills both temporary and permanent positions. Acumen offers permanent placement in the following fields: accounting, business management, computers, and more. **Specializes in the areas of:** Accounting/Auditing; Computer Hardware/Software.

MANAGEMENT RECRUITERS OF RENO
1025 Ridgeview Drive, Suite 100, Reno NV 89509. 702/826-5243. **Fax:** 702/826-8329. **Contact:** J. Edward Trapp, Owner/Manager. **Description:** An executive search firm operating on a contingency basis. A franchised office of Management Recruiters International. Founded in 1979. Company pays fee. **Specializes in the areas of:** Accounting/Auditing; Administration/MIS/EDP; Computer Hardware/Software; Engineering; Health/Medical; Insurance; Manufacturing; Sales and

Marketing; Transportation. **Positions commonly filled include:** Accountant/Auditor; Actuary; Buyer; Chemical Engineer; Computer Programmer; Cost Estimator; EEG Technologist; EKG Technician; Electrical/Electronics Engineer; General Manager; Hotel Manager; Mathematician; Mechanical Engineer; MIS Specialist; Operations/Production Manager; Pharmacist; Physician; Software Engineer; Systems Analyst; Underwriter/Assistant Underwriter. **Corporate headquarters location:** This Location. **Average salary range of placements:** $30,000 - $50,000. **Number of placements per year:** 1 - 49.

EXECUTIVE SEARCH FIRMS OF NEW HAMPSHIRE

BARCLAY PERSONNEL SYSTEMS, INC.
One Executive Park Drive, Bedford NH 03110. 603/669-2011. **Contact:** Human Resources Manager. **Description:** An executive search firm. **Specializes in the areas of:** Accounting/ Auditing; Administration/MIS/EDP; Automation/Robotics; Banking; Computer Hardware/ Software; Engineering; Finance; Food Industry; Manufacturing; Real Estate; Sales and Marketing; Technical and Scientific. **Number of placements per year:** 50 - 99.

CHAUCER GROUP
55 Morrill Road, Canterbury NH 03224. 603/783-9500. **Fax:** 603/783-9229. **Contact:** Robert L. Harmon, CPC/Manager. **Description:** An executive search firm. Company pays fee. **Specializes in the areas of:** Computer Science/Software; Engineering; Manufacturing; Technical and Scientific. **Positions commonly filled include:** Chemical Engineer; Electrical/Electronics Engineer; Industrial Engineer; Mechanical Engineer; Software Engineer. **Number of placements per year:** 50 - 99.

DUBOIS & COMPANY
30 Centre Road, Somersworth NH 03878. 603/692-4471. **Fax:** 603/692-6015. **Contact:** Paulette Dubois, Principal/Owner. **Description:** An executive search firm operating on both retainer and contingency bases. **Specializes in the areas of:** Accounting/Auditing; Administration/MIS/EDP; Banking; Biology; Computer Science/Software; Engineering; General Management; Health/ Medical; Manufacturing; Personnel/Labor Relations; Printing/Publishing; Retail; Sales and Marketing; Technical and Scientific. **Positions commonly filled include:** Accountant/Auditor; Administrative Services Manager; Aerospace Engineer; Architect; Bank Officer/Manager; Biochemist; Biological Scientist; Biomedical Engineer; Budget Analyst; Buyer; Chemical Engineer; Chemist; Civil Engineer; Clerical Supervisor; Clinical Lab Technician; Computer Programmer; Credit Manager; Customer Service Representative; Design Engineer; Designer; Economist; Education Administrator; Electrical/Electronics Engineer; Electrician; Emergency Medical Technician; Environmental Engineer; Financial Analyst; Food Scientist/Technologist; General Manager; Health Services Manager; Hotel Manager; Human Resources Specialist; Industrial Engineer; Industrial Production Manager; Internet Services Manager; Manufacturer's/ Wholesaler's Sales Rep.; Market Research Analyst; Mathematician; Mechanical Engineer; Medical Records Technician; Metallurgical Engineer; MIS Specialist; Multimedia Designer; Nuclear Engineer; Operations/Production Manager; Public Relations Specialist; Purchasing Agent/ Manager; Quality Control Supervisor; Science Technologist; Securities Sales Representative; Services Sales Representative; Software Engineer; Stationary Engineer; Statistician; Strategic Relations Manager; Structural Engineer; Systems Analyst; Telecommunications Manager; Transportation/Traffic Specialist. **Corporate headquarters location:** This Location. **Average salary range of placements:** More than $50,000. **Number of placements per year:** 1 - 49.

ENTERPRISE TECHNOLOGIES
198 Main Street, Salem NH 03079. 603/890-3700. **Fax:** 603/890-8701. **Contact:** Linda Bonvie, Executive Director. **E-mail address:** ent@inc-net.com. **Description:** An executive search firm. **Specializes in the areas of:** Computer Science/Software; Engineering; Manufacturing;

Personnel/Labor Relations; Technical and Scientific. **Positions commonly filled include:** Computer Programmer; Design Engineer; Electrical/Electronics Engineer; MIS Specialist; Quality Control Supervisor; Software Engineer; Systems Analyst; Technical Writer/Editor; Telecommunications Manager. **Average salary range of placements:** More than $50,000. **Number of placements per year:** 1 - 49.

LLOYD PERSONNEL CONSULTANTS
7 Medallion Center, Merrimack NH 03054. 603/424-0020. **Fax:** 603/424-8207. **Contact:** Paul D. Smith, President. **Description:** An executive search firm operating on both retainer and contingency bases. Company pays fee. **Specializes in the areas of:** Computer Science/Software; Manufacturing; Personnel/Labor Relations; Sales and Marketing; Technical and Scientific. **Positions commonly filled include:** Biochemist; Biological Scientist; Biomedical Engineer; Customer Service Representative; General Manager; Human Resources Specialist; Internet Services Manager; Operations/Production Manager; Public Relations Specialist; Quality Control Supervisor; Services Sales Representative; Software Engineer; Systems Analyst. **Other U.S. locations:** NJ; NY. **Average salary range of placements:** More than $50,000. **Number of placements per year:** 50 - 99.

MANAGEMENT RECRUITERS INTERNATIONAL OF BEDFORD
Cold Stream Office Park, 116-C South River Road, Bedford NH 03110. 603/669-9800. **Fax:** 603/623-8609. **Contact:** Mike Bacon, Manager. **Description:** An executive search firm. **Specializes in the areas of:** Accounting/Auditing; Administration/MIS/EDP; Advertising; Architecture/Construction/Real Estate; Banking; Communications; Computer Hardware/Software; Construction; Electrical; Engineering; Finance; Food Industry; General Management; Health/Medical; Insurance; Legal; Manufacturing; Operations Management; Personnel/Labor Relations; Printing/Publishing; Procurement; Retail; Sales and Marketing; Technical and Scientific; Transportation.

PREFERRED RESOURCES GROUP
P.O. Box 6370, Nashua NH 03063. 603/889-0112. **Fax:** 603/598-4915. **Contact:** Thomas Shiber, Principal. **Description:** An executive search firm operating on both retainer and contingency bases. Company pays fee. **Specializes in the areas of:** Administration/MIS/EDP; Computer Science/Software; Engineering; General Management; Industrial; Manufacturing; Technical and Scientific. **Positions commonly filled include:** Buyer; Chemical Engineer; Chemist; Computer Programmer; Design Engineer; Electrical/Electronics Engineer; General Manager; Industrial Engineer; Industrial Production Manager; Internet Services Manager; Mechanical Engineer; Metallurgical Engineer; MIS Specialist; Multimedia Designer; Operations/Production Manager; Purchasing Agent/Manager; Quality Control Supervisor; Software Engineer; Structural Engineer; Systems Analyst; Technical Writer/Editor; Telecommunications Manager. **Corporate headquarters location:** This Location. **Other U.S. locations:** Nationwide. **Average salary range of placements:** More than $50,000. **Number of placements per year:** 1 - 49.

R.G.T. ASSOCIATES
P.O. Box 1032, Portsmouth NH 03802-1032. 603/431-9500. **Fax:** 603/431-6984. **Contact:** Bob Thiboutot, CPC, President. **Description:** An executive search firm. Company pays fee. **Specializes in the areas of:** Accounting/Auditing; Administration/MIS/EDP; Banking; Clerical; Computer Hardware/Software; Engineering; Finance; Manufacturing; Sales and Marketing; Technical and Scientific. **Positions commonly filled include:** Accountant/Auditor; Administrative Assistant; Aerospace Engineer; Bank Officer/Manager; Bookkeeper; Computer Operator; Computer Programmer; Credit Manager; Data Entry Clerk; Draftsperson; EDP Specialist; Electrical/Electronics Engineer; Financial Analyst; General Manager; Industrial Engineer; Legal Secretary; Marketing Specialist; Mechanical Engineer; Medical Secretary; MIS Specialist; Operations/Production Manager; Purchasing Agent/Manager; Quality Control Supervisor; Receptionist; Sales Representative; Secretary; Systems Analyst; Typist/Word Processor. **Average salary range of placements:** More than $50,000. **Number of placements per year:** 1 - 49.

SALES CONSULTANTS OF NASHUA-MANCHESTER
Six Medallion Center, Merrimack NH 03054. 603/424-3282. **Fax:** 603/424-3286. **Contact:** Sheldon Baron, Manager/Owner. **Description:** An executive search firm that also operates as an

employment agency. Company pays fee. **Specializes in the areas of:** Advertising; Banking; Biology; Computer Science/Software; Engineering; Fashion; Finance; Health/Medical; Industrial; Insurance; Operations Management; Procurement; Retail; Sales and Marketing; Technical and Scientific. **Positions commonly filled include:** Bank Officer/Manager; Biological Scientist; Branch Manager; Chemical Engineer; Chemist; Design Engineer; Dietician/Nutritionist; Electrical/ Electronics Engineer; Environmental Engineer; Financial Analyst; Health Services Manager; Industrial Engineer; Industrial Production Manager; Insurance Agent/Broker; Mechanical Engineer; Pharmacist; Physical Therapist; Physician; Registered Nurse; Respiratory Therapist; Science Technologist; Software Engineer; Telecommunications Manager. **Average salary range of placements:** $30,000 - $50,000. **Number of placements per year:** 50 - 99.

PERMANENT EMPLOYMENT AGENCIES OF NEW HAMPSHIRE

ACCESS CONSULTING, INC.
54A Oyster River Road, Durham NH 03824-3029. 603/868-7884. **Contact:** Consultant. **Description:** An employment agency. Company pays fee. **Specializes in the areas of:** Administration/MIS/EDP; Communications; Computer Operations; Computer Science/Software; Data Processing; Engineering; Technical and Scientific. **Positions commonly filled include:** Computer Programmer; Designer; Electrical/Electronics Engineer; Management Analyst/ Consultant; Software Engineer; Systems Analyst; Technical Writer/Editor. **Number of placements per year:** 50 - 99.

BARROS ASSOCIATES
199 Wheelwright Road, Hampstead NH 03841. 603/329-1100. **Fax:** 603/329-1111. **Contact:** Daniel Barros, President. **Description:** An employment agency. Company pays fee. **Specializes in the areas of:** Communications; Computer Science/Software; Network Administration. **Positions commonly filled include:** Software Engineer. **Number of placements per year:** 1 - 49.

CENTRAL NEW HAMPSHIRE EMPLOYMENT SERVICES
67 Water Street, Room 210, Laconia NH 03246. 603/528-2828. **Toll-free phone:** 800/256-2482. **Fax:** 603/528-6625. **Contact:** Abraham Dadian, President. **Description:** An employment agency. Company pays fee. **Specializes in the areas of:** Accounting/Auditing; Administration/MIS/EDP; Banking; Computer Science/Software; Engineering; Finance; Health/Medical; Insurance; Legal; Manufacturing; Personnel/Labor Relations; Sales and Marketing; Secretarial. **Positions commonly filled include:** Accountant/Auditor; Aerospace Engineer; Architect; Bank Officer/Manager; Bookkeeper; Civil Engineer; Clerk; Computer Programmer; Credit Manager; Customer Service Representative; Data Entry Clerk; Draftsperson; EDP Specialist; Electrical/Electronics Engineer; Hotel Manager; Human Resources Manager; Industrial Designer; Legal Secretary; Mechanical Engineer; Medical Secretary; Metallurgical Engineer; Receptionist; Sales Representative; Secretary; Systems Analyst; Typist/Word Processor; Underwriter/Assistant Underwriter. **Corporate headquarters location:** This Location. **Other U.S. locations:** Concord NH. **Average salary range of placements:** $20,000 - $29,999. **Number of placements per year:** 500 - 999.

EXETER 2100
Computer Park, P.O. Box 2120, Hampton NH 03842. 603/926-6712. **Fax:** 603/926-0536. **Contact:** Bruce Montville, Managing Partner. **Description:** An employment agency. Company pays fee. **Specializes in the areas of:** Computer Science/Software; Information Systems. **Positions commonly filled include:** Computer Programmer; EDP Specialist; Software Engineer.

KENDA SYSTEMS
One Stiles Road, Suite 106, Salem NH 03079. 603/898-7884. **Contact:** Manager. **Description:** An employment agency. Kenda Systems focuses on the placement of computer consultants and

software engineers on a permanent and temporary basis. **Specializes in the areas of:** Computer Science/Software.

NATIONAL EMPLOYMENT SERVICE CORPORATION

95 Albany Street, Suite 3, Portsmouth NH 03801. **Toll-free phone:** 800/EMPLOYMENT. **Fax:** 603/427-1411. **Contact:** Contract Manager. **World Wide Web address:** http://www. nationalemployment.com. **Description:** An employment agency focusing on permanent and temporary placements in the areas of engineering, finance, and information technologies. Founded in 1992. Company pays fee. **Specializes in the areas of:** Accounting/Auditing; Administration/ MIS/EDP; Architecture/Construction/Real Estate; Banking; Computer Science/Software; Economics; Engineering; Finance; Food Industry; General Management; Health/Medical; Industrial; Manufacturing; Personnel/Labor Relations; Retail; Sales and Marketing; Transportation. **Corporate headquarters location:** This Location. **Other U.S. locations:** Nashua NH. **Average salary range of placements:** $30,000 - $50,000. **Number of placements per year:** 1000+.

PROFESSIONAL RECRUITERS INC.

5 Coliseum Avenue, Nashua NH 03063. 603/886-3909x316. **Fax:** 603/886-4205. **Recorded jobline:** 603/886-3909x304. **Contact:** Timothy E. Moran, CPC/CEO. **Description:** An employment agency. Company pays fee. **Specializes in the areas of:** Accounting/Auditing; Banking; Computer Science/Software; Network Administration; Personnel/Labor Relations; Technical and Scientific. **Positions commonly filled include:** Accountant/Auditor; Bank Officer/ Manager; Branch Manager; Budget Analyst; Computer Programmer; Credit Manager; Financial Analyst; Software Engineer; Systems Analyst. **Average salary range of placements:** More than $50,000. **Number of placements per year:** 1000+.

RESOURCE RECRUITING
CONTEMPORARY ACCOUNTANTS

400 Amherst Street, Nashua NH 03063. 603/595-2822. **Fax:** 603/889-0259. **Contact:** Robert C. Harrington, Executive Vice President. **Description:** An employment agency. Founded in 1992. Company pays fee. **Specializes in the areas of:** Accounting/Auditing; Administration/MIS/EDP; Banking; Computer Science/Software; Legal; Manufacturing; Operations Management; Sales and Marketing; Secretarial. **Positions commonly filled include:** Accountant/Auditor; Bookkeeper; Chief Financial Officer; Clerk; Collections Agent; Controller; Credit Manager; Manager of Information Systems; Payroll Clerk; Secretary. **Average salary range of placements:** $30,000 - $50,000. **Number of placements per year:** 50 - 99.

SOFTWARE NETWORKS INC.

125 Main Street, Suite A, Newmarket NH 03857. 603/659-1000. **Fax:** 603/359-1005. **Contact:** Doug Cotton, Recruitment Manager. **Description:** An employment agency that focuses on permanent placements in the software development and communications industries. Founded in 1990. Company pays fee. **Specializes in the areas of:** Computer Science/Software. **Positions commonly filled include:** Computer Programmer; Design Engineer; Electrical/Electronics Engineer; Software Engineer. **Average salary range of placements:** More than $50,000. **Number of placements per year:** 50 - 99.

TECH/AID OF NEW HAMPSHIRE

71 Spit Brook Road, Suite 102, Nashua NH 03060. 603/891-4100. **Contact:** Recruiter. **Description:** An employment agency. Company pays fee. **Specializes in the areas of:** Architecture/Construction/Real Estate; Cable TV; Computer Hardware/Software; Construction; Engineering; Manufacturing; Technical and Scientific. **Positions commonly filled include:** Aerospace Engineer; Architectural Engineer; Buyer; Ceramics Engineer; Chemical Engineer; Civil Engineer; Draftsperson; Electrical/Electronics Engineer; Estimator; Industrial Designer; Mechanical Engineer; Metallurgical Engineer; Mining Engineer; Operations/Production Manager; Petroleum Engineer; Purchasing Agent/Manager; Quality Control Supervisor; Technical Writer/Editor; Technician. **Number of placements per year:** 1000+.

TECHNICAL DIRECTIONS, INC. (TDI)

78 Northeastern Boulevard, Suite 2B, Nashua NH 03062. 603/880-6720. **Fax:** 603/880-7859. **Contact:** Bill Wright, Recruiter. **Description:** An employment agency. Company pays fee.

Specializes in the areas of: Administration/MIS/EDP; Computer Hardware/Software; Engineering; Sales and Marketing; Technical and Scientific. **Positions commonly filled include:** Aerospace Engineer; Biomedical Engineer; Computer Programmer; EDP Specialist; Electrical/Electronics Engineer; Industrial Engineer; Marketing Specialist; Mechanical Engineer; Metallurgical Engineer; MIS Specialist; Systems Analyst; Technical Writer/Editor.

EXECUTIVE SEARCH FIRMS OF NEW JERSEY

ALTA ASSOCIATES, INC.
Eight Bartles Corner Road, Suite 21, Flemington NJ 08822. 908/806-8442. **Fax:** 908/806-8443. **Contact:** Joyce Brocaglia, Vice President. **Description:** An executive search firm. Company pays fee. **Specializes in the areas of:** Computer Hardware/Software. **Positions commonly filled include:** Computer Programmer; Database Manager; EDP Specialist; MIS Specialist. **Number of placements per year:** 50 - 99.

BLAKE & ASSOCIATES EXECUTIVE SEARCH
213 East Verona Boulevard, Suite 200, Pleasantville NJ 08232. 609/645-3330. **Fax:** 609/383-0320. **Contact:** Ed Blake, President. **Description:** An executive search firm. Company pays fee. **Specializes in the areas of:** Accounting/Auditing; Administration/MIS/EDP; Advertising; Architecture/Construction/Real Estate; Art/Design; Banking; Biology; Computer Science/Software; Economics; Engineering; Finance; Food Industry; General Management; Health/Medical; Industrial; Insurance; Legal; Manufacturing; Personnel/Labor Relations; Printing/Publishing; Retail; Sales and Marketing; Secretarial. **Positions commonly filled include:** Accountant/Auditor; Actuary; Adjuster; Administrative Services Manager; Agricultural Engineer; Agricultural Scientist; Architect; Attorney; Bank Officer/Manager; Biological Scientist; Biomedical Engineer; Branch Manager; Brokerage Clerk; Budget Analyst; Buyer; Chemical Engineer; Chemist; Civil Engineer; Claim Representative; Clerical Supervisor; Collector; Computer Programmer; Construction and Building Inspector; Construction Contractor; Cost Estimator; Credit Manager; Customer Service Representative; Dental Lab Technician; Dentist; Dietician/Nutritionist; Draftsperson; EEG Technologist; EKG Technician; Electrical/Electronics Engineer; Electrician; Financial Analyst; Food Scientist/Technologist; General Manager; Geologist/Geophysicist; Health Services Manager; Human Resources Manager; Human Service Worker; Industrial Engineer; Industrial Production Manager; Insurance Agent/Broker; Investigator; Landscape Architect; Management Trainee; Manufacturer's/Wholesaler's Sales Rep.; Materials Engineer; Mechanical Engineer; Medical Records Technician; Metallurgical Engineer; Mining Engineer; Nuclear Engineer; Nuclear Medicine Technologist; Occupational Therapist; Paralegal; Petroleum Engineer; Physical Therapist; Property and Real Estate Manager; Public Relations Specialist; Purchasing Agent/ Manager; Quality Control Supervisor; Radiologic Technologist; Real Estate Agent; Recreational Therapist; Respiratory Therapist; Restaurant/Food Service Manager; Science Technologist; Securities Sales Representative; Services Sales Representative; Software Engineer; Speech-Language Pathologist; Stationary Engineer; Statistician; Structural Engineer; Surgical Technician; Surveyor; Systems Analyst; Technical Writer/Editor; Transportation/Traffic Specialist; Travel Agent; Underwriter/Assistant Underwriter; Urban/Regional Planner; Wholesale and Retail Buyer. **Number of placements per year:** 50 - 99.

CARTER McKENZIE INC.
200 Executive Drive, West Orange NJ 07052-3303. 201/736-7100. **Fax:** 201/736-9416. **Contact:** John Capo, Vice President. **E-mail address:** carter@media-ware.com. **World Wide Web address:** http://www.media-ware.com/carter. **Description:** An executive search firm operating on both retainer and contingency bases that specializes in the placement of MIS professionals. Company pays fee. **Specializes in the areas of:** Administration/MIS/EDP; Computer Science/Software. **Positions commonly filled include:** Computer Programmer; MIS Specialist; Software Engineer; Systems Analyst; Telecommunications Manager. **Corporate headquarters location:** This

Location. **Average salary range of placements:** More than $50,000. **Number of placements per year:** 100 - 199.

L. CAVALIERE & ASSOCIATES

2300 State Route 27, North Brunswick NJ 08902. 908/940-3100. **Fax:** 908/940-2266. **Contact:** Louis Cavaliere, Managing Director. **Description:** An executive search firm operating on both retainer and contingency bases and specializing in placing MIS professionals. Company pays fee. **Specializes in the areas of:** Computer Science/Software. **Positions commonly filled include:** Computer Programmer; Management Analyst/Consultant; Software Engineer; Systems Analyst. **Corporate headquarters location:** This Location. **Average salary range of placements:** More than $50,000. **Number of placements per year:** 1 - 49.

CHURCHILL & HARRIMAN, INC.

601 Ewing Street, Suite B7, Princeton NJ 08540. 609/921-3551. **Fax:** 609/921-1061. **Contact:** Kenneth J. Peterson, President. **Description:** An executive search firm that provides professional search and information technology solutions. The firm also provides per diem information technology consulting services. Company pays fee. **Specializes in the areas of:** Administration/MIS/EDP; Computer Science/Software. **Positions commonly filled include:** Computer Programmer; Systems Analyst. **Corporate headquarters location:** This Location. **Average salary range of placements:** More than $50,000. **Number of placements per year:** 50 - 99.

DATA HUNTERS, INC.

P.O. Box 884, Ramsey NJ 07446-0884. 201/825-1368. **Fax:** 201/327-4234. **Contact:** Bette Rosenfeld, President. **E-mail address:** datahunt@nis.net. **Description:** An executive search firm and employment agency that specializes in placing computer and data processing professionals. Company pays fee. **Specializes in the areas of:** Computer Science/Software; Data Processing. **Positions commonly filled include:** Computer Programmer; Internet Services Manager; MIS Specialist; Software Engineer; Systems Analyst; Telecommunications Manager. **Corporate headquarters location:** This Location. **Number of placements per year:** 1 - 49.

THE DATAFINDERS GROUP, INC.

25 East Spring Valley Avenue, Maywood NJ 07607. 201/845-7700. **Fax:** 201/845-7365. **Contact:** Thomas J. Credidio, Vice President. **Description:** An executive search firm. Company pays fee. **Specializes in the areas of:** Administration/MIS/EDP; Computer Science/Software; Sales and Marketing. **Positions commonly filled include:** Computer Programmer; EDP Specialist; Manufacturer's/Wholesaler's Sales Rep.; Services Sales Representative; Software Engineer; Systems Analyst. **Number of placements per year:** 200 - 499.

EXECUTIVE HEALTHCARE PLACEMENT, INC.

128 Sussex Way, Suite A, Jamesburg NJ 08831. 609/655-8686. **Toll-free phone:** 800/585-7001. **Fax:** 609/655-8081. **Contact:** Steven H. Horowitz, President. **Description:** An executive search firm. **Specializes in the areas of:** Administration/MIS/EDP; Computer Science/Software; Health/Medical; Personnel/Labor Relations. **Positions commonly filled include:** Computer Programmer; Financial Analyst; Health Services Manager; Human Resources Specialist; Management Analyst/Consultant; MIS Specialist; Occupational Therapist; Pharmacist; Physical Therapist; Physician; Quality Control Supervisor; Recreational Therapist; Registered Nurse; Respiratory Therapist; Speech-Language Pathologist; Surgical Technician; Systems Analyst. **Average salary range of placements:** More than $50,000. **Number of placements per year:** 1 - 49.

EXPRESS PERSONNEL SERVICES

2569 State Route 10, Morris Plains NJ 07950-3410. 201/898-1001. **Fax:** 201/898-1005. **Contact:** Nicole Simons, Personnel Supervisor. **Description:** An executive search firm. **Specializes in the areas of:** Accounting/Auditing; Administration/MIS/EDP; Advertising; Computer Hardware/Software; Engineering; General Management; Industrial; Legal; Manufacturing; Printing/Publishing; Sales and Marketing; Secretarial; Technical and Scientific. **Positions commonly filled include:** Accountant/Auditor; Administrative Services Manager; Advertising Clerk; Blue-Collar Worker Supervisor; Claim Representative; Clerical Supervisor; Computer Programmer; Customer Service Representative; Design Engineer; Designer; Draftsperson; Editor; Electrical/Electronics Engineer; Electrician; General Manager; Human Resources Specialist; Management Trainee;

Manufacturer's/Wholesaler's Sales Rep.; MIS Specialist; Paralegal; Quality Control Supervisor; Software Engineer; Typist/Word Processor. **Number of placements per year:** 200 - 499.

GIBSON MARTIN CONSULTING
694 Route 15 South, Suite 205B, Lake Hopatcong NJ 07849. 201/663-3300. **Fax:** 201/663-3316. **Contact:** Robert Lee, Principal. **Description:** An executive search firm that specializes in placing accounting, finance, and computer professionals. Company pays fee. **Specializes in the areas of:** Accounting/Auditing; Administration/MIS/EDP; Computer Science/Software; Finance. **Positions commonly filled include:** Accountant/Auditor; Architect; Budget Analyst; Chemist; Computer Programmer; Financial Analyst; Systems Analyst.

HOLM PERSONNEL CONSULTANTS
333 Route 46 West, Suite 202, Fairfield NJ 07004. 201/808-1933. **Contact:** Personnel. **Description:** An executive search firm that also operates as an employment agency. Company pays fee. **Specializes in the areas of:** Accounting/Auditing; Administration/MIS/EDP; Computer Science/Software; General Management; Personnel/Labor Relations; Sales and Marketing; Secretarial. **Positions commonly filled include:** Accountant/Auditor; Administrative Services Manager; Budget Analyst; Buyer; Computer Programmer; Credit Manager; Financial Analyst; Human Resources Specialist; Management Trainee; MIS Specialist; Purchasing Agent/Manager; Software Engineer; Systems Analyst; Telecommunications Manager; Transportation/Traffic Specialist. **Average salary range of placements:** $30,000 - $50,000. **Number of placements per year:** 50 - 99.

HRESHKO CONSULTING GROUP
850 U.S. Highway 1, North Brunswick NJ 08902. 908/545-9000. **Fax:** 908/545-0800. **Contact:** Frank Hreshko, Managing Director. **Description:** An executive search firm. Company pays fee. **Specializes in the areas of:** Accounting/Auditing; Banking; Computer Science/Software; Finance; General Management; Insurance; Manufacturing; Personnel/Labor Relations; Sales and Marketing. **Positions commonly filled include:** Accountant/Auditor; Computer Programmer; Financial Analyst; General Manager; Human Resources Specialist; Management Analyst/Consultant; MIS Specialist; Software Engineer; Systems Analyst; Telecommunications Manager. **Corporate headquarters location:** This Location. **Average salary range of placements:** More than $50,000. **Number of placements per year:** 100 - 199.

J.M. JOSEPH ASSOCIATES
P.O. Box 104, High Bridge NJ 08829-0104. 908/638-6877. **Fax:** 908/638-8220. **Contact:** C. Russell Ditzel, Managing Director. **E-mail address:** cditzel@vcx.net. **Description:** An executive search firm. Company pays fee. **Specializes in the areas of:** Administration/MIS/EDP; Computer Science/Software; Engineering; Food Industry; Health/Medical; Manufacturing; Personnel/Labor Relations; Sales and Marketing; Technical and Scientific. **Positions commonly filled include:** Financial Analyst; Food Scientist/Technologist; Human Resources Specialist; Industrial Engineer; Industrial Production Manager; MIS Specialist; Operations/Production Manager; Strategic Relations Manager. **Average salary range of placements:** More than $50,000. **Number of placements per year:** 1 - 49.

T.J. KOELLHOFFER & ASSOCIATES
250 State Route 28, Suite 206, Bridgewater NJ 08807. 908/526-6880. **Fax:** 908/725-2653. **Contact:** Tom Koellhoffer, Principal. **E-mail address:** tomkoell@aol.com. **Description:** An executive search firm. Company pays fee. **Specializes in the areas of:** Broadcasting; Computer Science/Software; Engineering; Manufacturing; Technical and Scientific. **Positions commonly filled include:** Aerospace Engineer; Biochemist; Biological Scientist; Biomedical Engineer; Electrical/Electronics Engineer; Materials Engineer; Mechanical Engineer; Metallurgical Engineer; Multimedia Designer; Physicist/Astronomer; Software Engineer. **Average salary range of placements:** More than $50,000. **Number of placements per year:** 1 - 49.

PAUL KULL & COMPANY
121 Center Grove Road, Randolph NJ 07869. 201/361-7440. **Contact:** Paul Kull, Owner. **Description:** An executive search firm. Founded in 1966. Company pays fee. **Specializes in the areas of:** Computer Hardware/Software; Engineering; General Management; Manufacturing; Sales

and Marketing; Technical and Scientific. **Positions commonly filled include:** Aerospace Engineer; Biochemist; Biomedical Engineer; Chemical Engineer; Computer Programmer; Electrical/ Electronics Engineer; Industrial Engineer; Manufacturing Engineer; Marketing Specialist; Mechanical Engineer; Software Engineer. **Number of placements per year:** 1 - 49.

L&K ASSOCIATES
179 West Broadway, Salem NJ 08079-1328. 609/935-3070. **Contact:** Gene Lank, President. **Description:** An executive search firm placing 10 percent of its clients on a retainer basis and 90 percent of its clients on a contingency basis. Company pays fee. **Specializes in the areas of:** Computer Science/Software; Legal; Technical and Scientific; Telecommunications. **Positions commonly filled include:** Attorney; Computer Programmer; MIS Specialist; Systems Analyst; Telecommunications Manager. **Corporate headquarters location:** This Location. **Other U.S. locations:** Nationwide. **Average salary range of placements:** More than $50,000. **Number of placements per year:** 1 - 49.

LANCASTER ASSOCIATES
94 Grove Street, Somerville NJ 08876. 908/526-5440. **Fax:** 908/526-1992. **Contact:** Ray Lancaster, President. **Description:** A retainer and contingency search firm. Company pays fee. **Specializes in the areas of:** Computer Science/Software; Technical and Scientific. **Positions commonly filled include:** MIS Specialist; Telecommunications Manager. **Corporate headquarters location:** This Location. **Other U.S. locations:** Nationwide. **Average salary range of placements:** More than $50,000. **Number of placements per year:** 1 - 49.

JONATHAN LAWRENCE ASSOCIATES
103 Washington Street, Morristown NJ 07960. 201/285-1988. **Contact:** Manager. **Description:** An executive search firm. Company pays fee. **Specializes in the areas of:** Administration/MIS/EDP; Computer Science/Software.

MIS SEARCH
450 Harmon Meadow Boulevard, First Floor, Secaucus NJ 07094. 201/330-0080. **Fax:** 201/330-8729. **Contact:** Maryanne McGuire, Technical Recruiter. **Description:** An executive search firm. Company pays fee. **Specializes in the areas of:** Administration/MIS/EDP; Computer Science/Software. **Positions commonly filled include:** Computer Programmer; Systems Analyst.

MANAGEMENT RECRUITERS OF BAY HEAD
106 Bridge Avenue, Bay Head NJ 08742. 908/714-1300. **Fax:** 908/714-1311. **Contact:** Bob Ceresi, General Manager. **E-mail address:** mribayhead@aol.com. **Description:** An executive search firm. Company pays fee. **Specializes in the areas of:** Computer Science/Software; Electrical; Electronics; Engineering; Industrial; Manufacturing; Sales and Marketing; Technical and Scientific. **Positions commonly filled include:** Buyer; Chemical Engineer; Computer Programmer; Customer Service Representative; Draftsperson; Electrical/Electronics Engineer; Industrial Engineer; Industrial Production Manager; Management Trainee; Manufacturer's/ Wholesaler's Sales Rep.; Mechanical Engineer; Metallurgical Engineer; Purchasing Agent/ Manager; Quality Control Supervisor; Services Sales Representative; Software Engineer; Systems Analyst; Wholesale and Retail Buyer. **Average salary range of placements:** More than $50,000. **Number of placements per year:** 50 - 99.

MANAGEMENT RECRUITERS OF PASSAIC COUNTY
750 Hamburg Turnpike, Pompton Lakes NJ 07442. 201/831-7778. **Contact:** David Zawicki, Manager. **Description:** An executive search firm. **Specializes in the areas of:** Accounting/ Auditing; Administration/MIS/EDP; Advertising; Architecture/Construction/Real Estate; Banking; Communications; Computer Science/Software; Construction; Electrical; Engineering; Finance; Food Industry; General Management; Health/Medical; Personnel/Labor Relations; Printing/ Publishing; Procurement; Retail; Sales and Marketing; Technical and Scientific; Textiles; Transportation.

MANAGEMENT RECRUITERS OF SPARTA
191 Woodport Road, Suite 201, Sparta NJ 07871. 201/729-1888. **Fax:** 201/729-1620. **Contact:** Lance Incitti, Manager. **Description:** An executive search firm. **Specializes in the areas of:**

Accounting/Auditing; Administration/MIS/EDP; Advertising; Architecture/Construction/Real Estate; Banking; Communications; Computer Science/Software; Construction; Electrical; Engineering; Finance; Food Industry; General Management; Health/Medical; Industrial; Insurance; Legal; Manufacturing; Operations Management; Personnel/Labor Relations; Printing/Publishing; Procurement; Retail; Sales and Marketing; Technical and Scientific; Textiles; Transportation.

PERSONNEL ASSOCIATES INC.
239 U.S. Highway 22, Green Brook NJ 08812-1916. 908/968-8866. **Fax:** 908/968-9437. **Contact:** Thomas C. Wood, President. **Description:** An executive search firm operating on a contingency basis and specializing in the placement of programmer analysts, systems analysts, system administrators, and project managers. Company pays fee. **Specializes in the areas of:** Administration/MIS/EDP; Computer Science/Software. **Positions commonly filled include:** Computer Programmer; Internet Services Manager; MIS Specialist; Systems Analyst. **Corporate headquarters location:** This Location. **Other U.S. locations:** Nationwide. **Number of placements per year:** 50 - 99.

PRINCETON EXECUTIVE SEARCH
2667 Nottingham Way, Hamilton NJ 08619. 609/584-1100. **Fax:** 609/584-1141. **Contact:** Andrew B. Barkocy, CPC, President. **Description:** An executive search firm. Company pays fee. **Specializes in the areas of:** Accounting/Auditing; Administration/MIS/EDP; Banking; Computer Science/Software; Engineering; Film Production; Personnel/Labor Relations. **Positions commonly filled include:** Accountant/Auditor; Aerospace Engineer; Agricultural Engineer; Bank Officer/Manager; Biomedical Engineer; Budget Analyst; Chemical Engineer; Civil Engineer; Credit Manager; Electrical/Electronics Engineer; Financial Analyst; General Manager; Human Resources Manager; Industrial Engineer; Mechanical Engineer; Nuclear Engineer; Petroleum Engineer; Purchasing Agent/Manager; Software Engineer; Systems Analyst. **Number of placements per year:** 1 - 49.

PRO PLACEMENTS, INC.
1158 Raritan Road, Clark NJ 07066. 908/499-0999. **Fax:** 908/499-0707. **Contact:** Personnel. **Description:** An executive search firm. Company pays fee. **Specializes in the areas of:** Accounting/Auditing; Administration/MIS/EDP; Computer Science/Software; Engineering; Finance; General Management; Manufacturing; Personnel/Labor Relations; Sales and Marketing; Secretarial; Technical and Scientific. **Average salary range of placements:** $20,000 - $29,999. **Number of placements per year:** 1 - 49.

ROCHESTER SYSTEMS INC.
227 East Bergen Place, Red Bank NJ 07701. 908/747-7474. **Fax:** 908/747-7055. **Contact:** Peter Gotch, Account Executive. **Description:** An executive search firm. Company pays fee. **Specializes in the areas of:** Accounting/Auditing; Computer Science/Software; Logistics; Manufacturing; MIS/EDP. **Positions commonly filled include:** Computer Programmer; Financial Analyst; Industrial Production Manager; Management Analyst/Consultant; Manager of Information Systems; Operations/Production Manager; Software Engineer; Systems Analyst. **Number of placements per year:** 50 - 99.

R.S. SADOW ASSOCIATES
24 Heather Drive, Somerset NJ 08873. 908/545-4550. **Fax:** 908/545-0797. **Contact:** Ray Sadow, President. **Description:** An executive search firm operating on a contingency basis. Company pays fee. **Specializes in the areas of:** Accounting/Auditing; Administration/MIS/EDP; Banking; Computer Science/Software; Engineering; Finance; Industrial; Manufacturing; Printing/Publishing. **Positions commonly filled include:** Accountant/Auditor; Aerospace Engineer; Agricultural Engineer; Bank Officer/Manager; Biological Scientist; Biomedical Engineer; Budget Analyst; Ceramics Engineer; Chemical Engineer; Chemist; Civil Engineer; Clerical Supervisor; Computer Programmer; Credit Manager; Customer Service Representative; Draftsperson; Editor; Electrical/Electronics Engineer; Financial Analyst; Industrial Engineer; Industrial Production Manager; Materials Engineer; Mechanical Engineer; Metallurgical Engineer; Mining Engineer; Nuclear Engineer; Petroleum Engineer; Purchasing Agent/Manager; Quality Control Supervisor; Software Engineer; Stationary Engineer; Structural Engineer; Systems Analyst; Technical Writer/Editor.

Average salary range of placements: More than $50,000. **Number of placements per year:** 1 - 49.

SALES CONSULTANTS OF MORRIS COUNTY
364 Parsippany Road, Parsippany NJ 07054. 201/887-3838. **Fax:** 201/887-2304. **Contact:** Ernest Bivona, Manager. **E-mail address:** scmorris.manager@mrinet.com. **Description:** An executive search firm operating on a contingency basis. **Specializes in the areas of:** Computer Science/Software; Engineering; Finance; Food Industry; General Management; Health/Medical; Industrial; Insurance; Printing/Publishing; Sales and Marketing; Technical and Scientific. **Positions commonly filled include:** Bank Officer/Manager; Biological Scientist; Biomedical Engineer; Chemical Engineer; Civil Engineer; Mechanical Engineer; Metallurgical Engineer; Sales Representative; Software Engineer; Telecommunications Manager. **Other U.S. locations:** Nationwide. **Number of placements per year:** 50 - 99.

SALES CONSULTANTS OF SPARTA
376 Route 15, Suite 200, Sparta NJ 07871. 201/579-5555. **Fax:** 201/579-2220. **Contact:** Harvey Bass, Manager. **Description:** An executive search firm. **Specializes in the areas of:** Accounting/Auditing; Administration/MIS/EDP; Advertising; Architecture/Construction/Real Estate; Banking; Communications; Computer Science/Software; Design; Electrical; Engineering; Finance; Food Industry; General Management; Health/Medical; Industrial; Insurance; Legal; Manufacturing; Operations Management; Printing/Publishing; Procurement; Retail; Sales and Marketing; Technical and Scientific; Textiles; Transportation.

SELECT PROFESSIONALS, INC.
P.O. Box 580, Denville NJ 07834. 201/625-2856. **Fax:** 201/625-1779. **Contact:** Ray Farrant, Vice President. **Description:** Select Professionals, Inc. provides placements for technical professionals. **Specializes in the areas of:** Computer Science/Software; Engineering; Technical and Scientific. **Positions commonly filled include:** Aerospace Engineer; Computer Programmer; Electrical/ Electronics Engineer; MIS Specialist; Software Engineer; Systems Analyst; Technical Writer/ Editor. **Benefits available to temporary workers:** Bonuses. **Corporate headquarters location:** This Location. **Average salary range of placements:** $30,000 - $50,000. **Number of placements per year:** 1 - 49.

SKUPPSEARCH, INC.
580 Sylvan Avenue, Englewood Cliffs NJ 07632. 201/894-1824. **Fax:** 201/894-1324. **Contact:** Holly Skupp, President. **Description:** A placement firm specializing in technical communications occupations such as technical writers, documentation specialists, designers, developers, and trainers. Company pays fee. **Specializes in the areas of:** Banking; Computer Science/Software; Finance; Health/Medical; Insurance. **Positions commonly filled include:** Editor; Management Analyst/Consultant; Multimedia Designer; Systems Analyst; Technical Writer/Editor. **Average salary range of placements:** $30,000 - $50,000. **Number of placements per year:** 1 - 49.

TENEK CORPORATION
250 Carter Drive, Edison NJ 08817. 908/248-1600. **Fax:** 908/248-1665. **Contact:** Peter Scocchi, Vice President. **Description:** An executive search firm. Company pays fee. **Specializes in the areas of:** Administration/MIS/EDP; Computer Science/Software; Engineering; Government; Industrial; Manufacturing; Secretarial; Technical and Scientific; Telecommunications. **Positions commonly filled include:** Biomedical Engineer; Chemical Engineer; Civil Engineer; Electrical/ Electronics Engineer; Geologist/Geophysicist; Industrial Designer; Mechanical Engineer; Nuclear Engineer; Petroleum Engineer; Software Engineer; Stationary Engineer; Structural Engineer; Transportation/Traffic Specialist. **Number of placements per year:** 50 - 99.

ALLEN THOMAS ASSOCIATES
518 Prospect Avenue, Little Silver NJ 07739. **Contact:** Tom Benoit, President. **Description:** An executive search firm. **Specializes in the areas of:** Computer Hardware/Software; Computer Science/Software; Health/Medical. **Positions commonly filled include:** Computer Programmer; Electrical/Electronics Engineer; Health Services Manager; Human Resources Manager; MIS Specialist; Occupational Therapist; Physical Therapist; Physician; Systems Analyst.

ULTIMATE SOLUTIONS, INC.
121 West Passaic Street, Rochelle Park NJ 07662. 201/909-3717. **Fax:** 201/587-0772. **Contact:** Caryn S. Reiman, Owner. **Description:** An executive search firm and employment agency, operating on a contingency basis. Founded in 1986. Company pays fee. **Specializes in the areas of:** Administration/MIS/EDP; Computer Science/Software; Technical and Scientific. **Positions commonly filled include:** Computer Programmer; Internet Services Manager; MIS Specialist; Software Engineer; Systems Analyst; Telecommunications Manager. **Corporate headquarters location:** This Location. **Average salary range of placements:** More than $50,000. **Number of placements per year:** 100 - 199.

WORLCO COMPUTER RESOURCES, INC.
901 Route 38, Cherry Hill NJ 08002. 609/665-4700. **Fax:** 609/665-8142. **Contact:** Frank Parisi, Managing Partner. **Description:** An executive search firm that also operates as a temporary and permanent employment agency. Company pays fee. **Specializes in the areas of:** Administration/MIS/EDP; Computer Hardware/Software; Sales and Marketing. **Positions commonly filled include:** Computer Programmer; Internet Services Manager; Marketing Specialist; MIS Specialist; Sales Representative; Systems Analyst; Technical Writer/Editor; Telecommunications Manager. **Number of placements per year:** 100 - 199.

PERMANENT EMPLOYMENT AGENCIES OF NEW JERSEY

A+ PERSONNEL
1017 Broadway, Bayonne NJ 07002. 201/437-5594. **Fax:** 201/437-2914. **Contact:** Jill G. Rowland, Vice President. **Description:** An employment agency. Company pays fee. **Specializes in the areas of:** Accounting/Auditing; Administration/MIS/EDP; Computer Science/Software; Finance; Legal; Manufacturing; Personnel/Labor Relations; Secretarial. **Positions commonly filled include:** Accountant/Auditor; Advertising Clerk; Bank Officer/Manager; Blue-Collar Worker Supervisor; Brokerage Clerk; Budget Analyst; Buyer; Chemical Engineer; Civil Engineer; Claim Representative; Clerical Supervisor; Computer Programmer; Credit Manager; Customer Service Representative; Financial Analyst; General Manager; Human Resources Manager; Industrial Engineer; Industrial Production Manager; Insurance Agent/Broker; Management Analyst/ Consultant; Management Trainee; Mechanical Engineer; Operations/Production Manager; Paralegal; Public Relations Specialist; Purchasing Agent/Manager; Quality Control Supervisor; Restaurant/Food Service Manager; Securities Sales Representative; Software Engineer; Systems Analyst; Travel Agent; Underwriter/Assistant Underwriter; Wholesale and Retail Buyer. **Number of placements per year:** 50 - 99.

ADEL - LAWRENCE ASSOCIATES
300 Highway 34, Suite 18, Aberdeen NJ 07747. 908/566-4914. **Fax:** 908/566-9326. **Contact:** Larry Radzely, President. **Description:** An employment agency. Adel - Lawrence Associates recruits engineering/technical personnel in the areas of management, field service, software, biomedical engineering, telecommunications, and manufacturing. Company pays fee. **Specializes in the areas of:** Computer Science/Software; Engineering; Health/Medical; Manufacturing; Technical and Scientific. **Positions commonly filled include:** Biomedical Engineer; Clinical Lab Technician; Computer Programmer; Design Engineer; Electrical/Electronics Engineer; Mechanical Engineer; MIS Specialist; Software Engineer; Systems Analyst. **Average salary range of placements:** $30,000 - $50,000. **Number of placements per year:** 200 - 499.

ANDREW PERSONNEL SERVICES
P.O. Box 790, Chester NJ 07930. 908/879-2995. **Fax:** 908/879-8482. **Contact:** Tina Bennett, Regional Sales/Marketing Manager. **Description:** An employment agency. Company pays fee. **Specializes in the areas of:** Accounting/Auditing; Administration/MIS/EDP; Computer

Science/Software; Engineering; Finance; Industrial; Insurance; Legal; Retail; Sales and Marketing; Secretarial; Technical and Scientific. **Positions commonly filled include:** Bank Officer/Manager; Blue-Collar Worker Supervisor; Chemical Engineer; Claim Representative; Clerical Supervisor; Computer Programmer; Counselor; Credit Manager; Customer Service Representative; Design Engineer; Draftsperson; Editor; Environmental Engineer; General Manager; Human Resources Specialist; Industrial Engineer; Industrial Production Manager; Internet Services Manager; Manufacturer's/Wholesaler's Sales Rep.; Mechanical Engineer; Multimedia Designer; Operations/ Production Manager; Paralegal; Purchasing Agent/Manager; Restaurant/Food Service Manager; Services Sales Representative; Software Engineer; Structural Engineer; Systems Analyst; Technical Writer/Editor; Telecommunications Manager; Travel Agent; Typist/Word Processor. **Benefits available to temporary workers:** Medical Insurance. **Corporate headquarters location:** This Location. **Other area locations:** Clinton NJ; Little Falls NJ; Parsippany NJ. **Average salary range of placements:** $20,000 - $29,999.

BAI PERSONNEL SOLUTIONS INC.

P.O. Box 2023, Princeton NJ 08542. 609/734-9631. **Fax:** 609/734-9619. **Contact:** Margaret Spells, Branch Manager. **Description:** An employment agency that also offers contract services. Company pays fee. **NOTE:** The company's physical address is One Independence Way, Princeton NJ. **Specializes in the areas of:** Accounting/Auditing; Administration/MIS/EDP; Banking; Computer Hardware/Software; Finance; Insurance; Legal; Personnel/Labor Relations; Sales and Marketing; Secretarial. **Positions commonly filled include:** Accountant/Auditor; Brokerage Clerk; Budget Analyst; Claim Representative; Clerical Supervisor; Computer Programmer; Credit Manager; Customer Service Representative; Financial Analyst; Human Resources Specialist; Management Analyst/Consultant; MIS Specialist; Operations/Production Manager; Paralegal; Reporter; Services Sales Representative; Technical Writer/Editor; Telecommunications Manager; Typist/Word Processor. **Corporate headquarters location:** This Location. **Other U.S. locations:** Albany NY; Schenectady NY. **Number of placements per year:** 100 - 199.

BERMAN & LARSON

140 Route 17 North, Paramus NJ 07652. 201/262-9200. **Toll-free phone:** 800/640-0126. **Fax:** 201/262-7060. **Contact:** Bob Larson, CPC, President. **E-mail address:** jobsbl@tribeca.ios.com. **World Wide Web address:** http://www.tribeca.ios.com/~jobsbl/. **Description:** Berman & Larson is an employment agency specializing in placing information systems professionals. Company pays fee. **Specializes in the areas of:** Computer Science/Software. **Positions commonly filled include:** Computer Programmer; Internet Services Manager; MIS Specialist; Software Engineer; Systems Analyst; Technical Writer/Editor. **Benefits available to temporary workers:** 401(k); Medical Insurance. **Corporate headquarters location:** This Location. **Average salary range of placements:** More than $50,000. **Number of placements per year:** 200 - 499.

CAREERS FIRST, INC.

305 U.S. Route 130, Cinnaminson NJ 08077-3398. 609/786-0004. **Contact:** Gail Duncan, President. **Description:** An employment agency. **Specializes in the areas of:** Administration/ MIS/EDP; Computer Hardware/Software; Technical and Scientific.

CAREERS USA

533 North Evergreen Avenue, Woodbury NJ 08096. 609/384-1600. **Fax:** 609/384-1310. **Contact:** Carla Janoff, President/Owner. **Description:** An employment agency. Company pays fee. **Specializes in the areas of:** Accounting/Auditing; Administration/MIS/EDP; Computer Science/Software; General Management; Industrial; Manufacturing; Personnel/Labor Relations; Secretarial; Technical and Scientific. **Positions commonly filled include:** Administrative Services Manager; Blue-Collar Worker Supervisor; Claim Representative; Clerical Supervisor; Customer Service Representative; Draftsperson; Editor; Financial Analyst; Human Resources Specialist; Industrial Production Manager; MIS Specialist; Operations/Production Manager; Paralegal; Purchasing Agent/Manager; Quality Control Supervisor; Typist/Word Processor.

CITIZENS EMPLOYMENT SERVICES, INC.

One Magnolia Avenue, Montvale NJ 07645. 201/391-5144. **Fax:** 201/391-4477. **Contact:** Elaine Larfier, Manager. **Description:** An employment agency. Company pays fee. **Specializes in the areas of:** Banking; Clerical; Computer Science/Software; Industrial; Insurance; Manufacturing;

Retail; Sales and Marketing. **Positions commonly filled include:** Accountant/Auditor; Actuary; Bank Officer/Manager; Bookkeeper; Chemical Engineer; Claim Representative; Computer Operator; Computer Programmer; Credit Manager; Customer Service Representative; Data Entry Clerk; Draftsperson; EDP Specialist; Electrical/Electronics Engineer; Industrial Production Manager; Insurance Agent/Broker; Legal Secretary; Mechanical Engineer; Operations/Production Manager; Purchasing Agent/Manager; Quality Control Supervisor; Sales Representative; Secretary; Stenographer; Systems Analyst; Technician; Travel Agent; Typist/Word Processor; Underwriter/Assistant Underwriter. **Corporate headquarters location:** This Location. **Other U.S. locations:** Parsippany NJ. **Average salary range of placements:** $30,000 - $50,000. **Number of placements per year:** 1000+.

GLENN DAVIS ASSOCIATES
124 Morris Turnpike, Randolph NJ 07869-2976. 201/895-4242. **Contact:** Manager. **Description:** An employment agency. **Specializes in the areas of:** Computer Science/Software. **Positions commonly filled include:** MIS Specialist. **Number of placements per year:** 50 - 99.

EXECUTIVE NETWORK, INC.
147 Columbia Turnpike, Florham Park NJ 07932-2145. 201/966-5400. **Fax:** 201/966-0304. **Contact:** Elissa Marcus, President. **Description:** A temporary and permanent employment agency. Company pays fee. **Specializes in the areas of:** Accounting/Auditing; Administration/MIS/EDP; Computer Science/Software; Finance; Insurance; Personnel/Labor Relations; Retail; Sales and Marketing; Secretarial; Technical and Scientific. **Positions commonly filled include:** Accountant/Auditor; Administrative Services Manager; Advertising Clerk; Branch Manager; Budget Analyst; Claim Representative; Clerical Supervisor; Computer Programmer; Customer Service Representative; Financial Analyst; General Manager; Human Resources Specialist; Management Trainee; MIS Specialist; Purchasing Agent/Manager; Services Sales Representative; Software Engineer; Systems Analyst; Underwriter/Assistant Underwriter. **Average salary range of placements:** $30,000 - $50,000. **Number of placements per year:** 50 - 99.

EXECUTIVE SOFTWARE PLUS
24 Lyons Place, Westwood NJ 07675. 201/666-5484. **Fax:** 201/664-0693. **Contact:** Claire Monte, Vice President. **Description:** An employment agency specializing in placing software professionals. **Specializes in the areas of:** Computer Science/Software. **Positions commonly filled include:** Computer Programmer; Systems Analyst.

IMPACT PERSONNEL, INC.
1901 North Olden Avenue, Suite 26A, Trenton NJ 08618. 609/406-1200. **Contact:** Manager. **Description:** An employment agency. Company pays fee. **Specializes in the areas of:** Accounting/Auditing; Administration/MIS/EDP; Advertising; Banking; Computer Hardware/Software; Engineering; Fashion; Finance; Food Industry; General Management; Health/Medical; Industrial; Insurance; Legal; Manufacturing; Printing/Publishing; Retail; Sales and Marketing; Secretarial; Technical and Scientific. **Positions commonly filled include:** Accountant/Auditor; Administrative Assistant; Bookkeeper; Chemical Engineer; Claim Representative; Clerk; Commercial Artist; Computer Operator; Computer Programmer; Customer Service Representative; Data Entry Clerk; Driver; Electrical/Electronics Engineer; Factory Worker; Hotel Manager; Industrial Designer; Industrial Engineer; Legal Secretary; Light Industrial Worker; Mechanical Engineer; Medical Secretary; Quality Control Supervisor; Receptionist; Recruiter; Sales Representative; Secretary; Software Engineer; Technician; Typist/Word Processor. **Number of placements per year:** 500 - 999.

INFOSYSTEMS PLACEMENT SERVICE
17 Holmes Lane, Marlton NJ 08053-1911. 609/596-7770. **Fax:** 609/596-7772. **Contact:** Joseph A. Dougherty, Owner/Consultant. **Description:** An employment agency specializing in the recruitment and placement of information systems professionals. Company pays fee. **Specializes in the areas of:** Administration/MIS/EDP; Computer Science/Software. **Positions commonly filled include:** Computer Programmer; Internet Services Manager; Software Engineer; Systems Analyst; Telecommunications Manager. **Corporate headquarters location:** This Location. **Average salary range of placements:** More than $50,000. **Number of placements per year:** 1 - 49.

MAYFAIR SERVICES

372 Buffalo Avenue, Paterson NJ 07503. 201/742-0990. **Fax:** 201/742-0991. **Contact:** Mary Costello, Owner. **Description:** An employment and contract services agency. **Specializes in the areas of:** Administration/MIS/EDP; Art/Design; Biology; Computer Hardware/Software; Engineering; General Management; Health/Medical; Industrial; Manufacturing; Personnel/Labor Relations; Printing/Publishing; Technical and Scientific; Transportation. **Positions commonly filled include:** Aerospace Engineer; Agricultural Engineer; Aircraft Mechanic/Engine Specialist; Architect; Biochemist; Biological Scientist; Biomedical Engineer; Buyer; Chemical Engineer; Chemist; Civil Engineer; Clinical Lab Technician; Construction and Building Inspector; Design Engineer; Designer; Draftsperson; Electrical/Electronics Engineer; Electrician; Environmental Engineer; General Manager; Health Services Manager; Human Resources Specialist; Industrial Engineer; Industrial Production Manager; Library Technician; Management Trainee; Mathematician; Mechanical Engineer; Medical Records Technician; Metallurgical Engineer; MIS Specialist; Multimedia Designer; Nuclear Engineer; Nuclear Medicine Technologist; Occupational Therapist; Operations/Production Manager; Pharmacist; Quality Control Supervisor; Science Technologist; Software Engineer; Stationary Engineer; Structural Engineer; Typist/Word Processor; Veterinarian. **Corporate headquarters location:** This Location. **Average salary range of placements:** $20,000 - $29,999. **Number of placements per year:** 200 - 499.

POMERANTZ PERSONNEL

1375 Plainfield Avenue, Watchung NJ 07060. 908/754-3092. **Toll-free phone:** 800/754-7000. **Fax:** 908/757-0298. **Contact:** Keith Grude, Corporate Director. **Description:** An employment agency. **Specializes in the areas of:** Accounting/Auditing; Administration/MIS/EDP; Banking; Computer Hardware/Software; Finance; Food Industry; General Management; Health/Medical; Industrial; Legal; Personnel/Labor Relations; Retail; Sales and Marketing; Secretarial. **Other U.S. locations:** Nationwide.

RSVP SERVICES

P.O. Box 8369, Cherry Hill NJ 08002-0369. 609/667-4488. **Contact:** Howard Levine, Director. **E-mail address:** hl@rsvpjobs.com. **Description:** An employment agency. **Specializes in the areas of:** Administration/MIS/EDP; Computer Hardware/Software; Electrical; Electronics.

SCIENTIFIC SEARCH, INC.

560 Fellowship Road, Suite 309, Mount Laurel NJ 08054. 609/866-0200. **Fax:** 609/722-5307. **Contact:** Robert I. Greensberg, President. **Description:** An employment agency. Company pays fee. **Specializes in the areas of:** Computer Science/Software; Health/Medical. **Positions commonly filled include:** Computer Programmer; Health Services Manager; Nuclear Medicine Technologist; Occupational Therapist; Physical Therapist; Physician; Registered Nurse; Respiratory Therapist; Systems Analyst. **Number of placements per year:** 50 - 99.

SELECTIVE PERSONNEL

288 Summerhill Road, East Brunswick NJ 08816. 609/497-2900. **Contact:** Manager. **Description:** An employment agency. **Specializes in the areas of:** Accounting/Auditing; Administration/MIS/EDP; Banking; Clerical; Computer Hardware/Software; Engineering; Finance; Health/Medical; Insurance; Legal; Manufacturing; Personnel/Labor Relations; Sales and Marketing; Technical and Scientific.

SOURCE SERVICES CORPORATION

15 Essex Road, Paramus NJ 07652. 201/843-2020. **Fax:** 201/843-7705. **Contact:** Jackie Finestone, Branch Manager. **Description:** An employment agency. Company pays fee. **Specializes in the areas of:** Accounting/Auditing; Administration/MIS/EDP; Banking; Computer Science/Software; Engineering; Finance; Health/Medical; Legal; Manufacturing. **Positions commonly filled include:** Accountant/Auditor; Attorney; Budget Analyst; Computer Programmer; Credit Manager; Financial Analyst; Licensed Practical Nurse; MIS Specialist; Registered Nurse; Software Engineer. **Benefits available to temporary workers:** Profit Sharing. **Number of placements per year:** 1000+.

CLAIRE WRIGHT ASSOCIATION

1280 U.S. Highway 46, Parsippany NJ 07054-4911. 201/402-8400. **Fax:** 201/402-8519. **Contact:** K. Kelley, Counselor. **Description:** An employment agency. Company pays fee. **Specializes in the

areas of: Accounting/Auditing; Administration/MIS/EDP; Advertising; Computer Hardware/Software; Engineering; Finance; Insurance; Legal; Manufacturing; Sales and Marketing; Technical and Scientific. **Positions commonly filled include:** Accountant/Auditor; Advertising Clerk; Chemical Engineer; Claim Representative; Computer Programmer; Draftsperson; Electrical/Electronics Engineer; Environmental Engineer; Financial Analyst; Human Resources Specialist; Industrial Engineer; Insurance Agent/Broker; Internet Services Manager; Market Research Analyst; Mechanical Engineer; Metallurgical Engineer; MIS Specialist; Nuclear Engineer; Paralegal; Property and Real Estate Manager; Purchasing Agent/Manager; Securities Sales Representative; Software Engineer; Technical Writer/Editor; Telecommunications Manager; Typist/Word Processor; Underwriter/Assistant Underwriter. **Average salary range of placements:** $30,000 - $50,000. **Number of placements per year:** 1 - 49.

PERMANENT EMPLOYMENT AGENCIES OF NEW MEXICO

ALBUQUERQUE PERSONNEL, INC.
6011 Osuna Road NE, Suite B, Albuquerque NM 87109. 505/888-3555. **Fax:** 505/883-9022. **Contact:** Mark R. Pyszkowski, Vice President. **Description:** An employment agency. **Specializes in the areas of:** Accounting/Auditing; Administration/MIS/EDP; Advertising; Architecture/Construction/Real Estate; Banking; Broadcasting; Computer Science/Software; Engineering; Finance; Food Industry; General Management; Health/Medical; Industrial; Legal; Manufacturing; Personnel/Labor Relations; Printing/Publishing; Retail; Sales and Marketing; Secretarial; Technical and Scientific; Transportation. **Positions commonly filled include:** Accountant/Auditor; Adjuster; Administrative Services Manager; Advertising Clerk; Automotive Mechanic; Bank Officer/Manager; Blue-Collar Worker Supervisor; Branch Manager; Budget Analyst; Buyer; Civil Engineer; Claim Representative; Clerical Supervisor; Computer Programmer; Construction and Building Inspector; Construction Contractor; Cost Estimator; Counselor; Credit Manager; Customer Service Representative; Dental Assistant/Dental Hygienist; Designer; Draftsperson; Economist; Electrician; Emergency Medical Technician; Financial Analyst; General Manager; Health Services Manager; Hotel Manager; Human Resources Manager; Industrial Engineer; Industrial Production Manager; Insurance Agent/Broker; Management Analyst/Consultant; Management Trainee; Manufacturer's/Wholesaler's Sales Rep.; Mechanical Engineer; Operations/Production Manager; Paralegal; Property and Real Estate Manager; Public Relations Specialist; Purchasing Agent/Manager; Quality Control Supervisor; Radio/TV Announcer/Broadcaster; Real Estate Agent; Restaurant/Food Service Manager; Services Sales Representative; Software Engineer; Systems Analyst; Technical Writer/Editor; Transportation/Traffic Specialist; Travel Agent; Urban/Regional Planner; Veterinarian; Wholesale and Retail Buyer. **Number of placements per year:** 50 - 99.

CDI CORPORATION-MIDWEST
2730 San Pedro Drive NE, Suite H, Albuquerque NM 87110. 505/888-4544. **Toll-free phone:** 800/354-7314. **Fax:** 505/888-3233. **Contact:** Shawn Murphy, Technical Manager. **Description:** An employment agency. Founded in 1945, this is one of the largest full-service contract engineering, design, scientific, and technical services agencies in the world, with over 450 offices nationwide. Company pays fee. **Specializes in the areas of:** Administration/MIS/EDP; Art/Design; Biology; Computer Science/Software; Engineering; Finance; General Management; Health/Medical; Industrial; Manufacturing; Personnel/Labor Relations; Printing/Publishing; Secretarial; Technical and Scientific; Transportation. **Positions commonly filled include:** Accountant/Auditor; Aerospace Engineer; Agricultural Engineer; Architect; Biological Scientist; Biomedical Engineer; Blue-Collar Worker Supervisor; Budget Analyst; Buyer; Chemical Engineer; Chemist; Civil Engineer; Clerical Supervisor; Clinical Lab Technician; Computer Programmer; Construction Contractor; Cost Estimator; Customer Service Representative; Design Engineer; Designer; Draftsperson; Editor; EEG Technologist; EKG Technician; Electrical/Electronics Engineer;

Environmental Engineer; Financial Analyst; Food Scientist/Technologist; General Manager; Geographer; Geologist/Geophysicist; Human Resources Manager; Human Service Worker; Industrial Engineer; Industrial Production Manager; Internet Services Manager; Landscape Architect; Librarian; Library Technician; Management Analyst/Consultant; Management Trainee; Market Research Analyst; Mathematician; Mechanical Engineer; Medical Records Technician; Metallurgical Engineer; Mining Engineer; MIS Specialist; Multimedia Designer; Nuclear Engineer; Nuclear Medicine Technologist; Operations/Production Manager; Petroleum Engineer; Pharmacist; Physical Therapist; Physician; Physicist/Astronomer; Public Relations Specialist; Quality Control Supervisor; Radiologic Technologist; Science Technologist; Software Engineer; Stationary Engineer; Statistician; Strategic Relations Manager; Structural Engineer; Surveyor; Systems Analyst; Technical Writer/Editor; Telecommunications Manager; Transportation/Traffic Specialist; Typist/Word Processor; Urban/Regional Planner; Video Production Coordinator. **Corporate headquarters location:** Philadelphia PA. **International locations:** Worldwide. **Average salary range of placements:** $30,000 - $50,000. **Number of placements per year:** 200 - 499.

SANDERSON EMPLOYMENT SERVICE INC.
500 Chama Boulevard NE, Albuquerque NM 87108. 505/265-8827. **Contact:** Bill Sanderson, President. **Description:** An employment agency. Company pays fee. **Specializes in the areas of:** Accounting/Auditing; Clerical; Computer Hardware/Software; Engineering; Sales and Marketing; Secretarial; Technical and Scientific. **Positions commonly filled include:** Accountant/Auditor; Aerospace Engineer; Bookkeeper; Civil Engineer; Computer Programmer; Data Entry Clerk; EDP Specialist; Electrical/Electronics Engineer; Legal Secretary; Mechanical Engineer; Medical Secretary; Physicist/Astronomer; Receptionist; Sales Representative; Secretary; Stenographer; Systems Analyst; Typist/Word Processor. **Number of placements per year:** 50 - 99.

SNELLING PERSONNEL SERVICES
2601 Wyoming Boulevard NE, Suite 106, Albuquerque NM 87112-1031. 505/293-7800. **Contact:** Sue Lane, Manager. **Description:** An employment agency. Company pays fee. **Specializes in the areas of:** Accounting/Auditing; Computer Science/Software; Engineering; Health/Medical; Retail; Sales and Marketing; Secretarial. **Positions commonly filled include:** Accountant/Auditor; Adjuster; Administrative Services Manager; Advertising Clerk; Architect; Bank Officer/Manager; Branch Manager; Buyer; Clerical Supervisor; Computer Programmer; Credit Manager; Customer Service Representative; Dental Assistant/Dental Hygienist; Design Engineer; Draftsperson; Electrical/Electronics Engineer; General Manager; Hotel Manager; Human Resources Manager; Landscape Architect; Management Trainee; Mechanical Engineer; Metallurgical Engineer; MIS Specialist; Occupational Therapist; Operations/Production Manager; Paralegal; Physician; Purchasing Agent/Manager; Registered Nurse; Respiratory Therapist; Restaurant/Food Service Manager; Services Sales Representative; Software Engineer; Systems Analyst; Telecommunications Manager; Typist/Word Processor; Underwriter/Assistant Underwriter. **Corporate headquarters location:** Dallas TX. **International locations:** Worldwide. **Average salary range of placements:** $20,000 - $29,999. **Number of placements per year:** 50 - 99.

EXECUTIVE SEARCH FIRMS OF NEW YORK

AJC SEARCH
119 North Park Avenue, Rockville Centre NY 11570. 516/766-1699. **Fax:** 516/766-3889. **Contact:** Jay Cohen, Principal. **Description:** An executive search firm. Company pays fee. **Specializes in the areas of:** Accounting/Auditing; Computer Science/Software; Economics; Finance; Health/Medical; Insurance; Legal; Sales and Marketing. **Positions commonly filled include:** Accountant/Auditor; Actuary; Attorney; Computer Programmer; Consultant; Dental Assistant/Dental Hygienist; Dentist; Dietician/Nutritionist; Financial Analyst; Insurance Agent/Broker; Mathematician; Paralegal; Pharmacist; Physical Therapist; Physician; Registered Nurse; Securities Sales Representative; Statistician; Systems Analyst; Underwriter/Assistant Underwriter. **Average salary range of placements:** More than $50,000. **Number of placements per year:** 1 - 49.

AZR INC.

245 Fifth Avenue, New York NY 10016. 212/545-7796. **Contact:** Richard Silverman, President. **Description:** An executive search firm. **Specializes in the areas of:** Computer Science/Software. **Positions commonly filled include:** Computer Programmer; Internet Services Manager; MIS Specialist; Multimedia Designer; Software Engineer; Systems Analyst; Technical Writer/Editor; Telecommunications Manager. **Average salary range of placements:** More than $50,000. **Number of placements per year:** 50 - 99.

ADEPT TECH RECRUITING

219 Glendale Road, Scarsdale NY 10583. 914/725-8583. **Contact:** Fred Press, President. **Description:** An executive search firm. Company pays fee. **Specializes in the areas of:** Administration/MIS/EDP; Computer Science/Software; Finance; Health/Medical; Information Systems. **Average salary range of placements:** $30,000 - $50,000. **Number of placements per year:** 1 - 49.

ALITE ASSOCIATES, INC.

56 Beaver Street, Suite 403, New York NY 10004. 212/809-1400. **Fax:** 212/809-1730. **Contact:** Maria Alite, President. **Description:** A full-service contingency executive recruiting firm specializing in the personnel needs of the banking and brokerage industries. **Specializes in the areas of:** Accounting/Auditing; Banking; Computer Science/Software; Finance. **Positions commonly filled include:** Telecommunications Manager.

AMES O'NEILL ASSOCIATES INC.

330 Venderbilt Motor Parkway, Hauppauge NY 11788. 516/582-4800. **Fax:** 516/234-6094. **Contact:** George C. Ames, President. **Description:** An executive search firm. Company pays fee. **Specializes in the areas of:** Computer Science/Software; Engineering; Technical and Scientific. **Positions commonly filled include:** Aerospace Engineer; Biomedical Engineer; Chemical Engineer; Computer Programmer; Electrical/Electronics Engineer; Industrial Engineer; Mechanical Engineer; Metallurgical Engineer; Software Engineer; Systems Analyst; Technical Writer/Editor. **Number of placements per year:** 1 - 49.

BERKEL ASSOCIATES, INC.

477 Madison Avenue, Suite 707, New York NY 10022-4503. 212/826-3000. **Fax:** 212/826-3006. **Contact:** Carol Bernstein, President. **Description:** A contingency search firm and employment agency. **Specializes in the areas of:** Administration/MIS/EDP; Computer Science/Software; Legal; Personnel/Labor Relations; Secretarial. **Positions commonly filled include:** Accountant/Auditor; Administrative Services Manager; Attorney; Clerical Supervisor; Computer Programmer; Counselor; Human Resources Specialist; Internet Services Manager; Legal Secretary; Library Technician; MIS Specialist; Paralegal; Typist/Word Processor. **Corporate headquarters location:** This Location.

BOS BUSINESS CONSULTANTS

4211 North Buffalo Street, Orchard Park NY 14127. 716/662-0800. **Fax:** 716/662-0623. **Contact:** John Bos, Owner. **E-mail address:** careersbbc@microagewny.com. **Description:** An executive search firm that specializes in the placement of personnel in the engineering, marketing of medical devices, computer, and scientific fields. Bos Business Consultants operates on both contingency and retainer bases. Company pays fee. **Specializes in the areas of:** Computer Science/Software; Engineering; Industrial; Manufacturing. **Positions commonly filled include:** Aerospace Engineer; Biomedical Engineer; Chemical Engineer; Chemist; Computer Programmer; Design Engineer; Designer; Electrical/Electronics Engineer; Financial Analyst; Food Scientist/Technologist; Human Resources Manager; Industrial Engineer; Mechanical Engineer; Metallurgical Engineer; MIS Specialist; Operations/Production Manager; Quality Control Supervisor; Software Engineer; Systems Analyst. **Corporate headquarters location:** This Location. **Average salary range of placements:** More than $50,000. **Number of placements per year:** 1 - 49.

BRANTHOVER ASSOCIATES

51 East 42nd Street, Suite 500, New York NY 10017. 212/949-9400. **Fax:** 212/949-5905. **Contact:** Jeanne Branthover, President. **E-mail address:** branthover@aol.com. **Description:** An executive search firm. **Specializes in the areas of:** Accounting/Auditing; Administration/MIS/EDP;

Advertising; Banking; Computer Science/Software; Finance; General Management; Manufacturing; Personnel/Labor Relations; Printing/Publishing; Sales and Marketing. **Positions commonly filled include:** Accountant/Auditor; Bank Officer/Manager; Budget Analyst; Computer Programmer; Credit Manager; Customer Service Representative; Financial Analyst; General Manager; Human Resources Manager; Systems Analyst. **Number of placements per year:** 50 - 99.

BRUML ASSOCIATES INC.

306 Birchwood Park Drive, Jericho NY 11753-2307. 516/822-7940. **Contact:** Michael Bruml, Owner. **Description:** Founded in 1970, Bruml Associates, Inc. is an executive search firm operating on a contingency basis that specializes in the technical industry. Company pays fee. **Specializes in the areas of:** Computer Science/Software; Engineering; Technical and Scientific. **Positions commonly filled include:** Biomedical Engineer; Computer Engineer; Design Engineer; Electrical/Electronics Engineer; Software Engineer. **Corporate headquarters location:** This Location. **Average salary range of placements:** $30,000 - $50,000. **Number of placements per year:** 1 - 49.

BURNS PERSONNEL

3300 Monroe Avenue, Rochester NY 14618. 716/385-6300. **Contact:** Jackie Tedesco, President. **World Wide Web address:** http://www.burnspersonnel.com. **Description:** An executive search firm that also provides temporary placement. Company pays fee. **Specializes in the areas of:** Computer Science/Software; Engineering; Industrial; Manufacturing; Personnel/Labor Relations; Secretarial; Technical and Scientific. **Positions commonly filled include:** Biochemist; Biological Scientist; Biomedical Engineer; Chemical Engineer; Chemist; Computer Programmer; Draftsperson; EKG Technician; Electrical/Electronics Engineer; Emergency Medical Technician; Human Resources Specialist; Industrial Engineer; Materials Engineer; Mechanical Engineer; MIS Specialist; Software Engineer; Systems Analyst; Typist/Word Processor. **Average salary range of placements:** $20,000 - $29,999. **Number of placements per year:** 500 - 999.

CAVAN SYSTEMS LTD.

10 Cuttermill Road, Suite 403A, Great Neck NY 11021. 516/487-7777. **Fax:** 516/487-7857. **Contact:** Manager of Recruiting. **Description:** An executive search firm. **Specializes in the areas of:** Computer Science/Software; Information Systems; Multimedia. **Positions commonly filled include:** Communications Engineer; Computer Programmer; Media Specialist; Network Engineer; Systems Analyst; Telecommunications Analyst. **Number of placements per year:** 50 - 99.

COLTON PARTNERSHIP

63 Wall Street, Suite 2901, New York NY 10005. 212/248-9700. **Fax:** 212/509-1633. **Contact:** Daniel Alzapiedi, President. **E-mail address:** citi63i@aol.com. **World Wide Web address:** http://www.colt.nexel.com. **Description:** A retainer search firm. Colton Partnership specializes in investment banking placements. The firm's subdivision, Colton Information Technology, offers technical recruitment and placement. Company pays fee. **Specializes in the areas of:** Computer Science/Software; Engineering; General Management; Insurance; Investment; Manufacturing; Personnel/Labor Relations; Technical and Scientific. **Positions commonly filled include:** Bank Officer/Manager; Computer Programmer; Design Engineer; Designer; Electrical/Electronics Engineer; Financial Analyst; Human Resources Manager; Internet Services Manager; Management Analyst/Consultant; MIS Specialist; Multimedia Designer; Quality Control Supervisor; Science Technologist; Securities Sales Representative; Software Engineer. **Corporate headquarters location:** This Location. **Other U.S. locations:** Nationwide. **Number of placements per year:** 1 - 49.

COMPU-TECH PERSONNEL AGENCY

33 Walt Whitman Road, Huntington Station NY 11746-3627. **Contact:** Freda Frankel, President. **Description:** An executive search firm operating on a contingency basis. **Specializes in the areas of:** Accounting/Auditing; Administration/MIS/EDP; Computer Science/Software; Finance. **Positions commonly filled include:** Accountant/Auditor; Budget Analyst; Computer Programmer; Credit Manager; Financial Analyst; Systems Analyst. **Corporate headquarters location:** This Location. **Average salary range of placements:** $30,000 - $50,000. **Number of placements per year:** 50 - 99.

CONSORTIUM

One Times Square Plaza, 13th Floor, New York NY 10036-6560. 212/221-1544. **Contact:** Alexander C. Valcic, Recruiter. **Description:** An executive search and contract services firm. Consortium operates on both retainer and contingency bases. **Specializes in the areas of:** Accounting/Auditing; Administration/MIS/EDP; Banking; Computer Hardware/Software; Engineering; Finance; Health/Medical; Legal; Personnel/Labor Relations; Technical and Scientific. **Positions commonly filled include:** Accountant/Auditor; Aerospace Engineer; Aircraft Mechanic/Engine Specialist; Attorney; Bank Officer/Manager; Chemical Engineer; Computer Programmer; Design Engineer; Designer; Electrical/Electronics Engineer; Emergency Medical Technician; Financial Analyst; Health Services Manager; Human Resources Manager; Human Service Worker; Management Analyst/Consultant; MIS Specialist; Multimedia Designer; Nuclear Engineer; Occupational Therapist; Physical Therapist; Physician; Registered Nurse; Respiratory Therapist; Securities Sales Representative; Software Engineer; Speech-Language Pathologist; Systems Analyst; Technical Writer/Editor; Telecommunications Manager. **Corporate headquarters location:** This Location. **Other U.S. locations:** Nationwide. **Average salary range of placements:** More than $50,000. **Number of placements per year:** 100 - 199.

CORPORATE CAREERS, INC.

188 East Post Road, White Plains NY 10601. 914/946-2003. **Fax:** 914/946-2019. **Contact:** Richard Birnbaum, President. **Description:** An executive search firm. **Specializes in the areas of:** Accounting/Auditing; Administration/MIS/EDP; Advertising; Biology; Computer Science/Software; Engineering; Finance; Food Industry; General Management; Industrial; Manufacturing; Personnel/Labor Relations; Printing/Publishing; Sales and Marketing; Secretarial; Technical and Scientific. **Positions commonly filled include:** Accountant/Auditor; Biochemist; Biological Scientist; Biomedical Engineer; Budget Analyst; Chemical Engineer; Chemist; Civil Engineer; Clinical Lab Technician; Computer Programmer; Credit Manager; Customer Service Representative; Design Engineer; Designer; Draftsperson; Electrical/Electronics Engineer; Environmental Engineer; Financial Analyst; Food Scientist/Technologist; Human Resources Specialist; Industrial Engineer; Industrial Production Manager; Management Analyst/Consultant; Market Research Analyst; Mechanical Engineer; Metallurgical Engineer; MIS Specialist; Multimedia Designer; Nuclear Engineer; Operations/Production Manager; Public Relations Specialist; Purchasing Agent/Manager; Quality Control Supervisor; Software Engineer; Strategic Relations Manager; Structural Engineer; Systems Analyst; Telecommunications Manager; Transportation/Traffic Specialist; Typist/Word Processor. **Benefits available to temporary workers:** 401(k); Paid Holidays; Paid Vacations. **Corporate headquarters location:** This Location. **Average salary range of placements:** $20,000 - $29,999.

DAPEXS CONSULTANTS INC.

5320 West Genesee Street, Camillus NY 13031-2203. 315/484-9300. **Fax:** 315/484-9330. **Contact:** Peter J. Leofsky, President. **Description:** An executive search firm founded in 1970 for computer professionals, accounting/financial executives, and manufacturing/engineering personnel. Dapexs Consultants Inc. operates on both retainer and contingency bases. Company pays fee. **Specializes in the areas of:** Administration/MIS/EDP; Computer Science/Software. **Positions commonly filled include:** Computer Programmer; Internet Services Manager; Management Analyst/Consultant; MIS Specialist; Software Engineer; Systems Analyst; Technical Writer/Editor; Telecommunications Manager.

SETH DIAMOND ASSOCIATES INC.

45 West 45th Street, Suite 801, New York NY 10036. 212/944-6190. **Fax:** 212/944-6197. **Contact:** Seth Diamond, CPC, President. **Description:** An executive search and placement agency specializing in permanent and temporary recruitment at all levels. Company pays fee. **Specializes in the areas of:** Accounting/Auditing; Administration/MIS/EDP; Advertising; Architecture/Construction/Real Estate; Banking; Broadcasting; Computer Science/Software; Economics; Fashion; Finance; General Management; Health/Medical; Industrial; Legal; Manufacturing; Personnel/Labor Relations; Printing/Publishing; Retail; Sales and Marketing; Secretarial. **Positions commonly filled include:** Accountant/Auditor; Actuary; Administrative Services Manager; Advertising Clerk; Bank Officer/Manager; Brokerage Clerk; Budget Analyst; Buyer; Clerical Supervisor; Computer Programmer; Credit Manager; Customer Service Representative; Dental Assistant/Dental Hygienist; Dental Lab Technician; Economist; Financial Analyst; General

Manager; Human Resources Manager; Industrial Engineer; Industrial Production Manager; Management Trainee; Manufacturer's/Wholesaler's Sales Rep.; Mechanical Engineer; Operations/Production Manager; Property and Real Estate Manager; Purchasing Agent/Manager; Software Engineer; Statistician; Systems Analyst; Wholesale and Retail Buyer. **Corporate headquarters location:** This Location. **Number of placements per year:** 200 - 499.

ETC SEARCH, INC.
226 East 54th Street, Suite 308, New York NY 10022. 212/371-3880. **Fax:** 212/754-4877. **Contact:** Marlene Eskenazie, Managing Director. **Description:** An executive search firm. Company pays fee. **Specializes in the areas of:** Administration/MIS/EDP; Computer Science/Software; Information Technology. **Positions commonly filled include:** Computer Programmer; Database Manager; Network Engineer; Software Engineer; Systems Analyst. **Number of placements per year:** 50 - 99.

DAVID M. ELLNER ASSOCIATES
Two Penn Plaza, New York NY 10121. 212/279-0665. **Fax:** 212/714-0672. **Contact:** David Ellner, CEO. **Description:** An executive search firm. Company pays fee. **Specializes in the areas of:** Administration/MIS/EDP; Computer Science/Software; Legal; Printing/Publishing; Technical and Scientific. **Positions commonly filled include:** Aerospace Engineer; Attorney; Computer Programmer; Editor; Electrical/Electronics Engineer; Systems Analyst. **Average salary range of placements:** More than $50,000.

ERIC ROBERT ASSOCIATES
350 7th Avenue, New York NY 10001-5013. 212/695-5900. **Fax:** 212/695-5809. **Contact:** Eric Silverman, President. **E-mail address:** ericrob@nyc.pipeline.com. **Description:** An executive search firm that places full-time employees in all areas of information technology. Company pays fee. **Specializes in the areas of:** Computer Science/Software. **Positions commonly filled include:** Computer Programmer; Internet Services Manager; MIS Specialist; Technical Writer/Editor; Telecommunications Manager. **Corporate headquarters location:** This Location. **Average salary range of placements:** More than $50,000. **Number of placements per year:** 100 - 199.

ETHAN ALLEN PERSONNEL PLACEMENT
30 West State Street, Binghamton NY 13901-2332. 607/772-1560. **Fax:** 607/772-1564. **Contact:** Michael H. Houlihan, President. **Description:** An executive search firm. Company pays fee. **Specializes in the areas of:** Accounting/Auditing; Administration/MIS/EDP; Banking; Computer Science/Software; Engineering; Finance; General Management; Industrial; Manufacturing; Personnel/Labor Relations; Sales and Marketing; Technical and Scientific. **Positions commonly filled include:** Accountant/Auditor; Aerospace Engineer; Agricultural Engineer; Bank Officer/Manager; Budget Analyst; Buyer; Chemical Engineer; Computer Programmer; Cost Estimator; Credit Manager; Design Engineer; Designer; Draftsperson; Electrical/Electronics Engineer; Environmental Engineer; Financial Analyst; General Manager; Human Resources Manager; Industrial Engineer; Industrial Production Manager; Internet Services Manager; Management Analyst/Consultant; Manufacturer's/Wholesaler's Sales Rep.; Market Research Analyst; Mechanical Engineer; Metallurgical Engineer; MIS Specialist; Nuclear Engineer; Operations/Production Manager; Purchasing Agent/Manager; Quality Control Supervisor; Software Engineer; Stationary Engineer; Structural Engineer; Systems Analyst; Technical Writer/Editor; Telecommunications Manager; Transportation/Traffic Specialist. **Corporate headquarters location:** This Location. **Number of placements per year:** 100 - 199.

EXECUTIVE DIRECTIONS INC.
Two Penn Plaza, Suite 1185, New York NY 10121-1185. 212/594-5775. **Fax:** 212/594-4183. **Contact:** Gus Oakes, CPC, President. **Description:** An executive search firm. **Specializes in the areas of:** Administration/MIS/EDP; Computer Science/Software. **Positions commonly filled include:** Computer Programmer; Management Analyst/Consultant; Systems Analyst. **Number of placements per year:** 100 - 199.

FANNING PERSONNEL
507 5th Avenue, Suite 800, New York NY 10017-4906. 212/867-1725. **Fax:** 212/867-1338. **Contact:** Dave Cowen, President. **E-mail address:** resume@fanning.com. **World Wide Web**

address: http://www.fanning.com. **Description:** A contingency search firm and employment agency. Company pays fee. **Specializes in the areas of:** Administration/MIS/EDP; Banking; Broadcasting; Computer Science/Software; Entertainment; Fashion; Finance; Personnel/Labor Relations; Secretarial. **Positions commonly filled include:** Computer Programmer; Human Resources Specialist; Internet Services Manager; MIS Specialist; Systems Analyst; Typist/Word Processor. **Corporate headquarters location:** This Location. **Average salary range of placements:** $30,000 - $50,000. **Number of placements per year:** 100 - 199.

FILCRO MEDIA STAFFING
275 Madison Avenue, New York NY 10016. 212/599-0909. **Fax:** 212/599-1024. **Contact:** Tony Filson, President. **Description:** An employment agency and an executive search firm. Company pays fee. **Specializes in the areas of:** Advertising; Broadcasting; Computer Science/Software; Personnel/Labor Relations; Sales and Marketing; Secretarial. **Positions commonly filled include:** Broadcast Technician; Editor; Human Resources Manager; Multimedia Designer; Public Relations Specialist; Technical Writer/Editor; Typist/Word Processor. **Corporate headquarters location:** This Location. **Average salary range of placements:** $30,000 - $50,000. **Number of placements per year:** 200 - 499.

GRUEN RESOURCES, INC.
323 Clocktower Commons, Brewster NY 10509. 914/279-8827. **Fax:** 914/279-8845. **Contact:** Connie Gruen, President. **E-mail address:** cmgruen@aol.com. **Description:** An executive search firm that recruits commercial real estate and financial services personnel. The company places candidates on both retainer and contingency bases. **Specializes in the areas of:** Banking; Computer Science/Software; Finance. **Positions commonly filled include:** Financial Analyst; Management Analyst/Consultant; MIS Specialist; Property and Real Estate Manager. **Average salary range of placements:** More than $50,000. **Number of placements per year:** 1 - 49.

THE HAAS ASSOCIATES, INC.
443 West 24th Street, New York NY 10011. 212/741-2457. **Contact:** Margaret Haas, President. **E-mail address:** mhaas@pipeline.com. **Description:** An executive search firm. The firm's focus is on high-tech recruiting for Western companies in Asia. Company pays fee. **Specializes in the areas of:** Banking; Computer Science/Software; Engineering; Finance; Industrial; Manufacturing. **Number of placements per year:** 1 - 49.

F.P. HEALY & COMPANY INC.
230 Park Avenue, Suite 232, New York NY 10169-0083. **Toll-free phone:** 800/FP-HEALY. **Fax:** 212/661-0383. **Contact:** Frank P. Healy, President. **Description:** An executive search firm operating on a retainer basis. Company pays fee. **Specializes in the areas of:** Administration/MIS/EDP; Banking; Computer Science/Software; Engineering; General Management; Manufacturing; Sales and Marketing. **Positions commonly filled include:** Accountant/Auditor; Actuary; Aerospace Engineer; Electrical/Electronics Engineer; General Manager; Science Technologist; Software Engineer; Systems Analyst. **Corporate headquarters location:** This Location. **Average salary range of placements:** More than $50,000. **Number of placements per year:** 100 - 199.

HESSEL ASSOCIATES
420 Lexington Avenue, Suite 300, New York NY 10170-0399. 212/297-6105. **Fax:** 212/682-1029. **Contact:** Jeffrey Hessel, President. **Description:** An executive search firm operating on both retainer and contingency bases. **Specializes in the areas of:** Accounting/Auditing; Banking; Computer Science/Software; Consulting; Finance. **Average salary range of placements:** More than $50,000.

HUNTINGTON PERSONNEL CONSULTANTS, INC.
P.O. Box 1077, Huntington NY 11743-0640. 516/549-8888. **Fax:** 516/549-3012. **Contact:** Jeannette A. Henry, CPC, President. **E-mail address:** jahenry@i-2000.com. **Description:** An executive search firm that also operates as an employment agency. Client companies are located primarily in New York City and on Long Island. Company pays fee. **Specializes in the areas of:** Administration/MIS/EDP; Business Systems Analysis; Computer Operations; Computer Programming; Consulting; Data Security; Data Communications; Information Technology;

Network Administration; Software Development; Software Documentation; Software Engineering; Software Quality Assurance; Software Training; Systems Administration; Systems Design; Systems Programming; Technical Writing; Telecommunications; Telephone Technical Support. **Positions commonly filled include:** Computer Programmer; Software Engineer; Systems Analyst; Technical Writer/Editor.

INTERSPACE INTERACTIVE INC.
50 East 42nd Street, Suite 2400, New York NY 10017. 212/867-6661. **Fax:** 212/867-6682. **Contact:** Bill Ellis, President. **Description:** An executive search firm. Company pays fee. **Specializes in the areas of:** Accounting/Auditing; Computer Science/Software; Engineering; Finance; General Management; Personnel/Labor Relations; Sales and Marketing. **Positions commonly filled include:** Accountant/Auditor; Manufacturer's/Wholesaler's Sales Rep.; Marketing Specialist; Public Relations Specialist; Services Sales Representative; Software Engineer; Systems Analyst. **Number of placements per year:** 100 - 199.

JOSEPH ASSOCIATES INC.
229 Main Street, Huntington NY 11743-6955. **Fax:** 516/642-1413. **Contact:** Joe Nakelski, President. **E-mail address:** joeassoc@aol.com. **Description:** An executive search firm. **Specializes in the areas of:** Administration/MIS/EDP; Computer Hardware/Software; Market Research. **Positions commonly filled include:** Actuary; Computer Programmer; Human Resources Manager; Market Research Analyst; Mathematician; MIS Specialist; Software Engineer; Statistician; Systems Analyst; Technical Writer/Editor; Telecommunications Manager. **Corporate headquarters location:** This Location. **Other U.S. locations:** Nationwide. **Average salary range of placements:** More than $50,000. **Number of placements per year:** 1 - 49.

IRENE KANE PERSONNEL
27 West Neck Road, Huntington NY 11743. 516/351-1800. **Fax:** 516/351-1626. **Contact:** Ellen West, Owner/Manager. **Description:** An executive search firm. Company pays fee. **Specializes in the areas of:** Computer Science/Software; Engineering. **Positions commonly filled include:** Aerospace Engineer; Computer Programmer; Design Engineer; Electrical/ Electronics Engineer; MIS Specialist; Software Engineer; Systems Analyst; Technical Writer/ Editor. **Corporate headquarters location:** This Location. **Average salary range of placements:** $30,000 - $50,000. **Number of placements per year:** 50 - 99.

DAVID LAWRENCE ASSOCIATES
60 East 42nd Street, New York NY 11710. 212/883-1100. **Fax:** 212/883-0838. **Contact:** Larry Rheingold, President. **Description:** An executive search firm. **Specializes in the areas of:** Computer Hardware/Software. **Positions commonly filled include:** Computer Programmer; MIS Specialist; Software Engineer; Technician.

LAWRENCE EXECUTIVE SEARCH
32 Reni Road, Manhasset NY 11030-1223. 516/627-5361. **Fax:** 516/627-5536. **Contact:** Lawrence Kamisher, President. **E-mail address:** lkamisher@aol.com. **Description:** An executive search firm. Company pays fee. **Specializes in the areas of:** Advertising; Art/Design; Computer Science/ Software; Sales and Marketing. **Corporate headquarters location:** This Location. **Other U.S. locations:** Nationwide. **Average salary range of placements:** More than $50,000. **Number of placements per year:** 1 - 49.

THE MVP GROUP
150 Broadway, 21st Floor, New York NY 10038. 212/571-1830. **Fax:** 212/393-1048. **Contact:** Charlie Otersen, Vice President. **Description:** An executive search firm. **Specializes in the areas of:** Administration/MIS/EDP; Banking; Computer Science/Software; Finance; Personnel/Labor Relations. **Positions commonly filled include:** Accountant/Auditor; Economist; Financial Analyst; Human Resources Manager; Securities Sales Representative. **Number of placements per year:** 500 - 999.

MANAGEMENT RECRUITERS OF MANHATTAN
370 Lexington Avenue, Suite 1412, New York NY 10017. 212/972-7300. **Fax:** 212/972-7309. **Contact:** Jeff Heath, Manager. **Description:** An executive search firm. **Specializes in the areas of:**

Accounting/Auditing; Administration/MIS/EDP; Advertising; Architecture/Construction/Real Estate; Banking; Communications; Computer Science/Software; Electrical; Engineering; Finance; Food Industry; General Management; Health/Medical; Insurance; Legal; Manufacturing; Operations Management; Personnel/Labor Relations; Printing/Publishing; Procurement; Retail; Sales and Marketing; Technical and Scientific; Textiles; Transportation.

MANAGEMENT RECRUITERS OF ORANGE COUNTY
P.O. Box 1530, Greenwood Lake NY 10925. 914/477-9509. **Fax:** 914/477-3016. **Contact:** Carolyn Chermak, President. **Description:** An executive search firm that also operates as a temporary agency. The agency is a division of Management Recruiters International. Company pays fee. **Specializes in the areas of:** Administration/MIS/EDP; Architecture/Construction/Real Estate; Biology; Computer Science/Software; Engineering; Food Industry; General Management; Industrial; Legal; Manufacturing; Nonprofit; Retail; Sales and Marketing; Transportation. **Average salary range of placements:** More than $50,000. **Number of placements per year:** 1 - 49.

MANAGEMENT RECRUITERS OF WOODBURY
COMPUSEARCH
100 Crossways Park West, Suite 208, Woodbury NY 11797. 516/364-9290. **Fax:** 516/364-4478. **Contact:** Bill Jose, Manager. **Description:** An executive search firm. **Specializes in the areas of:** Accounting/Auditing; Administration/MIS/EDP; Architecture/Construction/Real Estate; Banking; Communications; Computer Science/Software; Electrical; Engineering; Finance; Food Industry; General Management; Health/Medical; Industrial; Insurance; Legal; Manufacturing; Personnel/ Labor Relations; Printing/Publishing; Procurement; Retail; Sales and Marketing; Technical and Scientific; Textiles; Transportation.

OPTIMAL RESOURCES
18 East 48th Street, New York NY 10017. 212/486-7713. **Fax:** 212/486-8042. **Contact:** Frank J. Marzi, Vice President. **Description:** An executive search firm operating on a contingency basis. Company pays fee. **Specializes in the areas of:** Accounting/Auditing; Administration/MIS/EDP; Advertising; Banking; Computer Science/Software; Fashion; Finance; Personnel/Labor Relations; Printing/Publishing; Sales and Marketing; Secretarial. **Positions commonly filled include:** Accountant/Auditor; Brokerage Clerk; Financial Analyst; Human Resources Specialist; Management Trainee; Paralegal; Public Relations Specialist; Securities Sales Representative; Services Sales Representative. **Corporate headquarters location:** This Location. **Average salary range of placements:** $30,000 - $50,000. **Number of placements per year:** 200 - 499.

PC DATA INC.
462 7th Avenue, 4th Floor, New York NY 10018. 212/736-5870x212. **Fax:** 212/736-9046. **Contact:** Joe Sabrin, Executive Vice President. **Description:** An executive search firm. Company pays fee. **Specializes in the areas of:** Administration/MIS/EDP; Computer Science/Software; Personnel/Labor Relations; Sales and Marketing. **Positions commonly filled include:** Human Resources Specialist; Internet Services Manager; MIS Specialist; Multimedia Designer; Services Sales Representative; Software Engineer; Technical Writer/Editor; Telecommunications Manager. **Number of placements per year:** 50 - 99.

PATHWAY EXECUTIVE SEARCH, INC.
60 East 42nd Street, Suite 405, New York NY 10165. 212/557-2650. **Fax:** 212/682-1743. **Contact:** Jay Berger, President. **Description:** An executive search firm. Company pays fee. **Specializes in the areas of:** Banking; Computer Science/Software; Finance; Technical and Scientific. **Positions commonly filled include:** Computer Programmer; Economist; Management Analyst/Consultant; Mathematician; Science Technologist; Securities Sales Representative; Software Engineer; Systems Analyst. **Number of placements per year:** 50 - 99.

P.G. PRAGER SEARCH ASSOCIATES, LTD.
1461 Franklin Avenue, Garden City NY 11530. 516/294-4400. **Fax:** 516/294-4443. **Contact:** Paul Gershon Prager, President. **Description:** A contingency search firm specializing in legal, financial, insurance, human resources, marketing, management, office support, and MIS placements. Company pays fee. **Specializes in the areas of:** Accounting/Auditing; Administration/MIS/EDP; Banking; Computer Science/Software; Finance; Insurance; Legal; Personnel/Labor Relations; Sales

and Marketing; Secretarial. **Positions commonly filled include:** Accountant/Auditor; Attorney; Bank Officer/Manager; Budget Analyst; Buyer; Credit Manager; Customer Service Representative; General Manager; Human Resources Manager; Industrial Production Manager; Insurance Agent/ Broker; Management Analyst/Consultant; Manufacturer's/Wholesaler's Sales Rep.; Paralegal; Property and Real Estate Manager; Public Relations Specialist; Purchasing Agent/ Manager; Quality Control Supervisor; Statistician; Systems Analyst; Underwriter/Assistant Underwriter. **Corporate headquarters location:** This Location. **Other U.S. locations:** Nationwide. **Average salary range of placements:** More than $50,000. **Number of placements per year:** 1 - 49.

PRO SEARCH ASSOCIATES, INC.
15 North Mill Street, Nyack NY 10960-3015. 914/353-2260. **Fax:** 914/353-2366. **Contact:** Edwin Kahn, Principal. **Description:** An executive search firm that operates on both retainer and contingency bases and specializes in finance, information systems, and computer professions. Company pays fee. **Specializes in the areas of:** Administration/MIS/EDP; Computer Science/Software; Finance. **Positions commonly filled include:** Computer Programmer; Financial Analyst; Internet Services Manager; Systems Analyst; Telecommunications Manager. **Corporate headquarters location:** This Location. **Average salary range of placements:** More than $50,000. **Number of placements per year:** 1 - 49.

RESPONSE STAFFING SERVICES
271 Madison Avenue, 18th floor, New York NY 10016. 212/983-8870. **Fax:** 212/983-9492. **Contact:** Allen Gutterman, President. **Description:** An executive search firm operating on both retainer and contingency bases. Response Staffing Services is also a specialty recruiting and temporary placement firm. Company pays fee. **NOTE:** This is also the location of Career Advisors, Inc. **Specializes in the areas of:** Accounting/Auditing; Architecture/Construction/Real Estate; Banking; Computer Hardware/Software; Finance; Health/Medical; Insurance; Secretarial. **Positions commonly filled include:** Accountant/Auditor; Actuary; Bank Officer/Manager; Biological Scientist; Biomedical Engineer; Branch Manager; Brokerage Clerk; Budget Analyst; Claim Representative; Clinical Lab Technician; Computer Programmer; Construction and Building Inspector; Credit Manager; Customer Service Representative; EEG Technologist; EKG Technician; Emergency Medical Technician; Financial Analyst; Health Services Manager; Insurance Agent/Broker; Licensed Practical Nurse; Medical Records Technician; Nuclear Medicine Technologist; Occupational Therapist; Paralegal; Pharmacist; Physical Therapist; Physician; Property and Real Estate Manager; Recreational Therapist; Registered Nurse; Respiratory Therapist; Securities Sales Representative; Software Engineer; Surgical Technician; Systems Analyst; Underwriter/Assistant Underwriter. **Benefits available to temporary workers:** Discount Health Plan; Free Bank Account. **Corporate headquarters location:** This Location. **Average salary range of placements:** $30,000 - $50,000. **Number of placements per year:** 200 - 499.

BOB ROSS EXECUTIVE SEARCH CORPORATION
150 West 51st Street, Suite 1811, New York NY 10019. 212/969-9030. **Fax:** 212/969-9067. **Contact:** Bob Ross, President. **E-mail address:** ross14@ix.netcom.com. **Description:** An executive search firm. Bob Ross Executive Search places candidates on both retainer and contingency bases. Company pays fee. **Specializes in the areas of:** Computer Operations; Software Quality Assurance. **Positions commonly filled include:** Computer Programmer; Systems Analyst. **Corporate headquarters location:** This Location. **Other U.S. locations:** Nationwide. **Average salary range of placements:** $30,000 - $50,000. **Number of placements per year:** 100 - 199.

SALES CONSULTANTS OF WESTCHESTER
Nine Skyline Drive, Hawthorne NY 10532. 914/592-1290. **Fax:** 914/592-1258. **Contact:** Bob Penney, Manager. **Description:** An executive search firm operating on a retained and contingency basis. Sales Consultants is a sister company of Management Recruiters, a worldwide recruiting firm. **Specializes in the areas of:** Accounting/Auditing; Administration/MIS/EDP; Architecture/Construction/Real Estate; Banking; Communications; Computer Science/Software; Construction; Electrical; Engineering; Finance; Food Industry; General Management; Health/Medical; Insurance; Legal; Manufacturing; Personnel/Labor Relations; Printing/Publishing; Procurement; Retail; Sales and Marketing; Technical and Scientific; Textiles; Transportation. **Other U.S. locations:** Nationwide.

SHARP PLACEMENT PROFESSIONALS
55 Post Avenue, Westbury NY 11590. 516/876-9222. **Fax:** 516/876-9080. **Contact:** Donald Levine, CPC, President. **Description:** An executive recruiting firm. Company pays fee. **Specializes in the areas of:** Advertising; Computer Science/Software; Engineering; Personnel/Labor Relations; Sales and Marketing. **Positions commonly filled include:** General Manager; Human Resources Specialist; Internet Services Manager; Market Research Analyst; Multimedia Designer; Services Sales Representative; Systems Analyst; Telecommunications Manager. **Corporate headquarters location:** This Location. **Other U.S. locations:** Nationwide. **Average salary range of placements:** More than $50,000. **Number of placements per year:** 1 - 49.

TAYLOR JORDAN ASSOCIATES, INC.
10818 Queens Boulevard, Flushing NY 11375-4789. 718/793-4400. **Contact:** Mr. Pace Langsam, Director of Marketing. **Description:** A contingency search firm. Company pays fee. **Specializes in the areas of:** Accounting/Auditing; Computer Science/Software; General Management; Manufacturing. **Positions commonly filled include:** Accountant/Auditor; Human Resources Manager; Industrial Engineer; Industrial Production Manager; MIS Specialist; Operations/Production Manager. **Average salary range of placements:** More than $50,000.

TECH OPTIONS, INC.
P.O. Box 386, Lenox Hill Station NY 10021. 212/988-3087. **Fax:** 212/744-3977. **Contact:** Martha Kellner, President. **Description:** An executive search and consulting firm specializing in the placement of client/server and computer systems experts. The clientele are primarily financial firms in the metropolitan area. Company pays fee. **Specializes in the areas of:** Computer Science/Software; Finance. **Positions commonly filled include:** Computer Programmer; MIS Specialist; Software Engineer; Systems Analyst. **Corporate headquarters location:** This Location. **Other U.S. locations:** Montclair NJ. **Average salary range of placements:** More than $50,000. **Number of placements per year:** 1 - 49.

TECHNO-TRAC SYSTEMS, INC.
251 Central Park West, New York NY 10024. 212/769-TRAC. **Fax:** 212/873-1596. **Contact:** Mort Trachtenberg, President. **E-mail address:** technomt@nyc.pipeline.com. **Description:** An executive search firm that places information technology, financial, and research professionals. Company pays fee. **Specializes in the areas of:** Administration/MIS/EDP; Banking; Computer Science/Software. **Positions commonly filled include:** Budget Analyst; Computer Operator; Computer Programmer; Database Manager; Financial Analyst; Management Analyst/Consultant; MIS Specialist; Software Engineer; Systems Analyst; Systems Manager. **Corporate headquarters location:** This Location. **Average salary range of placements:** More than $50,000. **Number of placements per year:** 1 - 49.

TRAYNOR CONFIDENTIAL LTD.
P.O. Box 189, Pittsford NY 14534. 716/387-0383. **Fax:** 716/387-0384. **Contact:** Tom Traynor, President. **Description:** An executive search firm operating on a contingency basis. Company pays fee. **Specializes in the areas of:** Accounting/Auditing; Architecture/Construction/Real Estate; Computer Science/Software; Finance; Telecommunications. **Positions commonly filled include:** Accountant/Auditor; Architect; Budget Analyst; Civil Engineer; Computer Programmer; Construction and Building Inspector; Construction Contractor; Cost Estimator; Financial Analyst; Internet Services Manager; Landscape Architect; Market Research Analyst; MIS Specialist; Multimedia Designer; Property and Real Estate Manager; Software Engineer; Statistician. **Corporate headquarters location:** This Location. **Average salary range of placements:** $30,000 - $50,000. **Number of placements per year:** 1 - 49.

WEHN ASSOCIATES, INC.
71 Vanderbilt Avenue, New York NY 10017. 212/675-3224. **Fax:** 212/986-4245. **Contact:** President. **Description:** An executive search firm. **Specializes in the areas of:** Administration/MIS/EDP; Computer Science/Software. **Positions commonly filled include:** Architect; Computer Programmer; Designer; Software Engineer; Systems Analyst; Technical Writer/Editor. **Number of placements per year:** 50 - 99.

PERMANENT EMPLOYMENT AGENCIES OF NEW YORK

ACCOUNTING & COMPUTER PERSONNEL
200 Salina Meadows Parkway, Suite 180, Syracuse NY 13212. 315/457-8000. **Fax:** 315/457-0029. **Contact:** William E. Winnewisser, President. **Description:** An employment agency. Company pays fee. **Specializes in the areas of:** Accounting/Auditing; Administration/MIS/EDP; Advertising; Architecture/Construction/Real Estate; Art/Design; Banking; Biology; Broadcasting; Computer Science/Software; Economics; Engineering; Finance; Food Industry; General Management; Health/Medical; Industrial; Insurance; Legal; Manufacturing; Nonprofit; Personnel/ Labor Relations; Printing/Publishing; Retail; Sales and Marketing; Technical and Scientific. **Positions commonly filled include:** Accountant/Auditor; Bank Officer/Manager; Budget Analyst; Computer Programmer; Cost Estimator; Credit Manager; Financial Analyst; Software Engineer; Systems Analyst; Technical Writer/Editor. **Number of placements per year:** 50 - 99.

AMES GROUP
928 Broadway, Suite 1101A, New York NY 10010-6008. 212/475-5900. **Fax:** 212/674-2401. **Contact:** Max Sabrin, Managing Director. **E-mail address:** amesgroup@aol.com. **Description:** An employment agency. Company pays fee. **Specializes in the areas of:** Administration/MIS/EDP; Banking; Broadcasting; Computer Hardware/Software; Finance; Health/Medical; Insurance; Legal; Personnel/Labor Relations; Printing/Publishing; Sales and Marketing; Secretarial. **Positions commonly filled include:** Accountant/Auditor; Administrative Assistant; Bookkeeper; Computer Operator; Computer Programmer; Customer Service Representative; EDP Specialist; Legal Secretary; Marketing Specialist; MIS Specialist; Public Relations Specialist; Secretary; Software Engineer; Systems Analyst; Technical Writer/Editor; Typist/Word Processor. **Corporate headquarters location:** This Location. **Average salary range of placements:** More than $50,000. **Number of placements per year:** 1 - 49.

ANALYTIC RECRUITING INC.
21 East 40th Street, Suite 500, New York NY 10016. 212/545-8511. **Fax:** 212/545-8520. **Contact:** Rita Raz, Principal. **Description:** An employment agency. Company pays fee. **Specializes in the areas of:** Administration/MIS/EDP; Banking; Computer Science/Software; Economics; Finance; Sales and Marketing. **Positions commonly filled include:** Computer Programmer; Economist; Financial Analyst; Management Analyst/Consultant; Mathematician; Statistician; Systems Analyst. **Number of placements per year:** 100 - 199.

APRIL TECHNICAL RECRUITING
P.O. Box 40303, Rochester NY 14604-0803. 716/325-5220. **Fax:** 716/546-4870. **Contact:** Tom Miller, Technical Recruiter. **Description:** Founded in 1972, April Technical Recruiting is a full-service employment agency. Company pays fee. **Specializes in the areas of:** Administration/MIS/EDP; Computer Science/Software; Engineering; Manufacturing; Technical and Scientific. **Positions commonly filled include:** Accountant/Auditor; Chemical Engineer; Civil Engineer; Computer Programmer; Cost Estimator; Design Engineer; Draftsperson; Electrical/ Electronics Engineer; Environmental Engineer; Industrial Engineer; Industrial Production Manager; Mechanical Engineer; MIS Specialist; Quality Control Supervisor; Software Engineer; Structural Engineer; Systems Analyst. **Corporate headquarters location:** This Location. **Other U.S. locations:** Nationwide. **Average salary range of placements:** $30,000 - $50,000.

C.C. BURKE LIMITED
60 East 42nd Street, New York NY 10165-0999. 212/286-0092. **Fax:** 212/286-0396. **Contact:** Charlene Burke, President. **E-mail address:** burkecc@msn.com. **Description:** An employment agency. Company pays fee. **Specializes in the areas of:** Accounting/Auditing; Computer Science/ Software; Fashion; Personnel/Labor Relations; Sales and Marketing; Secretarial. **Positions commonly filled include:** Accountant/Auditor; Computer Programmer; Human Resources Manager; Management Trainee; Public Relations Specialist; Systems Analyst. **Average salary range of placements:** $30,000 - $50,000. **Number of placements per year:** 50 - 99.

CARLILE PERSONNEL AGENCY INC.

Three Ellingwood Court, Suite 202, New Hartford NY 13413. 315/736-3083. **Fax:** 315/736-5340. **Contact:** Doug Manning, Recruiter. **Description:** An employment agency. Company pays fee. **Specializes in the areas of:** Accounting/Auditing; Administration/MIS/EDP; Computer Science/Software; Engineering; Finance; Industrial; Manufacturing; Personnel/Labor Relations; Sales and Marketing; Technical and Scientific. **Positions commonly filled include:** Accountant/ Auditor; Chemical Engineer; Civil Engineer; Computer Programmer; Designer; Draftsperson; Electrical/Electronics Engineer; Human Resources Manager; Industrial Engineer; Industrial Production Manager; Management Analyst/Consultant; Manufacturer's/Wholesaler's Sales Rep.; Mechanical Engineer; Metallurgical Engineer; Operations/Production Manager; Purchasing Agent/Manager; Quality Control Supervisor; Software Engineer; Structural Engineer; Systems Analyst. **Number of placements per year:** 1 - 49.

IRENE COHEN TEMPS

475 Fifth Avenue, 2nd Floor, New York NY 10017-6220. 212/725-1666. **Fax:** 212/889-6746. **Contact:** Diane Cohen, Vice President. **Description:** An employment agency. Company pays fee. **Specializes in the areas of:** Accounting/Auditing; Administration/MIS/EDP; Advertising; Computer Science/Software; Finance; Personnel/Labor Relations; Secretarial. **Positions commonly filled include:** Accountant/Auditor; Administrative Services Manager; Customer Service Representative; Financial Analyst; Human Resources Manager; Internet Services Manager; Systems Analyst; Typist/Word Processor. **Corporate headquarters location:** This Location. **Average salary range of placements:** $30,000 - $50,000. **Number of placements per year:** 500 - 999.

EDEN PERSONNEL, INC.

280 Madison Avenue, New York NY 10016. 212/685-8600. **Contact:** Randie Rice, Office Manager. **Description:** An employment agency. Company pays fee. **Specializes in the areas of:** Accounting/Auditing; Administration/MIS/EDP; Advertising; Computer Hardware/Software; Fashion; Finance; Food Industry; Health/Medical; Industrial; Legal; Nonprofit; Personnel/Labor Relations; Retail; Secretarial. **Positions commonly filled include:** Administrative Assistant; Clerk; Computer Operator; Data Entry Clerk; Legal Secretary; Light Industrial Worker; Medical Secretary; Nurse; Receptionist; Secretary; Stenographer; Typist/Word Processor.

EMPLOYMENT RECRUITERS AGENCY

11821 Queens Boulevard, Suite 609, Flushing NY 11375. 718/263-2300. **Fax:** 718/263-9668. **Contact:** Brian Moran, President. **Description:** An employment services agency. Company pays fee. **Specializes in the areas of:** Accounting/Auditing; Administration/MIS/EDP; Advertising; Architecture/Construction/Real Estate; Banking; Computer Science/Software; Fashion; Finance; General Management; Health/Medical; Insurance; Legal; Manufacturing; Personnel/Labor Relations; Secretarial. **Positions commonly filled include:** Accountant/Auditor; Administrative Services Manager; Clerical Supervisor; Credit Manager; Customer Service Representative; Financial Analyst; General Manager; Human Resources Manager; Management Analyst/ Consultant; Management Trainee; Paralegal. **Number of placements per year:** 200 - 499.

EXTRA HELP EMPLOYMENT SERVICE

950 New Loudon Road, Troy NY 12180. 518/272-2080. **Fax:** 518/272-3643. **Contact:** Genevieve Wolk, Manager. **Description:** An employment agency. Company pays fee. **Specializes in the areas of:** Accounting/Auditing; Computer Science/Software; Engineering; General Management; Industrial; Legal. **Positions commonly filled include:** Administrative Services Manager; Clerical Supervisor; Computer Programmer; Customer Service Representative; Draftsperson; General Manager; Industrial Engineer; Insurance Agent/Broker; Manufacturer's/Wholesaler's Sales Rep.; Mechanical Engineer; Paralegal. **Corporate headquarters location:** Rochester NY. **Average salary range of placements:** $20,000 - 50,000. **Number of placements per year:** 500 - 999.

ROBERT HALF INTERNATIONAL INC.
INFORMATION SYSTEMS DIVISION

565 Fifth Avenue, 12th Floor, New York NY 10017. 212/290-2700. **Fax:** 212/290-8100. **Contact:** Director. **Description:** A division of the worldwide employment agency. Company pays fee. **Specializes in the areas of:** Accounting/Auditing; Administration/MIS/EDP; Advertising;

Architecture/Construction/Real Estate; Banking; Broadcasting; Computer Science/Software; Engineering; Finance; General Management; Industrial; Insurance; Legal; Manufacturing; Nonprofit; Personnel/Labor Relations; Sales and Marketing; Technical and Scientific. **Positions commonly filled include:** Computer Programmer; Financial Analyst; Management; Science Technologist; Software Engineer; Statistician; Systems Analyst; Technical Writer/Editor. **Number of placements per year:** 200 - 499.

HUNTER MAC & ASSOCIATES

139 Fulton Street, New York NY 10038. 212/267-2790. **Fax:** 212/962-2339. **Contact:** Patrick McKeown, Vice President. **Description:** An employment agency. Company pays fee. **Specializes in the areas of:** Accounting/Auditing; Administration/MIS/EDP; Banking; Computer Science/ Software; Finance; Insurance; Personnel/Labor Relations; Secretarial. **Positions commonly filled include:** Accountant/Auditor; Administrative Services Manager; Brokerage Clerk; Computer Programmer; Financial Analyst; Human Resources Manager; Librarian; Management Trainee; Securities Sales Representative; Systems Analyst. **Number of placements per year:** 200 - 499.

INFORMATION SYSTEMS STAFFING, INC.

5730 Commons Park Drive, East Syracuse NY 13057. 315/449-1838. **Toll-free phone:** 800/466-1939. **Fax:** 315/449-1939. **Contact:** Manager. **E-mail address:** issi@dreamscape.com. **Description:** An employment agency specializing in the placement of contract and permanent computer professionals in central New York. Company pays fee. **Specializes in the areas of:** Computer Science/Software. **Positions commonly filled include:** Computer Programmer; MIS Specialist; Software Engineer; Systems Analyst; Technical Writer/Editor; Telecommunications Manager. **Corporate headquarters location:** This Location. **Average salary range of placements:** $30,000 - $50,000. **Number of placements per year:** 1 - 49.

INNOVATIONS ASSOCIATES

627 Field Street, Johnson City NY 13790-1057. 607/798-9376. **Fax:** 607/797-3485. **Contact:** Gene P. George, Director of Personnel Services. **Description:** An employment agency that specializes in long-term contract assignments, permanent placement, and temp-to-perm positions. Company pays fee. **Specializes in the areas of:** Accounting/Auditing; Administration/MIS/EDP; Advertising; Art/Design; Computer Science/Software; Engineering; Sales and Marketing. **Positions commonly filled include:** Accountant/Auditor; Advertising Clerk; Civil Engineer; Clerical Supervisor; Computer Programmer; Customer Service Representative; Draftsperson; Electrical/Electronics Engineer; Industrial Engineer; Library Technician; Market Research Analyst; Mechanical Engineer; MIS Specialist; Multimedia Designer; Public Relations Specialist; Quality Control Supervisor; Software Engineer; Structural Engineer; Surveyor; Systems Analyst; Technical Writer/Editor; Telecommunications Manager. **Benefits available to temporary workers:** 401(k); Medical Insurance. **Corporate headquarters location:** This Location.

JDC ASSOCIATES

330 Vanderbilt-Motor Parkway, Suite 101, Hauppauge NY 11788. 516/231-8581. **Fax:** 516/231-8011. **Contact:** Lori Boyle, President. **Description:** An employment agency. Company pays fee. **Specializes in the areas of:** Accounting/Auditing; Administration/MIS/EDP; Computer Science/Software; Health/Medical; Sales and Marketing; Secretarial. **Positions commonly filled include:** Accountant/Auditor; Clerical Supervisor; Computer Programmer; Customer Service Representative; Occupational Therapist; Systems Analyst; Typist/Word Processor.

KTECH SYSTEMS GROUP, INC.

150 Broadway, New York NY 10038. 212/227-0800. **Fax:** 212/227-1412. **Contact:** Dick Lewis, President. **E-mail address:** ktechsys@interport.net. **World Wide Web address:** http://www.ktech.com. **Description:** An employment agency that recruits and places computer technology professionals with *Fortune* 1000 companies, banks, investment firms, management consulting firms, and software/hardware manufacturing companies. Company pays fee. **Specializes in the areas of:** Administration/MIS/EDP; Banking; Computer Science/Software; Finance; Personnel/Labor Relations; Technical and Scientific. **Positions commonly filled include:** Computer Programmer; Financial Analyst; Human Resources Manager; Internet Services Manager; MIS Specialist; Science Technologist; Software Engineer; Systems Analyst; Technical Writer/

Editor; Telecommunications Manager. **Other U.S. locations:** Short Hills NJ. **Average salary range of placements:** More than $50,000. **Number of placements per year:** 200 - 499.

LYNN MARSHALL PERSONNEL AGENCY, INC.

91-31 Queens Boulevard, Rego Park NY 11373. 718/446-5200. **Fax:** 718/446-5202. **Contact:** Lynn Marshall, Owner/President. **Description:** An employment agency. Company pays fee. **Specializes in the areas of:** Accounting/Auditing; Banking; Computer Science/Software; Legal; Secretarial. **Positions commonly filled include:** Accountant/Auditor; Administrative Services Manager; Bank Officer/Manager; Clerical Supervisor; Computer Programmer; Credit Manager; Customer Service Representative; Human Resources Manager; Management Trainee; Services Sales Representative; Systems Analyst; Typist/Word Processor. **Number of placements per year:** 50 - 99.

MOGUL CONSULTANTS INC.

380 North Broadway, Suite 208, Jericho NY 11753. 516/822-4363. **Fax:** 516/822-4364. **Contact:** Gene Mogul, President. **Description:** An employment agency. Company pays fee. **Specializes in the areas of:** Computer Science/Software; Engineering; General Management; Sales and Marketing; Technical and Scientific. **Positions commonly filled include:** Broadcast Technician; Computer Programmer; Electrical/Electronics Engineer; Software Engineer; Systems Analyst. **Number of placements per year:** 1 - 49.

NATIONAL EMPLOYMENT DATABASE, INC.

50 Broad Street, Suite 421, New York NY 10004. 212/269-1994. **Fax:** 212/269-1995. **Contact:** Ray Bloom, President. **World Wide Web address:** http://www.supertek.com. **Description:** National Employment Database, Inc. is a database employment service placing professionals in office support, accounting, bookkeeping, and computer-related positions. Company pays fee. **Specializes in the areas of:** Accounting/Auditing; Administration/MIS/EDP; Computer Science/Software; Finance; Printing/Publishing. **Positions commonly filled include:** Accountant/Auditor; Administrative Services Manager; Brokerage Clerk; Computer Programmer; Economist; Electrical/Electronics Engineer; Human Resources Manager; Management Trainee; Metallurgical Engineer; Operations/Production Manager; Paralegal; Software Engineer; Systems Analyst. **Corporate headquarters location:** This Location. **Other U.S. locations:** Nationwide. **Average salary range of placements:** $30,000 - $50,000. **Number of placements per year:** 500 - 999.

NATIONWIDE PERSONNEL GROUP

474 Elmwood Avenue, Suite 201, Buffalo NY 14222. 716/881-2144. **Fax:** 716/881-0711. **Contact:** Mark Gademsky, President. **E-mail address:** gademsky@localnet.com. **Description:** An employment agency focusing on the placement of computer and engineering personnel. Company pays fee. **Specializes in the areas of:** Administration/MIS/EDP; Computer Science/Software; Engineering. **Positions commonly filled include:** Computer Programmer; Design Engineer; Electrical/Electronics Engineer; Internet Services Manager; Mathematician; Mechanical Engineer; MIS Specialist; Software Engineer; Systems Analyst; Telecommunications Manager. **Corporate headquarters location:** This Location. **Average salary range of placements:** $30,000 - $50,000. **Number of placements per year:** 50 - 99.

NORRELL SERVICES

420 Lexington Street, Suite 2516, New York NY 10017. 212/697-5240. **Fax:** 212/557-0034. **Contact:** Cristi Riccio, Recruiter. **Description:** A national employment service founded in 1961. Company pays fee. **Specializes in the areas of:** Accounting/Auditing; Administration/MIS/EDP; Computer Science/Software; Finance; Personnel/Labor Relations; Sales and Marketing; Secretarial; Technical and Scientific. **Positions commonly filled include:** Accountant/Auditor; Customer Service Representative; Financial Analyst; Human Resources Manager; Paralegal; Services Sales Representative; Typist/Word Processor. **Benefits available to temporary workers:** Dental Insurance; Medical Insurance; Paid Holidays; Paid Vacation; Stock Option Plan. **Corporate headquarters location:** Atlanta GA. **Other U.S. locations:** Nationwide. **Average salary range of placements:** $30,000 - $50,000. **Number of placements per year:** 1000+.

PARSONS, ANDERSON AND GEE, INC.
642 Kreag Road, Pittsford NY 14534-3705. 716/586-8679. **Fax:** 716/586-0247. **Contact:** Art Fandel, General Manager. **Description:** An employment agency. Company pays fee. **Specializes in the areas of:** Accounting/Auditing; Administration/MIS/EDP; Computer Science/Software; Engineering; Finance; Food Industry; General Management; Industrial; Insurance; Manufacturing; Personnel/Labor Relations; Printing/Publishing; Sales and Marketing; Technical and Scientific. **Positions commonly filled include:** Accountant/Auditor; Biomedical Engineer; Buyer; Chemical Engineer; Computer Programmer; Designer; Electrical/Electronics Engineer; General Manager; Human Resources Manager; Industrial Engineer; Industrial Production Manager; Manufacturer's/Wholesaler's Sales Rep.; Mechanical Engineer; Metallurgical Engineer; Purchasing Agent/Manager; Quality Control Supervisor; Securities Sales Representative; Software Engineer; Stationary Engineer; Structural Engineer; Systems Analyst; Technical Writer/Editor. **Number of placements per year:** 1 - 49.

QUANTUM PERSONNEL AGENCY INC.
17 East 45th Street, New York NY 10017. 212/286-0111. **Fax:** 212/808-5279. **Contact:** Stu Howard, Executive Recruiter. **Description:** An employment agency. **Specializes in the areas of:** Computer Science/Software. **Positions commonly filled include:** Computer Programmer; Systems Analyst. **Number of placements per year:** 500 - 999.

REM RESOURCES, INC.
507 Fifth Avenue, Suite 1101, New York NY 10017. 212/661-0090. **Fax:** 212/661-0107. **Contact:** Gabrielle Rem, President. **Description:** An employment agency operating on a contingency basis. The firm also provides temporary placements. Company pays fee. **Specializes in the areas of:** Accounting/Auditing; Administration/MIS/EDP; Advertising; Banking; Computer Science/ Software; Fashion; Finance; Health/Medical; Legal; Manufacturing; Nonprofit; Personnel/Labor Relations; Printing/Publishing; Sales and Marketing; Secretarial. **Positions commonly filled include:** Accountant/Auditor; Adjuster; Administrative Services Manager; Advertising Clerk; Bank Officer/Manager; Branch Manager; Brokerage Clerk; Budget Analyst; Claim Representative; Clerical Supervisor; Computer Programmer; Economist; Editor; Financial Analyst; Health Services Manager; Human Resources Manager; Human Service Worker; Operations/Production Manager; Public Relations Specialist; Securities Sales Representative; Systems Analyst; Technical Writer/Editor; Typist/Word Processor. **Average salary range of placements:** $30,000 - $50,000. **Number of placements per year:** 100 - 199.

FRAN ROGERS PERSONNEL
One Huntington Quadrangle, Suite 2S09, Melville NY 11747. 516/752-8888. **Contact:** Fran Rogers, President. **Description:** An employment agency. Company pays fee. **Specializes in the areas of:** Accounting/Auditing; Administration/MIS/EDP; Banking; Computer Science/Software; Engineering; Finance; Technical and Scientific. **Positions commonly filled include:** Accountant/Auditor; Administrative Assistant; Bookkeeper; EDP Specialist; Electrical/Electronics Engineer; Financial Analyst; MIS Specialist; Receptionist; Secretary; Systems Analyst; Typist/Word Processor. **Number of placements per year:** 200 - 499.

SIGMA SEARCH
535 Broad Hollow Road, Melville NY 11747. 516/694-7707. **Contact:** Thea Linker, President. **Description:** An employment agency. Company pays fee. **Specializes in the areas of:** Accounting/Auditing; Administration/MIS/EDP; Computer Science/Software; Finance; Technical and Scientific. **Positions commonly filled include:** Accountant/Auditor; Bookkeeper; Computer Programmer; EDP Specialist; Financial Analyst; MIS Specialist; Systems Analyst. **Number of placements per year:** 50 - 99.

STAMM PERSONNEL AGENCY INC.
27 Whitehall Street, New York NY 10004. 212/509-6600. **Fax:** 212/509-3773. **Contact:** Ava Hall, Executive Recruiter. **Description:** An employment agency. Company pays fee. **Specializes in the areas of:** Administration/MIS/EDP; Computer Science/Software; Consulting; Finance; Investment; Sales and Marketing; Secretarial. **Positions commonly filled include:** Accountant/Auditor; Attorney; Brokerage Clerk; Computer Programmer; Customer Service Representative; Economist;

Editor; Financial Analyst; Securities Sales Representative; Systems Analyst; Technical Writer/Editor. **Number of placements per year:** 100 - 199.

UNITED PERSONNEL AGENCY, INC.

51 East 42nd Street, Suite 417, New York NY 10017. 212/490-2197. **Fax:** 212/661-2767. **Contact:** Michael P. Williams, President. **Description:** An employment agency. Company pays fee. **Specializes in the areas of:** Accounting/Auditing; Advertising; Banking; Broadcasting; Computer Science/Software; Legal; Nonprofit; Personnel/Labor Relations; Printing/Publishing; Secretarial; Word Processing. **Positions commonly filled include:** Accountant/Auditor; Budget Analyst; Claim Representative; Paralegal; Software Engineer; Systems Analyst. **Number of placements per year:** 1000+.

H.L. YOH COMPANY

301 Exchange Boulevard, Rochester NY 14608. 716/454-5400. **Fax:** 716/454-2105. **Contact:** Johnna Ufret, Technical Recruiter. **E-mail address:** yohit@vivanet.com. **Description:** A full-service employment agency founded in 1942. Company pays fee. **Specializes in the areas of:** Administration/MIS/EDP; Art/Design; Computer Science/Software; Engineering; Industrial; Manufacturing; Technical and Scientific. **Positions commonly filled include:** Accountant/Auditor; Aerospace Engineer; Agricultural Engineer; Aircraft Mechanic/Engine Specialist; Biological Scientist; Biomedical Engineer; Blue-Collar Worker Supervisor; Chemical Engineer; Chemist; Civil Engineer; Clinical Lab Technician; Computer Programmer; Customer Service Representative; Design Engineer; Designer; Draftsperson; Electrical/Electronics Engineer; Environmental Engineer; Human Resources Manager; Industrial Engineer; Industrial Production Manager; Internet Services Manager; Management Analyst/Consultant; Mechanical Engineer; Metallurgical Engineer; MIS Specialist; Operations/Production Manager; Quality Control Supervisor; Software Engineer; Stationary Engineer; Structural Engineer; Systems Analyst; Technical Writer/Editor; Telecommunications Manager. **Benefits available to temporary workers:** Paid Holidays; Paid Vacation. **Corporate headquarters location:** Radnor PA. **Other U.S. locations:** Nationwide. **Average salary of placements:** $30,000 - $50,000. **Number of placements per year:** 100 - 199.

EXECUTIVE SEARCH FIRMS OF NORTH CAROLINA

ACCURATE STAFFING CONSULTANTS, INC.

1328-D Starbrook Drive, Charlotte NC 28210. 704/554-9675. **Fax:** 704/554-5914. **Contact:** Catherine Wall, President. **Description:** An executive search firm. Company pays fee. **Specializes in the areas of:** Accounting/Auditing; Administration/MIS/EDP; Banking; Computer Science/Software; Engineering; Industrial; Manufacturing; Personnel/Labor Relations; Secretarial; Technical and Scientific. **Average salary range of placements:** $30,000 - $50,000. **Number of placements per year:** 50 - 99.

ADVANCED PERSONNEL RESOURCES

20 Oak Branch Drive, Suite D, Greensboro NC 27407-2145. 910/855-6664. **Fax:** 910/299-8746. **Contact:** Diane Z. Gaines, President. **Description:** An executive search firm. Company pays fee. **Specializes in the areas of:** Accounting/Auditing; Administration/MIS/EDP; Computer Science/Software; Finance; General Management; Insurance; Legal; Manufacturing; Personnel/Labor Relations; Printing/Publishing; Sales and Marketing; Secretarial.

AMERIPRO SEARCH, INC.

20468-A Chartwell Centre Drive, Cornelius NC 28031. 704/896-8991. **Fax:** 704/896-8855. **Contact:** Elaine C. Brauninger, President. **Description:** An executive search firm. Company pays fee. **Specializes in the areas of:** Accounting/Auditing; Administration/MIS/EDP; Computer

Science/Software; Data Processing; Engineering; Finance; General Management; Industrial; Manufacturing; MIS/EDP; Personnel/Labor Relations; Sales and Marketing; Technical and Scientific. **Positions commonly filled include:** Accountant/Auditor; Administrative Services Manager; Advertising Clerk; Aerospace Engineer; Biochemist; Biological Scientist; Branch Manager; Designer; Engineer; Financial Analyst; General Manager; Industrial Production Manager; Internet Services Manager; MIS Specialist; Operations Research Analyst; Science Technologist; Scientist; Software Engineer; Technical Writer/Editor; Telecommunications Manager. **Average salary range of placements:** More than $50,000. **Number of placements per year:** 100 - 199.

AMOS & ASSOCIATES
633-B Chapel Hill Road, Burlington NC 27215. 910/222-0231. **Fax:** 910/222-1214. **Contact:** Diane Amos, President. **E-mail address:** famsamos@netpath.net. **World Wide Web address:** http://www.resourcecenter.com/amos. **Description:** An executive search firm. Company pays fee. **Specializes in the areas of:** Computer Science/Software. **Positions commonly filled include:** Computer Programmer; Software Engineer; Systems Analyst. **Average salary range of placements:** $30,000 - $50,000. **Number of placements per year:** 50 - 99.

EASTERN SEARCH GROUP
P.O. Box 4655, Wilmington NC 28406. 910/799-7700. **Fax:** 910/392-6266. **Contact:** Fred Wells, President. **Description:** An executive search firm. Company pays fee. **Specializes in the areas of:** Apparel; Computer Science/Software; Engineering; Industrial; Manufacturing; Textiles. **Positions commonly filled include:** Accountant/Auditor; Chemical Engineer; Computer Programmer; Electrical/Electronics Engineer; Industrial Engineer; Industrial Production Manager; Mechanical Engineer; Quality Control Supervisor; Systems Analyst. **Number of placements per year:** 1 - 49.

EXECUTIVE RECRUITMENT SPECIALISTS, INC.
6407 Idlewild Road, Building 1, Suite 103, Charlotte NC 28212. 704/536-8830. **Fax:** 704/536-8893. **Contact:** Eric Sklut, President. **Description:** An executive search firm. Company pays fee. **Specializes in the areas of:** Computer Science/Software; Engineering; Finance; General Management; Government; Health/Medical; Manufacturing. **Positions commonly filled include:** Accountant/Auditor; Biomedical Engineer; Electrical/Electronics Engineer; General Manager; Health Services Manager; Licensed Practical Nurse; Mechanical Engineer; Medical Records Technician; Nuclear Medicine Technologist; Occupational Therapist; Physical Therapist; Quality Control Supervisor; Radiologic Technologist; Recreational Therapist; Registered Nurse; Respiratory Therapist; Software Engineer; Speech-Language Pathologist. **Number of placements per year:** 50 - 99.

ROBERT HALF INTERNATIONAL, INC.
300 North Greene Street, Suite 275, Greensboro NC 27401. 919/274-4253. **Fax:** 919/273-2882. **Contact:**. Human Resources. **Description:** An executive search firm. Company pays fee. **Specializes in the areas of:** Accounting/Auditing; Administration/MIS/EDP; Banking; Computer Hardware/Software; Finance. **Positions commonly filled include:** Accountant/Auditor; Bookkeeper; Clerk; Computer Operator; Computer Programmer; CPA; Credit Manager; Data Entry Clerk; EDP Specialist; Software Engineer; Systems Analyst.

INFORMATION SYSTEMS PROFESSIONALS, INC.
5904 Castlebrook Drive, Raleigh NC 27604. 919/954-9100. **Fax:** 919/954-1947. **Contact:** Brad Moses, President. **E-mail address:** ispros@nando.net. **Description:** An executive search firm. **Specializes in the areas of:** Administration/MIS/EDP; Computer Science/Software; Technical and Scientific. **Positions commonly filled include:** Computer Programmer; Management Analyst/Consultant; Software Engineer; Systems Analyst; Technical Writer/Editor. **Average salary range of placements:** $30,000 - $50,000. **Number of placements per year:** 1 - 49.

KEY PERSONNEL SERVICES
1400 Battleground Avenue, Suite 211, Greensboro NC 27408. 910/272-1333. **Fax:** 910/272-9397. **Contact:** Chris Sparks, Vice President. **Description:** An executive search firm. Company pays fee. **Specializes in the areas of:** Accounting/Auditing; Administration/MIS/EDP; Advertising; Banking; Computer Science/Software; Engineering; Finance; General Management; Industrial;

Manufacturing; Personnel/Labor Relations; Sales and Marketing; Secretarial; Textiles. **Positions commonly filled include:** Accountant/Auditor; Bank Officer/Manager; Blue-Collar Worker Supervisor; Branch Manager; Buyer; Claim Representative; Computer Programmer; Credit Manager; Customer Service Representative; Electrical/Electronics Engineer; Financial Analyst; General Manager; Human Resources Specialist; Industrial Engineer; Industrial Production Manager; Insurance Agent/Broker; Management Analyst/Consultant; Management Trainee; Mechanical Engineer; MIS Specialist; Purchasing Agent/Manager; Securities Sales Representative; Software Engineer; Systems Analyst; Travel Agent. **Average salary range of placements:** $30,000 - $50,000. **Number of placements per year:** 50 - 99.

KILGO & COMPANY

8318 Pineville-Matthews Road, Charlotte NC 28226. 704/544-0342. **Fax:** 704/542-6353. **Contact:** Don Kilgo, Owner. **Description:** An executive search firm. Company pays fee. **NOTE:** The firm places experienced technical personnel only. **Specializes in the areas of:** Computer Science/Software. **Positions commonly filled include:** Sales and Marketing Representative; Sales Manager. **Number of placements per year:** 1 - 49.

MANAGEMENT RECRUITERS

305 South Main Street, Kannapolis NC 28081. 704/938-6144. **Fax:** 704/938-3480. **Contact:** Tom Whitley, President. **Description:** An executive search firm. Company pays fee. **Specializes in the areas of:** Computer Science/Software. **Positions commonly filled include:** Computer Programmer; Systems Analyst. **Number of placements per year:** 1 - 49.

MANAGEMENT RECRUITERS

22 South Pack Square, Suite 302, Asheville NC 28801. 704/258-9646. **Fax:** 704/252-0866. **Contact:** Paul Rumson, President. **Description:** An executive search firm operating on both retainer and contingency bases. Company pays fee. **Specializes in the areas of:** Computer Science/Software; Engineering; Industrial; Manufacturing. **Positions commonly filled include:** Chemical Engineer; Chemist; Computer Programmer; Designer; Draftsperson; Electrical/Electronics Engineer; Hotel Manager; Industrial Engineer; Industrial Production Manager; Materials Engineer; Mechanical Engineer; Mining Engineer; Purchasing Agent/Manager; Quality Control Supervisor; Restaurant/Food Service Manager; Software Engineer; Structural Engineer. **Corporate headquarters location:** Cleveland OH. **International locations:** Worldwide. **Average salary range of placements:** More than $50,000. **Number of placements per year:** 1 - 49.

MANAGEMENT RECRUITERS

5509 Creedmoor Road, Suite 206, Raleigh NC 27612-2812. 919/781-0400. **Contact:** Phillip Stanley, Manager. **Description:** An executive search firm. **Specializes in the areas of:** Accounting/Auditing; Administration/MIS/EDP; Advertising; Architecture/Construction/Real Estate; Banking; Communications; Computer Hardware/Software; Design; Electrical; Engineering; Food Industry; General Management; Health/Medical; Insurance; Legal; Manufacturing; Operations Management; Personnel/Labor Relations; Printing/Publishing; Procurement; Retail; Sales and Marketing; Technical and Scientific; Textiles; Transportation.

MANAGEMENT RECRUITERS

P.O. Box 6077, Hickory NC 28603. 704/495-8233. **Contact:** Byron King, Owner. **Description:** An executive search firm. **Specializes in the areas of:** Accounting/Auditing; Administration/MIS/EDP; Advertising; Architecture/Construction/Real Estate; Banking; Communications; Computer Hardware/Software; Design; Electrical; Engineering; Food Industry; General Management; Health/Medical; Insurance; Legal; Manufacturing; Operations Management; Personnel/Labor Relations; Printing/Publishing; Procurement; Retail; Sales and Marketing; Technical and Scientific; Textiles; Transportation.

MANAGEMENT RECRUITERS

P.O. Box 629, Cedar Mountain NC 28718. 704/884-4118. **Fax:** 704/884-3512. **Contact:** Frank Schoff, President. **Description:** An executive search firm. **Specializes in the areas of:** Accounting/Auditing; Administration/MIS/EDP; Advertising; Architecture/Construction/Real Estate; Banking; Communications; Computer Hardware/Software; Design; Electrical; Engineering; Food Industry; General Management; Health/Medical; Insurance; Legal; Manufacturing;

Operations Management; Personnel/Labor Relations; Printing/Publishing; Procurement; Retail; Sales and Marketing; Technical and Scientific; Textiles; Transportation.

MANAGEMENT RECRUITERS
835 Highland Avenue SE, Hickory NC 28602. 704/324-2020. **Contact:** Scott Volz, Manager. **Description:** An executive search firm. **Specializes in the areas of:** Accounting/Auditing; Administration/MIS/EDP; Advertising; Architecture/Construction/Real Estate; Banking; Communications; Computer Hardware/Software; Design; Electrical; Engineering; Food Industry; General Management; Health/Medical; Insurance; Legal; Manufacturing; Operations Management; Personnel/Labor Relations; Printing/Publishing; Procurement; Technical and Scientific; Textiles; Transportation.

MANAGEMENT RECRUITERS
P.O. Box 1186, Rocky Mount NC 27802-1186. 919/442-8000. **Contact:** Bob Manning, Manager. **Description:** An executive search firm. **Specializes in the areas of:** Accounting/Auditing; Administration/MIS/EDP; Advertising; Architecture/Construction/Real Estate; Banking; Communications; Computer Hardware/Software; Design; Electrical; Engineering; Food Industry; General Management; Health/Medical; Insurance; Legal; Manufacturing; Operations Management; Personnel/Labor Relations; Printing/Publishing; Procurement; Retail; Sales and Marketing; Technical and Scientific; Textiles; Transportation.

MANAGEMENT RECRUITERS INTERNATIONAL
SALES CONSULTANTS
107 Edinburgh South, Suite 210, Cary NC 27511. 919/460-9595. **Fax:** 919/460-0642. **Contact:** Donna Durkin, Administrative Assistant. **Description:** An executive search firm. Company pays fee. **Specializes in the areas of:** Accounting/Auditing; Administration/MIS/EDP; Advertising; Architecture/Construction/Real Estate; Banking; Communications; Computer Hardware/Software; Design; Electrical; Engineering; Food Industry; General Management; Health/Medical; Insurance; Legal; Manufacturing; Operations Management; Personnel/Labor Relations; Printing/Publishing; Procurement; Retail; Sales and Marketing; Technical and Scientific; Textiles; Transportation. **Positions commonly filled include:** Aerospace Engineer; Branch Manager; Chemical Engineer; Claim Representative; Design Engineer; Electrical/Electronics Engineer; General Manager; Mechanical Engineer; Nuclear Engineer; Software Engineer; Technical Writer/Editor; Telecommunications Manager. **Average salary range of placements:** $30,000 - $50,000. **Number of placements per year:** 100 - 199.

MANAGEMENT RECRUITERS INTERNATIONAL
5701 Westpark Drive, Suite 110, Charlotte NC 28217. 704/525-9270. **Fax:** 704/527-0070. **Contact:** Mr. Ev Fuller, General Manager. **Description:** An executive search firm. Company pays fee. **Specializes in the areas of:** Art/Design; Computer Science/Software; Engineering; Finance; Food Industry; General Management; Health/Medical; Industrial; Insurance; Manufacturing; Personnel/Labor Relations; Printing/Publishing; Sales and Marketing; Technical and Scientific; Transportation. **Positions commonly filled include:** Administrative Services Manager; Bank Officer/Manager; Biomedical Engineer; Branch Manager; Buyer; Chemical Engineer; Chiropractor; Computer Programmer; Construction Contractor; Customer Service Representative; Dentist; Designer; EEG Technologist; EKG Technician; Financial Analyst; General Manager; Health Services Manager; Industrial Engineer; Industrial Production Manager; Management Analyst/Consultant; Manufacturer's/Wholesaler's Sales Rep.; Occupational Therapist; Operations/Production Manager; Pharmacist; Physical Therapist; Physician; Purchasing Agent/Manager; Quality Control Supervisor; Recreational Therapist; Registered Nurse; Respiratory Therapist; Securities Sales Representative; Software Engineer; Surveyor; Systems Analyst; Transportation/Traffic Specialist; Underwriter/Assistant Underwriter; Veterinarian. **Number of placements per year:** 200 - 499.

MANAGEMENT RECRUITERS OF DURHAM
5102 Durham/Chapel Hill Boulevard, Durham NC 27707. 919/489-6521. **Contact:** Ann Phillips, Manager. **Description:** An executive search firm. **Specializes in the areas of:** Accounting/Auditing; Administration/MIS/EDP; Advertising; Architecture/Construction/Real Estate; Banking; Communications; Computer Hardware/Software; Design; Electrical; Engineering;

Food Industry; General Management; Health/Medical; Insurance; Legal; Manufacturing; Operations Management; Personnel/Labor Relations; Printing/Publishing; Procurement; Retail; Sales and Marketing; Technical and Scientific; Textiles; Transportation.

MANAGEMENT RECRUITERS OF FAYETTEVILLE
951 South McPherson Church Road, Suite 105, Fayetteville NC 28303. 910/483-2555. **Fax:** 910/483-6524. **Contact:** John Semmes, Manager. **Description:** An executive search firm. **Specializes in the areas of:** Accounting/Auditing; Administration/MIS/EDP; Advertising; Architecture/Construction/Real Estate; Banking; Communications; Computer Hardware/Software; Design; Electrical; Engineering; Food Industry; General Management; Health/Medical; Insurance; Legal; Manufacturing; Operations Management; Personnel/Labor Relations; Printing/Publishing; Procurement; Retail; Sales and Marketing; Technical and Scientific; Textiles; Transportation.

MANAGEMENT RECRUITERS OF WINSTON-SALEM
P.O. Box 17054, Winston-Salem NC 27116-7054. 910/723-0484. **Contact:** Mike Jones, Manager. **Description:** An executive search firm. **Specializes in the areas of:** Accounting/Auditing; Administration/MIS/EDP; Advertising; Architecture/Construction/Real Estate; Banking; Communications; Computer Hardware/Software; Design; Electrical; Engineering; Food Industry; General Management; Health/Medical; Insurance; Legal; Manufacturing; Operations Management; Personnel/Labor Relations; Printing/Publishing; Procurement; Retail; Sales and Marketing; Technical and Scientific; Textiles; Transportation.

NATIONAL SERVICES, INC.
P.O. Box 6505, Raleigh NC 27628-6505. 919/787-8000. **Contact:** Bill Poole, President. **Description:** An executive search firm. **Specializes in the areas of:** Accounting/Auditing; Computer Hardware/Software; Engineering; General Management; Manufacturing; Personnel/Labor Relations; Technical and Scientific. **Number of placements per year:** 1 - 49.

PARENICA & COMPANY
19250 Stableford Lane, Huntersville NC 28078. 704/896-0060. **Fax:** 704/896-0240. **Contact:** James Parenica, President. **Description:** An executive search firm. Company pays fee. **Specializes in the areas of:** Banking; Computer Science/Software; General Management; Personnel/Labor Relations. **Positions commonly filled include:** General Manager; Human Resources Specialist; MIS Specialist. **Average salary range of placements:** More than $50,000. **Number of placements per year:** 1 - 49.

SALES CONSULTANTS OF CONCORD, INC.
254 Church Street, Concord NC 28025. 704/786-0700. **Fax:** 704/782-1356. **Contact:** Anna Lee Pearson, President. **Description:** An executive search firm. Company pays fee. **Specializes in the areas of:** Computer Science/Software; General Management; Health/Medical; Sales and Marketing; Technical and Scientific; Telecommunications. **Positions commonly filled include:** Branch Manager; Customer Service Representative; Economist; Electrical/Electronics Engineer; General Manager; Health Services Manager; Management Analyst/Consultant; Manufacturer's/Wholesaler's Sales Rep.; Marketing Manager; Physical Therapist; Physician; Product Manager; Psychologist; Sales Manager; Services Sales Representative; Software Engineer. **Number of placements per year:** 1 - 49.

SALES CONSULTANTS OF HIGH POINT
2411 Penny Road, Suite 101, High Point NC 27265. 910/883-4433. **Contact:** Manager. **Description:** An executive search firm. **Specializes in the areas of:** Accounting/Auditing; Administration/MIS/EDP; Advertising; Architecture/Construction/Real Estate; Banking; Communications; Computer Hardware/Software; Design; Electrical; Engineering; Food Industry; General Management; Health/Medical; Insurance; Legal; Manufacturing; Operations Management; Personnel/Labor Relations; Printing/Publishing; Procurement; Retail; Sales and Marketing; Technical and Scientific; Textiles; Transportation.

SANFORD ROSE ASSOCIATES
P.O. Box 13490, Charlotte NC 28270. 704/366-0730. **Fax:** 704/365-0620. **Contact:** James L. Downs, CEO. **Description:** An executive search firm operating on a contingency basis. Company

pays fee. **Specializes in the areas of:** Banking; Computer Science/Software. **Positions commonly filled include:** Bank Officer/Manager; Credit Manager; Economist; Financial Analyst; General Manager; Internet Services Manager; Management Analyst/Consultant; Market Research Analyst; MIS Specialist; Securities Sales Representative; Software Engineer; Statistician; Systems Analyst; Technical Writer/Editor; Telecommunications Manager. **Corporate headquarters location:** This Location. **Average salary range of placements:** More than $50,000. **Number of placements per year:** 1 - 49.

SOURCE EDP

100 North Tryon Street, Charlotte NC 28202-4000. 704/333-8311. **Toll-free phone:** 800/334-3617. **Contact:** Dan Buttrey, Managing Director. **Description:** An executive search firm that also operates as a temporary and permanent employment agency. Company pays fee. **Specializes in the areas of:** Accounting/Auditing; Administration/MIS/EDP; Banking; Computer Science/Software; Finance; Technical and Scientific. **Positions commonly filled include:** Accountant/Auditor; Bank Officer/Manager; Branch Manager; Budget Analyst; Computer Programmer; Credit Manager; Financial Analyst; Human Resources Specialist; MIS Manager; Systems Analyst; Technical Writer/Editor. **Average salary range of placements:** More than $50,000. **Number of placements per year:** 100 - 199.

DAVID WEINFELD GROUP

6512 Six Forks Road, Suite 603B, Raleigh NC 27615. 919/676-7828. **Fax:** 919/676-7399. **Contact:** David Weinfeld, President. **Description:** An executive search firm. Company pays fee. **Specializes in the areas of:** Broadcasting; Computer Science/Software; Engineering; Sales and Marketing; Telecommunications. **Positions commonly filled include:** Computer Programmer; Design Engineer; Electrical/Electronics Engineer; Multimedia Designer; Project Manager; Services Sales Representative; Software Engineer; Systems Analyst; Telecommunications Manager. **Average salary range of placements:** More than $50,000. **Number of placements per year:** 50 - 99.

PERMANENT EMPLOYMENT AGENCIES OF NORTH CAROLINA

ANDERSON & DANIEL PERSONNEL

4900 Randall Parkway, Suite F, Wilmington NC 28403-2831. 910/799-8500. **Fax:** 910/791-0706. **Contact:** Elma B. Daniel, President. **Description:** An employment agency offering permanent, temporary, and temp-to-perm placements. Company pays fee. **Specializes in the areas of:** Accounting/Auditing; Administration/MIS/EDP; Computer Science/Software; Engineering; Secretarial. **Positions commonly filled include:** Accountant/Auditor; Bank Officer/Manager; Blue-Collar Worker Supervisor; Buyer; Ceramics Engineer; Chemical Engineer; Chemist; Civil Engineer; Clerical Supervisor; Clinical Lab Technician; Computer Programmer; Cost Estimator; Credit Manager; Dental Assistant/Dental Hygienist; Design Engineer; Draftsperson; Electrical/Electronics Engineer; Environmental Engineer; Financial Analyst; Human Resources Specialist; Industrial Engineer; Industrial Production Manager; Licensed Practical Nurse; Management Trainee; Materials Engineer; Mechanical Engineer; Medical Records Technician; Metallurgical Engineer; MIS Specialist; Operations/Production Manager; Property and Real Estate Manager; Purchasing Agent/Manager; Quality Control Supervisor; Registered Nurse; Software Engineer; Statistician; Structural Engineer; Systems Analyst; Technical Writer/Editor; Typist/Word Processor. **Average salary range of placements:** $20,000 - $29,999. **Number of placements per year:** 1 - 49.

CAREER STAFFING

800 Clanton Road, Suite W, Charlotte NC 28217. 704/525-8400. **Fax:** 704/525-8682. **Contact:** Jim Chambers, President. **Description:** An employment agency. Company pays fee. **Specializes in**

the areas of: Banking; Clerical; Computer Hardware/Software; Secretarial. **Positions commonly filled include:** Administrative Assistant; Bank Officer/Manager; Bookkeeper; Clerk; Computer Programmer; Construction Trade Worker; Credit Manager; Data Entry Clerk; Draftsperson; Driver; EDP Specialist; Factory Worker; Financial Analyst; Legal Secretary; Light Industrial Worker; Medical Secretary; Receptionist; Secretary; Stenographer; Systems Analyst; Typist/Word Processor. **Number of placements per year:** 200 - 499.

CAREERS UNLIMITED INC.
1911 Hillandale Road, Suite 1210, Durham NC 27705. 919/383-7431. **Fax:** 919/383-5706. **Contact:** Manager. **Description:** An employment agency. **Specializes in the areas of:** Accounting/Auditing; Computer Science/Software; Finance; Office Support; Sales and Marketing; Temporary Assignments.

CORPORATE STAFFING CONSULTANTS, INC.
P.O. Box 221739, Charlotte NC 28222. 704/366-1800. **Fax:** 704/366-0070. **Contact:** Alan W. Madsen, CPC, President. **Description:** An employment agency. **Specializes in the areas of:** Computer Science/Software; Engineering; Environmental.

DATA MASTERS
P.O. Box 14548, Greensboro NC 27415-4548. 910/373-1461. **Toll-free phone:** 800/DataMasters. **Fax:** 910/373-1501. **Contact:** Manager. **E-mail addresses:** datamast@vnet.net; datamas@aol.com; 70702.62@compuserve.com. **World Wide Web address:** http://www. datamasters.com/dm. **Description:** An employment agency. Company pays fee. **Specializes in the areas of:** Computer Science/Software. **Positions commonly filled include:** Computer Programmer; Science Technologist; Systems Analyst. **Average salary range of placements:** $30,000 - $50,000. **Number of placements per year:** 100 - 199.

F-O-R-T-U-N-E PERSONNEL CONSULTANTS OF RALEIGH, INC.
P.O. Box 98388, Raleigh NC 27624-8388. 919/848-9929. **Fax:** 919/848-9666. **Contact:** Rick Deckelbaum, Vice President. **Description:** An employment agency. Company pays fee. **Specializes in the areas of:** Accounting/Auditing; Banking; Computer Science/Software; Engineering; Finance; Food Industry; Manufacturing; Personnel/Labor Relations. **Positions commonly filled include:** Accountant/Auditor; Aerospace Engineer; Biomedical Engineer; Ceramics Engineer; Chemical Engineer; Chemist; Civil Engineer; Computer Programmer; Electrical/Electronics Engineer; Financial Analyst; General Manager; Human Resources Manager; Industrial Engineer; Materials Engineer; Mechanical Engineer; Metallurgical Engineer; Petroleum Engineer; Software Engineer; Statistician; Systems Analyst. **Number of placements per year:** 200 - 499.

GRAHAM & ASSOCIATES
2100-J West Cornwallis Drive, Greensboro NC 27408. 910/288-9330. **Contact:** Gary Graham, CPC, President. **Description:** An employment agency. Company pays fee. **Specializes in the areas of:** Accounting/Auditing; Banking; Clerical; Computer Hardware/Software; Engineering; Legal; Manufacturing; MIS/EDP; Personnel/Labor Relations; Technical and Scientific. **Positions commonly filled include:** Accountant/Auditor; Administrative Assistant; Aerospace Engineer; Agricultural Engineer; Attorney; Bank Officer/Manager; Bookkeeper; Buyer; Ceramics Engineer; Chemical Engineer; Chemist; Civil Engineer; Clerk; Computer Operator; Computer Programmer; Credit Manager; Customer Service Representative; Data Entry Clerk; Draftsperson; Economist; EDP Specialist; Electrical/Electronics Engineer; Factory Worker; Financial Analyst; General Manager; Human Resources Manager; Industrial Designer; Industrial Engineer; Legal Secretary; Light Industrial Worker; Marketing Specialist; Mechanical Engineer; Medical Secretary; Metallurgical Engineer; Mining Engineer; Operations/Production Manager; Petroleum Engineer; Physicist/Astronomer; Purchasing Agent/Manager; Quality Control Supervisor; Receptionist; Secretary; Stenographer; Systems Analyst; Technical Writer/Editor; Technician; Typist/Word Processor. **Number of placements per year:** 200 - 499.

PROFESSIONAL PERSONNEL ASSOCIATES INC.
7520 East Independence Boulevard, Suite 160, Charlotte NC 28227. 704/532-2599. **Fax:** 704/536-8192. **Contact:** Gregory Whitt, President. **Description:** An employment agency that specializes in

technical engineering and managerial placements and also offers contract services. **Specializes in the areas of:** Computer Science/Software; Engineering; Industrial; Manufacturing; Personnel/Labor Relations. **Positions commonly filled include:** Aerospace Engineer; Buyer; Chemical Engineer; Chemist; Civil Engineer; Computer Programmer; Design Engineer; Designer; Electrical/Electronics Engineer; Environmental Engineer; Human Resources Specialist; Industrial Engineer; Industrial Production Manager; Mechanical Engineer; Metallurgical Engineer; MIS Specialist; Nuclear Engineer; Purchasing Agent/Manager; Quality Control Supervisor; Software Engineer; Structural Engineer. **Average salary range of placements:** $30,000 - $50,000. **Number of placements per year:** 50 - 99.

SUMMIT COMPUTER SERVICES INC.

7520 Independence Boulevard, Charlotte NC 28227-9405. 704/568-8095. **Fax:** 704/568-8098. **Contact:** Robert A. Moeller, Recruiter. **Description:** An employment agency that also offers contract technical services. Company pays fee. **Specializes in the areas of:** Computer Science/Software. **Positions commonly filled include:** Computer Scientist; Software Engineer; Technical Writer/Editor. **Corporate headquarters location:** This Location. **Average salary range of placements:** $30,000 - $50,000. **Number of placements per year:** 50 - 99.

THE UNDERWOOD GROUP/ACCREDITED PERSONNEL

2840 Plaza Place, Suite 211, Raleigh NC 27612. 919/782-3024. **Contact:** Manager. **Description:** An employment agency that also provides contract services. Company pays fee. **Specializes in the areas of:** Administration/MIS/EDP; Computer Science/Software; Data Processing; Engineering. **Positions commonly filled include:** Computer Programmer; Design Engineer; Electrical/ Electronics Engineer; MIS Specialist; Software Engineer; Systems Analyst. **Number of placements per year:** 50 - 99.

WOODS-HOYLE, INC.

P.O. Box 9902, Greensboro NC 27429. 910/273-4557. **Contact:** Anne Marie Woods, President. **Description:** An employment agency. **Specializes in the areas of:** Computer Science/Software; Data Processing.

EXECUTIVE SEARCH FIRMS OF NORTH DAKOTA

CAREER CONNECTION

1621 South University, Fargo ND 58103. 701/232-4614. **Fax:** 701/241-9822. **Contact:** Human Resources. **Description:** An executive search firm operating on a contingency basis. Company pays fee. **Specializes in the areas of:** Administration/MIS/EDP; Computer Science/Software; Engineering; Industrial; Manufacturing; Personnel/Labor Relations; Technical and Scientific. **Positions commonly filled include:** Accountant/Auditor; Agricultural Engineer; Civil Engineer; Computer Programmer; Design Engineer; Designer; Draftsperson; Electrical/Electronics Engineer; Environmental Engineer; Manufacturer's/Wholesaler's Sales Rep.; Mechanical Engineer; MIS Specialist; Operations/Production Manager; Purchasing Agent/Manager; Quality Control Supervisor; Services Sales Representative; Software Engineer; Structural Engineer; Surveyor; Systems Analyst; Technical Writer/Editor; Telecommunications Manager. **Corporate headquarters location:** This Location. **Average salary range of placements:** $30,000 - $50,000. **Number of placements per year:** 100 - 199.

PROFESSIONAL MANAGEMENT ASSOCIATION

109 1/2 Broadway, Fargo ND 58102. 701/237-9262. **Toll-free phone:** 800/473-2512. **Contact:** Kent Hochgraber, Recruiter. **Description:** An executive search firm. **Specializes in the areas of:** Accounting/Auditing; Banking; Computer Science/Software; Engineering. **Positions commonly filled include:** Accountant/Auditor; Chemical Engineer; Chemist; Civil Engineer; Clinical Lab

Technician; Dietician/Nutritionist; Health Services Manager; Mechanical Engineer; Occupational Therapist; Pharmacist; Physical Therapist; Physician; Recreational Therapist; Systems Analyst. **Corporate headquarters location:** This Location. **Other U.S. locations:** Nationwide. **Average salary range of placements:** $30,000 - $50,000. **Number of placements per year:** 1 - 49.

PERMANENT EMPLOYMENT AGENCIES OF NORTH DAKOTA

PERSONNEL SERVICES, INC.
401 South Main Street, Minot ND 58701. 701/852-2038. **Toll-free phone:** 800/PSI-2038. **Contact:** Manager. **Description:** A full-service employment agency engaged in executive search, permanent placement, and temporary placement of workers. The firm also provides contract services and offers career/outplacement counseling. **Specializes in the areas of:** Accounting/Auditing; Administration/MIS/EDP; Advertising; Banking; Computer Science/Software; Finance; General Management; Personnel/Labor Relations; Retail; Sales and Marketing; Secretarial. **Positions commonly filled include:** Accountant/Auditor; Administrative Services Manager; Advertising Clerk; Architect; Bank Officer/Manager; Branch Manager; Broadcast Technician; Budget Analyst; Claim Representative; Computer Programmer; Construction and Building Inspector; Customer Service Representative; Dental Assistant/Dental Hygienist; Editor; Financial Analyst; General Manager; Health Services Manager; Hotel Manager; Human Resources Manager; Human Service Worker; Management Analyst/Consultant; Management Trainee; Manufacturer's/Wholesaler's Sales Rep.; Market Research Analyst; MIS Specialist; Multimedia Designer; Public Relations Specialist; Purchasing Agent/Manager; Radio/TV Announcer/Broadcaster; Reporter; Restaurant/ Food Service Manager; Securities Sales Representative; Services Sales Representative; Surveyor; Technical Writer/Editor. **Average salary range of placements:** $20,000 - $29,999. **Number of placements per year:** 100 - 199.

EXECUTIVE SEARCH FIRMS OF OHIO

ACCOUNTANTS ON CALL
700 Ackerman Road, Suite 390, Columbus OH 43202. 614/267-7200. **Fax:** 614/267-7595. **Contact:** Russell Sheets, President. **Description:** An executive search firm operating on both retainer and contingency bases. Company pays fee. **Specializes in the areas of:** Accounting/Auditing; Administration/MIS/EDP; Banking; Computer Science/Software; Finance. **Positions commonly filled include:** Accountant/Auditor; Budget Analyst; Computer Programmer; Credit Manager; Financial Analyst; Internet Services Manager; Management Analyst/Consultant; MIS Specialist; Systems Analyst; Telecommunications Manager. **Other U.S. locations:** Nationwide. **Number of placements per year:** 200 - 499.

J.B. BROWN & ASSOCIATES
50 Public Square, Terminal Tower, Suite 1114, Cleveland OH 44113. 216/696-2525. **Fax:** 216/696-5825. **Contact:** Jeffrey B. Brown, President. **Description:** An executive search firm that also operates as an employment agency and contract services firm. Company pays fee. **Specializes in the areas of:** Accounting/Auditing; Administration/MIS/EDP; Advertising; Banking; Computer Science/Software; Engineering; Finance; Insurance; Manufacturing; Personnel/Labor Relations; Sales and Marketing. **Positions commonly filled include:** Accountant/Auditor; Actuary; Attorney; Bank Officer/Manager; Budget Analyst; Buyer; Chemical Engineer; Computer Programmer; Credit Manager; Design Engineer; Designer; Electrical/Electronics Engineer; Environmental Engineer; Financial Analyst; General Manager; Human Resources Specialist; Industrial Engineer; Industrial

Production Manager; Insurance Agent/Broker; Management Trainee; Market Research Analyst; Mechanical Engineer; MIS Specialist; Pharmacist; Purchasing Agent/Manager; Software Engineer; Structural Engineer; Telecommunications Manager; Underwriter/Assistant Underwriter. **Corporate headquarters location:** This Location. **Average salary range of placements:** $30,000 - $50,000. **Number of placements per year:** 200 - 499.

BURKS GROUP

23811 Chagrin Boulevard, Suite 280, Beachwood OH 44122. 216/595-8765. **Fax:** 216/595-8770. **Contact:** Human Resources. **Description:** Burks Group is an executive search firm that primarily recruits females and minorities. Clients include Cornell University, Digital Equipment, General Electric, General Motors, Ford Motor Company, Hewlett-Packard, 3M, Merck, and Xerox, among others. Company pays fee. **Specializes in the areas of:** Accounting/Auditing; Administration/MIS/EDP; Automotive; Banking; Computer Science/Software; Engineering; Finance; General Management; Industrial; Manufacturing; Personnel/Labor Relations. **Positions commonly filled include:** Accountant/ Auditor; Administrative Services Manager; Buyer; Chemical Engineer; Chemist; Civil Engineer; Computer Programmer; Credit Manager; Design Engineer; Electrical/Electronics Engineer; Environmental Engineer; Financial Analyst; General Manager; Human Resources Manager; Industrial Engineer; Management Analyst/Consultant; Market Research Analyst; Materials Engineer; Mechanical Engineer; MIS Manager; Purchasing Agent/Manager; Quality Control Supervisor; Software Engineer; Systems Analyst; Technical Writer/Editor. **Average salary range of placements:** More than $50,000. **Number of placements per year:** 1 - 49.

COMBINED RESOURCES INC.

14701 Detroit Avenue, Suite 750, Lakewood OH 44107. 216/221-1161. **Fax:** 216/221-1186. **Contact:** Gil Sherman, President. **Description:** An executive search firm. Company pays fee. **Specializes in the areas of:** Administration/MIS/EDP; Advertising; Art/Design; Computer Science/Software; Economics; Engineering; General Management; Industrial; Manufacturing; Personnel/Labor Relations; Retail; Sales and Marketing. **Positions commonly filled include:** Administrative Services Manager; Branch Manager; Buyer; Computer Programmer; Credit Manager; Customer Service Representative; Designer; Economist; Electrical/Electronics Engineer; Financial Analyst; General Manager; Human Resources Manager; Industrial Engineer; Management Analyst/Consultant; Manufacturer's/Wholesaler's Sales Rep.; Mechanical Engineer; Operations/Production Manager; Purchasing Agent/Manager; Services Sales Representative; Software Engineer; Systems Analyst; Transportation/Traffic Specialist; Wholesale and Retail Buyer.

COMPUTER & TECHNICAL ASSOCIATES

200 Westview Towers, 21010 Center Ridge Road, Suite 200, Rocky River OH 44116. **Toll-free phone:** 800/752-3674. **Contact:** Dean Flood, Manager. **Description:** An executive search firm and employment agency specializing in computer disciplines. Founded in 1965. Company pays fee. **Specializes in the areas of:** Administration/MIS/EDP; Computer Science/Software; Engineering. **Positions commonly filled include:** Computer Programmer; Electrical/Electronics Engineer; Software Engineer; Systems Analyst; Technical Writer/Editor; Telecommunications Manager. **Corporate headquarters location:** This Location. **Other U.S. locations:** Nationwide. **Average salary range of placements:** More than $50,000. **Number of placements per year:** 100 - 199.

DATA BANK CORPORATION

635 West 7th Street, Suite 100, Cincinnati OH 45203-1546. 513/241-9955. **Toll-free phone:** 800/733-0020. **Fax:** 513/333-6364. **Contact:** Wayne Ivey, President. **E-mail address:** jobs@databankcorp.com. **World Wide Web address:** http://www.databankcorp.com. **Description:** An executive search firm and employment agency. Data Bank is an information systems recruiter that places experienced professionals, mid-level managers, and executives. Founded in 1975. Company pays fee. **Specializes in the areas of:** Administration/MIS/EDP; Computer Science/Software; Information Systems. **Positions commonly filled include:** Computer Programmer; Internet Services Manager; Management Analyst/Consultant; MIS Specialist; Software Engineer; Systems Analyst; Technical Writer/Editor; Telecommunications Manager. **Corporate headquarters location:** This Location. **Number of placements per year:** 100 - 199.

DUNHILL PROFESSIONAL SEARCH OF COLUMBUS
1166 Goodale Boulevard, Suite 200, Columbus OH 43212. 614/421-0111. **Contact:** John Salzman, Owner. **Description:** An executive search firm. **Specializes in the areas of:** Computer Hardware/Software; Sales and Marketing.

EXECUTECH CONSULTANTS, INC.
P.O. Box 29385, Cincinnati OH 45229. 513/281-6416. **Fax:** 513/281-6674. **Contact:** Howard Bond, Publisher. **Description:** An executive search firm. **Specializes in the areas of:** Accounting/Auditing; Advertising; Broadcasting; Computer Science/Software; Economics; Education; Engineering; Finance; Food Industry; General Management; Manufacturing; Personnel/Labor Relations; Printing/Publishing; Sales and Marketing. **Positions commonly filled include:** Accountant/Auditor; Advertising Clerk; Bank Officer/Manager; Buyer; Economist; Editor; Management Analyst/Consultant; Management Trainee; Manufacturer's/Wholesaler's Sales Rep.

EXECUTIVE SEARCH LTD.
4830 Interstate Drive, Cincinnati OH 45246. 513/874-6901. **Fax:** 513/870-6348. **Contact:** Jim Cimino, Vice President. **Description:** An executive search firm operating on both retainer and contingency bases. Company pays fee. **Specializes in the areas of:** Accounting/Auditing; Administration/MIS/EDP; Computer Science/Software; Engineering; Finance; General Management; Industrial; Manufacturing; Personnel/Labor Relations; Sales and Marketing; Technical and Scientific. **Positions commonly filled include:** Accountant/Auditor; Biomedical Engineer; Budget Analyst; Buyer; Chemical Engineer; Chemist; Civil Engineer; Computer Programmer; Design Engineer; Designer; Electrical/Electronics Engineer; Environmental Engineer; Financial Analyst; General Manager; Health Services Manager; Human Resources Specialist; Industrial Engineer; Industrial Production Manager; Manufacturer's/Wholesaler's Sales Rep.; Mechanical Engineer; Metallurgical Engineer; Mining Engineer; MIS Specialist; Nuclear Engineer; Pharmacist; Physical Therapist; Physician; Purchasing Agent/Manager; Quality Control Supervisor; Respiratory Therapist; Software Engineer; Stationary Engineer; Structural Engineer; Systems Analyst; Transportation/Traffic Specialist. **Corporate headquarters location:** This Location. **Average salary range of placements:** More than $50,000. **Number of placements per year:** 100 - 199.

FENZEL MILAR ASSOCIATES
602 Quincy Street, Ironton OH 45638. 614/532-6409. **Fax:** 614/533-0813. **Contact:** John Milar, Owner. **Description:** An executive search firm operating on a contingency basis. Company pays fee. **Specializes in the areas of:** Administration/MIS/EDP; Computer Science/Software; Engineering; Manufacturing. **Positions commonly filled include:** Ceramics Engineer; Chemical Engineer; Electrical/Electronics Engineer; Industrial Engineer; Materials Engineer; Mechanical Engineer; Metallurgical Engineer; MIS Specialist; Software Engineer; Systems Analyst. **Average salary range of placements:** $30,000 - $50,000. **Number of placements per year:** 1 - 49.

FLOWERS & ASSOCIATES
ASSOCIATED TEMPORARIES
1446 South Reynolds Road, Maumee OH 43537-1634. 419/893-4816. **Fax:** 419/891-0779. **Contact:** William J. Ross, President. **Description:** An executive search firm operating on a contingency basis. Flowers & Associates also offers career/outplacement counseling. Associated Temporaries, also at this location, is a temporary placement agency. Company pays fee. **Specializes in the areas of:** Accounting/Auditing; Computer Science/Software; Engineering; Finance; General Management; Industrial; Manufacturing; Technical and Scientific; Transportation. **Positions commonly filled include:** Accountant/Auditor; Aerospace Engineer; Blue-Collar Worker Supervisor; Buyer; Ceramics Engineer; Chemical Engineer; Chemist; Civil Engineer; Clerical Supervisor; Clinical Lab Technician; Computer Programmer; Customer Service Representative; Design Engineer; Designer; Draftsperson; Electrical/Electronics Engineer; Environmental Engineer; Financial Analyst; General Manager; Human Resources Specialist; Industrial Engineer; Industrial Production Manager; Management Analyst/Consultant; Management Trainee; Materials Engineer; Mechanical Engineer; Metallurgical Engineer; MIS Specialist; Operations/Production Manager; Purchasing Agent/Manager; Quality Control Supervisor; Services Sales Representative; Software Engineer; Statistician; Structural Engineer;

Systems Analyst; Transportation/Traffic Specialist; Typist/Word Processor. **Benefits available to temporary workers:** Paid Holidays; Paid Vacation. **Average salary range of placements:** More than $50,000. **Number of placements per year:** 50 - 99.

GAYHART & ASSOCIATES
1250 Old River Road, Second Floor, Cleveland OH 44113. 216/861-7010. **Contact:** Richard F. Albertini, President. **Description:** An executive search firm. Company pays fee. **Specializes in the areas of:** Architecture/Construction/Real Estate; Computer Science/Software; Engineering; General Management; Industrial; Manufacturing; Personnel/Labor Relations; Technical and Scientific. **Positions commonly filled include:** Aerospace Engineer; Aircraft Mechanic/Engine Specialist; Architect; Biological Scientist; Biomedical Engineer; Buyer; Ceramics Engineer; Chemical Engineer; Chemist; Civil Engineer; Computer Programmer; Construction and Building Inspector; Construction Contractor; Cost Estimator; Designer; Draftsperson; Electrical/Electronics Engineer; Food Scientist/Technologist; General Manager; Geologist/Geophysicist; Industrial Engineer; Landscape Architect; Materials Engineer; Mechanical Engineer; Metallurgical Engineer; Mining Engineer; Nuclear Engineer; Petroleum Engineer; Quality Control Supervisor; Software Engineer; Stationary Engineer; Structural Engineer; Systems Analyst; Technical Writer/Editor. **Number of placements per year:** 100 - 199.

RUSS HADICK & ASSOCIATES
7100 Corporate Way, Suite B, Centerville OH 45459. 937/439-7700. **Contact:** Russ Hadick, President. **Description:** An executive search and recruitment agency. Company pays fee. **Specializes in the areas of:** Accounting/Auditing; Administration/MIS/EDP; Computer Hardware/Software; Engineering; Finance; Manufacturing; Personnel/Labor Relations; Technical and Scientific. **Positions commonly filled include:** Accountant/Auditor; Aerospace Engineer; Agricultural Engineer; Attorney; Buyer; Chemical Engineer; Chemist; Civil Engineer; Computer Programmer; Credit Manager; Draftsperson; Economist; EDP Specialist; Electrical/Electronics Engineer; Financial Analyst; General Manager; Human Resources Manager; Industrial Engineer; Marketing Specialist; Mechanical Engineer; Metallurgical Engineer; MIS Specialist; Purchasing Agent/Manager; Quality Control Supervisor; Sales Representative; Systems Analyst; Technical Writer/Editor; Technician. **Number of placements per year:** 50 - 99.

HAMMANN & ASSOCIATES
3540 Blue Rock Road, Cincinnati OH 45239. 513/385-2528. **Contact:** Ed Hammann, Owner. **E-mail address:** ed.hamman@ccc-bbs.com. **Description:** An executive search firm. Company pays fee. **Specializes in the areas of:** Administration/MIS/EDP; Computer Science/Software. **Positions commonly filled include:** Computer Programmer; Systems Analyst. **Average salary range of placements:** $30,000 - $50,000. **Number of placements per year:** 1 - 49.

J.D. HERSEY & ASSOCIATES
1695 Old Henderson Road, Columbus OH 43220. 614/459-4555. **Fax:** 614/459-4544. **Contact:** Jeff Hersey, President. **Description:** An executive search firm. Company pays fee. **Specializes in the areas of:** Architecture/Construction/Real Estate; Computer Science/Software; Retail; Sales and Marketing. **Positions commonly filled include:** Buyer; Computer Programmer; Construction Contractor; Cost Estimator; Manufacturer's/Wholesaler's Sales Rep.; Systems Analyst. **Number of placements per year:** 1 - 49.

HITE EXECUTIVE SEARCH
THE HITE COMPANIES, INC.
P.O. Box 43217, Cleveland OH 44143-0217. 216/461-1600. **Contact:** William A. Hite, III, President. **Description:** An executive search firm. **Specializes in the areas of:** Accounting/Auditing; Administration/MIS/EDP; Advertising; Architecture/Construction/Real Estate; Banking; Computer Hardware/Software; Engineering; Finance; General Management; Health/Medical; Legal; Manufacturing; Nonprofit; Personnel/Labor Relations; Sales and Marketing; Technical and Scientific. **Number of placements per year:** 50 - 99.

ITS TECHNICAL STAFFING
911 Madison Avenue, Toledo OH 43624. 419/259-3656. **Fax:** 419/255-0519. **Contact:** Roger Radelhoff, President. **Description:** An executive search firm operating on a contingency basis.

Specializes in the areas of: Computer Science/Software; Engineering; Industrial; Manufacturing; Personnel/Labor Relations; Technical and Scientific. **Positions commonly filled include:** Chemical Engineer; Civil Engineer; Computer Programmer; Construction and Building Inspector; Construction Contractor; Cost Estimator; Design Engineer; Designer; Draftsperson; Electrical/ Electronics Engineer; Electrician; Environmental Engineer; Human Resources Specialist; Industrial Engineer; Industrial Production Manager; Mechanical Engineer; Petroleum Engineer; Software Engineer; Strategic Relations Manager; Structural Engineer; Surveyor; Technical Writer/Editor. **Corporate headquarters location:** This Location. **Average salary range of placements:** $30,000 - $50,000. **Number of placements per year:** 100 - 199.

ICON MANAGEMENT GROUP
621 West Broad Street, Suite 2-D, Pataskala OH 43062. 614/927-4404. **Fax:** 614/927-0392. **Contact:** Robert Bremer, President. **Description:** An executive search firm. Company pays fee. **Specializes in the areas of:** Administration/MIS/EDP; Computer Science/Software; Engineering; Industrial; Manufacturing; Personnel/Labor Relations. **Positions commonly filled include:** Aerospace Engineer; Agricultural Engineer; Biomedical Engineer; Ceramics Engineer; Chemical Engineer; Chemist; Civil Engineer; Computer Programmer; Designer; Electrical/Electronics Engineer; General Manager; Human Resources Manager; Industrial Engineer; Industrial Production Manager; Manufacturer's/Wholesaler's Sales Rep.; Materials Engineer; Mechanical Engineer; Metallurgical Engineer; Mining Engineer; Nuclear Engineer; Operations/Production Manager; Petroleum Engineer; Purchasing Agent/Manager; Quality Control Supervisor; Software Engineer; Stationary Engineer; Structural Engineer; Systems Analyst. **Number of placements per year:** 50 - 99.

JAEGER INTERNATIONAL INC.
4889 Sinclair Road, Suite 112, Columbus OH 43229. 614/885-0364. **Fax:** 614/885-0415. **Contact:** Ted Langley, President. **E-mail address:** ilcook2@aol.com. **World Wide Web address:** http://www.jaegerint.com. **Description:** An executive search firm specializing in technical, professional, and managerial placements for manufacturing, telecommunications, and technology-based companies. Company pays fee. **Specializes in the areas of:** Computer Science/Software; Engineering; Industrial; Manufacturing; Technical and Scientific. **Positions commonly filled include:** Chemical Engineer; Computer Programmer; Industrial Engineer; Industrial Production Manager; Quality Control Supervisor. **Corporate headquarters location:** This Location. **Average salary range of placements:** $30,000 - $50,000. **Number of placements per year:** 1 - 49.

M&M PERSONNEL INC.
812 Huron Road E, Cleveland OH 44115. 216/436-2436. **Fax:** 216/436-2441. **Contact:** Gary Gardiner, President. **E-mail address:** 76351.2662@compuserve.com. **Description:** An executive search firm. The firm also operates as an employment agency. Company pays fee. **Specializes in the areas of:** Accounting/Auditing; Administration/MIS/EDP; Architecture/Construction/Real Estate; Computer Science/Software; Engineering; Health/Medical; Secretarial. **Positions commonly filled include:** Accountant/Auditor; Aerospace Engineer; Budget Analyst; Claim Representative; Computer Programmer; Construction and Building Inspector; Construction Contractor; Data Processor; Designer; Engineer; Licensed Practical Nurse; Medical Records Technician; MIS Specialist; Occupational Therapist; Physical Therapist; Registered Nurse; Software Engineer; Systems Analyst. **Average salary range of placements:** $30,000 - $50,000. **Number of placements per year:** 50 - 99.

MANAGEMENT RECRUITERS INTERNATIONAL
1900 East Dublin-Granville Road, Suite 110B, Columbus OH 43229. 614/794-3200. **Fax:** 614/794-3233. **Contact:** Dick Stoltz, Business Manager. **E-mail address:** columbus! manager@mrinet.com. **Description:** An executive search firm specializing in the fields of manufacturing, logistics, construction, office and administrative operations, and information technologies. Company pays fee. **Specializes in the areas of:** Administration/MIS/EDP; Architecture/Construction/Real Estate; Computer Science/Software; Engineering; Food Industry; General Management; Industrial; Manufacturing; Personnel/Labor Relations; Transportation. **Positions commonly filled include:** Agricultural Engineer; Biochemist; Buyer; Chemical Engineer; Computer Programmer; Construction Manager; Cost Estimator; Credit Manager; Customer Service Representative; Electrical/Electronics Engineer; Environmental Engineer; Food

Scientist/Technologist; General Manager; Human Resources Specialist; Industrial Engineer; Industrial Production Manager; Internet Services Manager; Metallurgical Engineer; MIS Specialist; Multimedia Designer; Operations/Production Manager; Purchasing Agent/Manager; Quality Control Supervisor; Software Engineer; Systems Analyst; Telecommunications Manager. **Number of placements per year:** 200 - 499.

MANAGEMENT RECRUITERS INTERNATIONAL
3450 West Central Avenue, Suite 360, Toledo OH 43606. **Fax:** 419/537-8730. **Contact:** Branch Manager. **Description:** An executive search firm. Company pays fee. **Specializes in the areas of:** Computer Science/Software; Engineering; General Management; Industrial; Manufacturing; Personnel/Labor Relations; Sales and Marketing. **Positions commonly filled include:** Chemical Engineer; Computer Programmer; Electrical/Electronics Engineer; General Manager; Human Resources Manager; Industrial Engineer; Industrial Production Manager; Mechanical Engineer; Metallurgical Engineer; Operations/Production Manager; Purchasing Agent/Manager; Quality Control Supervisor; Systems Analyst. **Number of placements per year:** 50 - 99.

MANAGEMENT RECRUITERS OF BOARDMAN
8090 Market Street, Suite 2, Boardman OH 44512. 330/726-6656. **Contact:** Manager. **Description:** An executive search firm. **Specializes in the areas of:** Accounting/Auditing; Administration/MIS/EDP; Advertising; Architecture/Construction/Real Estate; Banking; Communications; Computer Hardware/Software; Construction; Design; Electrical; Engineering; Finance; Food Industry; General Management; Health/Medical; Industrial; Insurance; Legal; Manufacturing; Operations Management; Personnel/Labor Relations; Printing/Publishing; Retail; Sales and Marketing; Technical and Scientific; Textiles; Transportation.

MANAGEMENT RECRUITERS OF CINCINNATI
36 East 4th Street, Suite 800, Cincinnati OH 45202. 513/651-5500. **Fax:** 513/651-3298. **Contact:** Joe McCullough, Co-Owner/Manager. **Description:** An executive search firm. **Specializes in the areas of:** Accounting/Auditing; Administration/MIS/EDP; Advertising; Architecture/Construction/Real Estate; Banking; Communications; Computer Hardware/Software; Construction; Design; Electrical; Engineering; Finance; Food Industry; General Management; Health/Medical; Industrial; Insurance; Legal; Manufacturing; Operations Management; Personnel/Labor Relations; Printing/Publishing; Procurement; Retail; Sales and Marketing; Technical and Scientific; Textiles; Transportation.

MANAGEMENT RECRUITERS OF CLEVELAND (LAKE)
Euclid Office Plaza, 26250 Euclid Avenue, Suite 811, Cleveland OH 44132-3674. 216/261-7696. **Fax:** 216/261-7699. **Contact:** Terry Wesley, Manager. **Description:** An executive search firm. **Specializes in the areas of:** Accounting/Auditing; Administration/MIS/EDP; Advertising; Architecture/Construction/Real Estate; Banking; Communications; Computer Hardware/Software; Construction; Electrical; Engineering; Finance; Food Industry; General Management; Health/Medical; Industrial; Insurance; Legal; Manufacturing; Operations Management; Printing/Publishing; Procurement; Retail; Sales and Marketing; Technical and Scientific; Textiles; Transportation.

MANAGEMENT RECRUITERS OF CLEVELAND (SOUTH)
9700 Rockside Road, Suite 100, Cleveland OH 44125-6264. 216/642-5788. **Fax:** 216/642-5933. **Contact:** Paul Montigny, Owner. **Description:** An executive search firm. **Specializes in the areas of:** Accounting/Auditing; Administration/MIS/EDP; Advertising; Architecture/Construction/Real Estate; Banking; Communications; Computer Hardware/Software; Construction; Design; Electrical; Engineering; Finance; Food Industry; General Management; Health/Medical; Legal; Manufacturing; Operations Management; Personnel/Labor Relations; Printing/Publishing; Procurement; Retail; Sales and Marketing; Technical and Scientific; Textiles; Transportation.

MANAGEMENT RECRUITERS OF CLEVELAND (SOUTHWEST)
P.O. Box 178, Brunswick OH 44212-0178. 330/273-4300. **Fax:** 330/273-2862. **Contact:** Bob Boal, Manager. **Description:** An executive search firm. **Specializes in the areas of:** Accounting/Auditing; Administration/MIS/EDP; Advertising; Architecture/Construction/Real Estate; Banking; Communications; Computer Hardware/Software; Construction; Design;

Electrical; Engineering; Finance; Food Industry; General Management; Health/Medical; Industrial; Insurance; Legal; Manufacturing; Operations Management; Personnel/Labor Relations; Printing/Publishing; Procurement; Retail; Sales and Marketing; Technical and Scientific; Textiles; Transportation.

MANAGEMENT RECRUITERS OF COLUMBUS (WEST)
800 East Broad Street, Columbus OH 43205. 614/252-6200. **Contact:** Cheryl Clossman, Office Manager. **Description:** An executive search firm. **Specializes in the areas of:** Accounting/Auditing; Administration/MIS/EDP; Advertising; Architecture/Construction/Real Estate; Banking; Communications; Computer Hardware/Software; Construction; Design; Electrical; Engineering; Finance; Food Industry; General Management; Health/Medical; Industrial; Insurance; Legal; Manufacturing; Operations Management; Personnel/Labor Relations; Printing/Publishing; Procurement; Retail; Sales and Marketing; Technical and Scientific; Textiles; Transportation.

MANAGEMENT RECRUITERS OF DAYTON
333 West First Street, Suite 304, Dayton OH 45402-1831. 937/228-8271. **Fax:** 937/228-2620. **Contact:** Manager. **Description:** An executive search firm. **Specializes in the areas of:** Accounting/Auditing; Administration/MIS/EDP; Advertising; Architecture/Construction/Real Estate; Banking; Communications; Computer Hardware/Software; Construction; Design; Electrical; Engineering; Finance; Food Industry; General Management; Health/Medical; Industrial; Insurance; Legal; Manufacturing; Operations Management; Personnel/Labor Relations; Printing/Publishing; Procurement; Retail; Sales and Marketing; Technical and Scientific; Textiles; Transportation.

MANAGEMENT RECRUITERS OF NORTH CANTON
P.O. Box 2970, North Canton OH 44720. 330/497-0122. **Fax:** 330/497-9730. **Contact:** Roger Bascom, Manager. **E-mail address:** 75030.3611@compuserve.com. **World Wide Web address:** http://www.mrnc.com. **Description:** An executive search firm. Company pays fee. **Specializes in the areas of:** Administration/MIS/EDP; Computer Hardware/Software; Information Technology; Manufacturing; Packaging; Printing/Publishing. **Positions commonly filled include:** Computer Programmer; Design Engineer; Designer; General Manager; Mechanical Engineer; MIS Specialist; Operations/Production Manager; Production Manager; Quality Control Supervisor; Systems Analyst. **Average salary range of placements:** $30,000 - $50,000.

MANAGEMENT RECRUITERS OF SOLON
P.O. Box 39361, Solon OH 44139. 216/248-7300. **Contact:** Kim Barnett, Manager. **Description:** An executive search firm. **Specializes in the areas of:** Accounting/Auditing; Administration/MIS/EDP; Advertising; Architecture/Construction/Real Estate; Banking; Communications; Computer Hardware/Software; Construction; Design; Electrical; Engineering; Finance; Food Industry; General Management; Health/Medical; Industrial; Insurance; Legal; Manufacturing; Operations Management; Personnel/Labor Relations; Printing/Publishing; Procurement; Retail; Sales and Marketing; Technical and Scientific; Textiles; Transportation.

MESSINA MANAGEMENT SYSTEMS
4770 Duke Drive, Suite 140, Mason OH 45040. 513/398-3331. **Fax:** 513/398-0496. **Contact:** Vincent Messina, CPC, President. **Description:** An executive search firm that also operates as a temporary agency. Company pays fee. **Specializes in the areas of:** Accounting/Auditing; Art/Design; Computer Science/Software; Engineering; Finance; Manufacturing; Personnel/Labor Relations. **Average salary range of placements:** $30,000 - $50,000. **Number of placements per year:** 100 - 199.

MIDLAND CONSULTANTS
4715 State Road, Cleveland OH 44109. 216/398-9330. **Fax:** 216/398-0879. **Contact:** David Sgro, President. **E-mail address:** midland@bright.net. **Description:** An executive search firm. **Specializes in the areas of:** Administration/MIS/EDP; Biology; Computer Science/Software; Engineering; General Management; Industrial; Manufacturing; Plastics; Printing/Publishing; Rubber; Sales and Marketing. **Positions commonly filled include:** Bank Officer/Manager; Biochemist; Biological Scientist; Chemist; Computer Programmer; Design Engineer; Designer;

Draftsperson; Electrical/Electronics Engineer; Environmental Engineer; General Manager; Human Resources Specialist; Industrial Engineer; Mechanical Engineer; MIS Manager; Operations/Production Manager; Software Engineer; Systems Analyst. **Average salary range of placements:** $30,000 - $50,000. **Number of placements per year:** 100 - 199.

MIDWEST SEARCH CONSULTANTS
471 East Broad Street, Suite 1201, Columbus OH 43215. 614/224-3600. **Fax:** 614/224-5585. **Contact:** Roosevelt Tabb, Principal. **Description:** An executive search firm. Company pays fee. **Specializes in the areas of:** Computer Science/Software; Engineering; Food Industry; General Management; Industrial; Manufacturing; Personnel/Labor Relations; Sales and Marketing. **Positions commonly filled include:** Chemical Engineer; Computer Programmer; Electrical/Electronics Engineer; Environmental Engineer; General Manager; Human Service Worker; Industrial Engineer; Industrial Production Manager; Internet Services Manager; Materials Engineer; Mechanical Engineer; Metallurgical Engineer; Purchasing Agent/Manager; Quality Control Supervisor; Restaurant/Food Service Manager; Software Engineer; Systems Analyst. **Corporate headquarters location:** This Location. **Average salary range of placements:** More than $50,000. **Number of placements per year:** 1 - 49.

MYERS & ASSOCIATES
4571 Stephen Circle, Canton OH 44718. 330/494-3274. **Fax:** 330/489-0790. **Contact:** George Myers, President. **Description:** An executive search firm operating on a contingency basis. Company pays fee. **Specializes in the areas of:** Computer Science/Software; Engineering; Manufacturing. **Positions commonly filled include:** Accountant/Auditor; Attorney; Computer Programmer; Design Engineer; Designer; Draftsperson; Electrical/Electronics Engineer; Electrician; Industrial Engineer; Industrial Production Manager; Mechanical Engineer; Metallurgical Engineer; Systems Analyst. **Corporate headquarters location:** This Location. **Average salary range of placements:** Less than $20,000. **Number of placements per year:** 1 - 49.

NEWCOMB - DESMOND & ASSOCIATES
73 Powhatton Drive, Milford OH 45150. 513/831-9522. **Fax:** 513/831-9557. **Contact:** Michael J. Desmond, Chief Operating Officer. **E-mail address:** mdesmond@ix.netcom.com. **Description:** An executive search firm that also operates as a temporary agency and contract services firm. The firm handles interim contract staffing and executive leasing. Founded in 1979. Company pays fee. **Specializes in the areas of:** Accounting/Auditing; Administration/MIS/EDP; Advertising; Banking; Biology; Computer Science/Software; Engineering; Finance; Food Industry; General Management; Industrial; Manufacturing; Personnel/Labor Relations; Printing/Publishing; Sales and Marketing; Technical and Scientific. **Positions commonly filled include:** Accountant/Auditor; Administrative Services Manager; Advertising Clerk; Aerospace Engineer; Aircraft Mechanic/Engine Specialist; Bank Officer/Manager; Biochemist; Biological Scientist; Biomedical Engineer; Chemical Engineer; Chemist; Computer Programmer; Customer Service Representative; Design Engineer; Designer; Draftsperson; Electrical/Electronics Engineer; Electrician; Environmental Engineer; Human Resources Specialist; Human Service Worker; Industrial Engineer; Management Analyst/Consultant; Manufacturer's/Wholesaler's Sales Rep.; Market Research Analyst; Mechanical Engineer; MIS Specialist; Operations/Production Manager; Purchasing Agent/Manager; Quality Control Supervisor; Services Sales Representative; Software Engineer; Stationary Engineer; Structural Engineer; Systems Analyst; Technical Writer/Editor; Telecommunications Manager. **Corporate headquarters location:** This Location. **Other locations:** Worldwide. **Number of placements per year:** 100 - 199.

NORTH PEAK GROUP
812 Huron Road, Suite 308, Cleveland OH 44115. 216/621-1070. **Fax:** 216/621-0825. **Contact:** Matthew Bruns, President. **E-mail address:** mbruns6108@aol.com. **Description:** An executive search firm. The firm focuses on placements in information systems and computer fields. Founded in 1993. Company pays fee. **Specializes in the areas of:** Computer Science/Software; Engineering; Environmental; Food Industry. **Positions commonly filled include:** Biochemist; Computer Programmer; Environmental Engineer; Hotel Manager; MIS Specialist; Restaurant/Food Service Manager; Software Engineer; Telecommunications Manager. **Corporate headquarters location:**

This Location. **Other U.S. locations:** Nationwide. **Average salary range of placements:** More than $50,000. **Number of placements per year:** 50 - 99.

OLSTEN PROFESSIONAL & TECHNICAL STAFFING SERVICES
3515 Michigan Avenue, Cincinnati OH 45208. 513/321-4313. **Fax:** 513/533-6757. **Contact:** Sharon Nichols, Branch Manager. **E-mail address:** olsten@megalinx.net. **Description:** An executive search firm that also operates as an employment agency and contract services firm. Placement is done on a contingency basis. This franchise was founded in 1968. Company pays fee. **Specializes in the areas of:** Accounting/Auditing; Administration/MIS/EDP; Art/Design; Banking; Computer Science/Software; Engineering; Finance; General Management; Manufacturing; Personnel/Labor Relations; Printing/Publishing; Sales and Marketing; Technical and Scientific. **Positions commonly filled include:** Accountant/Auditor; Bank Officer/Manager; Branch Manager; Budget Analyst; Ceramics Engineer; Chemical Engineer; Civil Engineer; Claim Representative; Clinical Lab Technician; Computer Programmer; Customer Service Representative; Draftsperson; Electrical/Electronics Engineer; Environmental Engineer; Financial Analyst; Human Resources Specialist; Industrial Engineer; Industrial Production Manager; Internet Services Manager; Manufacturer's/Wholesaler's Sales Rep.; Market Research Analyst; Materials Engineer; Mechanical Engineer; Metallurgical Engineer; MIS Specialist; Operations/Production Manager; Quality Control Supervisor; Securities Sales Representative; Services Sales Representative; Software Engineer; Stationary Engineer; Structural Engineer; Surveyor; Systems Analyst; Technical Writer/Editor; Telecommunications Manager; Typist/Word Processor. **Benefits available to temporary workers:** Medical Insurance; Paid Vacations. **Average salary range of placements:** $30,000 - $50,000.

OPPORTUNITIES CONSULTANTS INC.
435 Elm Street, Suite 810, Cincinnati OH 45202. 513/241-8675. **Fax:** 513/241-6285. **Contact:** Jay Richardson, Vice President. **Description:** An executive search firm. Company pays fee. **Specializes in the areas of:** Accounting/Auditing; Administration/MIS/EDP; Computer Science/Software; Engineering; Finance; Industrial; Manufacturing; Personnel/Labor Relations; Sales and Marketing; Technical and Scientific. **Positions commonly filled include:** Accountant/Auditor; Branch Manager; Buyer; Chemical Engineer; Chemist; Computer Programmer; Credit Analyst; Design Engineer; Designer; Electrical/Electronics Engineer; Financial Analyst; Human Resources Specialist; Industrial Engineer; Industrial Production Manager; Internet Services Manager; Market Research Analyst; Mechanical Engineer; MIS Specialist; Multimedia Designer; Operations/Production Manager; Purchasing Agent/Manager; Quality Control Supervisor; Systems Analyst. **Average salary range of placements:** $30,000 - $50,000. **Number of placements per year:** 50 - 99.

PARAGON RECRUITING OFFICIALS
2000 Henderson Road, Columbus OH 43220. 614/457-1211. **Contact:** Vince Procopio, President. **Description:** An executive search firm specializing in all aspects of data processing. **Specializes in the areas of:** Computer Science/Software. **Positions commonly filled include:** Computer Programmer; MIS Specialist; Systems Analyst. **Number of placements per year:** 1 - 49.

PLACEMENT SERVICES LIMITED, INC.
265 South Main Street, Akron OH 44308. 330/762-3838. **Toll-free phone:** 800/860-2252. **Contact:** Manager. **Description:** An executive search firm that also operates as a temporary agency and contract services firm. Placement Services Limited operates on a retainer basis. **Specializes in the areas of:** Computer Science/Software; Engineering; Food Industry; Health/Medical; Industrial; Manufacturing; Personnel/Labor Relations. **Positions commonly filled include:** Blue-Collar Worker Supervisor; Chemical Engineer; Chemist; Civil Engineer; Computer Programmer; Designer; Electrical/Electronics Engineer; Human Resources Manager; Industrial Engineer; Mechanical Engineer; Metallurgical Engineer; Mining Engineer; Nuclear Engineer; Petroleum Engineer; Purchasing Agent/Manager; Software Engineer; Structural Engineer; Systems Analyst. **Average salary range of placements:** $30,000 - $50,000. **Number of placements per year:** 100 - 199.

RECRUITMASTERS OF CINCINNATI
5237 Traverse Court, West Chester OH 45069-5587. 513/860-1717. **Fax:** 513/860-1717. **Contact:** Frank J. Watson, President. **Description:** An executive search firm operating on a contingency basis. **Specializes in the areas of:** Computer Science/Software; Personnel/Labor Relations; Sales and Marketing. **Positions commonly filled include:** Accountant/Auditor; Computer Programmer; Sales Representative; Systems Analyst. **Average salary range of placements:** $30,000 - $50,000. **Number of placements per year:** 1 - 49.

W.R. RENNER & ASSOCIATES
6998 Dublin Court, West Chester OH 45069. 513/777-7000. **Fax:** 513/777-4942. **Contact:** Bill Renner, Owner. **Description:** An executive search firm that recruits engineering and computer professionals for the manufacturing and service industries. Company pays fee. **Specializes in the areas of:** Accounting/Auditing; Computer Science/Software; Engineering; General Management; Industrial; Manufacturing; Technical and Scientific. **Positions commonly filled include:** Accountant/Auditor; Administrative Services Manager; Aerospace Engineer; Agricultural Engineer; Aircraft Mechanic/Engine Specialist; Buyer; Chemical Engineer; Chemist; Civil Engineer; Computer Programmer; Cost Estimator; Design Engineer; Designer; Electrical/Electronics Engineer; Environmental Engineer; Financial Analyst; Industrial Engineer; Industrial Production Manager; Mechanical Engineer; Metallurgical Engineer; Mining Engineer; MIS Specialist; Nuclear Engineer; Petroleum Engineer; Purchasing Agent/Manager; Quality Control Supervisor; Software Engineer; Structural Engineer; Systems Analyst; Telecommunications Manager. **Corporate headquarters location:** This Location. **Average salary range of placements:** $30,000 - $50,000. **Number of placements per year:** 1 - 49.

SACHS ASSOCIATES
568 Liberty Avenue, Huron OH 44839. 419/433-3837. **Contact:** Scott Sachs, President. **Description:** An executive search and contract services firm. Company pays fee. **Specializes in the areas of:** Administration/MIS/EDP; Computer Science/Software; Engineering; Personnel/Labor Relations. **Positions commonly filled include:** Aerospace Engineer; Computer Programmer; Design Engineer; Electrical/Electronics Engineer; Human Resources Specialist; Internet Services Manager; Management Analyst/Consultant; MIS Specialist; Multimedia Designer; Software Engineer; Systems Analyst. **Number of placements per year:** 1 - 49.

SALES CONSULTANTS OF CINCINNATI
11311 Cornell Park Drive, Suite 404, Cincinnati OH 45242. 513/247-0707. **Fax:** 513/247-2575. **Contact:** Brittany Crist, Office Coordinator. **Description:** An executive search firm. **Specializes in the areas of:** Accounting/Auditing; Administration/MIS/EDP; Advertising; Architecture/Construction/Real Estate; Banking; Communications; Computer Hardware/Software; Construction; Design; Electrical; Engineering; Finance; Food Industry; General Management; Health/Medical; Industrial; Insurance; Legal; Manufacturing; Operations Management; Personnel/Labor Relations; Printing/Publishing; Procurement; Retail; Sales and Marketing; Technical and Scientific; Textiles; Transportation.

SANFORD ROSE ASSOCIATES OF CLEVELAND
26250 Euclid Avenue, Suite 211, Cleveland OH 44132. 216/731-0005. **Contact:** Ralph Orkin, Owner. **Description:** A contingency search firm specializing in information systems. Company pays fee. **Specializes in the areas of:** Administration/MIS/EDP; Computer Science/Software. **Positions commonly filled include:** Computer Programmer; MIS Specialist; Software Engineer; Systems Analyst. **Average salary range of placements:** $30,000 - $50,000. **Number of placements per year:** 1 - 49.

SANFORD ROSE ASSOCIATES OF COLUMBUS
6230 Busch Boulevard, Columbus OH 43229. 614/436-3778. **Fax:** 614/436-8157. **Contact:** Bill Earhart, Owner/President. **Description:** An executive search firm specializing in computer hardware and software, manufacturing, and data processing. Company pays fee. **Specializes in the areas of:** Administration/MIS/EDP; Computer Science/Software; Engineering; Manufacturing. **Positions commonly filled include:** Computer Programmer; Mechanical Engineer; MIS Specialist; Software Engineer; Systems Analyst; Telecommunications Manager. **Corporate headquarters**

location: This Location. **Average salary range of placements:** More than $50,000. **Number of placements per year:** 1 - 49.

SANFORD ROSE ASSOCIATES OF YOUNGSTOWN

545 North Broad Street, Suite 2, Canfield OH 44406-9204. 330/533-9270. **Fax:** 330/533-9272. **Contact:** Richard H. Ellison, CPC, President. **E-mail address:** ellisor@aol.com. **Description:** An executive search firm. The company primarily recruits for information systems employers. Founded in 1970. Company pays fee. **Specializes in the areas of:** Administration/MIS/EDP; Computer Science/Software. **Positions commonly filled include:** Computer Programmer; Internet Services Manager; MIS Specialist; Multimedia Designer; Software Engineer; Systems Analyst; Telecommunications Manager. **Corporate headquarters location:** This Location. **Other U.S. locations:** Nationwide. **Average salary range of placements:** $30,000 - $50,000. **Number of placements per year:** 50 - 99.

SHAFER JONES ASSOCIATES

P.O. Box 405, Troy OH 45373-0405. 937/335-1885. **Fax:** 937/335-2237. **Contact:** Paul Jones, Owner. **Description:** An executive search firm operating on a contingency basis. Founded in 1992. Company pays fee. **Specializes in the areas of:** Accounting/Auditing; Computer Science/Software; Engineering; Finance; Health/Medical; Manufacturing; Personnel/Labor Relations. **Positions commonly filled include:** Accountant/Auditor; Buyer; Computer Programmer; Design Engineer; Electrical/Electronics Engineer; Financial Analyst; Human Resources Specialist; Mechanical Engineer; MIS Specialist; Pharmacist; Purchasing Agent/Manager; Quality Control Supervisor; Software Engineer; Technical Writer/Editor. **Average salary range of placements:** More than $50,000.

SOURCE SERVICES CORPORATION

525 Vine Street, Suite 2250, Cincinnati OH 45202. 513/651-4044. **Fax:** 513/651-3512. **Contact:** Greg Johnson, Managing Director. **E-mail address:** ssccio@sourcesvc.com. **Description:** An executive search firm operating on a contingency basis. **Specializes in the areas of:** Accounting/Auditing; Administration/MIS/EDP; Computer Science/Software; Economics; Engineering; Finance; Health/Medical; Manufacturing. **Positions commonly filled include:** Accountant/Auditor; Aerospace Engineer; Agricultural Engineer; Bank Officer/Manager; Biomedical Engineer; Budget Analyst; Chemical Engineer; Civil Engineer; Clerical Supervisor; Computer Programmer; Cost Estimator; Credit Manager; Customer Service Representative; Design Engineer; Economist; Electrical/Electronics Engineer; Environmental Engineer; Financial Analyst; Industrial Engineer; Industrial Production Manager; Internet Services Manager; Licensed Practical Nurse; Management Analyst/Consultant; Market Research Analyst; Mechanical Engineer; Metallurgical Engineer; Mining Engineer; MIS Specialist; Nuclear Engineer; Operations/Production Manager; Paralegal; Petroleum Engineer; Purchasing Agent/Manager; Quality Control Supervisor; Registered Nurse; Software Engineer; Strategic Relations Manager; Structural Engineer; Systems Analyst; Transportation/Traffic Specialist. **Corporate headquarters location:** Dallas TX. **Other U.S. locations:** Nationwide. **Average salary range of placements:** $30,000 - $50,000. **Number of placements per year:** 200 - 499.

TEKNON EMPLOYMENT RESOURCES, INC.

17 South Saint Clair Street, Suite 300, Dayton OH 45402. 937/222-5300. **Fax:** 937/222-6311. **Contact:** Bill Gaffney, Vice President of Recruiting. **E-mail address:** 102152.735@ compuserve.com. **Description:** An executive search firm. Company pays fee. **Specializes in the areas of:** Computer Science/Software; Data Communications; Telecommunications. **Positions commonly filled include:** Branch Manager; Computer Programmer; General Manager; Sales and Marketing Representative; Software Engineer; Systems Analyst; Telecommunications Manager. **Average salary range of placements:** $30,000 - $50,000. **Number of placements per year:** 50 - 99.

THOMAS - SCHADE & ASSOCIATES

15110 Foltz Parkway, Suite 104, Strongsville OH 44136. 216/846-0011. **Fax:** 216/572-5001. **Contact:** Danielle Schade, Partner. **Description:** An executive search firm. Company pays fee. **Specializes in the areas of:** Computer Science/Software; Engineering; Fire Protection Engineering; Manufacturing; Sales and Marketing; Security. **Positions commonly filled include:**

Branch Manager; Computer Programmer; Construction Contractor; Electrical/Electronics Engineer; General Manager; Librarian; Library Technician; MIS Manager; Services Sales Representative; Software Engineer; Systems Analyst. **Average salary range of placements:** More than $50,000. **Number of placements per year:** 50 - 99.

WEIPER RECRUITING SERVICES
1672 McCabe Lane, Cincinnati OH 45255-3095. 513/232-4300. **Contact:** Dave Weiper, Recruiter. **E-mail address:** coec@cin.ix.net. **World Wide Web address:** http://www.occ.com/wrs. **Description:** Weiper Recruiting Services is an executive search firm operating on a contingency basis. Company pays fee. **Specializes in the areas of:** Computer Science/Software. **Positions commonly filled include:** Computer Programmer; Internet Services Manager; Systems Analyst. **Average salary range of placements:** $30,000 - $50,000. **Number of placements per year:** 50 - 99.

PERMANENT EMPLOYMENT AGENCIES OF OHIO

ALLTECH RESOURCES, INC.
6000 Lombardo Center, Suite 310, Cleveland OH 44131. 216/642-5689. **Fax:** 216/642-9419. **Contact:** David A. Zupan, Chief Operations Officer. **Description:** An employment agency. **Specializes in the areas of:** Administration/MIS/EDP; Computer Science/Software; Technical and Scientific. **Positions commonly filled include:** Computer Programmer; Management Analyst/Consultant; MIS Specialist; Science Technologist; Software Engineer; Systems Analyst; Technical Writer/Editor; Telecommunications Manager. **Benefits available to temporary workers:** Dental Insurance; Medical Insurance; Paid Holidays; Paid Vacation. **Corporate headquarters location:** Chicago IL. **Average salary range of placements:** More than $50,000. **Number of placements per year:** 200 - 499.

AMERICAN BUSINESS PERSONNEL SERVICES, INC.
11499 Chester Road, Suite 2610, Cincinnati OH 45246. 513/772-1200. **Fax:** 513/326-2278. **Contact:** Chad Johnson, MIS Manager. **Description:** An employment agency focusing on the computer and engineering industries. Company pays fee. **Specializes in the areas of:** Computer Hardware/Software; Computer Programming; Computer Science/Software; Engineering. **Positions commonly filled include:** Computer Programmer; Engineer; Systems Analyst; Technical Support Representative. **Average salary range of placements:** $30,000 - $50,000. **Number of placements per year:** 1000+.

N.L. BENKE & ASSOCIATES, INC.
1422 Euclid Avenue, Suite 956, Cleveland OH 44115. 216/771-6822. **Contact:** Norman L. Benke, President. **Description:** N.L. Benke & Associates is an employment agency. Company pays fee. **Specializes in the areas of:** Accounting/Auditing; Banking; Computer Science/Software; Finance; Insurance; Personnel/Labor Relations. **Positions commonly filled include:** Accountant/Auditor; Adjuster; Administrative Services Manager; Attorney; Bank Officer/Manager; Branch Manager; Budget Analyst; Buyer; Clerical Supervisor; Computer Programmer; Credit Manager; Customer Service Representative; Economist; Financial Analyst; General Manager; Human Resources Manager; Management Analyst/Consultant; Management Trainee; Property and Real Estate Manager; Purchasing Agent/Manager; Quality Control Supervisor; Securities Sales Representative; Software Engineer; Systems Analyst; Underwriter/Assistant Underwriter. **Number of placements per year:** 200 - 499.

CBS COMPANIES
435 Elm Street, Suite 700, Cincinnati OH 45202. 513/651-1111. **Contact:** Wayne J. Vinson, Manager. **Description:** An employment agency. **Specializes in the areas of:** Bookkeeping; Clerical; Computer Science/Software; Finance.

CBS COMPANIES
130 West Second Street, Suite 1910, Dayton OH 45402. 937/222-2525. **Contact:** Robert L. Brown, CPC, President. **Description:** An employment agency. **Specializes in the areas of:** Clerical; Computer Science/Software; Finance; Office Support.

CAREER CONNECTIONS
35 Elliott Street, Athens OH 45701. 614/594-4941. **Fax:** 614/592-6289. **Contact:** Valerie Kinnard, General Manager. **Description:** An employment agency. Company pays fee. **Specializes in the areas of:** Computer Science/Software; Education; Industrial; Legal; Manufacturing; Nonprofit; Secretarial. **Positions commonly filled include:** Automotive Mechanic; Computer Programmer; Customer Service Representative; Paralegal; Social Worker; Technical Writer/Editor. **Number of placements per year:** 50 - 99.

CLOPTON'S PLACEMENT SERVICE
23241 Shurmer Drive, Cleveland OH 44128-4927. 216/292-4830. **Fax:** 216/292-4830. **Contact:** Fred Clopton, President. **Description:** An employment agency. **Specializes in the areas of:** Computer Science/Software; Engineering; Technical and Scientific. **Positions commonly filled include:** Biomedical Engineer; Electrical/Electronics Engineer; Mechanical Engineer; Quality Control Supervisor; Software Engineer; Systems Analyst; Technical Writer/Editor. **Number of placements per year:** 1 - 49.

ALAN N. DAUM AND ASSOCIATES
6241 Riverside Drive, Dublin OH 43017. 614/793-1200. **Contact:** Alan N. Daum, President. **E-mail address:** aldaum@aol.com. **Description:** An employment agency. Company pays fee. **Specializes in the areas of:** Computer Hardware/Software; Engineering; Food Industry; Manufacturing; Technical and Scientific. **Positions commonly filled include:** Computer Programmer; Electrical/Electronics Engineer; Software Engineer. **Number of placements per year:** 1 - 49.

ENTERPRISE SEARCH ASSOCIATES
77 West Elmwood Drive, Dayton OH 45459. 937/438-8774. **Contact:** Jeff Linck, Owner. **Description:** An employment agency. Company pays fee. **Specializes in the areas of:** Administration/MIS/EDP; Computer Hardware/Software. **Positions commonly filled include:** Computer Programmer; EDP Specialist; MIS Specialist; Software Engineer; Systems Analyst.

R.E. LOWE ASSOCIATES, INC.
8080 Ravines Edge Court, Worthington OH 43235. 614/436-6650. **Fax:** 614/436-2789. **Contact:** Dave Deringer, Recruiter. **Description:** An employment agency. **Specializes in the areas of:** Accounting/Auditing; Computer Science/Software; Engineering; Finance; Health/Medical; Insurance. **Positions commonly filled include:** Accountant/Auditor; Budget Analyst; Chemical Engineer; Claim Representative; Consultant; Credit Manager; Data Analyst; EDP Specialist; Electrical/Electronics Engineer; Environmental Engineer; Financial Analyst; Industrial Engineer; Manufacturing Engineer; Nurse; Packaging Engineer; Physician; Tax Specialist; Underwriter/Assistant Underwriter; Water/Wastewater Engineer. **Number of placements per year:** 200 - 499.

JERRY PAUL ASSOCIATES
1662 State Road, Cuyahoga Falls OH 44223. 330/923-2345. **Contact:** Jerry Paul, President. **Description:** An employment agency. **Specializes in the areas of:** Administration/MIS/EDP; Banking; Computer Hardware/Software; Hotel/Restaurant; Sales and Marketing. **Positions commonly filled include:** Hotel Manager; Restaurant/Food Service Manager; Sales Representative. **Number of placements per year:** 50 - 99.

PROVIDENCE PERSONNEL CONSULTANTS
2404 Fourth Street, Suite 1, Cuyahoga Falls OH 44221-2659. 330/929-6431. **Contact:** Donna Early, CPC, President. **Description:** An employment agency. Company pays fee. **Specializes in the areas of:** Accounting/Auditing; Administration/MIS/EDP; Advertising; Banking; Computer

Hardware/Software; Construction; Engineering; Finance; Health/Medical; Legal; Manufacturing; Sales and Marketing; Technical and Scientific; Transportation. **Positions commonly filled include:** Accountant/Auditor; Advertising Account Executive; Aerospace Engineer; Agricultural Engineer; Attorney; Bank Officer/Manager; Biological Scientist; Biomedical Engineer; Ceramics Engineer; Chemist; Civil Engineer; Computer Programmer; Credit Manager; Customer Service Representative. **Number of placements per year:** 50 - 99.

S&P SOLUTIONS
35000 Chardon Road, Suite 100, Cleveland OH 44094. 216/646-9111. **Fax:** 216/646-1429. **Contact:** Gary Bates, President. **Description:** S&P Solutions is an employment agency specializing in placements in the computer field. **Specializes in the areas of:** Computer Science/Software. **Positions commonly filled include:** Computer Programmer; Systems Analyst. **Number of placements per year:** 50 - 99.

SELECTIVE SEARCH ASSOCIATES
1206 North Main Street, Suite 112, North Canton OH 44720. 330/494-5584. **Fax:** 330/494-8911. **Contact:** Michael E. Ziarko, President. **Description:** Selective Search Associates is an employment agency specializing in computer and engineering placements. Company pays fee. **Specializes in the areas of:** Computer Science/Software; Engineering. **Positions commonly filled include:** Computer Programmer; Electrical/Electronics Engineer; Mechanical Engineer; Nuclear Engineer; Software Engineer; Systems Analyst. **Number of placements per year:** 1 - 49.

TAD RESOURCES
6555 Bush Boulevard, Suite 230, Columbus OH 43229. 614/863-2481. **Contact:** Ms. Bobbie Gallo, Branch Manager. **Description:** An employment agency. Company pays fee. **Specializes in the areas of:** Computer Hardware/Software; Engineering; Manufacturing; Technical and Scientific. **Positions commonly filled include:** Administrative Assistant; Aerospace Engineer; Agricultural Scientist; Architect; Biological Scientist; Biomedical Engineer; Bookkeeper; Ceramics Engineer; Civil Engineer; Clerk; Commercial Artist; Computer Programmer; Data Entry Clerk; Dietician/Nutritionist; Draftsperson; Electrical/Electronics Engineer; Factory Worker; General Manager; Industrial Designer; Industrial Engineer; Legal Secretary; Light Industrial Worker; Mechanical Engineer; Medical Secretary; Metallurgical Engineer; Mining Engineer; Nurse; Petroleum Engineer; Purchasing Agent/Manager; Receptionist; Secretary; Statistician; Systems Analyst; Technical Writer/Editor; Technician; Typist/Word Processor. **Number of placements per year:** 1000+.

TECH/AID OF OHIO
34950 Chardon Road, Suite 101, Willoughby Hills OH 44094. 216/749-3060. **Contact:** Manager. **Description:** An employment agency. Company pays fee. **Specializes in the areas of:** Accounting/Auditing; Banking; Computer Hardware/Software; Engineering; Finance; Insurance; Manufacturing; MIS/EDP; Nonprofit; Personnel/Labor Relations; Printing/Publishing; Technical and Scientific. **Positions commonly filled include:** Computer Operator; Computer Programmer; EDP Specialist; MIS Specialist; Systems Analyst; Technical Writer/Editor. **Number of placements per year:** 1000+.

VECTOR TECHNICAL
7911 Enterprise Drive, Second Floor, Mentor OH 44060. 216/946-8800. **Fax:** 216/946-8808. **Contact:** Timothy Bleich, Vice President. **Description:** An employment agency. Company pays fee. **Specializes in the areas of:** Biology; Computer Science/Software; Engineering; Industrial; Manufacturing; Technical and Scientific. **Positions commonly filled include:** Aerospace Engineer; Agricultural Engineer; Architect; Biological Scientist; Biomedical Engineer; Blue-Collar Worker Supervisor; Ceramics Engineer; Chemical Engineer; Chemist; Civil Engineer; Computer Programmer; Cost Estimator; Designer; Draftsperson; Electrical/Electronics Engineer; Electrician; Industrial Engineer; Industrial Production Manager; Materials Engineer; Mechanical Engineer; Metallurgical Engineer; Mining Engineer; Nuclear Engineer; Operations/Production Manager; Petroleum Engineer; Software Engineer; Stationary Engineer; Structural Engineer; Systems Analyst. **Number of placements per year:** 100 - 199.

EXECUTIVE SEARCH FIRMS OF OKLAHOMA

AMERI RESOURCE
2525 Northwest Expressway, Suite 532, Oklahoma City OK 73112. 405/842-5900. **Toll-free phone:** 800/583-7823. **Fax:** 405/843-9879. **Contact:** Len Branch, National Account Manager. **Description:** An executive search firm operating on both retainer and contingency bases. Company pays fee. **Specializes in the areas of:** Accounting/Auditing; Administration/MIS/EDP; Architecture/Construction/Real Estate; Banking; Computer Science/Software; Engineering; Finance; General Management; Health/Medical; Industrial; Manufacturing; Personnel/Labor Relations; Printing/Publishing; Secretarial; Technical and Scientific; Telecommunications; Transportation. **Positions commonly filled include:** Accountant/Auditor; Actuary; Administrative Services Manager; Aerospace Engineer; Agricultural Engineer; Aircraft Mechanic/Engine Specialist; Architect; Automotive Mechanic; Bank Officer/Manager; Blue-Collar Worker Supervisor; Branch Manager; Civil Engineer; Clerical Supervisor; Computer Programmer; Construction and Building Inspector; Construction Contractor; Cost Estimator; Credit Manager; Customer Service Representative; Dental Assistant/Dental Hygienist; Design Engineer; Designer; Draftsperson; Electrical/Electronics Engineer; Electrician; Financial Analyst; General Manager; Geographer; Geologist/Geophysicist; Human Resources Specialist; Industrial Engineer; Industrial Production Manager; Internet Services Manager; Licensed Practical Nurse; Management Analyst/Consultant; Management Trainee; Manufacturer's/Wholesaler's Sales Rep.; Market Research Analyst; Mechanical Engineer; Metallurgical Engineer; Mining Engineer; MIS Specialist; Multimedia Designer; Nuclear Engineer; Occupational Therapist; Operations/Production Manager; Petroleum Engineer; Purchasing Agent/Manager; Quality Control Supervisor; Recreational Therapist; Registered Nurse; Respiratory Therapist; Software Engineer; Structural Engineer; Systems Analyst; Technical Writer/Editor; Telecommunications Manager; Typist/Word Processor. **Corporate headquarters location:** This Location. **Other U.S. locations:** Tulsa OK. **Number of placements per year:** 1000+.

EXPRESS PERSONNEL SERVICES
7321 Southwestern, Oklahoma City OK 73139. 405/634-6600. **Contact:** P.J. Jackson, Regional Manager. **Description:** An executive search firm. Company pays fee. **Specializes in the areas of:** Accounting/Auditing; Computer Science/Software; Engineering; Health/Medical; Industrial; Legal. **Positions commonly filled include:** Accountant/Auditor; Administrative Services Manager; Advertising Clerk; Automotive Mechanic; Blue-Collar Worker Supervisor; Branch Manager; Clerical Supervisor; Computer Programmer; Customer Service Representative; Draftsperson; General Manager; Human Resources Specialist; Industrial Production Manager; Registered Nurse; Typist/Word Processor. **Other U.S. locations:** Nationwide. **Average salary range of placements:** Less than $20,000.

EXPRESS PERSONNEL SERVICES
6300 NW Expressway, Oklahoma City OK 73132. 405/840-5000. **Fax:** 405/720-9390. **Contact:** Harvey H. H. Homsey, Research & Development Manager. **Description:** An executive search firm. **Specializes in the areas of:** Accounting/Auditing; Advertising; Computer Science/Software; Food Industry; General Management; Industrial; Insurance; Legal; Manufacturing; Personnel/Labor Relations; Printing/Publishing; Sales and Marketing; Secretarial; Technical and Scientific. **Positions commonly filled include:** Administrative Services Manager; Advertising Clerk; Architect; Automotive Mechanic; Blue-Collar Worker Supervisor; Branch Manager; Buyer; Claim Representative; Computer Programmer; Customer Service Representative; Dental Assistant/Dental Hygienist; Designer; Dietician/Nutritionist; Draftsperson; Editor; Education Administrator; EEG Technologist; EKG Technician; Electrical/Electronics Engineer; Electrician; Emergency Medical Technician; Food Scientist/Technologist; General Manager; Human Resources Specialist; Human Service Worker; Librarian; Manufacturer's/Wholesaler's Sales Rep.; Market Research Analyst; Medical Records Technician; Multimedia Designer; Paralegal; Public Relations Specialist; Quality Control Supervisor; Restaurant/Food Service Manager; Services Sales Representative; Software Engineer; Systems Analyst; Technical Writer/Editor; Travel Agent. **Corporate headquarters location:** This Location. **Average salary range of placements:** $30,000 - $50,000. **Number of placements per year:** 1000+.

THE HENWOOD GROUP
4600 SE 29th Street, Oklahoma City OK 73115. 405/670-9070. **Toll-free phone:** 800/430-6760. **Fax:** 405/670-9071. **Contact:** Bill Henwood, President. **E-mail address:** henwood@icon.net. **Description:** The Henwood Group is an executive search firm specializing in information systems and telecommunications placements. Company pays fee. **Specializes in the areas of:** Administration/MIS/EDP; Computer Science/Software; Information Systems; Technical and Scientific; Telecommunications. **Positions commonly filled include:** Computer Programmer; Internet Services Manager; Management Analyst/Consultant; MIS Specialist; Multimedia Designer; Software Engineer; Systems Analyst; Telecommunications Manager. **Average salary range of placements:** More than $50,000. **Number of placements per year:** 1 - 49.

MANAGEMENT RECRUITERS OF OKLAHOMA CITY
3441 West Memorial Road, Suite 4, Oklahoma City OK 73134. 405/752-8848. **Fax:** 405/752-8783. **Contact:** Gary Roy, Manager. **Description:** Management Recruiters of Oklahoma City is an executive search firm. **Specializes in the areas of:** Accounting/Auditing; Administration/MIS/EDP; Advertising; Architecture/Construction/Real Estate; Banking; Communications; Computer Hardware/Software; Finance; Food Industry; General Management; Health/Medical; Insurance; Legal; Manufacturing; Printing/Publishing; Retail; Sales and Marketing; Technical and Scientific; Textiles; Transportation.

TERRY NEESE PERSONNEL AGENCY
2709 NW 39th, Oklahoma City OK 73112. 405/942-8551. **Contact:** Terry Neese, CEO. **Description:** An executive search firm operating on a contingency basis. Company pays fee. **Specializes in the areas of:** Accounting/Auditing; Administration/MIS/EDP; Computer Science/Software; Engineering; Finance; Health/Medical; Legal; Personnel/Labor Relations; Sales and Marketing; Secretarial. **Positions commonly filled include:** Accountant/Auditor; Administrative Services Manager; Advertising Clerk; Bank Officer/Manager; Blue-Collar Worker Supervisor; Brokerage Clerk; Budget Analyst; Chemical Engineer; Computer Operator; Computer Programmer; Construction Trade Worker; Credit Manager; Customer Service Representative; Data Entry Clerk; Dental Assistant/Dental Hygienist; Draftsperson; Electrical/Electronics Engineer; Emergency Medical Technician; General Manager; Human Resources Manager; Internet Services Manager; Management Analyst/Consultant; Management Trainee; Manufacturer's/Wholesaler's Sales Rep.; Medical Records Technician; Paralegal; Public Relations Specialist; Receptionist; Registered Nurse; Sales Representative; Secretary; Stenographer; Systems Analyst; Technical Writer/Editor; Typist/Word Processor. **Corporate headquarters location:** This Location. **Number of placements per year:** 200 - 499.

JOHN WYLIE ASSOCIATES, INC.
1727 East 71st Street, Tulsa OK 74136. 918/496-2100. **Contact:** John L. Wylie, President. **Description:** An executive search firm. **Specializes in the areas of:** Computer Hardware/Software; Engineering; Manufacturing; Technical and Scientific. **Number of placements per year:** 1 - 49.

PERMANENT EMPLOYMENT AGENCIES OF OKLAHOMA

LLOYD RICHARDS PERSONNEL SERVICE
507 South Main Street, Suite 502, Tulsa OK 74103. 918/582-5251. **Fax:** 918/582-5250. **Contact:** Lloyd Richards, CPC. **Description:** An employment agency. **Specializes in the areas of:** Accounting/Auditing; Bookkeeping; Chemical; Clerical; Computer Science/Software; Engineering; Finance; Health/Medical; Legal; Manufacturing; Retail; Sales and Marketing; Temporary Assignments.

EXECUTIVE SEARCH FIRMS OF OREGON

THE BRENTWOOD GROUP LIMITED
9 Monroe Parkway, Suite 230, Lake Oswego OR 97035. 503/697-8136. **Fax:** 503/697-8161. **Contact:** Manager. **E-mail address:** brentwood@transport.com. **Description:** An executive search firm. Company pays fee. **Specializes in the areas of:** Computer Science/Software; Health/Medical; Insurance; Personnel/Labor Relations. **Positions commonly filled include:** Adjuster; Claim Representative; Credit Manager; Human Resources Specialist; Physical Therapist; Physician; Software Engineer; Telecommunications Manager; Underwriter/Assistant Underwriter. **Average salary range of placements:** More than $50,000. **Number of placements per year:** 50 - 99.

MANAGEMENT RECRUITERS INTERNATIONAL
OFFICEMATES5 OF PORTLAND
2020 Lloyd Center, Portland OR 97232-1376. 503/287-8701. **Contact:** Manager. **Description:** An executive search firm. **Specializes in the areas of:** Accounting/Auditing; Administration/MIS/EDP; Advertising; Architecture/Construction/Real Estate; Banking; Communications; Computer Hardware/Software; Design; Electrical; Engineering; Food Industry; General Management; Health/Medical; Insurance; Legal; Manufacturing; Operations Management; Personnel/Labor Relations; Printing/Publishing; Procurement; Retail; Sales and Marketing; Technical and Scientific; Textiles; Transportation.

MASUT, HALE & ASSOCIATES UK, LTD
506 SW 6th Avenue, Fourth Floor, Portland OR 97204. 503/223-9190. **Fax:** 503/223-7472. **Contact:** Dominic Masut, Senior Managing Partner. **E-mail address:** mha@globalpo.com. **Description:** An executive search firm that provides various search, selection, and placement services primarily to the electronics, computer, software, and telecommunications industries. The company also provides mergers and acquisitions, venture capital, and management advisory services. Company pays fee. **Specializes in the areas of:** Accounting/Auditing; Computer Science/Software; Engineering; Sales and Marketing. **Positions commonly filled include:** Accountant/Auditor; Attorney; Computer Programmer; Design Engineer; Electrical/Electronics Engineer; General Manager; Software Engineer; Systems Analyst; Technical Writer/Editor. **Corporate headquarters location:** London, England. **Average salary range of placements:** More than $50,000.

NPRC/NATIONWIDE PERSONNEL RECRUITING & CONSULTING INC.
20834 SW Martinazzi Avenue, Tualatin OR 97062-9327. 503/692-4925. **Fax:** 503/692-6764. **Contact:** Barbara Bodle, President. **Description:** An executive search firm. Company pays fee. **Specializes in the areas of:** Architecture/Construction/Real Estate; Computer Science/Software; Engineering; General Management; Industrial; Manufacturing; Sales and Marketing; Technical and Scientific. **Positions commonly filled include:** Ceramics Engineer; Chemical Engineer; Civil Engineer; Design Engineer; Electrical/Electronics Engineer; Environmental Engineer; Financial Analyst; General Manager; Industrial Engineer; Industrial Production Manager; Manufacturer's/Wholesaler's Sales Rep.; Mechanical Engineer; MIS Specialist; Nuclear Engineer; Operations/Production Manager; Quality Control Supervisor; Services Sales Representative; Software Engineer; Stationary Engineer; Structural Engineer. **Average salary range of placements:** More than $50,000. **Number of placements per year:** 100 - 199.

SALES CONSULTANTS OF PORTLAND
5100 SW Macadam Avenue, Suite 325, Portland OR 97201. 503/241-1230. **Contact:** Manager. **Description:** An executive search firm. **Specializes in the areas of:** Accounting/Auditing; Administration/MIS/EDP; Advertising; Architecture/Construction/Real Estate; Banking; Communications; Computer Hardware/Software; Design; Electrical; Engineering; Food Industry; General Management; Health/Medical; Insurance; Legal; Manufacturing; Operations Management; Personnel/Labor Relations; Printing/Publishing; Procurement; Retail; Sales and Marketing; Technical and Scientific; Textiles; Transportation.

SEARCH NORTHWEST ASSOCIATES (SNA)
10117 SE Sunnyside Road, Suite F-727, Clackamas OR 97015. 503/654-1487. **Fax:** 503/654-9110. **Contact:** Douglas L. Jansen, CPC, President. **Description:** Search Northwest Associates is an executive search firm primarily serving the foundry, metal fabrication, chemicals, and high-tech industries. Founded in 1977. Company pays fee. **Specializes in the areas of:** Computer Science/Software; Engineering; General Management; Manufacturing. **Positions commonly filled include:** Attorney; Biochemist; Biomedical Engineer; Ceramics Engineer; Chemical Engineer; Civil Engineer; Electrical/ Electronics Engineer; Environmental Engineer; General Manager; Geologist/Geophysicist; Industrial Engineer; Materials Engineer; Mechanical Engineer; Metallurgical Engineer; Mining Engineer; Nuclear Engineer; Petroleum Engineer; Purchasing Agent/Manager; Quality Control Supervisor; Software Engineer. **Average salary range of placements:** More than $50,000. **Number of placements per year:** 100 - 199.

WOODWORTH INTERNATIONAL GROUP
620 SW 5th Avenue, Suite 1225, Portland OR 97204-1426. 503/225-5000. **Fax:** 503/225-5005. **Contact:** Gail Woodworth, President. **Description:** Woodworth International Group is a full-service executive search and management consulting organization. Company pays fee. **Specializes in the areas of:** Architecture/Construction/Real Estate; Computer Science/Software; Engineering; Finance; General Management; Health/Medical; Insurance; Manufacturing; Personnel/Labor Relations; Retail; Sales and Marketing; Transportation. **Positions commonly filled include:** Accountant/Auditor; Administrative Services Manager; Aerospace Engineer; Bank Officer/Manager; Biomedical Engineer; Branch Manager; Budget Analyst; Buyer; Chemical Engineer; Chemist; Civil Engineer; Computer Programmer; Credit Manager; Customer Service Representative; Design Engineer; Designer; Editor; Electrical/Electronics Engineer; Environmental Engineer; Financial Analyst; General Manager; Human Resources Manager; Industrial Engineer; Industrial Production Manager; Internet Services Manager; Management Analyst/Consultant; Mechanical Engineer; Metallurgical Engineer; MIS Specialist; Multimedia Designer; Nuclear Engineer; Nuclear Medicine Technologist; Operations/Production Manager; Physician; Property and Real Estate Manager; Purchasing Agent/Manager; Quality Control Supervisor; Software Engineer; Structural Engineer; Systems Analyst; Technical Writer/Editor; Telecommunications Manager; Transportation/Traffic Specialist; Video Producer. **Corporate headquarters location:** This Location. **Average salary range of placements:** More than $50,000. **Number of placements per year:** 200 - 499.

PERMANENT EMPLOYMENT AGENCIES OF OREGON

FIRST CHOICE PERSONNEL
11330 SW Ambiance Place, Tigard OR 97223. 503/620-0717. **Fax:** 503/244-1544. **Contact:** Randall B. Carrier, CPC, President. **Description:** An employment agency. Company pays fee. **Specializes in the areas of:** Computer Science/Software. **Positions commonly filled include:** Computer Programmer; Software Engineer; Systems Analyst. **Number of placements per year:** 1 - 49.

TRIAD TECHNOLOGY GROUP
10200 SW Greenburg Road, Suite 350, Portland OR 97223. 503/293-9545. **Fax:** 503/293-9546. **Contact:** Bruno Amicci, President. **Description:** Triad Technology Group is an employment agency. Company pays fee. **Specializes in the areas of:** Administration/MIS/EDP; Computer Science/Software; Engineering. **Positions commonly filled include:** Computer Programmer; Designer; Draftsperson; Software Engineer; Systems Analyst. **Number of placements per year:** 50 - 99.

EXECUTIVE SEARCH FIRMS OF PENNSYLVANIA

ADVANCED TECHNOLOGY RESOURCES
239 4th Avenue, Suite 212, Pittsburgh PA 15222. 412/281-9930. **Fax:** 412/281-5353. **Contact:** Bernie Flynn, Principle. **E-mail address:** bflynn@atrinc.com. **World Wide Web address:** http://www.industry.net.advanced.tech. **Description:** An executive search firm operating on both retainer and contingency bases. Advanced Technology Resources also offers contract services. The firm concentrates on five recruiting areas: Information Systems, Software Engineering, Telecommunications, Engineering, and Technology Sales and Marketing. Company pays fee. **NOTE:** Interested applicants should send resumes to 350 Saxonburg Road, Butler PA 16001-3621. **Specializes in the areas of:** Administration/MIS/EDP; Computer Science/Software; Engineering; Manufacturing; Sales and Marketing; Technical and Scientific. **Positions commonly filled include:** Computer Programmer; Electrical/Electronics Engineer; Financial Analyst; Internet Services Manager; Management Analyst/Consultant; Software Engineer; Statistician; Systems Analyst; Technical Writer/Editor. **Other U.S. locations:** Nationwide. **Average salary range of placements:** More than $50,000. **Number of placements per year:** 100 - 199.

BARTON PERSONNEL SYSTEMS, INC.
121 North Cedar Crest Boulevard, Allentown PA 18104-4664. 610/439-8751. **Fax:** 610/439-1207. **Contact:** Malcolm Singerman, Manager. **Description:** Barton Personnel Systems, Inc. is an executive search firm. Founded in 1972. Company pays fee. **Specializes in the areas of:** Accounting/Auditing; Computer Science/Software; Engineering; Health/Medical; Manufacturing. **Positions commonly filled include:** Accountant/Auditor; Buyer; Computer Programmer; Electrical/Electronics Engineer; Human Resources Manager; Management Analyst/Consultant; MIS Specialist; Physical Therapist; Software Engineer; Systems Analyst. **Number of placements per year:** 200 - 499.

BASILONE-OLIVER EXECUTIVE SEARCH
2987 Babcock Boulevard, Pittsburgh PA 15237. 412/931-9501. **Fax:** 412/931-9741. **Contact:** Larry S. Basilone, Partner/Owner. **Description:** Basilone-Oliver Executive Search is an executive search firm operating on a contingency basis. Company pays fee. **Specializes in the areas of:** Accounting/Auditing; Banking; Computer Science/Software; Economics; Finance; Industrial; Manufacturing; Personnel/Labor Relations; Retail. **Positions commonly filled include:** Accountant/Auditor; Cost Estimator; Credit Manager; Economist; Financial Analyst; Human Resources Manager; Market Research Analyst; Statistician. **Corporate headquarters location:** This Location. **Average salary range of placements:** $30,000 - $50,000. **Number of placements per year:** 50 - 99.

CMIS
24 Hagerty Boulevard, Suite 9, West Chester PA 19382. 610/430-0013. **Fax:** 610/696-5430. **Contact:** Peter DiNicola, President. **Description:** An executive search firm. Company pays fee. **Specializes in the areas of:** Computer Science/Software; Sales and Marketing. **Positions commonly filled include:** Computer Programmer; Software Engineer; Systems Analyst; Telecommunications Manager. **Average salary range of placements:** More than $50,000. **Number of placements per year:** 1 - 49.

CAREER CONCEPTS STAFFING SERVICES
4504 Peach Street, Erie PA 16509. 814/868-2333. **Fax:** 814/868-3238. **Contact:** Joseph A. DiGiorgio, Vice President. **Description:** An executive search firm. Company pays fee. **Specializes in the areas of:** Accounting/Auditing; Administration/MIS/EDP; Computer Science/Software; Engineering; Finance; Food Industry; General Management; Industrial; Manufacturing; Personnel/Labor Relations; Plastics; Sales and Marketing; Technical and Scientific. **Positions commonly filled include:** Accountant/Auditor; Aerospace Engineer; Agricultural Engineer; Biological Scientist; Biomedical Engineer; Buyer; Chemical Engineer; Chemist; Civil Engineer; Electrical/Electronics Engineer; General Manager; Health Services Manager; Human Resources

Manager; Industrial Engineer; Industrial Production Manager; Mechanical Engineer; Purchasing Agent/Manager; Quality Control Supervisor; Science Technologist; Technical Writer/Editor. **Number of placements per year:** 50 - 99.

CLIFFORD ASSOCIATES, INC.
306 Corporate Drive East, Langhorne PA 19047. 215/968-1980. **Fax:** 215/968-6686. **Contact:** Cliff Milles, Owner. **Description:** An executive search firm operating on a contingency basis. Company pays fee. **Specializes in the areas of:** Computer Hardware/Software. **Positions commonly filled include:** Computer Programmer; MIS Specialist; Systems Analyst. **Corporate headquarters location:** This Location. **Number of placements per year:** 1 - 49.

GARRICK HALL & ASSOCIATES
260 South Broad Street, Suite 1600, Philadelphia PA 19102-5021. 215/546-0030. **Fax:** 215/546-4920. **Contact:** William J. Yamarick, President. **Description:** An executive search firm. Company pays fee. **Specializes in the areas of:** Computer Science/Software; Health/Medical; Industrial; Manufacturing; Sales and Marketing. **Positions commonly filled include:** Management Trainee; Services Sales Representative. **Number of placements per year:** 100 - 199.

HUMAN RESOURCE SOLUTIONS
200 Penn Street, Suite 203-206, Reading PA 19602-1000. 610/371-9505. **Fax:** 610/373-8618. **Contact:** Thomas N. Dondore, President. **E-mail address:** tomhrs@aol.com. **Description:** An executive search firm engaged in human resource consulting offering a full range of services including executive and management search, training and development, contract recruiting, and electronic staff sourcing. Company pays fee. **Specializes in the areas of:** Accounting/Auditing; Administration/MIS/EDP; Banking; Computer Science/Software; Engineering; Fashion; Finance; Food Industry; General Management; Industrial; Insurance; Manufacturing; Personnel/Labor Relations; Retail; Sales and Marketing; Technical and Scientific. **Positions commonly filled include:** Accountant/Auditor; Administrative Services Manager; Aerospace Engineer; Agricultural Engineer; Attorney; Bank Officer/Manager; Biological Scientist; Biomedical Engineer; Branch Manager; Budget Analyst; Buyer; Chemical Engineer; Chemist; Civil Engineer; Computer Programmer; Credit Manager; Design Engineer; Designer; Economist; Editor; Electrical/Electronics Engineer; Environmental Engineer; Financial Analyst; General Manager; Health Services Manager; Human Resources Manager; Industrial Engineer; Industrial Production Manager; Management Analyst/Consultant; Management Trainee; Manufacturer's/Wholesaler's Sales Rep.; Market Research Analyst; Materials Engineer; Mechanical Engineer; MIS Specialist; Multimedia Designer; Nuclear Engineer; Petroleum Engineer; Purchasing Agent/Manager; Quality Control Supervisor; Science Technologist; Securities Sales Representative; Services Sales Representative; Software Engineer; Stationary Engineer; Strategic Relations Manager; Structural Engineer; Systems Analyst; Telecommunications Manager; Transportation/Traffic Specialist.

J-RAND SEARCH
Two Bethlehem Plaza, Bethlehem PA 18018. 610/867-4649. **Fax:** 610/867-9750. **Contact:** Michael P. Watts, President. **Description:** A search and recruitment firm. Company pays fee. **Specializes in the areas of:** Accounting/Auditing; Administration/MIS/EDP; Banking; Biology; Computer Science/Software; Engineering; Finance; Food Industry; General Management; Health/Medical; Industrial; Manufacturing; Personnel/Labor Relations; Technical and Scientific; Transportation. **Positions commonly filled include:** Accountant/Auditor; Aerospace Engineer; Agricultural Engineer; Biological Scientist; Biomedical Engineer; Ceramics Engineer; Chemical Engineer; Chemist; Civil Engineer; Computer Programmer; Cost Estimator; Designer; Electrical/Electronics Engineer; Financial Analyst; Food Scientist/Technologist; General Manager; Industrial Engineer; Industrial Production Manager; Materials Engineer; Mechanical Engineer; Metallurgical Engineer; Meteorologist; Mining Engineer; Operations/Production Manager; Petroleum Engineer; Pharmacist; Physician; Purchasing Agent/Manager; Quality Control Supervisor; Software Engineer; Stationary Engineer; Statistician; Structural Engineer; Systems Analyst. **Number of placements per year:** 50 - 99.

NANCY JACKSON INC.
343 North Washington Avenue, Scranton PA 18503. 717/346-8711. **Fax:** 717/346-9940. **Contact:** Nancy Jackson, President. **Description:** An executive search firm operating on a contingency basis.

Company pays fee. **Specializes in the areas of:** Accounting/Auditing; Administration/MIS/EDP; Banking; Computer Science/Software; Engineering; Finance; General Management; Health/Medical; Manufacturing; Personnel/Labor Relations; Printing/Publishing; Sales and Marketing; Secretarial. **Positions commonly filled include:** Accountant/Auditor; Administrative Services Manager; Bank Officer/Manager; Biomedical Engineer; Buyer; Ceramics Engineer; Chemical Engineer; Chemist; Civil Engineer; Claim Representative; Clerical Supervisor; Computer Programmer; Credit Manager; Customer Service Representative; Dental Assistant/Dental Hygienist; Draftsperson; Economist; Electrical/Electronics Engineer; Financial Analyst; General Manager; Health Services Manager; Human Resources Manager; Industrial Engineer; Industrial Production Manager; Management Analyst/Consultant; Materials Engineer; Mechanical Engineer; Metallurgical Engineer; Occupational Therapist; Operations/Production Manager; Paralegal; Physical Therapist; Public Relations Specialist; Purchasing Agent/Manager; Quality Control Supervisor; Software Engineer. **Corporate headquarters location:** This Location. **Number of placements per year:** 100 - 199.

CLIFTON JOHNSON ASSOCIATES INC.
One Monroeville Center, Suite 450, Monroeville PA 15146. 412/856-8000. **Fax:** 412/856-8026. **Contact:** Clifton Johnson, President. **E-mail address:** clifton@nb.net. **Description:** An executive search firm established in 1969. The firm's clients include many *Fortune* 500 companies. The firm belongs to the NIS (Nationwide Interchange Service) Network, one of the largest computerized networks of professional recruiters. **Specializes in the areas of:** Computer Science/Software; Engineering; Industrial; Manufacturing. **Positions commonly filled include:** Aerospace Engineer; Chemical Engineer; Civil Engineer; Computer Programmer; Design Engineer; Electrical/Electronics Engineer; Environmental Engineer; Food Scientist/Technologist; Geologist/Geophysicist; Industrial Engineer; Industrial Production Manager; Mechanical Engineer; Metallurgical Engineer; Mining Engineer; Petroleum Engineer; Quality Control Supervisor; Software Engineer; Structural Engineer; Systems Analyst; Telecommunications Manager. **Corporate headquarters location:** This Location. **Average salary range of placements:** $30,000 - $50,000.

LAWRENCE PERSONNEL
1000 Valley Forge Circle, Suite 110, King of Prussia PA 19406-1111. 610/783-5400. **Fax:** 610/783-6008. **Contact:** Larry Goldberg, CPC, General Manager. **Description:** An executive search firm. Company pays fee. **Specializes in the areas of:** Computer Science/Software; Data Communications; Engineering; Telecommunications. **Positions commonly filled include:** Broadcast Technician; Designer; Electrical/Electronics Engineer; Internet Services Manager; Mechanical Engineer; MIS Specialist; Software Engineer; Systems Analyst; Technical Writer/Editor; Telecommunications Manager. **Number of placements per year:** 1 - 49.

MICHAEL LUSKIN ASSOCIATES
100 West Avenue, Jenkintown PA 19046-2625. 215/571-1410. **Contact:** Michael Luskin, President. **E-mail address:** luskin@aol.com. **Description:** An executive search firm that specializes in mid- and senior-level placements for universities, high-technology companies, and medical centers. Company pays fee. **Specializes in the areas of:** Computer Science/Software; Education; Engineering; General Management; Health/Medical; Nonprofit. **Positions commonly filled include:** Aerospace Engineer; Biological Scientist; Education Administrator; General Manager; MIS Specialist; Nuclear Engineer. **Average salary of placements:** More than $50,000. **Number of placements per year:** 1 - 49.

MANAGEMENT RECRUITERS OF DELAWARE COUNTY, INC.
COMPUSEARCH
90 South Newtown Street Road, Suite 9, Newtown Square PA 19073. 610/356-8360. **Fax:** 610/356-8731. **Contact:** Sandy Bishop, Manager. **Description:** An executive search firm. **Specializes in the areas of:** Accounting/Auditing; Administration/MIS/EDP; Advertising; Architecture/Construction/Real Estate; Banking; Communications; Computer Hardware/Software; Design; Electrical; Engineering; Finance; Food Industry; General Management; Health/Medical; Insurance; Legal; Manufacturing; Operations Management; Personnel/Labor Relations; Printing/Publishing; Procurement; Retail; Sales and Marketing; Technical and Scientific; Textiles; Transportation.

MANAGEMENT RECRUITERS OF LEHIGH VALLEY, INC.
COMPUSEARCH
1414 Millard Street, Suite 102, Bethlehem PA 18018. 610/974-9770. **Contact:** Fred Meyer, Manager. **Description:** An executive search firm. **Specializes in the areas of:** Accounting/Auditing; Administration/MIS/EDP; Advertising; Architecture/Construction/Real Estate; Banking; Communications; Computer Hardware/Software; Design; Electrical; Engineering; Finance; Food Industry; General Management; Health/Medical; Insurance; Legal; Manufacturing; Operations Management; Personnel/Labor Relations; Printing/Publishing; Procurement; Retail; Sales and Marketing; Technical and Scientific; Textiles; Transportation.

MANAGEMENT RECRUITERS OF MANAYUNK/CHESTNUT HILL, INC.
COMPUSEARCH
161 Leverington Avenue, Suite 102, Philadelphia PA 19127. 215/482-6881. **Fax:** 215/482-7518. **Contact:** Recruiter. **Description:** An executive search firm. Company pays fee. **Specializes in the areas of:** Biology; Biotechnology; Computer Science/Software; Legal; Pharmaceutical; Technical and Scientific. **Positions commonly filled include:** Attorney; Biological Scientist; Clinical Lab Technician; Computer Programmer; Science Technologist; Systems Analyst. **Number of placements per year:** 50 - 99.

MANAGEMENT RECRUITERS OF PHILADELPHIA, INC.
COMPUSEARCH
325 Chestnut Street, Chestnut Place, Suite 1106, Philadelphia PA 19106. 215/829-1900. **Contact:** Manager. **Description:** An executive search firm. **Specializes in the areas of:** Accounting/Auditing; Administration/MIS/EDP; Advertising; Architecture/Construction/Real Estate; Banking; Communications; Computer Hardware/Software; Design; Electrical; Engineering; Finance; Food Industry; General Management; Health/Medical; Insurance; Legal; Manufacturing; Operations Management; Personnel/Labor Relations; Printing/Publishing; Procurement; Retail; Sales and Marketing; Technical and Scientific; Textiles; Transportation.

GEORGE R. MARTIN EXECUTIVE SEARCH
P.O. Box 673, Doylestown PA 18901. 215/348-8146. **Contact:** George R. Martin, Owner/Manager. **Description:** An executive search firm. **Specializes in the areas of:** Computer Hardware/Software; Engineering; Insurance; Manufacturing; Personnel/Labor Relations; Sales and Marketing; Technical and Scientific.

LaMONTE OWENS & COMPANY
P.O. Box 27742, Philadelphia PA 19118. 215/248-0500. **Fax:** 215/233-3737. **Contact:** LaMonte Owens, President/Owner. **Description:** An executive search firm. **NOTE:** The physical address of the firm is 805 East Willow Grove Avenue, Philadelphia PA. **Specializes in the areas of:** Accounting/Auditing; Administration/MIS/EDP; Architecture/Construction/Real Estate; Banking; Biology; Computer Science/Software; Engineering; Finance; Health/Medical; Personnel/Labor Relations; Sales and Marketing; Technical and Scientific. **Positions commonly filled include:** Accountant/Auditor; Administrative Services Manager; Aerospace Engineer; Architect; Bank Officer/Manager; Biological Scientist; Biomedical Engineer; Branch Manager; Budget Analyst; Buyer; Ceramics Engineer; Chemical Engineer; Chemist; Civil Engineer; Clerical Supervisor; Computer Programmer; Credit Manager; Economist; Electrical/Electronics Engineer; Financial Analyst; Health Services Manager; Human Resources Manager; Industrial Engineer; Management Analyst/Consultant; Materials Engineer; Mechanical Engineer; Metallurgical Engineer; Nuclear Engineer; Purchasing Agent/Manager; Registered Nurse; Securities Sales Representative; Software Engineer; Statistician; Systems Analyst. **Number of placements per year:** 1 - 49.

RHA EXECUTIVE PERSONNEL SERVICES
J. ALLEN ENTERPRISES
33 West Lancaster Avenue, Ardmore PA 19003. 610/642-3092. **Fax:** 610/642-8347. **Contact:** Joel M. Allen, Owner. **Description:** An executive search firm. Company pays fee. **Specializes in the areas of:** Accounting/Auditing; Administration/MIS/EDP; Advertising; Computer Hardware/Software; Engineering; Food Industry; General Management; Health/Medical; Industrial; Insurance; Legal; Manufacturing; Nonprofit; Personnel/Labor Relations; Printing/Publishing;

Retail; Sales and Marketing; Secretarial; Technical and Scientific; Transportation. **Positions commonly filled include:** Accountant/Auditor; Administrative Assistant; Aerospace Engineer; Architect; Biological Scientist; Biomedical Engineer; Bookkeeper; Buyer; Ceramics Engineer; Chemical Engineer; Chemist; Civil Engineer; Claim Representative; Clerk; Commercial Artist; Computer Operator; Computer Programmer; Construction Trade Worker; Credit Manager; Customer Service Representative; Data Entry Clerk; Draftsperson; Driver; Editor; EDP Specialist; Electrical/Electronics Engineer; Factory Worker; Hotel Manager; Industrial Designer; Industrial Engineer; Legal Secretary; Light Industrial Worker; Manufacturing Engineer; Marketing Specialist; Mechanical Engineer; Medical Secretary; Metallurgical Engineer; MIS Specialist; Operations/Production Manager; Public Relations Specialist; Purchasing Agent/Manager; Quality Control Supervisor; Receptionist; Recruiter; Reporter; Sales Representative; Secretary; Software Engineer; Systems Analyst; Technical Illustrator; Technical Writer/Editor; Technician; Typist/Word Processor. **Number of placements per year:** 50 - 99.

THE RICHARDS GROUP

1608 Walnut Street, Suite 1702, Philadelphia PA 19103. 215/735-9450. **Fax:** 215/735-9430. **Contact:** Larry Winitsky, President. **Description:** An executive search firm. Founded in 1984. Company pays fee. **Specializes in the areas of:** Accounting/Auditing; Administration/MIS/EDP; Advertising; Banking; Computer Science/Software; Finance; General Management; Health/Medical; Insurance; Nonprofit; Personnel/Labor Relations; Retail; Sales and Marketing; Secretarial. **Positions commonly filled include:** Accountant/Auditor; Actuary; Administrative Services Manager; Bank Officer/Manager; Branch Manager; Brokerage Clerk; Buyer; Claim Representative; Clerical Supervisor; Computer Programmer; Credit Manager; Customer Service Representative; Dietician/Nutritionist; EEG Technologist; EKG Technician; Financial Analyst; General Manager; Health Services Manager; Human Resources Manager; Insurance Agent/Broker; Management Trainee; Manufacturer's/Wholesaler's Sales Rep.; Market Research Analyst; Medical Records Technician; Occupational Therapist; Operations/Production Manager; Recreational Therapist; Registered Nurse; Respiratory Therapist; Services Sales Representative; Systems Analyst; Technical Writer/Editor; Telecommunications Manager; Transportation/Traffic Specialist; Typist/Word Processor; Underwriter/Assistant Underwriter. **Corporate headquarters location:** This Location.

JASON ROBERTS ASSOCIATES INC.

One Belmont Avenue, Suite 526, Bala Cynwyd PA 19004. 610/667-1440. **Fax:** 610/667-1573. **Contact:** Robert Kirschner, President. **Description:** An executive search firm operating on both retainer and contingency bases. Company pays fee. **Specializes in the areas of:** Administration/MIS/EDP; Banking; Computer Science/Software; Finance; Food Industry; General Management; Insurance; Retail; Sales and Marketing; Secretarial; Transportation. **Positions commonly filled include:** Adjuster; Aerospace Engineer; Agricultural Scientist; Bank Officer/Manager; Biomedical Engineer; Ceramics Engineer; Chemical Engineer; Civil Engineer; Claim Representative; Clerical Supervisor; Collector; Computer Programmer; Credit Manager; Electrical/Electronics Engineer; Financial Analyst; Hotel Manager; Industrial Engineer; Investigator; Management Trainee; Manufacturer's/Wholesaler's Sales Rep.; Materials Engineer; Mechanical Engineer; Metallurgical Engineer; Mining Engineer; Nuclear Engineer; Petroleum Engineer; Restaurant/Food Service Manager; Services Sales Representative; Software Engineer; Stationary Engineer; Structural Engineer; Systems Analyst; Underwriter/Assistant Underwriter. **Corporate headquarters location:** This Location. **Average salary range of placements:** $30,000 - $50,000. **Number of placements per year:** 100 - 199.

S-H-S INTERNATIONAL OF WILKES-BARRE

216 North River Street, Suite 550, Wilkes-Barre PA 18702-2594. 717/825-3411. **Fax:** 717/825-7790. **Contact:** Chris Hackett, Search Consultant. **E-mail address:** chrish@headhunt. microserve.com. **World Wide Web address:** http://www.shsint.com. **Description:** An executive search firm founded in 1968. Company pays fee. **Specializes in the areas of:** Accounting/Auditing; Banking; Computer Science/Software; Engineering; Finance; Personnel/Labor Relations; Printing/Publishing. **Positions commonly filled include:** Accountant/Auditor; Aerospace Engineer; Biological Scientist; Budget Analyst; Ceramics

Engineer; Chemical Engineer; Chemist; Civil Engineer; Clinical Lab Technician; Computer Programmer; Credit Manager; Designer; Draftsperson; Editor; Financial Analyst; Human Resources Manager; Industrial Engineer; Landscape Architect; Materials Engineer; Mechanical Engineer; Medical Records Technician; Metallurgical Engineer; Nuclear Medicine Technologist; Occupational Therapist; Physical Therapist; Purchasing Agent/Manager; Radiologic Technologist; Respiratory Therapist; Software Engineer; Systems Analyst; Technical Writer/Editor. **Average salary range of placements:** $30,000 - $50,000. **Number of placements per year:** 50 - 99.

SALES CONSULTANTS OF PHILADELPHIA

301 Oxford Valley Road, Suite 1506-A, Yardley PA 19067. 215/321-4100. **Contact:** Manager. **Description:** An executive search firm. **Specializes in the areas of:** Accounting/Auditing; Administration/MIS/EDP; Advertising; Architecture/Construction/Real Estate; Banking; Communications; Computer Hardware/Software; Design; Electrical; Engineering; Finance; Food Industry; General Management; Health/Medical; Insurance; Legal; Manufacturing; Operations Management; Personnel/Labor Relations; Printing/Publishing; Procurement; Retail; Sales and Marketing; Technical and Scientific; Textiles; Transportation.

KENN SPINRAND INC.

P.O. Box 4095, Reading PA 19606. 610/779-0944. **Fax:** 610/779-8338. **Contact:** Kenn Spinrand, President. **Description:** An executive search firm. Company pays fee. **Specializes in the areas of:** Administration/MIS/EDP; Apparel; Computer Science/Software; Engineering; Manufacturing; Textiles. **Positions commonly filled include:** Accountant/Auditor; Aerospace Engineer; Agricultural Engineer; Apparel Worker; Biomedical Engineer; Ceramics Engineer; Chemical Engineer; Chemist; Civil Engineer; Computer Programmer; Draftsperson; Electrical/Electronics Engineer; Human Resources Manager; Industrial Engineer; Materials Engineer; Mechanical Engineer; Metallurgical Engineer; Mining Engineer; Nuclear Engineer; Petroleum Engineer; Quality Control Supervisor; Software Engineer; Stationary Engineer; Structural Engineer; Systems Analyst; Textile Manager.

SUBURBAN PLACEMENT SERVICE

21 North York Road, Willow Grove PA 19090-3420. 215/657-6262. **Fax:** 215/657-6431. **Contact:** Ed Fort, Manager. **Description:** An executive search firm founded in 1971. Company pays fee. **Specializes in the areas of:** Computer Science/Software. **Positions commonly filled include:** Aerospace Engineer; Computer Programmer; Customer Service Representative; Electrical/ Electronics Engineer; MIS Specialist; Software Engineer; Systems Analyst; Telecommunications Analyst. **Number of placements per year:** 50 - 99.

W.G. TUCKER & ASSOCIATES

2908 McKelvey Road, Suite 2, Pittsburgh PA 15221-4569. 412/244-9309. **Fax:** 412/244-9195. **Contact:** Manager. **Description:** An executive search firm. Company pays fee. **Specializes in the areas of:** Accounting/Auditing; Banking; Biology; Computer Science/Software; Engineering; Finance; Health/Medical; Insurance; Manufacturing; Personnel/Labor Relations; Technical and Scientific. **Positions commonly filled include:** Accountant/Auditor; Actuary; Aerospace Engineer; Attorney; Bank Officer/Manager; Biological Scientist; Biomedical Engineer; Branch Manager; Budget Analyst; Buyer; Ceramics Engineer; Chemical Engineer; Chemist; Civil Engineer; Claim Representative; Clinical Lab Technician; Computer Programmer; Cost Estimator; Credit Manager; Customer Service Representative; Dietician/Nutritionist; Draftsperson; Editor; Education Administrator; Electrical/Electronics Engineer; Financial Analyst; Food Scientist/Technologist; General Manager; Geologist/Geophysicist; Health Services Manager; Human Resources Manager; Industrial Engineer; Insurance Agent/Broker; Library Technician; Management Analyst/ Consultant; Management Trainee; Materials Engineer; Mechanical Engineer; Metallurgical Engineer; Mining Engineer; Nuclear Engineer; Occupational Therapist; Paralegal; Petroleum Engineer; Pharmacist; Physical Therapist; Physician; Public Relations Specialist; Purchasing Agent/Manager; Quality Control Supervisor; Restaurant/Food Service Manager; Science Technologist; Services Sales Representative; Software Engineer; Structural Engineer; Systems Analyst; Technical Writer/Editor; Underwriter/Assistant Underwriter; Urban/Regional Planner. **Number of placements per year:** 50 - 99.

PERMANENT EMPLOYMENT AGENCIES OF PENNSYLVANIA

ACROPOLIS SERVICES INC.
P.O. Box 425, Springhouse PA 19477. 215/542-9520. **Fax:** 215/542-9530. **Contact:** Gus Mechalas, President. **Description:** An employment agency. **Specializes in the areas of:** Accounting/Auditing; Administration/MIS/EDP; Banking; Biology; Computer Science/Software; Engineering; Finance; Health/Medical; Industrial; Insurance; Legal; Manufacturing; Personnel/ Labor Relations; Sales and Marketing; Secretarial; Technical and Scientific. **Positions commonly filled include:** Accountant/Auditor; Aerospace Engineer; Agricultural Engineer; Attorney; Bank Officer/Manager; Biological Scientist; Biomedical Engineer; Budget Analyst; Buyer; Ceramics Engineer; Chemical Engineer; Chemist; Civil Engineer; Computer Programmer; Credit Manager; Data Analyst; Data Entry Clerk; Dentist; Designer; Dietician/Nutritionist; Electrical/Electronics Engineer; Emergency Medical Technician; Financial Analyst; Health Services Manager; Industrial Engineer; Industrial Production Manager; Licensed Practical Nurse; Materials Engineer; Mathematician; Mechanical Engineer; Medical Records Technician; Metallurgical Engineer; Mining Engineer; Nuclear Engineer; Nuclear Medicine Technologist; Occupational Therapist; Paralegal; Petroleum Engineer; Pharmacist; Physical Therapist; Physician; Physicist/Astronomer; Psychologist; Purchasing Agent/Manager; Quality Control Supervisor; Radiologic Technologist; Recreational Therapist; Registered Nurse; Science Technologist; Software Engineer; Stationary Engineer; Statistician; Structural Engineer; Systems Analyst; Veterinarian. **Number of placements per year:** 50 - 99.

ADVANCE RECRUITING SERVICES
1250 Wall Avenue, Clairton PA 15025. 412/233-8808. **Fax:** 412/233-8814. **Contact:** Joe Giansante, Owner. **Description:** An employment agency that also offers contract services, temporary placement, and conducts executive searches on a contingency basis. Advance Recruiting Services specializes in the placement of telecommunications, data processing, and technical sales professionals. Company pays fee. **Specializes in the areas of:** Computer Science/Software; Sales and Marketing; Telephone Technical Support. **Positions commonly filled include:** MIS Specialist; Multimedia Designer; Services Sales Representative; Software Engineer; Technical Writer/Editor; Telecommunications Manager; Video Production Coordinator. **Benefits available to temporary workers:** Medical Insurance. **Average salary range of placements:** $30,000 - $50,000. **Number of placements per year:** 1 - 49.

ALL STAFFING INC.
100 West Ridge Street, P.O. Box 219, Lansford PA 18232. 717/645-8883. **Fax:** 717/645-9771. **Contact:** Stan Costello, Jr., President. **Description:** All Staffing Inc. is a permanent employment agency. Company pays fee. **Specializes in the areas of:** Accounting/Auditing; Administration/MIS/EDP; Advertising; Computer Science/Software; Finance; Health/Medical; Industrial; Manufacturing; Personnel/Labor Relations; Retail; Sales and Marketing; Secretarial. **Positions commonly filled include:** Accountant/Auditor; Computer Programmer; Dental Assistant/Dental Hygienist; Dental Lab Technician; Dentist; Draftsperson; Economist; EEG Technologist; EKG Technician; Financial Analyst; General Manager; Health Services Manager; Human Resources Manager; Insurance Agent/Broker; Management Trainee; Manufacturer's/ Wholesaler's Sales Rep.; Medical Records Technician; Nuclear Medicine Technologist; Occupational Therapist; Pharmacist; Physical Therapist; Physician; Physicist/Astronomer; Psychologist; Radiologic Technologist; Registered Nurse; Respiratory Therapist; Services Sales Representative; Surgical Technician; Systems Analyst; Wholesale and Retail Buyer. **Number of placements per year:** 50 - 99.

BRADLEY PROFESSIONAL SERVICES
440 East Swedesford Road, Suite 1070, Wayne PA 19087. 610/254-9995. **Fax:** 610/971-9480. **Contact:** Manager. **Description:** An employment agency. Company pays fee. **Specializes in the areas of:** Accounting/Auditing; Biology; Computer Science/Software; Engineering; Health/Medical; Personnel/Labor Relations; Technical and Scientific. **Positions commonly filled**

include: Accountant/Auditor; Aerospace Engineer; Agricultural Scientist; Architect; Biological Scientist; Biomedical Engineer; Buyer; Ceramics Engineer; Chemical Engineer; Chemist; Civil Engineer; Clinical Lab Technician; Computer Programmer; Designer; Draftsperson; Electrical/Electronics Engineer; Human Resources Manager; Industrial Engineer; Management Analyst/Consultant; Materials Engineer; Mechanical Engineer; Metallurgical Engineer; Mining Engineer; Nuclear Engineer; Petroleum Engineer; Pharmacist; Physician; Science Technologist; Software Engineer; Stationary Engineer; Statistician; Structural Engineer; Systems Analyst; Technical Writer/Editor. **Number of placements per year:** 100 - 199.

COMPUTER PROFESSIONALS UNLIMITED

5000 McKnight Road, Suite 302, Pittsburgh PA 15237. 412/367-4191. **Fax:** 412/367-1152. **Contact:** Personnel Manager. **Description:** An employment agency. **Specializes in the areas of:** Banking; Computer Science/Software. **Positions commonly filled include:** Bank Officer/Manager; Computer Programmer; Human Resources Manager; Software Engineer; Systems Analyst. **Number of placements per year:** 100 - 199.

ROBERT J. DePAUL & ASSOCIATES INC.

71 McMurray Road, Pittsburgh PA 15241. 412/561-0417. **Contact:** David J. Shopf, Placement Counselor. **Description:** Robert J. DePaul & Associates is an employment agency specializing in information management. The company offers both permanent and temporary placement, contracting, and management consulting. Consulting services include Third Generation Programming Languages (COBOL, CICS, C); Database Design (IMS, Oracle, Rdb, DB2); PC Support & Training Services; Microcomputer Based Languages (DBase, Clipper, Paradox); and Networking & Client Server Technology (Novell). Company pays fee. **Specializes in the areas of:** Administration/MIS/EDP; Banking; Computer Science/Software; Health/Medical; Retail. **Positions commonly filled include:** Computer Programmer; Network Engineer; Systems Analyst; Technical Writer/Editor; Telecommunications Manager. **Benefits available to temporary workers:** Dental Insurance; Medical Insurance. **Corporate headquarters location:** This Location. **Average salary range of placements:** $30,000 - $50,000.

DiCENZO PERSONNEL SPECIALISTS

428 Forbes Avenue, Suite 110, Pittsburgh PA 15219. 412/281-6207. **Fax:** 412/281-9326. **Contact:** Carmela DiCenzo, Owner. **Description:** An employment agency. Company pays fee. **Specializes in the areas of:** Accounting/Auditing; Administration/MIS/EDP; Computer Science/Software; Engineering; Finance; Food Industry; General Management; Industrial; Manufacturing; Personnel/Labor Relations; Sales and Marketing; Secretarial; Technical and Scientific. **Positions commonly filled include:** Accountant/Auditor; Adjuster; Administrative Services Manager; Aerospace Engineer; Agricultural Engineer; Aircraft Mechanic/Engine Specialist; Bank Officer/Manager; Biological Scientist; Biomedical Engineer; Blue-Collar Worker Supervisor; Branch Manager; Budget Analyst; Buyer; Chemical Engineer; Chemist; Civil Engineer; Claim Representative; Computer Programmer; Credit Manager; Customer Service Representative; Draftsperson; Economist; Electrical/Electronics Engineer; Electrician; Financial Analyst; General Manager; Industrial Engineer; Industrial Production Manager; Management Analyst/Consultant; Mechanical Engineer; Metallurgical Engineer; Mining Engineer; Nuclear Engineer; Paralegal; Petroleum Engineer; Public Relations Specialist; Purchasing Agent/Manager; Quality Control Supervisor; Services Sales Representative; Software Engineer; Stationary Engineer; Structural Engineer; Technical Writer/Editor.

GENERAL EMPLOYMENT & TRIAD PERSONNEL

1617 JFK Boulevard, Suite 930, Philadelphia PA 19103. 215/569-3226. **Fax:** 215/569-8164. **Contact:** Bill Gouldey, Manager. **Description:** An employment agency. **Specializes in the areas of:** Administration/MIS/EDP; Computer Science/Software. **Positions commonly filled include:** Computer Programmer; Internet Services Manager; MIS Specialist; Systems Analyst; Telecommunications Manager.

HOSKINS HAINS ASSOCIATES

3835 Walnut Street, Harrisburg PA 17109. 717/657-8444. **Contact:** Patricia Hoskins, Proprietor. **Description:** An employment agency. Founded in 1980. Company pays fee. **Specializes in the areas of:** Accounting/Auditing; Administration/MIS/EDP; Banking; Clerical; Computer

Hardware/Software; Engineering; Finance; Manufacturing; Personnel/Labor Relations; Sales and Marketing; Technical and Scientific. **Positions commonly filled include:** Accountant/Auditor; Administrative Assistant; Bank Officer/Manager; Computer Operator; Computer Programmer; Credit Manager; Data Entry Clerk; Draftsperson; EDP Specialist; Electrical/Electronics Engineer; Financial Analyst; Human Resources Manager; Industrial Engineer; Mechanical Engineer; Metallurgical Engineer; MIS Specialist; Purchasing Agent/Manager; Quality Control Supervisor; Receptionist; Sales Representative; Secretary; Statistician; Systems Analyst; Technical Writer/Editor; Typist/Word Processor. **Number of placements per year:** 1 - 49.

MAIN LINE PERSONNEL SERVICE
100 Presidential Boulevard, Suite 200, Bala Cynwyd PA 19004. 610/667-1820. **Fax:** 610/668-5000. **Contact:** Bart Marshall, Vice President. **Description:** An employment agency. Company pays fee. **Specializes in the areas of:** Computer Science/Software; Engineering; Personnel/Labor Relations; Technical and Scientific. **Positions commonly filled include:** Aerospace Engineer; Agricultural Scientist; Biological Scientist; Biomedical Engineer; Ceramics Engineer; Chemical Engineer; Civil Engineer; Computer Programmer; Electrical/Electronics Engineer; Geologist/Geophysicist; Human Resources Manager; Industrial Engineer; Materials Engineer; Mechanical Engineer; Metallurgical Engineer; Mining Engineer; Nuclear Engineer; Petroleum Engineer; Software Engineer; Stationary Engineer; Structural Engineer; Systems Analyst. **Number of placements per year:** 1000+.

Q-SOURCE
2950 Felton Road, Suite 102, Norristown PA 19401. 610/278-7993. **Fax:** 610/278-7985. **Contact:** John Quigley, President. **Description:** An employment agency. Company pays fee. **Specializes in the areas of:** Administration/MIS/EDP; Computer Science/Software; Engineering; Industrial; Manufacturing; Technical and Scientific. **Positions commonly filled include:** Ceramics Engineer; Chemical Engineer; Chemist; Computer Programmer; Design Engineer; Electrical/Electronics Engineer; Food Scientist/Technologist; Management Analyst/Consultant; Materials Engineer; Mechanical Engineer; Metallurgical Engineer; MIS Specialist; Quality Control Supervisor; Software Engineer; Structural Engineer; Systems Analyst. **Number of placements per year:** 1 - 49.

QUEST SYSTEMS, INC.
1150 First Avenue, Suite 255, King of Prussia PA 19406. 610/265-8100. **Fax:** 610/265-6974. **Contact:** Charles Lagana, Manager. **E-mail address:** questsyst@aol.com. **World Wide Web address:** http://www.questsyst.com. **Description:** An employment agency specializing in the fields of computer technologies and information systems. Company pays fee. **Specializes in the areas of:** Administration/MIS/EDP; Computer Science/Software. **Positions commonly filled include:** Computer Programmer; MIS Specialist; Software Engineer; Systems Analyst. **Corporate headquarters location:** Bethesda MD. **Other U.S. locations:** Atlanta GA; Baltimore MD; Bethesda MD. **Average salary range of placements:** $30,000 - $50,000. **Number of placements per year:** 1000+.

S-H-S INTERNATIONAL
101 East Lancaster Avenue, Wayne PA 19087. 610/687-6104. **Fax:** 610/687-6102. **Contact:** Paul Reitman, General Manager. **Description:** An employment agency. Company pays fee. **Specializes in the areas of:** Accounting/Auditing; Banking; Computer Science/Software; Engineering; Finance; Insurance; Legal; Sales and Marketing; Secretarial; Technical and Scientific. **Positions commonly filled include:** Accountant/Auditor; Actuary; Adjuster; Administrative Services Manager; Architect; Attorney; Bank Officer/Manager; Biochemist; Biological Scientist; Biomedical Engineer; Branch Manager; Buyer; Chemical Engineer; Chemist; Civil Engineer; Claim Representative; Clerical Supervisor; Computer Programmer; Customer Service Representative; Dental Assistant/Dental Hygienist; Design Engineer; Draftsperson; Editor; EKG Technician; Electrical/Electronics Engineer; Emergency Medical Technician; Environmental Engineer; Financial Analyst; Food Scientist/Technologist; Industrial Engineer; Insurance Agent/Broker; Internet Services Manager; Management Analyst/Consultant; Management Trainee; Market Research Analyst; Mechanical Engineer; Medical Records Technician; Metallurgical Engineer; MIS Specialist; Multimedia Designer; Nuclear Engineer; Occupational Therapist; Operations/Production Manager; Paralegal; Pharmacist; Physical Therapist; Physician; Purchasing

Agent/Manager; Quality Control Supervisor; Radiologic Technologist; Registered Nurse; Respiratory Therapist; Science Technologist; Securities Sales Representative; Services Sales Representative; Software Engineer; Statistician; Structural Engineer; Technical Writer/Editor; Telecommunications Manager; Travel Agent; Typist/Word Processor; Underwriter/Assistant Underwriter. **Corporate headquarters location:** This Location. **Other U.S. locations:** Nationwide. **Number of placements per year:** 50 - 99.

SELECT PERSONNEL, INC.
3070 Bristol Pike, Building 2, Suite 205, Bensalem PA 19020. 215/245-4800. **Fax:** 215/245-4990. **Contact:** Marjorie Stilwell, President. **Description:** An employment agency. Company pays fee. **Specializes in the areas of:** Computer Hardware/Software; Electronics; Engineering; Finance; Industrial; Manufacturing; Personnel/Labor Relations; Sales and Marketing; Technical and Scientific. **Positions commonly filled include:** Accountant/Auditor; Aerospace Engineer; Architect; Biological Scientist; Bookkeeper; Buyer; Ceramics Engineer; Chemical Engineer; Chemist; Civil Engineer; Computer Programmer; Customer Service Representative; Electrical/ Electronics Engineer; Industrial Engineer; Manufacturing Engineer; Marketing Specialist; Mechanical Engineer; Metallurgical Engineer; MIS Specialist; Operations/Production Manager; Purchasing Agent/Manager; Quality Control Supervisor; Sales Representative; Software Engineer; Systems Analyst; Technical Illustrator; Technical Writer/Editor; Technician. **Number of placements per year:** 100 - 199.

SOURCE EDP
150 South Warner Road, Suite 238, King of Prussia PA 19406. 610/341-1960. **Contact:** Manager. **Description:** An employment agency. **Specializes in the areas of:** Administration/MIS/EDP; Computer Hardware/Software; Technical and Scientific.

EXECUTIVE SEARCH FIRMS OF RHODE ISLAND

DORRA SEARCH INC.
One Richmond Square, Providence RI 02906. 401/453-1555. **Fax:** 401/453-1566. **Contact:** Bethany Gold, Managing Director. **Description:** An executive search firm. Company pays fee. **Specializes in the areas of:** Accounting/Auditing; Administration/MIS/EDP; Computer Science/Software. **Positions commonly filled include:** Accountant/Auditor; Computer Programmer; MIS Specialist; Systems Analyst. **Average salary range of placements:** More than $50,000. **Number of placements per year:** 1 - 49.

MANAGEMENT RECRUITERS
101 Dyer Street, Providence RI 02903. 401/274-2810. **Contact:** Manager. **Description:** An executive search firm. **Specializes in the areas of:** Accounting/Auditing; Administration/ MIS/EDP; Advertising; Architecture/Construction/Real Estate; Banking; Communications; Computer Hardware/Software; Design; Electrical; Engineering; Food Industry; General Management; Health/Medical; Insurance; Legal; Manufacturing; Operations Management; Personnel/Labor Relations; Printing/Publishing; Procurement; Retail; Sales and Marketing; Technical and Scientific; Textiles; Transportation.

PKS ASSOCIATES, INC.
P.O. Box 5021, Greene RI 02827. 401/397-6154. **Fax:** 401/397-6722. **Contact:** Paul Spremulli, Operations Manager. **Description:** An executive search firm operating on a contingency basis. Company pays fee. **Specializes in the areas of:** Computer Science/Software; Engineering; General Management; Industrial; Manufacturing; Personnel/Labor Relations; Technical and Scientific. **Positions commonly filled include:** Accountant/Auditor; Attorney; Blue-Collar Worker Supervisor; Computer Programmer; Dental Assistant/Dental Hygienist; Dentist; Design Engineer;

Designer; Draftsperson; Electrical/Electronics Engineer; Environmental Engineer; General Manager; Human Resources Specialist; Industrial Engineer; Industrial Production Manager; Licensed Practical Nurse; Mechanical Engineer; MIS Specialist; Occupational Therapist; Operations/Production Manager; Physical Therapist; Quality Control Supervisor; Recreational Therapist; Registered Nurse; Respiratory Therapist; Software Engineer; Structural Engineer; Systems Analyst. **Corporate headquarters location:** This Location. **Other U.S. locations:** Warwick RI. **Average salary range of placements:** $30,000 - $50,000. **Number of placements per year:** 200 - 499.

SALES CONSULTANTS OF WARWICK
349 Centerville Road, Warwick RI 02886-4324. 401/737-3200. **Contact:** Peter C. Cotton, Owner/President. **E-mail address:** slscnsltnt@aol.com. **World Wide Web address:** http://www.mrinet.com. **Description:** An executive search firm. Company pays fee. **Specializes in the areas of:** Accounting/Auditing; Administration/MIS/EDP; Advertising; Architecture/ Construction/Real Estate; Banking; Communications; Computer Hardware/Software; Design; Electrical; Engineering; Food Industry; General Management; Health/Medical; Insurance; Legal; Manufacturing; Operations Management; Personnel/Labor Relations; Printing/Publishing; Procurement; Retail; Sales and Marketing; Technical and Scientific; Textiles; Transportation. **Number of placements per year:** 1 - 49.

STORTI ASSOCIATES
4042 Post Road, Unit #8, Warwick RI 02886. 401/885-3100. **Fax:** 401/885-3107. **Contact:** Michael Storti, President. **Description:** An executive search firm. Company pays fee. **Specializes in the areas of:** Accounting/Auditing; Administration/MIS/EDP; Affirmative Action; Banking; Computer Hardware/Software; Engineering; Finance; Health/Medical; Manufacturing; Sales and Marketing. **Positions commonly filled include:** Accountant/Auditor; Biochemist; Biological Scientist; Biomedical Engineer; Buyer; Ceramics Engineer; Chemical Engineer; Chemist; Computer Programmer; EDP Specialist; Electrical/Electronics Engineer; Marketing Specialist; Mechanical Engineer; Metallurgical Engineer; MIS Specialist; Purchasing Agent/Manager; Quality Control Supervisor; Sales Representative; Software Engineer; Systems Analyst.

PERMANENT EMPLOYMENT AGENCIES OF RHODE ISLAND

NEW ENGLAND CONSULTANTS, INC.
156 Centerville Road, Warwick RI 02886. 401/732-4650. **Fax:** 401/732-4654. **Contact:** Mary Shaw, President. **Description:** An employment agency. Company pays fee. **Specializes in the areas of:** Computer Hardware/Software; Engineering; Manufacturing; Technical and Scientific. **Positions commonly filled include:** Computer Operator; Computer Programmer; EDP Specialist; Electrical/Electronics Engineer; Industrial Designer; Industrial Engineer; Manufacturing Engineer; Mechanical Engineer; Metallurgical Engineer; MIS Specialist; Operations/Production Manager; Quality Control Supervisor; Sales Representative; Software Engineer; Systems Analyst. **Number of placements per year:** 1 - 49.

NORRELL STAFFING SERVICES
400 Reservoir Avenue, Providence RI 02907-3553. 401/785-2077. **Fax:** 401/467-3930. **Contact:** Michael Oliver, President. **Description:** An employment agency. Founded in 1987. **Specializes in the areas of:** Accounting/Auditing; Administration/MIS/EDP; Banking; Computer Science/Software; Economics; Engineering; Finance; Food Industry; General Management; Industrial; Insurance; Legal; Manufacturing; Nonprofit; Personnel/Labor Relations; Printing/Publishing; Retail; Sales and Marketing; Secretarial; Technical and Scientific; Transportation. **Positions commonly filled include:** Accountant/Auditor; Administrative Services Manager; Aerospace Engineer; Bank Officer/Manager; Biomedical Engineer; Branch Manager;

Budget Analyst; Buyer; Chemical Engineer; Chemist; Civil Engineer; Claim Representative; Clerical Supervisor; Computer Programmer; Cost Estimator; Credit Manager; Customer Service Representative; Design Engineer; Designer; Dietician/Nutritionist; Draftsperson; Electrical/Electronics Engineer; Environmental Engineer; Financial Analyst; Food Scientist/Technologist; General Manager; Hotel Manager; Human Resources Specialist; Industrial Engineer; Industrial Production Manager; Management Analyst/Consultant; Management Trainee; Market Research Analyst; Mechanical Engineer; Metallurgical Engineer; MIS Specialist; Nuclear Engineer; Operations/Production Manager; Paralegal; Public Relations Specialist; Purchasing Agent/Manager; Quality Control Supervisor; Restaurant/Food Service Manager; Services Sales Representative; Software Engineer; Stationary Engineer; Strategic Relations Manager; Structural Engineer; Systems Analyst; Technical Writer/Editor; Telecommunications Manager; Typist/Word Processor. **Benefits available to temporary workers:** 401(k); Medical Insurance; Paid Holidays; Paid Vacation. **Corporate headquarters location:** Atlanta GA. **Other U.S. locations:** Nationwide. **Average salary range of placements:** $20,000 - $29,999. **Number of placements per year:** 50 - 99.

ON LINE TEMP INC.
One Richmond Square, Providence RI 02906. 401/274-1500. **Fax:** 401/274-1803. **Contact:** Susan Reid, Owner. **Description:** An employment agency that fills both temporary and permanent positions. Company pays fee. **Specializes in the areas of:** Computer Science/Software; Engineering; Personnel/Labor Relations; Technical and Scientific. **Positions commonly filled include:** Biochemist; Biological Scientist; Chemist; Computer Programmer; Design Engineer; Designer; Draftsperson; Editor; Engineer; Internet Services Manager; Mathematician; MIS Specialist; Multimedia Designer; Science Technologist; Systems Analyst; Technical Writer/Editor; Telecommunications Manager. **Number of placements per year:** 100 - 199.

EXECUTIVE SEARCH FIRMS OF SOUTH CAROLINA

DUNHILL PROFESSIONAL SEARCH
Six Village Square, 231 Hampton Street, Greenwood SC 29646. 864/229-5251. **Contact:** Hal Freese, President. **Description:** An executive search firm. **Specializes in the areas of:** Accounting/Auditing; Computer Hardware/Software; Engineering; Manufacturing.

EASTERN PERSONNEL SERVICES
8410 Rivers Avenue, Suite B, North Charleston SC 29406. 803/863-9111. **Fax:** 803/863-0710. **Contact:** Paul Day, President. **Description:** An executive search firm operating on a contingency basis. The firm also operates as a temporary agency, a contract services firm, and a career/outplacement counseling agency. Company pays fee. **Specializes in the areas of:** Administration/MIS/EDP; Computer Science/Software; General Management; Insurance; Legal; Personnel/Labor Relations; Sales and Marketing; Secretarial. **Positions commonly filled include:** Accountant/Auditor; Administrative Services Manager; Advertising Clerk; Architect; Attorney; Bank Officer/Manager; Branch Manager; Buyer; Claim Representative; Clerical Supervisor; Credit Manager; Customer Service Representative; Designer; Draftsperson; Electrical/Electronics Engineer; Electrician; Financial Analyst; General Manager; Human Resources Specialist; Human Service Worker; Industrial Engineer; Industrial Production Manager; Landscape Architect; Management Analyst/Consultant; Manufacturer's/Wholesaler's Sales Rep.; Market Research Analyst; Medical Records Technician; MIS Specialist; Operations/Production Manager; Property and Real Estate Manager; Purchasing Agent/Manager; Quality Control Supervisor; Services Sales Representative; Systems Analyst; Typist/Word Processor; Underwriter/Assistant Underwriter. **Benefits available to temporary workers:** Medical Insurance; Vacation Pay. **Corporate headquarters location:** This Location. **Other U.S. locations:** Savannah GA. **Average salary range of placements:** $20,000 - $29,999. **Number of placements per year:** 500 - 999.

MANAGEMENT RECRUITERS OF AIKEN
P.O. Box 730, Aiken SC 29802-0730. 803/648-1361. **Contact:** Michael Hardwick, Manager. **Description:** An executive search firm. Company pays fee. **Specializes in the areas of:** Apparel; Computer Hardware/Software; Electronics; Health/Medical; Medical Software; Metals; Plastics; Textiles. **Number of placements per year:** 500 - 999.

MANAGEMENT RECRUITERS OF ANDERSON
P.O. Box 2874, Anderson SC 29622. 864/225-1258. **Fax:** 864/225-2332. **Contact:** Rod Pagan, Owner. **Description:** Management Recruiters of Anderson is an executive search firm. Company pays fee. **Specializes in the areas of:** Automotive; Computer Science/Software; Data Processing; Health/Medical; Pharmaceutical. **Positions commonly filled include:** Accountant/Auditor; Biological Scientist; Biomedical Engineer; Buyer; Chemical Engineer; Chemist; Computer Programmer; Electrical/Electronics Engineer; General Manager; Human Resources Manager; Industrial Engineer; Mechanical Engineer; Purchasing Agent/Manager; Systems Analyst. **Number of placements per year:** 50 - 99.

MANAGEMENT RECRUITERS OF COLUMBIA
P.O. Box 58785, Columbia SC 29250. 803/254-1334. **Fax:** 803/254-1527. **Contact:** Bob Keen, Manager. **Description:** An executive search firm. **Specializes in the areas of:** Accounting/Auditing; Administration/MIS/EDP; Advertising; Architecture/Construction/Real Estate; Banking; Communications; Computer Hardware/Software; Design; Electrical; Engineering; Food Industry; General Management; Health/Medical; Insurance; Legal; Manufacturing; Operations Management; Personnel/Labor Relations; Printing/Publishing; Procurement; Retail; Sales and Marketing; Technical and Scientific; Textiles; Transportation.

MANAGEMENT RECRUITERS OF GREENVILLE
330 Pelham Road, Suite 109B, Greenville SC 29615. 864/370-1341. **Contact:** Office Manager. **Description:** An executive search firm. **Specializes in the areas of:** Accounting/Auditing; Administration/MIS/EDP; Advertising; Architecture/Construction/Real Estate; Banking; Communications; Computer Hardware/Software; Design; Electrical; Engineering; Food Industry; General Management; Health/Medical; Insurance; Legal; Manufacturing; Operations Management; Personnel/Labor Relations; Printing/Publishing; Procurement; Retail; Sales and Marketing; Technical and Scientific; Textiles; Transportation.

MANAGEMENT RECRUITERS OF ORANGEBURG
2037 Saint Matthews Road, Orangeburg SC 29118. 803/531-4101. **Fax:** 803/536-3714. **Contact:** Dick Crawford/Ed Chewning, Co-Managers. **Description:** An executive search firm. **Specializes in the areas of:** Accounting/Auditing; Administration/MIS/EDP; Advertising; Architecture/Construction/Real Estate; Banking; Communications; Computer Hardware/Software; Design; Electrical; Engineering; Food Industry; General Management; Health/Medical; Insurance; Legal; Manufacturing; Operations Management; Personnel/Labor Relations; Printing/Publishing; Procurement; Retail; Sales and Marketing; Technical and Scientific; Textiles; Transportation.

MANAGEMENT RECRUITERS OF ROCK HILL
1925 Ebenezer Road, Rock Hill SC 29732. 803/324-5181. **Fax:** 803/324-3431. **Contact:** Herman Smith, Manager. **Description:** An executive search firm. **Specializes in the areas of:** Accounting/Auditing; Administration/MIS/EDP; Advertising; Architecture/Construction/Real Estate; Banking; Communications; Computer Hardware/Software; Design; Electrical; Engineering; Food Industry; General Management; Health/Medical; Insurance; Legal; Manufacturing; Operations Management; Personnel/Labor Relations; Printing/Publishing; Procurement; Retail; Sales and Marketing; Technical and Scientific; Textiles; Transportation.

MILLER & ASSOCIATES
1852 Wallace School Road, Suite E, Charleston SC 29407-4887. 803/571-6630. **Fax:** 803/571-0230. **Contact:** Al E. Miller Jr., Owner/Manager. **Description:** An executive search firm that handles recruitment for various manufacturing industries and engineering professions. Company pays fee. **Specializes in the areas of:** Accounting/Auditing; Administration/MIS/EDP; Computer Science/Software; Engineering; Finance; Industrial; Manufacturing; Personnel/Labor Relations; Sales and Marketing; Technical and Scientific. **Positions commonly filled include:**

Accountant/Auditor; Ceramics Engineer; Civil Engineer; Design Engineer; Designer; Draftsperson; Electrical/Electronics Engineer; Environmental Engineer; Financial Analyst; General Manager; Human Resources Specialist; Industrial Engineer; Industrial Production Manager; Materials Engineer; Mechanical Engineer; Metallurgical Engineer; MIS Specialist; Purchasing Agent/Manager; Quality Control Supervisor; Software Engineer; Systems Analyst. **Average salary range of placements:** More than $50,000. **Number of placements per year:** 1 - 49.

SOUTHERN RECRUITERS
P.O. Box 2745, Aiken SC 29802-2745. 803/648-7834. **Contact:** Ray Fehrenbach, President. **Description:** An executive search firm. Company pays fee. **Specializes in the areas of:** Accounting/Auditing; Administration/MIS/EDP; Computer Science/Software; Engineering; General Management; Health/Medical; Manufacturing; MIS/EDP; Personnel/Labor Relations; Technical and Scientific. **Positions commonly filled include:** Accountant/Auditor; Bank Officer/Manager; Biochemist; Biomedical Engineer; Buyer; Civil Engineer; Computer Programmer; Design Engineer; Designer; Draftsperson; Electrical/Electronics Engineer; Environmental Engineer; General Manager; Human Resources Manager; Industrial Engineer; Mechanical Engineer; Metallurgical Engineer; MIS Specialist; Purchasing Agent/Manager; Quality Control Supervisor; Software Engineer; Structural Engineer; Systems Analyst. **Corporate headquarters location:** This Location. **Other U.S. locations:** Nationwide. **Average salary range of placements:** $30,000 - $50,000. **Number of placements per year:** 50 - 99.

PERMANENT EMPLOYMENT AGENCIES OF SOUTH CAROLINA

COMPANION EMPLOYMENT SERVICES
400 Germays Street, Columbia SC 29201. 803/771-6454. **Fax:** 803/765-1431. **Contact:** Personnel Consultant. **Description:** An employment agency. **Specializes in the areas of:** Computer Science/Software; Data Processing.

HARVEY PERSONNEL, INC.
P.O. Box 1931, Spartanburg SC 29304. 864/582-5616. **Fax:** 864/582-3588. **Contact:** Howard L. Harvey, CPC, President. **Description:** An employment agency. Company pays fee. **Specializes in the areas of:** Accounting/Auditing; Administration/MIS/EDP; Computer Hardware/Software; Engineering; General Management; Industrial; Manufacturing; Personnel/Labor Relations; Technical and Scientific. **Positions commonly filled include:** Accountant/Auditor; Biological Scientist; Biomedical Engineer; Buyer; Ceramics Engineer; Chemical Engineer; Chemist; Civil Engineer; EDP Specialist; Electrical/Electronics Engineer; Industrial Engineer; Manufacturing Engineer; Manufacturing Manager; Mechanical Engineer; Metallurgical Engineer; MIS Specialist; Operations/Production Manager; Plastics Engineer; Purchasing Agent/Manager; Quality Control Supervisor; Software Engineer; Systems Analyst. **Number of placements per year:** 1 - 49.

SEARCH AND RECRUIT INTERNATIONAL
2501 Northforest Drive, North Charleston SC 29420. 803/572-4040. **Fax:** 803/572-4045. **Contact:** Les Callahan, Southeast Regional Manager. **Description:** An employment agency that places personnel in high technology fields such as engineering, manufacturing, electronics, and computer hardware/software. Company pays fee. **Specializes in the areas of:** Accounting/Auditing; Administration/MIS/EDP; Computer Science/Software; Engineering; Finance; Food Industry; Health/Medical; Industrial; Manufacturing; Nuclear Power; Technical and Scientific. **Positions commonly filled include:** Accountant/Auditor; Biological Scientist; Biomedical Engineer; Ceramics Engineer; Chemical Engineer; Civil Engineer; Computer Programmer; Electrical/ Electronics Engineer; Electrician; Financial Analyst; Geologist/Geophysicist; Industrial Engineer; Industrial Production Manager; Materials Engineer; Mechanical Engineer; Medical Records Technician; Metallurgical Engineer; Nuclear Engineer; Nuclear Medicine Technologist;

Occupational Therapist; Operations/Production Manager; Pharmacist; Physical Therapist; Physician; Quality Control Supervisor; Respiratory Therapist; Software Engineer; Speech-Language Pathologist; Stationary Engineer; Structural Engineer; Systems Analyst. **Corporate headquarters location:** This Location. **Average salary range of placements:** $30,000 - $50,000. **Number of placements per year:** 100 - 199.

STAFFING RESOURCES
1755 St. Julian Place, Columbia SC 29204. 803/765-0820. **Contact:** Frank Staley, Owner. **Description:** An employment agency. **Specializes in the areas of:** Banking; Computer Hardware/Software; Health/Medical; Insurance; Manufacturing; Office Support; Temporary Assignments. **Positions commonly filled include:** Accountant/Auditor; Administrative Assistant; Computer Operator; Computer Programmer; Controller; CPA; Financial Manager; Receptionist; Secretary; Systems Analyst; Typist/Word Processor.

EXECUTIVE SEARCH FIRMS OF SOUTH DAKOTA

REGENCY RECRUITING, INC.
P.O. Box 77, North Sioux City SD 57049. 605/232-3205. **Fax:** 605/232-3159. **Contact:** Brad Moore, CPC, President. **E-mail address:** rgncyrctg@aol.com. **Description:** An executive search firm focusing on placement of computer and computer management personnel in the Midwest. Company pays fee. **Specializes in the areas of:** Computer Science/Software. **Positions commonly filled include:** Computer Programmer; Management Analyst/Consultant; MIS Specialist; Systems Analyst. **Number of placements per year:** 50 - 99.

PERMANENT EMPLOYMENT AGENCIES OF SOUTH DAKOTA

AVAILABILITY EMPLOYMENT INC.
1521 South Minnesota Avenue, Sioux Falls SD 57105. 605/336-0353. **Contact:** Recruiting. **Description:** Fills both temporary and permanent positions. Availability Employment covers a variety of fields including the computer industry. **Specializes in the areas of:** Administration/MIS/EDP; Computer Programming.

EXECUTIVE SEARCH FIRMS OF TENNESSEE

ACADEMY GRADUATES CAREER CONSULTANTS
1764 Doc Terry Road, Dandridge TN 37725-6248. 423/397-3300. **Contact:** Tom Karpick, President. **Description:** An executive search firm. Company pays fee. **Specializes in the areas of:** Computer Science/Software; Economics; Engineering; Industrial; Manufacturing; Personnel/Labor Relations; Technical and Scientific. **Positions commonly filled include:** Administrative Services Manager; Aerospace Engineer; Bank Officer/Manager; Branch Manager; Civil Engineer;

Construction Contractor; Electrical/Electronics Engineer; General Manager; Industrial Engineer; Industrial Production Manager; Management Analyst/Consultant; Management Trainee; Mechanical Engineer; MIS Specialist; Nuclear Engineer; Operations/Production Manager; Quality Control Supervisor; Software Engineer; Structural Engineer; Systems Analyst; Telecommunications Manager. **Number of placements per year:** 1 - 49.

ANDERSON McINTYRE PERSONNEL SERVICES
6148 Lee Highway, Suite 100, Chattanooga TN 37421. 423/894-9571. **Fax:** 423/892-7413. **Contact:** Maureen McIntyre, Owner. **Description:** An executive search firm. The firm is also a full-service employment agency offering career counseling and contract services. **Specializes in the areas of:** Administration/MIS/EDP; Advertising; Architecture/Construction/Real Estate; Computer Science/Software; Finance; General Management; Health/Medical; Legal; Manufacturing; Sales and Marketing; Secretarial. **Positions commonly filled include:** Accountant/Auditor; Administrative Services Manager; Advertising Clerk; Branch Manager; Claim Representative; Clerical Supervisor; Computer Programmer; Credit Manager; Draftsperson; Management Trainee; Medical Records Technician; Paralegal; Physical Therapist; Physician; Systems Analyst; Travel Agent; Typist/Word Processor. **Average salary range of placements:** $30,000 - $50,000. **Number of placements per year:** 100 - 199.

COOK ASSOCIATES INTERNATIONAL, INC.
P.O. Box 962, Brentwood TN 37024. 615/373-8263. **Fax:** 615/371-8215. **Contact:** Juli C. Wells, Regional Manager. **Description:** An executive search firm operating on both retainer and contingency bases. **Specializes in the areas of:** Accounting/Auditing; Administration/MIS/EDP; Computer Science/Software; Engineering; Health/Medical; Industrial; Insurance; Legal; Manufacturing; Personnel/Labor Relations; Printing/Publishing. **Positions commonly filled include:** Accountant/Auditor; Actuary; Attorney; Biomedical Engineer; Budget Analyst; Buyer; Chemical Engineer; Computer Programmer; Cost Estimator; Design Engineer; Dietician/Nutritionist; Electrical/Electronics Engineer; Electrician; Environmental Engineer; Financial Analyst; General Manager; Health Services Manager; Hotel Manager; Human Resources Specialist; Industrial Engineer; Mathematician; Mechanical Engineer; MIS Specialist; Occupational Therapist; Paralegal; Physical Therapist; Quality Control Supervisor; Registered Nurse; Software Engineer; Stationary Engineer; Statistician; Systems Analyst. **Corporate headquarters location:** This Location. **Other U.S. locations:** Hopkinsville KY; Greenville SC. **Number of placements per year:** 100 - 199.

HAMILTON RYKER COMPANY
P.O Box 1068, Martin TN 38237. 901/587-3161. **Toll-free phone:** 800/644-9449. **Fax:** 901/588-0810. **Contact:** Lisa Moss Hester, Professional Staffing Manager. **Description:** An executive search firm operating on both retainer and contingency bases. Company pays fee. **Specializes in the areas of:** Accounting/Auditing; Administration/MIS/EDP; Architecture/Construction/Real Estate; Computer Science/Software; Engineering; General Management; Industrial; Manufacturing; Personnel/Labor Relations; Technical and Scientific. **Positions commonly filled include:** Accountant/Auditor; Architect; Budget Analyst; Buyer; Chemical Engineer; Civil Engineer; Computer Programmer; Cost Estimator; Credit Manager; Design Engineer; Designer; Draftsperson; Electrical/Electronics Engineer; Environmental Engineer; Financial Analyst; General Manager; Human Resources Specialist; Industrial Engineer; Industrial Production Manager; Management Analyst/Consultant; Management Trainee; Market Research Analyst; Mechanical Engineer; Metallurgical Engineer; Mining Engineer; MIS Specialist; Operations/Production Manager; Purchasing Agent/Manager; Quality Control Supervisor; Software Engineer; Structural Engineer; Systems Analyst. **Corporate headquarters location:** This Location. **Other U.S. locations:** Memphis TN; Nashville TN.

INFORMATION SYSTEMS GROUP, INC.
6363 Poplar Avenue, Suite 336, Memphis TN 38119. 901/684-1030. **Fax:** 901/684-1068. **Contact:** Harold Lepman, President. **Description:** An executive search firm. Company pays fee. **Specializes in the areas of:** Computer Science/Software. **Positions commonly filled include:** Computer Programmer; Systems Analyst. **Number of placements per year:** 1 - 49.

J&D RESOURCES (JDR)
6555 Quince Road, Suite 425, Memphis TN 38119. 901/753-0500. **Fax:** 901/753-0550. **Contact:** Jill T. Herrin, President. **Description:** An executive search firm. J&D Resources (JDR) provides permanent and contract information systems positions. Founded in 1987. Company pays fee. **Specializes in the areas of:** Administration/MIS/EDP; Computer Science/Software. **Positions commonly filled include:** Computer Programmer; MIS Specialist; Software Engineer; Systems Analyst. **Average salary range of placements:** $30,000 - $50,000. **Number of placements per year:** 50 - 99.

MANAGEMENT RECRUITERS OF CHATTANOOGA
SALES CONSULTANTS OF CHATTANOOGA
7405 Shallowford Road, Suite 520, Chattanooga TN 37421. 423/894-5500. **Fax:** 423/894-1177. **Contact:** Bill Cooper, General Manager. **Description:** An executive search firm operating on a contingency basis. Management Recruiters and Sales Consultants are two divisions of the same company. Founded in 1980. Company pays fee. **NOTE:** A third division, Compusearch, is also based at this location. **Specializes in the areas of:** Administration/MIS/EDP; Computer Science/Software; Engineering; General Management; Manufacturing; Sales and Marketing; Technical and Scientific. **Positions commonly filled include:** Chemical Engineer; Chemist; Computer Programmer; Design Engineer; Environmental Engineer; General Manager; Industrial Engineer; Information Systems Consultant; Internet Services Manager; Manufacturer's/Wholesaler's Sales Rep.; Mechanical Engineer; MIS Specialist; Quality Control Supervisor; Services Sales Representative; Software Engineer; Systems Analyst. **Corporate headquarters location:** Cleveland OH. **Other U.S. locations:** Nationwide. **Average salary range of placements:** More than $50,000. **Number of placements per year:** 100 - 199.

MANAGEMENT RECRUITERS OF KNOXVILLE
9050B Executive Park Drive, Suite 16, Knoxville TN 37923. 423/694-1628. **Contact:** Jim Kline, Manager. **Description:** An executive search firm. **Specializes in the areas of:** Accounting/Auditing; Administration/MIS/EDP; Advertising; Architecture/Construction/Real Estate; Banking; Communications; Computer Hardware/Software; Finance; Food Industry; General Management; Health/Medical; Insurance; Legal; Manufacturing; Personnel/Labor Relations; Printing/Publishing; Retail; Technical and Scientific; Textiles; Transportation.

MANAGEMENT RECRUITERS OF LENOIR CITY
530 Highway 321 North, Suite 303, Lenoir City TN 37771. 423/986-3000. **Contact:** Mr. R.S. Strobo, Manager. **Description:** An executive search firm. **Specializes in the areas of:** Accounting/Auditing; Administration/MIS/EDP; Advertising; Architecture/Construction/Real Estate; Banking; Communications; Computer Hardware/Software; Finance; Food Industry; General Management; Health/Medical; Insurance; Legal; Manufacturing; Personnel/Labor Relations; Printing/Publishing; Retail; Sales and Marketing; Technical and Scientific; Textiles; Transportation.

GENE MURPHY & ASSOCIATES
1102 Murray Creek Lane, Franklin TN 37069-4727. 615/790-6990. **Fax:** 615/790-6990. **Contact:** Gene Murphy, President. **Description:** An executive search firm operating on a contingency basis. Founded in 1993. Company pays fee. **Specializes in the areas of:** Accounting/Auditing; Banking; Computer Science/Software; Food Industry; Health/Medical; Hotel/Restaurant; Manufacturing; Personnel/Labor Relations; Retail. **Positions commonly filled include:** Accountant/Auditor; Computer Programmer; Financial Analyst; Human Resources Specialist; MIS Specialist; Systems Analyst. **Average salary range of placements:** $30,000 - $50,000. **Number of placements per year:** 1 - 49.

SALES CONSULTANTS OF NASHVILLE
7003 Chadwick Drive, Suite 331, Brentwood TN 37027. 615/373-1111. **Contact:** Branch Manager. **Description:** An executive search firm. **Specializes in the areas of:** Accounting/Auditing; Administration/MIS/EDP; Advertising; Architecture/Construction/Real Estate; Banking; Communications; Computer Hardware/Software; Electrical; Engineering; Finance; Food Industry; General Management; Health/Medical; Insurance; Legal; Manufacturing; Operations Management;

Personnel/Labor Relations; Printing/Publishing; Retail; Sales and Marketing; Technical and Scientific; Textiles; Transportation.

SNELLING PERSONNEL SERVICE

6100 Building 3300, Chattanooga TN 37411. 423/894-1500. **Toll-free phone:** 800/891-1505. **Fax:** 423/894-1507. **Contact:** John C. Parham, Manager. **Description:** An executive search firm operating on a contingency basis. Company pays fee. **Specializes in the areas of:** Accounting/Auditing; Administration/MIS/EDP; Banking; Computer Science/Software; Engineering; Finance; General Management; Health/Medical; Industrial; Insurance; Manufacturing; Personnel/Labor Relations. **Positions commonly filled include:** Accountant/ Auditor; Architect; Bank Officer/Manager; Chemical Engineer; Chemist; Civil Engineer; Computer Programmer; Credit Manager; Customer Service Representative; Design Engineer; Designer; Draftsperson; Education Administrator; Electrical/Electronics Engineer; Environmental Engineer; Human Resources Specialist; Industrial Engineer; Management Analyst/Consultant; Management Trainee; Manufacturer's/Wholesaler's Sales Rep.; Mechanical Engineer; MIS Specialist; Operations/Production Manager; Physician; Purchasing Agent/Manager; Quality Control Supervisor; Restaurant/Food Service Manager; Software Engineer; Structural Engineer; Systems Analyst; Technical Writer/Editor; Telecommunications Manager; Typist/Word Processor; Underwriter/Assistant Underwriter. **Corporate headquarters location:** Dallas TX. **Other U.S. locations:** Nationwide. **Average salary range of placements:** $30,000 - $50,000. **Number of placements per year:** 50 - 99.

SOUTHERN PERSONNEL
dba SNELLING PERSONNEL

800 South Gay Street, Knox TN 37929. 423/637-5779. **Fax:** 423/523-3180. **Contact:** Greg O'Connor, General Manager. **Description:** An executive search firm. The firm also provides career/outplacement counseling and temporary services. Company pays fee. **Specializes in the areas of:** Accounting/Auditing; Computer Science/Software; Engineering; Finance; Food Industry; Health/Medical; Manufacturing; Personnel/Labor Relations. **Positions commonly filled include:** Accountant/Auditor; Chemical Engineer; Civil Engineer; Computer Programmer; Design Engineer; Electrical/Electronics Engineer; Environmental Engineer; Financial Analyst; General Manager; Human Resources Specialist; Industrial Engineer; Industrial Production Manager; Manufacturer's/Wholesaler's Sales Rep.; Mechanical Engineer; Physical Therapist; Physician; Purchasing Agent/Manager; Software Engineer; Systems Analyst. **Average salary range of placements:** $30,000 - $50,000. **Number of placements per year:** 100 - 199.

PERMANENT EMPLOYMENT AGENCIES OF TENNESSEE

ENGINEER ONE, INC.

P.O. Box 23037, Knoxville TN 37933. 423/690-2611. **Contact:** George Chaney, President. **Description:** An employment agency. Company pays fee. **Specializes in the areas of:** Computer Science/Software; Construction; Engineering; Food Industry; Industrial; Manufacturing; MIS/EDP; Technical and Scientific. **Positions commonly filled include:** Aerospace Engineer; Agricultural Engineer; Ceramics Engineer; Civil Engineer; Computer Programmer; Electrical/Electronics Engineer; General Manager; Industrial Engineer; Mechanical Engineer; Petroleum Engineer; Systems Engineer. **Number of placements per year:** 50 - 99.

EXPRESS PERSONNEL SERVICES, INC.

8807 Kingston Pike, Knoxville TN 37923. 423/531-1720. **Fax:** 423/531-3267. **Contact:** Celia Spinner, Owner. **Description:** An employment agency. Company pays fee. **Specializes in the areas of:** Computer Science/Software; Engineering; General Management; Industrial; Manufacturing; Personnel/Labor Relations; Sales and Marketing; Technical and Scientific;

Transportation. **Positions commonly filled include:** Administrative Services Manager; Biomedical Engineer; Branch Manager; Chemical Engineer; Clerical Supervisor; Computer Programmer; Construction Contractor; Customer Service Representative; Electrical/Electronics Engineer; Emergency Medical Technician; General Manager; Human Resources Manager; Industrial Engineer; Industrial Production Manager; Management Analyst/Consultant; Mechanical Engineer; Metallurgical Engineer; Operations/Production Manager; Purchasing Agent/Manager; Quality Control Supervisor; Sociologist; Software Engineer; Systems Analyst; Technical Writer/Editor; Transportation/Traffic Specialist. **Number of placements per year:** 1 - 49.

STAFFING SOLUTIONS
1801 Downtown West Boulevard, Knoxville TN 37919. 423/690-2311. **Contact:** Manager. **Description:** An employment agency. **Specializes in the areas of:** Accounting/Auditing; Banking; Computer Science/Software; Finance; Office Support; Technical and Scientific. **Positions commonly filled include:** Accountant/Auditor; Administrative Assistant; Computer Operator; Computer Programmer; Data Entry Clerk; Engineer; Industrial Designer.

EXECUTIVE SEARCH FIRMS OF TEXAS

ACKERMAN JOHNSON CONSULTANTS, INC.
333 North Sam Houston Parkway East, Suite 1210, Houston TX 77060-2417. 713/999-8879. **Fax:** 713/999-7570. **Contact:** Frederick W. Stang, President. **Description:** An executive search firm specializing in the placement of sales, management, and engineering technical personnel in a wide variety of fields. Company pays fee. **Specializes in the areas of:** Advertising; Computer Science/Software; Engineering; Food Industry; General Management; Industrial; Manufacturing; Personnel/Labor Relations; Sales and Marketing. **Corporate headquarters location:** This Location. **Other U.S. locations:** Nationwide. **Average salary range of placements:** More than $50,000. **Number of placements per year:** 100 - 199.

R. GAINES BATY ASSOCIATES, INC.
6360 Lyndon B. Johnson Freeway, Suite 100, Dallas TX 75240. 214/386-7900. **Fax:** 214/934-8476. **Contact:** R. Gaines Baty, President. **E-mail address:** gbaty@onramp.net. **World Wide Web address:** http://www.rampages.onramp.net/~gbaty. **Description:** R. Gaines Baty Associates, Inc. is a worldwide executive search firm for MIS management and information technology consulting positions, as well as bilingual accounting and auditing positions. Company pays fee. **Specializes in the areas of:** Computer Science/Software; Finance; MIS/EDP. **Positions commonly filled include:** Accountant/Auditor; Management Analyst/Consultant; MIS Specialist; Systems Analyst; Telecommunications Manager. **Corporate headquarters location:** This Location. **Other U.S. locations:** Atlanta GA. **Average salary range of placements:** More than $50,000. **Number of placements per year:** 1 - 49.

BEST/WORLD ASSOCIATES
505 West Abram Street, 3rd Floor, Arlington TX 76010. 817/861-0000. **Toll-free phone:** 800/749-2846. **Fax:** 817/459-2378. **Contact:** G. Tim Best, President. **Description:** An executive search firm operating on a retainer basis. Company pays fee. **Specializes in the areas of:** Banking; Computer Science/Software; Engineering; Finance; Food Industry; Manufacturing; Personnel/Labor Relations; Sales and Marketing. **Positions commonly filled include:** Accountant/Auditor; Chemical Engineer; Economist; Electrical/Electronics Engineer; Environmental Engineer; Financial Analyst; Food Scientist/Technologist; Human Resources Specialist; Management Analyst/Consultant; Market Research Analyst; Mechanical Engineer; MIS Specialist; Quality Control Supervisor; Software Engineer; Statistician; Systems Analyst. **Corporate headquarters location:** This Location. **Other U.S. locations:** Phoenix AZ; Torrance CA; Orlando FL; Piscataway NJ. **Average salary range of placements:** More than $50,000. **Number of placements per year:** 50 - 99.

BILSON & HAZEN INTERNATIONAL

1231 Greenway Drive, Suite 390, Irving TX 75034. 214/753-1193. **Fax:** 214/753-0969. **Contact:** Frederick Sagoe, President. **Description:** An executive search firm that also offers temporary and contract services. Company pays fee. **Specializes in the areas of:** Computer Science/Software; Personnel/Labor Relations; Sales and Marketing. **Positions commonly filled include:** Administrative Services Manager; Branch Manager; Claim Representative; Computer Programmer; Design Engineer; Electrical/Electronics Engineer; Human Resources Specialist; Manufacturer's/Wholesaler's Sales Rep.; Market Research Analyst; MIS Specialist; Software Engineer; Technical Writer/Editor; Telecommunications Manager. **Benefits available to temporary workers:** Dental Insurance; Medical Insurance. **Average salary range of placements:** More than $50,000. **Number of placements per year:** 50 - 99.

BIOSOURCE INTERNATIONAL

1878 Hilltop Drive, Suite 100, Lewisville TX 75067-2114. 214/317-7060. **Fax:** 214/317-0500. **Contact:** Ric J. Favors, Principal. **Description:** An executive search firm. Company pays fee. **Specializes in the areas of:** Biology; Biotechnology; Computer Science/Software; Engineering; Pharmaceutical; Technical and Scientific. **Positions commonly filled include:** Biochemist; Biological Scientist; Biomedical Engineer; Chemical Engineer; Chemist; Computer Programmer; Electrical/Electronics Engineer; General Manager; Management Analyst/Consultant; Mechanical Engineer; MIS Specialist; Multimedia Designer; Physician; Quality Control Supervisor; Science Technologist; Software Engineer; Statistician; Systems Analyst; Technical Writer/Editor. **Corporate headquarters location:** This Location. **Other U.S. locations:** Carlsbad CA; Sarasota FL; Greensboro NC. **Average salary range of placements:** More than $50,000. **Number of placements per year:** 1 - 49.

BRIDGE PERSONNEL

8350 North Central Expressway, Suite M1226, Dallas TX 75206. 214/692-8273. **Fax:** 214/369-6070. **Contact:** Jim Peeler, CPA, Owner. **Description:** An executive search firm operating on both retainer and contingency bases. The firm specializes in accounting and information systems placements. Company pays fee. **Specializes in the areas of:** Accounting/Auditing; Administration/MIS/EDP; Computer Science/Software; Finance. **Positions commonly filled include:** Accountant/Auditor; Computer Programmer; Financial Analyst; Software Engineer; Systems Analyst; Telecommunications Manager. **Corporate headquarters location:** This Location.

BUNDY-STEWART ASSOCIATES, INC.

13601 Preston Road, Suite 107W, Dallas TX 75240. 214/458-0626. **Fax:** 214/661-2670. **Contact:** Carolyn Stewart, Owner. **Description:** An executive search firm operating on a contingency basis. **Specializes in the areas of:** Accounting/Auditing; Administration/MIS/EDP; Computer Science/Software; Engineering; Industrial; Insurance; Manufacturing; Personnel/Labor Relations; Real Estate; Sales and Marketing; Telecommunications. **Positions commonly filled include:** Accountant/Auditor; Aircraft Mechanic/Engine Specialist; Attorney; Buyer; Computer Programmer; Credit Manager; Customer Service Representative; Design Engineer; Draftsperson; Electrical/Electronics Engineer; Human Resources Specialist; Industrial Engineer; Industrial Production Manager; Market Research Analyst; Mechanical Engineer; MIS Specialist; Operations/Production Manager; Purchasing Agent/Manager; Quality Control Supervisor; Securities Sales Representative; Software Engineer; Systems Analyst; Telecommunications Manager.

C.G. & COMPANY

5050 East University, 9B, Odessa TX 79762. 915/362-7681. **Fax:** 915/362-3578. **Contact:** Cathy George, CPC, Owner. **Description:** An executive search firm operating on both retainer and contingency bases. **Specializes in the areas of:** Computer Science/Software; Engineering; Manufacturing; Technical and Scientific. **Positions commonly filled include:** Accountant/Auditor; Administrative Services Manager; Advertising Clerk; Aerospace Engineer; Agricultural Engineer; Aircraft Mechanic/Engine Specialist; Architect; Attorney; Bank Officer/Manager; Blue-Collar Worker Supervisor; Branch Manager; Brokerage Clerk; Budget Analyst; Buyer; Chemical Engineer; Chemist; Civil Engineer; Clerical Supervisor; Computer Programmer; Construction Contractor; Cost Estimator; Counselor; Credit Manager; Customer Service Representative; Design

Engineer; Designer; Draftsperson; Electrical/Electronics Engineer; Electrician; Environmental Engineer; Financial Analyst; General Manager; Geologist/Geophysicist; Human Resources Specialist; Industrial Engineer; Industrial Production Manager; Management Analyst/Consultant; Management Trainee; Manufacturer's/Wholesaler's Sales Rep.; Market Research Analyst; Mechanical Engineer; Medical Records Technician; Metallurgical Engineer; Mining Engineer; MIS Specialist; Multimedia Designer; Occupational Therapist; Operations/Production Manager; Paralegal; Petroleum Engineer; Physical Therapist; Physician; Public Relations Specialist; Purchasing Agent/Manager; Quality Control Supervisor; Radio/TV Announcer/Broadcaster; Radiologic Technologist; Recreational Therapist; Registered Nurse; Respiratory Therapist; Restaurant/Food Service Manager; Securities Sales Representative; Services Sales Representative; Software Engineer; Speech-Language Pathologist; Systems Analyst; Technical Writer/Editor; Telecommunications Manager; Travel Agent; Typist/Word Processor. **Corporate headquarters location:** This Location. **Average salary range of placements:** More than $50,000. **Number of placements per year:** 100 - 199.

CAD TECHNOLOGY, INC.

1111 Wilcrest Green, Suite 450, Houston TX 77042. 713/785-2411. **Fax:** 713/785-1625. **Contact:** Jeani DeSisto, Staffing Coordinator. **Description:** An executive search firm operating on a contingency basis. Company pays fee. **Specializes in the areas of:** Architecture/Construction/Real Estate; Computer Science/Software; Engineering. **Positions commonly filled include:** Architect; Buyer; Chemical Engineer; Civil Engineer; Computer Programmer; Construction and Building Inspector; Cost Estimator; Design Engineer; Designer; Draftsperson; Electrical/Electronics Engineer; Environmental Engineer; Geologist/Geophysicist; Industrial Engineer; Mechanical Engineer; Mining Engineer; MIS Specialist; Operations/Production Manager; Petroleum Engineer; Quality Control Supervisor; Software Engineer; Structural Engineer; Systems Analyst; Technical Writer/Editor; Telecommunications Manager; Transportation/Traffic Specialist. **Corporate headquarters location:** This Location. **Number of placements per year:** 100 - 199.

CHERBONNIER GROUP

3050 Post Oak Boulevard, Houston TX 77056-6527. **Contact:** L.M. Cherbonnier, President. **Description:** An executive search firm operating on a retainer basis. Founded in 1969. Company pays fee. **Specializes in the areas of:** Accounting/Auditing; Administration/MIS/EDP; Architecture/Construction/Real Estate; Banking; Computer Science/Software; Engineering; Finance; General Management; Health/Medical; Legal; Personnel/Labor Relations. **Positions commonly filled include:** Attorney; Biomedical Engineer; Chemical Engineer; Chemist; Civil Engineer; Design Engineer; Electrical/Electronics Engineer; Environmental Engineer; General Manager; Geologist/Geophysicist; Mechanical Engineer; MIS Specialist; Occupational Therapist; Operations/Production Manager; Petroleum Engineer; Physician; Software Engineer; Telecommunications Manager. **Corporate headquarters location:** This Location. **Other U.S. locations:** Jackson MS; Seattle WA. **Number of placements per year:** 1 - 49.

COMPUTER PROFESSIONALS UNLIMITED

13612 Midway Road, Suite 333, Dallas TX 75244. 972/233-1773. **Fax:** 972/233-9619. **Contact:** V.J. Zapotocky, Owner/President. **E-mail address:** zipzap@onramp.net. **Description:** An executive search firm that also provides contract services, focusing on placement in computer sciences, information technology, and engineering professions. Founded in 1978. Company pays fee. **Specializes in the areas of:** Computer Science/Software; Engineering. **Positions commonly filled include:** Computer Programmer; Electrical/Electronics Engineer; Internet Services Manager; MIS Specialist; Software Engineer; Systems Analyst; Technical Writer/Editor; Telecommunications Manager. **Average salary range of placements:** More than $50,000. **Number of placements per year:** 50 - 99.

ELSWORTH GROUP

10127 Morocco Street, Suite 116, San Antonio TX 78216. 210/341-9197. **Fax:** 210/341-9413. **Contact:** Beverly O'Daniel, Director. **Description:** An executive search firm. Company pays fee. **Specializes in the areas of:** Computer Science/Software; Engineering; General Management; Industrial; Manufacturing; Sales and Marketing; Technical and Scientific; Transportation. **Positions commonly filled include:** Aerospace Engineer; Aircraft Mechanic/Engine Specialist; Biomedical Engineer; Chemical Engineer; Civil Engineer; Computer Programmer; Cost Estimator;

Designer; Draftsperson; Electrical/Electronics Engineer; General Manager; Industrial Engineer; Management Analyst/Consultant; Manufacturer's/Wholesaler's Sales Rep.; Mechanical Engineer; Metallurgical Engineer; Nuclear Engineer; Operations/Production Manager; Petroleum Engineer; Physicist/Astronomer; Production Manager; Purchasing Agent/Manager; Quality Control Supervisor; Software Engineer; Structural Engineer; Systems Analyst; Technical Writer/Editor. **Number of placements per year:** 1 - 49.

ABEL M. GONZALES & ASSOCIATES
P.O. Box 681845, San Antonio TX 78268. 210/695-5555. **Fax:** 210/695-8955. **Contact:** Abel Gonzales, General Manager. **Description:** An executive search firm operating on both retainer and contingency bases. Founded in 1978. Company pays fee. **Specializes in the areas of:** Accounting/Auditing; Advertising; Banking; Computer Science/Software; Engineering; Food Industry; General Management; Industrial; Manufacturing; Personnel/Labor Relations; Sales and Marketing; Technical and Scientific. **Positions commonly filled include:** Accountant/Auditor; Bank Officer/Manager; Biological Scientist; Buyer; Chemical Engineer; Chemist; Civil Engineer; Computer Programmer; Construction Contractor; Electrical/Electronics Engineer; Environmental Engineer; Financial Analyst; Food Scientist/Technologist; General Manager; Human Resources Specialist; Industrial Engineer; Industrial Production Manager; Management Analyst/Consultant; Management Trainee; Manufacturer's/Wholesaler's Sales Rep.; Market Research Analyst; Mechanical Engineer; Metallurgical Engineer; Mining Engineer; MIS Specialist; Operations/ Production Manager; Petroleum Engineer; Public Relations Specialist; Quality Control Supervisor; Restaurant/Food Service Manager; Services Sales Representative; Software Engineer; Strategic Relations Manager. **Corporate headquarters location:** This Location. **Average salary range of placements:** More than $50,000.

INSIDE TRACK
504 Hilltop Drive, Weatherford TX 76086. 817/599-7094. **Fax:** 817/596-0807. **Contact:** Matthew DiLorenzo, Senior Technical Recruiter. **Description:** An executive search firm. Inside Track focuses on high-tech positions, particularly in wireless telecommunications. Founded in 1989. Company pays fee. **Specializes in the areas of:** Administration/MIS/EDP; Computer Science/Software; Engineering; Industrial; Manufacturing; Sales and Marketing; Telecommunications. **Positions commonly filled include:** Aerospace Engineer; Biomedical Engineer; Ceramics Engineer; Chemical Engineer; Chemist; Civil Engineer; Computer Programmer; Design Engineer; Electrical/Electronics Engineer; Environmental Engineer; General Manager; Industrial Engineer; Materials Engineer; Mechanical Engineer; MIS Specialist; Purchasing Agent/Manager; Quality Control Supervisor; Software Engineer; Structural Engineer; Systems Analyst; Telecommunications Manager. **Average salary range of placements:** More than $50,000. **Number of placements per year:** 1 - 49.

INTRATECH RESOURCE GROUP, INC.
6565 West Loop South, Suite 540, Bellaire TX 77401. 713/669-1733. **Fax:** 713/667-5507. **Contact:** James B. Lewis, President. **Description:** An executive search firm. Company pays fee. **Specializes in the areas of:** Computer Science/Software. **Positions commonly filled include:** Computer Programmer; Management Analyst/Consultant; Software Engineer; Systems Analyst. **Number of placements per year:** 50 - 99.

KELLY TECHNICAL SERVICES
801 West Freeway Street, Suite 710, Grand Prairie TX 75051-1468. 214/263-1467. **Fax:** 214/264-6962. **Contact:** Jeff Hilman, Technical Branch Manager. **E-mail address:** ktps@aol.com. **World Wide Web address:** http://www.kellyservices.com. **Description:** An executive search firm. Kelly Technical Services is a division of Kelly Services, Inc. **Specializes in the areas of:** Accounting/Auditing; Administration/MIS/EDP; Art/Design; Computer Science/Software; Engineering; Finance; Personnel/Labor Relations. **Positions commonly filled include:** Accountant/Auditor; Actuary; Aerospace Engineer; Architect; Budget Analyst; Buyer; Clinical Lab Technician; Computer Programmer; Cost Estimator; Draftsperson; Editor; Electrical/Electronics Engineer; Environmental Engineer; Financial Analyst; Human Resources Manager; Industrial Engineer; Industrial Production Manager; Internet Services Manager; Management Analyst/ Consultant; Mechanical Engineer; MIS Specialist; Multimedia Designer; Purchasing Agent/Manager; Quality Control Supervisor; Software Engineer; Structural Engineer; Systems

Analyst; Technical Writer/Editor. **Corporate headquarters location:** Troy MI. **Average salary range of placements:** $30,000 - $50,000.

MANAGEMENT ALLIANCE CORPORATION
COMPUTER TECHNOLOGY SEARCH
2600 South Gessner Road, Suite 325, Houston TX 77063. 713/785-2005. **Fax:** 713/785-5179. **Contact:** Karl F. Decker, Manager. **Description:** An executive search firm operating on a contingency basis. Company pays fee. **Specializes in the areas of:** Computer Science/Software. **Positions commonly filled include:** Computer Programmer; Internet Services Manager; MIS Specialist; Multimedia Designer; Software Engineer; Statistician; Systems Analyst; Technical Writer/Editor; Telecommunications Manager. **Corporate headquarters location:** Dallas TX. **Other U.S. locations:** Los Angeles CA; Atlanta GA; Chicago IL; Kansas City KS. **Number of placements per year:** 100 - 199.

MANAGEMENT RECRUITERS INTERNATIONAL
1360 Post Oak Boulevard, Suite 2110, Houston TX 77056. 713/850-9850. **Fax:** 713/850-1429. **Contact:** Rich Bolls, Manager. **Description:** An executive search firm. **Specializes in the areas of:** Accounting/Auditing; Administration/MIS/EDP; Advertising; Architecture/Construction/Real Estate; Banking; Communications; Computer Hardware/Software; Construction; Design; Electrical; Engineering; Finance; Food Industry; General Management; Health/Medical; Industrial; Insurance; Legal; Manufacturing; Personnel/Labor Relations; Printing/Publishing; Procurement; Retail; Sales and Marketing; Technical and Scientific; Textiles; Transportation.

MANAGEMENT RECRUITERS INTERNATIONAL, INC.
8131 LBJ Freeway, Suite 800, Dallas TX 75251. 972/907-1010. **Fax:** 972/680-1919. **Contact:** George Buntrock, General Manager. **Description:** An executive search firm. Company pays fee. **Specializes in the areas of:** Accounting/Auditing; Administration/MIS/EDP; Computer Science/Software; Engineering; Food Industry; General Management; Health/Medical; Paper; Retail; Technical and Scientific. **Positions commonly filled include:** Ceramics Engineer; Chemical Engineer; Computer Programmer; Customer Service Representative; Electrical/Electronics Engineer; General Manager; Health Services Manager; Industrial Engineer; Industrial Production Manager; Materials Engineer; Mechanical Engineer; Metallurgical Engineer; Operations/Production Manager; Pharmacist; Physical Therapist; Purchasing Agent/Manager; Quality Control Supervisor; Registered Nurse; Software Engineer; Speech-Language Pathologist; Systems Analyst; Transportation/Traffic Specialist. **Number of placements per year:** 1 - 49.

MANAGEMENT RECRUITERS INTERNATIONAL, INC.
1009 West Randol Mill Road, Suite 209, Arlington TX 76012. 817/469-6161. **Contact:** Bob Stoessel, Manager. **Description:** An executive search firm. **Specializes in the areas of:** Accounting/Auditing; Administration/MIS/EDP; Advertising; Architecture/Construction/Real Estate; Banking; Communications; Computer Hardware/Software; Design; Electrical; Engineering; Finance; Food Industry; General Management; Health/Medical; Insurance; Legal; Manufacturing; Operations Management; Personnel/Labor Relations; Printing/Publishing; Procurement; Retail; Sales and Marketing; Technical and Scientific; Textiles; Transportation.

MANAGEMENT RECRUITERS OF DALLAS
13101 Preston Road, Suite 560, Dallas TX 75240. 214/788-1515. **Fax:** 214/701-8242. **Contact:** Robert S. Lineback, General Manager. **Description:** An executive search firm operating on both retainer and contingency bases. Company pays fee. **Specializes in the areas of:** Accounting/Auditing; Administration/MIS/EDP; Advertising; Architecture/Construction/Real Estate; Banking; Communications; Computer Hardware/Software; Design; Electrical; Engineering; Finance; Food Industry; General Management; Health/Medical; Insurance; Legal; Manufacturing; Operations Management; Personnel/Labor Relations; Printing/Publishing; Procurement; Retail; Sales and Marketing; Technical and Scientific; Transportation. **Positions commonly filled include:** Accountant/Auditor; Actuary; Administrative Services Manager; Aerospace Engineer; Agricultural Engineer; Bank Officer/Manager; Biochemist; Biological Scientist; Biomedical Engineer; Branch Manager; Chemical Engineer; Chemist; Civil Engineer; Clinical Lab Technician; Computer Programmer; Design Engineer; Designer; Dietician/Nutritionist; EEG Technologist; EKG Technician; Electrical/Electronics Engineer; Emergency Medical Technician; Environmental

Engineer; Financial Analyst; Food Scientist/Technologist; General Manager; Health Services Manager; Human Resources Specialist; Industrial Engineer; Licensed Practical Nurse; Management Analyst/Consultant; Management Trainee; Manufacturer's/Wholesaler's Sales Rep.; Mechanical Engineer; Medical Records Technician; Metallurgical Engineer; Mining Engineer; MIS Specialist; Multimedia Designer; Nuclear Engineer; Nuclear Medicine Technologist; Occupational Therapist; Operations/Production Manager; Petroleum Engineer; Pharmacist; Physical Therapist; Physician; Physicist/Astronomer; Purchasing Agent/Manager; Quality Control Supervisor; Radiologic Technologist; Registered Nurse; Respiratory Therapist; Restaurant/Food Service Manager; Science Technologist; Securities Sales Representative; Services Sales Representative; Software Engineer; Structural Engineer; Surgical Technician; Systems Analyst; Telecommunications Manager; Transportation/Traffic Specialist; Underwriter/Assistant Underwriter. **Corporate headquarters location:** Cleveland OH. **Average salary range of placements:** More than $50,000. **Number of placements per year:** 200 - 499.

MANAGEMENT RECRUITERS OF NASSAU BAY, INC.
1335 Regents Park Dirve, Suite 150, Houston TX 77058. 713/286-9977. **Fax:** 713/286-9922. **Contact:** Julia Smith, Office Manager. **Description:** An executive search firm. **Specializes in the areas of:** Computer Science/Software; Information Systems; Systems Programming. **Positions commonly filled include:** Aerospace Engineer; Branch Manager; Computer Programmer; Electrical/Electronics Engineer; Management Analyst/Consultant; MIS Specialist; Software Engineer; Systems Analyst; Technical Writer/Editor.

McDUFFY-EDWARDS
3117 Medina Drive, Garland TX 75041. 214/864-1174. **Fax:** 214/864-8559. **Contact:** Tom Edwards, Partner. **Description:** An executive search firm. McDuffy-Edwards also provides consulting services and seminars. Founded in 1980. Company pays fee. **Specializes in the areas of:** Computer Science/Software; Sales and Marketing; Technical and Scientific. **Positions commonly filled include:** Computer Programmer; Customer Service Representative; Electrical/Electronics Engineer; General Manager; Software Engineer; Systems Analyst; Telecommunications Manager. **Average salary range of placements:** More than $50,000. **Number of placements per year:** 50 - 99.

NATIONAL HUMAN RESOURCE GROUP
100 Capital of Texas Highway South, Building L, Suite 100, Austin TX 78746. 512/328-4448. **Fax:** 512/328-1696. **Contact:** Vicki Volick, President. **E-mail address:** nbrg@zilker.net. **Description:** An executive search firm operating on both retainer and contingency bases. The firm also offers technical consulting and contract services. Company pays fee. **Specializes in the areas of:** Computer Science/Software. **Positions commonly filled include:** Computer Programmer; MIS Specialist; Multimedia Designer; Systems Analyst. **Corporate headquarters location:** This Location. **Average salary range of placements:** More than $50,000. **Number of placements per year:** 1 - 49.

NOLL HUMAN RESOURCE SERVICES
5720 Lyndon B. Johnson Freeway, Suite 610, Dallas TX 75240. 214/392-2900. **Toll-free phone:** 800/536-7600. **Fax:** 214/934-3600. **Contact:** Edward J. Bunn, Manager. **Description:** An executive search firm. **Specializes in the areas of:** Administration/MIS/EDP; Computer Science/Software; Health/Medical; Transportation. **Positions commonly filled include:** Computer Programmer; Database Manager; Industrial Engineer; Management Analyst/Consultant; Physical Therapist; Physician; Systems Analyst; Transportation/Traffic Specialist. **Other U.S. locations:** Omaha NE. **Number of placements per year:** 50 - 99.

OPPORTUNITY UNLIMITED PERSONNEL CONSULTANTS
2720 West Mockingbird Lane, Dallas TX 75235. 214/357-9196. **Toll-free phone:** 800/969-0888. **Fax:** 214/357-0140. **Contact:** John T. Kearley, President. **E-mail address:** oui@onramp.net. **Description:** An executive search firm operating on a contingency basis. The firm focuses on the placement of engineers and computer scientists. Founded in 1959. Company pays fee. **Specializes in the areas of:** Computer Science/Software; Engineering; Personnel/Labor Relations; Technical and Scientific. **Positions commonly filled include:** Aerospace Engineer; Biomedical Engineer; Computer Programmer; Design Engineer; Electrical/Electronics Engineer; Human Resources

Manager; Mechanical Engineer; Multimedia Designer; Software Engineer; Systems Analyst; Telecommunications Manager. **Number of placements per year:** 100 - 199.

THE PAILIN GROUP
PROFESSIONAL SEARCH CONSULTANTS
8500 North Stemmons Freeway, Suite 6070, LB #55, Dallas TX 75247-3819. 214/630-1703. **Fax:** 214/630-1704. **Contact:** David L. Pailin, Sr., Senior Partner. **Description:** An executive search firm. Founded in 1989. Company pays fee. **Specializes in the areas of:** Accounting/Auditing; Administration/MIS/EDP; Advertising; Banking; Computer Science/Software; Economics; Engineering; Environmental; Finance; Food Industry; General Management; Health/Medical; Industrial; Insurance; Legal; Manufacturing; Nonprofit; Personnel/Labor Relations; Retail; Sales and Marketing; Transportation. **Positions commonly filled include:** Accountant/Auditor; Administrative Services Manager; Aerospace Engineer; Architect; Attorney; Bank Officer/ Manager; Budget Analyst; Ceramics Engineer; Civil Engineer; Computer Programmer; Construction Contractor; Cost Estimator; Credit Manager; Customer Service Representative; Design Engineer; Environmental Engineer; Financial Analyst; General Manager; Health Services Manager; Human Service Worker; Industrial Engineer; Materials Engineer; Mechanical Engineer; Metallurgical Engineer; Mining Engineer; MIS Specialist; Nuclear Engineer; Petroleum Engineer; Pharmacist; Physician; Quality Control Supervisor; Securities Sales Representative; Services Sales Representative; Software Engineer; Statistician; Systems Analyst; Technical Writer/Editor; Telecommunications Manager. **Corporate headquarters location:** This Location. **Other U.S. locations:** Philadelphia PA. **Average salary range of placements:** $30,000 - $50,000. **Number of placements per year:** 100 - 199.

THE PERSONNEL OFFICE
24127 Boerne Stage Road, San Antonio TX 78255-9517. 210/698-0300. **Fax:** 210/698-3299. **Contact:** F. Carl Hensley, President & CEO. **Description:** An executive search firm operating on both retainer and contingency bases. The firm also offers contract staffing, temporary staffing, outplacement, and custom services. The firm specializes in programmer and analyst placements. Company pays fee. **Specializes in the areas of:** Administration/MIS/EDP; Banking; Computer Science/Software; Engineering; Sales and Marketing. **Positions commonly filled include:** Accountant/Auditor; Bank Officer/Manager; Branch Manager; Computer Programmer; Customer Service Representative; Design Engineer; Electrical/Electronics Engineer; Environmental Engineer; General Manager; Health Services Manager; Human Resources Manager; Industrial Engineer; Manufacturer's/Wholesaler's Sales Rep.; MIS Specialist; Systems Analyst; Telecommunications Manager; Typist/Word Processor. **Average salary of placements:** $30,000 - $50,000. **Number of placements per year:** 1000+.

FRED PIERCE & ASSOCIATES
1406 Stonehollow, Suite 700, Kingwood TX 77339. 713/359-3100. **Fax:** 713/359-5100. **Contact:** Director. **E-mail address:** fred@sccsi.com. **Description:** An executive search firm operating on a retainer basis. **Specializes in the areas of:** Computer Science/Software. **Positions commonly filled include:** Computer Programmer; Human Resources Specialist; Software Engineer; Systems Analyst. **Corporate headquarters location:** This Location. **Other U.S. locations:** Nationwide. **Average salary range of placements:** More than $50,000. **Number of placements per year:** 1 - 49.

QUALITY INFORMATION SERVICE (QIS)
P.O. Box 1559, Whitney TX 76692. 817/694-6319. **Fax:** 817/694-6434. **Contact:** Betty Schatz, Senior Account Manager. **Description:** An executive search firm. Company pays fee. **Specializes in the areas of:** Computer Science/Software. **Positions commonly filled include:** Computer Programmer; Database Manager; Electrical/Electronics Engineer; MIS Specialist; Software Engineer; Systems Analyst; Technical Writer/Editor. **Number of placements per year:** 50 - 99.

RECRUITING ASSOCIATES
P.O. Box 8473, Amarillo TX 79114. 806/353-9548. **Fax:** 806/353-9540. **Contact:** Mike Rokey, CPC, Owner/Manager. **E-mail address:** mikedr@arn.net. **Description:** An executive search firm operating on a contingency basis. The firm provides permanent and contract opportunities for technical and data processing candidates. Founded in 1978. Company pays fee. **Specializes in the**

areas of: Accounting/Auditing; Computer Science/Software; Engineering. **Positions commonly filled include:** Accountant/Auditor; Chemical Engineer; Computer Programmer; Mechanical Engineer; MIS Specialist; Software Engineer; Systems Analyst. **Benefits available to temporary workers:** 401(k); Medical Insurance. **Average salary range of placements:** $30,000 - $50,000. **Number of placements per year:** 1 - 49.

RICCIONE & ASSOCIATES

16415 Addison Road, Suite 404, Dallas TX 75248. 214/380-6432. **Fax:** 214/407-0659. **Contact:** Nick Riccione, Owner. **E-mail address:** riccione@onramp.net. **World Wide Web address:** http://www.riccione.com. **Description:** An executive search and contract services firm operating on a contingency basis. **Specializes in the areas of:** Computer Science/Software. **Positions commonly filled include:** Computer Programmer; Software Engineer; Systems Analyst. **Average salary range of placements:** More than $50,000. **Number of placements per year:** 1 - 49.

SALES CONSULTANTS OF HOUSTON

5075 Westheimer, Suite 790, Houston TX 77056. 713/627-0809. **Fax:** 713/622-7285. **Contact:** Jim DeForest, General Manager. **Description:** An executive search firm. **Specializes in the areas of:** Accounting/Auditing; Administration/MIS/EDP; Advertising; Architecture/Construction/Real Estate; Banking; Communications; Computer Hardware/Software; Construction; Design; Electrical; Engineering; Finance; Food Industry; General Management; Health/Medical; Industrial; Insurance; Legal; Manufacturing; Operations Management; Personnel/Labor Relations; Printing/Publishing; Procurement; Retail; Sales and Marketing; Technical and Scientific; Textiles; Transportation.

SEARCH NETWORK INTERNATIONAL

12801 North Central Expressway, Suite 460, Dallas TX 75243. 214/980 4991. **Fax:** 214/980-8917. **Contact:** Alan Butz, Director of Marketing & Research. **E-mail address:** resumes@snint.com. **World Wide Web address:** http://www.snint.com. **Description:** An executive search firm operating on a contingency basis. Founded in 1976. Company pays fee. **Specializes in the areas of:** Accounting/Auditing; Computer Science/Software; Engineering; Food Industry; Industrial; Manufacturing. **Positions commonly filled include:** Accountant/Auditor; Aerospace Engineer; Architect; Biochemist; Biomedical Engineer; Buyer; Chemical Engineer; Chemist; Civil Engineer; Computer Programmer; Cost Estimator; Design Engineer; Designer; Draftsperson; Environmental Engineer; Financial Analyst; Food Scientist/Technologist; Geologist/Geophysicist; Industrial Engineer; Industrial Production Manager; Internet Services Manager; Mathematician; Mechanical Engineer; Metallurgical Engineer; Mining Engineer; MIS Specialist; Multimedia Designer; Operations/Production Manager; Petroleum Engineer; Public Relations Specialist; Purchasing Agent/Manager; Quality Control Supervisor; Software Engineer; Statistician; Structural Engineer; Systems Analyst; Telecommunications Manager. **Corporate headquarters location:** This Location. **Number of placements per year:** 500 - 999.

SNELLING PERSONNEL SERVICES

5151 Belt Line Road, Suite 365, Dallas TX 75240. 214/934-9030. **Fax:** 214/934-3639. **Contact:** Sam D. Bingham, CPC, Owner. **Description:** An executive search firm operating on a contingency basis. Company pays fee. **Specializes in the areas of:** Accounting/Auditing; Biology; Computer Science/Software; Engineering; Food Industry; Industrial; Manufacturing; Sales and Marketing; Secretarial; Technical and Scientific. **Positions commonly filled include:** Accountant/Auditor; Chemical Engineer; Chemist; Computer Programmer; Credit Manager; Customer Service Representative; Electrical/Electronics Engineer; Environmental Engineer; Food Scientist/Technologist; Industrial Engineer; Industrial Production Manager; Mechanical Engineer; Metallurgical Engineer; Mining Engineer; Nuclear Engineer; Operations/Production Manager; Petroleum Engineer; Restaurant/Food Service Manager; Services Sales Representative; Software Engineer; Systems Analyst. **Other U.S. locations:** Nationwide. **Average salary range of placements:** $30,000 - $50,000. **Number of placements per year:** 100 - 199.

STAFF EXTENSION INC.

5050 Quorum Drive, Suite 337, Dallas TX 75240-6723. 214/991-4737. **Fax:** 214/991-5325. **Contact:** Jack R. Williams, President. **E-mail address:** staffing@staffext.com. **World Wide Web address:** http://www.staffext.com. **Description:** An executive search firm. The firm also operates as a temporary agency and contract services firm. Founded in 1990. Company pays fee. **Specializes in the areas of:** Accounting/Auditing; Administration/MIS/EDP; Computer Science/Software; Engineering; Finance; General Management; Health/Medical; Manufacturing; Personnel/Labor Relations; Sales and Marketing; Technical and Scientific. **Corporate headquarters location:** This Location. **Other U.S. locations:** Denver CO; Houston TX. **Number of placements per year:** 50 - 99.

STRATEGIC OUTSOURCING CORPORATION

1201 Richardson Drive, Richardson TX 75080. 214/437-2220. **Fax:** 214/437-2310. **Contact:** Brandt Hamby, Recruiting Manager. **E-mail address:** 50C@why.net. **Description:** An executive search firm operating on a retainer basis. Company pays fee. **Specializes in the areas of:** Computer Science/Software; Engineering; General Management; Printing/Publishing; Sales and Marketing; Technical and Scientific. **Positions commonly filled include:** Branch Manager; Computer Programmer; General Manager; Software Engineer; Strategic Relations Manager; Systems Analyst. **Corporate headquarters location:** Dallas TX. **Average salary range of placements:** More than $50,000. **Number of placements per year:** 1 - 49.

TGA COMPANY

P.O. Box 331121, Fort Worth TX 76163. 817/370-0865. **Fax:** 817/292-6451. **Contact:** Tom Green, President. **Description:** An executive search firm focusing on the placement of financial and information systems professionals. Company pays fee. **Specializes in the areas of:** Accounting/Auditing; Computer Science/Software; Finance; Technical and Scientific. **Positions commonly filled include:** Accountant/Auditor; Computer Programmer; Credit Manager; Financial Analyst; MIS Specialist; Software Engineer; Systems Analyst. **Average salary range of placements:** More than $50,000. **Number of placements per year:** 50 - 99.

TECH-NET

14785 Preston Road, Dallas TX 75240-7876. 214/789-5126. **Fax:** 214/789-5186. **Contact:** Chris Cole, Owner. **Description:** An executive search firm specializing in the placement of engineers in sales positions utilizing UNIX-based design tools. Founded in 1989. Company pays fee. **Specializes in the areas of:** Computer Science/Software; Engineering; Sales and Marketing; Technical and Scientific. **Positions commonly filled include:** Aerospace Engineer; Design Engineer; Electrical/Electronics Engineer; Mechanical Engineer; MIS Specialist; Software Engineer; Technical Representative. **Average salary range of placements:** More than $50,000. **Number of placements per year:** 1 - 49.

TEXAS PERSONNEL

985 Interstate 10 North, Beaumont TX 77706. 409/892-5000. **Fax:** 409/892-5068. **Contact:** Cliff Heubel, Owner. **Description:** Founded in 1992, Texas Personnel is a search firm that operates on a contingency basis. Company pays fee. **Specializes in the areas of:** Banking; Computer Science/Software; Insurance; Sales and Marketing; Secretarial. **Positions commonly filled include:** Accountant/Auditor; Automotive Mechanic; Bank Officer/Manager; Computer Programmer; Dental Assistant/Dental Hygienist; Manufacturer's/Wholesaler's Sales Rep.; MIS Specialist; Services Sales Representative; Systems Analyst; Typist/Word Processor. **Number of placements per year:** 100 - 199.

TOTAL PERSONNEL

P.O. Box 28975, Dallas TX 75228. 214/327-1165. **Fax:** 214/328-3061. **Contact:** Sherry Phillips, President. **Description:** An executive search firm operating on both retainer and contingency bases. The firm specializes in the recruitment of data processing personnel. Company pays fee. **Specializes in the areas of:** Administration/MIS/EDP; Computer Science/Software. **Positions commonly filled include:** Computer Programmer; Education Administrator; Software Engineer; Systems Analyst; Teacher/Professor; Technical Writer/Editor; Telecommunications Manager. **Corporate headquarters location:** This Location. **Average salary range of placements:** More than $50,000. **Number of placements per year:** 1 - 49.

DICK VAN VLIET & ASSOCIATES

2401 Fountain View Drive, Suite 322, Houston TX 77057. 713/952-0371. **Contact:** Dick Van Vliet, President. **Description:** An executive search firm operating on a contingency basis. Founded in 1985. Company pays fee. **Specializes in the areas of:** Accounting/Auditing; Administration/MIS/EDP; Computer Science/Software; Engineering; Finance; General Management; Industrial; Manufacturing; Personnel/Labor Relations; Sales and Marketing. **Positions commonly filled include:** Accountant/Auditor; Budget Analyst; Chemical Engineer; Civil Engineer; Computer Programmer; Credit Manager; Design Engineer; Electrical/Electronics Engineer; Financial Analyst; Human Resources Specialist; Industrial Engineer; Industrial Production Manager; Mechanical Engineer; MIS Specialist; Operations/Production Manager; Software Engineer; Systems Analyst. **Average salary range of placements:** $30,000 - $50,000. **Number of placements per year:** 50 - 99.

VICK & ASSOCIATES
RECRUITERS ONLINE NETWORK
3325 Landershire Lane, Suite 1001, Plano TX 75023-6218. 972/612-8425. **Fax:** 972/612-1924. **Contact:** Bill Vick, Owner. **World Wide Web address:** http://www.ipa.com. **Description:** An executive search firm. Company pays fee. **Specializes in the areas of:** Computer Science/Software; General Management; Sales and Marketing; Technical and Scientific. **Positions commonly filled include:** Regional Manager; Sales Manager. **Number of placements per year:** 50 - 99.

WYMAN & ASSOCIATES, INC.
P.O. Box 13253, Arlington TX 76094. 817/572-5212. **Fax:** 817/483-5550. **Contact:** David Wyman, President. **Description:** An executive search firm operating on both retainer and contingency bases. Company pays fee. **Specializes in the areas of:** Computer Science/Software; Food Industry; General Management; Personnel/Labor Relations. **Positions commonly filled include:** Attorney; Claim Representative; Computer Programmer; Dietician/Nutritionist; EEG Technologist; EKG Technician; Food Scientist/Technologist; Health Services Manager; Hotel Manager; Human Resources Specialist; Insurance Agent/Broker; Internet Services Manager; Licensed Practical Nurse; Market Research Analyst; MIS Specialist; Multimedia Designer; Occupational Therapist; Physical Therapist; Registered Nurse; Respiratory Therapist; Restaurant/Food Service Manager; Software Engineer; Surgical Technician; Systems Analyst; Telecommunications Manager; Underwriter/Assistant Underwriter. **Corporate headquarters location:** This Location. **Average salary range of placements:** $30,000 - $50,000. **Number of placements per year:** 1 - 49.

PERMANENT EMPLOYMENT AGENCIES OF TEXAS

BABICH & ASSOCIATES, INC.
6060 North Central Expressway, Suite 544, Dallas TX 75206. 214/361-5735. **Contact:** Anthony Beshara, President. **Description:** An employment agency. Company pays fee. **Specializes in the areas of:** Accounting/Auditing; Administration/MIS/EDP; Clerical; Computer Hardware/Software; Engineering; Finance; Manufacturing; Sales and Marketing; Technical and Scientific. **Positions commonly filled include:** Accountant/Auditor; Administrative Assistant; Agricultural Engineer; Bookkeeper; Ceramics Engineer; Civil Engineer; Computer Programmer; Customer Service Representative; Data Entry Clerk; EDP Specialist; Electrical/Electronics Engineer; Financial Analyst; General Manager; Human Resources Manager; Industrial Engineer; Mechanical Engineer; Medical Secretary; Metallurgical Engineer; Receptionist; Secretary; Systems Analyst; Typist/Word Processor. **Number of placements per year:** 500 - 999.

BABICH & ASSOCIATES, INC.
One Summit Avenue, Suite 602, Fort Worth TX 76102. 817/336-7261. **Contact:** Anthony Beshara, President. **Description:** An employment agency. Company pays fee. **Specializes in the areas of:** Accounting/Auditing; Administration/MIS/EDP; Clerical; Computer Hardware/Software;

Engineering; Manufacturing; Sales and Marketing; Technical and Scientific. **Positions commonly filled include:** Accountant/Auditor; Administrative Assistant; Agricultural Engineer; Bookkeeper; Ceramics Engineer; Civil Engineer; Computer Programmer; Customer Service Representative; Data Entry Clerk; EDP Specialist; Electrical/Electronics Engineer; Financial Analyst; General Manager; Human Resources Manager; Industrial Engineer; Mechanical Engineer; Medical Secretary; Metallurgical Engineer; Receptionist; Sales Representative; Secretary; Stenographer; Systems Analyst; Typist/Word Processor. **Number of placements per year:** 500 - 999.

DATAPRO PERSONNEL CONSULTANTS
13355 Noel Road, Suite 2001, Dallas TX 75240. 972/661-8600. **Fax:** 972/661-1309. **Contact:** Jack Kallison, Owner. **Description:** An employment agency. Company pays fee. **Specializes in the areas of:** Computer Programming; Computer Science/Software. **Positions commonly filled include:** Computer Programmer; EDP Specialist; Project Manager; Software Engineer; Systems Analyst; Technical Writer/Editor.

EXPRESS PERSONNEL SERVICES
P.O. Box 8136, Waco TX 76714-8136. 817/776-3300. **Contact:** J.G. Scofield, Owner. **Description:** An employment agency. **Specializes in the areas of:** Accounting/Auditing; Administration/MIS/EDP; Architecture/Construction/Real Estate; Banking; Clerical; Computer Hardware/Software; Engineering; Finance; Food Industry; Health/Medical; Insurance; Legal; Manufacturing; Physician Executive; Sales and Marketing; Secretarial.

EXPRESS PERSONNEL SERVICES
3701 South Cooper Street, Suite 250, Arlington TX 76015. 817/468-9118. **Fax:** 817/468-9211. **Contact:** Gary Gibson, Owner/Manager. **Description:** A permanent and temporary employment agency operating on a contingency basis. Company pays fee. **Specializes in the areas of:** Accounting/Auditing; Banking; Computer Science/Software; Finance; General Management; Health/Medical; Industrial; Insurance; Personnel/Labor Relations; Printing/Publishing; Sales and Marketing; Secretarial. **Positions commonly filled include:** Accountant/Auditor; Administrative Services Manager; Advertising Clerk; Bank Officer/Manager; Blue-Collar Worker Supervisor; Branch Manager; Clerical Supervisor; Cost Estimator; Customer Service Representative; General Manager; Human Resources Specialist; Industrial Production Manager; Management Trainee; Paralegal; Public Relations Specialist; Quality Control Supervisor; Technical Writer/Editor; Typist/Word Processor. **Benefits available to temporary workers:** Medical Insurance; Paid Holidays; Paid Vacation. **Other U.S. locations:** Nationwide. **Average salary range of placements:** $30,000 - $50,000. **Number of placements per year:** 1 - 49.

GULCO INTERNATIONAL RECRUITING SERVICES
15710 John F. Kennedy Boulevard, Suite 110, Houston TX 77032. 713/590-9001. **Fax:** 713/590-1503. **Contact:** Rod Gullo, President. **Description:** An employment agency. Company pays fee. **Specializes in the areas of:** Accounting/Auditing; Administration/MIS/EDP; Banking; Computer Science/Software; Engineering; Finance; General Management; Health/Medical; Industrial; Manufacturing; Operations Management; Personnel/Labor Relations. **Positions commonly filled include:** Accountant/Auditor; Aerospace Engineer; Budget Analyst; Buyer; Ceramics Engineer; Chemical Engineer; Civil Engineer; Computer Programmer; Cost Estimator; Designer; Electrical/Electronics Engineer; Financial Analyst; Geologist/Geophysicist; Health Services Manager; Human Resources Manager; Licensed Practical Nurse; Materials Engineer; Mechanical Engineer; Mergers/Acquisitions Specialist; Mining Engineer; Nuclear Engineer; Occupational Therapist; Operations/Production Manager; Petroleum Engineer; Physical Therapist; Physician; Purchasing Agent/Manager; Quality Control Supervisor; Registered Nurse; Respiratory Therapist; Science Technologist; Software Engineer; Stationary Engineer; Structural Engineer; Surgical Technician; Systems Analyst; Technical Writer/Editor. **Number of placements per year:** 50 - 99.

KEYPEOPLE RESOURCES, INC.
520 Post Oak Boulevard, Suite 830, Houston TX 77027. 713/877-1427. **Fax:** 713/877-1826. **Contact:** Betty Thompson, President. **Description:** An employment agency. Company pays fee. **Specializes in the areas of:** Accounting/Auditing; Computer Science/Software; Secretarial. **Positions commonly filled include:** Accountant/Auditor; Administrative Assistant; Bookkeeper;

Clerk; Computer Programmer; EDP Specialist; Systems Analyst. **Number of placements per year:** 1 - 49.

LUSK & ASSOCIATES
P.O. Box 7500-331, Dallas TX 75209-0500. 214/528-9966. **Contact:** B.J. Lusk, President. **Description:** An employment agency. **Specializes in the areas of:** Administration/MIS/EDP; Clerical; Computer Hardware/Software; Secretarial.

O'KEEFE & ASSOCIATES
3420 Executive Center Drive, Suite 114, Austin TX 78731. 512/343-1134. **Fax:** 512/343-0142. **Contact:** Recruiter. **Description:** An employment agency. Company pays fee. **Specializes in the areas of:** Administration/MIS/EDP; Computer Hardware/Software; Engineering. **Positions commonly filled include:** Computer Programmer; Electrical/Electronics Engineer; Software Engineer; Systems Analyst. **Number of placements per year:** 50 - 99.

P&P PERSONNEL SERVICE
1711 East Central Texas Expressway, Killeen TX 76540. 817/526-9962. **Contact:** Gordon Plumlee, Owner. **Description:** An employment agency. **Specializes in the areas of:** Accounting/Auditing; Banking; Computer Science/Software; Finance; General Management; Insurance; Legal; Secretarial. **Positions commonly filled include:** Accountant/Auditor; Claim Representative; Clerical Supervisor; Dental Assistant/Dental Hygienist; Insurance Agent/Broker; Licensed Practical Nurse; Management Trainee; Medical Records Technician; Paralegal; Registered Nurse; Social Worker. **Number of placements per year:** 50 - 99.

PERSONNEL ONE, INC.
One Lincoln Center, 5400 LBJ Freeway, Suite 120, Dallas TX 75240. 214/982-8500. **Fax:** 214/982-8505. **Contact:** Manager. **Description:** An employment agency. Company pays fee. **Specializes in the areas of:** Accounting/Auditing; Banking; Computer Science/Software; Engineering; General Management; Legal; Personnel/Labor Relations; Sales and Marketing; Secretarial; Technical and Scientific. **Positions commonly filled include:** Accountant/Auditor; Aerospace Engineer; Agricultural Engineer; Bank Officer/Manager; Biological Scientist; Biomedical Engineer; Branch Manager; Ceramics Engineer; Chemical Engineer; Chemist; Civil Engineer; Clerical Supervisor; Clinical Lab Technician; Computer Programmer; Construction and Building Inspector; Construction Contractor; Counselor; Credit Manager; Electrical/Electronics Engineer; Financial Analyst; Human Resources Manager; Human Service Worker; Industrial Engineer; Manufacturer's/Wholesaler's Sales Rep.; Materials Engineer; Mechanical Engineer; Metallurgical Engineer; Mining Engineer; Nuclear Engineer; Paralegal; Petroleum Engineer; Property and Real Estate Manager; Purchasing Agent/Manager; Services Sales Representative; Software Engineer; Stationary Engineer; Structural Engineer; Systems Analyst. **Number of placements per year:** 200 - 499.

PROFESSIONS TODAY
2811 South Loop 289, Suite 20, Lubbock TX 79423. 806/745-8595. **Fax:** 806/748-0571. **Contact:** Gebrell Ward, Owner. **Description:** An employment agency. Company pays fee. **Specializes in the areas of:** Accounting/Auditing; Administration/MIS/EDP; Computer Hardware/Software; Engineering; General Management; Health/Medical; Industrial; Sales and Marketing; Secretarial. **Positions commonly filled include:** Accountant/Auditor; Administrative Assistant; Bookkeeper; Clerk; Computer Programmer; Customer Service Representative; Data Entry Clerk; Legal Secretary; Marketing Specialist; Medical Secretary; Receptionist; Sales Representative; Secretary; Typist/Word Processor. **Number of placements per year:** 100 - 199.

RESOURCE RECRUITERS INC.
4100 Spring Valley Road, Suite 800, Dallas TX 75244. 972/851-5408. **Contact:** Ms. Terese Scribner, CPC, President. **Description:** An employment agency. Company pays fee. **Specializes in the areas of:** Accounting/Auditing; Advertising; Architecture/Construction/Real Estate; Computer Science/Software; Finance; Food Industry; General Management; Insurance; Legal; Manufacturing; Paper; Printing/Publishing; Sales and Marketing; Secretarial. **Positions commonly filled include:** Accountant/Auditor; Adjuster; Clerical Supervisor; Collector; Credit Manager; Customer Service Representative; Human Resources Manager; Investigator; Management Trainee;

Paralegal; Restaurant/Food Service Manager; Services Sales Representative; Underwriter/Assistant Underwriter. **Number of placements per year:** 50 - 99.

SALINAS & ASSOCIATES PERSONNEL SERVICE
1700 Commerce Street, Suite 1660, Dallas TX 75201. 214/747-7878. **Fax:** 214/747-7877. **Contact:** Gerry Salinas, Owner/Recruiter. **Description:** An employment agency. **Specializes in the areas of:** Accounting/Auditing; Administration/MIS/EDP; Clerical; Computer Hardware/Software; Engineering; Health/Medical; Legal; Manufacturing; Sales and Marketing; Secretarial.

STEHOUWER & ASSOCIATES
2939 Mossrock, San Antonio TX 78230-5118. 210/349-4995. **Fax:** 210/349-4996. **Contact:** Personnel. **E-mail address:** ronsteh@connect.com. **Description:** An employment agency that provides organizational design consulting and psychological profiling. **Specializes in the areas of:** Administration/MIS/EDP; Art/Design; Computer Science/Software; Engineering; Technical and Scientific. **Positions commonly filled include:** Aerospace Engineer; Chemical Engineer; Chemist; Civil Engineer; Computer Programmer; Design Engineer; Mechanical Engineer; MIS Specialist; Software Engineer; Structural Engineer; Systems Analyst. **Average salary range of placements:** More than $50,000.

TAD TECHNICAL SERVICES
4300 Alpha Road, Suite 100, Dallas TX 75244. 972/980-0510. **Contact:** Tom Lash, Area Manager. **Description:** An employment agency. **Specializes in the areas of:** Administration/MIS/EDP; Aerospace; Clerical; Computer Hardware/Software; Engineering; Manufacturing; Technical and Scientific.

EXECUTIVE SEARCH FIRMS OF UTAH

EXEC-U-SOURCE
4746 South 900 East, Suite 250, Salt Lake City UT 84117-4903. 801/262-3179. **Fax:** 801/261-3480. **Contact:** Mike Kennedy, President. **Description:** An executive search firm. Exec-U-Source specializes in MIS recruiting. The firm primarily handles client/server technology and positions related to the IBM AS400 mid-range computer, covering all levels of technical and management placements within that field. Company pays fee. **Specializes in the areas of:** Computer Science/Software. **Positions commonly filled include:** Computer Programmer; MIS Specialist; Operations/Production Manager; Systems Analyst; Telecommunications Manager. **Corporate headquarters location:** This Location. **Average salary range of placements:** $30,000 - $50,000. **Number of placements per year:** 1 - 49.

MANAGEMENT RECRUITERS OF PROVO
2230 North University Parkway, Building 11-I, Provo UT 84604-1509. 801/375-0777. **Fax:** 801/375-5757. **Contact:** Larry J. Massung, General Manager. **E-mail address:** gen_mgr@recruitr.com. **Description:** An executive search firm. Management Recruiters of Provo is a division of Management Recruiters International, Inc., one of the nation's largest search and recruiting firms, with nearly 650 offices and 3,000 professional recruiters. The organization completes more than 30,000 placements each year with thousands of companies throughout the world. Management Recruiters fills staffing needs for permanent positions, interim assignments, and outsourcing of staffing projects. Other services include outplacements of displaced employees, videoconferencing, and sales staffing for short-term projects and objectives. Company pays fee. **Specializes in the areas of:** Automation/Robotics; Computer Science/Software; Engineering; Finance; General Management; Information Technology; Manufacturing; Technical and Scientific. **Positions commonly filled include:** Computer Programmer; Controller; Design Engineer; Electrical/Electronics Engineer; Financial Analyst; General Manager; Industrial Engineer; Market Research Analyst; Mechanical Engineer; MIS Specialist; Operations/Production Manager; Quality

Control Supervisor; Software Engineer; Systems Analyst. **Corporate headquarters location:** Cleveland OH. **Other locations:** Worldwide. **Average salary range of placements:** More than $50,000. **Number of placements per year:** 50 - 99.

TROUT & ASSOCIATES, INC.
15 Gatehouse Lane, Sandy UT 84092-4846. 801/576-1547. **Fax:** 801/576-1541. **Contact:** Thomas L. Trout, President. **Description:** Trout & Associates is an executive search firm operating on a retainer basis. Trout & Associates also provides outplacement and career counseling for client companies and individuals. Founded in 1973. Company pays fee. **Specializes in the areas of:** Accounting/Auditing; Computer Science/Software; Engineering; Finance; Health/Medical; Manufacturing; Technical and Scientific. **Positions commonly filled include:** Accountant/Auditor; Aerospace Engineer; Chemical Engineer; Computer Programmer; Design Engineer; Financial Analyst; General Manager; Materials Engineer; Mechanical Engineer; MIS Specialist; Software Engineer; Systems Analyst. **Average salary range of placements:** More than $50,000. **Number of placements per year:** 1 - 49.

PERMANENT EMPLOYMENT AGENCIES OF UTAH

DEECO INTERNATIONAL
P.O. Box 57033, Salt Lake City UT 84157. 801/261-3326. **Fax:** 801/261-3955. **Contact:** Dee McBride, Manager. **Description:** Deeco International is an employment agency operating on both retainer and contingency bases. Deeco provides medical manufacturers with qualified candidates in marketing, sales, and engineering. Company pays fee. **Specializes in the areas of:** Computer Science/Software; Health/Medical; Sales and Marketing; Technical and Scientific. **Positions commonly filled include:** Biomedical Engineer; Chemical Engineer; Design Engineer; Mechanical Engineer; Software Engineer. **Corporate headquarters location:** This Location. **Average salary range of placements:** More than $50,000. **Number of placements per year:** 1 - 49.

PROFESSIONAL RECRUITERS AND TEMPORARIES
220 East 3900 South, Suite 9, Salt Lake City UT 84107. 801/268-9940. **Fax:** 801/268-9940. **Contact:** Lora Lea Mock, Owner. **Description:** Professional Recruiters and Temporaries is an employment agency. **Specializes in the areas of:** Accounting/Auditing; Administration/MIS/EDP; Computer Hardware/Software; Electrical; Engineering; General Management; Manufacturing; Medical Sales and Marketing; Technical and Scientific. **Number of placements per year:** 500 - 999.

EXECUTIVE SEARCH FIRMS OF VERMONT

EPN/ECKLER PERSONNEL NETWORK
P.O. Box 549, Woodstock VT 05091. 802/457-1605. **Fax:** 802/457-1606. **Contact:** G.K. Eckler, President. **E-mail address:** epn@sover.net. **Description:** EPN/Eckler Personnel Network is an executive search firm that operates on a contingency basis. The firm focuses on business software fields including information technology services and software product support and development. Company pays fee. **Specializes in the areas of:** Administration/MIS/EDP; Computer Science/Software. **Positions commonly filled include:** Computer Programmer; Internet Services Manager; MIS Specialist; Multimedia Designer; Software Engineer; Systems Analyst. **Average salary range of placements:** $30,000 - $50,000. **Number of placements per year:** 50 - 99.

MANAGEMENT RECRUITERS OF BURLINGTON
187 St. Paul Street, Suite 4, Burlington VT 05401. 802/865-0541. **Contact:** Michael Connor, Recruiting. **Description:** This executive search firm covers the areas of finance, banking, and computers. **Specializes in the areas of:** Banking; Computer Programming; Finance; Investment.

PERMANENT EMPLOYMENT AGENCIES OF VERMONT

TECHNICAL CONNECTION
P.O. Box 1402, Burlington VT 05401. 802/658-8324. **Contact:** Dorothy Coe, Technical Recruiter. **Description:** This employment agency places applicants in permanent positions in a variety of technical fields including computers, mechanical engineering, and surveying. **Specializes in the areas of:** Computer Hardware/Software; Computer Programming; Electrical; Engineering.

EXECUTIVE SEARCH FIRMS OF VIRGINIA

ABILITY RESOURCES, INC.
716 Church Street, Alexandria VA 22314. 703/548-6400. **Contact:** Noel L. Ruppert, President. **Description:** An executive search firm. Company pays fee. **Specializes in the areas of:** Administration/MIS/EDP; Computer Science/Software; Defense Industry; Economics; Engineering; Finance; General Management; Nonprofit; Personnel/Labor Relations; Sales and Marketing; Technical and Scientific. **Positions commonly filled include:** Accountant/Auditor; Administrative Services Manager; Computer Programmer; Economist; Engineer; Financial Analyst; General Manager; Management Analyst/Consultant; Mathematician; Operations/ Production Manager; Public Relations Specialist; Statistician; Systems Analyst; Technical Writer/Editor. **Number of placements per year:** 1 - 49.

CONTEC SEARCH, INC.
5803 Stone Ridge Drive, Centreville VA 22020-2888. 703/968-0477. **Fax:** 703/968-0064. **Contact:** Brian Canatsey, President. **Description:** An executive search firm operating on a contingency basis. Company pays fee. **Specializes in the areas of:** Computer Science/Software; Finance; Health/Medical; Manufacturing. **Positions commonly filled include:** Computer Programmer; MIS Specialist; Systems Analyst. **Corporate headquarters location:** This Location. **Average salary range of placements:** More than $50,000. **Number of placements per year:** 1 - 49.

CORPORATE CONNECTION LTD.
7204 Glen Forest Drive, Richmond VA 23226-3778. 804/288-8844. **Contact:** Marshall W. Rotella, Jr., President. **Description:** Founded in 1980, Corporate Connection is a full-service recruiting firm. **Specializes in the areas of:** Accounting/Auditing; Administration/MIS/EDP; Banking; Computer Science/Software; Engineering; Finance; General Management; Industrial; Insurance; Legal; Manufacturing; Nonprofit; Personnel/Labor Relations; Printing/Publishing; Retail; Sales and Marketing; Secretarial; Technical and Scientific. **Positions commonly filled include:** Accountant/Auditor; Adjuster; Administrative Services Manager; Advertising Clerk; Aerospace Engineer; Agricultural Engineer; Aircraft Mechanic/Engine Specialist; Bank Officer/Manager; Biological Scientist; Biomedical Engineer; Blue-Collar Worker Supervisor; Branch Manager; Brokerage Clerk; Budget Analyst; Buyer; Chemical Engineer; Chemist; Civil Engineer; Claim Representative; Clerical Supervisor; Clinical Lab Technician; Computer Programmer; Construction Contractor; Cost Estimator; Credit Manager; Customer Service Representative; Design Engineer;

Designer; Draftsperson; Electrical/Electronics Engineer; Environmental Engineer; Financial Analyst; General Manager; Human Resources Manager; Industrial Engineer; Industrial Production Manager; Insurance Agent/Broker; Librarian; Library Technician; Management Analyst/Consultant; Management Trainee; Manufacturer's/Wholesaler's Sales Rep.; Market Research Analyst; Mechanical Engineer; Medical Records Technician; Metallurgical Engineer; Mining Engineer; Property and Real Estate Manager; Public Relations Specialist; Purchasing Agent/Manager; Quality Control Supervisor; Restaurant/Food Service Manager; Securities Sales Representative; Services Sales Representative; Software Engineer; Strategic Relations Manager; Structural Engineer; Systems Analyst; Transportation/Traffic Specialist; Travel Agent; Typist/Word Processor; Underwriter/Assistant Underwriter; Urban/Regional Planner. **Corporate headquarters location:** This Location. **Average salary range of placements:** $20,000 - $50,000. **Number of placements per year:** 200 - 499.

CAROL DAY AND ASSOCIATES
2105 Electric Road SW, Roanoke VA 24018. 540/989-2831. **Fax:** 540/989-5910. **Contact:** Carol Day, Owner. **Description:** An executive search firm and employment agency. Company pays fee. **Specializes in the areas of:** Accounting/Auditing; Administration/MIS/EDP; Computer Science/Software; Engineering; Health/Medical; Manufacturing; Retail; Sales and Marketing; Secretarial. **Positions commonly filled include:** Accountant/Auditor; Buyer; Chemical Engineer; Civil Engineer; Clerical Supervisor; Computer Programmer; Customer Service Representative; Electrical/Electronics Engineer; Hotel Manager; Industrial Engineer; Management Trainee; Mechanical Engineer; MIS Specialist; Paralegal; Quality Control Supervisor; Software Engineer. **Corporate headquarters location:** This Location. **Average salary range of placements:** $20,000 - $29,999. **Number of placements per year:** 200 - 499.

DONMAC ASSOCIATES
P.O. Box 2541, Reston VA 22090. 703/620-2866. **Fax:** 703/620-2867. **Contact:** Connie Andersen, President. **Description:** An executive search firm. Company pays fee. **Specializes in the areas of:** Computer Science/Software. **Positions commonly filled include:** Computer Programmer; Electrical/Electronics Engineer; Software Engineer; Systems Analyst. **Number of placements per year:** 50 - 99.

EFFECTIVE STAFFING INC.
209 Elden Street, Suite 208, Herndon VA 20170-4815. 703/742-9300. **Fax:** 703/742-9747. **Contact:** President. **Description:** An executive search firm operating on a contingency basis. The firm also provides temporary and contract services. Company pays fee. **Specializes in the areas of:** Accounting/Auditing; Administration/MIS/EDP; Banking; Computer Science/Software; Finance; General Management; Legal; Personnel/Labor Relations; Printing/Publishing; Sales and Marketing; Secretarial; Technical and Scientific. **Positions commonly filled include:** Accountant/Auditor; Administrative Services Manager; Bank Officer/Manager; Branch Manager; Broadcast Technician; Budget Analyst; Computer Programmer; Counselor; Customer Service Representative; Design Engineer; Economist; Education Administrator; Financial Analyst; General Manager; Health Services Manager; Human Resources Specialist; Human Service Worker; Management Analyst/Consultant; Market Research Analyst; Medical Records Technician; MIS Specialist; Multimedia Designer; Occupational Therapist; Operations/Production Manager; Paralegal; Physical Therapist; Quality Control Supervisor; Registered Nurse; Services Sales Representative; Technical Writer/Editor; Telecommunications Manager. **Average salary range of placements:** More than $50,000. **Number of placements per year:** 1 - 49.

ENGINEERING & MIS GUILD
8260 Greensboro Drive, Suite 460, McLean VA 22102. 703/761-4023. **Fax:** 703/761-4024. **Contact:** William J. Joyce, Principal. **E-mail address:** resumes@guildcorp.com. **Description:** An executive search firm operating on both retainer and contingency bases. Company pays fee. **Specializes in the areas of:** Administration/MIS/EDP; Computer Science/Software; Engineering. **Positions commonly filled include:** Computer Programmer; Customer Service Representative; Design Engineer; Electrical/Electronics Engineer; Internet Services Manager; Management Analyst/Consultant; MIS Specialist; Software Engineer; Telecommunications Manager. **Average salary range of placements:** More than $50,000. **Number of placements per year:** 200 - 499.

THE GEMINI GROUP
9749 Ashworth Drive, Richmond VA 23236. 804/276-3091. **Toll-free phone:** 800/258-1120. **Fax:** 804/320-2880. **Contact:** Mr. Terry Lee Stacy, M.A., President. **E-mail address:** tstacy@erols.com. **Description:** An executive search firm that also offers career/outplacement counseling. Company pays fee. **Specializes in the areas of:** Administration/MIS/EDP; Art/Design; Computer Science/Software; Food Industry; General Management; Health/Medical; Manufacturing. **Positions commonly filled include:** Accountant/Auditor; Administrative Services Manager; Attorney; Blue-Collar Worker Supervisor; Buyer; Chemist; Claim Representative; Computer Programmer; Customer Service Representative; Editor; Electrical/Electronics Engineer; Financial Analyst; General Manager; Human Resources Specialist; Industrial Production Manager; Management Analyst/Consultant; Manufacturer's/Wholesaler's Sales Rep.; Operations/Production Manager; Property and Real Estate Manager; Purchasing Agent/Manager; Quality Control Supervisor; Real Estate Agent; Restaurant/Food Service Manager; Securities Sales Representative. **Other U.S. locations:** DC; MD; NC. **Average salary range of placements:** More than $50,000. **Number of placements per year:** 100 - 199.

HALBRECHT & COMPANY, INC.
10195 Main Street, Suite L, Fairfax VA 22031. 703/359-2880. **Contact:** Thomas J. Maltby, Director. **Description:** An executive search firm operating on both retainer and contingency bases. Company pays fee. **Specializes in the areas of:** Administration/MIS/EDP; Computer Science/Software; Technical and Scientific. **Positions commonly filled include:** Actuary; Computer Programmer; EDP Specialist; Electrical/Electronics Engineer; Internet Services Manager; Management Analyst/Consultant; Mathematician; Software Engineer; Statistician; Systems Analyst. **Other U.S. locations:** Greenwich CT. **Number of placements per year:** 50 - 99.

INFORMATION SPECIALISTS COMPANY, INC.
P.O. Box 55313, Virginia Beach VA 23455. 757/460-7790. **Fax:** 757/460-7886. **Contact:** Hugo E. Schluter, CPC, Senior Vice President. **Description:** An executive search firm. Company pays fee. **Specializes in the areas of:** Computer Science/Software; Engineering; Industrial; Manufacturing; Technical and Scientific. **Positions commonly filled include:** Aerospace Engineer; Biomedical Engineer; Chemical Engineer; Civil Engineer; Construction and Building Inspector; Electrical/Electronics Engineer; Geologist/Geophysicist; Industrial Engineer; Mechanical Engineer; Metallurgical Engineer; Nuclear Engineer; Quality Control Supervisor; Science Technologist; Software Engineer; Stationary Engineer; Structural Engineer; Systems Analyst; Technical Writer/Editor. **Average salary range of placements:** $30,000 - $50,000. **Number of placements per year:** 1 - 49.

M.S.I.
1593 Spring Hill Road, Vienna VA 22182-2245. 703/893-5660. **Toll-free phone:** 800/347-5660. **Contact:** Edward Hughes, General Manager. **Description:** An executive search firm operating on a contingency basis. Company pays fee. **Specializes in the areas of:** Computer Science/Software; Health/Medical; Sales and Marketing. **Corporate headquarters location:** Atlanta GA. **Other U.S. locations:** Nationwide.

MANAGEMENT RECRUITERS OF MANASSAS
8807 Sudley Road, Suite 208, Manassas VA 20110-4719. 703/330-1830. **Contact:** Professional Placement. **Description:** An executive search firm. **Specializes in the areas of:** Accounting/Auditing; Administration/MIS/EDP; Advertising; Architecture/Construction/Real Estate; Banking; Communications; Computer Hardware/Software; Design; Electrical; Engineering; Finance; Food Industry; General Management; Health/Medical; Insurance; Legal; Manufacturing; Operations Management; Personnel/Labor Relations; Printing/Publishing; Procurement; Retail; Sales and Marketing; Technical and Scientific; Textiles; Transportation.

MANAGEMENT RECRUITERS OF McLEAN
6849 Old Dominion Drive, Suite 225, McLean VA 22101. 703/442-4842. **Contact:** Howard Reitkopp, Manager. **Description:** An executive search firm. **Specializes in the areas of:** Accounting/Auditing; Administration/MIS/EDP; Advertising; Architecture/Construction/Real Estate; Banking; Communications; Computer Hardware/Software; Design; Electrical; Engineering; Finance; Food Industry; General Management; Health/Medical; Insurance; Legal; Manufacturing;

Operations Management; Personnel/Labor Relations; Printing/Publishing; Procurement; Retail; Sales and Marketing; Technical and Scientific; Textiles; Transportation.

MANAGEMENT RECRUITERS OF ROANOKE
1960 Electric Road, Suite B, Roanoke VA 24018. 540/989-1676. **Contact:** Paul Sharp, Manager. **Description:** An executive search firm. **Specializes in the areas of:** Accounting/Auditing; Administration/MIS/EDP; Advertising; Architecture/Construction/Real Estate; Banking; Communications; Computer Hardware/Software; Design; Electrical; Engineering; Finance; Food Industry; General Management; Health/Medical; Insurance; Legal; Manufacturing; Operations Management; Personnel/Labor Relations; Printing/Publishing; Procurement; Retail; Sales and Marketing; Technical and Scientific; Textiles; Transportation.

THE McCORMICK GROUP
4024 Plank Road, Fredericksburg VA 22409. 703/786-9777. **Fax:** 703/786-9355. **Contact:** William J. McCormick, President. **Description:** An executive search and consulting firm. Company pays fee. **Specializes in the areas of:** Biology; Computer Science/Software; Engineering; Health/Medical; Insurance; Legal; Personnel/Labor Relations; Retail; Sales and Marketing; Technical and Scientific. **Positions commonly filled include:** Accountant/Auditor; Architect; Attorney; Biological Scientist; Civil Engineer; Computer Programmer; Cost Estimator; Design Engineer; Designer; Editor; EEG Technologist; Electrical/Electronics Engineer; Environmental Engineer; Financial Analyst; Health Services Manager; Human Resources Specialist; Licensed Practical Nurse; Management Analyst/Consultant; MIS Specialist; Multimedia Designer; Physical Therapist; Physician; Public Relations Specialist; Services Sales Representative; Software Engineer; Systems Analyst; Technical Writer/Editor. **Corporate headquarters location:** This Location. **Other U.S. locations:** Jacksonville FL; Boston MA; Kansas City MO; Arlington VA. **Average salary range of placements:** More than $50,000. **Number of placements per year:** 1 - 49.

NETWORK COMPANIES
1595 Spring Hill Road, Suite 220, Vienna VA 22182. 703/790-1100. **Fax:** 703/790-1123. **Contact:** Ron Sall, Director. **Description:** An executive search firm focusing on accounting, finance, and information technology placements. The firm also provides temporary and contract services. Founded in 1985. **Specializes in the areas of:** Accounting/Auditing; Computer Science/Software; Finance; Personnel/Labor Relations; Sales and Marketing; Technical and Scientific. **Positions commonly filled include:** Accountant/Auditor; Budget Analyst; Computer Programmer; Customer Service Representative; Financial Analyst; General Manager; Human Resources Specialist; Management Analyst/Consultant; MIS Specialist; Services Sales Representative; Software Engineer; Systems Analyst; Technical Writer/Editor; Telecommunications Manager; Typist/Word Processor. **Corporate headquarters location:** This Location. **Average salary range of placements:** More than $50,000. **Number of placements per year:** 500 - 999.

PLACEMENT PROFESSIONALS INC.
P.O. Box 29772, Richmond VA 23242. 804/741-1246. **Contact:** Manager. **Description:** An executive search firm. Company pays fee. **Specializes in the areas of:** Accounting/Auditing; Advertising; Banking; Computer Hardware/Software; Economics; Engineering; Finance; Health/Medical; Insurance; Legal; Manufacturing; Personnel/Labor Relations; Sales and Marketing; Technical and Scientific. **Positions commonly filled include:** Accountant/Auditor; Aerospace Engineer; Agricultural Engineer; Attorney; Bank Officer/Manager; Biological Scientist; Biomedical Engineer; Budget Analyst; Buyer; Chemical Engineer; Civil Engineer; Computer Programmer; Controller; Credit Manager; Economist; Electrical/Electronics Engineer; Financial Analyst; Human Resources Manager; Industrial Engineer; Mechanical Engineer; Metallurgical Engineer; Nuclear Engineer; Occupational Therapist; Petroleum Engineer; Pharmacist; Physical Therapist; Public Relations Specialist; Purchasing Agent/Manager; Registered Nurse; Software Engineer; Speech-Language Pathologist; Stationary Engineer; Structural Engineer; Systems Analyst; Technical Writer/Editor; Underwriter/Assistant Underwriter; Wholesale and Retail Buyer.

PROFESSIONAL SEARCH PERSONNEL
4900 Leesburg Pike, Suite 402, Alexandria VA 22302-1103. **Contact:** Chuck Cherel, Owner/Manager. **Description:** An executive search firm operating on both retainer and

contingency bases. Company pays fee. **Specializes in the areas of:** Accounting/Auditing; Administration/MIS/EDP; Architecture/Construction/Real Estate; Banking; Biology; Computer Hardware/Software; Engineering; Finance; Health/Medical; Industrial; Personnel/Labor Relations; Technical and Scientific. **Positions commonly filled include:** Accountant/Auditor; Actuary; Aerospace Engineer; Architect; Bank Officer/Manager; Biological Scientist; Biomedical Engineer; Buyer; Chemical Engineer; Chemist; Computer Programmer; Cost Estimator; Credit Manager; Design Engineer; Designer; Draftsperson; Electrical/Electronics Engineer; Environmental Engineer; Financial Analyst; Geologist/Geophysicist; Human Resources Manager; Industrial Engineer; Internet Services Manager; Manufacturer's/Wholesaler's Sales Rep.; Mechanical Engineer; Metallurgical Engineer; MIS Specialist; Nuclear Engineer; Occupational Therapist; Petroleum Engineer; Physical Therapist; Quality Control Supervisor; Respiratory Therapist; Securities Sales Representative; Software Engineer; Systems Analyst. **Corporate headquarters location:** This Location. **Average salary range of placements:** More than $50,000. **Number of placements per year:** 50 - 99.

DON RICHARD ASSOCIATION OF RICHMOND
7275 Glen Forest Drive, Suite 200, Richmond VA 23226-3772. 804/282-6300. **Fax:** 804/282-6792. **Recorded Jobline:** 804/282-1177. **Contact:** Mike Beck, MIS Director. **E-mail address:** dranet@i2020.net. **World Wide Web address:** http://www.donrichard.com. **Description:** Established in 1978, this is the third of 12 nationwide offices on the East Coast. The company specializes in permanent and temporary placement of accounting and information systems professionals. Company pays fee. **Specializes in the areas of:** Accounting/Auditing; Administration/MIS/EDP; Computer Science/Software; Finance; Secretarial. **Positions commonly filled include:** Accountant/Auditor; Administrative Services Manager; Budget Analyst; Clerical Supervisor; Computer Programmer; Credit Manager; Design Engineer; Financial Analyst; Internet Services Manager; Management Analyst/Consultant; MIS Specialist; Multimedia Designer; Operations/Production Manager; Quality Control Supervisor; Services Sales Representative; Software Engineer; Systems Analyst; Technical Writer/Editor; Telecommunications Manager; Typist/Word Processor. **Corporate headquarters location:** This Location. **Other U.S. locations:** Atlanta, Charlotte, Norfolk, Tampa, and Washington DC. **Average salary range of placements:** $30,000 - $50,000. **Number of placements per year:** 50 - 99.

STRATEGIC SEARCH, INC.
5206 Markel Road, Suite 302, Richmond VA 23230. 804/285-6100. **Fax:** 804/285-6182. **Contact:** Dorrie Steinberg, President. **Description:** An executive search firm focusing primarily on the permanent placement of MIS professionals. The firm also offers placement in the personnel, purchasing, and manufacturing fields. Founded in 1985. Company pays fee. **Specializes in the areas of:** Computer Science/Software; Manufacturing; Personnel/Labor Relations; Technical and Scientific. **Positions commonly filled include:** Computer Programmer; Human Resources Specialist; MIS Specialist; Purchasing Agent/Manager; Software Engineer; Systems Analyst; Technical Writer/Editor; Telecommunications Manager. **Corporate headquarters location:** This Location. **Other U.S. locations:** Yorktown PA. **Average salary range of placements:** $30,000 - $50,000. **Number of placements per year:** 50 - 99.

TRC
8300 Boone Boulevard, Suite 500, Vienna VA 22182. 703/821-0009. **Fax:** 703/761-6763. **Contact:** Ramon Rahbar, Vice President. **Description:** An executive search firm and temporary agency focusing on the placement of clerical, computer, and communications personnel. Founded in 1994. **Specializes in the areas of:** Accounting/Auditing; Administration/MIS/EDP; Computer Science/Software; Personnel/Labor Relations; Secretarial. **Positions commonly filled include:** Clerical Supervisor; Computer Programmer; Financial Analyst; Paralegal; Systems Analyst. **Average salary range of placements:** $30,000 - $50,000. **Number of placements per year:** 50 - 99.

BILL YOUNG & ASSOCIATES
8381 Old Courthouse Road, Suite 300, Vienna VA 22182. 703/573-0200. **Fax:** 703/573-3612. **Contact:** Angela C. Berkman, Resource Center Manager. **E-mail address:** applicant@ billyoung.com. **World Wide Web address:** http://www.billyoung.com. **Description:** An executive search firm focusing on placement of technical and computer personnel. Company pays fee.

Specializes in the areas of: Administration/MIS/EDP; Computer Science/Software; Telecommunications. **Positions commonly filled include:** Computer Programmer; Human Resources Specialist; Internet Services Manager; Management Analyst/Consultant; Multimedia Designer; Operations/Production Manager; Software Engineer; Systems Analyst; Technical Writer/Editor; Telecommunications Manager. **Average salary range of placements:** More than $50,000. **Number of placements per year:** 50 - 99.

PERMANENT EMPLOYMENT AGENCIES OF VIRGINIA

ALPHA OMEGA RESOURCES INC.
Briarwood Business Center, Suite 14, Route 221, Forest VA 24551. 804/385-8640. **Fax:** 804/385-0192. **Contact:** Ben Liveray, President. **Description:** An employment agency that also provides temporary placement. Company pays fee. **Specializes in the areas of:** Accounting/Auditing; Administration/MIS/EDP; Computer Science/Software; Engineering; Finance; Food Industry; General Management; Health/Medical; Industrial; Insurance; Manufacturing; Personnel/Labor Relations; Printing/Publishing; Sales and Marketing; Secretarial; Technical and Scientific; Transportation. **Positions commonly filled include:** Accountant/Auditor; Administrative Services Manager; Architect; Attorney; Branch Manager; Buyer; Computer Programmer; Customer Service Representative; Design Engineer; Electrical/Electronics Engineer; Environmental Engineer; General Manager; Human Resources Specialist; Industrial Engineer; Industrial Production Manager; Insurance Agent/Broker; Internet Services Manager; Manufacturer's/Wholesaler's Sales Rep.; Mechanical Engineer; MIS Specialist; Operations/Production Manager; Paralegal; Physician; Public Relations Specialist; Quality Control Supervisor; Services Sales Representative; Software Engineer; Typist/Word Processor. **Benefits available to temporary workers:** Medical Insurance; Paid Holidays; Paid Vacation. **Number of placements per year:** 1000+.

AMERICAN TECHNICAL RESOURCES
1651 Old Meadow Road, Sixth Floor, McLean VA 22102-4308. 703/917-7800. **Contact:** Manager of Technical Recruiting. **Description:** An employment agency. Company pays fee. **Specializes in the areas of:** Administration/MIS/EDP; Computer Hardware/Software; Defense Industry; Engineering; Military; Technical and Scientific. **Positions commonly filled include:** Account Representative; Computer Operator; Computer Programmer; Computer Scientist; Customer Service Representative; Database Manager; Electrical/Electronics Engineer; MIS Specialist; Operations/Production Manager; Software Engineer; Systems Analyst; Technical Writer/Editor. **Number of placements per year:** 200 - 499.

CORE PERSONNEL
8201 Greensboro Drive, Suite 100, McLean VA 22102. 703/556-9610. **Contact:** Harvey Silver, President. **Description:** An employment agency. Company pays fee. **Specializes in the areas of:** Computer Hardware/Software. **Positions commonly filled include:** Administrative Assistant; Clerk; Computer Programmer; Customer Service Representative; EDP Specialist; Legal Secretary; Medical Secretary; Receptionist; Sales Representative; Secretary; Stenographer; Systems Analyst; Typist/Word Processor.

CAROL McNEW EMPLOYMENT SERVICE
300 Moore Street, Bristol VA 24201. 540/466-3318. **Fax:** 540/466-6894. **Contact:** Carol McNew, Owner. **Description:** An employment agency. Founded in 1981. Company pays fee. **Specializes in the areas of:** Accounting/Auditing; Banking; Computer Science/Software; Fashion; Finance; General Management; Health/Medical; Legal; Personnel/Labor Relations; Printing/Publishing; Retail; Sales and Marketing; Secretarial. **Positions commonly filled include:** Accountant/Auditor; Administrative Assistant; Bank Officer/Manager; Bookkeeper; Clerk; Computer Operator; Computer Programmer; Credit Manager; Customer Service Representative; Data Entry Clerk;

Factory Worker; General Manager; Insurance Agent/Broker; Medical Secretary; Nurse; Operations/Production Manager; Purchasing Agent/Manager; Quality Control Supervisor; Receptionist; Sales Representative; Secretary; Stenographer; Typist/Word Processor. **Number of placements per year:** 200 - 499.

PAUL-TITTLE ASSOCIATES, INC.
1485 Chain Bridge Road, Suite 304, McLean VA 22101-4501. 703/442-0500. **Fax:** 703/893-3871. **Contact:** Manager. **E-mail address:** pta@paul-tittle.com. **Description:** An employment agency. **Specializes in the areas of:** Computer Science/Software; Electronics; Engineering; Sales and Marketing; Telecommunications.

SELECT STAFFING SERVICES
240 Corporate, Suite 120, Norfolk VA 23502. 757/461-1582. **Fax:** 757/461-2835. **Contact:** Jeanie Hurrell, Manager. **Description:** A long-term, short-term, and temporary to full-time staffing service founded in 1960. Select Staffing Services specializes in administrative, clerical, and technical job assignments. This is a regional firm with 14 offices. **Specializes in the areas of:** Accounting/Auditing; Administration/MIS/EDP; Computer Science/Software; Industrial; Legal; Personnel/Labor Relations; Sales and Marketing; Secretarial. **Positions commonly filled include:** Accountant/Auditor; Claim Representative; Clerical Supervisor; Credit Manager; Customer Service Representative; Draftsperson; Human Resources Manager; MIS Specialist; Paralegal; Typist/Word Processor. **Benefits available to temporary workers:** 401(k); Holidays; Life Insurance; Training; Vacation. **Corporate headquarters location:** Reston VA. **Average salary range of placements:** $20,000 - $29,999. **Number of placements per year:** 1000+.

TECHNICAL SEARCH CORPORATION
7400 Beaufont Springs Drive, Suite 425, Richmond VA 23225. 804/323-3000. **Fax:** 804/330-9378. **Contact:** Mr. W.J. Kymmell, President. **Description:** An employment agency. **Specializes in the areas of:** Computer Science/Software. **Positions commonly filled include:** Computer Programmer; Software Engineer; Systems Analyst. **Number of placements per year:** 50 - 99.

VANTAGE PERSONNEL, INC.
2300 Clarendon Boulevard, Suite 1109, Arlington VA 22201. 703/247-4100. **Fax:** 703/247-4102. **Contact:** Mary Ann Wilkinson, CPC, President. **Description:** An employment agency. **Specializes in the areas of:** Administration/MIS/EDP; Computer Science/Software; Office Support; Telecommunications. **Positions commonly filled include:** Administrator; Management; Secretary; Typist/Word Processor.

EXECUTIVE SEARCH FIRMS OF WASHINGTON

BERKANA INTERNATIONAL
3417 Fremont Avenue North, Suite 225, Seattle WA 98103. 206/547-3226. **Fax:** 206/547-3843. **Contact:** Paul Allen, Vice President. **E-mail address:** berkana@headhunters.com. **World Wide Web address:** http://www.headhunters.com. **Description:** An executive search firm specializing in high-tech industry placement. Company pays fee. **Specializes in the areas of:** Computer Science/Software. **Positions commonly filled include:** Computer Programmer; Electrical/ Electronics Engineer; General Manager; Internet Services Manager; Multimedia Designer; Software Engineer; Strategic Relations Manager; Technical Writer/Editor; Telecommunications Manager; Video Production Coordinator. **Average salary range of placements:** More than $50,000. **Number of placements per year:** 1 - 49.

THE CAREER CLINIC, INC.
9725 3rd Avenue NE, Suite 509, Seattle WA 98115. 206/524-9831. **Fax:** 206/524-4125. **Contact:** Jane Ray, President. **E-mail address:** career@careerseanet.com. **Description:** An executive search firm operating on a contingency basis. The firm also provides temporary services and

career/outplacement counseling. Founded in 1967. Company pays fee. **Specializes in the areas of:** Administration/MIS/EDP; Architecture/Construction/Real Estate; Banking; Computer Science/ Software; Engineering; Food Industry; General Management; Industrial; Insurance; Legal; Manufacturing; Retail; Sales and Marketing; Secretarial; Technical and Scientific; Transportation. **Positions commonly filled include:** Accountant/Auditor; Actuary; Adjuster; Administrative Services Manager; Aircraft Mechanic/Engine Specialist; Attorney; Bank Officer/Manager; Budget Analyst; Chemical Engineer; Civil Engineer; Claim Representative; Computer Programmer; Construction Contractor; Cost Estimator; Credit Manager; Customer Service Representative; Design Engineer; Draftsperson; Economist; Editor; Electrical/Electronics Engineer; Environmental Engineer; Financial Analyst; General Manager; Health Services Manager; Hotel Manager; Human Resources Specialist; Industrial Engineer; Insurance Agent/Broker; Internet Services Manager; Landscape Architect; Management Analyst/Consultant; Manufacturer's/Wholesaler's Sales Rep.; Market Research Analyst; Mechanical Engineer; MIS Specialist; Occupational Therapist; Paralegal; Property and Real Estate Manager; Purchasing Agent/Manager; Quality Control Supervisor; Radio/TV Announcer/Broadcaster; Recreational Therapist; Restaurant/Food Service Manager; Securities Sales Representative; Software Engineer; Stationary Engineer; Structural Engineer; Surgical Technician; Surveyor; Systems Analyst; Teacher/Professor; Technical Writer/Editor; Typist/Word Processor; Urban/Regional Planner. **Number of placements per year:** 50 - 99.

J.F. CHURCH ASSOCIATES

P.O. Box 6128, Bellevue WA 98008-0128. 206/644-3278. **Contact:** Jim Church, President. **E-mail address:** jfchurch@scn.org. **Description:** An executive search firm. Company pays fee. **Specializes in the areas of:** Computer Science/Software; Sales and Marketing. **Number of placements per year:** 1 - 49.

COMPUTER PERSONNEL

720 Olide Way, Suite 510, Seattle WA 98101. 206/340-2722. **Fax:** 206/340-8845. **Contact:** Ron Meints, President. **Description:** An executive search firm. **Specializes in the areas of:** Computer Science/Software. **Positions commonly filled include:** Computer Programmer; Software Engineer; Systems Analyst.

EXPRESS PERSONNEL SERVICES

222 South Washington Street, Spokane WA 99204. 509/747-6011. **Fax:** 509/747-8930. **Contact:** Manager. **Description:** An executive search firm that also provides temporary and contract services. Company pays fee. **Specializes in the areas of:** Accounting/Auditing; Computer Science/Software; General Management; Insurance; Personnel/Labor Relations. **Positions commonly filled include:** Accountant/Auditor; Adjuster; Architect; Attorney; Blue-Collar Worker Supervisor; Branch Manager; Credit Manager; Designer; Draftsperson. **Corporate headquarters location:** Oklahoma City OK. **Average salary range of placements:** $20,000 - $29,999. **Number of placements per year:** 1000+.

HALL KINION ASSOCIATES

3005 112th Avenue NE, Suite 102, Bellevue WA 98004. 206/889-5003. **Toll-free phone:** 800/234-1136. **Fax:** 206/889-5985. **Contact:** Debbie Allen Oberbillig, Technical Recruiting Manager. **E-mail address:** hallkinb@ix.netcom.com. **World Wide Web address:** http://www.hallkinion.com. **Description:** An executive search firm. Company pays fee. **Specializes in the areas of:** Computer Hardware/Software; Computer Science/Software; Sales and Marketing; Technical and Scientific. **Positions commonly filled include:** Computer Programmer; Design Engineer; Editor; Engineer; Software Engineer; Systems Analyst; Technical Writer/Editor. **Number of placements per year:** 100 - 199.

N.G. HAYES

401 Park Place, Suite 207, Kirkland WA 98033. 206/889-1522. **Fax:** 206/889-1524. **Contact:** Nelia Hayes, President. **E-mail address:** nghayes@aol.com. **Description:** An executive search firm. Company pays fee. **Specializes in the areas of:** Computer Science/Software. **Positions commonly filled include:** Computer Programmer; Software Engineer; Systems Analyst. **Number of placements per year:** 1 - 49.

THE JOBS COMPANY

8900 East Sprague Avenue, Spokane WA 99212-2927. 509/928-3151. **Fax:** 509/928-3168. **Contact:** Gloria Hager, Manager. **Description:** An executive search firm and employment agency operating on a contingency basis. Founded in 1973. **Specializes in the areas of:** Accounting/Auditing; Administration/MIS/EDP; Computer Science/Software; Engineering; General Management; Health/Medical; Personnel/Labor Relations; Printing/Publishing; Retail; Sales and Marketing; Secretarial; Technical and Scientific. **Positions commonly filled include:** Accountant/Auditor; Adjuster; Administrative Services Manager; Advertising Clerk; Automotive Mechanic; Biological Scientist; Biomedical Engineer; Blue-Collar Worker Supervisor; Branch Manager; Broadcast Technician; Brokerage Clerk; Budget Analyst; Buyer; Chemist; Claim Representative; Clerical Supervisor; Clinical Lab Technician; Computer Programmer; Construction and Building Inspector; Construction Contractor; Cost Estimator; Counselor; Credit Manager; Dental Assistant/Dental Hygienist; Dental Lab Technician; Designer; Draftsperson; Editor; EEG Technologist; EKG Technician; Electrical/Electronics Engineer; Electrician; Emergency Medical Technician; Financial Analyst; Financial Services Sales Rep.; General Manager; Health Services Manager; Hotel Manager; Human Resources Manager; Industrial Engineer; Industrial Production Manager; Licensed Practical Nurse; Management Trainee; Manufacturer's/Wholesaler's Sales Rep.; Mechanical Engineer; Medical Records Technician; Metallurgical Engineer; Nuclear Medicine Technologist; Occupational Therapist; Paralegal; Purchasing Agent/Manager; Quality Control Supervisor; Registered Nurse; Reporter; Respiratory Therapist; Restaurant/Food Service Manager; Securities Sales Representative; Services Sales Representative; Software Engineer; Stationary Engineer; Structural Engineer; Surgical Technician; Systems Analyst; Technical Writer/Editor; Telecommunications Manager; Typist/Word Processor.

KOSSUTH & ASSOCIATES

800 Bellevue Way NE, Suite 400, Bellevue WA 98004. 206/450-9050. **Fax:** 206/450-0513. **Contact:** Jane Kossuth, President. **Description:** An executive search firm. Company pays fee. **Specializes in the areas of:** Communications; Computer Science/Software; General Management; Sales and Marketing; Technical and Scientific. **Positions commonly filled include:** Computer Programmer; General Manager; Public Relations Specialist; Software Engineer; Systems Analyst; Technical Writer/Editor. **Number of placements per year:** 50 - 99.

MACROSEARCH

13353 Bel-Red Road, Suite 206, Bellevue WA 98005. 206/641-7252. **Fax:** 206/641-0969. **Contact:** Vickie Stovall, Manager, Recruiting. **Description:** An executive search firm. Company pays fee. **Specializes in the areas of:** Computer Science/Software. **Positions commonly filled include:** Computer Programmer; Software Engineer; Systems Analyst; Teacher/Professor; Technical Writer/Editor. **Number of placements per year:** 50 - 99.

MANAGEMENT RECRUITERS OF MERCER ISLAND

9725 SE 36th Street, Suite 312, Mercer Island WA 98040-3896. 206/232-0204. **Contact:** James J. Dykeman, Manager. **E-mail address:** mercer!jjd@mrinet.com. **Description:** An executive search firm. **Specializes in the areas of:** Accounting/Auditing; Administration/MIS/EDP; Advertising; Architecture/Construction/Real Estate; Banking; Communications; Computer Science/Software; Design; Electrical; Engineering; Finance; Food Industry; General Management; Health/Medical; Insurance; Legal; Manufacturing; MIS/EDP; Operations Management; Personnel/Labor Relations; Printing/Publishing; Procurement; Retail; Sales and Marketing; Technical and Scientific; Textiles; Transportation. **Number of placements per year:** 100 - 199.

MANAGEMENT RECRUITERS OF NORTH TACOMA

535 Dock Street, Suite 111, Tacoma WA 98402. 206/572-7542. **Fax:** 206/572-7872. **Contact:** Bill Saylor, President/Manager. **Description:** An executive search firm. Company pays fee. **Specializes in the areas of:** Accounting/Auditing; Computer Hardware/Software; Computer Science/Software; Engineering; Finance; Manufacturing; Sales and Marketing; Technical and Scientific; Telecommunications. **Positions commonly filled include:** Accountant/Auditor; Actuary; Architect; Buyer; Computer Programmer; Economist; Engineer; Financial Analyst; Systems Analyst. **Number of placements per year:** 50 - 99.

MANAGEMENT RECRUITERS OF SEATTLE

2510 Fairview Avenue East, Seattle WA 98102-3216. 206/328-0936. **Toll-free phone:** 800/237-6562. **Fax:** 206/328-3256. **Contact:** Dan Jilka, Manager/Co-Owner. **Description:** An executive search firm that operates on retainer and contingency bases. **Specializes in the areas of:** Administration/MIS/EDP; Computer Science/Software; Engineering; Food Industry; General Management; Health/Medical; Manufacturing; Retail; Sales and Marketing; Technical and Scientific. **Number of placements per year:** 1 - 49.

MANAGEMENT RECRUITERS OF TACOMA

2709 Jahn Avenue NW, Suite H-11, Gig Harbor WA 98335. 206/858-9991. **Contact:** Dennis Johnson, Manager. **Description:** An executive search firm. **Specializes in the areas of:** Accounting/Auditing; Administration/MIS/EDP; Advertising; Architecture/Construction/Real Estate; Banking; Communications; Computer Science/Software; Design; Electrical; Engineering; Finance; Food Industry; General Management; Health/Medical; Insurance; Legal; Manufacturing; Operations Management; Personnel/Labor Relations; Printing/Publishing; Procurement; Retail; Sales and Marketing; Technical and Scientific; Textiles; Transportation. **Number of placements per year:** 1 - 49.

JOHN MASON & ASSOCIATES

2135 112th Avenue NE, Bellevue WA 98004-2950. 206/453-1608. **Fax:** 206/451-9214. **Contact:** John Mason/Duff Mason, Human Resources. **E-mail address:** masonsail@aol.com. **Description:** An executive search firm that places mid- to senior-level management and provides human resources consulting for small- to mid-size companies. The firm operates on both retainer and contingency bases. **Specializes in the areas of:** Computer Science/Software; Engineering; Finance; General Management; Manufacturing; Personnel/Labor Relations; Technical and Scientific. **Positions commonly filled include:** Computer Programmer, Credit Manager; Design Engineer; Electrical/Electronics Engineer; General Manager; Human Resources Manager; Manufacturer's/Wholesaler's Sales Rep.; Mechanical Engineer; MIS Specialist; Software Engineer; Structural Engineer; Systems Analyst; Telecommunications Manager. **Average salary range of placements:** More than $50,000. **Number of placements per year:** 50 - 99.

PERSONNEL UNLIMITED INC.

West 25 Nora, Spokane WA 99205. 509/326-8880. **Fax:** 509/326-0112. **Contact:** Gary P. Desgrosellier, President. **Description:** An executive search firm. Company pays fee. **Specializes in the areas of:** Accounting/Auditing; Administration/MIS/EDP; Clerical; Computer Science/Software; Engineering; Finance; Food Industry; General Management; Health/Medical; Insurance; Legal; Manufacturing; MIS/EDP; Personnel/Labor Relations; Sales and Marketing; Secretarial. **Positions commonly filled include:** Accountant/Auditor; Adjuster; Administrative Worker/Clerk; Agricultural Engineer; Bank Officer/Manager; Bookkeeper; Buyer; Ceramics Engineer; Chemical Engineer; Civil Engineer; Claim Representative; Clerical Supervisor; Collector; Computer Operator; Computer Programmer; Credit Manager; Customer Service Representative; Data Entry Clerk; Draftsperson; EDP Specialist; Electrical/Electronics Engineer; Financial Analyst; Food Scientist/Technologist; General Manager; Health Services Manager; Hotel Manager; Industrial Engineer; Investigator; Legal Secretary; Management Trainee; Manufacturer's/Wholesaler's Sales Rep.; Marketing Specialist; Materials Engineer; Mechanical Engineer; Medical Records Technician; Medical Secretary; Metallurgical Engineer; MIS Specialist; Operations/Production Manager; Public Relations Specialist; Purchasing Agent/Manager; Quality Control Supervisor; Receptionist; Registered Nurse; Secretary; Services Sales Representative; Statistician; Stenographer; Structural Engineer; Systems Analyst; Technical Writer/Editor; Technician; Travel Agent; Typist/Word Processor; Underwriter/Assistant Underwriter. **Number of placements per year:** 500 - 999.

RIGEL COMPUTER RESOURCES

1611 116th Avenue NE, Bellevue WA 98004. 206/646-4990. **Fax:** 206/646-3058. **Contact:** Rita Ashley, President. **Description:** An executive search firm. Company pays fee. **Specializes in the areas of:** Computer Science/Software. **Average salary range of placements:** More than $50,000. **Number of placements per year:** 50 - 99.

SMALL BUSINESS SOLUTIONS INC.
4511 100th Street East, Tacoma WA 98446. 206/537-1040. **Fax:** 206/531-7323. **Contact:** President. **Description:** Small Business Solutions is an executive search firm. Founded in 1986. **Specializes in the areas of:** Accounting/Auditing; Computer Science/Software. **Positions commonly filled include:** Accountant/Auditor. **Average salary range of placements:** $20,000 - $29,999. **Number of placements per year:** 1 - 49.

SNELLING PERSONNEL SERVICES
2101 4th Avenue, Suite 1330, Seattle WA 98121. 206/441-8895. **Fax:** 206/448-5373. **Contact:** Sue and Tom Truscott, Owners/Managers. **Description:** An executive search firm and employment agency operating on a contingency basis. Founded in 1966. Company pays fee. **Specializes in the areas of:** Accounting/Auditing; Administration/MIS/EDP; Computer Science/Software; Engineering; Finance; General Management; Insurance; Personnel/Labor Relations; Retail; Sales and Marketing; Secretarial. **Positions commonly filled include:** Accountant/Auditor; Administrative Services Manager; Administrative Worker/Clerk; Bookkeeper; Branch Manager; Brokerage Clerk; Budget Analyst; Buyer; Claim Representative; Clerical Supervisor; Collector; Computer Operator; Credit Manager; Customer Service Representative; Data Entry Clerk; Financial Analyst; General Manager; Health Services Manager; Insurance Agent/Broker; Investigator; Legal Secretary; Management Trainee; Manufacturer's/Wholesaler's Sales Rep.; Medical Records Technician; Medical Secretary; Property and Real Estate Manager; Purchasing Agent/Manager; Receptionist; Secretary; Securities Sales Representative; Services Sales Representative; Technical Writer/Editor; Telecommunications Manager; Typist/Word Processor. **Corporate headquarters location:** Dallas TX. **Average salary range of placements:** $30,000 - $50,000. **Number of placements per year:** 200 - 499.

STRAIN PERSONNEL SPECIALISTS
801 Pine Street, Suite 2000, Seattle WA 98101-1807. 206/382-1588. **Fax:** 206/622-1572. **Contact:** Joe Strain, CPC, Partner. **E-mail address:** joestrain@msn.com. **Description:** Strain Personnel Specialists is an executive search firm operating on a retainer basis. The company also offers contract services. Company pays fee. **Specializes in the areas of:** Administration/MIS/EDP; Computer Science/Software; Engineering; Manufacturing; Personnel/Labor Relations; Technical and Scientific. **Positions commonly filled include:** Computer Programmer; Design Engineer; Human Resources Manager; Internet Services Manager; Mathematician; MIS Specialist; Multimedia Designer; Science Technologist; Software Engineer; Systems Analyst; Technical Writer/Editor; Telecommunications Manager. **Average salary range of placements:** More than $50,000. **Number of placements per year:** 50 - 99.

THE WASHINGTON FIRM
Two Nickerson Street, Courtyard Suite, Seattle WA 98109. 206/284-4800. **Fax:** 206/284-8844. **Contact:** Al Battson, Principal. **Description:** The Washington Firm is an executive search firm. Company pays fee. **Specializes in the areas of:** Accounting/Auditing; Administration/MIS/EDP; Architecture/ Construction/Real Estate; Biology; Biotechnology; Computer Science/Software; Engineering; Finance; General Management; Health/Medical; Personnel/Labor Relations; Telecommunications. **Positions commonly filled include:** Accountant/Auditor; Biological Scientist; Biomedical Engineer; Budget Analyst; Chemical Engineer; Civil Engineer; Computer Programmer; Construction Contractor; Cost Estimator; Electrical/Electronics Engineer; Financial Analyst; Human Resources Manager; Property and Real Estate Manager; Software Engineer; Structural Engineer; Systems Analyst; Technical Writer/Editor. **Number of placements per year:** 50 - 99.

WHITTALL MANAGEMENT GROUP
720 South 333rd Street, Suite 102, Federal Way WA 98003. 206/874-0710. **Fax:** 206/952-2918. **Contact:** Geoff Whittall, Vice President. **Description:** Whittall Management Group is an executive search firm that operates on retainer and contingency bases. Company pays fee. **Specializes in the areas of:** Computer Science/Software; Engineering; Food Industry; General Management; Industrial; Manufacturing; Personnel/Labor Relations; Sales and Marketing; Technical and Scientific. **Positions commonly filled include:** Computer Programmer; Construction Contractor; Design Engineer; Electrical/Electronics Engineer; Electrician; Environmental Engineer;

Forester/Conservation Scientist; General Manager; Human Resources Manager; Industrial Engineer; Industrial Production Manager; Mechanical Engineer; MIS Specialist; Operations/Production Manager; Quality Control Supervisor; Software Engineer; Structural Engineer; Systems Analyst. **Average salary range of placements:** More than $50,000. **Number of placements per year:** 50 - 99.

PERMANENT EMPLOYMENT AGENCIES OF WASHINGTON

HOUSER, MARTIN, MORRIS & ASSOCIATES
P.O. Box 90015, Bellevue WA 98009. 206/453-2700. **Contact:** Bob Holert, President. **Description:** Houser, Martin, Morris & Associates is an employment agency. Company pays fee. **Specializes in the areas of:** Computer Science/Software; Engineering; Finance; Insurance; Manufacturing; MIS/EDP; Sales and Marketing. **Positions commonly filled include:** Accountant/Auditor; Actuary; Aerospace Engineer; Bank Officer/Manager; Computer Programmer; Credit Manager; EDP Specialist; Electrical/Electronics Engineer; Financial Analyst; General Manager; Industrial Engineer; Insurance Agent/Broker; Mechanical Engineer; Metallurgical Engineer; MIS Specialist; Purchasing Agent/Manager; Underwriter/Assistant Underwriter. **Number of placements per year:** 1 - 49.

NELSON, COULSON & ASSOCIATES INC.
14450 NE 29th Place, Suite 115, Bellevue WA 98007-3697. 206/883-6612. **Toll-free phone:** 888/883-6612. **Fax:** 206/883-6916. **Recorded jobline:** 206/883-6612. **Contact:** Jennifer Ellis, Manager. **World Wide Web address:** http://www.ncainc.com/ncainc. **Description:** Nelson, Coulson & Associates is an employment agency. Company pays fee. **Specializes in the areas of:** Computer Science/Software; Personnel/Labor Relations. **Positions commonly filled include:** Buyer; Computer Programmer; Designer; Draftsperson; Editor; Engineer; Financial Analyst; Systems Analyst; Technical Writer/Editor. **Number of placements per year:** 200 - 499.

STAFFING RESOURCES
1000 2nd Avenue, Suite 1700, Seattle WA 98104. 206/583-2711. **Fax:** 206/583-2725. **Contact:** Dave Martin, President. **Description:** An employment agency. Company pays fee. **Specializes in the areas of:** Accounting/Auditing; Administration/MIS/EDP; Computer Science/Software; Finance; Legal; Personnel/Labor Relations; Secretarial. **Positions commonly filled include:** Accountant/Auditor; Adjuster; Administrative Services Manager; Brokerage Clerk; Budget Analyst; Buyer; Claim Representative; Clerical Supervisor; Collector; Credit Manager; Customer Service Representative; Financial Analyst; Investigator; Management Trainee; Paralegal; Property and Real Estate Manager; Services Sales Representative. **Number of placements per year:** 500 - 999.

THOMAS COMPANY
15434 SE 167th Place, Renton WA 98058. 206/255-7637. **Contact:** Thomas J. Yankowski, Executive Director. **Description:** An employment agency. Company pays fee. **Specializes in the areas of:** Banking; Computer Science/Software; Finance; Insurance; MIS/EDP; Sales and Marketing. **Positions commonly filled include:** Accountant/Auditor; Actuary; Administrative Worker/Clerk; Advertising Account Executive; Attorney; Bookkeeper; Claim Representative; Computer Programmer; Customer Service Representative; Data Entry Clerk; Economist; EDP Specialist; Financial Analyst; General Manager; Human Resources Manager; Insurance Agent/Broker; Management Analyst/Consultant; Marketing Specialist; Statistician; Systems Analyst; Technical Writer/Editor; Technician; Underwriter/Assistant Underwriter. **Number of placements per year:** 50 - 99.

PERMANENT EMPLOYMENT AGENCIES OF WEST VIRGINIA

KEY PERSONNEL, INC.
1124 Fourth Avenue, Suite 300, Huntington WV 25701. 304/529-3377. **Contact:** Recruiter. **Description:** A full-service employment agency. Founded in 1975. Company pays fee. **Specializes in the areas of:** Accounting/Auditing; Banking; Computer Science/Software; Engineering; Finance; Food Industry; General Management; Health/Medical; Industrial; Manufacturing; Personnel/Labor Relations; Sales and Marketing; Technical and Scientific; Transportation. **Positions commonly filled include:** Accountant/Auditor; Adjuster; Bank Officer/Manager; Branch Manager; Chemical Engineer; Claim Representative; Cost Estimator; Credit Manager; Customer Service Representative; Health Services Manager; Human Resources Specialist; Industrial Engineer; Manufacturer's/Wholesaler's Sales Rep.; Mechanical Engineer; MIS Specialist; Purchasing Agent/Manager; Quality Control Supervisor; Restaurant/Food Service Manager; Services Sales Representative; Systems Analyst; Transportation/Traffic Specialist; Underwriter/Assistant Underwriter. **Average salary range of placements:** $30,000 - $50,000. **Number of placements per year:** 50 - 99.

ONSITE COMMERCIAL STAFFING
1430-1 Edwin Miller Boulevard, Martinsburg WV 25401. 304/267-7363. **Contact:** Recruiting. **Description:** An employment agency that provides both temporary and permanent placements. Onsite Commercial Staffing focuses on technical and business placements. **Specializes in the areas of:** Computer Hardware/Software; Computer Programming; Environmental; Telecommunications.

QUANTUM RESOURCES
P.O. Box 1751, Parkersburg WV 26102. 304/428-8028. **Contact:** Cindy Miller, Technical Recruiter. **Description:** An employment agency that provides both permanent and temporary placements. **Specializes in the areas of:** Clerical; Computer Hardware/Software; Engineering; Industrial.

SNELLING PERSONNEL SERVICES
P.O. Box 4522, Charleston WV 25364. 304/344-0101. **Contact:** Recruiting. **Description:** Offers both temporary and permanent placements. **Specializes in the areas of:** Accounting/Auditing; Computer Hardware/Software; Sales and Marketing.

EXECUTIVE SEARCH FIRMS OF WISCONSIN

CAREER RECRUITERS
15850 West Bluemond Road, Suite 201, Brookfield WI 53005-6007. 414/784-0595. **Fax:** 414/797-0853. **Contact:** Don Schoberg, Owner. **Description:** Career Recruiters focuses on permanent placement of MIS professionals. Company pays fee. **Specializes in the areas of:** Computer Science/Software. **Positions commonly filled include:** Computer Programmer; MIS Specialist; Software Engineer; Systems Analyst. **Corporate headquarters location:** This Location. **Average salary range of placements:** More than $50,000. **Number of placements per year:** 1 - 49.

CAREER RESOURCES
757 Sand Lake Road, Onalaska WI 54650. 608/783-6307. **Fax:** 608/783-6302. **Contact:** Chris M. Jansson, CPC, Owner. **Description:** An executive search firm. **Specializes in the areas of:** Accounting/Auditing; Administration/MIS/EDP; Computer Science/Software; Finance; Personnel/Labor Relations. **Positions commonly filled include:** Accountant/Auditor; Bank

Officer/Manager; Computer Programmer; Financial Analyst; Human Resources Manager; MIS Manager; Systems Analyst.

COMPUSEARCH OF WAUSAU
3205 Terrace Court, Suite 1A, Wausau WI 54401-4915. 715/842-1750. **Fax:** 715/842-1741. **Contact:** Laurie L. Prochnow, President. **Description:** An executive search firm specializing in the information systems industry. The firm fills positions of all levels, with a primary geographic concentration on Wisconsin and surrounding states. Company pays fee. **Specializes in the areas of:** Computer Science/Software; Fashion; General Management; Health/Medical; Insurance; Manufacturing; Retail; Sales and Marketing. **Positions commonly filled include:** Computer Programmer; MIS Specialist; Software Engineer; Systems Analyst; Telecommunications Manager. **Corporate headquarters location:** This Location. **Average salary range of placements:** More than $50,000. **Number of placements per year:** 1 - 49.

J.M. EAGLE PARTNERS
11514 North Port Washington Road, Suite 105, Mequon WI 53092. 414/241-1113. **Fax:** 414/241-4745. **Contact:** Jerry Moses, President. **Description:** An executive search firm. Company pays fee. **Specializes in the areas of:** Computer Science/Software; Engineering; Finance; General Management; Health/Medical; Manufacturing; Personnel/Labor Relations; Sales and Marketing; Technical and Scientific. **Number of placements per year:** 1000+.

HUNTER MIDWEST
11101 West Janesville Road, Hales Corners WI 53130-2530. 414/529-3930x3100. **Toll-free phone:** 800/236-3930. **Fax:** 414/529-0394. **Contact:** Michael Certalic, Office Manager. **E-mail address:** hmidwest@execpc.com. **Description:** An executive search firm that specializes in placing AS400 IS professionals, PC LAN, and UNIX technical specialists. The firm's client base consists of over 7,000 companies in the upper Midwest. Company pays fee. **NOTE:** A minimum of one year of experience is required. The firm does not place recent college graduates. **Specializes in the areas of:** Administration/MIS/EDP; Computer Science/Software; Network Administration; Systems Administration. **Positions commonly filled include:** Computer Programmer; MIS Specialist; Systems Analyst. **Number of placements per year:** 50 - 99.

INTERNATIONAL SEARCH
P.O. Box 381, Green Bay WI 54305-0381. 414/437-8055. **Toll-free phone:** 800/276-8913. **Fax:** 414/437-0343. **Contact:** Michael Wingers, Owner. **Description:** An executive search firm. Company pays fee. **Specializes in the areas of:** Accounting/Auditing; Computer Science/Software; Finance. **Positions commonly filled include:** Accountant/Auditor; Budget Analyst; Human Resources Manager; MIS Specialist. **Corporate headquarters location:** This Location. **Other U.S. locations:** Nationwide. **Average salary range of placements:** More than $50,000. **Number of placements per year:** 50 - 99.

MANAGEMENT RECRUITERS OF APPLETON
COMPUSEARCH
911 North Lynndale Drive, Appleton WI 54914. 414/731-5221. **Contact:** Russ Hanson, Manager. **Description:** An executive search firm. **Specializes in the areas of:** Accounting/Auditing; Administration/MIS/EDP; Advertising; Architecture/Construction/Real Estate; Banking; Communications; Computer Hardware/Software; Electrical; Engineering; Finance; Food Industry; General Management; Health/Medical; Insurance; Legal; Manufacturing; Operations Management; Personnel/Labor Relations; Printing/Publishing; Technical and Scientific; Textiles; Transportation.

MANAGEMENT RECRUITERS OF ELM GROVE
13000 West Bluemound Road, Elm Grove WI 53122-2655. 414/797-7500. **Fax:** 414/797-7500. **Contact:** Peder Medtlie, Manager. **Description:** An executive search firm that recruits for specialized professional and middle management positions. Company pays fee. **Specializes in the areas of:** Accounting/Auditing; Computer Science/Software; Engineering; Finance; General Management; Manufacturing. **Positions commonly filled include:** Accountant/Auditor; Bank Officer/Manager; Biomedical Engineer; Budget Analyst; Chemical Engineer; Computer Programmer; Environmental Engineer; General Manager; Industrial Engineer; Industrial Production Manager; Mechanical Engineer; Metallurgical Engineer; MIS Specialist; Quality

Control Supervisor; Software Engineer; Systems Analyst. **Other U.S. locations:** Nationwide. **Average salary range of placements:** $30,000 - $50,000. **Number of placements per year:** 50 - 99.

MANAGEMENT RECRUITERS OF GREEN BAY

444 South Adams Street, Green Bay WI 54301. 414/437-4353. **Contact:** Mr. Garland Ross, Manager. **Description:** An executive search firm. **Specializes in the areas of:** Accounting/Auditing; Administration/MIS/EDP; Advertising; Architecture/Construction/Real Estate; Banking; Communications; Computer Hardware/Software; Electrical; Engineering; Finance; Food Industry; General Management; Health/Medical; Insurance; Legal; Manufacturing; Operations Management; Personnel/Labor Relations; Printing/Publishing; Technical and Scientific; Transportation.

MANAGEMENT RECRUITERS OF MILWAUKEE (SOUTH)

5307 South 92nd Street, Suite 125, Hales Corners WI 53130. 414/529-8020. **Contact:** Office Manager. **Description:** An executive search firm. **Specializes in the areas of:** Accounting/Auditing; Administration/MIS/EDP; Advertising; Architecture/Construction/Real Estate; Banking; Communications; Computer Hardware/Software; Electrical; Engineering; Finance; Food Industry; General Management; Health/Medical; Insurance; Legal; Manufacturing; Operations Management; Personnel/Labor Relations; Printing/Publishing; Technical and Scientific; Transportation.

PRAIRIE ENGINEERING

P.O. Box 165, DeForest WI 53532-0165. 608/846-7600. **Fax:** 608/846-7601. **Contact:** Human Resources. **Description:** An executive search firm. Company pays fee. **Specializes in the areas of:** Computer Science/Software; Engineering. **Positions commonly filled include:** Computer Programmer; Software Engineer; Systems Analyst; Technical Writer/Editor. **Average salary range of placements:** $30,000 - $50,000. **Number of placements per year:** 1 - 49.

PROFESSIONAL ENGINEERING

11941 West Rawson Avenue, Franklin WI 53132. 414/427-1700. **Fax:** 414/427-8080. **Contact:** Patty Wiza, President. **E-mail address:** proeng33@aol.com. **Description:** An executive search firm. Company pays fee. **Specializes in the areas of:** Computer Science/Software; Engineering; Industrial; Manufacturing. **Positions commonly filled include:** Buyer; Chemical Engineer; Design Engineer; Designer; Draftsperson; Electrical/Electronics Engineer; Industrial Engineer; Industrial Production Manager; Mechanical Engineer; Metallurgical Engineer; Purchasing Agent/Manager; Quality Control Supervisor; Software Engineer. **Number of placements per year:** 50 - 99.

PROFESSIONAL RESOURCE SERVICES

1825 Lone Oak Circle West, Brookfield WI 53045-5017. 414/782-6901. **Fax:** 414/785-1813. **Contact:** Sandra DeChant, Principal. **Description:** An executive search firm specializing in information systems staffing. Company pays fee. **Specializes in the areas of:** Computer Science/Software. **Positions commonly filled include:** Computer Programmer; Software Engineer; Systems Analyst. **Corporate headquarters location:** This Location. **Other U.S. locations:** Nationwide. **Average salary range of placements:** $30,000 - $50,000. **Number of placements per year:** 1 - 49.

ROWBOTTOM & ASSOCIATES

7707 West Menomonee River Parkway, Milwaukee WI 53213-2632. 414/475-1974. **Fax:** 414/475-5038. **Contact:** Mark Rowbottom, Owner. **E-mail address:** robottom@execpc.com. **Description:** An executive search firm. Company pays fee. **Specializes in the areas of:** Computer Science/Software; Information Systems. **Positions commonly filled include:** Computer Programmer; Internet Services Manager; MIS Manager; Software Engineer; Systems Analyst; Telecommunications Manager. **Average salary range of placements:** $30,000 - $50,000. **Number of placements per year:** 100 - 199.

SALES CONSULTANTS

601 East Henry Clay, Milwaukee WI 53217-5646. 414/963-2520. **Contact:** Tim Lawler, Manager. **Description:** An executive search firm. **Specializes in the areas of:** Accounting/Auditing;

Administration/MIS/EDP; Advertising; Architecture/Construction/Real Estate; Banking; Communications; Computer Hardware/Software; Electrical; Engineering; Finance; Food Industry; General Management; Health/Medical; Insurance; Legal; Manufacturing; Operations Management; Personnel/Labor Relations; Printing/Publishing; Technical and Scientific; Transportation.

TECHTRONIX TECHNICAL EMPLOYMENT

5401 North 76th Street, Milwaukee WI 53218. 414/466-3100. **Fax:** 414/466-3598. **Contact:** Louis Beauchamp, Vice President. **Description:** An executive search firm. Company pays fee. **Specializes in the areas of:** Administration/MIS/EDP; Computer Science/Software; Engineering; Manufacturing; Technical and Scientific. **Positions commonly filled include:** Aerospace Engineer; Biomedical Engineer; Chemical Engineer; Computer Programmer; Electrical/Electronics Engineer; Human Resources Manager; Industrial Engineer; Industrial Production Manager; Mechanical Engineer; Metallurgical Engineer; Quality Control Supervisor; Software Engineer; Systems Analyst. **Number of placements per year:** 50 - 99.

PERMANENT EMPLOYMENT AGENCIES OF WISCONSIN

ARGUS TECHNICAL SERVICES

2339 West Wisconsin Avenue, Appleton WI 54914. 414/731-7703. **Fax:** 414/731-1886. **Contact:** John LaFay, District Sales Manager. **Description:** An employment agency. Company pays fee. **Specializes in the areas of:** Architecture/Construction/Real Estate; Art/Design; Computer Science/Software; Engineering; Industrial; Manufacturing; Personnel/Labor Relations; Technical and Scientific. **Positions commonly filled include:** Aerospace Engineer; Agricultural Engineer; Architect; Biomedical Engineer; Chemical Engineer; Chemist; Civil Engineer; Computer Programmer; Designer; Draftsperson; Electrical/Electronics Engineer; Electrician; General Manager; Industrial Engineer; Industrial Production Manager; Landscape Architect; Mechanical Engineer; Metallurgical Engineer; Mining Engineer; Nuclear Engineer; Petroleum Engineer; Quality Control Supervisor; Software Engineer; Stationary Engineer; Structural Engineer; Surveyor; Technical Writer/Editor; Urban/Regional Planner.

DUNHILL & CONCORD
DUNHILL STAFFING SERVICES

735 North Water Street, Milwaukee WI 53202. 414/272-4860. **Fax:** 414/272-3852. **Contact:** Bradley Brin, President. **Description:** An employment agency with both temporary and permanent divisions including permanent divisions for clerical, physician, and data processing professions. **Specializes in the areas of:** Accounting/Auditing; Administration/MIS/EDP; Computer Science/Software; Health/Medical; Insurance; Secretarial. **Positions commonly filled include:** Accountant/Auditor; Budget Analyst; Butcher; Clerical Supervisor; Computer Programmer; Credit Manager; Customer Service Representative; Dietician/Nutritionist; EEG Technologist; EKG Technician; Financial Analyst; Human Resources Manager; Insurance Agent/Broker; MIS Specialist; Occupational Therapist; Paralegal; Pharmacist; Physical Therapist; Physician; Recreational Therapist; Registered Nurse; Respiratory Therapist; Software Engineer; Speech-Language Pathologist; Systems Analyst; Typist/Word Processor; Underwriter/Assistant Underwriter. **Number of placements per year:** 200 - 499.

NORTHERN TECHNICAL SERVICES

8899 North 60th Street, Milwaukee WI 53223. **Toll-free phone:** 800/686-2819. **Fax:** 414/362-8880. **Contact:** Peter Ryan, Sales Manager. **Description:** An employment agency and contract services firm specializing in the hourly rental of engineering and information systems personnel. Founded in 1975. Company pays fee. **Specializes in the areas of:** Computer Science/Software; Engineering; Personnel/Labor Relations. **Positions commonly filled include:** Aerospace Engineer; Agricultural Engineer; Architect; Ceramics Engineer; Chemical Engineer; Civil Engineer;

Computer Programmer; Design Engineer; Designer; Draftsperson; Electrical/Electronics Engineer; Industrial Engineer; Industrial Production Manager; Landscape Architect; Materials Engineer; Mechanical Engineer; Metallurgical Engineer; Mining Engineer; MIS Specialist; Petroleum Engineer; Quality Control Supervisor; Software Engineer; Structural Engineer; Systems Analyst; Technical Writer/Editor; Transportation/Traffic Specialist. **Benefits available to temporary workers:** 401(k); Paid Holidays. **Corporate headquarters location:** This Location. **Other U.S. locations:** Green Bay WI; Mosinee WI. **Average salary range of placements:** $30,000 - $50,000. **Number of placements per year:** 200 - 499.

PLACEMENTS OF RACINE INC.
222 Main Street, Suite 101, Racine WI 53403. 414/637-9355. **Contact:** Office Manager. **Description:** An employment agency. Company pays fee. **Specializes in the areas of:** Accounting/Auditing; Administration/MIS/EDP; Clerical; Computer Hardware/Software; Engineering; Finance; Manufacturing; Sales and Marketing. **Positions commonly filled include:** Accountant/Auditor; Agricultural Engineer; Bookkeeper; Buyer; Computer Programmer; Credit Manager; Customer Service Representative; Draftsperson; EDP Specialist; Electrical/Electronics Engineer; Financial Analyst; Industrial Engineer; Legal Secretary; Marketing Specialist; Mechanical Engineer; Medical Secretary; Metallurgical Engineer; MIS Specialist; Purchasing Agent/Manager; Quality Control Supervisor; Receptionist; Sales Representative; Secretary; Stenographer; Systems Analyst; Technical Writer/Editor; Technician; Typist/Word Processor. **Number of placements per year:** 50 - 99.

TECHNOLOGY CONSULTING CORPORATION
N-16 W23233, Stone Ridge Drive, Waukesha WI 53188. 414/650-6500. **Contact:** Human Resources. **Description:** An employment agency that places high-tech professionals and computer consultants. **Specializes in the areas of:** Computer Hardware/Software.

WALLENS, BECK AND ASSOCIATES, INC.
5626 North 91st Street, Suite 306, Milwaukee WI 53225-2745. 414/527-2400. **Fax:** 414/461-3757. **Contact:** Charles N. Wallens, President. **Description:** An employment agency. Company pays fee. **Specializes in the areas of:** Accounting/Auditing; Administration/MIS/EDP; Architecture/ Construction/Real Estate; Banking; Computer Science/Software; Engineering; Finance; General Management; Industrial; Insurance; Legal; Manufacturing; Personnel/Labor Relations; Printing/Publishing; Sales and Marketing; Transportation. **Positions commonly filled include:** Accountant/Auditor; Actuary; Agricultural Engineer; Architect; Attorney; Bank Officer/Manager; Biological Scientist; Biomedical Engineer; Branch Manager; Budget Analyst; Buyer; Chemical Engineer; Chemist; Civil Engineer; Claim Representative; Computer Programmer; Construction Contractor; Cost Estimator; Credit Manager; Customer Service Representative; Designer; Economist; Electrical/Electronics Engineer; Environmental Engineer; Financial Analyst; Food Scientist/Technologist; General Manager; Geologist/Geophysicist; Health Services Manager; Hotel Manager; Human Resources Manager; Industrial Engineer; Industrial Production Manager; Insurance Agent/Broker; Librarian; Management Analyst/Consultant; Management Trainee; Manufacturer's/Wholesaler's Sales Rep.; Mechanical Engineer; Metallurgical Engineer; Mining Engineer; Operations/Production Manager; Paralegal; Property and Real Estate Manager; Purchasing Agent/Manager; Quality Control Supervisor; Real Estate Agent; Science Technologist; Securities Sales Representative; Services Sales Representative; Software Engineer; Stationary Engineer; Statistician; Structural Engineer; Systems Analyst; Technical Writer/Editor; Transportation/Traffic Specialist. **Number of placements per year:** 1 - 49.

EXECUTIVE SEARCH FIRMS OF WYOMING

MANAGEMENT RECRUITERS OF CHEYENNE
1008 East 21st Street, Cheyenne WY 82001. 307/635-8731. **Fax:** 307/635-6653. **Contact:** Manager. **Description:** An executive search firm operating on a contingency basis. Founded in

1978. Company pays fee. **Specializes in the areas of:** Accounting/Auditing; Administration/MIS/EDP; Advertising; Architecture/Construction/Real Estate; Banking; Communications; Computer Hardware/Software; Design; Electrical; Engineering; Food Industry; General Management; Health/Medical; Insurance; Legal; Manufacturing; Operations Management; Personnel/Labor Relations; Printing/Publishing; Procurement; Retail; Sales and Marketing; Technical and Scientific; Textiles; Transportation. **Positions commonly filled include:** Ceramics Engineer; Design Engineer; Industrial Engineer; Materials Engineer; Mechanical Engineer; Metallurgical Engineer; Quality Control Supervisor. **Average salary range of placements:** More than $50,000. **Number of placements per year:** 1 - 49.

PROFESSIONAL ASSOCIATIONS

ACADEMY OF INTERACTIVE ARTS AND SCIENCES
500 South Buena Vista Street, Burbank CA 91501. 818/623-3730. E-mail address: academy@interactive.org. World Wide Web address: http://www.interactive.org. Provides a forum for members of the interactive technology community.

AMERICAN ASSOCIATION FOR ARTIFICIAL INTELLIGENCE
445 Burgess Drive, Menlo Park CA 94025-3442. 415/328-3123. E-mail address: info@aaai.org. World Wide Web address: http://www.aaai.org. A nonprofit scientific society promoting the advancement of artificial intelligence research.

AMERICAN INTERNET ASSOCIATION
World Wide Web address: http://www.amernet.org. A nonprofit association providing assistance in the use of the Internet. Membership required.

ASSOCIATION FOR COMPUTING MACHINERY
1515 Broadway, 17th Floor, New York NY 10036. 212/869-7440. World Wide Web address: http://www.acm.org. Membership required.

ASSOCIATION FOR INFORMATION SYSTEMS
222 Mervis Hall, Katz Graduate School of Business, University of Pittsburgh, Pittsburgh PA 15260. 412/648-1588. World Wide Web address: http://www1.pitt.edu/~ais. A professional association serving academicians in the information systems field.

ASSOCIATION FOR MULTIMEDIA COMMUNICATIONS
P.O. Box 10645, Chicago IL 60610. 312/409-1032. E-mail address: amc@amcomm.org. World Wide Web address: http://www.amcomm.org. A multimedia and Internet association.

ASSOCIATION FOR MULTI-MEDIA INTERNATIONAL, INC.
813/960-1692. World Wide Web address: http://www.ami.org. An international nonprofit association serving professionals in the Internet and multimedia industries.

ASSOCIATION FOR WOMEN IN COMPUTING
41 Sutter Street, Suite 1006, San Francisco CA 94104. 415/905-4663. E-mail address: awc@acm.org. World Wide Web address: http://www.awc-hq.org/awc. A nonprofit organization promoting women in computing professions.

ASSOCIATION OF INTERNET PROFESSIONALS
1301 Fifth Avenue, Suite 3300, Seattle WA 98101. E-mail address: info@associp.org. World Wide Web address: http://www.associp.org. A nonprofit trade association providing a forum for Internet users and professionals.

ASSOCIATION OF ONLINE PROFESSIONALS
6096 Franconia Road, Suite D, Alexandria VA 22310. 703/803-0455. E-mail address: aop@cris.com. World Wide Web address: http://www.aop.org. An international association for computer-based communications systems professionals.

ASSOCIATION OF PERSONAL COMPUTER USER GROUPS
4020 McEwen, Suite 105, Dallas TX 75244-5019. 972/233-9107. E-mail address: office@apcug.org. World Wide Web address: http://www.apcug.org. An international platform-independent, nonprofit corporation concerned with computer user groups.

ASSOCIATION OF SHAREWARE PROFESSIONALS
545 Grover Road, Muskegon MI 49442. 616/788-5131. World Wide Web address: http://www.asp-shareware.org. A user information site focusing on shareware software.

BLACK DATA PROCESSING ASSOCIATES
1111 14th Street NW, Suite 700, Washington DC 20005-5603. Toll-free phone: 800/727-BDPA. E-mail address: nbdpa@bdpa.org. World Wide Web address: http://www.bdpa.org. An organization of information technology professionals serving the minority community.

CFI, INC.
4030 West Braker Lane, Suite 550, Austin TX 78759. 512/342-2244. E-mail address: cfi@cfi.org. World Wide Web address: http://www.cfi.org. A nonprofit consortium concerned with the promotion of electronic design automation (EDA) technology.

THE CENTER FOR SOFTWARE DEVELOPMENT
111 West St. John, Suite 200, San Jose CA 95113. 408/494-8378. E-mail address: info@center.org. World Wide Web address: http://www.center.org. A nonprofit organization providing technical and business resources for software developers.

COALITION FOR NETWORKED INFORMATION
21 Dupont Circle NW, Suite 800, Washington DC 20036. 202/296-5098. World Wide Web address: http://www.cni.org. Promotes high-performance computers and networks.

COMMERCIAL INTERNET EXCHANGE ASSOCIATION (CIX)
4875 Eisenhower Avenue, Alexandria VA 22304-0797. 703/824-9249. E-mail address: helpdesk@cix.org. World Wide Web address: http://www.cix.org. A nonprofit trade association of data internetworking service providers.

COMPUTER GAME DEVELOPERS' ASSOCIATION
960 North San Antonio Road, Suite 125, Los Altos CA 94022. 415/948-2432. World Wide Web address: http://www.cgda.org. A professional society for the interactive entertainment, educational software, and multimedia industries.

COMPUTER INDUSTRY CONTRACT PROFESSIONALS ASSOCIATION
925 South Mason Road, Suite 401, Katy TX 77450. 713/346-2664. World Wide Web address: http://www.cicpa.org. A national nonprofit organization providing professional and personal services to its members.

THE COMPUTING RESEARCH ASSOCIATION
1875 Connecticut Avenue NW, Suite 718, Washington DC 20009-5728. 202/234-2111. E-mail address: info@cra.org. An association of academic departments of computer science and engineering, industry laboratories, and affiliated professional societies.

ELECTRONIC DESIGN AUTOMATION COMPANIES (EDAC)
111 West Saint John Street, San Jose CA 95113. 408/287-EDAC. E-mail address: information@edac.org. World Wide Web address: http://www.edac.org. An international association of companies concerned with the electronic design automation industry.

THE ENTERPRISE COMPUTER TELEPHONY FORUM
303 Vintage Park Drive, Foster City CA 94404-1138. 415/578-6852. World Wide Web address: http://www.ectf.org. A nonprofit corporation concerned with promoting open, competitive markets for computer telephony integration (CTI).

FABLESS SEMICONDUCTOR ASSOCIATION
Galleria Tower I, 13355 Noel Road, Dallas TX 75240-6636. 972/239-5119. World Wide Web address: http://www.computek.net/info. A semiconductor industry association.

FEDERAL NETWORKING COUNCIL
World Wide Web address: http://www.fnc.gov. A chartered federal council concerned with networking technology.

FEDERATION OF AMERICAN RESEARCH NETWORKS (FARNET)
E-mail address: info@farnet.org. World Wide Web address: http://www.farnet.org. A nonprofit association of internetworking technology companies.

GRAPHIC COMMUNICATIONS ASSOCIATION

100 Daingerfield Road, Alexandria VA 22314-2888. 703/519-8160. E-mail address: info@gca.org. World Wide Web address: http://www.gca.org. A nonprofit association concerned with computer technology in the graphics and communications industries. Membership required.

HTML WRITERS GUILD

World Wide Web address: http://www.hwg.org. An international organization of Web page writers and Internet professionals.

THE IPG SOCIETY

World Wide Web address: http://www.ipgnet.com. A professional trade association representing programmers internationally.

INDEPENDENT COMPUTER CONSULTANTS ASSOCIATION

11131 South Towne Square, Suite F, St. Louis MO 63123. Toll-free phone: 800/774-4222. World Wide Web address: http://www.icca.org. A nonprofit organization providing professional development and business support for independent computer consultants.

INFORMATION TECHNOLOGY ASSOCIATION OF AMERICA

1616 North Fort Myer Drive, Suite 1300, Arlington VA 22209. 703/522-5055. World Wide Web address: http://www.itaa.org.

INTERACTIVE MULTIMEDIA ASSOCIATION

48 Maryland Avenue, Suite 202, Annapolis MD 21401-8011. 410/626-1380. E-mail address: info@ima.org. World Wide Web address: http://www.ima.org. A multimedia and Internet association.

INTERNATIONAL COMMUNICATIONS INDUSTRIES ASSOCIATION, INC.

11242 Waples Mill Road, Suite 200, Fairfax VA 22030. 800/659-7469. E-mail address: icia@icia.org. World Wide Web address: http://www.icia.org. A nonprofit organization supporting the multimedia and audio-visual industries. Membership required.

INTERNATIONAL COUNCIL ON SYSTEMS ENGINEERING

2033 Sixth Avenue, Suite 804, Seattle WA 98121. 206/441-8262. World Wide Web address: http://www.incose.org. An organization concerned with the systems engineering industry.

INTERNATIONAL DISK DRIVE EQUIPMENT AND MATERIALS ASSOCIATION

710 Lakeway, Suite 140, Sunnyvale CA 94086. E-mail address: jlpinder@aol.com. World Wide Web address: http://www.idema.org. An organization which promotes the disk drive industry. Membership required.

THE INTERNATIONAL INTERACTIVE COMMUNICATIONS SOCIETY

10160 SW Nimbus Avenue, Suite F2, Portland OR 97223. 503/620-3604. World Wide Web address: http://www.iics.org. A society of multimedia and Internet professionals with 34 chapters worldwide.

THE INTERNATIONAL SOCIETY OF INTERNET PROFESSIONALS, INC.

1318 South Westcliff Place, Suite 93, Spokane WA 99204. World Wide Web address: http://www.webpro.org. An international nonprofit professional and trade organization for Internet professionals.

INTERNET DEVELOPERS ASSOCIATION

World Wide Web address: http://www.association.org. A trade association concerned with content development for the Internet.

INTERNET SERVICE PROVIDERS' CONSORTIUM

P.O. Box 88, St. Peters MO 63376. E-mail address: ispc-list@ispc.org. World Wide Web address: http://www.ispc.org. An Internet consortium.

INTERNET SOCIETY
12020 Sunrise Valley Drive, Reston VA 20191-3429. World Wide Web address: http://www.isoc.org. A non-governmental, Internet-oriented organization promoting global cooperation and coordination.

THE MICROELECTRONICS AND COMPUTER TECHNOLOGY CORPORATION (MCC)
3500 West Balcones Center Drive, Austin TX 78759. 512/338-3421. E-mail address: ask@mcc.com. World Wide Web address: http://www.mcc.com. A consortial research and development organization serving the high-tech industry. Membership required.

MULTIBUS MANUFACTURERS GROUP
P.O. Box 6208, Aloha OR 97007. World Wide Web address: http://www.multibus.org. A nonprofit trade organization concerned with open bus technology in the computer industry.

MULTIMEDIA DEVELOPMENT GROUP
2601 Mariposa Street, San Francisco CA 94110. 415/553-2300. Fax: 415/553-2403. E-mail address: info@mdg.org. A nonprofit trade association dedicated to the business and market development of multimedia companies.

NATIONAL MULTIMEDIA ASSOCIATION OF AMERICA
4920 Niagra Road, 3rd Floor, College Park MD 20740. 800/819-1335. World Wide Web address: http://www.nmaa.org. A nonprofit corporation concerned with the multimedia industry.

THE NETWORK PROFESSIONALS ASSOCIATION
151 East 1700 South, Provo UT 84606-7380. 801/379-0330. World Wide Web address: http://www.npa.org. An association promoting the network computing industry

NORTH AMERICAN INTERNET SERVICE PROVIDERS' ASSOCIATION
World Wide Web address: http://www.naispa.org. An association promoting standardization of the Internet industry.

NORTH AMERICAN NETWORK OPERATORS' GROUP
World Wide Web address: http://www.merit.edu/~nanog. A forum for technical and operations engineers and managers of Internet service providers.

OBJECT MANAGEMENT GROUP
492 Old Connecticut Path, Framingham MA 01701. 508/820-4300. E-mail address: info@omg.org. World Wide Web address: http://www.omg.org. A consortium promoting object technology (OT).

THE OPEN GROUP
11 Cambridge Center, Cambridge MA 02142-1405. 617/621-8700. World Wide Web address: http://www.osf.org. A consortium concerned with open systems technology in the information systems industry. Membership required.

SEMICONDUCTOR EQUIPMENT AND MATERIALS INTERNATIONAL
805 East Middlefield Road, Mountain View CA 94043-4080. 415/964-5111. E-mail address: semihq@semi.org. World Wide Web address: http://www.semi.org. An international trade association concerned with the semiconductor and flat-panel display industries. Membership required.

SOCIETY FOR DOCUMENTATION PROFESSIONALS
World Wide Web address: http://www.sdpro.com. A professional organization for technical communicators.

SOCIETY FOR INFORMATION MANAGEMENT
401 North Michigan Avenue, Chicago IL 60611-4267. 312/644-6610. E-mail address: info@simnet.org. World Wide Web address: http://www.simnet.org. A forum for information technology professionals.

THE SOCIETY OF COMPUTER PROFESSIONALS

20 Acorn Road, Secaucus NJ 07094. 201/865-4669. World Wide Web address: http://www.comprof.com. Requires annual dues.

SOFTWARE FORUM

P.O. Box 61031, Palo Alto CA 94306. 415/854-7219. E-mail address: 73771.1176@ compuserve.com. World Wide Web address: http://www.softwareforum.org. An independent, nonprofit organization for software industry professionals.

SOFTWARE PUBLISHERS ASSOCIATION

1730 M Street NW, Suite 700, Washington DC 20036. 202/452-1600. World Wide Web address: http://www.spa.org.

SOFTWARE SUPPORT PROFESSIONALS ASSOCIATION

11858 Bernardo Plaza Court, Suite 101C, San Diego CA 92128. 619/674-4864. World Wide Web address: http://www.sspa-online.com. A forum for service and support professionals in the software industry.

TECHNOLOGY ASSOCIATION LEADERS' COUNCIL (TALC)

13 Crescent Road, Livingston NJ 07039. 201/716-9457. World Wide Web address: http://www.thevine.com/talc/wow.htm. A consortium of executives of user groups and technology associations.

UNIFORUM

2901 Tasman Drive, Suite 205, Santa Clara CA 95054-1100. 408/986-8840. E-mail address: barbaram@uniforum.org. World Wide Web address: http://www.uniforum.org. An international, vendor-independent, nonprofit association of information systems professionals.

USENIX ASSOCIATION

2560 Ninth Street, Berkeley CA 94710. 510/528-8649. World Wide Web address: http://www.usenix.org. An advanced computing systems professional association for engineers, systems administrators, scientists, and technicians.

WORLD WIDE WEB CONSORTIUM

545 Technology Square, Cambridge MA 02139. 617/253-2613. E-mail address: admin@w3.org. World Wide Web address: http://www.w3.org. An academic consortium concerned with World Wide Web technology, located at Massachusetts Institute of Technology (MIT).

WORLD WIDE WEB TRADE ASSOCIATION

World Wide Web address: http://www.web-star.com/wwwta.html. An association promoting responsible use of the World Wide Web.

INDEXES

GEOGRAPHICAL INDEX

NOTE: *Below is an alphabetical index of primary employer listings included in this book. Those employers in each industry that fall under the headings "Additional employers" are not indexed here.*

COLORADO

GEORGIA

NEC TECHNOLOGIES, INC., 224
THE NETWORK CONNECTION, 224
PEACHTREE SOFTWARE, 149
PLATINUM TECHNOLOGY, 151
QUANTRA CORPORATION, 155
S2, INC., 331
SAMPO CORP. OF AMERICA, 401
SEQUENT COMPUTER SYSTEMS, 332
SOFTLAB INC., 163
SOFTWARE AG, 163
SOFTWARE SOLUTIONS INC., 164
SOUTHERN ELECTRONICS
    DISTRIBUTORS, 401
STERLING SOFTWARE, INC., 167
STRATUS COMPUTER, INC., 233
SUNGARD INSURANCE SYSTEMS, 169
SYSTEMS & PROGRAMMING
    CONSULTANTS, 355
SYSTEMS MAINTENANCE INC., 386
TSG TECHNICAL SERVICES, 356
TSW INTERNATIONAL, 170
TRECOM BUSINESS SYSTEMS, INC., 357
UNICOMP, 174
UNISYS CORPORATION, 339
UNIVERSAL DATA CONSULTANTS, 386
VSI ENTERPRISES, INC., 175
XCELLENET, INC., 180

HAWAII

VERIFONE, INC., 175

IDAHO

ADAGER CORPORATION, 193
DIGITAL EQUIPMENT CORP., 203
EXTENDED SYSTEMS, INC., 206
HEWLETT-PACKARD COMPANY, 210
INSURANCE SOLUTIONS, 124
MICRON ELECTRONICS, 221
MICRON TECHNOLOGY, 278
SPUR PRODUCTS CORPORATION, 334
ZILOG INC., 285

ILLINOIS

ASAP, 71
ALPHA MICROSYSTEMS, 381
AMBASSADOR BUSINESS SOLUTIONS,
    390
ANALYSTS INTERNATIONAL CORP., 343
ANICOM, INC., 194
APPLIED SYSTEMS, INC., 296
BUSINESS SOLUTIONS INTERNATL., 345
COMARK, INC., 393
COMPDISK, 369
COMPUTER ASSOCIATES
    INTERNATIONAL, INC., 93
COMPUTER HORIZONS CORP., 305
COMPUTER IDENTICS ID MATRIX, 305

COMPUTER MANAGEMENT SCIENCES,
    INC., 347
CORPORATE DISK COMPANY, 370
CYBERTEK CORPORATION, 98
CYBORG SYSTEMS INC., 98
DARTEK COMPUTER SUPPLY, 394
DATA COMMUNICATION FOR BUSINESS
    INC., 309
DATACRON INC., 100
DATAIR EMPLOYEE BENEFITS
    SYSTEMS, 101
DATALOGICS INC., 101
DELPHI INFORMATION SYSTEMS, 103
DOMINO AMJET INC., 205
EDGE SYSTEMS, INC., 313
J.D. EDWARDS & COMPANY, 105
ENTERPRISE SYSTEMS INC., 106
ENTEX INFORMATION SERVICES, 313
FUTURESOURCE, 364
GALILEO INTERNATIONAL, 112
GREENBRIER & RUSSEL INC., 351
HEALTH SYSTEMS ARCHITECTS, 118
HURLETRON INC., 267
IBM CORPORATION, 212
INFORMATION RESOURCES, INC., 372
INFORMIX SOFTWARE, INC., 123
INSO CORPORATION, 124
INTERIM TECHNOLOGY, 351
INTERLINK COMPUTER SCIENCES, 319
INTERNET SYSTEMS CORPORATION, 127
JBA INTERNATIONAL INC., 129
KINETIC SYSTEMS CORPORATION, 217
KLEINSCHMIDT INC., 320
MEDICUS SYSTEMS CORPORATION, 137
MERCURY INTERACTIVE CORP., 138
MICRO PROFESSIONALS INC., 399
MICRO SOLUTIONS COMPUTER
    PRODUCTS INC., 221
NCR CORPORATION, 223
PERLE SYSTEMS INC., 328
PETRO VEND INC., 328
PITNEY BOWES SOFTWARE SYSTEMS,
    151
PLATINUM TECHNOLOGY, 151
QUILL CORPORATION, 401
RIM SYSTEMS INC., 156
RISC MANAGEMENT, 330
SPSS INC., 159
SEAGATE SOFTWARE, 160
SEQUENT COMPUTER SYSTEMS, 332
SERVITECH INC., 386
SILVON SOFTWARE INC., 162
SOFTWARE ARCHITECTS, 355
SPYGLASS, INC., 166
STORAGETEK TERIS, 167
STREAMLOGIC, 271
SYNTRONIC INSTRUMENTS INC., 271
SYSTEM SOFTWARE ASSOCIATES, 171
TECHNIUM, INC., 356
TRI-COR INDUSTRIES, INC., 173
U.S. ROBOTICS, INC., 237
UNISYS CORPORATION, 339

MASSACHUSETTS

NEW JERSEY

NEW MEXICO

NEW YORK

NORTH CAROLINA

NORTH DAKOTA

OHIO

OKLAHOMA

OREGON

UTAH

VERMONT

VIRGINIA

WASHINGTON

WEST VIRGINIA

WISCONSIN

WYOMING

ALPHABETICAL INDEX

NOTE: *Below is an alphabetical index of primary employer listings included in this book. Those employers in each industry that fall under the headings "Additional employers" are not indexed here.*

O

P

Q

R

T

Your Job Hunt
Your Feedback

Comments, questions, or suggestions? We want to hear from you. Please complete this questionnaire and mail it to:

The JobBank Staff
Adams Media Corporation
260 Center Street
Holbrook, MA 02343

Did this book provide helpful advice and valuable information which you used in your job search? Was the information easy to access?

Recommendations for improvements. How could we improve this book to help in your job search? No suggestion is too small or too large.

Would you recommend this book to a friend beginning a job hunt?

Name: _____

Occupation: _____

Which JobBank did you use? _____

Address: _____

Daytime phone: _____

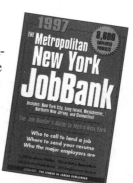